STUDENT'S SOLUTIONS MANUAL

DANIEL S. MILLER
Niagara County Community College

ALGEBRA & TRIGONOMETRY
FIFTH EDITION

Robert Blitzer
Miami Dade College

PEARSON

Boston Columbus Indianapolis New York San Francisco Upper Saddle River
Amsterdam Cape Town Dubai London Madrid Milan Munich Paris Montreal Toronto
Delhi Mexico City São Paulo Sydney Hong Kong Seoul Singapore Taipei Tokyo

Reproduced by Pearson from electronic files supplied by the author.

ISBN-13: 978-0-321-83736-3
ISBN-10: 0-321-83736-3

7 8 9 10 v092 17 16 15 14

www.pearsonhighered.com

PEARSON

TABLE OF CONTENTS for STUDENT SOLUTIONS

ALGEBRA AND TRIGONOMETRY 5E

Chapter P
Fundamental Concepts of Algebra

Section P.1

Check Point Exercises

1. $8 + 6(x-3)^2 = 8 + 6(13-3)^2$

$$= 8 + 6(10)^2$$
$$= 8 + 6(100)$$
$$= 8 + 600$$
$$= 608$$

2. Since 2015 is 15 years after 2000, substitute 15 for x.
$T = 4x^2 + 341x + 3194$

$$= 4(15)^2 + 341(15) + 3194$$
$$= 9209$$

If trends continue, the tuition and fees will be $9209.

3. The elements common to {3, 4, 5, 6, 7} and {3, 7, 8, 9} are 3 and 7.
$\{3,4,5,6,7\} \cap \{3,7,8,9\} = \{3,7\}$

4. The union is the set containing all the elements of either set.
$\{3,4,5,6,7\} \cup \{3,7,8,9\} = \{3,4,5,6,7,8,9\}$

5. $\left\{ -9, \ -1.3, \ 0, \ 0.\overline{3}, \ \dfrac{\pi}{2}, \ \sqrt{9}, \ \sqrt{10} \right\}$

   a. Natural numbers: $\sqrt{9}$ because $\sqrt{9} = 3$

   b. Whole numbers: $0, \ \sqrt{9}$

   c. Integers: $-9, 0, \sqrt{9}$

   d. Rational numbers: $-9, \ -1.3, \ 0, \ 0.\overline{3}, \ \sqrt{9}$

   e. Irrational numbers: $\dfrac{\pi}{2}, \ \sqrt{10}$

   f. Real numbers: $-9, \ -1.3, \ 0, \ 0.\overline{3}, \ \dfrac{\pi}{2}, \ \sqrt{9}, \ \sqrt{10}$

6. a. $\left| 1 - \sqrt{2} \right|$

   Because $\sqrt{2} \approx 1.4$, the number inside the absolute value bars is negative. The absolute value of x when $x < 0$ is $-x$. Thus,
   $$\left| 1 - \sqrt{2} \right| = -\left(1 - \sqrt{2} \right) = \sqrt{2} - 1$$

   b. $\left| \pi - 3 \right|$

   Because $\pi \approx 3.14$, the number inside the absolute value bars is positive. The absolute value of a positive number is the number itself. Thus,
   $$\left| \pi - 3 \right| = \pi - 3.$$

   c. $\dfrac{|x|}{x}$

   Because $x > 0$, $\ |x| = x$.

   Thus, $\dfrac{|x|}{x} = \dfrac{x}{x} = 1$

7. $|-4 - (5)| = |-9| = 9$

   The distance between -4 and 5 is 9.

8. $7(4x^2 + 3x) + 2(5x^2 + x)$

   $$= 7(4x^2 + 3x) + 2(5x^2 + x)$$
   $$= 28x^2 + 21x + 10x^2 + 2x$$
   $$= 38x^2 + 23x$$

9. $6 + 4[7 - (x-2)]$

   $$= 6 + 4[7 - x + 2)]$$
   $$= 6 + 4[9 - x]$$
   $$= 6 + 36 - 4x$$
   $$= 42 - 4x$$

Concept and Vocabulary Check P.1

1. expression

2. b to the nth power; base; exponent

3. formula; modeling; models

4. intersection; $A \cap B$

5. union; $A \cup B$

6. natural

7. whole

8. integers

9. rational

10. irrational

11. rational; irrational

12. absolute value; x, $-x$

13. $b+a$; ba

14. $a+(b+c)$; $(ab)c$

15. $ab+ac$

16. 0; inverse; 0; identity

17. inverse; 1; identity

18. simplified

19. a

Exercise Set P.1

1. $7+5(10)=7+50=57$

3. $6(3)-8=18-8=10$

5. $8^2+3(8)=64+24=88$

7. $7^2-6(7)+3=49-42+3=7+3=10$

9. $4+5(9-7)^3=4+5(2)^3$
$=4+5(8)=4+40=44$

11. $8^2-3(8-2)=64-3(6)$
$=64-18=46$

13. $\dfrac{5(x+2)}{2x-14}=\dfrac{5(10+2)}{2(10)-14}$
$=\dfrac{5(12)}{6}$
$=5\cdot2$
$=10$

15. $\dfrac{2x+3y}{x+1}; x=-2, y=4$
$=\dfrac{2(-2)+3(4)}{-2+1}=\dfrac{-4+12}{-1}=\dfrac{8}{-1}=-8$

17. $C=\dfrac{5}{9}(50-32)=\dfrac{5}{9}(18)=10$
$50°\mathrm{F}$ is equivalent to $10°\mathrm{C}$.

19. $h=4+60t-16t^2=4+60(2)-16(2)^2$
$=4+120-16(4)=4+120-64$
$=124-64=60$
Two seconds after it is kicked, the ball's height is 60 feet.

21. $\{1,2,3,4\}\cap\{2,4,5\}=\{2,4\}$

23. $\{s,e,t\}\cap\{t,e,s\}=\{s,e,t\}$

25. $\{1,3,5,7\}\cap\{2,4,6,8,10\}=\{\ \ \}$
The empty set is also denoted by $\varnothing$.

27. $\{a,b,c,d\}\cap\varnothing=\varnothing$

29. $\{1,2,3,4\}\cup\{2,4,5\}=\{1,2,3,4,5\}$

31. $\{1,3,5,7\}\cup\{2,4,6,8,10\}$
$=\{1,2,3,4,5,6,7,8,10\}$

33. $\{a,e,i,o,u\}\cup\varnothing=\{a,e,i,o,u\}$

35. a. $\sqrt{100}$

b. $0,\sqrt{100}$

c. $-9,0,\sqrt{100}$

d. $-9,-\dfrac{4}{5},0,0.25,9.2,\sqrt{100}$

e. $\sqrt{3}$

f. $-9,-\dfrac{4}{5},0,0.25,\sqrt{3},9.2,\sqrt{100}$

37. a. $\sqrt{64}$

b. $0,\sqrt{64}$

c. $-11,0,\sqrt{64}$

d. $-11,-\dfrac{5}{6},0,0.75,\sqrt{64}$

e. $\sqrt{5},\pi$

f. $-11,-\dfrac{5}{6},0,0.75,\sqrt{5},\pi,\sqrt{64}$

39. 0

41. Answers will vary. An example is 2.

43. true; −13 is to the left of −2 on the number line.

45. true; 4 is to the right of −7 on the number line.

47. true; $-\pi = -\pi$

49. true; 0 is to the right of −6 on the number line.

51. $|300| = 300$

53. $|12 - \pi| = 12 - \pi$

55. $\left|\sqrt{2} - 5\right| = 5 - \sqrt{2}$

57. $\dfrac{-3}{|-3|} = \dfrac{-3}{3} = -1$

59. $\left||-3| - |-7|\right| = |3 - 7| = |-4| = 4$

61. $|x + y| = |2 + (-5)| = |-3| = 3$

63. $|x| + |y| = |2| + |-5| = 2 + 5 = 7$

65. $\dfrac{y}{|y|} = \dfrac{-5}{|-5|} = \dfrac{-5}{5} = -1$

67. The distance is $|2 - 17| = |-15| = 15$.

69. The distance is $|-2 - 5| = |-7| = 7$.

71. The distance is $|-19 - (-4)| = |-19 + 4| = |-15| = 15$.

73. The distance is
$|-3.6 - (-1.4)| = |-3.6 + 1.4| = |-2.2| = 2.2$.

75. $6 + (-4) = (-4) + 6$;
commutative property of addition

77. $6 + (2 + 7) = (6 + 2) + 7$;
associative property of addition

79. $(2 + 3) + (4 + 5) = (4 + 5) + (2 + 3)$;
commutative property of addition

81. $2(-8 + 6) = -16 + 12$;
distributive property of multiplication over addition

83. $\dfrac{1}{x+3}(x+3) = 1; x \neq -3$,
inverse property of multiplication

85. $5(3x + 4) - 4 = 5 \cdot 3x + 5 \cdot 4 - 4$
$= 15x + 20 - 4$
$= 15x + 16$

87. $5(3x - 2) + 12x = 5 \cdot 3x - 5 \cdot 2 + 12x$
$= 15x - 10 + 12x$
$= 27x - 10$

89. $7(3y - 5) + 2(4y + 3)$
$= 7 \cdot 3y - 7 \cdot 5 + 2 \cdot 4y + 2 \cdot 3$
$= 21y - 35 + 8y + 6$
$= 29y - 29$

91. $5(3y - 2) - (7y + 2) = 15y - 10 - 7y - 2$
$= 8y - 12$

93. $7 - 4[3 - (4y - 5)] = 7 - 4[3 - 4y + 5]$
$= 7 - 4[8 - 4y]$
$= 7 - 32 + 16y$
$= 16y - 25$

95. $18x^2 + 4 - \left[6(x^2 - 2) + 5\right]$
$= 18x^2 + 4 - \left[6x^2 - 12 + 5\right]$
$= 18x^2 + 4 - \left[6x^2 - 7\right]$
$= 18x^2 + 4 - 6x^2 + 7$
$= 18x^2 - 6x^2 + 4 + 7$
$= (18 - 6)x^2 + 11 = 12x^2 + 11$

97. $-(-14x) = 14x$

99. $-(2x - 3y - 6) = -2x + 3y + 6$

101. $\dfrac{1}{3}(3x) + [(4y) + (-4y)] = x + 0 = x$

103. $|-6| \;\square\; |-3|$
$6 \;\square\; 3$
$6 > 3$
Since $6 > 3$, $|-6| > |-3|$.

105.
$$\left|\frac{3}{5}\right| \ \square \ |-0.6|$$
$$|0.6| \ \square \ |-0.6|$$
$$0.6 \ \square \ 0.6$$
$$0.6 \ = \ 0.6$$
Since $0.6 = 0.6$, $\left|\frac{3}{5}\right| = |-0.6|$.

107.
$$\frac{30}{40} - \frac{3}{4} \ \square \ \frac{14}{15} \cdot \frac{15}{14}$$
$$\frac{30}{40} - \frac{30}{40} \ \square \ \frac{\cancel{14}}{\cancel{15}} \cdot \frac{\cancel{15}}{\cancel{14}}$$
$$0 \ \square \ 1$$
$$0 < 1$$
Since $0 < 1$, $\frac{30}{40} - \frac{3}{4} < \frac{14}{15} \cdot \frac{15}{14}$.

109.
$$\frac{8}{13} \div \frac{8}{13} \ \square \ |-1|$$
$$\frac{8}{13} \cdot \frac{13}{8} \ \square \ 1$$
$$1 \ \square \ 1$$
$$1 = 1$$
Since $1 = 1$, $\frac{8}{13} \div \frac{8}{13} = |-1|$.

111. $8^2 - 16 \div 2^2 \cdot 4 - 3 = 64 - 16 \div 4 \cdot 4 - 3$
$$= 64 - 4 \cdot 4 - 3$$
$$= 64 - 16 - 3$$
$$= 48 - 3$$
$$= 45$$

113.
$$\frac{5 \cdot 2 - 3^2}{[3^2 - (-2)]^2} = \frac{5 \cdot 2 - 9}{[9 - (-2)]^2}$$
$$= \frac{10 - 9}{[9 + 2]^2}$$
$$= \frac{10 - 9}{11^2}$$
$$= \frac{1}{121}$$

115. $8 - 3[-2(2 - 5) - 4(8 - 6)] = 8 - 3[-2(-3) - 4(2)]$
$$= 8 - 3[6 - 8]$$
$$= 8 - 3[-2]$$
$$= 8 + 6$$
$$= 14$$

117.
$$\frac{2(-2) - 4(-3)}{5 - 8} = \frac{-4 + 12}{-3}$$
$$= \frac{8}{-3}$$
$$= -\frac{8}{3}$$

119.
$$\frac{(5-6)^2 - 2|3 - 7|}{89 - 3 \cdot 5^2} = \frac{(-1)^2 - 2|-4|}{89 - 3 \cdot 25}$$
$$= \frac{1 - 2(4)}{89 - 75}$$
$$= \frac{1 - 8}{14}$$
$$= \frac{-7}{14}$$
$$= -\frac{1}{2}$$

121. $x - (x + 4) = x - x - 4 = -4$

123. $6(-5x) = -30x$

125. $5x - 2x = 3x$

127. $8x - (3x + 6) = 8x - 3x - 6 = 5x - 6$

129. a. $H = \frac{7}{10}(220 - a)$
$$H = \frac{7}{10}(220 - 20)$$
$$= \frac{7}{10}(200)$$
$$= 140$$
The lower limit of the heart rate for a 20-year-old with this exercise goal is 140 beats per minute.

b. $H = \frac{4}{5}(220 - a)$
$$H = \frac{4}{5}(220 - 20)$$
$$= \frac{4}{5}(200)$$
$$= 160$$
The upper limit of the heart rate for a 20-year-old with this exercise goal is 160 beats per minute.

131. a. $T = 26x^2 + 819x + 15,527$

$= 26(10)^2 + 819(10) + 15,527$

$= 26,317$

The formula estimates the cost to have been $26,317 in 2010.

b. This overestimates the value in the graph by $44.

c. $T = 26x^2 + 819x + 15,527$

$= 26(13)^2 + 819(13) + 15,527$

$= 30,568$

The formula projects the cost to be $30,568 in 2013.

133. a. $0.05x + 0.12(10,000 - x)$

$= 0.05x + 1200 - 0.12x$

$= 1200 - 0.07x$

b. $1200 - 0.07x = 1200 - 0.07(6000)$

$= \$780$

135. – 143. Answers will vary.

145. does not make sense; Explanations will vary. Sample explanation: To use the model, substitute 0 for x.

147. does not make sense; Explanations will vary. Sample explanation: The commutative property changes order and the associative property changes groupings.

149. false; Changes to make the statement true will vary. A sample change is: All whole numbers are integers.

151. false; Changes to make the statement true will vary. A sample change is: Some irrational numbers are negative.

153. false; Changes to make the statement true will vary. A sample change is: $5 + 3(x - 4) = 5 + 3x - 12 = 3x - 7$.

155. true

157. $-\pi > -3.5$

159. a. $b^4 \cdot b^3 = (b \cdot b \cdot b \cdot b)(b \cdot b \cdot b) = b^7$

b. $b^5 \cdot b^5 = (b \cdot b \cdot b \cdot b \cdot b)(b \cdot b \cdot b \cdot b \cdot b) = b^{10}$

c. add the exponents

160. a. $\dfrac{b^7}{b^3} = \dfrac{b \cdot b \cdot b \cdot b \cdot b \cdot b \cdot b}{b \cdot b \cdot b} = b^4$

b. $\dfrac{b^8}{b^2} = \dfrac{b \cdot b \cdot b \cdot b \cdot b \cdot b \cdot b \cdot b}{b \cdot b} = b^6$

c. subtract the exponents

161. $6.2 \times 10^3 = 6.2 \times 10 \times 10 \times 10 = 6200$

It moves the decimal point 3 places to the right.

Section P.2

Check Point Exercises

1. a. $3^3 3^2 = 3^{3+2} = 3^5$ or 243

b. $\left(4x^3 y^4\right)\left(10x^2 y^6\right) = 4 \cdot 10 \cdot x^3 \cdot x^2 \cdot y^4 \cdot y^6$

$= 40x^{3+2} \cdot y^{4+6}$

$= 40x^5 y^{10}$

2. a. $\dfrac{(-3)^6}{(-3)^3} = (-3)^3 = -27$

b. $\dfrac{27x^{14} y^8}{3x^3 y^5} = \dfrac{27}{3} \cdot \dfrac{x^{14}}{x^3} \cdot \dfrac{y^8}{y^5} = 9x^{14-3} y^{8-5} = 9x^{11} y^3$

3. a. $5^{-2} = \dfrac{1}{5^2} = \dfrac{1}{25}$

b. $(-3)^{-3} = \dfrac{1}{(-3)^3} = \dfrac{1}{-27} = -\dfrac{1}{27}$

c. $\dfrac{1}{4^{-2}} = \dfrac{1}{\frac{1}{4^2}} = 1 \cdot \dfrac{4^2}{1} = 4^2 = 16$

d. $3x^{-6} y^4 = 3 \cdot \dfrac{1}{x^6} \cdot y^4 = \dfrac{3y^4}{x^6}$

4. a. $\left(3^3\right)^2 = 3^{3 \cdot 2} = 3^6$ or 729

b. $\left(y^7\right)^{-2} = y^{7(-2)} = y^{-14} = \dfrac{1}{y^{14}}$

c. $\left(b^{-3}\right)^{-4} = b^{-3(-4)} = b^{12}$

5. $(-4x)^3 = (-4)^3(x)^3 = -64x^3$

6. a. $\left(-\dfrac{2}{y}\right)^5 = \dfrac{(-2)^5}{y^5} = \dfrac{-32}{y^5}$

b. $\left(\dfrac{x^5}{3}\right)^3 = \dfrac{(x^5)^3}{3^3} = \dfrac{x^{15}}{27}$

7. a. $\left(2x^3 y^6\right)^4 = (2)^4\left(x^3\right)^4\left(y^6\right)^4 = 16x^{12} y^{24}$

b. $\left(-6x^2 y^5\right)\left(3xy^3\right) = (-6)\cdot 3\cdot x^2 \cdot x \cdot y^5 \cdot y^3$
$$= -18x^3 y^8$$

c. $\dfrac{100x^{12} y^2}{20x^{16} y^{-4}} = \left(\dfrac{100}{20}\right)\left(\dfrac{x^{12}}{x^{16}}\right)\left(\dfrac{y^2}{y^{-4}}\right)$
$$= 5x^{12-16} y^{2-(-4)}$$
$$= 5x^{-4} y^6$$
$$= \dfrac{5y^6}{x^4}$$

d. $\left(\dfrac{5x}{y^4}\right)^{-2} = \dfrac{(5)^{-2}(x)^{-2}}{\left(y^4\right)^{-2}}$
$$= \dfrac{(5)^{-2}(x)^{-2}}{\left(y^4\right)^{-2}}$$
$$= \dfrac{5^{-2} x^{-2}}{y^{-8}}$$
$$= \dfrac{y^8}{5^2 x^2}$$
$$= \dfrac{y^8}{25x^2}$$

8. a. $-2.6\times 10^9 = -2,600,000,000$

b. $3.017\times 10^{-6} = 0.000003017$

9. a. $5,210,000,000 = 5.21\times 10^9$

b. $-0.00000006893 = -6.893\times 10^{-8}$

10. $410\times 10^7 = \left(4.1\times 10^2\right)\times 10^7$
$$= 4.1\times\left(10^2 \times 10^7\right)$$
$$= 4.1\times 10^9$$

11 a. $\left(7.1\times 10^5\right)\left(5\times 10^{-7}\right)$
$$= 7.1\cdot 5\times 10^5 \cdot 10^{-7}$$
$$= 35.5\times 10^{-2}$$
$$= \left(3.55\times 10^1\right)\times 10^{-2}$$
$$= 3.55\times\left(10^1 \times 10^{-2}\right)$$
$$= 3.55\times 10^{-1}$$

b. $\dfrac{1.2\times 10^6}{3\times 10^{-3}} = \dfrac{1.2}{3}\cdot\dfrac{10^6}{10^{-3}}$
$$= 0.4\times 10^{6-(-3)}$$
$$= 0.4\times 10^9$$
$$= 4\times 10^8$$

12. $\dfrac{2.6\times 10^{12}}{3.12\times 10^8} = \dfrac{2.6}{3.12}\cdot\dfrac{10^{12}}{10^8}$
$$\approx 0.83\cdot 10^4$$
$$\approx 8300$$
The cost was \$8300 per citizen.

Concept and Vocabulary Check P.2

1. b^{m+n}; add

2. b^{m-n}; subtract

3. 1

4. $\dfrac{1}{b^n}$

5. false

6. b^n

7. true

8. a number greater than or equal to 1 and less than 10; integer

9. true

10. false

Exercise Set P.2

1. $5^2 \cdot 2 = (5 \cdot 5) \cdot 2 = 25 \cdot 2 = 50$

3. $(-2)^6 = (-2)(-2)(-2)(-2)(-2)(-2) = 64$

5. $-2^6 = -2 \cdot 2 \cdot 2 \cdot 2 \cdot 2 \cdot 2 = -64$

7. $(-3)^0 = 1$

9. $-3^0 = -1$

11. $4^{-3} = \dfrac{1}{4^3} = \dfrac{1}{4 \cdot 4 \cdot 4} = \dfrac{1}{64}$

13. $2^2 \cdot 2^3 = 2^{2+3} = 2^5 = 2 \cdot 2 \cdot 2 \cdot 2 \cdot 2 = 32$

15. $(2^2)^3 = 2^{2 \cdot 3} = 2^6 = 2 \cdot 2 \cdot 2 \cdot 2 \cdot 2 \cdot 2 = 64$

17. $\dfrac{2^8}{2^4} = 2^{8-4} = 2^4 = 2 \cdot 2 \cdot 2 \cdot 2 = 16$

19. $3^{-3} \cdot 3 = 3^{-3+1} = 3^{-2} = \dfrac{1}{3^2} = \dfrac{1}{3 \cdot 3} = \dfrac{1}{9}$

21. $\dfrac{2^3}{2^7} = 2^{3-7} = 2^{-4} = \dfrac{1}{2^4} = \dfrac{1}{2 \cdot 2 \cdot 2 \cdot 2} = \dfrac{1}{16}$

23. $x^{-2}y = \dfrac{1}{x^2} \cdot y = \dfrac{y}{x^2}$

25. $x^0 y^5 = 1 \cdot y^5 = y^5$

27. $x^3 \cdot x^7 = x^{3+7} = x^{10}$

29. $x^{-5} \cdot x^{10} = x^{-5+10} = x^5$

31. $(x^3)^7 = x^{3 \cdot 7} = x^{21}$

33. $(x^{-5})^3 = x^{-5 \cdot 3} = x^{-15} = \dfrac{1}{x^{15}}$

35. $\dfrac{x^{14}}{x^7} = x^{14-7} = x^7$

37. $\dfrac{x^{14}}{x^{-7}} = x^{14-(-7)} = x^{14+7} = x^{21}$

39. $(8x^3)^2 = 8^2(x^3)^2 = 8^2 x^{3 \cdot 2} = 64x^6$

41. $\left(-\dfrac{4}{x}\right)^3 = \dfrac{(-4)^3}{x^3} = -\dfrac{64}{x^3}$

43. $(-3x^2 y^5)^2 = (-3)^2(x^2)^2 \cdot (y^5)^2$
$$= 9x^{2 \cdot 2} y^{5 \cdot 2}$$
$$= 9x^4 y^{10}$$

45. $(3x^4)(2x^7) = 3 \cdot 2 x^4 \cdot x^7 = 6x^{4+7} = 6x^{11}$

47. $(-9x^3 y)(-2x^6 y^4) = (-9)(-2)x^3 x^6 yy^4$
$$= 18x^{3+6} y^{1+4}$$
$$= 18x^9 y^5$$

49. $\dfrac{8x^{20}}{2x^4} = \left(\dfrac{8}{2}\right)\left(\dfrac{x^{20}}{x^4}\right) = 4x^{20-4} = 4x^{16}$

51. $\dfrac{25a^{13} \cdot b^4}{-5a^2 \cdot b^3} = \left(\dfrac{25}{-5}\right)\left(\dfrac{a^{13}}{a^2}\right)\left(\dfrac{b^4}{b^3}\right)$
$$= -5a^{13-2}b^{4-3}$$
$$= -5a^{11}b$$

53. $\dfrac{14b^7}{7b^{14}} = \left(\dfrac{14}{7}\right)\left(\dfrac{b^7}{b^{14}}\right) = 2 \cdot b^{7-14} = 2b^{-7} = \dfrac{2}{b^7}$

55. $(4x^3)^{-2} = (4^{-2})(x^3)^{-2}$
$$= 4^{-2} x^{-6}$$
$$= \dfrac{1}{4^2 x^6}$$
$$= \dfrac{1}{16x^6}$$

57. $\dfrac{24x^3 \cdot y^5}{32x^7 y^{-9}} = \dfrac{3}{4}x^{3-7} y^{5-(-9)}$
$$= \dfrac{3}{4}x^{-4} y^{14}$$
$$= \dfrac{3y^{14}}{4x^4}$$

59. $\left(\dfrac{5x^3}{y}\right)^{-2} = \dfrac{5^{-2} x^{-6}}{y^{-2}} = \dfrac{y^2}{25x^6}$

61. $\left(\dfrac{-15a^4b^2}{5a^{10}b^{-3}}\right)^3 = \left(\dfrac{-3b^{2-(-3)}}{a^{10-4}}\right)^3$

$\qquad = \left(\dfrac{-3b^5}{a^6}\right)^3$

$\qquad = \dfrac{-27b^{15}}{a^{18}}$

63. $\left(\dfrac{3a^{-5}b^2}{12a^3b^{-4}}\right)^0 = 1$

65. $3.8 \times 10^2 = 380$

67. $6 \times 10^{-4} = 0.0006$

69. $-7.16 \times 10^6 = -7,160,000$

71. $7.9 \times 10^{-1} = 0.79$

73. $-4.15 \times 10^{-3} = -0.00415$

75. $-6.00001 \times 10^{10} = -60,000,100,000$

77. $32,000 = 3.2 \times 10^4$

79. $638,000,000,000,000,000$
$\qquad = 6.38 \times 10^{17}$

81. $-5716 = -5.716 \times 10^3$

83. $0.0027 = 2.7 \times 10^{-3}$

85. $-0.00000000504 = -5.04 \times 10^{-9}$

87. $\left(3 \times 10^4\right)\left(2.1 \times 10^3\right) = (3 \times 2.1)\left(10^4 \times 10^3\right)$

$\qquad\qquad = 6.3 \times 10^{4+3} = 6.3 \times 10^7$

89. $\left(1.6 \times 10^{15}\right)\left(4 \times 10^{-11}\right) = (1.6 \times 4)\left(10^{15} \times 10^{-11}\right)$

$\qquad\qquad = 6.4 \times 10^{15+(-11)}$

$\qquad\qquad = 6.4 \times 10^4$

91. $\left(6.1 \times 10^{-8}\right)\left(2 \times 10^{-4}\right) = (6.1 \times 2)\left(10^{-8} \times 10^{-4}\right)$

$\qquad\qquad = 12.2 \times 10^{-8+(-4)}$

$\qquad\qquad = 12.2 \times 10^{-12}$

$\qquad\qquad = 1.22 \times 10^{-11}$

93. $\left(4.3 \times 10^8\right)\left(6.2 \times 10^4\right)$

$\qquad = (4.3 \times 6.2)\left(10^8 \times 10^4\right)$

$\qquad = 26.66 \times 10^{8+4}$

$\qquad = 26.66 \times 10^{12}$

$\qquad = 2.666 \times 10^{13} \approx 2.67 \times 10^{13}$

95. $\dfrac{8.4 \times 10^8}{4 \times 10^5} = \dfrac{8.4}{4} \times \dfrac{10^8}{10^5}$

$\qquad\qquad = 2.1 \times 10^{8-5} = 2.1 \times 10^3$

97. $\dfrac{3.6 \times 10^4}{9 \times 10^{-2}} = \dfrac{3.6}{9} \times \dfrac{10^4}{10^{-2}}$

$\qquad\qquad = 0.4 \times 10^{4-(-2)}$

$\qquad\qquad = 0.4 \times 10^6 = 4 \times 10^5$

99. $\dfrac{4.8 \times 10^{-2}}{2.4 \times 10^6} = \dfrac{4.8}{2.4} \times \dfrac{10^{-2}}{10^6}$

$\qquad\qquad = 2 \times 10^{-2-6} = 2 \times 10^{-8}$

101. $\dfrac{2.4 \times 10^{-2}}{4.8 \times 10^{-6}} = \dfrac{2.4}{4.8} \times \dfrac{10^{-2}}{10^{-6}}$

$\qquad\qquad = 0.5 \times 10^{-2-(-6)}$

$\qquad\qquad = 0.5 \times 10^4 = 5 \times 10^3$

103. $\dfrac{480,000,000,000}{0.00012} = \dfrac{4.8 \times 10^{11}}{1.2 \times 10^{-4}}$

$\qquad\qquad = \dfrac{4.8}{1.2} \times \dfrac{10^{11}}{10^{-4}}$

$\qquad\qquad = 4 \times 10^{11-(-4)}$

$\qquad\qquad = 4 \times 10^{15}$

105. $\dfrac{0.00072 \times 0.003}{0.00024}$

$\qquad = \dfrac{\left(7.2 \times 10^{-4}\right)\left(3 \times 10^{-3}\right)}{2.4 \times 10^{-4}}$

$\qquad = \dfrac{7.2 \times 3}{2.4} \times \dfrac{10^{-4} \cdot 10^{-3}}{10^{-4}} = 9 \times 10^{-3}$

107. $\dfrac{\left(x^{-2}y\right)^{-3}}{\left(x^2y^{-1}\right)^3} = \dfrac{x^6y^{-3}}{x^6y^{-3}}$

$\qquad\qquad = x^{6-6}y^{-3-(-3)} = x^0y^0 = 1$

109. $\left(2x^{-3}yz^{-6}\right)(2x)^{-5} = 2x^{-3}yz^{-6} \cdot 2^{-5}x^{-5}$

$= 2^{-4}x^{-8}yz^{-6} = \dfrac{y}{2^4 x^8 z^6} = \dfrac{y}{16x^8 z^6}$

111. $\left(\dfrac{x^3 y^4 z^5}{x^{-3}y^{-4}z^{-5}}\right)^{-2} = \left(x^6 y^8 z^{10}\right)^{-2}$

$= x^{-12}y^{-16}z^{-20} = \dfrac{1}{x^{12}y^{16}z^{20}}$

113. $\dfrac{\left(2^{-1}x^{-2}y^{-1}\right)^{-2}\left(2x^{-4}y^3\right)^{-2}\left(16x^{-3}y^3\right)^0}{\left(2x^{-3}y^{-5}\right)^2}$

$= \dfrac{\left(2^2 x^2 y^2\right)\left(2^{-2}x^8 y^{-6}\right)(1)}{\left(2^2 x^{-6}y^{-10}\right)}$

$= \dfrac{x^{18}y^6}{4}$

115. **a.** 2.17×10^{12}

b. 3.09×10^8

c. $\dfrac{2.17 \times 10^{12}}{3.09 \times 10^8} = \dfrac{2.17}{3.09} \times \dfrac{10^{12}}{10^8}$

$\approx 0.7023 \times 10^4$

≈ 7023

$\$7023$ per American

117. $1340 \times 10^6 \cdot 7.90 = 1.34 \times 10^9 \cdot 7.9$

$= 1.34 \cdot 7.9 \times 10^9$

$= 10.586 \times 10^9$

$= 1.0586 \times 10^{10}$

Box-office receipts were $\$1.0586 \times 10^{10}$ in 2010.

119. $5.3 \times 10^{-23} \cdot 20,000 = 5.3 \times 10^{-23} \cdot 2 \times 10^4$

$= 5.3 \cdot 2 \times 10^{-23} \cdot 10^4$

$= 10.6 \times 10^{-19}$

$= 1.06 \times 10^1 \cdot 10^{-19}$

$= 1.06 \times 10^{-18}$

The mass is 1.06×10^{-18} gram.

121. $3.2 \times 10^7 \cdot 127 = 3.2 \times 10^7 \cdot 1.27 \times 10^2$

$= 3.2 \cdot 1.27 \times 10^7 \cdot 10^2$

$= 4.064 \times 10^9$

Americans eat 4.064×10^9 chickens per year.

123. – 131. Answers will vary.

133. makes sense

135. makes sense

137. true

139. false; Changes to make the statement true will vary. A sample change is: $5^2 \cdot 5^{-2} = 2^5 \cdot 2^{-5}$.

141. false; Changes to make the statement true will vary. A sample change is:

$\dfrac{8 \times 10^{30}}{2 \times 10^{-5}} = 4 \times 10^{30-(-5)} = 4 \times 10^{35}$.

143. true

145. $b^A = MN, b^C = M, b^D = N$

$b^A = b^C b^D$

$A = C + D$

147. Answers will vary.

148. **a.** $\sqrt{16} \cdot \sqrt{4} = 4 \cdot 2 = 8$

b. $\sqrt{16 \cdot 4} = \sqrt{64} = 8$

c. $\sqrt{16} \cdot \sqrt{4} = \sqrt{16 \cdot 4}$

149. **a.** $\sqrt{300} \approx 17.32$

b. $10\sqrt{3} \approx 17.32$

c. $\sqrt{300} = 10\sqrt{3}$

150. **a.** $21x + 10x = 31x$

b. $21\sqrt{2} + 10\sqrt{2} = 31\sqrt{2}$

Section P.3

Check Point Exercises

1. a. $\sqrt{81} = 9$

    b. $-\sqrt{9} = -3$

    c. $\sqrt{\dfrac{1}{25}} = \dfrac{1}{5}$

    d. $\sqrt{36+64} = \sqrt{100} = 10$

    e. $\sqrt{36} + \sqrt{64} = 6 + 8 = 14$

2. a. $\sqrt{75} = \sqrt{25 \cdot 3} = \sqrt{25}\sqrt{3} = 5\sqrt{3}$

    b. $\sqrt{5x} \cdot \sqrt{10x} = \sqrt{5x \cdot 10x}$
        $= \sqrt{50x^2}$
        $= \sqrt{25 \cdot 2x^2}$
        $= \sqrt{25x^2} \cdot \sqrt{2}$
        $= 5x\sqrt{2}$

3. a. $\sqrt{\dfrac{25}{16}} = \dfrac{\sqrt{25}}{\sqrt{16}} = \dfrac{5}{4}$

    b. $\dfrac{\sqrt{150x^3}}{\sqrt{2x}} = \sqrt{\dfrac{150x^3}{2x}}$
        $= \sqrt{75x^2}$
        $= \sqrt{25x^2} \cdot \sqrt{3}$
        $= 5x\sqrt{3}$

4. a. $8\sqrt{13} + 9\sqrt{13} = (8+9)\sqrt{3}$
        $= 17\sqrt{13}$

    b. $\sqrt{17x} - 20\sqrt{17x}$
        $= 1\sqrt{17x} - 20\sqrt{17x}$
        $= (1-20)\sqrt{17x}$
        $= -19\sqrt{17x}$

5. a. $5\sqrt{27} + \sqrt{12}$
        $= 5\sqrt{9 \cdot 3} + \sqrt{4 \cdot 3}$
        $= 5 \cdot 3\sqrt{3} + 2\sqrt{3}$
        $= 15\sqrt{3} + 2\sqrt{3}$
        $= (15+2)\sqrt{3}$
        $= 17\sqrt{3}$

    b. $6\sqrt{18x} - 4\sqrt{8x}$
        $= 6\sqrt{9 \cdot 2x} - 4\sqrt{4 \cdot 2x}$
        $= 6 \cdot 3\sqrt{2x} - 4 \cdot 2\sqrt{2x}$
        $= 18\sqrt{2x} - 8\sqrt{2x}$
        $= (18-8)\sqrt{2x}$
        $= 10\sqrt{2x}$

6. a. If we multiply numerator and denominator by $\sqrt{3}$, the denominator becomes $\sqrt{3} \cdot \sqrt{3} = \sqrt{9} = 3$. Therefore, multiply by 1, choosing $\dfrac{\sqrt{3}}{\sqrt{3}}$ for 1.

        $\dfrac{5}{\sqrt{3}} = \dfrac{5}{\sqrt{3}} \cdot \dfrac{\sqrt{3}}{\sqrt{3}} = \dfrac{5\sqrt{3}}{\sqrt{9}} = \dfrac{5\sqrt{3}}{3}$

    b. The *smallest* number that will produce a perfect square in the denominator of $\dfrac{6}{\sqrt{12}}$ is $\sqrt{3}$ because $\sqrt{12} \cdot \sqrt{3} = \sqrt{36} = 6$. So multiply by 1, choosing $\dfrac{\sqrt{3}}{\sqrt{3}}$ for 1.

        $\dfrac{6}{\sqrt{12}} = \dfrac{6}{\sqrt{12}} \cdot \dfrac{\sqrt{3}}{\sqrt{3}} = \dfrac{6\sqrt{3}}{\sqrt{36}} = \dfrac{6\sqrt{3}}{6} = \sqrt{3}$

7. Multiply by $\dfrac{4-\sqrt{5}}{4-\sqrt{5}}$.

    $\dfrac{8}{4+\sqrt{5}} = \dfrac{8}{4+\sqrt{5}} \cdot \dfrac{4-\sqrt{5}}{4-\sqrt{5}}$
    $= \dfrac{8(4-\sqrt{5})}{4^2 - (\sqrt{5})^2}$
    $= \dfrac{8(4-\sqrt{5})}{16-5}$
    $= \dfrac{8(4-\sqrt{5})}{11}$ or $\dfrac{32-8\sqrt{5}}{11}$

8. a. $\sqrt[3]{40} = \sqrt[3]{8 \cdot 5} = \sqrt[3]{8} \cdot \sqrt[3]{5} = 2\sqrt[3]{5}$

b. $\sqrt[5]{8} \cdot \sqrt[5]{8} = \sqrt[5]{64} = \sqrt[5]{32} \cdot \sqrt[5]{2} = 2\sqrt[5]{2}$

c. $\sqrt[3]{\dfrac{125}{27}} = \dfrac{\sqrt[3]{125}}{\sqrt[3]{27}} = \dfrac{5}{3}$

9. $3\sqrt[3]{81} - 4\sqrt[3]{3}$
$= 3\sqrt[3]{27 \cdot 3} - 4\sqrt[3]{3}$
$= 3 \cdot 3\sqrt[3]{3} - 4\sqrt[3]{3}$
$= 9\sqrt[3]{3} - 4\sqrt[3]{3}$
$= (9 - 4)\sqrt[3]{3}$
$= 5\sqrt[3]{3}$

10. a. $25^{\frac{1}{2}} = \sqrt{25} = 5$

b. $8^{\frac{1}{3}} = \sqrt[3]{8} = 2$

c. $-81^{\frac{1}{4}} = -\sqrt[4]{81} = -3$

d. $(-8)^{\frac{1}{3}} = \sqrt[3]{-8} = -2$

e. $27^{-\frac{1}{3}} = \dfrac{1}{27^{\frac{1}{3}}} = \dfrac{1}{\sqrt[3]{27}} = \dfrac{1}{3}$

11. a. $27^{\frac{4}{3}} = \left(\sqrt[3]{27}\right)^4 = (3)^4 = 81$

b. $4^{\frac{3}{2}} = \left(\sqrt[2]{4}\right)^3 = (2)^3 = 8$

c. $32^{-\frac{2}{5}} = \dfrac{1}{32^{\frac{2}{5}}} = \dfrac{1}{\left(\sqrt[5]{32}\right)^2} = \dfrac{1}{2^2} = \dfrac{1}{4}$

12. a. $\left(2x^{4/3}\right)\left(5x^{8/3}\right)$
$= 2 \cdot 5x^{4/3} \cdot x^{8/3}$
$= 10x^{(4/3)+(8/3)}$
$= 10x^{12/3}$
$= 10x^4$

b. $\dfrac{20x^4}{5x^{3/2}} = \left(\dfrac{20}{5}\right)\left(\dfrac{x^4}{x^{3/2}}\right)$
$= 4x^{4-(3/2)}$
$= 4x^{(8/2)-(3/2)}$
$= 4x^{5/2}$

13. $\sqrt[6]{x^3} = x^{3/6} = x^{1/2} = \sqrt{x}$

Concept and Vocabulary Check P.3

1. principal

2. 8^2

3. $|a|$

4. $\sqrt{a} \cdot \sqrt{b}$

5. $\dfrac{\sqrt{a}}{\sqrt{b}}$

6. $18\sqrt{3}$

7. $5;\ 6\sqrt{3}$

8. $7 - \sqrt{3}$

9. $\sqrt{10} + \sqrt{2}$

10. index; radicand

11. $(-2)^5$

12. $a;\ |a|$

13. $\sqrt[n]{a}$

14. $2;\ 8$

Exercise Set P.3

1. $\sqrt{36} = \sqrt{6^2} = 6$

3. $-\sqrt{36} = -\sqrt{6^2} = -6$

5. $\sqrt{-36}$, The square root of a negative number is not real.

7. $\sqrt{25-16} = \sqrt{9} = 3$

9. $\sqrt{25} - \sqrt{16} = 5 - 4 = 1$

11. $\sqrt{(-13)^2} = \sqrt{169} = 13$

13. $\sqrt{50} = \sqrt{25 \cdot 2} = \sqrt{25}\sqrt{2} = 5\sqrt{2}$

15. $\sqrt{45x^2} = \sqrt{9x^2 \cdot 5}$
$\qquad = \sqrt{9x^2}\sqrt{5}$
$\qquad = \sqrt{9}\sqrt{x^2}\sqrt{5}$
$\qquad = 3|x|\sqrt{5}$

17. $\sqrt{2x} \cdot \sqrt{6x} = \sqrt{2x \cdot 6x}$
$\qquad = \sqrt{12x^2}$
$\qquad = \sqrt{4x^2} \cdot \sqrt{3}$
$\qquad = 2x\sqrt{3}$

19. $\sqrt{x^3} = \sqrt{x^2} \cdot \sqrt{x} = x\sqrt{x}$

21. $\sqrt{2x^2} \cdot \sqrt{6x} = \sqrt{2x^2 \cdot 6x}$
$\qquad = \sqrt{12x^3}$
$\qquad = \sqrt{4x^2} \cdot \sqrt{3x}$
$\qquad = 2x\sqrt{3x}$

23. $\sqrt{\dfrac{1}{81}} = \dfrac{\sqrt{1}}{\sqrt{81}} = \dfrac{1}{9}$

25. $\sqrt{\dfrac{49}{16}} = \dfrac{\sqrt{49}}{\sqrt{16}} = \dfrac{7}{4}$

27. $\dfrac{\sqrt{48x^3}}{\sqrt{3x}} = \sqrt{\dfrac{48x^3}{3x}} = \sqrt{16x^2} = 4x$

29. $\dfrac{\sqrt{150x^4}}{\sqrt{3x}} = \sqrt{\dfrac{150x^4}{3x}}$
$\qquad = \sqrt{50x^3}$
$\qquad = \sqrt{25x^2} \cdot \sqrt{2x}$
$\qquad = 5x\sqrt{2x}$

31. $\dfrac{\sqrt{200x^3}}{\sqrt{10x^{-1}}} = \sqrt{\dfrac{200x^3}{10x^{-1}}}$
$\qquad = \sqrt{20x^{3-(-1)}}$
$\qquad = \sqrt{20x^4}$
$\qquad = \sqrt{4 \cdot 5x^4}$
$\qquad = 2x^2\sqrt{5}$

33. $7\sqrt{3} + 6\sqrt{3} = (7+6)\sqrt{3} = 13\sqrt{3}$

35. $6\sqrt{17x} - 8\sqrt{17x} = (6-8)\sqrt{17x} = -2\sqrt{17x}$

37. $\sqrt{8} + 3\sqrt{2} = \sqrt{4 \cdot 2} + 3\sqrt{2}$
$\qquad = 2\sqrt{2} + 3\sqrt{2}$
$\qquad = (2+3)\sqrt{2}$
$\qquad = 5\sqrt{2}$

39. $\sqrt{50x} - \sqrt{8x} = \sqrt{25 \cdot 2x} - \sqrt{4 \cdot 2x}$
$\qquad = 5\sqrt{2x} - 2\sqrt{2x}$
$\qquad = (5-2)\sqrt{2x}$
$\qquad = 3\sqrt{2x}$

41. $3\sqrt{18} + 5\sqrt{50} = 3\sqrt{9 \cdot 2} + 5\sqrt{25 \cdot 2}$
$\qquad = 3 \cdot 3\sqrt{2} + 5 \cdot 5\sqrt{2}$
$\qquad = 9\sqrt{2} + 25\sqrt{2}$
$\qquad = (9+25)\sqrt{2}$
$\qquad = 34\sqrt{2}$

43. $3\sqrt{8} - \sqrt{32} + 3\sqrt{72} - \sqrt{75}$
$\qquad = 3\sqrt{4 \cdot 2} - \sqrt{16 \cdot 2} + 3\sqrt{36 \cdot 2} - \sqrt{25 \cdot 3}$
$\qquad = 3 \cdot 2\sqrt{2} - 4\sqrt{2} + 3 \cdot 6\sqrt{2} - 5\sqrt{3}$
$\qquad = 6\sqrt{2} - 4\sqrt{2} + 18\sqrt{2} - 5\sqrt{3}$
$\qquad = 20\sqrt{2} - 5\sqrt{3}$

45. $\dfrac{1}{\sqrt{7}} = \dfrac{1}{\sqrt{7}} \cdot \dfrac{\sqrt{7}}{\sqrt{7}} = \dfrac{\sqrt{7}}{7}$

47. $\dfrac{\sqrt{2}}{\sqrt{5}} = \dfrac{\sqrt{2}}{\sqrt{5}} \cdot \dfrac{\sqrt{5}}{\sqrt{5}} = \dfrac{\sqrt{10}}{5}$

49.
$$\frac{13}{3+\sqrt{11}} = \frac{13}{3+\sqrt{11}} \cdot \frac{3-\sqrt{11}}{3-\sqrt{11}}$$
$$= \frac{13(3-\sqrt{11})}{3^2 - (\sqrt{11})^2}$$
$$= \frac{13(3-\sqrt{11})}{9-11}$$
$$= \frac{13(3-\sqrt{11})}{-2}$$

51.
$$\frac{7}{\sqrt{5}-2} = \frac{7}{\sqrt{5}-2} \cdot \frac{\sqrt{5}+2}{\sqrt{5}+2}$$
$$= \frac{7(\sqrt{5}+2)}{(\sqrt{5})^2 - 2^2}$$
$$= \frac{7(\sqrt{5}+2)}{5-4}$$
$$= 7(\sqrt{5}+2)$$

53.
$$\frac{6}{\sqrt{5}+\sqrt{3}} = \frac{6}{\sqrt{5}+\sqrt{3}} \cdot \frac{\sqrt{5}-\sqrt{3}}{\sqrt{5}-\sqrt{3}}$$
$$= \frac{6(\sqrt{5}-\sqrt{3})}{(\sqrt{5})^2 - (\sqrt{3})^2}$$
$$= \frac{6(\sqrt{5}-\sqrt{3})}{5-3}$$
$$= \frac{6(\sqrt{5}-\sqrt{3})}{2}$$
$$= 3(\sqrt{5}-\sqrt{3})$$

55. $\sqrt[3]{125} = \sqrt[3]{5^3} = 5$

57. $\sqrt[3]{-8} = \sqrt[3]{(-2)^3} = -2$

59. $\sqrt[4]{-16}$ is not a real number.

61. $\sqrt[4]{(-3)^4} = |-3| = 3$

63. $\sqrt[5]{(-3)^5} = -3$

65. $\sqrt[5]{-\frac{1}{32}} = \sqrt[5]{-\frac{1}{2^5}} = -\frac{1}{2}$

67. $\sqrt[3]{32} = \sqrt[3]{8 \cdot 4} = \sqrt[3]{8}\sqrt[3]{4} = 2 \cdot \sqrt[3]{4}$

69. $\sqrt[3]{x^4} = \sqrt[3]{x^3 \cdot x} = x \cdot \sqrt[3]{x}$

71. $\sqrt[3]{9} \cdot \sqrt[3]{6} = \sqrt[3]{54} = \sqrt[3]{27 \cdot 2} = \sqrt[3]{27}\sqrt[3]{2} = 3\sqrt[3]{2}$

73. $\dfrac{\sqrt[5]{64x^6}}{\sqrt[5]{2x}} = \sqrt[5]{\dfrac{64x^6}{2x}} = \sqrt[5]{32x^5} = 2x$

75. $4\sqrt[5]{2} + 3\sqrt[5]{2} = 7\sqrt[5]{2}$

77.
$$5\sqrt[3]{16} + \sqrt[3]{54} = 5\sqrt[3]{8 \cdot 2} + \sqrt[3]{27 \cdot 2}$$
$$= 5 \cdot 2\sqrt[3]{2} + 3\sqrt[3]{2}$$
$$= 10\sqrt[3]{2} + 3\sqrt[3]{2}$$
$$= 13\sqrt[3]{2}$$

79.
$$\sqrt[3]{54xy^3} - y\sqrt[3]{128x}$$
$$= \sqrt[3]{27 \cdot 2xy^3} - y\sqrt[3]{64 \cdot 2x}$$
$$= 3y\sqrt[3]{2x} - 4y\sqrt[3]{2x}$$
$$= -y\sqrt[3]{2x}$$

81. $\sqrt{2} + \sqrt[3]{8} = \sqrt{2} + 2$

83. $36^{1/2} = \sqrt{36} = 6$

85. $8^{1/3} = \sqrt[3]{8} = 2$

87. $125^{2/3} = \left(\sqrt[3]{125}\right)^2 = 5^2 = 25$

89. $32^{-4/5} = \dfrac{1}{32^{4/5}} = \dfrac{1}{2^4} = \dfrac{1}{16}$

91.
$$\left(7x^{1/3}\right)\left(2x^{1/4}\right) = 7 \cdot 2x^{1/3} \cdot x^{1/4}$$
$$= 14 \cdot x^{1/3+1/4}$$
$$= 14x^{7/12}$$

93.
$$\frac{20x^{1/2}}{5x^{1/4}} = \left(\frac{20}{5}\right)\left(\frac{x^{1/2}}{x^{1/4}}\right)$$
$$= 4 \cdot x^{1/2-1/4}$$
$$= 4x^{1/4}$$

95. $\left(x^{2/3}\right)^3 = x^{2/3 \cdot 3} = x^2$

97. $(25x^4y^6)^{1/2} = 25^{1/2}x^{4 \cdot 1/2}y^{6 \cdot 1/2} = 5x^2|y|^3$

99. $\dfrac{\left(3y^{\frac{1}{4}}\right)^3}{y^{\frac{1}{12}}} = \dfrac{27y^{\frac{3}{4}}}{y^{\frac{1}{12}}} = 27y^{\frac{3}{4}-\frac{1}{12}}$

$= 27y^{\frac{8}{12}} = 27y^{\frac{2}{3}}$

101. $\sqrt[4]{5^2} = 5^{2/4} = 5^{1/2} = \sqrt{5}$

103. $\sqrt[3]{x^6} = x^{6/3} = x^2$

105. $\sqrt[6]{x^4} = \sqrt[6/2]{x^{4/2}} = \sqrt[3]{x^2}$

107. $\sqrt[9]{x^6 y^3} = x^{\frac{6}{9}} y^{\frac{3}{9}} = x^{\frac{2}{3}} y^{\frac{1}{3}} = \sqrt[3]{x^2 y}$

109. $\sqrt[3]{\sqrt[4]{16} + \sqrt{625}} = \sqrt[3]{2 + 25} = \sqrt[3]{27} = 3$

111. $\left(49x^{-2}y^4\right)^{-1/2}\left(xy^{1/2}\right)$

$= (49)^{-1/2}\left(x^{-2}\right)^{-1/2}\left(y^4\right)^{-1/2}\left(xy^{1/2}\right)$

$= \dfrac{1}{49^{1/2}} x^{(-2)(-1/2)} y^{(4)(-1/2)} \left(xy^{1/2}\right)$

$= \dfrac{1}{7} x^1 y^{-2} \cdot xy^{1/2} = \dfrac{1}{7} x^{1+1} y^{-2+(1/2)}$

$= \dfrac{1}{7} x^2 y^{-3/2} = \dfrac{x^2}{7y^{3/2}}$

113. $\left(\dfrac{x^{-5/4} y^{1/3}}{x^{-3/4}}\right)^{-6} = \left(x^{(-5/4)-(-3/4)} y^{1/3}\right)^{-6}$

$= \left(x^{-2/4} y^{1/3}\right)^{-6} = x^{(-2/4)(-6)} y^{(1/3)(-6)}$

$= x^3 y^{-2} = \dfrac{x^3}{y^2}$

115. The message is "Paige Fox is bad at math."

117. $\dfrac{2}{\sqrt{5}-1} \cdot \dfrac{\sqrt{5}+1}{\sqrt{5}+1} = \dfrac{2(\sqrt{5}+1)}{5-1}$

$= \dfrac{2(\sqrt{5}+1)}{4}$

$= \dfrac{\sqrt{5}+1}{2}$

≈ 1.62

About 1.62 to 1.

119. Perimeter:
$P = 2l + 2w$

$= 2\cdot\sqrt{125} + 2\cdot 2\sqrt{20}$

$= 2\cdot\sqrt{25\cdot 5} + 4\sqrt{4\cdot 5}$

$= 2\cdot 5\sqrt{5} + 4\cdot 2\sqrt{5}$

$= 10\sqrt{5} + 8\sqrt{5}$

$= 18\sqrt{5} \text{ feet}$

Area:
$A = lw$

$= \sqrt{125} \cdot 2\sqrt{20}$

$= 2\sqrt{125\cdot 20}$

$= 2\sqrt{2500}$

$= 2\cdot 50$

$= 100 \text{ square feet}$

121. – 127. Answers will vary.

129. does not make sense; Explanations will vary. Sample explanation: The denominator is rationalized correctly.

131. does not make sense; Explanations will vary. Sample explanation: $2\sqrt{20} + 4\sqrt{75}$ simplifies to $4\sqrt{5} + 20\sqrt{3}$ and thus the radical terms are not common.

133. false; Changes to make the statement true will vary. A sample change is: $7^{\frac{1}{2}} \cdot 7^{\frac{1}{2}} = 7^1 = 7$.

135. false; Changes to make the statement true will vary. The cube root of –8 is the real number –2.

137. $\left(5+\sqrt{\boxed{3}}\right)\left(5-\sqrt{\boxed{3}}\right) = 22$

$25 - \boxed{3} = 22$

$\boxed{3} = 3$

139. $\sqrt{13+\sqrt{2}+\dfrac{7}{3+\sqrt{2}}}$

$= \sqrt{13+\sqrt{2}+\dfrac{7}{3+\sqrt{2}}\cdot\dfrac{3-\sqrt{2}}{3-\sqrt{2}}}$

$= \sqrt{13+\sqrt{2}+\dfrac{21-7\sqrt{2}}{9-2}}$

$= \sqrt{13+\sqrt{2}+\dfrac{21-7\sqrt{2}}{7}}$

$= \sqrt{13+\sqrt{2}+3-\sqrt{2}}$

$= \sqrt{16}$

$= 4$

141. a.

$$\frac{ab}{a^2+ab+b^2}+\left(\frac{ac-ad-bc+bd}{ac-ad+bc-bd}\div\frac{a^3-b^3}{a^3+b^3}\right)=\frac{ab}{a^2+ab+b^2}+\left(\frac{a(c-d)-b(c-d)}{a(c-d)+b(c-d)}\cdot\frac{a^3+b^3}{a^3-b^3}\right)$$

$$=\frac{ab}{a^2+ab+b^2}+\left(\frac{(c-d)(a-b)}{(c-d)(a+b)}\cdot\frac{(a+b)(a^2-ab+b^2)}{(a-b)(a^2+ab+b^2)}\right)=\frac{ab}{a^2+ab+b^2}+\frac{a^2-ab+b^2}{a^2+ab+b^2}$$

$$=\frac{ab+a^2-ab+b^2}{a^2+ab+b^2}=\frac{a^2+b^2}{a^2+ab+b^2}$$

Her son is 8 years old.

b. Son's portion:

$$\frac{8^{-\frac{4}{3}}+2^{-2}}{16^{-\frac{3}{4}}+2^{-1}}=\frac{\dfrac{1}{\left(\sqrt[3]{8}\right)^4}+\dfrac{1}{2^2}}{\dfrac{1}{\left(\sqrt[4]{16}\right)^3}+\dfrac{1}{2}}$$

$$=\frac{\dfrac{1}{2^4}+\dfrac{1}{4}}{\dfrac{1}{2^3}+\dfrac{1}{2}}$$

$$=\frac{\dfrac{1}{16}+\dfrac{1}{4}}{\dfrac{1}{8}+\dfrac{1}{2}}$$

$$=\frac{\dfrac{5}{16}}{\dfrac{5}{8}}$$

$$=\frac{8}{16}$$

$$=\frac{1}{2}$$

Mom's portion:

$$\frac{1}{2}\left(1-\frac{1}{2}\right)=\frac{1}{2}\left(\frac{1}{2}\right)=\frac{1}{4}$$

142. $(2x^3y^2)(5x^4y^7)=10x^7y^9$

143. $2x^4(8x^4+3x)=2x^4(8x^4)+2x^4(3x)$
$$=16x^8+6x^5$$

144. $2x(x^2+4x+5)+3(x^2+4x+5)$
$$=2x^3+8x^2+10x+3x^2+12x+15$$
$$=2x^3+8x^2+3x^2+10x+12x+15$$
$$=2x^3+11x^2+22x+15$$

Section P.4

Check Point Exercises

1. **a.** $(-17x^3 + 4x^2 - 11x - 5) + (16x^3 - 3x^2 + 3x - 15)$

$= (-17x^3 + 16x^3) + (4x^2 - 3x^2) + (-11x + 3x) + (-5 - 15)$

$= -x^3 + x^2 - 8x - 20$

  b. $(13x^2 - 9x^2 - 7x + 1) - (-7x^3 + 2x^2 - 5x + 9)$

$= (13x^3 - 9x^2 - 7x + 1) + (7x^3 - 2x^2 + 5x - 9)$

$= (13x^3 + 7x^3) + (-9x^2 - 2x^2) + (-7x + 5x) + (1 - 9)$

$= 20x^3 - 11x^2 - 2x - 8$

2. $(5x - 2)(3x^2 - 5x + 4)$

$= 5x(3x^2 - 5x + 4) - 2(3x^2 - 5x + 4)$

$= 5x \cdot 3x^2 - 5x \cdot 5x + 5x \cdot 4 - 2 \cdot 3x^2 + 2 \cdot 5x - 2 \cdot 4$

$= 15x^3 - 25x^2 + 20x - 6x^2 + 10x - 8$

$= 15x^3 - 31x^2 + 30x - 8$

3. $(7x - 5)(4x - 3) = 7x \cdot 4x + 7x(-3) + (-5)4x + (-5)(-3)$

$= 28x^2 - 21x - 20x + 15$

$= 28x^2 - 41x + 15$

4. **a.** Use the special-product formula shown.

$(A + B)(A - B) = A^2 - B^2$

$(7x + 8)(7x - 8) = (7x)^2 - (8)^2$

$= 49x^2 - 64$

  b. Use the special-product formula shown.

$(A + B)(A - B) = A^2 - B^2$

$(2y^3 - 5)(2y^3 + 5) = (2y^3 + 5)(2y^3 - 5)$

$= (2y^3)^2 - (5)^2$

$= 4y^6 - 25$

5. **a.** Use the special-product formula shown.

$(A + B)^2 = A^2 + 2AB + B^2$

$(x + 10)^2 = x^2 + 2(x)(10) + 10^2$

$= x^2 + 20x + 100$

  b. Use the special-product formula shown.

$(A + B)^2 = A^2 + 2AB + B^2$

$(5x + 4)^2 = (5x)^2 + 2(5x)(4) + 4^2$

$= 25x^2 + 40x + 16$

6. **a.** Use the special-product formula shown.

$$(A-B)^2 = A^2 - 2AB + B^2$$

$$(x-9)^2 = x^2 - 2(x)(9) + 9^2$$

$$= x^2 - 18x + 81$$

b. Use the special-product formula shown.

$$(A-B)^2 = A^2 - 2AB + B^2$$

$$(7x-3)^2 = (7x)^2 - 2(7x)(3) + 3^2$$

$$= 49x^2 - 42x + 9$$

7. $(x^3 - 4x^2y + 5xy^2 - y^3) - (x^3 - 6x^2y + y^3)$

$$= (x^3 - 4x^2y + 5xy^2 - y^3) + (-x^3 + 6x^2y - y^3)$$

$$= (x^3 - x^3) + (-4x^2y + 6x^2y) + (5xy^2) + (-y^3 - y^3)$$

$$= 2x^2y + 5xy^2 - 2y^3$$

8. **a.** $(7x - 6y)(3x - y) = (7x)(3x) + (7x)(-y) + (-6y)(3x) + (-6y)(-y)$

$$= 21x^2 - 7xy - 18xy + 6y^2$$

$$= 21x^2 - 25xy + 6y^2$$

b. $(2x + 4y)^2 = (2x)^2 + 2(2x)(4y) + (4y)^2$

$$= 4x^2 + 16xy + 16y^2$$

Concept and Vocabulary Check P.4

1. whole

2. standard

3. monomial

4. binomial

5. trinomial

6. n

7. like;

8. distributive; $4x^3 - 8x^2 + 6$; $7x^3$

9. $5x$; 3; like

10. $3x^2$; $5x$; $21x$; 35

11. $A^2 - B^2$; minus

12. $A^2 + 2AB + B^2$; squared; product of the terms; squared

13. $A^2 - 2AB + B^2$; minus; product of the terms; plus

14. $n + m$

Exercise Set P.4

1. yes; $2x + 3x^2 - 5 = 3x^2 + 2x - 5$

3. no; The form of a polynomial involves addition and subtraction, not division.

5. $3x^2$ has degree 2
$-5x$ has degree 1
4 has degree 0
$3x^2 - 5x + 4$ has degree 2.

7. x^2 has degree 2
$-4x^3$ has degree 3
$9x$ has degree 1
$-12x^4$ has degree 4
63 has degree 0
$x^2 - 4x^3 + 9x - 12x^4 + 63$ has degree 4.

9. $(-6x^3 + 5x^2 - 8x + 9) + (17x^3 + 2x^2 - 4x - 13) = (-6x^3 + 17x^3) + (5x^2 + 2x^2) + (-8x - 4x) + (9 - 13)$
$$= 11x^3 + 7x^2 - 12x - 4$$
The degree is 3.

11. $(17x^3 - 5x^2 + 4x - 3) - (5x^3 - 9x^2 - 8x + 11) = (17x^3 - 5x^2 + 4x - 3) + (-5x^3 + 9x^2 + 8x - 11)$
$$= (17x^3 - 5x^3) + (-5x^2 + 9x^2) + (4x + 8x) + (-3 - 11)$$
$$= 12x^3 + 4x^2 + 12x - 14$$
The degree is 3.

13. $(5x^2 - 7x - 8) + (2x^2 - 3x + 7) - (x^2 - 4x - 3) = (5x^2 - 7x - 8) + (2x^2 - 3x + 7) + (-x^2 + 4x + 3)$
$$= (5x^2 + 2x^2 - x^2) + (-7x - 3x + 4x) + (-8 + 7 + 3)$$
$$= 6x^2 - 6x + 2$$
The degree is 2.

15. $(x + 1)(x^2 - x + 1) = x(x^2) - x \cdot x + x \cdot 1 + 1(x^2) - 1 \cdot x + 1 \cdot 1$
$$= x^3 - x^2 + x + x^2 - x + 1$$
$$= x^3 + 1$$

17. $(2x - 3)(x^2 - 3x + 5) = (2x)(x^2) + (2x)(-3x) + (2x)(5) + (-3)(x^2) + (-3)(-3x) + (-3)(5)$
$$= 2x^3 - 6x^2 + 10x - 3x^2 + 9x - 15$$
$$= 2x^3 - 9x^2 + 19x - 15$$

19. $(x + 7)(x + 3) = x^2 + 3x + 7x + 21 = x^2 + 10x + 21$

21. $(x-5)(x+3) = x^2 + 3x - 5x - 15 = x^2 - 2x - 15$

23. $(3x+5)(2x+1) = (3x)(2x) + 3x(1) + 5(2x) + 5 = 6x^2 + 3x + 10x + 5 = 6x^2 + 13x + 5$

25. $(2x-3)(5x+3) = (2x)(5x) + (2x)(3) + (-3)(5x) + (-3)(3) = 10x^2 + 6x - 15x - 9 = 10x^2 - 9x - 9$

27. $(5x^2 - 4)(3x^2 - 7) = (5x^2)(3x^2) + (5x^2)(-7) + (-4)(3x^2) + (-4)(-7) = 15x^4 - 35x^2 - 12x^2 + 28 = 15x^4 - 47x^2 + 28$

29. $\left(8x^3 + 3\right)\left(x^2 - 5\right) = \left(8x^3\right)\left(x^2\right) + \left(8x^3\right)(-5) + (3)\left(x^2\right) + (3)(-5) = 8x^5 - 40x^3 + 3x^2 - 15$

31. $(x+3)(x-3) = x^2 - 3^2 = x^2 - 9$

33. $(3x+2)(3x-2) = (3x)^2 - 2^2 = 9x^2 - 4$

35. $(5-7x)(5+7x) = 5^2 - (7x)^2 = 25 - 49x^2$

37. $(4x^2 + 5x)(4x^2 - 5x) = (4x^2)^2 - (5x)^2 = 16x^4 - 25x^2$

39. $\left(1 - y^5\right)\left(1 + y^5\right) = (1)^2 - \left(y^5\right)^2 = 1 - y^{10}$

41. $(x+2)^2 = x^2 + 2 \cdot x \cdot 2 + 2^2 = x^2 + 4x + 4$

43. $(2x+3)^2 = (2x)^2 + 2(2x)(3) + 3^2 = 4x^2 + 12x + 9$

45. $(x-3)^2 = x^2 - 2 \cdot x \cdot 3 + 3^2 = x^2 - 6x + 9$

47. $(4x^2 - 1)^2 = (4x^2)^2 - 2(4x^2)(1) + 1^2 = 16x^4 - 8x^2 + 1$

49. $(7-2x)^2 = 7^2 - 2(7)(2x) + (2x)^2 = 49 - 28x + 4x^2 = 4x^2 - 28x + 49$

51. $(x+1)^3 = x^3 + 3 \cdot x^2 \cdot 1 + 3x \cdot 1^2 + 1^3 = x^3 + 3x^2 + 3x + 1$

53. $(2x+3)^3 = (2x)^3 + 3 \cdot (2x)^2 \cdot 3 + 3(2x) \cdot 3^2 + 3^3 = 8x^3 + 36x^2 + 54x + 27$

55. $(x-3)^3 = x^3 - 3 \cdot x^3 \cdot 3 + 3 \cdot x \cdot 3^2 - 3^3 = x^3 - 9x^2 + 27x - 27$

57. $(3x-4)^3 = (3x)^3 - 3(3x)^2 \cdot 4 + 3(3x) \cdot 4^2 - 4^3 = 27x^3 - 108x^2 + 144x - 64$

59. $(5x^2y - 3xy) + (2x^2y - xy) = (5x^2y + 2x^2y) + (-3xy - xy)$
$$= (5+2)x^2y + (-3-1)xy$$
$$= 7x^2y - 4xy \text{ is of degree 3.}$$

61. $(4x^2y + 8xy + 11) + (-2x^2y + 5xy + 2) = (4x^2y - 2x^2y) + (8xy + 5xy) + (11+2)$
$$= (4-2)x^2y + (8+5)xy + 13$$
$$= 2x^2y + 13xy + 13 \text{ is of degree 3.}$$

63. $(x^3 + 7xy - 5y^2) - (6x^3 - xy + 4y^2) = (x^3 + 7xy - 5y^2)$

$$= (x^3 - 6x^3) + (7xy + xy) + (-5y^2 - 4y^2)$$
$$= (1 - 6)x^3 + (7 + 1)xy + (-5 - 4)y^2$$
$$= -5x^3 + 8xy - 9y^2 \text{ is of degree 3.}$$

65. $(3x^4y^2 + 5x^3y - 3y) - (2x^4y^2 - 3x^3y - 4y + 6x) = (3x^4y^2 + 5x^3y - 3y) + (-2x^4y^2 + 3x^3y + 4y - 6x)$

$$= (3x^4y^2 - 2x^4y^2) + (5x^3y + 3x^3y) + (-3y + 4y) - 6x$$
$$= (3 - 2)x^4y^2 + (5 + 3)x^3y + (-3 + 4)y - 6x$$
$$= x^4y^2 + 8x^3y + y - 6x \text{ is of degree 6.}$$

67. $(x + 5y)(7x + 3y) = x(7x) + x(3y) + (5y)(7x) + (5y)(3y)$

$$= 7x^2 + 3xy + 35xy + 15y^2$$
$$= 7x^2 + 38xy + 15y^2$$

69. $(x - 3y)(2x + 7y) = x(2x) + x(7y) + (-3y)(2x) + (-3y)(7y)$

$$= 2x^2 + 7xy - 6xy - 21y^2$$
$$= 2x^2 + xy - 21y^2$$

71. $(3xy - 1)(5xy + 2) = (3xy)(5xy) + (3xy)(2) + (-1)(5xy) + (-1)(2)$

$$= 15x^2y^2 + 6xy - 5xy - 2$$
$$= 15x^2y^2 + xy - 2$$

73. $(7x + 5y)^2 = (7x)^2 + 2(7x)(5y) + (5y)^2 = 49x^2 + 70xy + 25y^2$

75. $(x^2y^2 - 3)^2 = (x^2y^2)^2 - 2(x^2y^2)(3) + 3^2 = x^4y^4 - 6x^2y^2 + 9$

77. $(x - y)(x^2 + xy + y^2) = x(x^2) + x(xy) + x(y^2) + (-y)(x^2) + (-y)(xy) + (-y)(y^2)$

$$= x^3 + x^2y + xy^2 - x^2y - xy^2 - y^3$$
$$= x^3 - y^3$$

79. $(3x + 5y)(3x - 5y) = (3x)^2 - (5y)^2 = 9x^2 - 25y^2$

81. $(7xy^2 - 10y)(7xy^2 + 10y) = (7xy^2)^2 - (10y)^2 = 49x^2y^4 - 100y^2$

83. $(3x + 4y)^2 - (3x - 4y)^2 = \left[(3x)^2 + 2(3x)(4y) + (4y)^2\right] - \left[(3x)^2 - 2(3x)(4y) + (4y)^2\right]$

$$= (9x^2 + 24xy + 16y^2) - (9x^2 - 24xy + 16y^2)$$
$$= 9x^2 + 24xy + 16y^2 - 9x^2 + 24xy - 16y^2$$
$$= 48xy$$

85. $(5x-7)(3x-2)-(4x-5)(6x-1)$

$= \left[15x^2 - 10x - 21x + 14\right] - \left[24x^2 - 4x - 30x + 5\right]$

$= \left(15x^2 - 31x + 14\right) - \left(24x^2 - 34x + 5\right)$

$= 15x^2 - 31x + 14 - 24x^2 + 34x - 5$

$= -9x^2 + 3x + 9$

87. $(2x+5)(2x-5)\left(4x^2+25\right)$

$= \left[(2x)^2 - 5^2\right]\left(4x^2 + 25\right)$

$= \left(4x^2 - 25\right)\left(4x^2 + 25\right)$

$= \left(4x^2\right)^2 - (25)^2$

$= 16x^4 - 625$

89. $\dfrac{(2x-7)^5}{(2x-7)^3} = (2x-7)^{5-3}$

$= (2x-7)^2$

$= (2x)^2 - 2(2x)(7) + (7)^2$

$= 4x^2 - 28x + 49$

91. **a.** $M = 312x^2 - 2615x + 16,615$

$M = 312(16)^2 - 2615(16) + 16,615 = 54,647$

The model estimates the median annual income for a man with 16 years of education to be $54,647.
The model overestimates the actual value of $54,091 shown in the bar graph by $556.

b. $M - W = (0.6x^3 + 285x^2 - 2256x + 15,112) - (-1.2x^3 + 367x^2 - 4900x + 26,561)$

$M - W = 0.6x^3 + 285x^2 - 2256x + 15,112 + 1.2x^3 - 367x^2 + 4900x - 26,561$

$M - W = 0.6x^3 + 1.2x^3 + 285x^2 - 367x^2 - 2256x + 4900x + 15,112 - 26,561$

$M - W = 1.8x^3 - 82x^2 + 2644x - 11,449$

c. $M - W = 1.8x^3 - 82x^2 + 2644x - 11,449$

$M - W = 1.8(14)^3 - 82(14)^2 + 2644(14) - 11,449 = 14,434$

The difference in the median income between men and women with 14 years experience is $14,434.

d. $42,163 - 27,027 = 15,136$

The actual difference displayed in the graph in the median income between men and women with 14 years experience is $15,136.
The model underestimates this difference by $15,136 - \$14,434 = \$702$.

93. $x(8-2x)(10-2x) = x\left(80 - 36x + 4x^2\right)$

$= 80x - 36x^2 + 4x^3$

$= 4x^3 - 36x^2 + 80x$

95. $(x+9)(x+3)-(x+5)(x+1)$

$\quad = x^2+12x+27-\left(x^2+6x+5\right)$

$\quad = x^2+12x+27-x^2-6x-5$

$\quad = 6x+22$

97. – 101. Answers will vary.

103. makes sense

105. makes sense

107. false; Changes to make the statement true will vary. A sample change is: $(3x^3+2)(3x^3-2)=9x^6-4$

109. false; Changes to make the statement true will vary. A sample change is: $(x+1)^2=x^2+2x+1$

111. $[(7x+5)+4y][(7x+5)-4y]=(7x+5)^2-4y^2$

$\qquad\qquad\qquad\qquad\qquad = (7x)^2+2(7x)(5)+5^2-16y^2$

$\qquad\qquad\qquad\qquad\qquad = 49x^2+70x+25-16y^2$

113. $(x^n+2)(x^n-2)-(x^n-3)^2$

$\quad (x^n+2)(x^n-2)-(x^n-3)^2$

$\quad = (x^{2n}-4)-(x^{2n}-6x^n+9)$

$\quad = x^{2n}-4-x^{2n}+6x^n-9$

$\quad = 6x^n-13$

115. $(x+3)(x+\boxed{4})=x^2+7x+12$

116. $(x-\boxed{2})(x-12)=x^2-14x+24$

117. $(4x+1)(2x-\boxed{3})=8x^2-10x-3$

Mid-Chapter P Check Point

1. $(3x+5)(4x-7)=(3x)(4x)+(3x)(-7)+(5)(4x)+(5)(-7)$

$\qquad\qquad\qquad\quad = 12x^2-21x+20x-35$

$\qquad\qquad\qquad\quad = 12x^2-x-35$

2. $(3x+5)-(4x-7)=3x+5-4x+7$

$\qquad\qquad\qquad\quad = 3x-4x+5+7$

$\qquad\qquad\qquad\quad = -x+12$

3. $\sqrt{6}+9\sqrt{6}=10\sqrt{6}$

4. $3\sqrt{12}-\sqrt{27}=3\cdot2\sqrt{3}-3\sqrt{3}=6\sqrt{3}-3\sqrt{3}=3\sqrt{3}$

5. $7x+3[9-(2x-6)]=7x+3[9-2x+6]=7x+3[15-2x]=7x+45-6x=x+45$

Copyright © 2014 Pearson Education, Inc.

6. $(8x-3)^2 = (8x)^2 - 2(8x)(3) + (3)^2 = 64x^2 - 48x + 9$

7. $\left(x^{\frac{1}{3}} y^{-\frac{1}{2}}\right)^6 = x^{\frac{1}{3} \cdot 6} y^{-\frac{1}{2} \cdot 6} = x^2 y^{-3} = \dfrac{x^2}{y^3}$

8. $\left(\dfrac{2}{7}\right)^0 - 32^{-\frac{2}{5}} = 1 - \dfrac{1}{\left(\sqrt[5]{32}\right)^2} = 1 - \dfrac{1}{(2)^2} = 1 - \dfrac{1}{4} = \dfrac{3}{4}$

9. $(2x-5) - (x^2 - 3x + 1) = 2x - 5 - x^2 + 3x - 1 = -x^2 + 5x - 6$

10. $\begin{aligned}(2x-5)(x^2 - 3x + 1) &= 2x(x^2 - 3x + 1) - 5(x^2 - 3x + 1)\\ &= 2x(x^2 - 3x + 1) - 5(x^2 - 3x + 1)\\ &= 2x^3 - 6x^2 + 2x - 5x^2 + 15x - 5\\ &= 2x^3 - 6x^2 - 5x^2 + 2x + 15x - 5\\ &= 2x^3 - 11x^2 + 17x - 5\end{aligned}$

11. $x^3 + x^3 - x^3 \cdot x^3 = 2x^3 - x^6 = -x^6 + 2x^3$

12. $\begin{aligned}(9a - 10b)(2a + b) &= (9a)(2a) + (9a)(b) + (-10b)(2a) + (-10b)(b)\\ &= (9a)(2a) + (9a)(b) + (-10b)(2a) + (-10b)(b)\\ &= 18a^2 + 9ab - 20ab - 10b^2\\ &= 18a^2 - 11ab - 10b^2\end{aligned}$

13. $\{a,c,d,e\} \cup \{c,d,f,h\} = \{a,c,d,e,f,h\}$

14. $\{a,c,d,e\} \cap \{c,d,f,h\} = \{c,d\}$

15. $\begin{aligned}\left(3x^2 y^3 - xy + 4y^2\right) - \left(-2x^2 y^3 - 3xy + 5y^2\right) &= 3x^2 y^3 - xy + 4y^2 + 2x^2 y^3 + 3xy - 5y^2\\ &= 3x^2 y^3 - xy + 4y^2 + 2x^2 y^3 + 3xy - 5y^2\\ &= 3x^2 y^3 + 2x^2 y^3 - xy + 3xy + 4y^2 - 5y^2\\ &= 5x^2 y^3 + 2xy - y^2\end{aligned}$

16. $\dfrac{24x^2 y^{13}}{-2x^5 y^{-2}} = -12x^{2-5} y^{13-(-2)} = -12x^{-3} y^{15} = -\dfrac{12y^{15}}{x^3}$

17. $\left(\dfrac{1}{3} x^{-5} y^4\right)\left(18x^{-2} y^{-1}\right) = 6x^{-5-2} y^{4-1} = \dfrac{6y^3}{x^7}$

18. $\sqrt[12]{x^4} = x^{\frac{4}{12}} = \left|x^{\frac{1}{3}}\right| = \left|\sqrt[3]{x}\right|$

19. $\dfrac{24 \times 10^3}{2 \times 10^6} = \dfrac{24}{2} \cdot \dfrac{10^3}{10^6} = 12 \times 10^{-3} = \left(1.2 \times 10^1\right) \times 10^{-3} = 1.2 \times \left(10^1 \times 10^{-3}\right) = 1.2 \times 10^{-2}$

20. $\dfrac{\sqrt[3]{32}}{\sqrt[3]{2}} = \sqrt[3]{\dfrac{32}{2}} = \sqrt[3]{16} = \sqrt[3]{2^4} = 2\sqrt[3]{2}$

21. $(x^3 + 2)(x^3 - 2) = x^6 - 4$

22. $(x^2 + 2)^2 = (x^2)^2 + 2(x^2)(2) + (2)^2 = x^4 + 4x^2 + 4$

23. $\sqrt{50} \cdot \sqrt{6} = 5\sqrt{2} \cdot \sqrt{6} = 5\sqrt{2 \cdot 6} = 5\sqrt{12} = 5 \cdot 2\sqrt{3} = 10\sqrt{3}$

24. $\dfrac{11}{7 - \sqrt{3}} = \dfrac{11}{7 - \sqrt{3}} \cdot \dfrac{7 + \sqrt{3}}{7 + \sqrt{3}} = \dfrac{77 + 11\sqrt{3}}{49 - 3} = \dfrac{77 + 11\sqrt{3}}{46}$

25. $\dfrac{11}{\sqrt{3}} = \dfrac{11}{\sqrt{3}} \cdot \dfrac{\sqrt{3}}{\sqrt{3}} = \dfrac{11\sqrt{3}}{3}$

26. $\left\{ -11, \ -\dfrac{3}{7}, \ 0, \ 0.45, \ \sqrt{25} \right\}$

27. Since $2 - \sqrt{13} < 0$ then $\left| 2 - \sqrt{13} \right| = \sqrt{13} - 2$

28. Since $x < 0$ then $|x| = -x$. Thus $x^2 |x| = -x^2 x = -x^3$

29. $4.6 \cdot 3.0 \times 10^8 = 4.6 \times 10^8 = 13.8 \times 10^8 = 1.38 \times 10^9$
The U.S. produces 1.38×10^9 pounds of garbage per day.

30. $\dfrac{3 \times 10^{10}}{7.5 \times 10^9} = \dfrac{3}{7.5} \cdot \dfrac{10^{10}}{10^9} = 0.4 \times 10 = 4$
A human brain has 4 times as many neurons as a gorilla brain.

31. **a.**   Model 1:
$$P = -86x + 890$$
$$P = -86(0) + 890$$
$$P = 890$$
Model 2:
$$P = 18x^2 - 175x + 950$$
$$P = 18(0)^2 - 175(0) + 950$$
$$P = 950$$

Model 2 best describes the data in 2007.

b. $P = 18x^2 - 175x + 950$
$$P = 18(5)^2 - 175(5) + 950$$
$$P = 525$$
Model 2 overestimates the average price in 2012 by $5.

Section P.5

Check Point Exercises

1. a. $10x^3 - 4x^2$
    $$= 2x^2(5x) - 2x^2(2)$$
    $$= 2x^2(5x - 2)$$

    b. $2x(x-7) + 3(x-7)$
    $$= (x-7)(2x+3)$$

2. $x^3 + 5x^2 - 2x - 10$
    $$= (x^3 + 5x^2) - (2x + 10)$$
    $$= x^2(x+5) - 2(x+5)$$
    $$= (x+5)(x^2 - 2)$$

3. Find two numbers whose product is 40 and whose sum is 13. The required integers are 8 and 5. Thus,
    $x^2 + 13x + 40 = (x+8)(x+5)$ or $(x+5)(x+8)$.

4. Find two numbers whose product is –14 and whose sum is –5. The required integers are –7 and 2. Thus,
    $x^2 - 5x - 14 = (x-7)(x+2)$ or $(x+2)(x-7)$.

5. Find two First terms whose product is $6x^2$.
    $$6x^2 + 19x - 7 = (6x \quad)(x \quad)$$
    $$6x^2 + 19x - 7 = (3x \quad)(2x \quad)$$

    Find two Last terms whose product is –7.
    The possible factors are 1(–7) and –1(7).

    Try various combinations of these factors to find the factorization in which the sum of the Outside and Inside products is 19x.

Possible Factors of $6x^2 + 19x - 7$	Sum of Outside and Inside Products (Should Equal 19x)
$(6x+1)(x-7)$	$-42x + x = -41x$
$(6x-7)(x+1)$	$6x - 7x = -x$
$(6x-1)(x+7)$	$42x - x = 41x$
$(6x+7)(x-1)$	$-6x + 7x = x$
$(3x+1)(2x-7)$	$-21x + 2x = -19x$
$(3x-7)(2x+1)$	$3x - 14x = -11x$
$(3x-1)(2x+7)$	$21x - 2x = 19x$
$(3x+7)(2x-1)$	$-3x + 14x = 11x$

    Thus, $6x^2 + 19x - 7 = (3x-1)(2x+7)$ or $(2x+7)(3x-1)$.

6. Find two First terms whose product is $3x^2$.

$$3x^2 - 13xy + 4y^2 = (3x \quad)(x \quad)$$

Find two Last terms whose product is $4y^2$.
The possible factors are $(2y)(2y)$, $(-2y)(-2y)$, $(4y)(y)$, and $(-4y)(-y)$.

Try various combinations of these factors to find the factorization in which the sum of the Outside and Inside products is $-13xy$.

$$3x^2 - 13xy + y^2 = (3x - y)(x - 4y) \text{ or } (x - 4y)(3x - y).$$

7. Express each term as the square of some monomial. Then use the formula for factoring $A^2 - B^2$.

    a. $\quad x^2 - 81 = x^2 - 9^2 = (x+9)(x-9)$

    b. $\quad 36x^2 - 25 = (6x)^2 - 5^2 = (6x+5)(6x-5)$

8. Express $81x^4 - 16$ as the difference of two squares and use the formula for factoring $A^2 - B^2$.
$$81x^4 - 16 = (9x^2)^2 - 4^2 = (9x^2 + 4)(9x^2 - 4)$$

The factor $9x^2 - 4$ is the difference of two squares and can be factored. Express $9x^2 - 4$ as the difference of two squares and again use the formula for factoring $A^2 - B^2$.
$$(9x^2 + 4)(9x^2 - 4) = (9x^2 + 4)\left[(3x)^2 - 2^2\right] = (9x^2 + 4)(3x + 2)(3x - 2)$$

Thus, factored completely,
$$81x^4 - 16 = (9x^2 + 4)(3x + 2)(3x - 2).$$

9.   **a.** $\quad x^2 + 14x + 49 = x^2 + 2 \cdot x \cdot 7 + 7^2 = (x+7)^2$

    b. Since $16x^2 = (4x)^2$ and $49 = 7^2$, check to see if the middle term can be expressed as twice the product of $4x$ and 7. Since $2 \cdot 4x \cdot 7 = 56x$, $16x^2 - 56x + 49$ is a perfect square trinomial. Thus, $16x^2 - 56x + 49 = (4x)^2 - 2 \cdot 4x \cdot 7 + 7^2$
$$= (4x - 7)^2$$

10.   **a.** $\quad x^3 + 1 = x^3 + 1^3$
$$= (x+1)(x^2 - x \cdot 1 + 1^2)$$
$$= (x+1)(x^2 - x + 1)$$

    b. $\quad 125x^3 - 8 = (5x)^3 - 2^3$
$$= (5x - 2)\left[(5x)^2 + (5x)(2) + 2^2\right]$$
$$= (5x - 2)(25x^2 + 10x + 4)$$

11. Factor out the greatest common factor.
$$3x^3 - 30x^2 + 75x = 3x\left(x^2 - 10x + 25\right)$$
Factor the perfect square trinomial.
$$3x\left(x^2 - 10x + 25\right) = 3x\left(x - 5\right)^2$$

12. Reorder to write as a difference of squares.

$x^2 - 36a^2 + 20x + 100$

$= x^2 + 20x + 100 - 36a^2$

$= \left(x^2 + 20x + 100\right) - 36a^2$

$= \left(x + 10\right)^2 - 36a^2$

$= \left(x + 10 + 6a\right)\left(x + 10 - 6a\right)$

13. $x\left(x-1\right)^{-\frac{1}{2}} + \left(x-1\right)^{\frac{1}{2}}$

$= \left(x-1\right)^{-\frac{1}{2}}\left[x + \left(x-1\right)^{\frac{1}{2} - \left(-\frac{1}{2}\right)}\right]$

$= \left(x-1\right)^{-\frac{1}{2}}\left[x + \left(x-1\right)\right]$

$= \left(x-1\right)^{-\frac{1}{2}}\left(2x - 1\right)$

$= \dfrac{2x-1}{\left(x-1\right)^{\frac{1}{2}}}$

Concept and Vocabulary Check P.5

1. d

2. g

3. b

4. c

5. c

6. a

7. f

8. $(x+1)^{\frac{1}{2}}$

Exercise Set P.5

1. $18x + 27 = 9 \cdot 2x + 9 \cdot 3 = 9(2x + 3)$

3. $3x^2 + 6x = 3x \cdot x + 3x \cdot 2 = 3x(x + 2)$

5. $9x^4 - 18x^3 + 27x^2$

$= 9x^2(x^2) + 9x^2(-2x) + 9x^2(3)$

$= 9x^2(x^2 - 2x + 3)$

7. $x(x + 5) + 3(x + 5) = (x + 5)(x + 3)$

9. $x^2(x - 3) + 12(x - 3) = (x - 3)(x^2 + 12)$

11. $x^3 - 2x^2 + 5x - 10 = x^2(x-2) + 5(x-2)$
$$= (x^2 + 5)(x - 2)$$

13. $x^3 - x^2 + 2x - 2 = x^2(x-1) + 2(x-1)$
$$= (x-1)(x^2 + 2)$$

15. $3x^3 - 2x^2 - 6x + 4 = x^2(3x-2) - 2(3x-2)$
$$= (3x-2)(x^2 - 2)$$

17. $x^2 + 5x + 6 = (x+2)(x+3)$

19. $x^2 - 2x - 15 = (x-5)(x+3)$

21. $x^2 - 8x + 15 = (x-5)(x-3)$

23. $3x^2 - x - 2 = (3x+2)(x-1)$

25. $3x^2 - 25x - 28 = (3x-28)(x+1)$

27. $6x^2 - 11x + 4 = (2x-1)(3x-4)$

29. $4x^2 + 16x + 15 = (2x+3)(2x+5)$

31. $9x^2 - 9x + 2 = (3x-1)(3x-2)$

33. $20x^2 + 27x - 8 = (5x+8)(4x-1)$

35. $2x^2 + 3xy + y^2 = (2x+y)(x+y)$

37. $6x^2 - 5xy - 6y^2 = (3x+2y)(2x-3y)$

39. $x^2 - 100 = x^2 - 10^2 = (x+10)(x-10)$

41. $36x^2 - 49 = (6x)^2 - 7^2 = (6x+7)(6x-7)$

43. $9x^2 - 25y^2 = (3x)^2 - (5y)^2$
$$= (3x+5y)(3x-5y)$$

45. $x^4 - 16 = (x^2)^2 - 4^2$
$$= (x^2 + 4)(x^2 - 4)$$
$$= (x^2 + 4)(x+2)(x-2)$$

47. $16x^4 - 81 = (4x^2)^2 - 9^2$
$$= (4x^2 + 9)(4x^2 - 9)$$
$$= (4x^2 + 9)[(2x)^2 - 3^2]$$
$$= (4x^2 + 9)(2x+3)(2x-3)$$

49. $x^2 + 2x + 1 = x^2 + 2 \cdot x \cdot 1 + 1^2 = (x+1)^2$

51. $x^2 - 14x + 49 = x^2 - 2 \cdot x \cdot 7 + 7^2$
$$= (x-7)^2$$

53. $4x^2 + 4x + 1 = (2x)^2 + 2 \cdot 2x \cdot 1 + 1^2$
$$= (2x+1)^2$$

55. $9x^2 - 6x + 1 = (3x)^2 - 2 \cdot 3x \cdot 1 + 1^2$
$$= (3x-1)^2$$

57. $x^3 + 27 = x^3 + 3^3$
$$= (x+3)(x^2 - x \cdot 3 + 3^2)$$
$$= (x+3)(x^2 - 3x + 9)$$

59. $x^3 - 64 = x^3 - 4^3$
$$= (x-4)(x^2 + x \cdot 4 + 4^2)$$
$$= (x-4)(x^2 + 4x + 16)$$

61. $8x^3 - 1 = (2x)^3 - 1^3$
$$= (2x-1)[(2x)^2 + (2x)(1) + 1^2]$$
$$= (2x-1)(4x^2 + 2x + 1)$$

63. $64x^3 + 27 = (4x)^3 + 3^3$
$$= (4x+3)[(4x)^2 - (4x)(3) + 3^2]$$
$$= (4x+3)(16x^2 - 12x + 9)$$

65. $3x^3 - 3x = 3x(x^2 - 1) = 3x(x+1)(x-1)$

67. $4x^2 - 4x - 24 = 4(x^2 - x - 6)$
$$= 4(x+2)(x-3)$$

69. $2x^4 - 162 = 2(x^4 - 81)$
$$= 2[(x^2)^2 - 9^2]$$
$$= 2(x^2 + 9)(x^2 - 9)$$
$$= 2(x^2 + 9)(x^2 - 3^2)$$
$$= 2(x^2 + 9)(x+3)(x-3)$$

71. $x^3 + 2x^2 - 9x - 18 = (x^3 + 2x^2) - (9x + 18)$
$$= x^2(x+2) - 9(x+2)$$
$$= (x^2 - 9)(x+2)$$
$$= (x^2 - 3^2)(x+2)$$
$$= (x-3)(x+3)(x+2)$$

73. $2x^2 - 2x - 112 = 2(x^2 - x - 56) = 2(x-8)(x+7)$

75. $x^3 - 4x = x(x^2 - 4)$
$$= x(x^2 - 2^2)$$
$$= x(x-2)(x+2)$$

77. $x^2 + 64$ is prime.

79. $x^3 + 2x^2 - 4x - 8 = (x^3 + 2x^2) + (-4x - 8)$
$$= x^2(x+2) - 4(x+2) = (x^2 - 4)(x+2) = (x^2 - 2^2)(x+2) = (x-2)(x+2)(x+2) = (x-2)(x+2)^2$$

81. $y^5 - 81y$
$$= y(y^4 - 81) = y[(y^2)^2 - 9^2] = y(y^2 + 9)(y^2 - 9) = y(y^2 + 9)(y^2 - 3^2) = y(y^2 + 9)(y+3)(y-3)$$

83. $20y^4 - 45y^2 = 5y^2(4y^2 - 9) = 5y^2[(2y)^2 - 3^2] = 5y^2(2y+3)(2y-3)$

85. $x^2 - 12x + 36 - 49y^2 = (x^2 - 12x + 36) - 49y^2 = (x-6)^2 - 49y^2 = (x-6+7y)(x-6-7y)$

87. $9b^2x - 16y - 16x + 9b^2y$
$$= (9b^2x + 9b^2y) + (-16x - 16y) = 9b^2(x+y) - 16(x+y) = (x+y)(9b^2 - 16) = (x+y)(3b+4)(3b-4)$$

89. $x^2y - 16y + 32 - 2x^2$
$$= (x^2y - 16y) + (-2x^2 + 32) = y(x^2 - 16) - 2(x^2 - 16) = (x^2 - 16)(y-2) = (x+4)(x-4)(y-2)$$

91. $2x^3 - 8a^2x + 24x^2 + 72x$
$$= 2x(x^2 - 4a^2 + 12x + 36) = 2x\left[(x^2 + 12x + 36) - 4a^2\right] = 2x\left[(x+6)^2 - 4a^2\right] = 2x(x+6-2a)(x+6+2a)$$

93. $x^{\frac{3}{2}} - x^{\frac{1}{2}} = x^{\frac{1}{2}}\left(x^{\frac{3}{2} - \frac{1}{2}}\right) - 1 = x^{\frac{1}{2}}(x-1)$

95. $4x^{-\frac{2}{3}} + 8x^{\frac{1}{3}} = 4x^{-\frac{2}{3}}\left(1 + 2x^{\frac{1}{3} - \left(-\frac{2}{3}\right)}\right) = 4x^{-\frac{2}{3}}(1+2x) = \dfrac{4(1+2x)}{x^{\frac{2}{3}}}$

97. $(x+3)^{\frac{1}{2}}-(x+3)^{\frac{3}{2}}=(x+3)^{\frac{1}{2}}\left[1-(x+3)^{\frac{3}{2}-\frac{1}{2}}\right]=(x+3)^{\frac{1}{2}}\left[1-(x+3)\right]=(x+3)^{\frac{1}{2}}(-x-2)=-(x+3)^{\frac{1}{2}}(x+2)$

99. $(x+5)^{-\frac{1}{2}}-(x+5)^{-\frac{3}{2}}=(x+5)^{-\frac{3}{2}}\left[(x+5)^{-\frac{1}{2}-\left(-\frac{3}{2}\right)}-1\right]=(x+5)^{-\frac{3}{2}}\left[(x+5)-1\right]=(x+5)^{-\frac{3}{2}}(x+4)=\dfrac{x+4}{(x+5)^{\frac{3}{2}}}$

101. $(4x-1)^{\frac{1}{2}}-\dfrac{1}{3}(4x-1)^{\frac{3}{2}}$

$=(4x-1)^{\frac{1}{2}}\left[1-\dfrac{1}{3}(4x-1)^{\frac{3}{2}-\frac{1}{2}}\right]=(4x-1)^{\frac{1}{2}}\left[1-\dfrac{1}{3}(4x-1)\right]=(4x-1)^{\frac{1}{2}}\left[1-\dfrac{4}{3}x+\dfrac{1}{3}\right]$

$=(4x-1)^{\frac{1}{2}}\left(\dfrac{4}{3}-\dfrac{4}{3}x\right)=(4x-1)^{\frac{1}{2}}\dfrac{4}{3}(1-x)=\dfrac{-4(4x-1)^{\frac{1}{2}}(x-1)}{3}$

103. $10x^2(x+1)-7x(x+1)-6(x+1)=(x+1)\left(10x^2-7x-6\right)=(x+1)(5x-6)(2x+1)$

105. $6x^4+35x^2-6=\left(x^2+6\right)\left(6x^2-1\right)$

107. $y^7+y=y\left(y^6+1\right)=y\left[\left(y^2\right)^3+1^3\right]=y\left(y^2+1\right)\left(y^4-y^2+1\right)$

109. $x^4-5x^2y^2+4y^4=\left(x^2-4y^2\right)\left(x^2-y^2\right)=(x+2y)(x-2y)(x+y)(x-y)$

111. $(x-y)^4-4(x-y)^2$

$=(x-y)^2\left((x-y)^2-4\right)=(x-y)^2\left((x-y)+2\right)\left((x-y)-2\right)=(x-y)^2(x-y+2)(x-y-2)$

113. $2x^2-7xy^2+3y^4=\left(2x-y^2\right)\left(x-3y^2\right)$

115. a. $(x-0.4x)-0.4(x-0.4x)=(x-0.4x)(1-0.4)=(0.6x)(0.6)=0.36x$

  b. No, the computer is selling at 36% of its original price.

117. a. $(3x)^2-4\cdot2^2=9x^2-16$

  b. $9x^2-16=(3x+4)(3x-4)$

119. a. $x(x+y)-y(x+y)$

  b. $x(x+y)-y(x+y)=(x+y)(x-y)$

121. $V_{shaded} = V_{outside} - V_{inside}$

$\qquad = a \cdot a \cdot 4a - b \cdot b \cdot 4a$

$\qquad = 4a^3 - 4ab^2$

$\qquad = 4a\left(a^2 - b^2\right)$

$\qquad = 4a\left(a+b\right)\left(a-b\right)$

123. – 129. Answers will vary.

131. makes sense

133. makes sense

135. true

137. false; Changes to make the statement true will vary. A sample change is: $x^3 - 64 = \left(x-4\right)\left(x+4x+16\right)$

139. $-x^2 - 4x + 5 = -1\left(x^2 + 4x - 5\right) = -1\left(x+5\right)\left(x-1\right) = -\left(x+5\right)\left(x-1\right)$

141. $(x-5)^{-\frac{1}{2}}(x+5)^{-\frac{1}{2}} - (x+5)^{\frac{1}{2}}(x-5)^{-\frac{3}{2}} = (x-5)^{-\frac{3}{2}}(x+5)^{-\frac{1}{2}}\left[(x-5)^{\frac{1}{2}-\left(\frac{3}{2}\right)} - (x+5)^{\frac{1}{2}-\left(-\frac{1}{2}\right)}\right]$

$$= (x-5)^{-\frac{3}{2}}(x+5)^{-\frac{1}{2}}\left[(x-5)-(x+5)\right]$$

$$= (x-5)^{-\frac{3}{2}}(x+5)^{-\frac{1}{2}}(-10) = \frac{-10}{(x-5)^{\frac{3}{2}}(x+5)^{\frac{1}{2}}}$$

143. $b = 0, 3, 4,$ or $-c(c+4),$ where $c > 0$ is an integer.

144. $\dfrac{x^2 + 6x + 5}{x^2 - 25} = \dfrac{(x+5)(x+1)}{(x+5)(x-5)} = \dfrac{x+1}{x-5}$

145. $\dfrac{5}{4} \cdot \dfrac{8}{15} = \dfrac{5}{4} \cdot \dfrac{4 \cdot 2}{5 \cdot 3} = \dfrac{1}{1} \cdot \dfrac{2}{3} = \dfrac{2}{3}$

146. $\dfrac{1}{2} + \dfrac{2}{3} = \dfrac{3}{6} + \dfrac{4}{6} = \dfrac{7}{6}$

Section P.6

Check Point Exercises

1. **a.** The denominator would equal zero if $x = -5$, so -5 must be excluded from the domain.

    b. $x^2 - 36 = (x+6)(x-6)$
    The denominator would equal zero if $x = -6$ or $x = 6$, so -6 and 6 must both must be excluded from the domain.

    c. $x^2 - 5x - 14 = (x+2)(x-7)$
    The denominator would equal zero if $x = -2$ or $x = 7$, so -2 and 7 must both must be excluded from the domain.

2. **a.** $\dfrac{x^3 + 3x^2}{x+3} = \dfrac{x^2(x+3)}{x+3}$

    $= \dfrac{x^2(x+3)}{x+3}$

    $= x^2, \; x \neq -3$

    Because the denominator is $x + 3$, $x \neq -3$

    b. $\dfrac{x^2 - 1}{x^2 + 2x + 1} = \dfrac{(x-1)(x+1)}{(x+1)(x+1)} = \dfrac{x-1}{x+1}, x \neq -1$

    Because the denominator is
    $(x+1)(x+1), x \neq -1$

3. $\dfrac{x+3}{x^2-4} \cdot \dfrac{x^2-x-6}{x^2+6x+9}$

    $= \dfrac{x+3}{(x+2)(x-2)} \cdot \dfrac{(x-3)(x+2)}{(x+3)(x+3)}$

    $= \dfrac{x+3}{(x+2)(x-2)} \cdot \dfrac{(x-3)(x+2)}{(x+3)(x+3)}$

    $= \dfrac{x-3}{(x-2)(x+3)}, \; x \neq -2, \; x \neq 2, \; x \neq -3$

    Because the denominator has factors of
    $x+2, \; x-2, \text{ and } x+3, \; x \neq -2, \; x \neq 2, \text{ and } x \neq -3$.

4. $\dfrac{x^2 - 2x + 1}{x^3 + x} \div \dfrac{x^2 + x - 2}{3x^2 + 3}$

    $= \dfrac{x^2 - 2x + 1}{x^3 + x} \cdot \dfrac{3x^2 + 3}{x^2 + x - 2}$

    $= \dfrac{(x-1)(x-1)}{x(x^2+1)} \cdot \dfrac{3(x^2+1)}{(x+2)(x-1)}$

    $= \dfrac{3(x-1)}{x(x+2)}, \; x \neq 1, \; x \neq 0, \; x \neq -2$

5. $\dfrac{x}{x+1} - \dfrac{3x+2}{x+1} = \dfrac{x - 3x - 2}{x+1}$

    $= \dfrac{-2x - 2}{x+1}$

    $= \dfrac{-2(x+1)}{x+1}$

    $= -2, \; x \neq -1$

6. $\dfrac{3}{x+1} + \dfrac{5}{x-1}$

    $= \dfrac{3x(x-1) + 5(x+1)}{(x+1)(x-1)}$

    $= \dfrac{3x - 3 + 5x + 5}{(x+1)(x-1)}$

    $= \dfrac{8x + 2}{(x+1)(x-1)}$

    $= \dfrac{2(4x+1)}{(x+1)(x-1)}$

    $= \dfrac{2(4x+1)}{(x+1)(x-1)}, \; x \neq -1 \text{ and } x \neq 1.$

7. Factor each denominator completely.
    $x^2 - 6x + 9 = (x-3)^2$

    $x^2 - 9 = (x+3)(x-3)$
    List the factors of the first denominator.
    $x-3, \; x-3$
    Add any unlisted factors from the second denominator.
    $x-3, \; x-3, \; x+3$
    The least common denominator is the product of all factors in the final list.
    $(x-3)(x-3)(x+3)$ or $(x-3)^2(x+3)$ is the least common denominator.

8. Find the least common denominator.
    $x^2 - 10x + 25 = (x-5)^2$
    $2x - 10 = 2(x-5)$

    The least common denominator is $2(x-5)(x-5)$.
    Write all rational expressions in terms of the least common denominator.

    $\dfrac{x}{x^2 - 10x + 25} - \dfrac{x-4}{2x-10}$

    $= \dfrac{x}{(x-5)(x-5)} - \dfrac{x-4}{2(x-5)}$

    $= \dfrac{2x}{2(x-5)(x-5)} - \dfrac{(x-4)(x-5)}{2(x-5)(x-5)}$

Add numerators, putting this sum over the least common denominator.

$$= \frac{2x - (x-4)(x-5)}{2(x-5)(x-5)}$$

$$= \frac{2x - (x^2 - 5x - 4x + 20)}{2(x-5)(x-5)}$$

$$= \frac{2x - x^2 + 5x + 4x - 20}{2(x-5)(x-5)}$$

$$= \frac{2x - x^2 + 5x + 4x - 20}{2(x-5)(x-5)}$$

$$= \frac{-x^2 + 11x - 20}{2(x-5)(x-5)}$$

$$= \frac{-x^2 + 11x - 20}{2(x-5)^2}, \ x \ne 5$$

9.
$$\frac{\dfrac{1}{x} - \dfrac{3}{2}}{\dfrac{1}{x} + \dfrac{3}{4}} = \frac{\dfrac{2}{2x} - \dfrac{3x}{2x}}{\dfrac{4}{4x} + \dfrac{3x}{4x}}, \ x \ne 0$$

$$= \frac{\dfrac{2-3x}{2x}}{\dfrac{4+3x}{4x}}, \ x \ne \frac{-4}{3}$$

$$= \frac{2-3x}{2x} \div \frac{4+3x}{4x}$$

$$= \frac{2-3x}{2x} \cdot \frac{4x}{4+3x}$$

$$= \frac{2-3x}{4+3x} \cdot \frac{4}{2}$$

$$= \frac{2-3x}{4+3x} \cdot \frac{2}{1}$$

$$= \frac{2(2-3x)}{4+3x}, \ x \ne 0, \ x \ne \frac{-4}{3}$$

10. Multiply each of the three terms, $\dfrac{1}{x+7}$, $\dfrac{1}{x}$, and 7 by the least common denominator of $x(x+7)$.

$$\frac{\dfrac{1}{x+7} - \dfrac{1}{x}}{7} = \frac{x(x+7)\left(\dfrac{1}{x+7}\right) - x(x+7)\left(\dfrac{1}{x}\right)}{7x(x+7)}$$

$$= \frac{x - (x+7)}{7x(x+7)}$$

$$= \frac{-7}{7x(x+7)}$$

$$= -\frac{1}{x(x+7)}, \ x \ne 0, \ x \ne -7$$

Concept and Vocabulary Check P.6

1. polynomials

2. domain; 0

3. factoring; common factors

4. $\dfrac{x^2}{15}$

5. $\dfrac{3}{5}$

6. $\dfrac{x^2 - x + 4}{3}$

7. $x+3$ and $x-2$; $x+3$ and $x+1$; $(x+3)(x-2)(x+1)$

8. $3x+4$

9. complex; complex

10. x; $x+3$; -3; $\dfrac{1}{x(x+3)}$

Exercise Set P.6

1. $\dfrac{7}{x-3}, \ x \ne 3$

3. $\dfrac{x+5}{x^2 - 25} = \dfrac{x+5}{(x+5)(x-5)}, \ x \ne 5, -5$

5. $\dfrac{x-1}{x^2 + 11x + 10} = \dfrac{x-1}{(x+1)(x+10)}, \ x \ne -1, -10$

7. $\dfrac{3x-9}{x^2 - 6x + 9} = \dfrac{3(x-3)}{(x-3)(x-3)}$
$$= \dfrac{3}{x-3}, \ x \ne 3$$

9. $\dfrac{x^2 - 12x + 36}{4x - 24} = \dfrac{(x-6)(x-6)}{4(x-6)} = \dfrac{x-6}{4}.$
$x \ne 6$

11. $\dfrac{y^2 + 7y - 18}{y^2 - 3y + 2} = \dfrac{(y+9)(y-2)}{(y-2)(y-1)} = \dfrac{y+9}{y-1},$
$y \ne 1, 2$

13. $\dfrac{x^2+12x+36}{x^2-36} = \dfrac{(x+6)^2}{(x+6)(x-6)} = \dfrac{x+6}{x-6}$,

$x \neq 6, -6$

15. $\dfrac{x-2}{3x+9} \cdot \dfrac{2x+6}{2x-4} = \dfrac{x-2}{3(x+3)} \cdot \dfrac{2(x+3)}{2(x-2)}$

$= \dfrac{2}{6} = \dfrac{1}{3}, x \neq 2, -3$

17. $\dfrac{x^2-9}{x^2} \cdot \dfrac{x^2-3x}{x^2+x-12}$

$= \dfrac{(x-3)(x+3)}{x^2} \cdot \dfrac{x(x-3)}{(x+4)(x-3)}$

$= \dfrac{(x-3)(x+3)}{x(x+4)}, x \neq 0, -4, 3$

19. $\dfrac{x^2-5x+6}{x^2-2x-3} \cdot \dfrac{x^2-1}{x^2-4}$

$= \dfrac{(x-3)(x-2)}{(x-3)(x+1)} \cdot \dfrac{(x+1)(x-1)}{(x-2)(x+2)}$

$= \dfrac{x-1}{x+2}, \; x \neq -2, -1, 2, 3$

21. $\dfrac{x^3-8}{x^2-4} \cdot \dfrac{x+2}{3x} = \dfrac{(x-2)(x^2+2x+4)}{(x-2)(x+2)} \cdot \dfrac{x+2}{3x}$

$= \dfrac{x^2+2x+4}{3x}, x \neq -2, 0, 2$

23. $\dfrac{x+1}{3} \div \dfrac{3x+3}{7} = \dfrac{x+1}{3} \div \dfrac{3(x+1)}{7}$

$= \dfrac{x+1}{3} \cdot \dfrac{7}{3(x+1)}$

$= \dfrac{7}{9}, x \neq -1$

25. $\dfrac{x^2-4}{x} \div \dfrac{x+2}{x-2} = \dfrac{(x-2)(x+2)}{x} \cdot \dfrac{x-2}{x+2}$

$= \dfrac{(x-2)^2}{x}; x \neq 0, -2, 2$

27. $\dfrac{4x^2+10}{x-3} \div \dfrac{6x^2+15}{x^2-9}$

$= \dfrac{2(2x^2+5)}{x-3} \div \dfrac{3(2x^2+5)}{(x-3)(x+3)}$

$= \dfrac{2(2x^2+5)}{x-3} \cdot \dfrac{(x-3)(x+3)}{3(2x^2+5)}$

$= \dfrac{2(x+3)}{3}, x \neq 3, -3$

29. $\dfrac{x^2-25}{2x-2} \div \dfrac{x^2+10x+25}{x^2+4x-5}$

$= \dfrac{(x-5)(x+5)}{2(x-1)} \div \dfrac{(x+5)^2}{(x+5)(x-1)}$

$= \dfrac{(x-5)(x+5)}{2(x-1)} \cdot \dfrac{(x+5)(x-1)}{(x+5)^2}$

$= \dfrac{x-5}{2}, x \neq 1, -5$

31. $\dfrac{x^2+x-12}{x^2+x-30} \cdot \dfrac{x^2+5x+6}{x^2-2x-3} \div \dfrac{x+3}{x^2+7x+6}$

$= \dfrac{(x+4)(x-3)}{(x+6)(x-5)} \cdot \dfrac{(x+2)(x+3)}{(x+1)(x-3)} \cdot \dfrac{(x+6)(x+1)}{x+3}$

$= \dfrac{(x+4)(x+2)}{x-5}$

$x \neq -6, -3, -1, 3, 5$

33. $\dfrac{4x+1}{6x+5} + \dfrac{8x+9}{6x+5} = \dfrac{4x+1+8x+9}{6x+5}$

$= \dfrac{12x+10}{6x+5}$

$= \dfrac{2(6x+5)}{6x+5} = 2, x \neq -\dfrac{5}{6}$

35. $\dfrac{x^2-2x}{x^2+3x} + \dfrac{x^2+x}{x^2+3x} = \dfrac{x^2-2x+x^2+x}{x^2+3x}$

$= \dfrac{2x^2-x}{x^2+3x}$

$= \dfrac{x(2x-1)}{x(x+3)}$

$= \dfrac{2x-1}{x+3}, x \neq 0, -3$

37. $\dfrac{4x-10}{x-2}-\dfrac{x-4}{x-2}=\dfrac{4x-10-(x-4)}{x-2}$

$\qquad\qquad\qquad=\dfrac{4x-10-x+4}{x-2}$

$\qquad\qquad\qquad=\dfrac{3x-6}{x-2}$

$\qquad\qquad\qquad=\dfrac{3(x-2)}{x-2}$

$\qquad\qquad\qquad=3,\ x\neq 2$

39. $\dfrac{x^2+3x}{x^2+x-12}-\dfrac{x^2-12}{x^2+x-12}$

$\qquad=\dfrac{x^2+3x-(x^2-12)}{x^2+x-12}$

$\qquad=\dfrac{x^2+3x-x^2+12}{x^2+x-12}$

$\qquad=\dfrac{3x+12}{x^2+x-12}$

$\qquad=\dfrac{3(x+4)}{(x+4)(x-3)}$

$\qquad=\dfrac{3}{x-3},\ x\neq 3,-4$

41. $\dfrac{3}{x+4}+\dfrac{6}{x+5}=\dfrac{3(x+5)+6(x+4)}{(x+4)(x+5)}$

$\qquad=\dfrac{3x+15+6x+24}{(x+4)(x+5)}$

$\qquad=\dfrac{9x+39}{(x+4)(x+5)},\ x\neq -4,-5$

43. $\dfrac{3}{x+1}-\dfrac{3}{x}=\dfrac{3x-3(x+1)}{x(x+1)}$

$\qquad=\dfrac{3x-3x-3}{x(x+1)}=-\dfrac{3}{x(x+1)},\ x\neq -1,0$

45. $\dfrac{2x}{x+2}+\dfrac{x+2}{x-2}=\dfrac{2x(x-2)+(x+2)(x+2)}{(x+2)(x-2)}$

$\qquad=\dfrac{2x^2-4x+x^2+4x+4}{(x+2)(x-2)}$

$\qquad=\dfrac{3x^2+4}{(x+2)(x-2)},\ x\neq -2,2$

47. $\dfrac{x+5}{x-5}+\dfrac{x-5}{x+5}$

$\qquad=\dfrac{(x+5)(x+5)+(x-5)(x-5)}{(x-5)(x+5)}$

$\qquad=\dfrac{x^2+10x+25+x^2-10x+25}{(x-5)(x+5)}$

$\qquad=\dfrac{2x^2+50}{(x-5)(x+5)},\ x\neq -5,5$

49. $\dfrac{3}{2x+4}+\dfrac{2}{3x+6}=\dfrac{3}{2(x+2)}+\dfrac{2}{3(x+2)}$

$\qquad\qquad\qquad=\dfrac{9}{6(x+2)}+\dfrac{4}{6(x+2)}$

$\qquad\qquad\qquad=\dfrac{9+4}{6(x+2)}$

$\qquad\qquad\qquad=\dfrac{13}{6(x+2)}$

$\quad x\neq -2$

51. $\dfrac{4}{x^2+6x+9}+\dfrac{4}{x+3}=\dfrac{4}{(x+3)^2}+\dfrac{4}{x+3}$

$\qquad=\dfrac{4+4(x+3)}{(x+3)^2}=\dfrac{4+4x+12}{(x+3)^2}=\dfrac{4x+16}{(x+3)^2},$

$\quad x\neq -3$

53. $\dfrac{3x}{x^2+3x-10}-\dfrac{2x}{x^2+x-6}$

$\qquad=\dfrac{3x}{(x+5)(x-2)}-\dfrac{2x}{(x+3)(x-2)}$

$\qquad=\dfrac{3x(x+3)-2x(x+5)}{(x+5)(x-2)(x+3)}$

$\qquad=\dfrac{3x^2+9x-2x^2-10x}{(x+5)(x-2)(x+3)}$

$\qquad=\dfrac{x^2-x}{(x+5)(x-2)(x+3)},\ x\neq -5,2,-3$

55. $\dfrac{x+3}{x^2-1}-\dfrac{x+2}{x-1}$

$=\dfrac{x+3}{(x+1)(x-1)}-\dfrac{x+2}{x-1}$

$=\dfrac{x+3}{(x+1)(x-1)}-\dfrac{(x+1)(x+2)}{(x+1)(x-1)}$

$=\dfrac{x+3}{(x+1)(x-1)}-\dfrac{x^2+3x+2}{(x+1)(x-1)}$

$=\dfrac{x+3-x^2-3x-2}{(x+1)(x-1)}$

$=\dfrac{-x^2-2x+1}{(x+1)(x-1)}$

$x\neq 1,-1$

57. $\dfrac{4x^2+x-6}{x^2+3x+2}-\dfrac{3x}{x+1}+\dfrac{5}{x+2}$

$=\dfrac{4x^2+x-6}{(x+1)(x+2)}+\dfrac{-3x}{x+1}+\dfrac{5}{x+2}$

$=\dfrac{4x^2+x-5}{(x+1)(x+2)}+\dfrac{-3x(x+2)}{(x+1)(x+2)}+\dfrac{5(x+1)}{(x+1)(x+2)}$

$=\dfrac{4x^2+x-6-3x^2-6x+5x+5}{(x+1)(x+2)}$

$=\dfrac{x^2-1}{(x+1)(x+2)}$

$=\dfrac{(x-1)(x+1)}{(x+1)(x+2)}$

$=\dfrac{x-1}{x+2};x\neq -2,-1$

59. $\dfrac{\frac{x}{3}-1}{x-3}=\dfrac{3\left[\frac{x}{3}-1\right]}{3\left[x-3\right]}=\dfrac{x-3}{3(x-3)}=\dfrac{1}{3},\ x\neq 3$

61. $\dfrac{1+\frac{1}{x}}{3-\frac{1}{x}}=\dfrac{x\left[1+\frac{1}{x}\right]}{x\left[3-\frac{1}{x}\right]}=\dfrac{x+1}{3x-1},x\neq 0,\dfrac{1}{3}$

63. $\dfrac{\frac{1}{x}+\frac{1}{y}}{x+y}=\dfrac{xy\left[\frac{1}{x}+\frac{1}{y}\right]}{xy\left[x+y\right]}=\dfrac{y+x}{xy(x+y)}=\dfrac{1}{xy},$

$x\neq 0,y\neq 0,x\neq -y$

65. $\dfrac{x-\frac{x}{x+3}}{x+2}=\dfrac{(x+3)\left[x-\frac{x}{x+3}\right]}{(x+3)(x+2)}=\dfrac{x(x+3)-x}{(x+3)(x+2)}$

$=\dfrac{x^2+3x-x}{(x+3)(x+2)}=\dfrac{x^2+2x}{(x+3)(x+2)}$

$=\dfrac{x(x+2)}{(x+3)(x+2)}=\dfrac{x}{x+3},x\neq -2,-3$

67. $\dfrac{\frac{3}{x-2}-\frac{4}{x+2}}{\frac{7}{x^2-4}}=\dfrac{\frac{3}{x-2}-\frac{4}{x+2}}{\frac{7}{(x-2)(x+2)}}$

$=\dfrac{\left[\frac{3}{x-2}-\frac{4}{x+2}\right](x-2)(x+2)}{\left[\frac{7}{(x-2)(x+2)}\right](x-2)(x+2)}$

$=\dfrac{3(x+2)-4(x-2)}{7}$

$=\dfrac{3x+6-4x+8}{7}=\dfrac{-x+14}{7}$

$=-\dfrac{x-14}{7}\ \ x\neq -2,2$

69. $\dfrac{\frac{1}{x+1}}{\frac{1}{x^2-2x-3}+\frac{1}{x-3}}=\dfrac{\frac{1}{x+1}}{\frac{1}{(x+1)(x-3)}+\frac{1}{x-3}}$

$=\dfrac{\frac{(x+1)(x-3)}{x+1}}{\frac{(x+1)(x-3)}{(x+1)(x-3)}+\frac{(x+1)(x-3)}{x-3}}$

$=\dfrac{x-3}{1+x+1}$

$=\dfrac{x-3}{x+2}\ \ x\neq -2,-1,3$

71.
$$\frac{\dfrac{1}{(x+h)^2}-\dfrac{1}{x^2}}{h}=\frac{\dfrac{x^2(x+h)^2}{(x+h)^2}-\dfrac{x^2(x+h)^2}{x^2}}{hx^2(x+h)^2}$$

$$=\frac{x^2-(x+h)^2}{hx^2(x+h)^2}$$

$$=\frac{x^2-(x^2+2hx+h^2)}{hx^2(x+h)^2}$$

$$=\frac{x^2-x^2-2hx-h^2}{hx^2(x+h)^2}$$

$$=\frac{-2hx-h^2}{hx^2(x+h)^2}$$

$$=\frac{-h(2x+h)}{hx^2(x+h)^2}$$

$$=-\frac{(2x+h)}{x^2(x+h)^2}$$

73.
$$\left(\frac{2x+3}{x+1}\cdot\frac{x^2+4x-5}{2x^2+x-3}\right)-\frac{2}{x+2}=\left(\frac{\cancel{(2x+3)}}{x+1}\cdot\frac{(x+5)\cancel{(x-1)}}{\cancel{(2x+3)}\cancel{(x-1)}}\right)-\frac{2}{x+2}=\frac{x+5}{x+1}-\frac{2}{x+2}$$

$$=\frac{(x+5)(x+2)}{(x+1)(x+2)}-\frac{2(x+1)}{(x+1)(x+2)}=\frac{(x+5)(x+2)-2(x+1)}{(x+1)(x+2)}=\frac{x^2+2x+5x+10-2x-2}{(x+1)(x+2)}=\frac{x^2+5x+8}{(x+1)(x+2)}$$

75.
$$\left(2-\frac{6}{x+1}\right)\left(1+\frac{3}{x-2}\right)=\left(\frac{2(x+1)}{(x+1)}-\frac{6}{(x+1)}\right)\left(\frac{(x-2)}{(x-2)}+\frac{3}{(x-2)}\right)$$

$$=\left(\frac{2x+2-6}{x+1}\right)\left(\frac{x-2+3}{x-2}\right)=\left(\frac{2x-4}{x+1}\right)\left(\frac{x+1}{x-2}\right)=\frac{2\cancel{(x-2)}\cancel{(x+1)}}{\cancel{(x+1)}\cancel{(x-2)}}=2$$

77.
$$\frac{y^{-1}-(y+5)^{-1}}{5}=\frac{\dfrac{1}{y}-\dfrac{1}{y+5}}{5}$$

$$\text{LCD}=y(y+5)$$

$$\frac{\dfrac{1}{y}-\dfrac{1}{y+5}}{5}=\frac{y(y+5)\left(\dfrac{1}{y}-\dfrac{1}{y+5}\right)}{y(y+5)(5)}=\frac{y+5-y}{5y(y+5)}=\frac{5}{5y(y+5)}=\frac{1}{y(y+5)}$$

79.
$$\left(\frac{1}{a^3-b^3}\cdot\frac{ac+ad-bc-bd}{1}\right)-\frac{c-d}{a^2+ab+b^2}=\left(\frac{1}{(a-b)(a^2+ab+b^2)}\cdot\frac{a(c+d)-b(c+d)}{1}\right)-\frac{c-d}{a^2+ab+b^2}$$

$$=\left(\frac{1}{\cancel{(a-b)}(a^2+ab+b^2)}\cdot\frac{(c+d)\cancel{(a-b)}}{1}\right)-\frac{c-d}{a^2+ab+b^2}=\frac{c+d}{a^2+ab+b^2}-\frac{c-d}{a^2+bd+b^2}$$

$$=\frac{c+d-c+d}{a^2+ab+b^2}=\frac{2d}{a^2+ab+b^2}$$

81. a. $\dfrac{130x}{100-x}$ is equal to

   1. $\dfrac{130\cdot 40}{100-40}=\dfrac{130\cdot 40}{60}=86.67$,
   when $x=40$

   2. $\dfrac{130\cdot 80}{100-80}=\dfrac{130\cdot 80}{20}=520$,
   when $x=80$

   3. $\dfrac{130\cdot 90}{100-90}=\dfrac{130\cdot 90}{10}=1170$,
   when $x=90$

   It costs $86,670,000 to inoculate 40% of the population against this strain of flu, and $520,000,000 to inoculate 80% of the population, and $1,170,000,000 to inoculate 90% of the population.

b.   For $x=100$, the function is not defined.

c.   As x approaches 100, the value of the function increases rapidly. So it costs an astronomical amount of money to inoculate almost all of the people, and it is impossible to inoculate 100% of the population.

83. a.   Substitute 4 for x in the model.

   $W=-66x^2+526x+1030$

   $W=-66(4)^2+526(4)+1030$

   $W=2078$

   According to the model, women between the ages of 19 and 30 with this lifestyle need 2078 calories per day. This underestimates the actual value shown in the bar graph by 22 calories.

b.   Substitute 4 for x in the model.

   $M=-120x^2+998x+590$

   $M=-120(4)^2+998(4)+590$

   $M=2662$

   According to the model, men between the ages of 19 and 30 with this lifestyle need 2662 calories per day. This underestimates the actual value shown in the bar graph by 38 calories.

c. $\dfrac{W}{M}=\dfrac{-66x^2+526x+1030}{-120x^2+998x+590}$

   $=\dfrac{2(-33x^2+263x+515)}{2(-60x^2+499x+295)}$

   $=\dfrac{-33x^2+263x+515}{-60x^2+499x+295}$

85. $P = 2L + 2W$

$$= 2\left(\frac{x}{x+5}\right) + 2\left(\frac{x}{x+6}\right)$$

$$= \frac{2x}{x+5} + \frac{2x}{x+6}$$

$$= \frac{2x(x+6)}{(x+5)(x+6)} + \frac{2x(x+5)}{(x+5)(x+6)}$$

$$= \frac{2x^2 + 12x + 2x^2 + 10x}{(x+5)(x+6)}$$

$$= \frac{4x^2 + 22x}{(x+5)(x+6)}$$

87. – 97. Answers will vary.

99. does not make sense; Explanations will vary. Sample explanation: The numerator and denominator of $\frac{7}{14+x}$ do not share a common factor.

101. makes sense

103. true

105. false; Changes to make the statement true will vary. A sample change is: $6 + \frac{1}{x} = \frac{6x}{x} + \frac{1}{x} = \frac{6x+1}{x}$

107. $\left(1 - \frac{1}{x}\right)\left(1 - \frac{1}{x+1}\right)\left(1 - \frac{1}{x+2}\right)\left(1 - \frac{1}{x+3}\right) = \left(\frac{x}{x} - \frac{1}{x}\right)\left(\frac{x+1}{x+1} - \frac{1}{x+1}\right)\left(\frac{x+2}{x+2} - \frac{1}{x+2}\right)\left(\frac{x+3}{x+3} - \frac{1}{x+3}\right)$

$$= \left(\frac{x-1}{x}\right)\left(\frac{(x+1)-1}{x+1}\right)\left(\frac{(x+2)-1}{x+2}\right)\left(\frac{(x+3)-1}{x+3}\right)$$

$$= \frac{x-1}{\cancel{x}} \cdot \frac{\cancel{x}}{\cancel{x+1}} \cdot \frac{\cancel{x+1}}{\cancel{x+2}} \cdot \frac{\cancel{x+2}}{x+3} = \frac{x-1}{x+3}$$

109. It cubes x.

$$\frac{\frac{1}{x} + \frac{1}{x^2} + \frac{1}{x^3}}{\frac{1}{x^4} + \frac{1}{x^5} + \frac{1}{x^6}} = \frac{\frac{x^6}{x} + \frac{x^6}{x^2} + \frac{x^6}{x^3}}{\frac{x^6}{x^4} + \frac{x^6}{x^5} + \frac{x^6}{x^6}} = \frac{x^5 + x^4 + x^3}{x^2 + x + 1} = \frac{x^3(x^2 + x + 1)}{x^2 + x + 1} = x^3$$

110. $y = 4 - x^2$

111. $y = 1 - x^2$

112. $y = |x+1|$

x	$y = |x+1|$
-4	$|-4+1| = 3$
-3	$|-3+1| = 2$
-2	$|-2+1| = 1$
-1	$|-1+1| = 0$
0	$|0+1| = 1$
1	$|1+1| = 2$
2	$|2+1| = 3$

Chapter P Review Exercises

1. $3 + 6(x-2)^3 = 3 + 6(4-2)^3$
$$= 3 + 6(2)^3$$
$$= 3 + 6(8)$$
$$= 3 + 48$$
$$= 51$$

2. $x^2 - 5(x-y) = 6^2 - 5(6-2)$
$$= 36 - 5(4)$$
$$= 36 - 20$$
$$= 16$$

3. $S = 0.015x^2 + x + 10$
$S = 0.015(60)^2 + (60) + 10$
$$= 0.015(3600) + 60 + 10$$
$$= 54 + 60 + 10$$
$$= 124$$

4. $A = \{a,b,c\} \quad B = \{a,c,d,e\}$
$\{a,b,c\} \cap \{a,c,d,e\} = \{a,c\}$

5. $A = \{a,b,c\} \quad B = \{a,c,d,e\}$
$\{a,b,c\} \cup \{a,c,d,e\} = \{a,b,c,d,e\}$

6. $A = \{a,b,c\} \quad C = \{a,d,f,g\}$
$\{a,b,c\} \cup \{a,d,f,g\} = \{a,b,c,d,f,g\}$

7. $A = \{a,b,c\} \quad C = \{a,d,f,g\}$
$\{a,d,f,g\} \cap \{a,b,c\} = \{a\}$

8. **a.** $\sqrt{81}$

 b. $0, \sqrt{81}$

 c. $-17, 0, \sqrt{81}$

 d. $-17, -\dfrac{9}{13}, 0, 0.75, \sqrt{81}$

 e. $\sqrt{2}, \pi$

 f. $-17, -\dfrac{9}{13}, 0, 0.75, \sqrt{2}, \pi, \sqrt{81}$

9. $|-103| = 103$

10. $|\sqrt{2} - 1| = \sqrt{2} - 1$

11. $|3 - \sqrt{17}| = \sqrt{17} - 3$ since $\sqrt{17}$ is greater than 3.

12. $|4 - (-17)| = |4 + 17| = |21| = 21$

13. $3 + 17 = 17 + 3$;
commutative property of addition.

14. $(6 \cdot 3) \cdot 9 = 6 \cdot (3 \cdot 9)$;
associative property of multiplication.

15. $\sqrt{3}(\sqrt{5}+\sqrt{3}) = \sqrt{15}+3$;

distributive property of multiplication over addition.

16. $(6 \cdot 9) \cdot 2 = 2 \cdot (6 \cdot 9)$;

commutative property of multiplication.

17. $\sqrt{3}(\sqrt{5}+\sqrt{3}) = (\sqrt{5}+\sqrt{3})\sqrt{3}$;

commutative property of multiplication.

18. $(3 \cdot 7)+(4 \cdot 7) = (4 \cdot 7)+(3 \cdot 7)$;

commutative property of addition.

19. $5(2x-3)+7x = 10x-15+7x = 17x-15$

20. $\dfrac{1}{5}(5x)+\left[(3y)+(-3y)\right]-(-x) = x+\left[0\right]+x = 2x$

21. $3(4y-5)-(7y+2) = 12y-15-7y-2 = 5y-17$

22. $8-2[3-(5x-1)] = 8-2[3-5x+1]$
$$= 8-2[4-5x]$$
$$= 8-8+10x$$
$$= 10x$$

23. $D = 0.005x^2+0.55x+34$

$D = 0.005(30)^2+0.55(30)+34$

$\quad = 55$

The U.S. diversity index was 55% in 2010.
This is the same as the value displayed in the bar graph.

24. $(-3)^3(-2)^2 = (-27) \cdot (4) = -108$

25. $2^{-4}+4^{-1} = \dfrac{1}{2^4}+\dfrac{1}{4}$
$$= \dfrac{1}{16}+\dfrac{1}{4}$$
$$= \dfrac{1}{16}+\dfrac{4}{16}$$
$$= \dfrac{5}{16}$$

26. $5^{-3} \cdot 5 = 5^{-3}5^1 = 5^{-3+1} = 5^{-2} = \dfrac{1}{5^2} = \dfrac{1}{25}$

27. $\dfrac{3^3}{3^6} = 3^{3-6} = 3^{-3} = \dfrac{1}{3^3} = \dfrac{1}{27}$

28. $(-2x^4y^3)^3 = (-2)^3(x^4)^3(y^3)^3$
$$= (-2)^3 x^{4 \cdot 3} y^{3 \cdot 3}$$
$$= -8x^{12}y^9$$

29. $(-5x^3y^2)(-2x^{-11}y^{-2})$
$$= (-5)(-2)x^3x^{-11}y^2y^{-2}$$
$$= 10 \cdot x^{3-11}y^{2-2}$$
$$= 10x^{-8}y^0$$
$$= \dfrac{10}{x^8}$$

30. $(2x^3)^{-4} = (2)^{-4}(x^3)^{-4}$
$$= 2^{-4}x^{-12}$$
$$= \dfrac{1}{2^4 x^{12}}$$
$$= \dfrac{1}{16x^{12}}$$

31. $\dfrac{7x^5y^6}{28x^{15}y^{-2}} = \left(\dfrac{7}{28}\right)(x^{5-15})(y^{6-(-2)})$
$$= \dfrac{1}{4}x^{-10}y^8$$
$$= \dfrac{y^8}{4x^{10}}$$

32. $3.74 \times 10^4 = 37,400$

33. $7.45 \times 10^{-5} = 0.0000745$

34. $3,590,000 = 3.59 \times 10^6$

35. $0.00725 = 7.25 \times 10^{-3}$

36. $(3 \times 10^3)(1.3 \times 10^2) = (3 \times 1.3) \times (10^3 \times 10^2)$
$$= 3.9 \times 10^5$$
$$= 390,000$$

37. $\dfrac{6.9 \times 10^3}{3 \times 10^5} = \left(\dfrac{6.9}{3}\right) \times 10^{3-5}$
$$= 2.3 \times 10^{-2}$$
$$= 0.023$$

38. 1.35×10^{12}

39. $32,000,000 = 3.2 \times 10^7$

40. $\dfrac{1.35\times10^{12}}{3.2\times10^{7}} = \dfrac{1.35}{3.2}\cdot\dfrac{10^{12}}{10^{7}} \approx 0.42188\times10^{5} = 42{,}188$

1.35×10^{12} seconds is approximately 42,188 years.

41. $\sqrt{300} = \sqrt{100\cdot3} = \sqrt{100}\cdot\sqrt{3} = 10\sqrt{3}$

42. $\sqrt{12x^2} = \sqrt{4x^2\cdot3} = \sqrt{4x^2}\cdot\sqrt{3} = 2|x|\sqrt{3}$

43. $\sqrt{10x}\cdot\sqrt{2x} = \sqrt{20x^2}$
$\qquad\qquad\quad = \sqrt{4x^2}\cdot\sqrt{5}$
$\qquad\qquad\quad = 2x\sqrt{5}$

44. $\sqrt{r^3} = \sqrt{r^2}\cdot\sqrt{r} = r\sqrt{r}$

45. $\sqrt{\dfrac{121}{4}} = \dfrac{\sqrt{121}}{\sqrt{4}} = \dfrac{11}{2}$

46. $\dfrac{\sqrt{96x^3}}{\sqrt{2x}} = \sqrt{\dfrac{96x^3}{2x}}$
$\qquad\quad = \sqrt{48x^2}$
$\qquad\quad = \sqrt{16x^2}\cdot\sqrt{3}$
$\qquad\quad = 4x\sqrt{3}$

47. $7\sqrt{5}+13\sqrt{5} = (7+13)\sqrt{5} = 20\sqrt{5}$

48. $2\sqrt{50}+3\sqrt{8} = 2\sqrt{25\cdot2}+3\sqrt{4\cdot2}$
$\qquad\qquad\quad = 2\cdot5\sqrt{2}+3\cdot2\sqrt{2}$
$\qquad\qquad\quad = 10\sqrt{2}+6\sqrt{2}$
$\qquad\qquad\quad = 16\sqrt{2}$

49. $4\sqrt{72}-2\sqrt{48} = 4\sqrt{36\cdot2}-2\sqrt{16\cdot3}$
$\qquad\qquad\quad = 4\cdot6\sqrt{2}-2\cdot4\sqrt{3}$
$\qquad\qquad\quad = 24\sqrt{2}-8\sqrt{3}$

50. $\dfrac{30}{\sqrt{5}} = \dfrac{30}{\sqrt{5}}\cdot\dfrac{\sqrt{5}}{\sqrt{5}} = \dfrac{30\sqrt{5}}{5} = 6\sqrt{5}$

51. $\dfrac{\sqrt{2}}{\sqrt{3}} = \dfrac{\sqrt{2}}{\sqrt{3}}\cdot\dfrac{\sqrt{3}}{\sqrt{3}} = \dfrac{\sqrt{6}}{3}$

52. $\dfrac{5}{6+\sqrt{3}} = \dfrac{5}{6+\sqrt{3}}\cdot\dfrac{6-\sqrt{3}}{6-\sqrt{3}}$
$\qquad\quad = \dfrac{5(6-\sqrt{3})}{36-3}$
$\qquad\quad = \dfrac{5(6-\sqrt{3})}{33}$

53.
$\dfrac{14}{\sqrt{7}-\sqrt{5}} = \dfrac{14}{\sqrt{7}-\sqrt{5}}\cdot\dfrac{\sqrt{7}+\sqrt{5}}{\sqrt{7}+\sqrt{5}}$
$\qquad\quad = \dfrac{14(\sqrt{7}+\sqrt{5})}{7-5}$
$\qquad\quad = \dfrac{14(\sqrt{7}+\sqrt{5})}{2}$
$\qquad\quad = 7(\sqrt{7}+\sqrt{5})$

54. $\sqrt[3]{125} = 5$

55. $\sqrt[5]{-32} = -2$

56. $\sqrt[4]{-125}$ is not a real number.

57. $\sqrt[4]{(-5)^4} = \sqrt[4]{625} = \sqrt[4]{5^4} = 5$

58. $\sqrt[3]{81} = \sqrt[3]{27\cdot3} = \sqrt[3]{27}\cdot\sqrt[3]{3} = 3\sqrt[3]{3}$

59. $\sqrt[3]{y^5} = \sqrt[3]{y^3 y^2} = y\sqrt[3]{y^2}$

60. $\sqrt[4]{8}\cdot\sqrt[4]{10} = \sqrt[4]{80} = \sqrt[4]{16\cdot5} = \sqrt[4]{16}\cdot\sqrt[4]{5} = 2\sqrt[4]{5}$

61. $4\sqrt[3]{16}+5\sqrt[3]{2} = 4\sqrt[3]{8\cdot2}+5\sqrt[3]{2}$
$\qquad\qquad\qquad = 4\cdot2\sqrt[3]{2}+5\sqrt[3]{2}$
$\qquad\qquad\qquad = 8\sqrt[3]{2}+5\sqrt[3]{2}$
$\qquad\qquad\qquad = 13\sqrt[3]{2}$

62. $\dfrac{\sqrt[4]{32x^5}}{\sqrt[4]{16x}} = \sqrt[4]{\dfrac{32x^5}{16x}} = \sqrt[4]{2x^4} = x\sqrt[4]{2}$

63. $16^{1/2} = \sqrt{16} = 4$

64. $25^{-1/2} = \dfrac{1}{25^{1/2}} = \dfrac{1}{\sqrt{25}} = \dfrac{1}{5}$

65. $125^{1/3} = \sqrt[3]{125} = 5$

66. $27^{-1/3} = \dfrac{1}{27^{1/3}} = \dfrac{1}{\sqrt[3]{27}} = \dfrac{1}{3}$

67. $64^{2/3} = (\sqrt[3]{64})^2 = 4^2 = 16$

68. $27^{-4/3} = \dfrac{1}{27^{4/3}} = \dfrac{1}{(\sqrt[3]{27})^4} = \dfrac{1}{3^4} = \dfrac{1}{81}$

69. $(5x^{2/3})(4x^{1/4}) = 5 \cdot 4x^{2/3+1/4} = 20x^{11/12}$

70. $\dfrac{15x^{3/4}}{5x^{1/2}} = \left(\dfrac{15}{5}\right)x^{3/4-1/2} = 3x^{1/4}$

71. $(125 \cdot x^6)^{2/3} = (\sqrt[3]{125x^6})^2$
$\qquad\qquad\quad = (5x^2)^2$
$\qquad\qquad\quad = 25x^4$

72. $\sqrt[6]{y^3} = (y^3)^{1/6} = y^{3 \cdot 1/6} = y^{1/2} = \sqrt{y}$

73. $(-6x^3 + 7x^2 - 9x + 3) + (14x^3 + 3x^2 - 11x - 7) = (-6x^3 + 14x^3) + (7x^2 + 3x^2) + (-9x - 11x) + (3 - 7)$
$\qquad\qquad\qquad\qquad\qquad\qquad\qquad\qquad\quad = 8x^3 + 10x^2 - 20x - 4$

The degree is 3.

74. $(13x^4 - 8x^3 + 2x^2) - (5x^4 - 3x^3 + 2x^2 - 6) = (13x^4 - 8x^3 + 2x^2) + (-5x^4 + 3x^3 - 2x^2 + 6)$
$\qquad\qquad\qquad\qquad\qquad\qquad\qquad\qquad\quad = (13x^4 - 5x^4) + (-8x^3 + 3x^3) + (2x^2 - 2x^2) + 6$
$\qquad\qquad\qquad\qquad\qquad\qquad\qquad\qquad\quad = 8x^4 - 5x^3 + 6$

The degree is 4.

75. $(3x - 2)(4x^2 + 3x - 5) = (3x)(4x^2) + (3x)(3x) + (3x)(-5) + (-2)(4x^2) + (-2)(3x) + (-2)(-5)$
$\qquad\qquad\qquad\qquad\qquad = 12x^3 + 9x^2 - 15x - 8x^2 - 6x + 10$
$\qquad\qquad\qquad\qquad\qquad = 12x^3 + x^2 - 21x + 10$

76. $(3x - 5)(2x + 1) = (3x)(2x) + (3x)(1) + (-5)(2x) + (-5)(1)$
$\qquad\qquad\qquad\qquad = 6x^2 + 3x - 10x - 5$
$\qquad\qquad\qquad\qquad = 6x^2 - 7x - 5$

77. $(4x + 5)(4x - 5) = (4x^2) - 5^2 = 16x^2 - 25$

78. $(2x + 5)^2 = (2x)^2 + 2(2x) \cdot 5 + 5^2 = 4x^2 + 20x + 25$

79. $(3x-4)^2 = (3x)^2 - 2(3x) \cdot 4 + (-4)^2 = 9x^2 - 24x + 16$

80. $(2x+1)^3 = (2x)^3 + 3(2x)^2(1) + 3(2x)(1)^2 + 1^3 = 8x^3 + 12x^2 + 6x + 1$

81. $(5x-2)^3 = (5x)^3 - 3(5x)^2(2) + 3(5x)(2)^2 - 2^3 = 125x^3 - 150x^2 + 60x - 8$

82. $(7x^2 - 8xy + y^2) + (-8x^2 - 9xy - 4y^2) = (7x^2 - 8x^2) + (-8xy - 9xy) + (y^2 - 4y^2)$
$$= -x^2 - 17xy - 3y^2$$

The degree is 2.

83. $(13x^3y^2 - 5x^2y - 9x^2) - (-11x^3y^2 - 6x^2y + 3x^2 - 4)$
$= (13x^3y^2 - 5x^2y - 9x^2) + (11x^3y^2 + 6x^2y - 3x^2 + 4)$
$= (13x^3y^2 + 11x^3y^2) + (-5x^2y + 6x^2y) + (-9x^2 - 3x^2) + 4$
$= 24x^3y^2 + x^2y - 12x^2 + 4$
The degree is 5.

84. $(x+7y)(3x-5y) = x(3x) + (x)(-5y) + (7y)(3x) + (7y)(-5y)$
$$= 3x^2 - 5xy + 21xy - 35y^2$$
$$= 3x^2 + 16xy - 35y^2$$

85. $(3x-5y)^2 = (3x)^2 - 2(3x)(5y) + (-5y)^2$
$$= 9x^2 - 30xy + 25y^2$$

86. $(3x^2 + 2y)^2 = (3x^2)^2 + 2(3x^2)(2y) + (2y)^2$
$$= 9x^4 + 12x^2y + 4y^2$$

87. $(7x+4y)(7x-4y) = (7x)^2 - (4y)^2$
$$= 49x^2 - 16y^2$$

88. $(a-b)(a^2 + ab + b^2)$
$= a(a^2) + a(ab) + a(b^2) + (-b)(a^2) + (-b)(ab) + (-b)(b^2)$
$= a^3 + a^2b + ab^2 - a^2b - ab^2 - b^3$
$= a^3 - b^3$

89. $15x^3 + 3x^2 = 3x^2 \cdot 5x + 3x^2 \cdot 1$
$$= 3x^2(5x+1)$$

90. $x^2 - 11x + 28 = (x - 4)(x - 7)$

91. $15x^2 - x - 2 = (3x + 1)(5x - 2)$

92. $64 - x^2 = 8^2 - x^2 = (8 - x)(8 + x)$

93. $x^2 + 16$ is prime.

94. $3x^4 - 9x^3 - 30x^2 = 3x^2(x^2 - 3x - 10)$
$$= 3x^2(x - 5)(x + 2)$$

95. $20x^7 - 36x^3 = 4x^3(5x^4 - 9)$

96. $x^3 - 3x^2 - 9x + 27 = x^2(x - 3) - 9(x - 3)$
$$= (x^2 - 9)(x - 3)$$
$$= (x + 3)(x - 3)(x - 3)$$
$$= (x + 3)(x - 3)^2$$

97. $16x^2 - 40x + 25 = (4x - 5)(4x - 5)$
$$= (4x - 5)^2$$

98. $x^4 - 16 = (x^2)^2 - 4^2$
$$= (x^2 + 4)(x^2 - 4)$$
$$= (x^2 + 4)(x + 2)(x - 2)$$

99. $y^3 - 8 = y^3 - 2^3 = (y - 2)(y^2 + 2y + 4)$

100. $x^3 + 64 = x^3 + 4^3 = (x + 4)(x^2 - 4x + 16)$

101. $3x^4 - 12x^2 = 3x^2(x^2 - 4)$
$$= 3x^2(x - 2)(x + 2)$$

102. $27x^3 - 125 = (3x)^3 - 5^3$
$$= (3x - 5)[(3x)^2 + (3x)(5) + 5^2]$$
$$= (3x - 5)(9x^2 + 15x + 25)$$

103. $x^5 - x = x(x^4 - 1)$
$$= x(x^2 - 1)(x^2 + 1)$$
$$= x(x - 1)(x + 1)(x^2 + 1)$$

104. $x^3 + 5x^2 - 2x - 10 = x^2(x + 5) - 2(x + 5)$
$$= (x^2 - 2)(x + 5)$$

105. $x^2 + 18x + 81 - y^2 = (x^2 + 18x + 81) - y^2$
$$= (x + 9)^2 - y^2$$
$$= (x + 9 - y)(x + 9 + y)$$

106. $16x^{-\frac{3}{4}} + 32x^{\frac{1}{4}} = 16x^{-\frac{3}{4}}\left(1 + 2x^{\frac{1}{4} - \left(-\frac{3}{4}\right)}\right)$
$$= 16x^{-\frac{3}{4}}(1 + 2x)$$
$$= \frac{16(1 + 2x)}{x^{\frac{3}{4}}}$$

107. $\left(x^2 - 4\right)\left(x^2 + 3\right)^{\frac{1}{2}} - \left(x^2 - 4\right)^2\left(x^2 + 3\right)^{\frac{3}{2}}$
$$= \left(x^2 - 4\right)\left(x^2 + 3\right)^{\frac{1}{2}}\left[1 - \left(x^2 - 4\right)\left(x^2 + 3\right)\right]$$
$$= (x - 2)(x + 2)\left(x^2 + 3\right)^{\frac{1}{2}}\left[1 - (x - 2)(x + 2)\left(x^2 + 3\right)\right]$$
$$= (x - 2)(x + 2)(x^2 + 3)^{\frac{1}{2}}(-x^4 + x^2 + 13)$$

108. $12x^{-\frac{1}{2}} + 6x^{-\frac{3}{2}} = 6x^{-\frac{3}{2}}(2x + 1) = \frac{6(2x + 1)}{x^{\frac{3}{2}}}$

109. $\dfrac{x^3 + 2x^2}{x + 2} = \dfrac{x^2(x + 2)}{x + 2} = x^2 , x \neq -2$

110. $\dfrac{x^2 + 3x - 18}{x^2 - 36} = \dfrac{(x + 6)(x - 3)}{(x + 6)(x - 6)} = \dfrac{x - 3}{x - 6},$
$x \neq -6, 6$

111. $\dfrac{x^2 + 2x}{x^2 + 4x + 4} = \dfrac{x(x + 2)}{(x + 2)^2} = \dfrac{x}{x + 2},$
$x \neq -2$

112. $\dfrac{x^2 + 6x + 9}{x^2 - 4} \cdot \dfrac{x + 3}{x - 2} = \dfrac{(x + 3)^2}{(x - 2)(x + 2)} \cdot \dfrac{x + 3}{x - 2}$
$$= \dfrac{(x + 3)^3}{(x - 2)^2(x + 2)},$$
$x \neq 2, -2$

113.

$$\frac{6x+2}{x^2-1} \div \frac{3x^2+x}{x-1}$$

$$= \frac{2(3x+1)}{(x-1)(x+1)} \div \frac{x(3x+1)}{x-1}$$

$$= \frac{2(3x+1)}{(x-1)(x+1)} \cdot \frac{x-1}{x(3x+1)}$$

$$= \frac{2}{x(x+1)},$$

$$x \neq 0, 1, -1, -\frac{1}{3}$$

114.

$$\frac{x^2-5x-24}{x^2-x-12} \div \frac{x^2-10x+16}{x^2+x-6}$$

$$= \frac{(x-8)(x+3)}{(x-4)(x+3)} \div \frac{(x-2)(x-8)}{(x+3)(x-2)}$$

$$= \frac{x-8}{x-4} \cdot \frac{x+3}{x-8}$$

$$= \frac{x+3}{x-4},$$

$$x \neq -3, 4, 2, 8$$

115. $\dfrac{2x-7}{x^2-9} - \dfrac{x-10}{x^2-9} = \dfrac{2x-7-(x-10)}{x^2-9}$

$$= \frac{x+3}{(x+3)(x-3)}$$

$$= \frac{1}{x-3},$$

$$x \neq 3, -3$$

116. $\dfrac{3x}{x+2} + \dfrac{x}{x-2} = \dfrac{3x}{x+2} \cdot \dfrac{x-2}{x-2} + \dfrac{x}{x-2} \cdot \dfrac{x+2}{x+2}$

$$= \frac{3x^2-6x+x^2+2x}{(x+2)(x-2)}$$

$$= \frac{4x^2-4x}{(x+2)(x-2)}$$

$$= \frac{4x(x-1)}{(x+2)(x-2)},$$

$$x \neq 2, -2$$

117.

$$\frac{x}{x^2-9} + \frac{x-1}{x^2-5x+6}$$

$$= \frac{x}{(x-3)(x+3)} + \frac{x-1}{(x-2)(x-3)}$$

$$= \frac{x}{(x-3)(x+3)} \cdot \frac{x-2}{x-2} + \frac{x-1}{(x-2)(x-3)} \cdot \frac{x+3}{x+3}$$

$$= \frac{x(x-2)+(x-1)(x+3)}{(x-3)(x+3)(x-2)}$$

$$= \frac{x^2-2x+x^2+2x-3}{(x-3)(x+3)(x-2)}$$

$$= \frac{2x^2-3}{(x-3)(x+3)(x-2)}$$

$$x \neq 3, -3, 2$$

118.

$$\frac{4x-1}{2x^2+5x-3} - \frac{x+3}{6x^2+x-2}$$

$$= \frac{4x-1}{(2x-1)(x+3)} - \frac{x+3}{(2x-1)(3x+2)}$$

$$= \frac{4x-1}{(2x-1)(x+3)} \cdot \frac{3x+2}{3x+2}$$

$$- \frac{x+3}{(2x-1)(3x+2)} \cdot \frac{x+3}{x+3}$$

$$= \frac{12x^2+8x-3x-2-x^2-6x-9}{(2x-1)(x+3)(3x+2)}$$

$$= \frac{11x^2-x-11}{(2x-1)(x+3)(3x+2)},$$

$$x \neq \frac{1}{2}, -3, -\frac{2}{3}$$

119. $\dfrac{\frac{1}{x}-\frac{1}{2}}{\frac{1}{3}-\frac{x}{6}} = \dfrac{\frac{1}{x}-\frac{1}{2}}{\frac{1}{3}-\frac{x}{6}} \cdot \dfrac{6x}{6x}$

$\qquad = \dfrac{6-3x}{2x-x^2}$

$\qquad = \dfrac{-3(x-2)}{-x(x-2)}$

$\qquad = \dfrac{3}{x},$

$\qquad x \neq 0,\, 2$

120. $\dfrac{3+\frac{12}{x}}{1-\frac{16}{x^2}} = \dfrac{3+\frac{12}{x}}{1-\frac{16}{x^2}} \cdot \dfrac{x^2}{x^2}$

$\qquad = \dfrac{3x^2+12x}{x^2-16}$

$\qquad = \dfrac{3x(x+4)}{(x+4)(x-4)}$

$\qquad = \dfrac{3x}{x-4},$

$\qquad x \neq 0,\, 4,\, -4$

121. $\dfrac{3-\frac{1}{x+3}}{3+\frac{1}{x+3}} = \dfrac{3-\frac{1}{x+3}}{3+\frac{1}{x+3}} \cdot \dfrac{x+3}{x+3}$

$\qquad = \dfrac{3(x+3)-1}{3(x+3)+1}$

$\qquad = \dfrac{3x+9-1}{3x+9+1}$

$\qquad = \dfrac{3x+8}{3x+10},$

$\qquad x \neq -3,\, -\dfrac{10}{3}$

Chapter P Test

1. $5(2x^2-6x)-(4x^2-3x) = 10x^2-30x-4x^2+3x$
$$= 6x^2-27x$$

2. $7+2[3(x+1)-2(3x-1)]$
$\qquad = 7+2[3x+3-6x+2]$
$\qquad = 7+2[-3x+5]$
$\qquad = 7-6x+10$
$\qquad = -6x+17$

3. $\{1,2,5\} \cap \{5,a\} = \{5\}$

4. $\{1,2,5\} \cup \{5,a\} = \{1,2,5,a\}$

5. $(2x^2y^3-xy+y^2)-(-4x^2y^3-5xy-y^2)$
$\qquad = 2x^2y^3-xy+y^2+4x^2y^3+5xy+y^2$
$\qquad = 2x^2y^3+4x^2y^3-xy+5xy+y^2+y^2$
$\qquad = 6x^2y^3+4xy+2y^2$

6. $\dfrac{30x^3y^4}{6x^9y^{-4}} = 5x^{3-9}y^{4-(-4)} = 5x^{-6}y^8 = \dfrac{5y^8}{x^6}$

7. $\sqrt{6r} \cdot \sqrt{3r} = \sqrt{18r^2} = \sqrt{9r^2} \cdot \sqrt{2} = 3r\sqrt{2}$

8. $4\sqrt{50}-3\sqrt{18} = 4\sqrt{25 \cdot 2}-3\sqrt{9 \cdot 2}$
$\qquad = 4 \cdot 5\sqrt{2}-3 \cdot 3\sqrt{2}$
$\qquad = 20\sqrt{2}-9\sqrt{2}$
$\qquad = 11\sqrt{2}$

9. $\dfrac{3}{5+\sqrt{2}} = \dfrac{3}{5+\sqrt{2}} \cdot \dfrac{5-\sqrt{2}}{5-\sqrt{2}}$
$\qquad = \dfrac{3(5-\sqrt{2})}{25-2}$
$\qquad = \dfrac{3(5-\sqrt{2})}{23}$

10. $\sqrt[3]{16x^4} = \sqrt[3]{8x^3 \cdot 2x}$
$= \sqrt[3]{8x^3} \cdot \sqrt[3]{2x}$
$= 2x\sqrt[3]{2x}$

11. $\dfrac{x^2 + 2x - 3}{x^2 - 3x + 2} = \dfrac{(x+3)(x-1)}{(x-2)(x-1)} = \dfrac{x+3}{x-2},$
$x \neq 2, 1$

12. $\dfrac{5 \times 10^{-6}}{20 \times 10^{-8}} = \dfrac{5}{20} \cdot \dfrac{10^{-6}}{10^{-8}} = 0.25 \times 10^2 = 2.5 \times 10^1$

13. $(2x - 5)(x^2 - 4x + 3)$
$= 2x^3 - 8x^2 + 6x - 5x^2 + 20x - 15$
$= 2x^3 - 13x^2 + 26x - 15$

14. $(5x + 3y)^2 = (5x)^2 + 2(5x)(3y) + (3y)^2$
$= 25x^2 + 30xy + 9y^2$

15. $\dfrac{2x + 8}{x - 3} \div \dfrac{x^2 + 5x + 4}{x^2 - 9}$
$= \dfrac{2(x+4)}{x-3} \div \dfrac{(x+1)(x+4)}{(x-3)(x+3)}$
$= \dfrac{2(x+4)}{x-3} \cdot \dfrac{(x-3)(x+3)}{(x+1)(x+4)}$
$= \dfrac{2(x+3)}{x+1},$
$x \neq 3, -1, -4, -3$

16. $\dfrac{x}{x+3} + \dfrac{5}{x-3}$
$= \dfrac{x}{x+3} \cdot \dfrac{x-3}{x-3} + \dfrac{5}{x-3} \cdot \dfrac{x+3}{x+3}$
$= \dfrac{x(x-3) + 5(x+3)}{(x+3)(x-3)}$
$= \dfrac{x^2 - 3x + 5x + 15}{(x+3)(x-3)}$
$= \dfrac{x^2 + 2x + 15}{(x+3)(x-3)}, x \neq 3, -3$

17. $\dfrac{2x+3}{x^2 - 7x + 12} - \dfrac{2}{x-3}$
$= \dfrac{2x+3}{(x-3)(x-4)} - \dfrac{2}{x-3}$
$= \dfrac{2x+3}{(x-3)(x-4)} - \dfrac{2}{x-3} \cdot \dfrac{x-4}{x-4}$
$= \dfrac{2x+3 - 2(x-4)}{(x-3)(x-4)}$
$= \dfrac{2x+3 - 2(x-4)}{(x-3)(x-4)}$
$= \dfrac{2x+3 - 2x + 8}{(x-3)(x-4)}$
$= \dfrac{11}{(x-3)(x-4)},$
$x \neq 3, 4$

18. $\dfrac{\frac{1}{x} - \frac{1}{3}}{\frac{1}{x}} = \dfrac{\frac{1}{x} - \frac{1}{3}}{\frac{1}{x}} \cdot \dfrac{3x}{3x} = \dfrac{3-x}{3},$
$x \neq 0$

19. $x^2 - 9x + 18 = (x-3)(x-6)$

20. $x^3 + 2x^2 + 3x + 6 = x^2(x+2) + 3(x+2)$
$= (x^2 + 3)(x+2)$

21. $25x^2 - 9 = (5x)^2 - 3^2 = (5x-3)(5x+3)$

22. $36x^2 - 84x + 49 = (6x)^2 - 2(6x) \cdot 7 + 7^2$
$= (6x - 7)^2$

23. $y^3 - 125 = y^3 - 5^3 = (y-5)(y^2 + 5y + 25)$

24. $(x^2 + 10x + 25) - 9y^2$
$= (x+5)^2 - 9y^2$
$= (x+5-3y)(x+5+3y)$

25. $x(x+3)^{-\frac{3}{5}} + (x+3)^{\frac{2}{5}}$
$= (x+3)^{-\frac{3}{5}} \left[x + (x+3) \right]$
$= (x+3)^{-\frac{3}{5}}(2x+3) = \dfrac{2x+3}{(x+3)^{\frac{3}{5}}}$

26. $-7, -\dfrac{4}{5}, 0, 0.25, \sqrt{4}, \dfrac{22}{7}$ are rational numbers.

27. $3(2 + 5) = 3(5 + 2)$;
commutative property of addition

28. $6(7 + 4) = 6 \cdot 7 + 6 \cdot 4$
distributive property of multiplication over addition

29. $0.00076 = 7.6 \times 10^{-4}$

30. $27^{-\frac{5}{3}} = \dfrac{1}{27^{\frac{5}{3}}} = \dfrac{1}{\left(\sqrt[3]{27}\right)^5} = \dfrac{1}{(3)^5} = \dfrac{1}{243}$

31. $2\left(6.6 \times 10^9\right) = 13.2 \times 10^9 = 1.32 \times 10^{10}$

32. **a.** 2003 is 14 years after 1989.
$M = -0.28n + 47$
$M = -0.28(14) + 47$
$\quad = 43.08$
In 2003, 43.08% of bachelor's degrees were awarded to men. This overestimates the actual percent shown by the bar graph by 0.08%.

b. $R = \dfrac{M}{W} = \dfrac{-0.28n + 47}{0.28n + 53}$

c. $R = \dfrac{-0.28n + 47}{0.28n + 53}$
$R = \dfrac{-0.28(25) + 47}{0.28(25) + 53}$
$\quad = \dfrac{2}{3}$
Three women will receive bachelor's degrees for every two men. This describes the projections exactly.

Chapter 1
Equations and Inequalities

Section 1.1

Check Point Exercises

1.

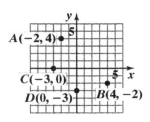

2.

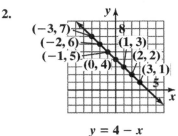

$$y = 4 - x$$

$x = -3, y = 7$

$x = -2, y = 6$

$x = -1, y = 5$

$x = 0, y = 4$

$x = 1, y = 3$

$x = 2, y = 2$

$x = 3, y = 1$

3.

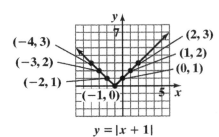

$$y = |x + 1|$$

$x = -4, y = 3$

$x = -3, y = 2$

$x = -2, y = 1$

$x = -1, y = 0$

$x = 0, y = 1$

$x = 1, y = 2$

$x = 2, y = 3$

4. The meaning of a $[-100, 100, 50]$ by $[-100, 100, 10]$ viewing rectangle is as follows:

$$[\ \overbrace{-100}^{\substack{\text{minimum} \\ x\text{-value}}}\ ,\ \overbrace{100}^{\substack{\text{maximum} \\ x\text{-value}}}\ ,\ \overbrace{50}^{\substack{\text{distance} \\ \text{between} \\ x\text{-axis} \\ \text{tick} \\ \text{marks}}}\ ]$$

by

$$[\ \overbrace{-100}^{\substack{\text{minimum} \\ y\text{-value}}}\ ,\ \overbrace{100}^{\substack{\text{maximum} \\ y\text{-value}}}\ ,\ \overbrace{10}^{\substack{\text{distance} \\ \text{between} \\ y\text{-axis} \\ \text{tick} \\ \text{marks}}}\ ]$$

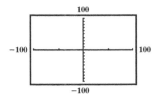

5. a. The graph crosses the x-axis at $(-3, 0)$.
Thus, the x-intercept is -3.
The graph crosses the y-axis at $(0, 5)$.
Thus, the y-intercept is 5.

   b. The graph does not cross the x-axis.
Thus, there is no x-intercept.
The graph crosses the y-axis at $(0, 4)$.
Thus, the y-intercept is 4.

   c. The graph crosses the x- and y-axes at the origin $(0, 0)$.
Thus, the x-intercept is 0 and the y-intercept is 0.

6. a. $d = 4n + 5$

   $d = 4(15) + 5 = 65$

   65% of marriages end in divorce after 15 years when the wife is under 18 at the time of marriage.

   b. According to the line graph, 60% of marriages end in divorce after 15 years when the wife is under 18 at the time of marriage.

   c. The mathematical model overestimates the actual percentage shown in the graph by 5%.

Concept and Vocabulary Check 1.1

1. *x*-axis

2. *y*-axis

3. origin

4. quadrants; four

5. *x*-coordinate; *y*-coordinate

6. solution; satisfies

7. *x*-intercept; zero

8. *y*-intercept; zero

Exercise Set 1.1

1.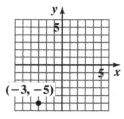

3.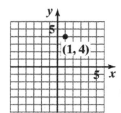

5.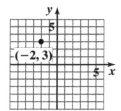

7.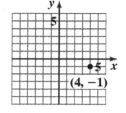

9.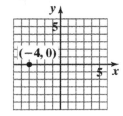

11.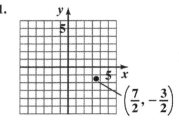

13.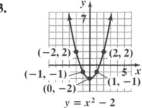

$x = -3, y = 7$

$x = -2, y = 2$

$x = -1, y = -1$

$x = 0, y = -2$

$x = 1, y = -1$

$x = 2, y = 2$

$x = 3, y = 7$

15.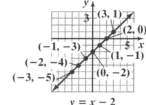

$x = -3, y = -5$

$x = -2, y = -4$

$x = -1, y = -3$

$x = 0, y = -2$

$x = 1, y = -1$

$x = 2, y = 0$

$x = 3, y = 1$

17.

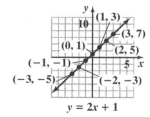

$x = -3, y = -5$

$x = -2, y = -3$

$x = -1, y = -1$

$x = 0, y = 1$

$x = 1, y = 3$

$x = 2, y = 5$

$x = 3, y = 7$

19.

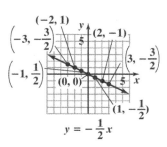

$x = -3, y = \dfrac{3}{2}$

$x = -2, y = 1$

$x = -1, y = \dfrac{1}{2}$

$x = 0, y = 0$

$x = 1, y = -\dfrac{1}{2}$

$x = 2, y = -1$

$x = 3, y = -\dfrac{3}{2}$

21.

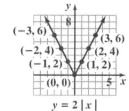

$x = -3, y = 6$

$x = -2, y = 4$

$x = -1, y = 2$

$x = 0, y = 0$

$x = 1, y = 2$

$x = 2, y = 4$

$x = 3, y = 6$

23.

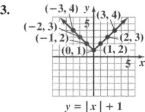

$x = -3, y = 4$

$x = -2, y = 3$

$x = -1, y = 2$

$x = 0, y = 1$

$x = 1, y = 2$

$x = 2, y = 3$

$x = 3, y = 4$

25.

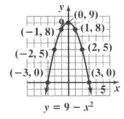

$$y = 9 - x^2$$

$x = -3, y = 0$

$x = -2, y = 5$

$x = -1, y = 8$

$x = 0, y = 9$

$x = 1, y = 8$

$x = 2, y = 5$

$x = 3, y = 0$

27.

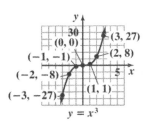

$$y = x^3$$

$x = -3, y = -27$

$x = -2, y = -8$

$x = -1, y = 1$

$x = 0, y = 0$

$x = 1, y = 1$

$x = 2, y = 8$

$x = 3, y = 27$

29. (c) *x*-axis tick marks –5, –4, –3, –2, –1, 0, 1, 2, 3, 4, 5; *y*-axis tick marks are the same.

31. (b); *x*-axis tick marks –20, –10, 0, 10, 20, 30, 40, 50, 60, 70, 80; *y*-axis tick marks –30, –20, –10, 0, 10, 20, 30, 40, 50, 60, 70

33. The equation that corresponds to Y_2 in the table is (c), $y_2 = 2 - x$. We can tell because all of the points $(-3, 5)$, $(-2, 4)$, $(-1, 3)$, $(0, 2)$, $(1, 1)$, $(2, 0)$, and $(3, -1)$ are on the line $y = 2 - x$, but all are not on any of the others.

35. No. It passes through the point $(0, 2)$.

37. $(2, 0)$

39. The graphs of Y_1 and Y_2 intersect at the points $(-2, 4)$ and $(1, 1)$.

41. a. 2; The graph intersects the *x*-axis at (2, 0).

   b. –4; The graph intersects the *y*-axis at (0,–4).

43. a. 1, –2; The graph intersects the *x*-axis at (1, 0) and (–2, 0).

   b. 2; The graph intersects the *y*-axis at (0, 2).

45. a. –1; The graph intersects the *x*-axis at (–1, 0).

   b. none; The graph does not intersect the *y*-axis.

47.

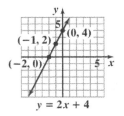

$$y = 2x + 4$$

49.

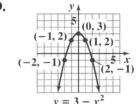

$$y = 3 - x^2$$

51.

x	(x, y)
–3	$(-3, 5)$
–2	$(-2, 5)$
–1	$(-1, 5)$
0	$(0, 5)$
1	$(1, 5)$
2	$(2, 5)$
3	$(3, 5)$

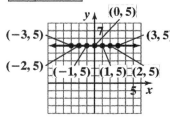

$$y = 5$$

53.

x	(x, y)
-2	$\left(-2, -\dfrac{1}{2}\right)$
-1	$(-1, -1)$
$-\dfrac{1}{2}$	$\left(-\dfrac{1}{2}, -2\right)$
$-\dfrac{1}{3}$	$\left(-\dfrac{1}{3}, -3\right)$
$\dfrac{1}{3}$	$\left(\dfrac{1}{3}, 3\right)$
$\dfrac{1}{2}$	$\left(\dfrac{1}{2}, 2\right)$
1	$(1, 1)$
2	$\left(2, \dfrac{1}{2}\right)$

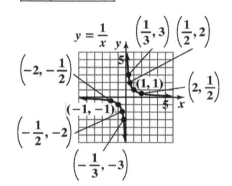

55.   **a.**   According to the line graph, 20% of seniors used marijuana in 2005.

   b.   2005 is 25 years after 1980.
   $M = -0.3n + 27$
   $M = -0.3(25) + 27 = 19.5$
   According to formula, 19.5% of seniors used marijuana in 2005. This underestimates the value in the graph by 0.5%.

   c.   According to the line graph, about 47% of seniors used alcohol in 2005.

   d.   2005 is 25 years after 1980.
   $A = -0.9n + 69$
   $A = -0.9(25) + 69 = 46.5$
   According to formula, 46.5% of seniors used alcohol in 2005. It is less than the estimate, although answers may vary.

   e.   The minimum for marijuana was reached in 1990. According to the line graph, about 14% of seniors used marijuana in 1990.

57.   At age 8, women have the least number of awakenings, averaging about 1 awakening per night.

59.   The difference between the number of awakenings for 25-year-old men and women is about 1.9.

61. – 65.   Answers will vary.

67.   makes sense

69.   does not make sense; Explanations will vary. Sample explanation: These three points are not collinear.

71.   false; Changes to make the statement true will vary. A sample change is: The product of the coordinates of a point in quadrant III is also positive.

73.   true

75.   I, III

77.   IV

79.   (a)

81.   (b)

83.   (b)

85.   (c)

87. $2(x-3) - 17 = 13 - 3(x+2)$
   $2(6-3) - 17 = 13 - 3(6+2)$
   $2(3) - 17 = 13 - 3(8)$
   $6 - 17 = 13 - 24$
   $-11 = -11, \text{ true}$

88. $12\left(\dfrac{x+2}{4} - \dfrac{x-1}{3}\right) = 12\left(\dfrac{x+2}{4}\right) - 12\left(\dfrac{x-1}{3}\right)$
   $= 3(x+2) - 4(x-1)$
   $= 3x + 6 - 4x + 4$
   $= -x + 10$

89. $(x-3)\ \dfrac{3}{x-3} + 9 = (x-3)\ \dfrac{3}{x-3} + (x-3)(9)$
   $= 3 + 9x - 27$
   $= 9x - 24$

Section 1.2

Check Point Exercises

1.
$$4x+5=29$$
$$4x+5-5=29-5$$
$$4x=24$$
$$\frac{4x}{4}=\frac{24}{4}$$
$$x=6$$
Check:
$$4x+5=29$$
$$4(6)+5=29$$
$$24+5=29$$
$$29=29 \ \text{ true}$$
The solution set is {6}.

2.
$$4(2x+1)-29=3(2x-5)$$
$$8x+4-29=6x-15$$
$$8x-25=6x-15$$
$$8x-25-6x=6x-15-6x$$
$$2x-25=-15$$
$$2x-25+25=-15+25$$
$$2x=10$$
$$\frac{2x}{2}=\frac{10}{2}$$
$$x=5$$
Check:
$$4(2x+1)-29=3(2x-5)$$
$$4[2(5)+1]-29=3[2(5)-5]$$
$$4[10+1]-29=3[10-5]$$
$$4[11]-29=3[5]$$
$$44-29=15$$
$$15=15 \ \text{ true}$$
The solution set is {5}.

3.
$$\frac{x-3}{4}=\frac{5}{14}-\frac{x+5}{7}$$
$$28\cdot\frac{x-3}{4}=28\left(\frac{5}{14}-\frac{x+5}{7}\right)$$
$$7(x-3)=2(5)-4(x+5)$$
$$7x-21=10-4x-20$$
$$7x-21=-4x-10$$
$$7x+4x=-10+21$$
$$11x=11$$
$$\frac{11x}{11}=\frac{11}{11}$$
$$x=1$$
Check:
$$\frac{x-3}{4}=\frac{5}{14}-\frac{x+5}{7}$$
$$\frac{1-3}{4}=\frac{5}{14}-\frac{1+5}{7}$$
$$\frac{-2}{4}=\frac{5}{14}-\frac{6}{7}$$
$$-\frac{1}{2}=-\frac{1}{2}$$
The solution set is {1}.

4.
$$\frac{5}{2x}=\frac{17}{18}-\frac{1}{3x}, \ x\neq 0$$
$$18x\cdot\frac{5}{2x}=18x\left(\frac{17}{18}-\frac{1}{3x}\right)$$
$$18\cdot\frac{5}{2x}=18x\cdot\frac{17}{18}-18x\cdot\frac{1}{3x}$$
$$45=17x-6$$
$$45+6=17x-6+6$$
$$51=17x$$
$$\frac{51}{17}=\frac{17x}{17}$$
$$3=x$$
The solution set is {3}.

5.
$$\frac{x}{x-2} = \frac{2}{x-2} - \frac{2}{3}, \quad x \neq 2$$

$$3(x-2) \cdot \frac{x}{x-2} = 3(x-2)\left[\frac{2}{x-2} - \frac{2}{3}\right]$$

$$3(x-2) \cdot \frac{x}{x-2} = (3x-2) \cdot \frac{2}{x-2} - 3(x-2) \cdot \frac{2}{3}$$

$$3x = 6 - (x-2) \cdot 2$$

$$3x = 6 - 2(x-2)$$

$$3x = 6 - 2x + 4$$

$$3x = 10 - 2x$$

$$3x + 2x = 10 - 2x + 2x$$

$$5x = 10$$

$$\frac{5x}{5} = \frac{10}{5}$$

$$x = 2$$

The solution set is the empty set, $\varnothing$.

6. Set $y_1 = y_2$.

$$\frac{1}{x+4} + \frac{1}{x-4} = \frac{22}{x^2-16}$$

$$\frac{1}{x+4} + \frac{1}{x-4} = \frac{22}{(x+4)(x-4)}$$

$$\frac{(x+4)(x-4)}{x+4} + \frac{(x+4)(x-4)}{x-4} = \frac{22(x+4)(x-4)}{(x+4)(x-4)}$$

$$(x-4) + (x+4) = 22$$

$$x - 4 + x + 4 = 22$$

$$2x = 22$$

$$x = 11$$

Check:

$$\frac{1}{x+4} + \frac{1}{x-4} = \frac{22}{x^2-16}$$

$$\frac{1}{11+4} + \frac{1}{11-4} = \frac{22}{11^2-16}$$

$$\frac{1}{15} + \frac{1}{7} = \frac{22}{105}$$

$$\frac{22}{105} = \frac{22}{105} \quad \text{true}$$

7.
$$4x - 7 = 4(x-1) + 3$$

$$4x - 7 = 4(x-1) + 3$$

$$4x - 7 = 4x - 4 + 3$$

$$4x - 7 = 4x - 1$$

$$-7 = -1$$

The original equation is equivalent to the statement $-7 = -1$, which is false for every value of x.
The solution set is the empty set, $\varnothing$.
The equation is an inconsistent equation.

8.
$$4x - 7 = 4(x-1) + 3$$

$$7x + 9 = 9(x+1) - 2x$$

$$7x + 9 = 9x + 9 - 2x$$

$$7x + 9 = 7x + 9$$

$$9 = 9$$

The original equation is equivalent to the statement $9 = 9$, which is true for every value of x.
The equation is an identity, and all real numbers are solutions. The solution set $\{x | x \text{ is a real number}\}$.

9.
$$D = \frac{10}{9}x + \frac{53}{9}$$

$$10 = \frac{10}{9}x + \frac{53}{9}$$

$$9 \cdot 10 = 9\left(\frac{10}{9}x + \frac{53}{9}\right)$$

$$90 = 10x + 53$$

$$90 - 53 = 10x + 53 - 53$$

$$37 = 10x$$

$$\frac{37}{10} = \frac{10x}{10}$$

$$3.7 = x$$

$$x = 3.7$$

The formula indicates that if the low-humor group averages a level of depression of 10 in response to a negative life event, the intensity of that event is 3.7. This is shown as the point whose corresponding value on the vertical axis is 10 and whose value on the horizontal axis is 3.7.

Concept and Vocabulary Check 1.2

1. linear

2. equivalent

3. apply the distributive property

4. least common denominator; 12

5. 0

6. $2x$

7. $(x+5)(x+1)$

8. $x \neq 2$; $x \neq 4$

9. $5(x+3) + 3(x+4) = 12x + 9$

10. identity

11. inconsistent

Exercise Set 1.2

1. $7x - 5 = 72$
  $7x = 77$
  $x = 11$
 Check:
  $7x - 5 = 72$
  $7(11) - 5 = 72$
  $77 - 5 = 72$
  $72 = 72$
 The solution set is $\{11\}$.

3. $11x - (6x - 5) = 40$
  $11x - 6x + 5 = 40$
  $5x + 5 = 40$
  $5x = 35$
  $x = 7$
 The solution set is $\{7\}$.

 Check:
  $11x - (6x - 5) = 40$
  $11(7) - [6(7) - 5] = 40$
  $77 - (42 - 5) = 40$
  $77 - (37) = 40$
  $40 = 40$

5. $2x - 7 = 6 + x$
  $x - 7 = 6$
  $x = 13$
 The solution set is $\{13\}$.

 Check:
  $2(13) - 7 = 6 + 13$
  $26 - 7 = 19$
  $19 = 19$

7. $7x + 4 = x + 16$
  $6x + 4 = 16$
  $6x = 12$
  $x = 2$
 The solution set is $\{2\}$.

 Check:
  $7(2) + 4 = 2 + 16$
  $14 + 4 = 18$
  $18 = 18$

9. $3(x - 2) + 7 = 2(x + 5)$
  $3x - 6 + 7 = 2x + 10$
  $3x + 1 = 2x + 10$
  $x + 1 = 10$
  $x = 9$
 The solution set is $\{9\}$.

 Check:
  $3(9 - 2) + 7 = 2(9 + 5)$
  $3(7) + 7 = 2(14)$
  $21 + 7 = 28$
  $28 = 28$

11. $3(x - 4) - 4(x - 3) = x + 3 - (x - 2)$
  $3x - 12 - 4x + 12 = x + 3 - x + 2$
  $-x = 5$
  $x = -5$
 The solution set is $\{-5\}$.

 Check:
  $3(-5 - 4) - 4(-5 - 3) = -5 + 3 - (-5 - 2)$
  $3(-9) - 4(-8) = -2 - (-7)$
  $-27 + 32 = -2 + 7$
  $5 = 5$

13. $16 = 3(x - 1) - (x - 7)$
  $16 = 3x - 3 - x + 7$
  $16 = 2x + 4$
  $12 = 2x$
  $6 = x$
 The solution set is $\{6\}$.

 Check:
  $16 = 3(6 - 1) - (6 - 7)$
  $16 = 3(5) - (-1)$
  $16 = 15 + 1$
  $16 = 16$

15. $25 - [2 + 5y - 3(y + 2)] = -3(2y - 5) - [5(y - 1) - 3y + 3]$

$25 - [2 + 5y - 3y - 6] = -6y + 15 - [5y - 5 - 3y + 3]$

$25 - [2y - 4] = -6y + 15 - [2y - 2]$

$25 - 2y + 4 = -6y + 15 - 2y + 2$

$-2y + 29 = -8y + 17$

$6y = -12$

$y = -2$

The solution set is $\{-2\}$.

Check:

$25 - [2 + 5y - 3(y + 2)] = -3(2y - 5) - [5(y - 1) - 3y + 3]$

$25 - [2 + 5(-2) - 3(-2 + 2)] = -3[2(-2) - 5] - [5(-2 - 1) - 3(-2) + 3]$

$25 - [2 - 10 - 3(0)] = -3[-4 - 5] - [5(-3) + 6 + 3]$

$25 - [-8] = -3(-9) - [-15 + 9]$

$25 + 8 = 27 - (-6)$

$33 = 27 + 6$

$33 = 33$

17. $\dfrac{x}{3} = \dfrac{x}{2} - 2$

$6\left[\dfrac{x}{3} = \dfrac{x}{2} - 2\right]$

$2x = 3x - 12$

$12 = 3x - 2x$

$x = 12$

The solution set is $\{12\}$.

19. $20 - \dfrac{x}{3} = \dfrac{x}{2}$

$6\left[20 - \dfrac{x}{3} = \dfrac{x}{2}\right]$

$120 - 2x = 3x$

$120 = 3x + 2x$

$120 = 5x$

$x = \dfrac{120}{5}$

$x = 24$

The solution set is $\{24\}$.

21. $\dfrac{3x}{5} = \dfrac{2x}{3} + 1$

$15\left[\dfrac{3x}{5} = \dfrac{2x}{3} + 1\right]$

$9x = 10x + 15$

$9x - 10x = 15$

$-x = 15$

$x = -15$

The solution set is $\{-15\}$.

23. $\dfrac{3x}{5} - x = \dfrac{x}{10} - \dfrac{5}{2}$

$10\left[\dfrac{3x}{5} - x = \dfrac{x}{10} - \dfrac{5}{2}\right]$

$6x - 10x = x - 25$

$-4x - x = -25$

$-5x = -25$

$x = 5$

The solution set is $\{5\}$.

25. $\dfrac{x+3}{6} = \dfrac{3}{8} + \dfrac{x-5}{4}$

$24\left[\dfrac{x+3}{6} = \dfrac{3}{8} + \dfrac{x-5}{4}\right]$

$4x + 12 = 9 + 6x - 30$

$4x - 6x = -21 - 12$

$-2x = -33$

$x = \dfrac{33}{2}$

The solution set is $\left\{\dfrac{33}{2}\right\}$.

27. $\dfrac{x}{4} = 2 + \dfrac{x-3}{3}$

$12\left[\dfrac{x}{4} = 2 + \dfrac{x-3}{3}\right]$

$3x = 24 + 4x - 12$

$3x - 4x = 12$

$-x = 12$

$x = -12$

The solution set is $\{-12\}$.

29. $\dfrac{x+1}{3} = 5 - \dfrac{x+2}{7}$

$21\left[\dfrac{x+1}{3} = 5 - \dfrac{x+2}{7}\right]$

$7x + 7 = 105 - 3x - 6$

$7x + 3x = 99 - 7$

$10x = 92$

$x = \dfrac{92}{10}$

$x = \dfrac{46}{5}$

The solution set is $\left\{\dfrac{46}{5}\right\}$.

31. a. $\dfrac{4}{x} = \dfrac{5}{2x} + 3 \,(x \neq 0)$

b. $\dfrac{4}{x} = \dfrac{5}{2x} + 3$

$8 = 5 + 6x$

$3 = 6x$

$\dfrac{1}{2} = x$

The solution set is $\left\{\dfrac{1}{2}\right\}$.

33. a. $\dfrac{2}{x} + 3 = \dfrac{5}{2x} + \dfrac{13}{4} \,(x \neq 0)$

b. $\dfrac{2}{x} + 3 = \dfrac{5}{2x} + \dfrac{13}{4}$

$8 + 12x = 10 + 13x$

$-x = 2$

$x = -2$

The solution set is $\{-2\}$.

35. a. $\dfrac{2}{3x} + \dfrac{1}{4} = \dfrac{11}{6x} - \dfrac{1}{3} \,(x \neq 0)$

b. $\dfrac{2}{3x} + \dfrac{1}{4} = \dfrac{11}{6x} - \dfrac{1}{3}$

$8 + 3x = 22 - 4x$

$7x = 14$

$x = 2$

The solution set is $\{2\}$.

37. a. $\dfrac{x-2}{2x} + 1 = \dfrac{x+1}{x} \quad (x \neq 0)$

b. $\dfrac{x-2}{2x} + 1 = \dfrac{x+1}{x}$

$x - 2 + 2x = 2x + 2$

$x - 2 = 2$

$x = 4$

The solution set is $\{4\}$.

39. a. $\dfrac{1}{x-1} + 5 = \dfrac{11}{x-1} \,(x \neq 1)$

b. $\dfrac{1}{x-1} + 5 = \dfrac{11}{x-1}$

$1 + 5(x-1) = 11$

$1 + 5x - 5 = 11$

$5x - 4 = 11$

$5x = 15$

$x = 3$

The solution set is $\{3\}$.

41. a. $\dfrac{8x}{x+1} = 4 - \dfrac{8}{x+1} \ (x \neq -1)$

b. $\dfrac{8x}{x+1} = 4 - \dfrac{8}{x+1}$

$8x = 4(x+1) - 8$

$8x = 4x + 4 - 8$

$4x = -4$

$x = -1 \Rightarrow$ no solution

The solution set is the empty set, $\varnothing$.

43. a. $\dfrac{3}{2x-2} + \dfrac{1}{2} = \dfrac{2}{x-1} \ (x \neq 1)$

b. $\dfrac{3}{2x-2} + \dfrac{1}{2} = \dfrac{2}{x-1}$

$\dfrac{3}{2(x-1)} + \dfrac{1}{2} = \dfrac{2}{x-1}$

$3 + 1(x-1) = 4$

$3 + x - 1 = 4$

$x = 2$

The solution set is $\{2\}$.

45. a. $\dfrac{3}{x+2} + \dfrac{2}{x-2} = \dfrac{8}{(x+2)(x-2)}; (x \neq -2, 2)$

b. $\dfrac{3}{x+2} + \dfrac{2}{x-2} = \dfrac{8}{(x+2)(x-2)}$

$(x \neq 2, x \neq -2)$

$3(x-2) + 2(x+2) = 8$

$3x - 6 + 2x + 4 = 8$

$5x = 10$

$x = 2 \Rightarrow$ no solution

The solution set is the empty set, $\varnothing$.

47. a. $\dfrac{2}{x+1} - \dfrac{1}{x-1} = \dfrac{2x}{x^2-1} \ (x \neq 1, x \neq -1)$

b.

$\dfrac{2}{x+1} - \dfrac{1}{x-1} = \dfrac{2x}{x^2-1}$

$\dfrac{2}{x+1} - \dfrac{1}{x-1} = \dfrac{2x}{(x+1)(x-1)}$

$2(x-1) - 1(x+1) = 2x$

$2x - 2 - x - 1 = 2x$

$-x = 3$

$x = -3$

The solution set is $\{-3\}$.

49. a. $\dfrac{1}{x-4} - \dfrac{5}{x+2} = \dfrac{6}{(x-4)(x+2)}; (x \neq -2, 4)$

b.

$\dfrac{1}{x-4} - \dfrac{5}{x+2} = \dfrac{6}{x^2-2x-8}$

$\dfrac{1}{x-4} - \dfrac{5}{x+2} = \dfrac{6}{(x-4)(x+2)}$

$(x \neq 4, x \neq -2)$

$1(x+2) - 5(x-4) = 6$

$x + 2 - 5x + 20 = 6$

$-4x = -16$

$x = 4 \Rightarrow$ no solution

The solution set is the empty set, $\varnothing$.

51. Set $y_1 = y_2$.

$5(2x-8) - 2 = 5(x-3) + 3$

$10x - 40 - 2 = 5x - 15 + 3$

$10x - 42 = 5x - 12$

$10x - 5x = -12 + 42$

$5x = 30$

$x = 6$

The solution set is $\{6\}$.

53. Set $y_1 - y_2 = 1$.

$\dfrac{x-3}{5} - \dfrac{x-5}{4} = 1$

$20 \cdot \dfrac{x-3}{5} - 20 \cdot \dfrac{x-5}{4} = 20 \cdot 1$

$4(x-3) - 5(x-5) = 20$

$4x - 12 - 5x + 25 = 20$

$-x + 13 = 20$

$-x = 7$

$x = -7$

The solution set is $\{-7\}$.

55. Set $y_1 + y_2 = y_3$.

$$\frac{5}{x+4} + \frac{3}{x+3} = \frac{12x+19}{x^2+7x+12}$$

$$\frac{5}{x+4} + \frac{3}{x+3} = \frac{12x+19}{(x+4)(x+3)}$$

$$(x+4)(x+3)\left(\frac{5}{x+4} + \frac{3}{x+3}\right) = (x+4)(x+3)\frac{12x+19}{(x+4)(x+3)}$$

$$5(x+3) + 3(x+4) = 12x+19$$

$$5x+15+3x+12 = 12x+19$$

$$8x+27 = 12x+19$$

$$-4x = -8$$

$$x = 2$$

The solution set is $\{2\}$.

57. $0 = 4[x-(3-x)] - 7(x+1)$

$0 = 4[x-3+x] - 7x - 7$

$0 = 4[2x-3] - 7x - 7$

$0 = 8x - 12 - 7x - 7$

$0 = x - 19$

$-x = -19$

$x = 19$

The solution set is $\{19\}$.

59.

$$0 = \frac{x+6}{3x-12} - \frac{5}{x-4} - \frac{2}{3}$$

$$0 = \frac{x+6}{3(x-4)} - \frac{5}{x-4} - \frac{2}{3}$$

$$3(x-4) \cdot 0 = 3(x-4)\left(\frac{x+6}{3(x-4)} - \frac{5}{x-4} - \frac{2}{3}\right)$$

$$0 = \frac{3(x-4)(x+6)}{3(x-4)} - \frac{5 \cdot 3(x-4)}{x-4} - \frac{2 \cdot 3(x-4)}{3}$$

$$0 = (x+6) - 15 - 2(x-4)$$

$$0 = x + 6 - 15 - 2x + 8$$

$$0 = -x - 1$$

$$x = -1$$

The solution set is $\{-1\}$.

61. $5x+9=9(x+1)-4x$

$5x+9=9x+9-4x$

$5x+9=5x+9$

$9=9$

The solution set $\{x|x \text{ is a real number}\}$.

The given equation is an identity.

63. $3(x+2)=7+3x$

$3x+6=7+3x$

$6=7$

The solution set $\varnothing$.

The given equation is an inconsistent equation.

65. $10x+3=8x+3$

$2x+3=3$

$2x=0$

$x=0$

The solution set $\{0\}$.

The given equation is a conditional equation.

67. $\dfrac{2x}{x-3}=\dfrac{6}{x-3}+4$

$2x=6+4(x-3)$

$2x=6+4x-12$

$-2x=-6$

$x=3 \Rightarrow$ no solution

The given equation is an inconsistent equation.

69. $\dfrac{x+5}{2}-4=\dfrac{2x-1}{3}$

$3(x+5)-24=2(2x-1)$

$3x+15-24=4x-2$

$-x=7$

$x=-7$

The solution set is $\{-7\}$.

The given equation is a conditional equation.

71. $\dfrac{2}{x-2}=3+\dfrac{x}{x-2}$

$2=3(x-2)+x$

$2=3x-6+x$

$-4x=-8$

$x=2 \Rightarrow$ no solution

The solution set is the empty set, $\varnothing$.

The given equation is an inconsistent equation.

73. $8x-(3x+2)+10=3x$

$8x-3x-2+10=3x$

$2x=-8$

$x=-4$

The solution set is $\{-4\}$.

The given equation is a conditional equation.

75. $\dfrac{2}{x}+\dfrac{1}{2}=\dfrac{3}{4}$

$8+2x=3x$

$-x=-8$

$x=8$

The solution set is $\{8\}$.

The given equation is a conditional equation.

77. $\dfrac{4}{x-2}+\dfrac{3}{x+5}=\dfrac{7}{(x+5)(x-2)}$

$4(x+5)+3(x-2)=7$

$4x+20+3x-6=7$

$7x=-7$

$x=-1$

The solution set is $\{-1\}$.

The given equation is a conditional equation.

79. $\dfrac{4x}{x+3}-\dfrac{12}{x-3}=\dfrac{4x^2+36}{x^2-9}; x\neq 3,-3$

$4x(x-3)-12(x+3)=4x^2+36$

$4x^2-12x-12x-36=4x^2+36$

$4x^2-24x-36=4x^2+36$

$-24x-36=36$

$-24x=72$

$x=-3 \quad$ No solution

The solution set is $\{\ \}$.

The given equation is an inconsistent equation.

81. The equation is $3(x-4)=3(2-2x)$, and the solution is $x=2$.

83. The equation is $-3(x-3)=5(2-x)$, and the solution is $x=0.5$.

85. Solve: $4(x-2)+2=4x-2(2-x)$

$$4x-8+2=4x-4+2x$$
$$4x-6=6x-4$$
$$-2x-6=-4$$
$$-2x=2$$
$$x=-1$$

Now, evaluate x^2-x for $x=-1$:

$$x^2-x=(-1)^2-(-1)$$
$$=1-(-1)=1+1=2$$

87. Solve for x: $\dfrac{3(x+3)}{5}=2x+6$

$$3(x+3)=5(2x+6)$$
$$3x+9=10x+30$$
$$-7x+9=30$$
$$-7x=21$$
$$x=-3$$

Solve for y: $-2y-10=5y+18$

$$-7y-10=18$$
$$-7y=28$$
$$y=-4$$

Now, evaluate $x^2-(xy-y)$ for $x=-3$ and $y=-4$:

$$x^2-(xy-y)$$
$$=(-3)^2-\left[-3(-4)-(-4)\right]$$
$$=(-3)^2-\left[12-(-4)\right]$$
$$=9-(12+4)=9-16=-7$$

89. $\left[(3+6)^2 \div 3\right]\cdot 4=-54x$

$$\left(9^2 \div 3\right)\cdot 4=-54x$$
$$(81\div 3)\cdot 4=-54x$$
$$27\cdot 4=-54x$$
$$108=-54x$$
$$-2=x$$

The solution set is $\{-2\}$.

91. $5-12x=8-7x-\left[6\div 3\left(2+5^3\right)+5x\right]$

$$5-12x=8-7x-\left[6\div 3\left(2+125\right)+5x\right]$$
$$5-12x=8-7x-\left[6\div 3\cdot 127+5x\right]$$
$$5-12x=8-7x-\left[2\cdot 127+5x\right]$$
$$5-12x=8-7x-\left[254+5x\right]$$
$$5-12x=8-7x-254-5x$$
$$5-12x=-12x-246$$
$$5=-246$$

The final statement is a contradiction, so the equation has no solution. The solution set is $\varnothing$.

93. $0.7x+0.4(20)=0.5(x+20)$

$$0.7x+8=0.5x+10$$
$$0.2x+8=10$$
$$0.2x=2$$
$$x=10$$

The solution set is $\{10\}$.

95. $4x+13-\left\{2x-\left[4(x-3)-5\right]\right\}=2(x-6)$

$$4x+13-\left\{2x-\left[4x-12-5\right]\right\}=2x-12$$
$$4x+13-\left\{2x-\left[4x-17\right]\right\}=2x-12$$
$$4x+13-\left\{2x-4x+17\right\}=2x-12$$
$$4x+13-\left\{-2x+17\right\}=2x-12$$
$$4x+13+2x-17=2x-12$$
$$6x-4=2x-12$$
$$4x-4=-12$$
$$4x=-8$$
$$x=-2$$

The solution set is $\{-2\}$.

97. a. $p+\dfrac{x}{2}=37$

$$p+\dfrac{40}{2}=37$$
$$p+20=37$$
$$p=17$$

According to the model, 17% of American adults smoked cigarettes in 2010. This underestimates the value shown in the bar graph by 2%.

b. $p + \dfrac{x}{2} = 37$

$7 + \dfrac{x}{2} = 37$

$\dfrac{x}{2} = 30$

$x = 60$

According to the model, only 7% of American adults will smoke cigarettes 60 years after 1970, or 2030.

99.   **a.**   What cost $10,000 in 1967 would cost about $52,000 in 2000.

b. $C = 1388x + 24{,}963$

$= 1388(20) + 24{,}963$

$= 52{,}723$

According to Model 1, what cost $10,000 in 1967 would cost about $52,723 in 2000. This describes the estimate from part (a) reasonably well.

c. $C = 3x^2 + 1308x + 25{,}268$

$= 3(20)^2 + 1308(20) + 25{,}268$

$= 52{,}628$

According to Model 2, what cost $10,000 in 1967 would cost about $52,628 in 2000. This describes the estimate from part (a) reasonably well.

101. $C = 1388x + 24{,}963$

$77{,}707 = 1388x + 24{,}963$

$52{,}744 = 1388x$

$38 = x$

Model 1 predicts the cost will be $77,707 38 years after 1980, or 2018.

103.   11 learning trials; represented by the point $(11, 0.95)$ on the graph.

105.

$$C = \dfrac{x + 0.1(500)}{x + 500}$$

$$0.28 = \dfrac{x + 0.1(500)}{x + 500}$$

$$0.28(x + 500) = x + 0.1(500)$$

$$0.28x + 140 = x + 50$$

$$-0.72x = -90$$

$$\dfrac{-0.72x}{-0.72} = \dfrac{-90}{-0.72}$$

$$x = 125$$

125 liters of pure peroxide must be added.

107. – 115.   Answers will vary.

117. $\{5\}$

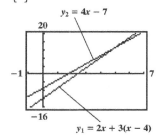

119. $\{-5\}$

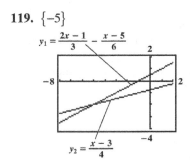

121.   makes sense

123.   makes sense

125.   false; Changes to make the statement true will vary. A sample change is: In the first equation, $x \neq 4$.

127.   false; Changes to make the statement true will vary. A sample change is: If $a = 0$, then $ax + b = 0$ is equivalent to $b = 0$, which either has no solution ($b \neq 0$) or infinitely many solutions ($b = 0$).

129.
$$\frac{7x+4}{b}+13=x$$
$$\frac{7(-6)+4}{b}+13=-6$$
$$\frac{-42+4}{b}+13=-6$$
$$\frac{-38}{b}+13=-6$$
$$\frac{-38}{b}=-19$$
$$-38=-19b$$
$$b=2$$

131. $x+150$

132. $20+0.05x$

133. $4x+400$

Section 1.3

Check Point Exercises

1. Let $x=$ the median starting salary, in thousands of dollars, of education majors.
Let $x+21=$ the median starting salary, in thousands of dollars, of computer science majors.
Let $x+14=$ the median starting salary, in thousands of dollars, of economics majors.
$$x+(x+21)+(x+14)=140$$
$$x+x+21+x+14=140$$
$$3x+35=140$$
$$3x=105$$
$$x=35$$
$$x+21=56$$
$$x+14=49$$
The median starting salary of education majors is $35 thousand, of computer science majors is $56 thousand, and of economics majors is $49 thousand.

2. Let $x=$ the number of years after 1969.
$$85-0.9x=25$$
$$-0.9x=25-85$$
$$-0.9x=-60$$
$$x=\frac{-60}{-0.9}$$
$$x\approx67$$
25% of freshmen will respond this way 67 years after 1969, or 2036.

3. Let $x=$ the number of text messages at which the costs of the two plans are the same.
$$\overbrace{15+0.08x}^{\text{Plan A}}=\overbrace{3+0.12x}^{\text{Plan B}}$$
$$15+0.08x-15=3+0.12x-15$$
$$0.08x=0.12x-12$$
$$0.08x-0.12x=0.12x-12-0.12x$$
$$-0.04x=-12$$
$$\frac{-0.04x}{-0.04}=\frac{-12}{-0.04}$$
$$x=300$$
The two plans are the same at 300 text messages.

4. Let $x=$ the computer's price before the reduction.
$$x-0.30x=840$$
$$0.70x=840$$
$$x=\frac{840}{0.70}$$
$$x=1200$$
Before the reduction the computer's price was $1200.

5. Let $x=$ the amount invested at 9%.
Let $5000-x=$ the amount invested at 11%.
$$0.09x+0.11(5000-x)=487$$
$$0.09x+550-0.11x=487$$
$$-0.02x+550=487$$
$$-0.02x=-63$$
$$x=\frac{-63}{-0.02}$$
$$x=3150$$
$$5000-x=1850$$
$3150 was invested at 9% and $1850 was invested at 11%.

6. Let $x=$ the width of the court.
Let $x+44=$ the length of the court.
$$2l+2w=P$$
$$2(x+44)+2x=288$$
$$2x+88+2x=288$$
$$4x+88=288$$
$$4x=200$$
$$x=\frac{200}{4}$$
$$x=50$$
$$x+44=94$$
The dimensions of the court are 50 feet by 94 feet.

7.
$$2l + 2w = P$$
$$2l + 2w - 2l = P - 2l$$
$$2w = P - 2l$$
$$\frac{2w}{2} = \frac{P - 2l}{2}$$
$$w = \frac{P - 2l}{2}$$

8.
$$P = C + MC$$
$$P = C(1 + M)$$
$$\frac{P}{1 + M} = \frac{C(1 + M)}{1 + M}$$
$$\frac{P}{1 + M} = C$$
$$C = \frac{P}{1 + M}$$

Concept and Vocabulary Check 1.3

1. $x + 658.6$

2. $31 + 2.4x$

3. $4 + 0.15x$

4. $x - 0.15x$ or $0.85x$

5. $0.12x + 0.09(30,000 - x)$

6. isolated on one side

7. factoring

Exercise Set 1.3

1. Let x = the number
$$5x - 4 = 26$$
$$5x = 30$$
$$x = 6$$
The number is 6.

3. Let x = the number
$$x - 0.20x = 20$$
$$0.80x = 20$$
$$x = 25$$
The number is 25.

5. Let x = the number
$$0.60x + x = 192$$
$$1.6x = 192$$
$$x = 120$$
The number is 120.

7. Let x = the number
$$0.70x = 224$$
$$x = 320$$
The number is 320.

9. Let x = the number
$$x + 26 = \text{the other number}$$
$$x + (x + 26) = 64$$
$$x + x + 26 = 64$$
$$2x + 26 = 64$$
$$2x = 38$$
$$x = 19$$
If $x = 19$, then $x + 26 = 45$.
The numbers are 19 and 45.

11.
$$y_1 - y_2 = 2$$
$$(13x - 4) - (5x + 10) = 2$$
$$13x - 4 - 5x - 10 = 2$$
$$8x - 14 = 2$$
$$8x = 16$$
$$\frac{8x}{8} = \frac{16}{8}$$
$$x = 2$$

13.
$$y_1 = 8y_2 + 14$$
$$10(2x - 1) = 8(2x + 1) + 14$$
$$20x - 10 = 16x + 8 + 14$$
$$20x - 10 = 16x + 22$$
$$4x = 32$$
$$\frac{4x}{4} = \frac{32}{4}$$
$$x = 8$$

15.
$$3y_1 - 5y_2 = y_3 - 22$$
$$3(2x + 6) - 5(x + 8) = (x) - 22$$
$$6x + 18 - 5x - 40 = x - 22$$
$$x - 22 = x - 22$$
$$x - x = -22 + 22$$
$$0 = 0$$
The solution set is the set of all real numbers.

17.
$$3y_1 + 4y_2 = 4y_3$$
$$3\left(\frac{1}{x}\right) + 4\left(\frac{1}{2x}\right) = 4\left(\frac{1}{x-1}\right)$$
$$\frac{3}{x} + \frac{2}{x} = \frac{4}{x-1}$$
$$\frac{5}{x} = \frac{4}{x-1}$$
$$\frac{5x(x-1)}{x} = \frac{4x(x-1)}{x-1}$$
$$5(x-1) = 4x$$
$$5x-5 = 4x$$
$$x = 5$$

19. Let $x =$ the number of years spent watching TV.
Let $x + 19 =$ the number of years spent sleeping.
$$x + (x+19) = 37$$
$$x + x + 19 = 37$$
$$2x + 19 = 37$$
$$2x = 18$$
$$x = 9$$
$$x + 19 = 28$$
Americans will spend 9 years watching TV and 28 years sleeping.

21. Let $x =$ the average salary, in thousands, for an American whose final degree is a bachelor's.
Let $2x - 49 =$ the average salary, in thousands, for an American whose final degree is a master's.
$$x + (2x - 49) = 116$$
$$x + 2x - 49 = 116$$
$$3x - 49 = 116$$
$$3x = 165$$
$$x = 55$$
$$2x - 49 = 61$$
The average salary for an American whose final degree is a bachelor's is $55 thousand and for an American whose final degree is a master's is $61 thousand.

23. Let $x =$ the number of years since 2000.
$$31 + 2.4x = 67$$
$$2.4x = 67 - 31$$
$$2.4x = 36$$
$$x = \frac{36}{2.4}$$
$$x = 15$$
67% of American adults will view college education as essential 15 years after 2000, or 2015.

25. a. $y = 24,000 - 3000x$

b.
$$y = 24,000 - 3000x$$
$$9000 = 24,000 - 3000x$$
$$9000 - 24,000 = -3000x$$
$$-15,000 = -3000x$$
$$x = \frac{-15,000}{-3000}$$
$$x = 5$$
The car's value will drop to $9000 after 5 years.

c. Graph:

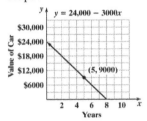

27. Let $x =$ the number of months.
The cost for Club A: $25x + 40$
The cost for Club B: $30x + 15$
$$25x + 40 = 30x + 15$$
$$-5x + 40 = 15$$
$$-5x = -25$$
$$x = 5$$
The total cost for the clubs will be the same at 5 months. The cost will be
$$25(5) + 40 = 30(5) + 15 = \$165$$

29. Let $x =$ the number of uses.
Cost without discount pass: $1.25x$
Cost with discount pass: $15 + 0.75x$
$$1.25x = 15 + 0.75x$$
$$0.50x = 15$$
$$x = 30$$
The bus must be used 30 times in a month for the costs to be equal.

31. a. Let x = the number of years (after 2010).
College A's enrollment: $13,300 + 1000x$

College B's enrollment: $26,800 - 500x$
$$13,300 + 1000x = 26,800 - 500x$$
$$13,300 + 1500x = 26,800$$
$$1500x = 13,500$$
$$x = 9$$
The two colleges will have the same enrollment 9 years after 2010, or 2019. That year the enrollment will be
$$13,300 + 1000(9)$$
$$= 26,800 - 500(9)$$
$$= 22,300 \text{ students}$$

b. Check points to determine that
$y_1 = 13,300 + 1000x$ and
$y_2 = 26,800 - 500x$.

33. Let x = the cost of the television set.
$$x - 0.20x = 336$$
$$0.80x = 336$$
$$x = 420$$
The television set's price is \$420.

35. Let x = the nightly cost
$$x + 0.08x = 162$$
$$1.08x = 162$$
$$x = 150$$
The nightly cost is \$150.

37. Let c = the dealer's cost
$$584 = c + 0.25c$$
$$584 = 1.25c$$
$$467.20 = c$$
The dealer's cost is \$467.20.

39. Let x = the amount invested at 6%.
Let $7000 - x$ = the amount invested at 8%.
$$0.06x + 0.08(7000 - x) = 520$$
$$0.06x + 560 - 0.08x = 520$$
$$-0.02x + 560 = 520$$
$$-0.02x = -40$$
$$x = \frac{-40}{-0.02}$$
$$x = 2000$$
$$7000 - x = 5000$$
\$2000 was invested at 6% and \$5000 was invested at 8%.

41. Let x = amount invested at 12%
$8000 - x$ = amount invested at 5% loss
$$.12x - .05(8000 - x) = 620$$
$$.12x - 400 + .05x = 620$$
$$.17x = 1020$$
$$x = 6000$$
$$8000 - x = 2000$$
\$6000 at 12%, \$2000 at 5% loss

43. Let w = the width of the field
Let $2w$ = the length of the field
$$P = 2(\text{length}) + 2(\text{width})$$
$$300 = 2(2w) + 2(w)$$
$$300 = 4w + 2w$$
$$300 = 6w$$
$$50 = w$$
If $w = 50$, then $2w = 100$. Thus, the dimensions are 50 yards by 100 yards.

45. Let w = the width of the field
Let $2w + 6$ = the length of the field
$$228 = 6w + 12$$
$$216 = 6w$$
$$36 = w$$
If $w = 36$, then $2w + 6 = 2(36) + 6 = 78$. Thus, the dimensions are 36 feet by 78 feet.

47. Let x = the width of the frame.
Total length: $16 + 2x$
Total width: $12 + 2x$
$$P = 2(\text{length}) + 2(\text{width})$$
$$72 = 2(16 + 2x) + 2(12 + 2x)$$
$$72 = 32 + 4x + 24 + 4x$$
$$72 = 8x + 56$$
$$16 = 8x$$
$$2 = x$$
The width of the frame is 2 inches.

49. Let x = number of hours
$35x$ = labor cost
$$35x + 63 = 448$$
$$35x = 385$$
$$x = 11$$
It took 11 hours.

51. Let x = inches over 5 feet
$$100 + 5x = 135$$
$$5x = 35$$
$$x = 7$$
A height of 5 feet 7 inches corresponds to 135 pounds.

53. Let x = the weight of unpeeled bananas.
$\dfrac{7}{8}x =$ weight of peeled bananas

$$x = \frac{7}{8}x + \frac{7}{8}$$
$$\frac{1}{8}x = \frac{7}{8}$$
$$x = 7$$
The banana with peel weighs 7 ounces.

55. $A = lw$
$$w = \frac{A}{l}$$
area of rectangle

57. $A = \dfrac{1}{2}bh$
$$2A = bh$$
$$b = \frac{2A}{h};$$
area of triangle

59. $I = Prt$
$$P = \frac{I}{rt};$$
interest

61. $E = mc^2$
$$m = \frac{E}{c^2};$$
Einstein's equation

63. $T = D + pm$
$$T - D = pm$$
$$\frac{T - D}{m} = \frac{pm}{m}$$
$$\frac{T - D}{m} = p$$
total of payment

65. $A = \dfrac{1}{2}h(a + b)$
$$2A = h(a + b)$$
$$\frac{2A}{h} = a + b$$
$$\frac{2A}{h} - b = a$$
area of trapezoid

67. $S = P + Prt$
$$S - P = Prt$$
$$\frac{S - P}{Pt} = r;$$
interest

69. $B = \dfrac{F}{S - V}$
$$B(S - V) = F$$
$$S - V = \frac{F}{B}$$
$$S = \frac{F}{B} + V$$

71. $IR + Ir = E$
$$I(R + r) = E$$
$$I = \frac{E}{R + r}$$
electric current

73. $\dfrac{1}{p} + \dfrac{1}{q} = \dfrac{1}{f}$
$$qf + pf = pq$$
$$f(q + p) = pq$$
$$f = \frac{pq}{p + q}$$
thin lens equation

75. – 79. Answers will vary.

81. does not make sense; Explanations will vary. Sample explanation: Though mathematical models can often provide excellent estimates about future attitudes, they cannot guaranty perfect precision.

83. does not make sense; Explanations will vary. Sample explanation: Solving a formula for one of its variables does not produce a numerical value for the variable.

85.
$$0.1x + .9(1000 - x) = 420$$
$$0.1 + 900 - 0.9x = 420$$
$$-0.8x = -480$$
$$x = 600$$

600 students at the north campus, 400 students at south campus.

87. Let x = woman's age
$3x$ = Coburn's age
$3x + 20 = 2(x + 20)$
$3x + 20 = 2x + 40$
$x + 20 = 40$
$x = 20$
Coburn is 60 years old the woman is 20 years old.

89. Let x = mother's amount
$2x$ = boy's amount
$\dfrac{x}{2}$ = girl's amount

$$x + 2x + \frac{x}{2} = 14,000$$
$$\frac{7}{2}x = 14,000$$
$$x = \$4,000$$

The mother received $4000, the boy received $8000, and the girl received $2000.

91.
$$V = C - \frac{C - S}{L}N$$
$$VL = CL - CN + SN$$
$$VL - SN = CL - CN$$
$$VL - SN = C(L - N)$$
$$\frac{VL - SN}{L - N} = C$$
$$C = \frac{VL - SN}{L - N}$$

93.
$$(7 - 3x)(-2 - 5x) = -14 - 35x + 6x + 15x^2$$
$$= -14 - 29x + 15x^2$$
$$\text{or}$$
$$= 15x^2 - 29x - 14$$

94.
$$\sqrt{18} - \sqrt{8} = \sqrt{9 \cdot 2} - \sqrt{4 \cdot 2}$$
$$= 3\sqrt{2} - 2\sqrt{2}$$
$$= \sqrt{2}$$

95.
$$\frac{7 + 4\sqrt{2}}{2 - 5\sqrt{2}} \cdot \frac{2 + 5\sqrt{2}}{2 + 5\sqrt{2}} = \frac{14 + 35\sqrt{2} + 8\sqrt{2} + 40}{4 + 10\sqrt{2} - 10\sqrt{2} - 50}$$
$$= \frac{54 + 43\sqrt{2}}{-46}$$
$$= -\frac{54 + 43\sqrt{2}}{46}$$

Section 1.4

Check Point Exercises

1. a.
$$(5 - 2i) + (3 + 3i)$$
$$= 5 - 2i + 3 + 3i$$
$$= (5 + 3) + (-2 + 3)i$$
$$= 8 + i$$

b.
$$(2 + 6i) - (12 - i)$$
$$= 2 + 6i - 12 + i$$
$$= (2 - 12) + (6 + 1)i$$
$$= -10 + 7i$$

2. a.
$$7i(2 - 9i) = 7i(2) - 7i(9i)$$
$$= 14i - 63i^2$$
$$= 14i - 63(-1)$$
$$= 63 + 14i$$

b.
$$(5 + 4i)(6 - 7i) = 30 - 35i + 24i - 28i^2$$
$$= 30 - 35i + 24i - 28(-1)$$
$$= 30 + 28 - 35i + 24i$$
$$= 58 - 11i$$

3.
$$\frac{5 + 4i}{4 - i} = \frac{5 + 4i}{4 - i} \cdot \frac{4 + i}{4 + i}$$
$$= \frac{20 + 5i + 16i + 4i^2}{16 + 4i - 4i - i^2}$$
$$= \frac{20 + 21i - 4}{16 + 1}$$
$$= \frac{16 + 21i}{17}$$
$$= \frac{16}{17} + \frac{21}{17}i$$

4. **a.** $\sqrt{-27} + \sqrt{-48} = i\sqrt{27} + i\sqrt{48}$
$$= i\sqrt{9 \cdot 3} + i\sqrt{16 \cdot 3}$$
$$= 3i\sqrt{3} + 4i\sqrt{3}$$
$$= 7i\sqrt{3}$$

b. $(-2 + \sqrt{-3})^2 = (-2 + i\sqrt{3})^2$
$$= (-2)^2 + 2(-2)(i\sqrt{3}) + (i\sqrt{3})^2$$
$$= 4 - 4i\sqrt{3} + 3i^2$$
$$= 4 - 4i\sqrt{3} + 3(-1)$$
$$= 1 - 4i\sqrt{3}$$

c. $\dfrac{-14 + \sqrt{-12}}{2} = \dfrac{-14 + i\sqrt{12}}{2}$
$$= \dfrac{-14 + 2i\sqrt{3}}{2}$$
$$= \dfrac{-14}{2} + \dfrac{2i\sqrt{3}}{2}$$
$$= -7 + i\sqrt{3}$$

Concept and Vocabulary Check 1.4

1. $\sqrt{-1}$; -1

2. complex; imaginary; real

3. $-6i$

4. $14i$

5. 18; $-15i$; $12i$; $-10i^2$; 10

6. $2 + 9i$

7. $2 + 5i$

8. i; $2i\sqrt{5}$

Exercise Set 1.4

1. $(7 + 2i) + (1 - 4i) = 7 + 2i + 1 - 4i$
$$= 7 + 1 + 2i - 4i$$
$$= 8 - 2i$$

3. $(3 + 2i) - (5 - 7i) = 3 - 5 + 2i + 7i$
$$= 3 + 2i - 5 + 7i$$
$$= -2 + 9i$$

5. $6 - (-5 + 4i) - (-13 - i) = 6 + 5 - 4i + 13 + i$
$$= 24 - 3i$$

7. $8i - (14 - 9i) = 8i - 14 + 9i$
$$= -14 + 8i + 9i$$
$$= -14 + 17i$$

9. $-3i(7i - 5) = -21i^2 + 15i$
$$= -21(-1) + 15i$$
$$= 21 + 15i$$

11. $(-5 + 4i)(3 + i) = -15 - 5i + 12i + 4i^2$
$$= -15 + 7i - 4$$
$$= -19 + 7i$$

13. $(7 - 5i)(-2 - 3i) = -14 - 21i + 10i + 15i^2$
$$= -14 - 15 - 11i$$
$$= -29 - 11i$$

15. $(3 + 5i)(3 - 5i) = 9 - 15i + 15i - 25i^2$
$$= 9 + 25$$
$$= 34$$

17. $(-5 + i)(-5 - i) = 25 + 5i - 5i - i^2$
$$= 25 + 1$$
$$= 26$$

19. $(2 + 3i)^2 = 4 + 12i + 9i^2$
$$= 4 + 12i - 9$$
$$= -5 + 12i$$

21. $\dfrac{2}{3-i} = \dfrac{2}{3-i} \cdot \dfrac{3+i}{3+i}$
$$= \dfrac{2(3+i)}{9+1}$$
$$= \dfrac{2(3+i)}{10}$$
$$= \dfrac{3+i}{5}$$
$$= \dfrac{3}{5} + \dfrac{1}{5}i$$

23. $\dfrac{2i}{1+i} = \dfrac{2i}{1+i} \cdot \dfrac{1-i}{1-i} = \dfrac{2i - 2i^2}{1+1} = \dfrac{2 + 2i}{2} = 1 + i$

25. $\dfrac{8i}{4-3i} = \dfrac{8i}{4-3i} \cdot \dfrac{4+3i}{4+3i}$

$= \dfrac{32i+24i^2}{16+9}$

$= \dfrac{-24+32i}{25}$

$= -\dfrac{24}{25} + \dfrac{32}{25}i$

27. $\dfrac{2+3i}{2+i} = \dfrac{2+3i}{2+i} \cdot \dfrac{2-i}{2-i}$

$= \dfrac{4+4i-3i^2}{4+1}$

$= \dfrac{7+4i}{5}$

$= \dfrac{7}{5} + \dfrac{4}{5}i$

29. $\sqrt{-64} - \sqrt{-25} = i\sqrt{64} - i\sqrt{25}$

$= 8i - 5i = 3i$

31. $5\sqrt{-16} + 3\sqrt{-81} = 5(4i) + 3(9i)$

$= 20i + 27i = 47i$

33. $\left(-2+\sqrt{-4}\right)^2 = (-2+2i)^2$

$= 4 - 8i + 4i^2$

$= 4 - 8i - 4$

$= -8i$

35. $\left(-3-\sqrt{-7}\right)^2 = \left(-3-i\sqrt{7}\right)^2$

$= 9 + 6i\sqrt{7} + i^2(7)$

$= 9 - 7 + 6i\sqrt{7}$

$= 2 + 6i\sqrt{7}$

37. $\dfrac{-8+\sqrt{-32}}{24} = \dfrac{-8+i\sqrt{32}}{24}$

$= \dfrac{-8+i\sqrt{16\cdot 2}}{24}$

$= \dfrac{-8+4i\sqrt{2}}{24}$

$= -\dfrac{1}{3} + \dfrac{\sqrt{2}}{6}i$

39. $\dfrac{-6-\sqrt{-12}}{48} = \dfrac{-6-i\sqrt{12}}{48}$

$= \dfrac{-6-i\sqrt{4\cdot 3}}{48}$

$= \dfrac{-6-2i\sqrt{3}}{48}$

$= -\dfrac{1}{8} - \dfrac{\sqrt{3}}{24}i$

41. $\sqrt{-8}\left(\sqrt{-3}-\sqrt{5}\right) = i\sqrt{8}(i\sqrt{3}-\sqrt{5})$

$= 2i\sqrt{2}\left(i\sqrt{3}-\sqrt{5}\right)$

$= -2\sqrt{6} - 2i\sqrt{10}$

43. $\left(3\sqrt{-5}\right)\left(-4\sqrt{-12}\right) = \left(3i\sqrt{5}\right)\left(-8i\sqrt{3}\right)$

$= -24i^2\sqrt{15}$

$= 24\sqrt{15}$

45. $(2-3i)(1-i) - (3-i)(3+i)$

$= \left(2-2i-3i+3i^2\right) - \left(3^2-i^2\right)$

$= 2 - 5i + 3i^2 - 9 + i^2$

$= -7 - 5i + 4i^2$

$= -7 - 5i + 4(-1)$

$= -11 - 5i$

47. $(2+i)^2 - (3-i)^2$

$= \left(4+4i+i^2\right) - \left(9-6i+i^2\right)$

$= 4 + 4i + i^2 - 9 + 6i - i^2$

$= -5 + 10i$

49. $5\sqrt{-16} + 3\sqrt{-81}$

$= 5\sqrt{16}\sqrt{-1} + 3\sqrt{81}\sqrt{-1}$

$= 5\cdot 4i + 3\cdot 9i$

$= 20i + 27i$

$= 47i$ or $0 + 47i$

51. $f(x) = x^2 - 2x + 2$

$f(1+i) = (1+i)^2 - 2(1+i) + 2$

$= 1 + 2i + i^2 - 2 - 2i + 2$

$= 1 + i^2$

$= 1 - 1$

$= 0$

53. $f(x) = \dfrac{x^2 + 19}{2 - x}$

$f(3i) = \dfrac{(3i)^2 + 19}{2 - 3i}$

$= \dfrac{9i^2 + 19}{2 - 3i}$

$= \dfrac{-9 + 19}{2 - 3i}$

$= \dfrac{10}{2 - 3i}$

$= \dfrac{10}{2 - 3i} \cdot \dfrac{2 + 3i}{2 + 3i}$

$= \dfrac{20 + 30i}{4 - 9i^2}$

$= \dfrac{20 + 30i}{4 + 9}$

$= \dfrac{20 + 30i}{13}$

$= \dfrac{20}{13} + \dfrac{30}{13}i$

55. $E = IR = (4 - 5i)(3 + 7i)$

$= 12 + 28i - 15i - 35i^2$

$= 12 + 13i - 35(-1)$

$= 12 + 35 + 13i = 47 + 13i$

The voltage of the circuit is $(47 + 13i)$ volts.

57. Sum:

$\left(5 + i\sqrt{15}\right) + \left(5 - i\sqrt{15}\right)$

$= 5 + i\sqrt{15} + 5 - i\sqrt{15}$

$= 5 + 5$

$= 10$

Product:

$\left(5 + i\sqrt{15}\right)\left(5 - i\sqrt{15}\right)$

$= 25 - 5i\sqrt{15} + 5i\sqrt{15} - 15i^2$

$= 25 + 15$

$= 40$

59. – 65. Answers will vary.

67. makes sense

69. does not make sense; Explanations will vary. Sample explanation: $i = \sqrt{-1}$; It is not a variable in this context.

71. false; Changes to make the statement true will vary. A sample change is: All irrational numbers are complex numbers.

73. false; Changes to make the statement true will vary. A sample change is:

$\dfrac{7 + 3i}{5 + 3i} = \dfrac{7 + 3i}{5 + 3i} \cdot \dfrac{5 - 3i}{5 - 3i} = \dfrac{44 - 6i}{34} = \dfrac{22}{17} - \dfrac{3}{17}i$

75. $\dfrac{4}{(2 + i)(3 - i)} = \dfrac{4}{6 - 2i + 3i - i^2}$

$= \dfrac{4}{6 + i + 1}$

$= \dfrac{4}{7 + i}$

$= \dfrac{4}{7 + i} \cdot \dfrac{7 - i}{7 - i}$

$= \dfrac{28 - 4i}{49 - i^2}$

$= \dfrac{28 - 4i}{49 + 1}$

$= \dfrac{28 - 4i}{50}$

$= \dfrac{28}{50} - \dfrac{4}{50}i$

$= \dfrac{14}{25} - \dfrac{2}{25}i$

77. $\dfrac{8}{1+\dfrac{2}{i}} = \dfrac{8}{\dfrac{i}{i}+\dfrac{2}{i}}$

$= \dfrac{8}{\dfrac{2+i}{i}}$

$= \dfrac{8i}{2+i}$

$= \dfrac{8i}{2+i} \cdot \dfrac{2-i}{2-i}$

$= \dfrac{16i - 8i^2}{4 - i^2}$

$= \dfrac{16i + 8}{4 + 1}$

$= \dfrac{8 + 16i}{5}$

$= \dfrac{8}{5} + \dfrac{16}{5}i$

78. $2x^2 + 7x - 4 = (2x - 1)(x + 4)$

79. $x^2 - 6x + 9 = (x - 3)(x - 3) = (x - 3)^2$

80. $\dfrac{-b - \sqrt{b^2 - 4ac}}{2a} = \dfrac{-(9) - \sqrt{(9)^2 - 4(2)(-5)}}{2(2)}$

$= \dfrac{-9 - \sqrt{81 + 40}}{4}$

$= \dfrac{-9 - \sqrt{121}}{4}$

$= \dfrac{-9 - 11}{4}$

$= -5$

Section 1.5

Check Point Exercises

1. **a.** $3x^2 - 9x = 0$

$3x(x - 3) = 0$

$3x = 0 \quad$ or $\quad x - 3 = 0$

$x = 0 \qquad\qquad x = 3$

The solution set is $\{0, 3\}$.

b. $2x^2 + x = 1$

$2x^2 + x - 1 = 0$

$(2x - 1)(x + 1) = 0$

$2x - 1 = 0 \quad$ or $\quad x + 1 = 0$

$2x = 1 \qquad\qquad x = -1$

$x = \dfrac{1}{2}$

The solution set is $\left\{-1, \dfrac{1}{2}\right\}$.

2. **a.** $3x^2 = 21$

$\dfrac{3x^2}{3} = \dfrac{21}{3}$

$x^2 = 7$

$x = \pm\sqrt{7}$

The solution set is $\left\{-\sqrt{7}, \sqrt{7}\right\}$.

b. $5x^2 + 45 = 0$

$5x^2 = -45$

$x^2 = -9$

$x = \pm\sqrt{-9}$

$x = \pm 3i$

The solution set is $\{-3i, 3i\}$.

c. $(x + 5)^2 = 11$

$x + 5 = \pm\sqrt{11}$

$x = -5 \pm\sqrt{11}$

The solution set is $\left\{-5 + \sqrt{11}, -5 - \sqrt{11}\right\}$.

3. **a.** The coefficient of the *x*-term is 6. Half of 6 is 3, and 3^2 is 9.

9 should be added to the binomial.

$x^2 + 6x + 9 = (x + 3)^2$

b. The coefficient of the *x*-term is –5.

Half of –5 is $-\dfrac{5}{2}$, and $\left(-\dfrac{5}{2}\right)^2$ is $\dfrac{25}{4}$.

$\dfrac{25}{4}$ should be added to the binomial.

$x^2 - 5x + \dfrac{25}{4} = \left(x - \dfrac{5}{2}\right)^2$

c. The coefficient of the x-term is $\dfrac{2}{3}$.

Half of $\dfrac{2}{3}$ is $\dfrac{1}{3}$, and $\left(\dfrac{1}{3}\right)^2$ is $\dfrac{1}{9}$.

$\dfrac{1}{9}$ should be added to the binomial.

$$x^2 + \frac{2}{3}x + \frac{1}{9} = \left(x + \frac{1}{3}\right)^2$$

4. $x^2 + 4x - 1 = 0$

$x^2 + 4x \quad = 1$

$x^2 + 4x + 4 = 1 + 4$

$(x+2)^2 = 5$

$x + 2 = \pm\sqrt{5}$

$x = -2 \pm \sqrt{5}$

The solution set is $\left\{-2 \pm \sqrt{5}\right\}$.

5. $2x^2 + 3x - 4 = 0$

$x^2 + \dfrac{3}{2}x - 2 = 0$

$x^2 + \dfrac{3}{2}x \quad = 2$

$x^2 + \dfrac{3}{2}x + \dfrac{9}{16} = 2 + \dfrac{9}{16}$

$\left(x + \dfrac{3}{4}\right)^2 = \dfrac{41}{16}$

$x + \dfrac{3}{4} = \pm\sqrt{\dfrac{41}{16}}$

$x + \dfrac{3}{4} = \pm\dfrac{\sqrt{41}}{4}$

$x = -\dfrac{3}{4} \pm \dfrac{\sqrt{41}}{4}$

$x = \dfrac{-3 \pm \sqrt{41}}{4}$

The solution set is $\left\{\dfrac{-3 \pm \sqrt{41}}{4}\right\}$.

6. $2x^2 + 2x - 1 = 0$

$a = 2, b = 2, c = -1$

$x = \dfrac{-b \pm \sqrt{b^2 - 4ac}}{2a}$

$= \dfrac{-2 \pm \sqrt{2^2 - 4(2)(-1)}}{2(2)}$

$= \dfrac{-2 \pm \sqrt{4 + 8}}{4}$

$= \dfrac{-2 \pm \sqrt{12}}{4}$

$= \dfrac{-2 \pm 2\sqrt{3}}{4}$

$= \dfrac{2(-1 \pm \sqrt{3})}{4}$

$= \dfrac{-1 \pm \sqrt{3}}{2}$

The solution set is $\left\{\dfrac{-1 \pm \sqrt{3}}{2}\right\}$.

7. $x^2 - 2x + 2 = 0$

$a = 1, b = -2, c = 2$

$x = \dfrac{-b \pm \sqrt{b^2 - 4ac}}{2a}$

$x = \dfrac{-(-2) \pm \sqrt{(-2)^2 - 4(1)(2)}}{2(1)}$

$x = \dfrac{2 \pm \sqrt{4 - 8}}{2}$

$x = \dfrac{2 \pm \sqrt{-4}}{2}$

$x = \dfrac{2 \pm 2i}{2}$

$x = 1 \pm i$

The solution set is $\{1 + i, 1 - i\}$.

8. a. $a = 1, \quad b = 6, \quad c = 9$

$b^2 - 4ac = (6)^2 - 4(1)(9)$

$\qquad = 36 - 36$

$\qquad = 0$

Since $b^2 - 4ac = 0$, the equation has one real solution that is rational.

b. $a = 2, \quad b = -7, \quad c = -4$

$$b^2 - 4ac = (-7)^2 - 4(2)(-4)$$
$$= 49 + 32$$
$$= 81$$

Since $b^2 - 4ac > 0$, the equation has two real solutions. Since 81 is a perfect square, the two solutions are rational.

c. $a = 3, \quad b = -2, \quad c = 4$

$$b^2 - 4ac = (-2)^2 - 4(3)(4)$$
$$= 4 - 48$$
$$= -44$$

Since $b^2 - 4ac < 0$, the equation has two imaginary solutions that are complex conjugates.

9.
$$P = 0.01A^2 + 0.05A + 107$$
$$115 = 0.01A^2 + 0.05A + 107$$
$$0 = 0.01A^2 + 0.05A - 8$$
$$a = 0.01, \quad b = 0.05, \quad c = -8$$

$$A = \frac{-b \pm \sqrt{b^2 - 4ac}}{2a}$$

$$A = \frac{-(0.05) \pm \sqrt{(0.05)^2 - 4(0.01)(-8)}}{2(0.01)}$$

$$A = \frac{-0.05 \pm \sqrt{0.3225}}{0.02}$$

$$A \approx \frac{-0.05 + \sqrt{0.3225}}{0.02} \qquad A \approx \frac{-0.05 - \sqrt{0.3225}}{0.02}$$

$$A \approx 26 \qquad\qquad A \approx -31$$

Age cannot be negative, reject the negative answer. Thus, a woman whose normal systolic blood pressure is 115 mm Hg is approximately 26 years old.

10. Let w = the screen's width.

$$w^2 + l^2 = d^2$$
$$w^2 + 15^2 = 25^2$$
$$w^2 + 225 = 625$$
$$w^2 = 400$$
$$w = \pm\sqrt{400}$$
$$w = \pm 20$$

Reject the negative value.
The width of the television is 20 inches.

Concept and Vocabulary Check 1.5

1. quadratic

2. $A = 0$ or $B = 0$

3. x-intercepts

4. $\pm\sqrt{d}$

5. $\pm\sqrt{7}$

6. $\dfrac{9}{4}$

7. $\dfrac{4}{25}$

8. 9

9. $\dfrac{1}{9}$

10. $\dfrac{-b \pm \sqrt{b^2 - 4ac}}{2a}$

11. $2; \; 9; \; -5$

12. $1; \; -4; \; -1$

13. $2 \pm \sqrt{2}$

14. $-1 \pm i\dfrac{\sqrt{6}}{2}$

15. $b^2 - 4ac$

16. no

17. two

18. the square root property

19. the quadratic formula

20. factoring and the zero-product principle

21. right; hypotenuse; legs

22. right; legs; the square of the length of the hypotenuse

Exercise Set 1.5

1. $x^2 - 3x - 10 = 0$
$(x + 2)(x - 5) = 0$
$x + 2 = 0$ or $x - 5 = 0$
$x = -2$ or $x = 5$
The solution set is $\{-2, 5\}$.

3. $x^2 = 8x - 15$
$x^2 - 8x + 15 = 0$
$(x - 3)(x - 5) = 0$
$x - 3 = 0$ or $x - 5 = 0$
$x = 3$ or $x = 5$
The solution set is $\{3, 5\}$.

5. $6x^2 + 11x - 10 = 0$
$(2x + 5)(3x - 2) = 0$
$2x + 5 = 0$ or $3x - 2 = 0$
$2x = -5$ $\qquad 3x = 2$
$x = -\dfrac{5}{2}$ or $x = \dfrac{2}{3}$
The solution set is $\left\{-\dfrac{5}{2}, \dfrac{2}{3}\right\}$.

7. $3x^2 - 2x = 8$
$3x^2 - 2x - 8 = 0$
$(3x + 4)(x - 2) = 0$
$3x + 4 = 0$ or $x - 2 = 0$
$3x = -4$
$x = -\dfrac{4}{3}$ or $x = 2$
The solution set is $\left\{-\dfrac{4}{3}, 2\right\}$.

9. $3x^2 + 12x = 0$
$3x(x + 4) = 0$
$3x = 0$ or $x + 4 = 0$
$x = 0$ or $x = -4$
The solution set is $\{-4, 0\}$.

11. $2x(x - 3) = 5x^2 - 7x$
$2x^2 - 6x - 5x^2 + 7x = 0$
$-3x^2 + x = 0$
$x(-3x + 1) = 0$
$x = 0$ or $-3x + 1 = 0$
$\qquad\qquad -3x = -1$
$\qquad\qquad x = \dfrac{1}{3}$
The solution set is $\left\{0, \dfrac{1}{3}\right\}$.

13. $7 - 7x = (3x + 2)(x - 1)$
$7 - 7x = 3x^2 - x - 2$
$7 - 7x - 3x^2 + x + 2 = 0$
$-3x^2 - 6x + 9 = 0$
$-3(x + 3)(x - 1) = 0$
$x + 3 = 0$ or $x - 1 = 0$
$x = -3$ or $x = 1$
The solution set is $\{-3, 1\}$.

15. $3x^2 = 27$
$x^2 = 9$
$x = \pm\sqrt{9} = \pm 3$
The solution set is $\{-3, 3\}$.

17. $5x^2 + 1 = 51$
$5x^2 = 50$
$x^2 = 10$
$x = \pm\sqrt{10}$
The solution set is $\left\{-\sqrt{10}, \sqrt{10}\right\}$.

19. $2x^2 - 5 = -55$
$2x^2 = -50$
$x^2 = -25$
$x = \pm\sqrt{-25} = \pm 5i$
The solution set is $\{5i, -5i\}$.

21. $(x + 2)^2 = 25$
$x + 2 = \pm\sqrt{25}$
$x + 2 = \pm 5$
$x = -2 \pm 5$
$x = -2 + 5$ or $x = -2 - 5$
$x = 3$ $\qquad\qquad x = -7$
The solution set is $\{-7, 3\}$.

23. $3(x-4)^2 = 15$

$\qquad (x-4)^2 = 5$

$\qquad x-4 = \pm\sqrt{5}$

$\qquad x = 4 \pm \sqrt{5}$

The solution set is $\left\{4+\sqrt{5}, 4-\sqrt{5}\right\}$.

25. $(x+3)^2 = -16$

$\qquad x+3 = \pm\sqrt{-16}$

$\qquad x+3 = \pm 4i$

$\qquad x = -3 \pm 4i$

The solution set is $\{-3+4i, -3-4i\}$.

27. $(x-3)^2 = -5$

$\qquad x-3 = \pm\sqrt{-5}$

$\qquad x-3 = \pm i\sqrt{5}$

$\qquad x = 3 \pm i\sqrt{5}$

The solution set is $\left\{3+i\sqrt{5}, 3-i\sqrt{5}\right\}$.

29. $(3x+2)^2 = 9$

$\qquad 3x+2 = \pm\sqrt{9} = \pm 3$

$\qquad 3x+2 = -3 \quad$ or $\quad 3x+2 = 3$

$\qquad 3x = -5 \qquad\qquad 3x = 1$

$\qquad x = -\dfrac{5}{3} \quad$ or $\quad x = \dfrac{1}{3}$

The solution set is $\left\{-\dfrac{5}{3}, \dfrac{1}{3}\right\}$.

31. $(5x-1)^2 = 7$

$\qquad 5x-1 = \pm\sqrt{7}$

$\qquad 5x = 1 \pm\sqrt{7}$

$\qquad x = \dfrac{1 \pm\sqrt{7}}{5}$

The solution set is $\left\{\dfrac{1-\sqrt{7}}{5}, \dfrac{1+\sqrt{7}}{5}\right\}$.

33. $(3x-4)^2 = 8$

$\qquad 3x-4 = \pm\sqrt{8} = \pm 2\sqrt{2}$

$\qquad 3x = 4 \pm 2\sqrt{2}$

$\qquad x = \dfrac{4 \pm 2\sqrt{2}}{3}$

The solution set is $\left\{\dfrac{4-2\sqrt{2}}{3}, \dfrac{4+2\sqrt{2}}{3}\right\}$.

35. $x^2 + 12x$

$\qquad \left(\dfrac{12}{2}\right)^2 = 6^2 = 36$

$\qquad x^2 + 12x + 36 = (x+6)^2$

37. $x^2 - 10x$

$\qquad \left(\dfrac{10}{2}\right)^2 = 5^2 = 25$

$\qquad x^2 - 10x + 25 = (x-5)^2$

39. $x^2 + 3x$

$\qquad \left(\dfrac{3}{2}\right)^2 = \dfrac{9}{4}$

$\qquad x^2 + 3x + \dfrac{9}{4} = \left(x+\dfrac{3}{2}\right)^2$

41. $x^2 - 7x$

$\qquad \left(\dfrac{7}{2}\right)^2 = \dfrac{49}{4}$

$\qquad x^2 - 7x + \dfrac{49}{4} = \left(x-\dfrac{7}{2}\right)^2$

43. $x^2 - \dfrac{2}{3}x$

$\qquad \left(\dfrac{\frac{2}{3}}{2}\right)^2 = \left(\dfrac{1}{3}\right)^2 = \dfrac{1}{9}$

$\qquad x^2 - \dfrac{2}{3}x + \dfrac{1}{9} = \left(x-\dfrac{1}{3}\right)^2$

45. $x^2 - \dfrac{1}{3}x$

$\qquad \left(\dfrac{\frac{1}{3}}{2}\right)^2 = \left(\dfrac{1}{6}\right)^2 = \dfrac{1}{36}$

$\qquad x^2 - \dfrac{1}{3}x + \dfrac{1}{36} = \left(x-\dfrac{1}{6}\right)^2$

47. $\qquad x^2 + 6x = 7$

$\qquad x^2 + 6x + 9 = 7 + 9$

$\qquad\qquad (x+3)^2 = 16$

$\qquad\qquad\quad x+3 = \pm 4$

$\qquad\qquad\qquad x = -3 \pm 4$

The solution set is $\{-7, 1\}$.

49.
$$x^2 - 2x = 2$$
$$x^2 - 2x + 1 = 2 + 1$$
$$(x-1)^2 = 3$$
$$x - 1 = \pm\sqrt{3}$$
$$x = 1 \pm \sqrt{3}$$
The solution set is $\left\{1+\sqrt{3}, 1-\sqrt{3}\right\}$.

51.
$$x^2 - 6x - 11 = 0$$
$$x^2 - 6x = 11$$
$$x^2 - 6x + 9 = 11 + 9$$
$$(x-3)^2 = 20$$
$$x - 3 = \pm\sqrt{20}$$
$$x = 3 \pm 2\sqrt{5}$$
The solution set is $\left\{3+2\sqrt{5}, 3-2\sqrt{5}\right\}$.

53.
$$x^2 + 4x + 1 = 0$$
$$x^2 + 4x = -1$$
$$x^2 + 4x + 4 = -1 + 4$$
$$(x+2)^2 = 3$$
$$x + 2 = \pm\sqrt{3}$$
$$x = -2 \pm \sqrt{3}$$
The solution set is $\left\{-2+\sqrt{3}, -2-\sqrt{3}\right\}$.

55.
$$x^2 - 5x + 6 = 0$$
$$x^2 - 5x = -6$$
$$x^2 - 5x + \frac{25}{4} = -6 + \frac{25}{4}$$
$$\left(x - \frac{5}{2}\right)^2 = \frac{1}{4}$$
$$x - \frac{5}{2} = \pm\sqrt{\frac{1}{4}}$$
$$x - \frac{5}{2} = \pm\frac{1}{2}$$
$$x = \frac{5}{2} \pm \frac{1}{2}$$
$$x = \frac{5}{2} + \frac{1}{2} \quad \text{or} \quad x = \frac{5}{2} - \frac{1}{2}$$
$$x = 3 \qquad\qquad x = 2$$
The solution set is {2, 3}.

57.
$$x^2 + 3x - 1 = 0$$
$$x^2 + 3x = 1$$
$$x^2 + 3x + \frac{9}{4} = 1 + \frac{9}{4}$$
$$\left(x + \frac{3}{2}\right)^2 = \frac{13}{4}$$
$$x + \frac{3}{2} = \pm\frac{\sqrt{13}}{2}$$
$$x = \frac{-3 \pm \sqrt{13}}{2}$$
The solution set is $\left\{\dfrac{-3+\sqrt{13}}{2}, \dfrac{-3-\sqrt{13}}{2}\right\}$.

59.
$$2x^2 - 7x + 3 = 0$$
$$x^2 - \frac{7}{2}x + \frac{3}{2} = 0$$
$$x^2 - \frac{7}{2}x = \frac{-3}{2}$$
$$x^2 - \frac{7}{2}x + \frac{49}{16} = -\frac{3}{2} + \frac{49}{16}$$
$$\left(x - \frac{7}{4}\right)^2 = \frac{25}{16}$$
$$x - \frac{7}{4} = \pm\frac{5}{4}$$
$$x = \frac{7}{4} \pm \frac{5}{4}$$
The solution set is $\left\{\dfrac{1}{2}, 3\right\}$.

61. $4x^2 - 4x - 1 = 0$

$$4x^2 - 4x - 1 = 0$$

$$x^2 - x - \frac{1}{4} = 0$$

$$x^2 - x = \frac{1}{4}$$

$$x^2 - x + \frac{1}{4} = \frac{1}{4} + \frac{1}{4}$$

$$\left(x - \frac{1}{2}\right)^2 = \frac{2}{4}$$

$$x - \frac{1}{2} = \frac{\pm\sqrt{2}}{2}$$

$$x = \frac{1 \pm \sqrt{2}}{2}$$

The solution set is $\left\{\dfrac{1+\sqrt{2}}{2}, \dfrac{1-\sqrt{2}}{2}\right\}$.

63. $3x^2 - 2x - 2 = 0$

$$x^2 - \frac{2}{3}x - \frac{2}{3} = 0$$

$$x^2 - \frac{2}{3}x = \frac{2}{3}$$

$$x^2 - \frac{2}{3}x + \frac{1}{9} = \frac{2}{3} + \frac{1}{9}$$

$$\left(x - \frac{1}{3}\right)^2 = \frac{7}{9}$$

$$x - \frac{1}{3} = \frac{\pm\sqrt{7}}{3}$$

$$x = \frac{1 \pm \sqrt{7}}{3}$$

The solution set is $\left\{\dfrac{1+\sqrt{7}}{3}, \dfrac{1-\sqrt{7}}{3}\right\}$.

65. $x^2 + 8x + 15 = 0$

$$x = \frac{-8 \pm \sqrt{8^2 - 4(1)(15)}}{2(1)}$$

$$x = \frac{-8 \pm \sqrt{64 - 60}}{2}$$

$$x = \frac{-8 \pm \sqrt{4}}{2}$$

$$x = \frac{-8 \pm 2}{2}$$

The solution set is $\{-5, -3\}$.

67. $x^2 + 5x + 3 = 0$

$$x = \frac{-5 \pm \sqrt{5^2 - 4(1)(3)}}{2(1)}$$

$$x = \frac{-5 \pm \sqrt{25 - 12}}{2}$$

$$x = \frac{-5 \pm \sqrt{13}}{2}$$

The solution set is $\left\{\dfrac{-5+\sqrt{13}}{2}, \dfrac{-5-\sqrt{13}}{2}\right\}$.

69. $3x^2 - 3x - 4 = 0$

$$x = \frac{3 \pm \sqrt{(-3)^2 - 4(3)(-4)}}{2(3)}$$

$$x = \frac{3 \pm \sqrt{9 + 48}}{6}$$

$$x = \frac{3 \pm \sqrt{57}}{6}$$

The solution set is $\left\{\dfrac{3+\sqrt{57}}{6}, \dfrac{3-\sqrt{57}}{6}\right\}$.

71. $\qquad 4x^2 = 2x + 7$

$$4x^2 - 2x - 7 = 0$$

$$x = \frac{2 \pm \sqrt{(-2)^2 - 4(4)(-7)}}{2(4)}$$

$$x = \frac{2 \pm \sqrt{4 + 112}}{8}$$

$$x = \frac{2 \pm \sqrt{116}}{8}$$

$$x = \frac{2 \pm 2\sqrt{29}}{8}$$

$$x = \frac{1 \pm \sqrt{29}}{4}$$

The solution set is $\left\{\dfrac{1+\sqrt{29}}{4}, \dfrac{1-\sqrt{29}}{4}\right\}$.

73. $x^2 - 6x + 10 = 0$

$$x = \frac{6 \pm \sqrt{(-6)^2 - 4(1)(10)}}{2(1)}$$

$$x = \frac{6 \pm \sqrt{36 - 40}}{2}$$

$$x = \frac{6 \pm \sqrt{-4}}{2}$$

$$x = \frac{6 \pm 2i}{2}$$

$$x = 3 \pm i$$

The solution set is $\{3 + i, 3 - i\}$.

75. $x^2 - 4x - 5 = 0$

$(-4)^2 - 4(1)(-5)$

$= 16 + 20$

$= 36;$ 2 unequal real solutions

77. $2x^2 - 11x + 3 = 0$

$(-11)^2 - 4(2)(3)$

$= 121 - 24$

$= 97;$ 2 unequal real solutions

79. $x^2 - 2x + 1 = 0$

$(-2)^2 - 4(1)(1)$

$= 4 - 4$

$= 0;$ 1 real solution

81. $x^2 - 3x - 7 = 0$

$(-3)^2 - 4(1)(-7)$

$= 9 + 28$

$= 37;$ 2 unequal real solutions

83. $2x^2 - x = 1$

$2x^2 - x - 1 = 0$

$(2x + 1)(x - 1) = 0$

$2x + 1 = 0$ or $x - 1 = 0$

$2x = -1$

$x = -\dfrac{1}{2}$ or $x = 1$

The solution set is $\left\{-\dfrac{1}{2}, 1\right\}$.

85. $5x^2 + 2 = 11x$

$5x^2 - 11x + 2 = 0$

$(5x - 1)(x - 2) = 0$

$5x - 1 = 0$ or $x - 2 = 0$

$5x = 1$

$x = \dfrac{1}{5}$ or $x = 2$

The solution set is $\left\{\dfrac{1}{5}, 2\right\}$.

87. $3x^2 = 60$

$x^2 = 20$

$x = \pm\sqrt{20}$

$x = \pm 2\sqrt{5}$

The solution set is $\left\{-2\sqrt{5}, 2\sqrt{5}\right\}$.

89. $x^2 - 2x = 1$

$x^2 - 2x + 1 = 1 + 1$

$(x - 1)^2 = 2$

$x - 1 = \pm\sqrt{2}$

$x = 1 \pm \sqrt{2}$

The solution set is $\left\{1 + \sqrt{2}, 1 - \sqrt{2}\right\}$.

91. $(2x + 3)(x + 4) = 1$

$2x^2 + 8x + 3x + 12 = 1$

$2x^2 + 11x + 11 = 0$

$$x = \frac{-11 \pm \sqrt{11^2 - 4(2)(11)}}{2(2)}$$

$$x = \frac{-11 \pm \sqrt{121 - 88}}{4}$$

$$x = \frac{-11 \pm \sqrt{33}}{4}$$

The solution set is $\left\{\dfrac{-11 + \sqrt{33}}{4}, \dfrac{-11 - \sqrt{33}}{4}\right\}$.

93. $(3x-4)^2 = 16$

$$3x-4 = \pm\sqrt{16}$$
$$3x-4 = \pm 4$$
$$3x = 4 \pm 4$$
$$3x = 8 \text{ or } 3x = 0$$
$$x = \frac{8}{3} \text{ or } x = 0$$

The solution set is $\left\{0, \frac{8}{3}\right\}$.

95. $3x^2 - 12x + 12 = 0$

$$x^2 - 4x + 4 = 0$$
$$(x-2)(x-2) = 0$$
$$x - 2 = 0$$
$$x = 2$$

The solution set is $\{2\}$.

97. $4x^2 - 16 = 0$

$$4x^2 = 16$$
$$x^2 = 4$$
$$x = \pm 2$$

The solution set is $\{-2, 2\}$.

99. $x^2 - 6x + 13 = 0$

$$x^2 - 6x = -13$$
$$x^2 - 6x + 9 = -13 + 9$$
$$(x-3)^2 = -4$$
$$x - 3 = \pm 2i$$
$$x = 3 \pm 2i$$

The solution set is $\{3+2i, 3-2i\}$.

101. $x^2 = 4x - 7$

$$x^2 - 4x = -7$$
$$x^2 - 4x + 4 = -7 + 4$$
$$(x-2)^2 = -3$$
$$x - 2 = \pm i\sqrt{3}$$
$$x = 2 \pm i\sqrt{3}$$

The solution set is $\left\{2+i\sqrt{3}, 2-i\sqrt{3}\right\}$.

103. $2x^2 - 7x = 0$

$$x(2x-7) = 0$$
$$x = 0 \text{ or } 2x - 7 = 0$$
$$2x = 7$$
$$x = 0 \text{ or } x = \frac{7}{2}$$

The solution set is $\left\{0, \frac{7}{2}\right\}$.

105. $\dfrac{1}{x} + \dfrac{1}{x+2} = \dfrac{1}{3}; x \neq 0, -2$

$$3x + 6 + 3x = x^2 + 2x$$
$$0 = x^2 - 4x - 6$$
$$x = \frac{-(-4) \pm \sqrt{(-4)^2 - 4(1)(-6)}}{2(1)}$$
$$x = \frac{4 \pm \sqrt{16+24}}{2}$$
$$x = \frac{4 \pm \sqrt{40}}{2}$$
$$x = \frac{4 \pm 2\sqrt{10}}{2}$$
$$x = 2 \pm \sqrt{10}$$

The solution set is $\{2+\sqrt{10}, \ 2-\sqrt{10}\}$.

107. $\dfrac{2x}{x-3} + \dfrac{6}{x+3} = \dfrac{-28}{x^2-9}; x \neq 3, -3$

$$2x(x+3) + 6(x-3) = -28$$
$$2x^2 + 6x + 6x - 18 = -28$$
$$2x^2 + 12x + 10 = 0$$
$$x^2 + 6x + 5 = 0$$
$$(x+1)(x+5) = 0$$

The solution set is $\{-5, \ -1\}$.

109. $x^2 - 4x - 5 = 0$

$$(x+1)(x-5) = 0$$
$$x + 1 = 0 \qquad x - 5 = 0$$
$$x = -1 \quad \text{or} \quad x = 5$$

This equation matches graph (d).

111.
$$0 = -(x+1)^2 + 4$$
$$(x+1)^2 = 4$$
$$x+1 = \pm 2$$
$$x = -1 \pm 2$$
$$x = -3, \quad x = 1$$
This equation matches graph (f).

113. $x^2 - 2x + 2 = 0$
$$a = 1, \quad b = -2, \quad c = 2$$
$$x = \frac{-b \pm \sqrt{b^2 - 4ac}}{2a}$$
$$x = \frac{-(-2) \pm \sqrt{(-2)^2 - 4(1)(2)}}{2(1)}$$
$$x = \frac{2 \pm \sqrt{-4}}{2}$$
$$x = \frac{2 \pm 2i}{2}$$
$$x = 1 \pm i$$
This equation has no real roots. Thus, its equation has no x-intercepts. This equation matches graph (b).

115. $y = 2x^2 - 3x$
$$2 = 2x^2 - 3x$$
$$0 = 2x^2 - 3x - 2$$
$$0 = (2x+1)(x-2)$$
$$x = -\frac{1}{2}, \quad x = 2$$

117.
$$y_1 y_2 = 14$$
$$(x-1)(x+4) = 14$$
$$x^2 + 3x - 4 = 14$$
$$x^2 + 3x - 18 = 0$$
$$(x+6)(x-3) = 0$$
$$x = -6, \quad x = 3$$

119.

$$y_1 + y_2 = 1$$

$$\frac{2x}{x+2} + \frac{3}{x+4} = 1$$

$$(x+2)(x+4)\left(\frac{2x}{x+2} + \frac{3}{x+4}\right) = 1(x+2)(x+4)$$

$$\frac{2x(x+2)(x+4)}{x+2} + \frac{3(x+2)(x+4)}{x+4} = (x+2)(x+4)$$

$$2x(x+4) + 3(x+2) = (x+2)(x+4)$$

$$2x^2 + 8x + 3x + 6 = x^2 + 6x + 8$$

$$x^2 + 5x - 2 = 0$$

$$x = \frac{-b \pm \sqrt{b^2 - 4ac}}{2a}$$

$$x = \frac{-(5) \pm \sqrt{(5)^2 - 4(1)(-2)}}{2(1)}$$

$$x = \frac{-5 \pm \sqrt{33}}{2}$$

The solution set is $\left\{\dfrac{-5+\sqrt{33}}{2}, \dfrac{-5-\sqrt{33}}{2}\right\}$.

121.

$$y_1 - y_2 = 0$$

$$(2x^2 + 5x - 4) - (-x^2 + 15x - 10) = 0$$

$$2x^2 + 5x - 4 + x^2 - 15x + 10 = 0$$

$$3x^2 - 10x + 6 = 0$$

$$x = \frac{-b \pm \sqrt{b^2 - 4ac}}{2a}$$

$$x = \frac{-(-10) \pm \sqrt{(-10)^2 - 4(3)(6)}}{2(3)}$$

$$x = \frac{10 \pm \sqrt{28}}{6}$$

$$x = \frac{10 \pm 2\sqrt{7}}{6}$$

$$x = \frac{5 \pm \sqrt{7}}{3}$$

The solution set is $\left\{\dfrac{5+\sqrt{7}}{3}, \dfrac{5-\sqrt{7}}{3}\right\}$.

123. Values that make the denominator zero must be excluded.

$$2x^2 + 4x - 9 = 0$$

$$x = \frac{-b \pm \sqrt{b^2 - 4ac}}{2a}$$

$$x = \frac{-(4) \pm \sqrt{(4)^2 - 4(2)(-9)}}{2(2)}$$

$$x = \frac{-4 \pm \sqrt{88}}{4}$$

$$x = \frac{-4 \pm 2\sqrt{22}}{4}$$

$$x = \frac{-2 \pm \sqrt{22}}{2}$$

125. $x^2 - (6 + 2x) = 0$

$$x^2 - 2x - 6 = 0$$

Apply the quadratic formula.

$$a = 1 \quad b = -2 \quad c = -6$$

$$x = \frac{-(-2) \pm \sqrt{(-2)^2 - 4(1)(-6)}}{2(1)}$$

$$= \frac{2 \pm \sqrt{4 - (-24)}}{2}$$

$$= \frac{2 \pm \sqrt{28}}{2}$$

$$= \frac{2 \pm \sqrt{4 \cdot 7}}{2} = \frac{2 \pm 2\sqrt{7}}{2} = 1 \pm \sqrt{7}$$

We disregard $1 - \sqrt{7}$ because it is negative, and we are looking for a positive number.

Thus, the number is $1 + \sqrt{7}$.

127.

$$\frac{1}{x^2 - 3x + 2} = \frac{1}{x + 2} + \frac{5}{x^2 - 4}$$

$$\frac{1}{(x - 1)(x - 2)} = \frac{1}{x + 2} + \frac{5}{(x + 2)(x - 2)}$$

Multiply both sides of the equation by the least common denominator, $(x - 1)(x - 2)(x + 2)$. This results in the following:

$$x + 2 = (x - 1)(x - 2) + 5(x - 1)$$

$$x + 2 = x^2 - 2x - x + 2 + 5x - 5$$

$$x + 2 = x^2 + 2x - 3$$

$$0 = x^2 + x - 5$$

Apply the quadratic formula:

$$a = 1 \quad b = 1 \quad c = -5.$$

$$x = \frac{-1 \pm \sqrt{1^2 - 4(1)(-5)}}{2(1)} = \frac{-1 \pm \sqrt{1 - (-20)}}{2}$$

$$= \frac{-1 \pm \sqrt{21}}{2}$$

The solutions are $\dfrac{-1 \pm \sqrt{21}}{2}$, and the solution set is

$$\left\{ \frac{-1 \pm \sqrt{21}}{2} \right\}.$$

129. $\sqrt{2}x^2 + 3x - 2\sqrt{2} = 0$

Apply the quadratic formula:

$$a = \sqrt{2} \quad b = 3 \quad c = -2\sqrt{2}$$

$$x = \frac{-3 \pm \sqrt{3^2 - 4(\sqrt{2})(-2\sqrt{2})}}{2(\sqrt{2})}$$

$$= \frac{-3 \pm \sqrt{9 - (-16)}}{2\sqrt{2}}$$

$$= \frac{-3 \pm \sqrt{25}}{2\sqrt{2}} = \frac{-3 \pm 5}{2\sqrt{2}}$$

Evaluate the expression to obtain two solutions.

$$x = \frac{-3 - 5}{2\sqrt{2}} \quad \text{or} \quad x = \frac{-3 + 5}{2\sqrt{2}}$$

$$= \frac{-8}{2\sqrt{2}} \cdot \frac{\sqrt{2}}{\sqrt{2}} \qquad = \frac{2}{2\sqrt{2}} \cdot \frac{\sqrt{2}}{\sqrt{2}}$$

$$= \frac{-8\sqrt{2}}{4} \qquad\qquad = \frac{2\sqrt{2}}{4}$$

$$= -2\sqrt{2} \qquad\qquad = \frac{\sqrt{2}}{2}$$

The solutions are $-2\sqrt{2}$ and $\dfrac{\sqrt{2}}{2}$, and the solution

set is $\left\{ -2\sqrt{2}, \dfrac{\sqrt{2}}{2} \right\}$.

131. $N = \dfrac{x^2 - x}{2}$

$21 = \dfrac{x^2 - x}{2}$

$42 = x^2 - x$

$0 = x^2 - x - 42$

$0 = (x + 6)(x - 7)$

$x + 6 = 0$ or $x - 7 = 0$

$\qquad x = -6 \qquad\qquad x = 7$

Reject the negative value.
There were 7 players.

133. This is represented on the graph as point (7, 21).

135. $f(x) = 0.013x^2 - 1.19x + 28.24$

$\qquad 3 = 0.013x^2 - 1.19x + 28.24$

$\qquad 0 = 0.013x^2 - 1.19x + 25.24$

Apply the quadratic formula.
$a = 0.013 \quad b = -1.19 \quad c = 25.24$

$x = \dfrac{-(-1.19) \pm \sqrt{(-1.19)^2 - 4(0.013)(25.24)}}{2(0.013)}$

$ = \dfrac{1.19 \pm \sqrt{1.4161 - 1.31248}}{0.026}$

$ = \dfrac{1.19 \pm \sqrt{0.10362}}{0.026}$

$ \approx \dfrac{1.19 \pm 0.32190}{0.026}$

$ \approx 58.15 \text{ or } 33.39$

The solutions are approximately 33.39 and 58.15.
Thus, 33 year olds and 58 year olds are expected to
be in 3 fatal crashes per 100 million miles driven.
The function models the actual data well.

137. Let $y_1 = -0.01x^2 + 0.7x + 6.1$

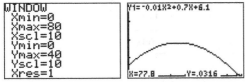

Using the TRACE feature, we find that the height
of the shot put is approximately 0 feet when the
distance is 77.8 feet. Graph (b) shows the shot'
path.

139. a. $\dfrac{1}{\Phi - 1}$

b. $\dfrac{\Phi}{1} = \dfrac{1}{\Phi - 1}$

$\qquad (\Phi - 1)\dfrac{\Phi}{1} = (\Phi - 1)\dfrac{1}{\Phi - 1}$

$\qquad\quad \Phi^2 - \Phi = 1$

$\qquad\quad \Phi^2 - \Phi - 1 = 0$

$\qquad\quad \Phi = \dfrac{-b \pm \sqrt{b^2 - 4ac}}{2a}$

$\qquad\quad \Phi = \dfrac{-(-1) \pm \sqrt{(-1)^2 - 4(1)(-1)}}{2(1)}$

$\qquad\quad \Phi = \dfrac{1 \pm \sqrt{1 + 4}}{2}$

$\qquad\quad \Phi = \dfrac{1 \pm \sqrt{5}}{2}, \text{ reject negative}$

$\qquad\quad \Phi = \dfrac{1 + \sqrt{5}}{2}$

c. The golden ratio is $\dfrac{1 + \sqrt{5}}{2}$ to 1.

141. $x^2 = 4^2 + 2^2$

$\quad x^2 = 16 + 4$

$\quad x^2 = 20$

$\quad\; x = \pm\sqrt{20}$

$\quad\; x = \pm 2\sqrt{5}$

We disregard $-2\sqrt{5}$ because we can't have a
negative measurement. The path is $2\sqrt{5}$ miles, or
approximately 4.5 miles.

143. $\quad 90^2 + 90^2 = x^2$

$\quad 8100 + 8100 = x^2$

$\qquad\quad 16200 = x^2$

$\qquad\qquad x \approx \pm 127.28$

The distance is 127.28 feet.

145. Let w = the width
Let $w + 3$ = the length

Area = lw

$$54 = (w+3)w$$

$$54 = w^2 + 3w$$

$$0 = w^2 + 3w - 54$$

$$0 = (w+9)(w-6)$$

Apply the zero product principle.

$$w+9 = 0 \qquad w-6 = 0$$

$$w = -9 \qquad w = 6$$

The solution set is $\{-9, 6\}$. Disregard -9 because we can't have a negative length measurement. The width is 6 feet and the length is $6 + 3 = 9$ feet.

147. Let x = the length of the side of the original square
Let $x + 3$ = the length of the side of the new, larger square

$$(x+3)^2 = 64$$

$$x^2 + 6x + 9 = 64$$

$$x^2 + 6x - 55 = 0$$

$$(x+11)(x-5) = 0$$

Apply the zero product principle.

$$x+11 = 0 \qquad x-5 = 0$$

$$x = -11 \qquad x = 5$$

The solution set is $\{-11, 5\}$. Disregard -11 because we can't have a negative length measurement. This means that x, the length of the side of the original square, is 5 inches.

149. Let x = the width of the path

$$(20+2x)(10+2x) = 600$$

$$200 + 40x + 20x + 4x^2 = 600$$

$$200 + 60x + 4x^2 = 600$$

$$4x^2 + 60x + 200 = 600$$

$$4x^2 + 60x - 400 = 0$$

$$4(x^2 + 15x - 100) = 0$$

$$4(x+20)(x-5) = 0$$

Apply the zero product principle.

$$4(x+20) = 0 \qquad x-5 = 0$$

$$x+20 = 0 \qquad x = 5$$

$$x = -20$$

The solution set is $\{-20, 5\}$. Disregard -20 because we can't have a negative width measurement. The width of the path is 5 meters.

151. $x(x)(2) = 200$

$$2x^2 = 200$$

$$x^2 = 100$$

$$x = \pm 10$$

The length and width are 10 inches.

153. $x(20 - 2x) = 13$

$$20x - 2x^2 = 13$$

$$0 = 2x^2 - 20x + 13$$

$$x = \frac{-(-20) \pm \sqrt{(-20)^2 - 4(2)(13)}}{2(2)}$$

$$x = \frac{20 \pm \sqrt{296}}{4}$$

$$x = \frac{10 \pm 17.2}{4}$$

$$x = 9.3, 0.7$$

9.3 in and 0.7 in

155. – 165. Answers will vary.

167. does not make sense; Explanations will vary. Sample explanation: Higher degree polynomial equations can have only one x-intercept.

169. makes sense

171. true

173. false; Changes to make the statement true will vary. A sample change is: The first step is to collect all the terms on one side and have 0 on the other.

175. $s = -16t^2 + v_0 t$

$$0 = -16t^2 + v_0 t - s$$

$$a = -16, \ b = v_0, \ c = -s$$

$$t = \frac{-v_0 \pm \sqrt{(v_0)^2 - 4(-16)(-s)}}{2(-16)}$$

$$t = \frac{-v_0 \pm \sqrt{(v_0)^2 - 64s}}{-32}$$

$$t = \frac{v_0 \pm \sqrt{v_0^2 - 64s}}{32}$$

177. $x^3 + x^2 - 4x - 4 = x^2(x+1) - 4(x+1)$

$$= (x+1)(x^2 - 4)$$
$$= (x+1)(x+2)(x-2)$$

178. $(\sqrt{x+4} + 1)^2 = \sqrt{x+4}^2 + 2(\sqrt{x+4})(1) + 1^2$

$$= x + 4 + 2\sqrt{x+4} + 1$$
$$= x + 5 + 2\sqrt{x+4}$$

179. $\quad 5x^{2/3} + 11x^{1/3} + 2 = 0$

$$5(-8)^{2/3} + 11(-8)^{1/3} + 2 = 0$$
$$5(-2)^2 + 11(-2)^1 + 2 = 0$$
$$5(4) + 11(-2) + 2 = 0$$
$$20 - 22 + 2 = 0$$
$$0 = 0, \text{ true}$$

The statement is true.

Mid-Chapter 1 Check Point

1. $\quad -5 + 3(x+5) = 2(3x-4)$

$$-5 + 3x + 15 = 6x - 8$$
$$3x + 10 = 6x - 8$$
$$-3x = -18$$
$$\frac{-3x}{-3} = \frac{-18}{-3}$$
$$x = 6$$

The solution set is $\{6\}$.

2. $\quad 5x^2 - 2x = 7$

$$5x^2 - 2x - 7 = 0$$
$$(5x - 7)(x + 1) = 0$$
$$5x - 7 = 0 \quad \text{or} \quad x + 1 = 0$$
$$5x = 7 \qquad\qquad x = -1$$
$$x = \frac{7}{5}$$

The solution set is $\left\{ -1, \dfrac{7}{5} \right\}$.

3. $\quad \dfrac{x-3}{5} - 1 = \dfrac{x-5}{4}$

$$20\left(\frac{x-3}{5} - 1 \right) = 20\left(\frac{x-5}{4} \right)$$
$$\frac{20(x-3)}{5} - 20(1) = \frac{20(x-5)}{4}$$
$$4(x-3) - 20 = 5(x-5)$$
$$4x - 12 - 20 = 5x - 25$$
$$4x - 32 = 5x - 25$$
$$-x = 7$$
$$x = -7$$

The solution set is $\{-7\}$.

4. $\quad 3x^2 - 6x - 2 = 0$

$$x = \frac{-b \pm \sqrt{b^2 - 4ac}}{2a}$$
$$x = \frac{-(-6) \pm \sqrt{(-6)^2 - 4(3)(-2)}}{2(3)}$$
$$x = \frac{6 \pm \sqrt{60}}{6}$$
$$x = \frac{6 \pm 2\sqrt{15}}{6}$$
$$x = \frac{3 \pm \sqrt{15}}{3}$$

The solution set is $\left\{ \dfrac{3+\sqrt{15}}{3}, \dfrac{3-\sqrt{15}}{3} \right\}$.

5. $\quad 4x - 2(1-x) = 3(2x+1) - 5$

$$4x - 2(1-x) = 3(2x+1) - 5$$
$$4x - 2 + 2x = 6x + 3 - 5$$
$$6x - 2 = 6x - 2$$
$$0 = 0$$

The equation is an identity.
The solution set is $\{x | x \text{ is a real number}\}$.

6. $5x^2 + 1 = 37$

$5x^2 = 36$

$\dfrac{5x^2}{5} = \dfrac{36}{5}$

$x^2 = \dfrac{36}{5}$

$x = \pm\sqrt{\dfrac{36}{5}}$

$x = \pm\dfrac{6}{\sqrt{5}}$

$x = \pm\dfrac{6}{\sqrt{5}} \cdot \dfrac{\sqrt{5}}{\sqrt{5}}$

$x = \pm\dfrac{6\sqrt{5}}{5}$

The solution set is $\left\{ -\dfrac{6\sqrt{5}}{5}, \dfrac{6\sqrt{5}}{5} \right\}$.

7. $x(2x - 3) = -4$

$2x^2 - 3x = -4$

$2x^2 - 3x + 4 = 0$

$x = \dfrac{-b \pm \sqrt{b^2 - 4ac}}{2a}$

$x = \dfrac{-(-3) \pm \sqrt{(-3)^2 - 4(2)(4)}}{2(2)}$

$x = \dfrac{3 \pm \sqrt{-23}}{4}$

$x = \dfrac{3 \pm i\sqrt{23}}{4}$

The solution set is $\left\{ \dfrac{3 + i\sqrt{23}}{4}, \dfrac{3 - i\sqrt{23}}{4} \right\}$.

8. $\dfrac{3x}{4} - \dfrac{x}{3} + 1 = \dfrac{4x}{5} - \dfrac{3}{20}$

$\dfrac{3x}{4} - \dfrac{x}{3} + 1 = \dfrac{4x}{5} - \dfrac{3}{20}$

$60\left(\dfrac{3x}{4} - \dfrac{x}{3} + 1 \right) = 60\left(\dfrac{4x}{5} - \dfrac{3}{20} \right)$

$\dfrac{60(3x)}{4} - \dfrac{60x}{3} + 60(1) = \dfrac{60(4x)}{5} - \dfrac{60(3)}{20}$

$45x - 20x + 60 = 48x - 9$

$25x + 60 = 48x - 9$

$-23x = -69$

$\dfrac{-23x}{-23} = \dfrac{-69}{-23}$

$x = 3$

The solution set is $\{3\}$.

9. $(x + 3)^2 = 24$

$x + 3 = \pm\sqrt{24}$

$x = -3 \pm 2\sqrt{6}$

The solution set is $\left\{ -3 + 2\sqrt{6}, -3 - 2\sqrt{6} \right\}$.

10. $\dfrac{1}{x^2} - \dfrac{4}{x} + 1 = 0$

$x^2\left(\dfrac{1}{x^2} - \dfrac{4}{x} + 1 \right) = x^2(0)$

$\dfrac{x^2}{x^2} - \dfrac{4x^2}{x} + x^2 = 0$

$1 - 4x + x^2 = 0$

$x^2 - 4x + 1 = 0$

$x = \dfrac{-b \pm \sqrt{b^2 - 4ac}}{2a}$

$x = \dfrac{-(-4) \pm \sqrt{(-4)^2 - 4(1)(1)}}{2(1)}$

$x = \dfrac{4 \pm \sqrt{12}}{2}$

$x = \dfrac{4 \pm 2\sqrt{3}}{2}$

$x = 2 \pm \sqrt{3}$

The solution set is $\left\{ 2 + \sqrt{3}, 2 - \sqrt{3} \right\}$.

11. $3x+1-(x-5) = 2x-4$

$2x+6 = 2x-4$

$6 = -4$

The solution set is $\varnothing$.

12. $\dfrac{2x}{x^2+6x+8} = \dfrac{x}{x+4} - \dfrac{2}{x+2}, \quad x \neq -2, x \neq -4$

$\dfrac{2x}{(x+4)(x+2)} = \dfrac{x}{x+4} - \dfrac{2}{x+2}$

$\dfrac{2x(x+4)(x+2)}{(x+4)(x+2)} = (x+4)(x+2)\left(\dfrac{x}{x+4} - \dfrac{2}{x+2}\right)$

$2x = \dfrac{x(x+4)(x+2)}{x+4} - \dfrac{2(x+4)(x+2)}{x+2}$

$2x = x(x+2) - 2(x+4)$

$2x = x^2+2x-2x-8$

$0 = x^2-2x-8$

$0 = (x+2)(x-4)$

$x+2=0 \quad \text{or} \quad x-4=0$

$x=-2 \qquad\qquad x=4$

-2 must be rejected.

The solution set is $\{4\}$.

13. Let $y=0$.

$0 = x^2+6x+2$

$x = \dfrac{-b \pm \sqrt{b^2-4ac}}{2a}$

$x = \dfrac{-(6) \pm \sqrt{(6)^2-4(1)(2)}}{2(1)}$

$x = \dfrac{-6 \pm \sqrt{28}}{2}$

$x = \dfrac{-6 \pm 2\sqrt{7}}{2}$

$x = -3 \pm \sqrt{7}$

x-intercepts: $-3+\sqrt{7}$ and $-3-\sqrt{7}$.

14. Let $y=0$.

$0 = 4(x+1)-3x-(6-x)$

$0 = 4x+4-3x-6+x$

$0 = 2x-2$

$-2x = -2$

$x = 1$

x-intercept: 1.

15. Let $y=0$.

$0 = 2x^2+26$

$-2x^2 = 26$

$x^2 = -13$

$x = \pm\sqrt{-13}$

$x = \pm i\sqrt{13}$

There are no x-intercepts.

16. Let $y=0$.

$0 = \dfrac{x^2}{3} + \dfrac{x}{2} - \dfrac{2}{3}$

$6(0) = 6\left(\dfrac{x^2}{3} + \dfrac{x}{2} - \dfrac{2}{3}\right)$

$0 = \dfrac{6 \cdot x^2}{3} + \dfrac{6 \cdot x}{2} - \dfrac{6 \cdot 2}{3}$

$0 = 2x^2+3x-4$

$x = \dfrac{-b \pm \sqrt{b^2-4ac}}{2a}$

$x = \dfrac{-(3) \pm \sqrt{(3)^2-4(2)(-4)}}{2(2)}$

$x = \dfrac{-3 \pm \sqrt{41}}{4}$

x-intercepts: $\dfrac{-3+\sqrt{41}}{4}$ and $\dfrac{-3-\sqrt{41}}{4}$.

17. Let $y=0$.

$0 = x^2-5x+8$

$x = \dfrac{-b \pm \sqrt{b^2-4ac}}{2a}$

$x = \dfrac{-(-5) \pm \sqrt{(-5)^2-4(1)(8)}}{2(1)}$

$x = \dfrac{5 \pm \sqrt{-7}}{2}$

$x = \dfrac{5 \pm i\sqrt{7}}{2}$

There are no x-intercepts.

18. $y_1 = y_2$

$3(2x-5)-2(4x+1) = -5(x+3)-2$

$6x-15-8x-2 = -5x-15-2$

$-2x-17 = -5x-17$

$3x = 0$

$x = 0$

The solution set is $\{0\}$.

19.

$$y_1 y_2 = 10$$
$$(2x+3)(x+2) = 10$$
$$2x^2 + 7x + 6 = 10$$
$$2x^2 + 7x - 4 = 0$$
$$(2x-1)(x+4) = 0$$
$$2x-1 = 0 \quad \text{or} \quad x+4 = 0$$
$$x = \frac{1}{2} \qquad x = -4$$

The solution set is $\left\{-4, \frac{1}{2}\right\}$.

20. $x^2 + 10x - 3 = 0$

$$x^2 + 10x \quad = 3$$

Since $b = 10$, we add $\left(\dfrac{10}{2}\right)^2 = 5^2 = 25$.

$$x^2 + 10x + 25 = 3 + 25$$
$$(x+5)^2 = 28$$

Apply the square root principle:

$$x + 5 = \pm\sqrt{28}$$
$$x + 5 = \pm\sqrt{4 \cdot 7} = \pm 2\sqrt{7}$$
$$x = -5 \pm 2\sqrt{7}$$

The solutions are $-5 \pm 2\sqrt{7}$, and the solution set is $\left\{-5 \pm 2\sqrt{7}\right\}$.

21. $2x^2 + 5x + 4 = 0$

$$a = 2 \quad b = 5 \quad c = 4$$
$$b^2 - 4ac = 5^2 - 4(2)(4)$$
$$= 25 - 32 = -7$$

Since the discriminant is negative, there are no real solutions. There are two imaginary solutions that are complex conjugates.

22.

$$10x(x+4) = 15x - 15$$
$$10x^2 + 40x = 15x - 15$$
$$10x^2 - 25x + 15 = 0$$
$$a = 10 \quad b = -25 \quad c = 15$$
$$b^2 - 4ac = (-25)^2 - 4(10)(15)$$
$$= 625 - 600 = 25$$

Since the discriminant is positive and a perfect square, there are two rational solutions.

23.

x	(x, y)
-2	-5
-1	-3
0	-1
1	1
2	3

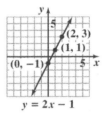

$$y = 2x - 1$$

24.

x	(x, y)
-3	-2
-2	-1
-1	0
0	1
1	0
2	-1
3	-2

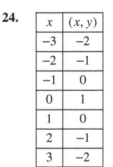

$$y = 1 - |x|$$

25.

x	(x, y)
-2	6
-1	3
0	2
1	3
2	6

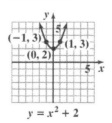

$$y = x^2 + 2$$

26.
$$L = a + (n-1)d$$
$$L = a + dn - d$$
$$-dn = a - d - L$$
$$\frac{-dn}{-d} = \frac{a}{-d} - \frac{d}{-d} - \frac{L}{-d}$$
$$n = -\frac{a}{d} + 1 + \frac{L}{d}$$
$$n = \frac{L}{d} - \frac{a}{d} + 1$$
$$n = \frac{L-a}{d} + 1$$

27.
$$A = 2lw + 2lh + 2wh$$
$$-2lw - 2lh = 2wh - A$$
$$l(-2w - 2h) = 2wh - A$$
$$l = \frac{2wh - A}{-2w - 2h}$$
$$l = \frac{A - 2wh}{2w + 2h}$$

28.
$$f = \frac{f_1 f_2}{f_1 + f_2}$$
$$(f_1 + f_2)(f) = (f_1 + f_2)\frac{f_1 f_2}{f_1 + f_2}$$
$$f_1 f + f_2 f = f_1 f_2$$
$$f_1 f - f_1 f_2 = -f_2 f$$
$$f_1(f - f_2) = -f_2 f$$
$$f_1 = \frac{-f_2 f}{f - f_2}$$
$$f_1 = -\frac{f f_2}{f - f_2} \text{ or } f_1 = \frac{f f_2}{f_2 - f}$$

29. Let x = the number of times "sorry" was used.
Let $x + 419$ = the number of times "love" was used.
Let $x + 32$ = the number of times "thanks" was used.
$$x + (x + 419) + (x + 32) = 1084$$
$$x + x + 419 + x + 32 = 1084$$
$$3x + 451 = 1084$$
$$3x = 633$$
$$x = 211$$
$$x + 419 = 630$$
$$x + 32 = 243$$
The word "sorry" was used 211 times, the word "love" was used 630 times, and the word "thanks" was used 243 times.

30. Let x = the number of years since 1960.
$$23 - 0.28x = 0$$
$$-0.28x = -23$$
$$\frac{-0.28x}{-0.28} = \frac{-23}{-0.28}$$
$$x \approx 82$$
If this trend continues, corporations will pay zero taxes 82 years after 1960, or 2042.

31. Let x = the amount invested at 8%.
Let $25,000 - x$ = the amount invested at 9%.
$$0.08x + 0.09(25,000 - x) = 2135$$
$$0.08x + 2250 - 0.09x = 2135$$
$$-0.01x + 2250 = 2135$$
$$-0.01x = -115$$
$$x = \frac{-115}{-0.01}$$
$$x = 11,500$$
$$25,000 - x = 13,500$$
$11,500 was invested at 8% and $13,500 was invested at 9%.

32. Let x = the number of text messages.
Plan A: $C = 15 + 0.05x$
Plan B: $C = 10 + 0.075x$
Set the costs equal to each other.
$$15 + 0.05x = 10 + 0.075x$$
$$15 = 10 + 0.025x$$
$$5 = 0.025x$$
$$200 = x$$
The cost will be the same for 200 text messages.

$$C = 15 + 0.05x$$
$$C = 15 + 0.05(200)$$
$$= 25$$
The cost for 200 text messages will be $25.

33. Let x = the price before the reduction.
$$x - 0.40x = 468$$
$$0.60x = 468$$
$$\frac{0.60x}{0.60} = \frac{468}{0.60}$$
$$x = 780$$
The price before the reduction was $780.

34. Let x = the amount invested at 4%.
Let $4000 - x$ = the amount invested that lost 3%.
$$0.04x - 0.03(4000 - x) = 55$$
$$0.04x - 120 + 0.03x = 55$$
$$0.07x - 120 = 55$$
$$0.07x = 175$$
$$x = \frac{175}{0.07}$$
$$x = 2500$$
$$4000 - x = 1500$$
$2500 was invested at 4% and $1500 lost 3%.

35. Let x = the width of the rectangle
Let $2x + 5$ = the length of the rectangle
$$2l + 2w = P$$
$$2(2x + 5) + 2x = 46$$
$$4x + 10 + 2x = 46$$
$$6x + 10 = 46$$
$$6x = 36$$
$$\frac{6x}{6} = \frac{36}{6}$$
$$x = 6$$
$$2x + 5 = 17$$
The dimensions of the rectangle are 6 ft by 17 ft.

36. Let x = the width of the rectangle
Let $2x - 1$ = the length of the rectangle
$$lw = A$$
$$(2x - 1)x = 28$$
$$2x^2 - x = 28$$
$$2x^2 - x - 28 = 0$$
$$(2x + 7)(x - 4) = 0$$
$$2x + 7 = 0 \quad \text{or} \quad x - 4 = 0$$
$$2x = -7 \qquad\qquad x = 4$$
$$x = -\frac{7}{2}$$
$-\frac{7}{2}$ must be rejected.
If $x = 4$, then $2x - 1 = 7$
The dimensions of the rectangle are 4 ft by 7 ft.

37. Let x = the height up the pole at which the wires are attached.
$$x^2 + 5^2 = 13^2$$
$$x^2 + 25 = 169$$
$$x^2 = 144$$
$$x = \pm 12$$
-12 must be rejected.
The wires are attached 12 yards up the pole.

38. **a.**
$$P = -10x^2 + 475x + 3500$$
$$5990 = -10x^2 + 475x + 3500$$
$$0 = -10x^2 + 475x - 2490$$
$$0 = 2x^2 - 95x + 498$$
$$0 = (x - 6)(2x - 83)$$
$$x - 6 = 0 \quad \text{or} \quad 2x - 83 = 0$$
$$x = 6 \qquad\qquad 2x = 83$$
$$x = 41.5$$
The population reached 5990 after 6 years.

b. This is represented by the point (6, 5990).

39. $P = 0.004x^2 - 0.37x + 14.1$
$$25 = 0.004x^2 - 0.37x + 14.1$$
$$0 = 0.004x^2 - 0.37x - 10.9$$
$$x = \frac{-b \pm \sqrt{b^2 - 4ac}}{2a}$$
$$x = \frac{-(-0.37) \pm \sqrt{(-0.37)^2 - 4(0.004)(-10.9)}}{2(0.004)}$$
$$x = \frac{0.37 \pm \sqrt{0.1369 + 0.1744}}{0.008}$$
$$x \approx 116, \quad x \approx -23 \text{ (rejected)}$$
The percentage of foreign born Americans will be 25% about 116 years after 1920, or 2036.

40. $(6 - 2i) - (7 - i) = 6 - 2i - 7 + i = -1 - i$

41. $3i(2 + i) = 6i + 3i^2 = -3 + 6i$

42. $(1 + i)(4 - 3i) = 4 - 3i + 4i - 3i^2$
$$= 4 + i + 3 = 7 + i$$

43. $\dfrac{1+i}{1-i} = \dfrac{1+i}{1-i} \cdot \dfrac{1+i}{1+i} = \dfrac{1+i+i+i^2}{1-i^2}$
$$= \frac{1 + 2i - 1}{1 + 1}$$
$$= \frac{2i}{2}$$
$$= i$$

44. $\sqrt{-75} - \sqrt{-12} = 5i\sqrt{3} - 2i\sqrt{3} = 3i\sqrt{3}$

45. $\left(2 - \sqrt{-3}\right)^2 = \left(2 - i\sqrt{3}\right)^2$
$$= 4 - 4i\sqrt{3} + 3i^2$$
$$= 4 - 4i\sqrt{3} - 3$$
$$= 1 - 4i\sqrt{3}$$

Section 1.6

Check Point Exercises

1.
$$4x^4 = 12x^2$$
$$4x^4 - 12x^2 = 0$$
$$4x^2(x^2 - 3) = 0$$
$$4x^2 = 0 \quad \text{or} \quad x^2 - 3 = 0$$
$$x^2 = 0 \qquad\qquad x^2 = 3$$
$$x = \pm\sqrt{0} \qquad\qquad x = \pm\sqrt{3}$$
$$x = 0 \qquad\qquad x = \pm\sqrt{3}$$
The solution set is $\left\{-\sqrt{3}, 0, \sqrt{3}\right\}$.

2.
$$2x^3 + 3x^2 = 8x + 12$$
$$x^2(2x+3) - 4(2x+3) = 10$$
$$(2x+3)(x^2 - 4) = 0$$
$$2x+3 = 0 \quad \text{or} \quad x^2 - 4 = 0$$
$$2x = -3 \qquad\qquad x^2 = 4$$
$$x = -\frac{3}{2} \qquad\qquad x = \pm 2$$
The solution set is $\left\{-2, -\dfrac{3}{2}, 2\right\}$.

3.
$$\sqrt{x+3} + 3 = x$$
$$\sqrt{x+3} = x - 3$$
$$\left(\sqrt{x+3}\right)^2 = (x-3)^2$$
$$x + 3 = x^2 - 6x + 9$$
$$0 = x^2 - 7x + 6$$
$$0 = (x-6)(x-1)$$
$$x - 6 = 0 \quad \text{or} \quad x - 1 = 0$$
$$x = 6 \qquad\qquad x = 1$$
1 does not check and must be rejected.
The solution set is $\{6\}$.

4.
$$\sqrt{x+5} - \sqrt{x-3} = 2$$
$$\sqrt{x+5} = 2 + \sqrt{x-3}$$
$$\left(\sqrt{x+5}\right)^2 = \left(2 + \sqrt{x-3}\right)^2$$
$$x+5 = (2)^2 + 2(2)\left(\sqrt{x-3}\right) + \left(\sqrt{x-3}\right)^2$$
$$x+5 = 4 + 4\sqrt{x-3} + x - 3$$
$$4 = 4\sqrt{x-3}$$
$$\frac{4}{4} = \frac{4\sqrt{x-3}}{4}$$
$$1 = \sqrt{x-3}$$
$$(1)^2 = \left(\sqrt{x-3}\right)^2$$
$$1 = x - 3$$
$$4 = x$$
The check indicates that 4 is a solution.
The solution set is $\{4\}$.

5. **a.**
$$5x^{3/2} - 25 = 0$$
$$5x^{3/2} = 25$$
$$x^{3/2} = 5$$
$$\left(x^{3/2}\right)^{2/3} = (5)^{2/3}$$
$$x = 5^{2/3} \quad \text{or} \quad \sqrt[3]{25}$$

Check:
$$5\left(5^{2/3}\right)^{3/2} - 25 = 0$$
$$5(5) - 25 = 0$$
$$25 - 25 = 0$$
$$0 = 0$$
The solution set is $\left\{5^{2/3}\right\}$ or $\left\{\sqrt[3]{25}\right\}$.

b.
$$x^{\frac{2}{3}} - 8 = -4$$
$$x^{2/3} = 4$$
$$\left(x^{2/3}\right)^{3/2} = 4^{3/2} \qquad \text{or}$$
$$x = \left(2^2\right)^{3/2} \qquad\qquad x = (-2)^3$$
$$x = 2^3 \qquad\qquad\qquad x = -8$$
$$x = 8$$
The solution set is $\{-8, 8\}$.

6.

$$x^4 - 5x^2 + 6 = 0$$

$$\left(x^2\right)^2 - 5x^2 + 6 = 0$$

Let $t = x^2$.

$$t^2 - 5t + 6 = 0$$

$$(t - 3)(t - 2) = 0$$

$t - 3 = 0$ or $t - 2 = 0$

$t = 3$ or $t = 2$

$x^2 = 3$ or $x^2 = 2$

$x = \pm\sqrt{3}$ or $x = \pm\sqrt{2}$

The solution set is $\left\{-\sqrt{3}, -\sqrt{2}, \sqrt{2}, \sqrt{3}\right\}$.

7. $3x^{2/3} - 11x^{1/3} - 4 = 0$

Let $t = x^{1/3}$.

$$3t^2 - 11t - 4 = 0$$

$$(3t + 1)(t - 4) = 0$$

$3t + 1 = 0$ or $t - 4 = 0$

$3t = -1$

$t = -\dfrac{1}{3}$ $t = 4$

$x^{1/3} = -\dfrac{1}{3}$ $x^{1/3} = 4$

$x = \left(-\dfrac{1}{3}\right)^3$ $x = 4^3$

$x = -\dfrac{1}{27}$ $x = 64$

The solution set is $\left\{-\dfrac{1}{27}, 64\right\}$.

8. $|2x - 1| = 5$

$2x - 1 = 5$ or $2x - 1 = -5$

$2x = 6$ $2x = -4$

$x = 3$ $x = -2$

The solution set is $\{-2, 3\}$.

9. $4|1 - 2x| - 20 = 0$

$$4|1 - 2x| = 20$$

$$|1 - 2x| = 5$$

$1 - 2x = 5$ or $1 - 2x = -5$

$-2x = 4$ $-2x = -6$

$x = -2$ $x = 3$

The solution set is $\{-2, 3\}$.

10.

$$H = -2.3\sqrt{I} + 67.6$$

$$33.1 = -2.3\sqrt{I} + 67.6$$

$$-34.5 = -2.3\sqrt{I}$$

$$\frac{-34.5}{-2.3} = \frac{-2.3\sqrt{I}}{-2.3}$$

$$15 = \sqrt{I}$$

$$15^2 = \left(\sqrt{I}\right)^2$$

$$225 = I$$

The model indicates that an annual income of 225 thousand dollars, or \$225,000, corresponds to 33.1 hours per week watching TV.

Concept and Vocabulary Check 1.6

1. subtract $8x$ and subtract 12 from both sides

2. radical

3. extraneous

4. $2x + 1$; $x^2 + 14x + 49$

5. $x + 2$; $x + 8 - 6\sqrt{x - 1}$

6. $5^{\frac{4}{3}}$

7. $\pm 5^{\frac{3}{2}}$

8. x^2; $u^2 - 13u + 36 = 0$

9. $x^{\frac{1}{3}}$; $u^2 + 2u - 3 = 0$

10. c; $-c$

11. $3x - 1 = 7$; $3x - 1 = -7$

Exercise Set 1.6

1.
$$3x^4 - 48x^2 = 0$$
$$3x^2(x^2 - 16) = 0$$
$$3x^2(x+4)(x-4) = 0$$
$$3x^2 = 0 \quad x+4 = 0 \quad x-4 = 0$$
$$x^2 = 0 \quad\quad x = -4 \quad\quad x = 4$$
$$x = 0$$
The solution set is $\{-4, 0, 4\}$.

3.
$$3x^3 + 2x^2 = 12x + 8$$
$$3x^3 + 2x^2 - 12x - 8 = 0$$
$$x^2(3x+2) - 4(3x+2) = 0$$
$$(3x+2)(x^2 - 4) = 0$$
$$3x+2 = 0 \quad\quad x^2 - 4 = 0$$
$$3x = -2 \quad\quad x^2 = 4$$
$$x = -\frac{2}{3} \quad\quad x = \pm 2$$
The solution set is $\left\{-2, -\frac{2}{3}, 2\right\}$.

5.
$$2x - 3 = 8x^3 - 12x^2$$
$$8x^3 - 12x^2 - 2x + 3 = 0$$
$$4x^2(2x-3) - (2x-3) = 0$$
$$(2x-3)(4x^2 - 1) = 0$$
$$2x - 3 = 0 \quad\quad 4x^2 - 1 = 0$$
$$2x = 3 \quad\quad 4x^2 = 1$$
$$\quad\quad\quad\quad x^2 = \frac{1}{4}$$
$$x = \frac{3}{2} \quad\quad x = \pm\frac{1}{2}$$
The solution set is $\left\{\frac{3}{2}, \frac{1}{2}, -\frac{1}{2}\right\}$.

7.
$$4y^3 - 2 = y - 8y^2$$
$$4y^3 + 8y^2 - y - 2 = 0$$
$$4y^2(y+2) - (y+2) = 0$$
$$(y+2)(4y^2 - 1) = 0$$
$$y + 2 = 0 \quad\quad 4y^2 - 1 = 0$$
$$\quad\quad\quad\quad 4y^2 = 1$$
$$\quad\quad\quad\quad y^2 = \frac{1}{4}$$
$$y = -2 \quad\quad y = \pm\frac{1}{2}$$
The solution set is $\left\{-2, \frac{1}{2}, -\frac{1}{2}\right\}$.

9.
$$2x^4 = 16x$$
$$2x^4 - 16x = 0$$
$$2x(x^3 - 8) = 0$$
$$2x = 0 \quad\quad\quad\quad x^3 - 8 = 0$$
$$x = 0 \quad\quad (x-2)(x^2 + 2x + 2) = 0$$
$$\quad\quad\quad x - 2 = 0 \quad x^2 + 2x + 4 = 0$$
$$x = 2 \quad x = \frac{-2 \pm \sqrt{2^2 - 4(1)(4)}}{2(1)}$$
$$x = \frac{-2 \pm \sqrt{-12}}{2}$$
$$x = \frac{-2 \pm 2i\sqrt{3}}{2}$$
$$x = -1 \pm i\sqrt{3}$$
The solution set is $\left\{0, 2, -1 \pm i\sqrt{3}\right\}$.

11.
$$\sqrt{3x+18} = x$$
$$3x + 18 = x^2$$
$$x^2 - 3x - 18 = 0$$
$$(x+3)(x-6) = 0$$
$$x + 3 = 0 \quad x - 6 = 0$$
$$x = -3 \quad\quad x = 6$$
$$\sqrt{3(-3)+18} = -3 \quad \sqrt{3(6)+18} = 6$$
$$\sqrt{-9+18} = -3 \quad\quad \sqrt{18+18} = 6$$
$$\sqrt{9} = -3 \text{ False} \quad \sqrt{36} = 6$$
The solution set is $\{6\}$.

13.
$$\sqrt{x+3} = x-3$$
$$x+3 = x^2 - 6x + 9$$
$$x^2 - 7x + 6 = 0$$
$$(x-1)(x-6) = 0$$
$$x-1 = 0 \quad x-6 = 0$$
$$x = 1 \qquad x = 6$$
$$\sqrt{1+3} = 1-3 \qquad \sqrt{6+3} = 6-3$$
$$\sqrt{4} = -2 \quad \text{False} \quad \sqrt{9} = 3$$
The solution set is $\{6\}$.

15.
$$\sqrt{2x+13} = x+7$$
$$2x+13 = (x+7)^2$$
$$2x+13 = x^2 + 14x + 49$$
$$x^2 + 12x + 36 = 0$$
$$(x+6)^2 = 0$$
$$x+6 = 0$$
$$x = -6$$
$$\sqrt{2(-6)+13} = -6+7$$
$$\sqrt{-12+13} = 1$$
$$\sqrt{1} = 1$$
The solution set is $\{-6\}$.

17.
$$x - \sqrt{2x+5} = 5$$
$$x-5 = \sqrt{2x+5}$$
$$(x-5)^2 = 2x+5$$
$$x^2 - 10x + 25 = 2x + 5$$
$$x^2 - 12x + 20 = 0$$
$$(x-2)(x-10) = 0$$
$$x-2 = 0 \quad x-10 = 0$$
$$x = 2 \qquad x = 10$$
$$2 - \sqrt{2(2)+5} = 5 \quad 10 - \sqrt{2(10)+5} = 5$$
$$2 - \sqrt{9} = 5 \qquad 10 - \sqrt{25} = 5$$
$$2-3 = 5 \quad \text{False} \quad 10-5 = 5$$
The solution set is $\{10\}$.

19.
$$\sqrt{2x+19} - 8 = x$$
$$\sqrt{2x+19} = x+8$$
$$\left(\sqrt{2x+19}\right)^2 = (x+8)^2$$
$$2x+19 = x^2 + 16x + 64$$
$$0 = x^2 + 14x + 45$$
$$0 = (x+9)(x+5)$$
$$x+9 = 0 \quad \text{or} \quad x+5 = 0$$
$$x = -9 \qquad x = -5$$
-9 does not check and must be rejected.
The solution set is $\{-5\}$.

21.
$$\sqrt{3x} + 10 = x+4$$
$$\sqrt{3x} = x-6$$
$$3x = (x-6)^2$$
$$3x = x^2 - 12x + 36$$
$$x^2 - 15x + 36 = 0$$
$$(x-12)(x-3) = 0$$
$$x-12 = 0 \quad x-3 = 0$$
$$x = 12 \qquad x = 3$$
$$\sqrt{3(12)} + 10 = 12+4 \quad \sqrt{3(3)} + 10 = 3+4$$
$$\sqrt{36} + 10 = 16 \qquad \sqrt{9} + 10 = 7$$
$$6+10 = 16 \qquad 3+10 = 7 \text{ False}$$
The solution set is $\{12\}$.

23.
$$\sqrt{x+8} - \sqrt{x-4} = 2$$
$$\sqrt{x+8} = \sqrt{x-4} + 2$$
$$x+8 = (\sqrt{x-4} + 2)^2$$
$$x+8 = x-4 + 4\sqrt{x-4} + 4$$
$$x+8 = x + 4\sqrt{x-4}$$
$$8 = 4\sqrt{x-4}$$
$$2 = \sqrt{x-4}$$
$$4 = x-4$$
$$x = 8$$
$$\sqrt{8+8} - \sqrt{8-4} = 2$$
$$\sqrt{16} - \sqrt{4} = 2$$
$$4-2 = 2$$
The solution set is $\{8\}$.

25. $\sqrt{x-5}-\sqrt{x-8}=3$

$\sqrt{x-5}=\sqrt{x-8}+3$

$x-5=(\sqrt{x-8}+3)^2$

$x-5=x-8+6\sqrt{x-8}+9$

$x-5=x+1+6\sqrt{x-8}$

$-6=6\sqrt{x-8}$

$-1=\sqrt{x-8}$

$1=x-8$

$x=9$

$\sqrt{9-5}-\sqrt{9-8}=3$

$\sqrt{4}-\sqrt{1}=3$

$2-1=3$ False

The solution set is the empty set, $\varnothing$.

27. $\sqrt{2x+3}+\sqrt{x-2}=2$

$\sqrt{2x+3}=2-\sqrt{x-2}$

$2x+3=(2-\sqrt{x-2})^2$

$2x+3=4-4\sqrt{x-2}+x-2$

$x+1=-4\sqrt{x-2}$

$(x+1)^2=16(x-2)$

$x^2+2x+1=16x-32$

$x^2-14x+33=0$

$(x-11)(x-3)=0$

$x-11=0 \quad x-3=0$

$x=11 \qquad x=3$

$\sqrt{2(11)+3}+\sqrt{11-2}=2$

$\sqrt{22+3}+\sqrt{9}=2$

$5+3=2$ False

$\sqrt{2(3)+3}+\sqrt{3-2}=2$

$\sqrt{6+3}+\sqrt{1}=2$

$3+1=2$ False

The solution set is the empty set, $\varnothing$.

29. $\sqrt{3\sqrt{x+1}}=\sqrt{3x-5}$

$3\sqrt{x+1}=3x-5$

$9(x+1)=9x^2-30x+25$

$9x^2-39x+16=0$

$$x=\frac{39\pm\sqrt{945}}{18}=\frac{13\pm\sqrt{105}}{6}$$

Check proposed solutions.

The solution set is $\left\{\dfrac{13+\sqrt{105}}{6}\right\}$.

31. $x^{3/2}=8$

$(x^{3/2})^{2/3}=8^{2/3}$

$x=\sqrt[3]{8}^{\,2}$

$x=2^2$

$x=4$

$4^{3/2}=8$

$\sqrt{4}^{\,3}=8$

$2^3=8$

The solution set is $\{4\}$.

33. $(x-4)^{3/2}=27$

$((x-4)^{3/2})^{2/3}=27^{2/3}$

$x-4=\sqrt[3]{27}^{\,2}$

$x-4=3^2$

$x-4=9$

$x=13$

$(13-4)^{3/2}=27$

$9^{3/2}=27$

$\sqrt{9}^{\,3}=27$

$3^3=27$

The solution set is $\{13\}$.

35. $6x^{5/2} - 12 = 0$

$6x^{5/2} = 12$

$x^{5/2} = 2$

$(x^{5/2})^{2/5} = 2^{2/5}$

$x = \sqrt[5]{2^2}$

$x = \sqrt[5]{4}$

$6(\sqrt[5]{4})^{5/2} - 12 = 0$

$6(4^{1/5})^{5/2} - 12 = 0$

$6(4^{1/2}) - 12 = 0$

$6(2) - 12 = 0$

The solution set is $\left\{\sqrt[5]{4}\right\}$.

37. $(x-4)^{2/3} = 16$

$\left[(x-4)^{2/3}\right]^{3/2} = (16)^{3/2}$

$x - 4 = \left(2^4\right)^{3/2}$

$x - 4 = 4^3 \qquad x - 4 = (-4)^3$

$x - 4 = 64 \qquad x - 4 = -64$

$x = 68 \qquad\qquad x = -60$

The solution set is $\{-60, 68\}$.

39. $(x^2 - x - 4)^{3/4} - 2 = 6$

$(x^2 - x - 4)^{3/4} = 8$

$((x^2 - x - 4)^{3/4})^{4/3} = 8^{4/3}$

$x^2 - x - 4 = \sqrt[3]{8}^4$

$x^2 - x - 4 = 2^4$

$x^2 - x - 4 = 16$

$x^2 - x - 20 = 0$

$(x-5)(x+4) = 0$

$x - 5 = 0 \quad x + 4 = 0$

$x = 5 \qquad x = -4$

$(5^2 - 5 - 4)^{3/4} - 2 = 6$

$(25 - 9)^{3/4} - 2 = 6$

$16^{3/4} - 2 = 6$

$\sqrt[4]{16}^3 - 2 = 6$

$2^3 - 2 = 6$

$8 - 2 = 6$

$((-4)^2 - (-4) - 4)^{3/4} - 2 = 6$

$(16 + 4 - 4)^{3/4} - 2 = 6$

$16^{3/4} - 2 = 6$

$\sqrt[4]{16}^3 - 2 = 6$

$2^3 - 2 = 6$

$8 - 2 = 6$

The solution set is $\{5, -4\}$.

41. $x^4 - 5x^2 + 4 = 0$ let $t = x^2$

$t^2 - 5t + 4 = 0$

$(t-1)(t-4) = 0$

$t - 1 = 0 \quad t - 4 = 0$

$t = 1 \qquad t = 4$

$x^2 = 1 \qquad x^2 = 4$

$x = \pm 1 \qquad x = \pm 2$

The solution set is $\{1, -1, 2, -2\}$.

43. $9x^4 = 25x^2 - 16$

$9x^4 - 25x^2 + 16 = 0$ let $t = x^2$

$9t^2 - 25t + 16 = 0$

$(9t - 16)(t - 1) = 0$

$9t - 16 = 0 \qquad t - 1 = 0$

$9t = 16 \qquad\quad t = 1$

$t = \dfrac{16}{9} \qquad x^2 = 1$

$\qquad\qquad\qquad x = \pm 1$

$x^2 = \dfrac{16}{9}$

$x = \pm\dfrac{4}{3}$

The solution set is $\left\{1, -1, \dfrac{4}{3}, -\dfrac{4}{3}\right\}$.

45. $x - 13\sqrt{x} + 40 = 0$ Let $t = \sqrt{x}$.

$t^2 - 13t + 40 = 0$

$(t - 8)(t - 5) = 0$

$t - 8 = 0$ $t - 5 = 0$

$\quad t = 8$ $t = 5$

$\quad \sqrt{x} = 8$ $\sqrt{x} = 5$

$\quad\quad x = 64$ $x = 25$

The solution set is $\{25, 64\}$.

47. $x^{-2} - x^{-1} - 20 = 0$ Let $t = x^{-1}$

$\quad t^2 - t - 20 = 0$

$(t - 5)(t + 4) = 0$

$t - 5 = 0 \; t + 4 = 0$

$t = 5$ $t = -4$

$x^{-1} = 5 \; x^{-1} = -4$

$\dfrac{1}{x} = 5 \quad \dfrac{1}{x} = -4$

$1 = 5x \quad 1 = -4x$

$\dfrac{1}{5} = x \quad -\dfrac{1}{4} = x$

The solution set is $\left\{ -\dfrac{1}{4}, \dfrac{1}{5} \right\}$.

49. $x^{2/3} - x^{1/3} - 6 = 0$ let $t = x^{1/3}$

$\quad t^2 - t - 6 = 0$

$(t - 3)(t + 2) = 0$

$t - 3 = 0$ $t + 2 = 0$

$\quad t = 3$ $t = -2$

$x^{1/3} = 3$ $x^{1/3} = -2$

$\quad x = 3^3$ $x = (-2)^3$

$\quad x = 27$ $x = -8$

The solution set is $\{27, -8\}$.

51. $x^{3/2} - 2x^{3/4} + 1 = 0$ let $t = x^{3/4}$

$\quad t^2 - 2t + 1 = 0$

$(t - 1)(t - 1) = 0$

$\quad t - 1 = 0$

$\quad\quad t = 1$

$\quad x^{3/4} = 1$

$\quad\quad x = 1^{4/3}$

$\quad\quad x = 1$

The solution set is $\{1\}$.

53. $2x - 3x^{1/2} + 1 = 0$ let $t = x^{1/2}$

$2t^2 - 3t + 1 = 0$

$(2t - 1)(t - 1) = 0$

$2t - 1 = 0$ $t - 1 = 0$

$2t = 1$

$t = \dfrac{1}{2}$ $t = 1$

$x^{1/2} = \dfrac{1}{2}$ $x^{1/2} = 1$

$x = \left(\dfrac{1}{2} \right)^2$ $x = 1^2$

$x = \dfrac{1}{4}$ $x = 1$

The solution set is $\left\{ \dfrac{1}{4}, 1 \right\}$.

55. $(x - 5)^2 - 4(x - 5) - 21 = 0$ let $t = x - 5$

$\quad t^2 - 4t - 21 = 0$

$\quad (t + 3)(t - 7) = 0$

$t + 3 = 0$ $t - 7 = 0$

$\quad t = -3$ $t = 7$

$x - 5 = -3$ $x - 5 = 7$

$\quad x = 2$ $x = 12$

The solution set is $\{2, 12\}$.

57. $\left(x^2 - x \right)^2 - 14\left(x^2 - x \right) + 24 = 0$

Let $t = x^2 - x$.

$t^2 - 14t + 24 = 0$

$(t - 2)(t - 12) = 0$

$t = 2$ or $t = 12$

$x^2 - x = 2$ or $x^2 - x = 12$

$x^2 - x - 2 = 0$ $x^2 - x - 12 = 0$

$(x - 2)(x + 1) = 0$ $(x - 4)(x + 3) = 0$

The solution set is $\{-3, -1, 2, 4\}$.

59. $\left(y - \dfrac{8}{y}\right)^2 + 5\left(y - \dfrac{8}{y}\right) - 14 = 0$

Let $t = y - \dfrac{8}{y}$.

$t^2 + 5t - 14 = 0$
$(t + 7)(t - 2) = 0$
$t = -7$ or $t = 2$

$y - \dfrac{8}{y} = -7$ or $y - \dfrac{8}{y} = 2$

$y^2 + 7y - 8 = 0$ $y^2 - 2y - 8 = 0$

$(y + 8)(y - 1) = 0$ $(y - 4)(y + 2) = 0$

The solution set is $\{-8, -2, 1, 4\}$.

61. $|x| = 8$

$x = 8, x = -8$
The solution set is $\{8, -8\}$.

63. $|x - 2| = 7$

$x - 2 = 7$ $x - 2 = -7$
$x = 9$ $x = -5$
The solution set is $\{9, -5\}$.

65. $|2x - 1| = 5$

$2x - 1 = 5$ $2x - 1 = -5$
$2x = 6$ $2x = -4$
$x = 3$ $x = -2$
The solution set is $\{3, -2\}$.

67. $2|3x - 2| = 14$

$|3x - 2| = 7$
$3x - 2 = 7$ $3x - 2 = -7$
$3x = 9$ $3x = -5$
$x = 3$ $x = -5/3$
The solution set is $\{3, -5/3\}$

69. $7|5x| + 2 = 16$

$7|5x| = 14$
$|5x| = 2$
$5x = 2$ $5x = -2$
$x = 2/5$ $x = -2/5$
The solution set is $\left\{\dfrac{2}{5}, -\dfrac{2}{5}\right\}$.

71. $2\left|4 - \dfrac{5}{2}x\right| + 6 = 18$

$2\left|4 - \dfrac{5}{2}x\right| = 12$

$\left|4 - \dfrac{5}{2}x\right| = 6$

$4 - \dfrac{5}{2}x = 6$ or $4 - \dfrac{5}{2}x = -6$

$-\dfrac{5}{2}x = 2$ $-\dfrac{5}{2}x = -10$

$-\dfrac{2}{5}\left(-\dfrac{5}{2}\right)x = -\dfrac{2}{5}(2)$ $-\dfrac{2}{5}\left(-\dfrac{5}{2}\right)x = -\dfrac{2}{5}(-10)$

$x = -\dfrac{4}{5}$ $x = 4$

The solution set is $\left\{-\dfrac{4}{5}, 4\right\}$.

73. $|x + 1| + 5 = 3$

$|x + 1| = -2$
No solution
The solution set is $\{\ \}$.

75. $|2x - 1| + 3 = 3$

$|2x - 1| = 0$
$2x - 1 = 0$
$2x = 1$
$x = \dfrac{1}{2}$

The solution set is $\left\{\dfrac{1}{2}\right\}$.

77. $|3x - 1| = |x + 5|$

$3x - 1 = x + 5$ $3x - 1 = -x - 5$
$2x - 1 = 5$ $4x - 1 = -5$
$2x = 6$ $4x = -4$
$x = 3$ $x = -1$
The solution set is $\{3, -1\}$.

79. Set $y = 0$ to find the x–intercept(s).

$$0 = \sqrt{x+2} + \sqrt{x-1} - 3$$
$$-\sqrt{x+2} = \sqrt{x-1} - 3$$
$$\left(-\sqrt{x+2}\right)^2 = \left(\sqrt{x-1} - 3\right)^2$$
$$x+2 = \left(\sqrt{x-1}\right)^2 - 2\left(\sqrt{x-1}\right)(3) + (3)^2$$
$$x+2 = x-1 - 6\sqrt{x-1} + 9$$
$$x+2 = x-1 - 6\sqrt{x-1} + 9$$
$$2 = 8 - 6\sqrt{x-1}$$
$$-6 = -6\sqrt{x-1}$$
$$\frac{-6}{-6} = \frac{-6\sqrt{x-1}}{-6}$$
$$1 = \sqrt{x-1}$$
$$(1)^2 = \left(\sqrt{x-1}\right)^2$$
$$1 = x-1$$
$$2 = x$$

The x-intercept is 2.
The corresponding graph is graph (c).

81. Set $y = 0$ to find the x–intercept(s).

$$0 = x^{\frac{1}{3}} + 2x^{\frac{1}{6}} - 3$$

Let $t = x^{\frac{1}{6}}$.

$$x^{\frac{1}{3}} + 2x^{\frac{1}{6}} - 3 = 0$$
$$\left(x^{\frac{1}{6}}\right)^2 + 2x^{\frac{1}{6}} - 3 = 0$$
$$t^2 + 2t - 3 = 0$$
$$(t+3)(t-1) = 0$$
$$t+3 = 0 \quad \text{or} \quad t-1 = 0$$
$$t = -3 \qquad\qquad t = 1$$

Substitute $x^{\frac{1}{6}}$ for t.

$$x^{\frac{1}{6}} = -3 \quad \text{or} \quad x^{\frac{1}{6}} = 1$$
$$\left(x^{\frac{1}{6}}\right)^6 = (-3)^6 \qquad \left(x^{\frac{1}{6}}\right)^6 = (1)^6$$
$$x = 729 \qquad\qquad x = 1$$

729 does not check and must be rejected.
The x-intercept is 1.
The corresponding graph is graph (e).

83. Set $y = 0$ to find the x–intercept(s).

$$(x+2)^2 - 9(x+2) + 20 = 0$$

Let $t = x+2$.

$$(x+2)^2 - 9(x+2) + 20 = 0$$
$$t^2 - 9t + 20 = 0$$
$$(t-5)(t-4) = 0$$
$$t-5 = 0 \quad \text{or} \quad t-4 = 0$$
$$t = 5 \qquad\qquad t = 4$$

Substitute $x+2$ for t.

$$x+2 = 5 \quad \text{or} \quad x+2 = 4$$
$$x = 3 \qquad\qquad x = 2$$

The x-intercepts are 2 and 3.
The corresponding graph is graph (f).

85. $|5 - 4x| = 11$

$$5 - 4x = 11 \qquad\qquad 5 - 4x = -11$$
$$-4x = 6 \quad \text{or} \quad -4x = -16$$
$$x = -\frac{3}{2} \qquad\qquad x = 4$$

The solution set is $\left\{-\dfrac{3}{2}, 4\right\}$.

87. $x + \sqrt{x+5} = 7$

$$\sqrt{x+5} = 7 - x$$
$$\left(\sqrt{x+5}\right)^2 = (7-x)^2$$
$$x+5 = 49 - 14x + x^2$$
$$0 = x^2 - 15x + 44$$
$$0 = (x-4)(x-11)$$
$$x-4 = 0 \quad \text{or} \quad x-11 = 0$$
$$x = 4 \qquad\qquad x = 11$$

11 does not check and must be rejected.
The solution set is $\{4\}$.

89.
$$2x^3 + x^2 - 8x + 2 = 6$$
$$2x^3 + x^2 - 8x - 4 = 0$$
$$x^2(2x+1) - 4(2x+1) = 0$$
$$(2x+1)(x^2 - 4) = 0$$
$$(2x+1)(x+2)(x-2) = 0$$
$$2x+1 = 0 \quad \text{or} \quad x+2 = 0 \quad \text{or} \quad x-2 = 0$$
$$x = -\frac{1}{2} \qquad\qquad x = -2 \qquad\qquad x = 2$$

The solution set is $\left\{-\dfrac{1}{2}, -2, 2\right\}$.

91. $(x+4)^{\frac{3}{2}} = 8$

$$\left((x+4)^{\frac{3}{2}}\right)^{\frac{2}{3}} = (8)^{\frac{2}{3}}$$

$$x+4 = \left(\sqrt[3]{8}\right)^2$$

$$x+4 = (2)^2$$

$$x+4 = 4$$

$$x = 0$$

The solution set is $\{0\}$.

93.
$$y_1 = y_2 + 3$$

$$\left(x^2-1\right)^2 = 2\left(x^2-1\right)+3$$

$$\left(x^2-1\right)^2 - 2\left(x^2-1\right)-3 = 0$$

Let $t = x^2 - 1$ and substitute.

$$t^2 - 2t - 3 = 0$$

$$(t+1)(t-3) = 0$$

$$t+1 = 0 \quad \text{or} \quad t-3 = 0$$

$$t = -1 \qquad\qquad t = 3$$

Substitute $x^2 - 1$ for t.

$$x^2 - 1 = -1 \quad \text{or} \quad x^2 - 1 = 3$$

$$x^2 = 0 \qquad\qquad x^2 = 4$$

$$x = 0 \qquad\qquad x = \pm 2$$

The solution set is $\{-2, 0, 2\}$.

95. $\left|x^2 + 2x - 36\right| = 12$

$$x^2 + 2x - 36 = 12 \qquad x^2 + 2x - 36 = -12$$

$$x^2 + 2x - 48 = 0 \quad \text{or} \quad x^2 + 2x - 24 = 0$$

$$(x+8)(x-6) = 0 \qquad (x+6)(x-4) = 0$$

Setting each of the factors above equal to zero gives
$x = -8$, $x = 6$, $x = -6$, and $x = 4$.

The solution set is $\{-8, -6, 4, 6\}$.

97. $x(x+1)^3 - 42(x+1)^2 = 0$

$$(x+1)^2\left(x(x+1)-42\right) = 0$$

$$(x+1)^2\left(x^2+x-42\right) = 0$$

$$(x+1)^2(x+7)(x-6) = 0$$

Setting each of the factors above equal to zero gives
$x = -7$, $x = -1$, and $x = 6$.

The solution set is $\{-7, -1, 6\}$.

99. Let x = the number.

$$\sqrt{5x-4} = x-2$$

$$\left(\sqrt{5x-4}\right)^2 = (x-2)^2$$

$$5x-4 = x^2 - 4x + 4$$

$$0 = x^2 - 9x + 8$$

$$0 = (x-8)(x-1)$$

$$x-8 = 0 \quad \text{or} \quad x-1 = 0$$

$$x = 8 \qquad\qquad x = 1$$

Check $x = 8$: $\sqrt{5(8)-4} = 8-2$

$$\sqrt{40-4} = 6$$

$$\sqrt{36} = 6$$

$$6 = 6$$

Check $x = 1$: $\sqrt{5(1)-4} = 1-2$

$$\sqrt{5-4} = -1$$

$$\sqrt{-1} \neq -1$$

Discard $x = 1$. The number is 8.

101.
$$r = \sqrt{\frac{3V}{\pi h}}$$

$$r^2 = \left(\sqrt{\frac{3V}{\pi h}}\right)^2$$

$$r^2 = \frac{3V}{\pi h}$$

$$\pi r^2 h = 3V$$

$$\frac{\pi r^2 h}{3} = V$$

$$V = \frac{\pi r^2 h}{3} \quad \text{or} \quad V = \frac{1}{3}\pi r^2 h$$

103. Exclude any value that causes the denominator to equal zero.

$$|x+2|-14=0$$
$$|x+2|=14$$

$$x+2=14 \quad \text{or} \quad x+2=-14$$
$$x=12 \qquad\qquad x=-16$$

-16 and 12 must be excluded from the domain.

105. $t=\dfrac{\sqrt{d}}{2}$

$$1.16=\dfrac{\sqrt{d}}{2}$$

$$2.32=\sqrt{d}$$

$$2.32^2=\left(\sqrt{d}\right)^2$$

$$d\approx 5.4$$

The vertical distance was about 5.4 feet.

107. It is represented by the point $(5.4, 1.16)$.

109. a. According to the line graph, about 48% $\pm 1\%$ of U.S. women participated in the labor force in 2000.

b. $p=2.2\sqrt{t}+36.2$

$$p=2.2\sqrt{28}+36.2\approx 47.8$$

According to the formula, 47.8% of U.S. women participated in the labor force in 2000.

c. $p=2.2\sqrt{t}+36.2$

$$52=2.2\sqrt{t}+36.2$$

$$15.8=2.2\sqrt{t}$$

$$\dfrac{15.8}{2.2}=\dfrac{2.2\sqrt{t}}{2.2}$$

$$\dfrac{15.8}{2.2}=\sqrt{t}$$

$$\left(\dfrac{15.8}{2.2}\right)^2=\left(\sqrt{t}\right)^2$$

$$52\approx t$$

According to the formula, 52% of U.S. women will participate in the labor force 52 years after 1972, or 2024.

111.
$$365=0.2x^{3/2}$$

$$\dfrac{365}{0.2}=\dfrac{0.2x^{3/2}}{0.2}$$

$$1825=x^{3/2}$$

$$1825^2=\left(x^{3/2}\right)^2$$

$$3,330,625=x^3$$

$$\sqrt[3]{3,330,625}=\sqrt[3]{x^3}$$

$$149.34\approx x$$

The average distance of the Earth from the sun is approximately 149 million kilometers.

113. $\sqrt{6^2+x^2}+\sqrt{8^2+(10-x)^2}=18$

$$\sqrt{36+x^2}=18-\sqrt{64+100-20x+x^2}$$

$$36+x^2=324-36\sqrt{x^2-20x+164}+x^2-20x+164$$

$$36\sqrt{x^2-20x+164}=-20x+452$$

$$9\sqrt{x^2-20x+164}=-5x+113$$

$$81(x^2-20x+164)=25x^2-1130x+12769$$

$$81x^2-1620x+13284=25x^2-1130x+12769$$

$$56x^2-490x+515=0$$

$$x=\dfrac{490\pm\sqrt{(-490)^2-4(56)(515)}}{2(56)}$$

$$x=\dfrac{490\pm 353.19}{112}$$

$$x\approx 1.2 \qquad x\approx 7.5$$

The point should be located approximately either 1.2 feet or 7.5 feet from the base of the 6-foot pole.

115. – 121. Answers will vary.

123. $-x^4+4x^3-4x^2=0$

The solution set is $\{0, 2\}$.

$$-(0)^4+4(0)^3-4(0)^2=0$$
$$0=0$$
$$-(2)^4+4(2)^3-4(2)^2=0$$
$$-16+32-16=0$$
$$0=0$$

125. does not make sense; Explanations will vary. Sample explanation: You should substitute into the original equation.

127. does not make sense; Explanations will vary. Sample explanation: Changing the order of the terms does not change the fact that this equation is quadratic in form.

129. false; Changes to make the statement true will vary. A sample change is: Squaring $x + 2$ results in $x^2 + 4x + 4.$

131. false; Changes to make the statement true will vary. A sample change is: To solve the equation, let $u^2 = x.$

133. $\sqrt{6x-2} = \sqrt{2x+3} - \sqrt{4x-1}$

$6x - 2 = 2x + 3 - 2\sqrt{(2x+3)(4x-1)} + 4x - 1$

$-4 = -2\sqrt{(2x+3)(4x-1)}$

$2 = \sqrt{8x^2 + 10x - 3}$

$4 = 8x^2 + 10x - 3$

$8x^2 + 10x - 7 = 0$

$x = \dfrac{-10 \pm \sqrt{10^2 - 4(8)(-7)}}{2(8)}$

$x = \dfrac{-10 \pm \sqrt{100 + 224}}{16}$

$x = \dfrac{-10 \pm \sqrt{324}}{16}$

$x = \dfrac{-10 \pm 18}{16}$

$x = \dfrac{-28}{26}, \dfrac{8}{16}$

$x = \dfrac{1}{2}$

The solution set is $\left\{ \dfrac{1}{2} \right\}.$

135. $\sqrt[3]{x\sqrt{x}} = 9$

$\sqrt[3]{x\sqrt{x}} = 9$

$\sqrt[3]{x^1 x^{\frac{1}{2}}} = 9$

$\left(x^1 x^{\frac{1}{2}} \right)^{\frac{1}{3}} = 9$

$\left(x^{\frac{3}{2}} \right)^{\frac{1}{3}} = 9$

$x^{\frac{1}{2}} = 9$

$\left(x^{\frac{1}{2}} \right)^2 = (9)^2$

$x = 81$

The solution set is $\{81\}.$

137. $3 - 2x \le 11$

$3 - 2(-1) \le 11$

$3 + 2 \le 11$

$5 \le 11, \text{ true}$

Yes, -1 is a solution.

138. $-2x - 4 = x + 5$

$-2x - x = 5 + 4$

$-3x = 9$

$x = \dfrac{9}{-3}$

$x = -3$

The solution set is $\{-3\}.$

139. $\dfrac{x+3}{4} = \dfrac{x-2}{3} + \dfrac{1}{4}$

$12 \left(\dfrac{x+3}{4} \right) = 12 \left(\dfrac{x-2}{3} + \dfrac{1}{4} \right)$

$3(x+3) = 4(x-2) + 3$

$3x + 9 = 4x - 8 + 3$

$3x + 9 = 4x - 5$

$3x - 4x = -5 - 9$

$-x = -14$

$x = 14$

The solution set is $\{14\}.$

Section 1.7

Check Point Exercises

1. **a.** $[-2, 5) = \{x | -2 \le x < 5\}$

 b. $[1, 3.5] = \{x | 1 \le x \le 3.5\}$

 c. $[-\infty, -1) = \{x | x < -1\}$

2. a. Graph $[1,3]$: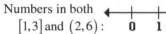

Graph $(2,6)$: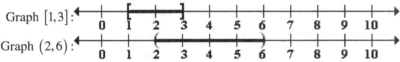

To find the intersection, take the portion of the number line that the two graphs have in common.

Numbers in both
$[1,3]$ and $(2,6)$: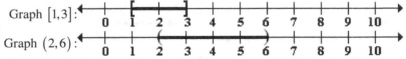

Thus, $[1,3] \cap (2,6) = (2,3]$.

b. Graph $[1,3]$: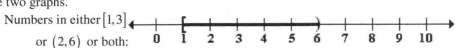

Graph $(2,6)$:

To find the union, take the portion of the number line representing the total collection of numbers in the two graphs.

Numbers in either $[1,3]$
or $(2,6)$ or both:

Thus, $[1,3] \cup (2,6) = [1,6)$.

3. $2 - 3x \le 5$

$-3x \le 3$

$x \ge -1$

The solution set is $\{x \mid x \ge -1\}$ or $[-1, \infty)$.

4. $3x + 1 > 7x - 15$

$-4x > -16$

$\dfrac{-4x}{-4} < \dfrac{-16}{-4}$

$x < 4$

The solution set is $\{x \mid x < 4\}$ or $(-\infty, 4)$.

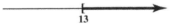

5. $\dfrac{x-4}{2} \ge \dfrac{x-2}{3} + \dfrac{5}{6}$

$6\left(\dfrac{x-4}{2}\right) \ge 6\left(\dfrac{x-2}{3} + \dfrac{5}{6}\right)$

$3(x-4) \ge 2(x-2) + 5$

$3x - 12 \ge 2x - 4 + 5$

$3x - 12 \ge 2x + 1$

$3x - 2x \ge 1 + 12$

$x \ge 13$

The solution set is $\{x \mid x \ge 13\}$ or $[13, \infty)$.

6. a. $3(x+1) > 3x+2$

$3x+3 > 3x+2$

$3 > 2$

$3 > 2$ is true for all values of x.

The solution set is $\{x \mid x \text{ is a real number}\}$ or $\Re$

or $(-\infty. \infty)$.

b. $x+1 \le x-1$

$1 \le -1$

$1 \le -1$ is false for all values of x.

The solution set is $\varnothing$.

7. $1 \le 2x+3 < 11$

$-2 \le 2x < 8$

$-1 \le x < 4$

The solution set is $\{x \mid -1 \le x < 4\}$ or $[-1, 4)$.

8. $|x-2| < 5$

$-5 < x-2 < 5$

$-3 < x < 7$

The solution set is $\{x \mid -3 < x < 7\}$ or $(-3, 7)$.

9. $-3|5x-2| + 20 \ge -19$

$-3|5x-2| \ge -39$

$\dfrac{-3|5x-2|}{-3} \le \dfrac{-39}{-3}$

$|5x-2| \le 13$

$-13 \le 5x-2 \le 13$

$-11 \le 5x \le 15$

$\dfrac{-11}{5} \le \dfrac{5x}{5} \le \dfrac{15}{5}$

$-\dfrac{11}{5} \le x \le 3$

The solution set is $\left\{x \mid -\dfrac{11}{5} \le x \le 3\right\}$ or $\left[-\dfrac{11}{5}, 3\right]$.

10. $18 < |6-3x|$

$6-3x < -18$ or $6-3x > 18$

$-3x < -24$ $\qquad$ $-3x > 12$

$\dfrac{-3x}{-3} > \dfrac{-24}{-3}$ $\qquad$ $\dfrac{-3x}{-3} < \dfrac{12}{-3}$

$x > 8$ $\qquad\qquad$ $x < -4$

The solution set is $\{x \mid x < -4 \text{ or } x > 8\}$

or $(-\infty, -4) \cup (8, \infty)$.

11. Let x = the number of miles driven in a week.

$260 < 80 + 0.25x$

$180 < 0.25x$

$720 < x$

Driving more than 720 miles in a week makes Basic the better deal.

Concept and Vocabulary Check 1.7

1. 2; 5; 2; 5

2. greater than

3. less than or equal to

4. $(-\infty, 9)$; intersection

5. $(-\infty, 12)$; union

6. adding 4; dividing; -3; direction; $>$; $<$

7. $\varnothing$

8. $(-\infty, \infty)$

9. middle

10. $-c$; c

11. $-c$; c

12. $-2 < x-7 < 2$

13. $x-7 < -2$ or $-7 > 2$

Exercise Set 1.7

1. $1 < x \le 6$

3. $-5 \le x < 2$

5. $-3 \le x \le 1$

7. $x > 2$

9. $x \ge -3$

11. $x < 3$

13. $x < 5.5$

15.   Graph $(-3, 0)$:

Graph $[-1, 2]$:

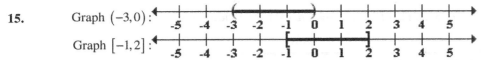

To find the intersection, take the portion of the number line that the two graphs have in common.

Numbers in both
$(-3, 0)$ and $[-1, 2]$:

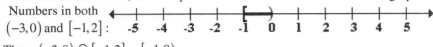

Thus, $(-3, 0) \cap [-1, 2] = [-1, 0)$.

17.   Graph $(-3, 0)$:

Graph $[-1, 2]$:

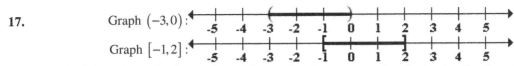

To find the union, take the portion of the number line representing the total collection of numbers in the two graphs.

Numbers in either $(-3, 0)$
or $[-1, 2]$ or both:

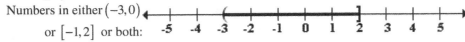

Thus, $(-3, 0) \cup [-1, 2] = (-3, 2]$.

19.   Graph $(-\infty, 5)$:

Graph $[1, 8]$:

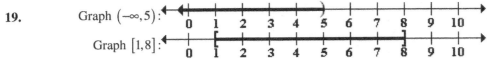

To find the intersection, take the portion of the number line that the two graphs have in common.

Numbers in both
$(-\infty, 5)$ and $[1, 8]$:

Thus, $(-\infty, 5) \cap [1, 8] = [1, 5)$.

21. Graph $(-\infty, 5)$: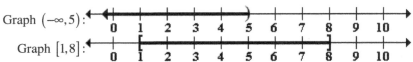

Graph $[1, 8]$: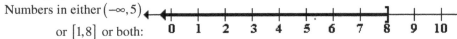

To find the union, take the portion of the number line representing the total collection of numbers in the two graphs.

Numbers in either $(-\infty, 5)$ or $[1, 8]$ or both: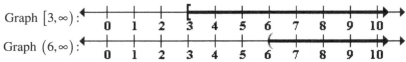

Thus, $(-\infty, 5) \cup [1, 8] = (-\infty, 8]$.

23. Graph $[3, \infty)$: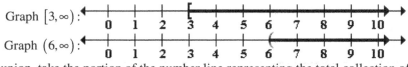

Graph $(6, \infty)$:

To find the intersection, take the portion of the number line that the two graphs have in common.

Numbers in both $[3, \infty)$ and $(6, \infty)$: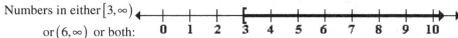

Thus, $[3, \infty) \cap (6, \infty) = (6, \infty)$.

25. Graph $[3, \infty)$:

Graph $(6, \infty)$:

To find the union, take the portion of the number line representing the total collection of numbers in the two graphs.

Numbers in either $[3, \infty)$ or $(6, \infty)$ or both:

Thus, $[3, \infty) \cup (6, \infty) = [3, \infty)$.

27. $5x + 11 < 26$
$5x < 15$
$x < 3$
The solution set is $\{x \mid x < 3\}$, or $(-\infty, 3)$.

29. $3x - 7 \geq 13$
$3x \geq 20$
$x \geq \dfrac{20}{3}$

The solution set is $\left\{ x \mid x > \dfrac{20}{3} \right\}$, or $\left[\dfrac{20}{3}, \infty \right)$.

31. $-9x \geq 36$
$x \leq -4$
The solution set is $\{x \mid x \leq -4\}$, or $(-\infty, -4]$.

33. $8x - 11 \le 3x - 13$
$8x - 3x \le -13 + 11$
$5x \le -2$
$x \le -\dfrac{2}{5}$
The solution set is $\left\{ x \middle| x \le -\dfrac{2}{5} \right\}$, or $\left(-\infty, -\dfrac{2}{5} \right]$.

35. $4(x + 1) + 2 \ge 3x + 6$
$4x + 4 + 2 \ge 3x + 6$
$4x + 6 \ge 3x + 6$
$4x - 3x \ge 6 - 6$
$x \ge 0$
The solution set is $\left\{ x \middle| x > 0 \right\}$, or $[0, \infty)$.

37. $2x - 11 < -3(x + 2)$
$2x - 11 < -3x - 6$
$5x < 5$
$x < 1$
The solution set is $\left\{ x \middle| x < 1 \right\}$, or $(-\infty, 1)$.

39. $1 - (x + 3) \ge 4 - 2x$
$1 - x - 3 \ge 4 - 2x$
$-x - 2 \ge 4 - 2x$
$x \ge 6$
The solution set is $\left\{ x \middle| x \ge 6 \right\}$, or $[6, \infty)$.

41. $\dfrac{x}{4} - \dfrac{3}{2} \le \dfrac{x}{2} + 1$
$\dfrac{4x}{4} - \dfrac{4 \cdot 3}{2} \le \dfrac{4 \cdot x}{2} + 4 \cdot 1$
$x - 6 \le 2x + 4$
$-x \le 10$
$x \ge -10$
The solution set is $\left\{ x \middle| x \ge -10 \right\}$, or $[-10, \infty)$.

43. $1 - \dfrac{x}{2} > 4$
$-\dfrac{x}{2} > 3$
$x < -6$
The solution set is $\left\{ x \middle| x, -6 \right\}$, or $(-\infty, -6)$.

45. $\dfrac{x-4}{6} \ge \dfrac{x-2}{9} + \dfrac{5}{18}$
$3(x - 4) \ge 2(x - 2) + 5$
$3x - 12 \ge 2x - 4 + 5$
$x \ge 13$
The solution set is $\left\{ x \middle| x \ge 13 \right\}$, or $[13, \infty)$.

47. $4(3x - 2) - 3x < 3(1 + 3x) - 7$
$12x - 8 - 3x < 3 + 9x - 7$
$9x - 8 < -4 + 9x$
$-8 < -4$
True for all x
The solution set is $\left\{ x \middle| x \text{ is any real number} \right\}$, or $(-\infty, \infty)$.

49. $5(x - 2) - 3(x + 4) \ge 2x - 20$
$5x - 10 - 3x - 12 \ge 2x - 20$
$2x - 22 \ge 2x - 20$
$-22 \ge -20$
Not true for any x.
The solution set is the empty set, $\varnothing$.

51. $6 < x + 3 < 8$
$6 - 3 < x + 3 - 3 < 8 - 3$
$3 < x < 5$
The solution set is $\left\{ x \middle| 3 < x < 5 \right\}$, or $(3, 5)$.

53. $-3 \le x - 2 < 1$
$-1 \le x < 3$
The solution set is $\left\{ x \middle| -1 \le x < 3 \right\}$, or $[-1, 3)$.

55. $-11 < 2x - 1 \le -5$
$-10 < 2x \le -4$
$-5 < x \le -2$
The solution set is $\left\{ x \middle| -5 < x \le -2 \right\}$, or $(-5, -2]$.

57. $-3 \le \dfrac{2}{3}x - 5 < -1$

$2 \le \dfrac{2}{3}x < 4$

$3 \le x < 6$

The solution set is $\left\{ x \mid 3 \le x < 6 \right\}$, or $[3, 6)$.

59. $|x| < 3$
$-3 < x < 3$
The solution set is $\left\{ x \mid -3 < x < 3 \right\}$, or $(-3, 3)$.

61. $|x - 1| \le 2$
$-2 \le x - 1 \le 2$
$-1 \le x \le 3$
The solution set is $\left\{ x \mid -1 \le x \le 3 \right\}$, or $[-1, 3]$.

63. $|2x - 6| < 8$
$-8 < 2x - 6 < 8$
$-2 < 2x < 14$
$-1 < x < 7$
The solution set is $\left\{ x \mid -1 < x < 7 \right\}$, or $(-1, 7)$.

65. $|2(x - 1) + 4| \le 8$
$-8 \le 2(x - 1) + 4 \le 8$
$-8 \le 2x - 2 + 4 \le 8$
$-8 \le 2x + 2 \le 8$
$-10 \le 2x \le 6$
$-5 \le x \le 3$
The solution set is $\left\{ x \mid -5 \le x \le 3 \right\}$, or $[-5, 3]$.

67. $\left| \dfrac{2y + 6}{3} \right| < 2$

$-2 < \dfrac{2y + 6}{3} < 2$

$-6 < 2y + 6 < 6$
$-12 < 2y < 0$
$-6 < y < 0$
The solution set is $\left\{ x \mid -6 < y < 0 \right\}$, or $(-6, 0)$.

69. $|x| > 3$
$x > 3 \text{ or } x < -3$
The solution set is $\left\{ x \mid x > 3 \text{ or } x < -3 \right\}$, that is,

$(-\infty, -3) \text{ or } (3, \infty)$.

71. $|x - 1| \ge 2$
$x - 1 \ge 2 \quad \text{ or } \quad x - 1 \le -2$
$x \ge 3 \qquad\qquad x \le -1$
The solution set is $\left\{ x \mid x \le -1 \text{ or } x \ge 3 \right\}$, that is,

$(-\infty, -1] \text{ or } [3, \infty)$.

73. $|3x - 8| > 7$
$3x - 8 > 7 \quad \text{ or } \quad 3x - 8 < -7$
$3x > 15 \qquad\qquad 3x < 1$

$x > 5 \qquad\qquad x < \dfrac{1}{3}$

The solution set is $\left\{ x \mid x < \dfrac{1}{3} \text{ or } x > 5 \right\}$, that is,

$\left(-\infty, \dfrac{1}{3} \right) \text{ or } (5, \infty)$.

75. $\left| \dfrac{2x + 2}{4} \right| \ge 2$

$\dfrac{2x + 2}{4} \ge 2 \quad \text{ or } \quad \dfrac{2x + 2}{4} \le -2$

$2x + 2 \ge 8 \qquad\qquad 2x + 2 \le -8$
$2x \ge 6 \qquad\qquad 2x \le -10$
$x \ge 3 \qquad\qquad x \le -5$
The solution set is $\left\{ x \mid x \le -5 \text{ or } x \ge 3 \right\}$, that is,

$(-\infty, -5] \text{ or } [3, \infty)$.

77. $\left| 3 - \dfrac{2}{3}x \right| > 5$

$3 - \dfrac{2}{3}x > 5 \quad \text{ or } \quad 3 - \dfrac{2}{3}x < -5$

$-\dfrac{2}{3}x > 2 \qquad\qquad -\dfrac{2}{3}x < -8$

$x < -3 \qquad\qquad x > 12$

The solution set is $\left\{ x \mid x < -3 \text{ or } x > 12 \right\}$, that is,

$(-\infty, -3) \text{ or } (12, \infty)$.

79. $3|x - 1| + 2 \ge 8$
$3|x - 1| \ge 6$
$|x - 1| \ge 2$
$x - 1 \ge 2 \quad \text{ or } \quad x - 1 \le -2$
$x \ge 3 \qquad\qquad x \le -1$
The solution set is $\left\{ x \mid x \le 1 \text{ or } x \ge 3 \right\}$, that is,

$(-\infty, -1] \text{ or } [3, \infty)$.

81. $-2|x-4| \geq -4$

$$\frac{-2|x-4|}{-2} \leq \frac{-4}{-2}$$

$$|x-4| \leq 2$$

$$-2 \leq x-4 \leq 2$$

$$2 \leq x \leq 6$$

The solution set is $\{x | 2 \leq x \leq 6\}$.

83. $-4|1-x| < -16$

$$\frac{-4|1-x|}{-4} > \frac{-16}{-4}$$

$$|1-x| > 4$$

$$1-x > 4 \qquad 1-x < -4$$

$$-x > 3 \quad \text{or} \quad -x < -5$$

$$x < -3 \qquad x > 5$$

The solution set is $\{x | x < -3 \text{ or } x > 5\}$.

85. $3 \leq |2x-1|$

$$2x-1 \geq 3 \qquad 2x-1 \leq -3$$

$$2x \geq 4 \quad \text{or} \quad 2x \leq -2$$

$$x \geq 2 \qquad x \leq -1$$

The solution set is $\{x | x \leq -1 \text{ or } x \geq 2\}$.

87. $5 > |4-x|$ is equivalent to $|4-x| < 5$.

$$-5 < 4-x < 5$$

$$-9 < -x < 1$$

$$\frac{-9}{-1} > \frac{-x}{-1} > \frac{1}{-1}$$

$$9 > x > -1$$

$$-1 < x < 9$$

The solution set is $\{x | -1 < x < 9\}$.

89. $1 < |2-3x|$ is equivalent to $|2-3x| > 1$.

$$2-3x > 1$$
$$-3x > -1$$
$$\frac{-3x}{-3} < \frac{-1}{-3} \quad \text{or}$$
$$x < \frac{1}{3}$$

$$2-3x < -1$$
$$-3x < -3$$
$$\frac{-3x}{-3} > \frac{-3}{-3}$$
$$x > 1$$

The solution set is $\left\{ x \middle| x < \frac{1}{3} \text{ or } x > 1 \right\}$.

91. $12 < \left| -2x + \frac{6}{7} \right| + \frac{3}{7}$

$$\frac{81}{7} < \left| -2x + \frac{6}{7} \right|$$

$$-2x + \frac{6}{7} > \frac{81}{7} \quad \text{or} \quad -2x + \frac{6}{7} < -\frac{81}{7}$$

$$-2x > \frac{75}{7} \qquad\qquad -2x < -\frac{87}{7}$$

$$x < -\frac{75}{14} \qquad\qquad x > \frac{87}{14}$$

The solution set is $\left\{ x \middle| x < -\frac{75}{14} \text{ or } x > \frac{87}{14} \right\}$, that is,

$$\left(-\infty, -\frac{75}{14} \right) \text{ or } \left(\frac{87}{14}, \infty \right).$$

93. $4 + \left| 3 - \frac{x}{3} \right| \geq 9$

$$\left| 3 - \frac{x}{3} \right| \geq 5$$

$$3 - \frac{x}{3} \geq 5 \quad \text{or} \quad 3 - \frac{x}{3} \leq -5$$

$$-\frac{x}{3} \geq 2 \qquad\qquad -\frac{x}{3} \leq -8$$

$$x \leq -6 \qquad\qquad x \geq 24$$

The solution set is $\{x | x \leq -6 \text{ or } x \geq 24\}$, that is,

$$(-\infty, -6] \text{ or } [24, \infty).$$

95. $\qquad y_1 \leq y_2$

$$\frac{x}{2} + 3 \leq \frac{x}{3} + \frac{5}{2}$$

$$6\left(\frac{x}{2} + 3 \right) \leq 6\left(\frac{x}{3} + \frac{5}{2} \right)$$

$$\frac{6x}{2} + 6(3) \leq \frac{6x}{3} + \frac{6(5)}{2}$$

$$3x + 18 \leq 2x + 15$$

$$x \leq -3$$

The solution set is $(-\infty, -3]$.

97. $\qquad y \geq 4$

$$1 - (x+3) + 2x \geq 4$$

$$1 - x - 3 + 2x \geq 4$$

$$x - 2 \geq 4$$

$$x \geq 6$$

The solution set is $[6, \infty)$.

99.
$$y < 8$$
$$|3x - 4| + 2 < 8$$
$$|3x - 4| < 6$$
$$-6 < 3x - 4 < 6$$
$$-2 < 3x < 10$$
$$\frac{-2}{3} < \frac{3x}{3} < \frac{10}{3}$$
$$\frac{-2}{3} < x < \frac{10}{3}$$

The solution set is $\left(\dfrac{-2}{3}, \dfrac{10}{3} \right)$.

101.
$$y \le 4$$
$$7 - \left| \frac{x}{2} + 2 \right| \le 4$$
$$-\left| \frac{x}{2} + 2 \right| \le -3$$
$$\left| \frac{x}{2} + 2 \right| \ge 3$$

$$\frac{x}{2} + 2 \ge 3 \quad \text{or} \quad \frac{x}{2} + 2 \le -3$$
$$x + 4 \ge 6 \qquad\qquad x + 4 \le -6$$
$$x \ge 2 \qquad\qquad\quad x \le -10$$

The solution set is $(-\infty, -10] \cup [2, \infty)$.

103. The graph's height is below 5 on the interval $(-1, 9)$.

105. The solution set is $\{ x \mid -1 \le x < 2 \}$ or $[-1, 2)$.

107. Let x be the number.
$$|4 - 3x| \ge 5 \quad \text{or} \quad |3x - 4| \ge 5$$
$$3x - 4 \ge 5 \qquad\qquad 3x - 4 \le -5$$
$$3x \ge 9 \quad \text{or} \quad 3x \le -1$$
$$x \ge 3 \qquad\qquad x \le -\frac{1}{3}$$

The solution set is $\left\{ x \mid x \le -\dfrac{1}{3} \text{ or } x \ge 3 \right\}$ or

109. $(0, 4)$

111. passion $\le$ intimacy or intimacy $\ge$ passion

113. passion<commitment or commitment > passion

115. 9, after 3 years

117. a. $I = \frac{1}{4}x + 26$
$$\frac{1}{4}x + 26 > 33$$
$$\frac{1}{4}x > 7$$
$$x > 28$$
More than 33% of U.S. households will have an interfaith marriage in years after 2016 (i.e. $1988 + 28$).

b. $N = \frac{1}{4}x + 6$
$$\frac{1}{4}x + 6 > 14$$
$$\frac{1}{4}x > 8$$
$$x > 32$$
More than 14% of U.S. households will have a person of faith married to someone with no religion in years after 2020 (i.e. $1988 + 32$).

c. More than 33% of U.S. households will have an interfaith marriage *and* more than 14% of U.S. households will have a person of faith married to someone with no religion in years after 2020.

d. More than 33% of U.S. households will have an interfaith marriage *or* more than 14% of U.S. households will have a person of faith married to someone with no religion in years after 2016.

119.
$$28 \le 20 + 0.40(x - 60) \le 40$$
$$28 \le 20 + 0.40x - 24 \le 40$$
$$28 \le 0.40x - 4 \le 40$$
$$32 \le 0.40x \le 44$$
$$80 \le x \le 110$$
Between 80 and 110 ten minutes, inclusive.

121.
$$\left| \frac{h - 50}{5} \right| \ge 1.645$$
$$\frac{h - 50}{5} \ge 1.645 \quad \text{or} \quad \frac{h - 50}{5} \le -1.645$$
$$h - 50 \ge 8.225 \qquad\qquad h - 50 \le -8.225$$
$$h \ge 58.225 \qquad\qquad\quad h \le 41.775$$
The number of outcomes would be 59 or more, or 41 or less.

123.
$$15 + 0.08x < 3 + .12x$$
$$12 < 0.04x$$
$$300 < x$$
Plan A is a better deal when texting more than 300 times per month.

125. $2 + 0.08x < 8 + 0.05x$

$0.03x < 6$

$x < 200$

The credit union is a better deal when writing less than 200 checks.

127. $3000 + 3x < 5.5x$

$3000 < 2.5x$

$1200 < x$

More then 1200 packets of stationary need to be sold each week to make a profit.

129. $245 + 95x \leq 3000$

$95x \leq 2755$

$x \leq 29$

29 bags or less can be lifted safely.

131. a. $\dfrac{86 + 88 + x}{3} \geq 90$

$\dfrac{174 + x}{3} \geq 90$

$174 + x \geq 270$

$x \geq 96$

You must get at least a 96.

b. $\dfrac{86 + 88 + x}{3} < 80$

$\dfrac{174 + x}{3} < 80$

$174 + x < 240$

$x < 66$

This will happen if you get a grade less than 66.

133. Let x = the number of times the bridge is crossed per three month period

The cost with the 3-month pass is $C_3 = 7.50 + 0.50x$.

The cost with the 6-month pass is $C_6 = 30$.

Because we need to buy two 3-month passes per 6-month pass, we multiply the cost with the 3-month pass by 2.

$2(7.50 + 0.50x) < 30$

$15 + x < 30$

$x < 15$

We also must consider the cost without purchasing a pass. We need this cost to be less than the cost with a 3-month pass.

$3x > 7.50 + 0.50x$

$2.50x > 7.50$

$x > 3$

The 3-month pass is the best deal when making more than 3 but less than 15 crossings per 3-month period.

135. – 141. Answers will vary.

143. $x < -3$

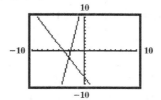

145. a. The cost of Plan A is $4 + 0.10x$;
The cost of Plan B is $2 + 0.15x$.

b. Graph:

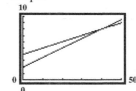

c. 41 or more checks make Plan A better.

d. $4 + 0.10x < 2 + 0.15x$

$2 < 0.05x$

$x > 40$

The solution set is $\{x \mid x > 40\}$ or $(40, \infty)$.

147. makes sense

149. makes sense

151. false; Changes to make the statement true will vary.
A sample change is: $(-\infty, 3) \cup (-\infty, -2) = (-\infty, 3)$

153. true

155. a. $|x - 4| < 3$

b. $|x - 4| \geq 3$

157. Set 1 has each x-coordinate paired with only one y-coordinate.

158.

x	$y = 2x$	(x, y)
–2	$y = 2(-2) = -4$	$(-2, -4)$
–1	$y = 2(-1) + 4 = 2$	$(-1, -2)$
0	$y = 2(0) = 0$	$(0, 0)$
1	$y = 2(1) = 2$	$(1, 2)$
2	$y = 2(2) = 4$	$(2, 4)$

x	$y = 2x + 4$	(x, y)
–2	$y = 2(-2) + 4 = 0$	$(-2, 0)$
–1	$y = 2(-1) + 4 = 2$	$(-1, 2)$
0	$y = 2(0) + 4 = 4$	$(0, 4)$
1	$y = 2(1) + 4 = 6$	$(1, 6)$
2	$y = 2(2) + 4 = 8$	$(2, 8)$

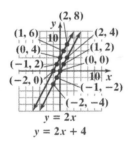

159. a. When the x-coordinate is 2, the y-coordinate is 3.

b. When the y-coordinate is 4, the x-coordinates are –3 and 3.

c. The x-coordinates are all real numbers.

d. The y-coordinates are all real numbers greater than or equal to 1.

Chapter 1 Review Exercises

1.

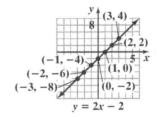

$x = -3, y = -8$
$x = -2, y = -6$
$x = -1, y = -4$
$x = 0, y = -2$
$x = 1, y = 0$
$x = 2, y = 2$
$x = 3, y = 4$

2.

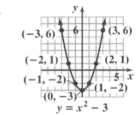

$x = -3, y = 6$
$x = -2, y = 1$
$x = -1, y = -2$
$x = 0, y = -3$
$x = 1, y = -2$
$x = 2, y = 1$
$x = 3, y = 6$

3.

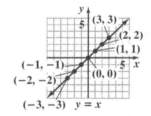

$x = -3, y = -3$
$x = -2, y = -2$
$x = -1, y = -1$
$x = 0, y = 0$
$x = 1, y = 1$
$x = 2, y = 2$
$x = 3, y = 3$

4.

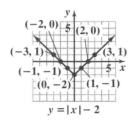

$$x = -3, y = 1$$
$$x = -2, y = 0$$
$$x = -1, y = -1$$
$$x = 0, y = -2$$
$$x = 1, y = -1$$
$$x = 2, y = 0$$
$$x = 3, y = 1$$

5.   A portion of Cartesian coordinate plane with minimum x-value equal to –20, maximum x-value equal to 40, x-scale equal to 10 and with minimum y-value equal to –5, maximum y-value equal to 5, and y-scale equal to 1.

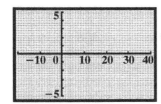

6. x-intercept: –2; The graph intersects the x-axis at $(-2, 0)$.

   y-intercept: 2; The graph intersects the y-axis at $(0, 2)$.

7. x-intercepts: 2, –2; The graph intersects the x-axis at $(-2, 0)$ and $(2, 0)$.

   y-intercept: –4; The graph intercepts the y-axis at $(0, -4)$.

8. x-intercept: 5; The graph intersects the x-axis at $(5, 0)$.

   y-intercept: None; The graph does not intersect the y-axis.

9.   The coordinates are (20, 8). This means that 8% of college students anticipated a starting salary of $20 thousand.

10.   The starting salary that was anticipated by the greatest percentage of college students was $30 thousand. 22% of students anticipated this salary.

11.   The starting salary that was anticipated by the least percentage of college students was $70 thousand. 2% of students anticipated this salary.

12.   Starting salaries of $25 thousand and $30 thousand were anticipated by more than 20% of college students

13.   14% of students anticipated a starting salary of $40 thousand.

14. $p = -0.01s^2 + 0.8s + 3.7$
   $p = -0.01(40)^2 + 0.8(40) + 3.7$
   $p = 19.7$
   This is greater than the estimate of the previous question.

15. $2x - 5 = 7$
   $2x = 12$
   $x = 6$
   The solution set is {6}.
   This is a conditional equation.

16. $5x + 20 = 3x$
   $2x = -20$
   $x = -10$
   The solution set is {–10}.
   This is a conditional equation.

17. $7(x - 4) = x + 2$
   $7x - 28 = x + 2$
   $6x = 30$
   $x = 5$
   The solution set is {5}.
   This is a conditional equation.

18. $1 - 2(6 - x) = 3x + 2$
   $1 - 12 + 2x = 3x + 2$
   $-11 - x = 2$
   $-x = 13$
   $x = -13$
   The solution set is {–13}.
   This is a conditional equation.

19. $2(x - 4) + 3(x + 5) = 2x - 2$
   $2x - 8 + 3x + 15 = 2x - 2$
   $5x + 7 = 2x - 2$
   $3x = -9$
   $x = -3$
   The solution set is {–3}.
   This is a conditional equation.

20. $2x - 4(5x + 1) = 3x + 17$

$2x - 20x - 4 = 3x + 17$

$-18x - 4 = 3x + 17$

$-21x = 21$

$x = -1$

The solution set is $\{-1\}$.

This is a conditional equation.

21. $7x + 5 = 5(x + 3) + 2x$

$7x + 5 = 5x + 15 + 2x$

$7x + 5 = 7x + 15$

$5 = 15$

The solution set is $\varnothing$.

This is an inconsistent equation.

22. $7x + 13 = 2(2x - 5) + 3x + 23$

$7x + 13 = 2(2x - 5) + 3x + 23$

$7x + 13 = 4x - 10 + 3x + 23$

$7x + 13 = 7x + 13$

$13 = 13$

The solution set is all real numbers.

This is an identity.

23. $\dfrac{2x}{3} = \dfrac{x}{6} + 1$

$2(2x) = x + 6$

$4x = x + 6$

$3x = 6$

$x = 2$

The solution set is $\{2\}$.

This is a conditional equation.

24. $\dfrac{x}{2} - \dfrac{1}{10} = \dfrac{x}{5} + \dfrac{1}{2}$

$5x - 1 = 2x + 5$

$3x = 6$

$x = 2$

The solution set is $\{2\}$.

This is a conditional equation.

25. $\dfrac{2x}{3} = 6 - \dfrac{x}{4}$

$4(2x) = 12(6) - 3x$

$8x = 72 - 3x$

$11x = 72$

$x = \dfrac{72}{11}$

The solution set is $\left\{\dfrac{72}{11}\right\}$.

This is a conditional equation.

26. $\dfrac{x}{4} = 2 - \dfrac{x-3}{3}$

$\dfrac{12 \cdot x}{4} = 12(2) - \dfrac{12(x-3)}{3}$

$3x = 24 - 4x + 12$

$7x = 36$

$x = \dfrac{36}{7}$

The solution set is $\left\{\dfrac{36}{7}\right\}$.

This is a conditional equation.

27. $\dfrac{3x+1}{3} - \dfrac{13}{2} = \dfrac{1-x}{4}$

$4(3x + 1) - 6(13) = 3(1 - x)$

$12x + 4 - 78 = 3 - 3x$

$12x - 74 = 3 - 3x$

$15x = 77$

$x = \dfrac{77}{15}$

The solution set is $\left\{\dfrac{77}{15}\right\}$.

This is a conditional equation.

28. $\dfrac{9}{4} - \dfrac{1}{2x} = \dfrac{4}{x}$

$9x - 2 = 16$

$9x = 18$

$x = 2$

The solution set is $\{2\}$.

This is a conditional equation.

29. $\dfrac{7}{x-5} + 2 = \dfrac{x+2}{x-5}$

$7 + 2(x - 5) = x + 2$

$7 + 2x - 10 = x + 2$

$2x - 3 = x + 2$

$x = 5$

5 does not check and must be rejected.

The solution set is the empty set, $\varnothing$.

This is an inconsistent equation.

30.
$$\frac{1}{x-1} - \frac{1}{x+1} = \frac{2}{x^2-1}$$
$$\frac{1}{x-1} - \frac{1}{x+1} = \frac{2}{(x+1)(x-1)}$$
$$x+1-(x-1) = 2$$
$$x+1-x+1 = 2$$
$$2 = 2$$
The solution set is all real numbers except -1 and 1.
This is a conditional equation.

31.
$$\frac{5}{x+3} + \frac{1}{x-2} = \frac{8}{x^2+x-6}$$
$$\frac{5}{x+3} + \frac{1}{x-2} = \frac{8}{(x+3)(x-2)}$$
$$\frac{5(x+3)(x-2)}{x+3} + \frac{(x+3)(x-2)}{x-2} = \frac{8(x+3)(x-2)}{(x+3)(x-2)}$$
$$5(x-2)+1(x+3) = 8$$
$$5x-10+x+3 = 8$$
$$6x-7 = 8$$
$$6x = 15$$
$$x = \frac{15}{6}$$
$$x = \frac{5}{2}$$

The solution set is $\left\{\frac{5}{2}\right\}$.

This is a conditional equation.

32.
$$\frac{1}{x+5} = 0$$
$$(x+5)\frac{1}{x+5} = (x+5)(0)$$
$$1 = 0$$
The solution set is the empty set, $\varnothing$.
This is an inconsistent equation.

33.
$$\frac{4}{x+2} + \frac{3}{x} = \frac{10}{x^2+2x}$$
$$\frac{4}{x+2} + \frac{3}{x} = \frac{10}{x(x+2)}$$
$$\frac{4 \cdot x(x+2)}{x+2} + \frac{3 \cdot x(x+2)}{x} = \frac{10 \cdot x(x+2)}{x(x+2)}$$
$$4x+3(x+2) = 10$$
$$4x+3x+6 = 10$$
$$7x+6 = 10$$
$$7x = 4$$
$$x = \frac{4}{7}$$

The solution set is $\left\{\frac{4}{7}\right\}$.

This is a conditional equation.

34.
$$3-5(2x+1)-2(x-4) = 0$$
$$3-5(2x+1)-2(x-4) = 0$$
$$3-10x-5-2x+8 = 0$$
$$-12x+6 = 0$$
$$-12x = -6$$
$$x = \frac{-6}{-12}$$
$$x = \frac{1}{2}$$

The solution set is $\left\{\frac{1}{2}\right\}$.

This is a conditional equation.

35.
$$\frac{x+2}{x+3} + \frac{1}{x^2+2x-3} - 1 = 0$$
$$\frac{x+2}{x+3} + \frac{1}{(x+3)(x-1)} - 1 = 0$$
$$\frac{x+2}{x+3} + \frac{1}{(x+3)(x-1)} = 1$$
$$\frac{(x+2)(x+3)(x-1)}{x+3} + 1 = (x+3)(x-1)$$
$$(x+2)(x-1)+1 = (x+3)(x-1)$$
$$x^2+x-2+1 = x^2+2x-3$$
$$x-1 = 2x-3$$
$$-x = -2$$
$$x = 2$$

The solution set is $\{2\}$.

This is a conditional equation.

36. Let x = the body count in *Scream*.
Let $x + 2$ = the body count in *Scream 2*.
Let $x + 3$ = the body count in *Scream 3*.
$$x + (x+2) + (x+3) + 10 = 33$$
$$x + x + 2 + x + 3 + 10 = 33$$
$$3x + 15 = 33$$
$$3x = 18$$
$$x = 6$$
$$x + 2 = 8$$
$$x + 3 = 9$$
The body count in *Scream*, *Scream 2*, and *Scream 3*, respectively, is 6, 8, and 9.

37. Let x = the number of years after 1980.
$$2.69 + 0.15x = 8.69$$
$$0.15x = 6$$
$$x = 40$$
The average price of a movie ticket will be \$8.69 40 years after 1980, or 2020.

38. $15 + .05x = 5 + .07x$
$$10 = .02x$$
$$500 = x$$
Both plans cost the same at 500 text messages.

39. Let x = the original price of the phone
$$48 = x - 0.20x$$
$$48 = 0.80x$$
$$60 = x$$
The original price is \$60.

40. Let x = the amount sold to earn \$800 in one week
$$800 = 300 + 0.05x$$
$$500 = 0.05x$$
$$10,000 = x$$
Sales must be \$10,000 in one week to earn \$800.

41. Let x = the amount invested at 4%
Let y = the amount invested at 7%
$$x + \quad y = 9000$$
$$0.04x + 0.07y = 555$$
Multiply the first equation by –0.04 and add.
$$-0.04x - 0.04y = -360$$
$$\underline{0.04x + 0.07y = 555}$$
$$0.03y = 195$$
$$y = 6500$$

Back-substitute 6500 for y in one of the original equations to find x.
$$x + y = 9000$$
$$x + 6500 = 9000$$
$$x = 2500$$
There was \$2500 invested at 4% and \$6500 invested at 7%.

42. Let x = the amount invested at 2%
Let $8000 - x$ = the amount invested at 5%.
$$0.05(8000 - x) = 0.02x + 85$$
$$400 - 0.05x = 0.02x + 85$$
$$-0.05x - 0.02x = 85 - 400$$
$$-0.07x = -315$$
$$\frac{-0.07x}{-0.07} = \frac{-315}{-0.07}$$
$$x = 4500$$
$$8000 - x = 3500$$
\$4500 was invested at 2% and \$3500 was invested at 5%.

43. Let w = the width of the playing field,
Let $3w - 6$ = the length of the playing field
$$P = 2(\text{length}) + 2(\text{width})$$
$$340 = 2(3w - 6) + 2w$$
$$340 = 6w - 12 + 2w$$
$$340 = 8w - 12$$
$$352 = 8w$$
$$44 = w$$
The dimensions are 44 yards by 126 yards.

44. a. Let x = the number of years after 2010.
College A's enrollment: $14,100 + 1500x$
College B's enrollment: $41,700 - 800x$
$$14,100 + 1500x = 41,700 - 800x$$

b. Check some points to determine that
$y_1 = 14,100 + 1500x$ and
$y_2 = 41,700 - 800x$. Since
$y_1 = y_2 = 32,100$ when $x = 12$, the two colleges will have the same enrollment in the year $2010 + 12 = 2022$. That year the enrollments will be 32,100 students.

45.
$$vt + gt^2 = s$$
$$gt^2 = s - vt$$
$$\frac{gt^2}{t^2} = \frac{s - vt}{t^2}$$
$$g = \frac{s - vt}{t^2}$$

46.
$$T = gr + gvt$$
$$T = g(r + vt)$$
$$\frac{T}{r + vt} = \frac{g(r + vt)}{r + vt}$$
$$\frac{T}{r + vt} = g$$
$$g = \frac{T}{r + vt}$$

47.
$$T = \frac{A - P}{Pr}$$
$$Pr(T) = Pr\frac{A - P}{Pr}$$
$$PrT = A - P$$
$$PrT + P = A$$
$$P(rT + 1) = A$$
$$P = \frac{A}{1 + rT}$$

48.
$$(8 - 3i) - (17 - 7i) = 8 - 3i - 17 + 7i$$
$$= -9 + 4i$$

49.
$$4i(3i - 2) = (4i)(3i) + (4i)(-2)$$
$$= 12i^2 - 8i$$
$$= -12 - 8i$$

50.
$$(7 - i)(2 + 3i)$$
$$= 7 \cdot 2 + 7(3i) + (-i)(2) + (-i)(3i)$$
$$= 14 + 21i - 2i + 3$$
$$= 17 + 19i$$

51.
$$(3 - 4i)^2 = 3^2 + 2 \cdot 3(-4i) + (-4i)^2$$
$$= 9 - 24i - 16$$
$$= -7 - 24i$$

52.
$$(7 + 8i)(7 - 8i) = 7^2 + 8^2 = 49 + 64 = 113$$

53.
$$\frac{6}{5 + i} = \frac{6}{5 + i} \cdot \frac{5 - i}{5 - i}$$
$$= \frac{30 - 6i}{25 + 1}$$
$$= \frac{30 - 6i}{26}$$
$$= \frac{15 - 3i}{13}$$
$$= \frac{15}{13} - \frac{3}{13}i$$

54.
$$\frac{3 + 4i}{4 - 2i} = \frac{3 + 4i}{4 - 2i} \cdot \frac{4 + 2i}{4 + 2i}$$
$$= \frac{12 + 6i + 16i + 8i^2}{16 - 4i^2}$$
$$= \frac{12 + 22i - 8}{16 + 4}$$
$$= \frac{4 + 22i}{20}$$
$$= \frac{1}{5} + \frac{11}{10}i$$

55.
$$\sqrt{-32} - \sqrt{-18} = i\sqrt{32} - i\sqrt{18}$$
$$= i\sqrt{16 \cdot 2} - i\sqrt{9 \cdot 2}$$
$$= 4i\sqrt{2} - 3i\sqrt{2}$$
$$= (4i - 3i)\sqrt{2}$$
$$= i\sqrt{2}$$

56.
$$(-2 + \sqrt{-100})^2 = (-2 + i\sqrt{100})^2$$
$$= (-2 + 10i)^2$$
$$= 4 - 40i + (10i)^2$$
$$= 4 - 40i - 100$$
$$= -96 - 40i$$

57.
$$\frac{4 + \sqrt{-8}}{2} = \frac{4 + i\sqrt{8}}{2} = \frac{4 + 2i\sqrt{2}}{2} = 2 + i\sqrt{2}$$

58.
$$2x^2 + 15x = 8$$
$$2x^2 + 15x - 8 = 0$$
$$(2x - 1)(x + 8) = 0$$
$$2x - 1 = 0 \quad x + 8 = 0$$
$$x = \frac{1}{2} \text{ or } x = -8$$

The solution set is $\left\{ \frac{1}{2}, -8 \right\}$.

59. $5x^2 + 20x = 0$

$5x(x+4) = 0$

$5x = 0 \quad x + 4 = 0$

$x = 0$ or $x = -4$

The solution set is $\{0, -4\}$.

60. $2x^2 - 3 = 125$

$2x^2 = 128$

$x^2 = 64$

$x = \pm 8$

The solution set is $\{8, -8\}$.

61. $\dfrac{x^2}{2} + 5 = -3$

$\dfrac{x^2}{2} = -8$

$x^2 = -16$

$\sqrt{x^2} = \pm\sqrt{-16}$

$x = \pm 4i$

62. $(x+3)^2 = -10$

$\sqrt{(x+3)^2} = \pm\sqrt{-10}$

$x + 3 = \pm i\sqrt{10}$

$x = -3 \pm i\sqrt{10}$

63. $(3x-4)^2 = 18$

$\sqrt{(3x-4)^2} = \pm\sqrt{18}$

$3x - 4 = \pm 3\sqrt{2}$

$3x = 4 \pm 3\sqrt{2}$

$\dfrac{3x}{3} = \dfrac{4 \pm 3\sqrt{2}}{3}$

$x = \dfrac{4 \pm 3\sqrt{2}}{3}$

64. $x^2 + 20x$

$\left(\dfrac{20}{2}\right)^2 = 10^2 = 100$

$x^2 + 20x + 100 = (x+10)^2$

65. $x^2 - 3x$

$\left(\dfrac{3}{2}\right)^2 = \dfrac{9}{4}$

$x^2 - 3x + \dfrac{9}{4} = \left(x - \dfrac{3}{2}\right)^2$

66. $x^2 - 12x = -27$

$x^2 - 12x + 36 = -27 + 36$

$(x-6)^2 = 9$

$x - 6 = \pm 3$

$x = 6 \pm 3$

$x = 9, 3$

The solution set is $\{9, 3\}$.

67. $3x^2 - 12x + 11 = 0$

$x^2 - 4x = -\dfrac{11}{3}$

$x^2 - 4x + 4 = -\dfrac{11}{3} + 4$

$(x-2)^2 = \dfrac{1}{3}$

$x - 2 = \pm\sqrt{\dfrac{1}{3}}$

$x = 2 \pm \dfrac{\sqrt{3}}{3}$

The solution set is $\left\{2 + \dfrac{\sqrt{3}}{3}, 2 - \dfrac{\sqrt{3}}{3}\right\}$.

68. $x^2 = 2x + 4$

$x^2 - 2x - 4 = 0$

$x = \dfrac{2 \pm \sqrt{(-2)^2 - 4(1)(-4)}}{2(1)}$

$x = \dfrac{2 \pm \sqrt{4+16}}{2}$

$x = \dfrac{2 \pm \sqrt{20}}{2}$

$x = \dfrac{2 \pm 2\sqrt{5}}{2}$

$x = 1 \pm \sqrt{5}$

The solution set is $\left\{1 + \sqrt{5}, 1 - \sqrt{5}\right\}$.

69. $x^2 - 2x + 19 = 0$

$$x = \frac{2 \pm \sqrt{(-2)^2 - 4(1)(19)}}{2(1)}$$

$$x = \frac{2 \pm \sqrt{4 - 76}}{2}$$

$$x = \frac{2 \pm \sqrt{-72}}{2}$$

$$x = \frac{2 \pm 6i\sqrt{2}}{2}$$

$$x = 1 \pm 3i\sqrt{2}$$

The solution set is $\left\{1 + 3i\sqrt{2}, 1 - 3i\sqrt{2}\right\}$.

70. $\quad 2x^2 = 3 - 4x$

$2x^2 + 4x - 3 = 0$

$$x = \frac{-4 \pm \sqrt{4^2 - 4(2)(-3)}}{2(2)}$$

$$x = \frac{-4 \pm \sqrt{16 + 24}}{4}$$

$$x = \frac{-4 \pm \sqrt{40}}{4}$$

$$x = \frac{-4 \pm 2\sqrt{10}}{4}$$

$$x = \frac{-2 \pm \sqrt{10}}{2}$$

The solution set is $\left\{\frac{-2 + \sqrt{10}}{2}, \frac{-2 - \sqrt{10}}{2}\right\}$.

71. $x^2 - 4x + 13 = 0$

$(-4)^2 - 4(1)(13)$

$= 16 - 52$

$= -36$; 2 complex imaginary solutions

72. $9x^2 = 2 - 3x$

$9x^2 + 3x - 2 = 0$

$3^2 - 4(9)(-2)$

$= 9 + 72$

$= 81$; 2 unequal real solutions

73. $2x^2 - 11x + 5 = 0$

$(2x - 1)(x - 5) = 0$

$2x - 1 = 0 \quad x - 5 = 0$

$x = \dfrac{1}{2} \ \text{ or } x = 5$

The solution set is $\left\{5, \dfrac{1}{2}\right\}$.

74. $\quad (3x + 5)(x - 3) = 5$

$3x^2 + 5x - 9x - 15 = 5$

$3x^2 - 4x - 20 = 0$

$$x = \frac{4 \pm \sqrt{(-4)^2 - 4(3)(-20)}}{2(3)}$$

$$x = \frac{4 \pm \sqrt{16 + 240}}{6}$$

$$x = \frac{4 \pm \sqrt{256}}{6}$$

$$x = \frac{4 \pm 16}{6}$$

$$x = \frac{20}{6}, \frac{-12}{6}$$

$$x = \frac{10}{3}, -2$$

The solution set is $\left\{-2, \dfrac{10}{3}\right\}$.

75. $3x^2 - 7x + 1 = 0$

$$x = \frac{7 \pm \sqrt{(-7)^2 - 4(3)(1)}}{2(3)}$$

$$x = \frac{7 \pm \sqrt{49 - 12}}{6}$$

$$x = \frac{7 \pm \sqrt{37}}{6}$$

The solution set is $\left\{\dfrac{7 + \sqrt{37}}{6}, \dfrac{7 - \sqrt{37}}{6}\right\}$.

76. $x^2 - 9 = 0$

$x^2 = 9$

$x = \pm 3$

The solution set is $\{-3, 3\}$.

77. $(x-3)^2 - 25 = 0$

$$(x-3)^2 = 25$$
$$x-3 = \pm 5$$
$$x = 3 \pm 5$$
$$x = 8, -2$$

The solution set is $\{8, -2\}$.

78. $3x^2 - x + 2 = 0$

$$x = \frac{1 \pm \sqrt{(-1)^2 - 4(3)(2)}}{2(3)}$$
$$x = \frac{1 \pm \sqrt{1 - 24}}{6}$$
$$x = \frac{1 \pm \sqrt{-23}}{6}$$
$$x = \frac{1 \pm i\sqrt{23}}{6}$$

The solution set is $\left\{ \frac{1 + i\sqrt{23}}{6}, \frac{1 - i\sqrt{23}}{6} \right\}$.

79. $3x^2 - 10x = 8$

$$3x^2 - 10x - 8 = 0$$
$$(3x+2)(x-4) = 0$$
$$3x + 2 = 0 \quad \text{or} \quad x - 4 = 0$$
$$3x = -2 \qquad\qquad x = 4$$
$$x = -\frac{2}{3}$$

The solution set is $\left\{ -\frac{2}{3}, 4 \right\}$.

80. $(x+2)^2 + 4 = 0$

$$(x+2)^2 = -4$$
$$\sqrt{(x+2)^2} = \pm\sqrt{-4}$$
$$x + 2 = \pm 2i$$
$$x = -2 \pm 2i$$

The solution set is $\{-2 + 2i, -2 - 2i\}$.

81.

$$\frac{5}{x+1} + \frac{x-1}{4} = 2$$

$$\frac{5 \cdot 4(x+1)}{x+1} + \frac{(x-1) \cdot 4(x+1)}{4} = 2 \cdot 4(x+1)$$

$$20 + (x-1)(x+1) = 8(x+1)$$
$$20 + x^2 - 1 = 8x + 8$$
$$x^2 - 8x - 11 = 0$$

$$x = \frac{-b \pm \sqrt{b^2 - 4ac}}{2a}$$

$$x = \frac{-(-8) \pm \sqrt{(-8)^2 - 4(1)(11)}}{2(1)}$$

$$x = \frac{8 \pm \sqrt{20}}{2}$$

$$x = \frac{8 \pm 2\sqrt{5}}{2}$$

$$x = 4 \pm \sqrt{5}$$

The solution set is $\left\{ 4 + \sqrt{5}, 4 - \sqrt{5} \right\}$.

82. $W(t) = 3t^2$

$$588 = 3t^2$$
$$196 = t^2$$

Apply the square root property.

$$t^2 = 196$$
$$t = \pm\sqrt{196}$$
$$t = \pm 14$$

The solutions are -14 and 14. We disregard -14, because we cannot have a negative time measurement. The fetus will weigh 588 grams after 14 weeks.

83. **a.** 2011 is 8 years after 2003.

$$B = 1.7x^2 + 6x + 26$$
$$B = 1.7(8)^2 + 6(8) + 26$$
$$= 182.8$$
$$\approx 183$$

According to the function, in 2011 there were 183 "Bicycle Friendly" communities. This overestimates the number shown in the graph by 3.

b.
$$B = 1.7x^2 + 6x + 26$$
$$826 = 1.7x^2 + 6x + 26$$
$$0 = 1.7x^2 + 6x - 800$$
$$x = \frac{-b \pm \sqrt{b^2 - 4ac}}{2a}$$
$$x = \frac{-(6) \pm \sqrt{(6)^2 - 4(1.7)(-800)}}{2(1.7)}$$
$$= \frac{-6 \pm \sqrt{36 + 5440}}{2(1.7)}$$
$$= \frac{-6 \pm \sqrt{5476}}{3.4}$$
$$= \frac{-6 \pm 74}{3.4}$$
$$x = 20 \text{ or } -23\tfrac{9}{17}$$

According to the function, there will be 826 "Bicycle Friendly" communities 20 years after 2003, or 2023.

84.
$$A = lw$$
$$15 = l(2l - 7)$$
$$15 = 2l^2 - 7l$$
$$0 = 2l^2 - 7l - 15$$
$$0 = (2l + 3)(l - 5)$$
$$l = 5$$
$$2l - 7 = 3$$
The length is 5 yards, the width is 3 yards.

85. Let x = height of building
$2x$ = shadow height
$$x^2 + (2x)^2 = 300^2$$
$$x^2 + 4x^2 = 90,000$$
$$5x^2 = 90,000$$
$$x^2 = 18,000$$
$$x \approx \pm 134.164$$
Discard negative height.
The building is approximately 134 meters high.

86.
$$2x^4 = 50x^2$$
$$2x^4 - 50x^2 = 0$$
$$2x^2(x^2 - 25) = 0$$
$$x = 0$$
$$x = \pm 5$$
The solution set is $\{-5, 0, 5\}$.

87.
$$2x^3 - x^2 - 18x + 9 = 0$$
$$x^2(2x - 1) - 9(2x - 1) = 0$$
$$(x^2 - 9)(2x - 1) = 0$$
$$x = \pm 3, \ x = \frac{1}{2}$$
The solution set is $\left\{-3, \dfrac{1}{2}, 3\right\}$.

88.
$$\sqrt{2x - 3} + x = 3$$
$$\sqrt{2x - 3} = 3 - x$$
$$2x - 3 = 9 - 6x + x^2$$
$$x^2 - 8x + 12 = 0$$
$$x^2 - 8x = -12$$
$$x^2 - 8x + 16 = -12 + 16$$
$$(x - 4)^2 = 4$$
$$x - 4 = \pm 2$$
$$x = 4 + 2$$
$$x = 6, 2$$
The solution set is $\{2\}$.

89.
$$\sqrt{x - 4} + \sqrt{x + 1} = 5$$
$$\sqrt{x - 4} = 5 - \sqrt{x + 1}$$
$$x - 4 = 25 - 10\sqrt{x + 1} + (x + 1)$$
$$x - 4 = 26 + x - 10\sqrt{x + 1}$$
$$-30 = -10\sqrt{x + 1}$$
$$3 = \sqrt{x + 1}$$
$$9 = x + 1$$
$$x = 8$$
The solution set is $\{8\}$.

90.
$$3x^{\frac{3}{4}} - 24 = 0$$
$$3x^{\frac{3}{4}} = 24$$
$$x^{\frac{3}{4}} = 8$$
$$\left(x^{\frac{3}{4}}\right)^{\frac{4}{3}} = (8)^{\frac{4}{3}}$$
$$x = 16$$
The solution set is $\{16\}$.

91. $(x-7)^{\frac{2}{3}} = 25$

$$\left[(x-7)^{\frac{2}{3}}\right]^{\frac{3}{2}} = 25^{\frac{3}{2}}$$

$$x-7 = \left(5^2\right)^{\frac{3}{2}}$$

$$x-7 = 5^3$$

$$x-7 = 125$$

$$x = 132$$

The solution set is $\{132\}$.

92. $x^4 - 5x^2 + 4 = 0$

Let $t = x^2$

$t^2 - 5t + 4 = 0$

$t = 4$ or $t = 1$

$x^2 = 4$ $\qquad$ $x^2 = 1$

$x = \pm 2$ $\qquad$ $x = \pm 1$

The solution set is $\{-2, -1, 1, 2\}$.

93. $x^{1/2} + 3x^{1/4} - 10 = 0$

Let $t = x^{1/4}$

$t^2 + 3t - 10 = 0$

$(t+5)(t-2) = 0$

$t = -5$ or $t = 2$

$x^{\frac{1}{4}} = -5$ $\qquad$ $x^{\frac{1}{4}} = 2$

$\left(x^{\frac{1}{4}}\right)^4 = (-5)^4$ $\qquad$ $\left(x^{\frac{1}{4}}\right)^4 = (2)^4$

$x = 625$ $\qquad$ $x = 16$

625 does not check and must be rejected.
The solution set is $\{16\}$.

94. $|2x+1| = 7$

$2x+1 = 7$ or $2x+1 = -7$

$2x = 6$ $\qquad$ $2x = -8$

$x = 3$ $\qquad$ $x = -8$

The solution set is $\{-4, 3\}$.

95. $2|x-3| - 6 = 10$

$2|x-3| = 16$

$|x-3| = 8$

$x-3 = 8$ or $x-3 = -8$

$x = 11$ $\qquad$ $x = -5$

The solution set is $\{-5, 11\}$.

96. $3x^{4/3} - 5x^{2/3} + 2 = 0$

Let $t = x^{\frac{2}{3}}$.

$3t^2 - 5t + 2 = 0$

$(3t-2)(t-1) = 0$

$3t - 2 = 0$ $\qquad$ or $\qquad$ $t - 1 = 0$

$3t = 2$ $\qquad\qquad\qquad$ $t = 1$

$t = \dfrac{2}{3}$ $\qquad\qquad\qquad$ $x^{\frac{2}{3}} = 1$

$x^{\frac{2}{3}} = \dfrac{2}{3}$ $\qquad\qquad$ $\left(x^{\frac{2}{3}}\right)^{\frac{3}{2}} = \pm(1)^{\frac{3}{2}}$

$\left(x^{\frac{2}{3}}\right)^{\frac{3}{2}} = \pm\left(\dfrac{2}{3}\right)^{\frac{3}{2}}$ $\qquad$ $x = \pm 1$

$x = \pm\sqrt[2]{\left(\dfrac{2}{3}\right)^3}$

$x = \pm\dfrac{2}{3}\sqrt{\dfrac{2}{3}}$

$x = \pm\dfrac{2}{3}\cdot\dfrac{\sqrt{2}}{\sqrt{3}}\cdot\dfrac{\sqrt{3}}{\sqrt{3}}$

$x = \pm\dfrac{2\sqrt{6}}{9}$

The solution set is $\left\{-\dfrac{2\sqrt{6}}{9}, \dfrac{2\sqrt{6}}{9}, -1, 1\right\}$.

97. $2\sqrt{x-1} = x$

$4(x-1) = x^2$

$4x - 4 = x^2$

$x^2 - 4x + 4 = 0$

$(x-2)^2 = 0$

$x = 2$

The solution set is $\{2\}$.

98. $|2x-5| - 3 = 0$

$2x-5 = 3$ or $2x-5 = -3$

$2x = 8$ $\qquad$ $2x = 2$

$x = 4$ $\qquad$ $x = 1$

The solution set is $\{4, 1\}$.

99. $x^3 + 2x^2 - 9x - 18 = 0$

$x^2(x+2) - 9(x+2) = 0$

$(x+2)(x^2 - 9) = 0$

$(x+2)(x+3)(x-3) = 0$

The solution set is $\{-3, -2, 3\}$.

100. $\sqrt{8-2x} - x = 0$

$\sqrt{8-2x} = x$

$\left(\sqrt{8-2x}\right)^2 = (x)^2$

$8 - 2x = x^2$

$0 = x^2 + 2x - 8$

$0 = (x+4)(x-2)$

$x + 4 = 0 \quad$ or $\quad x - 2 = 0$

$x = -4 \qquad\qquad x = 2$

-4 does not check.

The solution set is $\{2\}$.

101. $x^3 + 3x^2 - 2x - 6 = 0$

$x^2(x+3) - 2(x+3) = 0$

$(x+3)(x^2 - 2) = 0$

$x + 3 = 0 \quad$ or $\quad x^2 - 2 = 0$

$x = -3 \qquad\qquad x^2 = 2$

$\qquad\qquad\qquad x = \pm\sqrt{2}$

The solution set is $\left\{-3, -\sqrt{2}, \sqrt{2}\right\}$.

102. $-4|x+1| + 12 = 0$

$-4|x+1| = -12$

$|x+1| = 3$

$x + 1 = 3 \quad$ or $\quad x + 1 = -3$

$x = 2 \qquad\qquad x = -4$

The solution set is $\{-4, 2\}$.

103. $\quad p = -2.5\sqrt{t} + 17$

$7 = -2.5\sqrt{t} + 17$

$-10 = -2.5\sqrt{t}$

$4 = \sqrt{t}$

$16 = t$

The percentage will drop to 7% 16 years after 1993, or 2009.

104. $\left\{x \mid -3 \le x < 5\right\}$

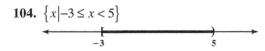

105. $\left\{x \mid x > -2\right\}$

106. $\left\{x \mid x \le 0\right\}$

107. Graph $(-2,1]$: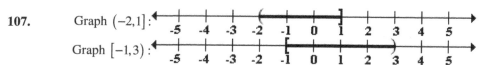

Graph $[-1,3)$: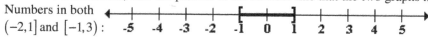

To find the intersection, take the portion of the number line that the two graphs have in common.

Numbers in both $(-2,1]$ and $[-1,3)$: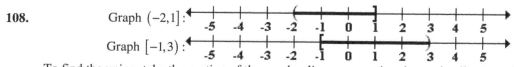

Thus, $(-2,1] \cap [-1,3) = [-1,1]$.

108. Graph $(-2,1]$: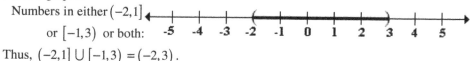

Graph $[-1,3)$: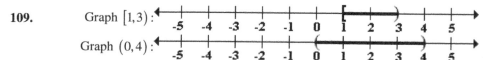

To find the union, take the portion of the number line representing the total collection of numbers in the two graphs.

Numbers in either $(-2,1]$ or $[-1,3)$ or both: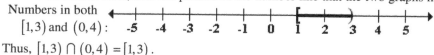

Thus, $(-2,1] \cup [-1,3) = (-2,3)$.

109. Graph $[1,3)$: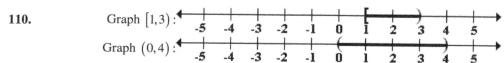

Graph $(0,4)$: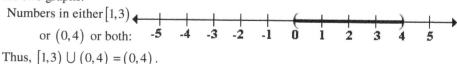

To find the intersection, take the portion of the number line that the two graphs have in common.

Numbers in both $[1,3)$ and $(0,4)$:

Thus, $[1,3) \cap (0,4) = [1,3)$.

110. Graph $[1,3)$:

Graph $(0,4)$:

To find the union, take the portion of the number line representing the total collection of numbers in the two graphs.

Numbers in either $[1,3)$ or $(0,4)$ or both:

Thus, $[1,3) \cup (0,4) = (0,4)$.

111. $-6x + 3 \le 15$
$-6x \le 12$
$x \ge 2$

The solution set is $[-2, \infty)$.

112. $6x - 9 \geq -4x - 3$

$ 10x \geq 6$

$ x \geq \dfrac{3}{5}$

The solution set is $\left[\dfrac{3}{5}, \infty\right)$.

113. $\dfrac{x}{3} - \dfrac{3}{4} - 1 > \dfrac{x}{2}$

$12\left(\dfrac{x}{3} - \dfrac{3}{4} - 1\right) > 12\left(\dfrac{x}{2}\right)$

$4x - 9 - 12 > 6x$

$-21 > 2x$

$-\dfrac{21}{2} > x$

The solution set is $\left(-\infty, -\dfrac{21}{2}\right)$.

114. $6x + 5 > -2(x - 3) - 25$

$6x + 5 > -2x + 6 - 25$

$8x + 5 > -19$

$8x > -24$

$x > -3$

The solution set is $(-3, \infty)$.

115. $3(2x - 1) - 2(x - 4) \geq 7 + 2(3 + 4x)$

$6x - 3 - 2x + 8 \geq 7 + 6 + 8x$

$4x + 5 \geq 8x + 13$

$-4x \geq 8$

$x \leq -2$

The solution set is $\left[-\infty, -2\right)$.

116. $5(x - 2) - 3(x + 4) \geq 2x - 20$

$5x - 10 - 3x - 12 \geq 2x - 20$

$2x - 22 \geq 2x - 20$

$-22 \geq -20$

The solution set is $\varnothing$.

117. $7 < 2x + 3 \leq 9$

$4 < 2x \leq 6$

$2 < x \leq 3$

$(2, 3]$

The solution set is $[2, 3)$.

118. $|2x + 3| \leq 15$

$-15 \leq 2x + 3 \leq 15$

$-18 \leq 2x \leq 12$

$-9 \leq x \leq 6$

The solution set is $[-9, 6]$.

119. $\left|\dfrac{2x + 6}{3}\right| > 2$

$\dfrac{2x + 6}{3} > 2 \qquad \dfrac{2x + 6}{3} < -2$

$2x + 6 > 6 \qquad 2x + 6 < -6$

$2x > 0 \qquad\quad 2x < -12$

$x > 0 \qquad\quad x < -6$

The solution set is $(-\infty, -6)$ or $(0, \infty)$.

120. $|2x + 5| - 7 \geq -6$

$|2x + 5| \geq 1$

$2x + 5 \geq 1$ or $2x + 5 \leq -1$

$2x \geq -4 \qquad 2x \leq -6$

$x \geq -2 \quad$ or $\quad x \leq -3$

The solution set is $(-\infty, -3]$ or $[-2, \infty)$.

121. $-4|x + 2| + 5 \leq -7$

$-4|x + 2| \leq -12$

$|x + 2| \geq 3$

$x + 2 \geq 3 \qquad x + 2 \leq -3$
$\qquad\qquad$ or
$x \geq 1 \qquad\quad x \leq -5$

The solution set is $(-\infty, -5] \cup [1, \infty)$.

122.
$$y_1 > y_2$$
$$-10 - 3(2x+1) > 8x+1$$
$$-10 - 6x - 3 > 8x+1$$
$$-6x - 13 > 8x+1$$
$$-14x > 14$$
$$\frac{-14x}{-14} < \frac{14}{-14}$$
$$x < -1$$
The solution set is $(-\infty, -1)$.

123. $3 - |2x-5| \geq -6$
$$-|2x-5| \geq -9$$
$$\frac{-|2x-5|}{-1} \leq \frac{-9}{-1}$$
$$|2x-5| \leq 9$$
$$-9 \leq 2x-5 \leq 9$$
$$-4 \leq 2x \leq 14$$
$$-2 \leq x \leq 7$$
The solution set is $[-2, 7]$.

124. $0.20x + 24 \leq 40$
$$0.20x \leq 16$$
$$\frac{0.20x}{0.20} \leq \frac{16}{0.20}$$
$$x \leq 80$$
A customer can drive no more than 80 miles.

125. $80 \leq \dfrac{95 + 79 + 91 + 86 + x}{5} < 90$
$$400 \leq 95 + 79 + 91 + 86 + x < 450$$
$$400 \leq 351 + x < 450$$
$$49 \leq x < 99$$
A grade of at least 49% but less than 99% will result in a B.

126. $0.075x \geq 9000$
$$\frac{0.075x}{0.075} \geq \frac{9000}{0.075}$$
$$x \geq 120,000$$
The investment must be at least \$120,000.

Chapter 1 Test

1. $7(x-2) = 4(x+1) - 21$
$$7x - 14 = 4x + 4 - 21$$
$$7x - 14 = 4x - 17$$
$$3x = -3$$
$$x = -1$$
The solution set is $\{-1\}$.

2. $-10 - 3(2x+1) - 8x - 1 = 0$
$$-10 - 6x - 3 - 8x - 1 = 0$$
$$-14x - 14 = 0$$
$$-14x = 14$$
$$x = -1$$
The solution set is $\{-1\}$.

3. $\dfrac{2x-3}{4} = \dfrac{x-4}{2} - \dfrac{x+1}{4}$
$$2x - 3 = 2(x-4) - (x+1)$$
$$2x - 3 = 2x - 8 - x - 1$$
$$2x - 3 = x - 9$$
$$x = -6$$
The solution set is $\{-6\}$.

4. $\dfrac{2}{x-3} - \dfrac{4}{x+3} = \dfrac{8}{(x-3)(x+3)}$
$$2(x+3) - 4(x-3) = 8$$
$$2x + 6 - 4x + 12 = 8$$
$$-2x + 18 = 8$$
$$-2x = -10$$
$$x = 5$$
The solution set is $\{5\}$.

5. $2x^2 - 3x - 2 = 0$
$(2x + 1)(x - 2) = 0$
$2x + 1 = 0 \quad \text{or} \quad x - 2 = 0$
$$x = -\frac{1}{2} \quad \text{or} \quad x = 2$$
The solution set is $\left\{ -\dfrac{1}{2}, 2 \right\}$.

6. $(3x-1)^2 = 75$
$$3x - 1 = \pm\sqrt{75}$$
$$3x = 1 \pm 5\sqrt{3}$$
$$x = \frac{1 \pm 5\sqrt{3}}{3}$$
The solution set is $\left\{ \dfrac{1 - 5\sqrt{3}}{3}, \dfrac{1 + 5\sqrt{3}}{3} \right\}$.

7. $(x+3)^2 + 25 = 0$

$(x+3)^2 = -25$

$x + 3 = \pm\sqrt{-25}$

$x = -3 \pm 5i$

The solution set is $\{-3+5i, -3-5i\}$.

8. $x(x-2) = 4$

$x^2 - 2x - 4 = 0$

$x = \dfrac{-b \pm \sqrt{b^2 - 4ac}}{2a}$

$x = \dfrac{2 \pm \sqrt{(-2)^2 - 4(1)(-4)}}{2}$

$x = \dfrac{2 \pm 2\sqrt{5}}{2}$

$x = 1 \pm \sqrt{5}$

The solution set is $\left\{1 - \sqrt{5}, 1 + \sqrt{5}\right\}$.

9. $4x^2 = 8x - 5$

$4x^2 - 8x + 5 = 0$

$x = \dfrac{-b \pm \sqrt{b^2 - 4ac}}{2a}$

$x = \dfrac{8 \pm \sqrt{(-8)^2 - 4(4)(5)}}{2(4)}$

$x = \dfrac{8 \pm \sqrt{-16}}{8}$

$x = \dfrac{8 \pm 4i}{8}$

$x = 1 \pm \dfrac{1}{2}i$

The solution set is $\left\{1 + \dfrac{1}{2}i, 1 - \dfrac{1}{2}i\right\}$.

10. $x^3 - 4x^2 - x + 4 = 0$

$x^2(x-4) - 1(x-4) = 0$

$(x^2 - 1)(x-4) = 0$

$(x-1)(x+1)(x-4) = 0$

$x = 1$ or $x = -1$ or $x = 4$

The solution set is $\{-1, 1, 4\}$.

11. $\sqrt{x-3} + 5 = x$

$\sqrt{x-3} = x - 5$

$x - 3 = x^2 - 10x + 25$

$x^2 - 11x + 28 = 0$

$x = \dfrac{11 \pm \sqrt{11^2 - 4(1)(28)}}{2(1)}$

$x = \dfrac{11 \pm \sqrt{121 - 112}}{2}$

$x = \dfrac{11 \pm \sqrt{9}}{2}$

$x = \dfrac{11 \pm 3}{2}$

$x = 7$ or $x = 4$

4 does not check and must be rejected.

The solution set is $\{7\}$.

12. $\sqrt{8 - 2x} - x = 0$

$\sqrt{8 - 2x} = x$

$\left(\sqrt{8 - 2x}\right)^2 = (x)^2$

$8 - 2x = x^2$

$0 = x^2 + 2x - 8$

$0 = (x+4)(x-2)$

$x + 4 = 0$ or $x - 2 = 0$

$x = -4 \qquad x = 2$

–4 does not check and must be rejected.

The solution set is $\{2\}$.

13. $\sqrt{x+4} + \sqrt{x-1} = 5$

$\sqrt{x+4} = 5 - \sqrt{x-1}$

$x + 4 = 25 - 10\sqrt{x-1} + (x-1)$

$x + 4 = 25 - 10\sqrt{x-1} + x - 1$

$-20 = -10\sqrt{x-1}$

$2 = \sqrt{x-1}$

$4 = x - 1$

$x = 5$

The solution set is $\{5\}$.

14. $5x^{3/2} - 10 = 0$

$5x^{3/2} = 10$

$x^{3/2} = 2$

$x = 2^{2/3}$

$x = \sqrt[3]{4}$

The solution set is $\left\{\sqrt[3]{4}\right\}$.

15. $x^{2/3} - 9x^{1/3} + 8 = 0$ let $t = x^{1/3}$

$$t^2 - 9t + 8 = 0$$
$$(t-1)(t-8) = 0$$
$$t = 1 \qquad t = 8$$
$$x^{1/3} = 1 \quad x^{1/3} = 8$$
$$x = 1 \qquad x = 512$$

The solution set is $\{1, 512\}$.

16. $\left|\dfrac{2}{3}x - 6\right| = 2$

$$\dfrac{2}{3}x - 6 = 2 \qquad \dfrac{2}{3}x - 6 = -2$$
$$\dfrac{2}{3}x = 8 \qquad \dfrac{2}{3}x = 4$$
$$x = 12 \qquad x = 6$$

The solution set is $\{6, 12\}$.

17. $-3|4x - 7| + 15 = 0$

$$-3|4x - 7| = -15$$
$$|4x - 7| = 5$$
$$4x - 7 = 5 \qquad 4x - 7 = -5$$
$$\qquad\quad \text{or}$$
$$4x = 12 \qquad 4x = 2$$
$$x = 3 \qquad\qquad x = \dfrac{1}{2}$$

The solution set is $\left\{\dfrac{1}{2}, 3\right\}$.

18. $\dfrac{1}{x^2} - \dfrac{4}{x} + 1 = 0$

$$\dfrac{x^2}{x^2} - \dfrac{4x^2}{x} + x^2 = 0$$
$$1 - 4x + x^2 = 0$$
$$x^2 - 4x + 1 = 0$$
$$x = \dfrac{-b \pm \sqrt{b^2 - 4ac}}{2a}$$
$$x = \dfrac{-(-4) \pm \sqrt{(-4)^2 - 4(1)(1)}}{2(1)}$$
$$x = \dfrac{4 \pm \sqrt{12}}{2}$$
$$x = \dfrac{4 \pm 2\sqrt{3}}{2}$$
$$x = 2 \pm \sqrt{3}$$

The solution set is $\left\{2 + \sqrt{3}, 2 - \sqrt{3}\right\}$.

19. $\dfrac{2x}{x^2 + 6x + 8} + \dfrac{2}{x+2} = \dfrac{x}{x+4}$

$$\dfrac{2x}{(x+4)(x+2)} + \dfrac{2}{x+2} = \dfrac{x}{x+4}$$
$$\dfrac{2x(x+4)(x+2)}{(x+4)(x+2)} + \dfrac{2(x+4)(x+2)}{x+2} = \dfrac{x(x+4)(x+2)}{x+4}$$
$$2x + 2(x+4) = x(x+2)$$
$$2x + 2x + 8 = x^2 + 2x$$
$$2x + 8 = x^2$$
$$0 = x^2 - 2x - 8$$
$$0 = (x-4)(x+2)$$
$$x - 4 = 0 \quad \text{or} \quad x + 2 = 0$$
$$x = 4 \qquad\qquad x = -2 \text{ (rejected)}$$

The solution set is $\{4\}$.

20. $3(x + 4) \geq 5x - 12$

$$3x + 12 \geq 5x - 12$$
$$-2x \geq -24$$
$$x \leq 12$$

The solution set is $(-\infty, 12]$.

21. $\dfrac{x}{6} + \dfrac{1}{8} \leq \dfrac{x}{2} - \dfrac{3}{4}$

$$4x + 3 \leq 12x - 18$$
$$-8x \leq -21$$
$$x \geq \dfrac{21}{8}$$

The solution set is $\left[\dfrac{21}{8}, \infty\right)$.

22. $-3 \leq \dfrac{2x + 5}{3} < 6$

$$-9 \leq 2x + 5 < 18$$
$$-14 \leq 2x < 13$$
$$-7 \leq x < \dfrac{13}{2}$$

The solution set is $\left[-7, \dfrac{13}{2}\right)$.

23. $|3x+2| \geq 3$

$3x+2 \geq 3$ or $3x+2 \leq -3$

$3x \geq 1$ $\qquad$ $3x \leq -5$

$x \geq \dfrac{1}{3}$ $\qquad$ $x \leq -\dfrac{5}{3}$

The solution set is $\left(-\infty, -\dfrac{5}{3}\right] \cup \left[\dfrac{1}{3}, \infty\right)$.

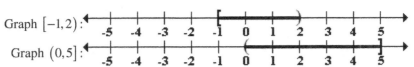

24. $-3 \leq y \leq 7$

$-3 \leq 2x-5 \leq 7$

$2 \leq 2x \leq 12$

$1 \leq x \leq 6$

The solution set is $[1,6]$.

25. $\qquad y \geq 1$

$\left|\dfrac{2-x}{4}\right| \geq 1$

$\dfrac{2-x}{4} \geq 1$ or $\dfrac{2-x}{4} \leq -1$

$2-x \geq 4$ $\qquad$ $2-x \leq -4$

$-x \geq 2$ $\qquad$ $-x \leq -6$

$x \leq -2$ $\qquad$ $x \geq 6$

The solution set is $(-\infty, -2] \cup [6, \infty)$.

26. Graph $[-1, 2)$:

Graph $(0, 5]$:

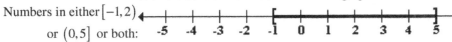

To find the union, take the portion of the number line representing the total collection of numbers in the two graphs.

Numbers in either $[-1, 2)$ or $(0, 5]$ or both:

Thus,

$[-1, 2) \cup (0, 5] = [-1, 5]$.

27. Graph $[-1, 2)$:

Graph $(0, 5]$:

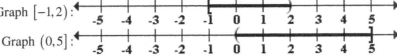

To find the intersection, take the portion of the number line that the two graphs have in common.

Numbers in both $[-1, 2)$ and $(0, 5]$:

Thus, $[-1, 2) \cap (0, 5] = (0, 2)$.

28.
$$V = \frac{1}{3}lwh$$
$$3V = lwh$$
$$\frac{3V}{lw} = \frac{lwh}{lw}$$
$$\frac{3V}{lw} = h$$
$$h = \frac{3V}{lw}$$

29.
$$y - y_1 = m(x - x_1)$$
$$y - y_1 = mx - mx_1$$
$$-mx = y_1 - mx_1 - y$$
$$\frac{-mx}{-m} = \frac{y_1 - mx_1 - y}{-m}$$
$$x = \frac{y - y_1}{m} + x_1$$

30.

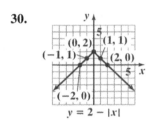

$$y = 2 - |x|$$

31.

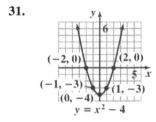

$$y = x^2 - 4$$

32.
$$(6 - 7i)(2 + 5i) = 12 + 30i - 14i - 35i^2$$
$$= 12 + 16i + 35$$
$$= 47 + 16i$$

33.
$$\frac{5}{2 - i} = \frac{5}{2 - i} \cdot \frac{2 + i}{2 + i}$$
$$= \frac{5(2 + i)}{4 + 1}$$
$$= \frac{5(2 + i)}{5}$$
$$= 2 + i$$

34.
$$2\sqrt{-49} + 3\sqrt{-64} = 2(7i) + 3(8i)$$
$$= 14i + 24i$$
$$= 38i$$

35.
$$43x + 575 = 1177$$
$$43x = 602$$
$$x = 14$$
The system's income will be $1177 billion 14 years after 2004, or 2018.

36.
$$B = 0.07x^2 + 47.4x + 500$$
$$1177 = 0.07x^2 + 47.4x + 500$$
$$0 = 0.07x^2 + 47.4x - 677$$
$$0 = 0.07x^2 + 47.4x - 677$$
$$x = \frac{-b \pm \sqrt{b^2 - 4ac}}{2a}$$
$$x = \frac{-(47.4) \pm \sqrt{(47.4)^2 - 4(0.07)(-677)}}{2(0.07)}$$
$$x \approx 14, \quad x \approx -691 \text{ (rejected)}$$
The system's income will be $1177 billion 14 years after 2004, or 2018.

37. The formulas model the data quite well.

38. Let $x =$ the percentage of strikingly-attractive men.
Let $x + 57 =$ the percentage of average-looking men.
Let $x + 25 =$ the percentage of good-looking men.
$$(x) + (x + 57) + (x + 25) = 88$$
$$x + x + 57 + x + 25 = 88$$
$$3x + 82 = 88$$
$$3x = 6$$
$$x = 2$$
$$x + 57 = 59$$
$$x + 25 = 27$$
2% of men are strikingly-attractive.
59% of men are average-looking.
27% of men are good-looking.

39.
$$29700 + 150x = 5000 + 1100x$$
$$24700 = 950x$$
$$26 = x$$
In 26 years, the cost will be $33,600.

40. Let x = amount invested at 8%
$10000 - x$ = amount invested at 10%

$$0.08x + 0.1(10000 - x) = 940$$
$$0.08x + 1000 - 0.1x = 940$$
$$-0.02x = -60$$
$$x = 3000$$
$$10000 - x = 7000$$

$3000 at 8%, $7000 at 10%

41.
$$l = 2w + 4$$
$$A = lw$$
$$48 = (2w + 4)w$$
$$48 = 2w^2 + 4w$$
$$0 = 2w^2 + 4w - 48$$
$$0 = w^2 + 2w - 24$$
$$0 = (w + 6)(w - 4)$$

$$w + 6 = 0 \qquad w - 4 = 0$$
$$w = -6 \qquad w = 4$$
$$2w + 4 = 2(4) + 4 = 12$$

width is 4 feet, length is 12 feet

42.
$$24^2 + x^2 = 26^2$$
$$576 + x^2 = 676$$
$$x^2 = 100$$
$$x = \pm 10$$

The wire should be attached 10 feet up the pole.

43. Let x = the original selling price
$$20 = x - 0.60x$$
$$20 = 0.40x$$
$$50 = x$$

The original price is $50.

44. Let x = the number text messages.
The monthly cost using Plan A is $C_A = 25$.
The monthly cost using Plan B is $C_B = 13 + 0.06x$.
For Plan A to be better deal, it must cost less than Plan B.

$$C_A < C_B$$
$$25 < 13 + 0.06x$$
$$12 < 0.06x$$
$$200 < x$$
$$x > 200$$

Plan A is a better deal when more than 200 text messages per month are sent/received.

Chapter 2
Functions and Graphs

Section 2.1

Check Point Exercises

1. The domain is the set of all first components: {0, 10, 20, 30, 40}. The range is the set of all second components: {9.1, 6.7, 10.7, 13.2, 21.2}.

2. **a.** The relation is not a function since the two ordered pairs (5, 6) and (5, 8) have the same first component but different second components.

   b. The relation is a function since no two ordered pairs have the same first component and different second components.

3. **a.** $2x + y = 6$

   $y = 6 - 2x$

   For each value of x, there is one and only one value for y, so the equation defines y as a function of x.

   b. $x^2 + y^2 = 1$

   $y^2 = 1 - x^2$

   $y = \pm\sqrt{1 - x^2}$

   Since there are values of x (all values between -1 and 1 exclusive) that give more than one value for y (for example, if $x = 0$, then $y = \pm\sqrt{1 - 0^2} = \pm 1$), the equation does not define y as a function of x.

4. **a.** $f(-5) = (-5)^2 - 2(-5) + 7$

   $= 25 - (-10) + 7$

   $= 42$

   b. $f(x + 4) = (x + 4)^2 - 2(x + 4) + 7$

   $= x^2 + 8x + 16 - 2x - 8 + 7$

   $= x^2 + 6x + 15$

   c. $f(-x) = (-x)^2 - 2(-x) + 7$

   $= x^2 - (-2x) + 7$

   $= x^2 + 2x + 7$

5.

x	$f(x) = 2x$	(x, y)
-2	-4	$(-2, -4)$
-1	-2	$(-1, -2)$
0	0	$(0, 0)$
1	2	$(1, 2)$
2	4	$(2, 4)$

x	$g(x) = 2x - 3$	(x, y)
-2	$g(-2) = 2(-2) - 3 = -7$	$(-2, -7)$
-1	$g(-1) = 2(-1) - 3 = -5$	$(-1, -5)$
0	$g(0) = 2(0) - 3 = -3$	$(0, -3)$
1	$g(1) = 2(1) - 3 = -1$	$(1, -1)$
2	$g(2) = 2(2) - 3 = 1$	$(2, 1)$

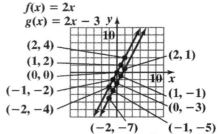

The graph of g is the graph of f shifted down 3 units.

6. The graph (a) passes the vertical line test and is therefore is a <u>function</u>.
   The graph (b) fails the vertical line test and is therefore <u>not a function</u>.
   The graph (c) passes the vertical line test and is therefore is a <u>function</u>.
   The graph (d) fails the vertical line test and is therefore <u>not a function</u>.

7. **a.** $f(5) = 400$

   b. $x = 9$, $f(9) = 100$

   c. The minimum T cell count in the asymptomatic stage is approximately 425.

8. a. domain: $\{x \mid -2 \le x \le 1\}$ or $[-2, 1]$.

range: $\{y \mid 0 \le y \le 3\}$ or $[0, 3]$.

b. domain: $\{x \mid -2 < x \le 1\}$ or $(-2, 1]$.

range: $\{y \mid -1 \le y < 2\}$ or $[-1, 2)$.

c. domain: $\{x \mid -3 \le x < 0\}$ or $[-3, 0)$.

range: $\{y \mid y = -3, -2, -1\}$.

Concept and Vocabulary Check 2.1

1. relation; domain; range

2. function

3. f; x

4. true

5. false

6. x; $x + 6$

7. ordered pairs

8. more than once; function

9. $[0, 3)$; domain

10. $[1, \infty)$; range

11. 0; 0; zeros

12. false

Exercise Set 2.1

1. The relation is a function since no two ordered pairs have the same first component and different second components. The domain is {1, 3, 5} and the range is {2, 4, 5}.

3. The relation is not a function since the two ordered pairs (3, 4) and (3, 5) have the same first component but different second components (the same could be said for the ordered pairs (4, 4) and (4, 5)). The domain is {3, 4} and the range is {4, 5}.

5. The relation is a function because no two ordered pairs have the same first component and different second components The domain is {3, 4, 5, 7} and the range is {−2, 1, 9}.

7. The relation is a function since there are no same first components with different second components. The domain is {−3, −2, −1, 0} and the range is {−3, −2, −1, 0}.

9. The relation is not a function since there are ordered pairs with the same first component and different second components. The domain is {1} and the range is {4, 5, 6}.

11. $x + y = 16$

$y = 16 - x$

Since only one value of y can be obtained for each value of x, y is a function of x.

13. $x^2 + y = 16$

$y = 16 - x^2$

Since only one value of y can be obtained for each value of x, y is a function of x.

15. $x^2 + y^2 = 16$

$y^2 = 16 - x^2$

$y = \pm\sqrt{16 - x^2}$

If $x = 0$, $y = \pm 4$.

Since two values, $y = 4$ and $y = -4$, can be obtained for one value of x, y is not a function of x.

17. $x = y^2$

$y = \pm\sqrt{x}$

If $x = 1$, $y = \pm 1$.

Since two values, $y = 1$ and $y = -1$, can be obtained for $x = 1$, y is not a function of x.

19. $y = \sqrt{x + 4}$

Since only one value of y can be obtained for each value of x, y is a function of x.

21. $x + y^3 = 8$

$y^3 = 8 - x$

$y = \sqrt[3]{8 - x}$

Since only one value of y can be obtained for each value of x, y is a function of x.

23. $xy + 2y = 1$

$y(x + 2) = 1$

$y = \dfrac{1}{x + 2}$

Since only one value of y can be obtained for each value of x, y is a function of x.

25. $|x| - y = 2$

$-y = -|x| + 2$

$y = |x| - 2$

Since only one value of y can be obtained for each value of x, y is a function of x.

27. a. $f(6) = 4(6) + 5 = 29$

b. $f(x + 1) = 4(x + 1) + 5 = 4x + 9$

c. $f(-x) = 4(-x) + 5 = -4x + 5$

29. a. $g(-1) = (-1)^2 + 2(-1) + 3$

$= 1 - 2 + 3$

$= 2$

b. $g(x + 5) = (x + 5)^2 + 2(x + 5) + 3$

$= x^2 + 10x + 25 + 2x + 10 + 3$

$= x^2 + 12x + 38$

c. $g(-x) = (-x)^2 + 2(-x) + 3$

$= x^2 - 2x + 3$

31. a. $h(2) = 2^4 - 2^2 + 1$

$= 16 - 4 + 1$

$= 13$

b. $h(-1) = (-1)^4 - (-1)^2 + 1$

$= 1 - 1 + 1$

$= 1$

c. $h(-x) = (-x)^4 - (-x)^2 + 1 = x^4 - x^2 + 1$

d. $h(3a) = (3a)^4 - (3a)^2 + 1$

$= 81a^4 - 9a^2 + 1$

33. a. $f(-6) = \sqrt{-6 + 6} + 3 = \sqrt{0} + 3 = 3$

b. $f(10) = \sqrt{10 + 6} + 3$

$= \sqrt{16} + 3$

$= 4 + 3$

$= 7$

c. $f(x - 6) = \sqrt{x - 6 + 6} + 3 = \sqrt{x} + 3$

35. a. $f(2) = \dfrac{4(2)^2 - 1}{2^2} = \dfrac{15}{4}$

b. $f(-2) = \dfrac{4(-2)^2 - 1}{(-2)^2} = \dfrac{15}{4}$

c. $f(-x) = \dfrac{4(-x)^2 - 1}{(-x)^2} = \dfrac{4x^2 - 1}{x^2}$

37. a. $f(6) = \dfrac{6}{|6|} = 1$

b. $f(-6) = \dfrac{-6}{|-6|} = \dfrac{-6}{6} = -1$

c. $f(r^2) = \dfrac{r^2}{|r^2|} = \dfrac{r^2}{r^2} = 1$

39.

x	$f(x) = x$	(x, y)
-2	$f(-2) = -2$	$(-2, -2)$
-1	$f(-1) = -1$	$(-1, -1)$
0	$f(0) = 0$	$(0, 0)$
1	$f(1) = 1$	$(1, 1)$
2	$f(2) = 2$	$(2, 2)$

x	$g(x) = x + 3$	(x, y)
-2	$g(-2) = -2 + 3 = 1$	$(-2, 1)$
-1	$g(-1) = -1 + 3 = 2$	$(-1, 2)$
0	$g(0) = 0 + 3 = 3$	$(0, 3)$
1	$g(1) = 1 + 3 = 4$	$(1, 4)$
2	$g(2) = 2 + 3 = 5$	$(2, 5)$

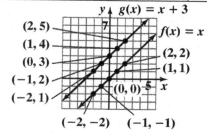

The graph of g is the graph of f shifted up 3 units.

41.

x	$f(x) = -2x$	(x, y)
–2	$f(-2) = -2(-2) = 4$	$(-2, 4)$
–1	$f(-1) = -2(-1) = 2$	$(-1, 2)$
0	$f(0) = -2(0) = 0$	$(0, 0)$
1	$f(1) = -2(1) = -2$	$(1, -2)$
2	$f(2) = -2(2) = -4$	$(2, -4)$

x	$g(x) = -2x - 1$	(x, y)
–2	$g(-2) = -2(-2) - 1 = 3$	$(-2, 3)$
–1	$g(-1) = -2(-1) - 1 = 1$	$(-1, 1)$
0	$g(0) = -2(0) - 1 = -1$	$(0, -1)$
1	$g(1) = -2(1) - 1 = -3$	$(1, -3)$
2	$g(2) = -2(2) - 1 = -5$	$(2, -5)$

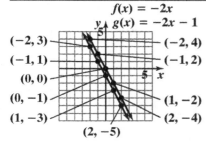

$f(x) = -2x$
$g(x) = -2x - 1$

The graph of g is the graph of f shifted down 1 unit.

43.

x	$f(x) = x^2$	(x, y)
–2	$f(-2) = (-2)^2 = 4$	$(-2, 4)$
–1	$f(-1) = (-1)^2 = 1$	$(-1, 1)$
0	$f(0) = (0)^2 = 0$	$(0, 0)$
1	$f(1) = (1)^2 = 1$	$(1, 1)$
2	$f(2) = (2)^2 = 4$	$(2, 4)$

x	$g(x) = x^2 + 1$	(x, y)
–2	$g(-2) = (-2)^2 + 1 = 5$	$(-2, 5)$
–1	$g(-1) = (-1)^2 + 1 = 2$	$(-1, 2)$
0	$g(0) = (0)^2 + 1 = 1$	$(0, 1)$
1	$g(1) = (1)^2 + 1 = 2$	$(1, 2)$
2	$g(2) = (2)^2 + 1 = 5$	$(2, 5)$

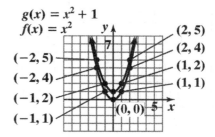

$g(x) = x^2 + 1$
$f(x) = x^2$

The graph of g is the graph of f shifted up 1 unit.

45.

x	$f(x) = |x|$	(x, y)
–2	$f(-2) = |-2| = 2$	$(-2, 2)$
–1	$f(-1) = |-1| = 1$	$(-1, 1)$
0	$f(0) = |0| = 0$	$(0, 0)$
1	$f(1) = |1| = 1$	$(1, 1)$
2	$f(2) = |2| = 2$	$(2, 2)$

x	$g(x) = |x| - 2$	(x, y)
–2	$g(-2) = |-2| - 2 = 0$	$(-2, 0)$
–1	$g(-1) = |-1| - 2 = -1$	$(-1, -1)$
0	$g(0) = |0| - 2 = -2$	$(0, -2)$
1	$g(1) = |1| - 2 = -1$	$(1, -1)$
2	$g(2) = |2| - 2 = 0$	$(2, 0)$

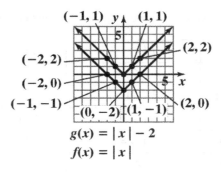

$g(x) = |x| - 2$
$f(x) = |x|$

The graph of g is the graph of f shifted down 2 units.

47.

x	$f(x) = x^3$	(x, y)
–2	$f(-2) = (-2)^3 = -8$	$(-2, -8)$
–1	$f(-1) = (-1)^3 = -1$	$(-1, -1)$
0	$f(0) = (0)^3 = 0$	$(0, 0)$
1	$f(1) = (1)^3 = 1$	$(1, 1)$
2	$f(2) = (2)^3 = 8$	$(2, 8)$

x	$g(x) = x^3 + 2$	(x, y)
–2	$g(-2) = (-2)^3 + 2 = -6$	$(-2, -6)$
–1	$g(-1) = (-1)^3 + 2 = 1$	$(-1, 1)$
0	$g(0) = (0)^3 + 2 = 2$	$(0, 2)$
1	$g(1) = (1)^3 + 2 = 3$	$(1, 3)$
2	$g(2) = (2)^3 + 2 = 10$	$(2, 10)$

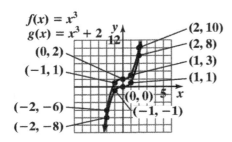

The graph of g is the graph of f shifted up 2 units.

49.

x	$f(x) = 3$	(x, y)
–2	$f(-2) = 3$	$(-2, 3)$
–1	$f(-1) = 3$	$(-1, 3)$
0	$f(0) = 3$	$(0, 3)$
1	$f(1) = 3$	$(1, 3)$
2	$f(2) = 3$	$(2, 3)$

x	$g(x) = 5$	(x, y)
–2	$g(-2) = 5$	$(-2, 5)$
–1	$g(-1) = 5$	$(-1, 5)$
0	$g(0) = 5$	$(0, 5)$
1	$g(1) = 5$	$(1, 5)$
2	$g(2) = 5$	$(2, 5)$

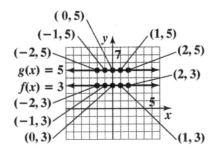

The graph of g is the graph of f shifted up 2 units.

51.

x	$f(x) = \sqrt{x}$	(x, y)
0	$f(0) = \sqrt{0} = 0$	$(0, 0)$
1	$f(1) = \sqrt{1} = 1$	$(1, 1)$
4	$f(4) = \sqrt{4} = 2$	$(4, 2)$
9	$f(9) = \sqrt{9} = 3$	$(9, 3)$

x	$g(x) = \sqrt{x} - 1$	(x, y)
0	$g(0) = \sqrt{0} - 1 = -1$	$(0, -1)$
1	$g(1) = \sqrt{1} - 1 = 0$	$(1, 0)$
4	$g(4) = \sqrt{4} - 1 = 1$	$(4, 1)$
9	$g(9) = \sqrt{9} - 1 = 2$	$(9, 2)$

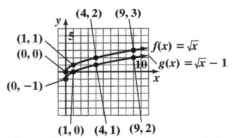

The graph of g is the graph of f shifted down 1 unit.

53.

x	$f(x) = \sqrt{x}$	(x, y)
0	$f(0) = \sqrt{0} = 0$	$(0, 0)$
1	$f(1) = \sqrt{1} = 1$	$(1, 1)$
4	$f(4) = \sqrt{4} = 2$	$(4, 2)$
9	$f(9) = \sqrt{9} = 3$	$(9, 3)$

x	$g(x) = \sqrt{x-1}$	(x, y)
1	$g(1) = \sqrt{1-1} = 0$	$(1, 0)$
2	$g(2) = \sqrt{2-1} = 1$	$(2, 1)$
5	$g(5) = \sqrt{5-1} = 2$	$(5, 2)$
10	$g(10) = \sqrt{10-1} = 3$	$(10, 3)$

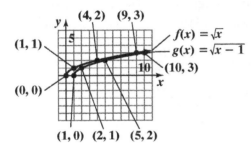

The graph of g is the graph of f shifted right 1 unit.

55. function

57. function

59. not a function

61. function

63. function

65. $f(-2) = -4$

67. $f(4) = 4$

69. $f(-3) = 0$

71. $g(-4) = 2$

73. $g(-10) = 2$

75. When $x = -2$, $g(x) = 1$.

77. **a.** domain: $(-\infty, \infty)$

b. range: $[-4, \infty)$

c. x-intercepts: -3 and 1

d. y-intercept: -3

e. $f(-2) = -3$ and $f(2) = 5$

79. **a.** domain: $(-\infty, \infty)$

b. range: $[1, \infty)$

c. x-intercept: none

d. y-intercept: 1

e. $f(-1) = 2$ and $f(3) = 4$

81. **a.** domain: $[0, 5)$

b. range: $[-1, 5)$

c. x-intercept: 2

d. y-intercept: -1

e. $f(3) = 1$

83. **a.** domain: $[0, \infty)$

b. range: $[1, \infty)$

c. x-intercept: none

d. y-intercept: 1

e. $f(4) = 3$

85. **a.** domain: $[-2, 6]$

b. range: $[-2, 6]$

c. x-intercept: 4

d. y-intercept: 4

e. $f(-1) = 5$

87. **a.** domain: $(-\infty, \infty)$

b. range: $(-\infty, -2]$

c. x-intercept: none

d. y-intercept: -2

e. $f(-4) = -5$ and $f(4) = -2$

89. **a.** domain: $(-\infty, \infty)$

b. range: $(0, \infty)$

c. x-intercept: none

d. y-intercept: 1.5

e. $f(4) = 6$

91. **a.** domain: $\{-5, -2, 0, 1, 3\}$

b. range: $\{2\}$

c. x-intercept: none

d. y-intercept: 2

e. $f(-5) + f(3) = 2 + 2 = 4$

93. $g(1) = 3(1) - 5 = 3 - 5 = -2$

$f(g(1)) = f(-2) = (-2)^2 - (-2) + 4$
$\qquad = 4 + 2 + 4 = 10$

95. $\sqrt{3 - (-1)} - (-6)^2 + 6 \div (-6) \cdot 4$

$= \sqrt{3 + 1} - 36 + 6 \div (-6) \cdot 4$

$= \sqrt{4} - 36 + -1 \cdot 4$

$= 2 - 36 + -4$

$= -34 + -4$

$= -38$

97. $f(-x) - f(x)$

$= (-x)^3 + (-x) - 5 - (x^3 + x - 5)$

$= -x^3 - x - 5 - x^3 - x + 5 = -2x^3 - 2x$

99. **a.** $\{(\text{Philippines, } 12), (\text{Spain, } 13),$

$(\text{Italy, } 14), (\text{Germany, } 14),$

$(\text{Russia, } 16)\}$

b. Yes, the relation is a function. Each country (element in the domain) corresponds to only one age (element in the range).

c. $\{(12, \text{ Philippines}), (13, \text{ Spain}),$

$(14, \text{ Italy}), (14, \text{ Germany}),$

$(16, \text{ Russia})\}$

d. No, the relation is not a function. 14 in the domain corresponds to two members in the range, Italy and Germany.

101. **a.** $f(70) = 83$ which means the chance that a 60-year old will survive to age 70 is 83%.

b. $g(70) = 76$ which means the chance that a 60-year old will survive to age 70 is 76%.

c. Function f is the better model.

103. **a.** $G(30) = -0.01(30)^2 + (30) + 60 = 81$

In 2010, the wage gap was 81%. This is represented as $(30, 81)$ on the graph.

b. $G(30)$ underestimates the actual data shown by the bar graph by 2%.

105. $C(x) = 100{,}000 + 100x$

$C(90) = 100{,}000 + 100(90) = \$109{,}000$

It will cost \$109,000 to produce 90 bicycles.

107. $T(x) = \dfrac{40}{x} + \dfrac{40}{x + 30}$

$T(30) = \dfrac{40}{30} + \dfrac{40}{30 + 30}$

$\qquad = \dfrac{80}{60} + \dfrac{40}{60}$

$\qquad = \dfrac{120}{60}$

$\qquad = 2$

If you travel 30 mph going and 60 mph returning, your total trip will take 2 hours.

109. – 117. Answers will vary.

119. does not make sense; Explanations will vary. Sample explanation: The parentheses used in function notation, such as $f(x)$, do not imply multiplication.

121. does not make sense; Explanations will vary. Sample explanation: This would not be a function because some elements in the domain would correspond to more than one age in the range.

123. false; Changes to make the statement true will vary. A sample change is: The range is $[-2, 2)$.

125. false; Changes to make the statement true will vary. A sample change is: $f(0) = 0.8$

127. Answers will vary.
An example is $\{(1,1),(2,1)\}$

129. $C(t) = 20 + 0.40(t - 60)$
$C(100) = 20 + 0.40(100 - 60)$
$\qquad = 20 + 0.40(40)$
$\qquad = 20 + 16$
$\qquad = 36$
For 100 calling minutes, the monthly cost is $36.

130. $f(x) = x + 2, \; x \leq 1$

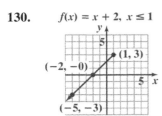

131. $2(x+h)^2 + 3(x+h) + 5 - (2x^2 + 3x + 5)$
$= 2(x^2 + 2xh + h^2) + 3x + 3h + 5 - 2x^2 - 3x - 5$
$= 2x^2 + 4xh + 2h^2 + 3x + 3h + 5 - 2x^2 - 3x - 5$
$= 2x^2 - 2x^2 + 4xh + 2h^2 + 3x - 3x + 3h + 5 - 5$
$= 4xh + 2h^2 + 3h$

Section 2.2

Check Point Exercises

1. The function is increasing on the interval $(-\infty, -1)$, decreasing on the interval $(-1, 1)$, and increasing on the interval $(1, \infty)$.

2. **a.** $f(-x) = (-x)^2 + 6 = x^2 + 6 = f(x)$
The function is even.

b. $g(-x) = 7(-x)^3 - (-x) = -7x^3 + x = -f(x)$
The function is odd.

c. $h(-x) = (-x)^5 + 1 = -x^5 + 1$
The function is neither even nor odd.

3. $C(t) = \begin{cases} 20 & \text{if } 0 \leq t \leq 60 \\ 20 + 0.40(t - 60) & \text{if } t > 60 \end{cases}$

b. Since $0 \leq 40 \leq 60$, $C(40) = 20$
With 40 calling minutes, the cost is $20.
This is represented by $(40, 20)$.

c. Since $80 > 60$,
$C(80) = 20 + 0.40(80 - 60) = 28$
With 80 calling minutes, the cost is $28.
This is represented by $(80, 28)$.

4.

$f(x) = \begin{cases} 3 & \text{if } x \leq -1 \\ x - 2 & \text{if } x > -1 \end{cases}$

5. **a.** $f(x) = -2x^2 + x + 5$
$f(x+h) = -2(x+h)^2 + (x+h) + 5$
$\qquad = -2(x^2 + 2xh + h^2) + x + h + 5$
$\qquad = -2x^2 - 4xh - 2h^2 + x + h + 5$

b. $\dfrac{f(x+h) - f(x)}{h}$
$= \dfrac{-2x^2 - 4xh - 2h^2 + x + h + 5 - \left(-2x^2 + x + 5\right)}{h}$
$= \dfrac{-2x^2 - 4xh - 2h^2 + x + h + 5 + 2x^2 - x - 5}{h}$
$= \dfrac{-4xh - 2h^2 + h}{h}$
$= \dfrac{h\left(-4x - 2h + 1\right)}{h}$
$= -4x - 2h + 1, \quad h \neq 0$

Concept and Vocabulary Check 2.2

1. $< f(x_2)$; $> f(x_2)$; $= f(x_2)$

2. maximum; minimum

3. $f(x)$; y-axis

4. $-f(x)$; origin

5. piecewise

6. less than or equal to x; 2; -3; 0

7. difference quotient; $x + h$; $f(x)$; h; h

8. false

9. false

Exercise Set 2.2

1. **a.** increasing: $(-1, \infty)$

    b. decreasing: $(-\infty, -1)$

    c. constant: none

3. **a.** increasing: $(0, \infty)$

    b. decreasing: none

    c. constant: none

5. **a.** increasing: none

    b. decreasing: $(-2, 6)$

    c. constant: none

7. **a.** increasing: $(-\infty, -1)$

    b. decreasing: none

    c. constant: $(-1, \infty)$

9. **a.** increasing: $(-\infty, 0)$ or $(1.5, 3)$

    b. decreasing: $(0, 1.5)$ or $(3, \infty)$

    c. constant: none

11. **a.** increasing: $(-2, 4)$

    b. decreasing: none

    c. constant: $(-\infty, -2)$ or $(4, \infty)$

13. **a.** $x = 0$, relative maximum $= 4$

    b. $x = -3, 3$, relative minimum $= 0$

15. **a.** $x = -2$, relative maximum $= 21$

    b. $x = 1$, relative minimum $= -6$

17. $f(x) = x^3 + x$
    $f(-x) = (-x)^3 + (-x)$
    $f(-x) = -x^3 - x = -(x^3 + x)$
    $f(-x) = -f(x)$, odd function

19. $g(x) = x^2 + x$
    $g(-x) = (-x)^2 + (-x)$
    $g(-x) = x^2 - x$, neither

21. $h(x) = x^2 - x^4$
    $h(-x) = (-x)^2 - (-x)^4$
    $h(-x) = x^2 - x^4$
    $h(-x) = h(x)$, even function

23. $f(x) = x^2 - x^4 + 1$
    $f(-x) = (-x)^2 - (-x)^4 + 1$
    $f(-x) = x^2 - x^4 + 1$
    $f(-x) = f(x)$, even function

25. $f(x) = \frac{1}{5}x^6 - 3x^2$
    $f(-x) = \frac{1}{5}(-x)^6 - 3(-x)^2$
    $f(-x) = \frac{1}{5}x^6 - 3x^2$
    $f(-x) = f(x)$, even function

27. $f(x) = x\sqrt{1 - x^2}$
    $f(-x) = -x\sqrt{1 - (-x)^2}$
    $f(-x) = -x\sqrt{1 - x^2}$
    $\quad\quad = -\left(x\sqrt{1 - x^2}\right)$
    $f(-x) = -f(x)$, odd function

29. The graph is symmetric with respect to the *y*-axis. The function is even.

31. The graph is symmetric with respect to the origin. The function is odd.

33. **a.** domain: $(-\infty, \infty)$

    b. range: $[-4, \infty)$

    c. *x*-intercepts: 1, 7

    d. *y*-intercept: 4

    e. $(4, \infty)$

    f. $(0, 4)$

    g. $(-\infty, 0)$

    h. $x = 4$

    i. $y = -4$

j. $f(-3) = 4$

k. $f(2) = -2$ and $f(6) = -2$

l. neither ; $f(-x) \neq x$, $f(-x) \neq -x$

35. a. domain: $(-\infty, 3]$

b. range: $(-\infty, 4]$

c. *x*-intercepts: –3, 3

d. $f(0) = 3$

e. $(-\infty, 1)$

f. $(1, 3)$

g. $(-\infty, -3]$

h. $f(1) = 4$

i. $x = 1$

j. positive; $f(-1) = +2$

37. a. $f(-2) = 3(-2) + 5 = -1$

b. $f(0) = 4(0) + 7 = 7$

c. $f(3) = 4(3) + 7 = 19$

39. a. $g(0) = 0 + 3 = 3$

b. $g(-6) = -(-6 + 3) = -(-3) = 3$

c. $g(-3) = -3 + 3 = 0$

41. a. $h(5) = \dfrac{5^2 - 9}{5 - 3} = \dfrac{25 - 9}{2} = \dfrac{16}{2} = 8$

b. $h(0) = \dfrac{0^2 - 9}{0 - 3} = \dfrac{-9}{-3} = 3$

c. $h(3) = 6$

43. a.

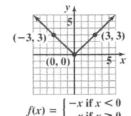

$f(x) = \begin{cases} -x \text{ if } x < 0 \\ x \text{ if } x \geq 0 \end{cases}$

b. range: $[0, \infty)$

45. a.

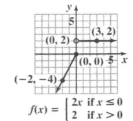

$f(x) = \begin{cases} 2x \text{ if } x \leq 0 \\ 2 \text{ if } x > 0 \end{cases}$

b. range: $(-\infty, 0] \cup \{2\}$

47. a.

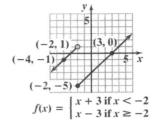

$f(x) = \begin{cases} x + 3 \text{ if } x < -2 \\ x - 3 \text{ if } x \geq -2 \end{cases}$

b. range: $(-\infty, \infty)$

49. a.

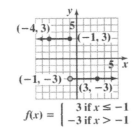

$f(x) = \begin{cases} 3 \text{ if } x \leq -1 \\ -3 \text{ if } x > -1 \end{cases}$

b. range: $\{-3, 3\}$

51. a.

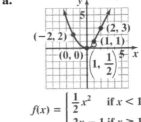

$f(x) = \begin{cases} \dfrac{1}{2}x^2 \quad \text{ if } x < 1 \\ 2x - 1 \text{ if } x \geq 1 \end{cases}$

b. range: $[0, \infty)$

53. a.

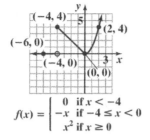

$$f(x) = \begin{cases} 0 & \text{if } x < -4 \\ -x & \text{if } -4 \le x < 0 \\ x^2 & \text{if } x \ge 0 \end{cases}$$

b. range: $[0, \infty)$

55.
$$\frac{f(x+h) - f(x)}{h}$$
$$= \frac{4(x+h) - 4x}{h}$$
$$= \frac{4x + 4h - 4x}{h}$$
$$= \frac{4h}{h}$$
$$= 4$$

57.
$$\frac{f(x+h) - f(x)}{h}$$
$$= \frac{3(x+h) + 7 - (3x + 7)}{h}$$
$$= \frac{3x + 3h + 7 - 3x - 7}{h}$$
$$= \frac{3h}{h}$$
$$= 3$$

59.
$$\frac{f(x+h) - f(x)}{h}$$
$$= \frac{(x+h)^2 - x^2}{h}$$
$$= \frac{x^2 + 2xh + h^2 - x^2}{h}$$
$$= \frac{2xh + h^2}{h}$$
$$= \frac{h(2x+h)}{h}$$
$$= 2x + h$$

61.
$$\frac{f(x+h) - f(x)}{h}$$
$$= \frac{(x+h)^2 - 4(x+h) + 3 - (x^2 - 4x + 3)}{h}$$
$$= \frac{x^2 + 2xh + h^2 - 4x - 4h + 3 - x^2 + 4x - 3}{h}$$
$$= \frac{2xh + h^2 - 4h}{h}$$
$$= \frac{h(2x + h - 4)}{h}$$
$$= 2x + h - 4$$

63.
$$\frac{f(x+h) - f(x)}{h}$$
$$= \frac{2(x+h)^2 + (x+h) - 1 - (2x^2 + x - 1)}{h}$$
$$= \frac{2x^2 + 4xh + 2h^2 + x + h - 1 - 2x^2 - x + 1}{h}$$
$$= \frac{4xh + 2h^2 + h}{h}$$
$$= \frac{h(4x + 2h + 1)}{h}$$
$$= 4x + 2h + 1$$

65.
$$\frac{f(x+h) - f(x)}{h}$$
$$= \frac{-(x+h)^2 + 2(x+h) + 4 - (-x^2 + 2x + 4)}{h}$$
$$= \frac{-x^2 - 2xh - h^2 + 2x + 2h + 4 + x^2 - 2x - 4}{h}$$
$$= \frac{-2xh - h^2 + 2h}{h}$$
$$= \frac{h(-2x - h + 2)}{h}$$
$$= -2x - h + 2$$

67. $\dfrac{f(x+h)-f(x)}{h}$

$= \dfrac{-2(x+h)^2+5(x+h)+7-(-2x^2+5x+7)}{h}$

$= \dfrac{-2x^2-4xh-2h^2+5x+5h+7+2x^2-5x-7}{h}$

$= \dfrac{-4xh-2h^2+5h}{h}$

$= \dfrac{h(-4x-2h+5)}{h}$

$= -4x-2h+5$

69. $\dfrac{f(x+h)-f(x)}{h}$

$= \dfrac{-2(x+h)^2-(x+h)+3-(-2x^2-x+3)}{h}$

$= \dfrac{-2x^2-4xh-2h^2-x-h+3+2x^2+x-3}{h}$

$= \dfrac{-4xh-2h^2-h}{h}$

$= \dfrac{h(-4x-2h-1)}{h}$

$= -4x-2h-1$

71. $\dfrac{f(x+h)-f(x)}{h} = \dfrac{6-6}{h} = \dfrac{0}{h} = 0$

73. $\dfrac{f(x+h)-f(x)}{h}$

$= \dfrac{\dfrac{1}{x+h}-\dfrac{1}{x}}{h}$

$= \dfrac{\dfrac{x}{x(x+h)}+\dfrac{-(x+h)}{x(x+h)}}{h}$

$= \dfrac{\dfrac{x-x-h}{x(x+h)}}{h}$

$= \dfrac{\dfrac{-h}{x(x+h)}}{h}$

$= \dfrac{-h}{x(x+h)} \cdot \dfrac{1}{h}$

$= \dfrac{-1}{x(x+h)}$

75. $\dfrac{f(x+h)-f(x)}{h}$

$= \dfrac{\sqrt{x+h}-\sqrt{x}}{h}$

$= \dfrac{\sqrt{x+h}-\sqrt{x}}{h} \cdot \dfrac{\sqrt{x+h}+\sqrt{x}}{\sqrt{x+h}+\sqrt{x}}$

$= \dfrac{x+h-x}{h\left(\sqrt{x+h}+\sqrt{x}\right)}$

$= \dfrac{h}{h\left(\sqrt{x+h}+\sqrt{x}\right)}$

$= \dfrac{1}{\sqrt{x+h}+\sqrt{x}}$

77. $\sqrt{f(-1.5)+f(-0.9)} - \left[f(\pi)\right]^2 + f(-3) \div f(1) \cdot f(-\pi)$

$= \sqrt{1+0} - \left[-4\right]^2 + 2 \div (-2) \cdot 3$

$= \sqrt{1} - 16 + (-1) \cdot 3$

$= 1 - 16 - 3$

$= -18$

79. $30 + 0.30(t-120) = 30 + 0.3t - 36 = 0.3t - 6$

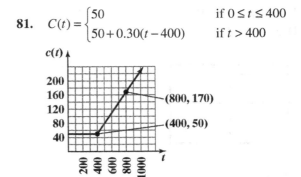

81. $C(t) = \begin{cases} 50 & \text{if } 0 \le t \le 400 \\ 50 + 0.30(t-400) & \text{if } t > 400 \end{cases}$

83. increasing: (25, 55); decreasing: (55, 75)

85. The percent body fat in women reaches a maximum at age 55. This maximum is 38%.

87. domain: [25, 75]; range: [34, 38]

89. This model describes percent body fat in men.

91. $T(20,000) = 850 + 0.15(20,000 - 8500)$
$= 2575$
A single taxpayer with taxable income of $20,000 owes $2575.

93. $42,449 + 0.33(x - 174,400)$

95. $f(3) = 0.85$
The cost of mailing a first-class letter weighing 3 ounces is $0.85.

97. The cost to mail a letter weighing 1.5 ounces is $0.65.

99.

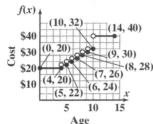

101. – 105. Answers will vary.

107.

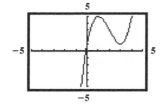

Increasing: $(-\infty, 1)$ or $(3, \infty)$
Decreasing: $(1, 3)$

109.

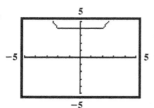

Increasing: $(2, \infty)$
Decreasing: $(-\infty, -2)$
Constant: $(-2, 2)$

111.

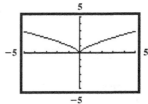

Increasing: $(0, \infty)$
Decreasing: $(-\infty, 0)$

113. a.

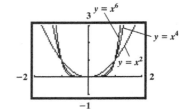

b.

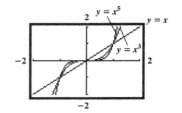

c. Increasing: $(0, \infty)$
Decreasing: $(-\infty, 0)$

d. $f(x) = x^n$ is increasing from $(-\infty, \infty)$ when n is odd.

e.

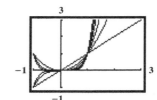

115. makes sense

117. makes sense

119. answers will vary

121. answers will vary

122. $\dfrac{y_2 - y_1}{x_2 - x_1} = \dfrac{4 - 1}{-2 - (-3)} = \dfrac{3}{1} = 3$

123. When $y = 0$:
$$4x - 3y - 6 = 0$$
$$4x - 3(0) - 6 = 0$$
$$4x - 6 = 0$$
$$4x = 6$$
$$x = \frac{3}{2}$$

The point is $\left(\frac{3}{2}, 0\right)$.

When $x = 0$:
$$4x - 3y - 6 = 0$$
$$4(0) - 3y - 6 = 0$$
$$-3y - 6 = 0$$
$$-3y = 6$$
$$x = -2$$

The point is $(0, -2)$.

124. $3x + 2y - 4 = 0$
$$2y = -3x + 4$$
$$y = \frac{-3x + 4}{2}$$
or
$$y = -\frac{3}{2}x + 2$$

Section 2.3

Check Point Exercises

1. a. $m = \dfrac{-2 - 4}{-4 - (-3)} = \dfrac{-6}{-1} = 6$

b. $m = \dfrac{5 - (-2)}{-1 - 4} = \dfrac{7}{-5} = -\dfrac{7}{5}$

2. Point-slope form:
$$y - y_1 = m(x - x_1)$$
$$y - (-5) = 6(x - 2)$$
$$y + 5 = 6(x - 2)$$

Slope-intercept form:
$$y + 5 = 6(x - 2)$$
$$y + 5 = 6x - 12$$
$$y = 6x - 17$$

3. $m = \dfrac{-6 - (-1)}{-1 - (-2)} = \dfrac{-5}{1} = -5$,

so the slope is –5.
Using the point (–2, –1), we get the following point-slope equation:
$$y - y_1 = m(x - x_1)$$
$$y - (-1) = -5[x - (-2)]$$
$$y + 1 = -5(x + 2)$$

Using the point (–1, –6), we get the following point-slope equation:
$$y - y_1 = m(x - x_1)$$
$$y - (-6) = -5[x - (-1)]$$
$$y + 6 = -5(x + 1)$$

Solve the equation for y:
$$y + 1 = -5(x + 2)$$
$$y + 1 = -5x - 10$$
$$y = -5x - 11.$$

4. The slope m is $\frac{3}{5}$ and the y-intercept is 1, so one point on the line is (1, 0). We can find a second point on the line by using the slope $m = \frac{3}{5} = \frac{\text{Rise}}{\text{Run}}$: starting at the point (0, 1), move 3 units up and 5 units to the right, to obtain the point (5, 4).

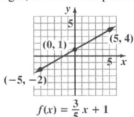

5. $y = 3$ is a horizontal line.

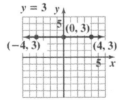

6. All ordered pairs that are solutions of $x = -3$ have a value of x that is always –3. Any value can be used for y.

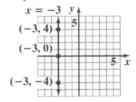

7. $3x + 6y - 12 = 0$

$$6y = -3x + 12$$

$$y = \frac{-3}{6}x + \frac{12}{6}$$

$$y = -\frac{1}{2}x + 2$$

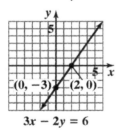

$3x + 6y - 12 = 0$

The slope is $-\dfrac{1}{2}$ and the y-intercept is 2.

8. Find the x-intercept:

$$3x - 2y - 6 = 0$$

$$3x - 2(0) - 6 = 0$$

$$3x - 6 = 0$$

$$3x = 6$$

$$x = 2$$

Find the y-intercept:

$$3x - 2y - 6 = 0$$

$$3(0) - 2y - 6 = 0$$

$$-2y - 6 = 0$$

$$-2y = 6$$

$$y = -3$$

$3x - 2y = 6$

9. First find the slope.

$$m = \frac{\text{Change in } y}{\text{Change in } x} = \frac{57.64 - 57.04}{354 - 317} = \frac{0.6}{37} \approx 0.016$$

Use the point-slope form and then find slope-intercept form.

$$y - y_1 = m(x - x_1)$$

$$y - 57.04 = 0.016(x - 317)$$

$$y - 57.04 = 0.016x - 5.072$$

$$y = 0.016x + 51.968$$

$$f(x) = 0.016x + 52.0$$

Find the temperature at a concentration of 600 parts per million.

$$f(x) = 0.016x + 52.0$$

$$f(600) = 0.016(600) + 52.0$$

$$= 61.6$$

The temperature at a concentration of 600 parts per million would be $61.6°$F.

Concept and Vocabulary Check 2.3

1. scatter plot; regression

2. $\dfrac{y_2 - y_1}{x_2 - x_1}$

3. positive

4. negative

5. zero

6. undefined

7. $y - y_1 = m(x - x_1)$

8. $y = mx + b$; slope; y-intercept

9. $(0, 3)$; 2; 5

10. horizontal

11. vertical

12. general

Exercise Set 2.3

1. $m = \dfrac{10 - 7}{8 - 4} = \dfrac{3}{4}$; rises

3. $m = \dfrac{2 - 1}{2 - (-2)} = \dfrac{1}{4}$; rises

5. $m = \dfrac{2 - (-2)}{3 - 4} = \dfrac{0}{-1} = 0$; horizontal

7. $m = \dfrac{-1 - 4}{-1 - (-2)} = \dfrac{-5}{1} = -5$; falls

9. $m = \dfrac{-2 - 3}{5 - 5} = \dfrac{-5}{0}$ undefined; vertical

11. $m = 2$, $x_1 = 3$, $y_1 = 5$;
point-slope form: $y - 5 = 2(x - 3)$;
slope-intercept form: $y - 5 = 2x - 6$
$$y = 2x - 1$$

13. $m = 6$, $x_1 = -2$, $y_1 = 5$;
point-slope form: $y - 5 = 6(x + 2)$;
slope-intercept form: $y - 5 = 6x + 12$
$$y = 6x + 17$$

15. $m = -3$, $x_1 = -2$, $y_1 = -3$;
point-slope form: $y + 3 = -3(x + 2)$;
slope-intercept form: $y + 3 = -3x - 6$
$$y = -3x - 9$$

17. $m = -4$, $x_1 = -4$, $y_1 = 0$;
point-slope form: $y - 0 = -4(x + 4)$;
slope-intercept form: $y = -4(x + 4)$
$$y = -4x - 16$$

19. $m = -1$, $x_1 = \dfrac{-1}{2}$, $y_1 = -2$;

point-slope form: $y + 2 = -1\left(x + \dfrac{1}{2}\right)$;

slope-intercept form: $y + 2 = -x - \dfrac{1}{2}$

$$y = -x - \dfrac{5}{2}$$

21. $m = \dfrac{1}{2}$, $x_1 = 0$, $y_1 = 0$;

point-slope form: $y - 0 = \dfrac{1}{2}(x - 0)$;

slope-intercept form: $y = \dfrac{1}{2}x$

23. $m = -\dfrac{2}{3}$, $x_1 = 6$, $y_1 = -2$;

point-slope form: $y + 2 = -\dfrac{2}{3}(x - 6)$;

slope-intercept form: $y + 2 = -\dfrac{2}{3}x + 4$

$$y = -\dfrac{2}{3}x + 2$$

25. $m = \dfrac{10 - 2}{5 - 1} = \dfrac{8}{4} = 2$;

point-slope form: $y - 2 = 2(x - 1)$ using
$(x_1, y_1) = (1, 2)$, or $y - 10 = 2(x - 5)$ using
$(x_1, y_1) = (5, 10)$;
slope-intercept form: $y - 2 = 2x - 2$ or
$$y - 10 = 2x - 10,$$
$$y = 2x$$

27. $m = \dfrac{3 - 0}{0 - (-3)} = \dfrac{3}{3} = 1$;

point-slope form: $y - 0 = 1(x + 3)$ using
$(x_1, y_1) = (-3, 0)$, or $y - 3 = 1(x - 0)$ using
$(x_1, y_1) = (0, 3)$; slope-intercept form: $y = x + 3$

29. $m = \dfrac{4 - (-1)}{2 - (-3)} = \dfrac{5}{5} = 1$;

point-slope form: $y + 1 = 1(x + 3)$ using
$(x_1, y_1) = (-3, -1)$, or $y - 4 = 1(x - 2)$ using
$(x_1, y_1) = (2, 4)$; slope-intercept form:
$y + 1 = x + 3$ or
$y - 4 = x - 2$
$$y = x + 2$$

31. $m = \dfrac{6 - (-2)}{3 - (-3)} = \dfrac{8}{6} = \dfrac{4}{3}$;

point-slope form: $y + 2 = \dfrac{4}{3}(x + 3)$ using

$(x_1, y_1) = (-3, -2)$, or $y - 6 = \dfrac{4}{3}(x - 3)$ using

$(x_1, y_1) = (3, 6)$;

slope-intercept form: $y + 2 = \dfrac{4}{3x} + 4$ or

$$y - 6 = \dfrac{4}{3}x - 4,$$

$$y = \dfrac{4}{3}x + 2$$

33. $m = \dfrac{-1 - (-1)}{4 - (-3)} = \dfrac{0}{7} = 0$;

point-slope form: $y + 1 = 0(x + 3)$ using
$(x_1, y_1) = (-3, -1)$, or $y + 1 = 0(x - 4)$ using
$(x_1, y_1) = (4, -1)$;
slope-intercept form: $y + 1 = 0$, so
$$y = -1$$

35. $m = \dfrac{0-4}{-2-2} = \dfrac{-4}{-4} = 1$;

point-slope form: $y - 4 = 1(x - 2)$ using
$(x_1, y_1) = (2, 4)$, or $y - 0 = 1(x + 2)$ using
$(x_1, y_1) = (-2, 0)$;

slope-intercept form: $y - 9 = x - 2,$ or
$$y = x + 2$$

37. $m = \dfrac{4-0}{0-\left(-\frac{1}{2}\right)} = \dfrac{4}{\frac{1}{2}} = 8$;

point-slope form: $y - 4 = 8(x - 0)$ using
$(x_1, y_1) = (0, 4)$, or $y - 0 = 8\left(x + \frac{1}{2}\right)$ using
$(x_1, y_1) = \left(-\frac{1}{2}, 0\right)$; or $y - 0 = 8\left(x + \frac{1}{2}\right)$

slope-intercept form: $y = 8x + 4$

39. $m = 2; b = 1$

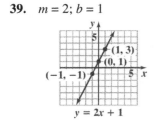

41. $m = -2; b = 1$

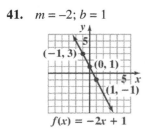

43. $m = \dfrac{3}{4}; b = -2$

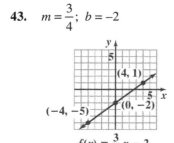

45. $m = -\dfrac{3}{5}; b = 7$

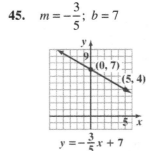

47. $m = -\dfrac{1}{2}; b = 0$

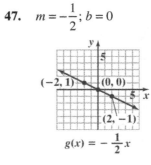

49.

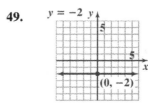

51.

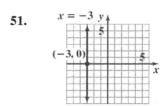

53.

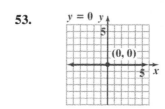

55.

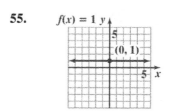

57. $3x - 18 = 0$

$3x = 18$

$x = 6$

59. a. $3x + y - 5 = 0$

$y - 5 = -3x$

$y = -3x + 5$

b. $m = -3; b = 5$

c.

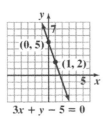

61. a. $2x + 3y - 18 = 0$

$2x - 18 = -3y$

$-3y = 2x - 18$

$y = \dfrac{2}{-3}x - \dfrac{18}{-3}$

$y = -\dfrac{2}{3}x + 6$

b. $m = -\dfrac{2}{3}; b = 6$

c.

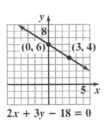

63. a. $8x - 4y - 12 = 0$

$8x - 12 = 4y$

$4y = 8x - 12$

$y = \dfrac{8}{4}x - \dfrac{12}{4}$

$y = 2x - 3$

b. $m = 2; b = -3$

c.

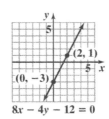

65. a. $3y - 9 = 0$

$3y = 9$

$y = 3$

b. $m = 0; b = 3$

c.

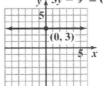

67. Find the *x*-intercept:
$$6x - 2y - 12 = 0$$
$$6x - 2(0) - 12 = 0$$
$$6x - 12 = 0$$
$$6x = 12$$
$$x = 2$$
Find the *y*-intercept:
$$6x - 2y - 12 = 0$$
$$6(0) - 2y - 12 = 0$$
$$-2y - 12 = 0$$
$$-2y = 12$$
$$y = -6$$

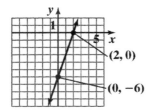

$6x - 2y - 12 = 0$

69. Find the *x*-intercept:
$$2x + 3y + 6 = 0$$
$$2x + 3(0) + 6 = 0$$
$$2x + 6 = 0$$
$$2x = -6$$
$$x = -3$$
Find the *y*-intercept:
$$2x + 3y + 6 = 0$$
$$2(0) + 3y + 6 = 0$$
$$3y + 6 = 0$$
$$3y = -6$$
$$y = -2$$

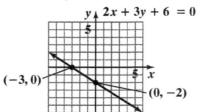

71. Find the *x*-intercept:
$$8x - 2y + 12 = 0$$
$$8x - 2(0) + 12 = 0$$
$$8x + 12 = 0$$
$$8x = -12$$
$$\frac{8x}{8} = \frac{-12}{8}$$
$$x = \frac{-3}{2}$$
Find the *y*-intercept:
$$8x - 2y + 12 = 0$$
$$8(0) - 2y + 12 = 0$$
$$-2y + 12 = 0$$
$$-2y = -12$$
$$y = -6$$

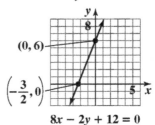

$8x - 2y + 12 = 0$

73.
$$m = \frac{0 - a}{b - 0} = \frac{-a}{b} = -\frac{a}{b}$$

Since *a* and *b* are both positive, $-\frac{a}{b}$ is

negative. Therefore, the line falls.

75.
$$m = \frac{(b + c) - b}{a - a} = \frac{c}{0}$$
The slope is undefined.
The line is vertical.

77. $Ax + By = C$
$$By = -Ax + C$$
$$y = -\frac{A}{B}x + \frac{C}{B}$$

The slope is $-\frac{A}{B}$ and the *y* – intercept is $\frac{C}{B}$.

79.
$$-3 = \frac{4-y}{1-3}$$
$$-3 = \frac{4-y}{-2}$$
$$6 = 4-y$$
$$2 = -y$$
$$-2 = y$$

81. $3x - 4f(x) = 6$
$$-4f(x) = -3x + 6$$
$$f(x) = \frac{3}{4}x - \frac{3}{2}$$

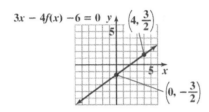

83. Using the slope-intercept form for the equation of a line:
$$-1 = -2(3) + b$$
$$-1 = -6 + b$$
$$5 = b$$

85. m_1, m_3, m_2, m_4

87. **a.** First, find the slope using $(20, 38.9)$ and $(10, 31.1)$.
$$m = \frac{38.9 - 31.1}{20 - 10} = \frac{7.8}{10} = 0.78$$
Then use the slope and one of the points to write the equation in point-slope form.
$$y - y_1 = m(x - x_1)$$
$$y - 31.1 = 0.78(x - 10)$$
or
$$y - 38.9 = 0.78(x - 20)$$

b. $y - 31.1 = 0.78(x - 10)$
$$y - 31.1 = 0.78x - 7.8$$
$$y = 0.78x + 23.3$$
$$f(x) = 0.78x + 23.3$$

c. $f(40) = 0.78(40) + 23.3 = 54.5$
The linear function predicts the percentage of never married American females, ages $25 - 29$, to be 54.5% in 2020.

89. **a.** **Life Expectancy for United States Males, by Year of Birth**

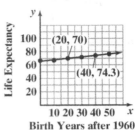

b. $m = \dfrac{\text{Change in } y}{\text{Change in } x} = \dfrac{74.3 - 70.0}{40 - 20} = 0.215$
$$y - y_1 = m(x - x_1)$$
$$y - 70.0 = 0.215(x - 20)$$
$$y - 70.0 = 0.215x - 4.3$$
$$y = 0.215x + 65.7$$
$$E(x) = 0.215x + 65.7$$

c. $E(x) = 0.215x + 65.7$
$$E(60) = 0.215(60) + 65.7$$
$$= 78.6$$
The life expectancy of American men born in 2020 is expected to be 78.6.

91. $(10, 230)$ $(60, 110)$ Points may vary.
$$m = \frac{110 - 230}{60 - 10} = -\frac{120}{50} = -2.4$$
$$y - 230 = -2.4(x - 10)$$
$$y - 230 = -2.4x + 24$$
$$y = -2.4x + 254$$
Answers will vary for predictions.

93. – 99. Answers will vary.

101. Two points are $(0, 6)$ and $(10, -24)$.
$$m = \frac{-24 - 6}{10 - 0} = \frac{-30}{10} = -3.$$
Check: $y = mx + b$: $y = -3x + 6$.

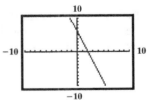

103. Two points are $(0, -2)$ and $(10, 5.5)$.

$$m = \frac{5.5 - (-2)}{10 - 0} = \frac{7.5}{10} = 0.75 \text{ or } \frac{3}{4}.$$

Check: $y = mx + b$: $y = \frac{3}{4}x - 2$.

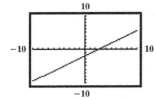

105. does not make sense; Explanations will vary. Sample explanation: Linear functions never change from increasing to decreasing.

107. does not make sense; Explanations will vary. Sample explanation: The slope of line's whose equations are in this form can be determined in several ways. One such way is to rewrite the equation in slope-intercept form.

109. false; Changes to make the statement true will vary. A sample change is: It is possible for m to equal b.

111. true

113. We are given that the x–intercept is -2 and the y–intercept is 4. We can use the points $(-2, 0)$ and $(0, 4)$ to find the slope.

$$m = \frac{4 - 0}{0 - (-2)} = \frac{4}{0 + 2} = \frac{4}{2} = 2$$

Using the slope and one of the intercepts, we can write the line in point-slope form.

$$y - y_1 = m(x - x_1)$$
$$y - 0 = 2(x - (-2))$$
$$y = 2(x + 2)$$
$$y = 2x + 4$$
$$-2x + y = 4$$

Find the x– and y–coefficients for the equation of the line with right-hand-side equal to 12. Multiply both sides of $-2x + y = 4$ by 3 to obtain 12 on the right-hand-side.

$$-2x + y = 4$$
$$3(-2x + y) = 3(4)$$
$$-6x + 3y = 12$$

Therefore, the coefficient of x is –6 and the coefficient of y is 3.

115. Answers will vary.

117. Answers will vary.

118. Since the slope is the same as the slope of $y = 2x + 1$, then $m = 2$.

$$y - y_1 = m(x - x_1)$$
$$y - 1 = 2(x - (-3))$$
$$y - 1 = 2(x + 3)$$
$$y - 1 = 2x + 6$$
$$y = 2x + 7$$

119. Since the slope is the negative reciprocal of $-\frac{1}{4}$, then $m = 4$.

$$y - y_1 = m(x - x_1)$$
$$y - (-5) = 4(x - 3)$$
$$y + 5 = 4x - 12$$
$$-4x + y + 17 = 0$$
$$4x - y - 17 = 0$$

120.
$$\frac{f(x_2) - f(x_1)}{x_2 - x_1} = \frac{f(4) - f(1)}{4 - 1}$$
$$= \frac{4^2 - 1^2}{4 - 1}$$
$$= \frac{15}{3}$$
$$= 5$$

Section 2.4

Check Point Exercises

1. The slope of the line $y = 3x + 1$ is 3.

$$y - y_1 = m(x - x_1)$$
$$y - 5 = 3(x - (-2))$$
$$y - 5 = 3(x + 2) \quad \text{point-slope}$$
$$y - 5 = 3x + 6$$
$$y = 3x + 11 \quad \text{slope-intercept}$$

2. a. Write the equation in slope-intercept form:

$$x + 3y - 12 = 0$$
$$3y = -x + 12$$
$$y = -\frac{1}{3}x + 4$$

The slope of this line is $-\frac{1}{3}$ thus the slope of any line perpendicular to this line is 3.

b. Use $m = 3$ and the point $(-2, -6)$ to write the equation.

$$y - y_1 = m(x - x_1)$$
$$y - (-6) = 3(x - (-2))$$
$$y + 6 = 3(x + 2)$$
$$y + 6 = 3x + 6$$
$$-3x + y = 0$$
$$3x - y = 0 \quad \text{general form}$$

3. $\quad m = \dfrac{\text{Change in } y}{\text{Change in } x} = \dfrac{14.7 - 9.0}{2008 - 1990} = \dfrac{5.7}{18} \approx 0.32$

The slope indicates that the number of U.S. men living alone increased at a rate of 0.32 million each year.
The rate of change is 0.32 million men per year.

4. **a.** $\quad \dfrac{f(x_2) - f(x_1)}{x_2 - x_1} = \dfrac{1^3 - 0^3}{1 - 0} = 1$

    b. $\quad \dfrac{f(x_2) - f(x_1)}{x_2 - x_1} = \dfrac{2^3 - 1^3}{2 - 1} = \dfrac{8 - 1}{1} = 7$

    c. $\quad \dfrac{f(x_2) - f(x_1)}{x_2 - x_1} = \dfrac{0^3 - (-2)^3}{0 - (-2)} = \dfrac{8}{2} = 4$

5. $\quad \dfrac{f(x_2) - f(x_1)}{x_2 - x_1} = \dfrac{f(3) - f(1)}{3 - 1}$

$$= \dfrac{0.05 - 0.03}{3 - 1}$$
$$= 0.01$$

The average rate of change in the drug's concentration between 1 hour and 3 hours is 0.01 mg per 100 mL per hour.

Concept and Vocabulary Check 2.4

1. the same

2. -1

3. $-\dfrac{1}{3}$; 3

4. -2 ; $\dfrac{1}{2}$

5. y; x

6. $\dfrac{f(x_2) - f(x_1)}{x_2 - x_1}$

Exercise Set 2.4

1. Since L is parallel to $y = 2x$, we know it will have slope $m = 2$. We are given that it passes through $(4, 2)$. We use the slope and point to write the equation in point-slope form.

$$y - y_1 = m(x - x_1)$$
$$y - 2 = 2(x - 4)$$

Solve for y to obtain slope-intercept form.

$$y - 2 = 2(x - 4)$$
$$y - 2 = 2x - 8$$
$$y = 2x - 6$$

In function notation, the equation of the line is $f(x) = 2x - 6$.

3. Since L is perpendicular to $y = 2x$, we know it will have slope $m = -\dfrac{1}{2}$. We are given that it passes through $(2, 4)$. We use the slope and point to write the equation in point-slope form.

$$y - y_1 = m(x - x_1)$$
$$y - 4 = -\dfrac{1}{2}(x - 2)$$

Solve for y to obtain slope-intercept form.

$$y - 4 = -\dfrac{1}{2}(x - 2)$$
$$y - 4 = -\dfrac{1}{2}x + 1$$
$$y = -\dfrac{1}{2}x + 5$$

In function notation, the equation of the line is $f(x) = -\dfrac{1}{2}x + 5$.

5. $m = -4$ since the line is parallel to $y = -4x + 3$; $x_1 = -8$, $y_1 = -10$; point-slope form:
$y + 10 = -4(x + 8)$
slope-intercept form: $y + 10 = -4x - 32$
$\qquad\qquad\qquad\qquad\qquad y = -4x - 42$

7. $m = -5$ since the line is perpendicular to $y = \dfrac{1}{5}x + 6$; $x_1 = 2$, $y_1 = -3$;
point-slope form: $y + 3 = -5(x - 2)$
slope-intercept form: $y + 3 = -5x + 10$
$\qquad\qquad\qquad\qquad\qquad y = -5x + 7$

9. $2x - 3y - 7 = 0$

$$-3y = -2x + 7$$

$$y = \frac{2}{3}x - \frac{7}{3}$$

The slope of the given line is $\frac{2}{3}$, so $m = \frac{2}{3}$ since the lines are parallel.

point-slope form: $y - 2 = \frac{2}{3}(x + 2)$

general form: $2x - 3y + 10 = 0$

11. $x - 2y - 3 = 0$

$$-2y = -x + 3$$

$$y = \frac{1}{2}x - \frac{3}{2}$$

The slope of the given line is $\frac{1}{2}$, so $m = -2$ since the lines are perpendicular.

point-slope form: $\quad y + 7 = -2(x - 4)$

general form: $2x + y - 1 = 0$

13. $\dfrac{15 - 0}{5 - 0} = \dfrac{15}{5} = 3$

15. $\dfrac{5^2 + 2 \cdot 5 - (3^2 + 2 \cdot 3)}{5 - 3} = \dfrac{25 + 10 - (9 + 6)}{2}$

$$= \frac{20}{2}$$

$$= 10$$

17. $\dfrac{\sqrt{9} - \sqrt{4}}{9 - 4} = \dfrac{3 - 2}{5} = \dfrac{1}{5}$

19. Since the line is perpendicular to $x = 6$ which is a vertical line, we know the graph of f is a horizontal line with 0 slope. The graph of f passes through $(-1, 5)$, so the equation of f is $f(x) = 5$.

21. First we need to find the equation of the line with x – intercept of 2 and y – intercept of –4. This line will pass through $(2, 0)$ and $(0, -4)$. We use these points to find the slope.

$$m = \frac{-4 - 0}{0 - 2} = \frac{-4}{-2} = 2$$

Since the graph of f is perpendicular to this line, it will have slope $m = -\dfrac{1}{2}$.

Use the point $(-6, 4)$ and the slope $-\dfrac{1}{2}$ to find the equation of the line.

$$y - y_1 = m(x - x_1)$$

$$y - 4 = -\frac{1}{2}(x - (-6))$$

$$y - 4 = -\frac{1}{2}(x + 6)$$

$$y - 4 = -\frac{1}{2}x - 3$$

$$y = -\frac{1}{2}x + 1$$

$$f(x) = -\frac{1}{2}x + 1$$

23. First put the equation $3x - 2y - 4 = 0$ in slope-intercept form.

$$3x - 2y - 4 = 0$$

$$-2y = -3x + 4$$

$$y = \frac{3}{2}x - 2$$

The equation of f will have slope $-\dfrac{2}{3}$ since it is perpendicular to the line above and the same y – intercept -2.

So the equation of f is $f(x) = -\dfrac{2}{3}x - 2$.

25. $p(x) = -0.25x + 22$

27. $m = \dfrac{1163 - 617}{1998 - 1994} = \dfrac{546}{4} \approx 137$

There was an average increase of approximately 137 discharges per year.

29. **a.** $f(x) = 1.1x^3 - 35x^2 + 264x + 557$

$f(0) = 1.1(0)^3 - 35(0)^2 + 264(0) + 557 = 557$

$f(4) = 1.1(4)^3 - 35(4)^2 + 264(4) + 557 = 1123.4$

$$m = \frac{1123.4 - 557}{4 - 0} \approx 142$$

b. This overestimates by 5 discharges per year.

31. – 35. Answers will vary.

37. $y = \frac{1}{3}x + 1$

$y = -3x - 2$

a. The lines are perpendicular because their slopes are negative reciprocals of each other. This is verified because product of their slopes is -1.

b. The lines do not appear to be perpendicular.

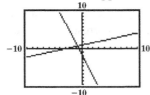

c. The lines appear to be perpendicular. The calculator screen is rectangular and does not have the same width and height. This causes the scale of the x–axis to differ from the scale on the y–axis despite using the same scale in the window settings. In part (b), this causes the lines not to appear perpendicular when indeed they are. The zoom square feature compensates for this and in part (c), the lines appear to be perpendicular.

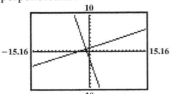

39. makes sense

41. makes sense

43. The slope of the line containing $(1, -3)$ and $(-2, 4)$

has slope $m = \dfrac{4 - (-3)}{-2 - 1} = \dfrac{4 + 3}{-3} = \dfrac{7}{-3} = -\dfrac{7}{3}$

Solve $Ax + y - 2 = 0$ for y to obtain slope-intercept form.

$Ax + y - 2 = 0$

$y = -Ax + 2$

So the slope of this line is $-A$.

This line is perpendicular to the line above so its

slope is $\dfrac{3}{7}$. Therefore, $-A = \dfrac{3}{7}$ so $A = -\dfrac{3}{7}$.

44. a.

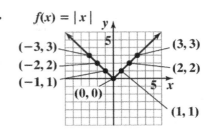

b.

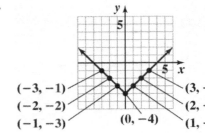

c. The graph in part (b) is the graph in part (a) shifted down 4 units.

45. a.

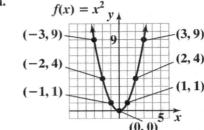

b.

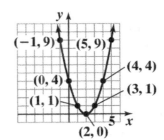

c. The graph in part (b) is the graph in part (a) shifted to the right 2 units.

46. a.

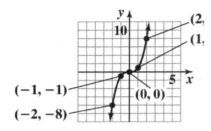

b.

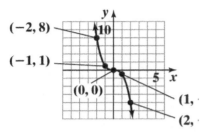

c. The graph in part (b) is the graph in part (a) reflected across the *y*-axis.

Mid-Chapter 2 Check Point

1. The relation is not a function.
   The domain is $\{1, 2\}$.
   The range is $\{-6, 4, 6\}$.

2. The relation is a function.
   The domain is $\{0, 2, 3\}$.
   The range is $\{1, 4\}$.

3. The relation is a function.
   The domain is $\{x \mid -2 \le x < 2\}$.
   The range is $\{y \mid 0 \le y \le 3\}$.

4. The relation is not a function.
   The domain is $\{x \mid -3 < x \le 4\}$.
   The range is $\{y \mid -1 \le y \le 2\}$.

5. The relation is not a function.
   The domain is $\{-2, -1, 0, 1, 2\}$.
   The range is $\{-2, -1, 1, 3\}$.

6. The relation is a function.
   The domain is $\{x \mid x \le 1\}$.
   The range is $\{y \mid y \ge -1\}$.

7. $x^2 + y = 5$
   $$y = -x^2 + 5$$
   For each value of *x*, there is one and only one value for *y*, so the equation defines *y* as a function of *x*.

8. $x + y^2 = 5$
   $$y^2 = 5 - x$$
   $$y = \pm\sqrt{5 - x}$$
   Since there are values of *x* that give more than one value for *y* (for example, if $x = 4$, then $y = \pm\sqrt{5 - 4} = \pm 1$), the equation does not define *y* as a function of *x*.

9. No vertical line intersects the graph in more than one point. Each value of *x* corresponds to exactly one value of *y*.

10. Domain: $(-\infty, \infty)$

11. Range: $(-\infty, 4]$

12. *x*-intercepts: –6 and 2

13. *y*-intercept: 3

14. increasing: $(-\infty, -2)$

15. decreasing: $(-2, \infty)$

16. $x = -2$

17. $f(-2) = 4$

18. $f(-4) = 3$

19. $f(-7) = -2$ and $f(3) = -2$

20. $f(-6) = 0$ and $f(2) = 0$

21. $(-6, 2)$

22. $f(100)$ is negative.

23. neither; $f(-x) \ne x$ and $f(-x) \ne -x$

24. $\dfrac{f(x_2) - f(x_1)}{x_2 - x_1} = \dfrac{f(4) - f(-4)}{4 - (-4)} = \dfrac{-5 - 3}{4 + 4} = -1$

25. $y = -2x$

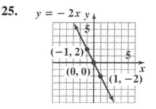

26. $y = -2$

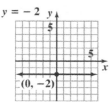

27. $x + y = -2$

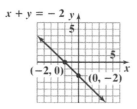

28. $y = \dfrac{1}{3}x - 2$

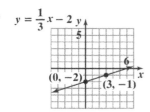

29. $x = 3.5$

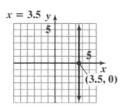

30.

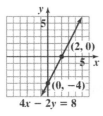

$4x - 2y = 8$

31.

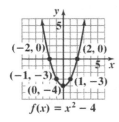

$f(x) = x^2 - 4$

32.

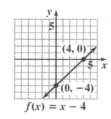

$f(x) = x - 4$

33.

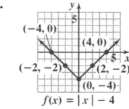

$f(x) = |x| - 4$

34. $5y = -3x$

$y = -\dfrac{3}{5}x$

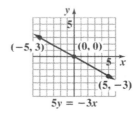

$5y = -3x$

35. $5y = 20$

$y = 4$

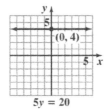

$5y = 20$

36.

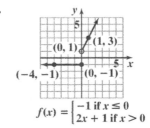

$f(x) = \begin{cases} -1 & \text{if } x \le 0 \\ 2x + 1 & \text{if } x > 0 \end{cases}$

37. a. $f(-x) = -2(-x)^2 - x - 5$

$\quad\quad = -2x^2 - x - 5$

neither; $f(-x) \neq x$ and $f(-x) \neq -x$

b. $\dfrac{f(x+h) - f(x)}{h}$

$= \dfrac{-2(x+h)^2 + (x+h) - 5 - (-2x^2 + x - 5)}{h}$

$= \dfrac{-2x^2 - 4xh - 2h^2 + x + h - 5 + 2x^2 - x + 5}{h}$

$= \dfrac{-4xh - 2h^2 + h}{h}$

$= \dfrac{h(-4x - 2h + 1)}{h}$

$= -4x - 2h + 1$

38. $C(x) = \begin{cases} 30 & \text{if } 0 \le t \le 200 \\ 30 + 0.40(t - 200) & \text{if } t > 200 \end{cases}$

a. $C(150) = 30$

b. $C(250) = 30 + 0.40(250 - 200) = 50$

39. $y - y_1 = m(x - x_1)$

$\quad y - 3 = -2(x - (-4))$

$\quad y - 3 = -2(x + 4)$

$\quad y - 3 = -2x - 8$

$\quad\quad y = -2x - 5$

$\quad f(x) = -2x - 5$

40. $m = \dfrac{\text{Change in } y}{\text{Change in } x} = \dfrac{1 - (-5)}{2 - (-1)} = \dfrac{6}{3} = 2$

$y - y_1 = m(x - x_1)$

$\quad y - 1 = 2(x - 2)$

$\quad y - 1 = 2x - 4$

$\quad\quad y = 2x - 3$

$\quad f(x) = 2x - 3$

41. $3x - y - 5 = 0$

$\quad -y = -3x + 5$

$\quad\quad y = 3x - 5$

The slope of the given line is 3, and the lines are parallel, so $m = 3$.

$y - y_1 = m(x - x_1)$

$y - (-4) = 3(x - 3)$

$\quad y + 4 = 3x - 9$

$\quad\quad y = 3x - 13$

$\quad f(x) = 3x - 13$

42. $2x - 5y - 10 = 0$

$\quad -5y = -2x + 10$

$\quad \dfrac{-5y}{-5} = \dfrac{-2x}{-5} + \dfrac{10}{-5}$

$\quad\quad y = \dfrac{2}{5}x - 2$

The slope of the given line is $\dfrac{2}{5}$, and the lines are perpendicular, so $m = -\dfrac{5}{2}$.

$y - y_1 = m(x - x_1)$

$y - (-3) = -\dfrac{5}{2}(x - (-4))$

$\quad y + 3 = -\dfrac{5}{2}x - 10$

$\quad\quad y = -\dfrac{5}{2}x - 13$

$\quad f(x) = -\dfrac{5}{2}x - 13$

43. $m_1 = \dfrac{\text{Change in } y}{\text{Change in } x} = \dfrac{0 - (-4)}{7 - 2} = \dfrac{4}{5}$

$m_2 = \dfrac{\text{Change in } y}{\text{Change in } x} = \dfrac{6 - 2}{1 - (-4)} = \dfrac{4}{5}$

The slope of the lines are equal thus the lines are parallel.

44. a. $m = \dfrac{\text{Change in } y}{\text{Change in } x} = \dfrac{42 - 26}{180 - 80} = \dfrac{16}{100} = 0.16$

b. For each minute of brisk walking, the percentage of patients with depression in remission increased by 0.16%. The rate of change is 0.16% per minute of brisk walking.

45. $\dfrac{f(x_2) - f(x_1)}{x_2 - x_1} = \dfrac{f(2) - f(-1)}{2 - (-1)}$

$\quad\quad = \dfrac{(3(2)^2 - 2) - (3(-1)^2 - (-1))}{2 + 1}$

$\quad\quad = 2$

Section 2.5

Check Point Exercises

1. Shift up vertically 3 units.

    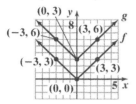

2. Shift to the right 4 units.

    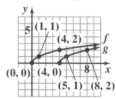

3. Shift to the right 1 unit and down 2 units.

    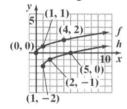

4. Reflect about the *x*-axis.

    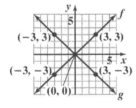

5. Reflect about the *y*-axis.

    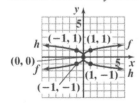

6. Vertically stretch the graph of $f(x) = |x|$.

    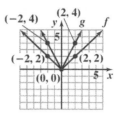

7. **a.** Horizontally shrink the graph of $y = f(x)$.

    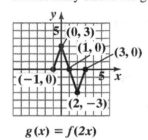

    $$g(x) = f(2x)$$

    b. Horizontally stretch the graph of $y = f(x)$.

    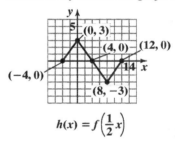

    $$h(x) = f\left(\frac{1}{2}x\right)$$

8. The graph of $y = f(x)$ is shifted 1 unit left, shrunk by a factor of $\frac{1}{3}$, reflected about the x-axis, then shifted down 2 units.

    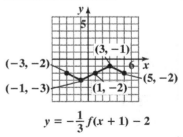

    $$y = -\frac{1}{3}f(x+1) - 2$$

9. The graph of $f(x) = x^2$ is shifted 1 unit right, stretched by a factor of 2, then shifted up 3 units.

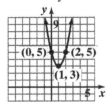

$$g(x) = 2(x-1)^2 + 3$$

Concept and Vocabulary Check 2.5

1. vertical; down

2. horizontal; to the right

3. x-axis

4. y-axis

5. vertical; y

6. horizontal; x

7. false

Exercise Set 2.5

1.

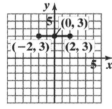

$$g(x) = f(x) + 1$$

3.

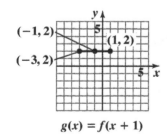

$$g(x) = f(x+1)$$

5.

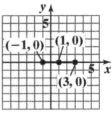

$$g(x) = f(x-1) - 2$$

7.

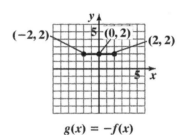

$$g(x) = -f(x)$$

9.

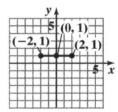

$$g(x) = -f(x) + 3$$

11.

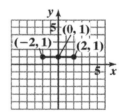

$$g(x) = \frac{1}{2}f(x)$$

13. $g(x) = f\left(\frac{1}{2}x\right)$

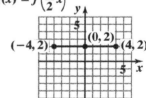

15.

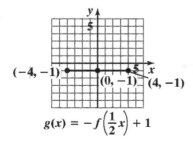

$$g(x) = -f\left(\frac{1}{2}x\right) + 1$$

17.

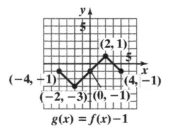

$$g(x) = f(x) - 1$$

19.

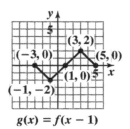

$$g(x) = f(x - 1)$$

21.

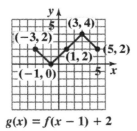

$$g(x) = f(x - 1) + 2$$

23.

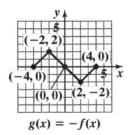

$$g(x) = -f(x)$$

25.

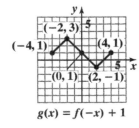

$$g(x) = f(-x) + 1$$

27.

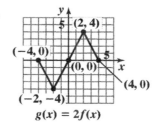

$$g(x) = 2f(x)$$

29.

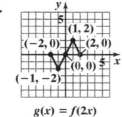

$$g(x) = f(2x)$$

31.

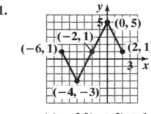

$$g(x) = 2f(x + 2) + 1$$

33.

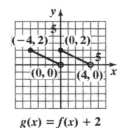

$$g(x) = f(x) + 2$$

35.

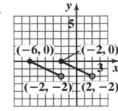

$$g(x) = f(x + 2)$$

37.

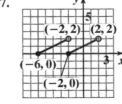

$$g(x) = -f(x + 2)$$

39.

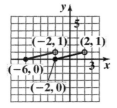

$$g(x) = -\frac{1}{2}f(x+2)$$

41.

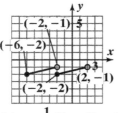

$$g(x) = -\frac{1}{2}f(x+2) - 2$$

43.

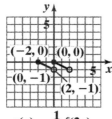

$$g(x) = \frac{1}{2}f(2x)$$

45.

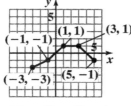

$$g(x) = f(x-1) - 1$$

47.

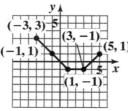

$$g(x) = -f(x-1) + 1$$

49.

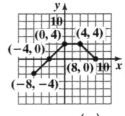

$$g(x) = 2f\left(\frac{1}{2}x\right)$$

51.

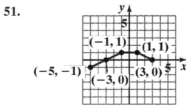

$$g(x) = \frac{1}{2}f(x+1)$$

53.

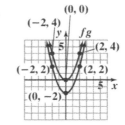

55.

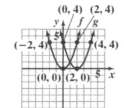

57.

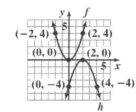

59.

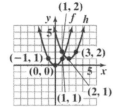

61.

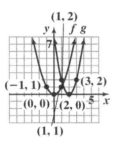

63.

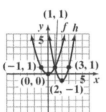

65.

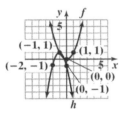

67.

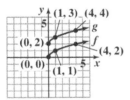

69.

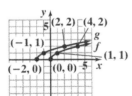

71.

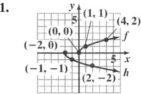

73.

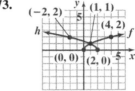

75.

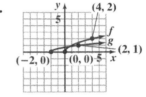

77.

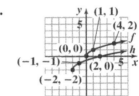

79.

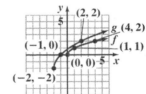

81.

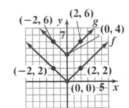

83.

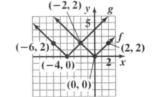

85.

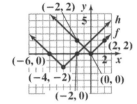

87.

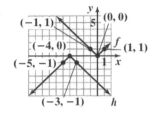

89.

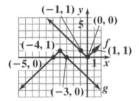

91.

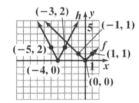

93.

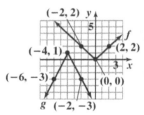

95.

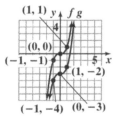

97.

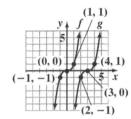

99.

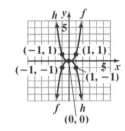

101.

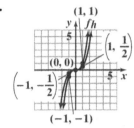

103.

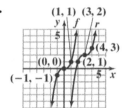

105.

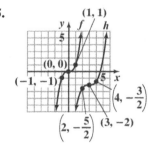

107.

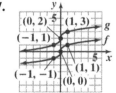

109.

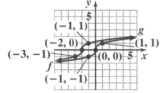

111.

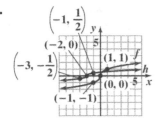

113.

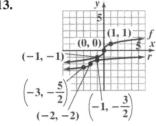

115.

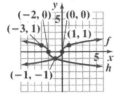

117.

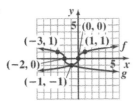

119.

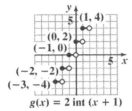

$g(x) = 2 \text{ int } (x+1)$

121.

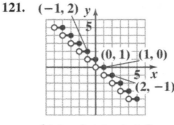

$h(x) = \text{int } (-x) + 1$

123. $y = \sqrt{x-2}$

125. $y = (x+1)^2 - 4$

127. a. First, vertically stretch the graph of $f(x) = \sqrt{x}$ by the factor 2.9; then shift the result up 20.1 units.

b. $f(x) = 2.9\sqrt{x} + 20.1$

$f(48) = 2.9\sqrt{48} + 20.1 \approx 40.2$

The model describes the actual data very well.

c. $\dfrac{f(x_2) - f(x_1)}{x_2 - x_1}$

$= \dfrac{f(10) - f(0)}{10 - 0}$

$= \dfrac{\left(2.9\sqrt{10} + 20.1\right) - \left(2.9\sqrt{0} + 20.1\right)}{10 - 0}$

$= \dfrac{29.27 - 20.1}{10}$

≈ 0.9

0.9 inches per month

d. $\dfrac{f(x_2) - f(x_1)}{x_2 - x_1}$

$= \dfrac{f(60) - f(50)}{60 - 50}$

$= \dfrac{\left(2.9\sqrt{60} + 20.1\right) - \left(2.9\sqrt{50} + 20.1\right)}{60 - 50}$

$= \dfrac{42.5633 - 40.6061}{10}$

≈ 0.2

This rate of change is lower than the rate of change in part (c). The relative leveling off of the curve shows this difference.

129. – 133. Answers will vary.

135. a.

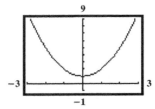

b.

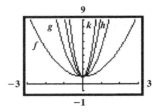

137. makes sense

139. does not make sense; Explanations will vary. Sample explanation: The reprogram should be $y = f(t+1)$.

141. false; Changes to make the statement true will vary. A sample change is: The graph of g is a translation of f three units to the <u>left</u> and three units upward.

143. false; Changes to make the statement true will vary. A sample change is: The stretch will be 5 units and the downward shift will be 10 units.

145. $g(x) = -(x+4)^2$

147. $g(x) = -\sqrt{x-2} + 2$

149. $(-a, b)$

151. $(a + 3, b)$

153. $(2x-1)(x^2 + x - 2) = 2x(x^2 + x - 2) - 1(x^2 + x - 2)$
$$= 2x^3 + 2x^2 - 4x - x^2 - x + 2$$
$$= 2x^3 + 2x^2 - x^2 - 4x - x + 2$$
$$= 2x^3 + x^2 - 5x + 2$$

154. $\left(f(x)\right)^2 - 2f(x) + 6 = (3x-4)^2 - 2(3x-4) + 6$
$$= 9x^2 - 24x + 16 - 6x + 8 + 6$$
$$= 9x^2 - 24x - 6x + 16 + 8 + 6$$
$$= 9x^2 - 30x + 30$$

155. $\dfrac{2}{\dfrac{3}{x} - 1} = \dfrac{2x}{\dfrac{3x}{x} - x} = \dfrac{2x}{3 - x}$

Section 2.6

Check Point Exercises

1. **a.** The function $f(x) = x^2 + 3x - 17$ contains neither division nor an even root. The domain of f is the set of all real numbers or $(-\infty, \infty)$.

b. The denominator equals zero when $x = 7$ or $x = -7$. These values must be excluded from the domain.
domain of $g = (-\infty, -7) \cup (-7, 7) \cup (7, \infty)$.

c. Since $h(x) = \sqrt{9x - 27}$ contains an even root; the quantity under the radical must be greater than or equal to 0.
$9x - 27 \geq 0$
$\quad 9x \geq 27$
$\quad\quad x \geq 3$
Thus, the domain of h is $\{x | x \geq 3\}$, or the interval $[3, \infty)$.

d. Since the denominator of $j(x)$ contains an even root; the quantity under the radical must be greater than or equal to 0. But that quantity must also not be 0 (because we cannot have division by 0). Thus, $24 - 3x$ must be strictly greater than 0.
$24 - 3x > 0$
$\quad -3x > -24$
$\quad\quad x < 8$
Thus, the domain of j is $\{x | x < 8\}$, or the interval $(-\infty, 8)$.

2. **a.** $(f + g)(x) = f(x) + g(x)$
$$= x - 5 + \left(x^2 - 1\right)$$
$$= x - 5 + x^2 - 1$$
$$= -x^2 + x - 6$$
domain: $(-\infty, \infty)$

b. $(f - g)(x) = f(x) - g(x)$
$$= x - 5 - \left(x^2 - 1\right)$$
$$= x - 5 - x^2 + 1$$
$$= -x^2 + x - 4$$
domain: $(-\infty, \infty)$

c. $(fg)(x) = (x-5)(x^2-1)$

$\qquad = x(x^2-1) - 5(x^2-1)$

$\qquad = x^3 - x - 5x^2 + 5$

$\qquad = x^3 - 5x^2 - x + 5$

domain: $(-\infty, \infty)$

d. $\left(\dfrac{f}{g}\right)(x) = \dfrac{f(x)}{g(x)}$

$\qquad = \dfrac{x-5}{x^2-1}, \; x \neq \pm 1$

domain: $(-\infty, -1) \cup (-1, 1) \cup (1, \infty)$

3. a. $(f+g)(x) = f(x) + g(x)$

$\qquad = \sqrt{x-3} + \sqrt{x+1}$

b. domain of f: $\quad x - 3 \geq 0$

$\qquad\qquad\qquad\quad x \geq 3$

$\qquad\qquad\qquad [3, \infty)$

domain of g: $\quad x + 1 \geq 0$

$\qquad\qquad\qquad\quad x \geq -1$

$\qquad\qquad\qquad [-1, \infty)$

The domain of $f + g$ is the set of all real numbers that are common to the domain of f and the domain of g. Thus, the domain of $f + g$ is $[3, \infty)$.

4. a. $(B+D)(x)$

$\qquad = B(x) + D(x)$

$\qquad = (-2.6x^2 + 49x + 3994) + (-0.6x^2 + 7x + 2412)$

$\qquad = -2.6x^2 + 49x + 3994 - 0.6x^2 + 7x + 2412$

$\qquad = -3.2x^2 + 56x + 6406$

b. $(B+D)(x) = -3.2x^2 + 56x + 6406$

$(B+D)(5) = -3.2(3)^2 + 56(3) + 6406$

$\qquad\qquad = 6545.2$

The number of births and deaths in the U.S. in 2003 was 6545.2 thousand.

c. $(B+D)(x)$ overestimates the actual number of births and deaths in 2003 by 7.2 thousand.

5. a. $(f \circ g)(x) = f(g(x))$

$\qquad = 5(2x^2 - x - 1) + 6$

$\qquad = 10x^2 - 5x - 5 + 6$

$\qquad = 10x^2 - 5x + 1$

b. $(g \circ f)(x) = g(f(x))$

$\qquad = 2(5x+6)^2 - (5x+6) - 1$

$\qquad = 2(25x^2 + 60x + 36) - 5x - 6 - 1$

$\qquad = 50x^2 + 120x + 72 - 5x - 6 - 1$

$\qquad = 50x^2 + 115x + 65$

c. $(f \circ g)(x) = 10x^2 - 5x + 1$

$(f \circ g)(-1) = 10(-1)^2 - 5(-1) + 1$

$\qquad\qquad = 10 + 5 + 1$

$\qquad\qquad = 16$

6. a. $(f \circ g)(x) = \dfrac{4}{\dfrac{1}{x} + 2} = \dfrac{4x}{1 + 2x}$

b. domain: $\left\{ x \,\middle|\, x \neq 0,\, x \neq -\dfrac{1}{2} \right\}$

or $\left(-\infty, -\dfrac{1}{2}\right) \cup \left(-\dfrac{1}{2}, 0\right) \cup (0, \infty)$

7. $h(x) = f \circ g$ where $f(x) = \sqrt{x}$; $g(x) = x^2 + 5$

Concept and Vocabulary Check 2.6

1. zero

2. negative

3. $f(x) + g(x)$

4. $f(x) - g(x)$

5. $f(x) \cdot g(x)$

6. $\dfrac{f(x)}{g(x)}$; $g(x)$

7. $(-\infty, \infty)$

8. $(2, \infty)$

9. $(0, 3)$; $(3, \infty)$

10. composition; $f(g(x))$

11. f; $g(x)$

12. composition; $g(f(x))$

13. g; $f(x)$

14. false

15. false

16. 2

Exercise Set 2.6

1. The function contains neither division nor an even root. The domain $= (-\infty, \infty)$

3. The denominator equals zero when $x = 4$. This value must be excluded from the domain.
   domain: $(-\infty, 4) \cup (4, \infty)$.

5. The function contains neither division nor an even root. The domain $= (-\infty, \infty)$

7. The values that make the denominator equal zero must be excluded from the domain.
   domain: $(-\infty, -3) \cup (-3, 5) \cup (5, \infty)$

9. The values that make the denominators equal zero must be excluded from the domain.
   domain: $(-\infty, -7) \cup (-7, 9) \cup (9, \infty)$

11. The first denominator cannot equal zero. The values that make the second denominator equal zero must be excluded from the domain.
    domain: $(-\infty, -1) \cup (-1, 1) \cup (1, \infty)$

13. Exclude x for $x = 0$.

    Exclude x for $\dfrac{3}{x} - 1 = 0$.

    $\dfrac{3}{x} - 1 = 0$

    $x\left(\dfrac{3}{x} - 1\right) = x(0)$

    $3 - x = 0$

    $-x = -3$

    $x = 3$

    domain: $(-\infty, 0) \cup (0, 3) \cup (3, \infty)$

15. Exclude x for $x - 1 = 0$.
    $x - 1 = 0$
    $x = 1$

Exclude x for $\dfrac{4}{x-1} - 2 = 0$.

$\dfrac{4}{x-1} - 2 = 0$

$(x-1)\left(\dfrac{4}{x-1} - 2\right) = (x-1)(0)$

$4 - 2(x-1) = 0$

$4 - 2x + 2 = 0$

$-2x + 6 = 0$

$-2x = -6$

$x = 3$

domain: $(-\infty, 1) \cup (1, 3) \cup (3, \infty)$

17. The expression under the radical must not be negative.
    $x - 3 \geq 0$
    $x \geq 3$
    domain: $[3, \infty)$

19. The expression under the radical must be positive.
    $x - 3 > 0$
    $x > 3$
    domain: $(3, \infty)$

21. The expression under the radical must not be negative.
    $5x + 35 \geq 0$
    $5x \geq -35$
    $x \geq -7$
    domain: $[-7, \infty)$

23. The expression under the radical must not be negative.
    $24 - 2x \geq 0$
    $-2x \geq -24$
    $\dfrac{-2x}{-2} \leq \dfrac{-24}{-2}$
    $x \leq 12$
    domain: $(-\infty, 12]$

25. The expressions under the radicals must not be negative.
    $x - 2 \geq 0$ and $x + 3 \geq 0$
    $x \geq 2$ $\qquad$ $x \geq -3$
    To make both inequalities true, $x \geq 2$.
    domain: $[2, \infty)$

27. The expression under the radical must not be negative.

$x - 2 \geq 0$

$x \geq 2$

The denominator equals zero when $x = 5$.

domain: $[2,5) \cup (5,\infty)$.

29. Find the values that make the denominator equal zero and must be excluded from the domain.

$x^3 - 5x^2 - 4x + 20$

$= x^2(x-5) - 4(x-5)$

$= (x-5)(x^2 - 4)$

$= (x-5)(x+2)(x-2)$

–2, 2, and 5 must be excluded.

domain: $(-\infty,-2) \cup (-2,2) \cup (2,5) \cup (5,\infty)$

31. $(f+g)(x) = 3x + 2$

domain: $(-\infty,\infty)$

$(f-g)(x) = f(x) - g(x)$

$= (2x+3) - (x-1)$

$= x + 4$

domain: $(-\infty,\infty)$

$(fg)(x) = f(x) \cdot g(x)$

$= (2x+3) \cdot (x-1)$

$= 2x^2 + x - 3$

domain: $(-\infty,\infty)$

$\left(\dfrac{f}{g}\right)(x) = \dfrac{f(x)}{g(x)} = \dfrac{2x+3}{x-1}$

domain: $(-\infty,1) \cup (1,\infty)$

33. $(f+g)(x) = 3x^2 + x - 5$

domain: $(-\infty,\infty)$

$(f-g)(x) = -3x^2 + x - 5$

domain: $(-\infty,\infty)$

$(fg)(x) = (x-5)(3x^2) = 3x^3 - 15x^2$

domain: $(-\infty,\infty)$

$\left(\dfrac{f}{g}\right)(x) = \dfrac{x-5}{3x^2}$

domain: $(-\infty,0) \cup (0,\infty)$

35. $(f+g)(x) = 2x^2 - 2$

domain: $(-\infty,\infty)$

$(f-g)(x) = 2x^2 - 2x - 4$

domain: $(-\infty,\infty)$

$(fg)(x) = (2x^2 - x - 3)(x+1)$

$= 2x^3 + x^2 - 4x - 3$

domain: $(-\infty,\infty)$

$\left(\dfrac{f}{g}\right)(x) = \dfrac{2x^2 - x - 3}{x+1}$

$= \dfrac{(2x-3)(x+1)}{(x+1)} = 2x - 3$

domain: $(-\infty,-1) \cup (-1,\infty)$

37. $(f+g)(x) = (3-x^2) + (x^2 + 2x - 15)$

$= 2x - 12$

domain: $(-\infty,\infty)$

$(f-g)(x) = (3-x^2) - (x^2 + 2x - 15)$

$= -2x^2 - 2x + 18$

domain: $(-\infty,\infty)$

$(fg)(x) = (3-x^2)(x^2 + 2x - 15)$

$= -x^4 - 2x^3 + 18x^2 + 6x - 45$

domain: $(-\infty,\infty)$

$\left(\dfrac{f}{g}\right)(x) = \dfrac{3-x^2}{x^2 + 2x - 15}$

domain: $(-\infty,-5) \cup (-5,3) \cup (3,\infty)$

39. $(f+g)(x) = \sqrt{x} + x - 4$

domain: $[0,\infty)$

$(f-g)(x) = \sqrt{x} - x + 4$

domain: $[0,\infty)$

$(fg)(x) = \sqrt{x}(x-4)$

domain: $[0,\infty)$

$\left(\dfrac{f}{g}\right)(x) = \dfrac{\sqrt{x}}{x-4}$

domain: $[0,4) \cup (4,\infty)$

41. $(f+g)(x) = 2 + \dfrac{1}{x} + \dfrac{1}{x} = 2 + \dfrac{2}{x} = \dfrac{2x+2}{x}$

domain: $(-\infty,0) \cup (0,\infty)$

$(f-g)(x) = 2 + \dfrac{1}{x} - \dfrac{1}{x} = 2$

domain: $(-\infty,0) \cup (0,\infty)$

$(fg)(x) = \left(2 + \dfrac{1}{x}\right) \cdot \dfrac{1}{x} = \dfrac{2}{x} + \dfrac{1}{x^2} = \dfrac{2x+1}{x^2}$

domain: $(-\infty,0) \cup (0,\infty)$

$\left(\dfrac{f}{g}\right)(x) = \dfrac{2 + \frac{1}{x}}{\frac{1}{x}} = \left(2 + \dfrac{1}{x}\right) \cdot x = 2x + 1$

domain: $(-\infty,0) \cup (0,\infty)$

43. $(f+g)(x) = f(x) + g(x)$

$$= \frac{5x+1}{x^2-9} + \frac{4x-2}{x^2-9}$$

$$= \frac{9x-1}{x^2-9}$$

domain: $(-\infty, -3) \cup (-3, 3) \cup (3, \infty)$

$(f-g)(x) = f(x) - g(x)$

$$= \frac{5x+1}{x^2-9} - \frac{4x-2}{x^2-9}$$

$$= \frac{x+3}{x^2-9}$$

$$= \frac{1}{x-3}$$

domain: $(-\infty, -3) \cup (-3, 3) \cup (3, \infty)$

$(fg)(x) = f(x) \cdot g(x)$

$$= \frac{5x+1}{x^2-9} \cdot \frac{4x-2}{x^2-9}$$

$$= \frac{(5x+1)(4x-2)}{\left(x^2-9\right)^2}$$

domain: $(-\infty, -3) \cup (-3, 3) \cup (3, \infty)$

$$\left(\frac{f}{g}\right)(x) = \frac{\dfrac{5x+1}{x^2-9}}{\dfrac{4x-2}{x^2-9}}$$

$$= \frac{5x+1}{x^2-9} \cdot \frac{x^2-9}{4x-2}$$

$$= \frac{5x+1}{4x-2}$$

The domain must exclude –3, 3, and any values that make $4x - 2 = 0$.

$4x - 2 = 0$

$4x = 2$

$x = \dfrac{1}{2}$

domain: $(-\infty, -3) \cup \left(-3, \frac{1}{2}\right) \cup \left(\frac{1}{2}, 3\right) \cup (3, \infty)$

45. $(f+g)(x) = f(x) + g(x)$

$$= \frac{8x}{x-2} + \frac{6}{x+3}$$

$$= \frac{8x(x+3)}{(x-2)(x+3)} + \frac{6(x-2)}{(x-2)(x+3)}$$

$$= \frac{8x^2+24x}{(x-2)(x+3)} + \frac{6x-12}{(x-2)(x+3)}$$

$$= \frac{8x^2+30x-12}{(x-2)(x+3)}$$

domain: $(-\infty, -3) \cup (-3, 2) \cup (2, \infty)$

$(f+g)(x) = f(x) - g(x)$

$$= \frac{8x}{x-2} - \frac{6}{x+3}$$

$$= \frac{8x(x+3)}{(x-2)(x+3)} - \frac{6(x-2)}{(x-2)(x+3)}$$

$$= \frac{8x^2+24x}{(x-2)(x+3)} - \frac{6x-12}{(x-2)(x+3)}$$

$$= \frac{8x^2+18x+12}{(x-2)(x+3)}$$

domain: $(-\infty, -3) \cup (-3, 2) \cup (2, \infty)$

$(fg)(x) = f(x) \cdot g(x)$

$$= \frac{8x}{x-2} \cdot \frac{6}{x+3}$$

$$= \frac{48x}{(x-2)(x+3)}$$

domain: $(-\infty, -3) \cup (-3, 2) \cup (2, \infty)$

$$\left(\frac{f}{g}\right)(x) = \frac{\dfrac{8x}{x-2}}{\dfrac{6}{x+3}}$$

$$= \frac{8x}{x-2} \cdot \frac{x+3}{6}$$

$$= \frac{4x(x+3)}{3(x-2)}$$

The domain must exclude –3, 2, and any values that make $3(x - 2) = 0$.

$3(x-2) = 0$

$3x - 6 = 0$

$3x = 6$

$x = 2$

domain: $(-\infty, -3) \cup (-3, 2) \cup (2, \infty)$

47. $(f+g)(x) = \sqrt{x+4} + \sqrt{x-1}$

domain: $[1, \infty)$

$(f-g)(x) = \sqrt{x+4} - \sqrt{x-1}$

domain: $[1, \infty)$

$(fg)(x) = \sqrt{x+4} \cdot \sqrt{x-1} = \sqrt{x^2+3x-4}$

domain: $[1, \infty)$

$\left(\dfrac{f}{g}\right)(x) = \dfrac{\sqrt{x+4}}{\sqrt{x-1}}$

domain: $(1, \infty)$

49. $(f+g)(x) = \sqrt{x-2} + \sqrt{2-x}$

domain: $\{2\}$

$(f-g)(x) = \sqrt{x-2} - \sqrt{2-x}$

domain: $\{2\}$

$(fg)(x) = \sqrt{x-2} \cdot \sqrt{2-x} = \sqrt{-x^2+4x-4}$

domain: $\{2\}$

$\left(\dfrac{f}{g}\right)(x) = \dfrac{\sqrt{x-2}}{\sqrt{2-x}}$

domain: $\varnothing$

51. $f(x) = 2x;\ g(x) = x + 7$

a. $(f \circ g)(x) = 2(x+7) = 2x+14$

b. $(g \circ f)(x) = 2x+7$

c. $(f \circ g)(2) = 2(2)+14 = 18$

53. $f(x) = x + 4;\ g(x) = 2x + 1$

a. $(f \circ g)(x) = (2x+1)+4 = 2x+5$

b. $(g \circ f)(x) = 2(x+4)+1 = 2x+9$

c. $(f \circ g)(2) = 2(2)+5 = 9$

55. $f(x) = 4x - 3;\ g(x) = 5x^2 - 2$

a. $(f \circ g)(x) = 4(5x^2-2)-3$
$= 20x^2 - 11$

b. $(g \circ f)(x) = 5(4x-3)^2 - 2$
$= 5(16x^2-24x+9)-2$
$= 80x^2 - 120x + 43$

c. $(f \circ g)(2) = 20(2)^2 - 11 = 69$

57. $f(x) = x^2 + 2;\ g(x) = x^2 - 2$

a. $(f \circ g)(x) = (x^2-2)^2 + 2$
$= x^4 - 4x^2 + 4 + 2$
$= x^4 - 4x^2 + 6$

b. $(g \circ f)(x) = (x^2+2)^2 - 2$
$= x^4 + 4x^2 + 4 - 2$
$= x^4 + 4x^2 + 2$

c. $(f \circ g)(2) = 2^4 - 4(2)^2 + 6 = 6$

59. $f(x) = 4 - x;\ g(x) = 2x^2 + x + 5$

a. $(f \circ g)(x) = 4 - \left(2x^2 + x + 5\right)$
$= 4 - 2x^2 - x - 5$
$= -2x^2 - x - 1$

b. $(g \circ f)(x) = 2(4-x)^2 + (4-x) + 5$
$= 2(16 - 8x + x^2) + 4 - x + 5$
$= 32 - 16x + 2x^2 + 4 - x + 5$
$= 2x^2 - 17x + 41$

c. $(f \circ g)(2) = -2(2)^2 - 2 - 1 = -11$

61. $f(x) = \sqrt{x};\ g(x) = x - 1$

a. $(f \circ g)(x) = \sqrt{x-1}$

b. $(g \circ f)(x) = \sqrt{x} - 1$

c. $(f \circ g)(2) = \sqrt{2-1} = \sqrt{1} = 1$

63. $f(x) = 2x - 3;\ g(x) = \dfrac{x+3}{2}$

a. $(f \circ g)(x) = 2\left(\dfrac{x+3}{2}\right) - 3$
$= x + 3 - 3$
$= x$

b. $(g \circ f)(x) = \dfrac{(2x-3)+3}{2} = \dfrac{2x}{2} = x$

c. $(f \circ g)(2) = 2$

65. $f(x) = \frac{1}{x}; \quad g(x) = \frac{1}{x}$

    a. $(f \circ g)(x) = \frac{1}{\frac{1}{x}} = x$

    b. $(g \circ f)(x) = \frac{1}{\frac{1}{x}} = x$

    c. $(f \circ g)(2) = 2$

67. **a.** $(f \circ g)(x) = f\left(\frac{1}{x}\right) = \frac{2}{\frac{1}{x} + 3}, x \neq 0$

$$= \frac{2(x)}{\left(\frac{1}{x} + 3\right)(x)}$$

$$= \frac{2x}{1 + 3x}$$

    b. We must exclude 0 because it is excluded from g.

We must exclude $-\frac{1}{3}$ because it causes the denominator of $f \circ g$ to be 0.

domain: $\left(-\infty, -\frac{1}{3}\right) \cup \left(-\frac{1}{3}, 0\right) \cup (0, \infty)$.

69. **a.** $(f \circ g)(x) = f\left(\frac{4}{x}\right) = \frac{\frac{4}{x}}{\frac{4}{x} + 1}$

$$= \frac{\left(\frac{4}{x}\right)(x)}{\left(\frac{4}{x} + 1\right)(x)}$$

$$= \frac{4}{4 + x}, x \neq -4$

    b. We must exclude 0 because it is excluded from g.
We must exclude -4 because it causes the denominator of $f \circ g$ to be 0.

domain: $(-\infty, -4) \cup (-4, 0) \cup (0, \infty)$.

71. **a.** $f \circ g(x) = f(x - 2) = \sqrt{x - 2}$

    b. The expression under the radical in $f \circ g$ must not be negative.

$x - 2 \geq 0$

$x \geq 2$

domain: $[2, \infty)$.

73. **a.** $(f \circ g)(x) = f(\sqrt{1 - x})$

$$= \left(\sqrt{1 - x}\right)^2 + 4$$

$$= 1 - x + 4$$

$$= 5 - x$$

    b. The domain of $f \circ g$ must exclude any values that are excluded from g.

$1 - x \geq 0$

$-x \geq -1$

$x \leq 1$

domain: $(-\bullet, 1]$.

75. $f(x) = x^4 \quad g(x) = 3x - 1$

77. $f(x) = \sqrt[3]{x} \quad g(x) = x^2 - 9$

79. $f(x) = |x| \quad g(x) = 2x - 5$

81. $f(x) = \frac{1}{x} \quad g(x) = 2x - 3$

83. $(f + g)(-3) = f(-3) + g(-3) = 4 + 1 = 5$

85. $(fg)(2) = f(2)g(2) = (-1)(1) = -1$

87. The domain of $f + g$ is $[-4, 3]$.

89. The graph of $f + g$

91. $(f \circ g)(-1) = f(g(-1)) = f(-3) = 1$

93. $(g \circ f)(0) = g(f(0)) = g(2) = -6$

95.
$$(f \circ g)(x) = 7$$
$$2\left(x^2 - 3x + 8\right) - 5 = 7$$
$$2x^2 - 6x + 16 - 5 = 7$$
$$2x^2 - 6x + 11 = 7$$
$$2x^2 - 6x + 4 = 0$$
$$x^2 - 3x + 2 = 0$$
$$(x-1)(x-2) = 0$$
$$x - 1 = 0 \quad \text{or} \quad x - 2 = 0$$
$$x = 1 \qquad\qquad x = 2$$

97. **a.**
$$(M + F)(x) = M(x) + F(x)$$
$$= (1.53x + 114.8) + (1.46x + 120.7)$$
$$= 2.99x + 235.5$$

b.
$$(M + F)(x) = 2.99x + 235.5$$
$$(M + F)(20) = 2.99(20) + 235.5$$
$$= 295.3$$
The total U.S. population in 2005 was 295.3 million.

c. The result in part (b) underestimates the actual total by 2.7 million.

99.
$$(R - C)(20,000)$$
$$= 65(20,000) - (600,000 + 45(20,000))$$
$$= -200,000$$
The company lost $200,000 since costs exceeded revenues.
$$(R - C)(30,000)$$
$$= 65(30,000) - (600,000 + 45(30,000))$$
$$= 0$$
The company broke even.

101. a. *f* gives the price of the computer after a $400 discount. *g* gives the price of the computer after a 25% discount.

b. $(f \circ g)(x) = 0.75x - 400$

This models the price of a computer after first a 25% discount and then a $400 discount.

c. $(g \circ f)(x) = 0.75(x - 400)$

This models the price of a computer after first a $400 discount and then a 25% discount.

d. The function $f \circ g$ models the greater discount, since the 25% discount is taken on the regular price first.

103. – 107. Answers will vary.

109.

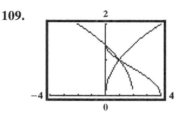

$$(f \circ g)(x) = \sqrt{2 - \sqrt{x}}$$
The domain of g is $[0, \infty)$.
The expression under the radical in $f \circ g$ must not be negative.
$$2 - \sqrt{x} \geq 0$$
$$-\sqrt{x} \geq -2$$
$$\sqrt{x} \leq 2$$
$$x \leq 4$$
domain: $[0, 4]$

111. makes sense

113. does not make sense; Explanations will vary. Sample explanation: The diagram illustrates
$$g\left(f(x)\right) = x^2 + 4.$$

115. false; Changes to make the statement true will vary. A sample change is:
$$f(x) = 2x; g(x) = 3x$$
$$(f \circ g)(x) = f\left(g(x)\right) = f(3x) = 2(3x) = 6x$$
$$(g \circ f)(x) = g\left(f(x)\right) = g(f(x)) = 3(2x) = 6x$$

117. true

119. Answers will vary.

120. $\{(4, -2), (1, -1), (1, 1), (4, 2)\}$
The element 1 in the domain corresponds to two elements in the range.
Thus, the relation is not a function.

121.

$$x = \frac{5}{y} + 4$$

$$y(x) = y\left(\frac{5}{y} + 4\right)$$

$$xy = 5 + 4y$$

$$xy - 4y = 5$$

$$y(x - 4) = 5$$

$$y = \frac{5}{x - 4}$$

122.

$$x = y^2 - 1$$

$$x + 1 = y^2$$

$$\sqrt{x+1} = \sqrt{y^2}$$

$$\sqrt{x+1} = y$$

$$y = \sqrt{x+1}$$

Section 2.7

Check Point Exercises

1.

$$f\left(g(x)\right) = 4\left(\frac{x+7}{4}\right) - 7$$

$$= x + 7 - 7$$

$$= x$$

$$g\left(f(x)\right) = \frac{(4x-7)+7}{4}$$

$$= \frac{4x-7+7}{4}$$

$$= \frac{4x}{4}$$

$$= x$$

$$f\left(g(x)\right) = g\left(f(x)\right) = x$$

2. $f(x) = 2x + 7$

Replace $f(x)$ with y:

$$y = 2x + 7$$

Interchange x and y:

$$x = 2y + 7$$

Solve for y:

$$x = 2y + 7$$

$$x - 7 = 2y$$

$$\frac{x-7}{2} = y$$

Replace y with $f^{-1}(x)$:

$$f^{-1}(x) = \frac{x-7}{2}$$

3. $f(x) = 4x^3 - 1$

Replace $f(x)$ with y:

$$y = 4x^3 - 1$$

Interchange x and y:

$$x = 4y^3 - 1$$

Solve for y:

$$x = 4y^3 - 1$$

$$x + 1 = 4y^3$$

$$\frac{x+1}{4} = y^3$$

$$\sqrt[3]{\frac{x+1}{4}} = y$$

Replace y with $f^{-1}(x)$:

$$f^{-1}(x) = \sqrt[3]{\frac{x+1}{4}}$$

Alternative form for answer:

$$f(x)^{-1} = \sqrt[3]{\frac{x+1}{4}} = \frac{\sqrt[3]{x+1}}{\sqrt[3]{4}}$$

$$= \frac{\sqrt[3]{x+1}}{\sqrt[3]{4}} \cdot \frac{\sqrt[3]{2}}{\sqrt[3]{2}} = \frac{\sqrt[3]{2x+2}}{\sqrt[3]{8}}$$

$$= \frac{\sqrt[3]{2x+2}}{2}$$

4. $f(x) = \frac{3}{x} - 1$

Replace $f(x)$ with y:

$$y = \frac{3}{x} - 1$$

Interchange x and y:

$$x = \frac{3}{y} - 1$$

Solve for y:

$$x = \frac{3}{y} - 1$$

$$xy = 3 - y$$

$$xy + y = 3$$

$$y(x + 1) = 3$$

$$y = \frac{3}{x+1}$$

Replace y with $f^{-1}(x)$:

$$f^{-1}(x) = \frac{3}{x+1}$$

5. The graphs of (b) and (c) pass the horizontal line test and thus have an inverse.

6. Find points of f^{-1}.

$f(x)$	$f^{-1}(x)$
$(-2,-2)$	$(-2,-2)$
$(-1,0)$	$(0,-1)$
$(1,2)$	$(2,1)$

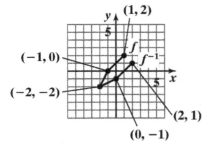

7. $f(x) = x^2 + 1$

Replace $f(x)$ with y:

$y = x^2 + 1$

Interchange x and y:

$x = y^2 + 1$

Solve for y:

$x = y^2 + 1$

$x - 1 = y^2$

$\sqrt{x-1} = y$

Replace y with $f^{-1}(x)$:

$f^{-1}(x) = \sqrt{x-1}$

Concept and Vocabulary Check 2.7

1. inverse

2. x; x

3. horizontal; one-to-one

4. $y = x$

Exercise Set 2.7

1. $f(x) = 4x;\ g(x) = \dfrac{x}{4}$

$f(g(x)) = 4\left(\dfrac{x}{4}\right) = x$

$g(f(x)) = \dfrac{4x}{4} = x$

f and *g* are inverses.

3. $f(x) = 3x + 8;\ g(x) = \dfrac{x-8}{3}$

$f(g(x)) = 3\left(\dfrac{x-8}{3}\right) + 8 = x - 8 + 8 = x$

$g(f(x)) = \dfrac{(3x+8)-8}{3} = \dfrac{3x}{3} = x$

f and *g* are inverses.

5. $f(x) = 5x - 9;\ g(x) = \dfrac{x+5}{9}$

$f(g(x)) = 5\left(\dfrac{x+5}{9}\right) - 9$

$= \dfrac{5x+25}{9} - 9$

$= \dfrac{5x-56}{9}$

$g(f(x)) = \dfrac{5x-9+5}{9} = \dfrac{5x-4}{9}$

f and *g* are not inverses.

7. $f(x) = \dfrac{3}{x-4};\ g(x) = \dfrac{3}{x} + 4$

$f(g(x)) = \dfrac{3}{\frac{3}{x}+4-4} = \dfrac{3}{\frac{3}{x}} = x$

$g(f(x)) = \dfrac{3}{\frac{3}{x-4}} + 4$

$= 3 \cdot \left(\dfrac{x-4}{3}\right) + 4$

$= x - 4 + 4$

$= x$

f and *g* are inverses.

9. $f(x) = -x;\ g(x) = -x$

$f(g(x)) = -(-x) = x$

$g(f(x)) = -(-x) = x$

f and *g* are inverses.

11. a.
$$f(x) = x + 3$$
$$y = x + 3$$
$$x = y + 3$$
$$y = x - 3$$
$$f^{-1}(x) = x - 3$$

b.
$$f(f^{-1}(x)) = x - 3 + 3 = x$$
$$f^{-1}(f(x)) = x + 3 - 3 = x$$

13. a.
$$f(x) = 2x$$
$$y = 2x$$
$$x = 2y$$
$$y = \frac{x}{2}$$
$$f^{-1}(x) = \frac{x}{2}$$

b.
$$f(f^{-1}(x)) = 2\left(\frac{x}{2}\right) = x$$
$$f^{-1}(f(x)) = \frac{2x}{2} = x$$

15. a.
$$f(x) = 2x + 3$$
$$y = 2x + 3$$
$$x = 2y + 3$$
$$x - 3 = 2y$$
$$y = \frac{x-3}{2}$$
$$f^{-1}(x) = \frac{x-3}{2}$$

b.
$$f(f^{-1}(x)) = 2\left(\frac{x-3}{2}\right) + 3$$
$$= x - 3 + 3$$
$$= x$$
$$f^{-1}(f(x)) = \frac{2x+3-3}{2} = \frac{2x}{2} = x$$

17. a.
$$f(x) = x^3 + 2$$
$$y = x^3 + 2$$
$$x = y^3 + 2$$
$$x - 2 = y^3$$
$$y = \sqrt[3]{x-2}$$
$$f^{-1}(x) = \sqrt[3]{x-2}$$

b.
$$f(f^{-1}(x)) = \left(\sqrt[3]{x-2}\right)^3 + 2$$
$$= x - 2 + 2$$
$$= x$$
$$f^{-1}(f(x)) = \sqrt[3]{x^3 + 2 - 2} = \sqrt[3]{x^3} = x$$

19. a.
$$f(x) = (x + 2)^3$$
$$y = (x + 2)^3$$
$$x = (y + 2)^3$$
$$\sqrt[3]{x} = y + 2$$
$$y = \sqrt[3]{x} - 2$$
$$f^{-1}(x) = \sqrt[3]{x} - 2$$

b.
$$f(f^{-1}(x)) = \left(\sqrt[3]{x} - 2 + 2\right)^3 = \left(\sqrt[3]{x}\right)^3 = x$$
$$f^{-1}(f(x)) = \sqrt[3]{(x+2)^3} - 2$$
$$= x + 2 - 2$$
$$= x$$

21. a.
$$f(x) = \frac{1}{x}$$
$$y = \frac{1}{x}$$
$$x = \frac{1}{y}$$
$$xy = 1$$
$$y = \frac{1}{x}$$
$$f^{-1}(x) = \frac{1}{x}$$

b.
$$f(f^{-1}(x)) = \frac{1}{\frac{1}{x}} = x$$
$$f^{-1}(f(x)) = \frac{1}{\frac{1}{x}} = x$$

23. a.
$$f(x) = \sqrt{x}$$
$$y = \sqrt{x}$$
$$x = \sqrt{y}$$
$$y = x^2$$
$$f^{-1}(x) = x^2, x \geq 0$$

b.
$$f(f^{-1}(x)) = \sqrt{x^2} = |x| = x \text{ for } x \geq 0.$$
$$f^{-1}(f(x)) = (\sqrt{x})^2 = x$$

25. a.

$$f(x) = \frac{7}{x} - 3$$

$$y = \frac{7}{x} - 3$$

$$x = \frac{7}{y} - 3$$

$$xy = 7 - 3y$$

$$xy + 3y = 7$$

$$y(x+3) = 7$$

$$y = \frac{7}{x+3}$$

$$f^{-1}(x) = \frac{7}{x+3}$$

b.

$$f\left(f^{-1}(x)\right) = \frac{7}{\dfrac{7}{x+3}} - 3 = x$$

$$f^{-1}\left(f(x)\right) = \frac{7}{\dfrac{7}{x} - 3 + 3} = x$$

27. a.

$$f(x) = \frac{2x+1}{x-3}$$

$$y = \frac{2x+1}{x-3}$$

$$x = \frac{2y+1}{y-3}$$

$$x(y-3) = 2y+1$$

$$xy - 3x = 2y+1$$

$$xy - 2y = 3x+1$$

$$y(x-2) = 3x+1$$

$$y = \frac{3x+1}{x-2}$$

$$f^{-1}(x) = \frac{3x+1}{x-2}$$

b.

$$f\left(f^{-1}(x)\right) = \frac{2\left(\dfrac{3x+1}{x-2}\right)+1}{\dfrac{3x+1}{x-2}-3}$$

$$= \frac{2(3x+1)+x-2}{3x+1-3(x-2)} = \frac{6x+2+x-2}{3x+1-3x+6}$$

$$= \frac{7x}{7} = x$$

$$f^{-1}\left(f(x)\right) = \frac{3\left(\dfrac{2x+1}{x-3}\right)+1}{\dfrac{2x+1}{x-3}-2}$$

$$= \frac{3(2x+1)+x-3}{2x+1-2(x-3)}$$

$$= \frac{6x+3+x-3}{2x+1-2x+6} = \frac{7x}{7} = x$$

29. The function fails the horizontal line test, so it does not have an inverse function.

31. The function fails the horizontal line test, so it does not have an inverse function.

33. The function passes the horizontal line test, so it does have an inverse function.

35.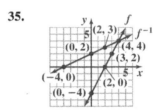

37.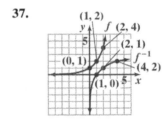

39. a.

$$f(x) = 2x-1$$

$$y = 2x-1$$

$$x = 2y-1$$

$$x+1 = 2y$$

$$\frac{x+1}{2} = y$$

$$f^{-1}(x) = \frac{x+1}{2}$$

b.

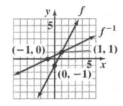

c. domain of f : $(-\infty, \infty)$

range of f : $(-\infty, \infty)$

domain of f^{-1} : $(-\infty, \infty)$

range of f^{-1} : $(-\infty, \infty)$

41. a. $f(x) = x^2 - 4$

$y = x^2 - 4$

$x = y^2 - 4$

$x + 4 = y^2$

$\sqrt{x+4} = y$

$f^{-1}(x) = \sqrt{x+4}$

b.

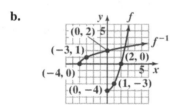

c. domain of f : $[0, \infty)$

range of f : $[-4, \infty)$

domain of f^{-1} : $[-4, \infty)$

range of f^{-1} : $[0, \infty)$

43. a. $f(x) = (x-1)^2$

$y = (x-1)^2$

$x = (y-1)^2$

$-\sqrt{x} = y - 1$

$-\sqrt{x} + 1 = y$

$f^{-1}(x) = 1 - \sqrt{x}$

b.

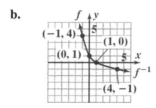

c. domain of f : $(-\infty, 1]$

range of f : $[0, \infty)$

domain of f^{-1} : $[0, \infty)$

range of f^{-1} : $(-\infty, 1]$

45. a. $f(x) = x^3 - 1$

$y = x^3 - 1$

$x = y^3 - 1$

$x + 1 = y^3$

$\sqrt[3]{x+1} = y$

$f^{-1}(x) = \sqrt[3]{x+1}$

b.

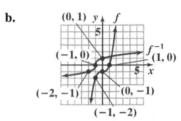

c. domain of f : $(-\infty, \infty)$

range of f : $(-\infty, \infty)$

domain of f^{-1} : $(-\infty, \infty)$

range of f^{-1} : $(-\infty, \infty)$

47. a. $f(x) = (x+2)^3$

$y = (x+2)^3$

$x = (y+2)^3$

$\sqrt[3]{x} = y + 2$

$\sqrt[3]{x} - 2 = y$

$f^{-1}(x) = \sqrt[3]{x} - 2$

b.

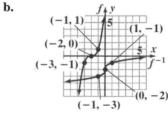

c. domain of f : $(-\infty, \infty)$

range of f : $(-\infty, \infty)$

domain of f^{-1} : $(-\infty, \infty)$

range of f^{-1} : $(-\infty, \infty)$

49. a.
$$f(x) = \sqrt{x-1}$$
$$y = \sqrt{x-1}$$
$$x = \sqrt{y-1}$$
$$x^2 = y-1$$
$$x^2+1 = y$$
$$f^{-1}(x) = x^2+1$$

b.

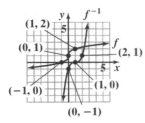

c. domain of f : $[1,\infty)$

range of f : $[0,\infty)$

domain of f^{-1} : $[0,\infty)$

range of f^{-1} : $[1,\infty)$

51. a.
$$f(x) = \sqrt[3]{x}+1$$
$$y = \sqrt[3]{x}+1$$
$$x = \sqrt[3]{y}+1$$
$$x-1 = \sqrt[3]{y}$$
$$(x-1)^3 = y$$
$$f^{-1}(x) = (x-1)^3$$

b.

c. domain of f : $(-\infty,\infty)$

range of f : $(-\infty,\infty)$

domain of f^{-1} : $(-\infty,\infty)$

range of f^{-1} : $(-\infty,\infty)$

53. $f(g(1)) = f(1) = 5$

55. $(g \circ f)(-1) = g(f(-1)) = g(1) = 1$

57. $f^{-1}(g(10)) = f^{-1}(-1) = 2$, since $f(2) = -1$.

59.
$$(f \circ g)(0) = f(g(0))$$
$$= f(4 \cdot 0 - 1)$$
$$= f(-1) = 2(-1) - 5 = -7$$

61. Let $f^{-1}(1) = x$. Then
$$f(x) = 1$$
$$2x - 5 = 1$$
$$2x = 6$$
$$x = 3$$
Thus, $f^{-1}(1) = 3$

63.
$$g(f[h(1)]) = g(f[1^2+1+2])$$
$$= g(f(4))$$
$$= g(2 \cdot 4 - 5)$$
$$= g(3)$$
$$= 4 \cdot 3 - 1 = 11$$

65. a. {(Zambia, 4.2), (Colombia, 4.5), (Poland, 3.3), (Italy, 3.3), (United States, 2.5)}

b. {(4.2, Zambia), (4.5 , Colombia), (3.3 , Poland), (3.3, Italy), (2.5, United States)}

f is not a one-to-one function because the inverse of f is not a function.

67. a. It passes the horizontal line test and is one-to-one.

b. $f^{-1}(0.25) = 15$ If there are 15 people in the room, the probability that 2 of them have the same birthday is 0.25.
$f^{-1}(0.5) = 21$ If there are 21 people in the room, the probability that 2 of them have the same birthday is 0.5.
$f^{-1}(0.7) = 30$ If there are 30 people in the room, the probability that 2 of them have the same birthday is 0.7.

69.
$$f(g(x)) = \frac{9}{5}\left[\frac{5}{9}(x-32)\right]+32$$
$$= x-32+32$$
$$= x$$
$$g(f(x)) = \frac{5}{9}\left[\left(\frac{9}{5}x+32\right)-32\right]$$
$$= x+32-32$$
$$= x$$
f and g are inverses.

71. – 75. Answers will vary.

77.

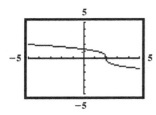

one-to-one

79.

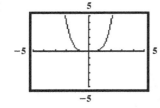

not one-to-one

81.

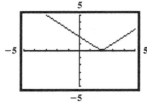

not one-to-one

83.

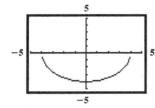

not one-to-one

85.

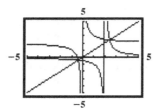

f and *g* are inverses

87. makes sense

89. makes sense

91. false; Changes to make the statement true will vary. A sample change is: The inverse is $\{(4,1), (7,2)\}$.

93. false; Changes to make the statement true will vary.

A sample change is: $f^{-1}(x) = \dfrac{x}{3}$.

95. $(f \circ g)(x) = 3(x+5) = 3x+15.$

$$y = 3x+15$$
$$x = 3y+15$$
$$y = \frac{x-15}{3}$$
$$(f \circ g)^{-1}(x) = \frac{x-15}{3}$$
$$g(x) = x+5$$
$$y = x+5$$
$$x = y+5$$
$$y = x-5$$
$$g^{-1}(x) = x-5$$
$$f(x) = 3x$$
$$y = 3x$$
$$x = 3y$$
$$y = \frac{x}{3}$$
$$f^{-1}(x) = \frac{x}{3}$$
$$\left(g^{-1} \circ f^{-1}\right)(x) = \frac{x}{3} - 5 = \frac{x-15}{3}$$

97. No, there will be 2 times when the spacecraft is at the same height, when it is going up and when it is coming down.

99. Answers will vary.

100. $\sqrt{(x_2 - x_1)^2 + (y_2 - y_1)^2} = \sqrt{(1-7)^2 + (-1-2)^2}$

$$= \sqrt{(-6)^2 + (-3)^2}$$
$$= \sqrt{36+9}$$
$$= \sqrt{45}$$
$$= 3\sqrt{5}$$

101.

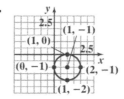

102. $y^2 - 6y - 4 = 0$

$$y^2 - 6y = 4$$
$$y^2 - 6y + 9 = 4 + 9$$
$$(y-3)^2 = 13$$
$$y - 3 = \pm\sqrt{13}$$
$$y = 3 \pm \sqrt{13}$$

Solution set: $\left\{3 \pm \sqrt{13}\right\}$

Section 2.8

Check Point Exercises

1. $d = \sqrt{(x_2 - x_1)^2 + (y_2 - y_1)^2}$

$d = \sqrt{(2 - (-1))^2 + (3 - (-3))^2}$

$= \sqrt{3^2 + 6^2}$

$= \sqrt{9 + 36}$

$= \sqrt{45}$

$= 3\sqrt{5}$

≈ 6.71

2. $\left(\dfrac{1+7}{2}, \dfrac{2+(-3)}{2} \right) = \left(\dfrac{8}{2}, \dfrac{-1}{2} \right) = \left(4, -\dfrac{1}{2} \right)$

3. $h = 0,\ k = 0,\ r = 4;$

$(x - 0)^2 + (y - 0)^2 = 4^2$

$x^2 + y^2 = 16$

4. $h = 0,\ k = -6,\ r = 10;$

$(x - 0)^2 + [y - (-6)]^2 = 10^2$

$(x - 0)^2 + (y + 6)^2 = 100$

$x^2 + (y + 6)^2 = 100$

5. a. $(x + 3)^2 + (y - 1)^2 = 4$

$[x - (-3)]^2 + (y - 1)^2 = 2^2$

So in the standard form of the circle's equation
$(x - h)^2 + (y - k)^2 = r^2$,

we have $h = -3,\ k = 1,\ r = 2.$

center: $(h, k) = (-3, 1)$

radius: $r = 2$

b.

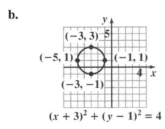

$(x + 3)^2 + (y - 1)^2 = 4$

c. domain: $[-5, -1]$

range: $[-1, 3]$

6. $x^2 + y^2 + 4x - 4y - 1 = 0$

$x^2 + y^2 + 4x - 4y - 1 = 0$

$(x^2 + 4x \quad) + (y^2 - 4y \quad) = 0$

$(x^2 + 4x + 4) + (y^2 + 4y + 4) = 1 + 4 + 4$

$(x + 2)^2 + (y - 2)^2 = 9$

$[x - (-x)]^2 + (y - 2)^2 = 3^2$

So in the standard form of the circle's equation
$(x - h)^2 + (y - k)^2 = r^2$, we have

$h = -2,\ k = 2,\ r = 3.$

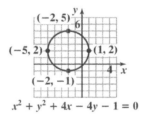

$x^2 + y^2 + 4x - 4y - 1 = 0$

Concept and Vocabulary Check 2.8

1. $\sqrt{(x_2 - x_1)^2 + (y_2 - y_1)^2}$

2. $\dfrac{x_1 + x_2}{2}$; $\dfrac{y_1 + y_2}{2}$

3. circle; center; radius

4. $(x - h)^2 + (y - k)^2 = r^2$

5. general

6. 4; 16

Exercise Set 2.8

1. $d = \sqrt{(14 - 2)^2 + (8 - 3)^2}$

$= \sqrt{12^2 + 5^2}$

$= \sqrt{144 + 25}$

$= \sqrt{169}$

$= 13$

3. $d = \sqrt{(-6-4)^2 + (3-(-1))^2}$

$\quad = \sqrt{(-10)^2 + (4)^2}$

$\quad = \sqrt{100+16}$

$\quad = \sqrt{116}$

$\quad = 2\sqrt{29}$

$\quad \approx 10.77$

5. $d = \sqrt{(-3-0)^2 + (4-0)^2}$

$\quad = \sqrt{3^2 + 4^2}$

$\quad = \sqrt{9+16}$

$\quad = \sqrt{25}$

$\quad = 5$

7. $d = \sqrt{[3-(-2)]^2 + [-4-(-6)]^2}$

$\quad = \sqrt{5^2 + 2^2}$

$\quad = \sqrt{25+4}$

$\quad = \sqrt{29}$

$\quad \approx 5.39$

9. $d = \sqrt{(4-0)^2 + [1-(-3)]^2}$

$\quad = \sqrt{4^2 + 4^2}$

$\quad = \sqrt{16+16}$

$\quad = \sqrt{32}$

$\quad = 4\sqrt{2}$

$\quad \approx 5.66$

11. $d = \sqrt{(-.5-3.5)^2 + (6.2-8.2)^2}$

$\quad = \sqrt{(-4)^2 + (-2)^2}$

$\quad = \sqrt{16+4}$

$\quad = \sqrt{20}$

$\quad = 2\sqrt{5}$

$\quad \approx 4.47$

13. $d = \sqrt{(\sqrt{5}-0)^2 + [0-(-\sqrt{3})]^2}$

$\quad = \sqrt{(\sqrt{5})^2 + (\sqrt{3})^2}$

$\quad = \sqrt{5+3}$

$\quad = \sqrt{8}$

$\quad = 2\sqrt{2}$

$\quad \approx 2.83$

15. $d = \sqrt{(-\sqrt{3}-3\sqrt{3})^2 + (4\sqrt{5}-\sqrt{5})^2}$

$\quad = \sqrt{(-4\sqrt{3})^2 + (3\sqrt{5})^2}$

$\quad = \sqrt{16(3)+9(5)}$

$\quad = \sqrt{48+45}$

$\quad = \sqrt{93}$

$\quad \approx 9.64$

17. $d = \sqrt{\left(\dfrac{1}{3}-\dfrac{7}{3}\right)^2 + \left(\dfrac{6}{5}-\dfrac{1}{5}\right)^2}$

$\quad = \sqrt{(-2)^2 + 1^2}$

$\quad = \sqrt{4+1}$

$\quad = \sqrt{5}$

$\quad \approx 2.24$

19. $\left(\dfrac{6+2}{2}, \dfrac{8+4}{2}\right) = \left(\dfrac{8}{2}, \dfrac{12}{2}\right) = (4,6)$

21. $\left(\dfrac{-2+(-6)}{2}, \dfrac{-8+(-2)}{2}\right)$

$\quad = \left(\dfrac{-8}{2}, \dfrac{-10}{2}\right) = (-4,-5)$

23. $\left(\dfrac{-3+6}{2}, \dfrac{-4+(-8)}{2}\right)$

$\quad = \left(\dfrac{3}{2}, \dfrac{-12}{2}\right) = \left(\dfrac{3}{2}, -6\right)$

25. $\left(\dfrac{\dfrac{-7}{2}+\left(-\dfrac{5}{2}\right)}{2}, \dfrac{\dfrac{3}{2}+\left(-\dfrac{11}{2}\right)}{2}\right)$

$\quad = \left(\dfrac{\dfrac{-12}{2}}{2}, \dfrac{\dfrac{-8}{2}}{2}\right) = \left(-\dfrac{6}{2}, \dfrac{-4}{2}\right) = (-3,-2)$

27. $\left(\dfrac{8+(-6)}{2}, \dfrac{3\sqrt{5}+7\sqrt{5}}{2}\right)$

$\quad = \left(\dfrac{2}{2}, \dfrac{10\sqrt{5}}{2}\right) = \left(1, 5\sqrt{5}\right)$

29. $\left(\dfrac{\sqrt{18}+\sqrt{2}}{2},\dfrac{-4+4}{2}\right)$

$=\left(\dfrac{3\sqrt{2}+\sqrt{2}}{2},\dfrac{0}{2}\right)=\left(\dfrac{4\sqrt{2}}{2},0\right)=(2\sqrt{2},0)$

31. $(x-0)^2+(y-0)^2=7^2$

$\qquad\qquad x^2+y^2=49$

33. $(x-3)^2+(y-2)^2=5^2$

$\quad(x-3)^2+(y-2)^2=25$

35. $[x-(-1)]^2+(y-4)^2=2^2$

$\qquad(x+1)^2+(y-4)^2=4$

37. $[x-(-3)]^2+[y-(-1)]^2=\left(\sqrt{3}\right)^2$

$\qquad(x+3)^2+(y+1)^2=3$

39. $[x-(-4)]^2+(y-0)^2=10^2$

$\qquad(x+4)^2+(y-0)^2=100$

41. $\qquad\qquad x^2+y^2=16$

$(x-0)^2+(y-0)^2=y^2$

$h=0,\ k=0,\ r=4;$

center $=(0,0);$ radius $=4$

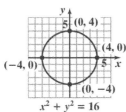

$x^2+y^2=16$

domain: $[-4,4]$

range: $[-4,4]$

43. $(x-3)^2+(y-1)^2=36$

$\quad(x-3)^2+(y-1)^2=6^2$

$h=3,\ k=1,\ r=6;$

center $=(3,1);$ radius $=6$

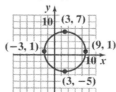

$(x-3)^2+(y-1)^2=36$

domain: $[-3,9]$

range: $[-5,7]$

45. $\qquad(x+3)^2+(y-2)^2=4$

$\quad[x-(-3)]^2+(y-2)^2=2^2$

$h=-3,\ k=2,\ r=2$

center $=(-3,2);$ radius $=2$

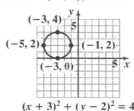

$(x+3)^2+(y-2)^2=4$

domain: $[-5,-1]$

range: $[0,4]$

47. $\qquad(x+2)^2+(y+2)^2=4$

$\quad[x-(-2)]^2+[y-(-2)]^2=2^2$

$h=-2,\ k=-2,\ r=2$

center $=(-2,-2);$ radius $=2$

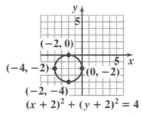

$(x+2)^2+(y+2)^2=4$

domain: $[-4,0]$

range: $[-4,0]$

49. $x^2 + (y-1)^2 = 1$

$h = 0, k = 1, r = 1;$

center = (0, 1); radius = 1

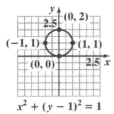

domain: $[-1, 1]$

range: $[0, 2]$

51. $(x+1)^2 + y^2 = 25$

$h = -1, k = 0, r = 5;$

center = (-1, 0); radius = 5

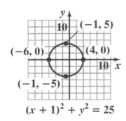

domain: $[-6, 4]$

range: $[-5, 5]$

53. $x^2 + y^2 + 6x + 2y + 6 = 0$

$(x^2 + 6x) + (y^2 + 2y) = -6$

$(x^2 + 6x + 9) + (y^2 + 2y + 1) = 9 + 1 - 6$

$(x+3)^2 + (y+1)^2 = 4$

$[x - (-3)]^2 + [9 - (-1)]^2 = 2^2$

center = (-3, -1); radius = 2

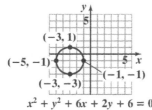

55. $x^2 + y^2 - 10x - 6y - 30 = 0$

$(x^2 - 10x) + (y^2 - 6y) = 30$

$(x^2 - 10x + 25) + (y^2 - 6y + 9) = 25 + 9 + 30$

$(x-5)^2 + (y-3)^2 = 64$

$(x-5)^2 + (y-3)^2 = 8^2$

center = (5, 3); radius = 8

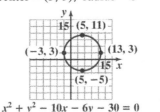

57. $x^2 + y^2 + 8x - 2y - 8 = 0$

$(x^2 + 8x) + (y^2 - 2y) = 8$

$(x^2 + 8x + 16) + (y^2 - 2y + 1) = 16 + 1 + 8$

$(x+4)^2 + (y-1)^2 = 25$

$[x - (-4)]^2 + (y-1)^2 = 5^2$

center = (-4, 1); radius = 5

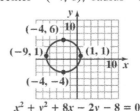

59. $x^2 - 2x + y^2 - 15 = 0$

$(x^2 - 2x) + y^2 = 15$

$(x^2 - 2x + 1) + (y - 0)^2 = 1 + 0 + 15$

$(x-1)^2 + (y-0)^2 = 16$

$(x-1)^2 + (y-0)^2 = 4^2$

center = (1, 0); radius = 4

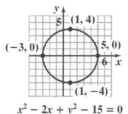

61.
$$x^2 + y^2 - x + 2y + 1 = 0$$
$$x^2 - x + y^2 + 2y = -1$$
$$x^2 - x + \frac{1}{4} + y^2 + 2y + 1 = -1 + \frac{1}{4} + 1$$
$$\left(x - \frac{1}{2}\right)^2 + (y + 1)^2 = \frac{1}{4}$$
center $= \left(\frac{1}{2}, -1\right)$; radius $= \frac{1}{2}$

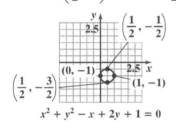

63.
$$x^2 + y^2 + 3x - 2y - 1 = 0$$
$$x^2 + 3x + y^2 - 2y = 1$$
$$x^2 + 3x + \frac{9}{4} + y^2 - 2y + 1 = 1 + \frac{9}{4} + 1$$
$$\left(x + \frac{3}{2}\right)^2 + (y - 1)^2 = \frac{17}{4}$$
center $= \left(-\frac{3}{2}, 1\right)$; radius $= \frac{\sqrt{17}}{2}$

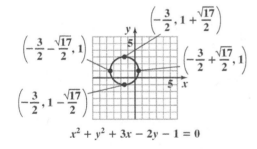

65. a. Since the line segment passes through the center, the center is the midpoint of the segment.
$$M = \left(\frac{x_1 + x_2}{2}, \frac{y_1 + y_2}{2}\right)$$
$$= \left(\frac{3 + 7}{2}, \frac{9 + 11}{2}\right) = \left(\frac{10}{2}, \frac{20}{2}\right)$$
$$= (5, 10)$$
The center is $(5, 10)$.

b. The radius is the distance from the center to one of the points on the circle. Using the point $(3, 9)$, we get:
$$d = \sqrt{(5 - 3)^2 + (10 - 9)^2}$$
$$= \sqrt{2^2 + 1^2} = \sqrt{4 + 1}$$
$$= \sqrt{5}$$
The radius is $\sqrt{5}$ units.

c.
$$(x - 5)^2 + (y - 10)^2 = \left(\sqrt{5}\right)^2$$
$$(x - 5)^2 + (y - 10)^2 = 5$$

67.
$$x^2 + y^2 = 16$$
$$x - y = 4$$

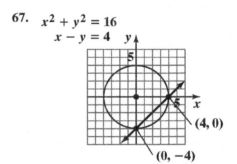

Intersection points: $(0, -4)$ and $(4, 0)$

Check $(0, -4)$:
$$0^2 + (-4)^2 = 16 \qquad 0 - (-4) = 4$$
$$16 = 16 \text{ true} \qquad 4 = 4 \text{ true}$$

Check $(4, 0)$:
$$4^2 + 0^2 = 16 \qquad 4 - 0 = 4$$
$$16 = 16 \text{ true} \qquad 4 = 4 \text{ true}$$
The solution set is $\{(0, -4), (4, 0)\}$.

69.
$$(x - 2)^2 + (y + 3)^2 = 4$$
$$y = x - 3$$

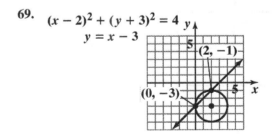

Intersection points: $(0, -3)$ and $(2, -1)$

Check $(0,-3)$:

$$(0-2)^2 + (-3+3)^2 = 9 \qquad -3 = 0-3$$
$$(-2)^2 + 0^2 = 4 \qquad -3 = -3 \text{ true}$$
$$4 = 4$$
$$\text{true}$$

Check $(2,-1)$:

$$(2-2)^2 + (-1+3)^2 = 4 \qquad -1 = 2-3$$
$$0^2 + 2^2 = 4 \qquad -1 = -1 \text{ true}$$
$$4 = 4$$
$$\text{true}$$

The solution set is $\{(0,-3),(2,-1)\}$.

71. $d = \sqrt{(8495-4422)^2 + (8720-1241)^2} \cdot \sqrt{0.1}$

$d = \sqrt{72,524,770} \cdot \sqrt{0.1}$

$d \approx 2693$

The distance between Boston and San Francisco is about 2693 miles.

73. If we place L.A. at the origin, then we want the equation of a circle with center at $(-2.4,-2.7)$ and radius 30.

$$(x-(-2.4))^2 + (y-(-2.7))^2 = 30^2$$
$$(x+2.4)^2 + (y+2.7)^2 = 900$$

75. – 81. Answers will vary.

83.

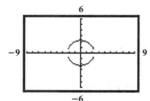

85.

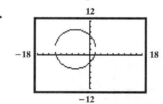

87. makes sense

89. makes sense

91. false; Changes to make the statement true will vary.
A sample change is: The center is at (3, –5).

93. false; Changes to make the statement true will vary.
A sample change is: Since $r^2 = -36$ this is not the equation of a circle.

95. a. d_1 is distance from (x_1,x_2) to midpoint

$$d_1 = \sqrt{\left(\frac{x_1+x_2}{2} - x_1\right)^2 + \left(\frac{y_1+y_2}{2} - y_1\right)^2}$$

$$d_1 = \sqrt{\left(\frac{x_1+x_2-2x_1}{2}\right)^2 + \left(\frac{y_1+y_2-2y_1}{2}\right)^2}$$

$$d_1 = \sqrt{\left(\frac{x_2-x_1}{2}\right)^2 + \left(\frac{y_2-y_1}{2}\right)^2}$$

$$d_1 = \sqrt{\frac{x_2-2x_1x_2+x_1^2}{4} + \frac{y_2^2-2y_2y_1+y_1^2}{4}}$$

$$d_1 = \sqrt{\frac{1}{4}\left(x_2-2x_1x_2+x_1+y_2^2-2y_2y_1+y_1^2\right)}$$

$$d_1 = \frac{1}{2}\sqrt{x_2-2x_1x_2+x_1+y_2^2-2y_2y_1+y_1^2}$$

d_2 is distance from midpoint to (x_2,y_2)

$$d_2 = \sqrt{\left(\frac{x_1+x_2}{2} - x_2\right)^2 + \left(\frac{y_1+y_2}{2} - y_2\right)^2}$$

$$d_2 = \sqrt{\left(\frac{x_1+x_2-2x_2}{2}\right)^2 + \left(\frac{y_1+y_2-2y_2}{2}\right)^2}$$

$$d_2 = \sqrt{\left(\frac{x_1-x_2}{2}\right)^2 + \left(\frac{y_1-y_2}{2}\right)^2}$$

$$d_2 = \sqrt{\frac{x_1^2-2x_1x_2+x_2^2}{4} + \frac{y_1^2-2y_2y_1+y_2^2}{4}}$$

$$d_2 = \sqrt{\frac{1}{4}\left(x_1^2-2x_1x_2+x_2^2+y_1^2-2y_2y_1+y_2^2\right)}$$

$$d_2 = \frac{1}{2}\sqrt{x_1^2-2x_1x_2+x_2^2+y_1^2-2y_2y_1+y_2^2}$$

$$d_1 = d_2$$

b. d_3 is the distance from (x_1,y_1) to $(x_2 y_2)$

$$d_3 = \sqrt{(x_2-x_1)^2 + (y_2-y_1)^2}$$

$$d_3 = \sqrt{x_2^2-2x_1x_2+x_1^2+y_2^2-2y_2y_1+y_1^2}$$

$d_1 + d_2 = d_3$ because $\dfrac{1}{2}\sqrt{a} + \dfrac{1}{2}\sqrt{a} = \sqrt{a}$

97. The circle is centered at (0,0). The slope of the radius with endpoints (0,0) and (3,–4) is

$m = -\dfrac{-4-0}{3-0} = -\dfrac{4}{3}$. The line perpendicular to the

radius has slope $\dfrac{3}{4}$. The tangent line has slope $\dfrac{3}{4}$ and

passes through (3,–4), so its equation is:

$y + 4 = \dfrac{3}{4}(x-3)$.

99. $-x^2 - 2x + 1 = 0$

$x^2 + 2x - 1 = 0$

$x = \dfrac{-b \pm \sqrt{b^2 - 4ac}}{2a}$

$x = \dfrac{-(-2) \pm \sqrt{(-2)^2 - 4(1)(-1)}}{2(1)}$

$= \dfrac{2 \pm \sqrt{8}}{2}$

$= \dfrac{2 \pm 2\sqrt{2}}{2}$

$= 1 \pm \sqrt{2}$

The solution set is $\{1 \pm \sqrt{2}\}$.

100. The graph of g is the graph of f shifted 1 unit up and 3 units to the left.

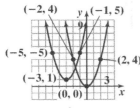

$f(x) = x^2$
$g(x) = (x+3)^2 + 1$

Chapter 2 Review Exercises

1. function
domain: {2, 3, 5}
range: {7}

2. function
domain: {1, 2, 13}
range: {10, 500, π}

3. not a function
domain: {12, 14}
range: {13, 15, 19}

4. $2x + y = 8$

$y = -2x + 8$

Since only one value of y can be obtained for each value of x, y is a function of x.

5. $3x^2 + y = 14$

$y = -3x^2 + 14$

Since only one value of y can be obtained for each value of x, y is a function of x.

6. $2x + y^2 = 6$

$y^2 = -2x + 6$

$y = \pm\sqrt{-2x + 6}$

Since more than one value of y can be obtained from some values of x, y is not a function of x.

7. $f(x) = 5 - 7x$

  a. $f(4) = 5 - 7(4) = -23$

  b. $f(x+3) = 5 - 7(x+3)$
          $= 5 - 7x - 21$
          $= -7x - 16$

  c. $f(-x) = 5 - 7(-x) = 5 + 7x$

8. $g(x) = 3x^2 - 5x + 2$

  a. $g(0) = 3(0)^2 - 5(0) + 2 = 2$

  b. $g(-2) = 3(-2)^2 - 5(-2) + 2$
          $= 12 + 10 + 2$
          $= 24$

  c. $g(x-1) = 3(x-1)^2 - 5(x-1) + 2$
          $= 3(x^2 - 2x + 1) - 5x + 5 + 2$
          $= 3x^2 - 11x + 10$

  d. $g(-x) = 3(-x)^2 - 5(-x) + 2$
          $= 3x^2 + 5x + 2$

9. **a.** $g(13) = \sqrt{13-4} = \sqrt{9} = 3$

  b. $g(0) = 4 - 0 = 4$

  c. $g(-3) = 4 - (-3) = 7$

10. **a.** $f(-2) = \dfrac{(-2)^2 - 1}{-2 - 1} = \dfrac{3}{-3} = -1$

b. $f(1) = 12$

c. $f(2) = \dfrac{2^2 - 1}{2 - 1} = \dfrac{3}{1} = 3$

11. The vertical line test shows that this is not the graph of a function.

12. The vertical line test shows that this is the graph of a function.

13. The vertical line test shows that this is the graph of a function.

14. The vertical line test shows that this is not the graph of a function.

15. The vertical line test shows that this is not the graph of a function.

16. The vertical line test shows that this is the graph of a function.

17. **a.** domain: $[-3, 5)$

b. range: $[-5, 0]$

c. x-intercept: -3

d. y-intercept: -2

e. increasing: $(-2, 0)$ or $(3, 5)$
decreasing: $(-3, -2)$ or $(0, 3)$

f. $f(-2) = -3$ and $f(3) = -5$

18. **a.** domain: $(-\infty, \infty)$

b. range: $(-\infty, 3]$

c. x-intercepts: -2 and 3

d. y-intercept: 3

e. increasing: $(-\infty, 0)$
decreasing: $(0, \infty)$

f. $f(-2) = 0$ and $f(6) = -3$

19. **a.** domain: $(-\infty, \infty)$

b. range: $[-2, 2]$

c. x-intercept: 0

d. y-intercept: 0

e. increasing: $(-2, 2)$
constant: $(-\infty, -2)$ or $(2, \infty)$

f. $f(-9) = -2$ and $f(14) = 2$

20. **a.** 0, relative maximum -2

b. -2, 3, relative minimum -3, -5

21. **a.** 0, relative maximum 3

b. none

22. $f(x) = x^3 - 5x$
$f(-x) = (-x)^3 - 5(-x)$
$\qquad = -x^3 + 5x$
$\qquad = -f(x)$
The function is odd. The function is symmetric with respect to the origin.

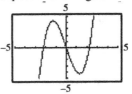

23. $f(x) = x^4 - 2x^2 + 1$
$f(-x) = (-x)^4 - 2(-x)^2 + 1$
$\qquad = x^4 - 2x^2 + 1$
$\qquad = f(x)$
The function is even. The function is symmetric with respect to the y-axis.

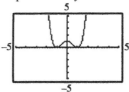

24.
$$f(x) = 2x\sqrt{1-x^2}$$
$$f(-x) = 2(-x)\sqrt{1-(-x)^2}$$
$$= -2x\sqrt{1-x^2}$$
$$= -f(x)$$

The function is odd. The function is symmetric with respect to the origin.

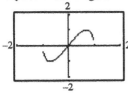

25. a.

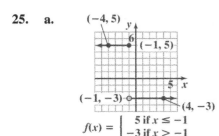

$$f(x) = \begin{cases} 5 \text{ if } x \le -1 \\ -3 \text{ if } x > -1 \end{cases}$$

b. range: $\{-3, 5\}$

26. a.

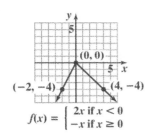

$$f(x) = \begin{cases} 2x \text{ if } x < 0 \\ -x \text{ if } x \ge 0 \end{cases}$$

b. range: $\{y \mid y \le 0\}$

27.
$$\frac{8(x+h)-11-(8x-11)}{h}$$
$$= \frac{8x+8h-11-8x+11}{h}$$
$$= \frac{8h}{8}$$
$$= 8$$

28.
$$\frac{-2(x+h)^2+(x+h)+10-\left(-2x^2+x+10\right)}{h}$$
$$= \frac{-2\left(x^2+2xh+h^2\right)+x+h+10+2x^2-x-10}{h}$$
$$= \frac{-2x^2-4xh-2h^2+x+h+10+2x^2-x-10}{h}$$
$$= \frac{-4xh-2h^2+h}{h}$$
$$= \frac{h\left(-4x-2h+1\right)}{h}$$
$$-4x-2h+1$$

29. a. Yes, the eagle's height is a function of time since the graph passes the vertical line test.

b. Decreasing: (3, 12)
The eagle descended.

c. Constant: (0, 3) or (12, 17)
The eagle's height held steady during the first 3 seconds and the eagle was on the ground for 5 seconds.

d. Increasing: (17, 30)
The eagle was ascending.

30.

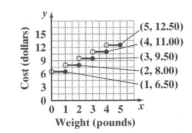

31. $m = \dfrac{1-2}{5-3} = \dfrac{-1}{2} = -\dfrac{1}{2}$; falls

32. $m = \dfrac{-4-(-2)}{-3-(-1)} = \dfrac{-2}{-2} = 1$; rises

33. $m = \dfrac{\frac{1}{4}-\frac{1}{4}}{6-(-3)} = \dfrac{0}{9} = 0$; horizontal

34. $m = \dfrac{10-5}{-2-(-2)} = \dfrac{5}{0}$ undefined; vertical

35. point-slope form: $y - 2 = -6(x + 3)$
slope-intercept form: $y = -6x - 16$

36. $m = \dfrac{2-6}{-1-1} = \dfrac{-4}{-2} = 2$

point-slope form: $y - 6 = 2(x - 1)$
or $y - 2 = 2(x + 1)$
slope-intercept form: $y = 2x + 4$

37. $3x + y - 9 = 0$
$y = -3x + 9$
$m = -3$
point-slope form:
$y + 7 = -3(x - 4)$
slope-intercept form:
$y = -3x + 12 - 7$
$y = -3x + 5$

38. perpendicular to $y = \dfrac{1}{3}x + 4$

$m = -3$
point-slope form:
$y - 6 = -3(x + 3)$
slope-intercept form:
$y = -3x - 9 + 6$
$y = -3x - 3$

39. Write $6x - y - 4 = 0$ in slope intercept form.
$6x - y - 4 = 0$
$-y = -6x + 4$
$y = 6x - 4$
The slope of the perpendicular line is 6, thus the
slope of the desired line is $m = -\dfrac{1}{6}$.

$$y - y_1 = m(x - x_1)$$
$$y - (-1) = -\tfrac{1}{6}\left(x - (-12)\right)$$
$$y + 1 = -\tfrac{1}{6}(x + 12)$$
$$y + 1 = -\tfrac{1}{6}x - 2$$
$$6y + 6 = -x - 12$$
$$x + 6y + 18 = 0$$

40. slope: $\dfrac{2}{5}$; y-intercept: -1

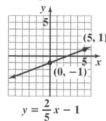

$y = \dfrac{2}{5}x - 1$

41. slope: -4; y-intercept: 5

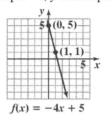

$f(x) = -4x + 5$

42. $2x + 3y + 6 = 0$
$$3y = -2x - 6$$
$$y = -\dfrac{2}{3}x - 2$$

slope: $-\dfrac{2}{3}$; y-intercept: -2

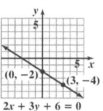

$2x + 3y + 6 = 0$

43. $2y - 8 = 0$
$$2y = 8$$
$$y = 4$$

slope: 0; y-intercept: 4

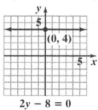

$2y - 8 = 0$

44. $2x - 5y - 10 = 0$

Find x-intercept:

$2x - 5(0) - 10 = 0$

$2x - 10 = 0$

$2x = 10$

$x = 5$

Find y-intercept:

$2(0) - 5y - 10 = 0$

$-5y - 10 = 0$

$-5y = 10$

$y = -2$

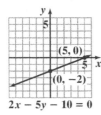

45. $2x - 10 = 0$

$2x = 10$

$x = 5$

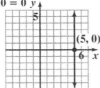

46. **a.** $m = \dfrac{11 - 2.3}{90 - 15} = \dfrac{8.7}{75} = 0.116$

$y - y_1 = m(x - x_1)$

$y - 11 = 0.116(x - 90)$

or

$y - 2.3 = 0.116(x - 15)$

b. $y - 11 = 0.116(x - 90)$

$y - 11 = 0.116x - 10.44$

$y = 0.116x + 0.56$

$f(x) = 0.116x + 0.56$

c. According to the graph, France has about 5 deaths per 100,000 persons.

d. $f(x) = 0.116x + 0.56$

$f(32) = 0.116(32) + 0.56$

$= 4.272$

≈ 4.3

According to the function, France has about 4.3 deaths per 100,000 persons. This underestimates the value in the graph by 0.7 deaths per 100,000 persons. The line passes below the point for France.

47. **a.** $m = \dfrac{52 - 64}{2010 - 1985} = \dfrac{-12}{25} = -0.48$

b. For each year from 1985 through 2010, the percentage of U.S. college freshmen rating their emotional health high or above average decreased by 0.48. The rate of change was -0.48% per year.

48. $\dfrac{f(x_2) - f(x_1)}{x_2 - x_1} = \dfrac{[9^2 - 4(9)] - [4^2 - 4 \cdot 5]}{9 - 5} = 10$

49.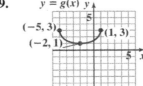

50.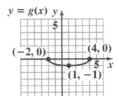

51.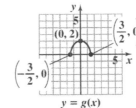

52.

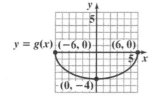

$y = g(x)$, $(-6, 0)$, $(6, 0)$, $(0, -4)$

53.

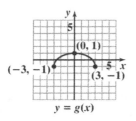

$(0, 1)$, $(-3, -1)$, $(3, -1)$, $y = g(x)$

54.

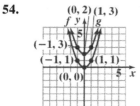

$(0, 2)$ $(1, 3)$, f g, $(-1, 3)$, $(-1, 1)$ $(1, 1)$, $(0, 0)$

55.

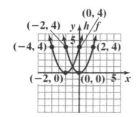

$(0, 4)$, $(-2, 4)$ h f, $(-4, 4)$ $(2, 4)$, $(-2, 0)$ $(0, 0)$

56.

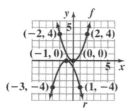

f, $(-2, 4)$ $(2, 4)$, $(-1, 0)$ $(0, 0)$, $(-3, -4)$ $(1, -4)$, r

57.

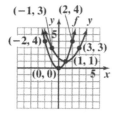

$(-1, 3)$ $(2, 4)$, f y, $(-2, 4)$, $(3, 3)$, $(1, 1)$, $(0, 0)$

58.

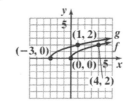

$(1, 2)$ g, $(-3, 0)$ f, $(0, 0)$, $(4, 2)$

59.

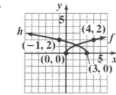

h, $(4, 2)$ f, $(-1, 2)$, $(0, 0)$, $(3, 0)$

60.

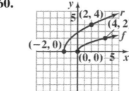

$(2, 4)$ r, $(4, 2)$, $(-2, 0)$ f, $(0, 0)$

61.

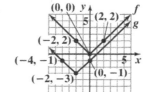

$(0, 0)$ f, $(2, 2)$ g, $(-2, 2)$, $(-4, -1)$, $(-2, -3)$ $(0, -1)$

62.

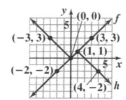

$(0, 0)$ f, $(-3, 3)$ $(3, 3)$, $(1, 1)$, $(-2, -2)$, $(4, -2)$ h

63.

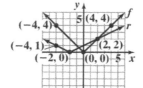

$(4, 4)$ f, $(-4, 4)$, $(-4, 1)$ r, $(2, 2)$, $(-2, 0)$ $(0, 0)$

64.

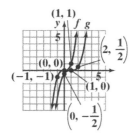

65.

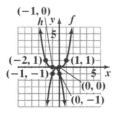

66.

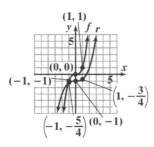

67.

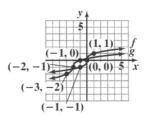

68.

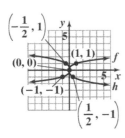

69.

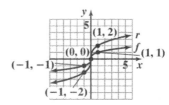

70. domain: $(-\infty, \infty)$

71. The denominator is zero when $x = 7$. The domain is $(-\infty, 7) \cup (7, \infty)$.

72. The expressions under each radical must not be negative.
$8 - 2x \geq 0$
$-2x \geq -8$
$x \leq 4$
domain: $(-\infty, 4]$.

73. The denominator is zero when $x = -7$ or $x = 3$.
domain: $(-\infty, -7) \cup (-7, 3) \cup (3, \infty)$

74. The expressions under each radical must not be negative. The denominator is zero when $x = 5$.
$x - 2 \geq 0$
$x \geq 2$
domain: $[2, 5) \cup (5, \infty)$

75. The expressions under each radical must not be negative.
$x - 1 \geq 0$ and $x + 5 \geq 0$
$x \geq 1$ $x \geq -5$
domain: $[1, \infty)$

76. $f(x) = 3x - 1;\ g(x) = x - 5$
$(f + g)(x) = 4x - 6$
domain: $(-\infty, \infty)$
$(f - g)(x) = (3x - 1) - (x - 5) = 2x + 4$
domain: $(-\infty, \infty)$
$(fg)(x) = (3x - 1)(x - 5) = 3x^2 - 16x + 5$
domain: $(-\infty, \infty)$
$\left(\dfrac{f}{g}\right)(x) = \dfrac{3x - 1}{x - 5}$
domain: $(-\infty, 5) \cup (5, \infty)$

77. $f(x) = x^2 + x + 1; g(x) = x^2 - 1$

$(f + g)(x) = 2x^2 + x$

domain: $(-\infty, \infty)$

$(f - g)(x) = (x^2 + x + 1) - (x^2 - 1) = x + 2$

domain: $(-\infty, \infty)$

$(fg)(x) = (x^2 + x + 1)(x^2 - 1)$

$= x^4 + x^3 - x - 1$

$\left(\dfrac{f}{g}\right)(x) = \dfrac{x^2 + x + 1}{x^2 - 1}$

domain: $(-\infty, -1) \cup (-1, 1) \cup (1, \infty)$

78. $f(x) = \sqrt{x + 7}; g(x) = \sqrt{x - 2}$

$(f + g)(x) = \sqrt{x + 7} + \sqrt{x - 2}$

domain: $[2, \infty)$

$(f - g)(x) = \sqrt{x + 7} - \sqrt{x - 2}$

domain: $[2, \infty)$

$(fg)(x) = \sqrt{x + 7} \cdot \sqrt{x - 2}$

$= \sqrt{x^2 + 5x - 14}$

domain: $[2, \infty)$

$\left(\dfrac{f}{g}\right)(x) = \dfrac{\sqrt{x + 7}}{\sqrt{x - 2}}$

domain: $(2, \infty)$

79. $f(x) = x^2 + 3; g(x) = 4x - 1$

    a. $(f \circ g)(x) = (4x - 1)^2 + 3$

        $= 16x^2 - 8x + 4$

    b. $(g \circ f)(x) = 4(x^2 + 3) - 1$

        $= 4x^2 + 11$

    c. $(f \circ g)(3) = 16(3)^2 - 8(3) + 4 = 124$

80. $f(x) = \sqrt{x}; \ g(x) = x + 1$

    a. $(f \circ g)(x) = \sqrt{x + 1}$

    b. $(g \circ f)(x) = \sqrt{x} + 1$

    c. $(f \circ g)(3) = \sqrt{3 + 1} = \sqrt{4} = 2$

81. **a.**

$(f \circ g)(x) = f\left(\dfrac{1}{x}\right)$

$= \dfrac{\dfrac{1}{x} + 1}{\dfrac{1}{x} - 2} = \dfrac{\left(\dfrac{1}{x} + 1\right)x}{\left(\dfrac{1}{x} - 2\right)x} = \dfrac{1 + x}{1 - 2x}$

    b. $\quad x \neq 0 \qquad\qquad 1 - 2x \neq 0$

$x \neq \dfrac{1}{2}$

$(-\infty, 0) \cup \left(0, \dfrac{1}{2}\right) \cup \left(\dfrac{1}{2}, \infty\right)$

82. **a.** $(f \circ g)(x) = f(x + 3) = \sqrt{x + 3 - 1} = \sqrt{x + 2}$

    b. $\quad x + 2 \geq 0$

$x \geq -2 \qquad [-2, \infty)$

83. $f(x) = x^4 \qquad g(x) = x^2 + 2x - 1$

84. $f(x) = \sqrt[3]{x} \qquad g(x) = 7x + 4$

85. $f(x) = \dfrac{3}{5}x + \dfrac{1}{2}; g(x) = \dfrac{5}{3}x - 2$

$f(g(x)) = \dfrac{3}{5}\left(\dfrac{5}{3}x - 2\right) + \dfrac{1}{2}$

$= x - \dfrac{6}{5} + \dfrac{1}{2}$

$= x - \dfrac{7}{10}$

$g(f(x)) = \dfrac{5}{3}\left(\dfrac{3}{5}x + \dfrac{1}{2}\right) - 2$

$= x + \dfrac{5}{6} - 2$

$= x - \dfrac{7}{6}$

f and *g* are not inverses of each other.

86. $f(x) = 2 - 5x; \; g(x) = \dfrac{2-x}{5}$

$f(g(x)) = 2 - 5\left(\dfrac{2-x}{5}\right)$

$\quad = 2 - (2 - x)$

$\quad = x$

$g(f(x)) = \dfrac{2-(2-5x)}{5} = \dfrac{5x}{5} = x$

f and *g* are inverses of each other.

87. a. $f(x) = 4x - 3$

$y = 4x - 3$

$x = 4y - 3$

$y = \dfrac{x+3}{4}$

$f^{-1}(x) = \dfrac{x+3}{4}$

b. $f(f^{-1}(x)) = 4\left(\dfrac{x+3}{4}\right) - 3$

$\quad = x + 3 - 3$

$\quad = x$

$f^{-1}(f(x)) = \dfrac{(4x-3)+3}{4} = \dfrac{4x}{4} = x$

88. a. $f(x) = 8x^3 + 1$

$y = 8x^3 + 1$

$x = 8y^3 + 1$

$x - 1 = 8y^3$

$\dfrac{x-1}{8} = y^3$

$\sqrt[3]{\dfrac{x-1}{8}} = y$

$\dfrac{\sqrt[3]{x-1}}{2} = y$

$f^{-1}(x) = \dfrac{\sqrt[3]{x-1}}{2}$

b. $f(f^{-1}(x)) = 8\left(\dfrac{\sqrt[3]{x-1}}{2}\right)^3 + 1$

$\quad = 8\left(\dfrac{x-1}{8}\right) + 1$

$\quad = x - 1 + 1$

$\quad = x$

$f^{-1}(f(x)) = \dfrac{\sqrt[3]{(8x^3+1)-1}}{2}$

$\quad = \dfrac{\sqrt[3]{8x^3}}{2}$

$\quad = \dfrac{2x}{2}$

$\quad = x$

89. a. $f(x) = \dfrac{2}{x} + 5$

$y = \dfrac{2}{x} + 5$

$x = \dfrac{2}{y} + 5$

$xy = 2 + 5y$

$xy - 5y = 2$

$y(x - 5) = 2$

$y = \dfrac{2}{x-5}$

$f^{-1}(x) = \dfrac{2}{x-5}$

b. $f(f^{-1}(x)) = \dfrac{2}{\frac{2}{x-5}} + 5$

$\quad = \dfrac{2(x-5)}{2} + 5$

$\quad = x - 5 + 5$

$\quad = x$

$f^{-1}(f(x)) = \dfrac{2}{\frac{2}{x}+5-5}$

$\quad = \dfrac{2}{\frac{2}{x}}$

$\quad = \dfrac{2x}{2}$

$\quad = x$

90. The inverse function exists.

91. The inverse function does not exist since it does not pass the horizontal line test.

92. The inverse function exists.

93. The inverse function does not exist since it does not pass the horizontal line test.

94.

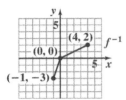

95. $f(x) = 1 - x^2$

$$y = 1 - x^2$$
$$x = 1 - y^2$$
$$y^2 = 1 - x$$
$$y = \sqrt{1-x}$$
$$f^{-1}(x) = \sqrt{1-x}$$

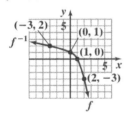

96. $f(x) = \sqrt{x} + 1$

$$y = \sqrt{x} + 1$$
$$x = \sqrt{y} + 1$$
$$x - 1 = \sqrt{y}$$
$$(x-1)^2 = y$$
$$f^{-1}(x) = (x-1)^2, \quad x \geq 1$$

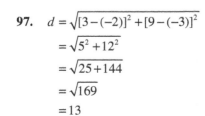

97. $d = \sqrt{[3-(-2)]^2 + [9-(-3)]^2}$

$$= \sqrt{5^2 + 12^2}$$
$$= \sqrt{25 + 144}$$
$$= \sqrt{169}$$
$$= 13$$

98. $d = \sqrt{[-2-(-4)]^2 + (5-3)^2}$

$$= \sqrt{2^2 + 2^2}$$
$$= \sqrt{4+4}$$
$$= \sqrt{8}$$
$$= 2\sqrt{2}$$
$$\approx 2.83$$

99. $\left(\dfrac{2+(-12)}{2}, \dfrac{6+4}{2}\right) = \left(\dfrac{-10}{2}, \dfrac{10}{2}\right) = (-5, 5)$

100. $\left(\dfrac{4+(-15)}{2}, \dfrac{-6+2}{2}\right) = \left(\dfrac{-11}{2}, \dfrac{-4}{2}\right) = \left(\dfrac{-11}{2}, -2\right)$

101. $x^2 + y^2 = 3^2$

$$x^2 + y^2 = 9$$

102. $(x-(-2))^2 + (y-4)^2 = 6^2$

$$(x+2)^2 + (y-4)^2 = 36$$

103. center: (0, 0); radius: 1

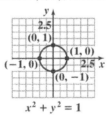

domain: $[-1, 1]$

range: $[-1, 1]$

104. center: (-2, 3); radius: 3

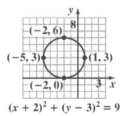

domain: $[-5, 1]$

range: $[0, 6]$

105.
$$x^2 + y^2 - 4x + 2y - 4 = 0$$
$$x^2 - 4x \quad + y^2 + 2y \quad = 4$$
$$x^2 - 4x + 4 + y^2 + 2y + 1 = 4 + 4 + 1$$
$$(x-2)^2 + (y+1)^2 = 9$$

center: (2, −1); radius: 3

$x^2 + y^2 - 4x + 2y - 4 = 0$

domain: $[-1, 5]$

range: $[-4, 2]$

Chapter 2 Test

1. (b), (c), and (d) are not functions.

2.
  a. $f(4) - f(-3) = 3 - (-2) = 5$

  b. domain: (−5, 6]

  c. range: [−4, 5]

  d. increasing: (−1, 2)

  e. decreasing: $(-5, -1)$ or $(2, 6)$

  f. $2, f(2) = 5$

  g. (−1, −4)

  h. *x*-intercepts: −4, 1, and 5.

  i. *y*-intercept: −3

3.
  a. −2, 2

  b. −1, 1

  c. 0

  d. even; $f(-x) = f(x)$

  e. no; f fails the horizontal line test

  f. $f(0)$ is a relative minimum.

g.

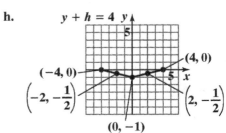

h.

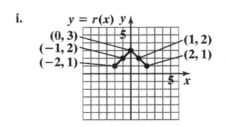

i.

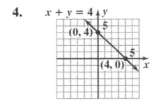

j. $\dfrac{f(x_2) - f(x_1)}{x_2 - x_1} = \dfrac{-1-0}{1-(-2)} = -\dfrac{1}{3}$

4.

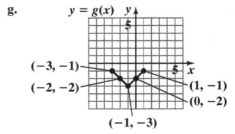

domain: $(-\infty, \infty)$

range: $(-\infty, \infty)$

5.

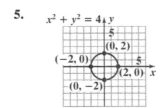

domain: $[-2, 2]$

range: $[-2, 2]$

6.

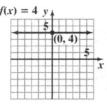

$f(x) = 4$

domain: $(-\infty, \infty)$

range: $\{4\}$

7.

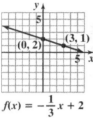

$f(x) = -\dfrac{1}{3}x + 2$

domain: $(-\infty, \infty)$

range: $(-\infty, \infty)$

8.

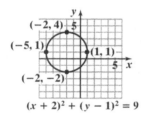

$(x + 2)^2 + (y - 1)^2 = 9$

domain: $[-5, 1]$

range: $[-2, 4]$

9.

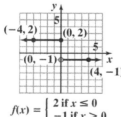

$f(x) = \begin{cases} 2 \text{ if } x \le 0 \\ -1 \text{ if } x > 0 \end{cases}$

domain: $(-\infty, \infty)$

range: $\{-1, 2\}$

10.

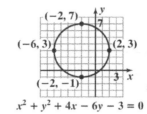

$x^2 + y^2 + 4x - 6y - 3 = 0$

domain: $[-6, 2]$

range: $[-1, 7]$

11.

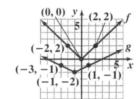

domain of f: $(-\infty, \infty)$

range of f: $[0, \infty)$

domain of g: $(-\infty, \infty)$

range of g: $[-2, \infty)$

12.

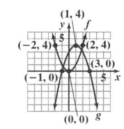

domain of f: $(-\infty, \infty)$

range of f: $[0, \infty)$

domain of g: $(-\infty, \infty)$

range of g: $(-\infty, 4]$

13.

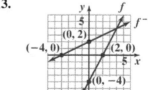

domain of f: $(-\infty, \infty)$

range of f: $(-\infty, \infty)$

domain of f^{-1}: $(-\infty, \infty)$

range of f^{-1}: $(-\infty, \infty)$

14.

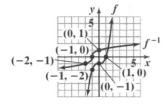

domain of f: $(-\infty, \infty)$

range of f: $(-\infty, \infty)$

domain of f^{-1}: $(-\infty, \infty)$

range of f^{-1}: $(-\infty, \infty)$

15.

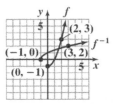

domain of f: $[0, \infty)$

range of f: $[-1, \infty)$

domain of f^{-1}: $[-1, \infty)$

range of f^{-1}: $[0, \infty)$

16. $f(x) = x^2 - x - 4$

$$f(x-1) = (x-1)^2 - (x-1) - 4$$
$$= x^2 - 2x + 1 - x + 1 - 4$$
$$= x^2 - 3x - 2$$

17. $\dfrac{f(x+h) - f(x)}{h}$

$$= \frac{(x+h)^2 - (x+h) - 4 - (x^2 - x - 4)}{h}$$
$$= \frac{x^2 + 2xh + h^2 - x - h - 4 - x^2 + x + 4}{h}$$
$$= \frac{2xh + h^2 - h}{h}$$
$$= \frac{h(2x + h - 1)}{h}$$
$$= 2x + h - 1$$

18. $(g - f)(x) = 2x - 6 - (x^2 - x - 4)$

$$= 2x - 6 - x^2 + x + 4$$
$$= -x^2 + 3x - 2$$

19. $\left(\dfrac{f}{g}\right)(x) = \dfrac{x^2 - x - 4}{2x - 6}$

domain: $(-\infty, 3) \cup (3, \infty)$

20. $(f \circ g)(x) = f(g(x))$

$$= (2x - 6)^2 - (2x - 6) - 4$$
$$= 4x^2 - 24x + 36 - 2x + 6 - 4$$
$$= 4x^2 - 26x + 38$$

21. $(g \circ f)(x) = g(f(x))$

$$= 2(x^2 - x - 4) - 6$$
$$= 2x^2 - 2x - 8 - 6$$
$$= 2x^2 - 2x - 14$$

22. $g(f(-1)) = 2((-1)^2 - (-1) - 4) - 6$

$$= 2(1 + 1 - 4) - 6$$
$$= 2(-2) - 6$$
$$= -4 - 6$$
$$= -10$$

23. $f(x) = x^2 - x - 4$

$$f(-x) = (-x)^2 - (-x) - 4$$
$$= x^2 + x - 4$$

f is neither even nor odd.

24. $m = \dfrac{-8 - 1}{-1 - 2} = \dfrac{-9}{-3} = 3$

point-slope form: $y - 1 = 3(x - 2)$
or $y + 8 = 3(x + 1)$
slope-intercept form: $y = 3x - 5$

25. $y = -\dfrac{1}{4}x + 5$ so $m = 4$

point-slope form: $y - 6 = 4(x + 4)$
slope-intercept form: $y = 4x + 22$

26. Write $4x + 2y - 5 = 0$ in slope intercept form.

$$4x + 2y - 5 = 0$$
$$2y = -4x + 5$$
$$y = -2x + \frac{5}{2}$$

The slope of the parallel line is –2, thus the slope of the desired line is $m = -2$.

$$y - y_1 = m(x - x_1)$$
$$y - (-10) = -2(x - (-7))$$
$$y + 10 = -2(x + 7)$$
$$y + 10 = -2x - 14$$
$$2x + y + 24 = 0$$

27. **a.** Find slope: $m = \dfrac{5870 - 4571}{4 - 1} = \dfrac{1299}{3} = 433$

point-slope form:
$$y - y_1 = m(x - x_1)$$
$$y - 4571 = 433(x - 1)$$

b. slope-intercept form:
$$y - 4571 = 433(x - 1)$$
$$y - 4571 = 433x - 433$$
$$y = 433x + 4138$$
$$f(x) = 433x + 4138$$

c. $f(x) = 433x + 4138$
$$= 433(10) + 4138$$
$$= 8468$$

According to the model, 8468 fatalities will involve distracted driving in 2014.

28. $\dfrac{3(10)^2 - 5 - [3(6)^2 - 5]}{10 - 6}$

$$= \frac{205 - 103}{4}$$

$$= \frac{192}{4}$$

$$= 48$$

29. $g(-1) = 3 - (-1) = 4$
$$g(7) = \sqrt{7 - 3} = \sqrt{4} = 2$$

30. The denominator is zero when $x = 1$ or $x = -5$.

domain: $(-\infty, -5) \cup (-5, 1) \cup (1, \infty)$

31. The expressions under each radical must not be negative.

$$x + 5 \geq 0 \quad \text{and} \quad x - 1 \geq 0$$
$$x \geq -5 \qquad\qquad x \geq 1$$

domain: $[1, \infty)$

32. $(f \circ g)(x) = \dfrac{7}{\dfrac{2}{x} - 4} = \dfrac{7x}{2 - 4x}$

$$x \neq 0, \quad 2 - 4x \neq 0$$
$$x \neq \frac{1}{2}$$

domain: $(-\infty, 0) \cup \left(0, \dfrac{1}{2}\right) \cup \left(\dfrac{1}{2}, \infty\right)$

33. $f(x) = x^7 \qquad g(x) = 2x + 3$

34. $d = \sqrt{(x_2 - x_1)^2 + (y_2 - y_1)^2}$

$$d = \sqrt{(x_2 - x_1)^2 + (y_2 - y_1)^2}$$
$$= \sqrt{(5 - 2)^2 + (2 - (-2))^2}$$
$$= \sqrt{3^2 + 4^2}$$
$$= \sqrt{9 + 16}$$
$$= \sqrt{25}$$
$$= 5$$

$$\left(\frac{x_1 + x_2}{2}, \frac{y_1 + y_2}{2}\right) = \left(\frac{2 + 5}{2}, \frac{-2 + 2}{2}\right)$$
$$= \left(\frac{7}{2}, 0\right)$$

The length is 5 and the midpoint is $\left(\dfrac{7}{2}, 0\right)$ or $(3.5, 0)$.

Cumulative Review Exercises (Chapters 1–2)

1. domain: $[0, 2)$

range: $[0, 2]$

2. $f(x) = 1$ at $\frac{1}{2}$ and $\frac{3}{2}$.

3. relative maximum: 2

4.

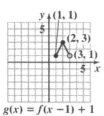

$g(x) = f(x-1) + 1$

5.

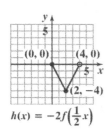

$h(x) = -2f\left(\frac{1}{2}x\right)$

6. $(x+3)(x-4) = 8$

$x^2 - x - 12 = 8$

$x^2 - x - 20 = 0$

$(x+4)(x-5) = 0$

$x + 4 = 0 \quad \text{or} \quad x - 5 = 0$

$x = -4 \quad \text{or} \quad\quad x = 5$

7. $3(4x - 1) = 4 - 6(x - 3)$

$12x - 3 = 4 - 6x + 18$

$18x = 25$

$x = \dfrac{25}{18}$

8. $\sqrt{x} + 2 = x$

$\sqrt{x} = x - 2$

$(\sqrt{x})^2 = (x - 2)^2$

$x = x^2 - 4x + 4$

$0 = x^2 - 5x + 4$

$0 = (x - 1)(x - 4)$

$x - 1 = 0 \quad \text{or} \quad x - 4 = 0$

$x = 1 \quad \text{or} \quad\quad x = 4$

A check of the solutions shows that $x = 1$ is an extraneous solution.

The solution set is $\{4\}$.

9. $x^{2/3} - x^{1/3} - 6 = 0$

Let $u = x^{1/3}$. Then $u^2 = x^{2/3}$.

$u^2 - u - 6 = 0$

$(u + 2)(u - 3) = 0$

$u = -2 \quad \text{or} \quad\quad u = 3$

$x^{1/3} = -2 \quad \text{or} \quad x^{1/3} = 3$

$x = (-2)^3 \quad \text{or} \quad\quad x = 3^3$

$x = -8 \quad \text{or} \quad\quad x = 27$

10. $\dfrac{x}{2} - 3 \le \dfrac{x}{4} + 2$

$4\left(\dfrac{x}{2} - 3\right) \le 4\left(\dfrac{x}{4} + 2\right)$

$2x - 12 \le x + 8$

$x \le 20$

The solution set is $(-\infty, 20]$.

11.

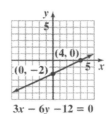

$3x - 6y - 12 = 0$

domain: $(-\infty, \infty)$

range: $(-\infty, \infty)$

12.

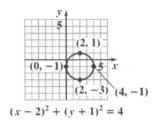

$(x - 2)^2 + (y + 1)^2 = 4$

domain: $[0, 4]$

range: $[-3, 1]$

13.

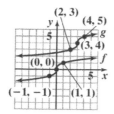

domain of f: $(-\infty, \infty)$

range of f: $(-\infty, \infty)$

domain of g: $(-\infty, \infty)$

range of g: $(-\infty, \infty)$

14.

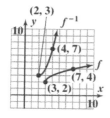

domain of f: $[3, \infty)$

range of f: $[2, \infty)$

domain of f^{-1}: $[2, \infty)$

range of f^{-1}: $[3, \infty)$

15. $\dfrac{f(x+h) - f(x)}{h}$

$= \dfrac{\left(4 - (x+h)^2\right) - \left(4 - x^2\right)}{h}$

$= \dfrac{4 - (x^2 + 2xh + h^2) - \left(4 - x^2\right)}{h}$

$= \dfrac{4 - x^2 - 2xh - h^2 - 4 + x^2}{h}$

$= \dfrac{-2xh - h^2}{h}$

$= \dfrac{h(-2x - h)}{h}$

$= -2x - h$

16. $(f \circ g)(x) = f\left(g(x)\right)$

$(f \circ g)(x) = f\left(x + 5\right)$

$0 = 4 - \left(x + 5\right)^2$

$0 = 4 - (x^2 + 10x + 25)$

$0 = 4 - x^2 - 10x - 25$

$0 = -x^2 - 10x - 21$

$0 = x^2 + 10x + 21$

$0 = (x + 7)(x + 3)$

The value of $(f \circ g)(x)$ will be 0 when $x = -3$ or $x = -7$.

17. $y = -\dfrac{1}{4}x + \dfrac{1}{3}$, so $m = 4$.

point-slope form: $y - 5 = 4(x + 2)$

slope-intercept form: $y = 4x + 13$

general form: $4x - y + 13 = 0$

18. $0.07x + 0.09(6000 - x) = 510$

$0.07x + 540 - 0.09x = 510$

$-0.02x = -30$

$x = 1500$

$6000 - x = 4500$

$1500 was invested at 7% and $4500 was invested at 9%.

19. $200 + 0.05x = .15x$

$200 = 0.10x$

$2000 = x$

For $2000 in sales, the earnings will be the same.

20. width = w

length = $2w + 2$

$2(2w + 2) + 2w = 22$

$4w + 4 + 2w = 22$

$6w = 18$

$w = 3$

$2w + 2 = 8$

The garden is 3 feet by 8 feet.

Chapter 3
Polynomial and Rational Functions

Section 3.1

Check Point Exercises

1. $f(x) = -(x-1)^2 + 4$

$$f(x) = \overset{a=-1}{-}\left(x - \overset{h=1}{1}\right)^2 + \overset{k=4}{4}$$

Step 1: The parabola opens down because $a < 0$.
Step 2: find the vertex: $(1, 4)$
Step 3: find the x-intercepts:

$$0 = -(x-1)^2 + 4$$
$$(x-1)^2 = 4$$
$$x-1 = \pm 2$$
$$x = 1 \pm 2$$
$$x = 3 \text{ or } x = -1$$

Step 4: find the y-intercept:

$$f(0) = -(0-1)^2 + 4 = 3$$

Step 5: The axis of symmetry is $x = 1$.

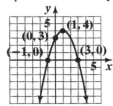

$f(x) = -(x - 1)^2 + 4$

2. $f(x) = (x-2)^2 + 1$

Step 1: The parabola opens up because $a > 0$.
Step 2: find the vertex: $(2, 1)$
Step 3: find the x-intercepts:

$$0 = (x-2)^2 + 1$$
$$(x-2)^2 = -1$$
$$x-2 = \sqrt{-1}$$
$$x = 2 \pm i$$

The equation has no real roots, thus the parabola has no x-intercepts.
Step 4: find the y-intercept:

$$f(0) = (0-2)^2 + 1 = 5$$

Step 5: The axis of symmetry is $x = 2$.

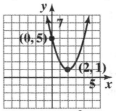

$f(x) = (x - 2)^2 + 1$

3. $f(x) = -x^2 + 4x + 1$

Step 1: The parabola opens down because $a < 0$.
Step 2: find the vertex:

$$x = -\frac{b}{2a} = -\frac{4}{2(-1)} = 2$$
$$f(2) = -2^2 + 4(2) + 1 = 5$$

The vertex is $(2, 5)$.
Step 3: find the x-intercepts:

$$0 = -x^2 + 4x + 1$$
$$x = \frac{-b \pm \sqrt{b^2 - 4ac}}{2a}$$
$$x = \frac{-4 \pm \sqrt{4^2 - 4(-1)(1)}}{2(-1)}$$
$$x = \frac{-4 \pm \sqrt{20}}{-2}$$
$$x = 2 \pm \sqrt{5}$$

The x-intercepts are $x \approx -0.2$ and $x \approx -4.2$.

Step 4: find the y-intercept: $f(0) = -0^2 + 4(0) + 1 = 1$

Step 5: The axis of symmetry is $x = 2$.

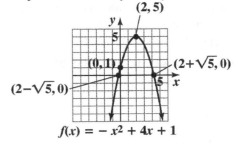

$f(x) = -x^2 + 4x + 1$

4. $f(x) = 4x^2 - 16x + 1000$

  a. $a = 4$. The parabola opens upward and has a minimum value.

  b. $x = \dfrac{-b}{2a} = \dfrac{16}{8} = 2$

    $f(2) = 4(2)^2 - 16(2) + 1000 = 984$

    The minimum point is 984 at $x = 2$.

  c. domain: $(-\infty, \infty)$ range: $[984, \infty)$

5. $f(x) = -0.005x^2 + 2x + 5$

  a. The information needed is found at the vertex.
    x-coordinate of vertex

    $x = \dfrac{-b}{2a} = \dfrac{-2}{2(-0.005)} = 200$

    y-coordinate of vertex

    $y = -0.005(200)^2 + 2(200) + 5 = 205$

    The vertex is (200,205).
    The maximum height of the arrow is 205 feet.
    This occurs 200 feet from its release.

  b. The arrow will hit the ground when the height reaches 0.

    $f(x) = -0.005x^2 + 2x + 5$

    $0 = -0.005x^2 + 2x + 5$

    $x = \dfrac{-b \pm \sqrt{b^2 - 4ac}}{2a}$

    $x = \dfrac{-2 \pm \sqrt{2^2 - 4(-0.005)(5)}}{2(-0.005)}$

    $x \approx -2$ or $x \approx 402$
    The arrow travels 402 feet before hitting the ground.

  c. The starting point occurs when $x = 0$. Find the corresponding y-coordinate.

    $f(x) = -0.005(0)^2 + 2(0) + 5 = 5$

    Plot $(0,5)$, $(402,0)$, and $(200,205)$, and connect them with a smooth curve.

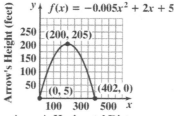

6. Let x = one of the numbers;
    $x - 8$ = the other number.
  The product is $f(x) = x(x - 8) = x^2 - 8x$
  The x-coordinate of the minimum is

  $x = -\dfrac{b}{2a} = -\dfrac{-8}{2(1)} = -\dfrac{-8}{2} = 4.$

  $f(4) = (4)^2 - 8(4)$
     $= 16 - 32 = -16$
  The vertex is $(4, -16)$.

  The minimum product is -16. This occurs when the two numbers are 4 and $4 - 8 = -4$.

7. Maximize the area of a rectangle constructed with 120 feet of fencing.
  Let x = the length of the rectangle. Let y = the width of the rectangle.
  Since we need an equation in one variable, use the perimeter to express y in terms of x.
  $2x + 2y = 120$

    $2y = 120 - 2x$

    $y = \dfrac{120 - 2x}{2} = 60 - x$

  We need to maximize $A = xy = x(60 - x)$. Rewrite A as a function of x.
  $A(x) = x(60 - x) = -x^2 + 60x$

  Since $a = -1$ is negative, we know the function opens downward and has a maximum at

  $x = -\dfrac{b}{2a} = -\dfrac{60}{2(-1)} = -\dfrac{60}{-2} = 30.$

  When the length x is 30, the width y is
  $y = 60 - x = 60 - 30 = 30$.
  The dimensions of the rectangular region with maximum area are 30 feet by 30 feet. This gives an area of $30 \cdot 30 = 900$ square feet.

Concept and Vocabulary Check 3.1

1. standard; parabola; (h, k); > 0; < 0

2. $-\dfrac{b}{2a}$; $f\left(-\dfrac{b}{2a}\right)$; $-\dfrac{b}{2a}$; $f\left(-\dfrac{b}{2a}\right)$

3. true

4. false

5. true

6. $x-8$; x^2-8x

7. $40-x$; $-x^2+40x$

Exercise Set 3.1

1.   vertex: (1, 1)
$$h(x)=(x-1)^2+1$$

3.   vertex: (1, −1)
$$j(x)=(x-1)^2-1$$

5.   The graph is $f(x)=x^2$ translated down one.
$$h(x)=x^2-1$$

7.   The point (1, 0) is on the graph and
$g(1)=0.$ $g(x)=x^2-2x+1$

9. $f(x)=2(x-3)^2+1$
$h=3, k=1$
The vertex is at (3, 1).

11. $f(x)=-2(x+1)^2+5$
$h=-1, k=5$
The vertex is at (−1, 5).

13. $f(x)=2x^2-8x+3$
$$x=\frac{-b}{2a}=\frac{8}{4}=2$$
$f(2)=2(2)^2-8(2)+3$
$=8-16+3=-5$
The vertex is at (2, −5).

15. $f(x)=-x^2-2x+8$
$$x=\frac{-b}{2a}=\frac{2}{-2}=-1$$
$f(-1)=-(-1)^2-2(-1)+8$
$=-1+2+8=9$
The vertex is at (−1, 9).

17. $f(x)=(x-4)^2-1$
vertex: (4, −1)
x-intercepts:
$$0=(x-4)^2-1$$
$$1=(x-4)^2$$
$\pm 1=x-4$
$x=3$ or $x=5$
y-intercept:
$$f(0)=(0-4)^2-1=15$$
The axis of symmetry is $x=4$.

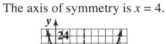

$f(x) = (x - 4)^2 - 1$

domain: $(-\infty,\infty)$

range: $[-1,\infty)$

19. $f(x)=(x-1)^2+2$
vertex: (1, 2)
x-intercepts:
$$0=(x-1)^2+2$$
$$(x-1)^2=-2$$
$x-1=\pm\sqrt{-2}$
$x=1\pm i\sqrt{2}$
No x-intercepts.
y-intercept:
$$f(0)=(0-1)^2+2=3$$
The axis of symmetry is $x=1$.

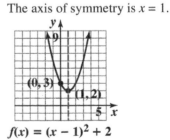

$f(x) = (x - 1)^2 + 2$

domain: $(-\infty,\infty)$

range: $[2,\infty)$

21. $y - 1 = (x-3)^2$

$y = (x-3)^2 + 1$

vertex: (3, 1)

x-intercepts:

$0 = (x-3)^2 + 1$

$(x-3)^2 = -1$

$x - 3 = \pm i$

$x = 3 \pm i$

No x-intercepts.

y-intercept: 10

$y = (0-3)^2 + 1 = 10$

The axis of symmetry is $x = 3$.

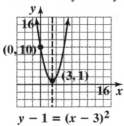

$y - 1 = (x - 3)^2$

domain: $(-\infty, \infty)$

range: $[1, \infty)$

23. $f(x) = 2(x+2)^2 - 1$

vertex: (−2, −1)

x-intercepts:

$0 = 2(x+2)^2 - 1$

$2(x+2)^2 = 1$

$(x+2)^2 = \dfrac{1}{2}$

$x + 2 = \pm \dfrac{1}{\sqrt{2}}$

$x = -2 \pm \dfrac{1}{\sqrt{2}} = -2 \pm \dfrac{\sqrt{2}}{2}$

y-intercept:

$f(0) = 2(0+2)^2 - 1 = 7$

The axis of symmetry is $x = -2$.

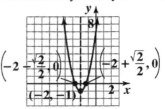

$\left(-2 - \dfrac{\sqrt{2}}{2}, 0\right)$ $\left(-2 + \dfrac{\sqrt{2}}{2}, 0\right)$

(−2, −1)

$f(x) = 2(x + 2)^2 - 1$

domain: $(-\infty, \infty)$

range: $[-1, \infty)$

25. $f(x) = 4 - (x-1)^2$

$f(x) = -(x-1)^2 + 4$

vertex: (1, 4)

x-intercepts:

$0 = -(x-1)^2 + 4$

$(x-1)^2 = 4$

$x - 1 = \pm 2$

$x = -1$ or $x = 3$

y-intercept:

$f(x) = -(0-1)^2 + 4 = 3$

The axis of symmetry is $x = 1$.

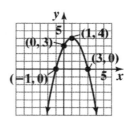

(0, 3) (1, 4)

(3, 0)

(−1, 0)

$f(x) = 4 - (x - 1)^2$

domain: $(-\infty, \infty)$

range: $(-\infty, 4]$

27. $f(x) = x^2 - 2x - 3$

$f(x) = (x^2 - 2x + 1) - 3 - 1$

$f(x) = (x-1)^2 - 4$

vertex: $(1, -4)$

x-intercepts:

$0 = (x-1)^2 - 4$

$(x-1)^2 = 4$

$x - 1 = \pm 2$

$x = -1$ or $x = 3$

y-intercept: -3

$f(0) = 0^2 - 2(0) - 3 = -3$

The axis of symmetry is $x = 1$.

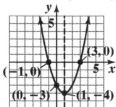

$f(x) = x^2 + 3x - 10$

domain: $(-\infty, \infty)$

range: $[-4, \infty)$

29. $f(x) = x^2 + 3x - 10$

$f(x) = \left(x^2 + 3x + \dfrac{9}{4} \right) - 10 - \dfrac{9}{4}$

$f(x) = \left(x + \dfrac{3}{2} \right)^2 - \dfrac{49}{4}$

vertex: $\left(-\dfrac{3}{2}, -\dfrac{49}{4} \right)$

x-intercepts:

$0 = \left(x + \dfrac{3}{2} \right)^2 - \dfrac{49}{4}$

$\left(x + \dfrac{3}{2} \right)^2 = \dfrac{49}{4}$

$x + \dfrac{3}{2} = \pm \dfrac{7}{2}$

$x = -\dfrac{3}{2} \pm \dfrac{7}{2}$

$x = 2$ or $x = -5$

y-intercept:

$f(x) = 0^2 + 3(0) - 10 = -10$

The axis of symmetry is $x = -\dfrac{3}{2}$.

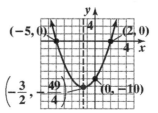

$f(x) = x^2 + 3x - 10$

domain: $(-\infty, \infty)$

range: $\left[-\dfrac{49}{4}, \infty \right)$

31. $f(x) = 2x - x^2 + 3$

$f(x) = -x^2 + 2x + 3$

$f(x) = -\left(x^2 - 2x + 1 \right) + 3 + 1$

$f(x) = -(x-1)^2 + 4$

vertex: $(1, 4)$

x-intercepts:

$0 = -(x-1)^2 + 4$

$(x-1)^2 = 4$

$x - 1 = \pm 2$

$x = -1$ or $x = 3$

y-intercept:

$f(0) = 2(0) - (0)^2 + 3 = 3$

The axis of symmetry is $x = 1$.

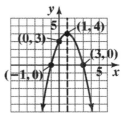

$f(x) = 2x - x^2 + 3$

domain: $(-\infty, \infty)$

range: $(-\infty, 4]$

33. $f(x) = x^2 + 6x + 3$

$f(x) = (x^2 + 6x + 9) + 3 - 9$

$f(x) = (x+3)^2 - 6$

vertex: $(-3, -6)$

x-intercepts:

$0 = (x+3)^2 - 6$

$(x+3)^2 = 6$

$x + 3 = \pm\sqrt{6}$

$x = -3 \pm \sqrt{6}$

y-intercept:

$f(0) = (0)^2 + 6(0) + 3$

$f(0) = 3$

The axis of symmetry is $x = -3$.

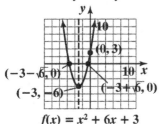

$f(x) = x^2 + 6x + 3$

domain: $(-\infty, \infty)$

range: $[-6, \infty)$

35. $f(x) = 2x^2 + 4x - 3$

$f(x) = 2(x^2 + 2x \quad) - 3$

$f(x) = 2(x^2 + 2x + 1) - 3 - 2$

$f(x) = 2(x+1)^2 - 5$

vertex: $(-1, -5)$

x-intercepts:

$0 = 2(x+1)^2 - 5$

$2(x+1)^2 = 5$

$(x+1)^2 = \dfrac{5}{2}$

$x + 1 = \pm\sqrt{\dfrac{5}{2}}$

$x = -1 \pm \dfrac{\sqrt{10}}{2}$

y-intercept:

$f(0) = 2(0)^2 + 4(0) - 3$

$f(0) = -3$

The axis of symmetry is $x = -1$.

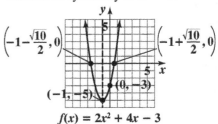

$f(x) = 2x^2 + 4x - 3$

domain: $(-\infty, \infty)$

range: $[-5, \infty)$

37. $f(x) = 2x - x^2 - 2$

$f(x) = -x^2 + 2x - 2$

$f(x) = -(x^2 - 2x + 1) - 2 + 1$

$f(x) = -(x-1)^2 - 1$

vertex: $(1, -1)$

x-intercepts:

$0 = -(x-1)^2 - 1$

$(x-1)^2 = -1$

$x - 1 = \pm i$

$x = 1 \pm i$

No x-intercepts.

y-intercept:

$f(0) = 2(0) - (0)^2 - 2 = -2$

The axis of symmetry is $x = 1$.

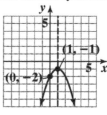

$f(x) = 2x - x^2 - 2$

domain: $(-\infty, \infty)$

range: $(-\infty, -1]$

39. $f(x) = 3x^2 - 12x - 1$

   a. $a = 3$. The parabola opens upward and has a minimum value.

   b. $x = \dfrac{-b}{2a} = \dfrac{12}{6} = 2$

       $f(2) = 3(2)^2 - 12(2) - 1$

       $= 12 - 24 - 1 = -13$

       The minimum is -13 at $x = 2$.

   c.   domain: $(-\infty, \infty)$ range: $[-13, \infty)$

41. $f(x) = -4x^2 + 8x - 3$

   a. $a = -4$. The parabola opens downward and has a maximum value.

   b. $x = \dfrac{-b}{2a} = \dfrac{-8}{-8} = 1$

       $f(1) = -4(1)^2 + 8(1) - 3$

       $= -4 + 8 - 3 = 1$

       The maximum is 1 at $x = 1$.

   c.   domain: $(-\infty, \infty)$ range: $(-\infty, 1]$

43. $f(x) = 5x^2 - 5x$

   a. $a = 5$. The parabola opens upward and has a minimum value.

   b. $x = \dfrac{-b}{2a} = \dfrac{5}{10} = \dfrac{1}{2}$

       $f\left(\dfrac{1}{2}\right) = 5\left(\dfrac{1}{2}\right)^2 - 5\left(\dfrac{1}{2}\right)$

       $= \dfrac{5}{4} - \dfrac{5}{2} = \dfrac{5}{4} - \dfrac{10}{4} = \dfrac{-5}{4}$

       The minimum is $\dfrac{-5}{4}$ at $x = \dfrac{1}{2}$.

   c.   domain: $(-\infty, \infty)$ range: $\left[\dfrac{-5}{4}, \infty\right)$

45. Since the parabola opens up, the vertex $(-1, -2)$ is a minimum point.

domain: $(-\infty, \infty)$. range: $[-2, \infty)$

47. Since the parabola has a maximum, it opens down from the vertex $(10, -6)$.

domain: $(-\infty, \infty)$. range: $(-\infty, -6]$

49. $(h, k) = (5, 3)$

$f(x) = 2(x - h)^2 + k = 2(x - 5)^2 + 3$

51. $(h, k) = (-10, -5)$

$f(x) = 2(x - h)^2 + k$

     $= 2[x - (-10)]^2 + (-5)$

     $= 2(x + 10)^2 - 5$

53. Since the vertex is a maximum, the parabola opens down and $a = -3$.

$(h, k) = (-2, 4)$

$f(x) = -3(x - h)^2 + k$

     $= -3[x - (-2)]^2 + 4$

     $= -3(x + 2)^2 + 4$

55. Since the vertex is a minimum, the parabola opens up and $a = 3$.

$(h, k) = (11, 0)$

$f(x) = 3(x - h)^2 + k$

     $= 3(x - 11)^2 + 0$

     $= 3(x - 11)^2$

57.   **a.** $y = -0.01x^2 + 0.7x + 6.1$

       $a = -0.01$, $b = 0.7$, $c = 6.1$

       x-coordinate of vertex

       $= \dfrac{-b}{2a} = \dfrac{-0.7}{2(-0.01)} = 35$

       y-coordinate of vertex

       $y = -0.01x^2 + 0.7x + 6.1$

       $y = -0.01(35)^2 + 0.7(35) + 6.1 = 18.35$

       The maximum height of the shot is about 18.35 feet. This occurs 35 feet from its point of release.

b. The ball will reach the maximum horizontal distance when its height returns to 0.

$$y = -0.01x^2 + 0.7x + 6.1$$

$$0 = -0.01x^2 + 0.7x + 6.1$$

$$a = -0.01, \ b = 0.7, \ c = 6.1$$

$$x = \frac{-b \pm \sqrt{b^2 - 4ac}}{2a}$$

$$x = \frac{-0.7 \pm \sqrt{0.7^2 - 4(-0.01)(6.1)}}{2(-0.01)}$$

$$x \approx 77.8 \text{ or } x \approx -7.8$$

The maximum horizontal distance is 77.8 feet.

c. The initial height can be found at $x = 0$.

$$y = -0.01x^2 + 0.7x + 6.1$$

$$y = -0.01(0)^2 + 0.7(0) + 6.1 = 6.1$$

The shot was released at a height of 6.1 feet.

59. $y = -0.8x^2 + 2.4x + 6$

a. The information needed is found at the vertex.
x-coordinate of vertex

$$x = \frac{-b}{2a} = \frac{-2.4}{2(-0.8)} = 1.5$$

y-coordinate of vertex

$$y = -0.8(1.5)^2 + 2.4(1.5) + 6 = 7.8$$

The vertex is (1.5, 7.8).
The maximum height of the ball is 7.8 feet.
This occurs 1.5 feet from its release.

b. The ball will hit the ground when the height reaches 0.

$$y = -0.8x^2 + 2.4x + 6$$

$$0 = -0.8x^2 + 2.4x + 6$$

$$x = \frac{-b \pm \sqrt{b^2 - 4ac}}{2a}$$

$$x = \frac{-2.4 \pm \sqrt{2.4^2 - 4(-0.8)(6)}}{2(-0.8)}$$

$$x \approx -1.6 \text{ or } x \approx 4.6$$

The ball travels 4.6 feet before hitting the ground.

c. The starting point occurs when $x = 0$. Find the corresponding y-coordinate.

$$y = -0.8(0)^2 + 2.4(0) + 6 = 6$$

Plot $(0, 6)$, $(1.5, 7.8)$, and $(4.7, 0)$, and connect them with a smooth curve.

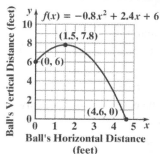

61. Let x = one of the numbers;
$16 - x$ = the other number.

The product is $f(x) = x(16 - x)$
$$= 16x - x^2 = -x^2 + 16x$$

The x-coordinate of the maximum is

$$x = -\frac{b}{2a} = -\frac{16}{2(-1)} = -\frac{16}{-2} = 8.$$

$$f(8) = -8^2 + 16(8) = -64 + 128 = 64$$

The vertex is (8, 64). The maximum product is 64.
This occurs when the two numbers are 8 and $16 - 8 = 8$.

63. Let x = one of the numbers;
$x - 16$ = the other number.

The product is $f(x) = x(x - 16) = x^2 - 16x$

The x-coordinate of the minimum is

$$x = -\frac{b}{2a} = -\frac{-16}{2(1)} = -\frac{-16}{2} = 8.$$

$$f(8) = (8)^2 - 16(8) = 64 - 128 = -64$$

The vertex is $(8, -64)$. The minimum product is -64.
This occurs when the two numbers are 8 and $8 - 16 = -8$.

65. Maximize the area of a rectangle constructed along a river with 600 feet of fencing.
Let x = the width of the rectangle;
$600 - 2x$ = the length of the rectangle
We need to maximize.

$A(x) = x(600 - 2x)$

$\qquad = 600x - 2x^2 = -2x^2 + 600x$

Since $a = -2$ is negative, we know the function opens downward and has a maximum at

$$x = -\frac{b}{2a} = -\frac{600}{2(-2)} = -\frac{600}{-4} = 150.$$

When the width is $x = 150$ feet, the length is
$600 - 2(150) = 600 - 300 = 300$ feet.

The dimensions of the rectangular plot with maximum area are 150 feet by 300 feet. This gives an area of $150 \cdot 300 = 45,000$ square feet.

67. Maximize the area of a rectangle constructed with 50 yards of fencing.
Let x = the length of the rectangle. Let y = the width of the rectangle.
Since we need an equation in one variable, use the perimeter to express y in terms of x.
$2x + 2y = 50$

$\qquad 2y = 50 - 2x$

$\qquad y = \dfrac{50 - 2x}{2} = 25 - x$

We need to maximize $A = xy = x(25 - x)$. Rewrite A as a function of x.

$A(x) = x(25 - x) = -x^2 + 25x$

Since $a = -1$ is negative, we know the function opens downward and has a maximum at

$$x = -\frac{b}{2a} = -\frac{25}{2(-1)} = -\frac{25}{-2} = 12.5.$$

When the length x is 12.5, the width y is
$y = 25 - x = 25 - 12.5 = 12.5$.

The dimensions of the rectangular region with maximum area are 12.5 yards by 12.5 yards. This gives an area of $12.5 \cdot 12.5 = 156.25$ square yards.

69. Maximize the area of the playground with 600 feet of fencing.
Let x = the length of the rectangle. Let y = the width of the rectangle.
Since we need an equation in one variable, use the perimeter to express y in terms of x.
$2x + 3y = 600$

$\qquad 3y = 600 - 2x$

$\qquad y = \dfrac{600 - 2x}{3}$

$\qquad y = 200 - \dfrac{2}{3}x$

We need to maximize $A = xy = x\left(200 - \dfrac{2}{3}x\right)$.

Rewrite A as a function of x.

$A(x) = x\left(200 - \dfrac{2}{3}x\right) = -\dfrac{2}{3}x^2 + 200x$

Since $a = -\dfrac{2}{3}$ is negative, we know the function opens downward and has a maximum at

$$x = -\frac{b}{2a} = -\frac{200}{2\left(-\dfrac{2}{3}\right)} = -\frac{200}{-\dfrac{4}{3}} = 150.$$

When the length x is 150, the width y is
$y = 200 - \dfrac{2}{3}x = 200 - \dfrac{2}{3}(150) = 100$.

The dimensions of the rectangular playground with maximum area are 150 feet by 100 feet. This gives an area of $150 \cdot 100 = 15,000$ square feet.

71. Maximize the cross-sectional area of the gutter:
$A(x) = x(20 - 2x)$

$\qquad = 20x - 2x^2 = -2x^2 + 20x.$

Since $a = -2$ is negative, we know the function opens downward and has a maximum at

$$x = -\frac{b}{2a} = -\frac{20}{2(-2)} = -\frac{20}{-4} = 5.$$

When the height x is 5, the width is
$20 - 2x = 20 - 2(5) = 20 - 10 = 10$.

$A(5) = -2(5)^2 + 20(5)$

$\qquad = -2(25) + 100 = -50 + 100 = 50$

The maximum cross-sectional area is 50 square inches. This occurs when the gutter is 5 inches deep and 10 inches wide.

73. a. $C(x) = 0.55x + 525$

b. $P(x) = R - C$

$$P(x) = -0.001x^2 + 3x - 0.55x - 525$$

$$P(x) = -0.001x^2 + 2.45x - 525$$

c. $x = \dfrac{-2.45}{2(-0.001)} = 1225$

$$P(x) = -0.001(1225)^2 + 2.45(1225) - 525$$

$$= 975.63$$

The maximum profit will be $975.63 per week obtained by selling 1225 sandwiches.

75. – 79. Answers will vary.

81. $y = 2x^2 - 82x + 720$

a.
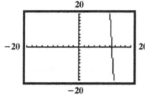

You can only see a little of the parabola.

b. $a = 2; b = -82$

$$x = -\frac{b}{2a} = -\frac{-82}{4} = 20.5$$

$$y = 2(20.5)^2 - 82(20.5) + 720$$

$$= 840.5 - 1681 + 720$$

$$= -120.5$$

vertex: $(20.5, -120.5)$

c. Ymax = 750

d. You can choose Xmin and Xmax so the x-value of the vertex is in the center of the graph. Choose Ymin to include the y-value of the vertex.

83. $y = -4x^2 + 20x + 160$

$$x = \frac{-b}{2a} = \frac{-20}{-8} = 2.5$$

$$y = -4(2.5)^2 + 20(2.5) + 160$$

$$= -2.5 + 50 + 160 = 185$$

The vertex is at $(2.5, 185)$.

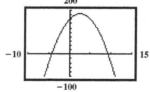

85. $y = 0.01x^2 + 0.6x + 100$

$$x = \frac{-b}{2a} = \frac{-0.6}{0.02} = -30$$

$$y = 0.01(-30)^2 + 0.6(-30) + 100$$

$$= 9 - 18 + 100 = 91$$

The vertex is at $(-30, 91)$.

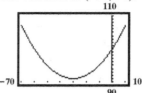

87. does not make sense; Explanations will vary. Sample explanation: Some parabolas have the y-axis as the axis of symmetry.

89. does not make sense; Explanations will vary. Sample explanation: If it is thrown vertically, its path will be a line segment.

91. true

93. false; Changes to make the statement true will vary. A sample change is: The graph has no x–intercepts. To find x–intercepts, set $y = 0$ and solve for x.

$$0 = -2(x + 4)^2 - 8$$

$$2(x + 4)^2 = -8$$

$$(x + 4)^2 = -4$$

Because the solutions to the equation are imaginary, we know that there are no x–intercepts.

95. $f(x) = 3(x + 2)^2 - 5; (-1, -2)$
axis: $x = -2$
$(-1, -2)$ is one unit right of $(-2, -2)$. One unit left of $(-2, -2)$ is $(-3, -2)$.
point: $(-3, -2)$

97. We start with the form $f(x) = a(x - h)^2 + k$.
Since we know the vertex is $(h, k) = (-3, -4)$, we have $f(x) = a(x + 3)^2 - 4$. We also know that the graph passes through the point $(1, 4)$, which allows us to solve for a.

$$4 = a(1 + 3)^2 - 4$$

$$8 = a(4)^2$$

$$8 = 16a$$

$$\frac{1}{2} = a$$

Therefore, the function is $f(x) = \dfrac{1}{2}(x + 3)^2 - 4$.

99. $3x + 4y = 1000$

$$4y = 1000 - 3x$$

$$y = 250 - 0.75x$$

$$A(x) = x(250 - 0.75x)$$

$$= 250x - 0.75x^2$$

$$x = \frac{-b}{2a} = \frac{-250}{-1.5} = 166\frac{2}{3}$$

$$y = 250 - 0.75\left(166\frac{2}{3}\right) = 125$$

The dimensions are $x = 166\frac{2}{3}$ ft, $y = 125$ ft.

The maximum area is about 20,833 ft^2.

101. Answers will vary.

102. $x^3 + 3x^2 - x - 3 = x^2(x+3) - 1(x+3)$

$$= (x+3)(x^2 - 1)$$

$$= (x+3)(x+1)(x-1)$$

103. $f(x) = x^3 - 2x - 5$

$$f(2) = (2)^3 - 2(2) - 5 = -1$$

$$f(3) = (3)^3 - 2(3) - 5 = 16$$

The graph passes through (2, –1), which is below the x-axis, and (3, 16), which is above the x-axis. Since the graph of f is continuous, it must cross the x-axis somewhere between 2 and 3 to get from one of these points to the other.

104. $f(x) = x^4 - 2x^2 + 1$

$$f(-x) = (-x)^4 - 2(-x)^2 + 1$$

$$= x^4 - 2x^2 + 1$$

Since $f(-x) = f(x)$, the function is even.

Thus, the graph is symmetric with respect to the y-axis.

Section 3.2

Check Point Exercises

1. Since n is even and $a_n > 0$, the graph rises to the left and to the right.

2. It is not necessary to multiply out the polynomial to determine its degree. We can find the degree of the polynomial by adding the degrees of each of its factors. $f(x) = 2 \overbrace{x^3}^{\text{degree 3}} \overbrace{(x-1)}^{\text{degree 1}} \overbrace{(x+5)}^{\text{degree 1}}$ has degree $3 + 1 + 1 = 5$.

$f(x) = 2x^3(x-1)(x+5)$ is of odd degree with a positive leading coefficient. Thus, the graph falls to the left and rises to the right.

3. Since n is odd and the leading coefficient is negative, the function falls to the right. Since the ratio cannot be negative, the model won't be appropriate.

4. The graph does not show the function's end behavior. Since $a_n > 0$ and n is odd, the graph should fall to the left but doesn't appear to do so.

5. $f(x) = x^3 + 2x^2 - 4x - 8$

$$0 = x^2(x+2) - 4(x+2)$$

$$0 = (x+2)(x^2 - 4)$$

$$0 = (x+2)^2(x-2)$$

$$x = -2 \text{ or } x = 2$$

The zeros are –2 and 2.

6. $f(x) = x^4 - 4x^2$

$$x^4 - 4x^2 = 0$$

$$x^2(x^2 - 4) = 0$$

$$x^2(x+2)(x-2) = 0$$

$$x = 0 \text{ or } x = -2 \text{ or } x = 2$$

The zeros are –2, 0, and 2.

7. $f(x) = -4\left(x + \dfrac{1}{2}\right)^2 (x-5)^3$

$-4\left(x + \dfrac{1}{2}\right)^2 (x-5)^3 = 0$

$x = -\dfrac{1}{2}$ or $x = 5$

The zeros are $-\dfrac{1}{2}$, with multiplicity 2, and 5, with multiplicity 3.

Because the multiplicity of $-\dfrac{1}{2}$ is even, the graph touches the *x*-axis and turns around at this zero. Because the multiplicity of 5 is odd, the graph crosses the *x*-axis at this zero.

8. $f(x) = 3x^3 - 10x + 9$

$f(-3) = 3(-3)^3 - 10(-3) + 9 = -42$

$f(-2) = 3(-2)^3 - 10(-2) + 9 = 5$

The sign change shows there is a zero between -3 and -2.

9. $f(x) = x^3 - 3x^2$

Since $a_n > 0$ and *n* is odd, the graph falls to the left and rises to the right.

$x^3 - 3x^2 = 0$

$x^2(x-3) = 0$

$x = 0$ or $x = 3$

The *x*-intercepts are 0 and 3.

$f(0) = 0^3 - 3(0)^2 = 0$

The *y*-intercept is 0.

$f(-x) = (-x)^3 - 3(-x)^2 = -x^3 - 3x^2$

No symmetry.

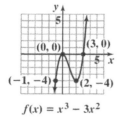

$f(x) = x^3 - 3x^2$

10. $f(x) = 2(x+2)^2 (x-3)$

The leading term is $2 \cdot x^2 \cdot x$, or $2x^3$.

Since $a_n > 0$ and *n* is odd, the graph falls to the left and rises to the right.

$2(x+2)^2(x-3) = 0$

$x = -2$ or $x = 3$

The *x*-intercepts are -2 and 3.

$f(0) = 2(0+2)^2(0-3) = -12$

The *y*-intercept is -12.

$f(-x) = 2\left((-x) + 2\right)^2 \left((-x) - 3\right)$

$\qquad = 2\left(-x + 2\right)^2 \left(-x - 3\right)$

No symmetry.

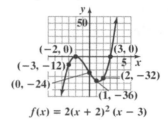

$f(x) = 2(x + 2)^2 (x - 3)$

Concept and Vocabulary Check 3.2

1. 5; -2

2. false

3. end; leading

4. falls; rises

5. rises; falls

6. rises; rises

7. falls; falls

8. true

9. true

10. *x*-intercept

11. turns around; crosses

12. 0; Intermediate Value

13. $n-1$

Exercise Set 3.2

1. polynomial function;
degree: 3

3. polynomial function;
degree: 5

5. not a polynomial function

7. not a polynomial function

9. not a polynomial function

11. polynomial function

13. Not a polynomial function because graph is not continuous.

15. (b)

17. (a)

19. $f(x) = 5x^3 + 7x^2 - x + 9$
Since $a_n > 0$ and n is odd, the graph of $f(x)$ falls to the left and rises to the right.

21. $f(x) = 5x^4 + 7x^2 - x + 9$
Since $a_n > 0$ and n is even, the graph of $f(x)$ rises to the left and to the right.

23. $f(x) = -5x^4 + 7x^2 - x + 9$
Since $a_n < 0$ and n is even, the graph of $f(x)$ falls to the left and to the right.

25. $f(x) = 2(x-5)(x+4)^2$
$x = 5$ has multiplicity 1;
The graph crosses the x-axis.
$x = -4$ has multiplicity 2;
The graph touches the x-axis and turns around.

27. $f(x) = 4(x-3)(x+6)^3$
x = 3 has multiplicity 1;
The graph crosses the x-axis.
$x = -6$ has multiplicity 3;
The graph crosses the x-axis.

29. $f(x) = x^3 - 2x^2 + x$
$$= x\left(x^2 - 2x + 1\right)$$
$$= x(x-1)^2$$
$x = 0$ has multiplicity 1;
The graph crosses the x-axis.
$x = 1$ has multiplicity 2;
The graph touches the x-axis and turns around.

31. $f(x) = x^3 + 7x^2 - 4x - 28$
$$= x^2(x+7) - 4(x+7)$$
$$= \left(x^2 - 4\right)(x+7)$$
$$= (x-2)(x+2)(x+7)$$
$x = 2$, $x = -2$ and $x = -7$ have multiplicity 1;
The graph crosses the x-axis.

33. $f(x) = x^3 - x - 1$
$f(1) = -1$
$f(2) = 5$
The sign change shows there is a zero between the given values.

35. $f(x) = 2x^4 - 4x^2 + 1$
$f(-1) = -1$
$f(0) = 1$
The sign change shows there is a zero between the given values.

37. $f(x) = x^3 + x^2 - 2x + 1$
$f(-3) = -11$
$f(-2) = 1$
The sign change shows there is a zero between the given values.

39. $f(x) = 3x^3 - 10x + 9$
$f(-3) = -42$
$f(-2) = 5$
The sign change shows there is a zero between the given values.

41. $f(x) = x^3 + 2x^2 - x - 2$

   a. Since $a_n > 0$ and n is odd, $f(x)$ rises to the right and falls to the left.

   b. $$x^3 + 2x^2 - x - 2 = 0$$
$$x^2(x+2) - (x+2) = 0$$
$$(x+2)(x^2 - 1) = 0$$
$$(x+2)(x-1)(x+1) = 0$$
$$x = -2, x = 1, x = -1$$
The zeros at -2, -1, and 1 have odd multiplicity so $f(x)$ crosses the x-axis at these points.

   c. $f(0) = (0)^3 + 2(0)^2 - 0 - 2$
$$= -2$$
The y-intercept is -2.

d. $f(-x) = (-x) + 2(-x)^2 - (-x) - 2$

$\qquad = -x^3 + 2x^2 + x - 2$

$-f(x) = -x^3 - 2x^2 + x + 2$

The graph has neither origin symmetry nor *y*-axis symmetry.

e. The graph has 2 turning points and $2 \le 3 - 1$.

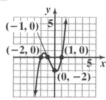

$f(x) = x^3 + 2x^2 - x - 2$

43. $f(x) = x^4 - 9x^2$

a. Since $a_n > 0$ and *n* is even, *f*(*x*) rises to the left and the right.

b. $x^4 - 9x^2 = 0$

$x^2(x^2 - 9) = 0$

$x^2(x - 3)(x + 3) = 0$

$x = 0, x = 3, x = -3$

The zeros at –3 and 3 have odd multiplicity, so *f*(*x*) crosses the *x*-axis at these points. The root at 0 has even multiplicity, so *f*(*x*) touches the *x*-axis at 0.

c. $f(0) = (0)^4 - 9(0)^2 = 0$

The *y*-intercept is 0.

d. $f(-x) = x^4 - 9x^2$

$f(-x) = f(x)$

The graph has *y*-axis symmetry.

e. The graph has 3 turning points and $3 \le 4 - 1$.

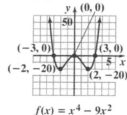

$f(x) = x^4 - 9x^2$

45. $f(x) = -x^4 + 16x^2$

a. Since $a_n < 0$ and *n* is even, *f*(*x*) falls to the left and the right.

b. $-x^4 + 16x^2 = 0$

$x^2(-x^2 + 16) = 0$

$x^2(4 - x)(4 + x) = 0$

$x = 0, x = 4, x = -4$

The zeros at –4 and 4 have odd multiplicity, so *f*(*x*) crosses the *x*-axis at these points. The root at 0 has even multiplicity, so *f*(*x*) touches the *x*-axis at 0.

c. $f(0) = (0)^4 - 9(0)^2 = 0$

The *y*-intercept is 0.

d. $f(-x) = -x^4 + 16x^2$

$f(-x) = f(x)$

The graph has *y*-axis symmetry.

e. The graph has 3 turning points and $3 \le 4 - 1$.

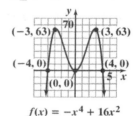

$f(x) = -x^4 + 16x^2$

47. $f(x) = x^4 - 2x^3 + x^2$

a. Since $a_n > 0$ and *n* is even, *f*(*x*) rises to the left and the right.

b. $x^4 - 2x^3 + x^2 = 0$

$x^2(x^2 - 2x + 1) = 0$

$x^2(x - 1)(x - 1) = 0$

$x = 0, x = 1$

The zeros at 1 and 0 have even multiplicity, so *f*(*x*) touches the *x*-axis at 0 and 1.

c. $f(0) = (0)^4 - 2(0)^3 + (0)^2 = 0$

The *y*-intercept is 0.

d. $f(-x) = x^4 + 2x^3 + x^2$

The graph has neither *y*-axis nor origin symmetry.

e. The graph has 3 turning points and $3 \le 4 - 1$.

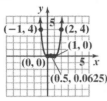

$$f(x) = x^4 - 2x^3 + x^2$$

49. $f(x) = -2x^4 + 4x^3$

a. Since $a_n < 0$ and n is even, $f(x)$ falls to the left and the right.

b. $-2x^4 + 4x^3 = 0$
$x^3(-2x + 4) = 0$
$x = 0, x = 2$
The zeros at 0 and 2 have odd multiplicity, so $f(x)$ crosses the x-axis at these points.

c. $f(0) = -2(0)^4 + 4(0)^3 = 0$
The y-intercept is 0.

d. $f(-x) = -2x^4 - 4x^3$
The graph has neither y-axis nor origin symmetry.

e. The graph has 1 turning point and $1 \le 4 - 1$.

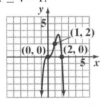

$$f(x) = -2x^4 + 4x^3$$

51. $f(x) = 6x^3 - 9x - x^5$

a. Since $a_n < 0$ and n is odd, $f(x)$ rises to the left and falls to the right.

b. $-x^5 + 6x^3 - 9x = 0$
$-x(x^4 - 6x^2 + 9) = 0$
$-x(x^2 - 3)(x^2 - 3) = 0$
$x = 0, x = \pm\sqrt{3}$
The root at 0 has odd multiplicity so $f(x)$ crosses the x-axis at $(0, 0)$. The zeros at $-\sqrt{3}$ and $\sqrt{3}$ have even multiplicity so $f(x)$ touches the x-axis at $\sqrt{3}$ and $-\sqrt{3}$.

c. $f(0) = -(0)^5 + 6(0)^3 - 9(0) = 0$
The y-intercept is 0.

d. $f(-x) = x^5 - 6x^3 + 9x$
$f(-x) = -f(x)$
The graph has origin symmetry.

e. The graph has 4 turning points and $4 \le 5 - 1$.

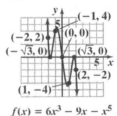

$$f(x) = 6x^3 - 9x - x^5$$

53. $f(x) = 3x^2 - x^3$

a. Since $a_n < 0$ and n is odd, $f(x)$ rises to the left and falls to the right.

b. $-x^3 + 3x^2 = 0$
$-x^2(x - 3) = 0$
$x = 0, x = 3$
The zero at 3 has odd multiplicity so $f(x)$ crosses the x-axis at that point. The root at 0 has even multiplicity so $f(x)$ touches the axis at $(0, 0)$.

c. $f(0) = -(0)^3 + 3(0)^2 = 0$
The y-intercept is 0.

d. $f(-x) = x^3 + 3x^2$
The graph has neither y-axis nor origin symmetry.

e. The graph has 2 turning points and $2 \le 3 - 1$.

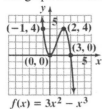

$$f(x) = 3x^2 - x^3$$

55. $f(x) = -3(x-1)^2 (x^2 - 4)$

a. Since $a_n < 0$ and n is even, $f(x)$ falls to the left and the right.

b. $-3(x-1)^2 (x^2 - 4) = 0$

$x = 1, x = -2, x = 2$
The zeros at −2 and 2 have odd multiplicity, so $f(x)$ crosses the x-axis at these points. The root at 1 has even multiplicity, so $f(x)$ touches the x-axis at (1, 0).

c. $f(0) = -3(0-1)^2 (0^2 - 4)^3$
$\quad = -3(1)(-4) = 12$
The y-intercept is 12.

d. $f(-x) = -3(-x-1)^2 (x^2 - 4)$
The graph has neither y-axis nor origin symmetry.

e. The graph has 1 turning point and $1 \le 4 - 1$.

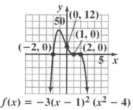

$$f(x) = -3(x-1)^2 (x^2 - 4)$$

57. $f(x) = x^2 (x-1)^3 (x+2)$

a. Since $a_n > 0$ and n is even, $f(x)$ rises to the left and the right.

b. $x = 0, x = 1, x = -2$
The zeros at 1 and −2 have odd multiplicity so $f(x)$ crosses the x-axis at those points. The root at 0 has even multiplicity so $f(x)$ touches the axis at (0, 0).

c. $f(0) = 0^2 (0-1)^3 (0+2) = 0$
The y-intercept is 0.

d. $f(-x) = x^2 (-x-1)^3 (-x+2)$
The graph has neither y-axis nor origin symmetry.

e. The graph has 2 turning points and $2 \le 6 - 1$.

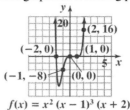

$$f(x) = x^2 (x-1)^3 (x+2)$$

59. $f(x) = -x^2 (x-1)(x+3)$

a. Since $a_n < 0$ and n is even, $f(x)$ falls to the left and the right.

b. $x = 0, x = 1, x = -3$
The zeros at 1 and −3 have odd multiplicity so $f(x)$ crosses the x-axis at those points. The root at 0 has even multiplicity so $f(x)$ touches the axis at (0, 0).

c. $f(0) = -0^2 (0-1)(0+3) = 0$
The y-intercept is 0.

d. $f(-x) = -x^2 (-x-1)(-x+3)$
The graph has neither y-axis nor origin symmetry.

e. The graph has 3 turning points and $3 \le 4 - 1$.

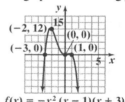

$$f(x) = -x^2 (x-1)(x+3)$$

61. $f(x) = -2x^3 (x-1)^2 (x+5)$

a. Since $a_n < 0$ and n is even, $f(x)$ falls to the left and the right.

b. $x = 0, x = 1, x = -5$
The roots at 0 and −5 have odd multiplicity so $f(x)$ crosses the x-axis at those points. The root at 1 has even multiplicity so $f(x)$ touches the axis at (1, 0).

c. $f(0) = -2(0)^3 (0-1)^2 (0+5) = 0$
The y-intercept is 0.

d. $f(-x) = 2x^3 (-x-1)^2 (-x+5)$
The graph has neither y-axis nor origin symmetry.

e. The graph has 2 turning points and $2 \le 6 - 1$.

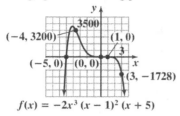

$f(x) = -2x^3 (x - 1)^2 (x + 5)$

63. $f(x) = (x-2)^2 (x+4)(x-1)$

a. Since $a_n > 0$ and n is even, $f(x)$ rises to the left and rises the right.

b. $x = 2$, $x = -4$, $x = 1$

The zeros at -4 and 1 have odd multiplicity so $f(x)$ crosses the x-axis at those points. The root at 2 has even multiplicity so $f(x)$ touches the axis at $(2, 0)$.

c. $f(0) = (0-2)^2 (0+4)(0-1) = -16$

The y-intercept is -16.

d. $f(-x) = (-x-2)^2 (-x+4)(-x-1)$

The graph has neither y-axis nor origin symmetry.

e. The graph has 3 turning points and $3 \le 4 - 1$.

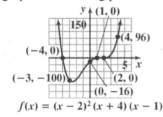

$f(x) = (x - 2)^2 (x + 4)(x - 1)$

65. a. The x-intercepts of the graph are -2, 1, and 4, so they are the zeros. Since the graph actually crosses the x-axis at all three places, all three have odd multiplicity.

b. Since the graph has two turning points, the function must be at least of degree 3. Since -2, 1, and 4 are the zeros, $x+2$, $x-1$, and $x-4$ are factors of the function. The lowest odd multiplicity is 1. From the end behavior, we can tell that the leading coefficient must be positive. Thus, the function is $f(x) = (x+2)(x-1)(x-4)$.

c. $f(0) = (0+2)(0-1)(0-4) = 8$

67. a. The x-intercepts of the graph are -1 and 3, so they are the zeros. Since the graph crosses the x-axis at -1, it has odd multiplicity. Since the graph touches the x-axis and turns around at 3, it has even multiplicity.

b. Since the graph has two turning points, the function must be at least of degree 3. Since -1 and 3 are the zeros, $x+1$ and $x-3$ are factors of the function. The lowest odd multiplicity is 1, and the lowest even multiplicity is 2. From the end behavior, we can tell that the leading coefficient must be positive. Thus, the function is $f(x) = (x+1)(x-3)^2$.

c. $f(0) = (0+1)(0-3)^2 = 9$

69. a. The x-intercepts of the graph are -3 and 2, so they are the zeros. Since the graph touches the x-axis and turns around at both -3 and 2, both have even multiplicity.

b. Since the graph has three turning points, the function must be at least of degree 4. Since -3 and 2 are the zeros, $x+3$ and $x-2$ are factors of the function. The lowest even multiplicity is 2. From the end behavior, we can tell that the leading coefficient must be negative. Thus, the function is $f(x) = -(x+3)^2 (x-2)^2$.

c. $f(0) = -(0+3)^2 (0-2)^2 = -36$

71. a. The x-intercepts of the graph are -2, -1, and 1, so they are the zeros. Since the graph crosses the x-axis at -1 and 1, they both have odd multiplicity. Since the graph touches the x-axis and turns around at -2, it has even multiplicity.

b. Since the graph has five turning points, the function must be at least of degree 6. Since -2, -1, and 1 are the zeros, $x+2$, $x+1$, and $x-1$ are factors of the function. The lowest even multiplicity is 2, and the lowest odd multiplicity is 1. However, to reach degree 6, one of the odd multiplicities must be 3. From the end behavior, we can tell that the leading coefficient must be positive. The function is

$f(x) = (x+2)^2 (x+1)(x-1)^3$.

c. $f(0) = (0+2)^2 (0+1)(0-1)^3 = -4$

73. a. $f(x) = 0.76x^3 - 30x^2 - 882x + 37,807$

$f(40) = 0.76(40)^3 - 30(40)^2 - 882(40) + 37807$

$= 3167$

The world tiger population in 2010 (40 years after 1970) was about 3167.
This is represented by the point $(40, 3167)$.

b. This underestimates the actual data shown in the bar graph by 33.

c. The leading coefficient is positive, thus the graph rises to the right.
No, if conservation efforts fail, the model will not be useful. The model indicates an increasing world tiger population that will actually decrease without conservation efforts.

75. a. The woman's heart rate was increasing from 1 through 4 minutes and from 8 through 10 minutes.

b. The woman's heart rate was decreasing from 4 through 8 minutes and from 10 through 12 minutes.

c. There were 3 turning points during the 12 minutes.

d. Since there were 3 turning points, a polynomial of degree 4 would provide the best fit.

e. The leading coefficient should be negative. The graph falls to the left and to the right.

f. The woman's heart rate reached a maximum of about 116 ± 1 beats per minute. This occurred after 10 minutes.

g. The woman's heart rate reached a minimum of about 64 ± 1 beats per minute. This occurred after 8 minutes.

77. – 93. Answers will vary.

95.

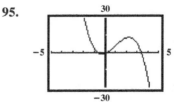

97.

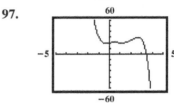

99.

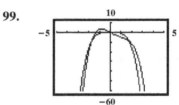

101. does not make sense; Explanations will vary. Sample explanation: Since $(x+2)$ is raised to an odd power, the graph crosses the *x*-axis at –2.

103. makes sense

105. false; Changes to make the statement true will vary. A sample change is: Such a function falls to the right and will eventually have negative values.

107. false; Changes to make the statement true will vary. A sample change is: A function with origin symmetry either falls to the left and rises to the right, or rises to the left and falls to the right.

109. $f(x) = x^3 - 2x^2$

110. $\dfrac{737}{21} = 35 + \dfrac{2}{21}$ or $35\dfrac{2}{21}$.

111. $6x^3 - x^2 - 5x + 4$

112. $2x^3 - x^2 - 11x + 6 = (x-3)(2x^2 + 3x - 2)$

$= (x-3)(2x-1)(x+2)$

Section 3.3

Check Point Exercises

1.
$$
\begin{array}{r}
x+5 \\
x+9{\overline{\smash{\big)}\,x^2+14x+45}} \\
\underline{x^2+9x} \\
5x+45 \\
\underline{5x+45} \\
0
\end{array}
$$

The answer is $x + 5$.

2.
$$
\begin{array}{r}
2x^2+3x-2 \\
x-3{\overline{\smash{\big)}\,2x^3-3x^2-11x+7}} \\
\underline{2x^3-6x^2} \\
3x^2-11x \\
\underline{3x^2-9x} \\
-2x+7 \\
\underline{-2x+6} \\
1
\end{array}
$$

The answer is $2x^2+3x-2+\dfrac{1}{x-3}$.

3.
$$
\begin{array}{r}
2x^2+7x+14 \\
x^2-2x{\overline{\smash{\big)}\,2x^4+3x^3+0x^2-7x-10}} \\
\underline{2x^4-4x^3} \\
7x^3+0x^2 \\
\underline{7x^3-14x^2} \\
14x^2-7x \\
\underline{14x^2-28x} \\
21x-10
\end{array}
$$

The answer is $2x^2+7x+14+\dfrac{21x-10}{x^2-2x}$.

4.
$$
\begin{array}{r|rrrr}
-2 & 1 & 0 & -7 & -6 \\
   &   & -2 & 4 & 6 \\
\hline
   & 1 & -2 & -3 & 0
\end{array}
$$

The answer is x^2-2x-3.

5.
$$
\begin{array}{r|rrrr}
-4 & 3 & 4 & -5 & 3 \\
   &   & -12 & 32 & -108 \\
\hline
   & 3 & -8 & 27 & -105
\end{array}
$$

$f(-4) = -105$

6.
$$
\begin{array}{r|rrrr}
-1 & 15 & 14 & -3 & -2 \\
   &   & -15 & 1 & 2 \\
\hline
   & 15 & -1 & -2 & 0
\end{array}
$$

$15x^2 - x - 2 = 0$

$(3x+1)(5x-2) = 0$

$x = -\dfrac{1}{3}$ or $x = \dfrac{2}{5}$

The solution set is $\left\{-1, -\dfrac{1}{3}, \dfrac{2}{5}\right\}$.

Concept and Vocabulary Check 3.3

1. $2x^3 + 0x^2 + 6x - 4$

2. $6x^3$; $3x$; $2x^2$; $7x^2$

3. $2x^2$; $5x-2$; $10x^3-4x^2$; $10x^3+6x^2$

4. $6x^2-10x$; $6x^2+8x$; $18x$; -4; $18x-4$

5. 9; $3x-5$; 9; $3x-5+\dfrac{9}{2x+1}$

6. divisor; quotient; remainder; dividend

7. 4; 1; 5; -7; 1

8. -5; 4; 0; -8; -2

9. true

10. $f(c)$

11. $x-c$

Exercise Set 3.3

1.
$$
\begin{array}{r}
x+3 \\
x+5{\overline{\smash{\big)}\,x^2+8x+15}} \\
\underline{x^2+5x} \\
3x+15 \\
\underline{3x+15} \\
0
\end{array}
$$

The answer is $x+3$.

3.

$$\begin{array}{r} x^2+3x+1 \\ x+2\overline{\smash{\big)}\,x^3+5x^2+7x+2} \\ \underline{x^3+2x^2} \\ 3x^2+7x \\ \underline{3x^2+6x} \\ x+2 \\ \underline{x+2} \\ 0 \end{array}$$

The answer is x^2+3x+1.

5.

$$\begin{array}{r} 2x^2+3x+5 \\ 3x-1\overline{\smash{\big)}\,6x^3+7x^2+12x-5} \\ \underline{6x^3-2x^2} \\ 9x^2+12x \\ \underline{9x^2-3x} \\ 15x-5 \\ \underline{15x-5} \\ 0 \end{array}$$

The answer is $2x^2+3x+5$.

7.

$$\begin{array}{r} 4x+3+\dfrac{2}{3x-2} \\ 3x-2\overline{\smash{\big)}\,12x^2+x-4} \\ \underline{12x^2-8x} \\ 9x-4 \\ \underline{9x-6} \\ 2 \end{array}$$

The answer is $4x+3+\dfrac{2}{3x-2}$.

9.

$$\begin{array}{r} 2x^2+x+6-\dfrac{38}{x+3} \\ x+3\overline{\smash{\big)}\,2x^3+7x^2+9x-20} \\ \underline{2x^3+6x^2} \\ x^2+9x \\ \underline{x^2+3x} \\ 6x-20 \\ \underline{6x+18} \\ -38 \end{array}$$

The answer is $2x^2+x+6-\dfrac{38}{x+3}$.

11.

$$\begin{array}{r} 4x^3+16x^2+60x+246+\dfrac{984}{x-4} \\ x-4\overline{\smash{\big)}\,4x^4-4x^2+6x} \\ \underline{4x^4-16x^3} \\ 16x^3-4x^2 \\ \underline{16x^3-64x^2} \\ 60x^2+6x \\ \underline{60x^2-240x} \\ 246x \\ \underline{246x-984} \\ 984 \end{array}$$

The answer is

$$4x^3+16x^2+60x+246+\dfrac{984}{x-4}.$$

13.

$$\begin{array}{r} 2x+5 \\ 3x^2-x-3\overline{\smash{\big)}\,6x^3+13x^2-11x-15} \\ \underline{6x^3-2x^2-6x} \\ 15x^2-5x-15 \\ \underline{15x^2-5x-15} \\ 0 \end{array}$$

The answer is $2x+5$.

15.

$$\begin{array}{r} 6x^2+3x-1 \\ 3x^2+1\overline{\smash{\big)}\,18x^4+9x^3+3x^2} \\ \underline{18x^4+6x^2} \\ 9x^3-3x^2 \\ \underline{9x^3+3x} \\ -3x^2-3x \\ \underline{-3x^2-1} \\ -3x+1 \end{array}$$

The answer is $6x^2+3x-1-\dfrac{3x-1}{3x^2+1}$.

17. $\left(2x^2+x-10\right)\div\left(x-2\right)$

$$\begin{array}{r|rrr} 2 & 2 & 1 & -10 \\ & & 4 & 10 \\ \hline & 2 & 5 & 0 \end{array}$$

The answer is $2x+5$.

19. $\left(3x^2 + 7x - 20\right) \div \left(x + 5\right)$

$$
\begin{array}{r|rrr}
-5 & 3 & 7 & -20 \\
   &   & -15 & 40 \\
\hline
   & 3 & -8 & 20
\end{array}
$$

The answer is $3x - 8 + \dfrac{20}{x+5}$.

21. $\left(4x^3 - 3x^2 + 3x - 1\right) \div \left(x - 1\right)$

$$
\begin{array}{r|rrrr}
1 & 4 & -3 & 3 & -1 \\
  &   & 4 & 1 & 4 \\
\hline
  & 4 & 1 & 4 & 3
\end{array}
$$

The answer is $4x^2 + x + 4 + \dfrac{3}{x-1}$.

23. $\left(6x^5 - 2x^3 + 4x^2 - 3x + 1\right) \div \left(x - 2\right)$

$$
\begin{array}{r|rrrrrr}
2 & 6 & 0 & -2 & 4 & -3 & 1 \\
  &   & 12 & 24 & 44 & 96 & 186 \\
\hline
  & 6 & 12 & 22 & 48 & 93 & 187
\end{array}
$$

The answer is

$6x^4 + 12x^3 + 22x^2 + 48x + 93 + \dfrac{187}{x-2}$.

25. $\left(x^2 - 5x - 5x^3 + x^4\right) \div \left(5 + x\right) \Rightarrow$

$\left(x^4 - 5x^3 + x^2 - 5x\right) \div \left(x + 5\right)$

$$
\begin{array}{r|rrrrr}
-5 & 1 & -5 & 1 & -5 & 0 \\
   &   & -5 & 50 & -255 & 1300 \\
\hline
   & 1 & -10 & 51 & -260 & 1300
\end{array}
$$

The answer is

$x^3 - 10x^2 + 51x - 260 + \dfrac{1300}{x+5}$.

27. $\dfrac{x^5 + x^3 - 2}{x - 1}$

$$
\begin{array}{r|rrrrrr}
1 & 1 & 0 & 1 & 0 & 0 & -2 \\
  &   & 1 & 1 & 2 & 2 & 2 \\
\hline
  & 1 & 1 & 2 & 2 & 2 & 0
\end{array}
$$

The answer is $x^4 + x^3 + 2x^2 + 2x + 2$.

29. $\dfrac{x^4 - 256}{x - 4}$

$$
\begin{array}{r|rrrrr}
4 & 1 & 0 & 0 & 0 & -256 \\
  &   & 4 & 16 & 64 & 256 \\
\hline
  & 1 & 4 & 16 & 64 & 0
\end{array}
$$

The answer is $x^3 + 4x^2 + 16x + 64$.

31. $\dfrac{2x^5 - 3x^4 + x^3 - x^2 + 2x - 1}{x + 2}$

$$
\begin{array}{r|rrrrrr}
-2 & 2 & -3 & 1 & -1 & 2 & -1 \\
   &   & -4 & 14 & -30 & 62 & -128 \\
\hline
   & 2 & -7 & 15 & -31 & 64 & -129
\end{array}
$$

The answer is

$2x^4 - 7x^3 + 15x^2 - 31x + 64 - \dfrac{129}{x+2}$.

33. $f(x) = 2x^3 - 11x^2 + 7x - 5$

$$
\begin{array}{r|rrrr}
4 & 2 & -11 & 7 & -5 \\
  &   & 8 & -12 & -20 \\
\hline
  & 2 & -3 & -5 & -25
\end{array}
$$

$f(4) = -25$

35. $f(x) = 3x^3 - 7x^2 - 2x + 5$

$$
\begin{array}{r|rrrr}
-3 & 3 & -7 & -2 & 5 \\
   &   & -9 & 48 & -138 \\
\hline
   & 3 & -16 & 46 & -133
\end{array}
$$

$f(-3) = -133$

37. $f(x) = x^4 + 5x^3 + 5x^2 - 5x - 6$

$$
\begin{array}{r|rrrrr}
3 & 1 & 5 & 5 & -5 & -6 \\
  &   & 3 & 24 & 87 & 246 \\
\hline
  & 1 & 8 & 29 & 82 & 240
\end{array}
$$

$f(3) = 240$

39. $f(x) = 2x^4 - 5x^3 - x^2 + 3x + 2$

$$
\begin{array}{r|rrrrr}
-\frac{1}{2} & 2 & -5 & -1 & 3 & 2 \\
             &   & -1 & 3 & -1 & -1 \\
\hline
             & 2 & -6 & 2 & 2 & 1
\end{array}
$$

$f\left(-\dfrac{1}{2}\right) = 1$

41. Dividend: $x^3 - 4x^2 + x + 6$
Divisor: $x + 1$

$$
\begin{array}{r|rrrr}
-1 & 1 & -4 & 1 & 6 \\
  &   & -1 & 5 & -6 \\
\hline
  & 1 & -5 & 6 & 0
\end{array}
$$

The quotient is $x^2 - 5x + 6$.
$(x+1)(x^2 - 5x + 6) = 0$
$(x + 1)(x - 2)(x - 3) = 0$
$x = -1, x = 2, x = 3$
The solution set is $\{-1, 2, 3\}$.

43. $2x^3 - 5x^2 + x + 2 = 0$

$$
\begin{array}{r|rrrr}
2 & 2 & -5 & 1 & 2 \\
  &   & 4 & -2 & -2 \\
\hline
  & 2 & -1 & -1 & 0
\end{array}
$$

$(x - 2)(2x^2 - x - 1) = 0$
$(x - 2)(2x + 1)(x - 1) = 0$
$x = 2, \ x = -\dfrac{1}{2}, x = 1$

The solution set is $\left\{-\dfrac{1}{2}, 1, 2\right\}$.

45. $12x^3 + 16x^2 - 5x - 3 = 0$

$$
\begin{array}{r|rrrr}
-\frac{3}{2} & 12 & 16 & -5 & -3 \\
  &   & -18 & 3 & 3 \\
\hline
  & 12 & -2 & -2 & 0
\end{array}
$$

$\left(x + \dfrac{3}{2}\right)(12x^2 - 2x - 2) = 0$

$\left(x + \dfrac{3}{2}\right)2\left(6x^2 - x - 1\right) = 0$

$\left(x + \dfrac{3}{2}\right)2(3x + 1)(2x - 1) = 0$

$x = -\dfrac{3}{2}, \ x = -\dfrac{1}{3}, \ x = \dfrac{1}{2}$

The solution set is $\left\{-\dfrac{3}{2}, -\dfrac{1}{3}, \dfrac{1}{2}\right\}$.

47. The graph indicates that 2 is a solution to the equation.

$$
\begin{array}{r|rrrr}
2 & 1 & 2 & -5 & -6 \\
  &   & 2 & 8 & 6 \\
\hline
  & 1 & 4 & 3 & 0
\end{array}
$$

The remainder is 0, so 2 is a solution.
$x^3 + 2x^2 - 5x - 6 = 0$
$(x - 2)\left(x^2 + 4x + 3\right) = 0$
$(x - 2)(x + 3)(x + 1) = 0$
The solutions are 2, -3, and -1, or $\{-3, -1, 2\}$.

49. The table indicates that 1 is a solution to the equation.

$$
\begin{array}{r|rrrr}
1 & 6 & -11 & 6 & -1 \\
  &   & 6 & -5 & 1 \\
\hline
  & 6 & -5 & 1 & 0
\end{array}
$$

The remainder is 0, so 1 is a solution.
$6x^3 - 11x^2 + 6x - 1 = 0$
$(x - 1)\left(6x^2 - 5x + 1\right) = 0$
$(x - 1)(3x - 1)(2x - 1) = 0$
The solutions are 1, $\dfrac{1}{3}$, and $\dfrac{1}{2}$, or $\left\{\dfrac{1}{3}, \dfrac{1}{2}, 1\right\}$.

51. a. $14x^3 - 17x^2 - 16x - 177 = 0$

$$
\begin{array}{r|rrrr}
3 & 14 & -17 & -16 & -177 \\
  &   & 42 & 75 & 177 \\
\hline
  & 14 & 25 & 59 & 0
\end{array}
$$

The remainder is 0 so 3 is a solution.
$14x^3 - 17x^2 - 16x - 177$
$= (x - 3)\left(14x^2 + 25x + 59\right)$

b. $f(x) = 14x^3 - 17x^2 - 16x + 34$

We need to find x when $f(x) = 211$.

$f(x) = 14x^3 - 17x^2 - 16x + 34$

$211 = 14x^3 - 17x^2 - 16x + 34$

$0 = 14x^3 - 17x^2 - 16x - 177$

This is the equation obtained in part **a.** One solution is 3. It can be used to find other solutions (if they exist).

$14x^3 - 17x^2 - 16x - 177 = 0$

$(x-3)(14x^2 + 25x + 59) = 0$

The polynomial $14x^2 + 25x + 59$ cannot be factored, so the only solution is $x = 3$. The female moth's abdominal width is 3 millimeters.

53. $A = l \cdot w$ so

$l = \dfrac{A}{w} = \dfrac{0.5x^3 - 0.3x^2 + 0.22x + 0.06}{x + 0.2}$

$$
\begin{array}{r|rrrr}
-0.2 & 0.5 & -0.3 & 0.22 & 0.06 \\
     &   & -0.1 & 0.08 & -0.06 \\
\hline
     & 0.5 & -0.4 & 0.3 & 0 \\
\end{array}
$$

Therefore, the length of the rectangle is $0.5x^2 - 0.4x + 0.3$ units.

55. a.

$f(30) = \dfrac{80(30) - 8000}{30 - 110} = 70$

(30, 70) At a 30% tax rate, the government tax revenue will be $70 ten billion.

b.

$$
\begin{array}{r|rr}
110 & 80 & -8000 \\
    &   & 8800 \\
\hline
    & 80 & 800 \\
\end{array}
$$

$f(x) = 80 + \dfrac{800}{x - 110}$

$f(30) = 80 + \dfrac{800}{80 - 110} = 70$

(30, 70) same answer as in **a.**

c. $f(x)$ is not a polynomial function. It is a rational function because it is the quotient of two linear polynomials.

57. – 65. Answers will vary.

67. makes sense

69. does not make sense; Explanations will vary. Sample explanation: The zeros of f are the same as the solutions of $f(x) = 0$.

71. true

73. false; Changes to make the statement true will vary. A sample change is: The divisor is a factor of the divided only if the remainder is the whole number 0.

75. $f(x) = d(x) \cdot q(x) + r(x)$

$2x^2 - 7x + 9 = d(x)(2x - 3) + 3$

$2x^2 - 7x + 6 = d(x)(2x - 3)$

$\dfrac{2x^2 - 7x + 6}{2x - 3} = d(x)$

$$
\begin{array}{r}
x - 2 \\
2x - 3 \overline{\smash{)}\, 2x^2 - 7x + 6} \\
\underline{2x^2 - 3x} \\
-4x + 6 \\
\underline{-4x + 6} \\
\end{array}
$$

The polynomial is $x - 2$.

77. $2x - 4 = 2(x - 2)$

Use synthetic division to divide by $x - 2$. Then divide the quotient by 2.

79. $x^2 + 4x - 1 = 0$

$$x = \frac{-b \pm \sqrt{b^2 - 4ac}}{2a}$$

$$x = \frac{-(4) \pm \sqrt{(4)^2 - 4(1)(-1)}}{2(1)}$$

$$x = \frac{-4 \pm \sqrt{20}}{2}$$

$$x = \frac{-4 \pm 2\sqrt{5}}{2}$$

$$x = -2 \pm \sqrt{5}$$

The solution set is $\left\{ -2 \pm \sqrt{5} \right\}$.

80. $x^2 + 4x + 6 = 0$

$$x = \frac{-b \pm \sqrt{b^2 - 4ac}}{2a}$$

$$x = \frac{-(4) \pm \sqrt{(4)^2 - 4(1)(6)}}{2(1)}$$

$$x = \frac{-4 \pm \sqrt{-8}}{2}$$

$$x = \frac{-4 \pm 2i\sqrt{2}}{2}$$

$$x = -2 \pm i\sqrt{2}$$

The solution set is $\left\{ -2 \pm i\sqrt{2} \right\}$.

81. $f(x) = a_n(x^4 - 3x^2 - 4)$

$$f(3) = -150$$

$$a_n\left((3)^4 - 3(3)^2 - 4\right) = -150$$

$$a_n(81 - 27 - 4) = -150$$

$$a_n(50) = -150$$

$$a_n = -3$$

Section 3.4

Check Point Exercises

1. $p : \pm 1, \ \pm 2, \ \pm 3, \ \pm 6$

$q : \pm 1$

$\dfrac{p}{q} : \pm 1, \ \pm 2, \ \pm 3, \ \pm 6$

are the possible rational zeros.

2. $p : \pm 1, \ \pm 3$

$q : \pm 1, \ \pm 2, \ \pm 4$

$\dfrac{p}{q} : \pm 1, \ \pm 3, \ \pm \dfrac{1}{2}, \ \pm \dfrac{1}{4}, \ \pm \dfrac{3}{2}, \ \pm \dfrac{3}{4}$

are the possible rational zeros.

3. $\pm 1, \ \pm 2, \ \pm 4, \ \pm 5, \ \pm 10, \ \pm 20$ are possible rational zeros

$$
\begin{array}{r|rrrr}
1 & 1 & 8 & 11 & -20 \\
  &   & 1 & 9 & 20 \\
\hline
  & 1 & 9 & 20 & 0
\end{array}
$$

1 is a zero.

$$x^2 + 9x + 20 = 0$$

$$(x + 4)(x + 5) = 0$$

$$x = -4 \quad \text{or} \quad x = -5$$

The zeros are $-5, -4,$ and 1.

4. $\pm 1, \ \pm 2$ are possible rational zeros

$$
\begin{array}{r|rrrr}
2 & 1 & 1 & -5 & -2 \\
  &   & 2 & 6 & 2 \\
\hline
  & 1 & 3 & 1 & 0
\end{array}
$$

2 is a zero.

$$x^2 + 3x + 1 = 0$$

$$x = \frac{-b \pm \sqrt{b^2 - 4ac}}{2a}$$

$$x = \frac{-3 \pm \sqrt{3^2 - 4(1)(1)}}{2(1)}$$

$$= \frac{-3 \pm \sqrt{5}}{2}$$

The zeros are $2, \dfrac{-3 - \sqrt{5}}{2},$ and $\dfrac{-3 + \sqrt{5}}{2}$..

5. $\pm 1,\ \pm 13$ are possible rational zeros.

$$
\begin{array}{r|rrrrr}
1 & 1 & -6 & 22 & -30 & 13 \\
  &   & 1 & -5 & 17 & -13 \\
\hline
  & 1 & -5 & 17 & -13 & 0
\end{array}
$$

1 is a zero.

$$
\begin{array}{r|rrrr}
1 & 1 & 5 & 17 & -13 \\
  &   & 1 & -4 & 13 \\
\hline
  & 1 & -4 & 13 & 0
\end{array}
$$

1 is a double root.

$x^2 - 4x + 13 = 0$

$x = \dfrac{4 \pm \sqrt{16 - 52}}{2} = \dfrac{4 \pm \sqrt{-36}}{2} = 2 + 3i$

The solution set is $\{1,\ 2 - 3i,\ 2 + 3i\}$.

6. $(x + 3)(x - i)(x + i) = (x + 3)(x^2 + 1)$

$f(x) = a_n(x + 3)(x^2 + 1)$

$f(1) = a_n(1 + 3)(1^2 + 1) = 8a_n = 8$

$a_n = 1$

$f(x) = (x + 3)(x^2 + 1)$

$f(x) = x^3 + 3x^2 + x + 3$

7. $f(x) = x^4 - 14x^3 + 71x^2 - 154x + 120$

$f(-x) = x^4 + 14x^3 + 71x^2 + 154x + 120$

Since $f(x)$ has 4 changes of sign, there are 4, 2, or 0 positive real zeros.
Since $f(-x)$ has no changes of sign, there are no negative real zeros.

Concept and Vocabulary Check 3.4

1. a_0; a_n

2. true

3. false

4. n

5. $a - bi$

6. -6; $(x + 6)(2x^2 - x - 1) = 0$

7. n; 1

8. false

9. true

10. true

Exercise Set 3.4

1. $f(x) = x^3 + x^2 - 4x - 4$

$p : \pm 1,\ \pm 2,\ \pm 4$

$q : \pm 1$

$\dfrac{p}{q} : \pm 1,\ \pm 2,\ \pm 4$

3. $f(x) = 3x^4 - 11x^3 - x^2 + 19x + 6$

$p : \pm 1,\ \pm 2,\ \pm 3,\ \pm 6$

$q : \pm 1,\ \pm 3$

$\dfrac{p}{q} : \pm 1,\ \pm 2,\ \pm 3,\ \pm 6,\ \pm \dfrac{1}{3},\ \pm \dfrac{2}{3}$

5. $f(x) = 4x^4 - x^3 + 5x^2 - 2x - 6$

$p : \pm 1,\ \pm 2,\ \pm 3,\ \pm 6$

$q : \pm 1,\ \pm 2,\ \pm 4$

$\dfrac{p}{q} : \pm 1,\ \pm 2,\ \pm 3,\ \pm 6,\ \pm \dfrac{1}{2},\ \pm \dfrac{1}{4},\ \pm \dfrac{3}{2},\ \pm \dfrac{3}{4}$

7. $f(x) = x^5 - x^4 - 7x^3 + 7x^2 - 12x - 12$

$p : \pm 1,\ \pm 2,\ \pm 3\ \pm 4\ \pm 6\ \pm 12$

$q : \pm 1$

$\dfrac{p}{q} : \pm 1,\ \pm 2,\ \pm 3\ \pm 4\ \pm 6\ \pm 12$

9. $f(x) = x^3 + x^2 - 4x - 4$

a. $p : \pm 1,\ \pm 2,\ \pm 4$

$q : \pm 1$

$\dfrac{p}{q} : \pm 1,\ \pm 2,\ \pm 4$

b.
$$
\begin{array}{r|rrrr}
2 & 1 & 1 & -4 & -4 \\
  &   & 2 & 6 & 4 \\
\hline
  & 1 & 3 & 2 & 0
\end{array}
$$

2 is a zero.
2, –2, –1 are rational zeros.

c. $x^3 + x^2 - 4x - 4 = 0$

$(x - 2)(x^2 + 3x + 2) = 0$

$(x - 2)(x + 2)(x + 1) = 0$

$x - 2 = 0\ \ x + 2 = 0\ \ x + 1 = 0$

$x = 2,\ x = -2,\ x = -1$

The solution set is $\{2, -2, -1\}$.

11. $f(x) = 2x^3 - 3x^2 - 11x + 6$

a. $p: \pm 1, \pm 2, \pm 3, \pm 6$

$q: \pm 1, \pm 2$

$\dfrac{p}{q}: \pm 1, \pm 2, \pm 3, \pm 6, \pm \dfrac{1}{2}, \pm \dfrac{3}{2}$

b.

$$
\begin{array}{r|rrrr}
3 & 2 & -3 & -11 & 6 \\
  &   & 6 & 9   & -6 \\
\hline
  & 2 & 3 & -2 & 0
\end{array}
$$

3 is a zero.

$3, \dfrac{1}{2}, -2$ are rational zeros.

c. $2x^3 - 3x^2 - 11x + 6 = 0$

$(x - 3)(2x^2 + 3x - 2) = 0$

$(x - 3)(2x - 1)(x + 2) = 0$

$x = 3, \; x = \dfrac{1}{2}, \; x = -2$

The solution set is $\left\{ 3, \dfrac{1}{2}, -2 \right\}$.

13. a. $f(x) = x^3 + 4x^2 - 3x - 6$

$p: \pm 1, \pm 2, \pm 3, \pm 6$

$q: \pm 1$

$\dfrac{p}{q}: \pm 1, \pm 2, \pm 3, \pm 6$

b.

$$
\begin{array}{r|rrrr}
-1 & 1 & 4 & -3 & -6 \\
   &   & -1 & -3 & 6 \\
\hline
   & 1 & 3 & -6 & 0
\end{array}
$$

−1 is a rational zero.

c. $x^2 + 3x - 6 = 0$

$x = \dfrac{-b \pm \sqrt{b^2 - 4ac}}{2a}$

$x = \dfrac{-3 \pm \sqrt{3^2 - 4(1)(-6)}}{2(1)}$

$= \dfrac{-3 \pm \sqrt{33}}{2}$

The solution set is $\left\{ -1, \dfrac{-3 + \sqrt{33}}{2}, \dfrac{-3 - \sqrt{33}}{2} \right\}$.

15. a. $f(x) = 2x^3 + 6x^2 + 5x + 2$

$p: \pm 1, \pm 2$

$q: \pm 1, \pm 2$

$\dfrac{p}{q}: \pm 1, \pm 2, \pm \dfrac{1}{2}$

b.

$$
\begin{array}{r|rrrr}
-2 & 2 & 6 & 5 & 2 \\
   &   & -4 & -4 & -2 \\
\hline
   & 2 & 2 & 1 & 0
\end{array}
$$

−2 is a rational zero.

c. $2x^2 + 2x + 1 = 0$

$x = \dfrac{-b \pm \sqrt{b^2 - 4ac}}{2a}$

$x = \dfrac{-2 \pm \sqrt{2^2 - 4(2)(1)}}{2(2)}$

$= \dfrac{-2 \pm \sqrt{-4}}{4}$

$= \dfrac{-2 \pm 2i}{4}$

$= \dfrac{-1 \pm i}{2}$

The solution set is $\left\{ -2, \dfrac{-1 + i}{2}, \dfrac{-1 - i}{2} \right\}$.

17. $x^3 - 2x^2 - 11x + 12 = 0$

a. $p: \pm 1, \pm 2, \pm 3, \pm 4, \pm 6, \pm 12$

$q: \pm 1$

$\dfrac{p}{q}: \pm 1, \pm 2, \pm 3, \pm 4, \pm 6, \pm 12$

b.

$$
\begin{array}{r|rrrr}
4 & 1 & -2 & -11 & 12 \\
  &   & 4 & 8   & -12 \\
\hline
  & 1 & 2 & -3 & 0
\end{array}
$$

4 is a root.

−3, 1, 4 are rational roots.

c. $x^3 - 2x^2 - 11x + 12$

$(x - 4)(x^2 + 2x - 3) = 0$

$(x - 4)(x + 3)(x - 1) = 0$

$x - 4 = 0 \quad x + 3 = 0 \quad x - 1 = 0$

$x = 4 \qquad x = -3 \qquad x = 1$

The solution set is {−3, 1, 4}.

19. $x^3 - 10x - 12 = 0$

 a. $p: \pm1, \pm2, \pm3, \pm4, \pm6, \pm12$

   $q: \pm1$

   $\dfrac{p}{q}: \pm1, \pm2, \pm3, \pm4, \pm6, \pm12$

 b.

-2	1	0	-10	-12
		-2	4	12
	1	-2	-6	0

   -2 is a rational root.

 c. $x^3 - 10x - 12 = 0$

   $(x+2)(x^2 - 2x - 6) = 0$

   $x = \dfrac{2 \pm \sqrt{4 + 24}}{2} = \dfrac{2 \pm \sqrt{28}}{2}$

   $\quad = \dfrac{2 \pm 2\sqrt{7}}{2} = 1 \pm \sqrt{7}$

   The solution set is $\left\{-2, 1 + \sqrt{7}, 1 - \sqrt{7}\right\}$.

21. $6x^3 + 25x^2 - 24x + 5 = 0$

 a. $p: \pm1, \pm5$

   $q: \pm1, \pm2, \pm3, \pm6$

   $\dfrac{p}{q}: \pm1, \pm5, \pm\dfrac{1}{2}, \pm\dfrac{5}{2}, \pm\dfrac{1}{3}, \pm\dfrac{5}{3}, \pm\dfrac{1}{6}, \pm\dfrac{5}{6}$

 b.

-5	6	25	-24	5
		-30	25	-5
	6	-5	1	0

   -5 is a root.

   $-5, \dfrac{1}{2}, \dfrac{1}{3}$ are rational roots.

 c. $6x^3 + 25x^2 - 24x + 5 = 0$

   $(x+5)(6x^2 - 5x + 1) = 0$

   $(x+5)(2x - 1)(3x - 1) = 0$

   $x + 5 = 0 \quad 2x - 1 = 0 \quad 3x - 1 = 0$

   $x = -5, \quad x = \dfrac{1}{2}, \quad x = \dfrac{1}{3}$

   The solution set is $\left\{-5, \dfrac{1}{2}, \dfrac{1}{3}\right\}$.

23. $x^4 - 2x^3 - 5x^2 + 8x + 4 = 0$

 a. $p: \pm1, \pm2, \pm4$

   $q: \pm1$

   $\dfrac{p}{q}: \pm1, \pm2, \pm4$

 b.

2	1	-2	-5	8	4
		2	0	-10	-4
	1	0	-5	-2	0

   2 is a root.
   $-2, 2$ are rational roots.

 c. $x^4 - 2x^3 - 5x^2 + 8x + 4 = 0$

   $(x-2)(x^3 - 5x - 2) = 0$

-2	1	0	-5	-2
		-2	4	2
	1	-2	-1	0

   -2 is a zero of $x^3 - 5x - 2 = 0$.

   $(x-2)(x+2)\left(x^2 - 2x - 1\right) = 0$

   $x = \dfrac{2 \pm \sqrt{4 + 4}}{2} = \dfrac{2 \pm \sqrt{8}}{2} = \dfrac{2 \pm 2\sqrt{2}}{2}$

   $\quad = 1 \pm \sqrt{2}$

   The solution set is
   $\left\{-2, 2, 1 + \sqrt{2}, 1 - \sqrt{2}\right\}$.

25. $(x-1)(x+5i)(x-5i)$

   $= (x-1)\left(x^2 + 25\right)$

   $= x^3 + 25x - x^2 - 25$

   $= x^3 - x^2 + 25x - 25$

   $f(x) = a_n\left(x^3 - x^2 + 25x - 25\right)$

   $f(-1) = a_n(-1 - 1 - 25 - 25)$

   $-104 = a_n(-52)$

   $a_n = 2$

   $f(x) = 2\left(x^3 - x^2 + 25x - 25\right)$

   $f(x) = 2x^3 - 2x^2 + 50x - 50$

27. $(x+5)(x-4-3i)(x-4+3i)$

$$= (x+5)\left(x^2 - 4x + 3ix - 4x + 16 - 12i\right.$$
$$\left. -3ix + 12i - 9i^2\right)$$
$$= (x+5)\left(x^2 - 8x + 25\right)$$
$$= \left(x^3 - 8x^2 + 25x + 5x^2 - 40x + 125\right)$$
$$= x^3 - 3x^2 - 15x + 125$$

$$f(x) = a_n(x^3 - 3x^2 - 15x + 125)$$
$$f(2) = a_n\left(2^3 - 3(2)^2 - 15(2) + 125\right)$$
$$91 = a_n(91)$$
$$a_n = 1$$
$$f(x) = 1\left(x^3 - 3x^2 - 15x + 125\right)$$
$$f(x) = x^3 - 3x^2 - 15x + 125$$

29. $(x-i)(x+i)(x-3i)(x+3i)$

$$= \left(x^2 - i^2\right)\left(x^2 - 9i^2\right)$$
$$= \left(x^2 + 1\right)\left(x^2 + 9\right)$$
$$= x^4 + 10x^2 + 9$$
$$f(x) = a_n(x^4 + 10x^2 + 9)$$
$$f(-1) = a_n((-1)^4 + 10(-1)^2 + 9)$$
$$20 = a_n(20)$$
$$a_n = 1$$
$$f(x) = x^4 + 10x^2 + 9$$

31. $(x+2)\ (x-5)\ (x-3+2i)\ (x-3-2i)$

$$= \left(x^2 - 3x - 10\right)\left(x^2 - 3x - 2ix - 3x + 9 + 6i + 2ix - 6i - 4i^2\right)$$
$$= \left(x^2 - 3x - 10\right)\left(x^2 - 6x + 13\right)$$
$$= x^4 - 6x + 13x^2 - 3x^3 + 18x^2 - 39x - 10x^2 + 60x - 130$$
$$= x^4 - 9x^3 + 21x^2 + 21x - 130$$
$$f(x) = a_n\left(x^4 - 9x^3 + 21x^2 + 21x - 130\right)$$
$$f(1) = a_n(1 - 9 + 21 + 21 - 130)$$
$$-96 = a_n(-96)$$
$$a_n = 1$$
$$f(x) = x^4 - 9x^3 + 21x^2 + 21x - 130$$

33. $f(x) = x^3 + 2x^2 + 5x + 4$

Since $f(x)$ has no sign variations, no positive real roots exist.

$f(-x) = -x^3 + 2x^2 - 5x + 4$

Since $f(-x)$ has 3 sign variations, 3 or 1 negative real roots exist.

35. $f(x) = 5x^3 - 3x^2 + 3x - 1$

Since $f(x)$ has 3 sign variations, 3 or 1 positive real roots exist.

$f(-x) = -5x^3 - 3x^2 - 3x - 1$

Since $f(-x)$ has no sign variations, no negative real roots exist.

37. $f(x) = 2x^4 - 5x^3 - x^2 - 6x + 4$

Since $f(x)$ has 2 sign variations, 2 or 0 positive real roots exist.

$f(-x) = 2x^4 + 5x^3 - x^2 + 6x + 4$

Since $f(-x)$ has 2 sign variations, 2 or 0 negative real roots exist.

39. $f(x) = x^3 - 4x^2 - 7x + 10$

$p : \pm 1, \pm 2, \pm 5, \pm 10$

$q : \pm 1$

$\dfrac{p}{q} : \pm 1, \pm 2, \pm 5, \pm 10$

Since $f(x)$ has 2 sign variations, 0 or 2 positive real zeros exist.

$f(-x) = -x^3 - 4x^2 + 7x + 10$

Since $f(-x)$ has 1 sign variation, exactly one negative real zeros exists.

$$
\begin{array}{r|rrr}
-2 & 1 & -4 & -7 & 10 \\
   &   & -2 & 12 & -10 \\
\hline
   & 1 & -6 & 5 & 0
\end{array}
$$

-2 is a zero.

$f(x) = (x + 2)(x^2 - 6x + 5)$

$\qquad = (x + 2)(x - 5)(x - 1)$

$x = -2,\ x = 5,\ x = 1$

The solution set is $\{-2, 5, 1\}$.

41. $2x^3 - x^2 - 9x - 4 = 0$

$p : \pm 1, \pm 2, \pm 4$

$q : \pm 1, \pm 2$

$\dfrac{p}{q} : \pm 1, \pm 2, \pm 4 \pm \dfrac{1}{2}$

1 positive real root exists.

$f(-x) = -2x^3 - x^2 + 9x - 4$ 2 or no negative real roots exist.

$$
\begin{array}{r|rrr}
-\dfrac{1}{2} & 2 & -1 & -9 & -4 \\
   &   & -1 & 1 & 4 \\
\hline
   & 2 & -2 & -8 & 0
\end{array}
$$

$-\dfrac{1}{2}$ is a root.

$\left(x + \dfrac{1}{2}\right)\left(2x^2 - 2x - 8\right) = 0$

$2\left(x + \dfrac{1}{2}\right)\left(x^2 - x - 4\right) = 0$

$x = \dfrac{1 \pm \sqrt{1 + 16}}{2} = \dfrac{1 \pm \sqrt{17}}{2}$

The solution set is $\left\{ -\dfrac{1}{2}, \dfrac{1 + \sqrt{17}}{2}, \dfrac{1 - \sqrt{17}}{2} \right\}$.

43. $f(x) = x^4 - 2x^3 + x^2 + 12x + 8$

$p : \pm 1,\ \pm 2,\ \pm 4,\ \pm 8$

$q : \pm 1$

$\dfrac{p}{q} : \pm 1,\ \pm 2,\ \pm 4,\ \pm 8$

Since $f(x)$ has 2 sign changes, 0 or 2 positive roots exist.

$f(-x) = (-x)^4 - 2(-x)^3 + (-x)^2 - 12x + 8$

$\qquad = x^4 + 2x^3 + x^2 - 12x + 8$

Since $f(-x)$ has 2 sign changes, 0 or 2 negative roots exist.

$$
\begin{array}{r|rrrr}
-1 & 1 & -2 & 1 & 12 & 8 \\
   &   & -1 & 4 & -4 & -8 \\
\hline
   & 1 & -3 & 4 & 8 & 0
\end{array}
$$

$$
\begin{array}{r|rrr}
-1 & 1 & -3 & 4 & 8 \\
   &   & -1 & 4 & -8 \\
\hline
   & 1 & -4 & 8 & 0
\end{array}
$$

$0 = x^2 - 4x + 8$

$x = \dfrac{-(-4) \pm \sqrt{(-4)^2 - 4(1)(8)}}{2(1)}$

$x = \dfrac{4 \pm \sqrt{16 - 32}}{2}$

$x = \dfrac{4 \pm \sqrt{-16}}{2}$

$x = \dfrac{4 \pm 4i}{2}$

$x = 2 \pm 2i$

The solution set is $\{ -1, -1, 2 + 2i, 2 - 2i \}$.

45. $x^4 - 3x^3 - 20x^2 - 24x - 8 = 0$

$p : \pm 1, \pm 2, \pm 4, \pm 8$

$q : \pm 1$

$\dfrac{p}{q} : \pm 1, \pm 2, \pm 4 \ \pm 8$

1 positive real root exists.

3 or 1 negative real roots exist.

$$\begin{array}{r|rrrrr} -1 & 1 & -3 & -20 & -24 & -8 \\ & & -1 & 4 & 16 & 8 \\ \hline & 1 & -4 & -16 & -8 & 0 \end{array}$$

$(x+1)\left(x^3 - 4x^2 - 16x - 8\right) = 0$

$$\begin{array}{r|rrrr} -2 & 1 & -4 & -16 & -8 \\ & & -2 & 12 & 8 \\ \hline & 1 & -6 & -4 & 0 \end{array}$$

$(x+1)(x+2)\left(x^2 - 6x - 4\right) = 0$

$x = \dfrac{6 \pm \sqrt{36 + 16}}{2} = \dfrac{6 \pm \sqrt{52}}{2}$

$= \dfrac{6 \pm 2\sqrt{13}}{2} = \dfrac{3 \pm \sqrt{13}}{2}$

The solution set is

$\left\{ -1, -2, 3 \pm \sqrt{13}, 3 - \sqrt{13} \right\}$.

47. $f(x) = 3x^4 - 11x^3 - x^2 + 19x + 6$

$p : \pm 1, \pm 2, \pm 3, \pm 6$

$q : \pm 1, \pm 3$

$\dfrac{p}{q} : \pm 1, \pm 2, \pm 3, \pm 6, \pm \dfrac{1}{3}, \pm \dfrac{2}{3}$

2 or no positive real zeros exists.

$f(-x) = 3x^4 + 11x^3 - x^2 - 19x + 6$

2 or no negative real zeros exist.

$$\begin{array}{r|rrrrr} -1 & 3 & -11 & -1 & 19 & 6 \\ & & -3 & 14 & -13 & -6 \\ \hline & 3 & -14 & 13 & 6 & 0 \end{array}$$

$f(x) = (x+1)\left(3x^3 - 14x^2 + 13x + 6\right)$

$$\begin{array}{r|rrrr} 2 & 3 & -14 & 13 & 6 \\ & & 6 & -16 & -6 \\ \hline & 3 & -8 & -3 & 0 \end{array}$$

$f(x) = (x+1)(x-2)\left(3x^2 - 8x - 3\right)$

$\quad = (x+1)(x-2)(3x+1)(x-3)$

$x = -1, \ x = 2 \ x = -\dfrac{1}{3}, \ x = 3$

The solution set is $\left\{ -1, 2, -\dfrac{1}{3}, 3 \right\}$.

49. $4x^4 - x^3 + 5x^2 - 2x - 6 = 0$

$p : \pm 1, \pm 2, \pm 3, \pm 6$

$q : \pm 1, \pm 2, \pm 4$

$\dfrac{p}{q} : \pm 1, \pm 2, \pm 3, \pm 6, \pm \dfrac{1}{2}, \pm \dfrac{3}{2}, \pm \dfrac{1}{4}, \pm \dfrac{3}{4}$

3 or 1 positive real roots exists.

1 negative real root exists.

$$\begin{array}{r|rrrrr} 1 & 4 & -1 & 5 & -2 & -6 \\ & & 4 & 3 & 8 & 6 \\ \hline & 4 & 3 & 8 & 6 & 0 \end{array}$$

$(x-1)(4x^3 + 3x^2 + 8x + 6) = 0$

$4x^3 + 3x^2 + 8x + 6 = 0$ has no positive real roots.

$$\begin{array}{r|rrrr} -\frac{3}{4} & 4 & 3 & 8 & 6 \\ & & -3 & 0 & -6 \\ \hline & 4 & 0 & 8 & 0 \end{array}$$

$$(x-1)\left(x+\frac{3}{4}\right)\left(4x^2+8\right)=0$$

$$4(x-1)\left(x+\frac{3}{4}\right)\left(x^2+2\right)=0$$

$$x^2+2=0$$

$$x^2=-2$$

$$x=\pm i\sqrt{2}$$

The solution set is $\left\{1,\,-\dfrac{3}{4},\,i\sqrt{2},\,-i\sqrt{2}\right\}$.

51. $2x^5+7x^4-18x^2-8x+8=0$

$p:\pm 1,\pm 2,\pm 4,\pm 8$

$q:\pm 1,\pm 2$

$\dfrac{p}{q}:\pm 1,\pm 2,\pm 4,\pm 8,\pm\dfrac{1}{2}$

2 or no positive real roots exists.
3 or 1 negative real root exist.

$$\begin{array}{r|rrrrrr} -2 & 2 & 7 & 0 & -18 & -8 & 8 \\ & & -4 & -6 & 12 & 12 & -8 \\ \hline & 2 & 3 & -6 & -6 & 4 & 0 \end{array}$$

$$(x+2)(2x^4+3x^3-6x^2-6x+4)=0$$

$$4x^3+3x^2+8x+6=0 \text{ has no positive real roots.}$$

$$\begin{array}{r|rrrrr} -2 & 2 & 3 & -6 & -6 & 4 \\ & & -4 & 2 & 8 & -4 \\ \hline & 2 & -1 & -4 & 2 & 0 \end{array}$$

$$(x+2)^2(2x^3-x^2-4x+2)$$

$$\begin{array}{r|rrrr} \frac{1}{2} & 2 & -1 & -4 & 2 \\ & & 1 & 0 & 2 \\ \hline & 2 & 0 & -4 & 0 \end{array}$$

$$(x+2)^2\left(x-\frac{1}{2}\right)\left(2x^2-4\right)=0$$

$$2(x+2)^2\left(x-\frac{1}{2}\right)\left(x^2-2\right)=0$$

$$x^2-2=0$$

$$x^2=2$$

$$x=\pm\sqrt{2}$$

The solution set is $\left\{-2,\,\dfrac{1}{2},\,\sqrt{2},\,-\sqrt{2}\right\}$.

53. $f(x)=-x^3+x^2+16x-16$

a. From the graph provided, we can see that -4 is an x-intercept and is thus a zero of the function. We verify this below:

$$\begin{array}{r|rrrr} -4 & -1 & 1 & 16 & -16 \\ & & 4 & -20 & 16 \\ \hline & -1 & 5 & -4 & 0 \end{array}$$

Thus, $-x^3+x^2+16x-16=0$

$$(x+4)(-x^2+5x-4)=0$$

$$-(x+4)(x^2-5x+4)=0$$

$$-(x+4)(x-1)(x-4)=0$$

$x+4=0$ or $x-1=0$ or $x-4=0$

$x=-4$ $\qquad x=1$ $\qquad x=4$

The zeros are -4, 1, and 4.

b.

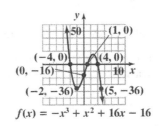

$$f(x)=-x^3+x^2+16x-16$$

55. $f(x) = 4x^3 - 8x^2 - 3x + 9$

a. From the graph provided, we can see that -1 is an x-intercept and is thus a zero of the function. We verify this below:

$$\begin{array}{r|rrrr} -1 & 4 & -8 & -3 & 9 \\ & & -4 & 12 & -9 \\ \hline & 4 & -12 & 9 & 0 \end{array}$$

Thus, $4x^3 - 8x^2 - 3x + 9 = 0$

$(x+1)(4x^2 - 12x + 9) = 0$

$(x+1)(2x-3)^2 = 0$

$x + 1 = 0$ or $(2x-3)^2 = 0$

$x = -1$ $2x - 3 = 0$

            $2x = 3$

            $x = \dfrac{3}{2}$

The zeros are -1 and $\dfrac{3}{2}$.

b.

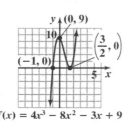

$f(x) = 4x^3 - 8x^2 - 3x + 9$

57. $f(x) = 2x^4 - 3x^3 - 7x^2 - 8x + 6$

a. From the graph provided, we can see that $\dfrac{1}{2}$ is an x-intercept and is thus a zero of the function. We verify this below:

$$\begin{array}{r|rrrrr} \frac{1}{2} & 2 & -3 & -7 & -8 & 6 \\ & & 1 & -1 & -4 & -6 \\ \hline & 2 & -2 & -8 & -12 & 0 \end{array}$$

Thus, $2x^4 - 3x^3 - 7x^2 - 8x + 6 = 0$

$\left(x - \dfrac{1}{2}\right)(2x^3 - 2x^2 - 8x - 12) = 0$

$2\left(x - \dfrac{1}{2}\right)(x^3 - x^2 - 4x - 6) = 0$

To factor $x^3 - x^2 - 4x - 6$, we use the Rational Zero Theorem to determine possible rational zeros.

Factors of the constant term -6:

$\pm 1,\ \pm 2,\ \pm 3,\ \pm 6$

Factors of the leading coefficient 1: ± 1

The possible rational zeros are:

$$\dfrac{\text{Factors of} -6}{\text{Factors of } 1} = \dfrac{\pm 1,\ \pm 2,\ \pm 3,\ \pm 6}{\pm 1}$$

$= \pm 1,\ \pm 2,\ \pm 3,\ \pm 6$

We test values from above until we find a zero. One possibility is shown next:

Test 3:

$$\begin{array}{r|rrrr} 3 & 1 & -1 & -4 & -6 \\ & & 3 & 6 & 6 \\ \hline & 1 & 2 & 2 & 0 \end{array}$$

The remainder is 0, so 3 is a zero of f.

$$2x^4 - 3x^3 - 7x^2 - 8x + 6 = 0$$

$\left(x - \dfrac{1}{2}\right)(2x^3 - 2x^2 - 8x - 12) = 0$

$2\left(x - \dfrac{1}{2}\right)(x^3 - x^2 - 4x - 6) = 0$

$2\left(x - \dfrac{1}{2}\right)(x - 3)(x^2 + 2x + 2) = 0$

Note that $x^2 + x + 1$ will not factor, so we use the quadratic formula:

$a = 1$ $b = 2$ $c = 2$

$$x = \dfrac{-2 \pm \sqrt{2^2 - 4(1)(2)}}{2(1)}$$

$$= \dfrac{-2 \pm \sqrt{-4}}{2} = \dfrac{-2 \pm 2i}{2} = -1 \pm i$$

The zeros are $\dfrac{1}{2}$, 3, and $-1 \pm i$.

b.

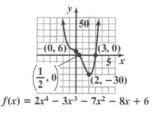

$f(x) = 2x^4 - 3x^3 - 7x^2 - 8x + 6$

59. $f(x) = 3x^5 + 2x^4 - 15x^3 - 10x^2 + 12x + 8$

a. From the graph provided, we can see that 1 and 2 are x-intercepts and are thus zeros of the function. We verify this below:

$$\underline{1|} \quad 3 \quad 2 \quad -15 \quad -10 \quad 12 \quad 8$$
$$\qquad\qquad 3 \quad 5 \quad -10 \quad -20 \quad -8$$
$$\overline{\qquad 3 \quad 5 \quad -10 \quad -20 \quad -8 \quad 0}$$

Thus, $3x^5 + 2x^4 - 15x^3 - 10x^2 + 12x + 8$
$$= (x-1)(3x^4 + 5x^3 - 10x^2 - 20x - 8)$$

$$\underline{2|} \quad 3 \quad 5 \quad -10 \quad -20 \quad -8$$
$$\qquad\qquad 6 \quad 22 \quad 24 \quad 8$$
$$\overline{\qquad 3 \quad 11 \quad 12 \quad 4 \quad 0}$$

Thus, $3x^5 + 2x^4 - 15x^3 - 10x^2 + 12x + 8$
$$= (x-1)(3x^4 + 5x^3 - 10x^2 - 20x - 8)$$
$$= (x-1)(x-2)(3x^3 + 11x^2 + 12x + 4)$$

To factor $3x^3 + 11x^2 + 12x + 4$, we use the Rational Zero Theorem to determine possible rational zeros.

Factors of the constant term 4: $\pm 1, \pm 2, \pm 4$
Factors of the leading coefficient 3: $\pm 1, \pm 3$

The possible rational zeros are:

$$\frac{\text{Factors of 4}}{\text{Factors of 3}} = \frac{\pm 1, \pm 2, \pm 4}{\pm 1, \pm 3}$$

$$= \pm 1, \pm 2, \pm 4, \pm\frac{1}{3}, \pm\frac{2}{3}, \pm\frac{4}{3}$$

We test values from above until we find a zero. One possibility is shown next:

Test -1:

$$\underline{-1|} \quad 3 \quad 11 \quad 12 \quad 4$$
$$\qquad\qquad -3 \quad -8 \quad -4$$
$$\overline{\qquad 3 \quad 8 \quad 4 \quad 0}$$

The remainder is 0, so -1 is a zero of f. We can now finish the factoring:

$$3x^5 + 2x^4 - 15x^3 - 10x^2 + 12x + 8 = 0$$
$$(x-1)(3x^4 + 5x^3 - 10x^2 - 20x - 8) = 0$$
$$(x-1)(x-2)(3x^3 + 11x^2 + 12x + 4) = 0$$
$$(x-1)(x-2)(x+1)(3x^2 + 8x + 4) = 0$$
$$(x-1)(x-2)(x+1)(3x+2)(x+2) = 0$$
$$x = 1, \; x = 2, \; x = -1, \; x = -\frac{2}{3}, \; x = -2$$

The zeros are -2, -1, $-\frac{2}{3}$, 1 and 2.

b.

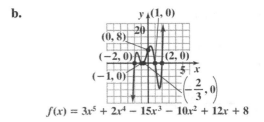

$$f(x) = 3x^5 + 2x^4 - 15x^3 - 10x^2 + 12x + 8$$

61. $V(x) = x(x+10)(30-2x)$
$$2000 = x(x+10)(30-2x)$$
$$2000 = -2x^3 + 10x^2 + 300x$$
$$2x^3 - 10x^2 - 300x + 2000 = 0$$
$$x^3 - 5x^2 - 150x + 1000 = 0$$

Find the roots.

$$\underline{10|} \quad 1 \quad -5 \quad -150 \quad 1000$$
$$\qquad\qquad 10 \quad 50 \quad -1000$$
$$\overline{\qquad 1 \quad 5 \quad -100 \quad 0}$$

Use the remaining quadratic to find the other 2 roots.

$$x = \frac{-b \pm \sqrt{b^2 - 4ac}}{2a}$$

$$x = \frac{-(5) \pm \sqrt{(5)^2 - 4(1)(-100)}}{2(1)}$$

$$x \approx -12.8, \; 7.8$$

Since the depth must be positive, reject the negative value.
The depth can be 10 inches or 7.8 inches to obtain a volume of 2000 cubic inches.

63. a. The answers correspond to the points $(7.8, 2000)$ and $(10, 2000)$.

b. The range is $(0, 15)$.

65. – 71. Answers will vary.

73. $6x^3 - 19x^2 + 16x - 4 = 0$
p: $\pm 1, \pm 2, \pm 4$
q: $\pm 1, \pm 2, \pm 3, \pm 6$

$$\frac{p}{q}: \pm 1, \pm 2, \pm 4, \pm\frac{1}{2}, \pm\frac{1}{3}, \pm\frac{2}{3}, \pm\frac{4}{3}, \pm\frac{1}{6}$$

From the graph, we see that the solutions are $\frac{1}{2}$, $\frac{2}{3}$ and 2.

75. $4x^4 + 4x^3 + 7x^2 - x - 2 = 0$

p: ± 1, ± 2

q: ± 1, ± 2, ± 4

$\dfrac{p}{q}$: ± 1, ± 2, $\pm \dfrac{1}{2}$, $\pm \dfrac{1}{4}$

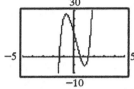

From the graph, we see that the solutions are

$-\dfrac{1}{2}$ and $\dfrac{1}{2}$.

77. $f(x) = x^5 - x^4 + x^3 - x^2 + x - 8$

$f(x)$ has 5 sign variations, so either 5, 3, or 1 positive real roots exist.

$f(-x) = -x^5 - x^4 - x^3 - x^2 - x - 8$

$f(-x)$ has no sign variations, so no negative real roots exist.

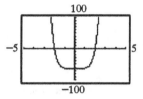

79. $f(x) = x^3 - 6x - 9$

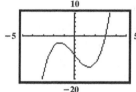

1 real zero

2 nonreal complex zeros

81. $f(x) = 3x^4 + 4x^3 - 7x^2 - 2x - 3$

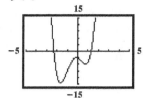

83. makes sense

85. makes sense

87. false; Changes to make the statement true will vary. A sample change is: The equation has 0 sign variations, so no positive roots exist.

89. true

91. $(2x+1)(x+5)(x+2) - 3x(x+5) = 208$

$(2x^2 + 11x + 5)(x+2) - 3x^2 - 15x = 208$

$2x^3 + 4x^2 + 11x^2 + 22x + 5x + 10 - 3x^2 - 15x = 208$

$2x^3 + 15x^2 + 27x - 3x^2 - 15x - 198 = 0$

$2x^3 + 12x^2 + 12x - 198 = 0$

$2(x^3 + 6x^2 + 6x - 99) = 0$

$$
\begin{array}{r|rrrr}
3 & 1 & 6 & 6 & -99 \\
  &   & 3 & 27 & 99 \\
\hline
  & 1 & 9 & 33 & 0
\end{array}
$$

$x^2 + 9x + 33 = 0$

$b^2 - 4ac = -51$

$x = 3$ in.

93. Because the polynomial has two obvious changes of direction; the smallest degree is 3.

95. Because the polynomial has two obvious changes of direction and two roots have multiplicity 2, the smallest degree is 5.

97. Answers will vary.

98. The function is undefined at $x = 1$ and $x = 2$.

99. The equation of the vertical asymptote is $x = 1$.

100. The equation of the horizontal asymptote is $y = 0$.

Mid-Chapter 3 Check Point

1. $f(x) = (x-3)^2 - 4$

The parabola opens up because $a > 0$.
The vertex is $(3, -4)$.
x-intercepts:

$0 = (x-3)^2 - 4$

$(x-3)^2 = 4$

$x - 3 = \pm\sqrt{4}$

$x = 3 \pm 2$

The equation has x-intercepts at $x = 1$ and $x = 5$.
y-intercept:

$f(0) = (0-3)^2 - 4 = 5$

domain: $(-\infty, \infty)$ range: $[-4, \infty)$

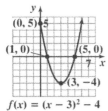

$f(x) = (x - 3)^2 - 4$

2. $f(x) = 5 - (x+2)^2$

The parabola opens down because $a < 0$.
The vertex is $(-2, 5)$.
x-intercepts:

$0 = 5 - (x+2)^2$

$(x+2)^2 = 5$

$x + 2 = \pm\sqrt{5}$

$x = -2 \pm \sqrt{5}$

y-intercept:

$f(0) = 5 - (0+2)^2 = 1$

domain: $(-\infty, \infty)$ range: $(-\infty, 5]$

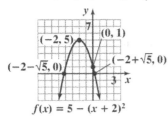

$f(x) = 5 - (x + 2)^2$

3. $f(x) = -x^2 - 4x + 5$

The parabola opens down because $a < 0$.

vertex: $x = -\dfrac{b}{2a} = -\dfrac{-4}{2(-1)} = -2$

$f(-2) = -(-2)^2 - 4(-2) + 5 = 9$

The vertex is $(-2, 9)$.
x-intercepts:

$0 = -x^2 - 4x + 5$

$x = \dfrac{-b \pm \sqrt{b^2 - 4ac}}{2a}$

$x = \dfrac{-(-4) \pm \sqrt{(-4)^2 - 4(-1)(5)}}{2(-1)}$

$x = \dfrac{4 \pm \sqrt{36}}{-2}$

$x = -2 \pm 3$

The x-intercepts are $x = 1$ and $x = -5$.
y-intercept:

$f(0) = -0^2 - 4(0) + 5 = 5$

domain: $(-\infty, \infty)$ range: $(-\infty, 9]$

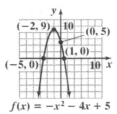

$f(x) = -x^2 - 4x + 5$

4. $f(x) = 3x^2 - 6x + 1$

The parabola opens up because $a > 0$.

vertex: $x = -\dfrac{b}{2a} = -\dfrac{-6}{2(3)} = 1$

$f(1) = 3(1)^2 - 6(1) + 1 = -2$

The vertex is $(1, -2)$.
x-intercepts:

$0 = 3x^2 - 6x + 1$

$x = \dfrac{-b \pm \sqrt{b^2 - 4ac}}{2a}$

$x = \dfrac{-(-6) \pm \sqrt{(-6)^2 - 4(3)(1)}}{2(3)}$

$x = \dfrac{6 \pm \sqrt{24}}{6}$

$x = \dfrac{3 \pm \sqrt{6}}{3}$

y-intercept:

$$f(0) = 3(0)^2 - 6(0) + 1 = 1$$

domain: $(-\infty, \infty)$ range: $[-2, \infty)$

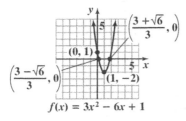

$$f(x) = 3x^2 - 6x + 1$$

5. $f(x) = (x-2)^2(x+1)^3$

$$(x-2)^2(x+1)^3 = 0$$

Apply the zero-product principle:

$(x-2)^2 = 0$ or $(x+1)^3 = 0$

$x - 2 = 0$ $x + 1 = 0$

$x = 2$ $x = -1$

The zeros are -1 and 2.

The graph of *f* crosses the *x*-axis at -1, since the zero has multiplicity 3. The graph touches the *x*-axis and turns around at 2 since the zero has multiplicity 2.

Since *f* is an odd-degree polynomial, degree 5, and since the leading coefficient, 1, is positive, the graph falls to the left and rises to the right.

Plot additional points as necessary and construct the graph.

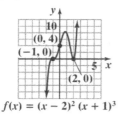

$$f(x) = (x-2)^2(x+1)^3$$

6. $f(x) = -(x-2)^2(x+1)^2$

$$-(x-2)^2(x+1)^2 = 0$$

Apply the zero-product principle:

$(x-2)^2 = 0$ or $(x+1)^2 = 0$

$x - 2 = 0$ $x + 1 = 0$

$x = 2$ $x = -1$

The zeros are -1 and 2.

The graph touches the *x*-axis and turns around both at -1 and 2 since both zeros have multiplicity 2.

Since *f* is an even-degree polynomial, degree 4, and since the leading coefficient, -1, is negative, the graph falls to the left and falls to the right.

Plot additional points as necessary and construct the graph.

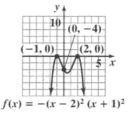

$$f(x) = -(x-2)^2(x+1)^2$$

7. $f(x) = x^3 - x^2 - 4x + 4$

$$x^3 - x^2 - 4x + 4 = 0$$

$$x^2(x-1) - 4(x-1) = 0$$

$$(x^2 - 4)(x-1) = 0$$

$$(x+2)(x-2)(x-1) = 0$$

Apply the zero-product principle:

$x + 2 = 0$ or $x - 2 = 0$ or $x - 1 = 0$

$x = -2$ $x = 2$ $x = 1$

The zeros are -2, 1, and 2.

The graph of *f* crosses the *x*-axis at all three zeros, -2, 1, and 2, since all have multiplicity 1.

Since *f* is an odd-degree polynomial, degree 3, and since the leading coefficient, 1, is positive, the graph falls to the left and rises to the right.

Plot additional points as necessary and construct the graph.

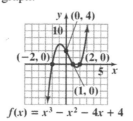

$$f(x) = x^3 - x^2 - 4x + 4$$

8. $f(x) = x^4 - 5x^2 + 4$

$$x^4 - 5x^2 + 4 = 0$$

$$(x^2 - 4)(x^2 - 1) = 0$$

$$(x + 2)(x - 2)(x + 1)(x - 1) = 0$$

Apply the zero-product principle,
$x = -2, \ x = 2, \ x = -1, \ x = 1$
The zeros are -2, -1, 1, and 2.

The graph crosses the x-axis at all four zeros, -2, -1, 1, and 2., since all have multiplicity 1.

Since f is an even-degree polynomial, degree 4, and since the leading coefficient, 1, is positive, the graph rises to the left and rises to the right.

Plot additional points as necessary and construct the graph.

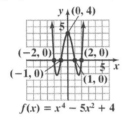

$f(x) = x^4 - 5x^2 + 4$

9. $f(x) = -(x + 1)^6$

$$-(x + 1)^6 = 0$$

$$(x + 1)^6 = 0$$

$$x + 1 = 0$$

$$x = -1$$

The zero is are -1.

The graph touches the x-axis and turns around at -1 since the zero has multiplicity 6.

Since f is an even-degree polynomial, degree 6, and since the leading coefficient, -1, is negative, the graph falls to the left and falls to the right.

Plot additional points as necessary and construct the graph.

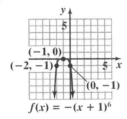

$f(x) = -(x + 1)^6$

10. $f(x) = -6x^3 + 7x^2 - 1$

To find the zeros, we use the Rational Zero Theorem:
List all factors of the constant term -1: ± 1
List all factors of the leading coefficient -6:
$\pm 1, \ \pm 2, \ \pm 3, \ \pm 6$

The possible rational zeros are:

$$\frac{\text{Factors of } -1}{\text{Factors of } -6} = \frac{\pm 1}{\pm 1, \ \pm 2, \ \pm 3, \ \pm 6}$$

$$= \pm 1, \ \pm \frac{1}{2}, \ \pm \frac{1}{3}, \ \pm \frac{1}{6}$$

We test values from the above list until we find a zero. One is shown next:

Test 1:

$$\begin{array}{r|rrrr} 1 & -6 & 7 & 0 & -1 \\ & & -6 & 1 & 1 \\ \hline & -6 & 1 & 1 & 0 \end{array}$$

The remainder is 0, so 1 is a zero. Thus,

$$-6x^3 + 7x^2 - 1 = 0$$

$$(x - 1)(-6x^2 + x + 1) = 0$$

$$-(x - 1)(6x^2 - x - 1) = 0$$

$$-(x - 1)(3x + 1)(2x - 1) = 0$$

Apply the zero-product property:

$$x = 1, \quad x = -\frac{1}{3}, \quad x = \frac{1}{2}$$

The zeros are $-\frac{1}{3}$, $\frac{1}{2}$, and 1.

The graph of f crosses the x-axis at all three zeros, $-\frac{1}{3}$, $\frac{1}{2}$, and 1, since all have multiplicity 1.

Since f is an odd-degree polynomial, degree 3, and since the leading coefficient, -6, is negative, the graph rises to the left and falls to the right.

Plot additional points as necessary and construct the graph.

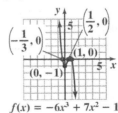

$f(x) = -6x^3 + 7x^2 - 1$

11. $f(x) = 2x^3 - 2x$

$$2x^3 - 2x = 0$$

$$2x(x^2 - 1) = 0$$

$$2x(x+1)(x-1) = 0$$

Apply the zero-product principle:

$x = 0, \qquad x = -1, \qquad x = 1$

The zeros are -1, 0, and 1.

The graph of f crosses the x-axis at all three zeros, -1, 0, and 1, since all have multiplicity 1.

Since f is an odd-degree polynomial, degree 3, and since the leading coefficient, 2, is positive, the graph falls to the left and rises to the right.

Plot additional points as necessary and construct the graph.

$f(x) = 2x^3 - 2x$

12. $f(x) = x^3 - 2x^2 + 26x$

$$x^3 - 2x^2 + 26x = 0$$

$$x(x^2 - 2x + 26) = 0$$

Note that $x^2 - 2x + 26$ does not factor, so we use the quadratic formula:

$x = 0 \quad$ or $\quad x^2 - 2x + 26 = 0$

$$a = 1, \ b = -2, \ c = 26$$

$$x = \frac{-(-2) \pm \sqrt{(-2)^2 - 4(1)(26)}}{2(1)}$$

$$= \frac{2 \pm \sqrt{-100}}{2} = \frac{2 \pm 10i}{2} = 1 \pm 5i$$

The zeros are 0 and $1 \pm 5i$.

The graph of f crosses the x-axis at 0 (the only real zero), since it has multiplicity 1.

Since f is an odd-degree polynomial, degree 3, and since the leading coefficient, 1, is positive, the graph falls to the left and rises to the right.

Plot additional points as necessary and construct the graph.

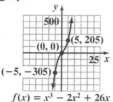

$f(x) = x^3 - 2x^2 + 26x$

13. $f(x) = -x^3 + 5x^2 - 5x - 3$

To find the zeros, we use the Rational Zero Theorem:

List all factors of the constant term -3: $\pm 1, \ \pm 3$

List all factors of the leading coefficient -1: ± 1

The possible rational zeros are:

$$\frac{\text{Factors of } -3}{\text{Factors of } -1} = \frac{\pm 1, \ \pm 3}{\pm 1} = \pm 1, \ \pm 3$$

We test values from the previous list until we find a zero. One is shown next:

Test 3:

$$\begin{array}{r|rrrr} 3 & -1 & 5 & -5 & -3 \\ & & -3 & 6 & 3 \\ \hline & -1 & 2 & 1 & 0 \end{array}$$

The remainder is 0, so 3 is a zero. Thus,

$$-x^3 + 5x^2 - 5x - 3 = 0$$

$$(x-3)(-x^2 + 2x + 1) = 0$$

$$-(x-3)(x^2 - 2x - 1) = 0$$

Note that $x^2 - 2x - 1$ does not factor, so we use the quadratic formula:

$x - 3 = 0 \quad$ or $\quad x^2 - 2x - 1 = 0$

$x = 3 \qquad\qquad a = 1, \ b = -2, \ c = -1$

$$x = \frac{-(-2) \pm \sqrt{(-2)^2 - 4(1)(-1)}}{2(1)}$$

$$= \frac{2 \pm \sqrt{8}}{2} = \frac{2 \pm 2\sqrt{2}}{2} = 1 \pm \sqrt{2}$$

The zeros are 3 and $1 \pm \sqrt{2}$.

The graph of f crosses the x-axis at all three zeros, 3 and $1 \pm \sqrt{2}$, since all have multiplicity 1.

Since f is an odd-degree polynomial, degree 3, and since the leading coefficient, -1, is negative, the graph rises to the left and falls to the right.

Plot additional points as necessary and construct the graph.

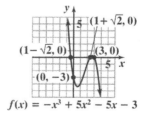

$f(x) = -x^3 + 5x^2 - 5x - 3$

14. $x^3 - 3x + 2 = 0$

We begin by using the Rational Zero Theorem to determine possible rational roots.

Factors of the constant term 2: $\pm 1, \pm 2$

Factors of the leading coefficient 1: ± 1

The possible rational zeros are:

$$\frac{\text{Factors of } 2}{\text{Factors of } 1} = \frac{\pm 1, \pm 2}{\pm 1} = \pm 1, \pm 2$$

We test values from above until we find a root. One is shown next:

Test 1:

$$\underline{1|} \quad 1 \quad 0 \quad -3 \quad 2$$
$$ \quad \quad 1 \quad 1 \quad -2$$
$$\overline{ \quad 1 \quad 1 \quad -2 \quad 0}$$

The remainder is 0, so 1 is a root of the equation. Thus,

$$x^3 - 3x + 2 = 0$$
$$(x-1)(x^2 + x - 2) = 0$$
$$(x-1)(x+2)(x-1) = 0$$
$$(x-1)^2(x+2) = 0$$

Apply the zero-product property:

$$(x-1)^2 = 0 \quad \text{or} \quad x + 2 = 0$$
$$x - 1 = 0 \qquad\qquad x = -2$$
$$x = 1$$

The solutions are -2 and 1, and the solution set is $\{-2, 1\}$.

15. $6x^3 - 11x^2 + 6x - 1 = 0$

We begin by using the Rational Zero Theorem to determine possible rational roots.

Factors of the constant term -1: ± 1

Factors of the leading coefficient 6: $\pm 1, \pm 2, \pm 3, \pm 6$

The possible rational zeros are:

$$\frac{\text{Factors of } -1}{\text{Factors of } 6} = \frac{\pm 1}{\pm 1, \pm 2, \pm 3, \pm 6}$$
$$= \pm 1, \pm \frac{1}{2}, \pm \frac{1}{3}, \pm \frac{1}{6}$$

We test values from above until we find a root. One is shown next:

Test 1:

$$\underline{1|} \quad 6 \quad -11 \quad 6 \quad -1$$
$$ \quad \quad 6 \quad -5 \quad 1$$
$$\overline{ \quad 6 \quad -5 \quad 1 \quad 0}$$

The remainder is 0, so 1 is a root of the equation. Thus,

$$6x^3 - 11x^2 + 6x - 1 = 0$$
$$(x-1)(6x^2 - 5x + 1) = 0$$
$$(x-1)(3x-1)(2x-1) = 0$$

Apply the zero-product property:

$$x - 1 = 0 \quad \text{or} \quad 3x - 1 = 0 \quad \text{or} \quad 2x - 1 = 0$$
$$x = 1 \qquad\qquad x = \frac{1}{3} \qquad\qquad x = \frac{1}{2}$$

The solutions are $\frac{1}{3}, \frac{1}{2}$ and 1, and the solution set is $\left\{ \frac{1}{3}, \frac{1}{2}, 1 \right\}$.

16. $(2x+1)(3x-2)^3(2x-7) = 0$

Apply the zero-product property:

$$2x + 1 = 0 \quad \text{or} \quad (3x-2)^3 = 0 \quad \text{or} \quad 2x - 7 = 0$$
$$x = -\frac{1}{2} \qquad\qquad 3x - 2 = 0 \qquad\qquad x = \frac{7}{2}$$
$$\phantom{x = -\frac{1}{2}} \qquad\qquad x = \frac{2}{3}$$

The solutions are $-\frac{1}{2}, \frac{2}{3}$ and $\frac{7}{2}$, and the solution set is $\left\{ -\frac{1}{2}, \frac{2}{3}, \frac{7}{2} \right\}$.

17. $2x^3 + 5x^2 - 200x - 500 = 0$

We begin by using the Rational Zero Theorem to determine possible rational roots.

Factors of the constant term -500:

$\pm 1, \pm 2, \pm 4, \pm 5, \pm 10, \pm 20, \pm 25,$

$\pm 50, \pm 100, \pm 125, \pm 250, \pm 500$

Factors of the leading coefficient 2: $\pm 1, \pm 2$

The possible rational zeros are:

$$\frac{\text{Factors of } 500}{\text{Factors of } 2} = \pm 1, \pm 2, \pm 4, \pm 5,$$

$\pm 10, \pm 20, \pm 25, \pm 50, \pm 100, \pm 125,$

$\pm 250, \pm 500, \pm \frac{1}{2}, \pm \frac{5}{2}, \pm \frac{25}{2}, \pm \frac{125}{2}$

We test values from above until we find a root. One is shown next:

Test 10:

$$\begin{array}{r|rrrr} 10 & 2 & 5 & -200 & -500 \\ & & 20 & 250 & 500 \\ \hline & 2 & 25 & 50 & 0 \end{array}$$

The remainder is 0, so 10 is a root of the equation. Thus,

$$2x^3 + 5x^2 - 200x - 500 = 0$$
$$(x-10)(2x^2 + 25x + 50) = 0$$
$$(x-10)(2x+5)(x+10) = 0$$

Apply the zero-product property:

$$x - 10 = 0 \quad \text{or} \quad 2x + 5 = 0 \quad \text{or} \quad x + 10 = 0$$
$$x = 10 \qquad\qquad x = -\frac{5}{2} \qquad\qquad x = -10$$

The solutions are -10, $-\frac{5}{2}$, and 10, and the solution set is $\left\{-10, \ -\frac{5}{2}, \ 10\right\}$.

18.
$$x^4 - x^3 - 11x^2 = x + 12$$
$$x^4 - x^3 - 11x^2 - x - 12 = 0$$

We begin by using the Rational Zero Theorem to determine possible rational roots.

Factors of the constant term -12:
$\pm 1, \ \pm 2, \ \pm 3, \ \pm 4, \ \pm 6, \ \pm 12$

Factors of the leading coefficient 1: ± 1

The possible rational zeros are:

$$\frac{\text{Factors of } -12}{\text{Factors of } 1}$$
$$= \frac{\pm 1, \ \pm 2, \ \pm 3, \ \pm 4, \ \pm 6, \ \pm 12}{\pm 1}$$
$$= \pm 1, \ \pm 2, \ \pm 3, \ \pm 4, \ \pm 6, \ \pm 12$$

We test values from this list we find a root. One possibility is shown next:

Test -3:

$$\begin{array}{r|rrrrr} -3 & 1 & -1 & -11 & -1 & -12 \\ & & -3 & 12 & -3 & 12 \\ \hline & 1 & -4 & 1 & -4 & 0 \end{array}$$

The remainder is 0, so -3 is a root of the equation. Using the Factor Theorem, we know that $x - 1$ is a factor. Thus,

$$x^4 - x^3 - 11x^2 - x - 12 = 0$$
$$(x+3)(x^3 - 4x^2 + x - 4) = 0$$
$$(x+3)\left[x^2(x-4) + 1(x-4)\right] = 0$$
$$(x+3)(x-4)(x^2 + 1) = 0$$

As this point we know that -3 and 4 are roots of the equation. Note that $x^2 + 1$ does not factor, so we use the square-root principle: $x^2 + 1 = 0$

$$x^2 = -1$$
$$x = \pm\sqrt{-1} = \pm i$$

The roots are -3, 4, and $\pm i$, and the solution set is $\{-3, \ 4, \ \pm i\}$.

19. $2x^4 + x^3 - 17x^2 - 4x + 6 = 0$

We begin by using the Rational Zero Theorem to determine possible rational roots.

Factors of the constant term 6: $\pm 1, \ \pm 2, \ \pm 3, \ \pm 6$

Factors of the leading coefficient 4: $\pm 1, \ \pm 2$

The possible rational roots are:

$$\frac{\text{Factors of } 6}{\text{Factors of } 2} = \frac{\pm 1, \ \pm 2, \ \pm 3, \ \pm 6}{\pm 1, \ \pm 2}$$
$$= \pm 1, \ \pm 2, \ \pm 3, \ \pm 6, \ \pm\frac{1}{2}, \ \pm\frac{3}{2}$$

We test values from above until we find a root. One possibility is shown next:

Test -3:

$$\begin{array}{r|rrrrr} -3 & 2 & 1 & -17 & -4 & 6 \\ & & -6 & 15 & 6 & -6 \\ \hline & 2 & -5 & -2 & 2 & 0 \end{array}$$

The remainder is 0, so -3 is a root. Using the Factor Theorem, we know that $x + 3$ is a factor of the polynomial. Thus,

$$2x^4 + x^3 - 17x^2 - 4x + 6 = 0$$
$$(x+3)(2x^3 - 5x^2 - 2x + 2) = 0$$

To solve the equation above, we need to factor $2x^3 - 5x^2 - 2x + 2$. We continue testing potential roots:

Test $\dfrac{1}{2}$:

$$\begin{array}{r|rrrr}
\frac{1}{2} & 2 & -5 & -2 & 2 \\
 & & 1 & -2 & -2 \\
\hline
 & 2 & -4 & -4 & 0
\end{array}$$

The remainder is 0, so $\dfrac{1}{2}$ is a zero and $x-\dfrac{1}{2}$ is a factor.

Summarizing our findings so far, we have

$$2x^4 + x^3 - 17x^2 - 4x + 6 = 0$$

$$(x+3)(2x^3 - 5x^2 - 2x + 2) = 0$$

$$(x+3)\left(x - \dfrac{1}{2}\right)(2x^2 - 4x - 4) = 0$$

$$2(x+3)\left(x - \dfrac{1}{2}\right)(x^2 - 2x - 2) = 0$$

At this point, we know that -3 and $\dfrac{1}{2}$ are roots of the equation. Note that $x^2 - 2x - 2$ does not factor, so we use the quadratic formula:

$$x^2 - 2x - 2 = 0$$

$$a = 1, \ b = -2, \ c = -2$$

$$x = \dfrac{-(-2) \pm \sqrt{(-2)^2 - 4(1)(-2)}}{2(1)}$$

$$= \dfrac{2 \pm \sqrt{4+8}}{2} = \dfrac{2 \pm \sqrt{12}}{2} = \dfrac{2 \pm 2\sqrt{3}}{2} = 1 \pm \sqrt{3}$$

The solutions are -3 , $\dfrac{1}{2}$, and $1 \pm \sqrt{3}$, and the

solution set is $\left\{-3, \ \dfrac{1}{2}, \ 1 \pm \sqrt{3}\right\}$.

20. $\quad P(x) = -x^2 + 150x - 4425$

Since $a = -1$ is negative, we know the function opens down and has a maximum at

$$x = -\dfrac{b}{2a} = -\dfrac{150}{2(-1)} = -\dfrac{150}{-2} = 75 \ .$$

$$P(75) = -75^2 + 150(75) - 4425$$

$$= -5625 + 11{,}250 - 4425 = 1200$$

The company will maximize its profit by manufacturing and selling 75 cabinets per day. The maximum daily profit is $1200.

21. Let $x =$ one of the numbers;
$-18 - x =$ the other number

The product is $f(x) = x(-18 - x) = -x^2 - 18x$

The x-coordinate of the maximum is

$$x = -\dfrac{b}{2a} = -\dfrac{-18}{2(-1)} = -\dfrac{-18}{-2} = -9.$$

$$f(-9) = -9\left[-18 - (-9)\right]$$

$$= -9(-18 + 9) = -9(-9) = 81$$

The vertex is $(-9, 81)$. The maximum product is 81. This occurs when the two numbers are -9 and $-18 - (-9) = -9$.

22. Let $x =$ height of triangle;
$40 - 2x =$ base of triangle

$$A = \dfrac{1}{2}bh = \dfrac{1}{2}x(40 - 2x)$$

$$A(x) = 20x - x^2$$

The height at which the triangle will have

maximum area is $x = -\dfrac{b}{2a} = -\dfrac{20}{2(-1)} = 10.$

$$A(10) = 20(10) - (10)^2 = 100$$

The maximum area is 100 squares inches.

23.

$$\begin{array}{r}
2x^2 - x - 3 \\
3x^2 - 1 \overline{\smash{\big)}\ 6x^4 - 3x^3 - 11x^2 + 2x + 4} \\
\underline{6x^4 - 2x^2 } \\
-3x^3 - 9x^2 + 2x \\
\underline{-3x^3 + x } \\
-9x^2 + x + 4 \\
\underline{-9x^2 + 3} \\
x + 1
\end{array}$$

$$2x^2 - x - 3 + \dfrac{x+1}{3x^2 - 1}$$

24. $\quad \left(2x^4 - 13x^3 + 17x^2 + 18x - 24\right) \div (x - 4)$

$$\begin{array}{r|rrrrr}
4 & 2 & -13 & 17 & 18 & -24 \\
 & & 8 & -20 & -12 & 24 \\
\hline
 & 2 & -5 & -3 & 6 & 0
\end{array}$$

The quotient is $2x^3 - 5x^2 - 3x + 6$.

25. $(x-1)(x-i)(x+i) = (x-1)(x^2+1)$

$f(x) = a_n(x-1)(x^2+1)$

$f(-1) = a_n(-1-1)\big((-1)^2+1\big) = -4a_n = 8$

$a_n = -2$

$f(x) = -2(x-1)(x^2+1)$ or $-2x^3 + 2x^2 - 2x + 2$

26. $(x-2)(x-2)(x-3i)(x+3i)$

$= (x-2)(x-2)(x^2+9)$

$f(x) = a_n(x-2)(x-2)(x^2+9)$

$f(0) = a_n(0-2)(0-2)(0^2+9)$

$36 = 36a_n$

$a_n = 1$

$f(x) = 1(x-2)(x-2)(x^2+9)$

$f(x) = x^4 - 4x^3 + 13x^2 - 36x + 36$

27. $f(x) = x^3 - x - 5$

$f(1) = 1^3 - 1 - 5 = -5$

$f(2) = 2^3 - 2 - 5 = 1$

Yes, the function must have a real zero between 1 and 2 because $f(1)$ and $f(2)$ have opposite signs.

Section 3.5

Check Point Exercises

1. Because division by 0 is undefined, we must exclude from the domain of each function values of x that cause the polynomial function in the denominator to be 0.

   a. $x - 5 = 0$

       $x = 5$

       $\{x \mid x \neq 5\}$ or $(-\infty, 5) \cup (5, \infty)$.

   b. $x^2 - 25 = 0$

       $x^2 = 25$

       $x = \pm 5$

       $\{x \mid x \neq 5, x \neq -5\}$ or $(-\infty, -5) \cup (-5, 5) \cup (5, \infty)$.

   c. The denominator cannot equal zero.
       All real numbers or $(-\infty, \infty)$. .

2.   **a.** $x^2 - 1 = 0$

       $x^2 = 1$

       $x = 1, x = -1$

  b. $g(x) = \dfrac{x-1}{x^2-1} = \dfrac{x-1}{(x-1)(x+1)} = \dfrac{1}{x+1}$

       $x = -1$

  c. The denominator cannot equal zero.
       No vertical asymptotes.

3.   **a.** The degree of the numerator, 2, is equal to the degree of the denominator, 2. Thus, the leading coefficients of the numerator and denominator, 9 and 3, are used to obtain the equation of the horizontal asymptote.

       $y = \dfrac{9}{3} = 3$

       $y = 3$ is a horizontal asymptote.

  b. The degree of the numerator, 1, is less than the degree of the denominator, 2. Thus, the graph has the x-axis as a horizontal asymptote $y = 0$ is a horizontal asymptote.

  c. The degree of the numerator, 3, is greater than the degree of the denominator, 2. Thus, the graph has no horizontal asymptote.

4. Begin with the graph of $f(x) = \dfrac{1}{x}$.

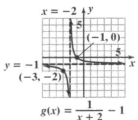

$$g(x) = \frac{1}{x+2} - 1$$

Shift the graph 2 units to the left by subtracting 2 from each x-coordinate. Shift the graph 1 unit down by subtracting 1 from each y-coordinate.

5. $f(x) = \dfrac{3x - 3}{x - 2}$

$f(-x) = \dfrac{3(-x) - 3}{-x - 2} = \dfrac{-3x - 3}{-x - 2} = \dfrac{3x + 3}{x + 2}$

no symmetry

$f(0) = \dfrac{3(0) - 3}{0 - 2} = \dfrac{3}{2}$

The y-intercept is $\dfrac{3}{2}$.

$3x - 3 = 0$

$\quad 3x = 3$

$\quad\ \ x = 1$

The x-intercept is 1.

Vertical asymptote:

$x - 2 = 0$

$\quad\ \ x = 2$

Horizontal asymptote:

$y = \dfrac{3}{1} = 3$

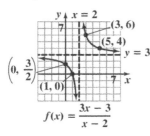

6. $f(x) = \dfrac{2x^2}{x^2 - 9}$

$f(-x) = \dfrac{2(-x)^2}{(-x)^2 - 9} = \dfrac{2x^2}{x^2 - 9} = f(x)$

The y-axis symmetry.

$f(0) = \dfrac{2(0)^2}{0^2 - 9} = 0$

The y-intercept is 0.

$2x^2 = 0$

$\quad\ x = 0$

The x-intercept is 0.

vertical asymptotes:

$x^2 - 9 = 0$

$x = 3, x = -3$

horizontal asymptote:

$y = \dfrac{2}{1} = 2$

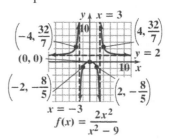

7. $f(x) = \dfrac{x^4}{x^2 + 2}$

$f(-x) = \dfrac{(-x)^4}{(-x)^2 + 2} = \dfrac{x^4}{x^2 + 2} = f(x)$

y-axis symmetry

$f(0) = \dfrac{0^4}{0^2 + 2} = 0$

The y-intercept is 0.

$x^4 = 0$

$\quad x = 0$

The x-intercept is 0.

vertical asymptotes:

$x^2 + 2 = 0$

$\quad\ x^2 = -2$

no vertical asymptotes

horizontal asymptote:

Since $n > m$, there is no horizontal asymptote.

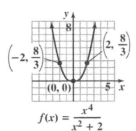

8.

$$\begin{array}{r|rrr}
2 & 2 & -5 & 7 \\
  &   & 4 & -2 \\
\hline
  & 2 & -1 & 5
\end{array}$$

the equation of the slant asymptote is
$y = 2x - 1$.

9. **a.** $C(x) = 500,000 + 400x$

b. $\bar{C}(x) = \dfrac{500,000 + 400x}{x}$

c. $\bar{C}(1000) = \dfrac{500,000 + 400(1000)}{1000}$

$= 900$

$\bar{C}(10,000) = \dfrac{500,000 + 400(10,000)}{10,000}$

$= 450$

$\bar{C}(100,000) = \dfrac{500,000 + 400(100,000)}{100,000}$

$= 405$

The average cost per wheelchair of producing 1000, 10,000, and 100,000 wheelchairs is $900, $450, and $405, respectively.

d. $y = \dfrac{400}{1} = 400$

The cost per wheelchair approaches $400 as more wheelchairs are produced.

Concept and Vocabulary Check 3.5

1. polynomial

2. false

3. true

4. vertical asymptote; $x = -5$

5. horizontal asymptote; $y = 0$; $y = \dfrac{1}{3}$

6. true

7. left; down

8. one more than

9. $y = 3x + 5$

Exercise Set 3.5

1. $f(x) = \dfrac{5x}{x-4}$

$\{x \mid x \neq 4\}$

3. $g(x) = \dfrac{3x^2}{(x-5)(x+4)}$

$\{x \mid x \neq 5, x \neq -4\}$

5. $h(x) = \dfrac{x+7}{x^2 - 49}$

$x^2 - 49 = (x-7)(x+7)$

$\{x \mid x \neq 7, x \neq -7\}$

7. $f(x) = \dfrac{x+7}{x^2 + 49}$

all real numbers

9. $-\infty$

11. $-\infty$

13. 0

15. $+\infty$

17. $-\infty$

19. 1

21. $f(x) = \dfrac{x}{x+4}$

$x + 4 = 0$

$x = -4$

vertical asymptote: $x = -4$

There are no holes.

23. $g(x) = \dfrac{x+3}{x(x+4)}$

$x(x+4) = 0$

$x = 0, x = -4$

vertical asymptotes: $x = 0$, $x = -4$

There are no holes.

25. $h(x) = \dfrac{x}{x(x+4)} = \dfrac{1}{x+4}$

$x + 4 = 0$

$x = -4$

vertical asymptote: $x = -4$

There is a hole at $x = 0$.

27. $r(x) = \dfrac{x}{x^2 + 4}$

$x^2 + 4$ has no real zeros
There are no vertical asymptotes.
There are no holes.

29. $f(x) = \dfrac{x^2 - 9}{x - 3}$

$\quad = \dfrac{(x+3)(x-3)}{x-3}$

$\quad = x + 3$

There are no vertical asymptotes.
There is a hole at $x = 3$.

31. $g(x) = \dfrac{x-3}{x^2 - 9}$

$\quad = \dfrac{x-3}{(x+3)(x-3)}$

$\quad = \dfrac{1}{x+3}$

vertical asymptote: $x = -3$
There is a hole at $x = 3$.

33. $h(x) = \dfrac{x+7}{x^2 + 4x - 21}$

$\quad = \dfrac{x+7}{(x+7)(x-3)}$

$\quad = \dfrac{1}{x-3}$

vertical asymptote: $x = 3$
There is a hole at $x = -7$.

35. $r(x) = \dfrac{x^2 + 4x - 21}{x+7}$

$\quad = \dfrac{(x+7)(x-3)}{x+7}$

$\quad = x - 3$

There are no vertical asymptotes.
There is a hole at $x = -7$.

37. $f(x) = \dfrac{12x}{3x^2 + 1}$

$n < m$
horizontal asymptote: $y = 0$

39. $g(x) = \dfrac{12x^2}{3x^2 + 1}$

$n = m$,

horizontal asymptote: $y = \dfrac{12}{3} = 4$

41. $h(x) = \dfrac{12x^3}{3x^2 + 1}$

$n > m$
no horizontal asymptote

43. $f(x) = \dfrac{-2x + 1}{3x + 5}$

$n = m$

horizontal asymptote: $y = -\dfrac{2}{3}$

45. $g(x) = \dfrac{1}{x - 1}$

Shift the graph of $f(x) = \dfrac{1}{x}$ 1 unit to the right.

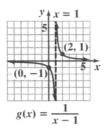

47. $h(x) = \dfrac{1}{x} + 2$

Shift the graph of $f(x) = \dfrac{1}{x}$ 2 units up.

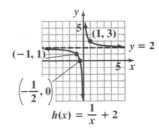

49. $g(x) = \dfrac{1}{x+1} - 2$

Shift the graph of $f(x) = \dfrac{1}{x}$ 1 unit left and 2 units down.

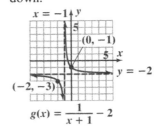

51. $g(x) = \dfrac{1}{(x+2)^2}$

Shift the graph of $f(x) = \dfrac{1}{x^2}$ 2 units left.

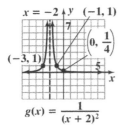

$g(x) = \dfrac{1}{(x+2)^2}$

53. $h(x) = \dfrac{1}{x^2} - 4$

Shift the graph of $f(x) = \dfrac{1}{x^2}$ 4 units down.

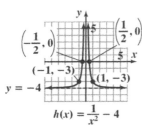

$h(x) = \dfrac{1}{x^2} - 4$

55. $h(x) = \dfrac{1}{(x-3)^2} + 1$

Shift the graph of $f(x) = \dfrac{1}{x^2}$ 3 units right and 1 unit up.

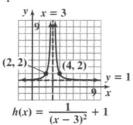

$h(x) = \dfrac{1}{(x-3)^2} + 1$

57. $f(x) = \dfrac{4x}{x-2}$

$f(-x) = \dfrac{4(-x)}{(-x)-2} = \dfrac{4x}{x+2}$

$f(-x) \neq f(x), f(-x) \neq -f(x)$

no symmetry

y-intercept: $y = \dfrac{4(0)}{0-2} = 0$

x-intercept: $4x = 0$

$x = 0$

vertical asymptote:

$x - 2 = 0$

$x = 2$

horizontal asymptote:

$n = m$, so $y = \dfrac{4}{1} = 4$

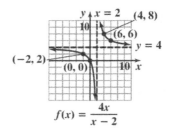

$f(x) = \dfrac{4x}{x-2}$

59. $f(x) = \dfrac{2x}{x^2 - 4}$

$f(-x) = \dfrac{2(-x)}{(-x)^2 - 4} = -\dfrac{2x}{x^2 - 4} = -f(x)$

Origin symmetry

y-intercept: $\dfrac{2(0)}{0^2 - 4} = \dfrac{0}{-4} = 0$

x-intercept:

$2x = 0$

$x = 0$

vertical asymptotes:

$x^2 - 4 = 0$

$x = \pm 2$

horizontal asymptote:

$n < m$ so $y = 0$

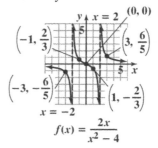

$f(x) = \dfrac{2x}{x^2 - 4}$

61. $f(x) = \dfrac{2x^2}{x^2 - 1}$

$f(-x) = \dfrac{2(-x)^2}{(-x)^2 - 1} = \dfrac{2x^2}{x^2 - 1} = f(x)$

y-axis symmetry

y-intercept: $y = \dfrac{2(0)^2}{0^2 - 1} = \dfrac{0}{1} = 0$

x-intercept:
$2x^2 = 0$
$x = 0$
vertical asymptote:
$x^2 - 1 = 0$
$x^2 = 1$
$x = \pm 1$
horizontal asymptote:

$n = m$, so $y = \dfrac{2}{1} = 2$

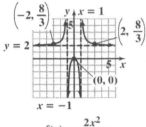

$$f(x) = \dfrac{2x^2}{x^2 - 1}$$

63. $f(x) = \dfrac{-x}{x + 1}$

$f(-x) = \dfrac{-(-x)}{(-x) + 1} = \dfrac{x}{-x + 1}$

$f(-x) \neq f(x), f(-x) \neq -f(x)$
no symmetry

y-intercept: $y = \dfrac{-(0)}{0 + 1} = \dfrac{0}{1} = 0$

x-intercept:
$-x = 0$
$x = 0$
vertical asymptote:
$x + 1 = 0$
$x = -1$

horizontal asymptote:

$n = m$, so $y = \dfrac{-1}{1} = -1$

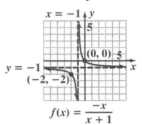

$$f(x) = \dfrac{-x}{x + 1}$$

65. $f(x) = -\dfrac{1}{x^2 - 4}$

$f(-x) = -\dfrac{1}{(-x)^2 - 4} = -\dfrac{1}{x^2 - 4} = f(x)$

y-axis symmetry

y-intercept: $y = -\dfrac{1}{0^2 - 4} = \dfrac{1}{4}$

x-intercept: $-1 \neq 0$
no *x*-intercept
vertical asymptotes:
$x^2 - 4 = 0$
$x^2 = 4$
$x = \pm 2$
horizontal asymptote:
$n < m$ or $y = 0$

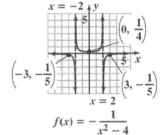

$$f(x) = -\dfrac{1}{x^2 - 4}$$

67. $f(x) = \dfrac{2}{x^2 + x - 2}$

$f(-x) = -\dfrac{2}{(-x)^2 - x - 2} = \dfrac{2}{x^2 - x - 2}$

$f(-x) \neq f(x), f(-x) \neq -f(x)$

no symmetry

y-intercept: $y = \dfrac{2}{0^2 + 0 - 2} = \dfrac{2}{-2} = -1$

x-intercept: none
vertical asymptotes:
$x^2 + x - 2 = 0$
$(x+2)(x-1) = 0$
$x = -2, x = 1$
horizontal asymptote:
$n < m$ so $y = 0$

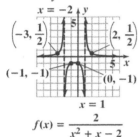

$f(x) = \dfrac{2}{x^2 + x - 2}$

69. $f(x) = \dfrac{2x^2}{x^2 + 4}$

$f(-x) = \dfrac{2(-x)^2}{(-x)^2 + 4} = \dfrac{2x^2}{x^2 + 4} = f(x)$

y axis symmetry

y-intercept: $y = \dfrac{2(0)^2}{0^2 + 4} = 0$

x-intercept: $2x^2 = 0$
$x = 0$
vertical asymptote: none
horizontal asymptote:

$n = m$, so $y = \dfrac{2}{1} = 2$

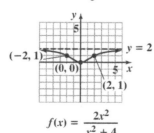

$f(x) = \dfrac{2x^2}{x^2 + 4}$

71. $f(x) = \dfrac{x+2}{x^2 + x - 6}$

$f(-x) = \dfrac{-x+2}{(-x)^2 - (-x) - 6} = \dfrac{-x+2}{x^2 + x - 6}$

$f(-x) \neq f(x), f(-x) \neq -f(x)$

no symmetry

y-intercept: $y = \dfrac{0+2}{0^2 + 0 - 6} = -\dfrac{2}{6} = -\dfrac{1}{3}$

x-intercept:
$x + 2 = 0$
$x = -2$
vertical asymptotes:
$x^2 + x - 6 = 0$
$(x+3)(x-2)$
$x = -3, x = 2$
horizontal asymptote:
$n < m$, so $y = 0$

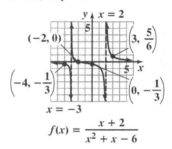

$f(x) = \dfrac{x+2}{x^2 + x - 6}$

73. $f(x) = \dfrac{x-2}{x^2 - 4}$

$f(-x) = \dfrac{-x-2}{(-x)^2 - 4} = \dfrac{-x-2}{x^2 - 4}$

$f(-x) \neq f(x), f(-x) \neq -f(x)$

no symmetry

y-intercept: $y = \dfrac{0-2}{0^2 - 4} = \dfrac{-2}{-4} = \dfrac{1}{2}$

x-intercept:
$x - 2 = 0, x = 2$
vertical asymptotes:

$f(x) = \dfrac{x-2}{x^2 - 4}$

$\quad = \dfrac{x-2}{(x-2)(x+2)}$

$\quad = \dfrac{1}{x+2}$

$x = -2$ is a vertical asymptote.

Furthermore, the value 2 causes the original denominator to be zero, but the reduced form of the function's equation does not cause the denominator to be zero. Thus, there is a hole at $x = 2$.

horizontal asymptote:

$n < m$, so $y = 0$

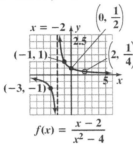

$$f(x) = \frac{x-2}{x^2-4}$$

75. $f(x) = \dfrac{x^4}{x^2+2}$

$f(-x) = \dfrac{(-x)^4}{(-x)^2+2} = \dfrac{x^4}{x^2+2} = f(x)$

y-axis symmetry

y-intercept: $y = \dfrac{0^4}{0^2+2} = 0$

x-intercept: $x^4 = 0$

$x = 0$

vertical asymptote: none
horizontal asymptote:

$n > m$, so none

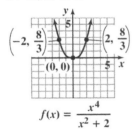

$$f(x) = \frac{x^4}{x^2+2}$$

77. $f(x) = \dfrac{x^2+x-12}{x^2-4}$

$f(-x) = \dfrac{(-x)^2-x-12}{(-x)^2-4} = \dfrac{x^2-x-12}{x^2-4}$

$f(-x) \ne f(x), f(-x) \ne -f(x)$

no symmetry

y-intercept: $y = \dfrac{0^2+0-12}{0^2-4} = 3$

x-intercept: $x^2+x-12 = 0$

$(x-3)(x+4) = 0$

$x = 3, x = -4$

vertical asymptotes:

$x^2-4 = 0$

$(x-2)(x+2) = 0$

$x = 2, x = -2$

horizontal asymptote:

$n = m$, so $y = \dfrac{1}{1} = 1$

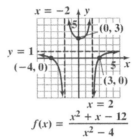

$$f(x) = \frac{x^2+x-12}{x^2-4}$$

79. $f(x) = \dfrac{3x^2+x-4}{2x^2-5x}$

$f(-x) = \dfrac{3(-x)^2-x-4}{2(-x)^2+5x} = \dfrac{3x^2-x-4}{2x^2+5x}$

$f(-x) \ne f(x), f(-x) \ne -f(x)$

no symmetry

y-intercept: $y = \dfrac{3(0)^2+0-4}{2(0)^2-5(0)} = \dfrac{-4}{0}$

no y-intercept
x-intercepts:

$3x^2+x-4 = 0$

$(3x+4)(x-1) = 0$

$3x+4 = 0 \quad x-1 = 0$

$3x = -4$

$x = -\dfrac{4}{3}, x = 1$

vertical asymptotes:

$2x^2-5x = 0$

$x(2x-5) = 0$

$x = 0, 2x = 5$

$x = \dfrac{5}{2}$

horizontal asymptote:

$n = m$, so $y = \dfrac{3}{2}$

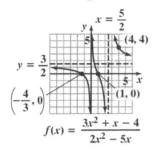

$$f(x) = \dfrac{3x^2 + x - 4}{2x^2 - 5x}$$

81. a. Slant asymptote:

$$f(x) = x - \dfrac{1}{x}$$

$$y = x$$

b. $f(x) = \dfrac{x^2 - 1}{x}$

$$f(-x) = \dfrac{(-x)^2 - 1}{(-x)} = \dfrac{x^2 - 1}{-x} = -f(x)$$

Origin symmetry

y-intercept: $y = \dfrac{0^2 - 1}{0} = \dfrac{-1}{0}$

no *y*-intercept

x-intercepts: $x^2 - 1 = 0$
$x = \pm 1$
vertical asymptote: $x = 0$

horizontal asymptote:
$n < m$, so none exist.

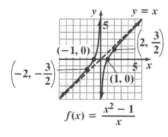

$$f(x) = \dfrac{x^2 - 1}{x}$$

83. a. Slant asymptote:

$$f(x) = x + \dfrac{1}{x}$$

$$y = x$$

b. $f(x) = \dfrac{x^2 + 1}{x}$

$$f(-x) = \dfrac{(-x)^2 + 1}{-x} = \dfrac{x^2 + 1}{-x} = -f(x)$$

Origin symmetry

y-intercept: $y = \dfrac{0^2 + 1}{0} = \dfrac{1}{0}$

no *y*-intercept
x-intercept:
$x^2 + 1 = 0$
$x^2 = -1$
no *x*-intercept
vertical asymptote: $x = 0$
horizontal asymptote:
$n > m$, so none exist.

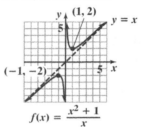

$$f(x) = \dfrac{x^2 + 1}{x}$$

85. a. Slant asymptote:

$$f(x) = x + 4 + \dfrac{6}{x - 3}$$

$$y = x + 4$$

b. $f(x) = \dfrac{x^2 + x - 6}{x - 3}$

$$f(-x) = \dfrac{(-x)^2 + (-x) - 6}{-x - 3} = \dfrac{x^2 - x - 6}{-x - 3}$$

$f(-x) \ne g(x)$, $g(-x) \ne -g(x)$
No symmetry

y-intercept: $y = \dfrac{0^2 + 0 - 6}{0 - 3} = \dfrac{-6}{-3} = 2$

x-intercept:
$x^2 + x - 6 = 0$
$(x + 3)(x - 2) = 0$
$x = -3$ and $x = 2$

vertical asymptote:
$x - 3 = 0$
$\quad x = 3$
horizontal asymptote:
$n > m$, so none exist.

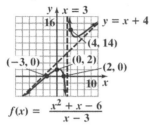

$$f(x) = \frac{x^2 + x - 6}{x - 3}$$

87. $f(x) = \dfrac{x^3 + 1}{x^2 + 2x}$

a. slant asymptote:

$$
\begin{array}{r}
x - 2 \\
x^2 + 2x \overline{\smash{)}\; x^3 \qquad\quad + 1} \\
\underline{x^3 + 2x^2} \\
-2x^2 \\
\underline{-2x^2 + 4x} \\
-4x + 1
\end{array}
$$

$y = x - 2$

b.
$$f(-x) = \frac{(-x)^3 + 1}{(-x)^2 + 2(-x)} = \frac{-x^3 + 1}{x^2 - 2x}$$

$f(-x) \neq f(x),\ f(-x) \neq -f(x)$

no symmetry

y-intercept: $y = \dfrac{0^3 + 1}{0^2 + 2(0)} = \dfrac{1}{0}$

no y-intercept
x-intercept: $x^3 + 1 = 0$
$\quad x^3 = -1$
$\quad\quad x = -1$
vertical asymptotes:
$\quad x^2 + 2x = 0$
$\quad x(x + 2) = 0$
$\quad\quad x = 0,\ x = -2$

horizontal asymptote:
$n > m$, so none

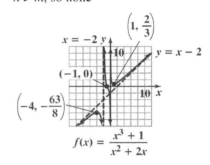

$$f(x) = \frac{x^3 + 1}{x^2 + 2x}$$

89. $\dfrac{5x^2}{x^2 - 4} \cdot \dfrac{x^2 + 4x + 4}{10x^3}$

$$= \frac{\cancel{5}\,\cancel{x^2}}{(\cancel{x+2})(x-2)} \cdot \frac{(x+2)^{\cancel{2}}}{\underset{2}{\cancel{10}\,x^{\cancel{3}1}}}$$

$$= \frac{x + 2}{2x(x - 2)}$$

So, $f(x) = \dfrac{x + 2}{2x(x - 2)}$

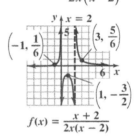

$$f(x) = \frac{x + 2}{2x(x - 2)}$$

91. $\dfrac{x}{2x + 6} - \dfrac{9}{x^2 - 9}$

$$\frac{x}{2x + 6} - \frac{9}{x^2 - 9} = \frac{x}{2(x + 3)} - \frac{9}{(x + 3)(x - 3)}$$

$$= \frac{x(x - 3) - 9(2)}{2(x + 3)(x - 3)}$$

$$= \frac{x^2 - 3x - 18}{2(x + 3)(x - 3)}$$

$$= \frac{(x - 6)(\cancel{x + 3})}{2(\cancel{x + 3})(x - 3)} = \frac{x - 6}{2(x - 3)}$$

So, $f(x) = \dfrac{x-6}{2(x-3)}$

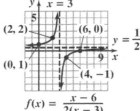

$$f(x) = \frac{x-6}{2(x-3)}$$

93. $\dfrac{1 - \dfrac{3}{x+2}}{1 + \dfrac{1}{x-2}} = \dfrac{1 - \dfrac{3}{x+2}}{1 + \dfrac{1}{x-2}} \cdot \dfrac{(x+2)(x-2)}{(x+2)(x-2)}$

$= \dfrac{(x+2)(x-2) - 3(x-2)}{(x+2)(x-2) + (x+2)}$

$= \dfrac{x^2 - 4 - 3x + 6}{x^2 - 4 + x + 2}$

$= \dfrac{x^2 - 3x + 2}{x^2 + x - 2}$

$= \dfrac{(x-2)\,\cancel{(x-1)}}{(x+2)\,\cancel{(x-1)}} = \dfrac{x-2}{x+2}$

So, $f(x) = \dfrac{x-2}{x+2}$

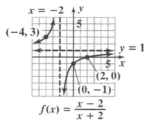

$$f(x) = \frac{x-2}{x+2}$$

95. $g(x) = \dfrac{2x+7}{x+3} = \dfrac{1}{x+3} + 2$

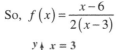

$$g(x) = \frac{1}{x+3} + 2$$

97. $g(x) = \dfrac{3x-7}{x-2} = \dfrac{-1}{x-2} + 3$

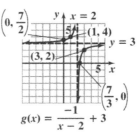

$$g(x) = \frac{-1}{x-2} + 3$$

99. a. $C(x) = 100x + 100,000$

  b. $\overline{C}(x) = \dfrac{100x + 100,000}{x}$

  c. $\overline{C}(500) = \dfrac{100(500) + 100,000}{500} = \$300$

  When 500 bicycles are manufactured, it costs $300 to manufacture each.

  $\overline{C}(1000) = \dfrac{100(1000) + 100,000}{1000} = \$200$

  When 1000 bicycles are manufactured, it costs $200 to manufacture each.

  $\overline{C}(2000) = \dfrac{100(2000) + 100,000}{2000} = \$150$

  When 2000 bicycles are manufactured, it costs $150 to manufacture each.

  $\overline{C}(4000) = \dfrac{100(4000) + 100,000}{4000} = \$125$

  When 4000 bicycles are manufactured, it costs $125 to manufacture each.
  The average cost decreases as the number of bicycles manufactured increases.

  d. $n = m$, so $y = \dfrac{100}{1} = 100$.

  As greater numbers of bicycles are manufactured, the average cost approaches $100.

101. a. From the graph the pH level of the human mouth 42 minutes after a person eats food containing sugar will be about 6.0.

  b. From the graph, the pH level is lowest after about 6 minutes.

  $$f(6) = \frac{6.5(6)^2 - 20.4(6) + 234}{6^2 + 36}$$

  $= 4.8$

  The pH level after 6 minutes (i.e. the lowest pH level) is 4.8.

c. From the graph, the pH level appears to approach 6.5 as time goes by. Therefore, the normal pH level must be 6.5.

d. $y = 6.5$

Over time, the pH level rises back to the normal level.

e. During the first hour, the pH level drops quickly below normal, and then slowly begins to approach the normal level.

103. $P(10) = \dfrac{100(10-1)}{10} = 90 \quad (10, 90)$

For a disease that smokers are 10 times more likely to contact than non-smokers, 90% of the deaths are smoking related.

105. $y = 100$ As incidence of the diseases increases, the percent of death approaches, but never gets to be, 100%.

107. a. $f(x) = \dfrac{p(x)}{q(x)} = \dfrac{1.75x^2 - 15.9x + 160}{2.1x^2 - 3.5x + 296}$

b. According to the graph, $\dfrac{2504.0}{3720.7} \approx 0.67$ or 67% of federal expenditures were spent on human resources in 2010.

c. According to the function,

$$f(x) = \dfrac{1.75(40)^2 - 15.9(40) + 160}{2.1(40)^2 - 3.5(40) + 296} \approx 0.66 \text{ or}$$

66% of federal expenditures were spent on human resources in 2010.

d. The degree of the numerator, 2, is equal to the degree of the denominator, 2. The leading coefficients of the numerator and denominator are 1.75 and 2.1, respectively. The equation of the horizontal asymptote is $y = \dfrac{1.75}{2.1}$ which is about 83%

Thus, about 83% of federal expenditures will be spent on human resources over time.

109. – 117. Answers will vary.

119.

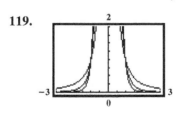

The graph approaches the horizontal asymptote faster and the vertical asymptote slower as *n* increases.

121. a. $f(x) = \dfrac{27725(x-14)}{x^2 + 9} - 5x$

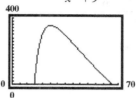

b. The graph increases from late teens until about the age of 25, and then the number of arrests decreases.

c. At age 25 the highest number arrests occurs. There are about 356 arrests for every 100,000 drivers.

123. does not make sense; Explanations will vary. Sample explanation: The function has one vertical asymptote, $x = 2$.

125. does not make sense; Explanations will vary. Sample explanation: As production level increases, the average cost for a company to produce each unit of its product decreases.

127. true

129. true

131. – 133. Answers will vary.

134.
$$2x^2 + x = 15$$
$$2x^2 + x - 15 = 0$$
$$(2x - 5)(x + 3) = 0$$
$$2x - 5 = 0 \quad \text{or} \quad x + 3 = 0$$
$$x = \dfrac{5}{2} \qquad\qquad x = -3$$

The solution set is $\left\{ -3, \dfrac{5}{2} \right\}$.

135.
$$x^3 + x^2 = 4x + 4$$
$$x^3 + x^2 - 4x - 4 = 0$$
$$x^2(x + 1) - 4(x + 1) = 0$$
$$(x + 1)(x^2 - 4) = 0$$
$$(x + 1)(x + 2)(x - 2) = 0$$

The solution set is $\{-2, -1, 2\}$.

136.
$$\frac{x+1}{x+3} - 2 = \frac{x+1}{x+3} - \frac{2(x+3)}{x+3}$$
$$= \frac{x+1}{x+3} - \frac{2x+6}{x+3}$$
$$= \frac{x+1-2x-6}{x+3}$$
$$= \frac{-x-5}{x+3} \text{ or } -\frac{x+5}{x+3}$$

Section 3.6

Check Point Exercises

1.
$$x^2 - x > 20$$
$$x^2 - x - 20 > 0$$
$$(x+4)(x-5) > 0$$

Solve the related quadratic equation.
$$(x+4)(x-5) = 0$$

Apply the zero product principle.
$$x + 4 = 0 \quad \text{or} \quad x - 5 = 0$$
$$x = -4 \qquad x = 5$$

The boundary points are –2 and 4.

Test Interval	Test Number	Test	Conclusion
$(-\infty, -4)$	–5	$(-5)^2 - (-5) > 20$ $30 > 20$, true	$(-\infty, -4)$ belongs to the solution set.
$(-4, 5)$	0	$(0)^2 - (0) > 20$ $0 > 20$, false	$(-4, 5)$ does not belong to the solution set.
$(5, \infty)$	10	$(10)^2 - (10) > 20$ $90 > 20$, true	$(5, \infty)$ belongs to the solution set.

The solution set is $(-\infty, -4) \cup (5, \infty)$ or $\{x \mid x < -4 \text{ or } x > 5\}$.

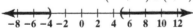

2. $2x^2 \le -6x - 1$

$2x^2 + 6x + 1 \le 0$

Solve the related quadratic equation to find the boundary points.

$2x^2 + 6x + 1 = 0$

$a = 2 \quad b = 6 \quad c = 1$

$x = \dfrac{-(6) \pm \sqrt{(6)^2 - 4(2)(1)}}{2(2)}$

$\quad = \dfrac{-6 \pm \sqrt{36 - 8}}{4}$

$\quad = \dfrac{-6 \pm \sqrt{28}}{4}$

$\quad = \dfrac{-6 \pm 2\sqrt{7}}{4}$

$\quad = \dfrac{-3 \pm \sqrt{7}}{2}$

$x = \dfrac{-3 + \sqrt{7}}{2} \quad \text{or} \quad x = \dfrac{-3 - \sqrt{7}}{2}$

$x \approx -0.2 \qquad\qquad x \approx -2.8$

Interval	Test Value	Test	Conclusion
$\left(-\infty, \dfrac{-3-\sqrt{7}}{2}\right)$	-10	$2(-10)^2 \le -6(-10)-1$ $200 \le 59$, false	$\left(-\infty, \dfrac{-3-\sqrt{7}}{2}\right)$ is not part of the solution set
$\left(\dfrac{-3-\sqrt{7}}{2}, \dfrac{-3+\sqrt{7}}{2}\right)$	-1	$2(-1)^2 \le -6(-1)-1$ $2 \le 5$, true	$\left(\dfrac{-3-\sqrt{7}}{2}, \dfrac{-3+\sqrt{7}}{2}\right)$ is part of the solution set
$\left(\dfrac{-3+\sqrt{7}}{2}, \infty\right)$	0	$2(0)^2 \le -6(0)-1$ $0 \le -1$, false	$\left(\dfrac{-3+\sqrt{7}}{2}, \infty\right)$ is not part of the solution set

The solution set is $\left(\dfrac{-3-\sqrt{7}}{2}, \dfrac{-3+\sqrt{7}}{2}\right)$.

3.
$$x^3 + 3x^2 \le x + 3$$
$$x^3 + 3x^2 - x - 3 \le 0$$
$$(x+1)(x-1)(x+3) \le 0$$
$$(x+1)(x-1)(x+3) = 0$$
$$x+1 = 0 \quad \text{or} \quad x-1 = 0 \quad \text{or} \quad x+3 = 0$$
$$x = -1 \qquad\qquad x = 1 \qquad\qquad x = -3$$

Test Interval	Test Number	Test	Conclusion
$(-\infty, -3)$	-4	$(-4)^3 + 3(-4)^2 \le (-4) + 3$ $-16 \le -1$ true	$(-\infty, -3)$ belongs to the solution set.
$(-3, -1]$	-2	$(-2)^3 + 3(-2)^2 \le (-2) + 3$ $4 \le 1$ false	$(-3, -1]$ does not belong to the solution set.
$[-1, 1]$	0	$(0)^3 + 3(0)^2 \le (0) + 3$ $0 \le 3$ true	$[-1, 1]$ belongs to the solution set.
$[1, \infty)$	2	$(6+3)(6-5) > 0$ true	$[1, \infty)$ does not belong to the solution set.

The solution set is $(-\infty, -3] \cup [-1, 1]$ or $\{x \mid x \le -3 \text{ or } -1 \le x \le 1\}$.

4.
$$\frac{2x}{x+1} \ge 1$$
$$\frac{2x}{x+1} - 1 \ge 0$$
$$\frac{x-1}{x+1} \ge 0$$
$$x - 1 = 0 \quad \text{or} \quad x + 1 = 0$$
$$x = 1 \qquad\qquad x = -1$$

Test Interval	Test Number	Test	Conclusion
$(-\infty, -1)$	-2	$\dfrac{2(-2)}{-2+1} \ge 1$ $4 \ge 1$, true	$(-\infty, -1)$ belongs to the solution set.
$(-1, 1]$	0	$\dfrac{2(0)}{0+1} \ge 1$ $0 \ge 1$, false	$(-1, 1]$ does not belong to the solution set.
$[1, \infty)$	2	$\dfrac{2(2)}{2+1} \ge 1$ $\dfrac{4}{3} \ge 1$, true	$[1, \infty)$ belongs to the solution set.

The solution set is $(-\infty, -1) \cup [1, \infty)$ or $\{x \mid x < -1 \text{ or } x \ge 1\}$.

5.

$$-16t^2 + 80t > 64$$

$$-16t^2 + 80t - 64 > 0$$

$$-16(t-1)(t-4) > 0$$

$$t - 1 = 0 \quad \text{or} \quad t - 4 = 0$$

$$t = 1 \qquad\qquad t = 4$$

Test Interval	Test Number	Test	Conclusion
$(-\infty, 1)$	0	$-16(0)^2 + 80(0) > 64$ $0 > 64$, false	$(-\infty, 1)$ does not belong to the solution set.
$(1, 4)$	2	$-16(2)^2 + 80(2) > 64$ $96 > 64$, true	$(1, 4)$ belongs to the solution set.
$(4, \infty)$	5	$-16(5)^2 + 80(5) > 64$ $0 > 64$, false	$(4, \infty)$ does not belong to the solution set.

The object will be more than 64 feet above the ground between 1 and 4 seconds.

Concept and Vocabulary Check 3.6

1. $x^2 + 8x + 15 = 0$; boundary

2. $(-\infty, -5)$; $(-5, -3)$; $(-3, \infty)$

3. true

4. true

5. $[-\infty, -2) \cup [1, \infty)$

Exercise Set 3.6

1. $(x - 4)(x + 2) > 0$
$x = 4$ or $x = -2$

$$
\begin{array}{c|c|c}
\text{T} & \text{F} & \text{T} \\
\hline
-2 & & 4
\end{array}
$$

Test -3: $(-3 - 4)(-3 + 2) > 0$
$\qquad\qquad 7 > 0$ True
Test 0: $(0 - 4)(0 + 2) > 0$
$\qquad\qquad -8 > 0$ False
Test 5: $(5 - 4)(5 + 2) > 0$
$\qquad\qquad 7 > 0$ True
$(-\infty, -2)$ or $(4, \infty)$

3. $(x-7)(x+3) \le 0$

$x = 7$ or $x = -3$

Test -4: $(-4-7)(-4+3) \le 0$

$11 \le 0$ False

Test 0: $(0-7)(0+3) \le 0$

$-21 \le 0$ True

Test 8: $(8-7)(8+3) \le 0$

$11 \le 0$ False

The solution set is $[-3, 7]$.

5. $x^2 - 5x + 4 > 0$

$(x-4)(x-1) > 0$

$x = 4$ or $x = 1$

Test 0: $0^2 - 5(0) + 4 > 0$

$4 > 0$ True

Test 2: $2^2 - 5(2) + 4 > 0$

$-2 > 0$ False

Test 5: $5^2 - 5(5) + 4 > 0$

$4 > 0$ True

The solution set is $(-\infty, 1)$ or $(4, \infty)$.

7. $x^2 + 5x + 4 > 0$

$(x+1)(x+4) > 0$

$x = -1$ or $x = -4$

Test -5: $(-5)^2 + 5(-5) + 4 > 0$

$4 > 0$ True

Test -3: $(-3)^2 + 5(-3) + 4 > 0$

$-2 > 0$ False

Test 0: $0^2 + 5(0) + 4 > 0$

$4 > 0$ True

The solution set is $(-\infty, -4)$ or $(-1, \infty)$.

9. $x^2 - 6x + 9 < 0$

$(x-3)(x-3) < 0$

$x = 3$

Test 0: $0^2 - 6(0) + 9 < 0$

$9 < 0$ False

Test 4: $4^2 - 6(4) + 9 < 0$

$1 < 0$ False

The solution set is the empty set, $\varnothing$.

11. $3x^2 + 10x - 8 \le 0$

$(3x-2)(x+4) \le 0$

$x = \dfrac{2}{3}$ or $x = -4$

Test -5: $3(-5)^2 + 10(-5) - 8 \le 0$

$17 \le 0$ False

Test 0: $3(0)^2 + 10(0) - 8 \le 0$

$8 \le 0$ True

Test 1: $3(1)^2 + 10(1) - 8 \le 0$

$5 \le 0$ False

The solution set is $\left[-4, \dfrac{2}{3} \right]$.

13.
$$2x^2 + x < 15$$
$$2x^2 + x - 15 < 0$$
$$(2x - 5)(x + 3) < 0$$
$$2x - 5 = 0 \quad \text{or} \quad x + 3 = 0$$
$$2x = 5$$

$$x = \frac{5}{2} \quad \text{or} \quad x = -3$$

F		T		F
	-3		$\frac{5}{2}$	

Test -4: $2(-4)^2 + (-4) < 15$
$$28 < 15 \text{ False}$$
Test 0: $2(0)^2 + 0 < 15$
$$0 < 15 \text{ True}$$
Test 3: $2(3)^2 + 3 < 15$
$$21 < 15 \text{ False}$$

The solution set is $\left(-3, \frac{5}{2}\right)$.

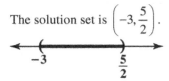

15.
$$4x^2 + 7x < -3$$
$$4x^2 + 7x + 3 < 0$$
$$(4x + 3)(x + 1) < 0$$
$$4x + 3 = 0 \quad \text{or} \quad x + 1 = 0$$
$$4x - 3 = 0$$

$$x = -\frac{3}{4} \quad \text{or} \quad x = -1$$

F		T		F
	-1		$-\frac{3}{4}$	

Test -2: $4(-2)^2 + 7(-2) < -3$
$$2 < -3 \text{ False}$$

Test $-\frac{7}{8}$: $4\left(-\frac{7}{8}\right)^2 + 7\left(-\frac{7}{8}\right) < -3$

$$\frac{49}{16} - \frac{49}{8} < -3$$

$$-\frac{49}{16} < -3 \text{ True}$$

Test 0: $4(0)^2 + 7(0) < -3$
$$0 < -3 \text{ False}$$

The solution set is $\left(-1, -\frac{3}{4}\right)$.

17.
$$5x \le 2 - 3x^2$$
$$3x^2 + 5x - 2 \le 0$$
$$(3x - 1)(x + 2) \le 0$$
$$3x - 1 = 0 \text{ or } x + 2 = 0$$
$$3x = 1$$
$$3x - 1 = 0 \quad \text{or} \quad x + 2 = 0$$
$$3x = 1$$

$$x = \frac{1}{3} \quad \text{or} \quad x = -2$$

F		T		F
	-2		$\frac{1}{3}$	

Test -3: $5(-3) \le 2 - 3(-3)^2$
$$-15 \le -25 \text{ False}$$
Test 0: $5(0) \le 2 - 3(0)^2$
$$0 \le 2 \text{ True}$$
Test 1: $5(1) \le 2 - 3(1)^2$
$$5 \le -1 \text{ False}$$

The solution set is $\left[-2, \frac{1}{3}\right]$.

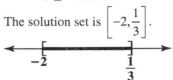

19. $x^2 - 4x \ge 0$
$$x(x - 4) \ge 0$$
$$x = 0 \text{ or } x - 4 = 0$$
$$x = 4$$

T		F		T
	0		4	

Test -1: $(-1)^2 - 4(-1) \ge 0$
$$5 \ge 0 \text{ True}$$
Test 1: $(1)^2 - 4(1) \ge 0$
$$-3 \ge 0 \text{ False}$$
$$0 \le 2 \text{ True}$$
Test 5: $5^2 - 4(5) \ge 0$
$$5 \ge 0 \text{ True}$$
The solution set is $(-\infty, 0]$ or $[4, \infty)$.

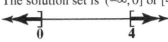

21. $2x^2 + 3x > 0$

$x(2x + 3) > 0$

$x = 0$ or $x = -\dfrac{3}{2}$

T	F	T
$-\frac{3}{2}$	0	

Test -2: $2(-2)^2 + 3(-2) > 0$

$2 > 0$ True

Test -1: $2(-1)^2 + 3(-1) > 0$

$-1 > 0$ False

Test 1: $2(1)^2 + 3(1) > 0$

$5 > 0$ True

The solution set is $\left(-\infty, -\dfrac{3}{2}\right)$ or $(0, \infty)$.

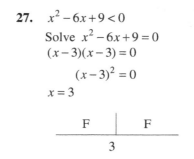

23. $-x^2 + x \geq 0$

$x^2 - x \leq 0$

$x(x - 1) \leq 0$

$x = 0$ or $x = 1$

F	T	F
0	1	

Test -1: $-(-1)^2 + (-1) \geq 0$

$-2 \geq 0$ False

Test $\dfrac{1}{2}$: $-\left(\dfrac{1}{2}\right)^2 + \left(\dfrac{1}{2}\right) \geq 0$

$\dfrac{1}{4} \geq 0$ True

Test 2: $-(2)^2 + 2 \geq 0$

$-2 \geq 0$ False

The solution set is [0, 1].

25. $x^2 \leq 4x - 2$

$x^2 - 4x + 2 \leq 0$

Solve $x^2 - 4x + 2 = 0$

$x = \dfrac{-b \pm \sqrt{b^2 - 4ac}}{2a}$

$x = \dfrac{-(-4) \pm \sqrt{(-4)^2 - 4(1)(2)}}{2(1)}$

$= \dfrac{4 \pm \sqrt{8}}{2}$

$= 2 \pm \sqrt{2}$

$x \approx 0.59$ or $x \approx 3.41$

F	T	F
0.59	3.41	

The solution set is $\left[2 - \sqrt{2}, 2 + \sqrt{2}\right]$ or $[0.59, 3.41]$.

27. $x^2 - 6x + 9 < 0$

Solve $x^2 - 6x + 9 = 0$

$(x - 3)(x - 3) = 0$

$(x - 3)^2 = 0$

$x = 3$

F	F
3	

The solution set is the empty set, $\varnothing$.

29. $(x - 1)(x - 2)(x - 3) \geq 0$

Boundary points: 1, 2, and 3

Test one value in each interval.

F	T	F	T
1	2	3	

The solution set is $[1, 2] \cup [3, \infty)$.

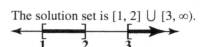

31. $x(3-x)(x-5) \le 0$

Boundary points: 0, 3, and 5
Test one value in each interval.

F		T		F		T
	0		3		5	

The solution set is $[0, 3] \cup [5, \infty)$.

33. $(2-x)^2 \left(x - \frac{7}{2}\right) < 0$

Boundary points: 2, and $\frac{7}{2}$
Test one value in each interval.

T		T		F
	2		$\frac{7}{2}$	

The solution set is $\left(-\infty, 2\right) \cup \left(2, \frac{7}{2}\right)$.

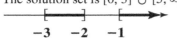

35. $x^3 + 2x^2 - x - 2 \ge 0$

$x^2(x+2) - 1(x+2) \ge 0$

$(x+2)(x^2-1) \ge 0$

$(x+2)(x-1)(x+1) \ge 0$

Boundary points: -2, -1, and 2
Test one value in each interval.

F		T		F		T
	-2		-1		2	

The solution set is $[-2, -1] \cup [1, \infty)$.

37. $x^3 + 2x^2 - x - 2 \ge 0$

$x^2(x-3) - 9(x-3) \ge 0$

$(x-3)(x^2-9) \ge 0$

$(x-3)(x+3)(x-3) \ge 0$

Boundary points: -3 and 3
Test one value in each interval.

T		F		F
	-3		3	

The solution set is $(-\infty, -3]$.

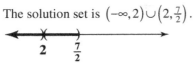

39. $x^3 + x^2 + 4x + 4 > 0$

$x^2(x+1) + 4(x+1) \ge 0$

$(x+1)(x^2+4) \ge 0$

Boundary point: -1
Test one value in each interval.

F		T
	-1	

The solution set is $(-1, \infty)$.

41. $x^3 - 9x^2 \ge 0$

$x^2(x-9) \ge 0$

Boundary points: 0 and 9
Test one value in each interval.

F		F		T
	0		9	

The solution set is $[0, 0] \cup [9, \infty)$.

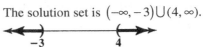

43. $\dfrac{x-4}{x+3} > 0$

$x - 4 = 0 \quad x + 3 = 0$

$x = 4 \qquad x = -3$

T		F		T
	-3		4	

The solution set is $(-\infty, -3) \cup (4, \infty)$.

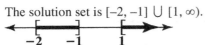

45. $\dfrac{x+3}{x+4} < 0$

$x = -3 \quad \text{or} \quad x = -4$

F		T		F
	-4		-3	

The solution set is $(-4, -3)$.

47. $\dfrac{-x+2}{x-4} \ge 0$

$x = 2 \text{ or } x = 4$

F		T		F
	2		4	

The solution set is $[2, 4)$.

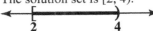

49. $\dfrac{4-2x}{3x+4} \le 0$

$x = 2 \text{ or } x = -\dfrac{4}{3}$

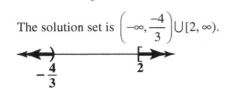

The solution set is $\left(-\infty, \dfrac{-4}{3}\right) \cup [2, \infty)$.

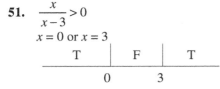

51. $\dfrac{x}{x-3} > 0$

$x = 0 \text{ or } x = 3$

T		F		T
	0		3	

The solution set is $(-\infty, 0) \cup (3, \infty)$.

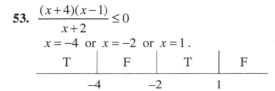

53. $\dfrac{(x+4)(x-1)}{x+2} \le 0$

$x = -4 \text{ or } x = -2 \text{ or } x = 1$.

T		F		T		F
	-4		-2		1	

Values of $x = -4$ or $x = 1$ result in $f(x) = 0$ and, therefore must be included in the solution set.
The solution set is $\left(-\infty, -4\right] \cup (-2, 1]$

T		F		T
	$\frac{1}{2}$		$\frac{7}{5}$	

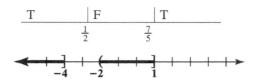

55. $\dfrac{x+1}{x+3} < 2$

$\dfrac{x+1}{x+3} - 2 < 0$

$\dfrac{x+1-2(x+3)}{x+3} < 0$

$\dfrac{x+1-2x-6}{x+3} < 0$

$\dfrac{-x-5}{x+3} < 0$

$x = \text{ or } x = -3$

	T		F		T
		-5		-3	

The solution set is $(-\infty, -5) \cup (-3, \infty)$.

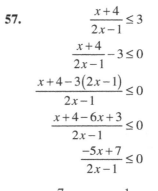

57. $\dfrac{x+4}{2x-1} \le 3$

$\dfrac{x+4}{2x-1} - 3 \le 0$

$\dfrac{x+4-3(2x-1)}{2x-1} \le 0$

$\dfrac{x+4-6x+3}{2x-1} \le 0$

$\dfrac{-5x+7}{2x-1} \le 0$

$x = \dfrac{7}{5} \quad \text{or} \quad x = \dfrac{1}{2}$

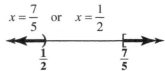

Copyright © 2014 Pearson Education, Inc.

59.
$$\frac{x-2}{x+2} \le 2$$

$$\frac{x-2}{x+2} - 2 \le 0$$

$$\frac{x-2-2(x+2)}{x+2} \le 0$$

$$\frac{x-2-2x-4}{x+2} \le 0$$

$$\frac{-x-6}{x+2} \le 0$$

$$x = -6 \quad \text{or} \quad x = -2$$

	T		F		T	
		-6		-2		

The solution set is $(-\infty, -6] \cup (-2, \infty)$.

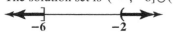

61. $f(x) = \sqrt{2x^2 - 5x + 2}$

The domain of this function requires that $2x^2 - 5x + 2 \ge 0$

Solve $2x^2 - 5x + 2 = 0$
$(x-2)(2x-1) = 0$

$x = \dfrac{1}{2} \text{ or } x = 2$

	T		F		T	
		$\frac{1}{2}$		2		

The domain is $\left(-\infty, \dfrac{1}{2}\right] \cup [2, \infty)$.

63. $f(x) = \sqrt{\dfrac{2x}{x+1} - 1}$

The domain of this function requires that $\dfrac{2x}{x+1} - 1 \ge 0$ or $\dfrac{x-1}{x+1} \ge 0$

$x = -1 \quad \text{or} \quad x = 1$

The value $x = 1$ results in 0 and, thus, it must be included in the domain.
The domain is $(-\infty, -1) \cup [1, \infty)$.

65. $\left|x^2+2x-36\right|>12$

Express the inequality without the absolute value symbol:

$x^2+2x-36<-12$ or $x^2+2x-36>12$

$x^2+2x-24<0$ $x^2+2x-48>0$

Solve the related quadratic equations.

$x^2+2x-24=0$ or $x^2+2x-48=0$

$(x+6)(x-4)=0$ $(x+8)(x-6)=0$

Apply the zero product principle.

$x+6=0$ or $x-4=0$ or $x+8=0$ or $x-6=0$

$x=-6$ $x=4$ $x=-8$ $x=6$

The boundary points are $-8,\ -6,\ 4$ and 6.

Test Interval	Test Number	Test	Conclusion
$(-\infty,-8)$	-9	$\left\|(-9)^2+2(-9)-36\right\|>12$ $27>12$, True	$(-\infty,-8)$ belongs to the solution set.
$(-8,-6)$	-7	$\left\|(-7)^2+2(-7)-36\right\|>12$ $1>12$, False	$(-8,-6)$ does not belong to the solution set.
$(-6,4)$	0	$\left\|0^2+2(0)-36\right\|>12$ $36>12$, True	$(-6,4)$ belongs to the solution set.
$(4,6)$	5	$\left\|5^2+2(5)-36\right\|>12$ $1>12$, False	$(4,6)$ does not belong to the solution set.
$(6,\infty)$	7	$\left\|7^2+2(7)-36\right\|>12$ $27>12$, True	$(6,\infty)$ belongs to the solution set.

The solution set is $(-\infty,-8)\cup(-6,4)\cup(6,\infty)$ or $\left\{x\,\middle|\,x<-8\ \text{or}\ -6<x<4\ \text{or}\ x>6\right\}$.

67. $\dfrac{3}{x+3} > \dfrac{3}{x-2}$

Express the inequality so that one side is zero.

$$\dfrac{3}{x+3} - \dfrac{3}{x-2} > 0$$

$$\dfrac{3(x-2)}{(x+3)(x-2)} - \dfrac{3(x+3)}{(x+3)(x-2)} > 0$$

$$\dfrac{3x-6-3x-9}{(x+3)(x-2)} < 0$$

$$\dfrac{-15}{(x+3)(x-2)} < 0$$

Find the values of x that make the denominator zero.

$x+3=0 \qquad x-2=0$

$\quad x=-3 \qquad\quad x=2$

The boundary points are -3 and 2.

Test Interval	Test Number	Test	Conclusion
$(-\infty,-3)$	-4	$\dfrac{3}{-4+3} > \dfrac{3}{-4-2}$ $-3 > \dfrac{1}{2}$, False	$(-\infty,-3)$ does not belong to the solution set.
$(-3,2)$	0	$\dfrac{3}{0+3} > \dfrac{3}{0-2}$ $1 > -\dfrac{3}{2}$, True	$(-3,2)$ belongs to the solution set.
$(2,\infty)$	3	$\dfrac{3}{3+3} > \dfrac{3}{3-2}$ $\dfrac{1}{2} > 3$, False	$(2,\infty)$ does not belong to the solution set.

The solution set is $(-3,2)$ or $\{x|-3 < x < 2\}$.

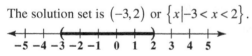

69.

$$\frac{x^2 - x - 2}{x^2 - 4x + 3} > 0$$

Find the values of x that make the numerator and denominator zero.

$x^2 - x - 2 = 0 \qquad x^2 - 4x + 3 = 0$

$(x - 2)(x + 1) = 0 \qquad (x - 3)(x - 1) = 0$

Apply the zero product principle.

$x - 2 = 0 \quad \text{or} \quad x + 1 = 0 \qquad x - 3 = 0 \quad \text{or} \quad x - 1 = 0$

$\qquad x = 2 \qquad\qquad x = -1 \qquad\qquad x = 3 \qquad\qquad x = 1$

The boundary points are -1, 1, 2 and 3.

Test Interval	Test Number	Test	Conclusion
$(-\infty, -1)$	-2	$\dfrac{(-2)^2 - (-2) - 2}{(-2)^2 - 4(-2) + 3} > 0$ $\dfrac{4}{15} > 0$, True	$(-\infty, -1)$ belongs to the solution set.
$(-1, 1)$	0	$\dfrac{0^2 - 0 - 2}{0^2 - 4(0) + 3} > 0$ $-\dfrac{2}{3} > 0$, False	$(-1, 1)$ does not belong to the solution set.
$(1, 2)$	1.5	$\dfrac{1.5^2 - 1.5 - 2}{1.5^2 - 4(1.5) + 3} > 0$ $\dfrac{5}{3} > 0$, True	$(1, 2)$ belongs to the solution set.
$(2, 3)$	2.5	$\dfrac{2.5^2 - 2.5 - 2}{2.5^2 - 4(2.5) + 3} > 0$ $-\dfrac{7}{3} > 0$, False	$(2, 3)$ does not belong to the solution set.
$(3, \infty)$	4	$\dfrac{4^2 - 4 - 2}{4^2 - 4(4) + 3} > 0$ $\dfrac{10}{3} > 0$, True	$(3, \infty)$ belongs to the solution set.

The solution set is $(-\infty, -1) \cup (1, 2) \cup (3, \infty)$ or $\{x \mid x < -1 \text{ or } 1 < x < 2 \text{ or } x > 3\}$.

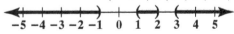

71.
$$2x^3 + 11x^2 \geq 7x + 6$$
$$2x^3 + 11x^2 - 7x - 6 \geq 0$$

The graph of $f(x) = 2x^3 + 11x^2 - 7x - 6$ appears to cross the x-axis at -6, $-\dfrac{1}{2}$, and 1. We verify this

numerically by substituting these values into the function:

$$f(-6) = 2(-6)^3 + 11(-6)^2 - 7(-6) - 6 = 2(-216) + 11(36) - (-42) - 6 = -432 + 396 + 42 - 6 = 0$$

$$f\left(-\frac{1}{2}\right) = 2\left(-\frac{1}{2}\right)^3 + 11\left(-\frac{1}{2}\right)^2 - 7\left(-\frac{1}{2}\right) - 6 = 2\left(-\frac{1}{8}\right) + 11\left(\frac{1}{4}\right) - \left(-\frac{7}{2}\right) - 6 = -\frac{1}{4} + \frac{11}{4} + \frac{7}{2} - 6 = 0$$

$$f(1) = 2(1)^3 + 11(1)^2 - 7(1) - 6 = 2(1) + 11(1) - 7 - 6 = 2 + 11 - 7 - 6 = 0$$

Thus, the boundaries are -6, $-\dfrac{1}{2}$, and 1. We need to find the intervals on which $f(x) \geq 0$. These intervals are

indicated on the graph where the curve is above the x-axis. Now, the curve is above the x-axis when

$-6 < x < -\dfrac{1}{2}$ and when $x > 1$. Thus, the solution set is $\left\{ x \,\middle|\, -6 \leq x \leq -\dfrac{1}{2} \text{ or } x \geq 1 \right\}$ or $\left[-6, -\dfrac{1}{2}\right] \cup [1, \infty)$.

73.
$$\frac{1}{4(x+2)} \leq -\frac{3}{4(x-2)}$$

$$\frac{1}{4(x+2)} + \frac{3}{4(x-2)} \leq 0$$

Simplify the left side of the inequality:

$$\frac{x-2}{4(x+2)} + \frac{3(x+2)}{4(x-2)} = \frac{x-2+3x+6}{4(x+2)(x-2)} = \frac{4x+4}{4(x+2)(x-2)} = \frac{4(x+1)}{4(x+2)(x-2)} = \frac{x+1}{x^2-4}.$$

The graph of $f(x) = \dfrac{x+1}{x^2-4}$ crosses the x-axis at -1, and has vertical asymptotes at $x = -2$ and $x = 2$. Thus,

the boundaries are -2, -1, and 1. We need to find the intervals on which $f(x) \leq 0$. These intervals are

indicated on the graph where the curve is below the x-axis. Now, the curve is below the x-axis when $x < -2$ and

when $-1 < x < 2$. Thus, the solution set is $\left\{ x \,|\, x < -2 \text{ or } -1 \leq x < 2 \right\}$ or $(-\infty, -2) \cup [-1, 2)$.

75. $s(t) = -16t^2 + 8t + 87$

The diver's height will exceed that of the cliff when $s(t) > 87$

$$-16t^2 + 8t + 87 > 87$$
$$-16t^2 + 8t > 0$$
$$-8t(2t - 1) > 0$$

The boundaries are 0 and $\dfrac{1}{2}$. Testing each interval shows that the diver will be higher than the cliff for the first half

second after beginning the jump. The interval is $\left(0, \dfrac{1}{2}\right)$.

77. $f(x) = 0.0875x^2 - 0.4x + 66.6$

$g(x) = 0.0875x^2 + 1.9x + 11.6$

a. $f(35) = 0.0875(35)^2 - 0.4(35) + 66.6 \approx 160$ feet

$g(35) = 0.0875(35)^2 + 1.9(35) + 11.6 \approx 185$ feet

b. Dry pavement: graph (b)
Wet pavement: graph (a)

c. The answers to part (a) model the actual stopping distances shown in the figure extremely well. The function values and the data are identical.

d. $0.0875x^2 - 0.4x + 66.6 > 540$

$0.0875x^2 - 0.4x + 473.4 > 0$
Solve the related quadratic equation.
$0.0875x^2 - 0.4x + 473.4 = 0$

$x = \dfrac{-b \pm \sqrt{b^2 - 4ac}}{2a}$

$x = \dfrac{-(-0.4) \pm \sqrt{(-0.4)^2 - 4(0.0875)(473.4)}}{2(0.0875)}$

$x \approx -71$ or 76

Since the function's domain is $x \geq 30$, we must test the following intervals.

Interval	Test Value	Test	Conclusion
$(30, 76)$	50	$0.0875(50)^2 - 0.4(50) + 66.6 > 540$ $265.35 > 540$, False	$(30, 76)$ does not belong to the solution set.
$(76, \infty)$	100	$0.0875(100)^2 - 0.4(100) + 66.6 > 540$ $901.6 > 540$, True	$(76, \infty)$ belongs to the solution set.

On dry pavement, stopping distances will exceed 540 feet for speeds exceeding 76 miles per hour. This is represented on graph (b) to the right of point (76, 540).

79. Let x = the length of the rectangle.
Since Perimeter $= 2(\text{length}) + 2(\text{width})$, we know

$50 = 2x + 2(\text{width})$

$50 - 2x = 2(\text{width})$

width $= \dfrac{50 - 2x}{2} = 25 - x$

Now, $A = (\text{length})(\text{width})$, so we have that

$A(x) \leq 114$

$x(25 - x) \leq 114$

$25x - x^2 \leq 114$
Solve the related equation
$25x - x^2 = 114$

$0 = x^2 - 25x + 114$

$0 = (x - 19)(x - 6)$

Apply the zero product principle:

$x - 19 = 0$ or $x - 6 = 0$

$x = 19$ $x = 6$

The boundary points are 6 and 19.

Test Interval	Test Number	Test	Conclusion
$(-\infty, 6)$	0	$25(0) - 0^2 \le 114$ $0 \le 114$, True	$(-\infty, 6)$ belongs to the solution set.
$(6, 19)$	10	$25(10) - 10^2 \le 114$ $150 \le 114$, False	$(6, 19)$ does not belong to the solution set.
$(19, \infty)$	20	$25(20) - 20^2 \le 114$ $100 \le 114$, True	$(19, \infty)$ belongs to the solution set.

If the length is 6 feet, then the width is 19 feet. If the length is less than 6 feet, then the width is greater than 19 feet. Thus, if the area of the rectangle is not to exceed 114 square feet, the length of the shorter side must be 6 feet or less.

81. – 85. Answers will vary.

87. Graph $y_1 = 2x^2 + 5x - 3$ in a standard window. The graph is below or equal to the *x*-axis for $-3 \le x \le \frac{1}{2}$.

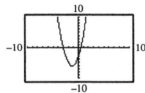

The solution set is $\left\{ x \middle| -3 \le x \le \frac{1}{2} \right\}$ or $\left[-3, \frac{1}{2} \right]$.

89. Graph $y_1 = \dfrac{x-4}{x-1}$ in a standard viewing window. The graph is below the *x*-axis for

$1 < x \le 4$.

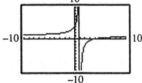

The solution set is $(1, 4]$.

91. Graph $y_1 = \dfrac{1}{x+1}$ and $y_2 = \dfrac{2}{x+4}$

y_1 less than or equal to y_2 for $-4 < x < -1$ or $x \ge 2$.

The solution set is $(-4, -1) \cup [2, \infty)$

93. **a.** $f(x) = 0.1375x^2 + 0.7x + 37.8$

    b. $0.1375x^2 + 0.7x + 37.8 > 446$

$0.1375x^2 + 0.7x + 408.2 > 0$
Solve the related quadratic equation.
$0.1375x^2 + 0.7x + 408.2 = 0$

$$x = \frac{-b \pm \sqrt{b^2 - 4ac}}{2a}$$

$$x = \frac{-(0.7) \pm \sqrt{(0.7)^2 - 4(0.1375)(408.2)}}{2(0.1375)}$$

$x \approx -57$ or 52
Since the function's domain must be $x \geq 0$, we must test the following intervals.

Interval	Test Value	Test	Conclusion
$(0, 52)$	10	$0.1375(10)^2 + 0.7(10) + 37.8 > 446$ $58.55 > 446$, False	$(0, 52)$ does not belong to the solution set.
$(52, \infty)$	100	$0.1375(100)^2 + 0.7(100) + 37.8 > 446$ $1482.8 > 446$, True	$(52, \infty)$ belongs to the solution set.

On wet pavement, stopping distances will exceed 446 feet for speeds exceeding 52 miles per hour.

95. does not make sense; Explanations will vary. Sample explanation: Polynomials are defined for all values.

97. does not make sense; Explanations will vary. Sample explanation: To solve this inequality you must first subtract 2 from both sides.

99. false; Changes to make the statement true will vary. A sample change is: The inequality cannot be solved by multiplying both sides by $x + 3$. We do not know if $x + 3$ is positive or negative. Thus, we would not know whether or not to reverse the order of the inequality.

101. true

103. One possible solution: $\dfrac{x-3}{x+4} \geq 0$

105. Because any number squared other than zero is positive, the solution includes only 2.

107. Because any number squared other than zero is positive, and the reciprocal of zero is undefined, the solution is all real numbers except 2.

109. $\sqrt{27-3x^2} \ge 0$

$27-3x^2 \ge 0$

$9-x^2 \ge 0$

$(3-x)(3+x) \ge 0$

$3-x=0 \qquad 3+x=0$

$x=3 \text{ or} \qquad x=-3$

Test -4: $\sqrt{27-3(-4)^2} \ge 0$

$\sqrt{27-48} \ge 0$

$\sqrt{-21} \ge 0$

no graph- imaginary

Test 0: $\sqrt{27-3(0)^2} \ge 0$

$\sqrt{27} \ge 0$ True

Test 4: $\sqrt{27-3(4)^2} \ge 0$

$\sqrt{27-48} \ge 0$

$\sqrt{-21} \ge 0$

no graph -imaginary

The solution set is $[-3, 3]$.

110. a. $y=kx^2$

$64=k \cdot 2^2$

$64=4k$

$16=k$

b. $y=kx^2$

$y=16x^2$

c. $y=kx^2$

$y=16x^2$

$y=16 \cdot 5^2$

$y=400$

111. a. $y=\dfrac{k}{x}$

$12=\dfrac{k}{8}$

$96=k$

b. $y=\dfrac{k}{x}$

$y=\dfrac{96}{x}$

c. $y=\dfrac{96}{x}$

$y=\dfrac{96}{3}$

$y=32$

112. $S=\dfrac{kA}{P}$

$12,000=\dfrac{k \cdot 60,000}{40}$

$\dfrac{12,000 \cdot 40}{60,000}=k$

$8=k$

Section 3.7

Check Point Exercises

1. y varies directly as x is expressed as $y=kx$.

The volume of water, W, varies directly as the time, t can be expressed as $W=kt$.

Use the given values to find k.

$W=kt$

$30=k(5)$

$6=k$

Substitute the value of k into the equation.

$W=kt$

$W=6t$

Use the equation to find W when $t=11$.

$W=6t$

$=6(11)$

$=66$

A shower lasting 11 minutes will use 66 gallons of water.

2. y varies directly as the cube of x is expressed as $y = kx^3$.

The weight, w, varies directly as the cube of the length, l can be expressed as $w = kl^2$.

Use the given values to find k.

$$w = kl^3$$

$$2025 = k(15)^3$$

$$0.6 = k$$

Substitute the value of k into the equation.

$$w = kl^3$$

$$w = 0.6l^3$$

Use the equation to find w when $l = 25$.

$$w = 0.6l^3$$

$$= 0.6(25)^3$$

$$= 9375$$

The 25-foot long shark was 9375 pounds.

3. y varies inversely as x is expressed as $y = \dfrac{k}{x}$.

The length, L, varies inversely as the frequency, f can be expressed as $L = \dfrac{k}{f}$.

Use the given values to find k.

$$L = \frac{k}{f}$$

$$8 = \frac{k}{640}$$

$$5120 = k$$

Substitute the value of k into the equation.

$$L = \frac{k}{f}$$

$$L = \frac{5120}{f}$$

Use the equation to find f when $L = 10$.

$$L = \frac{5120}{f}$$

$$10 = \frac{5120}{f}$$

$$10f = 5120$$

$$f = 512$$

A 10 inch violin string will have a frequency of 512 cycles per second.

4. let M represent the number of minutes

let Q represent the number of problems

let P represent the number of people

M varies directly as Q and inversely as P is expressed as $M = \dfrac{kQ}{P}$.

Use the given values to find k.

$$M = \frac{kQ}{P}$$

$$32 = \frac{k(16)}{4}$$

$$8 = k$$

Substitute the value of k into the equation.

$$M = \frac{kQ}{P}$$

$$M = \frac{8Q}{P}$$

Use the equation to find M when $P = 8$ and $Q = 24$.

$$M = \frac{8Q}{P}$$

$$M = \frac{8(24)}{8}$$

$$M = 24$$

It will take 24 minutes for 8 people to solve 24 problems.

5. V varies jointly with h and r^2 and can be modeled as $V = khr^2$.

Use the given values to find k.

$$V = khr^2$$

$$120\pi = k(10)(6)^2$$

$$\frac{\pi}{3} = k$$

Therefore, the volume equation is $V = \dfrac{1}{3}hr^2$.

$$V = \frac{\pi}{3}(2)(12)^2 = 96\pi \text{ cubic feet}$$

Concept and Vocabulary Check 3.7

1. $y - kx$; constant of variation

2. $y = kx^n$

3. $y = \dfrac{k}{x}$

4. $y = \dfrac{kx}{z}$

5. $y = kxz$

6. directly; inversely

7. jointly; inversely

Exercise Set 3.7

1. Use the given values to find k.
$y = kx$
$65 = k \cdot 5$
$\dfrac{65}{5} = \dfrac{k \cdot 5}{5}$
$13 = k$
The equation becomes $y = 13x$.
When $x = 12$, $y = 13x = 13 \cdot 12 = 156$.

3. Since y varies inversely with x, we have $y = \dfrac{k}{x}$.
Use the given values to find k.
$y = \dfrac{k}{x}$
$12 = \dfrac{k}{5}$
$5 \cdot 12 = 5 \cdot \dfrac{k}{5}$
$60 = k$
The equation becomes $y = \dfrac{60}{x}$.
When $x = 2$, $y = \dfrac{60}{2} = 30$.

5. Since y varies inversely as x and inversely as the square of z, we have $y = \dfrac{kx}{z^2}$.
Use the given values to find k.
$y = \dfrac{kx}{z^2}$
$20 = \dfrac{k(50)}{5^2}$
$20 = \dfrac{k(50)}{25}$
$20 = 2k$
$10 = k$
The equation becomes $y = \dfrac{10x}{z^2}$.
When $x = 3$ and $z = 6$,
$y = \dfrac{10x}{z^2} = \dfrac{10(3)}{6^2} = \dfrac{10(3)}{36} = \dfrac{30}{36} = \dfrac{5}{6}$.

7. Since y varies jointly as x and z, we have $y = kxz$.
Use the given values to find k.
$y = kxz$
$25 = k(2)(5)$
$25 = k(10)$
$\dfrac{25}{10} = \dfrac{k(10)}{10}$
$\dfrac{5}{2} = k$
The equation becomes $y = \dfrac{5}{2}xz$.
When $x = 8$ and $z = 12$, $y = \dfrac{5}{2}(8)(12) = 240$.

9. Since y varies jointly as a and b and inversely as the square root of c, we have $y = \dfrac{kab}{\sqrt{c}}$.

Use the given values to find k.

$$y = \frac{kab}{\sqrt{c}}$$

$$12 = \frac{k(3)(2)}{\sqrt{25}}$$

$$12 = \frac{k(6)}{5}$$

$$12(5) = \frac{k(6)}{5}(5)$$

$$60 = 6k$$

$$\frac{60}{6} = \frac{6k}{6}$$

$$10 = k$$

The equation becomes $y = \dfrac{10ab}{\sqrt{c}}$.

When $a = 5$, $b = 3$, $c = 9$,

$$y = \frac{10ab}{\sqrt{c}} = \frac{10(5)(3)}{\sqrt{9}} = \frac{150}{3} = 50.$$

11. $x = kyz$;

Solving for y:

$$x = kyz$$

$$\frac{x}{kz} = \frac{kyz}{yz}.$$

$$y = \frac{x}{kz}$$

13.

$$x = \frac{kz^3}{y};$$

Solving for y

$$x = \frac{kz^3}{y}$$

$$xy = y \cdot \frac{kz^3}{y}$$

$$xy = kz^3$$

$$\frac{xy}{x} = \frac{kz^3}{x}$$

$$y = \frac{kz^3}{x}$$

15.

$$x = \frac{kyz}{\sqrt{w}} ;$$

Solving for y:

$$x = \frac{kyz}{\sqrt{w}}$$

$$x\left(\sqrt{w}\right) = \left(\sqrt{w}\right)\frac{kyz}{\sqrt{w}}$$

$$x\sqrt{w} = kyz$$

$$\frac{x\sqrt{w}}{kz} = \frac{kyz}{kz}$$

$$y = \frac{x\sqrt{w}}{kz}$$

17. $x = kz(y + w)$;

Solving for y:

$$x = kz(y + w)$$

$$x = kzy + kzw$$

$$x - kzw = kzy$$

$$\frac{x - kzw}{kz} = \frac{kzy}{kz}$$

$$y = \frac{x - kzw}{kz}$$

19.

$$x = \frac{kz}{y - w} ;$$

Solving for y:

$$x = \frac{kz}{y - w}$$

$$(y - w)x = (y - w)\frac{kz}{y - w}$$

$$xy - wx = kz$$

$$xy = kz + wx$$

$$\frac{xy}{x} = \frac{kz + wx}{x}$$

$$y = \frac{xw + kz}{x}$$

21. Since T varies directly as B, we have $T = kB$.
Use the given values to find k.

$T = kB$

$3.6 = k(4)$

$\dfrac{3.6}{4} = \dfrac{k(4)}{4}$

$0.9 = k$

The equation becomes $T = 0.9B$.
When $B = 6$, $T = 0.9(6) = 5.4$.
The tail length is 5.4 feet.

23. Since B varies directly as D, we have $B = kD$.
Use the given values to find k.

$B = kD$

$8.4 = k(12)$

$\dfrac{8.4}{12} = \dfrac{k(12)}{12}$

$k = \dfrac{8.4}{12} = 0.7$

The equation becomes $B = 0.7D$.
When $B = 56$,

$56 = 0.7D$

$\dfrac{56}{0.7} = \dfrac{0.7D}{0.7}$

$D = \dfrac{56}{0.7} = 80$

It was dropped from 80 inches.

25. Since a man's weight varies directly as the cube of his height, we have $w = kh^3$.
Use the given values to find k.

$w = kh^3$

$170 = k(70)^3$

$170 = k(343{,}000)$

$\dfrac{170}{343{,}000} = \dfrac{k(343{,}000)}{343{,}000}$

$0.000496 = k$

The equation becomes $w = 0.000496h^3$.
When $h = 107$,

$w = 0.000496(107)^3$

$\quad = 0.000496(1{,}225{,}043) \approx 607$.

Robert Wadlow's weight was approximately 607 pounds.

27. Since the banking angle varies inversely as the turning radius, we have $B = \dfrac{k}{r}$.
Use the given values to find k.

$B = \dfrac{k}{r}$

$28 = \dfrac{k}{4}$

$28(4) = 28\left(\dfrac{k}{4}\right)$

$112 = k$

The equation becomes $B = \dfrac{112}{r}$.

When $r = 3.5$, $B = \dfrac{112}{r} = \dfrac{112}{3.5} = 32$.

The banking angle is $32°$ when the turning radius is 3.5 feet.

29. Since intensity varies inversely as the square of the distance, we have pressure, we have

$I = \dfrac{k}{d}$.

Use the given values to find k.

$I = \dfrac{k}{d^2}$.

$62.5 = \dfrac{k}{3^2}$

$62.5 = \dfrac{k}{9}$

$9(62.5) = 9\left(\dfrac{k}{9}\right)$

$562.5 = k$

The equation becomes $I = \dfrac{562.5}{d^2}$.

When $d = 2.5$, $I = \dfrac{562.5}{2.5^2} = \dfrac{562.5}{6.25} = 90$.

The intensity is 90 milliroentgens per hour.

31. Since index varies directly as weight and inversely as the square of one's height, we have $I = \dfrac{kw}{h^2}$.

Use the given values to find k.

$$I = \frac{kw}{h^2}$$

$$35.15 = \frac{k(180)}{60^2}$$

$$35.15 = \frac{k(180)}{3600}$$

$$(3600)35.15 = \frac{k(180)}{3600}$$

$$126540 = k(180)$$

$$k = \frac{126540}{180} = 703$$

The equation becomes $I = \dfrac{703w}{h^2}$.

When $w = 170$ and $h = 70$,

$$I = \frac{703(170)}{(70)^2} \approx 24.4.$$

This person has a BMI of 24.4 and is not overweight.

33. Since heat loss varies jointly as the area and temperature difference, we have $L = kAD$. Use the given values to find k.

$$L = kAD$$

$$1200 = k(3 \cdot 6)(20)$$

$$1200 = 360k$$

$$\frac{1200}{360} = \frac{360k}{360}$$

$$k = \frac{10}{3}$$

The equation becomes $L = \dfrac{10}{3}AD$

When $A = 6 \cdot 9 = 54$, $D = 10$,

$L = \dfrac{10}{3}(9 \cdot 6)(10) = 1800$.

The heat loss is 1800 Btu .

35. Since intensity varies inversely as the square of the distance from the sound source, we have $I = \dfrac{k}{d^2}$. If you move to a seat twice as far, then $d = 2d$. So we have

$$I = \frac{k}{(2d)^2} = \frac{k}{4d^2} = \frac{1}{4} \cdot \frac{k}{d^2}.$$ The intensity

will be multiplied by a factor of $\dfrac{1}{4}$. So the

sound intensity is $\dfrac{1}{4}$ of what it was originally.

37. a. Since the average number of phone calls varies jointly as the product of the populations and inversely as the square of the distance, we have

$$C = \frac{kP_1P_2}{d^2}.$$

b. Use the given values to find k.

$$C = \frac{kP_1P_2}{d^2}$$

$$326,000 = \frac{k(777,000)(3,695,000)}{(420)^2}$$

$$326,000 = \frac{k(2.87 \times 10^{12})}{176,400}$$

$$326,000 = 16269841.27k$$

$$0.02 \approx k$$

The equation becomes $C = \dfrac{0.02P_1P_2}{d^2}$.

c. $C = \dfrac{0.02(650,000)(490,000)}{(400)^2}$

$\approx 39,813$

There are approximately 39,813 daily phone calls.

39. a.

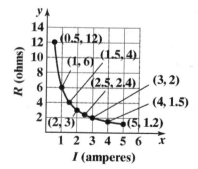

b. Current varies inversely as resistance. Answers will vary.

c. Since the current varies inversely as resistance we have $R = \dfrac{k}{I}$. Using one of the given ordered pairs to find k.

$$12 = \frac{k}{0.5}$$

$$12(0.5) = \frac{k}{0.5}(0.5)$$

$$k = 6$$

The equation becomes $R = \dfrac{6}{I}$.

41. – 47. Answers will vary.

49. does not make sense; Explanations will vary. Sample explanation: For an inverse variation, the independent variable can not be zero.

51. makes sense

53. Pressure, P, varies directly as the square of wind velocity, v, can be modeled as $P = kv^2$.

If $v = x$ then $P = k(x)^2 = kx^2$

If $v = 2x$ then $P = k(2x)^2 = 4kx^2$

If the wind speed doubles the pressure is 4 times more destructive.

55. The Heat, H, varies directly as the square of the voltage, v, and inversely as the resistance, r.

$$H = \frac{kv^2}{r}$$

If the voltage remains constant, to triple the heat the resistant must be reduced by a multiple of 3.

57. Answers will vary.

58.

$f(x) = 2^x$

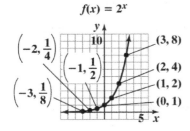

59.

$g(x) = f(-x) = 2^{-x}$

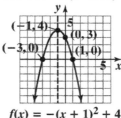

60.

$h(x) = f(x) + 1 = 2^x + 1$

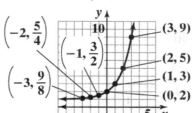

Chapter 3 Review Exercises

1. $f(x) = -(x+1)^2 + 4$

vertex: $(-1, 4)$

x-intercepts:

$$0 = -(x+1)^2 + 4$$

$$(x+1)^2 = 4$$

$$x + 1 = \pm 2$$

$$x = -1 \pm 2$$

$$x = -3 \text{ or } x = 1$$

y-intercept:

$$f(0) = -(0+1)^2 + 4 = 3$$

The axis of symmetry is $x = -1$.

$f(x) = -(x + 1)^2 + 4$

domain: $(-\infty, \infty)$ range: $(-\infty, 4]$

2. $f(x) = (x+4)^2 - 2$

vertex: $(-4, -2)$

x-intercepts:

$$0 = (x+4)^2 - 2$$

$$(x+4)^2 = 2$$

$$x + 4 = \pm\sqrt{2}$$

$$x = -4 \pm \sqrt{2}$$

y-intercept:

$$f(0) = (0+4)^2 - 2 = 14 = -1$$

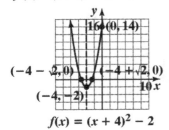

$$f(x) = (x+4)^2 - 2$$

The axis of symmetry is $x = -4$.
domain: $(-\infty, \infty)$ range: $[-2, \infty)$

3. $f(x) = -x^2 + 2x + 3$

$$= -\left(x^2 - 2x + 1\right) + 3 + 1$$

$$f(x) = -(x-1)^2 + 4$$

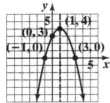

$$f(x) = -x^2 + 2x + 3$$

domain: $(-\infty, \infty)$ range: $(-\infty, 4]$

4. $f(x) = 2x^2 - 4x - 6$

$$f(x) = 2\left(x^2 - 2x + 1\right) - 6 - 2$$

$$2(x-1)^2 - 8$$

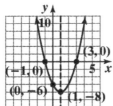

$$f(x) = 2x^2 - 4x - 6$$

axis of symmetry: $x = 1$
domain: $(-\infty, \infty)$ range: $[-8, \infty)$

5. $f(x) = -x^2 + 14x - 106$

a. Since $a < 0$ the parabola opens down with the maximum value occurring at

$$x = -\frac{b}{2a} = -\frac{14}{2(-1)} = 7 .$$

The maximum value is $f(7)$.

$$f(7) = -(7)^2 + 14(7) - 106 = -57$$

b. domain: $(-\infty, \infty)$ range: $(-\infty, -57]$

6. $f(x) = 2x^2 + 12x + 703$

a. Since $a > 0$ the parabola opens up with the minimum value occurring at

$$x = -\frac{b}{2a} = -\frac{12}{2(2)} = -3 .$$

The minimum value is $f(-3)$.

$$f(-3) = 2(-3)^2 + 12(-3) + 703 = 685$$

b. domain: $(-\infty, \infty)$ range: $[685, \infty)$

7. a. The maximum height will occur at the vertex.

$$f(x) = -0.025x^2 + x + 6$$

$$x = -\frac{b}{2a} = -\frac{1}{2(-0.025)} = 20$$

$$f(20) = -0.025(20)^2 + (20) + 6 = 16$$

The maximum height of 16 feet occurs when the ball is 20 yards downfield.

b. $f(x) = -0.025x^2 + x + 6$

$$f(0) = -0.025(0)^2 + (0) + 6 = 6$$

The ball was tossed at a height of 6 feet.

c. The ball is at a height of 0 when it hits the ground.

$$f(x) = -0.025x^2 + x + 6$$

$$0 = -0.025x^2 + x + 6$$

$$x = \frac{-b \pm \sqrt{b^2 - 4ac}}{2a}$$

$$x = \frac{-(1) \pm \sqrt{(1)^2 - 4(-0.025)(6)}}{2(-0.025)}$$

$$x \approx 45.3, \ -5.3 \text{(reject)}$$

The ball will hit the ground 45.3 yards downfield.

d. The football's path:

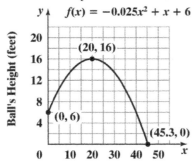

f(x) = -0.025x² + x + 6

Ball's Height (feet) vs Ball's Horizontal Distance (yards)

(20, 16)
(0, 6)
(45.3, 0)

8. Maximize the area using $A = lw$.

$A(x) = x(1000 - 2x)$

$A(x) = -2x^2 + 1000x$

Since $a = -2$ is negative, we know the function opens downward and has a maximum at

$$x = -\frac{b}{2a} = -\frac{1000}{2(-2)} = -\frac{1000}{-4} = 250.$$

The maximum area is achieved when the width is 250 yards. The maximum area is

$A(250) = 250(1000 - 2(250))$

$= 250(1000 - 500)$

$= 250(500) = 125,000.$

The area is maximized at 125,000 square yards when the width is 250 yards and the length is $1000 - 2 \cdot 250 = 500$ yards.

9. Let $x =$ one of the numbers
Let $14 + x =$ the other number

We need to minimize the function

$P(x) = x(14 + x)$

$= 14x + x^2$

$= x^2 + 14x.$

The minimum is at

$$x = -\frac{b}{2a} = -\frac{14}{2(1)} = -\frac{14}{2} = -7.$$

The other number is $14 + x = 14 + (-7) = 7$.

The numbers which minimize the product are 7 and -7. The minimum product is $-7 \cdot 7 = -49$.

10. $f(x) = -x^3 + 12x^2 - x$

The graph rises to the left and falls to the right and goes through the origin, so graph (c) is the best match.

11. $g(x) = x^6 - 6x^4 + 9x^2$

The graph rises to the left and rises to the right, so graph (b) is the best match.

12. $h(x) = x^5 - 5x^3 + 4x$

The graph falls to the left and rises to the right and crosses the y-axis at zero, so graph (a) is the best match.

13. $f(x) = -x^4 + 1$

$f(x)$ falls to the left and to the right so graph (d) is the best match.

14. a. Since n is odd and $a_n > 0$, the graph rises to the right.

b. No, the model will not be useful. The model indicates increasing deforestation despite a declining rate in which the forest is being cut down.

c. The graph of function g falls to the right.

d. No, the model will not be useful. The model indicates the amount of forest cleared, in square kilometers, will eventually be negative, which

15. In the polynomial, $f(x) = -x^4 + 21x^2 + 100$, the leading coefficient is -1 and the degree is 4. Applying the Leading Coefficient Test, we know that even-degree polynomials with negative leading coefficient will fall to the left and to the right. Since the graph falls to the right, we know that the elk population will die out over time.

16. $f(x) = -2(x-1)(x+2)^2(x+5)^3$

$x = 1$, multiplicity 1, the graph crosses the x-axis
$x = -2$, multiplicity 2, the graph touches the x-axis
$x = -5$, multiplicity 5, the graph crosses the x-axis

17. $f(x) = x^3 - 5x^2 - 25x + 125$

$= x^2(x - 5) - 25(x - 5)$

$= (x^2 - 25)(x - 5)$

$= (x + 5)(x - 5)^2$

$x = -5$, multiplicity 1, the graph crosses the x-axis
$x = 5$, multiplicity 2, the graph touches the x-axis

18. $f(x) = x^3 - 2x - 1$

$f(1) = (1)^3 - 2(1) - 1 = -2$

$f(2) = (2)^3 - 2(2) - 1 = 3$

The sign change shows there is a zero between the given values.

19. $f(x) = x^3 - x^2 - 9x + 9$

a. Since n is odd and $a_n > 0$, the graph falls to the left and rises to the right.

b. $f(-x) = (-x)^3 - (-x)^2 - 9(-x) + 9$

$= -x^3 - x^2 + 9x + 9$

$f(-x) \neq f(x), f(-x) \neq -f(x)$

no symmetry

c. $f(x) = (x - 3)(x + 3)(x - 1)$

zeros: $3, -3, 1$

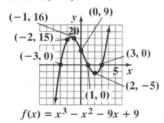

$f(x) = x^3 - x^2 - 9x + 9$

20. $f(x) = 4x - x^3$

a. Since n is odd and $a_n < 0$, the graph rises to the left and falls to the right.

b. $f(-x) = -4x + x^3$

$f(-x) = -f(x)$

origin symmetry

c. $f(x) = x(x^2 - 4) = x(x - 2)(x + 2)$

zeros: $x = 0, 2, -2$

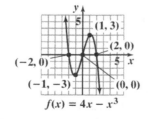

$f(x) = 4x - x^3$

21. $f(x) = 2x^3 + 3x^2 - 8x - 12$

a. Since h is odd and $a_n > 0$, the graph falls to the left and rises to the right.

b. $f(-x) = -2x^3 + 3x^2 + 8x - 12$

$f(-x) \neq f(x), f(-x) = -f(x)$

no symmetry

c. $f(x) = (x - 2)(x + 2)(2x + 3)$

zeros: $x = 2, -2, -\dfrac{3}{2}$

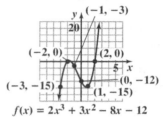

$f(x) = 2x^3 + 3x^2 - 8x - 12$

22. $g(x) = -x^4 + 25x^2$

a. The graph falls to the left and to the right.

b. $f(-x) = -(-x)^4 + 25(-x)^2$

$= -x^4 + 25x^2 = f(x)$

y-axis symmetry

c. $-x^4 + 25x^2 = 0$

$-x^2(x^2 - 25) = 0$

$-x^2(x - 5)(x + 5) = 0$

zeros: $x = -5, 0, 5$

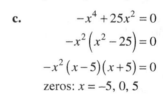

$f(x) = -x^4 + 25x^2$

23. $f(x) = -x^4 + 6x^3 - 9x^2$

a. The graph falls to the left and to the right.

b. $f(-x) = -(-x)^4 + 6(-x)^3 - 9(-x)$

$= -x^4 - 6x^3 - 9x^2 f(-x) \neq f(x)$

$f(-x) \neq -f(x)$

no symmetry

c. $= -x^2\left(x^2 - 6x + 9\right) = 0$

$-x^2(x-3)(x-3) = 0$

zeros: $x = 0, 3$

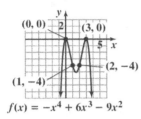

$f(x) = -x^4 + 6x^3 - 9x^2$

24. $f(x) = 3x^4 - 15x^3$

a. The graph rises to the left and to the right.

b. $f(-x) = 3(-x)^4 - 15(-x)^2 = 3x^4 + 15x^3$

$f(-x) \ne f(x),\ f(-x) \ne -f(x)$

no symmetry

c $3x^4 - 15x^3 = 0$

$3x^3(x-5) = 0$

zeros: $x = 0, 5$

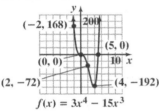

$f(x) = 3x^4 - 15x^3$

25. $f(x) = 2x^2(x-1)^3(x+2)$

Since $a_n > 0$ and n is even, $f(x)$ rises to the left and the right.

$x = 0,\ x = 1,\ x = -2$

The zeros at 1 and –2 have odd multiplicity so $f(x)$ crosses the x-axis at those points. The root at 0 has even multiplicity so $f(x)$ touches the axis at (0, 0)

$f(0) = 2(0)^2(0-1)^3(0+2) = 0$

The y-intercept is 0.

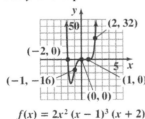

$f(x) = 2x^2(x-1)^3(x+2)$

26. $f(x) = -x^3(x+4)^2(x-1)$

Since $a_n < 0$ and n is even, $f(x)$ falls to the left and the right.

$x = 0,\ x = -4,\ x = 1$

The roots at 0 and 1 have odd multiplicity so $f(x)$ crosses the x-axis at those points. The root at –4 has even multiplicity so $f(x)$ touches the axis at (–4, 0)

$f(0) = -(0)^3(0+4)^2(0-1) = 0$

The y-intercept is 0.

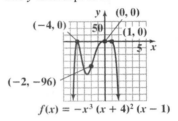

$f(x) = -x^3(x+4)^2(x-1)$

27.

$$\begin{array}{r} 4x^2 - 7x + 5 \\ x+1\overline{)4x^3 - 3x^2 - 2x + 1} \\ \underline{4x^3 + 4x^2} \\ -7x^2 - 2x \\ \underline{-7x^2 - 7x} \\ 5x + 1 \\ \underline{5x + 5} \\ -4 \end{array}$$

Quotient: $4x^2 - 7x + 5 - \dfrac{4}{x+1}$

28.

$$\begin{array}{r} 2x^2 - 4x + 1 \\ 5x-3\overline{)10x^3 - 26x^2 + 17x - 13} \\ \underline{10x^3 + 6x^2} \\ -20x^2 + 17x \\ \underline{-20x^2 + 12x} \\ 5x - 13 \\ \underline{5x - 3} \\ -10 \end{array}$$

Quotient: $2x^2 - 4x + 1 - \dfrac{10}{5x-3}$

29.

$$\begin{array}{r} 2x^2 + 3x - 1 \\ 2x^2 + 1 \overline{)\,4x^4 + 6x^3 + 3x - 1} \\ \underline{4x^2 + 2x^2} \\ 6x^3 - 2x^2 + 3x \\ \underline{6x^2 + 3x} \\ -2x^2 - 1 \\ \underline{-2x^2 - 1} \\ 0 \end{array}$$

30. $(3x^4 + 11x^3 - 20x^3 + 7x + 35) \div (x + 5)$

$$\begin{array}{r|rrrrr} -5 & 3 & 11 & -20 & 7 & 35 \\ & & -15 & 20 & 0 & -35 \\ \hline & 3 & -4 & 0 & 7 & 0 \end{array}$$

Quotient: $3x^3 - 4x^2 + 7$

31. $(3x^4 - 2x^2 - 10x) \div (x - 2)$

$$\begin{array}{r|rrrrr} 2 & 3 & 0 & -2 & -10 & 0 \\ & & 6 & 12 & 20 & 20 \\ \hline & 3 & 6 & 10 & 10 & 20 \end{array}$$

Quotient: $3x^3 + 6x^2 + 10x + 10 + \dfrac{20}{x-2}$

32. $f(x) = 2x^3 - 7x^2 + 9x - 3$

$$\begin{array}{r|rrrr} -13 & 2 & -7 & 9 & -3 \\ & & -26 & 429 & -5694 \\ \hline & 2 & -33 & 438 & -5697 \end{array}$$

Quotient: $f(-13) = -5697$

33. $f(x) = 2x^3 + x^2 - 13x + 6$

$$\begin{array}{r|rrrr} 2 & 2 & 1 & -13 & 6 \\ & & 4 & 10 & -6 \\ \hline & 2 & 5 & -3 & 0 \end{array}$$

$f(x) = (x-2)(2x^2 + 5x - 3)$
$\quad\;\; = (x-2)(2x-1)(x+3)$

Zeros: $x = 2, \dfrac{1}{2}, -3$

34. $x^3 - 17x + 4 = 0$

$$\begin{array}{r|rrrr} 4 & 1 & 0 & -17 & 4 \\ & & 4 & 16 & -4 \\ \hline & 1 & 4 & -1 & 0 \end{array}$$

$(x-4)(x^2 + 4x - 1) = 0$

$x = \dfrac{-4 \pm \sqrt{16+4}}{2} = \dfrac{-4 \pm 2\sqrt{5}}{2} = -2 \pm \sqrt{5}$

The solution set is $\left\{ 4, -2 + \sqrt{5}, -2 - \sqrt{5} \right\}$.

35. $f(x) = x^4 - 6x^3 + 14x^2 - 14x + 5$

$p: \pm 1, \pm 5$

$q: \pm 1$

$\dfrac{p}{q}: \pm 1, \pm 5$

36. $f(x) = 3x^5 - 2x^4 - 15x^3 + 10x^2 + 12x - 8$

$p: \pm 1, \pm 2, \pm 4, \pm 8$

$q: \pm 1, \pm 3$

$\dfrac{p}{q}: \pm 1, \pm 2, \pm 4, \pm 8, \pm \dfrac{8}{3}, \pm \dfrac{4}{3}, \pm \dfrac{2}{3}, \pm \dfrac{1}{3}$

37. $f(x) = 3x^4 - 2x^3 - 8x + 5$

$f(x)$ has 2 sign variations, so $f(x) = 0$ has 2 or 0 positive solutions.

$f(-x) = 3x^4 + 2x^3 + x + 5$

$f(-x)$ has no sign variations, so $f(x) = 0$ has no negative solutions.

38. $f(x) = 2x^5 - 3x^3 - 5x^2 + 3x - 1$

$f(x)$ has 3 sign variations, so $f(x) = 0$ has 3 or 1 positive real roots.

$f(-x) = -2x^5 + 3x^3 - 5x^2 - 3x - 1$

$f(-x)$ has 2 sign variations, so $f(x) = 0$ has 2 or 0 negative solutions.

39. $f(x) = f(-x) = 2x^4 + 6x^2 + 8$

No sign variations exist for either $f(x)$ or $f(-x)$, so no real roots exist.

40. $f(x) = x^3 + 3x^2 - 4$

   a. $p: \pm 1, \pm 2, \pm 4$

        $q: \pm 1$

        $\dfrac{p}{q}: \pm 1, \pm 2, \pm 4$

   b.   1 sign variation $\Rightarrow$ 1 positive real zero

        $f(-x) = -x^3 + 3x^2 - 4$

        2 sign variations $\Rightarrow$ 2 or no negative real zeros

   c.

$$\begin{array}{c|cccc} 1 & 1 & 3 & 0 & -4 \\ & & 1 & 4 & -4 \\ \hline & 1 & 4 & 4 & 0 \end{array}$$

        1 is a zero.

        1, –2 are rational zeros.

   d. $(x-1)(x^2 + 4x + 4) = 0$

        $(x-1)(x+2)^2 = 0$

        $x = 1$ or $x = -2$

        The solution set is $\{1, -2\}$.

41. $f(x) = 6x^3 + x^2 - 4x + 1$

   a. $p: \pm 1$

        $q: \pm 1, \pm 2, \pm 3, \pm 6$

        $\dfrac{p}{q}: \pm 1, \pm \dfrac{1}{2}, \pm \dfrac{1}{3}, \pm \dfrac{1}{6}$

   b. $f(x) = 6x^3 + x^2 - 4x + 1$

        2 sign variations; 2 or 0 positive real zeros.

        $f(-x) = -6x^3 + x^2 + 4x + 1$

        1 sign variation; 1 negative real zero.

   c.

$$\begin{array}{c|cccc} -1 & 6 & 1 & -4 & 1 \\ & & -6 & 5 & -1 \\ \hline & 6 & -5 & 1 & 0 \end{array}$$

        –1 is a zero.

        $-1, \dfrac{1}{3}, \dfrac{1}{2}$ are rational zeros.

   d. $6x^3 + x^2 - 4x + 1 = 0$

        $(x+1)(6x^2 - 5x + 1) = 0$

        $(x+1)(3x-1)(2x-1) = 0$

        $x = -1$ or $x = \dfrac{1}{3}$ or $x = \dfrac{1}{2}$

        The solution set is $\left\{ -1, \dfrac{1}{3}, \dfrac{1}{2} \right\}$.

42. $f(x) = 8x^3 - 36x^2 + 46x - 15$

   a. $p: \pm 1, \pm 3, \pm 5, \pm 15$

        $q: \pm 1, \pm 2, \pm 4, \pm 8$

        $\dfrac{p}{q}: \pm 1, \pm 3, \pm 5, \pm 15, \pm \dfrac{1}{2}, \pm \dfrac{1}{4}, \pm \dfrac{1}{8},$

        $\pm \dfrac{3}{2}, \pm \dfrac{3}{4}, \pm \dfrac{3}{8}, \pm \dfrac{5}{2}, \pm \dfrac{5}{4},$

        $\pm \dfrac{5}{8}, \pm \dfrac{15}{2}, \pm \dfrac{15}{4}, \pm \dfrac{15}{8}$

   b. $f(x) = 8x^3 - 36x^2 + 46x - 15$

        3 sign variations; 3 or 1 positive real solutions.

        $f(-x) = -8x^3 - 36x^2 - 46x - 15$

        0 sign variations; no negative real solutions.

   c.

$$\begin{array}{c|cccc} \frac{1}{2} & 8 & -36 & 46 & -15 \\ & & 4 & -16 & 15 \\ \hline & 8 & -32 & 30 & 0 \end{array}$$

        $\dfrac{1}{2}$ is a zero.

        $\dfrac{1}{2}, \dfrac{3}{2}, \dfrac{5}{2}$ are rational zeros.

   d.

$$8x^3 - 36x^2 + 46x - 15 = 0$$

$$\left(x - \dfrac{1}{2} \right)(8x^2 - 32x + 30) = 0$$

$$2\left(x - \dfrac{1}{2} \right)(4x^2 - 16x + 15) = 0$$

$$2\left(x - \dfrac{1}{2} \right)(2x-5)(2x-3) = 0$$

$$x = \dfrac{1}{2} \text{ or } x = \dfrac{5}{2} \text{ or } x = \dfrac{3}{2}$$

        The solution set is $\left\{ \dfrac{1}{2}, \dfrac{3}{2}, \dfrac{5}{2} \right\}$.

43. $2x^3 + 9x^2 - 7x + 1 = 0$

    a. $p: \pm 1$

          $q: \pm 1, \pm 2$

          $\dfrac{p}{q}: \pm 1, \pm \dfrac{1}{2}$

    b. $f(x) = 2x^3 + 9x^2 - 7x + 1$

        2 sign variations; 2 or 0 positive real zeros.

        $f(-x) = -2x^3 + 9x^2 + 7x + 1$

        1 sign variation; 1 negative real zero.

    c.

$\dfrac{1}{2}$	2	9	−7	1
		1	5	−1
	2	10	−2	0

        $\dfrac{1}{2}$ is a rational zero.

    d. $2x^3 + 9x^2 - 7x + 1 = 0$

        $\left(x - \dfrac{1}{2}\right)(2x^2 + 10x - 2) = 0$

        $2\left(x - \dfrac{1}{2}\right)(x^2 + 5x - 1) = 0$

        Solving $x^2 + 5x - 1 = 0$ using the quadratic

        formula gives $x = \dfrac{-5 \pm \sqrt{29}}{2}$

        The solution set is $\left\{ \dfrac{1}{2}, \dfrac{-5 + \sqrt{29}}{2}, \dfrac{-5 - \sqrt{29}}{2} \right\}$.

44. $x^4 - x^3 - 7x^2 + x + 6 = 0$

    a. $p = \dfrac{p}{q}: \pm 1, \pm 2, \pm 3, \pm 6$

    b. $f(x) = x^4 - x^3 - 7x^2 + x + 6$

        2 sign variations; 2 or 0 positive real zeros.

        $f(-x) = x^4 + x^3 - 7x^2 - x + 6$

        2 sign variations; 2 or 0 negative real zeros.

    c.

1	1	−1	−7	1	6
		1	0	−7	−6
	1	0	−7	−6	0

−1	1	0	−7	−6
		−1	1	6
	1	−1	−6	0

$-2, -1, 1, 3$ are rational zeros.

    d. $x^4 - x^3 - 7x^2 + x + 6 = 0$

        $(x - 1)(x + 1)(x^2 - x + 6) = 0$

        $(x - 1)(x + 1)(x - 3)(x + 2) = 0$

        The solution set is $\{-2, -1, 1, 3\}$.

45. $4x^4 + 7x^2 - 2 = 0$

    a. $p: \pm 1, \pm 2$

          $q: \pm 1, \pm 2, \pm 4$

          $\dfrac{p}{q}: \pm 1, \pm 2, \pm \dfrac{1}{2}, \pm \dfrac{1}{4}$

    b. $f(x) = 4x^4 + 7x^2 - 2$

        1 sign variation; 1 positive real zero.

        $f(-x) = 4x^4 + 7x^2 - 2$

        1 sign variation; 1 negative real zero.

    c.

$\dfrac{1}{2}$	4	0	7	0	−2
		2	1	4	2
	4	2	8	4	0

$-\dfrac{1}{2}$	4	2	8	4
		−2	0	−4
	4	0	8	0

$-\dfrac{1}{2}, \dfrac{1}{2}$ are rational zeros.

    d. $4x^4 + 7x^2 - 2 = 0$

        $\left(x - \dfrac{1}{2}\right)\left(x + \dfrac{1}{2}\right)(4x^2 + 8) = 0$

        $4\left(x - \dfrac{1}{2}\right)\left(x + \dfrac{1}{2}\right)(x^2 + 2) = 0$

        Solving $x^2 + 2 = 0$ using the quadratic formula

        gives $x = \pm 2i$

        The solution set is $\left\{ -\dfrac{1}{2}, \dfrac{1}{2}, 2i, -2i \right\}$.

46. $f(x) = 2x^4 + x^3 - 9x^2 - 4x + 4$

    a. $p: \pm 1, \pm 2, \pm 4$

      $q: \pm 1, \pm 2$

      $\dfrac{p}{q}: \pm 1, \pm 2, \pm 4, \pm \dfrac{1}{2}$

    b. $f(x) = 2x^4 + x^3 - 9x^2 - 4x + 4$

      2 sign variations; 2 or 0 positive real zeros.

      $f(-x) = 2x^4 - x^3 - 9x^2 + 4x + 4$

      2 sign variations; 2 or 0 negative real zeros.

    c.

2	2	1	−9	−4	4
		4	10	2	−4
	2	5	1	−2	0

−1	2	5	1	−2
		−2	−3	2
	2	3	−2	0

      $-2, -1, \dfrac{1}{2}, 2$ are rational zeros.

    d. $2x^2 + 3x - 2 = 0$

      $(2x - 1)(x + 2) = 0$

      $x = -2 \quad \text{or} \quad x = \dfrac{1}{2}$

      The solution set is $\left\{ -2, -1, \dfrac{1}{2}, 2 \right\}$.

47. $f(x) = a_n(x - 2)(x - 2 + 3i)(x - 2 - 3i)$

$f(x) = a_n(x - 2)(x^2 - 4x + 13)$

$f(1) = a_n(1 - 2)\left[1^2 - 4(1) + 13\right]$

$-10 = -10a_n$

$a_n = 1$

$f(x) = 1(x - 2)(x^2 - 4x + 13)$

$f(x) = x^3 - 4x^2 + 13x - 2x^2 + 8x - 26$

$f(x) = x^3 - 6x^2 + 21x - 26$

48. $f(x) = a_n(x - i)(x + i)(x + 3)^2$

$f(x) = a_n(x^2 + 1)(x^2 + 6x + 9)$

$f(-1) = a_n\left[(-1)^2 + 1\right]\left[(-1)^2 + 6(-1) + 9\right]$

$16 = 8a_n$

$a_n = 2$

$f(x) = 2(x^2 + 1)(x^2 + 6x + 9)$

$f(x) = 2(x^4 + 6x^3 + 9x^2 + x^2 + 6x + 9)$

$f(x) = 2x^4 + 12x^3 + 20x^2 + 12x + 18$

49. $f(x) = 2x^4 + 3x^3 + 3x - 2$

$p: \pm 1, \pm 2$

$q: \pm 1, \pm 2$

$\dfrac{p}{q}: \pm 1, \pm 2, \pm \dfrac{1}{2}$

−2	2	3	0	3	−2
		−4	2	−4	2
	2	−1	2	−1	0

$2x^4 + 3x^3 + 3x - 2 = 0$

$(x + 2)(2x^3 - x^2 + 2x - 1) = 0$

$(x + 2)[x^2(2x - 1) + (2x - 1)] = 0$

$(x + 2)(2x - 1)(x^2 + 1) = 0$

$x = -2, \; x = \dfrac{1}{2} \; \text{or} \; x = \pm i$

The zeros are $-2, \dfrac{1}{2}, \pm i$.

$f(x) = (x - i)(x + i)(x + 2)(2x - 1)$

50. $g(x) = x^4 - 6x^3 + x^2 + 24x + 16$

$p: \pm 1, \pm 2, \pm 4, \pm 8, \pm 16$

$q: \pm 1$

$\dfrac{p}{q}: \pm 1, \pm 2, \pm 4, \pm 8, \pm 16$

−1	1	−6	1	24	16
		−1	7	−8	−16
	1	−7	8	16	0

$x^4 - 6x^3 + x^2 + 24x + 16 = 0$

$(x + 1)(x^3 - 7x^2 + 8x + 16) = 0$

$$
\begin{array}{r|rrrr}
-1 & 1 & -7 & 8 & 16 \\
  &   & -1 & 8 & -16 \\
\hline
  & 1 & -8 & 16 & 0
\end{array}
$$

$$(x+1)^2(x^2-8x+16)=0$$
$$(x+1)^2(x-4)^2=0$$
$$x=-1 \text{ or } x=4$$
$$g(x)=(x+1)^2(x-4)^2$$

51. 4 real zeros, one with multiplicity two

52. 3 real zeros; 2 nonreal complex zeros

53. 2 real zeros, one with multiplicity two; 2 nonreal complex zeros

54. 1 real zero; 4 nonreal complex zeros

55. $g(x)=\dfrac{1}{(x+2)^2}-1$

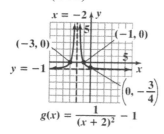

$$g(x)=\frac{1}{(x+2)^2}-1$$

56. $h(x)=\dfrac{1}{x-1}+3$

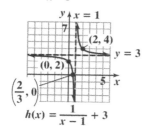

$$h(x)=\frac{1}{x-1}+3$$

57. $f(x)=\dfrac{2x}{x^2-9}$

Symmetry: $f(-x)=-\dfrac{2x}{x^2-9}=-f(x)$

origin symmetry
x-intercept:

$$0=\frac{2x}{x^2-9}$$
$$2x=0$$
$$x=0$$

y-intercept: $y=\dfrac{2(0)}{0^2-9}=0$

Vertical asymptote:

$$x^2-9=0$$
$$(x-3)(x+3)=0$$
$$x=3 \text{ and } x=-3$$

Horizontal asymptote:
$n<m$, so $y=0$

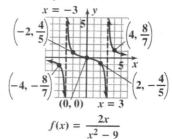

$$f(x)=\frac{2x}{x^2-9}$$

58. $g(x)=\dfrac{2x-4}{x+3}$

Symmetry: $g(-x)=\dfrac{-2x-4}{x+3}$

$g(-x)\neq g(x)$, $g(-x)\neq -g(x)$
No symmetry
x-intercept:

$$2x-4=0$$
$$x=2$$

y-intercept: $y=\dfrac{2(0)-4}{(0)+3}=-\dfrac{4}{3}$

Vertical asymptote:
$x+3=0$
$x=-3$

Horizontal asymptote:

$n=m$, so $y=\dfrac{2}{1}=2$

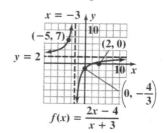

$$f(x)=\frac{2x-4}{x+3}$$

59. $h(x) = \dfrac{x^2 - 3x - 4}{x^2 - x - 6}$

Symmetry: $h(-x) = \dfrac{x^2 + 3x - 4}{x^2 + x - 6}$

$h(-x) \neq h(x), h(-x) \neq -h(x)$
No symmetry
x-intercepts:
$x^2 - 3x - 4 = 0$
$(x - 4)(x + 1)$
$x = 4 \quad x = -1$

y-intercept: $y = \dfrac{0^2 - 3(0) - 4}{0^2 - 0 - 6} = \dfrac{2}{3}$

Vertical asymptotes:
$x^2 - x - 6 = 0$
$(x - 3)(x + 2) = 0$
$x = 3, -2$

Horizontal asymptote:

$n = m$, so $y = \dfrac{1}{1} = 1$

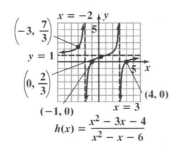

$h(x) = \dfrac{x^2 - 3x - 4}{x^2 - x - 6}$

60. $r(x) = \dfrac{x^2 + 4x + 3}{(x + 2)^2}$

Symmetry: $r(-x) = \dfrac{x^2 - 4x + 3}{(-x + 2)^2}$

$r(-x) \neq r(x), r(-x) \neq -r(x)$
No symmetry
x-intercepts:
$x^2 + 4x + 3 = 0$
$(x + 3)(x + 1) = 0$
$x = -3, -1$

y-intercept: $y = \dfrac{0^2 + 4(0) + 3}{(0 + 2)^2} = \dfrac{3}{4}$

Vertical asymptote:
$x + 2 = 0$
$x = -2$

Horizontal asymptote:

$n = m$, so $y = \dfrac{1}{1} = 1$

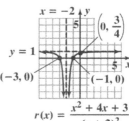

$r(x) = \dfrac{x^2 + 4x + 3}{(x + 2)^2}$

61. $y = \dfrac{x^2}{x + 1}$

Symmetry: $f(-x) = \dfrac{x^2}{-x + 1}$

$f(-x) \neq f(x), f(-x) \neq -f(x)$
No symmetry
x-intercept:
$x^2 = 0$
$x = 0$

y-intercept: $y = \dfrac{0^2}{0 + 1} = 0$

Vertical asymptote:
$x + 1 = 0$
$x = -1$
$n > m$, no horizontal asymptote.
Slant asymptote:

$y = x - 1 + \dfrac{1}{x + 1}$

$y = x - 1$

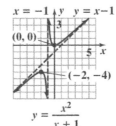

$y = \dfrac{x^2}{x + 1}$

62. $y = \dfrac{x^2 + 2x - 3}{x - 3}$

Symmetry: $f(-x) = \dfrac{x^2 - 2x - 3}{-x - 3}$

$f(-x) \neq f(x), f(-x) \neq -f(x)$
No symmetry
x-intercepts:
$x^2 + 2x - 3 = 0$
$(x + 3)(x - 1) = 0$
$x = -3, 1$

y-intercept: $y = \dfrac{0^2 + 2(0) - 3}{0 - 3} = \dfrac{-3}{-3} = 1$

Vertical asymptote:

$x - 3 = 0$

$x = 3$

Horizontal asymptote:

$n > m$, so no horizontal asymptote.

Slant asymptote:

$y = x + 5 + \dfrac{12}{x - 3}$

$y = x + 5$

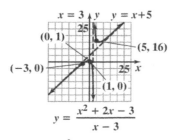

$$y = \dfrac{x^2 + 2x - 3}{x - 3}$$

63. $f(x) = \dfrac{-2x^3}{x^2 + 1}$

Symmetry: $f(-x) = \dfrac{2}{x^2 + 1} = -f(x)$

Origin symmetry

x-intercept:

$-2x^3 = 0$

$x = 0$

y-intercept: $y = \dfrac{-2(0)^3}{0^2 + 1} = \dfrac{0}{1} = 0$

Vertical asymptote:

$x^2 + 1 = 0$

$x^2 = -1$

No vertical asymptote.

Horizontal asymptote:

$n > m$, so no horizontal asymptote.

Slant asymptote:

$f(x) = -2x + \dfrac{2x}{x^2 + 1}$

$y = -2x$

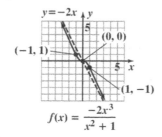

$$f(x) = \dfrac{-2x^3}{x^2 + 1}$$

64. $g(x) = \dfrac{4x^2 - 16x + 16}{2x - 3}$

Symmetry: $g(-x) = \dfrac{4x^2 + 16x + 16}{-2x - 3}$

$g(-x) \neq g(x)$, $g(-x) \neq -g(x)$

No symmetry

x-intercept:

$4x^2 - 16x + 16 = 0$

$4(x - 2)^2 = 0$

$x = 2$

y-intercept:

$y = \dfrac{4(0)^2 - 16(0) + 16}{2(0) - 3} = -\dfrac{16}{3}$

Vertical asymptote:

$2x - 3 = 0$

$x = \dfrac{3}{2}$

Horizontal asymptote:

$n > m$, so no horizontal asymptote.

Slant asymptote:

$g(x) = 2x - 5 + \dfrac{1}{2x - 3}$

$y = 2x - 5$

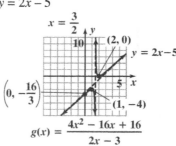

$$g(x) = \dfrac{4x^2 - 16x + 16}{2x - 3}$$

65. a. $C(x) = 50{,}000 + 25x$

b. $\overline{C}(x) = \dfrac{25x + 50{,}000}{x}$

c. $\overline{C}(50) = \dfrac{25(50) + 50{,}000}{50} = 1025$

When 50 calculators are manufactured, it costs $1025 to manufacture each.

$\overline{C}(100) = \dfrac{25(100) + 50{,}000}{100} = 525$

When 100 calculators are manufactured, it costs $525 to manufacture each.

$\overline{C}(1000) = \dfrac{25(1000) + 50{,}000}{1000} = 75$

When 1,000 calculators are manufactured, it costs \$75 to manufacture each.

$$\overline{C}(100,000) = \frac{25(100,000) + 50,000}{100,000} = 25.5 \text{ Wh}$$

en 100,000 calculators are manufactured, it costs \$25.50 to manufacture each.

d. $n = m$, so $y = \frac{25}{1} = 25$ is the horizontal

asymptote. Minimum costs will approach \$25.

66. $f(x) = \frac{150x + 120}{0.05x + 1}$

$n = m$, so $y = \frac{150}{0.05} = 3000$

The number of fish available in the pond approaches 3000.

67. $P(x) = \frac{72,900}{100x^2 + 729}$

$n < m$ so $y = 0$

As the number of years of education increases the percentage rate of unemployment approaches zero.

68. a. $P(x) = M(x) + F(x)$
$$= 1.53x + 114.8 + 1.46x + 120.7$$
$$= 2.99x + 235.5$$

b. $R(x) = \frac{M(x)}{P(x)} = \frac{1.53x + 114.8}{2.99x + 235.5}$

c. $y = \frac{1.53}{2.99} \approx 0.51$

Over time, the percentage of men in the U.S. population will approach 51%.

69. $2x^2 + 5x - 3 < 0$

Solve the related quadratic equation.

$$2x^2 + 5x - 3 = 0$$
$$(2x - 1)(x + 3) = 0$$

The boundary points are -3 and $\frac{1}{2}$.

Testing each interval gives a solution set of $\left(-3, \frac{1}{2}\right)$

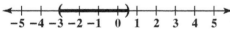

70. $2x^2 + 9x + 4 \geq 0$

Solve the related quadratic equation.

$$2x^2 + 9x + 4 = 0$$
$$(2x + 1)(x + 4) = 0$$

The boundary points are -4 and $-\frac{1}{2}$.

Testing each interval gives a solution set of

$$\left(-\infty, -4\right] \cup \left[-\frac{1}{2}, \infty\right)$$

71. $x^3 + 2x^2 > 3x$

Solve the related equation.

$$x^3 + 2x^2 = 3x$$
$$x^3 + 2x^2 - 3x = 0$$
$$x\left(x^2 + 2x - 3\right) = 0$$
$$x(x + 3)(x - 1) = 0$$

The boundary points are -3, 0, and 1.
Testing each interval gives a solution set of

$$(-3, 0) \cup (1, \infty)$$

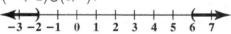

72. $\frac{x - 6}{x + 2} > 0$

Find the values of x that make the numerator and denominator zero.
The boundary points are -2 and 6.
Testing each interval gives a solution set of

$$(-\infty, -2) \cup (6, \infty).$$

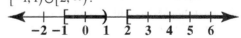

73. $\frac{(x + 1)(x - 2)}{x - 1} \geq 0$

Find the values of x that make the numerator and denominator zero.
The boundary points are -1, 1 and 2. We exclude 1 from the solution set, since this would make the denominator zero.
Testing each interval gives a solution set of

$$[-1, 1) \cup [2, \infty).$$

74. $\dfrac{x+3}{x-4} \le 5$

Express the inequality so that one side is zero.

$$\dfrac{x+3}{x-4} - 5 \le 0$$

$$\dfrac{x+3}{x-4} - \dfrac{5(x-4)}{x-4} \le 0$$

$$\dfrac{-4x+23}{x-4} \le 0$$

Find the values of x that make the numerator and denominator zero.

The boundary points are 4 and $\dfrac{23}{4}$. We exclude 4 from the solution set, since this would make the denominator zero.

Testing each interval gives a solution set of

$$(-\infty, 4) \cup \left[\dfrac{23}{4}, \infty\right).$$

75. **a.** $g(x) = 0.125x^2 + 2.3x + 27$

$g(35) = 0.125(35)^2 + 2.3(35) + 27 \approx 261$

The stopping distance on wet pavement for a motorcycle traveling 35 miles per hour is about 261 feet. This overestimates the distance shown in the graph by 1 foot.

b. $f(x) = 0.125x^2 - 0.8x + 99$

$0.125x^2 - 0.8x + 99 > 267$

$0.125x^2 - 0.8x - 168 > 0$

Solve the related quadratic equation.

$0.125x^2 - 0.8x - 168 = 0$

$$x = \dfrac{-b \pm \sqrt{b^2 - 4ac}}{2a}$$

$$x = \dfrac{-(-0.8) \pm \sqrt{(-0.8)^2 - 4(0.125)(-168)}}{2(0.125)}$$

$x = -33.6, \ 40$

Testing each interval gives a solution set of $(-\infty, -33.6) \cup (40, \infty)$.

Thus, speeds exceeding 40 miles per hour on dry pavement will require over 267 feet of stopping distance.

76. $s = -16t^2 + v_0 t + s_0$

$32 < -16t^2 + 48t + 0$

$0 < -16t^2 + 48t - 32$

$0 < -16\left(t^2 - 3t + 2\right)$

$0 < -16(t-2)(t-1)$

	F		T		F	
		1		2		

The projectile's height exceeds 32 feet during the time period from 1 to 2 seconds.

77. $w = ks$

$28 = k \cdot 250$

$0.112 = k$

Thus, $w = 0.112s$.

$w = 0.112(1200) = 134.4$

1200 cubic centimeters of melting snow will produce 134.4 cubic centimeters of water.

78. $d = kt^2$

$144 = k(3)^2$

$k = 16$

$d = 16t^2$

$d = 16(10)^2 = 1{,}600$ ft

79. $p = \dfrac{k}{w}$

$660 = \dfrac{k}{1.6}$

$1056 = k$

Thus, $p = \dfrac{1056}{w}$.

$p = \dfrac{1056}{2.4} = 440$

The pitch is 440 vibrations per second.

80. $l = \dfrac{k}{d^2}$

$28 = \dfrac{k}{8^2}$

$k = 1792$

$l = \dfrac{1792}{d^2}$

$l = \dfrac{1792}{4^2} = 112$ decibels

81.

$$t = \frac{kc}{w}$$

$$10 = \frac{k \cdot 30}{6}$$

$$10 = 5h$$

$$h = 2$$

$$t = \frac{2c}{w}$$

$$t = \frac{2(40)}{5} = 16 \text{ hours}$$

82.

$$V = khB$$

$$175 = k \cdot 15 \cdot 35$$

$$k = \frac{1}{3}$$

$$V = \frac{1}{3}hB$$

$$V = \frac{1}{3} \cdot 20 \cdot 120 = 800 \text{ ft}^3$$

83. **a.** Use $L = \dfrac{k}{R}$ to find k.

$$L = \frac{k}{R}$$

$$30 = \frac{k}{63}$$

$$63 \cdot 30 = 63 \cdot \frac{k}{63}$$

$$1890 = k$$

Thus, $L = \dfrac{1890}{R}$.

b. This is an approximate model.

c.

$$L = \frac{1890}{R}$$

$$L = \frac{1890}{27} = 70$$

The average life span of an elephant is 70 years.

Chapter 3 Test

1. $f(x) = (x+1)^2 + 4$

vertex: $(-1, 4)$
axis of symmetry: $x = -1$
x-intercepts:

$$(x+1)^2 + 4 = 0$$

$$x^2 + 2x + 5 = 0$$

$$x = \frac{-2 \pm \sqrt{4-20}}{2} = -1 \pm 2i$$

no x-intercepts
y-intercept:

$$f(0) = (0+1)^2 + 4 = 5$$

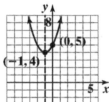

$f(x) = (x + 1)^2 + 4$
domain: $(-\infty, \infty)$; range: $[4, \infty)$

2. $f(x) = x^2 - 2x - 3$

$$x = \frac{-b}{2a} = \frac{2}{2} = 1$$

$$f(1) = 1^2 - 2(1) - 3 = -4$$

vertex: $(1, -4)$
axis of symmetry $x = 1$
x-intercepts:

$$x^2 - 2x - 3 = 0$$

$$(x-3)(x+1) = 0$$

$$x = 3 \text{ or } x = -1$$

y-intercept:

$$f(0) = 0^2 - 2(0) - 3 = -3$$

$f(x) = x^2 - 2x - 3$
domain: $(-\infty, \infty)$; range: $[-4, \infty)$

3. $f(x) = -2x^2 + 12x - 16$

Since the coefficient of x^2 is negative, the graph of $f(x)$ opens down and $f(x)$ has a maximum point.

$x = \dfrac{-12}{2(-2)} = 3$

$f(3) = -2(3)^2 + 12(3) - 16$

$\qquad = -18 + 36 - 16$

$\qquad = 2$

Maximum point: (3, 2)

domain: $(-\infty, \infty)$; range: $(-\infty, 2]$

4. $f(x) = -x^2 + 46x - 360$

$x = -\dfrac{b}{2a} = \dfrac{-46}{-2} = 23$

23 computers will maximize profit.

$f(23) = -(23)^2 + 46(23) - 360 = 169$

Maximum daily profit = \$16,900.

5. Let x = one of the numbers;
$14 - x$ = the other number.

The product is $f(x) = x(14 - x)$

$f(x) = x(14 - x) = -x^2 + 14x$

The x-coordinate of the maximum is

$x = -\dfrac{b}{2a} = -\dfrac{14}{2(-1)} = -\dfrac{14}{-2} = 7.$

$f(7) = -7^2 + 14(7) = 49$

The vertex is (7, 49). The maximum product is 49. This occurs when the two numbers are 7 and $14 - 7 = 7$.

6. a. $f(x) = x^3 - 5x^2 - 4x + 20$

$x^3 - 5x^2 - 4x + 20 = 0$

$x^2(x - 5) - 4(x - 5) = 0$

$(x - 5)(x - 2)(x + 2) = 0$

$x = 5, 2, -2$

The solution set is {5, 2, –2}.

b. The degree of the polynomial is odd and the leading coefficient is positive. Thus the graph falls to the left and rises to the right.

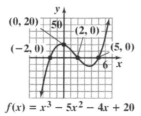

$f(x) = x^3 - 5x^2 - 4x + 20$

7. $f(x) = x^5 - x$

Since the degree of the polynomial is odd and the leading coefficient is positive, the graph of f should fall to the left and rise to the right. The x-intercepts should be –1 and 1.

8. a. The integral root is 2.

b.

2	6	–19	16	–4
		12	–14	4
	6	–7	2	0

$6x^2 - 7x + 2 = 0$

$(3x - 2)(2x - 1) = 0$

$x = \dfrac{2}{3}$ or $x = \dfrac{1}{2}$

The other two roots are $\dfrac{1}{2}$ and $\dfrac{2}{3}$.

9. $2x^3 + 11x^2 - 7x - 6 = 0$

$p: \pm 1, \pm 2, \pm 3, \pm 6$

$q: \pm 1, \pm 2$

$\dfrac{p}{q}: \pm 1, \pm 2, \pm 3, \pm 6, \pm \dfrac{1}{2}, \pm \dfrac{3}{2}$

10. $f(x) = 3x^5 - 2x^4 - 2x^2 + x - 1$

$f(x)$ has 3 sign variations.

$f(-x) = -3x^5 - 2x^4 - 2x^2 - x - 1$

$f(-x)$ has no sign variations.

There are 3 or 1 positive real solutions and no negative real solutions.

11. $x^3 + 9x^2 + 16x - 6 = 0$

Since the leading coefficient is 1, the possible rational zeros are the factors of 6

$p = \dfrac{p}{q}: \pm 1, \pm 2, \pm 3, \pm 6$

–3	1	9	16	–6
		–3	–18	6
	1	6	–2	0

Thus $x = 3$ is a root.

Solve the quotient $x^2 + 6x - 2 = 0$ using the quadratic formula to find the remaining roots.

$x = \dfrac{-b \pm \sqrt{b^2 - 4ac}}{2a}$

$$x = \frac{-(6) \pm \sqrt{(6)^2 - 4(1)(-2)}}{2(1)}$$

$$= \frac{-6 \pm \sqrt{44}}{2}$$

$$= -3 \pm \sqrt{11}$$

The zeros are -3, $-3 + \sqrt{11}$, and $-3 - \sqrt{11}$.

12. $f(x) = 2x^4 - x^3 - 13x^2 + 5x + 15$

a. Possible rational zeros are:

p: $\pm 1,\ \pm 3,\ \pm 5,\ \pm 15$

q: $\pm 1,\ \pm 2$

$\dfrac{p}{q}$: $\pm 1, \pm 3, \pm 5, \pm 15, \pm \dfrac{1}{2}, \pm \dfrac{3}{2}, \pm \dfrac{5}{2}, \pm \dfrac{15}{2}$

b. Verify that -1 and $\dfrac{3}{2}$ are zeros as it appears in the graph:

```
-1| 2  -1  -13   5   15
         -2   3  10  -15
   ─────────────────────
      2  -3  -10  15    0
```

```
 3|
 2| 2  -3  -10   15
          3   0  -15
   ─────────────────
      2   0  -10    0
```

Thus, -1 and $\dfrac{3}{2}$ are zeros, and the polynomial factors as follows:

$$2x^4 - x^3 - 13x^2 + 5x + 15 = 0$$

$$(x+1)(2x^3 - 3x^2 - 10x + 15) = 0$$

$$(x+1)\left(x - \frac{3}{2}\right)(2x^2 - 10) = 0$$

Find the remaining zeros by solving:

$$2x^2 - 10 = 0$$

$$2x^2 = 10$$

$$x^2 = 5$$

$$x = \pm\sqrt{5}$$

The zeros are -1, $\dfrac{3}{2}$, and $\pm\sqrt{5}$.

13. $f(x)$ has zeros at -2 and 1. The zero at -2 has multiplicity of 2.

$$x^3 + 3x^2 - 4 = (x - 1)(x + 2)^2$$

14. $f(x) = a_0(x+1)(x-1)(x+i)(x-i)$

$$= a_0(x^2 - 1)(x^2 + 1)$$

$$= a_0(x^4 - 1)$$

Since $f(3) = 160$, then

$$a_0(3^4 - 1) = 160$$

$$a_0(80) = 160$$

$$a_0 = \frac{160}{80}$$

$$a_0 = 2$$

$$f(x) = 2(x^4 - 1) = 2x^4 - 2$$

15. $f(x) = -3x^3 - 4x^2 + x + 2$

The graph shows a root at $x = -1$.
Use synthetic division to verify this root.

```
-1 | -3  -4   1   2
           3   1   4
     ─────────────────
     -3  -1   2   0
```

Factor the quotient to find the remaining zeros.

$$-3x^2 - x + 2 = 0$$

$$-(3x - 2)(x + 1) = 0$$

The zeros (*x*-intercepts) are -1 and $\dfrac{2}{3}$.

The *y*-intercept is $f(0) = 2$

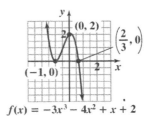

$$f(x) = -3x^3 - 4x^2 + x + 2$$

16. $f(x) = \dfrac{1}{(x+3)^2}$

domain: $\{x \mid x \neq -3\}$ or $(-\infty, -3) \cup (-3, \infty)$

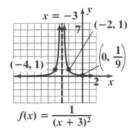

17. $f(x) = \dfrac{1}{x-1} + 2$

domain: $\{x \mid x \neq 1\}$ or $(-\infty, 1) \cup (1, \infty)$

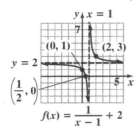

18. $f(x) = \dfrac{x}{x^2 - 16}$

domain: $\{x \mid x \neq 4, x \neq -4\}$

Symmetry: $f(-x) = \dfrac{-x}{x^2 - 16} = -f(x)$

y-axis symmetry

x-intercept: $x = 0$

y-intercept: $y = \dfrac{0}{0^2 - 16} = 0$

Vertical asymptotes:
$x^2 - 16 = 0$
$(x - 4)(x + 4) = 0$
$x = 4, -4$

Horizontal asymptote:
$n < m$, so $y = 0$ is the horizontal asymptote.

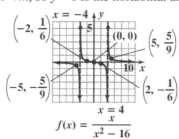

19. $f(x) = \dfrac{x^2 - 9}{x - 2}$

domain: $\{x \mid x \neq 2\}$

Symmetry: $f(-x) = \dfrac{x^2 - 9}{-x - 2}$

$f(-x) \neq f(x), f(-x) \neq -f(x)$
No symmetry

x-intercepts:
$x^2 - 9 = 0$
$(x - 3)(x + 3) = 0$
$x = 3, -3$

y-intercept: $y = \dfrac{0^2 - 9}{0 - 2} = \dfrac{9}{2}$

Vertical asymptote:
$x - 2 = 0$
$x = 2$

Horizontal asymptote:
$n > m$, so no horizontal asymptote exists.

Slant asymptote: $f(x) = x + 2 - \dfrac{5}{x - 2}$

$y = x + 2$

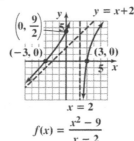

20. $f(x) = \dfrac{x + 1}{x^2 + 2x - 3}$

$x^2 + 2x - 3 = (x + 3)(x - 1)$

domain: $\{x \mid x \neq -3, x \neq 1\}$

Symmetry: $f(-x) = \dfrac{-x + 1}{x^2 - 2x - 3}$

$f(-x) \neq f(x), f(-x) \neq -f(x)$
No symmetry

x-intercept:
$x + 1 = 0$
$x = -1$

y-intercept: $y = \dfrac{0 + 1}{0^2 + 2(0) - 3} = -\dfrac{1}{3}$

Vertical asymptotes:
$x^2 + 2x - 3 = 0$
$(x + 3)(x - 1) = 0$
$x -3, 1$

Horizontal asymptote:

$n < m$, so $y = 0$ is the horizontal asymptote.

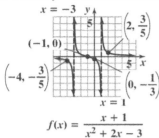

$$f(x) = \frac{x+1}{x^2 + 2x - 3}$$

21. $f(x) = \dfrac{4x^2}{x^2 + 3}$

domain: all real numbers

Symmetry: $f(-x) = \dfrac{4x^2}{x^2 + 3} = f(x)$

y-axis symmetry

x-intercept:

$4x^2 = 0$

$x = 0$

y-intercept: $y = \dfrac{4(0)^2}{0^2 + 3} = 0$

Vertical asymptote:

$x^2 + 3 = 0$

$x^2 = -3$

No vertical asymptote.

Horizontal asymptote:

$n = m$, so $y = \dfrac{4}{1} = 4$ is the horizontal asymptote.

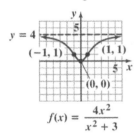

$$f(x) = \frac{4x^2}{x^2 + 3}$$

22. a. $\overline{C}(x) = \dfrac{300{,}000 + 10x}{x}$

b. Since the degree of the numerator equals the degree of the denominator, the horizontal asymptote is $x = \dfrac{10}{1} = 10$.

This represents the fact that as the number of satellite radio players produced increases, the production cost approaches $10 per radio.

23.
$$x^2 < x + 12$$
$$x^2 - x - 12 < 0$$
$$(x+3)(x-4) < 0$$
Boundary values: −3 and 4
Solution set: $(-3, 4)$

24. $\dfrac{2x+1}{x-3} \le 3$

$\dfrac{2x+1}{x-3} - 3 \le 0$

$\dfrac{10 - x}{x - 3} \le 0$

Boundary values: 3 and 10
Solution set: $(-\infty, 3) \cup [10, \infty)$

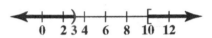

25. $i = \dfrac{k}{d^2}$

$20 = \dfrac{k}{15^2}$

$4500 = k$

$i = \dfrac{4500}{d^2} = \dfrac{4500}{10^2} = 45$ foot-candles

Cumulative Review Exercises (Chapters 1–3)

1. domain: $(-2, 2)$ range: $[0, \infty)$

2. The zero at −1 touches the x-axis at turns around so it must have a minimum multiplicity of 2.
The zero at 1 touches the x-axis at turns around so it must have a minimum multiplicity of 2.

3. There is a relative maximum at the point $(0, 3)$.

4. $(f \circ f)(-1) = f\big(f(-1)\big) = f(0) = 3$

5. $f(x) \to \infty$ as $\underline{x \to -2^+}$ or as $\underline{x \to 2^-}$

6.

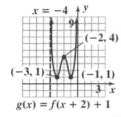

$g(x) = f(x + 2) + 1$

7. $|2x - 1| = 3$
$2x - 1 = 3$
$2x = 4$
$x = 2$
$2x - 1 = -3$
$2x = -2$
$x = -1$
The solution set is $\{-1, 2\}$.

8. $3x^2 - 5x + 1 = 0$
$x = \dfrac{5 \pm \sqrt{25 - 12}}{6} = \dfrac{5 \pm \sqrt{13}}{6}$
The solution set is $\left\{ \dfrac{5 + \sqrt{13}}{6}, \dfrac{5 - \sqrt{13}}{6} \right\}$.

9. $9 + \dfrac{3}{x} = \dfrac{2}{x^2}$
$9x^2 + 3x = 2$
$9x^2 + 3x - 2 = 0$
$(3x - 1)(3x + 2) = 0$
$3x - 1 = 0 \quad 3x + 2 = 0$
$x = \dfrac{1}{3} \quad \text{or} \quad x = -\dfrac{2}{3}$
The solution set is $\left\{ -\dfrac{2}{3}, \dfrac{1}{3} \right\}$.

10. $x^3 + 2x^2 - 5x - 6 = 0$
p: $\pm 1, \pm 2, \pm 3, \pm 6$
q: ± 1
$\dfrac{p}{q}$: $\pm 1, \pm 2, \pm 3, \pm 6$

$$
\begin{array}{r|rrrr}
-3 & 1 & 2 & -5 & -6 \\
   &   & -3 & 3 & 6 \\
\hline
   & 1 & -1 & -2 & 0
\end{array}
$$

$x^3 + 2x^2 - 5x - 6 = 0$
$(x + 3)(x^2 - x - 2) = 0$
$(x + 3)(x + 1)(x - 2) = 0$
$x = -3 \text{ or } x = -1 \text{ or } x = 2$
The solution set is $\{-3, -1, 2\}$.

11. $|2x - 5| > 3$
$2x - 5 > 3$
$2x > 8$
$x > 4$
$2x - 5 < -3$
$2x < 2$
$x < 1$
$(-\infty, 1) \text{ or } (4, \infty)$

12. $\qquad 3x^2 > 2x + 5$
$3x^2 - 2x - 5 > 0$
$3x^2 - 2x - 5 = 0$
$(3x - 5)(x + 1) = 0$
$x = \dfrac{5}{3} \text{ or } x = -1$

Test intervals are $(-\infty, -1)$, $\left(-1, \dfrac{5}{3} \right)$, $\left(\dfrac{5}{3}, \infty \right)$.

Testing points, the solution is $(-\infty, -1)$ or $\left(\dfrac{5}{3}, \infty \right)$.

13. $f(x) = x^3 - 4x^2 - x + 4$
x-intercepts:
$x^3 - 4x^2 - x + 4 = 0$
$x^2(x - 4) - 1(x - 4) = 0$
$(x - 4)(x^2 - 1) = 0$
$(x - 4)(x + 1)(x - 1) = 0$
$x = -1, 1, 4$
x-intercepts:
$f(0) = 0^3 - 4(0)^2 - 0 + 4 = 4$

The degree of the polynomial is odd and the leading coefficient is positive. Thus the graph falls to the left and rises to the right.

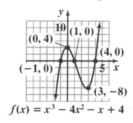

$f(x) = x^3 - 4x^2 - x + 4$

14. $f(x) = x^2 + 2x - 8$

$$x = \frac{-b}{2a} = \frac{-2}{2} = -1$$

$$f(-1) = (-1)^2 + 2(-1) - 8$$
$$= 1 - 2 - 8 = -9$$

vertex: $(-1, -9)$

x-intercepts:

$$x^2 + 2x - 8 = 0$$
$$(x+4)(x-2) = 0$$
$$x = -4 \text{ or } x = 2$$

y-intercept: $f(0) = -8$

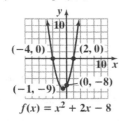

15. $f(x) = x^2(x-3)$

zeros: $x = 0$ (multiplicity 2) and $x = 3$

y-intercept: $y = 0$

$$f(x) = x^3 - 3x^2$$

$n = 3$, $a_n = 0$ so the graph falls to the left and rises to the right.

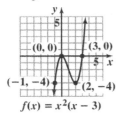

16. $f(x) = \dfrac{x-1}{x-2}$

vertical asymptote: $x = 2$

horizontal asymptote: $y = 1$

x-intercept: $x = 1$

y-intercept: $y = \dfrac{1}{2}$

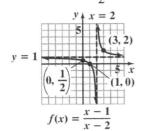

17.

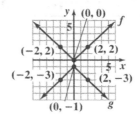

18.

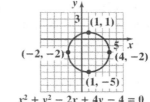

$$x^2 + y^2 - 2x + 4y - 4 = 0$$

19. $(f \circ g)(x) = f(g(x))$

$$= 2(4x-1)^2 - (4x-1) - 1$$
$$= 32x^2 - 20x + 2$$

20. $\dfrac{f(x+h) - f(x)}{h}$

$$= \frac{\left[2(x+h)^2 - (x+h) - 1\right] - \left[2x^2 - x - 1\right]}{h}$$

$$= \frac{2x^2 + 4hx - x + 2h^2 - h - 1 - 2x^2 + x + 1}{h}$$

$$= \frac{4hx + 2h^2 - h}{h}$$

$$= 4x + 2h - 1$$

Chapter 4
Exponential and Logarithmic Functions

Section 4.1

Check Point Exercises

1. $f(x) = 42.2(1.56)^x$

    $f(3) = 42.2(1.56)^3 \approx 160.20876 \approx 160$

    According to the function, the average amount spent after three hours of shopping at the mall is $160. This overestimates the actual amount shown by $11.

2. Begin by setting up a table of coordinates.

x	$f(x) = 3^x$
-3	$f(-3) = 3^{-3} = \frac{1}{27}$
-2	$f(-2) = 3^{-2} = \frac{1}{9}$
-1	$f(-1) = 3^{-1} = \frac{1}{3}$
0	$f(0) = 3^0 = 1$
1	$f(1) = 3^1 = 3$
2	$f(2) = 3^2 = 9$
3	$f(3) = 3^3 = 27$

    Plot a few of these points, connecting them with a continuous curve.

    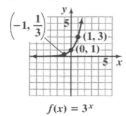

    $f(x) = 3^x$

3. Begin by setting up a table of coordinates.

x	$f(x) = \left(\frac{1}{3}\right)^x$
-2	$\left(\frac{1}{3}\right)^{-2} = 9$
-1	$\left(\frac{1}{3}\right)^{-1} = 3$
0	$\left(\frac{1}{3}\right)^0 = 1$
1	$\left(\frac{1}{3}\right)^1 = \frac{1}{3}$
2	$\left(\frac{1}{3}\right)^2 = \frac{1}{9}$

    Plot a few of these points, connecting them with a continuous curve.

    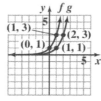

    $f(x) = \left(\frac{1}{3}\right)^x$

4. Note that the function $g(x) = 3^{x-1}$ has the general form $g(x) = b^{x+c}$ where $c = -1$. Because $c < 0$, we graph $g(x) = 3^{x-1}$ by shifting the graph of $f(x) = 3^x$ one unit to the right. Construct a table showing some of the coordinates for f and g.

x	$f(x) = 3^x$	$g(x) = 3^{x-1}$
-2	$3^{-2} = \frac{1}{9}$	$3^{-2-1} = 3^{-3} = \frac{1}{27}$
-1	$3^{-1} = \frac{1}{3}$	$3^{-1-1} = 3^{-2} = \frac{1}{9}$
0	$3^0 = 1$	$3^{0-1} = 3^{-1} = \frac{1}{3}$
1	$3^1 = 3$	$3^{1-1} = 3^0 = 1$
2	$3^2 = 9$	$3^{2-1} = 3^1 = 3$

    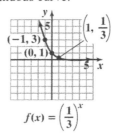

    $f(x) = 3^x$
    $g(x) = 3^{x-1}$

5. Note that the function $g(x) = 2^x + 1$ has the general form $g(x) = b^x + c$ where $c = 1$. Because $c > 0$, we graph $g(x) = 2^x + 1$ by shifting the graph of $f(x) = 2^x$ up one unit. Construct a table showing some of the coordinates for f and g.

x	$f(x) - 2^x$	$g(x) = 2^x + 1$
-2	$2^{-2} = \frac{1}{4}$	$2^{-2} + 1 = \frac{1}{4} + 1 = \frac{5}{4}$
-1	$2^{-1} = \frac{1}{2}$	$2^{-1} + 1 = \frac{1}{2} + 1 = \frac{3}{2}$
0	$2^0 = 1$	$2^0 + 1 = 1 + 1 = 2$
1	$2^1 = 2$	$2^1 + 1 = 2 + 1 = 3$
2	$2^2 = 4$	$2^2 + 1 = 4 + 1 = 5$

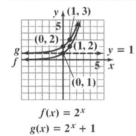

$$f(x) = 2^x$$
$$g(x) = 2^x + 1$$

6. 2012 is 34 years after 1978.
$$f(x) = 1066e^{0.042x}$$
$$f(34) = 1066e^{0.042(34)} \approx 4446$$
In 2012 the gray wolf population of the Western Great Lakes is projected to be about 4446.

7. **a.** $A = P\left(1 + \dfrac{r}{n}\right)^{nt}$

$A = 10,000\left(1 + \dfrac{0.08}{4}\right)^{4(5)}$

$= \$14,859.47$

b. $A = Pe^{rt}$

$A = 10,000e^{0.08(5)}$

$= \$14,918.25$

Concept and Vocabulary Check 4.1

1. b^x; $(-\infty, \infty)$; $(0, \infty)$

2. x; $y = 0$; horizontal

3. e; natural; 2.72

4. A; P; r; n

5. semiannually; quarterly; continuous

Exercise Set 4.1

1. $2^{3.4} \approx 10.556$

3. $3^{\sqrt{5}} \approx 11.665$

5. $4^{-1.5} = 0.125$

7. $e^{2.3} \approx 9.974$

9. $e^{-0.95} \approx 0.387$

11.

x	$f(x) = 4^x$
-2	$4^{-2} = \frac{1}{16}$
-1	$4^{-1} = \frac{1}{4}$
0	$4^0 = 1$
1	$4^1 = 4$
2	$4^2 = 16$

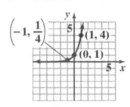

$$f(x) = 4^x$$

13.

x	$g(x) = \left(\frac{3}{2}\right)^x$
-2	$\left(\frac{3}{2}\right)^{-2} = \frac{4}{9}$
-1	$\left(\frac{3}{2}\right)^{-1} = \frac{2}{3}$
0	$\left(\frac{3}{2}\right)^0 = 1$
1	$\left(\frac{3}{2}\right)^1 = \frac{3}{2}$
2	$\left(\frac{3}{2}\right)^2 = \frac{9}{4}$

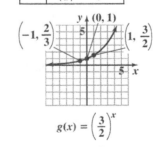

$$g(x) = \left(\frac{3}{2}\right)^x$$

15.

x	$h(x) = \left(\frac{1}{2}\right)^x$
-2	$\left(\frac{1}{2}\right)^{-2} = 4$
-1	$\left(\frac{1}{2}\right)^{-1} = 2$
0	$\left(\frac{1}{2}\right)^{0} = 1$
1	$\left(\frac{1}{2}\right)^{1} = \frac{1}{2}$
2	$\left(\frac{1}{2}\right)^{2} = \frac{1}{4}$

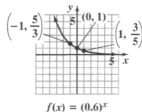

$$h(x) = \left(\frac{1}{2}\right)^x$$

17.

x	$f(x) = (0.6)^x$
-2	$(0.6)^{-2} = 2.\overline{7}$
-1	$(0.6)^{-1} = 1.\overline{6}$
0	$(0.6)^{0} = 1$
1	$(0.6)^{1} = 0.6$
2	$(0.6)^{2} = 0.36$

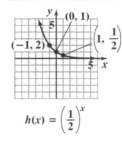

$$f(x) = (0.6)^x$$

19. This is the graph of $f(x) = 3^x$ reflected about the x-axis and about the y-axis, so the function is $H(x) = -3^{-x}$.

21. This is the graph of $f(x) = 3^x$ reflected about the x-axis, so the function is $F(x) = -3^x$.

23. This is the graph of $f(x) = 3^x$ shifted one unit downward, so the function is $h(x) = 3^x - 1$.

25. The graph of $g(x) = 2^{x+!}$ can be obtained by shifting the graph of $f(x) = 2^x$ one unit to the left.

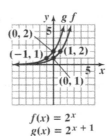

$$f(x) = 2^x$$
$$g(x) = 2^x + 1$$

asymptote: $y = 0$

domain: $(-\infty, \infty)$; range: $(0, \infty)$

27. The graph of $g(x) = 2^x - 1$ can be obtained by shifting the graph of $f(x) = 2^x$ downward one unit.

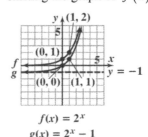

$$f(x) = 2^x$$
$$g(x) = 2^x - 1$$

asymptote: $y = -1$

domain: $(-\infty, \infty)$; range: $(-1, \infty)$

29. The graph of $h(x) = 2^{x+1} - 1$ can be obtained by shifting the graph of $f(x) = 2^x$ one unit to the left and one unit downward.

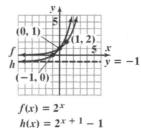

$$f(x) = 2^x$$
$$h(x) = 2^{x+1} - 1$$

asymptote: $y = -1$

domain: $(-\infty, \infty)$; range: $(-1, \infty)$

31. The graph of $g(x) = -2^x$ can be obtained by reflecting the graph of $f(x) = 2^x$ about the *x*-axis.

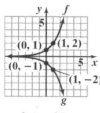

$f(x) = 2^x$
$g(x) = -2^x$

asymptote: $y = 0$
domain: $(-\infty, \infty)$; range: $(-\infty, 0)$

33. The graph of $g(x) = 2 \cdot 2^x$ can be obtained by vertically stretching the graph of $f(x) = 2^x$ by a factor of two.

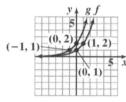

$f(x) = 2^x$
$g(x) = 2 \cdot 2^x$

asymptote: $y = 0$
domain: $(-\infty, \infty)$; range: $(0, \infty)$

35. The graph of $g(x) = e^{x-1}$ can be obtained by moving $f(x) = e^x$ 1 unit right.

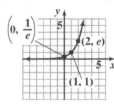

$g(x) = e^{x-1}$

asymptote: $y = 0$
domain: $(-\infty, \infty)$; range: $(0, \infty)$

37. The graph of $g(x) = e^x + 2$ can be obtained by moving $f(x) = e^x$ 2 units up.

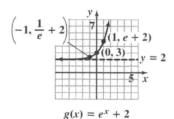

$g(x) = e^x + 2$

asymptote: $y = 2$
domain: $(-\infty, \infty)$; range: $(2, \infty)$

39. The graph of $h(x) = e^{x-1} + 2$ can be obtained by moving $f(x) = e^x$ 1 unit right and 2 units up.

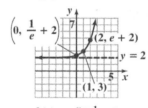

$h(x) = e^{x-1} + 2$

asymptote: $y = 2$
domain: $(-\infty, \infty)$; range: $(2, \infty)$

41. The graph of $h(x) = e^{-x}$ can be obtained by reflecting $f(x) = e^x$ about the *y*-axis.

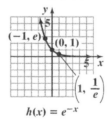

$h(x) = e^{-x}$

asymptote: $y = 0$
domain: $(-\infty, \infty)$; range: $(0, \infty)$

43. The graph of $g(x) = 2e^x$ can be obtained by stretching $f(x) = e^x$ vertically by a factor of 2.

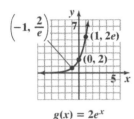

$$g(x) = 2e^x$$

asymptote: $y = 0$

domain: $(-\infty, \infty)$; range: $(0, \infty)$

45. The graph of $h(x) = e^{2x} + 1$ can be obtained by stretching $f(x) = e^x$ horizontally by a factor of 2 and then moving the graph up 1 unit.

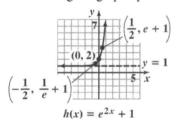

$$h(x) = e^{2x} + 1$$

asymptote: $y = 1$

domain: $(-\infty, \infty)$; range: $(1, \infty)$

47. The graph of $g(x)$ can be obtained by reflecting $f(x)$ about the y-axis.

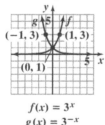

$$f(x) = 3^x$$
$$g(x) = 3^{-x}$$

asymptote of $f(x)$: $y = 0$
asymptote of $g(x)$: $y = 0$

49. The graph of $g(x)$ can be obtained by vertically shrinking $f(x)$ by a factor of $\frac{1}{3}$.

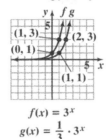

$$f(x) = 3^x$$
$$g(x) = \frac{1}{3} \cdot 3^x$$

asymptote of $f(x)$: $y = 0$
asymptote of $g(x)$: $y = 0$

51. The graph of $g(x)$ can be obtained by moving the graph of $f(x)$ one space to the right and one space up.

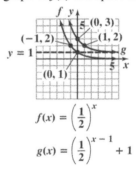

$$f(x) = \left(\frac{1}{2}\right)^x$$
$$g(x) = \left(\frac{1}{2}\right)^{x-1} + 1$$

asymptote of $f(x)$: $y = 0$
asymptote of $g(x)$: $y = 1$

53. a. $A = 10,000\left(1 + \dfrac{0.055}{2}\right)^{2(5)}$

$\approx \$13,116.51$

b. $A = 10,000\left(1 + \dfrac{0.055}{4}\right)^{4(5)}$

$\approx \$13,140.67$

c. $A = 10,000\left(1 + \dfrac{0.055}{12}\right)^{12(5)}$

$\approx \$13,157.04$

d. $A = 10,000e^{0.055(5)}$

$\approx \$13,165.31$

55. $A = 12,000\left(1 + \dfrac{0.07}{12}\right)^{12(3)}$

$\approx 14,795.11$ (7% yield)

$A = 12,000e^{0.0685(3)}$

$\approx 14,737.67$ (6.85% yield)

Investing $12,000 for 3 years at 7% compounded monthly yields the greater return.

57.

x	$f(x) = 2^x$	$g(x) = 2^{-x}$
-2	$\dfrac{1}{4}$	4
-1	$\dfrac{1}{2}$	2
0	1	1
1	2	$\dfrac{1}{2}$
2	4	$\dfrac{1}{4}$

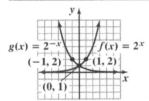

The point of intersection is $(0,1)$.

59.

x	$y = 2^x$
-2	$\dfrac{1}{4}$
-1	$\dfrac{1}{2}$
0	1
1	2
2	4

y	$x = 2^y$
-2	$\dfrac{1}{4}$
-1	$\dfrac{1}{2}$
0	1
1	2
2	4

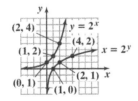

61. The graph is of the form $y = b^x$.

Substitute values from the point $(1, 4)$ to find b.

$y = b^x$

$4 = b^1$

$4 = b$

The equation of the graph is $y = 4^x$

63. The graph is of the form $y = -b^x$.

Substitute values from the point $(1, -e)$ to find b.

$y = -b^x$

$-e = -b^1$

$e = b$

The equation of the graph is $y = -e^x$

65. a. $f(0) = 574(1.026)^0$

$= 574(1) = 574$

India's population in 1974 was 574 million.

b. $f(27) = 574(1.026)^{27} \approx 1148$

India's population in 2001 will be 1148 million.

c. Since $2028 - 1974 = 54$, find

$f(54) = 574(1.026)^{54} \approx 2295$.

India's population in 2028 will be 2295 million.

d. $2055 - 1974 = 81$, find

$f(54) = 574(1.026)^{81} \approx 4590$.

India's population in 2055 will be 4590 million.

e. India's population appears to be doubling every 27 years.

67. $S = 465,000(1 + 0.06)^{10}$

$= 465,000(1.06)^{10} \approx \$832,744$

69. $2^{1.7} \approx 3.249009585$

$2^{1.73} \approx 3.317278183$

$2^{1.732} \approx 3.321880096$

$2^{1.73205} \approx 3.321995226$

$2^{1.7320508} \approx 3.321997068$

$2^{\sqrt{3}} \approx 3.321997085$

The closer the exponent is to $\sqrt{3}$, the closer the value is to $2^{\sqrt{3}}$.

71. a. 2011 is 11 years after 2000.

$f(x) = 782x + 6564$

$f(11) = 782(11) + 6564 = 15{,}166$

According to the linear model, the average cost of a family health insurance plan in 2011 was $15,166.

b. 2011 is 11 years after 2000.

$g(x) = 6875e^{0.077x}$

$g(11) = 6875e^{0.077(11)} \approx 16{,}037$

According to the exponential model, the average cost of a family health insurance plan in 2011 was $16,037.

c. The linear model is the better model for the data in 2011.

73. a. $f(0) = 80e^{-0.5(0)} + 20$

$= 80e^{0} + 20$

$= 80(1) + 20$

$= 100$

100% of the material is remembered at the moment it is first learned.

b. $f(1) = 80e^{-0.5(1)} + 20 \approx 68.5$

68.5% of the material is remembered 1 week after it is first learned.

c. $f(4) = 80e^{-0.5(4)} + 20 \approx 30.8$

30.8% of the material is remembered 4 weeks after it is first learned.

d. $f(52) = 80e^{-0.5(52)} + 20 \approx 20$

20% of the material is remembered 1 year after it is first learned.

75. $f(x) = 6.25(1.029)^{x}$

$f(59) = 6.25(1.029)^{59} \approx 33.8$

$g(x) = \dfrac{38.8}{1 + 6.3e^{-0.051x}}$

$g(59) = \dfrac{38.8}{1 + 6.3e^{-0.051(59)}} \approx 29.6$

Function $g(x)$ is a better model for the graph's value of 29.5 in 2009.

77. – 79. Answers will vary.

81. a. $A = 10{,}000\left(1 + \dfrac{0.05}{4}\right)^{4t}$

$A = 10{,}000\left(1 + \dfrac{0.045}{12}\right)^{12t}$

b. 5% compounded quarterly offers the better return.

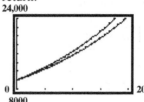

83. does not make sense; Explanations will vary. Sample explanation: The horizontal asymptote is $y = 0$.

85. does not make sense; Explanations will vary. Sample explanation: An exponential model appears better than a linear model.

87. false; Changes to make the statement true will vary. A sample change is: The amount of money will not increase without bound.

89. false; Changes to make the statement true will vary. A sample change is: If $f(x) = 2^{x}$ then $f(a+b) = f(a) \cdot f(b)$.

91. $y = 3^{x}$ is (d). y increases as x increases, but not as quickly as $y = 5^{x}$. $y = 5^{x}$ is (c). $y = \left(\dfrac{1}{3}\right)^{x}$ is (a).

$y = \left(\dfrac{1}{3}\right)^{x}$ is the same as $y = 3^{-x}$, so it is (d) reflected about the *y*-axis. $y = \left(\dfrac{1}{5}\right)^{x}$ is (b). $y = \left(\dfrac{1}{5}\right)^{x}$ is the same as $y = 5^{-x}$, so it is (c) reflected about the *y*-axis.

93. a.

$$\cosh(-x) = \frac{e^{-x} + e^{-(-x)}}{2}$$

$$= \frac{e^{-x} + e^{x}}{2}$$

$$= \frac{e^{x} + e^{-x}}{2}$$

$$= \cosh x$$

b.

$$\sinh(-x) = \frac{e^{-x} - e^{-(-x)}}{2}$$

$$= \frac{e^{-x} - e^{x}}{2}$$

$$= \frac{-\left(-e^{-x} + e^{x}\right)}{2}$$

$$= -\frac{e^{x} - e^{-x}}{2}$$

$$= -\sinh x$$

c.

$$(\cosh x)^2 - (\sinh x)^2 \overset{?}{=} 1$$

$$\left(\frac{e^{x} + e^{-x}}{2}\right)^2 - \left(\frac{e^{x} - e^{-x}}{2}\right)^2 \overset{?}{=} 1$$

$$\frac{e^{2x} + 2 + e^{-2x}}{4} - \frac{e^{2x} - 2 + e^{-2x}}{4} \overset{?}{=} 1$$

$$\frac{e^{2x} + 2 + e^{-2x} - e^{2x} + 2 - e^{-2x}}{4} \overset{?}{=} 1$$

$$\frac{4}{4} \overset{?}{=} 1$$

$$1 = 1$$

94. We do not know how to solve $x = 2^y$ for y.

95. $\dfrac{1}{2}$; i.e. $25^{1/2} = 5$

96. $(x-3)^2 > 0$

Solving the related equation, $(x-3)^2 = 0$, gives $x = 3$.
Note that the boundary value $x = 3$ does not satisfy the inequality.
Testing each interval gives a solution set of
$(-\infty, 3) \cup (3, \infty)$.

Section 4.2

Check Point Exercises

1. a. $3 = \log_7 x$ written in exponential form is
$7^3 = x$.

b. $2 = \log_b 25$ written in exponential form is
$b^2 = 25$.

c. $\log_4 26 = y$ written in exponential form is
$4^y = 26$.

2. a. $2^5 = x$ written in logarithmic form is
$5 = \log_2 x$.

b. $b^3 = 27$ written in logarithmic form
is $3 = \log_b 27$.

c. $e^y = 33$ written in logarithmic form
is $y = \log_e 33$.

3. a. Question: 10 to what power gives 100?
$\log_{10} 100 = 2$ because $10^2 = 100$.

b. Question: 5 to what power gives $\dfrac{1}{125}$?
$\log_5 \dfrac{1}{125} = -3$ because $5^{-3} = \dfrac{1}{5^3} = \dfrac{1}{125}$.

c. Question: 36 to what power gives 6?
$\log_{36} 6 = \dfrac{1}{2}$ because $36^{1/2} = \sqrt{36} = 6$

d. Question: 3 to what power gives $\sqrt[7]{3}$?
$\log_3 \sqrt[7]{3} = \dfrac{1}{7}$ because $3^{1/7} = \sqrt[7]{3}$.

4. a. Because $\log_b b = 1$, we conclude $\log_9 9 = 1$.

b. Because $\log_b 1 = 0$, we conclude $\log_8 1 = 0$.

5. a. Because $\log_b b^x = x$, we conclude $\log_7 7^8 = 8$.

b. Because $b^{\log_b x} = x$, we conclude $3^{\log_3 17} = 17$.

6. First, set up a table of coordinates for $f(x) = 3^x$.

x	-2	-1	0	1	2	3
$f(x) = 3^x$	$\frac{1}{9}$	$\frac{1}{3}$	1	3	9	27

Reversing these coordinates gives the coordinates for the inverse function $g(x) = \log_3 x$.

x	$\frac{1}{9}$	$\frac{1}{3}$	1	3	9	27
$g(x) = \log_3 x$	-2	-1	0	1	2	3

The graph of the inverse can also be drawn by reflecting the graph of $f(x) = 3^x$ about the line $y = x$.

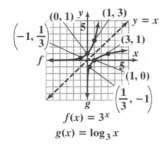

7. The domain of h consists of all x for which $x - 5 > 0$. Solving this inequality for x, we obtain $x > 5$. Thus, the domain of h is $(5, \infty)$.

8. Substitute the boy's age, 10, for x and evaluate the function at 10.
$$f(10) = 29 + 48.8 \log(10 + 1)$$
$$= 29 + 48.8 \log(11)$$
$$\approx 80$$
A 10-year-old boy is approximately 80% of his adult height.

9. Because $I = 10,000\, I_0$,
$$R = \log \frac{10,000 I_0}{I_0}$$
$$= \log 10,000$$
$$= 4$$
The earthquake registered 4.0 on the Richter scale.

10. a. The domain of f consists of all x for which $4 - x > 0$. Solving this inequality for x, we obtain $x < 4$.
The domain of f is $(-\infty, 4)$.

   b. The domain of g consists of all x for which $x^2 > 0$. Solving this inequality for x, we obtain $x < 0$ or $x > 0$.
The domain of g is $(-\infty, 0) \cup (0, \infty)$.

11. Find the temperature increase after 30 minutes by substituting 30 for x and evaluating the function at 30.
$$f(x) = 13.4 \ln x - 11.6$$
$$f(30) = 13.4 \ln 30 - 11.6$$
$$\approx 34$$
The temperature increase after 30 minutes is 34°. The function models the actual increase shown in the graph extremely well.

Concept and Vocabulary Check 4.2

1. $b^y = x$

2. logarithmic; b

3. 1

4. 0

5. x

6. x

7. $(0, \infty)$; $(-\infty, \infty)$

8. y; $x = 0$; vertical

9. up 5 units

10. to the left 5 units

11. x-axis

12. y-axis

13. $5 - x > 0$

14. common; $\log x$

15. natural; $\ln x$

Exercise Set 4.2

1. $2^4 = 16$

3. $3^2 = x$

5. $b^5 = 32$

7. $6^y = 216$

9. $\log_2 8 = 3$

11. $\log_2 \frac{1}{16} = -4$

13. $\log_8 2 = \frac{1}{3}$

15. $\log_{13} x = 2$

17. $\log_b 1000 = 3$

19. $\log_7 200 = y$

21. $\log_4 16 = 2$ because $4^2 = 16$.

23. $\log_2 64 = 6$ because $2^6 = 64$.

25. $\log_5 \frac{1}{5} = -1$ because $5^{-1} = \frac{1}{5}$.

27. $\log_2 \frac{1}{8} = -3$ because $2^{-3} = \frac{1}{8}$.

29. $\log_7 \sqrt{7} = \frac{1}{2}$ because $7^{\frac{1}{2}} = \sqrt{7}$.

31. $\log_2 \frac{1}{\sqrt{2}} = -\frac{1}{2}$ because $2^{-\frac{1}{2}} = \frac{1}{\sqrt{2}}$.

33. $\log_{64} 8 = \frac{1}{2}$ because $64^{1/2} = \sqrt{64} = 8$.

35. Because $\log_b b = 1$, we conclude $\log_5 5 = 1$.

37. Because $\log_b 1 = 0$, we conclude $\log_4 1 = 0$.

39. Because $\log_b b^x = x$, we conclude $\log_5 5^7 = 7$.

41. Because $b^{\log_b x} = x$, we conclude $8^{\log_8 19} = 19$.

43. First, set up a table of coordinates for $f(x) = 4^x$.

x	-2	-1	0	1	2	3
$f(x) = 4x$	$\frac{1}{16}$	$\frac{1}{4}$	1	4	16	64

Reversing these coordinates gives the coordinates for the inverse function $g(x) = \log_4 x$.

x	$\frac{1}{16}$	$\frac{1}{4}$	1	4	16	64
$g(x) = \log_{4x}$	-2	-1	0	1	2	3

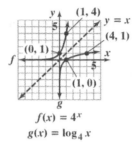

$f(x) = 4^x$
$g(x) = \log_4 x$

45. First, set up a table of coordinates for $f(x) = \left(\frac{1}{2}\right)^x$.

x	-2	-1	0	1	2	3
$f(x) = \left(\frac{1}{2}\right)^x$	4	2	1	$\frac{1}{2}$	$\frac{1}{4}$	$\frac{1}{8}$

Reversing these coordinates gives the coordinates for the inverse function $g(x) = \log_{1/2} x$.

x	4	2	1	$\frac{1}{2}$	$\frac{1}{4}$	$\frac{1}{8}$
$g(x) = \log_{1/2} x$	-2	-1	0	1	2	3

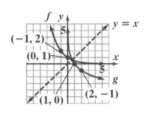

$f(x) = \left(\frac{1}{2}\right)^x$
$g(x) = \log_{1/2} x$

47. This is the graph of $f(x) = \log_3 x$ reflected about the x-axis and shifted up one unit, so the function is $H(x) = 1 - \log_3 x.$

49. This is the graph of $f(x) = \log_3 x$ shifted down one unit, so the function is $h(x) = \log_3 x - 1.$

51. This is the graph of $f(x) = \log_3 x$ shifted right one unit, so the function is $g(x) = \log_3(x - 1).$

53.

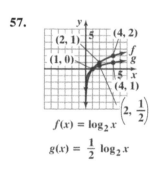

$$f(x) = \log_2 x$$
$$g(x) = \log_2(x+1)$$

vertical asymptote: $x = -1$
domain: $(-1, \infty)$; range: $(-\infty, \infty)$

55.

$$f(x) = \log_2 x$$
$$h(x) = 1 + \log_2 x$$

vertical asymptote: $x = 0$
domain: $(0, \infty)$; range: $(-\infty, \infty)$

57.

$$f(x) = \log_2 x$$

$$g(x) = \tfrac{1}{2} \log_2 x$$

vertical asymptote: $x = 0$
domain: $(0, \infty)$; range: $(-\infty, \infty)$

59.

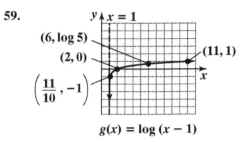

$$g(x) = \log(x - 1)$$

vertical asymptote: $x = 1$
domain: $(1, \infty)$; range: $(-\infty, \infty)$

61.

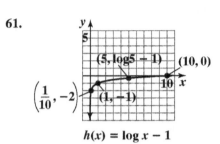

$$h(x) = \log x - 1$$

vertical asymptote: $x = 0$
domain: $(0, \infty)$; range: $(-\infty, \infty)$

63.

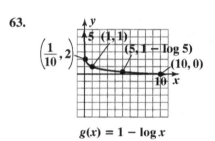

$$g(x) = 1 - \log x$$

vertical asymptote: $x = 0$
domain: $(0, \infty)$; range: $(-\infty, \infty)$

65.

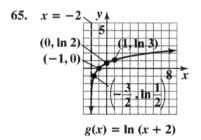

$$g(x) = \ln(x + 2)$$

vertical asymptote: $x = -2$
domain: $(-2, \infty)$; range: $(-\infty, \infty)$

67.

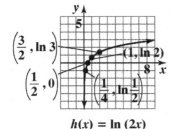

$$h(x) = \ln (2x)$$

vertical asymptote: $x = 0$
domain: $(0, \infty)$; range: $(-\infty, \infty)$

69.

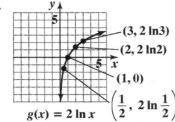

$$g(x) = 2 \ln x$$

vertical asymptote: $x = 0$
domain: $(0, \infty)$; range: $(-\infty, \infty)$

71.

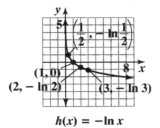

$$h(x) = -\ln x$$

vertical asymptote: $x = 0$
domain: $(0, \infty)$; range: $(-\infty, \infty)$

73.

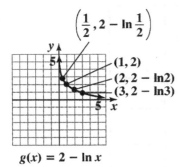

$$g(x) = 2 - \ln x$$

vertical asymptote: $x = 0$
domain: $(0, \infty)$; range: $(-\infty, \infty)$

75. The domain of f consists of all x for which
$x + 4 > 0$. Solving this inequality for x, we obtain
$x > -4$. Thus, the domain of f is $(-4, \infty)$.

77. The domain of f consists of all x for which
$2 - x > 0$. Solving this inequality for x, we obtain $x <$
2. Thus, the domain of f is $(-\infty, 2)$.

79. The domain of f consists of all x for which
$(x - 2)^2 > 0$. Solving this inequality for x, we obtain x
< 2 or $x > 2$. Thus, the domain of f is
$(-\infty, 2)$ or $(2, \infty)$.

81. $\log 100 = \log_{10} 100 = 2$
because $10^2 = 100$.

83. Because $\log 10^x = x$, we
conclude $\log 10^7 = 7$.

85. Because $10^{\log x} = x$, we
conclude $10^{\log 33} = 33$.

87. $\ln 1 = 0$ because $e^0 = 1$.

89. Because $\ln e^x = x$, we
conclude $\ln e^6 = 6$.

91. $\ln \dfrac{1}{e^6} = \ln e^{-6}$

Because $\ln e^x = x$ we conclude

$\ln e^{-6} = -6$, so $\ln \dfrac{1}{e^6} = -6$.

93. Because $e^{\ln x} = x$, we conclude $e^{\ln 125} = 125$.

95. Because $\ln e^x = x$, we conclude $\ln e^{9x} = 9x$.

97. Because $e^{\ln x} = x$, we conclude $e^{\ln 5x^2} = 5x^2$.

99. Because $10^{\log x} = x$, we conclude $10^{\log \sqrt{x}} = \sqrt{x}$.

101. $\log_3 (x - 1) = 2$
$$3^2 = x - 1$$
$$9 = x - 1$$
$$10 = x$$
The solution is 10, and the solution set is $\{10\}$.

103. $\log_4 x = -3$
$$4^{-3} = x$$
$$x = \dfrac{1}{4^3} = \dfrac{1}{64}$$

The solution is $\dfrac{1}{64}$, and the solution set is $\left\{\dfrac{1}{64}\right\}$.

105. $\log_3\left(\log_7 7\right) = \log_3 1 = 0$

107. $\log_2\left(\log_3 81\right) = \log_2\left(\log_3 3^4\right)$

$\qquad\qquad = \log_2 4 = \log_2 2^2 = 2$

109. For $f(x) = \ln(x^2 - x - 2)$ to be real, $x^2 - x - 2 > 0$.
Solve the related equation to find the boundary points:

$x^2 - x - 2 = 0$

$(x+1)(x-2) = 0$

The boundary points are –1 and 2. Testing each interval gives a domain of $(-\infty, -1) \cup (2, \infty)$.

111. For $f(x) = \ln\left(\dfrac{x+1}{x-5}\right)$ to be real, $\dfrac{x+1}{x-5} > 0$.

The boundary points are –1 and 5. Testing each interval gives a domain of $(-\infty, -1) \cup (5, \infty)$.

113. $f(13) = 62 + 35\log(13 - 4) \approx 95.4$
She is approximately 95.4% of her adult height.

115. a. 2008 is 39 years after 1969.

$f(x) = -7.52 \ln x + 53$

$f(39) = -7.52 \ln 39 + 53 \approx 25.5$

According to the function, 25.5% of first-year college men expressed antifeminist views in 2008. This underestimates the value in the graph by 0.7%.

b. 2015 is 46 years after 1969.

$f(x) = -7.52 \ln x + 53$

$f(46) = -7.52 \ln 46 + 53 \approx 24.2$

According to the function, 24.2% of first-year college men will express antifeminist views in 2015.

117. $D = 10\log\left[10^{12}(6.3 \times 10^6)\right] \approx 188$

Yes, the sound can rupture the human eardrum.

119. a. $f(0) = 88 - 15\ln(0 + 1) = 88$
The average score on the original exam was 88.

b. $f(2) = 88 - 15\ln(2 + 1) = 71.5$
$f(4) = 88 - 15\ln(4 + 1) = 63.9$
$f(6) = 88 - 15\ln(6 + 1) = 58.8$
$f(8) = 88 - 15\ln(8 + 1) = 55$

$f(10) = 88 - 15\ln(10 + 1) = 52$
$f(12) = 88 - 15\ln(12 + 1) = 49.5$

The average score after 2 months was about 71.5, after 4 months was about 63.9, after 6 months was about 58.8, after 8 months was about 55, after 10 months was about 52, and after one year was about 49.5.

121. – 127. Answers will vary.

129.

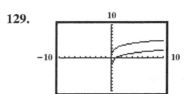

g(x) is *f(x)* shifted 3 units upward.

131.

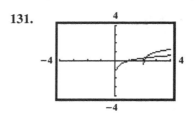

g(x) is *f(x)* shifted right 2 units and upward 1 unit.

133. a.

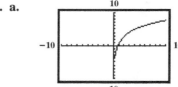

b.

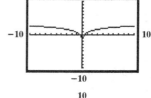

c.

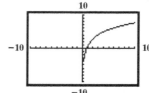

d They are the same.
$\log_b MN = \log_b M + \log_b N$

e. The sum of the logarithms of its factors.

135. makes sense

137. makes sense

139. false; Changes to make the statement true will vary.

A sample change is: $\dfrac{\log_2 8}{\log_2 4} = \dfrac{3}{2}$

141. false; Changes to make the statement true will vary. A sample change is: The domain of $f(x) = \log_2 x$ is $(0, \infty)$.

143. $\dfrac{\log_3 81 - \log_\pi 1}{\log_{2\sqrt{2}} 8 - \log 0.001} = \dfrac{4 - 0}{2 - (-3)} = \dfrac{4}{5}$

145. $\log_4 60 < \log_4 64 = 3$ so $\log_4 60 < 3$.

$\log_3 40 > \log_3 27 = 3$ so $\log_3 40 > 3$.

$\log_4 60 < 3 < \log_3 40$

$\log_3 40 > \log_4 60$

147. a. $\log_2 32 = \log_2 2^5 = 5$

b. $\log_2 8 + \log_2 4 = \log_2 2^3 + \log_2 2^2 = 3 + 2 = 5$

c. $\log_2(8 \cdot 4) = \log_2 8 + \log_2 4$

148. a. $\log_2 16 = \log_2 2^4 = 4$

b. $\log_2 32 - \log_2 2 = \log_2 2^5 - \log_2 2 = 5 - 1 = 4$

c. $\log_2\left(\dfrac{32}{2}\right) = \log_2 32 - \log_2 2$

149. a. $\log_3 81 = \log_3 3^4 = 4$

b. $2\log_3 9 = 2\log_3 3^2 = 2 \cdot 2 = 4$

c. $\log_3 9^2 = 2\log_3 9$

Section 4.3

Check Point Exercises

1. a. $\log_6(7 \cdot 11) = \log_6 7 + \log_6 11$

b. $\log(100x) = \log 100 + \log x$

$= 2 + \log x$

2. a. $\log_8\left(\dfrac{23}{x}\right) = \log_8 23 - \log_8 x$

b. $\ln\left(\dfrac{e^5}{11}\right) = \ln e^5 - \ln 11$

$= 5 - \ln 11$

3. a. $\log_6 3^9 = 9\log_6 3$

b. $\ln \sqrt[3]{x} = \ln x^{1/3} = \dfrac{1}{3}\ln x$

c. $\log(x+4)^2 = 2\log(x+4)$

4. a. $\log_b x^4 \sqrt[3]{y}$

$= \log_b x^4 y^{1/3}$

$= \log_b x^4 + \log_b y^{1/3}$

$= 4\log_b x + \dfrac{1}{3}\log_b y$

b. $\log_5 \dfrac{\sqrt{x}}{25y^3}$

$= \log_5 \dfrac{x^{1/2}}{25y^3}$

$= \log_5 x^{1/2} - \log_5 25y^3$

$= \log_5 x^{1/2} - \left(\log_5 5^2 + \log_5 y^3\right)$

$= \dfrac{1}{2}\log_5 x - \log_5 5^2 - \log_5 y^3$

$= \dfrac{1}{2}\log_5 x - 2\log_5 5 - 3\log_5 y$

$= \dfrac{1}{2}\log_5 x - 2 - 3\log_5 y$

5. a. $\log 25 + \log 4 = \log(25 \cdot 4) = \log 100 = 2$

b. $\log(7x+6) - \log x = \log\dfrac{7x+6}{x}$

6. a. $\ln x^2 + \dfrac{1}{3}\ln(x+5)$

$= \ln x^2 + \ln(x+5)^{1/3}$

$= \ln x^2 (x+5)^{1/3}$

$= \ln x^2 \sqrt[3]{x+5}$

b. $2\log(x-3) - \log x$

$= \log(x-3)^2 - \log x$

$= \log \dfrac{(x-3)^2}{x}$

c. $\dfrac{1}{4}\log_b x - 2\log_b 5 - 10\log_b y$

$= \log_b x^{1/4} - \log_b 5^2 - \log_b y^{10}$

$= \log_b x^{1/4} - \left(\log_b 25 - \log_b y^{10}\right)$

$= \log_b x^{1/4} - \log_b 25y^{10}$

$= \log_b \dfrac{x^{1/4}}{25y^{10}}$ or $\log_b \dfrac{\sqrt[4]{x}}{25y^{10}}$

7. $\log_7 2506 = \dfrac{\log 2506}{\log 7} \approx 4.02$

8. $\log_7 2506 = \dfrac{\ln 2506}{\ln 7} \approx 4.02$

Concept and Vocabulary Check 4.3

1. $\log_b M + \log_b N$; sum

2. $\log_b M - \log_b N$; difference

3. $p \log_b M$; product

4. $\log_a M$; $\log_a b$

Exercise Set 4.3

1. $\log_5(7\cdot 3) = \log_5 7 + \log_5 3$

3. $\log_7(7x) = \log_7 7 + \log_7 x = 1 + \log_7 x$

5. $\log(1000x) = \log 1000 + \log x = 3 + \log x$

7. $\log_7\left(\dfrac{7}{x}\right) = \log_7 7 - \log_7 x = 1 - \log_7 x$

9. $\log\left(\dfrac{x}{100}\right) = \log x - \log 100 = \log_x - 2$

11. $\log_4\left(\dfrac{64}{y}\right) = \log_4 64 - \log_4 y$

$= 3 - \log_4 y$

13. $\ln\left(\dfrac{e^2}{5}\right) = \ln e^2 - \ln 5 = 2\ln e - \ln 5 = 2 - \ln 5$

15. $\log_b x^3 = 3\log_b x$

17. $\log N^{-6} = -6\log N$

19. $\ln \sqrt[5]{x} = \ln x^{(1/5)} = \dfrac{1}{5}\ln x$

21. $\log_b x^2 y = \log_b x^2 + \log_b y = 2\log_b x + \log_b y$

23. $\log_4\left(\dfrac{\sqrt{x}}{64}\right) = \log_4 x^{1/2} - \log_4 64 = \dfrac{1}{2}\log_4 x - 3$

25. $\log_6\left(\dfrac{36}{\sqrt{x+1}}\right) = \log_6 36 - \log_6(x+1)^{1/2}$

$= 2 - \dfrac{1}{2}\log_6(x+1)$

27. $\log_b\left(\dfrac{x^2 y}{z^2}\right) = \log_b\left(x^2 y\right) - \log_b z^2$

$= \log_b x^2 + \log_b y - \log_b z^2$

$= 2\log_b x + \log_b y - 2\log_b z$

29. $\log \sqrt{100x} = \log(100x)^{1/2}$

$= \dfrac{1}{2}\log(100x)$

$= \dfrac{1}{2}(\log 100 + \log x)$

$= \dfrac{1}{2}(2 + \log x)$

$= 1 + \dfrac{1}{2}\log x$

31. $\log \sqrt[3]{\dfrac{x}{y}} = \log\left(\dfrac{x}{y}\right)^{1/3}$

$= \dfrac{1}{3}\left[\log\left(\dfrac{x}{y}\right)\right]$

$= \dfrac{1}{3}(\log x - \log y)$

$= \dfrac{1}{3}\log x - \dfrac{1}{3}\log y$

33. $\log_b \dfrac{\sqrt{x}\,y^3}{z^3}$

$= \log_b x^{1/2} + \log_b y^3 - \log_b z^3$

$= \dfrac{1}{2}\log_b x + 3\log_b y - 3\log_b z$

35. $\log_5 \sqrt[3]{\dfrac{x^2 y}{25}}$

$= \log_5 x^{2/3} + \log_5 y^{1/3} - \log_5 25^{1/3}$

$= \dfrac{2}{3}\log_5 x + \dfrac{1}{3}\log_5 y - \log_5 5^{2/3}$

$= \dfrac{2}{3}\log_5 x + \dfrac{1}{3}\log_5 y - \dfrac{2}{3}$

37. $\ln\left[\dfrac{x^3\sqrt{x^2+1}}{(x+1)^4}\right]$

$= \ln x^3 + \ln\sqrt{x^2+1} - \ln(x+1)^4$

$= 3\ln x + \dfrac{1}{2}\ln(x^2+1) - 4\ln(x+1)$

39. $\log\left[\dfrac{10x^2\sqrt[3]{1-x}}{7(x+1)^2}\right]$

$= \log 10 + \log x^2 + \log\sqrt[3]{1-x} - \log 7 - \log(x+1)^2$

$= 1 + 2\log x + \dfrac{1}{3}\log(1-x) - \log 7 - 2\log(x+1)$

41. $\log 5 + \log 2 = \log(5\cdot 2) = \log 10 = 1$

43. $\ln x + \ln 7 = \ln(7x)$

45. $\log_2 96 - \log_2 3 = \log_2\left(\dfrac{96}{3}\right) = \log_2 32 = 5$

47. $\log(2x+5) - \log x = \log\left(\dfrac{2x+5}{x}\right)$

49. $\log x + 3\log y = \log x + \log y^3 = \log(xy^3)$

51. $\dfrac{1}{2}\ln x + \ln y = \ln x^{1/2} + \ln y$

$= \ln\left(x^{\frac{1}{2}}y\right)$ or $\ln\left(y\sqrt{x}\right)$

53. $2\log_b x + 3\log_b y = \log_b x^2 + \log_b y^3$

$= \log_b(x^2 y^3)$

55. $5\ln x - 2\ln y = \ln x^5 - \ln y^2 = \ln\left(\dfrac{x^5}{y^2}\right)$

57. $3\ln x - \dfrac{1}{3}\ln y = \ln x^3 - \ln y^{1/3}$

$= \ln\left(\dfrac{x^3}{y^{1/3}}\right)$ or $\ln\left(\dfrac{x^3}{\sqrt[3]{y}}\right)$

59. $4\ln(x+6) - 3\ln x = \ln(x+6)^4 - \ln x^3$

$= \ln\dfrac{(x+6)^4}{x^3}$

61. $3\ln x + 5\ln y - 6\ln z$

$= \ln x^3 + \ln y^5 - \ln z^6$

$= \ln\dfrac{x^3 y^5}{z^6}$

63. $\dfrac{1}{2}\left(\log x + \log y\right)$

$= \dfrac{1}{2}(\log xy)$

$= \log(xy)^{1/2}$

$= \log\sqrt{xy}$

65. $\dfrac{1}{2}(\log_5 x + \log_5 y) - 2\log_5(x+1)$

$= \dfrac{1}{2}\log_5 xy - \log_5(x+1)^2$

$= \log_5(xy)^{1/2} - \log_5(x+1)^2$

$= \log_5\dfrac{(xy)^{1/2}}{(x+1)^2}$

$= \log_5\dfrac{\sqrt{xy}}{(x+1)^2}$

67. $\dfrac{1}{3}[2\ln(x+5)-\ln x-\ln(x^2-4)]$

$=\dfrac{1}{3}[\ln(x+5)^2-\ln x-\ln(x^2-4)]$

$=\dfrac{1}{3}\left[\ln\dfrac{(x+5)^2}{x(x^2-4)}\right]$

$=\ln\left[\dfrac{(x+5)^2}{x(x^2-4)}\right]^{1/3}$

$=\ln\sqrt[3]{\dfrac{(x+5)^2}{x(x^2-4)}}$

69. $\log x+\log(x^2-1)-\log 7-\log(x+1)$

$=\log x+\log(x^2-1)-(\log 7+\log(x+1))$

$=\log(x(x^2-1))-\log(7(x+1))$

$=\log\dfrac{x(x^2-1)}{7(x+1)}$

$=\log\dfrac{x(x+1)(x-1)}{7(x+1)}$

$=\log\dfrac{x(x-1)}{7}$

71. $\log_5 13=\dfrac{\log 13}{\log 5}\approx 1.5937$

73. $\log_{14}87.5=\dfrac{\ln 87.5}{\ln 14}\approx 1.6944$

75. $\log_{0.1}17=\dfrac{\log 17}{\log 0.1}\approx -1.2304$

77. $\log_\pi 63=\dfrac{\ln 63}{\ln \pi}\approx 3.6193$

79. $y=\log_3 x=\dfrac{\log x}{\log 3}$

81. $y=\log_2(x+2)=\dfrac{\log(x+2)}{\log 2}$

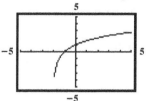

83. $\log_b\dfrac{3}{2}=\log_b 3-\log_b 2=C-A$

85. $\log_b 8=\log_b 2^3=3\log_b 2=3A$

87. $\log_b\sqrt{\dfrac{2}{27}}=\log_b\left(\dfrac{2}{27}\right)^{\frac{1}{2}}$

$=\dfrac{1}{2}\log_b\left(\dfrac{2}{3^3}\right)$

$=\dfrac{1}{2}\left(\log_b 2-\log_b 3^3\right)$

$=\dfrac{1}{2}\left(\log_b 2-3\log_b 3\right)$

$=\dfrac{1}{2}\log_b 2-\dfrac{3}{2}\log_b 3$

$=\dfrac{1}{2}A-\dfrac{3}{2}C$

89. false; $\ln e=1$

91. false; $\log_4(2x)^3=3\log_4(2x)$

93. true; $x\log 10^x=x\cdot x=x^2$

95. true; $\ln(5x)+\ln 1=\ln 5x+0=\ln 5x$

97. false; $\log(x+3)-\log(2x)=\log\dfrac{x+3}{2x}$

99. true; quotient rule

101. true; $\log_3 7=\dfrac{\log 7}{\log 3}=\dfrac{1}{\frac{\log 3}{\log 7}}=\dfrac{1}{\log_7 3}$

103. a. $D=10\log\dfrac{I}{I_0}$

b.
$$D_1 = 10\log\left(\frac{100I}{I_0}\right)$$
$$= 10\log\left(100I - I_0\right)$$
$$= 10\log 100 + 10\log I - 10\log I_0$$
$$= 10(2) + 10\log I - 10\log I_0$$
$$= 20 + 10\log\frac{I}{I_0}$$

This is 20 more than the loudness level of the softer sound. This means that the 100 times louder sound will be 20 decibels louder.

105. – 111. Answers will vary.

113. **a.** $y = \log_3 x = \dfrac{\ln x}{\ln 3}$

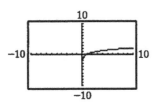

b.

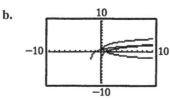

To obtain the graph of $y = 2 + \log_3 x$, shift the graph of $y = \log_3 x$ two units upward. To obtain the graph of $y = \log_3(x + 2)$, shift the graph of $y = \log_3 x$ two units left. To obtain the graph of $y = -\log_3 x$, reflect the graph of $y = \log_3 x$ about the x-axis.

115.
$$\log_3 x = \frac{\log x}{\log 3};$$
$$\log_{25} x = \frac{\log x}{\log 25};$$
$$\log_{100} x = \frac{\log x}{\log 100}$$

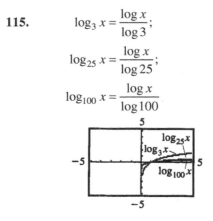

a. top graph: $y = \log_{100} x$
bottom graph: $y = \log_3 x$

b. top graph: $y = \log_3 x$
bottom graph: $y = \log_{100} x$

c. Comparing graphs of $\log_b x$ for $b > 1$, the graph of the equation with the largest b will be on the top in the interval $(0, 1)$ and on the bottom in the interval $(1, \infty)$.

117. – 119. Answers will vary.

121. makes sense

123. makes sense

125. true

127. false; Changes to make the statement true will vary. A sample change is: $\log_b\left(x^3 + y^3\right)$ cannot be simplified. If we were taking the logarithm of a product and not a sum, we would have been able to simplify as follows.
$$\log_b\left(x^3 y^3\right) = \log_b x^3 + \log_b y^3$$
$$= 3\log_b x + 3\log_b y$$

129. $\log e = \log_{10} e = \dfrac{\ln e}{\ln 10} = \dfrac{1}{\ln 10}$

131. $e^{\ln 8x^5 - \ln 2x^2} = e^{\ln\left(\frac{8x^5}{2x^2}\right)} = e^{\ln\left(4x^3\right)} = 4x^3$

133. Let $\log_b M = R$ and $\log_b N = S$.

Then $\log_b M = R$ means $b^R = M$ and

$\log_b N = S$ means $b^S = N$.

Thus, $\dfrac{M}{N} = \dfrac{b^R}{b^S}$

$\log_b \dfrac{M}{N} = \log_b b^{R-S}$

$\log_b \dfrac{M}{N} = R - S$

$\log_b \dfrac{M}{N} = \log_b M - \log_b N$

134. $a(x-2) = b(2x+3)$

$ax - 2a = 2bx + 3b$

$ax - 2bx = 2a + 3b$

$x(a - 2b) = 2a + 3b$

$x = \dfrac{2a + 3b}{a - 2b}$

135. $x(x-7) = 3$

$x^2 - 7x = 3$

$x^2 - 7x - 3 = 0$

$x = \dfrac{-b \pm \sqrt{b^2 - 4ac}}{2a}$

$x = \dfrac{-(-7) \pm \sqrt{(-7)^2 - 4(1)(-3)}}{2(1)}$

$x = \dfrac{7 \pm \sqrt{61}}{2}$

The solution set is $\left\{ \dfrac{7 \pm \sqrt{61}}{2} \right\}$.

136. $\dfrac{x+2}{4x+3} = \dfrac{1}{x}$

$x(4x+3)\left(\dfrac{x+2}{4x+3}\right) = x(4x+3)\left(\dfrac{1}{x}\right)$

$x(x+2) = 4x + 3$

$x^2 + 2x = 4x + 3$

$x^2 - 2x - 3 = 0$

$(x+1)(x-3) = 0$

$x + 1 = 0 \quad$ or $\quad x - 3 = 0$

$x = -1 \qquad\qquad x = 3$

The solution set is $\{-1, 3\}$.

Mid-Chapter 4 Check Point

1.

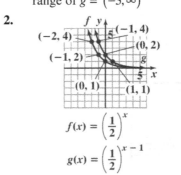

$f(x) = 2^x$
$g(x) = 2^x - 3$

asymptote of f: $\ y = 0$

asymptote of g: $\ y = -3$

domain of f = domain of $g = (-\infty, \infty)$

range of $f = \ (0, \infty)$

range of $g = (-3, \infty)$

2.

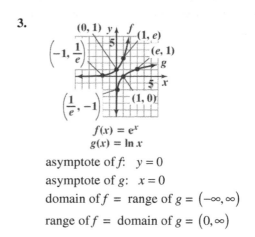

$f(x) = \left(\dfrac{1}{2}\right)^x$

$g(x) = \left(\dfrac{1}{2}\right)^{x-1}$

asymptote of f: $\ y = 0$

asymptote of g: $\ y = 0$

domain of f = domain of $g = (-\infty, \infty)$

range of f = range of $g = (0, \infty)$

3.

$f(x) = e^x$
$g(x) = \ln x$

asymptote of f: $\ y = 0$

asymptote of g: $\ x = 0$

domain of f = range of $g = (-\infty, \infty)$

range of f = domain of $g = (0, \infty)$

4.

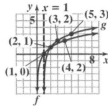

$$f(x) = \log_2 x$$
$$g(x) = \log_2 (x - 1) + 1$$

asymptote of f: $x = 0$
asymptote of g: $x = 1$
domain of $f = (0, \infty)$
domain of $g = (1, \infty)$
range of f = range of $g = (-\infty, \infty)$

5.

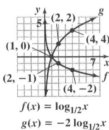

$$f(x) = \log_{1/2} x$$
$$g(x) = -2 \log_{1/2} x$$

asymptote of f: $x = 0$
asymptote of g: $x = 0$
domain of f = domain of $g = (0, \infty)$
range of f = range of $g = (-\infty, \infty)$

6. $f(x) = \log_3 (x + 6)$
The argument of the logarithm must be positive:
$x + 6 > 0$
$\quad x > -6$
domain: $\{x \mid x > -6\}$ or $(-6, \infty)$.

7. $f(x) = \log_3 x + 6$
The argument of the logarithm must be positive:
$x > 0$
domain: $\{x \mid x > 0\}$ or $(0, \infty)$.

8. $\log_3 (x + 6)^2$
The argument of the logarithm must be positive.
Now $(x + 6)^2$ is always positive, except when
$x = -6$
domain: $\{x \mid x \neq 0\}$ or $(-\infty, -6) \cup (-6, \infty)$.

9. $f(x) = 3^{x+6}$
domain: $\{x \mid x \text{ is a real number}\}$ or $(-\infty, \infty)$.

10. $\log_2 8 + \log_5 25 = \log_2 2^3 + \log_5 5^2$
$$= 3 + 2 = 5$$

11. $\log_3 \dfrac{1}{9} = \log_3 \dfrac{1}{3^2} = \log_3 3^{-2} = -2$

12. Let $\log_{100} 10 = y$
$$100^y = 10$$
$$\left(10^2\right)^y = 10^1$$
$$10^{2y} = 10^1$$
$$2y = 1$$
$$y = \frac{1}{2}$$

13. $\log \sqrt[3]{10} = \log 10^{\frac{1}{3}} = \dfrac{1}{3}$

14. $\log_2 \left(\log_3 81\right) = \log_2 \left(\log_3 3^4\right)$
$$= \log_2 4 = \log_2 2^2 = 2$$

15. $\log_3 \left(\log_2 \dfrac{1}{8}\right) = \log_3 \left(\log_2 \dfrac{1}{2^3}\right)$
$$= \log_3 \left(\log_2 2^{-3}\right)$$
$$= \log_3 (-3)$$
$$= \text{not possible}$$
This expression is impossible to evaluate because
$\log_3 (-3)$ is undefined.

16. $6^{\log_6 5} = 5$

17. $\ln e^{\sqrt{7}} = \sqrt{7}$

18. $10^{\log 13} = 13$

19. $\log_{100} 0.1 = y$
$$100^y = 0.1$$
$$\left(10^2\right)^y = \frac{1}{10}$$
$$10^{2y} = 10^{-1}$$
$$2y = -1$$
$$y = -\frac{1}{2}$$

20. $\log_\pi \pi^{\sqrt{\pi}} = \sqrt{\pi}$

21. $\log\left(\dfrac{\sqrt{xy}}{1000}\right) = \log\left(\sqrt{xy}\right) - \log 1000$

$$= \log\left(xy\right)^{\frac{1}{2}} - \log 10^3$$

$$= \frac{1}{2}\log\left(xy\right) - 3$$

$$= \frac{1}{2}\left(\log x + \log y\right) - 3$$

$$= \frac{1}{2}\log x + \frac{1}{2}\log y - 3$$

22. $\ln\left(e^{19} x^{20}\right) = \ln e^{19} + \ln x^{20}$

$$= 19 + 20\ln x$$

23. $8\log_7 x - \dfrac{1}{3}\log_7 y = \log_7 x^8 - \log_7 y^{\frac{1}{3}}$

$$= \log_7\left(\frac{x^8}{y^{\frac{1}{3}}}\right)$$

$$= \log_7\left(\frac{x^8}{\sqrt[3]{y}}\right)$$

24. $7\log_5 x + 2\log_5 x = \log_5 x^7 + \log_5 x^2$

$$= \log_5\left(x^7 \cdot x^2\right)$$

$$= \log_5 x^9$$

25. $\dfrac{1}{2}\ln x - 3\ln y - \ln\left(z-2\right)$

$$= \ln x^{\frac{1}{2}} - \ln y^3 - \ln\left(z-2\right)$$

$$= \ln\sqrt{x} - \left[\ln y^3 + \ln\left(z-2\right)\right]$$

$$= \ln\sqrt{x} - \ln\left[y^3\left(z-2\right)\right]$$

$$= \ln\left[\frac{\sqrt{x}}{y^3\left(z-2\right)}\right]$$

26. Continuously: $A = 8000e^{0.08(3)}$

$$\approx 10{,}170$$

Monthly: $A = 8000\left(1 + \dfrac{0.08}{12}\right)^{12\cdot 3}$

$$\approx 10{,}162$$

$10{,}170 - 10{,}162 = 8$

Interest returned will be $8 more if compounded continuously.

Section 4.4

Check Point Exercises

1. **a.** $5^{3x-6} = 125$

    $5^{3x-6} = 5^3$

    $3x - 6 = 3$

    $3x = 9$

    $x = 3$

    The solution set is $\{3\}$.

    b. $8^{x+2} = 4^{x-3}$

    $\left(2^3\right)^{x+2} = \left(2^2\right)^{x-3}$

    $2^{3x+6} = 2^{2x-6}$

    $3x + 6 = 2x - 6$

    $x = -12$

    The solution set is $\{-12\}$.

2. **a.** $5^x = 134$

    $\ln 5^x = \ln 134$

    $x \ln 5 = \ln 134$

    $x = \dfrac{\ln 134}{\ln 5} \approx 3.04$

    The solution set is $\left\{\dfrac{\ln 134}{\ln 5}\right\}$,

    approximately 3.04.

    b. $10^x = 8000$

    $\log 10^x = \log 8000$

    $x \log 10 = \log 8000$

    $x = \log 8000 \approx 3.90$

    The solution set is $\{\log 8000\}$, approximately 3.90.

3. $7e^{2x} = 63$

    $e^{2x} = 9$

    $\ln e^{2x} = \ln 9$

    $2x = \ln 9$

    $x = \dfrac{\ln 9}{2} \approx 1.10$

    The solution set is $\left\{\dfrac{\ln 9}{2}\right\}$,

    approximately 1.10.

4. $3^{2x-1} = 7^{x+1}$

    $\ln 3^{2x-1} = \ln 7^{x+1}$

    $(2x-1)\ln 3 = (x+1)\ln 7$

    $2x \ln 3 - \ln 3 = x \ln 7 + \ln 7$

    $2x \ln 3 - x \ln 7 = \ln 3 + \ln 7$

    $x(2\ln 3 - \ln 7) = \ln 3 + \ln 7$

    $x = \dfrac{\ln 3 + \ln 7}{2\ln 3 - \ln 7}$

    $x \approx 12.11$

    The solution set is $\left\{\dfrac{\ln 3 + \ln 7}{2\ln 3 - \ln 7}\right\}$,

    approximately 12.11.

5. $e^{2x} - 8e^x + 7 = 0$

    $\left(e^x - 7\right)\left(e^x - 1\right) = 0$

    $e^x - 7 = 0 \qquad \text{or} \quad e^x - 1 = 0$

    $\qquad e^x = 7 \qquad\qquad\quad e^x = 1$

    $\quad \ln e^x = \ln 7 \qquad\quad \ln e^x = \ln 1$

    $\qquad x = \ln 7 \qquad\qquad\quad x = 0$

    The solution set is $\{0, \ln 7\}$. The solutions are 0 and (approximately) 1.95.

6. **a.** $\log_2(x-4) = 3$

    $2^3 = x - 4$

    $8 = x - 4$

    $12 = x$

    Check:

    $\log_2(x-4) = 3$

    $\log_2(12-4) = 3$

    $\log_2 8 = 3$

    $3 = 3$

    The solution set is $\{12\}$.

b. $4\ln 3x = 8$

$\ln 3x = 2$

$e^{\ln 3x} = e^2$

$3x = e^2$

$x = \dfrac{e^2}{3} \approx 2.46$

Check

$4\ln 3x = 8$

$4\ln 3\left(\dfrac{e^2}{3}\right) = 8$

$4\ln e^2 = 8$

$4(2) = 8$

$8 = 8$

The solution set is $\left\{\dfrac{e^2}{3}\right\}$,

approximately 2.46.

7. $\log x + \log(x-3) = 1$

$\log x(x-3) = 1$

$10^1 = x(x-3)$

$10 = x^2 - 3x$

$0 = x^2 - 3x - 10$

$0 = (x-5)(x+2)$

$x - 5 = 0 \quad \text{or} \quad x + 2 = 0$

$x = 5 \quad \text{or} \qquad x = -2$

Check

Checking 5:

$\log 5 + \log(5-3) = 1$

$\log 5 + \log 2 = 1$

$\log(5 \cdot 2) = 1$

$\log 10 = 1$

$1 = 1$

Checking –2:

$\log x + \log(x-3) = 1$

$\log(-2) + \log(-2-3) \mathbf{\,0\,} 1$

Negative numbers do not have logarithms so
–2 does not check.
The solution set is $\{5\}$.

8. $\ln(x-3) = \ln(7x-23) - \ln(x+1)$

$\ln(x-3) = \ln\dfrac{7x-23}{x+1}$

$x - 3 = \dfrac{7x-23}{x+1}$

$(x-3)(x+1) = 7x - 23$

$x^2 - 2x - 3 = 7x - 23$

$x^2 - 9x + 20 = 0$

$(x-4)(x-5) = 0$

$x = 4 \quad \text{or} \quad x = 5$

Both values produce true statements.
The solution set is $\{4, 5\}$.

9.   For a risk of 7%, let $R = 7$ in

$R = 6e^{12.77x}$

$6e^{12.77x} = 7$

$e^{12.77x} = \dfrac{7}{6}$

$\ln e^{12.77x} = \ln\left(\dfrac{7}{6}\right)$

$12.77x = \ln\left(\dfrac{7}{6}\right)$

$x = \dfrac{\ln\left(\frac{7}{6}\right)}{12.77} \approx 0.01$

For a blood alcohol concentration of 0.01, the risk
of a car accident is 7%.

10. $A = P\left(1 + \dfrac{r}{n}\right)^{nt}$

$3600 = 1000\left(1 + \dfrac{0.08}{4}\right)^{4t}$

$1000\left(1 - \dfrac{0.08}{4}\right)^{4t} = 3600$

$1000(1 + 0.02)^{4t} = 3600$

$1000(1.02)^{4t} = 3600$

$(1.02)^{4t} = \ln 3.6$

$4t\ln(1.02) = \ln 3.6$

$t = \dfrac{\ln 3.6}{4\ln 1.02}$

≈ 16.2

After approximately 16.2 years, the $1000 will
grow to an accumulated value of $3600.

11. $f(x) = 62 + 35\log(x-4)$

Solve the equation when $f(x) = 97$.

$$62 + 35\log(x-4) = 97$$
$$35\log(x-4) = 35$$
$$\log(x-4) = \frac{35}{35}$$
$$\log(x-4) = 1$$
$$x - 4 = 10^1$$
$$x = 10 + 4$$
$$x = 14$$

At age 14, a girl will attain 97% of her adult height.

Concept and Vocabulary Check 4.4

1. $M = N$

2. $4x - 1$

3. $\dfrac{\ln 20}{\ln 9}$

4. $\ln 6$

5. 5^3

6. $(x^2 + x)$

7. $\dfrac{7x - 23}{x + 1}$

8. false

9. true

10. false

11. true

Exercise Set 4.4

1. $2^x = 64$
$$2^x = 2^6$$
$$x = 6$$
The solution is 6, and the solution set is $\{6\}$.

3. $5^x = 125$
$$5^x = 5^3$$
$$x = 3$$
The solution is 3, and the solution set is $\{3\}$.

5. $2^{2x-1} = 32$
$$2^{2x-1} = 2^5$$
$$2x - 1 = 5$$
$$2x = 6$$
$$x = 3$$
The solution is 3, and the solution set is $\{3\}$.

7. $4^{2x-1} = 64$
$$4^{2x-1} = 4^3$$
$$2x - 1 = 3$$
$$2x = 4$$
$$x = 2$$
The solution is 2, and the solution set is $\{2\}$.

9. $32^x = 8$
$$\left(2^5\right)^x = 2^3$$
$$2^{5x} = 2^3$$
$$5x = 3$$
$$x = \frac{3}{5}$$
The solution is $\dfrac{3}{5}$, and the solution set is $\left\{\dfrac{3}{5}\right\}$.

11. $9^x = 27$
$$\left(3^2\right)^x = 3^3$$
$$3^{2x} = 3^3$$
$$2x = 3$$
$$x = \frac{3}{2}$$
The solution is $\dfrac{3}{2}$, and the solution set is $\left\{\dfrac{3}{2}\right\}$.

13. $3^{1-x} = \dfrac{1}{27}$

$3^{1-x} = \dfrac{1}{3^3}$

$3^{1-x} = 3^{-3}$

$1 - x = -3$

$-x = -4$

$x = 4$

The solution set is $\{4\}$.

15. $6^{\frac{x-3}{4}} = \sqrt{6}$

$6^{\frac{x-3}{4}} = 6^{\frac{1}{2}}$

$\dfrac{x-3}{4} = \dfrac{1}{2}$

$2(x-3) = 4(1)$

$2x - 6 = 4$

$2x = 10$

$x = 5$

The solution is 5, and the solution set is $\{5\}$.

17. $4^x = \dfrac{1}{\sqrt{2}}$

$\left(2^2\right)^x = \dfrac{1}{2^{\frac{1}{2}}}$

$2^{2x} = 2^{-\frac{1}{2}}$

$2x = -\dfrac{1}{2}$

$x = \dfrac{1}{2}\left(-\dfrac{1}{2}\right) = -\dfrac{1}{4}$

The solution is $-\dfrac{1}{4}$, and the solution set is $\left\{-\dfrac{1}{4}\right\}$.

19. $8^{x+3} = 16^{x-1}$

$\left(2^3\right)^{x+3} = \left(2^4\right)^{x-1}$

$2^{3x+9} = 2^{4x-4}$

$3x + 9 = 4x - 4$

$13 = x$

The solution set is $\{13\}$.

21. $e^{x+1} = \dfrac{1}{e}$

$e^{x+1} = e^{-1}$

$x + 1 = -1$

$x = -2$

The solution set is $\{-2\}$.

23. $10^x = 3.91$

$\ln 10^x = \ln 3.91$

$x \ln 10 = \ln 3.91$

$x = \dfrac{\ln 3.91}{\ln 10} \approx 0.59$

25. $e^x = 5.7$

$\ln e^x = 5.7$

$x = \ln 5.7 \approx 1.74$

27. $5^x = 17$

$\ln 5^x = \ln 17$

$x \ln 5 = \ln 17$

$x = \dfrac{\ln 17}{\ln 5} \approx 1.76$

29. $5e^x = 23$

$e^x = \dfrac{23}{5}$

$\ln e^x = \ln \dfrac{23}{5}$

$x = \ln \dfrac{23}{5} \approx 1.53$

31. $3e^{5x} = 1977$

$e^{5x} = 659$

$\ln e^{5x} = \ln 659$

$x = \dfrac{\ln 659}{5} \approx 1.30$

33. $e^{1-5x} = 793$

$\ln e^{1-5x} = \ln 793$

$(1-5x)(\ln e) = \ln 793$

$1 - 5x = \ln 793$

$5x = 1 - \ln 793$

$x = \dfrac{1 - \ln 793}{5} \approx -1.14$

35.
$$e^{5x-3} - 2 = 10,476$$
$$e^{5x-3} = 10,478$$
$$\ln e^{5x-3} = \ln 10,478$$
$$(5x-3)\ln e = \ln 10,478$$
$$5x - 3 = \ln 10,478$$
$$5x = \ln 10,478 + 3$$
$$x = \frac{\ln 10,478 + 3}{5} \approx 2.45$$

37.
$$7^{x+2} = 410$$
$$\ln 7^{x+2} = \ln 410$$
$$(x+2)\ln 7 = \ln 410$$
$$x + 2 = \frac{\ln 410}{\ln 7}$$
$$x = \frac{\ln 410}{\ln 7} - 2 \approx 1.09$$

39.
$$7^{0.3x} = 813$$
$$\ln 7^{0.3x} = \ln 813$$
$$0.3x \ln 7 = \ln 813$$
$$x = \frac{\ln 813}{0.3 \ln 7} \approx 11.48$$

41.
$$5^{2x+3} = 3^{x-1}$$
$$\ln 5^{2x+3} = \ln 3^{x-1}$$
$$(2x+3)\ln 5 = (x-1)\ln 3$$
$$2x \ln 5 + 3\ln 5 = x \ln 3 - \ln 3$$
$$3\ln 5 + \ln 3 = x \ln 3 - 2x \ln 5$$
$$3\ln 5 + \ln 3 = x(\ln 3 - 2\ln 5)$$
$$\frac{3\ln 5 + \ln 3}{\ln 3 - 2\ln 5} = x$$
$$-2.80 \approx x$$

43.
$$e^{2x} - 3e^x + 2 = 0$$
$$\left(e^x - 2\right)\left(e^x - 1\right) = 0$$
$$e^x - 2 = 0 \quad \text{or} \quad e^x - 1 = 0$$
$$e^x = 2 \qquad\qquad e^x = 1$$
$$\ln e^x = \ln 2 \qquad \ln e^x = \ln 1$$
$$x = \ln 2 \qquad\qquad x = 0$$

The solution set is {0, ln 2}. The solutions are 0 and approximately 0.69.

45.
$$e^{4x} + 5e^{2x} - 24 = 0$$
$$\left(e^{2x} + 8\right)\left(e^{2x} - 3\right) = 0$$
$$e^{2x} + 8 = 0 \qquad \text{or} \quad e^{2x} - 3 = 0$$
$$e^{2x} = -8 \qquad\qquad e^{2x} = 3$$
$$\ln e^{2x} = \ln(-8) \qquad \ln e^{2x} = \ln 3$$
$$2x = \ln(-8) \qquad\qquad 2x = \ln 3$$
$$\ln(-8) \text{ does not exist} \qquad x = \frac{\ln 3}{2}$$

$$x = \frac{\ln 3}{2} \approx 0.55$$

47.
$$3^{2x} + 3^x - 2 = 0$$
$$(3^x + 2)(3^x - 1) = 0$$
$$3^x + 2 = 0 \qquad\qquad 3^x - 1 = 0$$
$$3^x = -2 \qquad\qquad 3^x = 1$$
$$\log 3^x = \log(-2) \qquad \log 3^x = \log 1$$
$$\text{does not exist} \qquad \log 3 = 0$$
$$x = \frac{0}{\log 3}$$
$$x = 0$$

The solution set is {0}.

49.
$$\log_3 x = 4$$
$$3^4 = x$$
$$81 = x$$

51.
$$\ln x = 2$$
$$e^2 = x$$
$$7.39 \approx x$$

53.
$$\log_4 (x+5) = 3$$
$$4^3 = x + 5$$
$$59 = x$$

55.
$$\log_2 (x+25) = 4$$
$$2^4 = x + 25$$
$$16 = x + 25$$
$$-9 = x$$

57. $\log_3(x+4) = -3$

$\quad 3^{-3} = x+4$

$\quad \dfrac{1}{27} = x+4$

$\quad \dfrac{1}{27} - 4 = x$

$\quad -\dfrac{107}{27} = x$

$\quad -3.96 \approx x$

59. $\log_4(3x+2) = 3$

$\quad 4^3 = 3x+2$

$\quad 64 = 3x+2$

$\quad 62 = 3x$

$\quad \dfrac{62}{3} = x$

$\quad 20.67 \approx x$

61. $5\ln(2x) = 20$

$\quad \ln(2x) = 4$

$\quad e^{\ln(2x)} = e^4$

$\quad 2x = e^4$

$\quad x = \dfrac{e^4}{2} \approx 27.30$

63. $6 + 2\ln x = 5$

$\quad 2\ln x = -1$

$\quad \ln x = -\dfrac{1}{2}$

$\quad e^{\ln x} = e^{-1/2}$

$\quad x = e^{-1/2} \approx 0.61$

65. $\ln\sqrt{x+3} = 1$

$\quad e^{\ln\sqrt{x+3}} = e^1$

$\quad \sqrt{x+3} = e$

$\quad x+3 = e^2$

$\quad x = e^2 - 3 \approx 4.39$

67. $\log_5 x + \log_5(4x-1) = 1$

$\quad \log_5(4x^2 - x) = 1$

$\quad 4x^2 - x = 5$

$\quad 4x^2 - x - 5 = 0$

$\quad (4x-5)(x+1) = 0$

$\quad x = \dfrac{5}{4}$ or $x = -1$

$x = -1$ does not check because $\log_5(-1)$ does not exist.

The solution set is $\left\{ \dfrac{5}{4} \right\}$.

69. $\log_3(x+6) + \log_3(x+4) = 1$

$\quad \log_3\left[(x+6)(x+4)\right] = 1$

$\quad (x+6)(x+4) = 3^1$

$\quad x^2 + 10x + 24 = 3$

$\quad x^2 + 10x + 21 = 0$

$\quad (x+7)(x+3) = 0$

$\quad x = -7$ or $x = -3$

$x = -7$ does not check because $\log_3(-7+6)$ does not exist. The solution set is $\{-3\}$.

71. $\log_2(x+2) - \log_2(x-5) = 3$

$\quad \log_2\left(\dfrac{x+2}{x-5}\right) = 3$

$\quad \dfrac{x+2}{x-5} = 2^3$

$\quad \dfrac{x+2}{x-5} = 8$

$\quad x+2 = 8(x-5)$

$\quad x+2 = 8x - 40$

$\quad 7x = 42$

$\quad x = 6$

73. $2\log_3(x+4) = \log_3 9 + 2$

$\quad 2\log_3(x+4) = 2 + 2$

$\quad 2\log_3(x+4) = 4$

$\quad \log_3(x+4) = 2$

$\quad 3^2 = x+4$

$\quad 9 = x+4$

$\quad 5 = x$

75. $\log_2(x-6)+\log_2(x-4)-\log_2 x = 2$

$$\log_2 \frac{(x-6)(x-4)}{x} = 2$$

$$\frac{(x-6)(x-4)}{x} = 2^2$$

$$x^2 - 10x + 24 = 4x$$

$$x^2 - 14x + 24 = 0$$

$$(x-12)(x-2) = 0$$

$x - 12 = 0 \quad$ or $\quad x - 2 = 0$

$x = 12 \qquad\qquad x = 2$

The solution set is $\{12\}$ since $\log_2(2-6) = \log_2(-4)$ is not possible.

77. $\log(x+4) = \log x + \log 4$

$\log(x+4) = \log 4x$

$x + 4 = 4x$

$4 = 3x$

$$x = \frac{4}{3}$$

This value is rejected. The solution set is $\left\{ \dfrac{4}{3} \right\}$.

79. $\log(3x-3) = \log(x+1) + \log 4$

$\log(3x-3) = \log(4x+4)$

$3x - 3 = 4x + 4$

$-7 = x$

This value is rejected. The solution set is $\{\ \}$.

81. $2\log x = \log 25$

$\log x^2 = \log 25$

$x^2 = 25$

$x = \pm 5$

-5 is rejected. The solution set is $\{5\}$.

83. $\log(x+4) - \log 2 = \log(5x+1)$

$$\log \frac{x+4}{2} = \log(5x+1)$$

$$\frac{x+4}{2} = 5x+1$$

$x + 4 = 10x + 2$

$-9x = -2$

$$x = \frac{2}{9}$$

$x \approx 0.22$

85. $2\log x - \log 7 = \log 112$

$\log x^2 - \log 7 = \log 112$

$$\log \frac{x^2}{7} = \log 112$$

$$\frac{x^2}{7} = 112$$

$x^2 = 784$

$x = \pm 28$

-28 is rejected. The solution set is $\{28\}$.

87. $\log x + \log(x+3) = \log 10$

$\log(x^2 + 3x) = \log 10$

$x^2 + 3x = 10$

$x^2 + 3x - 10 = 0$

$(x+5)(x-2) = 0$

$x = -5$ or $x = 2$

-5 is rejected. The solution set is $\{2\}$.

89. $\ln(x-4) + \ln(x+1) = \ln(x-8)$

$\ln(x^2 - 3x - 4) = \ln(x-8)$

$x^2 - 3x - 4 = x - 8$

$x^2 - 4x + 4 = 0$

$(x-2)(x-2) = 0$

$x = 2$

2 is rejected. The solution set is $\{\ \}$.

91. $\ln(x-2) - \ln(x+3) = \ln(x-1) - \ln(x+7)$

$$\ln \frac{x-2}{x+3} = \ln \frac{x-1}{x+7}$$

$$\frac{x-2}{x+3} = \frac{x-1}{x+7}$$

$(x-2)(x+7) = (x+3)(x-1)$

$x^2 + 5x - 14 = x^2 + 2x - 3$

$3x = 11$

$$x = \frac{11}{3}$$

$x \approx 3.67$

93. $5^{2x} \cdot 5^{4x} = 125$

$5^{2x+4x} = 5^3$

$5^{6x} = 5^3$

$6x = 3$

$$x = \frac{1}{2}$$

95. $2|\ln x| - 6 = 0$

$2|\ln x| = 6$

$|\ln x| = 3$

$\ln x = 3 \qquad$ or $\quad \ln x = -3$

$\quad x = e^3 \qquad\qquad x = e^{-3}$

$\quad x \approx 20.09 \qquad\quad x \approx 0.05$

97. $3^{x^2} = 45$

$\ln 3^{x^2} = \ln 45$

$x^2 \ln 3 = \ln 45$

$x^2 = \dfrac{\ln 45}{\ln 3}$

$x = \pm\sqrt{\dfrac{\ln 45}{\ln 3}} \approx \pm 1.86$

99. $\ln(2x+1) + \ln(x-3) - 2\ln x = 0$

$\ln(2x+1) + \ln(x-3) - \ln x^2 = 0$

$\ln \dfrac{(2x+1)(x-3)}{x^2} = 0$

$\dfrac{(2x+1)(x-3)}{x^2} = e^0$

$\dfrac{2x^2 - 5x - 3}{x^2} = 1$

$2x^2 - 5x - 3 = x^2$

$x^2 - 5x - 3 = 0$

$x = \dfrac{-b \pm \sqrt{b^2 - 4ac}}{2a}$

$x = \dfrac{-(-5) \pm \sqrt{(-5)^2 - 4(1)(-3)}}{2(1)}$

$x = \dfrac{5 \pm \sqrt{37}}{2}$

$x = \dfrac{5 + \sqrt{37}}{2} \approx 5.54$

$x = \dfrac{5 - \sqrt{37}}{2} \approx -0.54$ (rejected)

The solution set is $\left\{ \dfrac{5 + \sqrt{37}}{2} \right\}$.

101. $5^{x^2 - 12} = 25^{2x}$

$5^{x^2 - 12} = \left(5^2\right)^{2x}$

$5^{x^2 - 12} = 5^{4x}$

$x^2 - 12 = 4x$

$x^2 - 4x - 12 = 0$

$(x-6)(x+2) = 0$

Apply the zero product property:

$x - 6 = 0 \quad$ or $\quad x + 2 = 0$

$\quad x = 6 \qquad\qquad x = -2$

The solutions are -2 and 6, and the solution set is $\{-2, 6\}$.

103. a. 2010 is 0 years after 2010.

$A = 37.3e^{0.0095t}$

$A = 37.3e^{0.0095(0)} = 37.3$

The population of California was 37.3 million in 2010.

b. $A = 37.3e^{0.0095t}$

$40 = 37.3e^{0.0095t}$

$\dfrac{40}{37.3} = e^{0.0095t}$

$\ln \dfrac{40}{37.3} = \ln e^{0.0095t}$

$0.0095t = \ln \dfrac{40}{37.3}$

$t = \dfrac{\ln \dfrac{40}{37.3}}{0.0095} \approx 7$

The population of California will reach 40 million about 7 years after 2010, or 2017

105. $f(x) = 20(0.975)^x$

$1 = 20(0.975)^x$

$\dfrac{1}{20} = 0.975^x$

$\ln \dfrac{1}{20} = \ln 0.975^x$

$\ln \dfrac{1}{20} = x \ln 0.975$

$x = \dfrac{\ln \dfrac{1}{20}}{\ln 0.975}$

$x \approx 118$

There is 1% of surface sunlight at 118 feet. This is represented by the point (118,1).

107. $20,000 = 12,500\left(1 + \dfrac{0.0575}{4}\right)^{4t}$

$12,500(1.014375)^{4t} = 20,000$

$(1.014375)^{4t} = 1.6$

$\ln(1.014375)^{4t} = \ln 1.6$

$4t \ln(1.014375) = \ln 1.6$

$t = \dfrac{\ln 1.6}{4 \ln 1.014375} \approx 8.2$

8.2 years

109. $1400 = 1000\left(1 + \dfrac{0.168}{360}\right)^{360 \cdot t}$

$\dfrac{1400}{1000} = \left(1 + \dfrac{0.168}{360}\right)^{360 \cdot t}$

$\ln\left(1 + \dfrac{0.168}{360}\right)^{360 \cdot t} = \ln\left(\dfrac{1400}{1000}\right)$

$360 \cdot t \ln\left(1 + \dfrac{0.168}{360}\right) = \ln 1.4$

$t = \dfrac{\ln 1.4}{360 \ln\left(1 + \dfrac{0.168}{360}\right)}$

$t \approx 2.0$

111. accumulated amount $= 2(8000) = 16,000$

$16,000 = 8000e^{0.08t}$

$e^{0.08t} = 2$

$\ln e^{0.08t} = \ln 2$

$0.08t = \ln 2$

$t = \dfrac{\ln 2}{0.08}$

$t \approx 8.7$

The amount would double in 8.7 years.

113. accumulated amount $= 3(2350) = 7050$

$7050 = 2350e^{0.157t}$

$e^{0.157t} = \dfrac{7050}{2350}$

$\ln e^{0.157t} = \ln 3$

$0.157t = \ln 3$

$t = \dfrac{\ln 3}{0.157}$

$t \approx 7.0$

About 7 years.

115. a. 2009 is 3 years after 2006.

$f(x) = 1.2 \ln x + 15.7$

$f(3) = 1.2 \ln 3 + 15.7 \approx 17.0$

According to the function, 17.0% of the of the U.S. gross domestic product went toward healthcare in 2009. This underestimates the value shown in the graph by 0.3%.

b. $f(x) = 1.2 \ln x + 15.7$

$18.5 = 1.2 \ln x + 15.7$

$2.8 = 1.2 \ln x$

$\dfrac{2.8}{1.2} = \dfrac{1.2 \ln x}{1.2}$

$\dfrac{2.8}{1.2} = \ln x$

$x = e^{\frac{2.8}{1.2}}$

$x \approx 10$

If the trend continues, 18.5% of the U.S. gross domestic product will go toward healthcare 10 years after 2006, or 2016.

117. $P(x) = 95 - 30\log_2 x$

$40 = 95 - 30\log_2 x$

$30\log_2 x = 45$

$\log_2 x = 1.5$

$x = 2^{1.5} \approx 2.8$

Only half the students recall the important features of the lecture after 2.8 days.
This is represented by the point (2.8, 50).

119. a. $\text{pH} = -\log x$

$5.6 = -\log x$

$-5.6 = \log x$

$x = 10^{-5.6}$

The hydrogen ion concentration is $10^{-5.6}$ mole per liter.

b. $\text{pH} = -\log x$

$2.4 = -\log x$

$-2.4 = \log x$

$x = 10^{-2.4}$

The hydrogen ion concentration is $10^{-2.4}$ mole per liter.

c. $\dfrac{10^{-2.4}}{10^{-5.6}} = 10^{-2.4 - (-5.6)} = 10^{3.2}$

The concentration of the acidic rainfall in part (b) is $10^{3.2}$ times greater than the normal rainfall in part (a).

121. – 123. Answers will vary.

125.

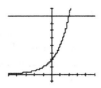

The intersection point is (2, 8).
Verify: $x = 2$

$2^{x+1} = 8$

$2^{2+1} = 2$

$2^3 = 8$

$8 = 8$

The solution set is $\{2\}$.

127.

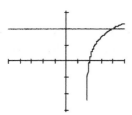

The intersection point is (4, 2).
Verify: $x = 4$

$\log_3(4 \cdot 4 - 7) = 2$

$\log_3 9 = 2$

$2 = 2$

The solution set is $\{4\}$.

129.

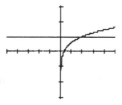

The intersection point is (2, 1).
Verify: $x = 2$

$\log(2+3) + \log 2 = 1$

$\log 5 + \log 2 = 1$

$\log(5 \cdot 2) = 1$

$\log 10 = 1$

$1 = 1$

The solution set is $\{2\}$.

131.

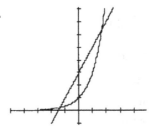

There are 2 points of intersection, approximately
$(-1.391606, 0.21678798)$ and
$(1.6855579, 6.3711158)$.
Verify $x \approx -1.391606$

$3^x = 2x + 3$

$3^{-1.391606} \approx 2(-1.391606) + 3$

$0.2167879803 \approx 0.216788$

Verify $x \approx 1.6855579$

$3^x = 2x + 3$

$3^{1.6855579} \approx 2(1.6855579) + 3$

$6.37111582 \approx 6.371158$

The solution set is $\{-1.391606, 1.6855579\}$.

133.

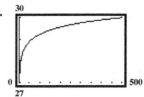

As the distance from the eye increases, barometric air pressure increases, leveling off at about 30 inches of mercury.

135.

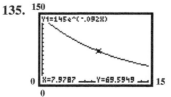

When $P = 70$, $t \approx 7.9$, so it will take about 7.9 minutes.
Verify:

$70 = 45e^{-0.092(7.9)}$

$70 \approx 70.10076749$

The runner's pulse will be 70 beats per minute after about 7.9 minutes.

137. does not make sense; Explanations will vary. Sample explanation: $2^x = 15$ requires logarithms. $2^x = 16$ can be solved by rewriting 16 as 2^4.

$$2^x = 15$$
$$\ln 2^x = \ln 15$$
$$x \ln 2 = \ln 15$$
$$x = \frac{\ln 15}{\ln 2}$$

$$2^x = 16$$
$$2^x = 2^4$$
$$x = 4$$

139. makes sense

141. false; Changes to make the statement true will vary. A sample change is: If $\log(x+3) = 2$, then $10^2 = x + 3$.

143. true

145. Account paying 3% interest:
$$A = 4000\left(1 + \frac{0.03}{1}\right)^{1 \cdot t}$$
Account paying 5% interest:
$$A = 2000\left(1 + \frac{0.05}{1}\right)^{1 \cdot t}$$
The two accounts will have the same balance when
$$4000(1.03)^t = 2000(1.05)^t$$
$$(1.03)^t = 0.5(1.05)^t$$
$$\left(\frac{1.03}{1.05}\right)^t = 0.5$$
$$\ln\left(\frac{1.03}{1.05}\right)^t = \ln 0.5$$
$$t \ln\left(\frac{1.03}{1.05}\right) = \ln 0.5$$
$$t = \frac{\ln 0.5}{\ln\left(\frac{1.03}{1.05}\right)} \approx 36$$

The accounts will have the same balance in about 36 years.

147. $(\log x)(2 \log x + 1) = 6$
$$2(\log x)^2 + \log x - 6 = 0$$
$$(2 \log x - 3)(\log x + 2) = 0$$
$$2 \log x - 3 = 0 \text{ or } \log x + 2 = 0$$
$$2 \log x = 3 \qquad \log x = -2$$
$$\log x = \frac{3}{2} \qquad x = 10^{-2}$$
$$x = 10^{3/2} \qquad x = \frac{1}{100}$$
$$x = 10\sqrt{10}$$

The solution set is $\left\{\dfrac{1}{100}, 10\sqrt{10}\right\}$.

Check by direct substitution:
$$\text{Check:} x = 10\sqrt{10} = 10^{3/2}$$
$$(\log x)(2 \log x + 1) = 6$$
$$\left(\log 10^{3/2}\right)(2 \log 10^{3/2} + 1) = 6$$
$$\left(\frac{3}{2}\right)\left(2 \cdot \frac{3}{2} + 1\right) = 6$$
$$\left(\frac{3}{2}\right)(3 + 1) = 6$$
$$\left(\frac{3}{2}\right)(4) = 6$$
$$6 = 6$$

149. Answers will vary.

150. $A = 10e^{-0.003t}$

a. 2006: $A = 10e^{-0.003(0)} = 10$ million
2007: $A = 10e^{-0.003(1)} \approx 9.97$ million
2008: $A = 10e^{-0.003(2)} \approx 9.94$ million
2009: $A = 10e^{-0.003(3)} \approx 9.91$ million

b. The population is decreasing.

151. a. $e^{\ln 3} = 3$

  b. $e^{\ln 3} = 3$

  $\left(e^{\ln 3}\right)^x = 3^x$

  $e^{(\ln 3)x} = 3^x$

152. An exponential function is the best choice.

Annual Retail Sales of
***Call of Duty* Games**

Section 4.5

Check Point Exercises

1. a. $A_0 = 807$. Since 2011 is 11 years after 2000, when $t = 11$, $A = 1052$.

$$A = A_0 e^{kt}$$

$$1052 = 807 e^{k(11)}$$

$$\frac{1052}{807} = e^{11k}$$

$$\ln\left(\frac{1052}{807}\right) = \ln e^{11k}$$

$$\ln\left(\frac{1052}{807}\right) = 11k$$

$$k = \frac{\ln\left(\frac{1052}{807}\right)}{11} \approx 0.024$$

Thus, the growth function is $A = 807 e^{0.024t}$.

b.

$$A = 807 e^{0.024t}$$

$$2000 = 807 e^{0.024t}$$

$$\frac{2000}{807} = e^{0.024t}$$

$$\ln\left(\frac{2000}{807}\right) = \ln e^{0.024t}$$

$$\ln\left(\frac{2000}{807}\right) = 0.024t$$

$$t = \frac{\ln\left(\frac{2000}{807}\right)}{0.024} \approx 38$$

Africa's population will reach 2000 million approximately 38 years after 2000, or 2038.

2. a. In the exponential decay model $A = A_0 e^{kt}$, substitute $\frac{A_0}{2}$ for A since the amount present after 28 years is half the original amount.

$$\frac{A_0}{2} = A_0 e^{k \cdot 28}$$

$$e^{28k} = \frac{1}{2}$$

$$\ln e^{28k} = \ln\frac{1}{2}$$

$$28k = \ln\frac{1}{2}$$

$$k = \frac{\ln^{1/2}}{28} \approx -0.0248$$

So the exponential decay model is $A = A_0 e^{-0.0248t}$

b. Substitute 60 for A_0 and 10 for A in the model from part (a) and solve for t.

$$10 = 60 e^{-0.0248t}$$

$$e^{-0.0248t} = \frac{1}{6}$$

$$\ln e^{-0.0248t} = \ln\frac{1}{6}$$

$$-0.0248t = \ln\frac{1}{6}$$

$$t = \frac{\ln\frac{1}{6}}{-0.0248} \approx 72$$

The strontium-90 will decay to a level of 10 grams about 72 years after the accident.

3. **a.** The time prior to learning trials corresponds to $t = 0$.

$$f(0) = \frac{0.8}{1 + e^{-0.2(0)}} = 0.4$$

The proportion of correct responses prior to learning trials was 0.4.

b. Substitute 10 for t in the model:

$$f(10) = \frac{0.8}{1 + e^{-0.2(10)}} \approx 0.7$$

The proportion of correct responses after 10 learning trials was 0.7.

c. In the logistic growth model, $f(t) = \dfrac{c}{1 + ae^{-bt}}$, the constant c represents the limiting size that $f(t)$ can attain. The limiting size of the proportion of correct responses as continued learning trials take place is 0.8.

4. A logarithmic function would be a good choice for modeling the data.

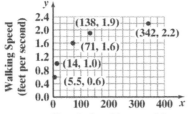

5. An exponential function would be a good choice for modeling the data although model choices may vary.

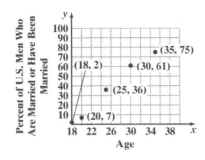

6. **a.** 1970 is 21 years after 1949.
$$f(x) = 0.074x + 2.294$$
$$f(21) = 0.074(21) + 2.294$$
$$\approx 3.8$$
$$g(x) = 2.577(1.017)^x$$
$$g(21) = 2.577(1.017)^{21}$$
$$\approx 3.7$$
The exponential function g serves as a better model for 1970.

b. 2050 is 101 years after 1949.
$$f(x) = 0.074x + 2.294$$
$$f(101) = 0.074(101) + 2.294$$
$$\approx 9.8$$
$$g(x) = 2.577(1.017)^x$$
$$g(101) = 2.577(1.017)^{101}$$
$$\approx 14.1$$
The linear function f serves as a better model for 2050.

7. $y = 4(7.8)^x$
$$= 4e^{(\ln 7.8)x}$$
Rounded to three decimal places:
$$y = 4e^{(\ln 7.8)x}$$
$$= 4e^{2.054x}$$

Concept and Vocabulary Check 4.5

1. > 0; < 0

2. A_0; A

3. A; c

4. logarithmic

5. exponential

6. linear

7. $\ln 5$

Exercise Set 4.5

1. Since 2010 is 0 years after 2010, find A when $t = 0$:

$A = 127.3e^{-0.006t}$

$A = 127.3e^{-0.006(0)}$

$A = 127.3e^0$

$A = 127.3(1)$

$A = 127.3$

In 2010, the population of Japan was 127.3 million.

3. Since $k = 0.019$, Iraq has the greatest growth rate at 1.9% per year.

5. Substitute $A = 1377$ into the model for India and solve for t:

$1377 = 1173.1e^{0.008t}$

$\dfrac{1377}{1173.1} = e^{0.008t}$

$\ln \dfrac{1377}{1173.1} = \ln e^{0.008t}$

$\ln \dfrac{1377}{1173.1} = 0.008t$

$t = \dfrac{\ln \dfrac{1377}{1173.1}}{0.008} \approx 20$

The population of India will be 1377 million approximately 20 years after 2010, or 2030.

7. **a.** $A_0 = 6.04$. Since 2050 is 50 years after 2000, when $t = 50$, $A = 10$.

$A = A_0 e^{kt}$

$10 = 6.04e^{k(50)}$

$\dfrac{10}{6.04} = e^{50k}$

$\ln \left(\dfrac{10}{6.04} \right) = \ln e^{50k}$

$\ln \left(\dfrac{10}{6.04} \right) = 50k$

$k = \dfrac{\ln \left(\dfrac{10}{6.04} \right)}{50} \approx 0.01$

Thus, the growth function is $A = 6.04e^{0.01t}$.

b. $9 = 6.04e^{0.01t}$

$\dfrac{9}{6.04} = e^{0.01t}$

$\ln \left(\dfrac{9}{6.04} \right) = \ln e^{0.01t}$

$\ln \left(\dfrac{9}{6.04} \right) = 0.01t$

$t = \dfrac{\ln \left(\dfrac{9}{6.04} \right)}{0.01} \approx 40$

Now, $2000 + 40 = 2040$, so the population will be 9 million is approximately the year 2040.

9. $P(x) = 99.9e^{0.0095t}$

$P(40) = 99.9e^{0.0095(40)}$

$P(40) = 99.9e^{0.0095(40)} \approx 146.1$

The population is projected to be 146.1 million in 2050.

11. $P(x) = 44.2e^{kt}$

$62.9 = 44.2e^{40k}$

$\dfrac{62.9}{44.2} = e^{40k}$

$\ln \left(\dfrac{62.9}{44.2} \right) = \ln e^{40k}$

$\ln \left(\dfrac{62.9}{44.2} \right) = 40k$

$\dfrac{\ln \left(\dfrac{62.9}{44.2} \right)}{40} = k$

$k \approx 0.0088$

The growth rate is 0.0088.

13. $P(x) = 82.3e^{kt}$

$70.5 = 82.3e^{40k}$

$\dfrac{70.5}{82.3} = e^{40k}$

$\ln \left(\dfrac{70.5}{82.3} \right) = \ln e^{40k}$

$\ln \left(\dfrac{70.5}{82.3} \right) = 40k$

$\dfrac{\ln \left(\dfrac{70.5}{82.3} \right)}{40} = k$

$k \approx -0.0039$

The growth rate is –0.0039.

15. $A = 16e^{-0.000121t}$

$A = 16e^{-0.000121(5715)}$

$A = 16e^{-0.691515}$

$A \approx 8.01$

Approximately 8 grams of carbon-14 will be present in 5715 years.

17. After 10 seconds, there will be $16 \cdot \dfrac{1}{2} = 8$ grams present. After 20 seconds, there will be $8 \cdot \dfrac{1}{2} = 4$ grams present. After 30 seconds, there will be $4 \cdot \dfrac{1}{2} = 2$ grams present. After 40 seconds, there will be $2 \cdot \dfrac{1}{2} = 1$ grams present. After 50 seconds, there will be $1 \cdot \dfrac{1}{2} = \dfrac{1}{2}$ gram present.

19. $A = A_0 e^{-0.000121t}$

$15 = 100e^{-0.000121t}$

$\dfrac{15}{100} = e^{-0.000121t}$

$\ln 0.15 = \ln e^{-0.000121t}$

$\ln 0.15 = -0.000121t$

$t = \dfrac{\ln 0.15}{-0.000121} \approx 15,679$

The paintings are approximately 15,679 years old.

21. $0.5 = e^{kt}$

$0.5 = e^{-0.055t}$

$\ln 0.5 = \ln e^{-0.055t}$

$\ln 0.5 = -0.055t$

$\dfrac{\ln 0.5}{-0.055} = t$

$t \approx 12.6$

The half-life is 12.6 years.

23. $0.5 = e^{kt}$

$0.5 = e^{1620k}$

$\ln 0.5 = \ln e^{1620k}$

$\ln 0.5 = 1620k$

$\dfrac{\ln 0.5}{1620} = k$

$k \approx -0.000428$

The decay rate is 0.0428% per year.

25. $0.5 = e^{kt}$

$0.5 = e^{17.5k}$

$\ln 0.5 = \ln e^{17.5k}$

$\ln 0.5 = 17.5k$

$\dfrac{\ln 0.5}{17.5} = k$

$k \approx -0.039608$

The decay rate is 3.9608% per day.

27. a. $\dfrac{1}{2} = 1e^{k1.31}$

$\ln \dfrac{1}{2} = \ln e^{1.31k}$

$\ln \dfrac{1}{2} = 1.31k$

$k = \dfrac{\ln \dfrac{1}{2}}{1.31} \approx -0.52912$

The exponential model is given by $A = A_0 e^{-0.52912t}$.

b. $A = A_0 e^{-0.52912t}$

$0.945 A_0 = A_0 e^{-0.52912t}$

$0.945 = e^{-0.52912t}$

$\ln 0.945 = \ln e^{-0.52912t}$

$\ln 0.945 = -0.52912t$

$t = \dfrac{\ln 0.945}{-0.52912} \approx 0.1069$

The age of the dinosaur ones is approximately 0.1069 billion or 106,900,000 years old.

29. First find the decay equation.

$0.5 = e^{kt}$

$0.5 = e^{22k}$

$\ln 0.5 = \ln e^{22k}$

$\ln 0.5 = 22k$

$\dfrac{\ln 0.5}{22} = k$

$k \approx -0.031507$

$A = e^{-0.031507t}$

Next use the decay equation answer question.

$$A = e^{-0.031507t}$$

$$0.8 = e^{-0.031507t}$$

$$\ln 0.8 = \ln e^{-0.031507t}$$

$$\ln 0.8 = -0.031507t$$

$$\frac{\ln 0.8}{-0.031507} = t$$

$$t \approx 7.1$$

It will take 7.1 years.

31. First find the decay equation.

$$0.5 = e^{kt}$$

$$0.5 = e^{36k}$$

$$\ln 0.5 = \ln e^{36k}$$

$$\ln 0.5 = 36k$$

$$\frac{\ln 0.5}{36} = k$$

$$k \approx -0.019254$$

$$A = e^{-0.019254t}$$

Next use the decay equation answer question.

$$A = e^{-0.019254t}$$

$$0.9 = e^{-0.019254t}$$

$$\ln 0.9 = \ln e^{-0.019254t}$$

$$\ln 0.9 = -0.019254t$$

$$\frac{\ln 0.9}{-0.019254} = t$$

$$t \approx 5.5$$

It will take 5.5 hours.

33.

$$2A_0 = A_0 e^{kt}$$

$$2 = e^{kt}$$

$$\ln 2 = \ln e^{kt}$$

$$\ln 2 = kt$$

$$t = \frac{\ln 2}{k}$$

The population will double in $t = \dfrac{\ln 2}{k}$ years.

35. $A = 4.3e^{0.01t}$

   a. $k = 0.01$, so New Zealand's growth rate is 1%.

   b.

$$A = 4.3e^{0.01t}$$

$$2 \cdot 4.3 = 4.3e^{0.01t}$$

$$2 = e^{0.01t}$$

$$\ln 2 = \ln e^{0.01t}$$

$$\ln 2 = 0.01t$$

$$t = \frac{\ln 2}{0.01} \approx 69$$

New Zealand's population will double in approximately 69 years.

37. **a.** When the epidemic began, $t = 0$.

$$f(0) = \frac{100,000}{1 + 5000e^0} \approx 20$$

Twenty people became ill when the epidemic began.

   b. $f(4) = \dfrac{100,000}{1 + 5,000e^{-4}} \approx 1080$

About 1080 people were ill at the end of the fourth week.

   c. In the logistic growth model,

$$f(t) = \frac{c}{1 + ae^{-bt}},$$

the constant c represents the limiting size that $f(t)$ can attain. The limiting size of the population that becomes ill is 100,000 people.

39. $f(x) = \dfrac{12.57}{1 + 4.11e^{-0.026(x)}}$

$$f(61) = \frac{12.57}{1 + 4.11e^{-0.026(61)}} \approx 6.8$$

The function models the data quite well.

41.

$$f(x) = \frac{12.57}{1 + 4.11e^{-0.026(x)}}$$

$$8 = \frac{12.57}{1 + 4.11e^{-0.026(x)}}$$

$$8\left(1 + 4.11e^{-0.026(x)}\right) = 12.57$$

$$8 + 32.88e^{-0.026(x)} = 12.57$$

$$32.88e^{-0.026(x)} = 4.57$$

$$e^{-0.026(x)} = \frac{4.57}{32.88}$$

$$\ln e^{-0.026(x)} = \ln \frac{4.57}{32.88}$$

$$-0.026x = \ln \frac{4.57}{32.88}$$

$$x = \frac{\ln \frac{4.57}{32.88}}{-0.026}$$

$$x \approx 76$$

The world population will reach 8 billion 76 years after 1949, or 2025.

43. $P(20) = \dfrac{90}{1 + 271e^{-0.122(20)}} \approx 3.7$

The probability that a 20-year-old has some coronary heart disease is about 3.7%.

45.

$$0.5 = \frac{0.9}{1 + 271e^{-0.122t}}$$

$$0.5\left(1 + 271e^{-0.122t}\right) = 0.9$$

$$1 + 271e^{-0.122t} = 1.8$$

$$271e^{-0.122t} = 0.8$$

$$e^{-0.122t} = \frac{0.8}{271}$$

$$\ln e^{-0.122t} = \ln \frac{0.8}{271}$$

$$-0.122t = \ln \frac{0.8}{271}$$

$$t = \frac{\ln \frac{0.8}{271}}{-0.122} \approx 48$$

The probability of some coronary heart disease is 50% at about age 48.

47. a. Scatter plot:

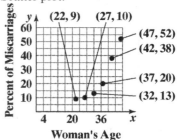

b. An exponential function appears to be the best choice for modeling the data.

49. a. Scatter plot:

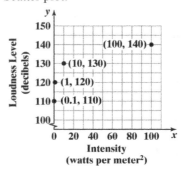

b. A logarithmic function appears to be the best choice for modeling the data.

51. a. Scatter plot:

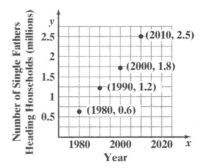

b. A linear function appears to be the best choice for modeling the data.

53. $y = 100(4.6)^x$ is equivalent to
$y = 100e^{(\ln 4.6)x}$;
Using $\ln 4.6 \approx 1.526$,
$y = 100e^{1.526x}$.

55. $y = 2.5(0.7)^x$ is equivalent to

$y = 2.5e^{(\ln 0.7)x}$;

Using $\ln 0.7 \approx -0.357$, $y = 2.5e^{-0.357x}$.

57. – 65. Answers will vary.

67. The logarithmic model is $y = 191.9 + 24.569 \ln x$.

Since $r = 0.870$ is fairly close to 1, the model fits the data okay, but not great.

69. The power regression model is $y = 195.05x^{0.100}$.

Since $r = 0.896$, the model fits the data fairly well.

71. a. Exponential Regression:

$y = 3.46(1.02)^x$; $r \approx 0.994$

Logarithmic Regression:

$y = 14.752 \ln x - 26.512$; $r \approx 0.673$

Linear Regression:

$y = 0.557x - 10.972$; $r \approx 0.947$

The exponential model has an r value closer to 1. Thus, the better model is $y = 3.46(1.02)^x$.

b. $y = 3.46(1.02)^x$

$y = 3.46e^{(\ln 1.02)x}$

$y = 3.46e^{0.02x}$

The 65-and-over population is increasing by approximately 2% each year.

73. does not make sense; Explanations will vary. Sample explanation: Since the car's value is decreasing (depreciating), the growth rate is negative.

75. makes sense

77. true

79. true

81. Use $T_0 = 210$, $C = 70$, $t = 30$, and $T = 140$ to determine the constant k:

$T = C + (T_0 - C)e^{-kt}$

$140 = 70 + (210 - 70)e^{-k(30)}$

$140 = 70 + 140e^{-30k}$

$70 = 140e^{-30k}$

$\dfrac{70}{140} = \dfrac{140e^{-30k}}{140}$

$0.5 = e^{-30k}$

$\ln 0.5 = \ln e^{-30k}$

$\ln 0.5 = -30k$

$k = \dfrac{\ln 0.5}{-30} \approx 0.0231$

Thus, the model for these conditions is

$T = 70 + 140e^{-0.0231t}$.

Evaluate the model for $t = 40$:

$T = 70 + 140e^{-0.0231(40)} \approx 126$

Thus, the temperature of the cake after 40 minutes will be approximately $126°F$.

83. $\dfrac{5\pi}{4} = 2\pi x$

$\dfrac{5\pi}{4 \cdot 2\pi} = \dfrac{2\pi x}{2\pi}$

$\dfrac{5}{8} = x$

The solution set is $\left\{ \dfrac{5}{8} \right\}$.

84. $\dfrac{17\pi}{6} - 2\pi = \dfrac{17\pi}{6} - \dfrac{12\pi}{6}$

$= \dfrac{17\pi - 12\pi}{6}$

$= \dfrac{5\pi}{6}$

85. $-\dfrac{\pi}{12} + 2\pi = -\dfrac{\pi}{12} + \dfrac{24\pi}{12}$

$= \dfrac{-\pi + 24\pi}{12}$

$= \dfrac{23\pi}{12}$

Chapter 4 Review Exercises

1. This is the graph of $f(x) = 4^x$ reflected about the *y*-axis, so the function is $g(x) = 4^{-x}$.

2. This is the graph of $f(x) = 4^x$ reflected about the *x*-axis and about the *y*-axis, so the function is $h(x) = -4^{-x}$.

3. This is the graph of $f(x) = 4^x$ reflected about the *x*-axis and about the *y*-axis then shifted upward 3 units, so the function is $r(x) = -4^{-x} + 3$.

4. This is the graph of $f(x) = 4^x$.

5. The graph of $g(x)$ shifts the graph of $f(x)$ one unit to the right.

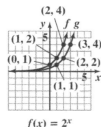

$$f(x) = 2^x$$
$$g(x) = 2^{x-1}$$

asymptote of *f*: $y = 0$
asymptote of *g*: $y = 0$
domain of f = domain of $g = (-\infty, \infty)$
range of f = range of $g = (0, \infty)$

6. The graph of $g(x)$ shifts the graph of $f(x)$ one unit down.

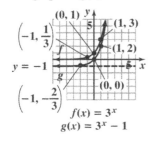

$$f(x) = 3^x$$
$$g(x) = 3^x - 1$$

asymptote of *f*: $y = 0$
asymptote of *g*: $y = -1$
domain of f = domain of $g = (-\infty, \infty)$
range of $f = (0, \infty)$
range of $g = (-1, \infty)$

7. The graph of $g(x)$ reflects the graph of *f*(x) about the *y* – axis.

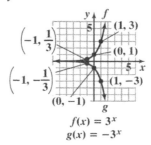

$$f(x) = 3^x$$
$$g(x) = -3^x$$

asymptote of *f*: $y = 0$
asymptote of *g*: $y = 0$
domain of f = domain of $g = (-\infty, \infty)$
range of $f = (0, \infty)$
range of $g = (-\infty, 0)$

8. The graph of $g(x)$ reflects the graph of *f*(x) about the *x* – axis.

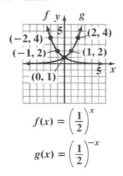

$$f(x) = \left(\frac{1}{2}\right)^x$$
$$g(x) = \left(\frac{1}{2}\right)^{-x}$$

asymptote of *f*: $y = 0$
asymptote of *g*: $y = 0$
domain of f = domain of $g = (-\infty, \infty)$
range of f = range of $g = (0, \infty)$

9. The graph of $g(x)$ vertically stretches the graph of $f(x)$ by a factor of 2.

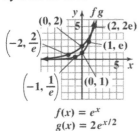

$f(x) = e^x$
$g(x) = 2e^{x/2}$

asymptote of f: $y = 0$
asymptote of g: $y = 0$
domain of f = domain of g = $(-\infty, \infty)$
range of f = range of g = $(0, \infty)$

10. 5.5% compounded semiannually:

$$A = 5000\left(1 + \frac{0.055}{2}\right)^{2 \cdot 5} \approx 6558.26$$

5.25% compounded monthly:

$$A = 5000\left(1 + \frac{0.0525}{12}\right)^{12 \cdot 5} \approx 6497.16$$

5.5% compounded semiannually yields the greater return.

11. 7% compounded monthly:

$$A = 14,000\left(1 + \frac{0.07}{12}\right)^{12 \cdot 10} \approx 28,135.26$$

6.85% compounded continuously:

$A = 14,000e^{0.0685(10)} \approx 27,772.81$
7% compounded monthly yields the greater return.

12. a. When first taken out of the microwave, the temperature of the coffee was 200°.

b. After 20 minutes, the temperature of the coffee was about 120°.
$T = 70 + 130e^{-0.04855(20)} \approx 119.23$
Using a calculator, the temperature is about 119°.

c. The coffee will cool to about 70°;
The temperature of the room is 70°.

13. $49^{1/2} = 7$

14. $4^3 = x$

15. $3^y = 81$

16. $\log_6 216 = 3$

17. $\log_b 625 = 4$

18. $\log_{13} 874 = y$

19. $\log_4 64 = 3$ because $4^3 = 64$.

20. $\log_5 \dfrac{1}{25} = -2$ because $5^{-2} = \dfrac{1}{25}$.

21. $\log_3(-9)$ is undefined and cannot be evaluated since $\log_b x$ is defined only for $x > 0$.

22. $\log_{16} 4 = \dfrac{1}{2}$ because $16^{1/2} = \sqrt{16} = 4$.

23. Because $\log_b b = 1$,
we conclude $\log_{17} 17 = 1$.

24. Because $\log_b b^x = x$,
we conclude $\log_3 3^8 = 8$.

25. Because $\ln e^x = x$,
we conclude $\ln e^5 = 5$.

26. $\log_3 \dfrac{1}{\sqrt{3}} = \log_3 \dfrac{1}{3^{\frac{1}{2}}} = \log_3 3^{-\frac{1}{2}} = -\dfrac{1}{2}$

27. $\ln \dfrac{1}{e^2} = \ln e^{-2} = -2$

28. $\log \dfrac{1}{1000} = \log \dfrac{1}{10^3} = \log 10^{-3} = -3$

29. Because $\log_b b = 1$,
we conclude $\log_8 8 = 1$.
So, $\log_3(\log_8 8) = \log_3 1$.
Because $\log_b 1 = 0$
we conclude $\log_3 1 = 0$.
Therefore, $\log_3(\log_8 8) = 0$.

30.

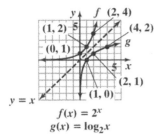

$$f(x) = 2^x$$
$$g(x) = \log_2 x$$

domain of f = range of g = $(-\infty, \infty)$

range of f = domain of g = $(0, \infty)$

31.

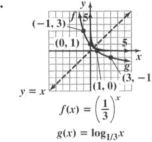

$$f(x) = \left(\frac{1}{3}\right)^x$$

$$g(x) = \log_{1/3} x$$

domain of f = range of g = $(-\infty, \infty)$

range of f = domain of g = $(0, \infty)$

32. This is the graph of $f(x) = \log x$ reflected about the y-axis, so the function is $g(x) = \log(-x)$.

33. This is the graph of $f(x) = \log x$ shifted left 2 units, reflected about the y-axis, then shifted upward one unit, so the function is $r(x) = 1 + \log(2 - x)$.

34. This is the graph of $f(x) = \log x$ shifted left 2 units then reflected about the y-axis, so the function is $h(x) = \log(2 - x)$.

35. This is the graph of $f(x) = \log x$.

36.

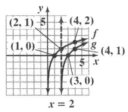

$$f(x) = \log_2 x$$
$$g(x) = \log_2(x - 2)$$

x-intercept: $(3, 0)$
vertical asymptote: $x = 2$
domain: $(2, \infty)$

range: $(-\infty, \infty)$

37.

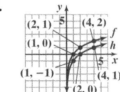

$$f(x) = \log_2 x$$
$$h(x) = -1 + \log_2 x$$

x-intercept: $(2, 0)$
vertical asymptote: $x = 0$
domain: $(0, \infty)$

range: $(-\infty, \infty)$

38.

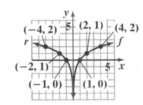

$$f(x) = \log_2 x$$
$$r(x) = \log_2(-x)$$

x-intercept: $(-1, 0)$
vertical asymptote: $x = 0$
domain: $(-\infty, 0)$

range: $(-\infty, \infty)$

39.

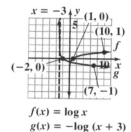

$f(x) = \log x$
$g(x) = -\log(x+3)$

asymptote of f: $x = 0$
asymptote of g: $x = -3$
domain of $f = (0, \infty)$

domain of $g = (-3, \infty)$

range of f = range of $g = (-\infty, \infty)$

40.

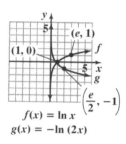

$f(x) = \ln x$
$g(x) = -\ln(2x)$

asymptote of f: $x = 0$
asymptote of g: $x = 0$
domain of f = domain of $g = (0, \infty)$

range of f = range of $g = (-\infty, \infty)$

41. The domain of f consists of all x for which $x + 5 > 0$.
Solving this inequality for x, we obtain $x > -5$.
Thus the domain of f is $(-5, \infty)$

42. The domain of f consists of all x for which $3 - x > 0$.
Solving this inequality for x, we obtain $x < 3$.
Thus, the domain of f is $(-\infty, 3)$.

43. The domain of f consists of all x for which
$(x-1)^2 > 0$.
Solving this inequality for x, we obtain
$x < 1$ or $x > 1$.
Thus, the domain of f is $(-\infty, 1) \cup (1, \infty)$.

44. Because $\ln e^x = x$, we conclude $\ln e^{6x} = 6x$.

45. Because $e^{\ln x} = x$, we conclude $e^{\ln \sqrt{x}} = \sqrt{x}$.

46. Because $10^{\log x} = x$, we conclude $10^{\log 4x^2} = 4x^2$.

47. $R = \log \dfrac{1000 I_0}{I_0} = \log 1000 = 3$
The Richter scale magnitude is 3.0.

48. a. $f(0) = 76 - 18\log(0+1) = 76$
When first given, the average score
was 76.

b. $f(2) = 76 - 18\log(2+1) \approx 67$
$f(4) = 76 - 18\log(4+1) \approx 63$
$f(6) = 76 - 18\log(6+1) \approx 61$
$f(8) = 76 - 18\log(8+1) \approx 59$
$f(12) = 76 - 18\log(12+1) \approx 56$
After 2, 4, 6, 8, and 12 months, the average
scores are about 67, 63, 61, 59, and 56,
respectively.

c.

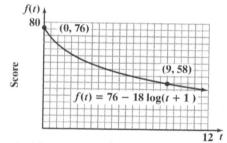

Retention decreases as time passes.

49. $t = \dfrac{1}{0.06}\ln\left(\dfrac{12}{12-5}\right) \approx 8.98$
It will take about 9 weeks.

50. $\log_6\left(36x^3\right)$
$= \log_6 36 + \log_6 x^3$
$= \log_6 36 + 3\log_6 x$
$= 2 + 3\log_6 x$

51. $\log_4 \dfrac{\sqrt{x}}{64} = \log_4 x^{1/2} - \log_4 64$
$= \dfrac{1}{2}\log_4 x - 3$

52. $\log_2 \dfrac{xy^2}{64} = \log_2 xy^2 - \log_2 64$
$= \log_2 x + \log_2 y^2 - \log_2 64$
$= \log_2 x + 2\log_2 y - 6$

53. $\ln \sqrt[3]{\dfrac{x}{e}}$

$= \ln\left(\dfrac{x}{e}\right)^{1/3}$

$= \dfrac{1}{3}\left[\ln x - \ln e\right]$

$= \dfrac{1}{3}\ln x - \dfrac{1}{3}\ln e$

$= \dfrac{1}{3}\ln x - \dfrac{1}{3}$

54. $\log_b 7 + \log_b 3$

$= \log_b (7 \cdot 3)$

$= \log_b 21$

55. $\log 3 - 3\log x$

$= \log 3 - \log x^3$

$= \log \dfrac{3}{x^3}$

56. $3\ln x + 4\ln y$

$= \ln x^3 + \ln y^4$

$= \ln\left(x^3 y^4\right)$

57. $\dfrac{1}{2}\ln x - \ln y$

$= \ln x^{1/2} - \ln y$

$= \ln \dfrac{\sqrt{x}}{y}$

58. $\log_6 72,348 = \dfrac{\log 72,348}{\log 6} \approx 6.2448$

59. $\log_4 0.863 = \dfrac{\ln 0.863}{\ln 4} \approx -0.1063$

60. true; $(\ln x)(\ln 1) = (\ln x)(0) = 0$

61. false; $\log(x+9) - \log(x+1) = \log\dfrac{(x+9)}{(x+1)}$

62. false; $\log_2 x^4 = 4\log_2 x$

63. true; $\ln e^x = x\ln e$

64. $2^{4x-2} = 64$

$2^{4x-2} = 2^6$

$4x - 2 = 6$

$4x = 8$

$x = 2$

65. $125^x = 25$

$\left(5^3\right)^x = 5^2$

$5^{3x} = 5^2$

$3x = 2$

$x = \dfrac{2}{3}$

66. $10^x = 7000$

$\log 10^x = \log 7000$

$x\log 10 = \log 7000$

$x = \log 7000$

$x \approx 3.85$

67. $9^{x+2} = 27^{-x}$

$\left(3^2\right)^{x+2} = \left(3^3\right)^{-x}$

$3^{2x+4} = 3^{-3x}$

$2x + 4 = -3x$

$5x = -4$

$x = -\dfrac{4}{5}$

68. $8^x = 12,143$

$\ln 8^x = \ln 12,143$

$x \ln 8 = \ln 12,143$

$x = \dfrac{\ln 12,143}{\ln 8} \approx 4.52$

69. $9e^{5x} = 1269$

$e^{5x} = 141$

$\ln e^{5x} = \ln 141$

$5x = \ln 141$

$x = \dfrac{\ln 141}{5}$

$x = \dfrac{1}{5}\ln 141 \approx 0.99$

70. $e^{12-5x} - 7 = 123$

$e^{12-5x} = 130$

$\ln e^{12-5x} = \ln 130$

$12 - 5x = \ln 130$

$5x = 12 - \ln 130$

$x = \dfrac{12 - \ln 130}{5} \approx 1.43$

71. $\qquad 5^{4x+2} = 37{,}500$

$\ln 5^{4x+2} = \ln 37{,}500$

$(4x+2)\ln 5 = \ln 37{,}500$

$4x\ln 5 + 2\ln 5 = \ln 37{,}500$

$4x\ln 5 = \ln 37{,}500 - 2\ln 5$

$x = \dfrac{\ln 37{,}500 - 2\ln 5}{4\ln 5} \approx 1.14$

72. $\qquad 3^{x+4} = 7^{2x-1}$

$\ln 3^{x+4} = \ln 7^{2x-1}$

$(x+4)\ln 3 = (2x-1)\ln 7$

$x\ln 3 + 4\ln 3 = 2x\ln 7 - \ln 7$

$x\ln 3 - 2x\ln 7 = -4\ln 3 - \ln 7$

$x(\ln 3 - 2\ln 7) = -4\ln 3 - \ln 7$

$x = \dfrac{-4\ln 3 - \ln 7}{\ln 3 - 2\ln 7}$

$x = \dfrac{4\ln 3 + \ln 7}{2\ln 7 - \ln 3}$

$x \approx 2.27$

73. $e^{2x} - e^x - 6 = 0$

$\left(e^x - 3\right)\left(e^x + 2\right) = 0$

$e^x - 3 = 0 \ $ or $\ e^x + 2 = 0$

$e^x = 3 \qquad\quad e^x = -2$

$\ln e^x = \ln 3 \quad \ln e^x - \ln(-2)$

$x = \ln 3 \qquad\quad x = \ln(-2)$

$x = \ln 3 \approx 1.099 \ \ \ln(-2)$ does not exist.

The solution set is $\{\ln 3\}$,

approximately 1.10.

74. $\log_4 (3x - 5) = 3$

$3x - 5 = 4^3$

$3x - 5 = 64$

$3x = 69$

$x = 23$

The solutions set is $\{23\}$.

75. $3 + 4\ln(2x) = 15$

$4\ln(2x) = 12$

$\ln(2x) = 3$

$2x = e^3$

$x = \dfrac{e^3}{2}$

$x \approx 10.04$

The solutions set is $\left\{ \dfrac{e^3}{2} \right\}$.

76. $\log_2 (x + 3) + \log_2 (x - 3) = 4$

$\log_2 (x+3)(x-3) = 4$

$\log_2 (x^2 - 9) = 4$

$x^2 - 9 = 2^4$

$x^2 - 9 = 16$

$x^2 = 25$

$x = \pm 5$

$x = -5$ does not check because $\log_2(-5+3)$ does not exist.

The solution set is $\{5\}$.

77. $\log_3 (x - 1) - \log_3 (x + 2) = 2$

$\log_3 \dfrac{x-1}{x+2} = 2$

$\dfrac{x-1}{x-2} = 3^2$

$\dfrac{x-1}{x+2} = 9$

$x - 1 = 9(x + 2)$

$x - 1 = 9x + 18$

$8x = -19$

$x = -\dfrac{19}{8}$

$x = -\dfrac{19}{8}$ does not check because $\log_3\left(-\dfrac{19}{8} - 1 \right)$

does not exist.

The solution set is $\varnothing$.

78. $\ln(x+4)-\ln(x+1)=\ln x$

$$\ln\frac{x+4}{x+1}=\ln x$$

$$\frac{x+4}{x+1}=x$$

$$x(x+1)=x+4$$

$$x^2+x=x+4$$

$$x^2=4$$

$$x=\pm 2$$

$x=-2$ does not check and must be rejected. The solution set is $\{2\}$.

79. $\log_4(2x+1)=\log_4(x-3)+\log_4(x+5)$

$$\log_4(2x+1)=\log_4(x-3)+\log_4(x+5)$$

$$\log_4(2x+1)=\log_4\left(x^2+2x-15\right)$$

$$2x+1=x^2+2x-15$$

$$16=x^2$$

$$x^2=16$$

$$x=\pm 4$$

$x=-4$ does not check and must be rejected. The solution set is $\{4\}$.

80. $P(x)=14.7e^{-0.21x}$

$$4.6=14.7e^{-0.21x}$$

$$\frac{4.6}{14.7}=e^{-0.21x}$$

$$\ln\frac{4.6}{14.7}=\ln e^{-0.21x}$$

$$\ln\frac{4.6}{14.7}=-0.21x$$

$$t=\frac{\ln\dfrac{4.6}{14.7}}{-0.21}\approx 5.5$$

The peak of Mt. Everest is about 5.5 miles above sea level.

81. $S(t)=21.4(1.56)^t$

$$1170.9=21.4(1.56)^t$$

$$\frac{1170.9}{21.4}=(1.56)^t$$

$$\ln\frac{1170.9}{21.4}=\ln(1.56)^t$$

$$\ln\frac{1170.9}{21.4}=t\ln 1.56$$

$$t=\frac{\ln\dfrac{1170.9}{21.4}}{\ln 1.56}\approx 9$$

This model projects that 1170.9 million smartphones will be sold 9 years after 2004, or 2013.

82. a. $f(x)=-6.2\ln x+40.5$

$$16=-6.2\ln x+40.5$$

$$-24.5=-6.2\ln x$$

$$\frac{-24.5}{-6.2}=\ln x$$

$$\frac{24.5}{6.2}=\ln x$$

$$e^{\frac{24.5}{6.2}}=e^{\ln x}$$

$$x=e^{\frac{24.5}{6.2}}\approx 52$$

This model projects that opposition to homosexual relationships among first-year college women will diminish to 16% 52 years after 1979, or 2031.

b. $g(x)=-7\ln x+59$

$$40=-7\ln x+59$$

$$-19=-7\ln x$$

$$\frac{-19}{-7}=\ln x$$

$$e^{\frac{19}{7}}=e^{\ln x}$$

$$x=e^{\frac{19}{7}}\approx 15$$

This model projects that opposition to homosexual relationships among first-year college men diminish to 40% 15 years after 1979, or 1994.

83. $20,000 = 12,500\left(1+\dfrac{0.065}{4}\right)^{4t}$

$12,500(1.01625)^{4t} = 20,000$

$(1.01625)^{4t} = 1.6$

$\ln(1.01625)^{4t} = \ln 1.6$

$4t\ln 1.01625 = \ln 1.6$

$t = \dfrac{\ln 1.6}{4\ln 1.01625} \approx 7.3$

It will take about 7.3 years.

84. $3 \cdot 50,000 = 50,000e^{0.075t}$

$50,000e^{0.075t} = 150,000$

$e^{0.075} = 3$

$\ln e^{0.075t} = \ln 3$

$0.075t = \ln 3$

$t = \dfrac{\ln 3}{0.075} \approx 14.6$

It will take about 14.6 years.

85. When an investment value triples, $A = 3P$.

$3P = Pe^{5r}$

$e^{5r} = 3$

$\ln e^{5r} = \ln 3$

$5r = \ln 3$

$r = \dfrac{\ln 3}{5} \approx 0.2197$

The interest rate would need to be about 22%

86. a. $50.5 = 35.3e^{k10}$

$\dfrac{50.5}{35.3} = e^{10k}$

$\ln \dfrac{50.5}{35.3} = \ln e^{10k}$

$\ln \dfrac{50.5}{35.3} = 10k$

$\dfrac{\ln \dfrac{50.5}{35.3}}{10} = k$

$0.036 \approx k$

$A = 35.3e^{0.036t}$

b. $A = 35.3e^{0.036(15)} \approx 60.6$

In 2015, the population will be about 60.6 million.

c. $70 = 35.3e^{0.036t}$

$\dfrac{70}{35.3} = e^{0.036t}$

$\ln \dfrac{70}{35.3} = \ln e^{0.036t}$

$\ln \dfrac{70}{35.3} = 0.036t$

$\dfrac{\ln \dfrac{70}{35.3}}{0.036} = t$

$19 \approx t$

The population will reach 70 million about 19 years after 2000, in 2019.

87.   Use the half-life of 140 days to find k.

$A = A_0 e^{kt}$

$\tfrac{1}{2} = e^{k \cdot 140}$

$\tfrac{1}{2} = e^{140k}$

$\ln \tfrac{1}{2} = \ln e^{140k}$

$\ln \tfrac{1}{2} = 140k$

$\dfrac{\ln \tfrac{1}{2}}{140} = k$

$k \approx -0.004951$

Use $A = A_0 e^{kt}$ to find t.

$A = A_0 e^{-0.004951t}$

$0.2 = e^{-0.004951t}$

$\ln 0.2 = \ln e^{-0.004951t}$

$\ln 0.2 = -0.004951t$

$t = \dfrac{\ln 0.2}{-0.004951}$

$t \approx 325$

It will take about 325 days for the substance to decay to 20% of its original amount.

88. a. $f(0) = \dfrac{500,000}{1+2499e^{-0.92(0)}} = 200$

200 people became ill when the epidemic began.

b. $f(6) = \dfrac{500,000}{1+2499e^{-0.92(6)}} = 45,411$

45,410 were ill after 6 weeks.

c.   500,000 people

89. a. Scatter plot:

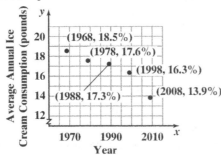

b. A linear function appears to be the better choice for modeling the data.

90. a. Scatter plot:

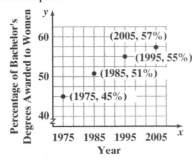

b. An logarithmic function appears to be the better choice for modeling the data.

91. a. Scatter plot:

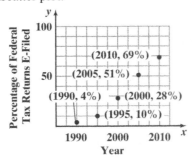

b. An exponential function appears to be the better choice for modeling the data.

92. $y = 73(2.6)^x$

$y = 73e^{(\ln 2.6)x}$

$y = 73e^{0.956x}$

93. $y = 6.5(0.43)^x$

$y = 6.5e^{(\ln 0.43)x}$

$y = 6.5e^{-0.844x}$

94. Answers will vary.

Chapter 4 Test

1.

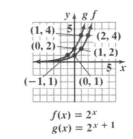

$f(x) = 2^x$
$g(x) = 2^x + 1$

2.

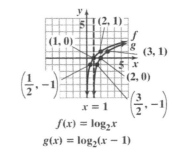

$f(x) = \log_2 x$
$g(x) = \log_2(x - 1)$

3. $5^3 = 125$

4. $\log_{36} 6 = \dfrac{1}{2}$

5. The domain of f consists of all x for which $3 - x > 0$. Solving this inequality for x, we obtain $x < 3$.
Thus, the domain of f is $(-\infty, 3)$.

6. $\log_4\left(64x^5\right) = \log_4 64 + \log_4 x^5$
$= 3 + 5\log_4 x$

7. $\log_3 \dfrac{\sqrt[3]{x}}{81} = \log_3 x^{\frac{1}{3}} - \log_3 81$
$= \dfrac{1}{3}\log_3 x - 4$

8. $6\log x + 2\log y = \log x^6 + \log y^2$
$= \log\left(x^6 y^2\right)$

9. $\ln 7 - 3\ln x = \ln 7 - \ln x^3$
$= \ln \dfrac{7}{x^3}$

10. $\log_{15} 71 = \dfrac{\log 71}{\log 15} \approx 1.5741$

11. $3^{x-2} = 9^{x+4}$

$3^{x-2} = \left(3^2\right)^{x+4}$

$3^{x-2} = 3^{2x+8}$

$x-2 = 2x+8$

$-x = 10$

$x = -10$

12. $5^x = 1.4$

$\ln 5^x = \ln 1.4$

$x \ln 5 = \ln 1.4$

$x = \dfrac{\ln 1.4}{\ln 5} \approx 0.2091$

13. $400e^{0.005x} = 1600$

$e^{0.005x} = 4$

$\ln e^{0.005x} = \ln 4$

$0.005x = \ln 4$

$x = \dfrac{\ln 4}{0.005} \approx 277.2589$

14. $e^{2x} - 6e^x + 5 = 0$

$\left(e^x - 5\right)\left(e^x - 1\right) = 0$

$e^x - 5 = 0 \qquad$ or $\qquad e^x - 1 = 0$

$ e^x = 5 \qquad\qquad\qquad e^x = 1$

$\ln e^x = \ln 5 \qquad\qquad \ln e^x = \ln 1$

$ x = \ln 5 \qquad\qquad\quad x = \ln 1$

$ x \approx 1.6094 \qquad\qquad\; x = 0$

The solution set is $\{0, \ln 5\}$; $\ln \approx 1.6094$.

15. $\log_6 (4x - 1) = 3$

$4x - 1 = 6^3$

$4x - 1 = 216$

$4x = 217$

$x = \dfrac{217}{4} = 54.25$

16. $2 \ln 3x = 8$

$\ln 3x = 4$

$3x = e^4$

$x = \dfrac{e^4}{3} \approx 18.1994$

17. $\log x + \log (x + 15) = 2$

$\log \left(x^2 + 15x\right) = 2$

$x^2 + 15x = 10^2$

$x^2 + 15x - 100 = 0$

$(x + 20)(x - 5) = 0$

$x + 20 = 0$ or $x - 5 = 0$

$x = -20 \qquad x = 5$

$x = -20$ does not check because $\log(-20)$ does not exist.

The solution set is $\{5\}$.

18. $\ln (x - 4) - \ln (x + 1) = \ln 6$

$\ln \dfrac{x-4}{x+1} = \ln 6$

$\dfrac{x-4}{x+1} = 6$

$6(x + 1) = x - 4$

$6x + 6 = x - 4$

$5x = -10$

$x = -2$

$x = -2$ does not check and must be rejected.

The solution set is $\varnothing$.

19. $D = 10 \log \dfrac{10^{12} I_0}{I_0}$

$ = 10 \log 10^{12}$

$ = 10 \cdot 12$

$ = 120$

The loudness of the sound is 120 decibels.

20. Since $\ln e^x = x$, $\ln e^{5x} = 5x$.

21. $\log_b b = 1$ because $b^1 = b$.

22. $\log_6 1 = 0$ because $6^0 = 1$.

23. 6.5% compounded semiannually:

$A = 3,000 \left(1 + \dfrac{0.065}{2}\right)^{2(10)} \approx \$5,687.51$

6% compounded continuously:

$A = 3,000 e^{0.06(10)} \approx \$5,466.36$

6.5% compounded semiannually yields about \$221 more than 6% compounded continuously.

24.

$$8000 = 4000\left(1+\frac{0.05}{4}\right)^{4t}$$

$$\frac{8000}{4000} = (1+0.0125)^{4t}$$

$$2 = (1.0125)^{4t}$$

$$\ln 2 = \ln(1.0125)^{4t}$$

$$\ln 2 = 4t\ln(1.0125)$$

$$\frac{\ln 2}{4\ln(1.0125)} = \frac{4t\ln(1.0125)}{4\ln(1.0125)}$$

$$t = \frac{\ln 2}{4\ln(1.0125)} \approx 13.9$$

It will take approximately 13.9 years for the money to grow to $8000.

25.

$$2 = 1e^{r10}$$

$$2 = e^{10r}$$

$$\ln 2 = \ln e^{10r}$$

$$\ln 2 = 10r$$

$$r = \frac{\ln 2}{10} \approx 0.069$$

The money will double in 10 years with an interest rate of approximately 6.9%.

26. a.

$$A = 82.3e^{-0.004(x)}$$

$$A = 82.3e^{-0.004(0)} \approx 82.3$$

In 2010, the population of Germany was 82.3 million.

b. The population of Germany is decreasing. We can tell because the model has a negative $k = -0.004$.

c.

$$79.1 = 82.3e^{-0.002t}$$

$$\frac{79.1}{82.3} = e^{-0.004t}$$

$$\ln\frac{79.1}{82.3} = \ln e^{-0.004t}$$

$$\ln\frac{79.1}{82.3} = -0.004t$$

$$t = \frac{\ln\frac{79.1}{82.3}}{-0.004} \approx 10$$

The population of Germany will be 79.1 million approximately 10 years after 2010 in the year 2020.

27. In 2010, $t = 0$ and $A_0 = 4121$
In 2050, $t = 2050 - 2010 = 40$ and $A = 5231$.

$$5231 = 4121e^{k40}$$

$$\frac{5231}{4121} = e^{40k}$$

$$\ln\frac{5231}{4121} = \ln e^{40k}$$

$$\ln\frac{5231}{4121} = 40k$$

$$\frac{\ln\frac{5231}{4121}}{40} = k$$

$$0.006 \approx k$$

The exponential growth function is
$A = 4121e^{0.006t}$.

28. First find the decay equation.

$$0.5 = e^{kt}$$

$$0.5 = e^{7.2k}$$

$$\ln 0.5 = \ln e^{7.2k}$$

$$\ln 0.5 = 7.2k$$

$$\frac{\ln 0.5}{7.2} = k$$

$$k \approx -0.096270$$

$$A = e^{-0.096270t}$$

Next use the decay equation answer question.

$$A = e^{-0.096270t}$$

$$0.3 = e^{-0.096270t}$$

$$\ln 0.3 = \ln e^{-0.096270t}$$

$$\ln 0.3 = -0.096270t$$

$$\frac{\ln 0.3}{-0.096270} = t$$

$$t \approx 12.5$$

It will take 12.5 days.

29. a.

$$f(0) = \frac{140}{1+9e^{-0.165(0)}} = 14$$

Fourteen elk were initially introduced to the habitat.

b.

$$f(10) = \frac{140}{1+9e^{-0.165(10)}} \approx 51$$

After 10 years, about 51 elk are expected.

c. In the logistic growth model,

$$f(t) = \frac{c}{1+ae^{-bt}},$$

the constant c represents the limiting size that $f(t)$ can attain. The limiting size of the elk population is 140 elk.

30. Plot the ordered pairs.

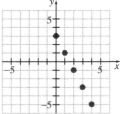

The values appear to belong to a linear function.

31. Plot the ordered pairs.

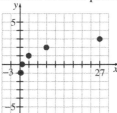

The values appear to belong to a logarithmic function.

32. Plot the ordered pairs.

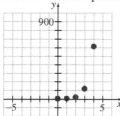

The values appear to belong to an exponential function.

33. Plot the ordered pairs.

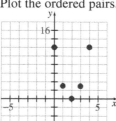

The values appear to belong to a quadratic function.

34. $y = 96(0.38)^x$

$y = 96e^{(\ln 0.38)x}$

$y = 96e^{-0.968x}$

Cumulative Review Exercises (Chapters 1–4)

1. $|3x-4| = 2$

$3x - 4 = 2$ or $3x - 4 = -2$

$3x = 6$ $\qquad 3x = 2$

$x = 2$ $\qquad x = \dfrac{2}{3}$

The solution set is $\left\{\dfrac{2}{3}, 2\right\}$.

2. $\sqrt{2x-5} - \sqrt{x-3} = 1$

$\sqrt{2x-5} = 1 + \sqrt{x-3}$

$\left(\sqrt{2x-5}\right)^2 = \left(1+\sqrt{x-3}\right)^2$

$2x - 5 = 1 + 2\sqrt{x-3} + x - 3$

$2x - 5 = x - 2 + 2\sqrt{x-3}$

$x - 3 = 2\sqrt{x-3}$

$(x-3)^2 = \left(2\sqrt{x-3}\right)^2$

$(x-3)^2 = 4(x-3)$

$x^2 - 6x + 9 = 4x - 12$

$x^2 - 10x + 21 = 0$

$(x-3)(x-7) = 0$

$x = 3$ or $x = 7$

Both solutions satisfy the original equation when checked.

The solution set is $\{3, 7\}$.

3. $x^4 + x^3 - 3x^2 - x + 2 = 0$

$p: \pm 1, \pm 2$

$q: \pm 1$

$\dfrac{p}{q}: \pm 1, \pm 2$

$$\begin{array}{r|rrrrr} -2 & 1 & 1 & -3 & -1 & 2 \\ & & -2 & 2 & 2 & -2 \\ \hline & 1 & -1 & -1 & 1 & 0 \end{array}$$

$(x+2)(x^3 - x^2 - x + 1) = 0$

$(x+2)[x^2(x-1) - (x-1)] = 0$

$(x+2)(x^2 - 1)(x-1) = 0$

$(x+2)(x+1)(x-1)(x-1) = 0$

$(x+2)(x+1)(x-1)^2 = 0$

$x+2 = 0 \quad$ or $\quad x+1 = 0 \quad$ or $\quad x-1 = 0$

$x = -2 \qquad\qquad x = -1 \qquad\qquad x = 1$

The solution set is $\{-2, -1, 1\}$.

4. $e^{5x} - 32 = 96$

$e^{5x} = 128$

$\ln e^{5x} = \ln 128$

$5x = \ln 128$

$x = \dfrac{\ln 128}{5} \approx 0.9704$

The solution set is $\left\{ \dfrac{\ln 128}{5} \right\}$, approximately 0.9704.

5. $\log_2(x+5) + \log_2(x-1) = 4$

$\log_2[(x+5)(x-1)] = 4$

$(x+5)(x-1) = 2^4$

$x^2 + 4x - 5 = 16$

$x^2 + 4x - 21 = 0$

$(x+7)(x-3) = 0$

$x+7 = 0 \quad$ or $\quad x-3 = 0$

$x = -7 \qquad\qquad x = 3$

$x = -7$ does not check because $\log_2(-7+5)$ does not exist.

The solution set is $\{3\}$.

6. $\ln(x+4) + \ln(x+1) = 2\ln(x+3)$

$\ln\big((x+4)(x+1)\big) = \ln(x+3)^2$

$(x+4)(x+1) = (x+3)^2$

$x^2 + 5x + 4 = x^2 + 6x + 9$

$5x + 4 = 6x + 9$

$-x = 5$

$x = -5$

$x = -5$ does not check and must be rejected.

The solution set is $\varnothing$.

7. $14 - 5x \geq -6$

$-5x \geq -20$

$x \leq 4$

The solution set is $(-\infty, 4]$.

8. $|2x - 4| \leq 2$

$2x - 4 \leq 2$ and $2x - 4 \geq -2$

$2x \leq 6 \qquad\qquad 2x \geq 2$

$x \leq 3 \qquad$ and $\quad x \geq 1$

The solution set is $[1, 3]$.

9. Circle with center: $(3, -2)$ and radius of 2

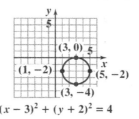

$(x-3)^2 + (y+2)^2 = 4$

10. Parabola with vertex: $(2, -1)$

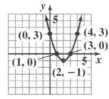

$f(x) = (x-2)^2 - 1$

11. *x*-intercepts:

$$x^2 - 1 = 0$$
$$x^2 = 1$$
$$x = \pm 1$$

The *x*-intercepts are $(1,0)$ and $(-1,0)$.

vertical asymptotes:

$$x^2 - 4 = 0$$
$$x^2 = 4$$
$$x = \pm 2$$

The vertical asymptotes are $x = 2$ and $x = -2$.

Horizontal asymptote: $y = 1$

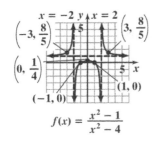

$$f(x) = \frac{x^2 - 1}{x^2 - 4}$$

12. *x*-intercepts:

$$x - 2 = 0 \quad \text{or} \quad x + 1 = 0$$
$$x = 2 \quad\quad \text{or} \quad x = -1$$

The *x*-intercepts are $(2,0)$ and $(-1,0)$.

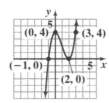

$$f(x) = (x - 2)^2(x + 1)$$

13.

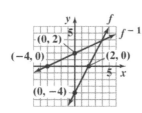

$$f(x) = 2x - 4$$
$$f^{-1}(x) = \frac{x + 4}{2}$$

14.

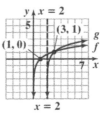

$$f(x) = \ln x$$
$$g(x) = \ln (x - 2) + 1$$

15. $m = \dfrac{3 - (-3)}{1 - 3} = \dfrac{6}{-2} = -3$

Using (1, 3) point-slope form:

$$y - 3 = -3(x - 1)$$

slope-intercept form:

$$y - 3 = -3(x - 1)$$
$$y - 3 = -3x + 3$$
$$y = -3x + 6$$

16. $(f \circ g)(x) = f(x + 2)$
$$= (x + 2)^2$$
$$= x^2 + 4x + 4$$

$(g \circ f)(x) = g(x^2)$
$$= x^2 + 2$$

17. *y* varies inversely as the square of *x* is expressed as

$$y = \frac{k}{x^2}.$$

The hours, *H* , vary inversely as the square of the number of cups of coffee, *C* can be expressed

as $H = \dfrac{k}{C^2}$.

Use the given values to find *k* .

$$H = \frac{k}{C^2}$$
$$8 = \frac{k}{2^2}$$
$$32 = k$$

Substitute the value of *k* into the equation.

$$H = \frac{k}{C^2}$$
$$H = \frac{32}{C^2}$$

Use the equation to find H when $C = 4$.

$$H = \frac{32}{C^2}$$

$$H = \frac{32}{4^2}$$

$$H = 2$$

If 4 cups of coffee are consumed you should expect to sleep 2 hours.

18. $s(t) = -16t^2 + 64t + 5$

The ball reaches its maximum height at

$$t = \frac{-b}{2a} = \frac{-(64)}{2(-16)} = 2 \text{ seconds.}$$

The maximum height is $s(2)$.

$$s(2) = -16(2)^2 + 64(2) + 5 = 69 \text{ feet.}$$

19. $s(t) = -16t^2 + 64t + 5$

Let $s(t) = 0$:

$$0 = -16t^2 + 64t + 5$$

Use the quadratic formula to solve.

$$t = \frac{-b \pm \sqrt{b^2 - 4ac}}{2a}$$

$$t = \frac{-(64) \pm \sqrt{(64)^2 - 4(-16)(5)}}{2(-16)}$$

$$t \approx 4.1, \quad t \approx -0.1$$

The negative value is rejected.
The ball hits the ground after about 4.1 seconds.

20. $40x + 10(1.5x) = 660$

$$40x + 15x = 660$$

$$55x = 660$$

$$x = 12$$

Your normal hourly salary is $12 per hour.

Section 5.1

Check Point Exercises

1. The radian measure of a central angle is the length of the intercepted arc, *s*, divided by the circle's radius, *r*. The length of the intercepted arc is 42 feet: $s = 42$ feet. The circle's radius is 12 feet: $r = 12$ feet. Now use the formula for radian measure to find the radian measure of θ.

$$\theta = \frac{s}{r} = \frac{42 \text{ feet}}{12 \text{ feet}} = 3.5$$

Thus, the radian measure of θ is 3.5

2. **a.** $60° = 60° \cdot \frac{\pi \text{ radians}}{180°} = \frac{60\pi}{180}$ radians

$= \frac{\pi}{3}$ radians

b. $270° = 270° \cdot \frac{\pi \text{ radians}}{180°} = \frac{270\pi}{180}$ radians

$= \frac{3\pi}{2}$ radians

c. $-300° = -300° \cdot \frac{\pi \text{ radians}}{180°} = \frac{-300\pi}{180}$ radians

$= -\frac{5\pi}{3}$ radians

3. **a.** $\frac{\pi}{4}$ radians $= \frac{\pi \text{ radians}}{4} \cdot \frac{180°}{\pi \text{ radians}}$

$= \frac{180°}{4} = 45°$

b. $-\frac{4\pi}{3}$ radians $= -\frac{4\pi \text{ radians}}{3} \cdot \frac{180°}{\pi}$

$= -\frac{4 \cdot 180°}{3} = -240°$

c. 6 radians $= 6$ radians $\cdot \frac{180°}{\pi \text{ radians}}$

$= \frac{6 \cdot 180°}{\pi} \approx 343.8°$

4. **a.**

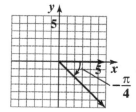

b.

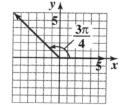

c.

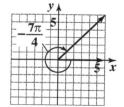

d.

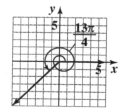

5. **a.** For a 400° angle, subtract 360° to find a positive coterminal angle.
$400° - 360° = 40°$

b. For a −135° angle, add 360° to find a positive coterminal angle.
$-135° + 360° = 225°$

6. **a.** $\frac{13\pi}{5} - 2\pi = \frac{13\pi}{5} - \frac{10\pi}{5} = \frac{3\pi}{5}$

b. $-\frac{\pi}{15} + 2\pi = -\frac{\pi}{15} + \frac{30\pi}{15} = \frac{29\pi}{15}$

7. **a.** $855° - 360° \cdot 2 = 855° - 720° = 135°$

b. $\frac{17\pi}{3} - 2\pi \cdot 2 = \frac{17\pi}{3} - 4\pi$

$= \frac{17\pi}{3} - \frac{12\pi}{3} = \frac{5\pi}{3}$

c. $-\dfrac{25\pi}{6}+2\pi\cdot3=-\dfrac{25\pi}{6}+6\pi$

$\qquad =-\dfrac{25\pi}{6}+\dfrac{36\pi}{6}=\dfrac{11\pi}{6}$

8. The formula $s=r\theta$ can only be used when θ is expressed in radians. Thus, we begin by converting $45°$ to radians. Multiply by $\dfrac{\pi\text{ radians}}{180°}$.

$45°=45°\cdot\dfrac{\pi\text{ radians}}{180°}=\dfrac{45}{180}\pi\text{ radians}$

$\qquad =\dfrac{\pi}{4}\text{ radians}$

Now we can use the formula $s=r\theta$ to find the length of the arc. The circle's radius is 6 inches : $r=6$ inches. The measure of the central angle in radians is $\dfrac{\pi}{4}:\theta=\dfrac{\pi}{4}$. The length of the arc intercepted by this central angle is

$s=r\theta$

$\qquad =(6\text{ inches})\left(\dfrac{\pi}{4}\right)$

$\qquad =\dfrac{6\pi}{4}\text{ inches}$

$\qquad =\dfrac{3\pi}{2}\text{ inches}$

$\qquad \approx4.71\text{ inches}.$

9. We are given ω, the angular speed.
$\omega=45$ revolutions per minute
We use the formula $v=r\omega$ to find v, the linear speed. Before applying the formula, we must express ω in radians per minute.

$\omega=\dfrac{45\text{ revolutions}}{1\text{ minute}}\cdot\dfrac{2\pi\text{ radians}}{1\text{ revolution}}$

$\qquad =\dfrac{90\pi\text{ radians}}{1\text{ minute}}$

The angular speed of the propeller is 90π radians per minute. The linear speed is
$v=r\omega$

$\qquad =1.5\text{ inches}\cdot\dfrac{90\pi}{1\text{ minute}}$

$\qquad =\dfrac{135\pi\text{ inches}}{\text{minute}}$

$\qquad =135\pi\dfrac{\text{inches}}{\text{minute}}$

$\qquad \approx424\dfrac{\text{inches}}{\text{minute}}$

The linear speed is 135π inches per minute, which is approximately 424 inches per minute.

Concept and Vocabulary Check 5.1

1. origin; *x*-axis

2. counterclockwise; clockwise

3. acute; right; obtuse; straight

4. $\dfrac{s}{r}$

5. $\dfrac{\pi}{180°}$

6. $\dfrac{180°}{\pi}$

7. coterminal; $360°$; 2π

8. $r\theta$

9. false

10. $r\omega$; angular

Exercise Set 5.1

1. obtuse

3. acute

5. straight

7. $\theta=\dfrac{s}{r}=\dfrac{40\text{ inches}}{10\text{ inches}}=4$ radians

9. $\theta=\dfrac{s}{r}=\dfrac{8\text{ yards}}{6\text{ yards}}=\dfrac{4}{3}$ radians

11. $\theta=\dfrac{s}{r}=\dfrac{400\text{ centimeters}}{100\text{ centimeters}}=4$ radians

13. $45°=45°\cdot\dfrac{\pi\text{ radians}}{180°}$

$\qquad =\dfrac{45\pi}{180}\text{ radians}$

$\qquad =\dfrac{\pi}{4}\text{ radians}$

15. $135° = 135° \cdot \dfrac{\pi \text{ radians}}{180°}$

$= \dfrac{135\pi}{180} \text{ radians}$

$= \dfrac{3\pi}{4} \text{ radians}$

17. $300° = 300° \cdot \dfrac{\pi \text{ radians}}{180°}$

$= \dfrac{300\pi}{180} \text{ radians}$

$= \dfrac{5\pi}{3} \text{ radians}$

19. $-225° = -225° \cdot \dfrac{\pi \text{ radians}}{180°}$

$= -\dfrac{225\pi}{180} \text{ radians}$

$= -\dfrac{5\pi}{4} \text{ radians}$

21. $\dfrac{\pi}{2} \text{ radians} = \dfrac{\pi \text{ radians}}{2} \cdot \dfrac{180°}{\pi \text{ radians}}$

$= \dfrac{180°}{2}$

$= 90°$

23. $\dfrac{2\pi}{3} \text{ radians} = \dfrac{2\pi \text{ radians}}{3} \cdot \dfrac{180°}{\pi \text{ radians}}$

$= \dfrac{2 \cdot 180°}{3}$

$= 120°$

25. $\dfrac{7\pi}{6} \text{ radians} = \dfrac{7\pi \text{ radians}}{6} \cdot \dfrac{180°}{\pi \text{ radians}}$

$= \dfrac{7 \cdot 180°}{6}$

$= 210°$

27. $-3\pi \text{ radians} = -3\pi \text{ radians} \cdot \dfrac{180°}{\pi \text{ radians}}$

$= -3 \cdot 180°$

$= -540°$

29. $18° = 18° \cdot \dfrac{\pi \text{ radians}}{180°}$

$= \dfrac{18\pi}{180} \text{ radians}$

$\approx 0.31 \text{ radians}$

31. $-40° = -40° \cdot \dfrac{\pi \text{ radians}}{180°}$

$= -\dfrac{40\pi}{180} \text{ radians}$

$\approx -0.70 \text{ radians}$

33. $200° = 200° \cdot \dfrac{\pi \text{ radians}}{180°}$

$= \dfrac{200\pi}{180} \text{ radians}$

$\approx 3.49 \text{ radians}$

35. $2 \text{ radians} = 2 \text{ radians} \cdot \dfrac{180°}{\pi \text{ radians}}$

$= \dfrac{2 \cdot 180°}{\pi}$

$\approx 114.59°$

37. $\dfrac{\pi}{13} \text{ radians} = \dfrac{\pi \text{ radians}}{13} \cdot \dfrac{180°}{\pi \text{ radians}}$

$= \dfrac{180°}{13}$

$\approx 13.85°$

39. $-4.8 \text{ radians} = -4.8 \text{ radians} \cdot \dfrac{180°}{\pi \text{ radians}}$

$= \dfrac{-4.8 \cdot 180°}{\pi}$

$\approx -275.02°$

41.

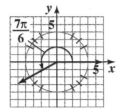

43.

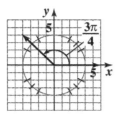

45.

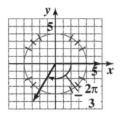

47.

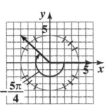

49.

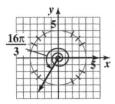

51.

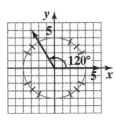

53.

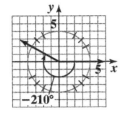

55.

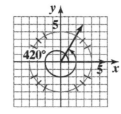

57. $395° - 360° = 35°$

59. $-150° + 360° = 210°$

61. $-765° + 360° \cdot 3 = -765° + 1080° = 315°$

63. $\dfrac{19\pi}{6} - 2\pi = \dfrac{19\pi}{6} - \dfrac{12\pi}{6} = \dfrac{7\pi}{6}$

65. $\dfrac{23\pi}{5} - 2\pi \cdot 2 = \dfrac{23\pi}{5} - 4\pi = \dfrac{23\pi}{5} - \dfrac{20\pi}{5} = \dfrac{3\pi}{5}$

67. $-\dfrac{\pi}{50} + 2\pi = -\dfrac{\pi}{50} + \dfrac{100\pi}{50} = \dfrac{99\pi}{50}$

69. $-\dfrac{31\pi}{7} + 2\pi \cdot 3 = -\dfrac{31\pi}{7} + 6\pi$

$= -\dfrac{31\pi}{7} + \dfrac{42\pi}{7} = \dfrac{11\pi}{7}$

71. $r = 12$ inches, $\theta = 45°$

Begin by converting $45°$ to radians, in order to use the formula $s = r\theta$.

$45° = 45° \cdot \dfrac{\pi \text{ radians}}{180°} = \dfrac{\pi}{4}$ radians

Now use the formula $s = r\theta$.

$s = r\theta = 12 \cdot \dfrac{\pi}{4} = 3\pi$ inches ≈ 9.42 inches

73. $r = 8$ feet, $\theta = 225°$

Begin by converting $225°$ to radians, in order to use the formula $s = r\theta$.

$225° = 225° \cdot \dfrac{\pi \text{ radians}}{180°} = \dfrac{5\pi}{4}$ radians

Now use the formula $s = r\theta$.

$s = r\theta = 8 \cdot \dfrac{5\pi}{4} = 10\pi$ feet ≈ 31.42 feet

75. 6 revolutions per second

$= \dfrac{6 \text{ revolutions}}{1 \text{ second}} \cdot \dfrac{2\pi \text{ radians}}{1 \text{ revolutions}} = \dfrac{12\pi \text{ radians}}{1 \text{ seconds}}$

$= 12\pi$ radians per second

77. $-\dfrac{4\pi}{3}$ and $\dfrac{2\pi}{3}$

79. $-\dfrac{3\pi}{4}$ and $\dfrac{5\pi}{4}$

81. $-\dfrac{\pi}{2}$ and $\dfrac{3\pi}{2}$

83. $\dfrac{55}{60}\cdot 2\pi = \dfrac{11\pi}{6}$

85. 3 minutes and 40 seconds equals 220 seconds.

$$\dfrac{220}{60}\cdot 2\pi = \dfrac{22\pi}{3}$$

87. First, convert to degrees.

$$\dfrac{1}{6}\text{ revolution} = \dfrac{1}{6}\text{ revolution}\cdot\dfrac{360°}{1\text{ revolution}}$$

$$= \dfrac{1}{6}\cdot 360° = 60°$$

Now, convert 60° to radians.

$$60° = 60°\cdot\dfrac{\pi\text{ radians}}{180°} = \dfrac{60\pi}{180}\text{ radians}$$

$$= \dfrac{\pi}{3}\text{ radians}$$

Therefore, $\dfrac{1}{6}$ revolution is equivalent to 60° or $\dfrac{\pi}{3}$ radians.

89. The distance that the tip of the minute hand moves is given by its arc length, s. Since $s = r\theta$, we begin by finding r and θ. We are given that $r = 8$ inches. The minute hand moves from 12 to 2 o'clock, or $\dfrac{1}{6}$ of a complete revolution. The formula $s = r\theta$ can only be used when θ is expressed in radians. We must convert $\dfrac{1}{6}$ revolution to radians.

$$\dfrac{1}{6}\text{ revolution} = \dfrac{1}{6}\text{ revolution}\cdot\dfrac{2\pi\text{ radians}}{1\text{ revolution}}$$

$$= \dfrac{\pi}{3}\text{ radians}$$

The distance the tip of the minute hand moves is

$$s = r\theta = (8\text{ inches})\left(\dfrac{\pi}{3}\right) = \dfrac{8\pi}{3}\text{ inches} \approx 8.38\text{ inches.}$$

91. The length of each arc is given by $s = r\theta$. We are given that $r = 24$ inches and $\theta = 90°$. The formula $s = r\theta$ can only be used when θ is expressed in radians.

$$90° = 90°\cdot\dfrac{\pi\text{ radians}}{180°} = \dfrac{90\pi}{180}\text{ radians}$$

$$= \dfrac{\pi}{2}\text{ radians}$$

The length of each arc is

$$s = r\theta = (24\text{ inches})\left(\dfrac{\pi}{2}\right) = 12\pi\text{ inches}$$

$$\approx 37.70\text{ inches.}$$

93. Recall that $\theta = \dfrac{s}{r}$. We are given that $s = 8000$ miles and $r = 4000$ miles.

$$\theta = \dfrac{s}{r} = \dfrac{8000\text{ miles}}{4000\text{ miles}} = 2\text{ radians}$$

Now, convert 2 radians to degrees.

$$2\text{ radians} = 2\text{ radians}\cdot\dfrac{180°}{\pi\text{ radians}} \approx 114.59°$$

95. Recall that $s = r\theta$. We are given that $r = 4000$ miles and $\theta = 30°$. The formula $s = r\theta$ can only be used when θ is expressed in radians.

$$30° = 30°\cdot\dfrac{\pi\text{ radians}}{180°} = \dfrac{30\pi}{180}\text{ radians}$$

$$= \dfrac{\pi}{6}\text{ radians}$$

$$s = r\theta = (4000\text{ miles})\left(\dfrac{\pi}{6}\right) \approx 2094\text{ miles}$$

To the nearest mile, the distance from A to B is

97. Linear speed is given by $v = r\omega$. We are given that $\omega = \dfrac{\pi}{12}$ radians per hour and $r = 4000$ miles. Therefore,

$$v = r\omega = (4000\text{ miles})\left(\dfrac{\pi}{12}\right)$$

$$= \dfrac{4000\pi}{12}\text{ miles per hour}$$

$$\approx 1047\text{ miles per hour}$$

The linear speed is about 1047 miles per hour.

99. Linear speed is given by $v = r\omega$. We are given that r = 12 feet and the wheel rotates at 20 revolutions per minute.

20 revolutions per minute

$= 20 \text{ revolutions per minute} \cdot \dfrac{2\pi \text{ radians}}{1 \text{ revolution}}$

$= 40\pi \text{ radians per minute}$

$v = r\omega = (12 \text{ feet})(40\pi)$

$\approx 1508 \text{ feet per minute}$

The linear speed of the wheel is about 1508 feet per minute.

101. – 111. Answers may vary.

113.

30.25°

115.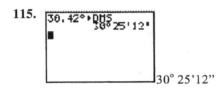

30° 25'12"

117. does not make sense; Explanations will vary. Sample explanation: Angles greater than π will exceed a straight angle.

119. makes sense

121. A right angle measures 90° and

$90° = \dfrac{\pi}{2}$ radians ≈ 1.57 radians.

If $\theta = \dfrac{3}{2}$ radians $= 1.5$ radians, θ is smaller than a right angle.

123. $s = r\theta$

Begin by changing $\theta = 26°$ to radians.

$26° = 26° \cdot \dfrac{\pi}{180°} = \dfrac{13\pi}{90}$ radians

$s = 4000 \cdot \dfrac{13\pi}{90}$

≈ 1815 miles

To the nearest mile, Miami, Florida is 1815 miles north of the equator.

124. First find the hypotenuse.

$c^2 = a^2 + b^2$

$c^2 = 5^2 + 12^2$

$c^2 = 25 + 144$

$c^2 = 169$

$c = 13$

Next write the ratio.

$\dfrac{a}{c} = \dfrac{5}{13}$

125. First find the hypotenuse.

$c^2 = a^2 + b^2$

$c^2 = 1^2 + 1^2$

$c^2 = 1 + 1$

$c^2 = 2$

$c = \sqrt{2}$

Next write the ratio and simplify.

$\dfrac{a}{c} = \dfrac{1}{\sqrt{2}}$

$= \dfrac{1}{\sqrt{2}} \cdot \dfrac{\sqrt{2}}{\sqrt{2}}$

$= \dfrac{\sqrt{2}}{2}$

126. $\left(\dfrac{a}{c}\right)^2 + \left(\dfrac{b}{c}\right)^2 = \dfrac{a^2}{c^2} + \dfrac{b^2}{c^2}$

$= \dfrac{a^2 + b^2}{c^2}$

Since $c^2 = a^2 + b^2$, continue simplifying by substituting c^2 for $a^2 + b^2$.

Section 5.2

Checkpoint Exercises

1. Use the Pythagorean Theorem, $c^2 = a^2 + b^2$, to find c.

   $a = 3, b = 4$

   $c^2 = a^2 + b^2 = 3^2 + 4^2 = 9 + 16 = 25$

   $c = \sqrt{25} = 5$

   Referring to these lengths as opposite, adjacent, and hypotenuse, we have

   $\sin\theta = \dfrac{\text{opposite}}{\text{hypotenuse}} = \dfrac{3}{5}$

   $\cos\theta = \dfrac{\text{adjacent}}{\text{hypotenuse}} = \dfrac{4}{5}$

   $\tan\theta = \dfrac{\text{opposite}}{\text{adjacent}} = \dfrac{3}{4}$

   $\csc\theta = \dfrac{\text{hypotenuse}}{\text{opposite}} = \dfrac{5}{3}$

   $\sec\theta = \dfrac{\text{hypotenuse}}{\text{adjacent}} = \dfrac{5}{4}$

   $\cot\theta = \dfrac{\text{adjacent}}{\text{opposite}} = \dfrac{4}{3}$

2. Use the Pythagorean Theorem, $c^2 = a^2 + b^2$, to find b.

   $a^2 + b^2 = c^2$

   $1^2 + b^2 = 5^2$

   $1 + b^2 = 25$

   $b^2 = 24$

   $b = \sqrt{24} = 2\sqrt{6}$

   Note that side a is opposite θ and side b is adjacent to θ.

   $\sin\theta = \dfrac{\text{opposite}}{\text{hypotenuse}} = \dfrac{1}{5}$

   $\cos\theta = \dfrac{\text{adjacent}}{\text{hypotenuse}} = \dfrac{2\sqrt{6}}{5}$

   $\tan\theta = \dfrac{\text{opposite}}{\text{adjacent}} = \dfrac{1}{2\sqrt{6}} = \dfrac{\sqrt{6}}{12}$

   $\csc\theta = \dfrac{\text{hypotenuse}}{\text{opposite}} = \dfrac{5}{1} = 5$

   $\sec\theta = \dfrac{\text{hypotenuse}}{\text{adjacent}} = \dfrac{5}{2\sqrt{6}} = \dfrac{5\sqrt{6}}{12}$

   $\cot\theta = \dfrac{\text{adjacent}}{\text{opposite}} = \dfrac{2\sqrt{6}}{1} = 2\sqrt{6}$

3. Apply the definitions of these three trigonometric functions.

   $\csc 45° = \dfrac{\text{length of hypotenuse}}{\text{length of side opposite } 45°}$

   $= \dfrac{\sqrt{2}}{1} = \sqrt{2}$

   $\sec 45° = \dfrac{\text{length of hypotenuse}}{\text{length of side adjacent to } 45°}$

   $= \dfrac{\sqrt{2}}{1} = \sqrt{2}$

   $\cot 45° = \dfrac{\text{length of side adjacent to } 45°}{\text{length of side opposite } 45°}$

   $= \dfrac{1}{1} = 1$

4. $\tan 60° = \dfrac{\text{length of side opposite } 60°}{\text{length of side adjacent to } 60°}$

   $= \dfrac{\sqrt{3}}{1} = \sqrt{3}$

   $\tan 30° = \dfrac{\text{length of side opposite } 30°}{\text{length of side adjacent to } 30°}$

   $= \dfrac{1}{\sqrt{3}} = \dfrac{1}{\sqrt{3}} \cdot \dfrac{\sqrt{3}}{3} = \dfrac{\sqrt{3}}{3}$

5. $\tan\theta = \dfrac{\sin\theta}{\cos\theta} = \dfrac{\dfrac{2}{3}}{\dfrac{\sqrt{5}}{3}}$

   $= \dfrac{2}{3} \cdot \dfrac{3}{\sqrt{5}} = \dfrac{2}{\sqrt{5}}$

   $= \dfrac{2}{\sqrt{5}} \cdot \dfrac{\sqrt{5}}{\sqrt{5}} = \dfrac{2\sqrt{5}}{5}$

   $\csc\theta = \dfrac{1}{\sin\theta} = \dfrac{1}{\dfrac{2}{3}} = \dfrac{3}{2}$

   $\sec\theta = \dfrac{1}{\cos\theta} = \dfrac{1}{\dfrac{\sqrt{5}}{3}} = \dfrac{3}{\sqrt{5}}$

   $= \dfrac{3}{\sqrt{5}} \cdot \dfrac{\sqrt{5}}{\sqrt{5}} = \dfrac{3\sqrt{5}}{5}$

   $\cot\theta = \dfrac{1}{\tan\theta} = \dfrac{1}{\dfrac{2}{\sqrt{5}}} = \dfrac{\sqrt{5}}{2}$

6. We can find the value of $\cos\theta$ by using the Pythagorean identity.

$$\sin^2\theta + \cos^2\theta = 1$$

$$\left(\frac{1}{2}\right)^2 + \cos^2\theta = 1$$

$$\frac{1}{4} + \cos^2\theta = 1$$

$$\cos^2\theta = 1 - \frac{1}{4}$$

$$\cos^2\theta = \frac{3}{4}$$

$$\cos\theta = \sqrt{\frac{3}{4}} = \frac{\sqrt{3}}{2}$$

7. a. $\sin 46^{\text{O}} = \cos(90^{\text{O}} - 46^{\text{O}}) = \cos 44^{\text{O}}$

  b. $\cot\dfrac{\pi}{12} = \tan\left(\dfrac{\pi}{2} - \dfrac{\pi}{12}\right)$

  $\qquad = \tan\left(\dfrac{6\pi}{12} - \dfrac{\pi}{12}\right)$

  $\qquad = \tan\dfrac{5\pi}{12}$

8.

Many Scientific Calculators			
Function	**Mode**	**Keystrokes**	**Display** (rounded to four places)
a. $\sin 72.8°$	Degree	72.8 SIN	0.9553
b. $\csc 1.5$	Radian	1.5 SIN 1/x	1.0025

Many Graphing Calculators			
Function	**Mode**	**Keystrokes**	**Display** (rounded to four places)
a. $\sin 72.8°$	Degree	SIN 72.8 ENTER	0.9553
b. $\csc 1.5$	Radian	(SIN 1.5) x^{-1} ENTER	1.0025

9. Because we have a known angle, an unknown opposite side, and a known adjacent side, we select the tangent function.

$$\tan 24^{\text{O}} = \frac{a}{750}$$

$$a = 750\tan 24^{\text{O}}$$

$$a \approx 750(0.4452) \approx 333.9$$

The distance across the lake is approximately 333.9 yards.

10. $\tan\theta = \dfrac{\text{side opposite}}{\text{side adjacent}} = \dfrac{14}{10}$

Use a calculator in degree mode to find θ.

Many Scientific Calculators	Many Graphing Calculators
$\boxed{\text{TAN}^{-1}}$ $\boxed{(}$ $\boxed{14}$ $\boxed{\div}$ $\boxed{10}$ $\boxed{)}$ $\boxed{\text{ENTER}}$	$\boxed{\text{TAN}}$ $\boxed{(}$ $14 \div 10$ $\boxed{)}$ $\boxed{\text{ENTER}}$

The display should show approximately 54. Thus, the angle of elevation of the sun is approximately $54°$.

Concept and Vocabulary Check 5.2

1. $\sin\theta = \dfrac{a}{c}$; $\csc\theta = \dfrac{c}{a}$; $\cos\theta = \dfrac{b}{c}$; $\sec\theta = \dfrac{c}{b}$' $\tan\theta = \dfrac{a}{b}$; $\cot\theta = \dfrac{b}{a}$

2. opposite; adjacent to; hypotenuse

3. true

4. $\sin\theta$; $\cos\theta$; $\tan\theta$;

5. $\tan\theta$; $\cot\theta$

6. 1; $\sec^2\theta$; $\csc^2\theta$

7. $\sin\theta$; $\tan\theta$; $\sec\theta$

Exercise Set 5.2

1. $c^2 = 9^2 + 12^2 = 225$

$c = \sqrt{225} = 15$

$\sin\theta = \dfrac{\text{opposite}}{\text{hypotenuse}} = \dfrac{9}{15} = \dfrac{3}{5}$

$\cos\theta = \dfrac{\text{adjacent}}{\text{hypotenuse}} = \dfrac{12}{15} = \dfrac{4}{5}$

$\tan\theta = \dfrac{\text{opposite}}{\text{adjacent}} = \dfrac{9}{12} = \dfrac{3}{4}$

$\csc\theta = \dfrac{\text{hypotenuse}}{\text{opposite}} = \dfrac{15}{9} = \dfrac{5}{3}$

$\sec\theta = \dfrac{\text{hypotenuse}}{\text{adjacent}} = \dfrac{15}{12} = \dfrac{5}{4}$

$\cot\theta = \dfrac{\text{adjacent}}{\text{opposite}} = \dfrac{12}{9} = \dfrac{4}{3}$

3. $a^2 + 21^2 = 29^2$

$\qquad a^2 = 841 - 441 = 400$

$\qquad a = \sqrt{400} = 20$

$\sin\theta = \dfrac{\text{opposite}}{\text{hypotenuse}} = \dfrac{20}{29}$

$\cos\theta = \dfrac{\text{adjacent}}{\text{hypotenuse}} = \dfrac{21}{29}$

$\tan\theta = \dfrac{\text{opposite}}{\text{adjacent}} = \dfrac{20}{21}$

$\csc\theta = \dfrac{\text{hypotenuse}}{\text{opposite}} = \dfrac{29}{20}$

$\sec\theta = \dfrac{\text{hypotenuse}}{\text{adjacent}} = \dfrac{29}{21}$

$\cot\theta = \dfrac{\text{adjacent}}{\text{opposite}} = \dfrac{21}{20}$

5. $10^2 + b^2 = 26^2$

$\qquad b^2 = 676 - 100 = 576$

$\qquad b = \sqrt{576} = 24$

$\sin\theta = \dfrac{\text{opposite}}{\text{hypotenuse}} = \dfrac{10}{26} = \dfrac{5}{13}$

$\cos\theta = \dfrac{\text{adjacent}}{\text{hypotenuse}} = \dfrac{24}{26} = \dfrac{12}{13}$

$\tan\theta = \dfrac{\text{opposite}}{\text{adjacent}} = \dfrac{10}{24} = \dfrac{5}{12}$

$\csc\theta = \dfrac{\text{hypotenuse}}{\text{opposite}} = \dfrac{26}{10} = \dfrac{13}{5}$

$\sec\theta = \dfrac{\text{hypotenuse}}{\text{adjacent}} = \dfrac{26}{24} = \dfrac{13}{12}$

$\cot\theta = \dfrac{\text{adjacent}}{\text{opposite}} = \dfrac{24}{10} = \dfrac{12}{5}$

7. $21^2 + b^2 = 35^2$

$\qquad b^2 = 1225 - 441 = 784$

$\qquad b = \sqrt{784} = 28$

$\sin\theta = \dfrac{\text{opposite}}{\text{hypotenuse}} = \dfrac{28}{35} = \dfrac{4}{5}$

$\cos\theta = \dfrac{\text{adjacent}}{\text{hypotenuse}} = \dfrac{21}{35} = \dfrac{3}{5}$

$\tan\theta = \dfrac{\text{opposite}}{\text{adjacent}} = \dfrac{28}{21} = \dfrac{4}{3}$

$\csc\theta = \dfrac{\text{hypotenuse}}{\text{opposite}} = \dfrac{35}{28} = \dfrac{5}{4}$

$\sec\theta = \dfrac{\text{hypotenuse}}{\text{adjacent}} = \dfrac{35}{21} = \dfrac{5}{3}$

$\cot\theta = \dfrac{\text{adjacent}}{\text{opposite}} = \dfrac{21}{28} = \dfrac{3}{4}$

9. $\cos 30° = \dfrac{\text{length of side adjacent to } 30°}{\text{length of hypotenuse}}$

$\qquad = \dfrac{\sqrt{3}}{2}$

11. $\sec 45° = \dfrac{\text{length of hypotenuse}}{\text{length of side adjacent to } 45°}$

$\qquad = \dfrac{\sqrt{2}}{1} = \sqrt{2}$

13. $\tan\dfrac{\pi}{3} = \tan 60°$

$\qquad = \dfrac{\text{length of side opposite } 60°}{\text{length of side adjacent to } 60°}$

$\qquad = \dfrac{\sqrt{3}}{1} = \sqrt{3}$

15. $\sin\dfrac{\pi}{4} - \cos\dfrac{\pi}{4} = \sin 45° - \cos 45°$

$\qquad = \dfrac{1}{\sqrt{2}} - \dfrac{1}{\sqrt{2}} = 0$

17. $\tan\theta = \dfrac{\sin\theta}{\cos\theta} = \dfrac{\dfrac{8}{17}}{\dfrac{15}{17}} = \dfrac{8}{15}$

$\csc\theta = \dfrac{1}{\sin\theta} = \dfrac{1}{\dfrac{8}{17}} = \dfrac{17}{8}$

$\sec\theta = \dfrac{1}{\cos\theta} = \dfrac{1}{\dfrac{15}{17}} = \dfrac{17}{15}$

$\cot\theta = \dfrac{\cos\theta}{\sin\theta} = \dfrac{\dfrac{15}{17}}{\dfrac{8}{17}} = \dfrac{15}{8}$

19. $\tan\theta = \dfrac{\sin\theta}{\cos\theta} = \dfrac{\dfrac{1}{3}}{\dfrac{2\sqrt{2}}{3}} = \dfrac{1}{2\sqrt{2}}$

$\quad = \dfrac{1}{2\sqrt{2}} \cdot \dfrac{\sqrt{2}}{\sqrt{2}} = \dfrac{\sqrt{2}}{4}$

$\csc\theta = \dfrac{1}{\sin\theta} = \dfrac{1}{\dfrac{1}{3}} = \dfrac{3}{1} = 3$

$\sec\theta = \dfrac{1}{\cos\theta} = \dfrac{1}{\dfrac{2\sqrt{2}}{3}} = \dfrac{3}{2\sqrt{2}}$

$\quad = \dfrac{3}{2\sqrt{2}} \cdot \dfrac{\sqrt{2}}{\sqrt{2}} = \dfrac{3\sqrt{2}}{4}$

$\cot\theta = \dfrac{\cos\theta}{\sin\theta} = \dfrac{\dfrac{2\sqrt{2}}{3}}{\dfrac{1}{3}} = \dfrac{2\sqrt{2}}{1} = 2\sqrt{2}$

21. $\sin^2\theta + \cos^2\theta = 1$

$\left(\dfrac{6}{7}\right)^2 + \cos^2\theta = 1$

$\dfrac{36}{49} + \cos^2\theta = 1$

$\cos^2\theta = 1 - \dfrac{36}{49}$

$\cos^2\theta = \dfrac{13}{49}$

$\cos\theta = \sqrt{\dfrac{13}{49}} = \dfrac{\sqrt{13}}{7}$

23. $\sin^2\theta + \cos^2\theta = 1$

$\left(\dfrac{\sqrt{39}}{8}\right)^2 + \cos^2\theta = 1$

$\dfrac{39}{64} + \cos^2\theta = 1$

$\cos^2\theta = 1 - \dfrac{39}{64}$

$\cos^2\theta = \dfrac{25}{64}$

$\cos\theta = \sqrt{\dfrac{25}{64}} = \dfrac{5}{8}$

25. $\sin 37^0 \csc 37^0 = \sin 37^0 \cdot \dfrac{1}{\sin 37^0} = 1$

27. $\sin^2\theta + \cos^2\theta = 1$

$\sin^2\dfrac{\pi}{9} + \cos^2\dfrac{\pi}{9} = 1$

29. $1 + \tan^2\theta = \sec^2\theta$

$1 + \tan^2 23^0 = \sec^2 23^0$

$1 = \sec^2 23^0 - \tan^2 23^0$

31. $\sin 7^\circ = \cos(90^\circ - 7^\circ) = \cos 83^\circ$

33. $\csc 25^\circ = \sec(90^\circ - 25^\circ) = \sec 65^\circ$

35. $\tan\dfrac{\pi}{9} = \cot\left(\dfrac{\pi}{2} - \dfrac{\pi}{9}\right)$

$\quad = \cot\left(\dfrac{9\pi}{18} - \dfrac{2\pi}{18}\right)$

$\quad = \cot\dfrac{7\pi}{18}$

37. $\cos\dfrac{2\pi}{5} = \sin\left(\dfrac{\pi}{2} - \dfrac{2\pi}{5}\right)$

$\quad = \sin\left(\dfrac{5\pi}{10} - \dfrac{4\pi}{10}\right)$

$\quad = \sin\dfrac{\pi}{10}$

39.

Many Scientific Calculators			
Function	**Mode**	**Keystrokes**	**Display** (rounded to four places)
sin 38°	Degree	38 SIN	0.6157

Many Graphing Calculators			
Function	**Mode**	**Keystrokes**	**Display** (rounded to four places)
sin 38°	Degree	SIN 38 ENTER	0.6157

41.

Many Scientific Calculators			
Function	**Mode**	**Keystrokes**	**Display** (rounded to four places)
tan 32.7°	Degree	32.7 TAN	0.6420

Many Graphing Calculators			
Function	**Mode**	**Keystrokes**	**Display** (rounded to four places)
tan 32.7°	Degree	TAN 32.7 ENTER	0.6420

43.

Many Scientific Calculators			
Function	**Mode**	**Keystrokes**	**Display** (rounded to four places)
csc 17°	Degree	17 SIN 1/x	3.4203

Many Graphing Calculators			
Function	**Mode**	**Keystrokes**	**Display** (rounded to four places)
csc 17°	Degree	(SIN 17) x^{-1} ENTER	3.4203

45.

Many Scientific Calculators			
Function	**Mode**	**Keystrokes**	**Display** (rounded to four places)
$\cos\dfrac{\pi}{10}$	Radian	π ÷ 10 = COS	0.9511

Many Graphing Calculators			
Function	**Mode**	**Keystrokes**	**Display** (rounded to four places)
$\cos\dfrac{\pi}{10}$	Radian	COS (π ÷ 10) ENTER	0.9511

47.

Many Scientific Calculators			
Function	**Mode**	**Keystrokes**	**Display** (rounded to four places)
$\cot\dfrac{\pi}{12}$	Radian	π ÷ 12 = TAN 1/x	3.7321

Many Graphing Calculators			
Function	**Mode**	**Keystrokes**	**Display** (rounded to four places)
$\cot\dfrac{\pi}{12}$	Radian	(TAN (π ÷ 12)) x^{-1} ENTER	3.7321

49. $\tan 37° = \dfrac{a}{250}$

$a = 250\tan 37°$

$a \approx 250(0.7536) \approx 188$ cm

51. $\cos 34° = \dfrac{b}{220}$

$b = 220\cos 34°$

$b \approx 220(0.8290) \approx 182$ in.

53. $\sin 23° = \dfrac{16}{c}$

$c = \dfrac{16}{\sin 23°} \approx \dfrac{16}{0.3907} \approx 41$ m

55.

Scientific Calculator	Graphing Calculator	Display (rounded to the nearest degree)
0.2974 $\boxed{\text{SIN}^{-1}}$	$\boxed{\text{SIN}^{-1}}$ 0.2974 $\boxed{\text{ENTER}}$	17

If $\sin \theta = 0.2974$, then $\theta \approx 17°$.

57.

Scientific Calculator	Graphing Calculator	Display (rounded to the nearest degree)
4.6252 $\boxed{\text{TAN}^{-1}}$	$\boxed{\text{TAN}^{-1}}$ 4.6252 $\boxed{\text{ENTER}}$	78

If $\tan \theta = 4.6252$, then $\theta \approx 78°$.

59.

Scientific Calculator	Graphing Calculator	Display (rounded to three places)
0.4112 $\boxed{\text{COS}^{-1}}$	$\boxed{\text{COS}^{-1}}$ 0.4112 $\boxed{\text{ENTER}}$	1.147

If $\cos \theta = 0.4112$, then $\theta \approx 1.147$ radians.

61.

Scientific Calculator	Graphing Calculator	Display (rounded to three places)
0.4169 $\boxed{\text{TAN}^{-1}}$	$\boxed{\text{TAN}^{-1}}$ 0.4169 $\boxed{\text{ENTER}}$	0.395

If $\tan \theta = 0.4169$, then $\theta \approx 0.395$ radians.

63.

$$\dfrac{\tan \dfrac{\pi}{3}}{2} - \dfrac{1}{\sec \dfrac{\pi}{6}} = \dfrac{\sqrt{3}}{2} - \dfrac{1}{\dfrac{1}{\cos \dfrac{\pi}{6}}}$$

$$= \dfrac{\sqrt{3}}{2} - \dfrac{1}{\dfrac{1}{\dfrac{\sqrt{3}}{2}}}$$

$$= \dfrac{\sqrt{3}}{2} - \dfrac{\sqrt{3}}{2}$$

$$= 0$$

65. $1 + \sin^2 40° + \sin^2 50°$

$= 1 + \sin^2 (90° - 50°) + \sin^2 50°$

$= 1 + \cos^2 50° + \sin^2 50°$

$= 1 + 1$

$= 2$

67. $\csc 37° \sec 53° - \tan 53° \cot 37°$

$= \sec 53° \sec 53° - \tan 53° \tan 53°$

$= \sec^2 53° - \tan^2 53°$

$= 1$

69. $f(\theta) = 2\cos\theta - \cos 2\theta$

$f\left(\dfrac{\pi}{6}\right) = 2\cos\dfrac{\pi}{6} - \cos\left(2 \cdot \dfrac{\pi}{6}\right)$

$\quad = 2\left(\dfrac{\sqrt{3}}{2}\right) - \cos\left(\dfrac{\pi}{3}\right)$

$\quad = \dfrac{2\sqrt{3}}{2} - \dfrac{1}{2}$

$\quad = \dfrac{2\sqrt{3} - 1}{2}$

71. $\tan\left(\dfrac{\pi}{2} - \theta\right) = \cot\theta = \dfrac{1}{4}$

73. $\tan 40° = \dfrac{a}{630}$

$a = 630\tan 40°$

$a \approx 630(0.8391) \approx 529$

The distance across the lake is approximately 529 yards.

75. $\tan\theta = \dfrac{125}{172}$

Use a calculator in degree mode to find θ.

Many Scientific Calculators	**Many Graphing Calculators**
125 $\boxed{\div}$ 172 $\boxed{=}$ $\boxed{\text{TAN}^{-1}}$	$\boxed{\text{TAN}^{-1}}$ $\boxed{(}$ 125 $\boxed{\div}$ 172 $\boxed{)}$ $\boxed{\text{ENTER}}$

The display should show approximately 36. Thus, the angle of elevation of the sun is approximately 36°.

77. $\sin 10° = \dfrac{500}{c}$

$c = \dfrac{500}{\sin 10°} \approx \dfrac{500}{0.1736} \approx 2880$

The plane has flown approximately 2880 feet.

79. $\cos\theta = \dfrac{60}{75}$

Use a calculator in degree mode to find θ.

Many Scientific Calculators	**Many Graphing Calculators**
60 $\boxed{\div}$ 75 $\boxed{=}$ $\boxed{\text{COS}^{-1}}$	$\boxed{\text{COS}^{-1}}$ $\boxed{(}$ 60 $\boxed{\div}$ 75 $\boxed{)}$ $\boxed{\text{ENTER}}$

The display should show approximately 37. Thus, the angle between the wire and the pole is approximately 37°.

81. – 91. Answers may vary.

93.

θ	0.4	0.3	0.2	0.1	0.01	0.001	0.0001	0.00001
$\cos\theta$	0.92106	0.95534	0.98007	0.99500	0.99995	0.9999995	0.999999995	1
$\dfrac{\cos\theta-1}{\theta}$	−0.19735	−0.148878	−0.099667	−0.04996	−0.005	−0.0005	−0.00005	0

$\dfrac{\cos\theta-1}{\theta}$ approaches 0 as θ approaches 0.

95. does not make sense; Explanations will vary. Sample explanation: This value is irrational. Irrational numbers are rounded on calculators.

97. makes sense

99. true

101. true

103. Use a calculator in degree mode to generate the following table. Then use the table to describe what happens to the tangent of an acute angle as the angle gets close to 90°.

θ	60	70	80	89	89.9	89.99	89.999	89.9999
$\tan\theta$	1.7321	2.7475	5.6713	57	573	5730	57,296	572,958

As θ approaches 90°, $\tan\theta$ increases without bound. At 90°, $\tan\theta$ is undefined.

105. a. $\dfrac{y}{r}$

b. First find r: $r = \sqrt{x^2 + y^2}$
$$r = \sqrt{(-3)^2 + 4^2}$$
$$r = 5$$
$\dfrac{y}{r} = \dfrac{4}{5}$, which is positive.

106. a. $\dfrac{x}{r}$

b. First find r: $r = \sqrt{x^2 + y^2}$
$$r = \sqrt{(-3)^2 + 5^2}$$
$$r = \sqrt{34}$$
$\dfrac{x}{r} = \dfrac{-3}{\sqrt{34}} = \dfrac{-3}{\sqrt{34}} \cdot \dfrac{\sqrt{34}}{\sqrt{34}} = \dfrac{-3\sqrt{34}}{34}$, which is negative.

107. a. $\theta' = 360^\circ - 345^\circ = 15^\circ$

b. $\theta' = \pi - \dfrac{5\pi}{6} = \dfrac{6\pi}{6} - \dfrac{5\pi}{6} = \dfrac{\pi}{6}$

Section 5.3

Checkpoint Exercises

1. $r = \sqrt{x^2 + y^2}$

   $r = \sqrt{1^2 + (-3)^2} = \sqrt{1 + 9} = \sqrt{10}$

   Now that we know x, y, and r, we can find the six trigonometric functions of θ.

   $\sin\theta = \dfrac{y}{r} = \dfrac{-3}{\sqrt{10}} = -\dfrac{3\sqrt{10}}{10}$

   $\cos\theta = \dfrac{x}{r} = \dfrac{1}{\sqrt{10}} = \dfrac{\sqrt{10}}{10}$

   $\tan\theta = \dfrac{y}{x} = \dfrac{-3}{1} = -3$

   $\csc\theta = \dfrac{r}{y} = \dfrac{\sqrt{10}}{-3} = -\dfrac{\sqrt{10}}{3}$

   $\sec\theta = \dfrac{r}{x} = \dfrac{\sqrt{10}}{1} = \sqrt{10}$

   $\cot\theta = \dfrac{x}{y} = \dfrac{1}{-3} = -\dfrac{1}{3}$

2. **a.** $\theta = 0° = 0$ radians

   The terminal side of the angle is on the positive x-axis. Select the point
   $P = (1,0)$: $x = 1$, $y = 0$, $r = 1$
   Apply the definitions of the cosine and cosecant functions.

   $\cos 0° = \cos 0 = \dfrac{x}{r} = \dfrac{1}{1} = 1$

   $\csc 0° = \csc 0 = \dfrac{r}{y} = \dfrac{1}{0}$, undefined

   b. $\theta = 90° = \dfrac{\pi}{2}$ radians

   The terminal side of the angle is on the positive y-axis. Select the point
   $P = (0,1)$: $x = 0$, $y = 1$, $r = 1$
   Apply the definitions of the cosine and cosecant functions.

   $\cos 90° = \cos\dfrac{\pi}{2} = \dfrac{x}{r} = \dfrac{0}{1} = 0$

   $\csc 90° = \csc\dfrac{\pi}{2} = \dfrac{r}{y} = \dfrac{1}{1} = 1$

   c. $\theta = 180° = \pi$ radians

   The terminal side of the angle is on the negative x-axis. Select the point
   $P = (-1,0)$: $x = -1$, $y = 0$, $r = 1$
   Apply the definitions of the cosine and cosecant functions.

   $\cos 180° = \cos\pi = \dfrac{x}{r} = \dfrac{-1}{1} = -1$

   $\csc 180° = \csc\pi = \dfrac{r}{y} = \dfrac{1}{0}$, undefined

   d. $\theta = 270° = \dfrac{3\pi}{2}$ radians

   The terminal side of the angle is on the negative y-axis. Select the point
   $P = (0,-1)$: $x = 0$, $y = -1$, $r = 1$
   Apply the definitions of the cosine and cosecant functions.

   $\cos 270° = \cos\dfrac{3\pi}{2} = \dfrac{x}{r} = \dfrac{0}{1} = 0$

   $\csc 270° = \csc\dfrac{3\pi}{2} = \dfrac{r}{y} = \dfrac{1}{-1} = -1$

3. Because $\sin\theta < 0$, θ cannot lie in quadrant I; all the functions are positive in quadrant I. Furthermore, θ cannot lie in quadrant II; $\sin\theta$ is positive in quadrant II. Thus, with $\sin\theta < 0$, θ lies in quadrant III or quadrant IV. We are also given that $\cos\theta < 0$. Because quadrant III is the only quadrant in which cosine is negative and the sine is negative, we conclude that θ lies in quadrant III.

4. Because the tangent is negative and the cosine is negative, θ lies in quadrant II. In quadrant II, x is negative and y is positive. Thus,

   $\tan\theta = -\dfrac{1}{3} = \dfrac{y}{x} = \dfrac{1}{-3}$

   $x = -3$, $y = 1$

   Furthermore,

   $r = \sqrt{x^2 + y^2} = \sqrt{(-3)^2 + 1^2} = \sqrt{9 + 1} = \sqrt{10}$

   Now that we know x, y, and r, we can find $\sin\theta$ and $\sec\theta$.

   $\sin\theta = \dfrac{y}{r} = \dfrac{1}{\sqrt{10}} = \dfrac{1}{\sqrt{10}} \cdot \dfrac{\sqrt{10}}{\sqrt{10}} = \dfrac{\sqrt{10}}{10}$

   $\sec\theta = \dfrac{r}{x} = \dfrac{\sqrt{10}}{-3} = -\dfrac{\sqrt{10}}{3}$

5. **a.** Because $210°$ lies between $180°$ and $270°$, it is in quadrant III. The reference angle is $\theta' = 210° - 180° = 30°$.

b. Because $\dfrac{7\pi}{4}$ lies between $\dfrac{3\pi}{2} = \dfrac{6\pi}{4}$ and $2\pi = \dfrac{8\pi}{4}$, it is in quadrant IV. The reference angle is $\theta' = 2\pi - \dfrac{7\pi}{4} = \dfrac{8\pi}{4} - \dfrac{7\pi}{4} = \dfrac{\pi}{4}$.

c. Because $-240°$ lies between $-180°$ and $-270°$, it is in quadrant II. The reference angle is $\theta = 240 - 180 = 60°$.

d. Because 3.6 lies between $\pi \approx 3.14$ and $\dfrac{3\pi}{2} \approx 4.71$, it is in quadrant III. The reference angle is $\theta' = 3.6 - \pi \approx 0.46$.

6. **a.** $665° - 360° = 305°$
This angle is in quadrant IV, thus the reference angle is $\theta' = 360° - 305° = 55°$.

b. $\dfrac{15\pi}{4} - 2\pi = \dfrac{15\pi}{4} - \dfrac{8\pi}{4} = \dfrac{7\pi}{4}$
This angle is in quadrant IV, thus the reference angle is $\theta' = 2\pi - \dfrac{7\pi}{4} = \dfrac{8\pi}{4} - \dfrac{7\pi}{4} = \dfrac{\pi}{4}$.

c. $-\dfrac{11\pi}{3} + 2 \cdot 2\pi = -\dfrac{11\pi}{3} + \dfrac{12\pi}{3} = \dfrac{\pi}{3}$
This angle is in quadrant I, thus the reference angle is $\theta' = \dfrac{\pi}{3}$.

7. **a.** $300°$ lies in quadrant IV. The reference angle is $\theta' = 360° - 300° = 60°$.

$\sin 60° = \dfrac{\sqrt{3}}{2}$

Because the sine is negative in quadrant IV,

$\sin 300° = -\sin 60° = -\dfrac{\sqrt{3}}{2}$.

b. $\dfrac{5\pi}{4}$ lies in quadrant III. The reference angle is

$\theta' = \dfrac{5\pi}{4} - \pi = \dfrac{5\pi}{4} - \dfrac{4\pi}{4} = \dfrac{\pi}{4}$.

$\tan \dfrac{\pi}{4} = 1$

Because the tangent is positive in quadrant III,

$\tan \dfrac{5\pi}{4} = +\tan \dfrac{\pi}{4} = 1$.

c. $-\dfrac{\pi}{6}$ lies in quadrant IV. The reference angle is

$\theta' = \dfrac{\pi}{6}$.

$\sec \dfrac{\pi}{6} = \dfrac{2\sqrt{3}}{3}$

Because the secant is positive in quadrant IV,

$\sec\left(-\dfrac{\pi}{6}\right) = +\sec \dfrac{\pi}{6} = \dfrac{2\sqrt{3}}{3}$.

8. **a.** $\dfrac{17\pi}{6} - 2\pi = \dfrac{17\pi}{6} - \dfrac{12\pi}{6} = \dfrac{5\pi}{6}$ lies in quadrant II. The reference angle is $\theta' = \pi - \dfrac{5\pi}{6} = \dfrac{\pi}{6}$.
The function value for the reference angle is

$\cos \dfrac{\pi}{6} = \dfrac{\sqrt{3}}{2}$.

Because the cosine is negative in quadrant II,

$\cos \dfrac{17\pi}{6} = \cos \dfrac{5\pi}{6} = -\cos \dfrac{\pi}{6} = -\dfrac{\sqrt{3}}{2}$.

b. $\dfrac{-22\pi}{3} + 8\pi = \dfrac{-22\pi}{3} + \dfrac{24\pi}{3} = \dfrac{2\pi}{3}$ lies in quadrant II. The reference angle is

$\theta' = \pi - \dfrac{2\pi}{3} = \dfrac{\pi}{3}$.

The function value for the reference angle is

$\sin \dfrac{\pi}{3} = \dfrac{\sqrt{3}}{2}$.

Because the sine is positive in quadrant II,

$\sin \dfrac{-22\pi}{3} = \sin \dfrac{2\pi}{3} = \sin \dfrac{\pi}{3} = \dfrac{\sqrt{3}}{2}$.

Concept and Vocabulary Check 5.3

1. $\sin\theta = \dfrac{y}{r}$; $\csc\theta = \dfrac{r}{y}$; $\cos\theta = \dfrac{x}{r}$; $\sec\theta = \dfrac{r}{x}$'

   $\tan\theta = \dfrac{y}{x}$; $\cot\theta = \dfrac{x}{y}$

2. $\tan\theta$; $\sec\theta$; $\cot\theta$; $\csc\theta$; $\tan\theta$; $\cot\theta$

3. $\sin\theta$; $\csc\theta$;

4. $\tan\theta$; $\cot\theta$;

5. $\cos\theta$; $\sec\theta$

6. terminal; x

7. **(a)** $180° - \theta$; **(b)** $\theta - 180°$; **(c)** $360° - \theta$

Exercise Set 5.3

1. We need values for x, y, and r. Because $P = (-4, 3)$ is a point on the terminal side of θ, $x = -4$ and $y = 3$. Furthermore,

   $r = \sqrt{x^2 + y^2} = \sqrt{(-4)^2 + 3^2} = \sqrt{16+9} = \sqrt{25} = 5$ Now that we know x, y, and r, we can find the six trigonometric functions of θ.

   $\sin\theta = \dfrac{y}{r} = \dfrac{3}{5}$

   $\cos\theta = \dfrac{x}{r} = \dfrac{-4}{5} = -\dfrac{4}{5}$

   $\tan\theta = \dfrac{y}{x} = \dfrac{3}{-4} = -\dfrac{3}{4}$

   $\csc\theta = \dfrac{r}{y} = \dfrac{5}{3}$

   $\sec\theta = \dfrac{r}{x} = \dfrac{5}{-4} = -\dfrac{5}{4}$

   $\cot\theta = \dfrac{x}{y} = \dfrac{-4}{3} = -\dfrac{4}{3}$

3. We need values for x, y, and r. Because $P = (2, 3)$ is a point on the terminal side of θ, $x = 2$ and $y = 3$. Furthermore,

   $r = \sqrt{x^2 + y^2} = \sqrt{2^2 + 3^2} = \sqrt{4+9} = \sqrt{13}$

   Now that we know x, y, and r, we can find the six trigonometric functions of θ.

   $\sin\theta = \dfrac{y}{r} = \dfrac{3}{\sqrt{13}} = \dfrac{3}{\sqrt{13}} \cdot \dfrac{\sqrt{13}}{\sqrt{13}} = \dfrac{3\sqrt{13}}{13}$

   $\cos\theta = \dfrac{x}{r} = \dfrac{2}{\sqrt{13}} = \dfrac{2}{\sqrt{13}} \cdot \dfrac{\sqrt{13}}{\sqrt{13}} = \dfrac{2\sqrt{13}}{13}$

   $\tan\theta = \dfrac{y}{x} = \dfrac{3}{2}$

   $\csc\theta = \dfrac{r}{y} = \dfrac{\sqrt{13}}{3}$

   $\sec\theta = \dfrac{r}{x} = \dfrac{\sqrt{13}}{2}$

   $\cot\theta = \dfrac{x}{y} = \dfrac{2}{3}$

5. We need values for x, y, and r. Because $P = (3, -3)$ is a point on the terminal side of θ, $x = 3$ and $y = -3$.

   Furthermore, $r = \sqrt{x^2 + y^2} = \sqrt{3^2 + (-3)^2} = \sqrt{9+9}$
   $= \sqrt{18} = 3\sqrt{2}$

   Now that we know x, y, and r, we can find the six trigonometric functions of θ.

   $\sin\theta = \dfrac{y}{r} = \dfrac{-3}{3\sqrt{2}} = -\dfrac{1}{\sqrt{2}} \cdot \dfrac{\sqrt{2}}{\sqrt{2}} = -\dfrac{\sqrt{2}}{2}$

   $\cos\theta = \dfrac{x}{r} = \dfrac{3}{3\sqrt{2}} = \dfrac{1}{\sqrt{2}} \cdot \dfrac{\sqrt{2}}{\sqrt{2}} = \dfrac{\sqrt{2}}{2}$

   $\tan\theta = \dfrac{y}{x} = \dfrac{-3}{3} = -1$

   $\csc\theta = \dfrac{r}{y} = \dfrac{3\sqrt{2}}{-3} = -\sqrt{2}$

   $\sec\theta = \dfrac{r}{x} = \dfrac{3\sqrt{2}}{3} = \sqrt{2}$

   $\cot\theta = \dfrac{x}{y} = \dfrac{3}{-3} = -1$

7. We need values for *x*, *y*, and *r*. Because $P = (-2, -5)$ is a point on the terminal side of θ, $x = -2$ and $y = -5$. Furthermore,

$r = \sqrt{x^2 + y^2} = \sqrt{(-2)^2 + (-5)^2} = \sqrt{4 + 25} = \sqrt{29}$ Now that we know *x*, *y*, and *r*, we can find the six trigonometric functions of θ.

$\sin \theta = \dfrac{y}{r} = \dfrac{-5}{\sqrt{29}} = \dfrac{-5}{\sqrt{29}} \cdot \dfrac{\sqrt{29}}{\sqrt{29}} = -\dfrac{5\sqrt{29}}{29}$

$\cos \theta = \dfrac{x}{r} = \dfrac{-2}{\sqrt{29}} = \dfrac{-2}{\sqrt{29}} \cdot \dfrac{\sqrt{29}}{\sqrt{29}} = -\dfrac{2\sqrt{29}}{29}$

$\tan \theta = \dfrac{y}{x} = \dfrac{-5}{-2} = \dfrac{5}{2}$

$\csc \theta = \dfrac{r}{y} = \dfrac{\sqrt{29}}{-5} = -\dfrac{\sqrt{29}}{5}$

$\sec \theta = \dfrac{r}{x} = \dfrac{\sqrt{29}}{-2} = -\dfrac{\sqrt{29}}{2}$

$\cot \theta = \dfrac{x}{y} = \dfrac{-2}{-5} = \dfrac{2}{5}$

9. $\theta = \pi$ radians
The terminal side of the angle is on the negative *x*-axis. Select the point $P = (-1, 0)$:
$x = -1$, $y = 0$, $r = 1$ Apply the definition of the cosine function.

$\cos \pi = \dfrac{x}{r} = \dfrac{-1}{1} = -1$

11. $\theta = \pi$ radians
The terminal side of the angle is on the negative *x*-axis. Select the point $P = (-1, 0)$:
$x = -1$, $y = 0$, $r = 1$ Apply the definition of the secant function.

$\sec \pi = \dfrac{r}{x} = \dfrac{1}{-1} = -1$

13. $\theta = \dfrac{3\pi}{2}$ radians
The terminal side of the angle is on the negative *y*-axis. Select the point $P = (0, -1)$:
$x = 0$, $y = -1$, $r = 1$ Apply the definition of the tangent function. $\tan \dfrac{3\pi}{2} = \dfrac{y}{x} = \dfrac{-1}{0}$, undefined

15. $\theta = \dfrac{\pi}{2}$ radians
The terminal side of the angle is on the positive *y*-axis. Select the point $P = (0, 1)$:
$x = 0$, $y = 1$, $r = 1$ Apply the definition of the cotangent function. $\cot \dfrac{\pi}{2} = \dfrac{x}{y} = \dfrac{0}{1} = 0$

17. Because $\sin \theta > 0$, θ cannot lie in quadrant III or quadrant IV; the sine function is negative in those quadrants. Thus, with $\sin \theta > 0$, θ lies in quadrant I or quadrant II. We are also given that $\cos \theta > 0$. Because quadrant I is the only quadrant in which the cosine is positive and sine is positive, we conclude that θ lies in quadrant I.

19. Because $\sin \theta < 0$, θ cannot lie in quadrant I or quadrant II; the sine function is positive in those two quadrants. Thus, with $\sin \theta < 0$, θ lies in quadrant III or quadrant IV. We are also given that $\cos \theta < 0$. Because quadrant III is the only quadrant in which the cosine is positive and the sine is negative, we conclude that θ lies in quadrant III.

21. Because $\tan \theta < 0$, θ cannot lie in quadrant I or quadrant III; the tangent function is positive in those quadrants. Thus, with $\tan \theta < 0$, θ lies in quadrant II or quadrant IV. We are also given that $\cos \theta < 0$. Because quadrant II is the only quadrant in which the cosine is negative and the tangent is negative, we conclude that θ lies in quadrant II.

23. In quadrant III *x* is negative and *y* is negative. Thus,

$\cos \theta = -\dfrac{3}{5} = \dfrac{x}{r} = \dfrac{-3}{5}$, $x = -3$, $r = 5$. Furthermore,

$r^2 = x^2 + y^2$

$5^2 = (-3)^2 + y^2$

$y^2 = 25 - 9 = 16$

$y = -\sqrt{16} = -4$

Now that we know *x*, *y*, and *r*, we can find the remaining trigonometric functions of θ.

$\sin \theta = \dfrac{y}{r} = \dfrac{-4}{5} = -\dfrac{4}{5}$

$\tan \theta = \dfrac{y}{x} = \dfrac{-4}{-3} = \dfrac{4}{3}$

$\csc \theta = \dfrac{r}{y} = \dfrac{5}{-4} = -\dfrac{5}{4}$

$\sec \theta = \dfrac{r}{x} = \dfrac{5}{-3} = -\dfrac{5}{3}$

$\cot \theta = \dfrac{x}{y} = \dfrac{-3}{-4} = \dfrac{3}{4}$

25. In quadrant II x is negative and y is positive. Thus,

$\sin\theta = \dfrac{5}{13} = \dfrac{y}{r}$, $y = 5$, $r = 13$. Furthermore,

$x^2 + y^2 = r^2$

$x^2 + 5^2 = 13^2$

$\qquad x^2 = 169 - 25 = 144$

$\qquad x = -\sqrt{144} = -12$

Now that we know x, y, and r, we can find the remaining trigonometric functions of θ.

$$\cos\theta = \frac{x}{r} = \frac{-12}{13} = -\frac{12}{13}$$

$$\tan\theta = \frac{y}{x} = \frac{5}{-12} = -\frac{5}{12}$$

$$\csc\theta = \frac{r}{y} = \frac{13}{5}$$

$$\sec\theta = \frac{r}{x} = \frac{13}{-12} = -\frac{13}{12}$$

$$\cot\theta = \frac{x}{y} = \frac{-12}{5} = -\frac{12}{5}$$

27. Because $270° < \theta < 360°$, θ is in quadrant IV. In quadrant IV x is positive and y is negative. Thus,

$\cos\theta = \dfrac{8}{17} = \dfrac{x}{r}$, $x = 8$,

$r = 17$. Furthermore

$x^2 + y^2 = r^2$

$8^2 + y^2 = 17^2$

$\qquad y^2 = 289 - 64 = 225$

$\qquad y = -\sqrt{225} = -15$

Now that we know x, y, and r, we can find the remaining trigonometric functions of θ.

$$\sin\theta = \frac{y}{r} = \frac{-15}{17} = -\frac{15}{17}$$

$$\tan\theta = \frac{y}{x} = \frac{-15}{8} = -\frac{15}{8}$$

$$\csc\theta = \frac{r}{y} = \frac{17}{-15} = -\frac{17}{15}$$

$$\sec\theta = \frac{r}{x} = \frac{17}{8}$$

$$\cot\theta = \frac{x}{y} = \frac{8}{-15} = -\frac{8}{15}$$

29. Because the tangent is negative and the sine is positive, θ lies in quadrant II. In quadrant II, x is negative and y is positive. Thus,

$\tan\theta = -\dfrac{2}{3} = \dfrac{y}{x} = \dfrac{2}{-3}$, $x = -3$, $y = 2$. Furthermore,

$r = \sqrt{x^2 + y^2} = \sqrt{(-3)^2 + 2^2} = \sqrt{9+4} = \sqrt{13}$

Now that we know x, y, and r, we can find the remaining trigonometric functions of θ.

$$\sin\theta = \frac{y}{r} = \frac{2}{\sqrt{13}} = \frac{2}{\sqrt{13}} \cdot \frac{\sqrt{13}}{\sqrt{13}} = \frac{2\sqrt{13}}{13}$$

$$\cos\theta = \frac{x}{r} = \frac{-3}{\sqrt{13}} = \frac{-3}{\sqrt{13}} \cdot \frac{\sqrt{13}}{\sqrt{13}} = -\frac{3\sqrt{13}}{13}$$

$$\csc\theta = \frac{r}{y} = \frac{\sqrt{13}}{2}$$

$$\sec\theta = \frac{r}{x} = \frac{\sqrt{13}}{-3} = -\frac{\sqrt{13}}{3}$$

$$\cot\theta = \frac{x}{y} = \frac{-3}{2} = -\frac{3}{2}$$

31. Because the tangent is positive and the cosine is negative, θ lies in quadrant III. In quadrant III, x is negative and y is negative. Thus, $\tan\theta = \dfrac{4}{3} = \dfrac{y}{x} = \dfrac{-4}{-3}$,

$x = -3$, $y = -4$. Furthermore,

$r = \sqrt{x^2 + y^2} = \sqrt{(-3)^2 + (-4)^2} = \sqrt{9+16}$

$\qquad = \sqrt{25} = 5$

Now that we know x, y, and r, we can find the remaining trigonometric functions of θ.

$$\sin\theta = \frac{y}{r} = \frac{-4}{5} = -\frac{4}{5}$$

$$\cos\theta = \frac{x}{r} = \frac{-3}{5} = -\frac{3}{5}$$

$$\csc\theta = \frac{r}{y} = \frac{5}{-4} = -\frac{5}{4}$$

$$\sec\theta = \frac{r}{x} = \frac{5}{-3} = -\frac{5}{3}$$

$$\cot\theta = \frac{x}{y} = \frac{-3}{-4} = \frac{3}{4}$$

33. Because the secant is negative and the tangent is positive, θ lies in quadrant III. In quadrant III, x is negative and y is negative. Thus,

$$\sec\theta = -3 = \frac{r}{x} = \frac{3}{-1}, \quad x = -1, \ r = 3 \ . \text{ Furthermore,}$$

$$x^2 + y^2 = r^2$$
$$(-1)^2 + y^2 = 3^2$$
$$y^2 = 9 - 1 = 8$$
$$y = -\sqrt{8} = -2\sqrt{2}$$

Now that we know x, y, and r, we can find the remaining trigonometric functions of θ.

$$\sin\theta = \frac{y}{r} = \frac{-2\sqrt{2}}{3} = -\frac{2\sqrt{2}}{3}$$

$$\cos\theta = \frac{x}{r} = \frac{-1}{3} = -\frac{1}{3}$$

$$\tan\theta = \frac{y}{x} = \frac{-2\sqrt{2}}{-1} = 2\sqrt{2}$$

$$\csc\theta = \frac{r}{y} = \frac{3}{-2\sqrt{2}} = \frac{3}{-2\sqrt{2}} \cdot \frac{\sqrt{2}}{\sqrt{2}} = -\frac{3\sqrt{2}}{4}$$

$$\cot\theta = \frac{x}{y} = \frac{-1}{-2\sqrt{2}} = \frac{1}{2\sqrt{2}} \cdot \frac{\sqrt{2}}{\sqrt{2}} = \frac{\sqrt{2}}{4}$$

35. Because $160°$ lies between $90°$ and $180°$, it is in quadrant II. The reference angle is $\theta' = 180° - 160° = 20°$.

37. Because $205°$ lies between $180°$ and $270°$, it is in quadrant III. The reference angle is $\theta' = 205° - 180° = 25°$.

39. Because $355°$ lies between $270°$ and $360°$, it is in quadrant IV. The reference angle is $\theta' = 360° - 355° = 5°$.

41. Because $\dfrac{7\pi}{4}$ lies between $\dfrac{3\pi}{2} = \dfrac{6\pi}{4}$ and $2\pi = \dfrac{8\pi}{4}$, it is in quadrant IV. The reference angle is

$$\theta' = 2\pi - \frac{7\pi}{4} = \frac{8\pi}{4} - \frac{7\pi}{4} = \frac{\pi}{4} \ .$$

43. Because $\dfrac{5\pi}{6}$ lies between $\dfrac{\pi}{2} = \dfrac{3\pi}{6}$ and $\pi = \dfrac{6\pi}{6}$, it is in quadrant II. The reference angle is

$$\theta' = \pi - \frac{5\pi}{6} = \frac{6\pi}{6} - \frac{5\pi}{6} = \frac{\pi}{6} \ .$$

45. $-150° + 360° = 210°$

Because the angle is in quadrant III, the reference angle is $\theta' = 210° - 180° = 30°$.

47. $-335° + 360° = 25°$

Because the angle is in quadrant I, the reference angle is $\theta' = 25°$.

49. Because 4.7 lies between $\pi \approx 3.14$ and $\dfrac{3\pi}{2} \approx 4.71$, it is in quadrant III. The reference angle is $\theta' = 4.7 - \pi \approx 1.56$.

51. $565° - 360° = 205°$

Because the angle is in quadrant III, the reference angle is $\theta' = 205° - 180° = 25°$.

53. $\dfrac{17\pi}{6} - 2\pi = \dfrac{17\pi}{6} - \dfrac{12\pi}{6} = \dfrac{5\pi}{6}$

Because the angle is in quadrant II, the reference angle is $\theta' = \pi - \dfrac{5\pi}{6} = \dfrac{\pi}{6}$.

55. $\dfrac{23\pi}{4} - 4\pi = \dfrac{23\pi}{4} - \dfrac{16\pi}{4} = \dfrac{7\pi}{4}$

Because the angle is in quadrant IV, the reference angle is $\theta' = 2\pi - \dfrac{7\pi}{4} = \dfrac{\pi}{4}$.

57. $-\dfrac{11\pi}{4} + 4\pi = -\dfrac{11\pi}{4} + \dfrac{16\pi}{4} = \dfrac{5\pi}{4}$

Because the angle is in quadrant III, the reference angle is $\theta' = \dfrac{5\pi}{4} - \pi = \dfrac{\pi}{4}$.

59. $-\dfrac{25\pi}{6} + 6\pi = -\dfrac{25\pi}{6} + \dfrac{36\pi}{6} = \dfrac{11\pi}{6}$

Because the angle is in quadrant IV, the reference angle is $\theta' = 2\pi - \dfrac{11\pi}{6} = \dfrac{\pi}{6}$.

61. $225°$ lies in quadrant III. The reference angle is $\theta' = 225° - 180° = 45°$.

$$\cos 45° = \frac{\sqrt{2}}{2}$$

Because the cosine is negative in quadrant III,

$$\cos 225° = -\cos 45° = -\frac{\sqrt{2}}{2} \ .$$

63. $210°$ lies in quadrant III. The reference angle is
$\theta' = 210° - 180° = 30°$.

$\tan 30° = \dfrac{\sqrt{3}}{3}$

Because the tangent is positive in quadrant III,

$\tan 210° = \tan 30° = \dfrac{\sqrt{3}}{3}$.

65. $420°$ lies in quadrant I. The reference angle is
$\theta' = 420° - 360° = 60°$.

$\tan 60° = \sqrt{3}$

Because the tangent is positive in quadrant I,
$\tan 420° = \tan 60° = \sqrt{3}$.

67. $\dfrac{2\pi}{3}$ lies in quadrant II. The reference angle is

$\theta' = \pi - \dfrac{2\pi}{3} = \dfrac{3\pi}{3} - \dfrac{2\pi}{3} = \dfrac{\pi}{3}$.

$\sin \dfrac{\pi}{3} = \dfrac{\sqrt{3}}{2}$

Because the sine is positive in quadrant II,

$\sin \dfrac{2\pi}{3} = \sin \dfrac{\pi}{3} = \dfrac{\sqrt{3}}{2}$.

69. $\dfrac{7\pi}{6}$ lies in quadrant III. The reference angle is

$\theta' = \dfrac{7\pi}{6} - \pi = \dfrac{7\pi}{6} - \dfrac{6\pi}{6} = \dfrac{\pi}{6}$.

$\csc \dfrac{\pi}{6} = 2$

Because the cosecant is negative in quadrant III,

$\csc \dfrac{7\pi}{6} = -\csc \dfrac{\pi}{6} = -2$.

71. $\dfrac{9\pi}{4}$ lies in quadrant I. The reference angle is

$\theta' = \dfrac{9\pi}{4} - 2\pi = \dfrac{9\pi}{4} - \dfrac{8\pi}{4} = \dfrac{\pi}{4}$.

$\tan \dfrac{\pi}{4} = 1$

Because the tangent is positive in quadrant I,
$\tan \dfrac{9\pi}{4} = \tan \dfrac{\pi}{4} = 1$

73. $-240°$ lies in quadrant II. The reference angle is
$\theta' = 240° - 180° = 60°$.

$\sin 60° = \dfrac{\sqrt{3}}{2}$

Because the sine is positive in quadrant II,

$\sin(-240°) = \sin 60° = \dfrac{\sqrt{3}}{2}$.

75. $-\dfrac{\pi}{4}$ lies in quadrant IV. The reference angle is

$\theta' = \dfrac{\pi}{4}$.

$\tan \dfrac{\pi}{4} = 1$

Because the tangent is negative in quadrant IV,

$\tan\left(-\dfrac{\pi}{4}\right) = -\tan \dfrac{\pi}{4} = -1$

77. $\sec 495° = \sec 135° = -\sqrt{2}$

79. $\cot \dfrac{19\pi}{6} = \cot \dfrac{7\pi}{6} = \sqrt{3}$

81. $\cos \dfrac{23\pi}{4} = \cos \dfrac{7\pi}{4} = \dfrac{\sqrt{2}}{2}$

83. $\tan\left(-\dfrac{17\pi}{6}\right) = \tan \dfrac{7\pi}{6} = \dfrac{\sqrt{3}}{3}$

85. $\sin\left(-\dfrac{17\pi}{3}\right) = \sin \dfrac{\pi}{3} = \dfrac{\sqrt{3}}{2}$

87. $\sin \dfrac{\pi}{3} \cos \pi - \cos \dfrac{\pi}{3} \sin \dfrac{3\pi}{2}$

$= \left(\dfrac{\sqrt{3}}{2}\right)(-1) - \left(\dfrac{1}{2}\right)(-1)$

$= -\dfrac{\sqrt{3}}{2} + \dfrac{1}{2}$

$= \dfrac{1 - \sqrt{3}}{2}$

89. $\sin\dfrac{11\pi}{4}\cos\dfrac{5\pi}{6} + \cos\dfrac{11\pi}{4}\sin\dfrac{5\pi}{6}$

$= \left(\dfrac{\sqrt{2}}{2}\right)\left(-\dfrac{\sqrt{3}}{2}\right) + \left(-\dfrac{\sqrt{2}}{2}\right)\left(\dfrac{1}{2}\right)$

$= -\dfrac{\sqrt{6}}{4} - \dfrac{\sqrt{2}}{4}$

$= -\dfrac{\sqrt{6}+\sqrt{2}}{4}$

91. $\sin\dfrac{3\pi}{2}\tan\left(-\dfrac{15\pi}{4}\right) - \cos\left(-\dfrac{5\pi}{3}\right)$

$= (-1)(1) - \left(\dfrac{1}{2}\right)$

$= -1 - \dfrac{1}{2}$

$= -\dfrac{2}{2} - \dfrac{1}{2}$

$= -\dfrac{3}{2}$

93. $f\left(\dfrac{4\pi}{3}+\dfrac{\pi}{6}\right) + f\left(\dfrac{4\pi}{3}\right) + f\left(\dfrac{\pi}{6}\right)$

$= \sin\left(\dfrac{4\pi}{3}+\dfrac{\pi}{6}\right) + \sin\dfrac{4\pi}{3} + \sin\dfrac{\pi}{6}$

$= \sin\dfrac{3\pi}{2} + \sin\dfrac{4\pi}{3} + \sin\dfrac{\pi}{6}$

$= (-1) + \left(-\dfrac{\sqrt{3}}{2}\right) + \left(\dfrac{1}{2}\right)$

$= -\dfrac{\sqrt{3}+1}{2}$

95. $(h\circ g)\left(\dfrac{17\pi}{3}\right) = h\left(g\left(\dfrac{17\pi}{3}\right)\right)$

$= 2\left(\cos\left(\dfrac{17\pi}{3}\right)\right)$

$= 2\left(\dfrac{1}{2}\right)$

$= 1$

97. The average rate of change is the slope of the line through the points $(x_1, f(x_1))$ and $(x_2, f(x_2))$

$m = \dfrac{f(x_2) - f(x_1)}{x_2 - x_1}$

$= \dfrac{\sin\left(\dfrac{3\pi}{2}\right) - \sin\left(\dfrac{5\pi}{4}\right)}{\dfrac{3\pi}{2} - \dfrac{5\pi}{4}}$

$= \dfrac{-1 - \left(-\dfrac{\sqrt{2}}{2}\right)}{\dfrac{\pi}{4}}$

$= \dfrac{-1 + \dfrac{\sqrt{2}}{2}}{\dfrac{\pi}{4}}$

$= \dfrac{4\left(-1+\dfrac{\sqrt{2}}{2}\right)}{4\left(\dfrac{\pi}{4}\right)}$

$= \dfrac{2\sqrt{2} - 4}{\pi}$

99. $\sin\theta = \dfrac{\sqrt{2}}{2}$ when the reference angle is $\dfrac{\pi}{4}$ and θ is in quadrants I or II.

QI	QII
$\theta = \dfrac{\pi}{4}$	$\theta = \pi - \dfrac{\pi}{4}$
	$= \dfrac{3\pi}{4}$

$\theta = \dfrac{\pi}{4}, \dfrac{3\pi}{4}$

101. $\sin\theta = -\dfrac{\sqrt{2}}{2}$ when the reference angle is $\dfrac{\pi}{4}$ and

θ is in quadrants III or IV.

QIII	QIV
$\theta = \pi + \dfrac{\pi}{4}$	$\theta = 2\pi - \dfrac{\pi}{4}$
$= \dfrac{5\pi}{4}$	$= \dfrac{7\pi}{4}$

$\theta = \dfrac{5\pi}{4}, \dfrac{7\pi}{4}$

103. $\tan\theta = -\sqrt{3}$ when the reference angle is $\dfrac{\pi}{3}$ and

θ is in quadrants II or IV.

QII	QIV
$\theta = \pi - \dfrac{\pi}{3}$	$\theta = 2\pi - \dfrac{\pi}{3}$
$= \dfrac{2\pi}{3}$	$= \dfrac{5\pi}{3}$

$\theta = \dfrac{2\pi}{3}, \dfrac{5\pi}{3}$

105. – 109. Answers may vary.

111. does not make sense; Explanations will vary. Sample explanation: Sine and cosecant have the same sign within any quadrant because they are reciprocals of each other.

113. makes sense

114.

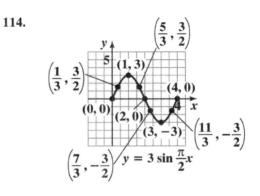

115. domain: $\{x \mid -1 \le x \le 1\}$ or $[-1,1]$

range: $\{y \mid -1 \le y \le 1\}$ or $[-1,1]$

116. a. $\sin\left(\dfrac{\pi}{4}\right) = \dfrac{\sqrt{2}}{2}$

$\sin\left(-\dfrac{\pi}{4}\right) = -\dfrac{\sqrt{2}}{2}$

$\sin\left(\dfrac{\pi}{3}\right) = \dfrac{\sqrt{3}}{2}$

$\sin\left(-\dfrac{\pi}{3}\right) = -\dfrac{\sqrt{3}}{2}$

The sine function is not even because $\sin(-\theta) \ne \sin\theta$

b. $\cos\left(\dfrac{\pi}{4}\right) = \dfrac{\sqrt{2}}{2}$

$\cos\left(-\dfrac{\pi}{4}\right) = \dfrac{\sqrt{2}}{2}$

$\cos\left(\dfrac{\pi}{3}\right) = \dfrac{1}{2}$

$\cos\left(-\dfrac{\pi}{3}\right) = \dfrac{1}{2}$

The cosine function is not odd because $\cos(-\theta) \ne -\cos\theta$

Section 5.4

Check Point Exercises

1. $P\left(\dfrac{\sqrt{3}}{2}, \dfrac{1}{2}\right)$

$\sin t = y = \dfrac{1}{2}$

$\cos t = x = \dfrac{\sqrt{3}}{2}$

$\tan t = \dfrac{y}{x} = \dfrac{\frac{1}{2}}{\frac{\sqrt{3}}{2}} = \dfrac{\sqrt{3}}{3}$

$\csc t = \dfrac{1}{y} = 2$

$\sec t = \dfrac{1}{x} = \dfrac{2\sqrt{3}}{3}$

$\cot t = \dfrac{x}{y} = \sqrt{3}$

2. The point P on the unit circle that corresponds to $t = \pi$ has coordinates $(-1, 0)$. Use $x = -1$ and $y = 0$ to find the values of the trigonometric functions.

$\sin \pi = y = 0$

$\cos \pi = x = -1$

$\tan \pi = \dfrac{y}{x} = \dfrac{0}{-1} = 0$

$\sec \pi = \dfrac{1}{x} = \dfrac{1}{-1} = -1$

$\cot \pi = \dfrac{x}{y} = \dfrac{-1}{0} = \text{undefined}$

$\csc \pi = \dfrac{1}{y} = \dfrac{1}{0} = \text{undefined}$

3. **a.** $\cos(-60°) = \cos 60° = \dfrac{1}{2}$

   b. $\tan\left(-\dfrac{\pi}{6}\right) = -\tan\left(\dfrac{\pi}{6}\right) = -\dfrac{\sqrt{3}}{3}$

4. **a.** $\cos 405° = \cos(360° + 45°)$

   $= \cos 45° = \dfrac{\sqrt{2}}{2}$

   b. $\sin \dfrac{7\pi}{3} = \sin\left(\dfrac{\pi}{3} + 2\pi\right) = \sin \dfrac{\pi}{3} = \dfrac{\sqrt{3}}{2}$

Concept and Vocabulary Check 5.4

1. intercepted arc

2. cosine; sine

3. sine; cosine; $(-\infty, \infty)$

4. 1; -1; $[-1, 1]$

5. $\cos t$; $\sec t$; even

6. $-\sin t$; $-\csc t$; $-\tan t$; $-\cot t$; odd

7. periodic; period

8. $\sin t$; $\cos t$; periodic; 2π

9. $\tan t$; $\cot t$; periodic; π

Exercise Set 5.4

1. The point P on the unit circle has coordinates $\left(-\dfrac{15}{17}, \dfrac{8}{17}\right)$. Use $x = -\dfrac{15}{17}$ and $y = \dfrac{8}{17}$ to find the values of the trigonometric functions.

$\sin t = y = \dfrac{8}{17}$

$\cos t = x = -\dfrac{15}{17}$

$\tan t = \dfrac{y}{x} = \dfrac{\frac{8}{17}}{-\frac{15}{17}} = -\dfrac{8}{15}$

$\csc t = \dfrac{1}{y} = \dfrac{17}{8}$

$\sec t = \dfrac{1}{x} = -\dfrac{17}{15}$

$\cot t = \dfrac{x}{y} = -\dfrac{15}{8}$

3. The point P on the unit circle that corresponds to $t = -\dfrac{\pi}{4}$ has coordinates $\left(\dfrac{\sqrt{2}}{2}, -\dfrac{\sqrt{2}}{2} \right)$. Use $x = \dfrac{\sqrt{2}}{2}$ and $y = -\dfrac{\sqrt{2}}{2}$ to find the values of the trigonometric functions.

$$\sin t = y = -\frac{\sqrt{2}}{2}$$

$$\cos t = x = \frac{\sqrt{2}}{2}$$

$$\tan t = \frac{y}{x} = \frac{-\frac{\sqrt{2}}{2}}{\frac{\sqrt{2}}{2}} = -1$$

$$\csc t = \frac{1}{y} = -\sqrt{2}$$

$$\sec t = \frac{1}{x} = \sqrt{2}$$

$$\cot t = \frac{x}{y} = -1$$

5. $\sin \dfrac{\pi}{6} = \dfrac{1}{2}$

7. $\cos \dfrac{5\pi}{6} = -\dfrac{\sqrt{3}}{2}$

9. $\tan \pi = \dfrac{0}{-1} = 0$

11. $\csc \dfrac{7\pi}{6} = \dfrac{1}{-\frac{1}{2}} = -2$

13. $\sec \dfrac{11\pi}{6} = \dfrac{1}{\frac{\sqrt{3}}{2}} = \dfrac{2\sqrt{3}}{3}$

15. $\sin \dfrac{3\pi}{2} = -1$

17. $\sec \dfrac{3\pi}{2} = $ undefined

19. a. $\cos \dfrac{\pi}{6} = \dfrac{\sqrt{3}}{2}$

   b. $\cos \left(-\dfrac{\pi}{6} \right) = \cos \dfrac{\pi}{6} = \dfrac{\sqrt{3}}{2}$

21. a. $\sin \dfrac{5\pi}{6} = \dfrac{1}{2}$

   b. $\sin \left(-\dfrac{5\pi}{6} \right) = -\sin \dfrac{5\pi}{6} = -\dfrac{1}{2}$

23. a. $\tan \dfrac{5\pi}{3} = \dfrac{-\frac{\sqrt{3}}{2}}{\frac{1}{2}} = -\sqrt{3}$

   b. $\tan \left(-\dfrac{5\pi}{3} \right) = -\tan \dfrac{5\pi}{3} = \sqrt{3}$

25. a. $\sin \dfrac{3\pi}{4} = \dfrac{\sqrt{2}}{2}$

   b. $\sin \dfrac{11\pi}{4} = \sin \left(\dfrac{3\pi}{4} + 2\pi \right) = \sin \dfrac{3\pi}{4} = \dfrac{\sqrt{2}}{2}$

27. a. $\cos \dfrac{\pi}{2} = 0$

   b. $\cos \dfrac{9\pi}{2} = \cos \left(\dfrac{\pi}{2} + 4\pi \right)$
   $$= \cos \left[\dfrac{\pi}{2} + 2(2\pi) \right]$$
   $$= \cos \dfrac{\pi}{2}$$
   $$= 0$$

29. a. $\tan \pi = \dfrac{0}{-1} = 0$

   b. $\tan 17\pi = \tan(\pi + 16\pi)$
   $$= \tan[\pi + 8(2\pi)]$$
   $$= \tan \pi$$
   $$= 0$$

31. a. $\sin\dfrac{7\pi}{4} = -\dfrac{\sqrt{2}}{2}$

    b. $\sin\dfrac{47\pi}{4} = \sin\left(\dfrac{7\pi}{4} + 10\pi\right)$

$$= \sin\left[\dfrac{7\pi}{4} + 5(2\pi)\right]$$

$$= \sin\dfrac{7\pi}{4}$$

$$= -\dfrac{\sqrt{2}}{2}$$

33. $\sin(-t) - \sin t = -\sin t - \sin t = -2\sin t = -2a$

35. $4\cos(-t) - \cos t = 4\cos t - \cos t = 3\cos t = 3b$

37. $\sin(t + 2\pi) - \cos(t + 4\pi) + \tan(t + \pi)$

$$= \sin(t) - \cos(t) + \tan(t)$$

$$= a - b + c$$

39. $\sin(-t - 2\pi) - \cos(-t - 4\pi) - \tan(-t - \pi)$

$$= -\sin(t + 2\pi) - \cos(t + 4\pi) + \tan(t + \pi)$$

$$= -\sin(t) - \cos(t) + \tan(t)$$

$$= -a - b + c$$

41. $\cos t + \cos(t + 1000\pi) - \tan t - \tan(t + 999\pi)$

$$\qquad\qquad - \sin t + 4\sin(t - 1000\pi)$$

$$= \cos t + \cos t - \tan t - \tan t - \sin t + 4\sin t$$

$$= 2\cos t - 2\tan t + 3\sin t$$

$$= 3a + 2b - 2c$$

43. a. $H = 12 + 8.3\sin\left[\dfrac{2\pi}{365}(80 - 80)\right]$

$$= 12 + 8.3\sin 0 = 12 + 8.3(0)$$

$$= 12$$

There are 12 hours of daylight in Fairbanks on March 21.

    b. $H = 12 + 8.3\sin\left[\dfrac{2\pi}{365}(172 - 80)\right]$

$$\approx 12 + 8.3\sin 1.5837$$

$$\approx 20.3$$

There are about 20.3 hours of daylight in Fairbanks on June 21.

    c. $H = 12 + 8.3\sin\left[\dfrac{2\pi}{365}(355 - 80)\right]$

$$\approx 12 + 8.3\sin 4.7339$$

$$\approx 3.7$$

There are about 3.7 hours of daylight in Fairbanks on December 21.

45. a. For $t = 7$,

$$E = \sin\dfrac{\pi}{14} \cdot 7 = \sin\dfrac{\pi}{2} = 1$$

For $t = 14$,

$$E = \sin\dfrac{\pi}{14} \cdot 14 = \sin\pi = 0$$

For $t = 21$,

$$E = \sin\dfrac{\pi}{14} \cdot 21 = \sin\dfrac{3\pi}{2} = -1$$

For $t = 28$,

$$E = \sin\dfrac{\pi}{14} \cdot 28 = \sin 2\pi = \sin 0 = 0$$

For $t = 35$,

$$E = \sin\dfrac{\pi}{14} \cdot 35 = \sin\dfrac{5\pi}{2} = \sin\dfrac{\pi}{2} = 1$$

Observations may vary.

    b. Because $E(35) = E(7) = 1$, the period is $35 - 7 = 28$ or 28 days.

47. – 51. Answers may vary.

53. makes sense

55. does not make sense; Explanations will vary. Sample explanation: Cosine is not an odd function.

57. $\cos 0° + \cos 1° + \cos 2° + \cos 3° + \cdots + \cos 179° + \cos 180°$

$= (\cos 0° + \cos 180°) + (\cos 1° + \cos 179°) + (\cos 2° + \cos 178°) + \cdots$

$= \overbrace{(\cos 0° + \cos 180°)}^{0} + \overbrace{(\cos 1° + \cos 179°)}^{0} + \overbrace{(\cos 2° + \cos 178°)}^{0} + \cdots$

$= 0 + 0 + 0 + \cdots + 0$

$= 0$

59. $f(a) + 2f(-a) = \sin(a) + 2\sin(-a)$

$\qquad\qquad\qquad = \sin(a) - 2\sin(a)$

$\qquad\qquad\qquad = f(a) - 2f(a)$

$\qquad\qquad\qquad = \dfrac{1}{4} - 2\left(\dfrac{1}{4}\right)$

$\qquad\qquad\qquad = -\dfrac{1}{4}$

61. $y = \dfrac{1}{2}\cos(4x + \pi)$

x	$-\dfrac{\pi}{4}$	$-\dfrac{\pi}{8}$	0	$\dfrac{\pi}{8}$	$\dfrac{\pi}{4}$
y	$\dfrac{1}{2}$	0	$-\dfrac{1}{2}$	0	$\dfrac{1}{2}$

62. $y = 4\sin\left(2x - \dfrac{2\pi}{3}\right)$

x	$\dfrac{\pi}{3}$	$\dfrac{7\pi}{12}$	$\dfrac{5\pi}{6}$	$\dfrac{13\pi}{12}$	$\dfrac{4\pi}{3}$
y	0	4	0	-4	0

63. $y = 3\sin\dfrac{\pi}{2}x$

x	0	$\dfrac{1}{3}$	1	$\dfrac{5}{3}$	2	$\dfrac{7}{3}$	3	$\dfrac{11}{3}$	4
y	0	$\dfrac{3}{2}$	3	$\dfrac{3}{2}$	0	$-\dfrac{3}{2}$	-3	$-\dfrac{3}{2}$	0

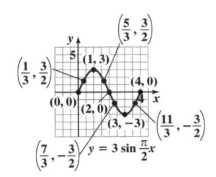

Mid-Chapter 5 Check Point

1. $10° = 10° \cdot \dfrac{\pi \text{ radians}}{180°} = \dfrac{10\pi}{180} \text{ radians}$

$\qquad = \dfrac{\pi}{18} \text{ radians}$

2. $-105° = -105° \cdot \dfrac{\pi \text{ radians}}{180°} = -\dfrac{105\pi}{180} \text{ radians}$

$\qquad = -\dfrac{7\pi}{12} \text{ radians}$

3. $\dfrac{5\pi}{12} \text{ radians} = \dfrac{5\pi \text{ radians}}{12} \cdot \dfrac{180^{\text{O}}}{\pi \text{ radians}} = 75^{\text{O}}$

4. $-\dfrac{13\pi}{20} \text{ radians} = -\dfrac{13\pi \text{ radians}}{20} \cdot \dfrac{180^{\text{O}}}{\pi \text{ radians}}$

$\qquad\qquad = -117^{\text{O}}$

5. **a.** $\dfrac{11\pi}{3} - 2\pi = \dfrac{11\pi}{3} - \dfrac{6\pi}{3} = \dfrac{5\pi}{3}$

b.

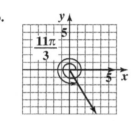

c. Since $\dfrac{5\pi}{3}$ is in quadrant IV, the reference angle

is $2\pi - \dfrac{5\pi}{3} = \dfrac{6\pi}{3} - \dfrac{5\pi}{3} = \dfrac{\pi}{3}$

6. **a.** $-\dfrac{19\pi}{4} + 6\pi = -\dfrac{19\pi}{4} + \dfrac{24\pi}{4} = \dfrac{5\pi}{4}$

b.

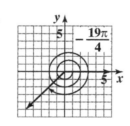

c. Since $\dfrac{5\pi}{4}$ is in quadrant III, the reference angle

is $\dfrac{5\pi}{4} - \pi = \dfrac{5\pi}{4} - \dfrac{4\pi}{4} = \dfrac{\pi}{4}$

7. **a.** $510° - 360° = 150°$

b.

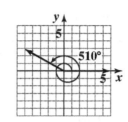

c. Since $150°$ is in quadrant II, the reference angle is $180° - 150° = 30°$

8. Use the Pythagorean theorem to find *b*.

$a^2 + b^2 = c^2$

$5^2 + b^2 = 6^2$

$25 + b^2 = 36$

$b^2 = 11$

$b = \sqrt{11}$

$\sin\theta = \dfrac{\text{opposite}}{\text{hypotenuse}} = \dfrac{5}{6}$

$\cos\theta = \dfrac{\text{adjacent}}{\text{hypotenuse}} = \dfrac{\sqrt{11}}{6}$

$\tan\theta = \dfrac{\text{opposite}}{\text{adjacent}} = \dfrac{5\sqrt{11}}{11}$

$\csc\theta = \dfrac{\text{hypotenuse}}{\text{opposite}} = \dfrac{6}{5}$

$\sec\theta = \dfrac{\text{hypotenuse}}{\text{adjacent}} = \dfrac{6}{\sqrt{11}} = \dfrac{6\sqrt{11}}{11}$

$\cot\theta = \dfrac{\text{adjacent}}{\text{opposite}} = \dfrac{\sqrt{11}}{5}$

9. $r = \sqrt{x^2 + y^2}$

$r = \sqrt{3^2 + (-2)^2} = \sqrt{9+4} = \sqrt{13}$

Now that we know *x*, *y*, and *r*, we can find the six trigonometric functions of θ.

$\sin \theta = \dfrac{y}{r} = \dfrac{-2}{\sqrt{13}} = -\dfrac{2\sqrt{13}}{13}$

$\cos \theta = \dfrac{x}{r} = \dfrac{3}{\sqrt{13}} = \dfrac{3\sqrt{13}}{13}$

$\tan \theta = \dfrac{y}{x} = \dfrac{-2}{3} = -\dfrac{2}{3}$

$\csc \theta = \dfrac{r}{y} = \dfrac{\sqrt{13}}{-2} = -\dfrac{\sqrt{13}}{2}$

$\sec \theta = \dfrac{r}{x} = \dfrac{\sqrt{13}}{3}$

$\cot \theta = \dfrac{x}{y} = \dfrac{3}{-2} = -\dfrac{3}{2}$

10. $r = \sqrt{x^2 + y^2}$

$r = \sqrt{\left(-\dfrac{3}{5}\right)^2 + \left(-\dfrac{4}{5}\right)^2} = \sqrt{\dfrac{9}{25} + \dfrac{16}{25}} = \sqrt{\dfrac{25}{25}} = 1$

Now that we know *x*, *y*, and *r*, we can find the six trigonometric functions of θ.

$\sin \theta = \dfrac{y}{r} = \dfrac{-\dfrac{4}{5}}{1} = -\dfrac{4}{5}$

$\cos \theta = \dfrac{x}{r} = \dfrac{-\dfrac{3}{5}}{1} = -\dfrac{3}{5}$

$\tan \theta = \dfrac{y}{x} = \dfrac{-\dfrac{4}{5}}{-\dfrac{3}{5}} = \dfrac{4}{3}$

$\csc \theta = \dfrac{r}{y} = \dfrac{1}{-\dfrac{4}{5}} = -\dfrac{5}{4}$

$\sec \theta = \dfrac{r}{x} = \dfrac{1}{-\dfrac{3}{5}} = -\dfrac{5}{3}$

$\cot \theta = \dfrac{x}{y} = \dfrac{-\dfrac{3}{5}}{-\dfrac{4}{5}} = \dfrac{3}{4}$

11. Because the tangent is negative and the cosine is negative, θ is in quadrant II. In quadrant II, *x* is negative and *y* is positive. Thus,

$\tan \theta = -\dfrac{3}{4} = \dfrac{x}{y}$, $\quad x = -4$, $y = 3$. Furthermore,

$r^2 = x^2 + y^2$

$r^2 = (-3)^2 + 4^2$

$r^2 = 9 + 16 = 25$

$r = 5$

Now that we know *x*, *y*, and *r*, we can find the remaining trigonometric functions of θ.

$\sin \theta = \dfrac{y}{r} = \dfrac{3}{5}$

$\cos \theta = \dfrac{x}{r} = \dfrac{-4}{5} = -\dfrac{4}{5}$

$\csc \theta = \dfrac{r}{y} = \dfrac{5}{3}$

$\sec \theta = \dfrac{r}{x} = \dfrac{5}{-3} = -\dfrac{5}{4}$

$\cot \theta = \dfrac{x}{y} = \dfrac{-3}{4} = -\dfrac{4}{3}$

12. Since $\cos \theta = \dfrac{3}{7} = \dfrac{x}{r}$, $\quad x = 3$, $r = 7$. Furthermore,

$x^2 + y^2 = r^2$

$3^2 + y^2 = 7^2$

$9 + y^2 = 49$

$y^2 = 40$

$y = \pm\sqrt{40} = \pm 2\sqrt{10}$

Because the cosine is positive and the sine is negative, θ is in quadrant IV. In quadrant IV, *x* is positive and *y* is negative.

Therefore $y = -2\sqrt{10}$

Use *x*, *y*, and *r* to find the remaining trigonometric functions of θ.

$\sin \theta = \dfrac{y}{r} = \dfrac{-2\sqrt{10}}{7} = -\dfrac{2\sqrt{10}}{7}$

$\tan \theta = \dfrac{y}{x} = \dfrac{-2\sqrt{10}}{3} = -\dfrac{2\sqrt{10}}{3}$

$\csc \theta = \dfrac{r}{y} = \dfrac{7}{-2\sqrt{10}} = -\dfrac{7\sqrt{10}}{20}$

$\sec \theta = \dfrac{r}{x} = \dfrac{7}{3}$

$\cot \theta = \dfrac{x}{y} = \dfrac{3}{-2\sqrt{10}} = -\dfrac{3\sqrt{10}}{20}$

13. $\tan \theta = \dfrac{\text{side opposite } \theta}{\text{side adjacent } \theta}$

$\tan 41° = \dfrac{a}{60}$

$\qquad a = 60 \tan 41°$

$\qquad a \approx 52 \text{ cm}$

14. $\cos \theta = \dfrac{\text{side adjacent } \theta}{\text{hypotenuse}}$

$\cos 72° = \dfrac{250}{c}$

$\qquad c = \dfrac{250}{\cos 72°}$

$\qquad c \approx 809 \text{ m}$

15. Since $\cos \theta = \dfrac{1}{6} = \dfrac{x}{r}$, $x = 1$, $r = 6$. Furthermore,

$x^2 + y^2 = r^2$

$1^2 + y^2 = 6^2$

$1 + y^2 = 36$

$\qquad y^2 = 35$

$\qquad y = \pm\sqrt{35}$

Since θ is acute, $y = +\sqrt{35} = \sqrt{35}$

$\cot\left(\dfrac{\pi}{2} - \theta\right) = \tan \theta = \dfrac{y}{x} = \dfrac{\sqrt{35}}{1} = \sqrt{35}$

16. $\tan 30° = \dfrac{\sqrt{3}}{3}$

17. $\cot 120° = \dfrac{1}{\tan 120°} = \dfrac{1}{-\tan 60°} = \dfrac{1}{-\sqrt{3}} = -\dfrac{\sqrt{3}}{3}$

18. $\cos 240° = -\cos 60° = -\dfrac{1}{2}$

19. $\sec \dfrac{11\pi}{6} = \dfrac{1}{\cos \dfrac{11\pi}{6}} = \dfrac{1}{\cos \dfrac{\pi}{6}} = \dfrac{1}{\dfrac{\sqrt{3}}{2}} = \dfrac{2}{\sqrt{3}} = \dfrac{2\sqrt{3}}{3}$

20. $\sin^2 \dfrac{\pi}{7} + \cos^2 \dfrac{\pi}{7} = 1$

21. $\sin\left(-\dfrac{2\pi}{3}\right) = \sin\left(-\dfrac{2\pi}{3} + 2\pi\right)$

$\qquad = \sin \dfrac{4\pi}{3} = -\sin \dfrac{\pi}{3}$

$\qquad = -\dfrac{\sqrt{3}}{2}$

22. $\csc\left(\dfrac{22\pi}{3}\right) = \csc\left(\dfrac{22\pi}{3} - 6\pi\right) = \csc \dfrac{4\pi}{3}$

$\qquad = \dfrac{1}{\sin \dfrac{4\pi}{3}} = \dfrac{1}{-\sin \dfrac{\pi}{3}} = \dfrac{1}{-\dfrac{\sqrt{3}}{2}}$

$\qquad = -\dfrac{2}{\sqrt{3}} = -\dfrac{2\sqrt{3}}{3}$

23. $\cos 495° = \cos\left(495° - 360°\right) = \cos 135°$

$\qquad = -\cos 45° = -\dfrac{\sqrt{2}}{2}$

24. $\tan\left(-\dfrac{17\pi}{6}\right) = \tan\left(-\dfrac{17\pi}{6} + 4\pi\right) = \tan \dfrac{7\pi}{6}$

$\qquad = \tan \dfrac{\pi}{6} = \dfrac{\sqrt{3}}{3}$

25. $\sin^2 \dfrac{\pi}{2} - \cos \pi = (1)^2 - (-1) = 1 + 1 = 2$

26. $\cos\left(\dfrac{5\pi}{6} + 2\pi n\right) + \tan\left(\dfrac{5\pi}{6} + n\pi\right)$

$\qquad = \cos \dfrac{5\pi}{6} + \tan \dfrac{5\pi}{6} = -\cos \dfrac{\pi}{6} - \tan \dfrac{\pi}{6}$

$\qquad = -\dfrac{\sqrt{3}}{2} - \dfrac{\sqrt{3}}{3} = -\dfrac{3\sqrt{3}}{6} - \dfrac{2\sqrt{3}}{6}$

$\qquad = -\dfrac{5\sqrt{3}}{6}$

27. Begin by converting from degrees to radians.

$36° = 36° \cdot \dfrac{\pi \text{ radians}}{180°} = \dfrac{\pi}{5} \text{ radians}$

$s = r\theta = 40 \cdot \dfrac{\pi}{5} = 8\pi \approx 25.13 \text{ cm}$

28. Linear speed is given by $v = r\omega$. It is given that $r = 10$ feet and the merry-go-round rotates at 8 revolutions per minute. Convert 8 revolutions per minute to radians per minute.

8 revolutions per minute

$= 8 \text{ revolutions per minute} \cdot \dfrac{2\pi \text{ radians}}{1 \text{ revolution}}$

$= 16\pi \text{ radians per minute}$

$v = r\omega = (10)(16\pi) = 160\pi \approx 502.7 \text{ feet per minute}$

The linear speed of the horse is about 502.7 feet per minute.

29. $\sin\theta = \dfrac{\text{side opposite } \theta}{\text{hypotenuse}}$

$\sin 6° = \dfrac{h}{5280}$

$h = 5280 \sin 6°$

$h \approx 551.9 \text{ feet}$

30. $\tan\theta = \dfrac{\text{side opposite } \theta}{\text{side adjacent } \theta}$

$\tan\theta = \dfrac{50}{60}$

$\theta = \tan^{-1}\left(\dfrac{50}{60}\right)$

$\theta \approx 40°$

Section 5.5

Checkpoint Exercises

1. The equation $y = 3\sin x$ is of the form $y = A\sin x$ with $A = 3$. Thus, the amplitude is $|A| = |3| = 3$ The period for both $y = 3\sin x$ and $y = \sin x$ is 2π. We find the three x–intercepts, the maximum point, and the minimum point on the interval $[0, 2\pi]$ by dividing the period, 2π, by 4,

$\dfrac{\text{period}}{4} = \dfrac{2\pi}{4} = \dfrac{\pi}{2}$, then by adding quarter-periods to generate x-values for each of the key points. The five x-values are

$x = 0$

$x = 0 + \dfrac{\pi}{2} = \dfrac{\pi}{2}$

$x = \dfrac{\pi}{2} + \dfrac{\pi}{2} = \pi$

$x = \pi + \dfrac{\pi}{2} = \dfrac{3\pi}{2}$

$x = \dfrac{3\pi}{2} + \dfrac{\pi}{2} = 2\pi$

Evaluate the function at each value of x.

x	$y = 3\sin x$	coordinates
0	$y = 3\sin 0 = 3 \cdot 0 = 0$	$(0, 0)$
$\dfrac{\pi}{2}$	$y = 3\sin\dfrac{\pi}{2} = 3 \cdot 1 = 3$	$\left(\dfrac{\pi}{2}, 3\right)$
π	$y = 3\sin x = 3 \cdot 0 = 0$	$(\pi, 0)$
$\dfrac{3\pi}{2}$	$y = 3\sin\dfrac{3\pi}{2}$ $= 3(-1) = -3$	$\left(\dfrac{3\pi}{2}, -3\right)$
2π	$y = 3\sin 2\pi = 3 \cdot 0 = 0$	$(2\pi, 0)$

Connect the five points with a smooth curve and graph one complete cycle of the given function with the graph of $y = \sin x$.

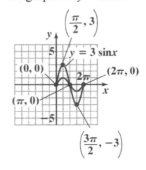

2. The equation $y = -\dfrac{1}{2}\sin x$ is of the form $y = A\sin x$ with $A = -\dfrac{1}{2}$. Thus, the amplitude is $|A| = \left|-\dfrac{1}{2}\right| = \dfrac{1}{2}$. The period for both $y = -\dfrac{1}{2}\sin x$ and $y = \sin x$ is 2π.

Find the *x*–values for the five key points by dividing the period, 2π, by 4, $\dfrac{\text{period}}{4} = \dfrac{2\pi}{4} = \dfrac{\pi}{2}$, then by adding quarter- periods. The five *x*-values are
$x = 0$

$x = 0 + \dfrac{\pi}{2} = \dfrac{\pi}{2}$

$x = \dfrac{\pi}{2} + \dfrac{\pi}{2} = \pi$

$x = \pi + \dfrac{\pi}{2} = \dfrac{3\pi}{2}$

$x = \dfrac{3\pi}{2} + \dfrac{\pi}{2} = 2\pi$

Evaluate the function at each value of *x*.

x	$y = -\dfrac{1}{2}\sin x$	coordinates
0	$y = -\dfrac{1}{2}\sin 0$ $= -\dfrac{1}{2} \cdot 0 = 0$	$(0, 0)$
$\dfrac{\pi}{2}$	$y = -\dfrac{1}{2}\sin \dfrac{\pi}{2}$ $= -\dfrac{1}{2} \cdot 1 = -\dfrac{1}{2}$	$\left(\dfrac{\pi}{2}, -\dfrac{1}{2}\right)$
π	$y = -\dfrac{1}{2}\sin \pi$ $= -\dfrac{1}{2} \cdot 0 = 0$	$(\pi, 0)$
$\dfrac{3\pi}{2}$	$y = -\dfrac{1}{2}\sin \dfrac{3\pi}{2}$ $= -\dfrac{1}{2}(-1) = \dfrac{1}{2}$	$\left(\dfrac{3\pi}{2}, \dfrac{1}{2}\right)$
2π	$y = -\dfrac{1}{2}\sin 2\pi$ $= -\dfrac{1}{2} \cdot 0 = 0$	$(2\pi, 0)$

Connect the five key points with a smooth curve and graph one complete cycle of the given function with the graph of $y = \sin x$. Extend the pattern of each graph to the left and right as desired.

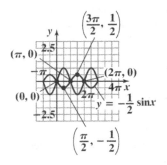

3. The equation $y = 2\sin \dfrac{1}{2}x$ is of the form

$y = A\sin Bx$ with $A = 2$ and $B = \dfrac{1}{2}$.

The amplitude is $|A| = |2| = 2$.

The period is $\dfrac{2\pi}{B} = \dfrac{2\pi}{\frac{1}{2}} = 4\pi$.

Find the *x*–values for the five key points by dividing the period, 4π, by 4, $\dfrac{\text{period}}{4} = \dfrac{4\pi}{4} = \pi$, then by adding quarter-periods.
The five *x*-values are
$x = 0$

$x = 0 + \pi = \pi$

$x = \pi + \pi = 2\pi$

$x = 2\pi + \pi = 3\pi$

$x = 3\pi + \pi = 4\pi$

Evaluate the function at each value of *x*.

x	$y = 2\sin \dfrac{1}{2}x$	coordinates
0	$y = 2\sin\left(\dfrac{1}{2} \cdot 0\right)$ $= 2\sin 0$ $= 2 \cdot 0 = 0$	$(0, 0)$
π	$y = 2\sin\left(\dfrac{1}{2} \cdot \pi\right)$ $= 2\sin \dfrac{\pi}{2} = 2 \cdot 1 = 2$	$(\pi, 2)$
2π	$y = 2\sin\left(\dfrac{1}{2} \cdot 2\pi\right)$ $= 2\sin \pi = 2 \cdot 0 = 0$	$(2\pi, 0)$

3π	$y = 2\sin\left(\dfrac{1}{2}\cdot 3\pi\right)$ $= 2\sin\dfrac{3\pi}{2}$ $= 2\cdot(-1) = -2$	$(3\pi, -2)$
4π	$y = 2\sin\left(\dfrac{1}{2}\cdot 4\pi\right)$ $= 2\sin 2\pi = 2\cdot 0 = 0$	$(4\pi, 0)$

Connect the five key points with a smooth curve and graph one complete cycle of the given function. Extend the pattern of the graph another full period to the right.

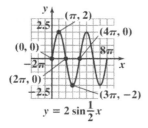

$y = 2\sin\dfrac{1}{2}x$

4. The equation $y = 3\sin\left(2x - \dfrac{\pi}{3}\right)$ is of the form

$y = A\sin(Bx - C)$ with $A = 3$, $B = 2$, and $C = \dfrac{\pi}{3}$.

The amplitude is $|A| = |3| = 3$.

The period is $\dfrac{2\pi}{B} = \dfrac{2\pi}{2} = \pi$.

The phase shift is $\dfrac{C}{B} = \dfrac{\frac{\pi}{3}}{2} = \dfrac{\pi}{3}\cdot\dfrac{1}{2} = \dfrac{\pi}{6}$.

Find the x-values for the five key points by dividing the period, π, by 4, $\dfrac{\text{period}}{4} = \dfrac{\pi}{4}$, then by adding quarter-periods to the value of x where the cycle begins, $x = \dfrac{\pi}{6}$.

The five x-values are

$x = \dfrac{\pi}{6}$

$x = \dfrac{\pi}{6} + \dfrac{\pi}{4} = \dfrac{2\pi}{12} + \dfrac{3\pi}{12} = \dfrac{5\pi}{12}$

$x = \dfrac{5\pi}{12} + \dfrac{\pi}{4} = \dfrac{5\pi}{12} + \dfrac{3\pi}{12} = \dfrac{8\pi}{12} = \dfrac{2\pi}{3}$

$x = \dfrac{2\pi}{3} + \dfrac{\pi}{4} = \dfrac{8\pi}{12} + \dfrac{3\pi}{12} = \dfrac{11\pi}{12}$

$x = \dfrac{11\pi}{12} + \dfrac{\pi}{4} = \dfrac{11\pi}{12} + \dfrac{3\pi}{12} = \dfrac{14\pi}{12} = \dfrac{7\pi}{6}$

Evaluate the function at each value of x.

x	$y = 3\sin\left(2x - \dfrac{\pi}{3}\right)$	coordinates
$\dfrac{\pi}{6}$	$y = 3\sin\left(2\cdot\dfrac{\pi}{6} - \dfrac{\pi}{3}\right)$ $= 3\sin 0 = 3\cdot 0 = 0$	$\left(\dfrac{\pi}{6}, 0\right)$
$\dfrac{5\pi}{12}$	$y = 3\sin\left(2\cdot\dfrac{5\pi}{12} - \dfrac{\pi}{3}\right)$ $= 3\sin\dfrac{3\pi}{6} = 3\sin\dfrac{\pi}{2}$ $= 3\cdot 1 = 3$	$\left(\dfrac{5\pi}{12}, 3\right)$
$\dfrac{2\pi}{3}$	$y = 3\sin\left(2\cdot\dfrac{2\pi}{3} - \dfrac{\pi}{3}\right)$ $= 3\sin\dfrac{3\pi}{3} = 3\sin\pi$ $= 3\cdot 0 = 0$	$\left(\dfrac{2\pi}{3}, 0\right)$
$\dfrac{11\pi}{12}$	$y = 3\sin\left(2\cdot\dfrac{11\pi}{12} - \dfrac{\pi}{3}\right)$ $= 3\sin\dfrac{9\pi}{6} = 3\sin\dfrac{3\pi}{2}$ $= 3(-1) = -3$	$\left(\dfrac{11\pi}{12}, -3\right)$
$\dfrac{7\pi}{6}$	$y = 3\sin\left(2\cdot\dfrac{7\pi}{6} - \dfrac{\pi}{3}\right)$ $= 3\sin\dfrac{6\pi}{3} = 3\sin 2\pi$ $= 3\cdot 0 = 0$	$\left(\dfrac{7\pi}{6}, 0\right)$

Connect the five key points with a smooth curve and graph one complete cycle of the given graph.

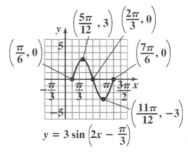

$y = 3\sin\left(2x - \dfrac{\pi}{3}\right)$

5. The equation $y = -4\cos \pi x$ is of the form
$y = A\cos Bx$ with $A = -4$, and $B = \pi$.
Thus, the amplitude is $|A| = |-4| = 4$.

The period is $\dfrac{2\pi}{B} = \dfrac{2\pi}{\pi} = 2$.

Find the x-values for the five key points by dividing
the period, 2, by 4, $\dfrac{\text{period}}{4} = \dfrac{2}{4} = \dfrac{1}{2}$, then by adding
quarter periods to the value of x where the cycle
begins. The five x-values are
$x = 0$

$x = 0 + \dfrac{1}{2} = \dfrac{1}{2}$

$x = \dfrac{1}{2} + \dfrac{1}{2} = 1$

$x = 1 + \dfrac{1}{2} = \dfrac{3}{2}$

$x = \dfrac{3}{2} + \dfrac{1}{2} = 2$

Evaluate the function at each value of x.

x	$y = -4\cos \pi x$	coordinates
0	$y = -4\cos(\pi \cdot 0)$ $= -4\cos 0 = -4$	$(0, -4)$
$\dfrac{1}{2}$	$y = -4\cos\left(\pi \cdot \dfrac{1}{2}\right)$ $= -4\cos\dfrac{\pi}{2} = 0$	$\left(\dfrac{1}{2}, 0\right)$
1	$y = -4\cos(\pi \cdot 1)$ $= -4\cos \pi = 4$	$(1, 4)$
$\dfrac{3}{2}$	$y = -4\cos\left(\pi \cdot \dfrac{3}{2}\right)$ $= -4\cos\dfrac{3\pi}{2} = 0$	$\left(\dfrac{3}{2}, 0\right)$
2	$y = -4\cos(\pi \cdot 2)$ $= -4\cos 2\pi = -4$	$(2, -4)$

Connect the five key points with a smooth curve and
graph one complete cycle of the given function. Extend
the pattern of the graph another full period to the left.

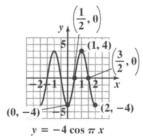

6. $y = \dfrac{3}{2}\cos(2x + \pi) = \dfrac{3}{2}\cos(2x - (-\pi))$

The equation is of the form $y = A\cos(Bx - C)$ with
$A = \dfrac{3}{2}$, $B = 2$, and $C = -\pi$.

Thus, the amplitude is $|A| = \left|\dfrac{3}{2}\right| = \dfrac{3}{2}$.

The period is $\dfrac{2\pi}{B} = \dfrac{2\pi}{2} = \pi$.

The phase shift is $\dfrac{C}{B} = \dfrac{-\pi}{2} = -\dfrac{\pi}{2}$.

Find the x-values for the five key points by dividing
the period, π, by 4, $\dfrac{\text{period}}{4} = \dfrac{\pi}{4}$, then by adding
quarter-periods to the value of x where the cycle
begins, $x = -\dfrac{\pi}{2}$.

The five x-values are
$x = -\dfrac{\pi}{2}$

$x = -\dfrac{\pi}{2} + \dfrac{\pi}{4} = -\dfrac{\pi}{4}$

$x = -\dfrac{\pi}{4} + \dfrac{\pi}{4} = 0$

$x = 0 + \dfrac{\pi}{4} = \dfrac{\pi}{4}$

$x = \dfrac{\pi}{4} + \dfrac{\pi}{4} = \dfrac{\pi}{2}$

Evaluate the function at each value of x.

x	$y = \dfrac{3}{2}\cos(2x+\pi)$	coordinates
$-\dfrac{\pi}{2}$	$y = \dfrac{3}{2}\cos(-\pi+\pi)$ $= \dfrac{3}{2}\cdot 1 = \dfrac{3}{2}$	$\left(-\dfrac{\pi}{2},\dfrac{3}{2}\right)$
$-\dfrac{\pi}{4}$	$y = \dfrac{3}{2}\cos\left(-\dfrac{\pi}{2}+\pi\right)$ $= \dfrac{3}{2}\cdot 0 = 0$	$\left(-\dfrac{\pi}{4},0\right)$
0	$y = \dfrac{3}{2}\cos(0+\pi)$ $= \dfrac{3}{2}\cdot -1 = -\dfrac{3}{2}$	$\left(0,-\dfrac{3}{2}\right)$
$\dfrac{\pi}{4}$	$y = \dfrac{3}{2}\cos\left(\dfrac{\pi}{2}+\pi\right)$ $= \dfrac{3}{2}\cdot 0 = 0$	$\left(\dfrac{\pi}{4},0\right)$
$\dfrac{\pi}{2}$	$y = \dfrac{3}{2}\cos(\pi+\pi)$ $= \dfrac{3}{2}\cdot 1 = \dfrac{3}{2}$	$\left(\dfrac{\pi}{2},\dfrac{3}{2}\right)$

Connect the five key points with a smooth curve and graph one complete cycle of the given graph.

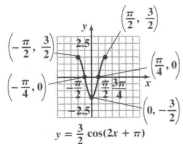

7. The graph of $y = 2\cos x + 1$ is the graph of $y = 2\cos x$ shifted one unit upwards. The period for both functions is 2π. The quarter-period is $\dfrac{2\pi}{4}$ or $\dfrac{\pi}{2}$. The cycle begins at $x = 0$. Add quarter-periods to generate x-values for the key points.

$x = 0$

$x = 0 + \dfrac{\pi}{2} = \dfrac{\pi}{2}$

$x = \dfrac{\pi}{2} + \dfrac{\pi}{2} = \pi$

$x = \pi + \dfrac{\pi}{2} = \dfrac{3\pi}{2}$

$x = \dfrac{3\pi}{2} + \dfrac{\pi}{2} = 2\pi$

Evaluate the function at each value of x.

x	$y = 2\cos x + 1$	coordinates
0	$y = 2\cos 0 + 1$ $= 2\cdot 1 + 1 = 3$	$(0, 3)$
$\dfrac{\pi}{2}$	$y = 2\cos\dfrac{\pi}{2} + 1$ $= 2\cdot 0 + 1 = 1$	$\left(\dfrac{\pi}{2}, 1\right)$
π	$y = 2\cos \pi + 1$ $= 2\cdot(-1)+1 = -1$	$(\pi, -1)$
$\dfrac{3\pi}{2}$	$y = 2\cos\dfrac{3\pi}{2} + 1$ $= 2\cdot 0 + 1 = 1$	$\left(\dfrac{3\pi}{2}, 1\right)$
2π	$y = 2\cos 2\pi + 1$ $= 2\cdot 1 + 1 = 3$	$(2\pi, 3)$

By connecting the points with a smooth curve, we obtain one period of the graph.

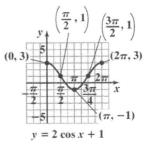

8. A, the amplitude, is the maximum value of y. The graph shows that this maximum value is 4, Thus, $A = 4$. The period is $\dfrac{\pi}{2}$, and period $= \dfrac{2\pi}{B}$. Thus,

$\dfrac{\pi}{2} = \dfrac{2\pi}{B}$

$\pi B = 4\pi$

$B = 4$

Substitute these values into $y = A\sin Bx$.

The graph is modeled by $y = 4\sin 4x$.

9. Because the hours of daylight ranges from a minimum of 10 hours to a maximum of 14 hours, the curve oscillates about the middle value, 12 hours. Thus, $D = 12$. The maximum number of hours is 2 hours above 12 hours. Thus, $A = 2$. The graph shows that one complete cycle occurs in 12–0, or 12 months. The period is 12. Thus, $12 = \dfrac{2\pi}{B}$

$$12B = 2\pi$$
$$B = \dfrac{2\pi}{12} = \dfrac{\pi}{6}$$

The graph shows that the starting point of the cycle is shifted from 0 to 3. The phase shift, $\dfrac{C}{B}$, is 3.

$$3 = \dfrac{C}{B}$$
$$3 = \dfrac{C}{\frac{\pi}{6}}$$
$$\dfrac{\pi}{2} = C$$

Substitute these values into $y = A\sin(Bx - C) + D$. The number of hours of daylight is modeled by

$$y = 2\sin\left(\dfrac{\pi}{6}x - \dfrac{\pi}{2}\right) + 12.$$

Concept and Vocabulary Check 5.5

1. $|A|$; $\dfrac{2\pi}{B}$

2. 3; 4π

3. π; 0; $\dfrac{\pi}{4}$; $\dfrac{\pi}{2}$; $\dfrac{3\pi}{4}$; π

4. $\dfrac{C}{B}$; right; left

5. $|A|$; $\dfrac{2\pi}{B}$

6. $\dfrac{1}{2}$; $\dfrac{2\pi}{3}$

7. false

8. true

9. true

10. true

Exercise Set 5.5

1. The equation $y = 4\sin x$ is of the form $y = A\sin x$ with $A = 4$. Thus, the amplitude is $|A| = |4| = 4$.

The period is 2π. The quarter-period is $\dfrac{2\pi}{4}$ or $\dfrac{\pi}{2}$.

The cycle begins at $x = 0$. Add quarter-periods to generate x-values for the key points.
$x = 0$

$$x = 0 + \dfrac{\pi}{2} = \dfrac{\pi}{2}$$
$$x = \dfrac{\pi}{2} + \dfrac{\pi}{2} = \pi$$
$$x = \pi + \dfrac{\pi}{2} = \dfrac{3\pi}{2}$$
$$x = \dfrac{3\pi}{2} + \dfrac{\pi}{2} = 2\pi$$

Evaluate the function at each value of x.

x	$y = 4\sin x$	coordinates
0	$y = 4\sin 0 = 4 \cdot 0 = 0$	$(0, 0)$
$\dfrac{\pi}{2}$	$y = 4\sin\dfrac{\pi}{2} = 4 \cdot 1 = 4$	$\left(\dfrac{\pi}{2}, 4\right)$
π	$y = 4\sin\pi = 4 \cdot 0 = 0$	$(\pi, 0)$
$\dfrac{3\pi}{2}$	$y = 4\sin\dfrac{3\pi}{2}$ $= 4(-1) = -4$	$\left(\dfrac{3\pi}{2}, -4\right)$
2π	$y = 4\sin 2\pi = 4 \cdot 0 = 0$	$(2\pi, 0)$

Connect the five key points with a smooth curve and graph one complete cycle of the given function with the graph of $y = \sin x$.

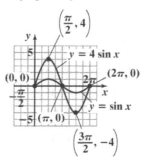

3. The equation $y = \dfrac{1}{3}\sin x$ is of the form $y = A\sin x$

with $A = \dfrac{1}{3}$. Thus, the amplitude is $|A| = \left|\dfrac{1}{3}\right| = \dfrac{1}{3}$.

The period is 2π. The quarter-period is $\dfrac{2\pi}{4}$ or $\dfrac{\pi}{2}$.

The cycle begins at $x = 0$. Add quarter-periods to generate x-values for the key points.

$x = 0$

$x = 0 + \dfrac{\pi}{2} = \dfrac{\pi}{2}$

$x = \dfrac{\pi}{2} + \dfrac{\pi}{2} = \pi$

$x = \pi + \dfrac{\pi}{2} = \dfrac{3\pi}{2}$

$x = \dfrac{3\pi}{2} + \dfrac{\pi}{2} = 2\pi$

Evaluate the function at each value of x.

x	$y = \dfrac{1}{3}\sin x$	coordinates
0	$y = \dfrac{1}{3}\sin 0 = \dfrac{1}{3}\cdot 0 = 0$	$(0, 0)$
$\dfrac{\pi}{2}$	$y = \dfrac{1}{3}\sin\dfrac{\pi}{2} = \dfrac{1}{3}\cdot 1 = \dfrac{1}{3}$	$\left(\dfrac{\pi}{2}, \dfrac{1}{3}\right)$
π	$y = \dfrac{1}{3}\sin \pi = \dfrac{1}{3}\cdot 0 = 0$	$(\pi, 0)$
$\dfrac{3\pi}{2}$	$y = \dfrac{1}{3}\sin\dfrac{3\pi}{2}$ $= \dfrac{1}{3}(-1) = -\dfrac{1}{3}$	$\left(\dfrac{3\pi}{2}, -\dfrac{1}{3}\right)$
2π	$y = \dfrac{1}{3}\sin 2\pi = \dfrac{1}{3}\cdot 0 = 0$	$(2\pi, 0)$

Connect the five key points with a smooth curve and graph one complete cycle of the given function with the graph of $y = \sin x$.

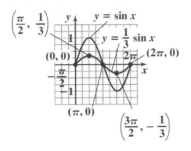

5. The equation $y = -3\sin x$ is of the form $y = A\sin x$ with $A = -3$. Thus, the amplitude is $|A| = |-3| = 3$.

The period is 2π. The quarter-period is $\dfrac{2\pi}{4}$ or $\dfrac{\pi}{2}$.

The cycle begins at $x = 0$. Add quarter-periods to generate x-values for the key points.

$x = 0$

$x = 0 + \dfrac{\pi}{2} = \dfrac{\pi}{2}$

$x = \dfrac{\pi}{2} + \dfrac{\pi}{2} = \pi$

$x = \pi + \dfrac{\pi}{2} = \dfrac{3\pi}{2}$

$x = \dfrac{3\pi}{2} + \dfrac{\pi}{2} = 2\pi$

Evaluate the function at each value of x.

x	$y = -3\sin x$	coordinates
0	$y = -3\sin x$ $= -3\cdot 0 = 0$	$(0, 0)$
$\dfrac{\pi}{2}$	$y = -3\sin\dfrac{\pi}{2}$ $= -3\cdot 1 = -3$	$\left(\dfrac{\pi}{2}, -3\right)$
π	$y = -3\sin \pi$ $= -3\cdot 0 = 0$	$(\pi, 0)$
$\dfrac{3\pi}{2}$	$y = -3\sin\dfrac{3\pi}{2}$ $= -3(-1) = 3$	$\left(\dfrac{3\pi}{2}, 3\right)$
2π	$y = -3\sin 2\pi$ $= -3\cdot 0 = 0$	$(2\pi, 0)$

Connect the five key points with a smooth curve and graph one complete cycle of the given function with the graph of $y = \sin x$.

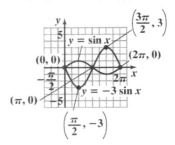

7. The equation $y = \sin 2x$ is of the form $y = A\sin Bx$ with $A = 1$ and $B = 2$. The amplitude is $|A| = |1| = 1$. The period is $\dfrac{2\pi}{B} = \dfrac{2\pi}{2} = \pi$. The quarter-period is $\dfrac{\pi}{4}$. The cycle begins at $x = 0$. Add quarter-periods to generate x-values for the key points.

$x = 0$

$x = 0 + \dfrac{\pi}{4}$

$x = \dfrac{\pi}{4} + \dfrac{\pi}{4} = \dfrac{\pi}{2}$

$x = \dfrac{\pi}{2} + \dfrac{\pi}{4} = \dfrac{3\pi}{4}$

$x = \dfrac{3\pi}{4} + \dfrac{\pi}{4} = \pi$

Evaluate the function at each value of x.

x	$y = \sin 2x$	coordinates
0	$y = \sin 2 \cdot 0 = \sin 0 = 0$	$(0, 0)$
$\dfrac{\pi}{4}$	$y = \sin\left(2 \cdot \dfrac{\pi}{4}\right)$ $= \sin\dfrac{\pi}{2} = 1$	$\left(\dfrac{\pi}{4}, 1\right)$
$\dfrac{\pi}{2}$	$y = \sin\left(2 \cdot \dfrac{\pi}{2}\right)$ $= \sin \pi = 0$	$\left(\dfrac{\pi}{2}, 0\right)$
$\dfrac{3\pi}{4}$	$y = \sin\left(2 \cdot \dfrac{3\pi}{4}\right)$ $= \sin\dfrac{3\pi}{2} = -1$	$\left(\dfrac{3\pi}{4}, -1\right)$
π	$y = \sin(2 \cdot \pi)$ $= \sin 2\pi = 0$	$(\pi, 0)$

Connect the five key points with a smooth curve and graph one complete cycle of the given function.

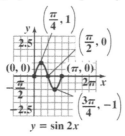

$y = \sin 2x$

9. The equation $y = 3\sin\dfrac{1}{2}x$ is of the form $y = A\sin Bx$ with $A = 3$ and $B = \dfrac{1}{2}$. The amplitude is $|A| = |3| = 3$. The period is $\dfrac{2\pi}{B} = \dfrac{2\pi}{\frac{1}{2}} = 2\pi \cdot 2 = 4\pi$. The quarter-period is $\dfrac{4\pi}{4} = \pi$. The cycle begins at $x = 0$. Add quarter-periods to generate x-values for the key points.

$x = 0$

$x = 0 + \pi = \pi$

$x = \pi + \pi = 2\pi$

$x = 2\pi + \pi = 3\pi$

$x = 3\pi + \pi = 4\pi$

Evaluate the function at each value of x.

x	$y = 3\sin\dfrac{1}{2}x$	coordinates
0	$y = 3\sin\left(\dfrac{1}{2} \cdot 0\right)$ $= 3\sin 0 = 3 \cdot 0 = 0$	$(0, 0)$
π	$y = 3\sin\left(\dfrac{1}{2} \cdot \pi\right)$ $= 3\sin\dfrac{\pi}{2} = 3 \cdot 1 = 3$	$(\pi, 3)$
2π	$y = 3\sin\left(\dfrac{1}{2} \cdot 2\pi\right)$ $= 3\sin \pi = 3 \cdot 0 = 0$	$(2\pi, 0)$
3π	$y = 3\sin\left(\dfrac{1}{2} \cdot 3\pi\right)$ $= 3\sin\dfrac{3\pi}{2}$ $= 3(-1) = -3$	$(3\pi, -3)$
4π	$y = 3\sin\left(\dfrac{1}{2} \cdot 4\pi\right)$ $= 3\sin 2\pi = 3 \cdot 0 = 0$	$(4\pi, 0)$

Connect the five points with a smooth curve and graph one complete cycle of the given function.

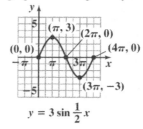

$y = 3\sin\dfrac{1}{2}x$

11. The equation $y = 4\sin \pi x$ is of the form $y = A\sin Bx$ with $A = 4$ and $B = \pi$. The amplitude is $|A| = |4| = 4$. The period is $\dfrac{2\pi}{B} = \dfrac{2\pi}{\pi} = 2$. The quarter-period is $\dfrac{2}{4} = \dfrac{1}{2}$. The cycle begins at $x = 0$. Add quarter-periods to generate x-values for the key points.

$x = 0$

$x = 0 + \dfrac{1}{2} = \dfrac{1}{2}$

$x = \dfrac{1}{2} + \dfrac{1}{2} = 1$

$x = 1 + \dfrac{1}{2} = \dfrac{3}{2}$

$x = \dfrac{3}{2} + \dfrac{1}{2} = 2$

Evaluate the function at each value of x.

x	$y = 4\sin \pi x$	coordinates
0	$y = 4\sin(\pi \cdot 0)$ $= 4\sin 0 = 4 \cdot 0 = 0$	$(0, 0)$
$\dfrac{1}{2}$	$y = 4\sin\left(\pi \cdot \dfrac{1}{2}\right)$ $= 4\sin\dfrac{\pi}{2} = 4(1) = 4$	$\left(\dfrac{1}{2}, 4\right)$
1	$y = 4\sin(\pi \cdot 1)$ $= 4\sin \pi = 4 \cdot 0 = 0$	$(1, 0)$
$\dfrac{3}{2}$	$y = 4\sin\left(\pi \cdot \dfrac{3}{2}\right)$ $= 4\sin\dfrac{3\pi}{2}$ $= 4(-1) = -4$	$\left(\dfrac{3}{2}, -4\right)$
2	$y = 4\sin(\pi \cdot 2)$ $= 4\sin 2\pi = 4 \cdot 0 = 0$	$(2, 0)$

Connect the five points with a smooth curve and graph one complete cycle of the given function.

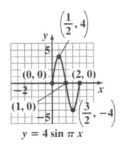

$y = 4 \sin \pi x$

13. The equation $y = -3\sin 2\pi x$ is of the form $y = A\sin Bx$ with $A = -3$ and $B = 2\pi$. The amplitude is $|A| = |-3| = 3$. The period is $\dfrac{2\pi}{B} = \dfrac{2\pi}{2\pi} = 1$. The quarter-period is $\dfrac{1}{4}$. The cycle begins at $x = 0$. Add quarter-periods to generate x-values for the key points.

$x = 0$

$x = 0 + \dfrac{1}{4} = \dfrac{1}{4}$

$x = \dfrac{1}{4} + \dfrac{1}{4} = \dfrac{1}{2}$

$x = \dfrac{1}{2} + \dfrac{1}{4} = \dfrac{3}{4}$

$x = \dfrac{3}{4} + \dfrac{1}{4} = 1$

Evaluate the function at each value of x.

x	$y = -3\sin 2\pi x$	coordinates
0	$y = -3\sin(2\pi \cdot 0)$ $= -3\sin 0$ $= -3 \cdot 0 = 0$	$(0, 0)$
$\dfrac{1}{4}$	$y = -3\sin\left(2\pi \cdot \dfrac{1}{4}\right)$ $= -3\sin\dfrac{\pi}{2}$ $= -3 \cdot 1 = -3$	$\left(\dfrac{1}{4}, -3\right)$
$\dfrac{1}{2}$	$y = -3\sin\left(2\pi \cdot \dfrac{1}{2}\right)$ $= -3\sin \pi$ $= -3 \cdot 0 = 0$	$\left(\dfrac{1}{2}, 0\right)$
$\dfrac{3}{4}$	$y = -3\sin\left(2\pi \cdot \dfrac{3}{4}\right)$ $= -3\sin\dfrac{3\pi}{2}$ $= -3(-1) = 3$	$\left(\dfrac{3}{4}, 3\right)$
1	$y = -3\sin(2\pi \cdot 1)$ $= -3\sin 2\pi$ $= -3 \cdot 0 = 0$	$(1, 0)$

Connect the five points with a smooth curve and graph one complete cycle of the given function.

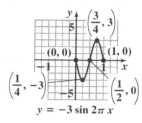

$$y = -3\sin 2\pi x$$

15. The equation $y = -\sin\dfrac{2}{3}x$ is of the form $y = A\sin Bx$

with $A = -1$ and $B = \dfrac{2}{3}$.

The amplitude is $|A| = |-1| = 1$.

The period is $\dfrac{2\pi}{B} = \dfrac{2\pi}{\frac{2}{3}} = 2\pi \cdot \dfrac{3}{2} = 3\pi$.

The quarter-period is $\dfrac{3\pi}{4}$. The cycle begins at $x = 0$.

Add quarter-periods to generate x-values for the key points.

$x = 0$

$x = 0 + \dfrac{3\pi}{4} = \dfrac{3\pi}{4}$

$x = \dfrac{3\pi}{4} + \dfrac{3\pi}{4} = \dfrac{3\pi}{2}$

$x = \dfrac{3\pi}{2} + \dfrac{3\pi}{4} = \dfrac{9\pi}{4}$

$x = \dfrac{9\pi}{4} + \dfrac{3\pi}{4} = 3\pi$

Evaluate the function at each value of x.

x	$y = -\sin\dfrac{2}{3}x$	coordinates
0	$y = -\sin\left(\dfrac{2}{3}\cdot 0\right)$ $= -\sin 0 = 0$	$(0, 0)$
$\dfrac{3\pi}{4}$	$y = -\sin\left(\dfrac{2}{3}\cdot\dfrac{3\pi}{4}\right)$ $= -\sin\dfrac{\pi}{2} = -1$	$\left(\dfrac{3\pi}{4}, -1\right)$
$\dfrac{3\pi}{2}$	$y = -\sin\left(\dfrac{2}{3}\cdot\dfrac{3\pi}{2}\right)$ $= -\sin\pi = 0$	$\left(\dfrac{3\pi}{2}, 0\right)$

$\dfrac{9\pi}{4}$	$y = -\sin\left(\dfrac{2}{3}\cdot\dfrac{9\pi}{4}\right)$ $= -\sin\dfrac{3\pi}{2}$ $= -(-1) = 1$	$\left(\dfrac{9\pi}{4}, 1\right)$
3π	$y = -\sin\left(\dfrac{2}{3}\cdot 3\pi\right)$ $= -\sin 2\pi = 0$	$(3\pi, 0)$

Connect the five points with a smooth curve and graph one complete cycle of the given function.

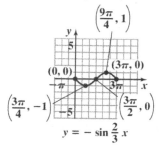

$$y = -\sin\dfrac{2}{3}x$$

17. The equation $y = \sin(x - \pi)$ is of the form $y = A\sin(Bx - C)$ with $A = 1$, $B = 1$, and $C = \pi$. The amplitude is $|A| = |1| = 1$. The period is $\dfrac{2\pi}{B} = \dfrac{2\pi}{1} = 2\pi$. The phase shift is $\dfrac{C}{B} = \dfrac{\pi}{1} = \pi$. The quarter-period is $\dfrac{2\pi}{4} = \dfrac{\pi}{2}$. The cycle begins at $x = \pi$. Add quarter-periods to generate x-values for the key points.

$x = \pi$

$x = \pi + \dfrac{\pi}{2} = \dfrac{3\pi}{2}$

$x = \dfrac{3\pi}{2} + \dfrac{\pi}{2} = 2\pi$

$x = 2\pi + \dfrac{\pi}{2} = \dfrac{5\pi}{2}$

$x = \dfrac{5\pi}{2} + \dfrac{\pi}{2} = 3\pi$

Evaluate the function at each value of x.

x	$y = \sin(x - \pi)$	coordinates
π	$y = \sin(\pi - \pi)$ $= \sin 0 = 0$	$(\pi, 0)$
$\dfrac{3\pi}{2}$	$y = \sin\left(\dfrac{3\pi}{2} - \pi\right)$ $= \sin\dfrac{\pi}{2} = 1$	$\left(\dfrac{3\pi}{2}, 1\right)$
2π	$y = \sin(2\pi - \pi)$ $= \sin \pi = 0$	$(2\pi, 0)$
$\dfrac{5\pi}{2}$	$y = \sin\left(\dfrac{5\pi}{2} - \pi\right)$ $= \sin\dfrac{3\pi}{2} = -1$	$\left(\dfrac{5\pi}{2}, -1\right)$
3π	$y = \sin(3\pi - \pi)$ $= \sin 2\pi = 0$	$(3\pi, 0)$

Connect the five points with a smooth curve and graph one complete cycle of the given function.

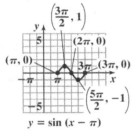

$y = \sin(x - \pi)$

19. The equation $y = \sin(2x - \pi)$ is of the form $y = A\sin(Bx - C)$ with $A = 1$, $B = 2$, and $C = \pi$. The amplitude is $|A| = |1| = 1$. The period is $\dfrac{2\pi}{B} = \dfrac{2\pi}{2} = \pi$. The phase shift is $\dfrac{C}{B} = \dfrac{\pi}{2}$. The quarter-period is $\dfrac{\pi}{4}$. The cycle begins at $x = \dfrac{\pi}{2}$. Add quarter-periods to generate x-values for the key points.

$x = \dfrac{\pi}{2}$

$x = \dfrac{\pi}{2} + \dfrac{\pi}{4} = \dfrac{3\pi}{4}$

$x = \dfrac{3\pi}{4} + \dfrac{\pi}{4} = \pi$

$x = \pi + \dfrac{\pi}{4} = \dfrac{5\pi}{4}$

$x = \dfrac{5\pi}{4} + \dfrac{\pi}{4} = \dfrac{3\pi}{2}$

Evaluate the function at each value of x.

x	$y = \sin(2x - \pi)$	coordinates
$\dfrac{\pi}{2}$	$y = \sin\left(2 \cdot \dfrac{\pi}{2} - \pi\right)$ $= \sin(\pi - \pi)$ $= \sin 0 = 0$	$\left(\dfrac{\pi}{2}, 0\right)$
$\dfrac{3\pi}{4}$	$y = \sin\left(2 \cdot \dfrac{3\pi}{4} - \pi\right)$ $= \sin\left(\dfrac{3\pi}{2} - \pi\right)$ $= \sin\dfrac{\pi}{2} = 1$	$\left(\dfrac{3\pi}{4}, 1\right)$
π	$y = \sin(2 \cdot \pi - \pi)$ $= \sin(2\pi - \pi)$ $= \sin \pi = 0$	$(\pi, 0)$
$\dfrac{5\pi}{4}$	$y = \sin\left(2 \cdot \dfrac{5\pi}{4} - \pi\right)$ $= \sin\left(\dfrac{5\pi}{2} - \pi\right)$ $= \sin\dfrac{3\pi}{2} = -1$	$\left(\dfrac{5\pi}{4}, -1\right)$
$\dfrac{3\pi}{2}$	$y = \sin\left(2 \cdot \dfrac{3\pi}{2} - \pi\right)$ $= \sin(3\pi - \pi)$ $= \sin 2\pi = 0$	$\left(\dfrac{3\pi}{2}, 0\right)$

Connect the five points with a smooth curve and graph one complete cycle of the given function.

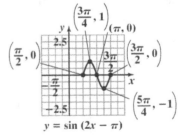

$y = \sin(2x - \pi)$

21. The equation $y = 3\sin(2x - \pi)$ is of the form $y = A\sin(Bx - C)$ with $A = 3$, $B = 2$, and $C = \pi$. The amplitude is $|A| = |3| = 3$. The period is $\dfrac{2\pi}{B} = \dfrac{2\pi}{2} = \pi$. The phase shift is $\dfrac{C}{B} = \dfrac{\pi}{2}$. The quarter-period is $\dfrac{\pi}{4}$. The cycle begins at $x = \dfrac{\pi}{2}$. Add quarter-periods to generate x-values for the key points.

$$x = \frac{\pi}{2}$$

$$x = \frac{\pi}{2} + \frac{\pi}{4} = \frac{3\pi}{4}$$

$$x = \frac{3\pi}{4} + \frac{\pi}{4} = \pi$$

$$x = \pi + \frac{\pi}{4} = \frac{5\pi}{4}$$

$$x = \frac{5\pi}{4} + \frac{\pi}{4} = \frac{3\pi}{2}$$

Evaluate the function at each value of x.

x	$y = 3\sin(2x - \pi)$	coordinates
$\dfrac{\pi}{2}$	$y = 3\sin\left(2 \cdot \dfrac{\pi}{2} - \pi\right)$ $= 3\sin(\pi - \pi)$ $= 3\sin 0 = 3 \cdot 0 = 0$	$\left(\dfrac{\pi}{2}, 0\right)$
$\dfrac{3\pi}{4}$	$y = 3\sin\left(2 \cdot \dfrac{3\pi}{4} - \pi\right)$ $= 3\sin\left(\dfrac{3\pi}{2} - \pi\right)$ $= 3\sin\dfrac{\pi}{2} = 3 \cdot 1 = 3$	$\left(\dfrac{3\pi}{4}, 3\right)$
π	$y = 3\sin(2 \cdot \pi - \pi)$ $= 3\sin(2\pi - \pi)$ $= 3\sin \pi = 3 \cdot 0 = 0$	$(\pi, 0)$
$\dfrac{5\pi}{4}$	$y = 3\sin\left(2 \cdot \dfrac{5\pi}{4} - \pi\right)$ $= 3\sin\left(\dfrac{5\pi}{2} - \pi\right)$ $= 3\sin\dfrac{3\pi}{2}$ $= 3(-1) = -3$	$\left(\dfrac{5\pi}{4}, -3\right)$
$\dfrac{3\pi}{2}$	$y = 3\sin\left(2 \cdot \dfrac{3\pi}{2} - \pi\right)$ $= 3\sin(3\pi - \pi)$ $= 3\sin 2\pi = 3 \cdot 0 = 0$	$\left(\dfrac{3\pi}{2}, 0\right)$

Connect the five points with a smooth curve and graph one complete cycle of the given function.

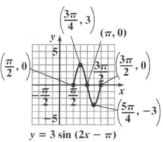

$$y = 3 \sin (2x - \pi)$$

23. $y = \dfrac{1}{2}\sin\left(x + \dfrac{\pi}{2}\right) = \dfrac{1}{2}\sin\left(x - \left(-\dfrac{\pi}{2}\right)\right)$

The equation $y = \dfrac{1}{2}\sin\left(x - \left(-\dfrac{\pi}{2}\right)\right)$ is of the form $y = A\sin(Bx - C)$ with $A = \dfrac{1}{2}$, $B = 1$, and $C = -\dfrac{\pi}{2}$.

The amplitude is $|A| = \left|\dfrac{1}{2}\right| = \dfrac{1}{2}$. The period is $\dfrac{2\pi}{B} = \dfrac{2\pi}{1} = 2\pi$. The phase shift is $\dfrac{C}{B} = \dfrac{-\frac{\pi}{2}}{1} = -\dfrac{\pi}{2}$.

The quarter-period is $\dfrac{2\pi}{4} = \dfrac{\pi}{2}$. The cycle begins at $x = -\dfrac{\pi}{2}$. Add quarter-periods to generate x-values for the key points.

$$x = -\frac{\pi}{2}$$

$$x = -\frac{\pi}{2} + \frac{\pi}{2} = 0$$

$$x = 0 + \frac{\pi}{2} = \frac{\pi}{2}$$

$$x = \frac{\pi}{2} + \frac{\pi}{2} = \pi$$

$$x = \pi + \frac{\pi}{2} = \frac{3\pi}{2}$$

Evaluate the function at each value of x.

x	$y = \dfrac{1}{2}\sin\left(x + \dfrac{\pi}{2}\right)$	coordinates
$-\dfrac{\pi}{2}$	$y = \dfrac{1}{2}\sin\left(-\dfrac{\pi}{2} + \dfrac{\pi}{2}\right)$ $= \dfrac{1}{2}\sin 0 = \dfrac{1}{2}\cdot 0 = 0$	$\left(-\dfrac{\pi}{2}, 0\right)$
0	$y = \dfrac{1}{2}\sin\left(0 + \dfrac{\pi}{2}\right)$ $= \dfrac{1}{2}\sin\dfrac{\pi}{2} = \dfrac{1}{2}\cdot 1 = \dfrac{1}{2}$	$\left(0, \dfrac{1}{2}\right)$
$\dfrac{\pi}{2}$	$y = \dfrac{1}{2}\sin\left(\dfrac{\pi}{2} + \dfrac{\pi}{2}\right)$ $= \dfrac{1}{2}\sin\pi = \dfrac{1}{2}\cdot 0 = 0$	$\left(\dfrac{\pi}{2}, 0\right)$
π	$y = \dfrac{1}{2}\sin\left(\pi + \dfrac{\pi}{2}\right)$ $= \dfrac{1}{2}\sin\dfrac{3\pi}{2}$ $= \dfrac{1}{2}\cdot(-1) = -\dfrac{1}{2}$	$\left(\pi, -\dfrac{1}{2}\right)$
$\dfrac{3\pi}{2}$	$y = \dfrac{1}{2}\sin\left(\dfrac{3\pi}{2} + \dfrac{\pi}{2}\right)$ $= \dfrac{1}{2}\sin 2\pi$ $= \dfrac{1}{2}\cdot 0 = 0$	$\left(\dfrac{3\pi}{2}, 0\right)$

Connect the five points with a smooth curve and graph one complete cycle of the given function.

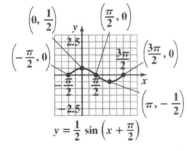

25. $y = -2\sin\left(2x + \dfrac{\pi}{2}\right) = -2\sin\left(2x - \left(-\dfrac{\pi}{2}\right)\right)$

The equation $y = -2\sin\left(2x - \left(-\dfrac{\pi}{2}\right)\right)$ is of the form $y = A\sin(Bx - C)$ with $A = -2$, $B = 2$, and $C = -\dfrac{\pi}{2}$. The amplitude is $|A| = |-2| = 2$. The period is $\dfrac{2\pi}{B} = \dfrac{2\pi}{2} = \pi$. The phase shift is $\dfrac{C}{B} = \dfrac{-\frac{\pi}{2}}{2} = -\dfrac{\pi}{2}\cdot\dfrac{1}{2} = -\dfrac{\pi}{4}$. The quarter-period is $\dfrac{\pi}{4}$. The cycle begins at $x = -\dfrac{\pi}{4}$. Add quarter-periods to generate x-values for the key points.

$x = -\dfrac{\pi}{4}$

$x = -\dfrac{\pi}{4} + \dfrac{\pi}{4} = 0$

$x = 0 + \dfrac{\pi}{4} = \dfrac{\pi}{4}$

$x = \dfrac{\pi}{4} + \dfrac{\pi}{4} = \dfrac{\pi}{2}$

$x = \dfrac{\pi}{2} + \dfrac{\pi}{4} = \dfrac{3\pi}{4}$

Evaluate the function at each value of x.

x	$y = -2\sin\left(2x + \dfrac{\pi}{2}\right)$	coordinates
$-\dfrac{\pi}{4}$	$y = -2\sin\left(2\cdot\left(-\dfrac{\pi}{4}\right) + \dfrac{\pi}{2}\right)$ $= -2\sin\left(-\dfrac{\pi}{2} + \dfrac{\pi}{2}\right)$ $= -2\sin 0 = -2\cdot 0 = 0$	$\left(-\dfrac{\pi}{4}, 0\right)$
0	$y = -2\sin\left(2\cdot 0 + \dfrac{\pi}{2}\right)$ $= -2\sin\left(0 + \dfrac{\pi}{2}\right)$ $= -2\sin\dfrac{\pi}{2}$ $= -2\cdot 1 = -2$	$(0, -2)$

$\dfrac{\pi}{4}$	$y = -2\sin\left(2\cdot\dfrac{\pi}{4}+\dfrac{\pi}{2}\right)$ $=-2\sin\left(\dfrac{\pi}{2}+\dfrac{\pi}{2}\right)$ $=-2\sin\pi$ $=-2\cdot 0 = 0$	$\left(\dfrac{\pi}{4},0\right)$
$\dfrac{\pi}{2}$	$y = -2\sin\left(2\cdot\dfrac{\pi}{2}+\dfrac{\pi}{2}\right)$ $=-2\sin\left(\pi+\dfrac{\pi}{2}\right)$ $=-2\sin\dfrac{3\pi}{2}$ $=-2(-1)=2$	$\left(\dfrac{\pi}{2},2\right)$
$\dfrac{3\pi}{4}$	$y = -2\sin\left(2\cdot\dfrac{3\pi}{4}+\dfrac{\pi}{2}\right)$ $=-2\sin\left(\dfrac{3\pi}{2}+\dfrac{\pi}{2}\right)$ $=-2\sin 2\pi$ $=-2\cdot 0 = 0$	$\left(\dfrac{3\pi}{4},0\right)$

Connect the five points with a smooth curve and graph one complete cycle of the given function.

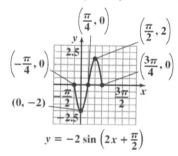

$y = -2\sin\left(2x+\dfrac{\pi}{2}\right)$

27. $y = 3\sin(\pi x + 2)$

The equation $y = 3\sin(\pi x - (-2))$ is of the form $y = A\sin(Bx - C)$ with $A = 3$, $B = \pi$, and $C = -2$. The amplitude is $|A| = |3| = 3$. The period is $\dfrac{2\pi}{B} = \dfrac{2\pi}{\pi} = 2$. The phase shift is $\dfrac{C}{B} = \dfrac{-2}{\pi} = -\dfrac{2}{\pi}$. The quarter-period is $\dfrac{2}{4} = \dfrac{1}{2}$. The cycle begins at $x = -\dfrac{2}{\pi}$. Add quarter-periods to generate x-values for the key points.

$x = -\dfrac{2}{\pi}$

$x = -\dfrac{2}{\pi}+\dfrac{1}{2}=\dfrac{\pi-4}{2\pi}$

$x = \dfrac{\pi-4}{2\pi}+\dfrac{1}{2}=\dfrac{\pi-2}{\pi}$

$x = \dfrac{\pi-2}{\pi}+\dfrac{1}{2}=\dfrac{3\pi-4}{2\pi}$

$x = \dfrac{3\pi-4}{2\pi}+\dfrac{1}{2}=\dfrac{2\pi-2}{\pi}$

Evaluate the function at each value of x.

x	$y = 3\sin(\pi x + 2)$	coordinates
$-\dfrac{2}{\pi}$	$y = 3\sin\left(\pi\left(-\dfrac{2}{\pi}\right)+2\right)$ $=3\sin(-2+2)$ $=3\sin 0 = 3\cdot 0 = 0$	$\left(-\dfrac{2}{\pi},0\right)$
$\dfrac{\pi-4}{2\pi}$	$y = 3\sin\left(\pi\left(\dfrac{\pi-4}{2\pi}\right)+2\right)$ $=3\sin\left(\dfrac{\pi-4}{2}+2\right)$ $=3\sin\left(\dfrac{\pi}{2}-2+2\right)$ $=3\sin\dfrac{\pi}{2}$ $=3\cdot 1 = 3$	$\left(\dfrac{\pi-4}{2\pi},3\right)$
$\dfrac{\pi-2}{\pi}$	$y = 3\sin\left(\pi\left(\dfrac{\pi-2}{\pi}\right)+2\right)$ $=3\sin(\pi-2+2)$ $=3\sin\pi = 3\cdot 0 = 0$	$\left(\dfrac{\pi-2}{\pi},0\right)$
$\dfrac{3\pi-4}{2\pi}$	$y = 3\sin\left(\pi\left(\dfrac{3\pi-4}{2\pi}\right)+2\right)$ $=3\sin\left(\dfrac{3\pi-4}{2}+2\right)$ $=3\sin\left(\dfrac{3\pi}{2}-2+2\right)$ $=3\sin\dfrac{3\pi}{2}$ $=3(-1)=-3$	$\left(\dfrac{5\pi}{4},-3\right)$

$\dfrac{2\pi-2}{\pi}$	$y = 3\sin\left(\pi\left(\dfrac{2\pi-2}{\pi}\right)\right)+2$	$\left(\dfrac{2\pi-2}{\pi}, 0\right)$
	$= 3\sin(2\pi-2+2)$	
	$= 3\sin 2\pi = 3\cdot 0 = 0$	

Connect the five points with a smooth curve and graph one complete cycle of the given function.

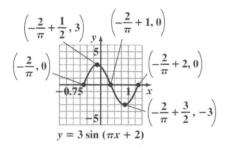

$y = 3 \sin (\pi x + 2)$

29. $y = -2\sin(2\pi x + 4\pi) = -2\sin(2\pi x - (-4\pi))$

The equation $y = -2\sin(2\pi x - (-4\pi))$ is of the form $y = A\sin(Bx - C)$ with $A = -2$, $B = 2\pi$, and $C = -4\pi$. The amplitude is $|A| = |-2| = 2$. The period is $\dfrac{2\pi}{B} = \dfrac{2\pi}{2\pi} = 1$. The phase shift is $\dfrac{C}{B} = \dfrac{-4\pi}{2\pi} = -2$. The quarter-period is $\dfrac{1}{4}$. The cycle begins at $x = -2$. Add quarter-periods to generate x-values for the key points.

$x = -2$

$x = -2 + \dfrac{1}{4} = -\dfrac{7}{4}$

$x = -\dfrac{7}{4} + \dfrac{1}{4} = -\dfrac{3}{2}$

$x = -\dfrac{3}{2} + \dfrac{1}{4} = -\dfrac{5}{4}$

$x = -\dfrac{5}{4} + \dfrac{1}{4} = -1$

Evaluate the function at each value of x.

x	$y = -2\sin(2\pi x + 4\pi)$	coordinates
-2	$y = -2\sin(2\pi(-2)+4\pi)$ $= -2\sin(-4\pi+4\pi)$ $= -2\sin 0$ $= -2\cdot 0 = 0$	$(-2, 0)$
$-\dfrac{7}{4}$	$y = -2\sin\left(2\pi\left(-\dfrac{7}{4}\right)+4\pi\right)$ $= -2\sin\left(-\dfrac{7\pi}{2}+4\pi\right)$ $= -2\sin\dfrac{\pi}{2} = -2\cdot 1 = -2$	$\left(-\dfrac{7}{4}, -2\right)$
$-\dfrac{3}{2}$	$y = -2\sin\left(2\pi\left(-\dfrac{3}{2}\right)+4\pi\right)$ $= -2\sin(-3\pi+4\pi)$ $= -2\sin\pi = -2\cdot 0 = 0$	$\left(-\dfrac{3}{2}, 0\right)$
$-\dfrac{5}{4}$	$y = -2\sin\left(2\pi\left(-\dfrac{5}{4}\right)+4\pi\right)$ $= -2\sin\left(-\dfrac{5\pi}{2}+4\pi\right)$ $= -2\sin\dfrac{3\pi}{2}$ $= -2(-1) = 2$	$\left(-\dfrac{5}{4}, 2\right)$
-1	$y = -2\sin(2\pi(-1)+4\pi)$ $= -2\sin(-2\pi+4\pi)$ $= -2\sin 2\pi$ $= -2\cdot 0 = 0$	$(-1, 0)$

Connect the five points with a smooth curve and graph one complete cycle of the given function.

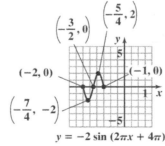

$y = -2 \sin (2\pi x + 4\pi)$

31. The equation $y = 2\cos x$ is of the form $y = A\cos x$ with $A = 2$. Thus, the amplitude is $\mid A \mid = \mid 2 \mid = 2$.

The period is 2π. The quarter-period is $\dfrac{2\pi}{4}$ or $\dfrac{\pi}{2}$.

The cycle begins at $x = 0$. Add quarter-periods to generate x-values for the key points.

$x = 0$

$x = 0 + \dfrac{\pi}{2} = \dfrac{\pi}{2}$

$x = \dfrac{\pi}{2} + \dfrac{\pi}{2} = \pi$

$x = \pi + \dfrac{\pi}{2} = \dfrac{3\pi}{2}$

$x = \dfrac{3\pi}{2} + \dfrac{\pi}{2} = 2\pi$

Evaluate the function at each value of x.

x	$y = 2\cos x$	coordinates
0	$y = 2\cos 0$ $= 2 \cdot 1 = 2$	$(0, 2)$
$\dfrac{\pi}{2}$	$y = 2\cos\dfrac{\pi}{2}$ $= 2 \cdot 0 = 0$	$\left(\dfrac{\pi}{2}, 0\right)$
π	$y = 2\cos \pi$ $= 2 \cdot (-1) = -2$	$(\pi, -2)$
$\dfrac{3\pi}{2}$	$y = 2\cos\dfrac{3\pi}{2}$ $= 2 \cdot 0 = 0$	$\left(\dfrac{3\pi}{2}, 0\right)$
2π	$y = 2\cos 2\pi$ $= 2 \cdot 1 = 2$	$(2\pi, 2)$

Connect the five points with a smooth curve and graph one complete cycle of the given function with the graph of $y = 2\cos x$.

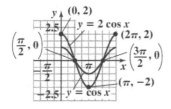

33. The equation $y = -2\cos x$ is of the form $y = A\cos x$ with $A = -2$. Thus, the amplitude is $\mid A \mid = \mid -2 \mid = 2$. The period is 2π. The quarter-period is $\dfrac{2\pi}{4}$ or $\dfrac{\pi}{2}$. The cycle begins at $x = 0$. Add quarter-periods to generate x-values for the key points.

$x = 0$

$x = 0 + \dfrac{\pi}{2} = \dfrac{\pi}{2}$

$x = \dfrac{\pi}{2} + \dfrac{\pi}{2} = \pi$

$x = \pi + \dfrac{\pi}{2} = \dfrac{3\pi}{2}$

$x = \dfrac{3\pi}{2} + \dfrac{\pi}{2} = 2\pi$

Evaluate the function at each value of x.

x	$y = -2\cos x$	coordinates
0	$y = -2\cos 0$ $= -2 \cdot 1 = -2$	$(0, -2)$
$\dfrac{\pi}{2}$	$y = -2\cos\dfrac{\pi}{2}$ $= -2 \cdot 0 = 0$	$\left(\dfrac{\pi}{2}, 0\right)$
π	$y = -2\cos \pi$ $= -2 \cdot (-1) = 2$	$(\pi, 2)$
$\dfrac{3\pi}{2}$	$y = -2\cos\dfrac{3\pi}{2}$ $= -2 \cdot 0 = 0$	$\left(\dfrac{3\pi}{2}, 0\right)$
2π	$y = -2\cos 2\pi$ $= -2 \cdot 1 = -2$	$(2\pi, -2)$

Connect the five points with a smooth curve and graph one complete cycle of the given function with the graph of $y = \cos x$.

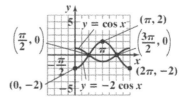

35. The equation $y = \cos 2x$ is of the form $y = A\cos Bx$ with $A = 1$ and $B = 2$. Thus, the amplitude is $|A| = |1| = 1$. The period is $\dfrac{2\pi}{B} = \dfrac{2\pi}{2} = \pi$. The quarter-period is $\dfrac{\pi}{4}$. The cycle begins at $x = 0$. Add quarter-periods to generate x-values for the key points.

$x = 0$

$x = 0 + \dfrac{\pi}{4} = \dfrac{\pi}{4}$

$x = \dfrac{\pi}{4} + \dfrac{\pi}{4} = \dfrac{\pi}{2}$

$x = \dfrac{\pi}{2} + \dfrac{\pi}{4} = \dfrac{3\pi}{4}$

$x = \dfrac{3\pi}{4} + \dfrac{\pi}{4} = \pi$

Evaluate the function at each value of x.

x	$y = \cos 2x$	coordinates
0	$y = \cos(2 \cdot 0)$ $= \cos 0 = 1$	$(0, 1)$
$\dfrac{\pi}{4}$	$y = \cos\left(2 \cdot \dfrac{\pi}{4}\right)$ $= \cos\dfrac{\pi}{2} = 0$	$\left(\dfrac{\pi}{4}, 0\right)$
$\dfrac{\pi}{2}$	$y = \cos\left(2 \cdot \dfrac{\pi}{2}\right)$ $= \cos\pi = -1$	$\left(\dfrac{\pi}{2}, -1\right)$
$\dfrac{3\pi}{4}$	$y = \cos\left(2 \cdot \dfrac{3\pi}{4}\right)$ $= \cos\dfrac{3\pi}{2} = 0$	$\left(\dfrac{3\pi}{4}, 0\right)$
π	$y = \cos(2 \cdot \pi)$ $= \cos 2\pi = 1$	$(\pi, 1)$

Connect the five points with a smooth curve and graph one complete cycle of the given function.

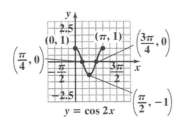

37. The equation $y = 4\cos 2\pi x$ is of the form $y = A\cos Bx$ with $A = 4$ and $B = 2\pi$. Thus, the amplitude is $|A| = |4| = 4$. The period is $\dfrac{2\pi}{B} = \dfrac{2\pi}{2\pi} = 1$. The quarter-period is $\dfrac{1}{4}$. The cycle begins at $x = 0$. Add quarter-periods to generate x-values for the key points.

$x = 0$

$x = 0 + \dfrac{1}{4} = \dfrac{1}{4}$

$x = \dfrac{1}{4} + \dfrac{1}{4} = \dfrac{1}{2}$

$x = \dfrac{1}{2} + \dfrac{1}{4} = \dfrac{3}{4}$

$x = \dfrac{3}{4} + \dfrac{1}{4} = 1$

Evaluate the function at each value of x.

x	$y = 4\cos 2\pi x$	coordinates
0	$y = 4\cos(2\pi \cdot 0)$ $= 4\cos 0$ $= 4 \cdot 1 = 4$	$(0, 4)$
$\dfrac{1}{4}$	$y = 4\cos\left(2\pi \cdot \dfrac{1}{4}\right)$ $= 4\cos\dfrac{\pi}{2}$ $= 4 \cdot 0 = 0$	$\left(\dfrac{1}{4}, 0\right)$
$\dfrac{1}{2}$	$y = 4\cos\left(2\pi \cdot \dfrac{1}{2}\right)$ $= 4\cos\pi$ $= 4 \cdot (-1) = -4$	$\left(\dfrac{1}{2}, -4\right)$
$\dfrac{3}{4}$	$y = 4\cos\left(2\pi \cdot \dfrac{3}{4}\right)$ $= 4\cos\dfrac{3\pi}{2}$ $= 4 \cdot 0 = 0$	$\left(\dfrac{3}{4}, 0\right)$
1	$y = 4\cos(2\pi \cdot 1)$ $= 4\cos 2\pi$ $= 4 \cdot 1 = 4$	$(1, 4)$

Connect the five points with a smooth curve and graph one complete cycle of the given function.

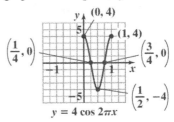

39. The equation $y = -4\cos\frac{1}{2}x$ is of the form

$y = A\cos Bx$ with $A = -4$ and $B = \frac{1}{2}$. Thus, the

amplitude is $|A| = |-4| = 4$. The period is

$\frac{2\pi}{B} = \frac{2\pi}{\frac{1}{2}} = 2\pi \cdot 2 = 4\pi$. The quarter-period is

$\frac{4\pi}{4} = \pi$. The cycle begins at $x = 0$. Add quarter-

periods to generate x-values for the key points.
$x = 0$
$x = 0 + \pi = \pi$
$x = \pi + \pi = 2\pi$
$x = 2\pi + \pi = 3\pi$
$x = 3\pi + \pi = 4\pi$

Evaluate the function at each value of x.

x	$y = -4\cos\frac{1}{2}x$	coordinates
0	$y = -4\cos\left(\frac{1}{2}\cdot 0\right)$ $= -4\cos 0$ $= -4\cdot 1 = -4$	$(0, -4)$
π	$y = -4\cos\left(\frac{1}{2}\cdot\pi\right)$ $= -4\cos\frac{\pi}{2}$ $= -4\cdot 0 = 0$	$(\pi, 0)$
2π	$y = -4\cos\left(\frac{1}{2}\cdot 2\pi\right)$ $= -4\cos\pi$ $= -4\cdot(-1) = 4$	$(2\pi, 4)$
3π	$y = -4\cos\left(\frac{1}{2}\cdot 3\pi\right)$ $= -4\cos\frac{3\pi}{2}$ $= -4\cdot 0 = 0$	$(3\pi, 0)$
4π	$y = -4\cos\left(\frac{1}{2}\cdot 4\pi\right)$ $= -4\cos 2\pi$ $= -4\cdot 1 = -4$	$(4\pi, -4)$

Connect the five points with a smooth curve and graph one complete cycle of the given function.

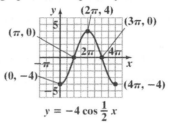

$y = -4\cos\frac{1}{2}x$

41. The equation $y = -\frac{1}{2}\cos\frac{\pi}{3}x$ is of the form

$y = A\cos Bx$ with $A = -\frac{1}{2}$ and $B = \frac{\pi}{3}$. Thus, the

amplitude is $|A| = \left|-\frac{1}{2}\right| = \frac{1}{2}$. The period is

$\frac{2\pi}{B} = \frac{2\pi}{\frac{\pi}{3}} = 2\pi \cdot \frac{3}{\pi} = 6$. The quarter-period is $\frac{6}{4} = \frac{3}{2}$.

The cycle begins at $x = 0$. Add quarter-periods to generate x-values for the key points.
$x = 0$

$x = 0 + \frac{3}{2} = \frac{3}{2}$

$x = \frac{3}{2} + \frac{3}{2} = 3$

$x = 3 + \frac{3}{2} = \frac{9}{2}$

$x = \frac{9}{2} + \frac{3}{2} = 6$

Evaluate the function at each value of x.

Copyright © 2014 Pearson Education, Inc.

x	$y = -\dfrac{1}{2}\cos\dfrac{\pi}{3}x$	coordinates
0	$y = -\dfrac{1}{2}\cos\left(\dfrac{\pi}{3}\cdot 0\right)$ $= -\dfrac{1}{2}\cos 0$ $= -\dfrac{1}{2}\cdot 1 = -\dfrac{1}{2}$	$\left(0, -\dfrac{1}{2}\right)$
$\dfrac{3}{2}$	$y = -\dfrac{1}{2}\cos\left(\dfrac{\pi}{3}\cdot\dfrac{3}{2}\right)$ $= -\dfrac{1}{2}\cos\dfrac{\pi}{2}$ $= -\dfrac{1}{2}\cdot 0 = 0$	$\left(\dfrac{3}{2}, 0\right)$
3	$y = -\dfrac{1}{2}\cos\left(\dfrac{\pi}{3}\cdot 3\right)$ $= -\dfrac{1}{2}\cos\pi$ $= -\dfrac{1}{2}\cdot(-1) = \dfrac{1}{2}$	$\left(3, \dfrac{1}{2}\right)$

x	$y = -\dfrac{1}{2}\cos\dfrac{\pi}{3}x$	coordinates
$\dfrac{9}{2}$	$y = -\dfrac{1}{2}\cos\left(\dfrac{\pi}{3}\cdot\dfrac{9}{2}\right)$ $= -\dfrac{1}{2}\cos\dfrac{3\pi}{2}$ $= -\dfrac{1}{2}\cdot 0 = 0$	$\left(\dfrac{9}{2}, 0\right)$
6	$y = -\dfrac{1}{2}\cos\left(\dfrac{\pi}{3}\cdot 6\right)$ $= -\dfrac{1}{2}\cos 2\pi$ $= -\dfrac{1}{2}\cdot 1 = -\dfrac{1}{2}$	$\left(6, -\dfrac{1}{2}\right)$

Connect the five points with a smooth curve and graph one complete cycle of the given function.

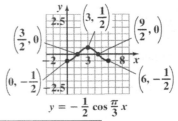

x	coordinates
$\dfrac{\pi}{2}$	$\left(\dfrac{\pi}{2}, 1\right)$
π	$(\pi, 0)$
$\dfrac{3\pi}{2}$	$\left(\dfrac{3\pi}{2}, -1\right)$
2π	$(2\pi, 0)$
$\dfrac{5\pi}{2}$	$\left(\dfrac{5\pi}{2}, 1\right)$

Connect the five points with a smooth curve and graph one complete cycle of the given function

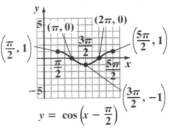

43. The equation $y = \cos\left(x - \dfrac{\pi}{2}\right)$ is of the form

$y = A\cos(Bx - C)$ with $A = 1$, and $B = 1$, and

$C = \dfrac{\pi}{2}$. Thus, the amplitude is $\mid A \mid = \mid 1 \mid = 1$. The

period is $\dfrac{2\pi}{B} = \dfrac{2\pi}{1} = 2\pi$. The phase shift is

$\dfrac{C}{B} = \dfrac{\dfrac{\pi}{2}}{1} = \dfrac{\pi}{2}$. The quarter-period is $\dfrac{2\pi}{4} = \dfrac{\pi}{2}$. The

cycle begins at $x = \dfrac{\pi}{2}$. Add quarter-periods to

generate x-values for the key points.

$x = \dfrac{\pi}{2}$

$x = \dfrac{\pi}{2} + \dfrac{\pi}{2} = \pi$

$x = \pi + \dfrac{\pi}{2} = \dfrac{3\pi}{2}$

$x = \dfrac{3\pi}{2} + \dfrac{\pi}{2} = 2\pi$

$x = 2\pi + \dfrac{\pi}{2} = \dfrac{5\pi}{2}$

Evaluate the function at each value of x.

45. The equation $y = 3\cos(2x - \pi)$ is of the form $y = A\cos(Bx - C)$ with $A = 3$, and $B = 2$, and $C = \pi$. Thus, the amplitude is $|A| = |3| = 3$. The period is $\dfrac{2\pi}{B} = \dfrac{2\pi}{2} = \pi$. The phase shift is $\dfrac{C}{B} = \dfrac{\pi}{2}$.

The quarter-period is $\dfrac{\pi}{4}$. The cycle begins at $x = \dfrac{\pi}{2}$. Add quarter-periods to generate x-values for the key points.

$x = \dfrac{\pi}{2}$

$x = \dfrac{\pi}{2} + \dfrac{\pi}{4} = \dfrac{3\pi}{4}$

$x = \dfrac{3\pi}{4} + \dfrac{\pi}{4} = \pi$

$x = \pi + \dfrac{\pi}{4} = \dfrac{5\pi}{4}$

$x = \dfrac{5\pi}{4} + \dfrac{\pi}{4} = \dfrac{3\pi}{2}$

Evaluate the function at each value of x.

x	coordinates
$\dfrac{\pi}{2}$	$\left(\dfrac{\pi}{2}, 3\right)$
$\dfrac{3\pi}{4}$	$\left(\dfrac{3\pi}{4}, 0\right)$
π	$(\pi, -3)$
$\dfrac{5\pi}{4}$	$\left(\dfrac{5\pi}{4}, 0\right)$
$\dfrac{3\pi}{2}$	$\left(\dfrac{3\pi}{2}, 3\right)$

Connect the five points with a smooth curve and graph one complete cycle of the given function

$y = 3\cos(2x - \pi)$

47. $y = \dfrac{1}{2}\cos\left(3x + \dfrac{\pi}{2}\right) = \dfrac{1}{2}\cos\left(3x - \left(-\dfrac{\pi}{2}\right)\right)$

The equation $y = \dfrac{1}{2}\cos\left(3x - \left(-\dfrac{\pi}{2}\right)\right)$ is of the form $y = A\cos(Bx - C)$ with $A = \dfrac{1}{2}$, and $B = 3$, and $C = -\dfrac{\pi}{2}$. Thus, the amplitude is $|A| = \left|\dfrac{1}{2}\right| = \dfrac{1}{2}$. The period is $\dfrac{2\pi}{B} = \dfrac{2\pi}{3}$. The phase shift is $\dfrac{C}{B} = \dfrac{-\frac{\pi}{2}}{3} = -\dfrac{\pi}{2} \cdot \dfrac{1}{3} = -\dfrac{\pi}{6}$. The quarter-period is $\dfrac{\frac{2\pi}{3}}{4} = \dfrac{2\pi}{3} \cdot \dfrac{1}{4} = \dfrac{\pi}{6}$. The cycle begins at $x = -\dfrac{\pi}{6}$. Add quarter-periods to generate x-values for the key points.

$x = -\dfrac{\pi}{6}$

$x = -\dfrac{\pi}{6} + \dfrac{\pi}{6} = 0$

$x = 0 + \dfrac{\pi}{6} = \dfrac{\pi}{6}$

$x = \dfrac{\pi}{6} + \dfrac{\pi}{6} = \dfrac{\pi}{3}$

$x = \dfrac{\pi}{3} + \dfrac{\pi}{6} = \dfrac{\pi}{2}$

Evaluate the function at each value of x.

x	coordinates
$-\dfrac{\pi}{6}$	$\left(-\dfrac{\pi}{6}, \dfrac{1}{2}\right)$
0	$(0, 0)$
$\dfrac{\pi}{6}$	$\left(\dfrac{\pi}{6}, -\dfrac{1}{2}\right)$
$\dfrac{\pi}{3}$	$\left(\dfrac{\pi}{3}, 0\right)$
$\dfrac{\pi}{2}$	$\left(\dfrac{\pi}{2}, \dfrac{1}{2}\right)$

Connect the five points with a smooth curve and graph one complete cycle of the given function

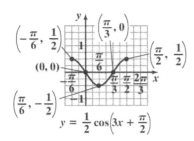

$$y = \tfrac{1}{2}\cos\left(3x + \tfrac{\pi}{2}\right)$$

49. The equation $y = -3\cos\left(2x - \dfrac{\pi}{2}\right)$ is of the form

$y = A\cos(Bx - C)$ with $A = -3$, and

$B = 2$, and $C = \dfrac{\pi}{2}$. Thus, the amplitude is

$|A| = |-3| = 3$. The period is $\dfrac{2\pi}{B} = \dfrac{2\pi}{2} = \pi$. The

phase shift is $\dfrac{C}{B} = \dfrac{\frac{\pi}{2}}{2} = \dfrac{\pi}{2}\cdot\dfrac{1}{2} = \dfrac{\pi}{4}$.

The quarter-period is $\dfrac{\pi}{4}$. The cycle begins at $x = \dfrac{\pi}{4}$.
Add quarter-periods to generate x-values for the key
points.

$x = \dfrac{\pi}{4}$

$x = \dfrac{\pi}{4} + \dfrac{\pi}{4} = \dfrac{\pi}{2}$

$x = \dfrac{\pi}{2} + \dfrac{\pi}{4} = \dfrac{3\pi}{4}$

$x = \dfrac{3\pi}{4} + \dfrac{\pi}{4} = \pi$

$x = \pi + \dfrac{\pi}{4} = \dfrac{5\pi}{4}$

Evaluate the function at each value of x.

x	coordinates
$\dfrac{\pi}{4}$	$\left(\dfrac{\pi}{4}, -3\right)$
$\dfrac{\pi}{2}$	$\left(\dfrac{\pi}{2}, 0\right)$
$\dfrac{3\pi}{4}$	$\left(\dfrac{3\pi}{4}, 3\right)$
π	$(\pi, 0)$
$\dfrac{5\pi}{4}$	$\left(\dfrac{5\pi}{4}, -3\right)$

Connect the five points with a smooth curve and
graph one complete cycle of the given function

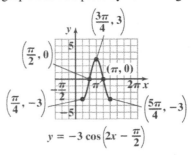

$$y = -3\cos\left(2x - \tfrac{\pi}{2}\right)$$

51. $y = 2\cos(2\pi x + 8\pi) = 2\cos(2\pi x - (-8\pi))$

The equation $y = 2\cos(2\pi x - (-8\pi))$ is of the form
$y = A\cos(Bx - C)$ with $A = 2$, $B = 2\pi$, and
$C = -8\pi$. Thus, the amplitude is $|A| = |2| = 2$.

The period is $\dfrac{2\pi}{B} = \dfrac{2\pi}{2\pi} = 1$. The phase shift is

$\dfrac{C}{B} = \dfrac{-8\pi}{2\pi} = -4$. The quarter-period is $\dfrac{1}{4}$. The cycle

begins at $x = -4$. Add quarter-periods to generate x-
values for the key points.
$x = -4$

$x = -4 + \dfrac{1}{4} = -\dfrac{15}{4}$

$x = -\dfrac{15}{4} + \dfrac{1}{4} = -\dfrac{7}{2}$

$x = -\dfrac{7}{2} + \dfrac{1}{4} = -\dfrac{13}{4}$

$x = -\dfrac{13}{4} + \dfrac{1}{4} = -3$

Evaluate the function at each value of x.

x	coordinates
-4	$(-4, 2)$
$-\dfrac{15}{4}$	$\left(-\dfrac{15}{4}, 0\right)$
$-\dfrac{7}{2}$	$\left(-\dfrac{7}{2}, -2\right)$
$-\dfrac{13}{4}$	$\left(-\dfrac{13}{4}, 0\right)$
-3	$(-3, 2)$

Connect the five points with a smooth curve and graph one complete cycle of the given function

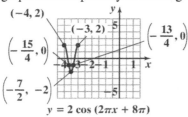

$y = 2\cos(2\pi x + 8\pi)$

By connecting the points with a smooth curve we obtain one period of the graph.

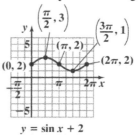

$y = \sin x + 2$

53. The graph of $y = \sin x + 2$ is the graph of $y = \sin x$ shifted up 2 units upward. The period for both functions is 2π. The quarter-period is $\dfrac{2\pi}{4}$ or $\dfrac{\pi}{2}$.

The cycle begins at $x = 0$. Add quarter-periods to generate x-values for the key points.

$x = 0$

$x = 0 + \dfrac{\pi}{2} = \dfrac{\pi}{2}$

$x = \dfrac{\pi}{2} + \dfrac{\pi}{2} = \pi$

$x = \pi + \dfrac{\pi}{2} = \dfrac{3\pi}{2}$

$x = \dfrac{3\pi}{2} + \dfrac{\pi}{2} = 2\pi$

Evaluate the function at each value of x.

x	$y = \sin x + 2$	coordinates
0	$y = \sin 0 + 2$ $= 0 + 2 = 2$	$(0, 2)$
$\dfrac{\pi}{2}$	$y = \sin \dfrac{\pi}{2} + 2$ $= 1 + 2 = 3$	$\left(\dfrac{\pi}{2}, 3\right)$
π	$y = \sin \pi + 2$ $= 0 + 2 = 2$	$(\pi, 2)$
$\dfrac{3\pi}{2}$	$y = \sin \dfrac{3\pi}{2} + 2$ $= -1 + 2 = 1$	$\left(\dfrac{3\pi}{2}, 1\right)$
2π	$y = \sin 2\pi + 2$ $= 0 + 2 = 2$	$(2\pi, 2)$

55. The graph of $y = \cos x - 3$ is the graph of $y = \cos x$ shifted 3 units downward. The period for both functions is 2π. The quarter-period is $\dfrac{2\pi}{4}$ or $\dfrac{\pi}{2}$.

The cycle begins at $x = 0$. Add quarter-periods to generate x-values for the key points.

$x = 0$

$x = 0 + \dfrac{\pi}{2} = \dfrac{\pi}{2}$

$x = \dfrac{\pi}{2} + \dfrac{\pi}{2} = \pi$

$x = \pi + \dfrac{\pi}{2} = \dfrac{3\pi}{2}$

$x = \dfrac{3\pi}{2} + \dfrac{\pi}{2} = 2\pi$

Evaluate the function at each value of x.

x	$y = \cos x - 3$	coordinates
0	$y = \cos 0 - 3$ $= 1 - 3 = -2$	$(0, -2)$
$\dfrac{\pi}{2}$	$y = \cos \dfrac{\pi}{2} - 3$ $= 0 - 3 = -3$	$\left(\dfrac{\pi}{2}, -3\right)$
π	$y = \cos \pi - 3$ $= -1 - 3 = -4$	$(\pi, -4)$
$\dfrac{3\pi}{2}$	$y = \cos \dfrac{3\pi}{2} - 3$ $= 0 - 3 = -3$	$\left(\dfrac{3\pi}{2}, -3\right)$
2π	$y = \cos 2\pi - 3$ $= 1 - 3 = -2$	$(2\pi, -2)$

By connecting the points with a smooth curve we obtain one period of the graph.

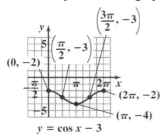

$$y = \cos x - 3$$

57. The graph of $y = 2\sin\frac{1}{2}x + 1$ is the graph

of $y = 2\sin\frac{1}{2}x$ shifted one unit upward. The

amplitude for both functions is $|2| = 2$. The period

for both functions is $\dfrac{2\pi}{\frac{1}{2}} = 2\pi \cdot 2 = 4\pi$. The quarter-

period is $\dfrac{4\pi}{4} = \pi$. The cycle begins at $x = 0$. Add

quarter-periods to generate x-values for the key
points.
$x = 0$

$x = 0 + \pi = \pi$

$x = \pi + \pi = 2\pi$

$x = 2\pi + \pi = 3\pi$

$x = 3\pi + \pi = 4\pi$

Evaluate the function at each value of x.

x	$y = 2\sin\dfrac{1}{2}x + 1$	coordinates
0	$y = 2\sin\left(\dfrac{1}{2}\cdot 0\right) + 1$ $= 2\sin 0 + 1$ $= 2\cdot 0 + 1 = 0 + 1 = 1$	$(0, 1)$
π	$y = 2\sin\left(\dfrac{1}{2}\cdot \pi\right) + 1$ $= 2\sin\dfrac{\pi}{2} + 1$ $= 2\cdot 1 + 1 = 2 + 1 = 3$	$(\pi, 3)$
2π	$y = 2\sin\left(\dfrac{1}{2}\cdot 2\pi\right) + 1$ $= 2\sin \pi + 1$ $= 2\cdot 0 + 1 = 0 + 1 = 1$	$(2\pi, 1)$

3π	$y = 2\sin\left(\dfrac{1}{2}\cdot 3\pi\right) + 1$ $= 2\sin\dfrac{3\pi}{2} + 1$ $= 2\cdot(-1) + 1$ $= -2 + 1 = -1$	$(3\pi, -1)$
4π	$y = 2\sin\left(\dfrac{1}{2}\cdot 4\pi\right) + 1$ $= 2\sin 2\pi + 1$ $= 2\cdot 0 + 1 = 0 + 1 = 1$	$(4\pi, 1)$

By connecting the points with a smooth curve we obtain one period of the graph.

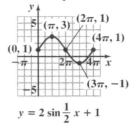

$$y = 2\sin\frac{1}{2}x + 1$$

59. The graph of $y = -3\cos 2\pi x + 2$ is the graph of

$y = -3\cos 2\pi x$ shifted 2 units upward. The

amplitude for both functions is $|-3| = 3$. The period

for both functions is $\dfrac{2\pi}{2\pi} = 1$. The quarter-period is

$\dfrac{1}{4}$. The cycle begins at $x = 0$. Add quarter-periods to

generate x-values for the key points.
$x = 0$

$x = 0 + \dfrac{1}{4} = \dfrac{1}{4}$

$x = \dfrac{1}{4} + \dfrac{1}{4} = \dfrac{1}{2}$

$x = \dfrac{1}{2} + \dfrac{1}{4} = \dfrac{3}{4}$

$x = \dfrac{3}{4} + \dfrac{1}{4} = 1$

Evaluate the function at each value of x.

x	$y = -3\cos 2\pi x + 2$	coordinates
0	$y = -3\cos(2\pi \cdot 0) + 2$ $= -3\cos 0 + 2$ $= -3 \cdot 1 + 2$ $= -3 + 2 = -1$	$(0, -1)$
$\dfrac{1}{4}$	$y = -3\cos\left(2\pi \cdot \dfrac{1}{4}\right) + 2$ $= -3\cos\dfrac{\pi}{2} + 2$ $= -3 \cdot 0 + 2$ $= 0 + 2 = 2$	$\left(\dfrac{1}{4}, 2\right)$
$\dfrac{1}{2}$	$y = -3\cos\left(2\pi \cdot \dfrac{1}{2}\right) + 2$ $= -3\cos\pi + 2$ $= -3 \cdot (-1) + 2$ $= 3 + 2 = 5$	$\left(\dfrac{1}{2}, 5\right)$
$\dfrac{3}{4}$	$y = -3\cos\left(2\pi \cdot \dfrac{3}{4}\right) + 2$ $= -3\cos\dfrac{3\pi}{2} + 2$ $= -3 \cdot 0 + 2$ $= 0 + 2 = 2$	$\left(\dfrac{3}{4}, 2\right)$
1	$y = -3\cos(2\pi \cdot 1) + 2$ $= -3\cos 2\pi + 2$ $= -3 \cdot 1 + 2$ $= -3 + 2 = -1$	$(1, -1)$

By connecting the points with a smooth curve we obtain one period of the graph.

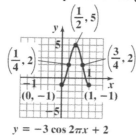

$$y = -3\cos 2\pi x + 2$$

61. Using $y = A\cos Bx$ the amplitude is 3 and $A = 3$,
The period is 4π and thus

$$B = \frac{2\pi}{\text{period}} = \frac{2\pi}{4\pi} = \frac{1}{2}$$

$$y = A\cos Bx$$

$$y = 3\cos\frac{1}{2}x$$

63. Using $y = A\sin Bx$ the amplitude is 2 and $A = -2$,
The period is π and thus

$$B = \frac{2\pi}{\text{period}} = \frac{2\pi}{\pi} = 2$$

$$y = A\sin Bx$$

$$y = -2\sin 2x$$

65. Using $y = A\sin Bx$ the amplitude is 2 and $A = 2$,
The period is 4 and thus

$$B = \frac{2\pi}{\text{period}} = \frac{2\pi}{4} = \frac{\pi}{2}$$

$$y = A\sin Bx$$

$$y = 2\sin\left(\frac{\pi}{2}x\right)$$

67.

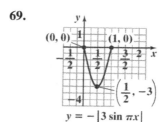

$$y = \left|2\cos\frac{x}{2}\right|$$

69.

$$y = -\left|3\sin \pi x\right|$$

71.

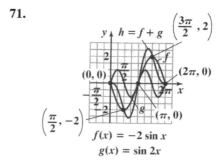

$$f(x) = -2\sin x$$
$$g(x) = \sin 2x$$

73.

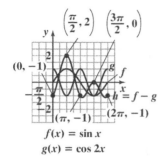

$$f(x) = \sin x$$
$$g(x) = \cos 2x$$

75. The period of the physical cycle is 33 days.

77. The period of the intellectual cycle is 23 days.

79. In the month of March, March 21 would be the best day to meet an on-line friend for the first time, because the emotional cycle is at a maximum.

81. Answers may vary.

83. The information gives the five key point of the graph.
(0, 14) corresponds to June,
(3, 12) corresponds to September,
(6, 10) corresponds to December,
(9, 12) corresponds to March,
(12, 14) corresponds to June
By connecting the five key points with a smooth curve we graph the information from June of one year to June of the following year.

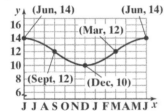

85. The function $y = 3\sin\dfrac{2\pi}{365}(x-79)+12$ is of the

form $y = A\sin B\left(x - \dfrac{C}{B}\right)+D$ with

$A = 3$ and $B = \dfrac{2\pi}{365}$.

a. The amplitude is $|A| = |3| = 3$.

b. The period is $\dfrac{2\pi}{B} = \dfrac{2\pi}{\frac{2\pi}{365}} = 2\pi \cdot \dfrac{365}{2\pi} = 365$.

c. The longest day of the year will have the most hours of daylight. This occurs when the sine function equals 1.

$$y = 3\sin\dfrac{2\pi}{365}(x-79)+12$$
$$y = 3(1)+12$$
$$y = 15$$

There will be 15 hours of daylight.

d. The shortest day of the year will have the least hours of daylight. This occurs when the sine function equals –1.

$$y = 3\sin\dfrac{2\pi}{365}(x-79)+12$$
$$y = 3(-1)+12$$
$$y = 9$$

There will be 9 hours of daylight.

e. The amplitude is 3. The period is 365. The phase shift is $\dfrac{C}{B} = 79$. The quarter-period is $\dfrac{365}{4} = 91.25$. The cycle begins at $x = 79$. Add quarter-periods to find the x-values of the key points.

$$x = 79$$
$$x = 79 + 91.25 = 170.25$$
$$x = 170.25 + 91.25 = 261.5$$
$$x = 261.5 + 91.25 = 352.75$$
$$x = 352.75 + 91.25 = 444$$

Because we are graphing for $0 \le x \le 365$, we will evaluate the function for the first four x-values along with $x = 0$ and $x = 365$. Using a calculator we have the following points.
(0, 9.1) (79, 12) (170.25, 15)
(261.5, 12) (352.75, 9) (365, 9.1)

By connecting the points with a smooth curve we obtain one period of the graph, starting on January 1.

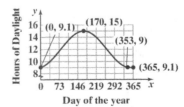

87. Because the depth of the water ranges from a minimum of 6 feet to a maximum of 12 feet, the curve oscillates about the middle value, 9 feet. Thus, $D = 9$. The maximum depth of the water is 3 feet above 9 feet. Thus, $A = 3$. The graph shows that one complete cycle occurs in 12-0, or 12 hours. The period is 12. Thus,

$$12 = \frac{2\pi}{B}$$

$$12B = 2\pi$$

$$B = \frac{2\pi}{12} = \frac{\pi}{6}$$

Substitute these values into $y = A\cos Bx + D$. The depth of the water is modeled by $y = 3\cos\frac{\pi x}{6} + 9$.

89. – 99. Answers may vary.

101. The function $y = 3\sin(2x + \pi) = 3\sin(2x - (-\pi))$ is of the form $y = A\sin(Bx - C)$ with $A = 3$, $B = 2$, and $C = -\pi$. The amplitude is $|A| = |3| = 3$. The period is $\frac{2\pi}{B} = \frac{2\pi}{2} = \pi$. The cycle begins at $x = \frac{C}{B} = \frac{-\pi}{2} = -\frac{\pi}{2}$. We choose $-\frac{\pi}{2} \le x \le \frac{3\pi}{2}$, and $-4 \le y \le 4$ for our graph.

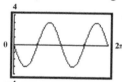

103. The function

$$y = 0.2\sin\left(\frac{\pi}{10}x + \pi\right) = 0.2\sin\left(\frac{\pi}{10}x - (-\pi)\right) \text{ is of}$$

the form $y = A\sin(Bx - C)$ with $A = 0.2$, $B = \frac{\pi}{10}$, and $C = -\pi$. The amplitude is $|A| = |0.2| = 0.2$.

The period is $\frac{2\pi}{B} = \frac{2\pi}{\frac{\pi}{10}} = 2\pi \cdot \frac{10}{\pi} = 20$. The cycle

begins at $x = \frac{C}{B} = \frac{-\pi}{\frac{\pi}{10}} = -\pi \cdot \frac{10}{\pi} = -10$. We choose

$-10 \le x \le 30$, and $-1 \le y \le 1$ for our graph.

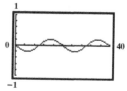

105.

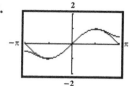

The graphs appear to be the same from $-\frac{\pi}{2}$ to $\frac{\pi}{2}$.

107.

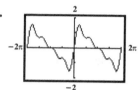

The graph is similar to $y = \sin x$, except the amplitude is greater and the curve is less smooth.

109. a.

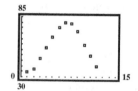

b. $y = 22.61\sin(0.50x - 2.04) + 57.17$

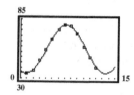

111. makes sense

113. makes sense

115. a. Since $A = 3$ and $D = -2$, the maximum will occur at $3 - 2 = 1$ and the minimum will occur at $-3 - 2 = -5$. Thus the range is $[-5, 1]$

Viewing rectangle: $\left[-\frac{\pi}{6}, \frac{23\pi}{6}, \frac{\pi}{6}\right]$ by $[-5, 1, 1]$

b. Since $A = 1$ and $D = -2$, the maximum will occur at $1 - 2 = -1$ and the minimum will occur at $-1 - 2 = -3$. Thus the range is $[-3, -1]$

Viewing rectangle: $\left[-\frac{\pi}{6}, \frac{7\pi}{6}, \frac{\pi}{6}\right]$ by $[-3, -1, 1]$

117. $y = \sin^2 x = \dfrac{1}{2} - \dfrac{1}{2}\cos 2x$

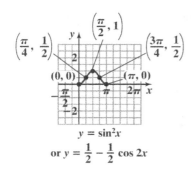

$$y = \sin^2 x$$
$$\text{or } y = \tfrac{1}{2} - \tfrac{1}{2}\cos 2x$$

119. Answers may vary.

120.

$$-\frac{\pi}{2} < x + \frac{\pi}{4} < \frac{\pi}{2}$$

$$-\frac{\pi}{2} - \frac{\pi}{4} < x + \frac{\pi}{4} - \frac{\pi}{4} < \frac{\pi}{2} - \frac{\pi}{4}$$

$$-\frac{2\pi}{4} - \frac{\pi}{4} < x < \frac{2\pi}{4} - \frac{\pi}{4}$$

$$-\frac{3\pi}{4} < x < \frac{\pi}{4}$$

$$\left\{ x \left| -\frac{3\pi}{4} < x < \frac{\pi}{4} \right. \right\} \text{ or } \left(-\frac{3\pi}{4}, \frac{\pi}{4} \right)$$

121. $\dfrac{-\dfrac{3\pi}{4} + \dfrac{\pi}{4}}{2} = \dfrac{-\dfrac{2\pi}{4}}{2} = \dfrac{-\dfrac{\pi}{2}}{2} = -\dfrac{\pi}{4}$

122. a.

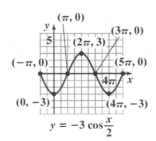

$$y = -3\cos\frac{x}{2}$$

b. The reciprocal function is undefined.

Section 5.6

Check Point Exercises

1. Solve the equations $2x = -\dfrac{\pi}{2}$ and $2x = \dfrac{\pi}{2}$

$$x = -\frac{\pi}{4} \qquad\qquad x = \frac{\pi}{4}$$

Thus, two consecutive asymptotes occur at $x = -\dfrac{\pi}{4}$
and $x = \dfrac{\pi}{4}$. Midway between these asymptotes is
$x = 0$. An x-intercept is 0 and the graph passes
through $(0, 0)$. Because the coefficient of the tangent
is 3, the points on the graph midway between an x-
intercept and the asymptotes have y-coordinates of
-3 and 3. Use the two asymptotes, the x-intercept,
and the points midway between to graph one period
of $y = 3\tan 2x$ from $-\dfrac{\pi}{4}$ to $\dfrac{\pi}{4}$. In order to graph for
$-\dfrac{\pi}{4} < x < \dfrac{3\pi}{4}$, Continue the pattern and extend the
graph another full period to the right.

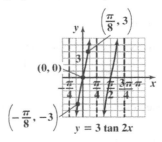

$$y = 3\tan 2x$$

2. Solve the equations

$$x - \frac{\pi}{2} = -\frac{\pi}{2} \qquad \text{and} \qquad x - \frac{\pi}{2} = \frac{\pi}{2}$$

$$x = \frac{\pi}{2} - \frac{\pi}{2} \qquad\qquad x = \frac{\pi}{2} + \frac{\pi}{2}$$

$$x = 0 \qquad\qquad\qquad x = \pi$$

Thus, two consecutive asymptotes occur at
$x = 0$ and $x = \pi$.

$$x\text{-intercept} = \frac{0 + \pi}{2} = \frac{\pi}{2}$$

An x-intercept is $\dfrac{\pi}{2}$ and the graph passes through
$\left(\dfrac{\pi}{2}, 0 \right)$. Because the coefficient of the tangent is 1,
the points on the graph midway between an x-
intercept and the asymptotes have y-coordinates of -1
and 1. Use the two consecutive asymptotes,
$x = 0$ and $x = \pi$, to graph one full period of

$y = \tan\left(x - \dfrac{\pi}{2}\right)$ from 0 to π. Continue the pattern and extend the graph another full period to the right.

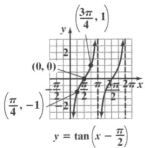

$$y = \tan\left(x - \dfrac{\pi}{2}\right)$$

3. Solve the equations

$$\dfrac{\pi}{2}x = 0 \quad \text{and} \quad \dfrac{\pi}{2}x = \pi$$

$$x = 0 \qquad\qquad x = \dfrac{\pi}{\frac{\pi}{2}}$$

$$x = 2$$

Two consecutive asymptotes occur at $x = 0$ and $x = 2$. Midway between $x = 0$ and $x = 2$ is $x = 1$. An x-intercept is 1 and the graph passes through (1, 0).

Because the coefficient of the cotangent is $\dfrac{1}{2}$, the points on the graph midway between an x-intercept and the asymptotes have y-coordinates of $-\dfrac{1}{2}$ and $\dfrac{1}{2}$. Use the two consecutive asymptotes, $x = 0$ and $x = 2$, to graph one full period of $y = \dfrac{1}{2}\cot\dfrac{\pi}{2}x$. The curve is repeated along the x-axis one full period as shown.

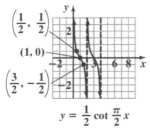

$$y = \dfrac{1}{2}\cot\dfrac{\pi}{2}x$$

4. The x-intercepts of $y = \sin\left(x + \dfrac{\pi}{4}\right)$ correspond to vertical asymptotes of $y = \csc\left(x + \dfrac{\pi}{4}\right)$.

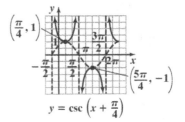

$$y = \csc\left(x + \dfrac{\pi}{4}\right)$$

5. Graph the reciprocal cosine function, $y = 2\cos 2x$. The equation is of the form $y = A\cos Bx$ with $A = 2$ and $B = 2$.

amplitude: $|A| = |2| = 2$

period: $\dfrac{2\pi}{B} = \dfrac{2\pi}{2} = \pi$

Use quarter-periods, $\dfrac{\pi}{4}$, to find x-values for the five key points. Starting with $x = 0$, the x-values are $0, \dfrac{\pi}{4}, \dfrac{\pi}{2}, \dfrac{3\pi}{4}$, and π. Evaluating the function at each value of x, the key points are

$(0, 2), \left(\dfrac{\pi}{4}, 0\right), \left(\dfrac{\pi}{2}, -2\right), \left(\dfrac{3\pi}{4}, 0\right), (\pi, 2)$. In order to graph for $-\dfrac{3\pi}{4} \le x \le \dfrac{3\pi}{4}$, Use the first four points and extend the graph $-\dfrac{3\pi}{4}$ units to the left. Use the graph to obtain the graph of the reciprocal function. Draw vertical asymptotes through the x-intercepts, and use them as guides to graph $y = 2\sec 2x$.

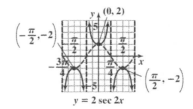

$$y = 2\sec 2x$$

Concept and Vocabulary Check 5.6

1. $\left(-\dfrac{\pi}{4}, \dfrac{\pi}{4}\right);\quad -\dfrac{\pi}{4};\quad \dfrac{\pi}{4}$

2. $(0, \pi);\quad 0;\quad \pi$

3. (0, 2); 0; 2

4. $\left(-\dfrac{\pi}{4}, \dfrac{3\pi}{4}\right)$; $-\dfrac{\pi}{4}$; $\dfrac{3\pi}{4}$

5. $3\sin 2x$

6. $y = 2\cos \pi x$

7. false

8. true

Exercise Set 5.6

1. The graph has an asymptote at $x = -\dfrac{\pi}{2}$.

The phase shift, $\dfrac{C}{B}$, from $\dfrac{\pi}{2}$ to $-\dfrac{\pi}{2}$ is $-\pi$ units.

Thus, $\dfrac{C}{B} = \dfrac{C}{1} = -\pi$

$C = -\pi$

The function with $C = -\pi$ is $y = \tan(x + \pi)$.

3. The graph has an asymptote at $x = \pi$.

$\pi = \dfrac{\pi}{2} + C$

$C = \dfrac{\pi}{2}$

The function is $y = -\tan\left(x - \dfrac{\pi}{2}\right)$.

5. Solve the equations $\dfrac{x}{4} = -\dfrac{\pi}{2}$ and $\dfrac{x}{4} = \dfrac{\pi}{2}$

$x = -\dfrac{\pi}{2} \cdot 4 \qquad x = \dfrac{\pi}{2} \cdot 4$

$x = -2\pi \qquad\qquad x = 2\pi$

Thus, two consecutive asymptotes occur at $x = -2\pi$ and $x = 2\pi$.

$x\text{-intercept} = \dfrac{-2\pi + 2\pi}{2} = \dfrac{0}{2} = 0$

An x-intercept is 0 and the graph passes through (0, 0). Because the coefficient of the tangent is 3, the points on the graph midway between an x-intercept and the asymptotes have y-coordinates of –3 and 3. Use the two consecutive asymptotes, $x = -2\pi$ and

$x = 2\pi$, to graph one full period of $y = 3\tan\dfrac{x}{4}$ from

-2π to 2π.

Continue the pattern and extend the graph another full period to the right.

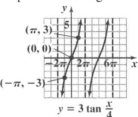

$y = 3\tan\dfrac{x}{4}$

7. Solve the equations $2x = -\dfrac{\pi}{2}$ and $2x = \dfrac{\pi}{2}$

$x = \dfrac{-\frac{\pi}{2}}{2} \qquad\qquad x = \dfrac{\frac{\pi}{2}}{2}$

$x = -\dfrac{\pi}{4} \qquad\qquad x = \dfrac{\pi}{4}$

Thus, two consecutive asymptotes occur at $x = -\dfrac{\pi}{4}$

and $x = \dfrac{\pi}{4}$.

$x\text{-intercept} = \dfrac{-\frac{\pi}{4} + \frac{\pi}{4}}{2} = \dfrac{0}{2} = 0$

An x-intercept is 0 and the graph passes through (0,

0). Because the coefficient of the tangent is $\dfrac{1}{2}$, the

points on the graph midway between an x-intercept

and the asymptotes have y-coordinates of $-\dfrac{1}{2}$ and

$\dfrac{1}{2}$. Use the two consecutive asymptotes, $x = -\dfrac{\pi}{4}$ and

$x = \dfrac{\pi}{4}$, to graph one full period of $y = \dfrac{1}{2}\tan 2x$ from

$-\dfrac{\pi}{4}$ to $\dfrac{\pi}{4}$. Continue the pattern and extend the

graph another full period to the right.

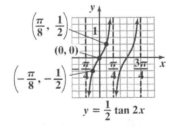

$y = \dfrac{1}{2}\tan 2x$

9. Solve the equations

$$\frac{1}{2}x = -\frac{\pi}{2} \qquad \text{and} \qquad \frac{1}{2}x = \frac{\pi}{2}$$

$$x = \left(-\frac{\pi}{2}\right)2 \qquad\qquad x = \left(\frac{\pi}{2}\right)2$$

$$x = -\pi \qquad\qquad\qquad x = \pi$$

Thus, two consecutive asymptotes occur at $x = -\pi$ and $x = \pi$.

$$x\text{-intercept} = \frac{-\pi + \pi}{2} = \frac{0}{2} = 0$$

An x-intercept is 0 and the graph passes through (0, 0). Because the coefficient of the tangent is –2, the points on the graph midway between an x-intercept and the asymptotes have y-coordinates of 2 and –2. Use the two consecutive asymptotes, $x = -\pi$ and $x = \pi$, to graph one full period of $y = -2\tan\frac{1}{2}x$ from $-\pi$ to π. Continue the pattern and extend the graph another full period to the right.

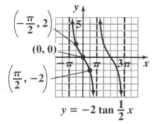

$$y = -2\tan\frac{1}{2}x$$

11. Solve the equations

$$x - \pi = -\frac{\pi}{2} \qquad \text{and} \qquad x - \pi = \frac{\pi}{2}$$

$$x = -\frac{\pi}{2} + \pi \qquad\qquad x = \frac{\pi}{2} + \pi$$

$$x = \frac{\pi}{2} \qquad\qquad\qquad x = \frac{3\pi}{2}$$

Thus, two consecutive asymptotes occur at $x = \frac{\pi}{2}$ and $x = \frac{3\pi}{2}$.

$$x\text{-intercept} = \frac{\frac{\pi}{2} + \frac{3\pi}{2}}{2} = \frac{\frac{4\pi}{2}}{2} = \frac{4\pi}{4} = \pi$$

An x-intercept is π and the graph passes through $(\pi,\ 0)$. Because the coefficient of the tangent is 1, the points on the graph midway between an x-intercept and the asymptotes have y-coordinates of –1 and 1. Use the two consecutive asymptotes, $x = \frac{\pi}{2}$ and $x = \frac{3\pi}{2}$, to graph one full period of

$y = \tan(x - \pi)$ from $\frac{\pi}{2}$ to $\frac{3\pi}{2}$. Continue the pattern and extend the graph another full period to the right.

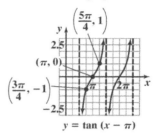

$$y = \tan(x - \pi)$$

13. There is no phase shift. Thus,

$$\frac{C}{B} = \frac{C}{1} = 0$$

$$C = 0$$

Because the points on the graph midway between an x-intercept and the asymptotes have y-coordinates of –1 and 1, $A = -1$. The function with $C = 0$ and $A = -1$ is $y = -\cot x$.

15. The graph has an asymptote at $-\frac{\pi}{2}$. The phase shift, $\frac{C}{B}$, from 0 to $-\frac{\pi}{2}$ is $-\frac{\pi}{2}$ units. Thus, $\frac{C}{B} = \frac{C}{1} = -\frac{\pi}{2}$

$$C = -\frac{\pi}{2}$$

The function with $C = -\frac{\pi}{2}$ is $y = \cot\left(x + \frac{\pi}{2}\right)$.

17. Solve the equations $x = 0$ and $x = \pi$. Two consecutive asymptotes occur at $x = 0$ and $x = \pi$.

$$x\text{-intercept} = \frac{0 + \pi}{2} = \frac{\pi}{2}$$

An x-intercept is $\frac{\pi}{2}$ and the graph passes through $\left(\frac{\pi}{2},\ 0\right)$. Because the coefficient of the cotangent is 2, the points on the graph midway between an x-intercept and the asymptotes have y-coordinates of 2 and –2. Use the two consecutive asymptotes, $x = 0$ and $x = \pi$, to graph one full period of $y = 2\cot x$. The curve is repeated along the x-axis one full period as shown.

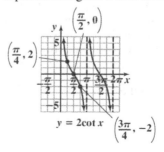

$$y = 2\cot x$$

19. Solve the equations $2x = 0$ and $2x = \pi$

$$x = 0 \qquad\qquad x = \frac{\pi}{2}$$

Two consecutive asymptotes occur at $x = 0$ and $x = \frac{\pi}{2}$.

x-intercept $= \dfrac{0 + \frac{\pi}{2}}{2} = \dfrac{\frac{\pi}{2}}{2} = \dfrac{\pi}{4}$

An x-intercept is $\dfrac{\pi}{4}$ and the graph passes through $\left(\dfrac{\pi}{4}, 0\right)$. Because the coefficient of the cotangent is $\dfrac{1}{2}$, the points on the graph midway between an x-intercept and the asymptotes have y-coordinates of $\dfrac{1}{2}$ and $-\dfrac{1}{2}$. Use the two consecutive asymptotes, $x = 0$ and $x = \dfrac{\pi}{2}$, to graph one full period of $y = \dfrac{1}{2}\cot 2x$.

The curve is repeated along the x-axis one full period as shown.

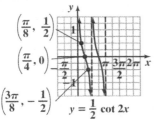

21. Solve the equations $\dfrac{\pi}{2}x = 0$ and $\dfrac{\pi}{2}x = \pi$

$$x = 0 \qquad\qquad x = \frac{\pi}{\frac{\pi}{2}}$$

$$\qquad\qquad\qquad\qquad x = 2$$

Two consecutive asymptotes occur at $x = 0$ and $x = 2$.

x-intercept $= \dfrac{0 + 2}{2} = \dfrac{2}{2} = 1$

An x-intercept is 1 and the graph passes through (1, 0). Because the coefficient of the cotangent is –3, the points on the graph midway between an x-intercept and the asymptotes have y-coordinates of –3 and 3. Use the two consecutive asymptotes, $x = 0$ and $x = 2$, to graph one full period of $y = -3\cot\dfrac{\pi}{2}x$. The curve is repeated along the x-axis

one full period as shown.

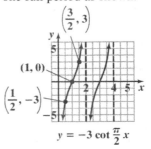

$y = -3\cot\dfrac{\pi}{2}x$

23. Solve the equations

$$x + \frac{\pi}{2} = 0 \qquad \text{and} \qquad x + \frac{\pi}{2} = \pi$$

$$x = 0 - \frac{\pi}{2} \qquad\qquad x = \pi - \frac{\pi}{2}$$

$$x = -\frac{\pi}{2} \qquad\qquad x = \frac{\pi}{2}$$

Two consecutive asymptotes occur at $x = -\dfrac{\pi}{2}$ and $x = \dfrac{\pi}{2}$.

x-intercept $= \dfrac{-\frac{\pi}{2} + \frac{\pi}{2}}{2} = \dfrac{0}{2} = 0$

An x-intercept is 0 and the graph passes through (0, 0). Because the coefficient of the cotangent is 3, the points on the graph midway between an x-intercept and the asymptotes have y-coordinates of 3 and –3.

Use the two consecutive asymptotes, $x = -\dfrac{\pi}{2}$ and $x = \dfrac{\pi}{2}$, to graph one full period of $y = 3\cot\left(x + \dfrac{\pi}{2}\right)$.

The curve is repeated along the x-axis one full period as shown.

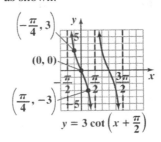

$y = 3\cot\left(x + \dfrac{\pi}{2}\right)$

25. The x-intercepts of $y = -\frac{1}{2}\sin\frac{x}{2}$ corresponds to vertical asymptotes of $y = -\frac{1}{2}\csc\frac{x}{2}$. Draw the vertical asymptotes, and use them as a guide to sketch the graph of $y = -\frac{1}{2}\csc\frac{x}{2}$.

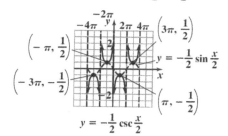

27. The x-intercepts of $y = \frac{1}{2}\cos 2\pi x$ corresponds to vertical asymptotes of $y = \frac{1}{2}\sec 2\pi x$. Draw the vertical asymptotes, and use them as a guide to sketch the graph of $y = \frac{1}{2}\sec 2\pi x$.

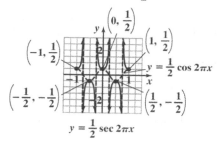

29. Graph the reciprocal sine function, $y = 3\sin x$. The equation is of the form $y = A\sin Bx$ with $A = 3$ and $B = 1$.

amplitude: $|A| = |3| = 3$

period: $\frac{2\pi}{B} = \frac{2\pi}{1} = 2\pi$

Use quarter-periods, $\frac{\pi}{2}$, to find x-values for the five key points. Starting with $x = 0$, the x-values are 0, $\frac{\pi}{2}$, π, $\frac{3\pi}{2}$, and 2π. Evaluating the function at each value of x, the key points are $(0, 0)$, $\left(\frac{\pi}{2}, 3\right)$, $(\pi, 0)$, $\left(\frac{3\pi}{2}, -3\right)$, and $(2\pi, 0)$. Use these key points to graph $y = 3\sin x$ from 0 to 2π. Extend the graph one cycle to the right. Use the graph to obtain the graph of the reciprocal

function. Draw vertical asymptotes through the x-intercepts, and use them as guides to graph $y = 3\csc x$.

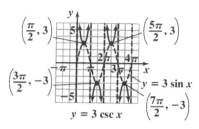

31. Graph the reciprocal sine function, $y = \frac{1}{2}\sin\frac{x}{2}$. The equation is of the form $y = A\sin Bx$ with $A = \frac{1}{2}$ and $B = \frac{1}{2}$.

amplitude: $|A| = \left|\frac{1}{2}\right| = \frac{1}{2}$

period: $\frac{2\pi}{B} = \frac{2\pi}{\frac{1}{2}} = 2\pi \cdot 2 = 4\pi$

Use quarter-periods, π, to find x-values for the five key points. Starting with $x = 0$, the x-values are 0, π, 2π, 3π, and 4π. Evaluating the function at each value of x, the key points are $(0, 0)$, $\left(\pi, \frac{1}{2}\right)$, $(2\pi, 0)$, $\left(3\pi, -\frac{1}{2}\right)$, and $(4\pi, 0)$. Use these key points to graph $y = \frac{1}{2}\sin\frac{x}{2}$ from 0 to 4π.

Extend the graph one cycle to the right. Use the graph to obtain the graph of the reciprocal function. Draw vertical asymptotes through the x-intercepts, and use them as guides to graph $y = \frac{1}{2}\csc\frac{x}{2}$.

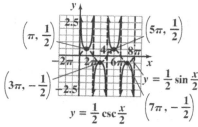

33. Graph the reciprocal cosine function, $y = 2\cos x$.
The equation is of the form $y = A\cos Bx$ with $A = 2$ and $B = 1$.
amplitude: $|A| = |2| = 2$

period: $\dfrac{2\pi}{B} = \dfrac{2\pi}{1} = 2\pi$

Use quarter-periods, $\dfrac{\pi}{2}$, to find x-values for the five key points. Starting with $x = 0$, the x-values are 0, $\dfrac{\pi}{2}$, π, $\dfrac{3\pi}{2}$, 2π. Evaluating the function at each value of x, the key points are $(0, 2)$, $\left(\dfrac{\pi}{2}, 0\right)$, $(\pi, -2)$, $\left(\dfrac{3\pi}{2}, 0\right)$, and $(2\pi, 2)$. Use these key points to graph $y = 2\cos x$ from 0 to 2π. Extend the graph one cycle to the right. Use the graph to obtain the graph of the reciprocal function. Draw vertical asymptotes through the x-intercepts, and use them as guides to graph $y = 2\sec x$.

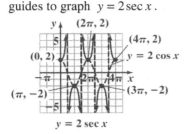

35. Graph the reciprocal cosine function, $y = \cos\dfrac{x}{3}$. The equation is of the form $y = A\cos Bx$ with $A = 1$ and $B = \dfrac{1}{3}$.
amplitude: $|A| = |1| = 1$

period: $\dfrac{2\pi}{B} = \dfrac{2\pi}{\frac{1}{3}} = 2\pi \cdot 3 = 6\pi$

Use quarter-periods, $\dfrac{6\pi}{4} = \dfrac{3\pi}{2}$, to find x-values for the five key points. Starting with $x = 0$, the x-values are 0, $\dfrac{3\pi}{2}$, 3π, $\dfrac{9\pi}{2}$, and 6π. Evaluating the function at each value of x, the key points are $(0, 1)$, $\left(\dfrac{3\pi}{2}, 0\right)$, $(3\pi, -1)$, $\left(\dfrac{9\pi}{2}, 0\right)$, and $(6\pi, 1)$. Use these key points to graph $y = \cos\dfrac{x}{3}$ from 0 to 6π. Extend the graph one cycle to the right. Use the graph to obtain

the graph of the reciprocal function. Draw vertical asymptotes through the x-intercepts, and use them as guides to graph $y = \sec\dfrac{x}{3}$.

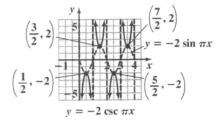

37. Graph the reciprocal sine function, $y = -2\sin \pi x$.
The equation is of the form $y = A\sin Bx$ with $A = -2$ and $B = \pi$.
amplitude: $|A| = |-2| = 2$

period: $\dfrac{2\pi}{B} = \dfrac{2\pi}{\pi} = 2$

Use quarter-periods, $\dfrac{2}{4} = \dfrac{1}{2}$, to find x-values for the five key points. Starting with $x = 0$, the x-values are 0, $\dfrac{1}{2}$, 1, $\dfrac{3}{2}$, and 2. Evaluating the function at each value of x, the key points are $(0, 0)$, $\left(\dfrac{1}{2}, -2\right)$, $(1, 0)$, $\left(\dfrac{3}{2}, 2\right)$, and $(2, 0)$. Use these key points to graph $y = -2\sin \pi x$ from 0 to 2. Extend the graph one cycle to the right. Use the graph to obtain the graph of the reciprocal function.

Draw vertical asymptotes through the x-intercepts, and use them as guides to graph $y = -2\csc \pi x$.

39. Graph the reciprocal cosine function, $y = -\dfrac{1}{2}\cos \pi x$.

The equation is of the form $y = A\cos Bx$ with

$A = -\dfrac{1}{2}$ and $B = \pi$.

amplitude: $|A| = \left|-\dfrac{1}{2}\right| = \dfrac{1}{2}$

period: $\dfrac{2\pi}{B} = \dfrac{2\pi}{\pi} = 2$

Use quarter-periods, $\dfrac{2}{4} = \dfrac{1}{2}$, to find x-values for the

five key points. Starting with $x = 0$, the x-values are

$0, \dfrac{1}{2}, 1, \dfrac{3}{2}$, and 2. Evaluating the function at each

value of x, the key points are $\left(0, -\dfrac{1}{2}\right)$,

$\left(\dfrac{1}{2}, 0\right), \left(1, \dfrac{1}{2}\right), \left(\dfrac{3}{2}, 0\right), \left(2, -\dfrac{1}{2}\right)$. Use these key

points to graph $y = -\dfrac{1}{2}\cos \pi x$ from 0 to 2. Extend

the graph one cycle to the right. Use the graph to
obtain the graph of the reciprocal function. Draw
vertical asymptotes through the
x-intercepts, and use them as guides to graph

$y = -\dfrac{1}{2}\sec \pi x$.

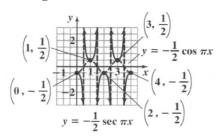

41. Graph the reciprocal sine function, $y = \sin(x - \pi)$.
The equation is of the form $y = A\sin(Bx - C)$ with A
= 1, and $B = 1$, and $C = \pi$.
amplitude: $|A| = |1| = 1$

period: $\dfrac{2\pi}{B} = \dfrac{2\pi}{1} = 2\pi$

phase shift: $\dfrac{C}{B} = \dfrac{\pi}{1} = \pi$

Use quarter-periods, $\dfrac{2\pi}{4} = \dfrac{\pi}{2}$, to find

x-values for the five key points. Starting with $x = \pi$,

the x-values are $\pi, \dfrac{3\pi}{2}, 2\pi, \dfrac{5\pi}{2}$, and 3π.

Evaluating the function at each value of x, the key

points are $(\pi, 0)$, $\left(\dfrac{3\pi}{2}, 1\right)$, $(2\pi, 0)$,

$\left(\dfrac{5\pi}{2}, -1\right)$, $(3\pi, 0)$. Use these key points to graph

$y = \sin(x - \pi)$ from π to 3π. Extend the graph one
cycle to the right. Use the graph to obtain the graph
of the reciprocal function.

Draw vertical asymptotes through the x-intercepts,
and use them as guides to graph $y = \csc(x - \pi)$.

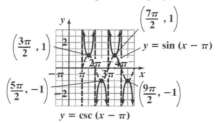

43. Graph the reciprocal cosine function,
$y = 2\cos(x + \pi)$. The equation is of the form
$y = A\cos(Bx + C)$ with $A = 2$, $B = 1$, and $C = -\pi$.
amplitude: $|A| = |2| = 2$

period: $\dfrac{2\pi}{B} = \dfrac{2\pi}{1} = 2\pi$

phase shift: $\dfrac{C}{B} = \dfrac{-\pi}{1} = -\pi$

Use quarter-periods, $\dfrac{2\pi}{4} = \dfrac{\pi}{2}$, to find x-values for the

five key points. Starting with $x = -\pi$, the x-values

are $-\pi, -\dfrac{\pi}{2}, 0, \dfrac{\pi}{2}$, and π. Evaluating the function

at each value of x, the key points are $(-\pi, 2)$,

$\left(-\dfrac{\pi}{2}, 0\right)$, $(0, -2)$, $\left(\dfrac{\pi}{2}, 0\right)$, and $(\pi, 2)$. Use these

key points to graph $y = 2\cos(x + \pi)$ from $-\pi$ to π.
Extend the graph one cycle to the right. Use the graph
to obtain the graph of the reciprocal function. Draw
vertical asymptotes through the x-intercepts, and use
them as guides to graph $y = 2\sec(x + \pi)$.

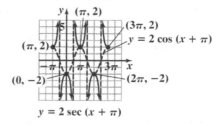

45.

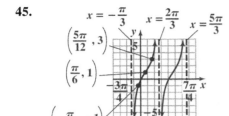

$$y = 2\tan\left(x - \frac{\pi}{6}\right) + 1$$

47.

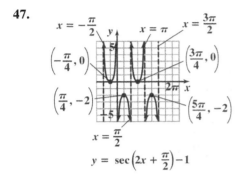

$$y = \sec\left(2x + \frac{\pi}{2}\right) - 1$$

49.

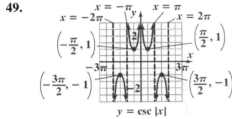

$$y = \csc|x|$$

51.

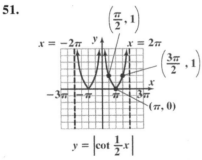

$$y = \left|\cot\frac{1}{2}x\right|$$

53. $y = (f \circ h)(x) = f(h(x)) = 2\sec\left(2x - \frac{\pi}{2}\right)$

$$y = 2\sec\left(2x - \frac{\pi}{2}\right)$$

55. Use a graphing utility with $y_1 = \tan x$ and $y_2 = -1$.
For the window use $\text{Xmin} = -2\pi$, $\text{Xmax} = 2\pi$,
$\text{Ymin} = -2$, and $\text{Ymax} = 2$.

$$x = -\frac{5\pi}{4}, \ -\frac{\pi}{4}, \ \frac{3\pi}{4}, \ \frac{7\pi}{4}$$

$$x \approx -3.93, \ -0.79, \ 2.36, \ 5.50$$

57. Use a graphing utility with $y_1 = 1/\sin x$ and $y_2 = 1$.
For the window use $\text{Xmin} = -2\pi$, $\text{Xmax} = 2\pi$,
$\text{Ymin} = -2$, and $\text{Ymax} = 2$.

$$x = -\frac{3\pi}{2}, \ \frac{\pi}{2}$$

$$x \approx -4.71, \ 1.57$$

59. $d = 12\tan 2\pi t$

a. Solve the equations

$$2\pi t = -\frac{\pi}{2} \quad \text{and} \quad 2\pi t = \frac{\pi}{2}$$

$$t = \frac{-\frac{\pi}{2}}{2\pi} \qquad\qquad t = \frac{\frac{\pi}{2}}{2\pi}$$

$$t = -\frac{1}{4} \qquad\qquad t = \frac{1}{4}$$

Thus, two consecutive asymptotes occur at

$$x = -\frac{1}{4} \quad \text{and} \quad x = \frac{1}{4}.$$

$$x\text{-intercept} = \frac{-\frac{1}{4} + \frac{1}{4}}{2} = \frac{0}{2} = 0$$

An x-intercept is 0 and the graph passes through (0, 0). Because the coefficient of the tangent is 12, the points on the graph midway between an x-intercept and the asymptotes have y-coordinates of –12 and 12. Use the two

consecutive asymptotes, $x = -\frac{1}{4}$ and $x = \frac{1}{4}$, to

graph one full period of $d = 12\tan 2\pi t$. To graph on [0, 2], continue the pattern and extend the graph to 2. (Do not use the left hand side of the first period of the graph on [0, 2].)

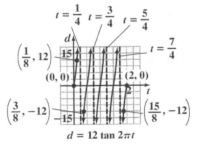

$$d = 12\tan 2\pi t$$

b. The function is undefined for $t = 0.25, 0.75, 1.25,$ and 1.75.
The beam is shining parallel to the wall at these times.

61. Use the function that relates the acute angle with the hypotenuse and the adjacent leg, the secant function.

$$\sec x = \frac{d}{10}$$

$$d = 10 \sec x$$

Graph the reciprocal cosine function, $y = 10 \cos x$.

The equation is of the form $y = A \cos Bx$ with $A = 10$ and $B = 1$.

amplitude: $|A| = |10| = 10$

period: $\dfrac{2\pi}{B} = \dfrac{2\pi}{1} = 2\pi$

For $-\dfrac{\pi}{2} < x < \dfrac{\pi}{2}$, use the *x*-values $-\dfrac{\pi}{2}$, 0, and $\dfrac{\pi}{2}$ to find the key points $\left(-\dfrac{\pi}{2}, 0\right)$, $(0, 10)$, and $\left(\dfrac{\pi}{2}, 0\right)$.

Connect these points with a smooth curve, then draw vertical asymptotes through the *x*-intercepts, and use them as guides to graph $d = 10 \sec x$ on $\left[-\dfrac{\pi}{2}, \dfrac{\pi}{2}\right]$.

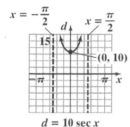

$d = 10 \sec x$

63.

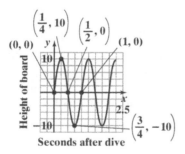

65. – 75. Answers may vary.

77. period: $\dfrac{\pi}{B} = \dfrac{\pi}{\frac{1}{4}} = \pi \cdot 4 = 4\pi$

Graph $y = \tan \dfrac{x}{4}$ for $0 \le x \le 8\pi$.

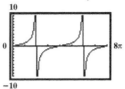

79. period: $\dfrac{\pi}{B} = \dfrac{\pi}{2}$

Graph $y = \cot 2x$ for $0 \le x \le \pi$.

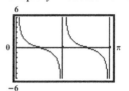

81. period: $\dfrac{\pi}{B} = \dfrac{\pi}{\pi} = 1$

Graph $y = \dfrac{1}{2} \tan \pi x$ for $0 \le x \le 2$.

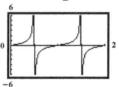

83. period: $\dfrac{2\pi}{B} = \dfrac{2\pi}{\frac{1}{2}} = 2\pi \cdot 2 = 4\pi$

Graph the functions for $0 \le x \le 8\pi$.

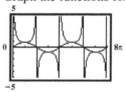

85. period: $\dfrac{2\pi}{B} = \dfrac{2\pi}{2} = \pi$

phase shift: $\dfrac{C}{B} = \dfrac{\frac{\pi}{6}}{2} = \dfrac{\pi}{12}$

Thus, we include $\dfrac{\pi}{12} \le x \le \dfrac{25\pi}{12}$ in our graph, and

graph for $0 \le x \le \dfrac{5\pi}{2}$.

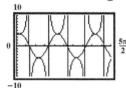

87.

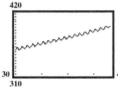

The graph shows that carbon dioxide concentration rises and falls each year, but over all the concentration increased from 1990 to 2008.

89. makes sense

91. does not make sense; Explanations will vary. Sample explanation: To obtain a cosecant graph, you can use a sine graph.

97. a. Since $A=1$, the range is $(-\infty, -1] \cup [1, \infty)$

Viewing rectangle: $\left[-\dfrac{\pi}{6}, \pi, \dfrac{7\pi}{6}\right]$ by $[-3, 3, 1]$

b. Since $A=3$, the range is $(-\infty, -3] \cup [3, \infty)$

Viewing rectangle: $\left[-\dfrac{1}{2}, \dfrac{7}{2}, 1\right]$ by $[-6, 6, 1]$

99. a.

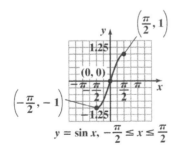

b. yes; Explanations will vary.

93. The graph has the shape of a cotangent function with consecutive asymptotes at

$x = 0$ and $x = \dfrac{2\pi}{3}$. The period is $\dfrac{2\pi}{3} - 0 = \dfrac{2\pi}{3}$. Thus,

$\dfrac{\pi}{B} = \dfrac{2\pi}{3}$

$2\pi B = 3\pi$

$B = \dfrac{3\pi}{2\pi} = \dfrac{3}{2}$

The points on the graph midway between an x-intercept and the asymptotes have y-coordinates of 1 and -1. Thus, $A = 1$. There is no phase shift. Thus, $C = 0$. An equation for this graph is $y = \cot\dfrac{3}{2}x$.

95. The range shows that $A = 2$.

Since the period is 3π, the coefficient of x is given by B where $\dfrac{2\pi}{B} = 3\pi$

$\dfrac{2\pi}{B} = 3\pi$

$3B\pi = 2\pi$

$B = \dfrac{2}{3}$

Thus, $y = 2\csc\dfrac{2x}{3}$

c. The angle is $-\dfrac{\pi}{6}$.

This is represented by the point $\left(-\dfrac{\pi}{6}, -\dfrac{1}{2}\right)$.

100. a.

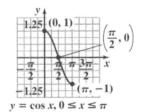

b. yes; Explanations will vary.

c. The angle is $\dfrac{5\pi}{6}$.

This is represented by the point $\left(\dfrac{5\pi}{6}, -\dfrac{\sqrt{3}}{2}\right)$.

101. a.

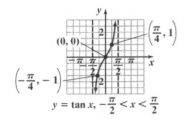

$$y = \tan x, -\frac{\pi}{2} < x < \frac{\pi}{2}$$

b. yes; Explanations will vary.

c. The angle is $-\frac{\pi}{3}$.

This is represented by the point $\left(-\frac{\pi}{3}, -\sqrt{3}\right)$.

Section 5.7

Check Point Exercises

1. Let $\theta = \sin^{-1}\frac{\sqrt{3}}{2}$, then $\sin\theta = \frac{\sqrt{3}}{2}$.

The only angle in the interval $\left[-\frac{\pi}{2}, \frac{\pi}{2}\right]$ that satisfies $\sin\theta = \frac{\sqrt{3}}{2}$ is $\frac{\pi}{3}$. Thus, $\theta = \frac{\pi}{3}$, or $\sin^{-1}\frac{\sqrt{3}}{2} = \frac{\pi}{3}$.

2. Let $\theta = \sin^{-1}\left(-\frac{\sqrt{2}}{2}\right)$, then $\sin\theta = -\frac{\sqrt{2}}{2}$.

The only angle in the interval $\left[-\frac{\pi}{2}, \frac{\pi}{2}\right]$ that satisfies $\cos\theta = -\frac{\sqrt{2}}{2}$ is $-\frac{\pi}{4}$. Thus $\theta = -\frac{\pi}{4}$, or $\sin^{-1}\left(-\frac{\sqrt{2}}{2}\right) = -\frac{\pi}{4}$.

3. Let $\theta = \cos^{-1}\left(-\frac{1}{2}\right)$, then $\cos\theta = -\frac{1}{2}$. The only angle in the interval $[0, \pi]$ that satisfies $\cos\theta = -\frac{1}{2}$ is $\frac{2\pi}{3}$. Thus,

$\theta = \frac{2\pi}{3}$, or $\cos^{-1}\left(-\frac{1}{2}\right) = \frac{2\pi}{3}$.

4. Let $\theta = \tan^{-1}(-1)$, then $\tan\theta = -1$. The only angle in the interval $\left(-\frac{\pi}{2}, \frac{\pi}{2}\right)$ that satisfies $\tan\theta = -1$ is $-\frac{\pi}{4}$. Thus

$\theta = -\frac{\pi}{4}$ or $\tan^{-1}\theta = -\frac{\pi}{4}$.

5.

Scientific Calculator Solution				
	Function	**Mode**	**Keystrokes**	**Display** (rounded to four places)
a.	$\cos^{-1}\left(\dfrac{1}{3}\right)$	Radian	1 ÷ 3 = $\boxed{\text{COS}^{-1}}$	1.2310
b.	$\tan^{-1}(-35.85)$	Radian	35.85 $\boxed{+/-}$ $\boxed{\text{TAN}^{-1}}$	−1.5429

Graphing Calculator Solution				
	Function	**Mode**	**Keystrokes**	**Display** (rounded to four places)
a.	$\cos^{-1}\left(\dfrac{1}{3}\right)$	Radian	$\boxed{\text{COS}^{-1}}$ $\boxed{(}$ 1 ÷ 3 $\boxed{)}$ $\boxed{\text{ENTER}}$	1.2310
b.	$\tan^{-1}(-35.85)$	Radian	$\boxed{\text{TAN}^{-1}}$ $\boxed{-}$ 35.85 $\boxed{\text{ENTER}}$	−1.5429

6. a. $\cos\left(\cos^{-1} 0.7\right)$

$x = 0.7$, x is in $[-1,1]$ so $\cos(\cos^{-1} 0.7) = 0.7$

b. $\sin^{-1}(\sin \pi)$

$x = \pi$, x is not in $\left[-\dfrac{\pi}{2}, \dfrac{\pi}{2}\right]$. x is in the domain of $\sin x$, so $\sin^{-1}(\sin \pi) = \sin^{-1}(0) = 0$

c. $\cos\left(\cos^{-1} \pi\right)$

$x = \pi$, x is not in $[-1,1]$ so $\cos\left(\cos^{-1} \pi\right)$ is not defined.

7. Let $\theta = \tan^{-1}\left(\dfrac{3}{4}\right)$, then $\tan \theta = \dfrac{3}{4}$. Because $\tan \theta$ is positive, θ is in the first quadrant.

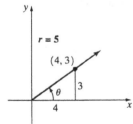

Use the Pythagorean Theorem to find r.

$r^2 = 3^2 + 4^2 = 9 + 16 = 25$

$r = \sqrt{25} = 5$

Use the right triangle to find the exact value.

$\sin\left(\tan^{-1} \dfrac{3}{4}\right) = \sin \theta = \dfrac{\text{side opposite } \theta}{\text{hypotenuse}} = \dfrac{3}{5}$

8. Let $\theta = \sin^{-1}\left(-\dfrac{1}{2}\right)$, then $\sin\theta = -\dfrac{1}{2}$. Because $\sin\theta$ is negative, θ is in quadrant IV.

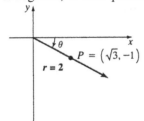

Use the Pythagorean Theorem to find x.

$$x^2 + (-1)^2 = 2^2$$
$$x^2 + 1 = 4$$
$$x^2 = 3$$
$$x = \sqrt{3}$$

Use values for x and r to find the exact value.

$$\cos\left[\sin^{-1}\left(-\dfrac{1}{2}\right)\right] = \cos\theta = \dfrac{x}{r} = \dfrac{\sqrt{3}}{2}$$

9. Let $\theta = \tan^{-1} x$, then $\tan\theta = x = \dfrac{x}{1}$.

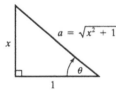

Use the Pythagorean Theorem to find the third side, a.

$$a^2 = x^2 + 1^2$$
$$a = \sqrt{x^2 + 1}$$

Use the right triangle to write the algebraic expression.

$$\sec\left(\tan^{-1} x\right) = \sec\theta = \dfrac{\sqrt{x^2 + 1}}{1} = \sqrt{x^2 + 1}$$

Concept and Vocabulary Check 5.7

1. $-\dfrac{\pi}{2} \le x \le \dfrac{\pi}{2}$; $\sin^{-1} x$

2. $0 \le x \le \pi$; $\cos^{-1} x$

3. $-\dfrac{\pi}{2} \le x \le \dfrac{\pi}{2}$; $\tan^{-1} x$

4. $[-1,1]$; $\left[-\dfrac{\pi}{2}, \dfrac{\pi}{2}\right]$

5. $[-1,1]$; $[0, \pi]$

6. $(-\infty, \infty)$; $\left(-\dfrac{\pi}{2}, \dfrac{\pi}{2}\right)$

7. $\left[-\dfrac{\pi}{2}, \dfrac{\pi}{2}\right]$

8. $[0, \pi]$

9. $\left(-\dfrac{\pi}{2}, \dfrac{\pi}{2}\right)$

10. false

Exercise Set 5.7

1. Let $\theta = \sin^{-1}\dfrac{1}{2}$, then $\sin\theta = \dfrac{1}{2}$. The only angle in the interval $\left[-\dfrac{\pi}{2}, \dfrac{\pi}{2}\right]$ that satisfies $\sin\theta = \dfrac{1}{2}$ is $\dfrac{\pi}{6}$. Thus, $\theta = \dfrac{\pi}{6}$, or $\sin^{-1}\dfrac{1}{2} = \dfrac{\pi}{6}$.

3. Let $\theta = \sin^{-1}\dfrac{\sqrt{2}}{2}$, then $\sin\theta = \dfrac{\sqrt{2}}{2}$. The only angle in the interval $\left[-\dfrac{\pi}{2}, \dfrac{\pi}{2}\right]$ that satisfies $\sin\theta = \dfrac{\sqrt{2}}{2}$ is $\dfrac{\pi}{4}$. Thus $\theta = \dfrac{\pi}{4}$, or $\sin^{-1}\dfrac{\sqrt{2}}{2} = \dfrac{\pi}{4}$.

5. Let $\theta = \sin^{-1}\left(-\dfrac{1}{2}\right)$, then $\sin\theta = -\dfrac{1}{2}$. The only angle in the interval $\left[-\dfrac{\pi}{2}, \dfrac{\pi}{2}\right]$ that satisfies $\sin\theta = -\dfrac{1}{2}$ is $-\dfrac{\pi}{6}$. Thus $\theta = -\dfrac{\pi}{6}$, or $\sin^{-1}\left(-\dfrac{1}{2}\right) = -\dfrac{\pi}{6}$.

7. Let $\theta = \cos^{-1}\dfrac{\sqrt{3}}{2}$, then $\cos\theta = \dfrac{\sqrt{3}}{2}$. The only angle in the interval $[0, \pi]$ that satisfies $\cos\theta = \dfrac{\sqrt{3}}{2}$ is $\dfrac{\pi}{6}$. Thus $\theta = \dfrac{\pi}{6}$, or $\cos^{-1}\dfrac{\sqrt{3}}{2} = \dfrac{\pi}{6}$.

9. Let $\theta = \cos^{-1}\left(-\dfrac{\sqrt{2}}{2}\right)$, then $\cos\theta = -\dfrac{\sqrt{2}}{2}$. The only angle in the interval $[0, \pi]$ that satisfies $\cos\theta = -\dfrac{\sqrt{2}}{2}$ is $\dfrac{3\pi}{4}$. Thus $\theta = \dfrac{3\pi}{4}$, or $\cos^{-1}\left(-\dfrac{\sqrt{2}}{2}\right) = \dfrac{3\pi}{4}$.

11. Let $\theta = \cos^{-1} 0$, then $\cos\theta = 0$. The only angle in the interval $[0, \pi]$ that satisfies $\cos\theta = 0$ is $\dfrac{\pi}{2}$. Thus $\theta = \dfrac{\pi}{2}$, or $\cos^{-1} 0 = \dfrac{\pi}{2}$.

13. Let $\theta = \tan^{-1}\dfrac{\sqrt{3}}{3}$, then $\tan\theta = \dfrac{\sqrt{3}}{3}$. The only angle in the interval $\left(-\dfrac{\pi}{2}, \dfrac{\pi}{2}\right)$ that satisfies $\tan\theta = \dfrac{\sqrt{3}}{3}$ is $\dfrac{\pi}{6}$. Thus $\theta = \dfrac{\pi}{6}$, or $\tan^{-1}\dfrac{\sqrt{3}}{3} = \dfrac{\pi}{6}$.

15. Let $\theta = \tan^{-1} 0$, then $\tan\theta = 0$. The only angle in the interval $\left(-\dfrac{\pi}{2}, \dfrac{\pi}{2}\right)$ that satisfies $\tan\theta = 0$ is 0. Thus $\theta = 0$, or $\tan^{-1} 0 = 0$.

17. Let $\theta = \tan^{-1}\left(-\sqrt{3}\right)$, then $\tan\theta = -\sqrt{3}$. The only angle in the interval $\left(-\dfrac{\pi}{2}, \dfrac{\pi}{2}\right)$ that satisfies $\tan\theta = -\sqrt{3}$ is $-\dfrac{\pi}{3}$. Thus $\theta = -\dfrac{\pi}{3}$, or $\tan^{-1}\left(-\sqrt{3}\right) = -\dfrac{\pi}{3}$.

19.

Scientific Calculator Solution			
Function	**Mode**	**Keystrokes**	**Display** (rounded to two places)
$\sin^{-1} 0.3$	Radian	0.3 SIN⁻¹	0.30

Graphing Calculator Solution			
Function	**Mode**	**Keystrokes**	**Display** (rounded to two places)
$\sin^{-1} 0.3$	Radian	SIN⁻¹ 0.3 ENTER	0.30

21.

Scientific Calculator Solution			
Function	**Mode**	**Keystrokes**	**Display** (rounded to two places)
$\sin^{-1}(-0.32)$	Radian	0.32 +/− SIN⁻¹	−0.33

Graphing Calculator Solution			
Function	**Mode**	**Keystrokes**	**Display** (rounded to two places)
$\sin^{-1}(-0.32)$	Radian	SIN⁻¹ − 0.32 ENTER	−0.33

23.

Scientific Calculator Solution			
Function	**Mode**	**Keystrokes**	**Display** (rounded to two places)
$\cos^{-1}\left(\dfrac{3}{8}\right)$	Radian	3 ÷ 8 = COS⁻¹	1.19

Graphing Calculator Solution			
Function	**Mode**	**Keystrokes**	**Display** (rounded to two places)
$\cos^{-1}\left(\dfrac{3}{8}\right)$	Radian	COS⁻¹ (3 ÷ 8) ENTER	1.19

25.

Scientific Calculator Solution			
Function	**Mode**	**Keystrokes**	**Display** (rounded to two places)
$\cos^{-1}\dfrac{\sqrt{5}}{7}$	Radian	5 √ ÷ 7 = COS⁻¹	1.25

Graphing Calculator Solution			
Function	**Mode**	**Keystrokes**	**Display** (rounded to two places)
$\cos^{-1}\dfrac{\sqrt{5}}{7}$	Radian	COS⁻¹ (√ 5 ÷ 7) ENTER	1.25

27.

Scientific Calculator Solution			
Function	**Mode**	**Keystrokes**	**Display** (rounded to two places)
$\tan^{-1}(-20)$	Radian	20 ⁺⁄₋ TAN⁻¹	−1.52

Graphing Calculator Solution			
Function	**Mode**	**Keystrokes**	**Display** (rounded to two places)
$\tan^{-1}(-20)$	Radian	TAN⁻¹ − 20 ENTER	−1.52

29.

Scientific Calculator Solution			
Function	**Mode**	**Keystrokes**	**Display** (rounded to two places)
$\tan^{-1}\left(-\sqrt{473}\right)$	Radian	473 √ ⁺⁄₋ TAN⁻¹	−1.52

Graphing Calculator Solution			
Function	**Mode**	**Keystrokes**	**Display** (rounded to two places)
$\tan^{-1}\left(-\sqrt{473}\right)$	Radian	TAN⁻¹ (− √ 473) ENTER	−1.52

31. $\sin\left(\sin^{-1}0.9\right)$

$x = 0.9$, x is in $[-1, 1]$, so $\sin(\sin^{-1}0.9) = 0.9$

33. $\sin^{-1}\left(\sin\dfrac{\pi}{3}\right)$

$x = \dfrac{\pi}{3}$, x is in $\left[-\dfrac{\pi}{2}, \dfrac{\pi}{2}\right]$, so $\sin^{-1}\left(\sin\dfrac{\pi}{3}\right) = \dfrac{\pi}{3}$

35. $\sin^{-1}\left(\sin\dfrac{5\pi}{6}\right)$

$x = \dfrac{5\pi}{6}$, x is not in $\left[-\dfrac{\pi}{2}, \dfrac{\pi}{2}\right]$, x is in the domain of

$\sin x$, so $\sin^{-1}\left(\sin\dfrac{5\pi}{6}\right) = \sin^{-1}\left(\dfrac{1}{2}\right) = \dfrac{\pi}{6}$

37. $\tan\left(\tan^{-1}125\right)$

$x = 125$, x is a real number, so $\tan\left(\tan^{-1}125\right) = 125$

39. $\tan^{-1}\left[\tan\left(-\dfrac{\pi}{6}\right)\right]$

$x = -\dfrac{\pi}{6}$, x is in $\left(-\dfrac{\pi}{2}, \dfrac{\pi}{2}\right)$, so

$\tan^{-1}\left[\tan\left(-\dfrac{\pi}{6}\right)\right] = -\dfrac{\pi}{6}$

41. $\tan^{-1}\left(\tan\dfrac{2\pi}{3}\right)$

$x = \dfrac{2\pi}{3}$, x is not in $\left(-\dfrac{\pi}{2}, \dfrac{\pi}{2}\right)$, x is in the domain of

$\tan x$, so $\tan^{-1}\left(\tan\dfrac{2\pi}{3}\right) = \tan^{-1}\left(-\sqrt{3}\right) = -\dfrac{\pi}{3}$

43. $\sin^{-1}(\sin\pi)$

$x = \pi$, x is not in $\left[-\dfrac{\pi}{2}, \dfrac{\pi}{2}\right]$,

x is in the domain of $\sin x$, so

$\sin^{-1}(\sin\pi) = \sin^{-1}0 = 0$

45. $\sin\left(\sin^{-1}\pi\right)$

$x = \pi$, x is not in $[-1, 1]$, so $\sin\left(\sin^{-1}\pi\right)$ is not

defined.

47. Let $\theta = \sin^{-1}\dfrac{4}{5}$, then $\sin\theta = \dfrac{4}{5}$. Because $\sin\theta$ is

positive, θ is in the first quadrant.

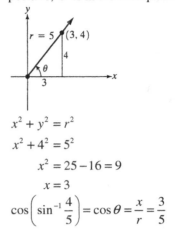

$x^2 + y^2 = r^2$

$x^2 + 4^2 = 5^2$

$x^2 = 25 - 16 = 9$

$x = 3$

$\cos\left(\sin^{-1}\dfrac{4}{5}\right) = \cos\theta = \dfrac{x}{r} = \dfrac{3}{5}$

49. Let $\theta = \cos^{-1}\dfrac{5}{13}$, then $\cos\theta = \dfrac{5}{13}$. Because $\cos\theta$ is

positive, θ is in the first quadrant.

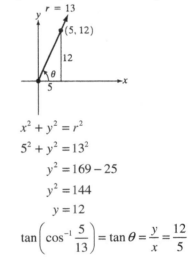

$x^2 + y^2 = r^2$

$5^2 + y^2 = 13^2$

$y^2 = 169 - 25$

$y^2 = 144$

$y = 12$

$\tan\left(\cos^{-1}\dfrac{5}{13}\right) = \tan\theta = \dfrac{y}{x} = \dfrac{12}{5}$

51. Let $\theta = \sin^{-1}\left(-\dfrac{3}{5}\right)$, then $\sin\theta = -\dfrac{3}{5}$. Because $\sin\theta$ is negative, θ is in quadrant IV.

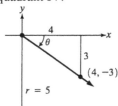

$$x^2 + y^2 = r^2$$
$$x^2 + (-3)^2 = 5^2$$
$$x^2 = 16$$
$$x = 4$$
$$\tan\left[\sin^{-1}\left(-\frac{3}{5}\right)\right] = \tan\theta = \frac{y}{x} = -\frac{3}{4}$$

53. Let, $\theta = \cos^{-1}\dfrac{\sqrt{2}}{2}$, then $\cos\theta = \dfrac{\sqrt{2}}{2}$. Because $\cos\theta$ is positive, θ is in the first quadrant.

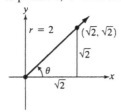

$$x^2 + y^2 = r^2$$
$$\left(\sqrt{2}\right)^2 + y^2 = 2^2$$
$$y^2 = 2$$
$$y = \sqrt{2}$$
$$\sin\left(\cos^{-1}\frac{\sqrt{2}}{2}\right) = \sin\theta = \frac{y}{r} = \frac{\sqrt{2}}{2}$$

55. Let $\theta = \sin^{-1}\left(-\dfrac{1}{4}\right)$, then $\sin\theta = -\dfrac{1}{4}$. Because $\sin\theta$ is negative, θ is in quadrant IV.

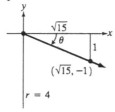

$$x^2 + y^2 = r^2$$
$$x^2 + (-1)^2 = 4^2$$
$$x^2 = 15$$
$$x = \sqrt{15}$$
$$\sec\left[\sin^{-1}\left(-\frac{1}{4}\right)\right] = \sec\theta = \frac{r}{x} = \frac{4}{\sqrt{15}} = \frac{4\sqrt{15}}{15}$$

57. Let $\theta = \cos^{-1}\left(-\dfrac{1}{3}\right)$, then $\cos\theta = -\dfrac{1}{3}$. Because $\cos\theta$ is negative, θ is in quadrant II.

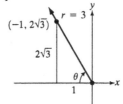

$$x^2 + y^2 = r^2$$
$$(-1)^2 + y^2 = 3^2$$
$$y^2 = 8$$
$$y = \sqrt{8}$$
$$y = 2\sqrt{2}$$

Use the right triangle to find the exact value.

$$\tan\left[\cos^{-1}\left(-\frac{1}{3}\right)\right] = \tan\theta = \frac{y}{x} = \frac{2\sqrt{2}}{-1} = -2\sqrt{2}$$

59. Let $\theta = \cos^{-1}\left(-\dfrac{\sqrt{3}}{2}\right)$, then $\cos\theta = -\dfrac{\sqrt{3}}{2}$. Because $\cos\theta$ is negative, θ is in quadrant II.

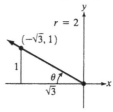

$$x^2 + y^2 = r^2$$
$$\left(-\sqrt{3}\right)^2 + y^2 = 2^2$$
$$y^2 = 1$$
$$y = 1$$
$$\csc\left[\cos^{-1}\left(-\frac{\sqrt{3}}{2}\right)\right] = \csc\theta = \frac{r}{y} = \frac{2}{1} = 2$$

61. Let $\theta = \tan^{-1}\left(-\dfrac{2}{3}\right)$, then $\tan\theta = -\dfrac{2}{3}$.

Because $\tan\theta$ is negative, θ is in quadrant IV.

$$r^2 = x^2 + y^2$$
$$r^2 = 3^2 + (-2)^2$$
$$r^2 = 9 + 4$$
$$r^2 = 13$$
$$r = \sqrt{13}$$

$$\cos\left[\tan^{-1}\left(-\frac{2}{3}\right)\right] = \cos\theta = \frac{x}{r} = \frac{3}{\sqrt{13}} = \frac{3\sqrt{13}}{13}$$

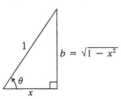

Use the Pythagorean Theorem to find the third side, b.
$$x^2 + b^2 = 1^2$$
$$b^2 = 1 - x^2$$
$$b = \sqrt{1-x^2}$$
Use the right triangle to write the algebraic expression.
$$\tan\left(\cos^{-1}x\right) = \tan\theta = \frac{\sqrt{1-x^2}}{x}$$

63. Let $\theta = \cos^{-1}x$, then $\cos\theta = x = \dfrac{x}{1}$.

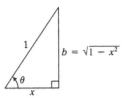

Use the Pythagorean Theorem to find the third side, b.
$$x^2 + b^2 = 1^2$$
$$b^2 = 1 - x^2$$
$$b = \sqrt{1-x^2}$$
Use the right triangle to write the algebraic expression.
$$\tan\left(\cos^{-1}x\right) = \tan\theta = \frac{\sqrt{1-x^2}}{x}$$

65. Let $\theta = \sin^{-1}2x$, then $\sin\theta = 2x$
$$y = 2x,\ r = 1$$
Use the Pythagorean Theorem to find x.
$$x^2 + (2x)^2 = 1^2$$
$$x^2 = 1 - 4x^2$$
$$x = \sqrt{1-4x^2}$$
$$\cos(\sin^{-1}2x) = \sqrt{1-4x^2}$$

67. Let $\theta = \sin^{-1}\dfrac{1}{x}$, then $\sin\theta = \dfrac{1}{x}$.

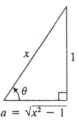

Use the Pythagorean Theorem to find the third side, a.
$$a^2 + 1^2 = x^2$$
$$a^2 = x^2 - 1$$
$$a = \sqrt{x^2 - 1}$$
Use the right triangle to write the algebraic expression.
$$\cos\left(\sin^{-1}\frac{1}{x}\right) = \cos\theta = \frac{\sqrt{x^2-1}}{x}$$

69. $\cot\left(\tan^{-1}\dfrac{x}{\sqrt{3}}\right) = \dfrac{\sqrt{3}}{x}$

71. Let $\theta = \sin^{-1}\dfrac{x}{\sqrt{x^2+4}}$, then $\sin\theta = \dfrac{x}{\sqrt{x^2+4}}$.

Use the Pythagorean Theorem to find the third side, a.
$$a^2 + x^2 = \left(\sqrt{x^2+4}\right)^2$$
$$a^2 = x^2 + 4 - x^2 = 4$$
$$a = 2$$
Use the right triangle to write the algebraic expression.
$$\sec\left(\sin^{-1}\frac{x}{\sqrt{x^2+4}}\right) = \sec\theta = \frac{\sqrt{x^2+4}}{2}$$

73. a. $y = \sec x$ is the reciprocal of $y = \cos x$. The x-values for the key points in the interval $[0, \pi]$ are $0, \dfrac{\pi}{4}, \dfrac{\pi}{2}, \dfrac{3\pi}{4}$, and π. The key points are

$(0, 1)$, $\left(\dfrac{\pi}{4}, \dfrac{\sqrt{2}}{2}\right)$, $\left(\dfrac{\pi}{2}, 0\right)$, $\left(\dfrac{3\pi}{4}, -\dfrac{\sqrt{2}}{2}\right)$, and

$(\pi, -1)$, Draw a vertical asymptote at $x = \dfrac{\pi}{2}$.

Now draw our graph from $(0, 1)$ through $\left(\dfrac{\pi}{4}, \sqrt{2}\right)$ to ∞ on the left side of the asymptote. From $-\infty$ on the right side of the asymptote through $\left(\dfrac{3\pi}{4}, -\sqrt{2}\right)$ to $(\pi, -1)$.

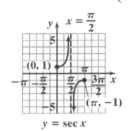

$y = \sec x$

b. With this restricted domain, no horizontal line intersects the graph of $y = \sec x$ more than once, so the function is one-to-one and has an inverse function.

c. Reflecting the graph of the restricted secant function about the line $y = x$, we get the graph of $y = \sec^{-1} x$.

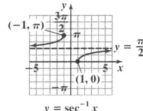

$y = \sec^{-1} x$

75.

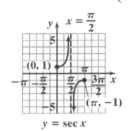

$f(x) = \sin^{-1} x + \dfrac{\pi}{2}$

domain: $[-1, 1]$;
range: $[0, \pi]$

77.

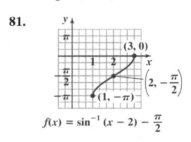

$g(x) = \cos^{-1}(x + 1)$

domain: $[-2, 0]$;
range: $[0, \pi]$

79.

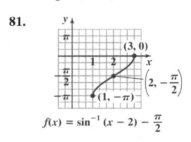

$f(x) = -2\tan^{-1} x$

domain: $(-\infty, \infty)$;
range: $(-\pi, \pi)$

81.

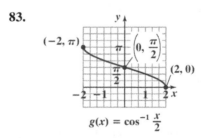

$f(x) = \sin^{-1}(x - 2) - \dfrac{\pi}{2}$

domain: $(1, 3)$;
range: $[-\pi, 0]$

83.

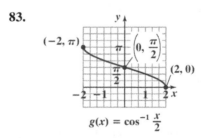

$g(x) = \cos^{-1}\dfrac{x}{2}$

domain: $[-2, 2]$;
range: $[0, \pi]$

85. The inner function, $\sin^{-1} x$, accepts values on the interval $[-1, 1]$. Since the inner and outer functions are inverses of each other, the domain and range are as follows.
domain: $[-1, 1]$; range: $[-1, 1]$

87. The inner function, $\cos x$, accepts values on the interval $(-\infty,\infty)$. The outer function returns values on the interval $[0,\pi]$

domain: $(-\infty,\infty)$; range: $[0,\pi]$

89. The inner function, $\cos x$, accepts values on the interval $(-\infty,\infty)$. The outer function returns values on the interval $\left[-\dfrac{\pi}{2},\dfrac{\pi}{2}\right]$

domain: $(-\infty,\infty)$; range: $\left[-\dfrac{\pi}{2},\dfrac{\pi}{2}\right]$

91. The functions $\sin^{-1}x$ and $\cos^{-1}x$ accept values on the interval $[-1,1]$. The sum of these values is always $\dfrac{\pi}{2}$.

domain: $[-1,1]$; range: $\left\{\dfrac{\pi}{2}\right\}$

93. $\theta = \tan^{-1}\dfrac{33}{x} - \tan^{-1}\dfrac{8}{x}$

x	θ
5	$\tan^{-1}\dfrac{33}{5} - \tan^{-1}\dfrac{8}{5} \approx 0.408$ radians
10	$\tan^{-1}\dfrac{33}{10} - \tan^{-1}\dfrac{8}{10} \approx 0.602$ radians
15	$\tan^{-1}\dfrac{33}{15} - \tan^{-1}\dfrac{8}{15} \approx 0.654$ radians
20	$\tan^{-1}\dfrac{33}{20} - \tan^{-1}\dfrac{8}{20} \approx 0.645$ radians
25	$\tan^{-1}\dfrac{33}{25} - \tan^{-1}\dfrac{8}{25} \approx 0.613$ radians

95. $\theta = 2\tan^{-1}\dfrac{21.634}{28} \approx 1.3157$ radians;

$1.3157\left(\dfrac{180}{\pi}\right) \approx 75.4°$

97. $\tan^{-1}b - \tan^{-1}a = \tan^{-1}2 - \tan^{-1}0$
≈ 1.1071 square units

99. – 109. Answers may vary.

111. The domain of $y = \cos^{-1}x$ is the interval $[-1, 1]$, and the range is the interval $[0, \pi]$. Because the second equation is the first equation with 1

subtracted from the variable, we will move our x max to π, and graph in a $\left[-\dfrac{\pi}{2},\ \pi,\ \dfrac{\pi}{4}\right]$ by [0, 4, 1] viewing rectangle.

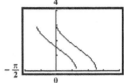

The graph of the second equation is the graph of the first equation shifted right 1 unit.

113. The domain of $y = \sin^{-1}x$ is the interval $[-1, 1]$, and the range is $\left[-\dfrac{\pi}{2},\dfrac{\pi}{2}\right]$. Because the second equation is the first equation plus 1, and with 2 added to the variable, we will move our y max to 3, and move our x min to $-\pi$, and graph in a $\left[-\pi,\ \dfrac{\pi}{2},\ \dfrac{\pi}{2}\right]$ by [–2, 3, 1] viewing rectangle.

The graph of the second equation is the graph of the first equation shifted left 2 units and up 1 unit.

115.

It seems $\sin^{-1}x + \cos^{-1}x = \dfrac{\pi}{2}$ for $-1 \le x \le 1$.

117. does not make sense; Explanations will vary. Sample explanation: Though this restriction works for tangent, it is not selected simply because it is easier to remember. Rather the restrictions are based on which intervals will have inverses.

119. does not make sense; Explanations will vary.
Sample explanation:

$$\sin^{-1}\left(\sin\frac{5\pi}{4}\right)=\sin^{-1}\left(-\frac{\sqrt{2}}{2}\right)=-\frac{\pi}{4}$$

121. $2\sin^{-1}x=\dfrac{\pi}{4}$

$$\sin^{-1}x=\frac{\pi}{8}$$

$$x=\sin\frac{\pi}{8}$$

123. Let α equal the acute angle in the smaller right
triangle.

$$\tan\alpha=\frac{8}{x}$$

so $\tan^{-1}\dfrac{8}{x}=\alpha$

$$\tan(\alpha+\theta)=\frac{33}{x}$$

so $\tan^{-1}\dfrac{33}{x}=\alpha+\theta$

$$\theta=\alpha+\theta-\alpha=\tan^{-1}\frac{33}{x}-\tan^{-1}\frac{8}{x}$$

124. $\tan A=\dfrac{a}{b}$

$$\tan 22.3°=\frac{a}{12.1}$$

$$a=12.1\tan 22.3°$$

$$a\approx 4.96$$

$$\cos A=\frac{b}{c}$$

$$\cos 22.3°=\frac{12.1}{c}$$

$$c=\frac{12.1}{\cos 22.3°}$$

$$c\approx 13.08$$

125. $\tan\theta=\dfrac{\text{opposite}}{\text{adjacent}}$

$$\tan\theta=\frac{18}{25}$$

$$\theta=\tan^{-1}\left(\frac{18}{25}\right)$$

$$\theta\approx 35.8°$$

126. $10\cos\left(\dfrac{\pi}{6}x\right)$

amplitude: $|10|=10$

period: $\dfrac{2\pi}{\frac{\pi}{6}}=2\pi\cdot\dfrac{6}{\pi}=12$

Section 5.8

Check Point Exercises

1. We begin by finding the measure of angle B. Because
    $C=90°$ and the sum of a triangle's angles is $180°$, we
    see that $A+B=90°$. Thus, $B=90°-A=90°-62.7°$
    $=27.3°$.
    Now we find b. Because we have a known angle, a
    known opposite side, and an unknown adjacent side,
    use the tangent function.

    $$\tan 62.7°=\frac{8.4}{b}$$

    $$b=\frac{8.4}{\tan 62.7°}\approx 4.34$$

    Finally, we need to find c. Because we have a known
    angle, a known opposite side and an unknown
    hypotenuse, use the sine function.

    $$\sin 62.7°=\frac{8.4}{c}$$

    $$c=\frac{8.4}{\sin 62.7}\approx 9.45$$

    In summary, $B=27.3°$, $b\approx 4.34$, and $c\approx 9.45$.

2. Using a right triangle, we have a known angle, an
    unknown opposite side, a, and a known adjacent side.
    Therefore, use the tangent function.

    $$\tan 85.4°=\frac{a}{80}$$

    $$a=80\tan 85.4°\approx 994$$

    The Eiffel tower is approximately 994 feet high.

3. Using a right triangle, we have an unknown angle, A,
    a known opposite side, and a known hypotenuse.
    Therefore, use the sine function.

    $$\sin A=\frac{6.7}{13.8}$$

    $$A=\sin^{-1}\frac{6.7}{13.8}\approx 29.0°$$

    The wire makes an angle of approximately $29.0°$ with
    the ground.

4. Using two right triangles, a smaller right triangle corresponding to the smaller angle of elevation drawn inside a larger right triangle corresponding to the larger angle of elevation, we have a known angle, an unknown opposite side, a in the smaller triangle, b in the larger triangle, and a known adjacent side in each triangle. Therefore, use the tangent function.

$$\tan 32° = \frac{a}{800}$$
$$a = 800\tan 32° \approx 499.9$$
$$\tan 35° = \frac{b}{800}$$
$$b = 800\tan 35° \approx 560.2$$

The height of the sculpture of Lincoln's face is 560.2 – 499.9, or approximately 60.3 feet.

5. a. We need the acute angle between ray OD and the north-south line through O. The measurement of this angle is given to be 25°. The angle is measured from the south side of the north-south line and lies east of the north-south line. Thus, the bearing from O to D is S 25°E.

b. We need the acute angle between ray OC and the north-south line through O. This angle measures $90° - 75° = 15°$. This angle is measured from the south side of the north-south line and lies west of the north-south line. Thus the bearing from O to C is S 15° W.

6. a. Your distance from the entrance to the trail system is represented by the hypotenuse, c, of a right triangle. Because we know the length of the two sides of the right triangle, we find c using the Pythagorean Theorem. We have

$$c^2 = a^2 + b^2 = (2.3)^2 + (3.5)^2 = 17.54$$
$$c = \sqrt{17.54} \approx 4.2$$

You are approximately 4.2 miles from the entrance to the trail system.

b. To find your bearing from the entrance to the trail system, consider a north-south line passing through the entrance. The acute angle from this line to the ray on which you lie is $31° + \theta$. Because we are measuring the angle from the south side of the line and you are west of the entrance, your bearing from the entrance is S $(31° + \theta)$ W. To find θ, Use a right triangle and the tangent function.

$$\tan\theta = \frac{3.5}{2.3}$$
$$\theta = \tan^{-1}\frac{3.5}{2.3} \approx 56.7°$$

Thus, $31° + \theta = 31° + 56.7° = 87.7°$. Your bearing from the entrance to the trail system is S 87.7° W.

7. When the object is released $(t = 0)$, the ball's distance, d, from its rest position is 6 inches down. Because it is down, d is negative: when $t = 0$, $d = -6$. Notice the greatest distance from rest position occurs at $t = 0$. Thus, we will use the equation with the cosine function, $y = a\cos\omega t$, to model the ball's motion. Recall that $|a|$ is the maximum distance. Because the ball initially moves down, $a = -6$. The value of ω can be found using the formula for the period.

$$\text{period} = \frac{2\pi}{\omega} = 4$$
$$2\pi = 4\omega$$
$$\omega = \frac{2\pi}{4} = \frac{\pi}{2}$$

Substitute these values into $d = a\cos wt$. The equation for the ball's simple harmonic motion is

$$d = -6\cos\frac{\pi}{2}t.$$

8. We begin by identifying values for a and ω.

$$d = 12\cos\frac{\pi}{4}t, \, a = 12 \text{ and } \omega = \frac{\pi}{4}.$$

a. The maximum displacement from the rest position is the amplitude. Because $a = 12$, the maximum displacement is 12 centimeters.

b. The frequency, f, is

$$f = \frac{\omega}{2\pi} = \frac{\frac{\pi}{4}}{2\pi} = \frac{\pi}{4}\cdot\frac{1}{2\pi} = \frac{1}{8}$$

The frequency is $\frac{1}{8}$ cm per second.

c. The time required for one cycle is the period.

$$\text{period} = \frac{2\pi}{\omega} = \frac{2\pi}{\frac{\pi}{4}} = 2\pi\cdot\frac{4}{\pi} = 8$$

The time required for one cycle is 8 seconds.

Concept and Vocabulary Check 5.8

1. sides; angles

2. north; south

3. simple harmonic; $|a|$; $\dfrac{2\pi}{\omega}$; $\dfrac{\omega}{2\pi}$

Exercise Set 5.8

1. Find the measure of angle B. Because
   $C = 90°$, $A + B = 90°$. Thus,
   $B = 90° - A = 90° - 23.5° = 66.5°$.
   Because we have a known angle, a known adjacent side, and an unknown opposite side, use the tangent function.

   $$\tan 23.5° = \frac{a}{10}$$
   $$a = 10 \tan 23.5° \approx 4.35$$

   Because we have a known angle, a known adjacent side, and an unknown hypotenuse, use the cosine function.

   $$\cos 23.5° = \frac{10}{c}$$
   $$c = \frac{10}{\cos 23.5°} \approx 10.90$$

3. Find the measure of angle B. Because
   $C = 90°$, $A + B = 90°$.
   Thus, $B = 90° - A = 90° - 52.6° = 37.4°$.
   Because we have a known angle, a known hypotenuse, and an unknown opposite side, use the sine function.

   $$\sin 52.6 = \frac{a}{54}$$
   $$a = 54 \sin 52.6° \approx 42.90$$

   Because we have a known angle, a known hypotenuse, and an unknown adjacent side, use the cosine function.

   $$\cos 52.6° = \frac{b}{54}$$
   $$b = 54 \cos 52.6° \approx 32.80$$

   In summary, $B = 37.4°$, $a \approx 42.90$, and $b \approx 32.80$.

5. Find the measure of angle A. Because
   $C = 90°$, $A + B = 90°$.
   Thus, $A = 90° - B = 90° - 16.8° = 73.2°$.
   Because we have a known angle, a known opposite side and an unknown adjacent side, use the tangent function.

   $$\tan 16.8° = \frac{30.5}{a}$$
   $$a = \frac{30.5}{\tan 16.8°} \approx 101.02$$

   Because we have a known angle, a known opposite side, and an unknown hypotenuse, use the sine function.

   $$\sin 16.8° = \frac{30.5}{c}$$
   $$c = \frac{30.5}{\sin 16.8°} \approx 105.52$$

   In summary, $A = 73.2°$, $a \approx 101.02$, and $c \approx 105.52$.

7. Find the measure of angle A. Because we have a known hypotenuse, a known opposite side, and an unknown angle, use the sine function.

   $$\sin A = \frac{30.4}{50.2}$$
   $$A = \sin^{-1}\left(\frac{30.4}{50.2}\right) \approx 37.3°$$

   Find the measure of angle B. Because
   $C = 90°$, $A + B = 90°$. Thus,
   $B = 90° - A \approx 90° - 37.3° = 52.7°$.
   Use the Pythagorean Theorem.

   $$a^2 + b^2 = c^2$$
   $$(30.4)^2 + b^2 = (50.2)^2$$
   $$b^2 = (50.2)^2 - (30.4)^2 = 1595.88$$
   $$b = \sqrt{1595.88} \approx 39.95$$

   In summary, $A \approx 37.3°$, $B \approx 52.7°$, and $b \approx 39.95$.

9. Find the measure of angle A. Because we have a known opposite side, a known adjacent side, and an unknown angle, use the tangent function.

   $$\tan A = \frac{10.8}{24.7}$$
   $$A = \tan^{-1}\left(\frac{10.8}{24.7}\right) \approx 23.6°$$

   Find the measure of angle B. Because
   $C = 90°$, $A + B = 90°$.
   Thus, $B = 90° - A \approx 90° - 23.6° = 66.4°$.
   Use the Pythagorean Theorem.

   $$c^2 = a^2 + b^2 = (10.8)^2 + (24.7)^2 = 726.73$$
   $$c = \sqrt{726.73} \approx 26.96$$

   In summary, $A \approx 23.6°$, $B \approx 66.4°$, and $c \approx 26.96$.

11. Find the measure of angle A. Because we have a known hypotenuse, a known adjacent side, and unknown angle, use the cosine function.

$$\cos A = \frac{2}{7}$$

$$A = \cos^{-1}\left(\frac{2}{7}\right) \approx 73.4°$$

Find the measure of angle B. Because
$C = 90°$, $A + B = 90°$.
Thus, $B = 90° - A \approx 90° - 73.4° = 16.6°$.
Use the Pythagorean Theorem.

$$a^2 + b^2 = c^2$$

$$a^2 + (2)^2 = (7)^2$$

$$a^2 = (7)^2 - (2)^2 = 45$$

$$a = \sqrt{45} \approx 6.71$$

In summary, $A \approx 73.4°$, $B \approx 16.6°$, and
$a \approx 6.71$.

13. We need the acute angle between ray OA and the north-south line through O. This angle measure $90° - 75° = 15°$. This angle is measured from the north side of the north-south line and lies east of the north-south line. Thus, the bearing from O and A is N $15°$ E.

15. The measurement of this angle is given to be $80°$. The angle is measured from the south side of the north-south line and lies west of the north-south line. Thus, the bearing from O to C is S $80°$ W.

17. When the object is released ($t = 0$), the object's distance, d, from its rest position is 6 centimeters down. Because it is down, d is negative: When $t = 0$, $d = -6$. Notice the greatest distance from rest position occurs at $t = 0$. Thus, we will use the equation with the cosine function, $y = a\cos\omega t$ to model the object's motion. Recall that $|a|$ is the maximum distance. Because the object initially moves down, $a = -6$. The value of ω can be found using the formula for the period.

$$\text{period} = \frac{2\pi}{\omega} = 4$$

$$2\pi = 4\omega$$

$$\omega = \frac{2\pi}{4} = \frac{\pi}{2}$$

Substitute these values into $d = a\cos\omega t$. The equation for the object's simple harmonic motion is

$$d = -6\cos\frac{\pi}{2}t.$$

19. When $t = 0$, $d = 0$. Therefore, we will use the equation with the sine function, $y = a\sin\omega t$, to model the object's motion. Recall that $|a|$ is the maximum distance. Because the object initially moves down, and has an amplitude of 3 inches, $a = -3$. The value of ω can be found using the formula for the period.

$$\text{period} = \frac{2\pi}{\omega} = 1.5$$

$$2\pi = 1.5\omega$$

$$\omega = \frac{2\pi}{1.5} = \frac{4\pi}{3}$$

Substitute these values into $d = a\sin\omega t$. The equation for the object's simple harmonic motion is

$$d = -3\sin\frac{4\pi}{3}t.$$

21. We begin by identifying values for a and ω.

$$d = 5\cos\frac{\pi}{2}t,\ a = 5 \text{ and } \omega = \frac{\pi}{2}$$

a. The maximum displacement from the rest position is the amplitude. Because $a = 5$, the maximum displacement is 5 inches.

b. The frequency, f, is

$$f = \frac{\omega}{2\pi} = \frac{\frac{\pi}{2}}{2\pi} = \frac{\pi}{2}\cdot\frac{1}{2\pi} = \frac{1}{4}.$$

The frequency is $\frac{1}{4}$ inch per second.

c. The time required for one cycle is the period.

$$\text{period} = \frac{2\pi}{\omega} = \frac{2\pi}{\frac{\pi}{2}} = 2\pi\cdot\frac{2}{\pi} = 4$$

The time required for one cycle is 4 seconds.

23. We begin by identifying values for a and ω.
$d = -6\cos 2\pi t$, $a = -6$ and $\omega = 2\pi$

a. The maximum displacement from the rest position is the amplitude.
Because $a = -6$, the maximum displacement is 6 inches.

b. The frequency, f, is

$$f = \frac{\omega}{2\pi} = \frac{2\pi}{2\pi} = 1.$$

The frequency is 1 inch per second.

c. The time required for one cycle is the period.

$$\text{period} = \frac{2\pi}{\omega} = \frac{2\pi}{2\pi} = 1$$

The time required for one cycle is
1 second.

25. We begin by identifying values for a and ω.

$$d = \frac{1}{2}\sin 2t, \ a = \frac{1}{2} \text{ and } \omega = 2$$

a. The maximum displacement from the rest position is the amplitude.

Because $a = \frac{1}{2}$, the maximum displacement is

$\frac{1}{2}$ inch.

b. The frequency, f, is

$$f = \frac{\omega}{2\pi} = \frac{2}{2\pi} = \frac{1}{\pi} \approx 0.32.$$

The frequency is approximately 0.32 cycle per second.

c. The time required for one cycle is the period.

$$\text{period} = \frac{2\pi}{\omega} = \frac{2\pi}{2} = \pi \approx 3.14$$

The time required for one cycle is approximately 3.14 seconds.

27. We begin by identifying values for a and ω.

$$d = -5\sin\frac{2\pi}{3}t, \ a = -5 \text{ and } \omega = \frac{2\pi}{3}$$

a. The maximum displacement from the rest position is the amplitude.
Because $a = -5$, the maximum displacement is 5 inches.

b. The frequency, f, is

$$f = \frac{\omega}{2\pi} = \frac{\frac{2\pi}{3}}{2\pi} = \frac{2\pi}{3} \cdot \frac{1}{2\pi} = \frac{1}{3}.$$

The frequency is $\frac{1}{3}$ cycle per second.

c. The time required for one cycle is the period.

$$\text{period} = \frac{2\pi}{\omega} = \frac{2\pi}{\frac{2\pi}{3}} = 2\pi \cdot \frac{3}{2\pi} = 3$$

The time required for one cycle is 3 seconds.

29. $x = 500\tan 40° + 500\tan 25°$

$x \approx 653$

31. $x = 600\tan 28° - 600\tan 25°$

$x \approx 39$

33. $x = \dfrac{300}{\tan 34°} - \dfrac{300}{\tan 64°}$

$x \approx 298$

35. $x = \dfrac{400\tan 40°\tan 20°}{\tan 40° - \tan 20°}$

$x \approx 257$

37. $d = 4\cos\left(\pi t - \dfrac{\pi}{2}\right)$

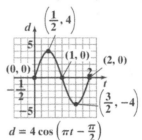

$$d = 4\cos\left(\pi t - \frac{\pi}{2}\right)$$

a. 4 in.

b. $\frac{1}{2}$ in. per sec

c. 2 sec

d. $\frac{1}{2}$

39. $d = -2\sin\left(\dfrac{\pi t}{4} + \dfrac{\pi}{2}\right)$

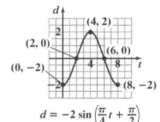

$$d = -2\sin\left(\frac{\pi}{4}t + \frac{\pi}{2}\right)$$

a. 2 in.

b. $\frac{1}{8}$ in. per sec

c. 8 sec

d. -2

41. Using a right triangle, we have a known angle, an unknown opposite side, a, and a known adjacent side. Therefore, use tangent function.

$$\tan 21.3° = \frac{a}{5280}$$

$$a = 5280\tan 21.3° \approx 2059$$

The height of the tower is approximately 2059 feet.

43. Using a right triangle, we have a known angle, a known opposite side, and an unknown adjacent side, a. Therefore, use the tangent function.

$$\tan 23.7° = \frac{305}{a}$$

$$a = \frac{305}{\tan 23.7°} \approx 695$$

The ship is approximately 695 feet from the statue's base.

45. The angle of depression from the helicopter to point P is equal to the angle of elevation from point P to the helicopter. Using a right triangle, we have a known angle, a known opposite side, and an unknown adjacent side, d. Therefore, use the tangent function.

$$\tan 36° = \frac{1000}{d}$$

$$d = \frac{1000}{\tan 36°} \approx 1376$$

The island is approximately 1376 feet off the coast.

47. Using a right triangle, we have an unknown angle, A, a known opposite side, and a known hypotenuse. Therefore, use the sine function.

$$\sin A = \frac{6}{23}$$

$$A = \sin^{-1}\left(\frac{6}{23}\right) \approx 15.1°$$

The ramp makes an angle of approximately 15.1° with the ground.

49. Using the two right triangles, we have a known angle, an unknown opposite side, a in the smaller triangle, b in the larger triangle, and a known adjacent side in each triangle. Therefore, use the tangent function.

$$\tan 19.2° = \frac{a}{125}$$

$$a = 125 \tan 19.2° \approx 43.5$$

$$\tan 31.7° = \frac{b}{125}$$

$$b = 125 \tan 31.7° \approx 77.2$$

The balloon rises approximately 77.2 – 43.5 or 33.7 feet.

51. Using a right triangle, we have a known angle, a known hypotenuse, and unknown sides. To find the opposite side, a, use the sine function.

$$\sin 53° = \frac{a}{150}$$

$$a = 150 \sin 53° \approx 120$$

To find the adjacent side, b, use the cosine function.

$$\cos 53° = \frac{b}{150}$$

$$b = 150 \cos 53° \approx 90$$

The boat has traveled approximately 90 miles north and 120 miles east.

53. The bearing from the fire to the second ranger is N 28° E. Using a right triangle, we have a known angle, a known opposite side, and an unknown adjacent side, b. Therefore, use the tangent function.

$$\tan 28° = \frac{7}{b}$$

$$b = \frac{7}{\tan 28°} \approx 13.2$$

The first ranger is 13.2 miles from the fire, to the nearest tenth of a mile.

55. Using a right triangle, we have a known adjacent side, a known opposite side, and an unknown angle, A. Therefore, use the tangent function.

$$\tan A = \frac{1.5}{2}$$

$$A = \tan\left(\frac{1.5}{2}\right) \approx 37°$$

We need the acute angle between the ray that runs from your house through your location, and the north-south line through your house. This angle measures approximately $90° - 37° = 53°$. This angle is measured from the north side of the north-south line and lies west of the north-south line. Thus, the bearing from your house to you is N 53° W.

57. To find the jet's bearing from the control tower, consider a north-south line passing through the tower. The acute angle from this line to the ray on which the jet lies is $35° + \theta$. Because we are measuring the angle from the north side of the line and the jet is east of the tower, the jet's bearing from the tower is N $(35° + \theta)$ E. To find θ, use a right triangle and the tangent function.

$$\tan \theta = \frac{7}{5}$$

$$\theta = \tan^{-1}\left(\frac{7}{5}\right) \approx 54.5°$$

Thus, $35° + \theta = 35° + 54.5° = 89.5°$.
The jet's bearing from the control tower is N 89.5° E.

59. The frequency, f, is $f = \dfrac{\omega}{2\pi}$, so

$$\frac{1}{2} = \frac{\omega}{2\pi}$$

$$\omega = \frac{1}{2} \cdot 2\pi = \pi$$

Because the amplitude is 6 feet, $a = 6$. Thus, the equation for the object's simple harmonic motion is $d = 6\sin \pi t$.

61. The frequency, f, is $f = \dfrac{\omega}{2\pi}$, so

$$264 = \frac{\omega}{2\pi}$$

$$\omega = 264 \cdot 2\pi = 528\pi$$

Thus, the equation for the tuning fork's simple harmonic motion is $d = \sin 528\pi t$.

63. – 69. Answers may vary.

71. $y = -6e^{-0.09x} \cos 2\pi x$

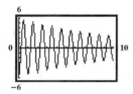

10 complete oscillations occur.

73. does not make sense; Explanations will vary. Sample explanation: When using bearings, the angle must be less than $90°$.

75. does not make sense; Explanations will vary. Sample explanation: Frequency and Period are inverses of each other. If the period is 10 seconds then the frequency is $\dfrac{1}{10} = 0.1$ oscillations per second.

77. Let d be the adjacent side to the $40°$ angle. Using the right triangles, we have a known angle and unknown sides in both triangles. Use the tangent function.

$$\tan 20° = \frac{h}{75 + d}$$

$$h = (75 + d)\tan 20°$$

Also, $\tan 40° = \dfrac{h}{d}$

$$h = d \tan 40°$$

Using the transitive property we have

$$(75 + d)\tan 20° = d \tan 40°$$

$$75 \tan 20° + d \tan 20° = d \tan 40°$$

$$d \tan 40° - d \tan 20° = 75 \tan 20°$$

$$d(\tan 40° - \tan 20°) = 75 \tan 20°$$

$$d = \frac{75 \tan 20°}{\tan 40° - \tan 20°}$$

Thus, $h = d \tan 40°$

$$= \frac{75 \tan 20°}{\tan 40° - \tan 20°}\tan 40° \approx 48$$

The height of the building is approximately 48 feet.

79. $\sec x \cot x = \dfrac{1}{\cos x} \cdot \dfrac{\cos x}{\sin x} = \dfrac{1}{\sin x}$ or $\csc x$

80. $\tan x \csc x \cos x = \dfrac{\sin x}{\cos x} \cdot \dfrac{1}{\sin x} \cdot \dfrac{\cos x}{1} = 1$

81. $\sec x + \tan x = \dfrac{1}{\cos x} + \dfrac{\sin x}{\cos x} = \dfrac{1 + \sin x}{\cos x}$

Chapter 5 Review Exercises

1. The radian measure of a central angle is the length of the intercepted arc divided by the circle's radius.

$$\theta = \frac{27}{6} = 4.5 \text{ radians}$$

2. $15° = 15° \cdot \dfrac{\pi \text{ radians}}{180°} = \dfrac{15\pi}{180}$ radian

$$= \frac{\pi}{12} \text{ radian}$$

3. $120° = 120° \cdot \dfrac{\pi \text{ radians}}{180°} = \dfrac{120\pi}{180}$ radians

$$= \frac{2\pi}{3} \text{ radians}$$

4. $315° = 315° \cdot \dfrac{\pi \text{ radians}}{180°} = \dfrac{315\pi}{180} \text{ radians}$

$\quad = \dfrac{7\pi}{4} \text{ radians}$

5. $\dfrac{5\pi}{3} \text{ radians} = \dfrac{5\pi}{3} \text{ radians} \cdot \dfrac{180°}{\pi \text{ radians}}$

$\quad\quad = \dfrac{5 \cdot 180°}{3} = 300°$

6. $\dfrac{7\pi}{5} \text{ radians} = \dfrac{7\pi}{5} \text{ radians} \cdot \dfrac{180°}{\pi \text{ radians}}$

$\quad\quad = \dfrac{7 \cdot 180°}{5} = 252°$

7. $-\dfrac{5\pi}{6} \text{ radians} = -\dfrac{5\pi}{6} \text{ radians} \cdot \dfrac{180°}{\pi \text{ radians}}$

$\quad\quad = -\dfrac{5 \cdot 180°}{6} = -150°$

8.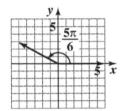

9.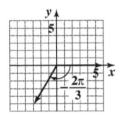

10.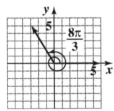

11.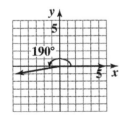

12.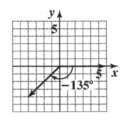

13. $400° - 360° = 40°$

14. $-445° + (2)360° = 275°$

15. $\dfrac{13\pi}{4} - 2\pi = \dfrac{13\pi}{4} - \dfrac{8\pi}{4} = \dfrac{5\pi}{4}$

16. $\dfrac{31\pi}{6} - (2)2\pi = \dfrac{31\pi}{6} - \dfrac{24\pi}{6} = \dfrac{7\pi}{6}$

17. $-\dfrac{8\pi}{3} + (2)2\pi = -\dfrac{8\pi}{3} + \dfrac{12\pi}{3} = \dfrac{4\pi}{3}$

18. $135° = 135° \cdot \dfrac{\pi \text{ radians}}{180°} = \dfrac{135 \cdot \pi}{180} \text{ radians}$

$\quad = \dfrac{3\pi}{4} \text{ radians}$

$s = r\theta$

$s = (10 \text{ ft})\left(\dfrac{3\pi}{4}\right) = \dfrac{15\pi}{2} \text{ ft} \approx 23.56 \text{ ft}$

19. $\dfrac{10.3 \text{ revolutions}}{1 \text{ minute}} \cdot \dfrac{2\pi \text{ radians}}{1 \text{ revolution}}$

$= \dfrac{20.6\pi \text{ radians}}{1 \text{ minute}} = 20.6\pi \text{ radians per minute}$

20. Use $v = r\omega$ where v is the linear speed and ω is the angular speed in radians per minute.

$\omega = \dfrac{2250 \text{ revolutions}}{1 \text{ minute}} \cdot \dfrac{2\pi \text{ radians}}{1 \text{ revolution}}$

$\quad = 4500\pi \text{ radians per minute}$

$v = 3 \text{ feet} \dfrac{4500\pi}{\text{minute}} = \dfrac{13,500\pi \text{ feet}}{\text{min}}$

$\quad \approx 42,412 \text{ ft per min}$

21. Use the Pythagorean Theorem to find the hypotenuse, c.

$$c^2 = a^2 + b^2$$

$$c = \sqrt{8^2 + 5^2} = \sqrt{64 + 25} = \sqrt{89}$$

$$\sin\theta = \frac{5}{\sqrt{89}} = \frac{5\sqrt{89}}{\sqrt{89}}$$

$$\cos\theta = \frac{8}{\sqrt{89}} = \frac{8\sqrt{89}}{\sqrt{89}}$$

$$\tan\theta = \frac{5}{8}$$

$$\csc\theta = \frac{\sqrt{89}}{5}$$

$$\sec\theta = \frac{\sqrt{89}}{8}$$

$$\cot\theta = \frac{3}{5}$$

22.
$$\sin\frac{\pi}{6} + \tan^2\frac{\pi}{3} = \frac{1}{2} + \left(\sqrt{3}\right)^2$$
$$= \frac{1}{2} + 3$$
$$= \frac{7}{2}$$

23.
$$\cos^2\frac{\pi}{4} + \tan^2\frac{\pi}{4} = \left(\frac{\sqrt{2}}{2}\right)^2 - (1)^2$$
$$= \frac{1}{2} - 1$$
$$= -\frac{1}{2}$$

24. $\sec^2\frac{\pi}{5} - \tan^2\frac{\pi}{5} = 1$

25. $\cos\frac{2\pi}{9}\sec\frac{2\pi}{9} = 1$

26. We can find the value of $\cos\theta$ by using the Pythagorean identity.

$$\sin^2\theta + \cos^2\theta = 1$$

$$\left(\frac{2\sqrt{7}}{7}\right)^2 + \cos^2\theta = 1$$

$$\frac{4}{7} + \cos^2\theta = 1$$

$$\cos^2\theta = 1 - \frac{4}{7}$$

$$\cos^2\theta = \frac{3}{7}$$

$$\cos\theta = \frac{\sqrt{3}}{\sqrt{7}} = \frac{\sqrt{3}}{\sqrt{7}} \cdot \frac{\sqrt{7}}{\sqrt{7}} = \frac{\sqrt{21}}{7}$$

Thus, $\cos\theta = \frac{\sqrt{21}}{7}$.

27. $\sin 70° = \cos(90° - 70°) = \cos 20°$

28. $\cos\frac{\pi}{2} = \sin\left(\frac{\pi}{2} - \frac{\pi}{2}\right) = \sin 0$

29.
$$\tan 23° = \frac{a}{100}$$
$$a = 100\tan 23°$$
$$a \approx 100(0.4245) \approx 42\,\text{mm}$$

30.
$$\sin 61° = \frac{20}{c}$$
$$c = \frac{20}{\sin 61°}$$
$$c \approx \frac{20}{0.8746} \approx 23\,\text{cm}$$

31.
$$\sin 48° = \frac{a}{50}$$
$$a = 50\sin 48°$$
$$a \approx 50(0.7431) \approx 37\,\text{in.}$$

32.
$$\sin\theta = \frac{y}{r} = \frac{1}{4}$$
$$x^2 + y^2 = r^2$$
$$x^2 + 1^2 = 4^2$$
$$x^2 = 15$$
$$x = \sqrt{15}$$
$$\tan\left(\frac{\pi}{2} - \theta\right) = \cot\theta = \frac{x}{y} = \frac{\sqrt{15}}{1} = \sqrt{15}$$

33. $\dfrac{1}{2}$ mi. $= \dfrac{1}{2} \cdot 5280$ ft $= 2640$ ft

$$\sin 17° = \frac{a}{2640}$$
$$a = 2640 \cdot \sin 17°$$
$$a \approx 2640(0.2924) \approx 772$$

The hiker gains 772 feet of altitude.

34. $\tan 32° = \dfrac{d}{50}$

$$d = 50 \tan 32°$$
$$d \approx 50(0.6249) \approx 31$$

The distance across the lake is about 31 meters.

35. $\tan \theta = \dfrac{6}{4}$

Use a calculator in degree mode to find θ.

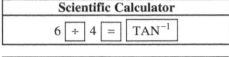

Scientific Calculator
6 ÷ 4 = TAN⁻¹

Graphing Calculator
TAN⁻¹ (6 ÷ 4) ENTER

The display should show approximately 56. Thus, the angle of elevation of the sun is approximately 56°.

36. We need values for x, y, and r. Because $P = (-1, -5)$ is a point on the terminal side of θ, $x = -1$ and $y = -5$. Furthermore,

$$r = \sqrt{(-1)^2 + (-5)^2}$$
$$= \sqrt{1 + 25} = \sqrt{26}$$

Now that we know x, y, and r, we can find the six trigonometric functions of θ.

$$\sin \theta = \frac{y}{r} = \frac{-5}{\sqrt{26}} = \frac{-5\sqrt{26}}{\sqrt{26} \cdot \sqrt{26}} = -\frac{5\sqrt{26}}{26}$$

$$\cos \theta = \frac{x}{r} = \frac{-1}{\sqrt{26}} = \frac{-1\sqrt{26}}{\sqrt{26} \cdot \sqrt{26}} = -\frac{\sqrt{26}}{26}$$

$$\tan \theta = \frac{y}{x} = \frac{-5}{-1} = 5$$

$$\csc \theta = \frac{r}{y} = \frac{\sqrt{26}}{-5} = -\frac{\sqrt{26}}{5}$$

$$\sec \theta = \frac{r}{x} = \frac{\sqrt{26}}{-1} = -\sqrt{26}$$

$$\cot \theta = \frac{x}{y} = \frac{-1}{-5} = \frac{1}{5}$$

37. We need values for x, y, and r. Because $P = (0, -1)$ is a point on the terminal side of θ, $x = 0$ and $y = -1$. Furthermore,

$$r = \sqrt{x^2 + y^2} = \sqrt{0^2 + (-1)^2}$$
$$= \sqrt{0+1} = \sqrt{1} = 1$$

Now that we know x, y, and r, we can find the six trigonometric functions of θ.

$$\sin \theta = \frac{y}{r} = \frac{-1}{1} = -1$$

$$\cos \theta = \frac{x}{r} = \frac{0}{1} = 0$$

$$\tan \theta = \frac{y}{x} = \frac{-1}{0}, \text{ undefined}$$

$$\csc \theta = \frac{r}{y} = \frac{1}{-1} = -1$$

$$\sec \theta = \frac{r}{x} = \frac{1}{0}, \text{ undefined}$$

$$\cot \theta = \frac{x}{y} = \frac{0}{-1} = 0$$

38. Because $\tan \theta > 0$, θ cannot lie in quadrant II and quadrant IV; the tangent function is negative in those two quadrants. Thus, with $\tan \theta > 0$, θ lies in quadrant I or quadrant III. We are also given that $\sec \theta > 0$. Because quadrant I is the only quadrant in which the tangent is positive and the secant is positive, we conclude that θ lies in quadrant I.

39. Because $\tan \theta > 0$, θ cannot lie in quadrant II and quadrant IV; the tangent function is negative in those two quadrants. Thus, with $\tan \theta > 0$, θ lies in quadrant I or quadrant III. We are also given that $\cos \theta < 0$. Because quadrant III is the only quadrant in which the tangent is positive and the cosine is negative, we conclude that θ lies in quadrant III.

40. Because the cosine is positive and the sine is negative, θ lies in quadrant IV. In quadrant IV, x is positive and y is negative. Thus, $\cos\theta = \dfrac{2}{5} = \dfrac{x}{r}$, $x = 2$, $r = 5$. Furthermore,

$$x^2 + y^2 = r^2$$
$$2^2 + y^2 = 5^2$$
$$y^2 = 25 - 4 = 21$$
$$y = -\sqrt{21}$$

Now that we know x, y, and r, we can find the six trigonometric functions of θ.

$$\sin\theta = \frac{y}{r} = \frac{-\sqrt{21}}{5} = -\frac{\sqrt{21}}{5}$$

$$\tan\theta = \frac{y}{x} = \frac{-\sqrt{21}}{2} = -\frac{\sqrt{21}}{2}$$

$$\csc\theta = \frac{r}{y} = \frac{5}{-\sqrt{21}} = -\frac{5\cdot\sqrt{21}}{\sqrt{21}\cdot\sqrt{21}} = -\frac{5\sqrt{21}}{21}$$

$$\sec\theta = \frac{r}{x} = \frac{5}{2}$$

$$\cot\theta = \frac{x}{y} = \frac{2}{-\sqrt{21}} = -\frac{2\sqrt{21}}{\sqrt{21}\cdot\sqrt{21}} = -\frac{2\sqrt{21}}{21}$$

41. Because the tangent is negative and the sine is positive, θ lies in quadrant II. In quadrant II x is negative and y is positive. Thus,

$$\tan\theta = -\frac{1}{3} = \frac{y}{x} = \frac{1}{-3}, \; x = -3, y = 1.$$

Furthermore,

$$r = \sqrt{x^2 + y^2} = \sqrt{(-3)^2 + 1^2} = \sqrt{9+1} = \sqrt{10}$$

Now that we know x, y, and r, we can find the six trigonometric functions of θ.

$$\sin\theta = \frac{y}{r} = \frac{1}{\sqrt{10}} = \frac{1\cdot\sqrt{10}}{\sqrt{10}\cdot\sqrt{10}} = \frac{\sqrt{10}}{10}$$

$$\cos\theta = \frac{x}{r} = \frac{-3}{\sqrt{10}} = -\frac{3\sqrt{10}}{\sqrt{10}\cdot\sqrt{10}} = -\frac{3\sqrt{10}}{10}$$

$$\csc\theta = \frac{r}{y} = \frac{\sqrt{10}}{1} = \sqrt{10}$$

$$\sec\theta = \frac{r}{x} = \frac{\sqrt{10}}{-3} = -\frac{\sqrt{10}}{3}$$

$$\cot\theta = \frac{x}{y} = \frac{-3}{1} = -3$$

42. Because the cotangent is positive and the cosine is negative, θ lies in quadrant III. In quadrant III x and y are both negative. Thus,

$$\cot\theta = \frac{3}{1} = \frac{x}{y} = \frac{-3}{-1}, \; x = -3, y = -1.$$

Furthermore,

$$r = \sqrt{x^2 + y^2} = \sqrt{(-3)^2 + (-1)^2} = \sqrt{9+1} = \sqrt{10}$$

Now that we know x, y, and r, we can find the six trigonometric functions of θ.

$$\sin\theta = \frac{y}{r} = \frac{-1}{\sqrt{10}} = -\frac{\sqrt{10}}{10}$$

$$\cos\theta = \frac{x}{r} = \frac{-3}{\sqrt{10}} = -\frac{3\sqrt{10}}{10}$$

$$\tan\theta = \frac{y}{x} = \frac{-1}{-3} = \frac{1}{3}$$

$$\csc\theta = \frac{r}{y} = \frac{\sqrt{10}}{-1} = -\sqrt{10}$$

$$\sec\theta = \frac{r}{x} = \frac{\sqrt{10}}{-3} = -\frac{\sqrt{10}}{3}$$

43. Because $265°$ lies between $180°$ and $270°$, it is in quadrant III.
The reference angle is $\theta' = 265° - 180° = 85°$.

44. Because $\dfrac{5\pi}{8}$ lies between $\dfrac{\pi}{2} = \dfrac{4\pi}{8}$ and $\pi = \dfrac{8\pi}{8}$, it is in quadrant II.

The reference angle is $\theta' = \pi - \dfrac{5\pi}{8} = \dfrac{8\pi}{8} - \dfrac{5\pi}{8} = \dfrac{3\pi}{8}$.

45. Find the coterminal angle: $-410° + (2)360° = 310°$
Find the reference angle: $360° - 310° = 50°$

46. Find the coterminal angle: $\dfrac{17\pi}{6} - 2\pi = \dfrac{5\pi}{6}$

Find the reference angle: $2\pi - \dfrac{5\pi}{6} = \dfrac{\pi}{6}$

47. Find the coterminal angle: $-\dfrac{11\pi}{3} + 4\pi = \dfrac{\pi}{3}$

Find the reference angle: $\dfrac{\pi}{3}$

48. 240° lies in quadrant III.
The reference angle is
$\theta' = 240° - 180° = 60°$.

$\sin 60° = \dfrac{\sqrt{3}}{2}$

In quadrant III, $\sin\theta < 0$, so

$\sin 240° = -\sin 60° = -\dfrac{\sqrt{3}}{2}$.

49. 120° lies in quadrant II.
The reference angle is
$\theta' = 180° - 120° = 60°$.

$\tan 60° = \sqrt{3}$

In quadrant II, $\tan\theta < 0$, so
$\tan 120° = -\tan 60° = -\sqrt{3}$.

50. $\dfrac{7\pi}{4}$ lies in quadrant IV.
The reference angle is

$\theta' = 2\pi - \dfrac{7\pi}{4} = \dfrac{8\pi}{4} - \dfrac{7\pi}{4} = \dfrac{\pi}{4}$.

$\sec \dfrac{\pi}{4} = \sqrt{2}$

In quadrant IV, $\sec\theta > 0$, so

$\sec \dfrac{7\pi}{4} = \sec \dfrac{\pi}{4} = \sqrt{2}$.

51. $\dfrac{11\pi}{6}$ lies in quadrant IV.
The reference angle is

$\theta' = 2\pi - \dfrac{11\pi}{6} = \dfrac{12\pi}{6} - \dfrac{11\pi}{6} = \dfrac{\pi}{6}$.

$\cos \dfrac{\pi}{6} = \dfrac{\sqrt{3}}{2}$

In quadrant IV, $\cos\theta > 0$, so $\cos \dfrac{11\pi}{6} = \cos \dfrac{\pi}{6} = \dfrac{\sqrt{3}}{2}$.

52. –210° lies in quadrant II.
The reference angle is
$\theta' = 210° - 180° = 30°$.

$\cot 30° = \sqrt{3}$

In quadrant II, $\cot\theta < 0$, so
$\cot(-210°) = -\cot 30° = -\sqrt{3}$.

53. $-\dfrac{2\pi}{3}$ lies in quadrant III.
The reference angle is

$\theta' = \pi + \dfrac{-2\pi}{3} = \dfrac{3\pi}{3} - \dfrac{2\pi}{3} = \dfrac{\pi}{3}$.

$\csc\left(\dfrac{\pi}{3}\right) = \dfrac{2\sqrt{3}}{3}$

In quadrant III, $\csc\theta < 0$, so

$\csc\left(-\dfrac{2\pi}{3}\right) = -\csc\left(\dfrac{\pi}{3}\right) = -\dfrac{2\sqrt{3}}{3}$.

54. $-\dfrac{\pi}{3}$ lies in quadrant IV.
The reference angle is

$\theta' = \dfrac{\pi}{3}$.

$\sin\left(\dfrac{\pi}{3}\right) = \dfrac{\sqrt{3}}{2}$

In quadrant IV, $\sin\theta < 0$, so

$\sin\left(-\dfrac{\pi}{3}\right) = -\sin\left(\dfrac{\pi}{3}\right) = -\dfrac{\sqrt{3}}{2}$.

55. 495° lies in quadrant II.
$495° - 360° = 135°$
The reference angle is
$\theta' = 180° - 135° = 45°$.

$\sin 45° = \dfrac{\sqrt{2}}{2}$

In quadrant II, $\sin\theta > 0$, so

$\sin 495° = \sin 45° = \dfrac{\sqrt{2}}{2}$.

56. $\dfrac{13\pi}{4}$ lies in quadrant III.

$\dfrac{13\pi}{4} - 2\pi = \dfrac{13\pi}{4} - \dfrac{8\pi}{4} = \dfrac{5\pi}{4}$

The reference angle is

$\theta' = \dfrac{5\pi}{4} - \pi = \dfrac{5\pi}{4} - \dfrac{4\pi}{4} = \dfrac{\pi}{4}$.

$\tan \dfrac{\pi}{4} = 1$

In quadrant III, $\tan\theta > 0$, so $\tan \dfrac{13\pi}{4} = \tan \dfrac{\pi}{4} = 1$.

57. $\sin \dfrac{22\pi}{3} = \sin\left(\dfrac{22\pi}{3} - 6\pi\right)$

$= \sin \dfrac{4\pi}{3}$

$= -\sin \dfrac{\pi}{3}$

$= -\dfrac{\sqrt{3}}{2}$

58. $\cos\left(-\dfrac{35\pi}{6}\right) = \cos\left(-\dfrac{35\pi}{6} + 6\pi\right)$

$= \cos \dfrac{\pi}{6}$

$= \dfrac{\sqrt{3}}{2}$

59. The equation $y = 3\sin 4x$ is of the form $y = A\sin Bx$ with $A = 3$ and $B = 4$. The amplitude is $|A| = |3| = 3$.

The period is $\dfrac{2\pi}{B} = \dfrac{2\pi}{4} = \dfrac{\pi}{2}$. The quarter-period is $\dfrac{\frac{\pi}{2}}{4} = \dfrac{\pi}{2} \cdot \dfrac{1}{4} = \dfrac{\pi}{8}$. The cycle begins at $x = 0$. Add quarter-periods to generate x-values for the key points.

$x = 0$

$x = 0 + \dfrac{\pi}{8} = \dfrac{\pi}{8}$

$x = \dfrac{\pi}{8} + \dfrac{\pi}{8} = \dfrac{\pi}{4}$

$x = \dfrac{\pi}{4} + \dfrac{\pi}{8} = \dfrac{3\pi}{8}$

$x = \dfrac{3\pi}{8} + \dfrac{\pi}{8} = \dfrac{\pi}{2}$

Evaluate the function at each value of x.

x	coordinates
0	$(0, 0)$
$\dfrac{\pi}{8}$	$\left(\dfrac{\pi}{8}, 3\right)$
$\dfrac{\pi}{4}$	$\left(\dfrac{\pi}{4}, 0\right)$
$\dfrac{3\pi}{8}$	$\left(\dfrac{3\pi}{8}, -3\right)$
$\dfrac{\pi}{2}$	$(2\pi, 0)$

Connect the five key points with a smooth curve and graph one complete cycle of the given function.

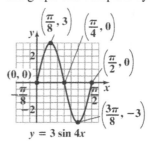

$y = 3\sin 4x$

60. The equation $y = -2\cos 2x$ is of the form $y = A\cos Bx$ with $A = -2$ and $B = 2$. The amplitude is $|A| = |-2| = 2$. The period is $\dfrac{2\pi}{B} = \dfrac{2\pi}{2} = \pi$. The quarter-period is $\dfrac{\pi}{4}$. The cycle begins at $x = 0$. Add quarter-periods to generate x-values for the key points.

$x = 0$

$x = 0 + \dfrac{\pi}{4} = \dfrac{\pi}{4}$

$x = \dfrac{\pi}{4} + \dfrac{\pi}{4} = \dfrac{\pi}{2}$

$x = \dfrac{\pi}{2} + \dfrac{\pi}{4} = \dfrac{3\pi}{4}$

$x = \dfrac{3\pi}{4} + \dfrac{\pi}{4} = \pi$

Evaluate the function at each value of x.

x	coordinates
0	$(0, -2)$
$\dfrac{\pi}{4}$	$\left(\dfrac{\pi}{4}, 0\right)$
$\dfrac{\pi}{2}$	$\left(\dfrac{\pi}{2}, 2\right)$
$\dfrac{3\pi}{4}$	$\left(\dfrac{3\pi}{4}, 0\right)$
π	$(\pi, -2)$

Connect the five key points with a smooth curve and graph one complete cycle of the given function.

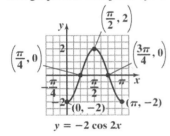

$y = -2\cos 2x$

61. The equation $y = 2\cos\dfrac{1}{2}x$ is of the form

$y = A\cos Bx$ with $A = 2$ and $B = \dfrac{1}{2}$. The amplitude

is $|A| = |2| = 2$. The period is $\dfrac{2\pi}{B} = \dfrac{2\pi}{\frac{1}{2}} = 2\pi \cdot 2 = 4\pi$.

The quarter-period is $\dfrac{4\pi}{4} = \pi$. The cycle begins at x

$= 0$. Add quarter-periods to generate x-values for the key points.
$x = 0$
$x = 0 + \pi = \pi$
$x = \pi + \pi = 2\pi$
$x = 2\pi + \pi = 3\pi$
$x = 3\pi + \pi = 4\pi$
Evaluate the function at each value of x.

x	coordinates
0	$(0, 2)$
π	$(\pi, 0)$
2π	$(2\pi, -2)$
3π	$(3\pi, 0)$
4π	$(4\pi, 2)$

Connect the five key points with a smooth curve and graph one complete cycle of the given function.

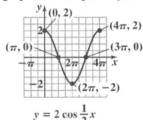

$y = 2\cos\dfrac{1}{2}x$

62. The equation $y = \dfrac{1}{2}\sin\dfrac{\pi}{3}x$ is of the form

$y = A\sin Bx$ with $A = \dfrac{1}{2}$ and $B = \dfrac{\pi}{3}$. The amplitude

is $|A| = \left|\dfrac{1}{2}\right| = \dfrac{1}{2}$. The period is

$\dfrac{2\pi}{B} = \dfrac{2\pi}{\frac{\pi}{3}} = 2\pi \cdot \dfrac{3}{\pi} = 6$. The quarter-period is

$\dfrac{6}{4} = \dfrac{3}{2}$. The cycle begins at $x = 0$. Add quarter-

periods to generate x-values for the key points.
$x = 0$
$x = 0 + \dfrac{3}{2} = \dfrac{3}{2}$
$x = \dfrac{3}{2} + \dfrac{3}{2} = 3$
$x = 3 + \dfrac{3}{2} = \dfrac{9}{2}$
$x = \dfrac{9}{2} + \dfrac{3}{2} = 6$
Evaluate the function at each value of x.

x	coordinates
0	(0, 0)
$\dfrac{3}{2}$	$\left(\dfrac{3}{2}, \dfrac{1}{2}\right)$
3	(3, 0)
$\dfrac{9}{2}$	$\left(\dfrac{9}{2}, -\dfrac{1}{2}\right)$
6	(6, 0)

Connect the five key points with a smooth curve and graph one complete cycle of the given function.

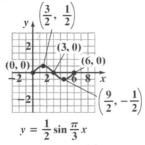

$$y = \tfrac{1}{2}\sin\tfrac{\pi}{3}x$$

63. The equation $y = -\sin \pi x$ is of the form
$y = A\sin Bx$ with $A = -1$ and $B = \pi$. The amplitude
is $|A| = |-1| = 1$. The period is $\dfrac{2\pi}{B} = \dfrac{2\pi}{\pi} = 2$. The

quarter-period is $\dfrac{2}{4} = \dfrac{1}{2}$. The cycle begins at $x = 0$.
Add quarter-periods to generate x-values for the key points.
$x = 0$

$x = 0 + \dfrac{1}{2} = \dfrac{1}{2}$

$x = \dfrac{1}{2} + \dfrac{1}{2} = 1$

$x = 1 + \dfrac{1}{2} = \dfrac{3}{2}$

$x = \dfrac{3}{2} + \dfrac{1}{2} = 2$

Evaluate the function at each value of x.

x	coordinates
0	(0, 0)
$\dfrac{1}{2}$	$\left(\dfrac{1}{2}, -1\right)$
1	(1, 0)
$\dfrac{3}{2}$	$\left(\dfrac{3}{2}, 1\right)$
2	(2, 0)

Connect the five key points with a smooth curve and graph one complete cycle of the given function.

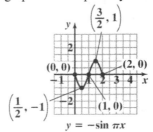

$$y = -\sin \pi x$$

64. The equation $y = 3\cos\dfrac{x}{3}$ is of the form

$y = A\cos Bx$ with $A = 3$ and $B = \dfrac{1}{3}$. The amplitude
is $|A| = |3| = 3$.

The period is $\dfrac{2\pi}{B} = \dfrac{2\pi}{\frac{1}{3}} = 2\pi \cdot 3 = 6\pi$. The quarter-

period is $\dfrac{6\pi}{4} = \dfrac{3\pi}{2}$. The cycle begins at $x = 0$. Add
quarter-periods to generate x-values for the key points.
$x = 0$

$x = 0 + \dfrac{3\pi}{2} = \dfrac{3\pi}{2}$

$x = \dfrac{3\pi}{2} + \dfrac{3\pi}{2} = 3\pi$

$x = 3\pi + \dfrac{3\pi}{2} = \dfrac{9\pi}{2}$

$x = \dfrac{9\pi}{2} + \dfrac{3\pi}{2} = 6\pi$

Evaluate the function at each value of x.

x	coordinates
0	$(0, 3)$
$\dfrac{3\pi}{2}$	$\left(\dfrac{3\pi}{2}, 0\right)$
3π	$(3\pi, -3)$
$\dfrac{9\pi}{2}$	$\left(\dfrac{9\pi}{2}, 0\right)$
6π	$(6\pi, 3)$

Connect the five key points with a smooth curve and graph one complete cycle of the given function.

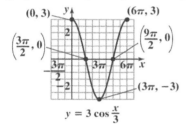

$y = 3\cos\dfrac{x}{3}$

65. The equation $y = 2\sin(x - \pi)$ is of the form
$y = A\sin(Bx - C)$ with $A = 2$, $B = 1$, and $C = \pi$. The
amplitude is $|A| = |2| = 2$. The period is

$\dfrac{2\pi}{B} = \dfrac{2\pi}{1} = 2\pi$. The phase shift is $\dfrac{C}{B} = \dfrac{\pi}{1} = \pi$. The

quarter-period is $\dfrac{2\pi}{4} = \dfrac{\pi}{2}$.

The cycle begins at $x = \pi$. Add quarter-periods to
generate x-values for the key points.
$x = \pi$

$x = \pi + \dfrac{\pi}{2} = \dfrac{3\pi}{2}$

$x = \dfrac{3\pi}{2} + \dfrac{\pi}{2} = 2\pi$

$x = 2\pi + \dfrac{\pi}{2} = \dfrac{5\pi}{2}$

$x = \dfrac{5\pi}{2} + \dfrac{\pi}{2} = 3\pi$

Evaluate the function at each value of x.

x	coordinates
π	$(\pi, 0)$
$\dfrac{3\pi}{2}$	$\left(\dfrac{3\pi}{2}, 2\right)$
2π	$(2\pi, 0)$
$\dfrac{5\pi}{2}$	$\left(\dfrac{5\pi}{2}, -2\right)$
3π	$(3\pi, 0)$

Connect the five key points with a smooth curve and graph one complete cycle of the given function.

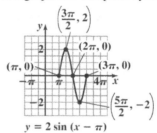

$y = 2\sin(x - \pi)$

66. $y = -3\cos(x + \pi) = -3\cos(x - (-\pi))$
The equation $y = -3\cos(x - (-\pi))$ is of the form
$y = A\cos(Bx - C)$ with $A = -3$, $B = 1$, and $C = -\pi$.
The amplitude is $|A| = |-3| = 3$.

The period is $\dfrac{2\pi}{B} = \dfrac{2\pi}{1} = 2\pi$. The phase shift is

$\dfrac{C}{B} = \dfrac{-\pi}{1} = -\pi$. The quarter-period is $\dfrac{2\pi}{4} = \dfrac{\pi}{2}$. The

cycle begins at $x = -\pi$. Add quarter-periods to
generate x-values for the key points.
$x = -\pi$

$x = -\pi + \dfrac{\pi}{2} = -\dfrac{\pi}{2}$

$x = -\dfrac{\pi}{2} + \dfrac{\pi}{2} = 0$

$x = 0 + \dfrac{\pi}{2} = \dfrac{\pi}{2}$

$x = \dfrac{\pi}{2} + \dfrac{\pi}{2} = \pi$

Evaluate the function at each value of x.

x	coordinates
$-\pi$	$(-\pi, -3)$
$-\dfrac{\pi}{2}$	$\left(-\dfrac{\pi}{2}, 0\right)$
0	$(0, 3)$
$\dfrac{\pi}{2}$	$\left(\dfrac{\pi}{2}, 0\right)$
π	$(\pi, -3)$

Connect the five key points with a smooth curve and graph one complete cycle of the given function.

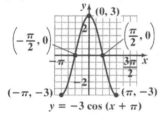

$$y = -3\cos(x + \pi)$$

67. $\quad y = \dfrac{3}{2}\cos\left(2x + \dfrac{\pi}{4}\right) = \dfrac{3}{2}\cos\left(2x - \left(-\dfrac{\pi}{4}\right)\right)$

The equation $y = \dfrac{3}{2}\cos\left(2x - \left(-\dfrac{\pi}{4}\right)\right)$ is of

the form $y = A\cos(Bx - C)$ with $A = \dfrac{3}{2}$,

$B = 2$, and $C = -\dfrac{\pi}{4}$. The amplitude is

$|A| = \left|\dfrac{3}{2}\right| = \dfrac{3}{2}$.

The period is $\dfrac{2\pi}{B} = \dfrac{2\pi}{2} = \pi$. The phase shift is

$\dfrac{C}{B} = \dfrac{-\frac{\pi}{4}}{2} = -\dfrac{\pi}{4} \cdot \dfrac{1}{2} = -\dfrac{\pi}{8}$. The quarter-period is $\dfrac{\pi}{4}$.

The cycle begins at $x = -\dfrac{\pi}{8}$. Add quarter-periods to generate x-values for the key points.

$x = -\dfrac{\pi}{8}$

$x = -\dfrac{\pi}{8} + \dfrac{\pi}{4} = \dfrac{\pi}{8}$

$x = \dfrac{\pi}{8} + \dfrac{\pi}{4} = \dfrac{3\pi}{8}$

$x = \dfrac{3\pi}{8} + \dfrac{\pi}{4} = \dfrac{5\pi}{8}$

$x = \dfrac{5\pi}{8} + \dfrac{\pi}{4} = \dfrac{7\pi}{8}$

Evaluate the function at each value of x.

x	coordinates
$-\dfrac{\pi}{8}$	$\left(-\dfrac{\pi}{8}, \dfrac{3}{2}\right)$
$\dfrac{\pi}{8}$	$\left(\dfrac{\pi}{8}, 0\right)$
$\dfrac{3\pi}{8}$	$\left(\dfrac{3\pi}{8}, -\dfrac{3}{2}\right)$
$\dfrac{5\pi}{8}$	$\left(\dfrac{5\pi}{8}, 0\right)$
$\dfrac{7\pi}{8}$	$\left(\dfrac{7\pi}{8}, \dfrac{3}{2}\right)$

Connect the five key points with a smooth curve and graph one complete cycle of the given function.

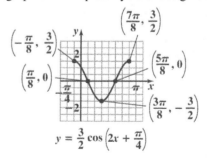

$$y = \dfrac{3}{2}\cos\left(2x + \dfrac{\pi}{4}\right)$$

68. $\quad y = \dfrac{5}{2}\sin\left(2x + \dfrac{\pi}{2}\right) = \dfrac{5}{2}\sin\left(2x - \left(-\dfrac{\pi}{2}\right)\right)$

The equation $y = \dfrac{5}{2}\sin\left(2x - \left(-\dfrac{\pi}{2}\right)\right)$ is of

the form $y = A\sin(Bx - C)$ with $A = \dfrac{5}{2}$,

$B = 2$, and $C = -\dfrac{\pi}{2}$. The amplitude is

$|A| = \left| \dfrac{5}{2} \right| = \dfrac{5}{2}.$

The period is $\dfrac{2\pi}{B} = \dfrac{2\pi}{2} = \pi.$ The phase shift is

$\dfrac{C}{B} = \dfrac{-\frac{\pi}{2}}{2} = -\dfrac{\pi}{2} \cdot \dfrac{1}{2} = -\dfrac{\pi}{4}.$ The quarter-period is $\dfrac{\pi}{4}.$

The cycle begins at $x = -\dfrac{\pi}{4}.$ Add quarter-periods to generate x-values for the key points.

$x = -\dfrac{\pi}{4}$

$x = -\dfrac{\pi}{4} + \dfrac{\pi}{4} = 0$

$x = 0 + \dfrac{\pi}{4} = \dfrac{\pi}{4}$

$x = \dfrac{\pi}{4} + \dfrac{\pi}{4} = \dfrac{\pi}{2}$

$x = \dfrac{\pi}{2} + \dfrac{\pi}{4} = \dfrac{3\pi}{4}$

Evaluate the function at each value of x.

x	coordinates
$-\dfrac{\pi}{4}$	$\left(-\dfrac{\pi}{4}, 0\right)$
0	$\left(0, \dfrac{5}{2}\right)$
$\dfrac{\pi}{4}$	$\left(\dfrac{\pi}{4}, 0\right)$
$\dfrac{\pi}{2}$	$\left(\dfrac{\pi}{2}, -\dfrac{5}{2}\right)$
$\dfrac{3\pi}{4}$	$\left(\dfrac{3\pi}{4}, 0\right)$

Connect the five key points with a smooth curve and graph one complete cycle of the given function.

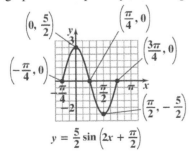

69. The equation $y = -3\sin\left(\dfrac{\pi}{3}x - 3\pi\right)$ is of the form $y = A\sin(Bx - C)$ with $A = -3$, $B = \dfrac{\pi}{3}$, and $C = 3\pi.$ The amplitude is $|A| = |-3| = 3.$ The period is $\dfrac{2\pi}{B} = \dfrac{2\pi}{\frac{\pi}{3}} = 2\pi \cdot \dfrac{3}{\pi} = 6.$ The phase shift is $\dfrac{C}{B} = \dfrac{3\pi}{\frac{\pi}{3}} = 3\pi \cdot \dfrac{3}{\pi} = 9.$ The quarter-period is $\dfrac{6}{4} = \dfrac{3}{2}.$ The cycle begins at $x = 9.$ Add quarter-periods to generate x-values for the key points.

$x = 9$

$x = 9 + \dfrac{3}{2} = \dfrac{21}{2}$

$x = \dfrac{21}{2} + \dfrac{3}{2} = 12$

$x = 12 + \dfrac{3}{2} = \dfrac{27}{2}$

$x = \dfrac{27}{2} + \dfrac{3}{2} = 15$

Evaluate the function at each value of x.

x	coordinates
9	$(9, 0)$
$\dfrac{21}{2}$	$\left(\dfrac{21}{2}, -3\right)$
12	$(12, 0)$
$\dfrac{27}{2}$	$\left(\dfrac{27}{2}, 3\right)$
15	$(15, 0)$

Connect the five key points with a smooth curve and graph one complete cycle of the given function.

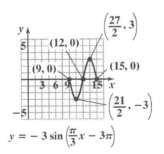

70. The graph of $y = \sin 2x + 1$ is the graph of $y = \sin 2x$ shifted one unit upward. The period for both functions is $\dfrac{2\pi}{2} = \pi$. The quarter-period is $\dfrac{\pi}{4}$. The cycle begins at $x = 0$. Add quarter-periods to generate x-values for the key points.

$x = 0$

$x = 0 + \dfrac{\pi}{4} = \dfrac{\pi}{4}$

$x = \dfrac{\pi}{4} + \dfrac{\pi}{4} = \dfrac{\pi}{2}$

$x = \dfrac{\pi}{2} + \dfrac{\pi}{4} = \dfrac{3\pi}{4}$

$x = \dfrac{3\pi}{4} + \dfrac{\pi}{4} = \pi$

Evaluate the function at each value of x.

x	coordinates
0	$(0, 1)$
$\dfrac{\pi}{4}$	$\left(\dfrac{\pi}{4}, 2\right)$
$\dfrac{\pi}{2}$	$\left(\dfrac{\pi}{2}, 1\right)$
$\dfrac{3\pi}{4}$	$\left(\dfrac{3\pi}{4}, 0\right)$
π	$(\pi, 1)$

By connecting the points with a smooth curve we obtain one period of the graph.

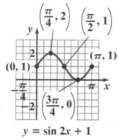

$y = \sin 2x + 1$

71. The graph of $y = 2\cos\dfrac{1}{3}x - 2$ is the graph of $y = 2\cos\dfrac{1}{3}x$ shifted two units downward. The period for both functions is $\dfrac{2\pi}{\frac{1}{3}} = 2\pi \cdot 3 = 6\pi$. The quarter-period is $\dfrac{6\pi}{4} = \dfrac{3\pi}{2}$. The cycle begins at $x = 0$. Add quarter-periods to generate x-values for the key points.

$x = 0$

$x = 0 + \dfrac{3\pi}{2} = \dfrac{3\pi}{2}$

$x = \dfrac{3\pi}{2} + \dfrac{3\pi}{2} = 3\pi$

$x = 3\pi + \dfrac{3\pi}{2} = \dfrac{9\pi}{2}$

$x = \dfrac{9\pi}{2} + \dfrac{3\pi}{2} = 6\pi$

Evaluate the function at each value of x.

x	coordinates
0	$(0, 0)$
$\dfrac{3\pi}{2}$	$\left(\dfrac{3\pi}{2}, -2\right)$
3π	$(3\pi, -4)$
$\dfrac{9\pi}{2}$	$\left(\dfrac{9\pi}{2}, -2\right)$
6π	$(6\pi, 0)$

By connecting the points with a smooth curve we obtain one period of the graph.

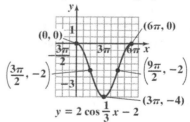

$y = 2\cos\dfrac{1}{3}x - 2$

72. a. At midnight $x = 0$. Thus,

$$y = 98.6 + 0.3\sin\left(\frac{\pi}{12} \cdot 0 - \frac{11\pi}{12}\right)$$

$$= 98.6 + 0.3\sin\left(-\frac{11\pi}{12}\right)$$

$$\approx 98.6 + 0.3(-0.2588) \approx 98.52$$

The body temperature is about 98.52°F.

b. period: $\dfrac{2\pi}{B} = \dfrac{2\pi}{\frac{\pi}{12}} = 2\pi \cdot \dfrac{12}{\pi} = 24$ hours

c. Solve the equation

$$\frac{\pi}{12}x - \frac{11\pi}{12} = \frac{\pi}{2}$$

$$\frac{\pi}{12}x = \frac{\pi}{2} + \frac{11\pi}{12} = \frac{6\pi}{12} + \frac{11\pi}{12} = \frac{17\pi}{12}$$

$$x = \frac{17\pi}{12} \cdot \frac{12}{\pi} = 17$$

The body temperature is highest for $x = 17$.

$$y = 98.6 + 0.3\sin\left(\frac{\pi}{12} \cdot 17 - \frac{11\pi}{12}\right)$$

$$= 98.6 + 0.3\sin\frac{\pi}{2} = 98.6 + 0.3 = 98.9$$

17 hours after midnight, which is
5 P.M., the body temperature is 98.9°F.

d. Solve the equation

$$\frac{\pi}{12}x - \frac{11\pi}{12} = \frac{3\pi}{2}$$

$$\frac{\pi}{12}x = \frac{3\pi}{2} + \frac{11\pi}{12} = \frac{18\pi}{12} + \frac{11\pi}{12} = \frac{29\pi}{12}$$

$$x = \frac{29\pi}{12} \cdot \frac{12}{\pi} = 29$$

The body temperature is lowest for $x = 29$.

$$y = 98.6 + 0.3\sin\left(\frac{\pi}{12} \cdot 29 - \frac{11\pi}{12}\right)$$

$$= 98.6 + 0.3\sin\left(\frac{3\pi}{2}\right)$$

$$= 98.6 + 0.3(-1) = 98.3°$$

29 hours after midnight or 5 hours after
midnight, at 5 A.M., the body temperature is
98.3°F.

e. The graph of $y = 98.6 + 0.3\sin\left(\dfrac{\pi}{12}x - \dfrac{11\pi}{12}\right)$ is

of the form $y = D + A\sin(Bx - C)$ with $A = 0.3$,

$B = \dfrac{\pi}{12}$, $C = \dfrac{11\pi}{12}$, and $D = 98.6$. The

amplitude is $|A| = |0.3| = 0.3$. The period
from part (b) is 24. The quarter-period is

$\dfrac{24}{4} = 6$. The phase shift is

$\dfrac{C}{B} = \dfrac{\frac{11\pi}{12}}{\frac{\pi}{12}} = \dfrac{11\pi}{12} \cdot \dfrac{12}{\pi} = 11$. The cycle begins at x

$= 11$. Add quarter-periods to generate x-values
for the key points.

$x = 11$

$x = 11 + 6 = 17$

$x = 17 + 6 = 23$

$x = 23 + 6 = 29$

$x = 29 + 6 = 35$

Evaluate the function at each value of x. The
key points are (11, 98.6), (17, 98.9), (23, 98.6),
(29, 98.3), (35, 98.6). Extend the pattern to the
left, and graph the function for $0 \le x \le 24$.

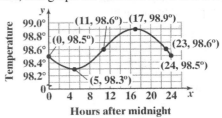

73. Blue:
This is a sine wave with a period of 480.
Since the amplitude is 1, $A = 1$.

$$B = \frac{2\pi}{\text{period}} = \frac{2\pi}{480} = \frac{\pi}{240}$$

The equation is $y = \sin\dfrac{\pi}{240}x$.

Red:
This is a sine wave with a period of 640.
Since the amplitude is 1, $A = 1$.

$$B = \frac{2\pi}{\text{period}} = \frac{2\pi}{640} = \frac{\pi}{320}$$

The equation is $y = \sin\dfrac{\pi}{320}x$.

74. Solve the equations

$$2x = -\frac{\pi}{2} \quad \text{and} \quad 2x = \frac{\pi}{2}$$

$$x = \frac{-\frac{\pi}{2}}{2} \qquad\qquad x = \frac{\frac{\pi}{2}}{2}$$

$$x = -\frac{\pi}{4} \qquad\qquad x = \frac{\pi}{4}$$

Thus, two consecutive asymptotes occur at

$$x = -\frac{\pi}{4} \text{ and } x = \frac{\pi}{4}.$$

$$x\text{-intercept} = \frac{-\frac{\pi}{4} + \frac{\pi}{4}}{2} = \frac{0}{2} = 0$$

An *x*-intercept is 0 and the graph passes through (0, 0). Because the coefficient of the tangent is 4, the points on the graph midway between an *x*-intercept and the asymptotes have *y*-coordinates of –4 and 4.

Use the two consecutive asymptotes. $x = -\frac{\pi}{4}$ and

$x = \frac{\pi}{4}$, to graph one full period of $y = 4\tan 2x$ from

$-\frac{\pi}{4}$ to $\frac{\pi}{4}$.

Continue the pattern and extend the graph another full period to the right.

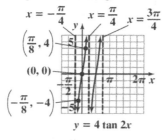

$y = 4\tan 2x$

75. Solve the equations

$$\frac{\pi}{4}x = -\frac{\pi}{2} \quad \text{and} \quad \frac{\pi}{4}x = \frac{\pi}{2}$$

$$x = -\frac{\pi}{2} \cdot \frac{4}{\pi} \qquad\qquad x = \frac{\pi}{2} \cdot \frac{4}{\pi}$$

$$x = -2 \qquad\qquad\qquad x = 2$$

Thus, two consecutive asymptotes occur at $x = -2$ and $x = 2$.

$$x\text{-intercept} = \frac{-2 + 2}{2} = \frac{0}{2} = 0$$

An *x*-intercept is 0 and the graph passes through (0, 0). Because the coefficient of the tangent is –2, the points on the graph midway between an *x*-intercept and the asymptotes have *y*-coordinates of 2 and –2.

Use the two consecutive asymptotes, $x = -2$ and $x = $

2, to graph one full period of $y = -2\tan\frac{\pi}{4}x$ from –2

to 2. Continue the pattern and extend the graph another full period to the right.

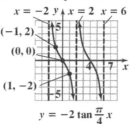

$y = -2\tan\frac{\pi}{4}x$

76. Solve the equations

$$x + \pi = -\frac{\pi}{2} \quad \text{and} \quad x + \pi = \frac{\pi}{2}$$

$$x = -\frac{\pi}{2} - \pi \qquad\qquad x = \frac{\pi}{2} - \pi$$

$$x = -\frac{3\pi}{2} \qquad\qquad\qquad x = -\frac{\pi}{2}$$

Thus, two consecutive asymptotes occur at

$$x = -\frac{3\pi}{2} \text{ and } x = -\frac{\pi}{2}.$$

$$x\text{-intercept} = \frac{-\frac{3\pi}{2} - \frac{\pi}{2}}{2} = \frac{-2\pi}{2} = -\pi$$

An *x*-intercept is $-\pi$ and the graph passes through $(-\pi, 0)$. Because the coefficient of the tangent is 1, the points on the graph midway between an *x*-intercept and the asymptotes have *y*-coordinates of –1 and 1. Use the two consecutive asymptotes,

$x = -\frac{3\pi}{2}$ and $x = -\frac{\pi}{2}$, to graph one full period of

$y = \tan(x + \pi)$ from $-\frac{3\pi}{2}$ to $-\frac{\pi}{2}$.

Continue the pattern and extend the graph another full period to the right.

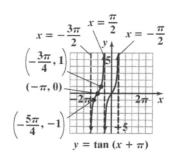

$y = \tan(x + \pi)$

77. Solve the equations

$$x - \frac{\pi}{4} = -\frac{\pi}{2} \quad \text{and} \quad x - \frac{\pi}{4} = \frac{\pi}{2}$$

$$x = -\frac{\pi}{2} + \frac{\pi}{4} \qquad\qquad x = \frac{\pi}{2} + \frac{\pi}{4}$$

$$x = -\frac{\pi}{4} \qquad\qquad\quad x = \frac{3\pi}{4}$$

Thus, two consecutive asymptotes occur at

$$x = -\frac{\pi}{4} \text{ and } x = -\frac{3\pi}{4}.$$

$$x\text{-intercept } = \frac{-\frac{\pi}{4} - \frac{3\pi}{4}}{2} = \frac{\frac{\pi}{2}}{2} = \frac{\pi}{4}$$

An x-intercept is $\frac{\pi}{4}$ and the graph passes through

$\left(\frac{\pi}{4}, 0\right)$. Because the coefficient of the tangent is -1, the points on the graph midway between an x-intercept and the asymptotes have y-coordinates of 1 and -1. Use the two consecutive asymptotes,

$x = -\frac{\pi}{4}$ and $x = \frac{3\pi}{4}$, to graph one full period of

$y = -\tan\left(x - \frac{\pi}{4}\right)$ from $-\frac{\pi}{4}$ to $\frac{3\pi}{4}$. Continue the pattern and extend the graph another full period to the right.

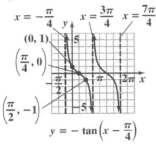

78. Solve the equations

$$3x = 0 \quad \text{and} \quad 3x = \pi$$

$$x = 0 \qquad\qquad x = \frac{\pi}{3}$$

Thus, two consecutive asymptotes occur at

$$x = 0 \text{ and } x = \frac{\pi}{3}.$$

$$x\text{-intercept } = \frac{0 + \frac{\pi}{3}}{2} = \frac{\frac{\pi}{3}}{2} = \frac{\pi}{6}$$

An x-intercept is $\frac{\pi}{6}$ and the graph passes through

$\left(\frac{\pi}{6}, 0\right)$.

Because the coefficient of the tangent is 2, the points on the graph midway between an x-intercept and the asymptotes have y-coordinates of 2 and -2. Use the two consecutive asymptotes, $x = 0$ and $x = \frac{\pi}{3}$, to

graph one full period of $y = 2\cot 3x$ from 0 to $\frac{\pi}{3}$.

Continue the pattern and extend the graph another full period to the right.

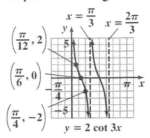

79. Solve the equations

$$\frac{\pi}{2}x = 0 \quad \text{and} \quad \frac{\pi}{2}x = \pi$$

$$x = 0 \qquad\qquad x = \pi \cdot \frac{2}{\pi}$$

$$x = 2$$

Thus, two consecutive asymptotes occur at $x = 0$ and $x = 2$.

$$x\text{-intercept } = \frac{0 + 2}{2} = \frac{2}{2} = 1$$

An x-intercept is 1 and the graph passes through $(1, 0)$. Because the coefficient of the cotangent is

$-\frac{1}{2}$, the points on the graph midway between an x-intercept and the asymptotes have y-coordinates of

$-\frac{1}{2}$ and $\frac{1}{2}$. Use the two consecutive asymptotes,

$x = 0$ and $x = 2$, to graph one full period of

$y = -\frac{1}{2}\cot\frac{\pi}{2}x$ from 0 to 2. Continue the pattern and

extend the graph another full period to the right.

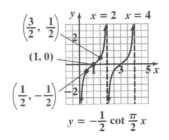

80. Solve the equations

$$x + \frac{\pi}{2} = 0 \qquad \text{and} \qquad x + \frac{\pi}{2} = \pi$$

$$x = 0 - \frac{\pi}{2} \qquad\qquad x = \pi - \frac{\pi}{2}$$

$$x = -\frac{\pi}{2} \qquad\qquad x = \frac{\pi}{2}$$

Thus, two consecutive asymptotes occur at

$$x = -\frac{\pi}{2} \text{ and } x = \frac{\pi}{2}.$$

$$x\text{-intercept} = \frac{-\frac{\pi}{2} + \frac{\pi}{2}}{2} = \frac{0}{2} = 0$$

An x-intercept is 0 and the graph passes through (0, 0). Because the coefficient of the cotangent is 2, the points on the graph midway between an x-intercept and the asymptotes have y-coordinates of 2 and –2.

Use the two consecutive asymptotes, $x = -\frac{\pi}{2}$ and

$x = \frac{\pi}{2}$, to graph one full period of $y = 2\cot\left(x + \frac{\pi}{2}\right)$

from $-\frac{\pi}{2}$ to $\frac{\pi}{2}$. Continue the pattern and extend the graph another full period to the right.

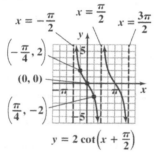

$$y = 2\cot\left(x + \frac{\pi}{2}\right)$$

81. Graph the reciprocal cosine function, $y = 3\cos 2\pi x$. The equation is of the form $y = A\cos Bx$ with $A = 3$ and $B = 2\pi$.

amplitude: $|A| = |3| = 3$

period: $\frac{2\pi}{B} = \frac{2\pi}{2\pi} = 1$

Use quarter-periods, $\frac{1}{4}$, to find x-values for the five

key points. Starting with $x = 0$, the x-values are 0, $\frac{1}{4}$,

$\frac{1}{2}, \frac{3}{4}, 1$. Evaluating the function at each value of x,

the key points are (0, 3),

$\left(\frac{1}{4}, 0\right), \left(\frac{1}{2}, -3\right), \left(\frac{3}{4}, 0\right)$, (1, 3).

Use these key points to graph $y = 3\cos 2\pi x$ from 0 to 1. Extend the graph one cycle to the right. Use the graph to obtain the graph of the reciprocal function. Draw vertical asymptotes through the x-intercepts, and use them as guides to graph $y = 3\sec 2\pi x$.

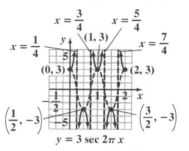

$$y = 3\sec 2\pi x$$

82. Graph the reciprocal sine function, $y = -2\sin \pi x$. The equation is of the form $y = A\sin Bx$ with $A = -2$ and $B = \pi$.

amplitude: $|A| = |-2| = 2$

period: $\frac{2\pi}{B} = \frac{2\pi}{\pi} = 2$

Use quarter-periods, $\frac{2}{4} = \frac{1}{2}$, to find

x-values for the five key points. Starting with

$x = 0$, the x-values are 0, $\frac{1}{2}$, 1, $\frac{3}{2}$, 2. Evaluating the

function at each value of x, the key points are (0, 0),

$\left(\frac{1}{2}, -2\right)$, (1, 0), $\left(\frac{3}{2}, 2\right)$, (2, 0). Use these key points

to graph $y = -2\sin \pi x$ from 0 to 2. Extend the graph one cycle to the right. Use the graph to obtain the graph of the reciprocal function. Draw vertical asymptotes through the x-intercepts, and use them as guides to graph $y = -2\csc \pi x$.

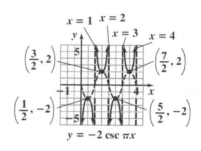

$$y = -2\csc \pi x$$

83. Graph the reciprocal cosine function,
$y = 3\cos(x + \pi)$. The equation is of the form
$y = A\cos(Bx - C)$ with $A = 3$, $B = 1$, and $C = -\pi$.

amplitude: $|A| = |3| = 3$

period: $\dfrac{2\pi}{B} = \dfrac{2\pi}{1} = 2\pi$

phase shift: $\dfrac{C}{B} = \dfrac{-\pi}{1} = -\pi$

Use quarter-periods, $\dfrac{2\pi}{4} = \dfrac{\pi}{2}$, to find

x-values for the five key points. Starting with

$x = -\pi$, the x-values are $-\pi, -\dfrac{\pi}{2},\ 0, \dfrac{\pi}{2}, \pi$.

Evaluating the function at each value of x, the key

points are $(-\pi, 3)$, $\left(-\dfrac{\pi}{2}, 0\right)$, $(0, -3)$,

$\left(\dfrac{\pi}{2}, 0\right)$, $(\pi, 3)$. Use these key points to graph

$y = 3\cos(x + \pi)$ from $-\pi$ to π. Extend the graph
one cycle to the right. Use the graph to obtain the
graph of the reciprocal function. Draw vertical
asymptotes through the x-intercepts, and use them as
guides to graph $y = 3\sec(x + \pi)$.

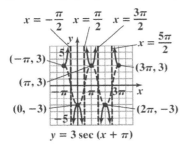

84. Graph the reciprocal sine function, $y = \dfrac{5}{2}\sin(x - \pi)$.
The equation is of the form $y = A\sin(Bx - C)$ with

$A = \dfrac{5}{2}$, $B = 1$, and $C = \pi$.

amplitude: $|A| = \left|\dfrac{5}{2}\right| = \dfrac{5}{2}$

period: $\dfrac{2\pi}{B} = \dfrac{2\pi}{1} = 2\pi$

phase shift: $\dfrac{C}{B} = \dfrac{\pi}{1} = \pi$

Use quarter-periods, $\dfrac{2\pi}{4} = \dfrac{\pi}{2}$, to find

x-values for the five key points. Starting with $x = \pi$,

the x-values are $\pi, \dfrac{3\pi}{2}, 2\pi, \dfrac{5\pi}{2}, 3\pi$. Evaluating the

function at each value of x, the key points

are $(\pi, 0)$, $\left(\dfrac{3\pi}{2}, \dfrac{5}{2}\right)$, $(2\pi, 0)$, $\left(\dfrac{5\pi}{2}, -\dfrac{5}{2}\right)$, $(3\pi, 0)$.

Use these key points to graph $y = \dfrac{5}{2}\sin(x - \pi)$ from

π to 3π. Extend the graph one cycle to the right.
Use the graph to obtain the graph of the reciprocal
function. Draw vertical asymptotes through the x-
intercepts, and use them as guides to graph

$y = \dfrac{5}{2}\csc(x - \pi)$.

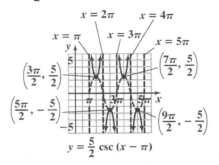

85. Let $\theta = \sin^{-1} 1$, then $\sin\theta = 1$.

The only angle in the interval $\left[-\dfrac{\pi}{2}, \dfrac{\pi}{2}\right]$ that satisfies

$\sin\theta = 1$ is $\dfrac{\pi}{2}$. Thus $\theta = \dfrac{\pi}{2}$, or $\sin^{-1} 1 = \dfrac{\pi}{2}$.

86. Let $\theta = \cos^{-1} 1$, then $\cos\theta = 1$.
The only angle in the interval $[0, \pi]$ that satisfies

$\cos\theta = 1$ is 0. Thus $\theta = 0$, or $\cos^{-1} 1 = 0$.

87. Let $\theta = \tan^{-1} 1$, then $\tan\theta = 1$.

The only angle in the interval $\left(-\dfrac{\pi}{2}, \dfrac{\pi}{2}\right)$ that satisfies

$\tan\theta = 1$ is $\dfrac{\pi}{4}$. Thus $\theta = \dfrac{\pi}{4}$, or $\tan^{-1} 1 = \dfrac{\pi}{4}$.

88. Let $\theta = \sin^{-1}\left(-\dfrac{\sqrt{3}}{2}\right)$, then $\sin\theta = -\dfrac{\sqrt{3}}{2}$.

The only angle in the interval $\left(-\dfrac{\pi}{2}, \dfrac{\pi}{2}\right)$ that satisfies

$\sin\theta = -\dfrac{\sqrt{3}}{2}$ is $-\dfrac{\pi}{3}$. Thus $\theta = -\dfrac{\pi}{3}$, or

$\sin^{-1}\left(-\dfrac{\sqrt{3}}{2}\right) = -\dfrac{\pi}{3}$.

89. Let $\theta = \cos^{-1}\left(-\dfrac{1}{2}\right)$, then $\cos\theta = -\dfrac{1}{2}$.

The only angle in the interval $[0, \pi]$ that satisfies

$\cos\theta = -\dfrac{1}{2}$ is $\dfrac{2\pi}{3}$. Thus $\theta = \dfrac{2\pi}{3}$, or

$\cos^{-1}\left(-\dfrac{1}{2}\right) = \dfrac{2\pi}{3}$.

90. Let $\theta = \tan^{-1}\left(-\dfrac{\sqrt{3}}{3}\right)$, then $\tan\theta = -\dfrac{\sqrt{3}}{3}$.

The only angle in the interval $\left(-\dfrac{\pi}{2}, \dfrac{\pi}{2}\right)$ that satisfies

$\tan\theta = -\dfrac{\sqrt{3}}{3}$ is $-\dfrac{\pi}{6}$.

Thus $\theta = -\dfrac{\pi}{6}$, or $\tan^{-1}\left(-\dfrac{\sqrt{3}}{3}\right) = -\dfrac{\pi}{6}$.

91. Let $\theta = \sin^{-1}\dfrac{\sqrt{2}}{2}$, then $\sin\theta = \dfrac{\sqrt{2}}{2}$. The only angle

in the interval $\left[-\dfrac{\pi}{2}, \dfrac{\pi}{2}\right]$ that satisfies $\sin\theta = \dfrac{\sqrt{2}}{2}$ is

$\dfrac{\pi}{4}$.

Thus, $\cos\left(\sin^{-1}\dfrac{\sqrt{2}}{2}\right) = \cos\dfrac{\pi}{4} = \dfrac{\sqrt{2}}{2}$.

92. Let $\theta = \cos^{-1}0$, then $\cos\theta = 0$. The only angle in

the interval $[0, \pi]$ that satisfies $\cos\theta = 0$ is $\dfrac{\pi}{2}$.

Thus, $\sin\left(\cos^{-1}0\right) = \sin\dfrac{\pi}{2} = 1$.

93. Let $\theta = \sin^{-1}\left(-\dfrac{1}{2}\right)$, then $\sin\theta = -\dfrac{1}{2}$. The only

angle in the interval $\left[-\dfrac{\pi}{2}, \dfrac{\pi}{2}\right]$ that satisfies

$\sin\theta = -\dfrac{1}{2}$ is $-\dfrac{\pi}{6}$.

Thus, $\tan\left[\sin^{-1}\left(-\dfrac{1}{2}\right)\right] = \tan\left(-\dfrac{\pi}{6}\right) = -\dfrac{\sqrt{3}}{3}$.

94. Let $\theta = \cos^{-1}\left(-\dfrac{\sqrt{3}}{2}\right)$, then $\cos\theta = -\dfrac{\sqrt{3}}{2}$. The only

angle in the interval $[0, \pi]$ that satisfies

$\cos\theta = -\dfrac{\sqrt{3}}{2}$ is $\dfrac{5\pi}{6}$.

Thus, $\tan\left[\cos^{-1}\left(-\dfrac{\sqrt{3}}{2}\right)\right] = \tan\dfrac{5\pi}{6} = -\dfrac{\sqrt{3}}{3}$.

95. Let $\theta = \tan^{-1}\dfrac{\sqrt{3}}{3}$, then $\tan\theta = \dfrac{\sqrt{3}}{3}$.

The only angle in the interval $\left(-\dfrac{\pi}{2}, \dfrac{\pi}{2}\right)$ that satisfies

$\tan\theta = \dfrac{\sqrt{3}}{3}$ is $\dfrac{\pi}{6}$.

Thus $\csc\left(\tan^{-1}\dfrac{\sqrt{3}}{3}\right) = \csc\dfrac{\pi}{6} = 2$.

96. Let $\theta = \tan^{-1}\dfrac{3}{4}$, then $\tan\theta = \dfrac{3}{4}$

Because $\tan\theta$ is positive, θ is in the first quadrant.

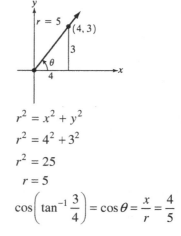

$r^2 = x^2 + y^2$

$r^2 = 4^2 + 3^2$

$r^2 = 25$

$r = 5$

$\cos\left(\tan^{-1}\dfrac{3}{4}\right) = \cos\theta = \dfrac{x}{r} = \dfrac{4}{5}$

97. Let $\theta = \cos^{-1}\dfrac{3}{5}$, then $\cos\theta = \dfrac{3}{5}$.

Because $\cos\theta$ is positive, θ is in the first quadrant.

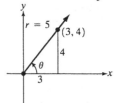

$$x^2 + y^2 = r^2$$
$$3^2 + y^2 = 5^2$$
$$y^2 = 25 - 9 = 16$$
$$y = \sqrt{16} = 4$$
$$\sin\left(\cos^{-1}\dfrac{3}{5}\right) = \sin\theta = \dfrac{y}{r} = \dfrac{4}{5}$$

98. Let $\theta = \sin^{-1}\left(-\dfrac{3}{5}\right)$, then $\sin\theta = -\dfrac{3}{5}$.

Because $\sin\theta$ is negative, θ is in quadrant IV.

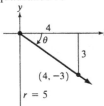

$$x^2 + (-3)^2 = 5^2$$
$$x^2 + y^2 = r^2$$
$$x^2 = 25 - 9 = 16$$
$$x = \sqrt{16} = 4$$
$$\tan\left[\sin^{-1}\left(-\dfrac{3}{5}\right)\right] = \tan\theta = \dfrac{y}{x} = -\dfrac{3}{4}$$

99. Let $\theta = \cos^{-1}\left(-\dfrac{4}{5}\right)$, then $\cos\theta = -\dfrac{4}{5}$.

Because $\cos\theta$ is negative, θ is in quadrant II.

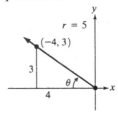

$$x^2 + y^2 = r^2$$
$$(-4)^2 + y^2 = 5^2$$
$$y^2 = 25 - 16 = 9$$
$$y = \sqrt{9} = 3$$

Use the right triangle to find the exact value.

$$\tan\left[\cos^{-1}\left(-\dfrac{4}{5}\right)\right] = \tan\theta = -\dfrac{3}{4}$$

100. Let $\theta = \tan^{-1}\left(-\dfrac{1}{3}\right)$,

Because $\tan\theta$ is negative, θ is in quadrant IV and $x = 3$ and $y = -1$.

$$r^2 = x^2 + y^2$$
$$r^2 = 3^2 + (-1)^2$$
$$r^2 = 10$$
$$r = \sqrt{10}$$
$$\sin\left[\tan^{-1}\left(-\dfrac{4}{5}\right)\right] = \sin\theta = \dfrac{y}{r} = \dfrac{-1}{\sqrt{10}} = -\dfrac{\sqrt{10}}{10}$$

101. $x = \dfrac{\pi}{3}$, x is in $\left[-\dfrac{\pi}{2}, \dfrac{\pi}{2}\right]$, so $\sin^{-1}\left(\sin\dfrac{\pi}{3}\right) = \dfrac{\pi}{3}$

102. $x = \dfrac{2\pi}{3}$, x is not in $\left[-\dfrac{\pi}{2}, \dfrac{\pi}{2}\right]$. x is in the domain of $\sin x$, so

$$\sin^{-1}\left(\sin\dfrac{2\pi}{3}\right) = \sin^{-1}\dfrac{\sqrt{3}}{2} = \dfrac{\pi}{3}$$

103. $\sin^{-1}\left(\cos\dfrac{2\pi}{3}\right) = \sin^{-1}\left(-\dfrac{1}{2}\right)$

Let $\theta = \sin^{-1}\left(-\dfrac{1}{2}\right)$, then $\sin\theta = -\dfrac{1}{2}$. The only

angle in the interval $\left[-\dfrac{\pi}{2}, \dfrac{\pi}{2}\right]$ that satisfies

$\sin\theta = -\dfrac{1}{2}$ is $-\dfrac{\pi}{6}$. Thus, $\theta = -\dfrac{\pi}{6}$, or

$\sin^{-1}\left(\cos\dfrac{2\pi}{3}\right) = \sin^{-1}\left(-\dfrac{1}{2}\right) = -\dfrac{\pi}{6}$.

104. Let $\theta = \tan^{-1}\dfrac{x}{2}$, then $\tan\theta = \dfrac{x}{2}$.

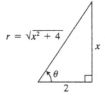

$r^2 = x^2 + 2^2$

$r^2 = x^2 + y^2$

$r = \sqrt{x^2 + 4}$

Use the right triangle to write the algebraic expression.

$\cos\left(\tan^{-1}\dfrac{x}{2}\right) = \cos\theta = \dfrac{2}{\sqrt{x^2 + 4}} = \dfrac{2\sqrt{x^2 + 4}}{x^2 + 4}$

105. Let $\theta = \sin^{-1}\dfrac{1}{x}$, then $\sin\theta = \dfrac{1}{x}$.

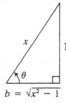

Use the Pythagorean theorem to find the third side, b.

$1^2 + b^2 = x^2$

$b^2 = x^2 - 1$

$b = \sqrt{x^2 - 1}$

Use the right triangle to write the algebraic expression.

$\sec\left(\sin^{-1}\dfrac{1}{x}\right) = \sec\theta = \dfrac{x}{\sqrt{x^2 - 1}} = \dfrac{x\sqrt{x^2 - 1}}{x^2 - 1}$

106. Find the measure of angle B. Because $C = 90°$, $A + B = 90°$. Thus,

$B = 90° - A = 90° - 22.3° = 67.7°$

We have a known angle, a known hypotenuse, and an unknown opposite side. Use the sine function.

$\sin 22.3° = \dfrac{a}{10}$

$a = 10\sin 22.3° \approx 3.79$

We have a known angle, a known hypotenuse, and an unknown adjacent side. Use the cosine function.

$\cos 22.3° = \dfrac{b}{10}$

$b = 10\cos 22.3° \approx 9.25$

In summary, $B = 67.7°$, $a \approx 3.79$, and $b \approx 9.25$.

107. Find the measure of angle A. Because $C = 90°$, $A + B = 90°$. Thus,

$A = 90° - B = 90° - 37.4° = 52.6°$

We have a known angle, a known opposite side, and an unknown adjacent side. Use the tangent function.

$\tan 37.4° = \dfrac{6}{a}$

$a = \dfrac{6}{\tan 37.4°} \approx 7.85$

We have a known angle, a known opposite side, and an unknown hypotenuse. Use the sine function.

$\sin 37.4° = \dfrac{6}{c}$

$c = \dfrac{6}{\sin 37.4°} \approx 9.88$

In summary, $A = 52.6°$, $a \approx 7.85$, and $c \approx 9.88$.

108. Find the measure of angle A. We have a known hypotenuse, a known opposite side, and an unknown angle. Use the sine function.

$\sin A = \dfrac{2}{7}$

$A = \sin^{-1}\left(\dfrac{2}{7}\right) \approx 16.6°$

Find the measure of angle B. Because $C = 90°$, $A + B = 90°$. Thus, $B = 90° - A \approx 90° - 16.6° = 73.4°$

We have a known hypotenuse, a known opposite side, and an unknown adjacent side. Use the Pythagorean theorem.

$a^2 + b^2 = c^2$

$2^2 + b^2 = 7^2$

$b^2 = 7^2 - 2^2 = 45$

$b = \sqrt{45} \approx 6.71$

In summary, $A \approx 16.6°$, $B \approx 73.4°$, and $b \approx 6.71$.

109. Find the measure of angle *A*. We have a known opposite side, a known adjacent side, and an unknown angle. Use the tangent function.

$$\tan A = \frac{1.4}{3.6}$$

$$A = \tan^{-1}\left(\frac{1.4}{3.6}\right) \approx 21.3°$$

Find the measure of angle *B*. Because $C = 90°$, $A + B = 90°$. Thus,

$B = 90° - A \approx 90° - 21.3° = 68.7°$

We have a known opposite side, a known adjacent side, and an unknown hypotenuse.

Use the Pythagorean theorem.

$$c^2 = a^2 + b^2 = (1.4)^2 + (3.6)^2 = 14.92$$

$$c = \sqrt{14.92} \approx 3.86$$

In summary, $A \approx 21.3°$, $B \approx 68.7°$, and $c \approx 3.86$.

110. Using a right triangle, we have a known angle, an unknown opposite side, *h*, and a known adjacent side. Therefore, use the tangent function.

$$\tan 25.6° = \frac{h}{80}$$

$$h = 80\tan 25.6°$$

$$\approx 38.3$$

The building is about 38 feet high.

111. Using a right triangle, we have a known angle, an unknown opposite side, *h*, and a known adjacent side. Therefore, use the tangent function.

$$\tan 40° = \frac{h}{60}$$

$$h = 60\tan 40° \approx 50 \text{ yd}$$

The second building is 50 yds taller than the first. Total height = $40 + 50 = 90$ yd .

112. Using two right triangles, a smaller right triangle corresponding to the smaller angle of elevation drawn inside a larger right triangle corresponding to the larger angle of elevation, we have a known angle, a known opposite side, and an unknown adjacent side, *d*, in the smaller triangle. Therefore, use the tangent function.

$$\tan 68° = \frac{125}{d}$$

$$d = \frac{125}{\tan 68°} \approx 50.5$$

We now have a known angle, a known adjacent side, and an unknown opposite side, *h*, in the larger triangle. Again, use the tangent function.

$$\tan 71° = \frac{h}{50.5}$$

$$h = 50.5\tan 71° \approx 146.7$$

The height of the antenna is $146.7 - 125$, or 21.7 ft, to the nearest tenth of a foot.

113. We need the acute angle between ray *OA* and the north-south line through *O*. This angle measures $90° - 55° = 35°$. This angle measured from the north side of the north-south line and lies east of the north-south line. Thus the bearing from *O* to *A* is N35°E.

114. We need the acute angle between ray *OA* and the north-south line through *O*. This angle measures $90° - 55° = 35°$. This angle measured from the south side of the north-south line and lies west of the north-south line. Thus the bearing from *O* to *A* is S35°W.

115. Using a right triangle, we have a known angle, a known adjacent side, and an unknown opposite side, *d*. Therefore, use the tangent function.

$$\tan 64° = \frac{d}{12}$$

$$d = 12\tan 64° \approx 24.6$$

The ship is about 24.6 miles from the lighthouse.

116.

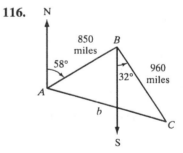

a. Using the figure,

$B = 58° + 32° = 90°$

Thus, use the Pythagorean Theorem to find the distance from city *A* to city *C*.

$$850^2 + 960^2 = b^2$$

$$b^2 = 722{,}500 + 921{,}600$$

$$b^2 = 1{,}644{,}100$$

$$b = \sqrt{1{,}644{,}100} \approx 1282.2$$

The distance from city A to city B is about 1282.2 miles.

b. Using the figure,

$$\tan A = \frac{\text{opposite}}{\text{adjacent}} = \frac{960}{850} \approx 1.1294$$

$$A \approx \tan^{-1}(1.1294) \approx 48°$$

$180° - 58° - 48° = 74°$

The bearing from city A to city C is S74°E.

117. $d = 20\cos\dfrac{\pi}{4}t$

$a = 20$ and $\omega = \dfrac{\pi}{4}$

a. maximum displacement:

$|a| = |20| = 20\,\text{cm}$

b. $f = \dfrac{\omega}{2\pi} = \dfrac{\frac{\pi}{4}}{2\pi} = \dfrac{\pi}{4}\cdot\dfrac{1}{2\pi} = \dfrac{1}{8}$

frequency: $\dfrac{1}{8}$ cm per second

c. period: $\dfrac{2\pi}{\omega} = \dfrac{2\pi}{\frac{\pi}{4}} = 2\pi\cdot\dfrac{4}{\pi} = 8$

The time required for one cycle is 8 seconds.

118. $d = \dfrac{1}{2}\sin 4t$

$a = \dfrac{1}{2}$ and $\omega = 4$

a. maximum displacement:

$|a| = \left|\dfrac{1}{2}\right| = \dfrac{1}{2}\,\text{cm}$

b. $f = \dfrac{\omega}{2\pi} = \dfrac{4}{2\pi} = \dfrac{2}{\pi} \approx 0.64$

frequency: 0.64 cm per second

c. period: $\dfrac{2\pi}{\omega} = \dfrac{2\pi}{4} = \dfrac{\pi}{2} \approx 1.57$

The time required for one cycle is about 1.57 seconds.

119. Because the distance of the object from the rest position at $t = 0$ is a maximum, use the form

$d = a\cos\omega t$. The period is $\dfrac{2\pi}{\omega}$ so,

$2 = \dfrac{2\pi}{\omega}$

$\omega = \dfrac{2\pi}{2} = \pi$

Because the amplitude is 30 inches, $|a| = 30$.
because the object starts below its rest position $a = -30$. the equation for the object's simple harmonic motion is $d = -30\cos\pi t$.

120. Because the distance of the object from the rest position at $t = 0$ is 0, use the form $d = a\sin\omega t$.

The period is $\dfrac{2\pi}{\omega}$ so

$5 = \dfrac{2\pi}{\omega}$

$\omega = \dfrac{2\pi}{5}$

Because the amplitude is $\dfrac{1}{4}$ inch, $|a| = \dfrac{1}{4}$. a is negative since the object begins pulled down. The equation for the object's simple harmonic motion is

$d = -\dfrac{1}{4}\sin\dfrac{2\pi}{5}t$.

Chapter 5 Test

1. $135° = 135°\cdot\dfrac{\pi \text{ radians}}{180°}$

$= \dfrac{135\pi}{180}$ radians

$= \dfrac{3\pi}{4}$ radians

2. $75° = 75°\cdot\dfrac{\pi \text{ radians}}{180°} = \dfrac{75\pi}{180}$ radians

$= \dfrac{5\pi}{12}$ radians

$s = r\theta$

$s = 20\left(\dfrac{5\pi}{12}\right) = \dfrac{25\pi}{3}$ ft ≈ 26.18 ft

3. a. $\dfrac{16\pi}{3} - 4\pi = \dfrac{16\pi}{3} - \dfrac{12\pi}{3} = \dfrac{4\pi}{3}$

b. $\dfrac{16\pi}{3}$ is coterminal with $\dfrac{4\pi}{3}$.

$\dfrac{4\pi}{3} - \pi = \dfrac{4\pi}{3} - \dfrac{3\pi}{3} = \dfrac{\pi}{3}$

4. $P = (-2, 5)$ is a point on the terminal side of θ, $x = -2$ and $y = 5$. Furthermore,

$r = \sqrt{x^2 + y^2} = \sqrt{(-2)^2 + (5)^2}$

$= \sqrt{4 + 25} = \sqrt{29}$

Use x, y, and r, to find the six trigonometric functions of θ.

$$\sin\theta = \frac{y}{r} = \frac{5}{\sqrt{29}} = \frac{5\sqrt{29}}{\sqrt{29}\sqrt{29}} = \frac{5\sqrt{29}}{29}$$

$$\cos\theta = \frac{x}{r} = \frac{-2}{\sqrt{29}} = -\frac{2\sqrt{29}}{\sqrt{29}\sqrt{29}} = -\frac{2\sqrt{29}}{29}$$

$$\tan\theta = \frac{y}{x} = \frac{5}{-2} = -\frac{5}{2}$$

$$\csc\theta = \frac{r}{y} = \frac{\sqrt{29}}{5}$$

$$\sec\theta = \frac{r}{x} = \frac{\sqrt{29}}{-2} = -\frac{\sqrt{29}}{2}$$

$$\cot\theta = \frac{x}{y} = \frac{-2}{5} = -\frac{2}{5}$$

5. Because $\cos\theta < 0$, θ cannot lie in quadrant I and quadrant IV; the cosine function is positive in those two quadrants. Thus, with $\cos\theta < 0$, θ lies in quadrant II or quadrant III. We are also given that $\cot\theta > 0$. Because quadrant III is the only quadrant in which the cosine is negative and the cotangent is positive, θ lies in quadrant III.

6. Because the cosine is positive and the tangent is negative, θ lies in quadrant IV. In quadrant IV x is positive and y is negative. Thus,

$$\cos\theta = \frac{1}{3} = \frac{x}{r},\ x = 1,\ r = 3.\ \text{Furthermore,}$$

$$x^2 + y^2 = r^3$$
$$1^2 + y^2 = 3^2$$
$$y^2 = 9 - 1 = 8$$
$$y = -\sqrt{8} = -2\sqrt{2}$$

Use x, y, and r, to find the six trigonometric functions of θ.

$$\sin\theta = \frac{y}{r} = \frac{-2\sqrt{2}}{3} = -\frac{2\sqrt{2}}{3}$$

$$\tan\theta = \frac{y}{x} = \frac{-2\sqrt{2}}{1} = -2\sqrt{2}$$

$$\csc\theta = \frac{r}{y} = \frac{3}{-2\sqrt{2}} = -\frac{3\sqrt{2}}{2\sqrt{2}\cdot\sqrt{2}} = -\frac{3\sqrt{2}}{4}$$

$$\sec\theta = \frac{r}{x} = \frac{3}{1} = 3$$

$$\cot\theta = \frac{x}{y} = \frac{1}{-2\sqrt{2}} = -\frac{1\cdot\sqrt{2}}{2\sqrt{2}\sqrt{2}} = -\frac{\sqrt{2}}{4}$$

7. $\tan\dfrac{\pi}{6}\cos\dfrac{\pi}{3} - \cos\dfrac{\pi}{2} = \dfrac{\sqrt{3}}{3}\cdot\dfrac{1}{2} - 0 = \dfrac{\sqrt{3}}{6}$

8. $300°$ lies in quadrant IV.
The reference angle is
$$\theta' = 360° - 300° = 60°$$
$$\tan 60° = \sqrt{3}$$
In quadrant IV, $\tan\theta < 0$, so
$$\tan 300° = -\tan 60 = -\sqrt{3}.$$

9. $\dfrac{7\pi}{4}$ lies in quadrant IV.
The reference angle is
$$\theta' = 2\pi - \frac{7\pi}{4} = \frac{8\pi}{4} - \frac{7\pi}{4} = \frac{\pi}{4}$$
$$\sin\frac{\pi}{4} = \frac{\sqrt{2}}{2}$$
In quadrant IV, $\sin\theta < 0$, so
$$\sin\frac{7\pi}{4} = -\sin\frac{\pi}{4} = -\frac{\sqrt{2}}{2}.$$

10. $\sec\dfrac{22\pi}{3} = \sec\dfrac{4\pi}{3} = -\sec\dfrac{\pi}{3}$
$$= \frac{1}{-\cos\dfrac{\pi}{3}} = \frac{1}{-\dfrac{1}{2}} = -2$$

11. $\cot\left(-\dfrac{8\pi}{3}\right) = \cot\left(\dfrac{4\pi}{3}\right) = \cot\dfrac{\pi}{3}$
$$= \frac{1}{\tan\dfrac{\pi}{3}} = \frac{1}{\sqrt{3}} = \frac{\sqrt{3}}{3}$$

12. $\tan\left(\dfrac{7\pi}{3} + n\pi\right) = \tan\dfrac{7\pi}{3} = \tan\dfrac{\pi}{3} = \sqrt{3}$

13. **a.** $\sin(-\theta) + \cos(-\theta) = -\sin(\theta) + \cos(\theta)$
$$= -a + b$$

   b. $\tan\theta - \sec\theta = \dfrac{\sin\theta}{\cos\theta} - \dfrac{1}{\cos\theta}$
$$= \frac{a}{b} - \frac{1}{b}$$
$$= \frac{a-1}{b}$$

14. The equation $y = 3\sin 2x$ is of the form $y = A\sin Bx$ with $A = 3$ and $B = 2$. The amplitude is $|A| = |3| = 3$.

The period is $\dfrac{2\pi}{B} = \dfrac{2\pi}{2} = \pi$. The quarter-period is $\dfrac{\pi}{4}$.

The cycle begins at $x = 0$. Add quarter-periods to generate x-values for the key points.

$x = 0$

$x = 0 + \dfrac{\pi}{4} = \dfrac{\pi}{4}$

$x = \dfrac{\pi}{4} + \dfrac{\pi}{4} = \dfrac{\pi}{2}$

$x = \dfrac{\pi}{2} + \dfrac{\pi}{4} = \dfrac{3\pi}{4}$

$x = \dfrac{3\pi}{4} + \dfrac{\pi}{4} = \pi$

Evaluate the function at each value of x.

x	coordinates
0	$(0, 0)$
$\dfrac{\pi}{4}$	$\left(\dfrac{\pi}{4}, 3\right)$
$\dfrac{\pi}{2}$	$\left(\dfrac{\pi}{2}, 0\right)$
$\dfrac{3\pi}{4}$	$\left(\dfrac{3\pi}{4}, -3\right)$
π	$(\pi, 0)$

Connect the five key points with a smooth curve and graph one complete cycle of the given function.

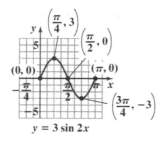

$y = 3\sin 2x$

15. The equation $y = -2\cos\left(x - \dfrac{\pi}{2}\right)$ is of the form $y = A\cos(Bx - C)$ with $A = -2$, $B = 1$, and $C = \dfrac{\pi}{2}$. The amplitude is $|A| = |-2| = 2$.

The period is $\dfrac{2\pi}{B} = \dfrac{2\pi}{1} = 2\pi$. The phase shift is $\dfrac{C}{B} = \dfrac{\frac{\pi}{2}}{1} = \dfrac{\pi}{2}$. The quarter-period is $\dfrac{2\pi}{4} = \dfrac{\pi}{2}$.

The cycle begins at $x = \dfrac{\pi}{2}$. Add quarter-periods to generate x-values for the key points.

$x = \dfrac{\pi}{2}$

$x = \dfrac{\pi}{2} + \dfrac{\pi}{2} = \pi$

$x = \pi + \dfrac{\pi}{2} = \dfrac{3\pi}{2}$

$x = \dfrac{3\pi}{2} + \dfrac{\pi}{2} = 2\pi$

$x = 2\pi + \dfrac{\pi}{2} = \dfrac{5\pi}{2}$

Evaluate the function at each value of x.

x	coordinates
$\dfrac{\pi}{2}$	$\left(\dfrac{\pi}{2}, -2\right)$
π	$(\pi, 0)$
$\dfrac{3\pi}{2}$	$\left(\dfrac{3\pi}{2}, 2\right)$
2π	$(2\pi, 0)$
$\dfrac{5\pi}{2}$	$\left(\dfrac{5\pi}{2}, -2\right)$

Connect the five key points with a smooth curve and graph one complete cycle of the given function.

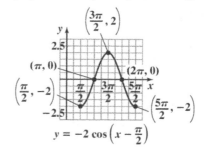

$y = -2\cos\left(x - \dfrac{\pi}{2}\right)$

16. Solve the equations

$$\frac{x}{2} = -\frac{\pi}{2} \quad \text{and} \quad \frac{x}{2} = \frac{\pi}{2}$$

$$x = -\frac{\pi}{2} \cdot 2 \qquad x = \frac{\pi}{2} \cdot 2$$

$$x = -\pi \qquad\qquad x = \pi$$

Thus, two consecutive asymptotes occur at $x = -\pi$ and $x = \pi$.

$$x\text{-intercept} = \frac{-\pi + \pi}{2} = \frac{0}{2} = 0$$

An x-intercept is 0 and the graph passes through (0, 0). Because the coefficient of the tangent is 2, the points on the graph midway between an x-intercept and the asymptotes have y-coordinates of -2 and 2. Use the two consecutive asymptotes, $x = -\pi$ and $x = \pi$, to graph one

full period of $y = 2\tan\dfrac{x}{2}$ from $-\pi$ to π.

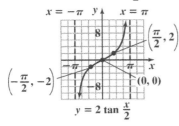

17. Graph the reciprocal sine function, $y = -\dfrac{1}{2}\sin\pi x$.

The equation is of the form $y = A\sin Bx$ with $A = -\dfrac{1}{2}$ and $B = \pi$.

amplitude: $|A| = \left|-\dfrac{1}{2}\right| = \dfrac{1}{2}$

period: $\dfrac{2\pi}{B} = \dfrac{2\pi}{\pi} = 2$

Use quarter-periods, $\dfrac{2}{4} = \dfrac{1}{2}$, to find x-values for the five key points. Starting with $x = 0$, the

x-values are $0, \dfrac{1}{2}, 1, \dfrac{3}{2}, 2$. Evaluating the function at each value of x, the key points are

$(0, 3)$, $\left(\dfrac{1}{2}, -\dfrac{1}{2}\right)$, $(1, 0)$, $\left(\dfrac{3}{2}, \dfrac{1}{2}\right)$, $(2, 0)$.

Use these key points to graph $y = -\dfrac{1}{2}\sin\pi x$ from 0 to 2. Use the graph to obtain the graph of the reciprocal function. Draw vertical asymptotes through the x-intercepts, and use them as guides to

graph $y = -\dfrac{1}{2}\csc\pi x$.

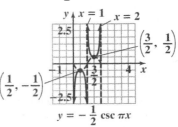

18. Let $\theta = \cos^{-1}\left(-\dfrac{1}{2}\right)$, then $\cos\theta = -\dfrac{1}{2}$.

Because $\cos\theta$ is negative, θ is in quadrant II.

$$x^2 + y^2 = r^2$$

$$(-1)^2 + y^2 = 2^2$$

$$y^2 = 4 - 1 = 3$$

$$y = \sqrt{3}$$

$$\tan\left[\cos^{-1}\left(-\dfrac{1}{2}\right)\right] = \tan\theta = \frac{y}{x} = \frac{\sqrt{3}}{-1} = -\sqrt{3}$$

19. Let $\theta = \cos^{-1}\left(\dfrac{x}{3}\right)$, then $\cos\theta = \dfrac{x}{3}$.

Because $\cos\theta$ is positive, θ is in quadrant I.

$$x^2 + y^2 = r^2$$

$$x^2 + y^2 = 3^2$$

$$y^2 = 9 - x^2$$

$$y = \sqrt{9 - x^2}$$

$$\sin\left[\cos^{-1}\left(\dfrac{x}{3}\right)\right] = \sin\theta = \frac{y}{r} = \frac{\sqrt{9 - x^2}}{3}$$

20. Find the measure of angle B. Because $C = 90°$, $A + B = 90°$.
Thus, $B = 90° - A = 90° - 21° = 69°$.
We have a known angle, a known hypotenuse, and an unknown opposite side. Use the sine function.

$$\sin 21° = \frac{a}{13}$$

$$a = 13\sin 21° \approx 4.7$$

We have a known angle, a known hypotenuse, and an unknown adjacent side. Use the cosine function.

$$\cos 21° = \frac{b}{13}$$

$$b = 13\cos 21° \approx 12.1$$

In summary, $B = 69°$, $a \approx 4.7$, and $b \approx 12.1$.

21. Using a right triangle, we have a known angle, an unknown opposite side, h, and a known adjacent side. Therefore, use the tangent function.

$$\tan 37° = \frac{h}{30}$$
$$h = 30 \tan 37° \approx 23$$

The building is about 23 yards high.

22. Using a right triangle, we have a known hypotenuse, a known opposite side, and an unknown angle. Therefore, use the sine function.

$$\sin \theta = \frac{43}{73}$$
$$\theta = \sin^{-1}\left(\frac{43}{73}\right) \approx 36.1°$$

The rope makes an angle of about $36.1°$ with the pole.

23. We need the acute angle between ray OP and the north-south line through O. This angle measures $90°$ $- 10°$. This angle is measured from the north side of the north-south line and lies west of the north-south line. Thus the bearing from O to P is N80°W.

24. $d = -6 \cos \pi t$

$a = -6$ and $\omega = \pi$

a. maximum displacement: $|a| = |-6| = 6$ in.

b. $f = \dfrac{\omega}{2\pi} = \dfrac{\pi}{2\pi} = \dfrac{1}{2}$

frequency: $\dfrac{1}{2}$ in. per second

c. period $= \dfrac{2\pi}{\omega} = \dfrac{2\pi}{\pi} = 2$

The time required for one cycle is 2 seconds.

25. Trigonometric functions are periodic.

Cumulative Review Exercises (Chapters 1-5)

1. $x^2 = 18 + 3x$

$x^2 - 3x - 18 = 0$

$(x-6)(x+3) = 0$

$x - 6 = 0$ or $x + 3 = 0$

$x = 6$ $\qquad$ $x = -3$

The solution set is $\{-3, 6\}$.

2. $x^3 + 5x^2 - 4x - 20 = 0$

$x^2(x+5) - 4(x+5) = 0$

$(x^2 - 4)(x+5) = 0$

$(x-2)(x+2)(x+5) = 0$

$x - 2 = 0$ or $x + 2 = 0$ or $x + 5 = 0$

$x = 2$ $\qquad$ $x = -2$ $\qquad$ $x = -5$

The solution set is $\{-5, -2, 2\}$.

3. $\log_2 x + \log_2(x-2) = 3$

$\log_2 x(x-2) = 3$

$x(x-2) = 2^3$

$x^2 - 2x = 2^3$

$x^2 - 2x - 8 = 0$

$(x-4)(x+2) = 0$

$x - 4 = 0$ or $x + 2 = 0$

$x = 4$ $\qquad$ $x = -2$

$x = -2$ is extraneous

The solution set is $\{4\}$

4. $\sqrt{x-3} + 5 = x$

$\sqrt{x-3} = x - 5$

$\left(\sqrt{x-3}\right)^2 = (x-5)^2$

$x - 3 = x^2 - 10x + 25$

$x^2 - 11x + 28 = 0$

$(x-4)(x-7) = 0$

$x - 4 = 0$ or $x - 7 = 0$

$x = 4$ $\qquad$ $x = 7$

$\sqrt{4-3} + 5 = 4$

$\sqrt{1} + 5 = 4$

$1 + 5 = 4$ $\quad$ false

$x = 4$ is not a solution

$\sqrt{7-3} + 5 = 7$

$\sqrt{4} + 5 = 7$

$2 + 5 = 7$ $\quad$ true

The solution set is $\{7\}$.

5. $x^3 - 4x^2 + x + 6 = 0$

$p: \pm 1, \pm 2, \pm 3, \pm 6$

$q: \pm 1$

$\dfrac{p}{q}: \pm 1, \pm 2, \pm 3, \pm 6$

$$
\begin{array}{r|rrrr}
2 & 1 & -4 & 1 & 6 \\
  &   & 2 & -4 & -6 \\
\hline
  & 1 & -2 & -3 & 0
\end{array}
$$

$x^3 - 4x^2 + x + 6 = (x - 2)\left(x^2 - 2x - 3\right)$

Thus,

$x^3 - 4x^2 + x + 6 = 0$

$(x - 2)\left(x^2 - 2x - 3\right) = 0$

$(x - 2)(x - 3)(x + 1) = 0$

$x - 2 = 0 \quad$ or $\quad x - 3 = 0 \quad$ or $\quad x + 1 = 0$

$\quad x = 2 \qquad\qquad x = 3 \qquad\qquad x = -1$

The solution set is $\{-1, 2, 3\}$

6. $|\,2x - 5\,| \le 11$

$-11 \le 2x - 5 \le 11$

$-6 \le 2x \le 16$

$-3 \le x \le 8$

The solution set is $\{x \mid -3 \le x \le 8\}$

7. $f(x) = \sqrt{x - 6}$

$x = \sqrt{y - 6}$

$x^2 = y - 6$

$y = x^2 + 6$

$f^{-1}(x) = x^2 + 6$

8.

$$
\require{enclose}
\begin{array}{r}
4x^2 - \dfrac{14}{5}x - \dfrac{17}{25} \\[2pt]
5x + 2 \enclose{longdiv}{20x^3 - 6x^2 - 9x + 10} \\[2pt]
\underline{20x^3 + 8x^2} \\[2pt]
-14x^2 - 9x \\[2pt]
\underline{-14x^2 - \dfrac{28}{5}x} \\[2pt]
-\dfrac{17}{5}x + 10 \\[2pt]
\underline{-\dfrac{17}{5}x - \dfrac{34}{25}} \\[2pt]
\dfrac{284}{25}
\end{array}
$$

The quotient is $4x^2 - \dfrac{14}{5}x - \dfrac{17}{25} + \dfrac{284}{125x + 50}$.

9. $\log 25 + \log 40 = \log(25 \cdot 40)$

$\qquad\qquad\qquad = \log 1000$

$\qquad\qquad\qquad = \log 10^3$

$\qquad\qquad\qquad = 3$

10. $\dfrac{14\pi}{9}$ radians $= \dfrac{14\pi}{9}$ radians $\cdot \dfrac{180°}{\pi \text{ radians}}$

$\qquad\qquad\qquad = \dfrac{14 \cdot 180°}{9} = 280°$

11. $3x^4 - 2x^3 + 5x^2 + x - 9 = 0$

The sign changes 3 times so the equation has at most 3 positive real roots;

$f(-x) = 3x^4 + 2x^3 + 5x^2 - x - 9$

The sign changes 1 time, so the equation has at most 1 negative real root.

12. $f(x) = \dfrac{x}{x^2 - 1}$

vertical asymptotes: $x^2 - 1 = 0$, $x = 1$ and $x = -1$

horizontal asymptote: $m = 1$ and $n = 2$ so $m < n$ and the x-axis is a horizontal asymptote.

x-intercept: $(0, 0)$

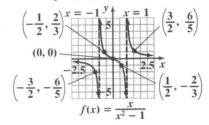

13. $(x-2)^2 + y^2 = 1$

The graph is a circle with center $(2,0)$ and $r = 1$.

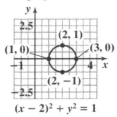

$(x-2)^2 + y^2 = 1$

14. $y = (x-1)(x+2)^2$

x-intercepts: $(1,0)$ and $(-2,0)$

y-intercept: $y = (-1)(2)^2 = -4$

$(0,-4)$

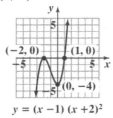

$y = (x-1)(x+2)^2$

15. $y = \sin\left(2x + \dfrac{\pi}{2}\right) = \sin\left(2x - \left(-\dfrac{\pi}{2}\right)\right)$

The equation $y = \sin\left(2x - \left(-\dfrac{\pi}{2}\right)\right)$ is of the form

$y = A\sin(Bx - C)$ with $A = 1$, $B = 2$, and $C = -\dfrac{\pi}{2}$.

The amplitude is $|A| = |1| = 1$

The period is $\dfrac{2\pi}{B} = \dfrac{2\pi}{2} = \pi$. The phase shift is

$\dfrac{C}{B} = \dfrac{-\frac{\pi}{2}}{2} = -\dfrac{\pi}{2}\cdot\dfrac{1}{2} = -\dfrac{\pi}{4}$. The quarter-period is $\dfrac{\pi}{4}$.

The cycle begins at $x = -\dfrac{\pi}{4}$. Add quarter-periods to generate x-values for the key points.

$x = -\dfrac{\pi}{4}$, $x = -\dfrac{\pi}{4} + \dfrac{\pi}{4} = 0$, $x = 0 + \dfrac{\pi}{4} = \dfrac{\pi}{4}$,

$x = \dfrac{\pi}{4} + \dfrac{\pi}{4} = \dfrac{\pi}{2}$, $x = \dfrac{\pi}{2} + \dfrac{\pi}{4} = \dfrac{3\pi}{4}$ To graph from 0 to

π, evaluate the function at the last four key points and at $x = \pi$.

x	coordinates
0	$(0, 1)$
$\dfrac{\pi}{4}$	$\left(\dfrac{\pi}{4}, 0\right)$
$\dfrac{\pi}{2}$	$\left(\dfrac{\pi}{2}, -1\right)$
$\dfrac{3\pi}{4}$	$\left(\dfrac{3\pi}{4}, 0\right)$
π	$(\pi, 1)$

Connect the points with a smooth curve and extend the graph one cycle to the right to graph from 0 to 2π.

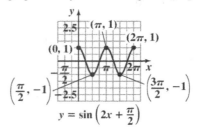

$y = \sin\left(2x + \dfrac{\pi}{2}\right)$

16. Solve the equations

$3x = -\dfrac{\pi}{2}$ and $3x = \dfrac{\pi}{2}$

$x = \dfrac{-\frac{\pi}{2}}{3}$ $\qquad$ $x = \dfrac{\frac{\pi}{2}}{3}$

$x = -\dfrac{\pi}{6}$ $\qquad$ $x = \dfrac{\pi}{6}$

Thus, two consecutive asymptotes occur at $x = -\dfrac{\pi}{6}$

and $x = \dfrac{\pi}{6}$.

$x\text{-intercept} = \dfrac{-\frac{\pi}{6} + \frac{\pi}{6}}{2} = \dfrac{0}{2} = 0$

An x-intercept is 0 and the graph passes through $(0,0)$.

Because the coefficient of the tangent is 2, the points on the graph midway between an x-intercept and the asymptotes have y-coordinates of -2 and 2. Use the two consecutive asymptotes, $x = -\dfrac{\pi}{6}$ and $x = \dfrac{\pi}{6}$, to graph one full period of $y = 2\tan 3x$ from $-\dfrac{\pi}{6}$ to $\dfrac{\pi}{6}$. Extend the pattern to the right to graph two complete cycles.

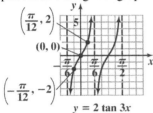

17. $C(p) = 30,000 + 2500p$

$R(p) = 3125p$

$30,000 + 2500p = 3125p$

$30,000 = 625p$

$p = 48$

48 performances must be played for you to break even.

18. **a.** Let t be the number of years after 2000.

$A = A_0 e^{kt}$

$A = 110e^{kt}$

$303 = 110e^{k(10)}$

$\dfrac{303}{110} = e^{k(10)}$

$\ln\dfrac{303}{110} = \ln e^{k(10)}$

$\ln\dfrac{303}{110} = 10k$

$\dfrac{\ln\dfrac{303}{110}}{10} = k$

$k \approx 0.1013$

Thus, $A = 110e^{0.1013t}$

b.

$A = 110e^{0.1013t}$

$400 = 110e^{0.1013t}$

$\dfrac{400}{110} = e^{0.1013t}$

$\ln\dfrac{400}{110} = \ln e^{0.1013t}$

$\ln\dfrac{400}{110} = 0.1013t$

$\dfrac{\ln\dfrac{400}{110}}{0.1013} = t$

$t \approx 13$

There will be 400 million cell phone subscribers in the United States 13 years after 2000, or 2013.

19. $2200 = \dfrac{k}{3.5}$

$k = 7700$

$h = \dfrac{7700}{5} = 1540$

The rate of heat loss is 1540 Btu per hour.

20. Using a right triangle, we have a known opposite side, a known adjacent side, and an unknown angle. Therefore, use the tangent function.

$\tan\theta = \dfrac{200}{50} = 4$

$\theta = \tan^{-1}(4) \approx 76°$

The angle of elevation is about $76°$.

Chapter 6
Analytic Trigonometry

Section 6.1

Check Point Exercises

1. $\csc x \tan x = \dfrac{1}{\sin x} \cdot \dfrac{\sin x}{\cos x}$

$= \dfrac{1}{\cos x}$

$= \sec x$

We worked with the left side and arrived at the right side. Thus, the identity is verified.

2. $\cos x \cot x + \sin x = \cos x \cdot \dfrac{\cos x}{\sin x} + \sin x$

$= \dfrac{\cos^2 x}{\sin x} + \sin x$

$= \dfrac{\cos^2 x}{\sin x} + \sin x \cdot \dfrac{\sin x}{\sin x}$

$= \dfrac{\cos^2 x}{\sin x} + \dfrac{\sin^2 x}{\sin x}$

$= \dfrac{\cos^2 x + \sin^2 x}{\sin x}$

$= \dfrac{1}{\sin x}$

$= \csc x$

We worked with the left side and arrived at the right side. Thus, the identity is verified.

3. $\sin x - \sin x \cos^2 x = \sin x \left(1 - \cos^2 x\right)$

$= \sin x \cdot \sin^2 x$

$= \sin^3 x$

We worked with the left side and arrived at the right side. Thus, the identity is verified.

4. $\dfrac{1 + \cos\theta}{\sin\theta} = \dfrac{1}{\sin\theta} + \dfrac{\cos\theta}{\sin\theta}$

$= \csc\theta + \cot\theta$

We worked with the left side and arrived at the right side. Thus, the identity is verified.

5. $\dfrac{\sin x}{1 + \cos x} + \dfrac{1 + \cos x}{\sin x}$

$= \dfrac{\sin x(\sin x)}{(1 + \cos x)\sin x} + \dfrac{(1 + \cos x)(1 + \cos x)}{\sin x(1 + \cos x)}$

$= \dfrac{\sin^2 x}{(1 + \cos x)\sin x} + \dfrac{1 + 2\cos x + \cos^2 x}{(1 + \cos x)\sin x}$

$= \dfrac{\sin^2 x + \cos^2 x + 2\cos x + 1}{(1 + \cos x)\sin x}$

$= \dfrac{1 + 1 + 2\cos x}{(1 + \cos x)\sin x}$

$= \dfrac{2 + 2\cos x}{(1 + \cos x)\sin x}$

$= \dfrac{2\left(1 + \cos x\right)}{(1 + \cos x)\sin x}$

$= \dfrac{2}{\sin x}$

$= 2\csc x$

We worked with the left side and arrived at the right side. Thus, the identity is verified.

6. $\dfrac{\cos x}{1 + \sin x} = \dfrac{\cos x}{(1 + \sin x)} \cdot \dfrac{1 - \sin x}{1 - \sin x}$

$= \dfrac{\cos x(1 - \sin x)}{1 - \sin^2 x}$

$= \dfrac{\cos x(1 - \sin x)}{\cos^2 x}$

$= \dfrac{1 - \sin x}{\cos x}$

We worked with the left side and arrived at the right side. Thus, the identity is verified.

7. $\dfrac{\sec x + \csc(-x)}{\sec x \csc x}$

$= \dfrac{\sec x - \csc x}{\sec x \csc x}$

$= \dfrac{\dfrac{1}{\cos x} - \dfrac{1}{\sin x}}{\dfrac{1}{\cos x}\dfrac{1}{\sin x}}$

$= \dfrac{\dfrac{\sin x}{\cos x \sin x} - \dfrac{\cos x}{\cos x \sin x}}{\dfrac{1}{\cos x \sin x}}$

$= \dfrac{\dfrac{\sin x - \cos x}{\cos x \sin x}}{\dfrac{1}{\cos x \sin x}}$

$= \dfrac{\sin x - \cos x}{\cos x \sin x}\dfrac{\cos x \sin x}{1}$

$= \sin x - \cos x$

We worked with the left side and arrived at the right side. Thus, the identity is verified.

8. Left side:

$\dfrac{1}{1 + \sin\theta} + \dfrac{1}{1 - \sin\theta}$

$= \dfrac{1(1 - \sin\theta)}{(1 + \sin\theta)(1 - \sin\theta)} + \dfrac{1(1 + \sin\theta)}{(1 - \sin\theta)(1 + \sin\theta)}$

$= \dfrac{1 - \sin\theta + 1 + \sin\theta}{(1 + \sin\theta)(1 - \sin\theta)}$

$= \dfrac{2}{(1 + \sin\theta)(1 - \sin\theta)}$

$= \dfrac{2}{1 - \sin^2\theta}$

Right side:

$2 + 2\tan^2\theta = 2 + 2\left(\dfrac{\sin^2\theta}{\cos^2\theta}\right)$

$= \dfrac{2\cos^2\theta}{\cos^2\theta} + \dfrac{2\sin^2\theta}{\cos^2\theta}$

$= \dfrac{2\cos^2\theta + 2\sin^2\theta}{\cos^2\theta}$

$= \dfrac{2}{\cos^2\theta} = \dfrac{2}{1 - \sin^2\theta}$

The identity is verified because both sides are equal to $\dfrac{2}{1 - \sin^2\theta}$.

Concept and Vocabulary Check 6.1

1. complicated; other

2. sines; cosines

3. false

4. $(\csc x - 1)(\csc x + 1)$

5. identical/the same

Exercise Set 6.1

1. $\sin x \sec x = \sin x \cdot \dfrac{1}{\cos x}$

$= \dfrac{\sin x}{\cos x}$

$= \tan x$

3. $\tan(-x) \cdot \cos x = -\tan x \cdot \cos x$

$= -\dfrac{\sin x}{\cos x} \cdot \cos x$

$= -\sin x$

5. $\tan x \csc x \cos x = \dfrac{\sin x}{\cos x} \cdot \dfrac{1}{\sin x}\cos x$

$= 1$

7. $\sec x - \sec x \sin^2 x = \sec x(1 - \sin^2 x)$

$= \dfrac{1}{\cos x} \cdot \cos^2 x$

$= \cos x$

9. $\cos^2 x - \sin^2 x = \left(1 - \sin^2 x\right) - \sin^2 x$

$= 1 - \sin^2 x - \sin^2 x$

$= 1 - 2\sin^2 x$

11. $\csc\theta - \sin\theta = \dfrac{1}{\sin\theta} - \sin\theta$

$\qquad = \dfrac{1}{\sin\theta} - \dfrac{\sin^2\theta}{\sin\theta}$

$\qquad = \dfrac{1 - \sin^2\theta}{\sin\theta}$

$\qquad = \dfrac{\cos^2\theta}{\sin\theta}$

$\qquad = \dfrac{\cos\theta}{\sin\theta}\cdot\cos\theta$

$\qquad = \cot\theta\cos\theta$

13. $\dfrac{\tan\theta\cot\theta}{\csc\theta} = \dfrac{\dfrac{\sin\theta}{\cos\theta}\cdot\dfrac{\cos\theta}{\sin\theta}}{\dfrac{1}{\sin\theta}}$

$\qquad = \dfrac{1}{\dfrac{1}{\sin\theta}}$

$\qquad = 1 \div \dfrac{1}{\sin\theta}$

$\qquad = 1 \cdot \dfrac{\sin\theta}{1}$

$\qquad = \sin\theta$

15. $\sin^2\theta(1 + \cot^2\theta) = \sin^2\theta(\csc^2\theta)$

$\qquad = \sin^2\theta\cdot\dfrac{1}{\sin^2\theta}$

$\qquad = 1$

17. $\dfrac{1 - \cos^2 t}{\cos t} = \dfrac{\sin^2 t}{\cos t}$

$\qquad = \sin t \cdot \dfrac{\sin t}{\cos t}$

$\qquad = \sin t \tan t$

19. $\dfrac{\csc^2 t}{\cot t} = \dfrac{\dfrac{1}{\sin^2 t}}{\dfrac{\cos t}{\sin t}}$

$\qquad = \dfrac{1}{\sin^2 t} \div \dfrac{\cos t}{\sin t}$

$\qquad = \dfrac{1}{\sin^2 t} \cdot \dfrac{\sin t}{\cos t}$

$\qquad = \dfrac{1}{\sin t} \cdot \dfrac{1}{\cos t}$

$\qquad = \csc t \sec t$

21. $\dfrac{\tan^2 t}{\sec t} = \dfrac{\sec^2 t - 1}{\sec t}$

$\qquad = \dfrac{\sec^2 t}{\sec t} - \dfrac{1}{\sec t}$

$\qquad = \sec t - \cos t$

23. $\dfrac{1 - \cos\theta}{\sin\theta} = \dfrac{1}{\sin\theta} - \dfrac{\cos\theta}{\sin\theta}$

$\qquad = \csc\theta - \cot\theta$

25. $\dfrac{\sin t}{\csc t} + \dfrac{\cos t}{\sec t} = \dfrac{\sin t}{\dfrac{1}{\sin t}} + \dfrac{\cos t}{\dfrac{1}{\cos t}}$

$\qquad = \sin t \div \dfrac{1}{\sin t} + \cos t \div \dfrac{1}{\cos t}$

$\qquad = \sin t \cdot \dfrac{\sin t}{1} + \cos t \cdot \dfrac{\cos t}{1}$

$\qquad = \sin^2 t + \cos^2 t$

$\qquad = 1$

27. $\tan t + \dfrac{\cos t}{1 + \sin t}$

$\qquad = \dfrac{\sin t}{\cos t} + \dfrac{\cos t}{1 + \sin t}$

$\qquad = \dfrac{\sin t}{\cos t}\cdot\dfrac{1 + \sin t}{1 + \sin t} + \dfrac{\cos t}{1 + \sin t}\cdot\dfrac{\cos t}{\cos t}$

$\qquad = \dfrac{\sin t + \sin^2 t}{\cos t(1 + \sin t)} + \dfrac{\cos^2 t}{\cos t(1 + \sin t)}$

$\qquad \dfrac{\sin t + \sin^2 t + \cos^2 t}{\cos t(1 + \sin t)}$

$\qquad = \dfrac{1 + \sin t}{\cos t(1 + \sin t)}$

$\qquad = \dfrac{1}{\cos t}$

$\qquad = \sec t$

29. $1 - \dfrac{\sin^2 x}{1 + \cos x} = 1 - \dfrac{\sin^2 x}{1 + \cos x} \cdot \dfrac{1 - \cos x}{1 - \cos x}$

$= 1 - \dfrac{\sin^2 x(1 - \cos x)}{1 - \cos^2 x}$

$= 1 - \dfrac{\sin^2 x(1 - \cos x)}{\sin^2 x}$

$= 1 - 1 + \cos x$

$= \cos x$

31. $\dfrac{\cos x}{1 - \sin x} + \dfrac{1 - \sin x}{\cos x}$

$= \dfrac{\cos x}{1 - \sin x} \cdot \dfrac{1 + \sin x}{1 + \sin x} + \dfrac{1 - \sin x}{\cos x}$

$= \dfrac{\cos x(1 + \sin x)}{1 - \sin^2 x} + \dfrac{1 - \sin x}{\cos x}$

$== \dfrac{\cos x(1 + \sin x)}{\cos^2 x} + \dfrac{1 - \sin x}{\cos x}$

$= \dfrac{1 + \sin x}{\cos x} + \dfrac{1 - \sin x}{\cos x}$

$= \dfrac{2}{\cos x}$

$= 2 \cdot \dfrac{1}{\cos x}$

$= 2 \sec x$

33. $\sec^2 x \csc^2 x = (1 + \tan^2 x) \csc^2 x$

$= \csc^2 x + \tan^2 x \csc^2 x$

$= \csc^2 x + \dfrac{\sin^2 x}{\cos^2 x} \cdot \dfrac{1}{\sin^2 x}$

$= \csc^2 x + \dfrac{1}{\cos^2 x}$

$= \csc^2 x + \sec^2 x$

$= \sec^2 x + \csc^2 x$

35. $\dfrac{\sec x - \csc x}{\sec x + \csc x} = \dfrac{\dfrac{1}{\cos x} - \dfrac{1}{\sin x}}{\dfrac{1}{\cos x} + \dfrac{1}{\sin x}}$

$= \dfrac{\dfrac{1}{\cos x} - \dfrac{1}{\sin x}}{\dfrac{1}{\cos x} + \dfrac{1}{\sin x}} \cdot \dfrac{\sin x}{\sin x}$

$= \dfrac{\dfrac{\sin x}{\cos x} - 1}{\dfrac{\sin x}{\cos x} + 1}$

$= \dfrac{\tan x - 1}{\tan x + 1}$

37. $\dfrac{\sin^2 x - \cos^2 x}{\sin x + \cos x} = \dfrac{(\sin x + \cos x)(\sin x - \cos x)}{\sin x + \cos x}$

$= \sin x - \cos x$

39. $\tan^2 2x + \sin^2 2x + \cos^2 2x = \tan^2 2x + 1$

$= \sec^2 2x$

41. $\dfrac{\tan 2\theta + \cot 2\theta}{\csc 2\theta} = \dfrac{\dfrac{\sin 2\theta}{\cos 2\theta} + \dfrac{\cos 2\theta}{\sin 2\theta}}{\dfrac{1}{\sin 2\theta}}$

$= \dfrac{\dfrac{\sin 2\theta}{\cos 2\theta} \cdot \dfrac{\sin 2\theta}{\sin 2\theta} + \dfrac{\cos 2\theta}{\sin 2\theta} \cdot \dfrac{\cos 2\theta}{\cos 2\theta}}{\dfrac{1}{\sin 2\theta}}$

$= \dfrac{\dfrac{\sin^2 2\theta + \cos 2\theta}{\cos 2\theta \sin 2\theta}}{\dfrac{1}{\sin 2\theta}}$

$= \dfrac{1}{\cos 2\theta \sin 2\theta} \div \dfrac{1}{\sin 2\theta}$

$= \dfrac{1}{\cos 2\theta \sin 2\theta} \cdot \dfrac{\sin 2\theta}{1}$

$= \dfrac{1}{\cos 2\theta} = \sec 2\theta$

43. $\dfrac{\tan x + \tan y}{1 - \tan x \tan y} = \dfrac{\dfrac{\sin x}{\cos x} + \dfrac{\sin y}{\cos y}}{1 - \dfrac{\sin x}{\cos y} \cdot \dfrac{\sin y}{\cos y}} \cdot \dfrac{\cos x \cos y}{\cos x \cos y}$

$= \dfrac{\sin x \cos y + \cos x \sin y}{\cos x \cos y - \sin x \sin y}$

45. Left side:

$$(\sec x - \tan x)^2 = \left(\frac{1}{\cos x} - \frac{\sin x}{\cos x}\right)^2$$

$$= \left(\frac{1 - \sin x}{\cos x}\right)^2$$

$$= \frac{(1 - \sin x)^2}{\cos^2 x}$$

Right side:

$$\frac{1 - \sin x}{1 + \sin x} = \frac{1 - \sin x}{1 + \sin x} \cdot \frac{1 - \sin x}{1 - \sin x}$$

$$= \frac{(1 - \sin x)^2}{1 - \sin^2 x}$$

$$= \frac{(1 - \sin x)^2}{\cos^2 x}$$

The identity is verified because both sides are equal

to $\dfrac{(1 - \sin x)^2}{\cos^2 x}$.

47. $\dfrac{\tan t}{\sec t - 1} = \dfrac{\tan t}{\sec t - 1} \cdot \dfrac{\sec t + 1}{\sec t + 1}$

$$= \frac{\tan t (\sec t + 1)}{\sec^2 t - 1}$$

$$= \frac{\tan t (\sec t + 1)}{\tan^2 t}$$

$$= \frac{\sec t + 1}{\tan t}$$

49. Left side:

$$\frac{1 + \cos t}{1 - \cos t} = \frac{1 + \cos t}{1 - \cos t} \cdot \frac{1 + \cos t}{1 + \cos t}$$

$$= \frac{(1 + \cos t)^2}{1 - \cos^2 t}$$

$$= \frac{(1 + \cos t)^2}{\sin^2 t}$$

Right side:

$$(\csc t + \cot t)^2 = \left(\frac{1}{\sin t} + \frac{\cos t}{\sin t}\right)^2$$

$$= \left(\frac{1 + \cos t}{\sin t}\right)^2$$

$$= \frac{(1 + \cos t)^2}{\sin^2 t}$$

The identity is verified because both sides are equal

to $\dfrac{(1 + \cos t)^2}{\sin^2 t}$.

51. $\cos^4 t - \sin^4 t = \left(\cos^2 t - \sin^2 t\right)\left(\cos^2 t + \sin^2 t\right)$

$$= \left(\cos^2 t - \sin^2 t\right) \cdot 1$$

$$= 1 - \sin^2 t - \sin^2 t$$

$$= 1 - 2\sin^2 t$$

53. $\dfrac{\sin \theta - \cos \theta}{\sin \theta} + \dfrac{\cos \theta - \sin \theta}{\cos \theta}$

$$= \frac{(\sin \theta - \cos \theta)\cos \theta}{\cos \theta \sin \theta} + \frac{(\cos \theta - \sin \theta)\sin \theta}{\cos \theta \sin \theta}$$

$$= \frac{\sin \theta \cos \theta - \cos^2 \theta + \sin \theta \cos \theta - \sin^2 \theta}{\sin \theta \cos \theta}$$

$$= \frac{2\sin \theta \cos \theta - \left(\cos^2 \theta + \sin^2 \theta\right)}{\sin \theta \cos \theta}$$

$$= \frac{2\sin \theta \cos \theta - 1}{\sin \theta \cos \theta}$$

$$= \frac{2\sin \theta \cos \theta}{\sin \theta \cos \theta} - \frac{1}{\sin \theta \cos \theta}$$

$$= 2 - \frac{1}{\sin \theta} \cdot \frac{1}{\cos \theta}$$

$$= 2 - \csc \theta \sec \theta$$

$$= 2 - \sec \theta \csc \theta$$

55. $\left(\tan^2 \theta + 1\right)\left(\cos^2 \theta + 1\right)$

$$= \tan^2 \theta \cos^2 \theta + \tan^2 \theta + \cos^2 \theta + 1$$

$$= \frac{\sin^2 \theta}{\cos^2 \theta} \cdot \cos^2 \theta + \tan^2 \theta + \cos^2 \theta + 1$$

$$= \sin^2 \theta + \tan^2 \theta + \cos^2 \theta + 1$$

$$= \sin^2 \theta + \cos^2 \theta + \tan^2 \theta + 1$$

$$= 1 + \tan^2 \theta + 1$$

$$= \tan^2 \theta + 2$$

57. $(\cos\theta - \sin\theta)^2 + (\cos\theta + \sin\theta)^2$

$= \cos^2\theta - 2\cos\theta\sin\theta + \sin^2\theta + \cos^2\theta + 2\cos\theta\sin\theta + \sin^2\theta$

$= \cos^2\theta + \sin^2\theta + \cos^2\theta + \sin^2\theta$

$= 1 + 1 = 2$

59. $\dfrac{\cos^2 x - \sin^2 x}{1 - \tan^2 x}$

$= \dfrac{\cos^2 x - \sin^2 x}{1 - \dfrac{\sin^2 x}{\cos^2 x}} = \dfrac{\cos^2 x - \sin^2 x}{\dfrac{\cos^2 x - \sin^2 x}{\cos^2 x}}$

$= \dfrac{\cos^2 x - \sin^2 x}{1} \div \dfrac{\cos^2 x - \sin^2 x}{\cos^2 x}$

$= \dfrac{\cos^2 x - \sin^2 x}{1} \cdot \dfrac{\cos^2 x}{\cos^2 x - \sin^2 x} = \cos^2 x$

61. Conjecture: left side is equal to $\cos x$

$\dfrac{(\sec x + \tan x)(\sec x - \tan x)}{\sec x} = \dfrac{\sec^2 x - \tan^2 x}{\sec x}$

$= \dfrac{1}{\sec x}$

$= \cos x$

63. Conjecture: left side is equal to $2\sin x$

$\dfrac{\cos x + \cot x \sin x}{\cot x} = \dfrac{\cos x}{\cot x} + \dfrac{\cot x \sin x}{\cot x}$

$= \dfrac{\cos x}{\frac{\cos x}{\sin x}} + \dfrac{\cot x \sin x}{\cot x}$

$= \dfrac{\cos x \sin x}{\cos x} + \sin x$

$= \sin x + \sin x$

$= 2\sin x$

65. Conjecture: left side is equal to $2\sec x$

$\dfrac{1}{\sec x + \tan x} + \dfrac{1}{\sec x - \tan x} = \dfrac{\sec^2 x - \tan^2 x}{\sec x + \tan x} + \dfrac{\sec^2 x - \tan^2 x}{\sec x - \tan x}$

$= \dfrac{(\sec x + \tan x)(\sec x - \tan x)}{\sec x + \tan x} + \dfrac{(\sec x + \tan x)(\sec x - \tan x)}{\sec x - \tan x}$

$= \sec x - \tan x + \sec x + \tan x$

$= 2\sec x$

67. $\dfrac{\tan x + \cot x}{\csc x} = \dfrac{\dfrac{\sin x}{\cos x} + \dfrac{\cos x}{\sin x}}{\dfrac{1}{\sin x}}$

$\phantom{\dfrac{\tan x + \cot x}{\csc x}} = \left(\dfrac{\sin x}{\cos x} + \dfrac{\cos x}{\sin x} \right) \dfrac{\sin x}{1}$

$\phantom{\dfrac{\tan x + \cot x}{\csc x}} = \dfrac{\sin^2 x}{\cos x} + \dfrac{\sin x \cos x}{\sin x}$

$\phantom{\dfrac{\tan x + \cot x}{\csc x}} = \dfrac{1 - \cos^2 x}{\cos x} + \dfrac{\cos x}{1}$

$\phantom{\dfrac{\tan x + \cot x}{\csc x}} = \dfrac{1 - \cos^2 x}{\cos x} + \dfrac{\cos^2 x}{\cos x}$

$\phantom{\dfrac{\tan x + \cot x}{\csc x}} = \dfrac{1}{\cos x}$

69. $\dfrac{\cos x}{1 + \sin x} + \tan x = \dfrac{\cos x}{1 + \sin x} \cdot \dfrac{\cos x}{\cos x} + \dfrac{\sin x}{\cos x} \cdot \dfrac{1 + \sin x}{1 + \sin x}$

$\phantom{\dfrac{\cos x}{1 + \sin x} + \tan x} = \dfrac{\cos^2 x}{(1 + \sin x)(\cos x)} + \dfrac{\sin x + \sin^2 x}{(1 + \sin x)(\cos x)}$

$\phantom{\dfrac{\cos x}{1 + \sin x} + \tan x} = \dfrac{\cos^2 x + \sin x + \sin^2 x}{(1 + \sin x)(\cos x)}$

$\phantom{\dfrac{\cos x}{1 + \sin x} + \tan x} = \dfrac{\sin x + \cos^2 x + \sin^2 x}{(1 + \sin x)(\cos x)}$

$\phantom{\dfrac{\cos x}{1 + \sin x} + \tan x} = \dfrac{\sin x + 1}{(1 + \sin x)(\cos x)}$

$\phantom{\dfrac{\cos x}{1 + \sin x} + \tan x} = \dfrac{1}{\cos x}$

71. $\dfrac{1}{1 - \cos x} - \dfrac{\cos x}{1 + \cos x} = \dfrac{1}{1 - \cos x} \cdot \dfrac{1 + \cos x}{1 + \cos x} - \dfrac{\cos x}{1 + \cos x} \cdot \dfrac{1 - \cos x}{1 - \cos x}$

$\phantom{\dfrac{1}{1 - \cos x} - \dfrac{\cos x}{1 + \cos x}} = \dfrac{1 + \cos x}{1 - \cos^2 x} - \dfrac{\cos x - \cos^2 x}{1 - \cos^2 x}$

$\phantom{\dfrac{1}{1 - \cos x} - \dfrac{\cos x}{1 + \cos x}} = \dfrac{1 + \cos x - \cos x + \cos^2 x}{1 - \cos^2 x}$

$\phantom{\dfrac{1}{1 - \cos x} - \dfrac{\cos x}{1 + \cos x}} = \dfrac{1 + \cos^2 x}{\sin^2 x}$

$\phantom{\dfrac{1}{1 - \cos x} - \dfrac{\cos x}{1 + \cos x}} = \dfrac{1}{\sin^2 x} + \dfrac{\cos^2 x}{\sin^2 x}$

$\phantom{\dfrac{1}{1 - \cos x} - \dfrac{\cos x}{1 + \cos x}} = \csc^2 x + \cot^2 x$

$\phantom{\dfrac{1}{1 - \cos x} - \dfrac{\cos x}{1 + \cos x}} = \csc^2 x + \csc^2 x - 1$

$\phantom{\dfrac{1}{1 - \cos x} - \dfrac{\cos x}{1 + \cos x}} = 2\csc^2 x - 1$

73. $\dfrac{1}{\csc x - \sin x} = \dfrac{1}{\dfrac{1}{\sin x} - \sin x}$

$= \dfrac{1}{\dfrac{1}{\sin x} - \sin x}$

$= \dfrac{1}{\dfrac{1}{\sin x} - \dfrac{\sin^2 x}{\sin x}}$

$= \dfrac{1}{\dfrac{1 - \sin^2 x}{\sin x}}$

$= \dfrac{1}{\dfrac{\cos^2 x}{\sin x}}$

$= \dfrac{\sin x}{\cos^2 x}$

$= \dfrac{1}{\cos x} \cdot \dfrac{\sin x}{\cos x}$

$= \sec x \tan x$

75. – 77. Answers may vary.

79.

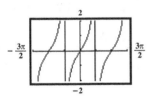

$\sec x(\sin x - \cos x) + 1 = \dfrac{1}{\cos x}(\sin x - \cos x) + 1$

$= \dfrac{\sin x}{\cos x} - \dfrac{\cos x}{\cos x} + 1$

$= \tan x - 1 + 1$

$= \tan x$

81.

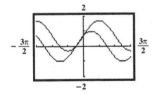

The graphs do not coincide.
Values for *x* may vary.

83.

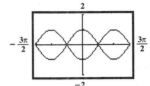

The graphs do not coincide.
Values for *x* may vary.

85.

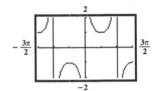

$\dfrac{\sin x}{1 - \cos^2 x} = \dfrac{\sin x}{\sin^2 x} = \dfrac{1}{\sin x} = \csc x$

87.

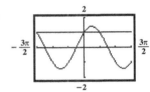

The graphs do not coincide.
Values for *x* may vary.

89. makes sense

91. does not make sense; Explanations will vary.
Sample explanation: The most efficient way to simplify the identity is to multiply out the numerator and then use a Pythagorean identity.

93. $\dfrac{\sin x - \cos x + 1}{\sin x + \cos x - 1}$

$= \dfrac{\sin x - \cos x + 1}{\sin x + \cos x - 1} \cdot \dfrac{\sin x - \cos x - 1}{\sin x - \cos x - 1}$

$= \dfrac{\sin^2 x - 2\cos x \sin x + \cos^2 x - 1}{\sin^2 x - 2\sin x - \cos^2 x + 1}$

$= \dfrac{\sin^2 x + \cos^2 x - 2\cos x \sin x - 1}{\sin x^2 - 2\sin x - (1 - \sin^2 x) + 1}$

$= \dfrac{1 - 2\cos x \sin x - 1}{\sin^2 x - 2\sin x + \sin x^2}$

$= \dfrac{-2\cos x \sin x}{2\sin^2 x - 2\sin x}$

$= \dfrac{-2\sin x \cos x}{2\sin x(\sin x - 1)}$

$= \dfrac{-\cos x}{\sin x - \cos x}$

$= \dfrac{-\cos x}{\sin x - 1} \cdot \dfrac{\sin x + 1}{\sin x + 1}$

$= \dfrac{-\cos x(\sin x + 1)}{\sin^2 x - 1}$

$= \dfrac{-\cos x(\sin x + 1)}{\cos^2 x - 1 - \cos^2 x}$

$= \dfrac{-\cos x(\sin x + 1)}{\cos^2 x}$

$= \dfrac{\sin x + 1}{\cos x}$

95. $\ln e^{\tan^2 x - \sec^2 x}$

$= \tan^2 x - \sec^2 x$

$= -(-\tan^2 x + \sec^2 x)$

$= -(\sec^2 x - \tan^2 x)$

$= -1$

97. Answers may vary.

98. $\cos 30° = \dfrac{\sqrt{3}}{2}$

$\sin 30° = \dfrac{1}{2}$

$\cos 60° = \dfrac{1}{2}$

$\sin 60° = \dfrac{\sqrt{3}}{2}$

$\cos 90° = 0$

$\sin 90° = 1$

99. **a.** No, they are not equal.

$\cos(30° + 60°) \neq \cos 30° + \cos 60°$

$\cos 90° \neq \dfrac{\sqrt{3}}{2} + \dfrac{1}{2}$

$0 \neq \dfrac{1 + \sqrt{3}}{2}$

 b. Yes, they are equal.

$\cos(30° + 60°) = \cos 30° \cos 60° - \sin 30° \sin 60°$

$\cos 90° = \left(\dfrac{\sqrt{3}}{2}\right)\left(\dfrac{1}{2}\right) - \left(\dfrac{1}{2}\right)\left(\dfrac{\sqrt{3}}{2}\right)$

$0 = \dfrac{\sqrt{3}}{4} - \dfrac{\sqrt{3}}{4}$

$0 = 0$

100. **a.** No, they are not equal.

$\sin(30° + 60°) \neq \sin 30° + \sin 60°$

$\sin 90° \neq \dfrac{1}{2} + \dfrac{\sqrt{3}}{2}$

$1 \neq \dfrac{1 + \sqrt{3}}{2}$

 b. Yes, they are equal.

$\sin(30° + 60°) = \sin 30° \cos 60° + \cos 30° \sin 60°$

$\sin 90° = \left(\dfrac{1}{2}\right)\left(\dfrac{1}{2}\right) + \left(\dfrac{\sqrt{3}}{2}\right)\left(\dfrac{\sqrt{3}}{2}\right)$

$1 = \dfrac{1}{4} + \dfrac{3}{4}$

$1 = 1$

Section 6.2

Check Point Exercises

1. $\cos 30° = \cos(90° - 60°)$

       $= \cos 90° \cos 60° + \sin 90° \sin 60°$

       $= 0 \cdot \dfrac{1}{2} + 1 \cdot \dfrac{\sqrt{3}}{2}$

       $= 0 + \dfrac{\sqrt{3}}{2}$

       $= \dfrac{\sqrt{3}}{2}$

2. $\cos 70° \cos 40° + \sin 70° \sin 40°$

       $= \cos(70 - 40°)$

       $= \cos 30°$

       $= \dfrac{\sqrt{3}}{2}$

3. $\dfrac{\cos(\alpha - \beta)}{\cos \alpha \cos \beta} = \dfrac{\cos \alpha \cos \beta + \sin \alpha \sin \beta}{\cos \alpha \cos \beta}$

       $= \dfrac{\cos \alpha}{\cos \alpha} \cdot \dfrac{\cos \beta}{\cos \beta} + \dfrac{\sin \alpha}{\cos \alpha} \cdot \dfrac{\sin \beta}{\cos \beta}$

       $= 1 \cdot 1 + \tan \alpha \cdot \tan \beta$

       $= 1 + \tan \alpha \tan \beta$

   We worked with the left side and arrived at the right side. Thus, the identity is verified.

4. $\sin \dfrac{5\pi}{12} = \sin\left(\dfrac{\pi}{6} + \dfrac{\pi}{4}\right)$

       $= \sin \dfrac{\pi}{6} \cos \dfrac{\pi}{4} + \cos \dfrac{\pi}{6} \sin \dfrac{\pi}{4}$

       $= \dfrac{1}{2} \cdot \dfrac{\sqrt{2}}{2} + \dfrac{\sqrt{3}}{2} \cdot \dfrac{\sqrt{2}}{2}$

       $= \dfrac{\sqrt{2}}{4} + \dfrac{\sqrt{6}}{4}$

       $= \dfrac{\sqrt{2} + \sqrt{6}}{4}$

5. **a.** $\sin \alpha = \dfrac{4}{5} = \dfrac{y}{r}$

       Find x:

       $x^2 + y^2 = r^2$

       $x^2 + 4^2 = 5^2$

       $x^2 + 16 = 25$

       $x^2 = 9$

Because α is in Quadrant II, x is negative.

$x = -\sqrt{9} = -3$

$\cos \alpha = \dfrac{x}{r} = \dfrac{-3}{5} = -\dfrac{3}{5}$

b. $\sin \beta = \dfrac{1}{2} = \dfrac{y}{r}$

       Find x:

       $x^2 + y^2 = r^2$

       $x^2 + 1^2 = 2^2$

       $x^2 + 1 = 4$

       $x^2 = 3$

       Because β is in Quadrant I, x is positive.

       $x = \sqrt{3}$

       $\cos \beta = \dfrac{x}{r} = \dfrac{\sqrt{3}}{2}$

c. $\cos(\alpha + \beta) = \cos \alpha \cos \beta - \sin \alpha \sin \beta$

       $= -\dfrac{3}{5} \cdot \dfrac{\sqrt{3}}{2} - \dfrac{4}{5} \cdot \dfrac{1}{2}$

       $= \dfrac{-3\sqrt{3}}{10} - \dfrac{4}{10}$

       $= \dfrac{-3\sqrt{3} - 4}{10}$

d. $\sin(\alpha + \beta) = \sin \alpha \cos \beta + \cos \alpha \sin \beta$

       $= \dfrac{4}{5} \cdot \dfrac{\sqrt{3}}{2} + \dfrac{-3}{5} \cdot \dfrac{1}{2}$

       $= \dfrac{4\sqrt{3}}{10} + \dfrac{-3}{10}$

       $= \dfrac{4\sqrt{3} - 3}{10}$

6. **a.**   The graph appears to be the sine curve, $y = \sin x$.

       It cycles through intercept, maximum, intercept, minimum and back to intercept. Thus, $y = \sin x$ also describes the graph.

b. $\cos\left(x + \dfrac{3\pi}{2}\right) = \cos x \cos \dfrac{3\pi}{2} - \sin x \sin \dfrac{3\pi}{2}$

                 $= \cos x \cdot 0 - \sin x \cdot (-1)$

                 $= \sin x$

       This verifies our observation that

       $y = \cos\left(x + \dfrac{3\pi}{2}\right)$ and $y = \sin x$ describe the same graph.

7.
$$\tan(x+\pi) = \frac{\tan x + \tan \pi}{1 - \tan x \tan \pi}$$
$$= \frac{\tan x + 0}{1 - \tan x \cdot 0}$$
$$= \frac{\tan x}{1}$$
$$= \tan x$$

Concept and Vocabulary Check 6.2

1. $\cos x \cos y - \sin x \sin y$

2. $\cos x \cos y + \sin x \sin y$

3. $\sin C \cos D + \cos C \sin D$

4. $\sin C \cos D - \cos C \sin D$

5. $\dfrac{\tan \theta + \tan \phi}{1 - \tan \theta \tan \phi}$

6. $\dfrac{\tan \theta - \tan \phi}{1 + \tan \theta \tan \phi}$

7. false

8. false

Exercise Set 6.2

1. $\cos(45° - 30°) = \cos 45° \cos 30° + \sin 45° \sin 30°$
$$= \frac{\sqrt{2}}{2} \cdot \frac{\sqrt{3}}{2} + \frac{\sqrt{2}}{2} \cdot \frac{1}{2}$$
$$= \frac{\sqrt{6}}{4} + \frac{\sqrt{2}}{4}$$
$$= \frac{\sqrt{6} + \sqrt{2}}{4}$$

3. $\cos\left(\dfrac{3\pi}{4} - \dfrac{\pi}{6}\right) = \cos\dfrac{3\pi}{4}\cos\dfrac{\pi}{6} + \sin\dfrac{3\pi}{4}\sin\dfrac{\pi}{6}$
$$= -\frac{\sqrt{2}}{2} \cdot \frac{\sqrt{3}}{2} + \frac{\sqrt{2}}{2} \cdot \frac{1}{2}$$
$$= -\frac{\sqrt{6}}{4} + \frac{\sqrt{2}}{4}$$
$$= \frac{\sqrt{2} - \sqrt{6}}{4}$$

5. **a.** $\cos 50° \cos 20° + \sin 50° \sin 20°$
$$= \cos\alpha\cos\beta + \sin\alpha\sin\beta$$
Thus, $\alpha = 50°$ and $\beta = 20°$.

b. $\cos 50° \cos 20° + \sin 50° \sin 20°$
$$= \cos(50° - 20°)$$
$$= \cos 30°$$

c. $\cos 30° = \dfrac{\sqrt{3}}{2}$

7. **a.** $\cos\dfrac{5\pi}{12}\cos\dfrac{\pi}{12} + \sin\dfrac{5\pi}{12}\sin\dfrac{\pi}{12}$
$$= \cos\alpha\cos\beta + \sin\alpha\sin\beta$$
Thus, $\alpha = \dfrac{5\pi}{12}$ and $\beta = \dfrac{\pi}{12}$.

b. $\cos\dfrac{5\pi}{12}\cos\dfrac{\pi}{12} + \sin\dfrac{5\pi}{12}\sin\dfrac{\pi}{12}$
$$= \cos\left(\frac{5\pi}{12} - \frac{\pi}{12}\right)$$
$$= \cos\frac{4\pi}{12}$$
$$= \cos\frac{\pi}{3}$$

c. $\cos\dfrac{\pi}{3} = \dfrac{1}{2}$

9. $\dfrac{\cos(\alpha - \beta)}{\cos\alpha\sin\beta} = \dfrac{\cos\alpha\cos\beta - \sin\alpha\sin\beta}{\cos\alpha\sin\beta}$
$$= \frac{\cos\alpha}{\cos\alpha} \cdot \frac{\cos\beta}{\sin\beta} - \frac{\sin\alpha}{\cos\alpha} \cdot \frac{\sin\beta}{\sin\beta}$$
$$= 1 \cdot \cot\beta + \tan\alpha \cdot 1$$
$$= \tan\alpha + \cot\beta$$

11. $\cos\left(x - \dfrac{\pi}{4}\right) = \cos x \cos\dfrac{\pi}{4} + \sin x \sin\dfrac{\pi}{4}$
$$= \cos x \cdot \frac{\sqrt{2}}{2} + \sin x \cdot \frac{\sqrt{2}}{2}$$
$$= \frac{\sqrt{2}}{2}(\cos x + \sin x)$$

13. $\sin(45° - 30°) = \sin 45° \cos 30° - \cos 45° \sin 30°$

$$= \frac{\sqrt{2}}{2} \cdot \frac{\sqrt{3}}{2} - \frac{\sqrt{2}}{2} \cdot \frac{1}{2}$$

$$= \frac{\sqrt{6}}{4} - \frac{\sqrt{2}}{4}$$

$$= \frac{\sqrt{6} - \sqrt{2}}{4}$$

15. $\sin(105°) = \sin(60° + 45°)$

$$= \sin 60° \cos 45° + \cos 60° \sin 45°$$

$$= \frac{\sqrt{3}}{2} \cdot \frac{\sqrt{2}}{2} + \frac{1}{2} \cdot \frac{\sqrt{2}}{2}$$

$$= \frac{\sqrt{6}}{4} + \frac{\sqrt{2}}{4}$$

$$= \frac{\sqrt{6} + \sqrt{2}}{4}$$

17. $\cos(135° + 30°) = \cos 135° \cos 30° - \sin 135° \sin 30°$

$$= \cos(90° + 45°) \cos 30° - \sin(90° + 45°) \sin 30°$$

$$= (\cos 90° \cos 45° - \sin 90° \sin 45°) \cos 30° - (\sin 90° \cos 45° + \cos 90° \sin 45°) \sin 30°$$

$$= \left(0 \cdot \frac{\sqrt{2}}{2} - 1 \cdot \frac{\sqrt{2}}{2} \right) \frac{\sqrt{3}}{2} - \left(1 \cdot \frac{\sqrt{2}}{2} + 0 \cdot \frac{\sqrt{2}}{2} \right) \frac{1}{2}$$

$$= \left(-\frac{\sqrt{2}}{2} \right) \frac{\sqrt{3}}{2} - \left(\frac{\sqrt{2}}{2} \right) \frac{1}{2}$$

$$= -\frac{\sqrt{6}}{4} - \frac{\sqrt{2}}{4}$$

$$= -\frac{\sqrt{6} + \sqrt{2}}{4}$$

19. $\cos 75° = \cos(45° + 30°)$

$$= \cos 45° \cos 30° - \sin 45° \sin 30°$$

$$= \frac{\sqrt{2}}{2} \cdot \frac{\sqrt{3}}{2} - \frac{\sqrt{2}}{2} \cdot \frac{1}{2}$$

$$= \frac{\sqrt{6}}{4} - \frac{\sqrt{2}}{4}$$

$$= \frac{\sqrt{6} - \sqrt{2}}{4}$$

21. $\tan\left(\dfrac{\pi}{6}+\dfrac{\pi}{4}\right)=\dfrac{\tan\dfrac{\pi}{6}+\tan\dfrac{\pi}{4}}{1-\tan\dfrac{\pi}{6}\tan\dfrac{\pi}{4}}$

$\qquad=\dfrac{\dfrac{\sqrt{3}}{3}+1}{1-\dfrac{\sqrt{3}}{3}\cdot 1}$

$\qquad=\dfrac{\dfrac{\sqrt{3}}{3}+\dfrac{3}{3}}{\dfrac{3}{3}-\dfrac{\sqrt{3}}{3}}$

$\qquad=\dfrac{\sqrt{3}+3}{3-\sqrt{3}}$

$\qquad=\dfrac{3+\sqrt{3}}{3-\sqrt{3}}\cdot\dfrac{3+\sqrt{3}}{3+\sqrt{3}}$

$\qquad=\dfrac{9+6\sqrt{3}+3}{9-3}$

$\qquad=\dfrac{12+6\sqrt{3}}{6}$

$\qquad=2+\sqrt{3}$

23. $\tan\left(\dfrac{4\pi}{3}-\dfrac{\pi}{4}\right)=\dfrac{\tan\dfrac{4\pi}{3}-\tan\dfrac{\pi}{4}}{1+\tan\dfrac{4\pi}{3}\tan\dfrac{\pi}{4}}$

$\qquad=\dfrac{\sqrt{3}-1}{1+\sqrt{3}\cdot 1}$

$\qquad=\dfrac{-1+\sqrt{3}}{1+\sqrt{3}}$

$\qquad=\dfrac{-1+\sqrt{3}}{1+\sqrt{3}}\cdot\dfrac{1-\sqrt{3}}{1-\sqrt{3}}$

$\qquad=\dfrac{-1+2\sqrt{3}-3}{1-3}$

$\qquad=\dfrac{-4+2\sqrt{3}}{-2}$

$\qquad=2-\sqrt{3}$

25. $\sin 25°\cos 5°+\cos 25°\sin 5°=\sin(25°+5°)$

$\qquad\qquad\qquad\qquad\qquad=\sin 30°$

$\qquad\qquad\qquad\qquad\qquad=\dfrac{1}{2}$

27. $\dfrac{\tan 10°+\tan 35°}{1-\tan 10°\tan 35°}=\tan(10°+35°)$

$\qquad\qquad\qquad\qquad=\tan 45°$

$\qquad\qquad\qquad\qquad=1$

29. $\sin\dfrac{5\pi}{12}\cos\dfrac{\pi}{4}-\cos\dfrac{5\pi}{12}\sin\dfrac{\pi}{4}=\sin\left(\dfrac{5\pi}{12}-\dfrac{\pi}{4}\right)$

$\qquad\qquad\qquad\qquad\qquad=\sin\left(\dfrac{2\pi}{12}\right)$

$\qquad\qquad\qquad\qquad\qquad=\sin\left(\dfrac{\pi}{6}\right)$

$\qquad\qquad\qquad\qquad\qquad=\dfrac{1}{2}$

31. $\dfrac{\tan\frac{\pi}{5}-\tan\frac{\pi}{30}}{1+\tan\frac{\pi}{5}\tan\frac{\pi}{30}}=\tan\left(\dfrac{\pi}{5}-\dfrac{\pi}{30}\right)$

$\qquad\qquad\qquad=\tan\left(\dfrac{5\pi}{30}\right)=\tan\left(\dfrac{\pi}{6}\right)$

$\qquad\qquad\qquad=\dfrac{\sqrt{3}}{3}$

33. $\sin\left(x+\dfrac{\pi}{2}\right)=\sin x\cos\dfrac{\pi}{2}+\cos x\sin\dfrac{\pi}{2}$

$\qquad\qquad\qquad=\sin x\cdot 0+\cos x\cdot 1$

$\qquad\qquad\qquad=\cos x$

35. $\cos\left(x-\dfrac{\pi}{2}\right)=\cos x\cos\dfrac{\pi}{2}+\sin x\sin\dfrac{\pi}{2}$

$\qquad\qquad\qquad=\cos x\cdot 0+\sin x\cdot 1$

$\qquad\qquad\qquad=\sin x$

37. $\tan(2\pi - x) = \dfrac{\tan 2\pi - \tan x}{1 + \tan 2\pi \tan x}$

$= \dfrac{0 - \tan x}{1 + 0 \cdot \tan x}$

$= -\tan x$

39. $\sin(\alpha + \beta) + \sin(\alpha - \beta)$

$= \sin \alpha \cos \beta + \cos \alpha \sin \beta$

$+ \sin \alpha \cos \beta - \cos \alpha \sin \beta$

$= 2 \sin \alpha \cos \beta$

41. $\dfrac{\sin(\alpha - \beta)}{\cos \alpha \cos \beta} = \dfrac{\sin \alpha \cos \beta - \cos \alpha \sin \beta}{\cos \alpha \cos \beta}$

$= \dfrac{\sin \alpha \cos \beta}{\cos \alpha \cos \beta} - \dfrac{\cos \alpha \sin \beta}{\cos \alpha \cos \beta}$

$= \tan \alpha \cdot 1 - 1 \cdot \tan \beta$

$= \tan \alpha - \tan \beta$

43. $\tan\left(\theta + \dfrac{\pi}{4}\right) = \dfrac{\tan \theta + \tan \frac{\pi}{4}}{1 - \tan \theta \tan \frac{\pi}{4}}$

$= \dfrac{\tan \theta + 1}{1 - \tan \theta}$

$= \dfrac{\frac{\sin \theta}{\cos \theta} + \frac{\cos \theta}{\cos \theta}}{\frac{\cos \theta}{\cos \theta} - \frac{\sin \theta}{\cos \theta}}$

$= \dfrac{\frac{\sin \theta + \cos \theta}{\cos \theta}}{\frac{\cos \theta - \sin \theta}{\cos \theta}}$

$= \dfrac{\sin \theta + \cos \theta}{\cos \theta} \cdot \dfrac{\cos \theta}{\cos \theta - \sin \theta}$

$= \dfrac{\sin \theta + \cos \theta}{\cos \theta - \sin \theta}$

$= \dfrac{\cos \theta + \sin \theta}{\cos \theta - \sin \theta}$

45. $\cos(\alpha + \beta)\cos(\alpha - \beta)$

$= (\cos \alpha \cos \beta - \sin \alpha \sin \beta)$

$\cdot (\cos \alpha \cos \beta + \sin \alpha \sin \beta)$

$= \cos^2 \alpha \cos^2 \beta - \sin^2 \alpha \sin^2 \beta$

$= \left(1 - \sin^2 \alpha\right)\cos^2 \beta - \sin^2 \alpha\left(1 - \cos^2 \beta\right)$

$= \cos^2 \beta - \sin^2 \alpha \cos^2 \beta$

$- \sin^2 \alpha + \sin^2 \alpha \cos^2 \beta$

$= \cos^2 \beta - \sin^2 \alpha$

47. $\dfrac{\sin(\alpha + \beta)}{\sin(\alpha - \beta)}$

$= \dfrac{\sin \alpha \cos \beta + \cos \alpha \sin \beta}{\sin \alpha \cos \beta - \cos \alpha \sin \beta}$

$= \dfrac{\sin \alpha \cos \beta + \cos \alpha \sin \beta}{\sin \alpha \cos \beta - \cos \alpha \sin \beta} \cdot \dfrac{\frac{1}{\cos \alpha \cos \beta}}{\frac{1}{\cos \alpha \cos \beta}}$

$= \dfrac{\frac{\sin \alpha \cos \beta + \cos \alpha \sin \beta}{\cos \alpha \cos \beta}}{\frac{\sin \alpha \cos \beta - \cos \alpha \sin \beta}{\cos \alpha \cos \beta}}$

$= \dfrac{\frac{\sin \alpha \cos \beta}{\cos \alpha \cos \beta} + \frac{\cos \alpha \sin \beta}{\cos \alpha \cos \beta}}{\frac{\sin \alpha \cos \beta}{\cos \alpha \cos \beta} - \frac{\cos \alpha \sin \beta}{\cos \alpha \cos \beta}}$

$= \dfrac{\tan \alpha \cdot 1 + 1 \cdot \tan \beta}{\tan \alpha \cdot 1 - 1 \cdot \tan \beta}$

$= \dfrac{\tan \alpha + \tan \beta}{\tan \alpha - \tan \beta}$

49. $\dfrac{\cos(x + h) - \cos x}{h}$

$= \dfrac{\cos x \cos h - \sin x \sin h - \cos x}{h}$

$= \dfrac{\cos x \cos h - \cos x - \sin x \sin h}{h}$

$= \dfrac{\cos x(\cos h - 1) - \sin x \sin h}{h}$

$= \cos x \cdot \dfrac{\cos h - 1}{h} - \sin x \cdot \dfrac{\sin h}{h}$

51. $\sin 2\alpha = \sin(\alpha + \alpha)$

$= \sin \alpha \cos \alpha + \cos \alpha \sin \alpha$

$= 2 \sin \alpha \cos \alpha$

53. $\tan 2\alpha = \tan(\alpha + \alpha)$

$= \dfrac{\tan \alpha + \tan \alpha}{1 - \tan \alpha \tan \alpha}$

$= \dfrac{2 \tan \alpha}{1 - \tan^2 \alpha}$

55. $\tan(\alpha + \beta) = \dfrac{\sin(\alpha + \beta)}{\cos(\alpha + \beta)}$

$= \dfrac{\sin\alpha\cos\beta + \cos\alpha\sin\beta}{\cos\alpha\cos\beta - \sin\alpha\sin\beta}$

$= \dfrac{\sin\alpha\cos\beta + \cos\alpha\sin\beta}{\cos\alpha\cos\beta - \sin\alpha\sin\beta} \cdot \dfrac{\frac{1}{\cos\alpha\cos\beta}}{\frac{1}{\cos\alpha\cos\beta}}$

$= \dfrac{\frac{\sin\alpha\cos\beta + \cos\alpha\sin\beta}{\cos\alpha\cos\beta}}{\frac{\cos\alpha\cos\beta - \sin\alpha\sin\beta}{\cos\alpha\cos\beta}}$

$= \dfrac{\frac{\sin\alpha\cos\beta}{\cos\alpha\cos\beta} + \frac{\cos\alpha\sin\beta}{\cos\alpha\cos\beta}}{\frac{\cos\alpha\cos\beta}{\cos\alpha\cos\beta} - \frac{\sin\alpha\sin\beta}{\cos\alpha\cos\beta}}$

$= \dfrac{\tan\alpha + \tan\beta}{1 - \tan\alpha\tan\beta}$

57. $\sin\alpha = \dfrac{3}{5} = \dfrac{y}{r}$

$x^2 + y^2 = r^2$

$x^2 + 3^2 = 5^2$

$x^2 + 9 = 25$

$x^2 = 16$

Because α lies in quadrant I, x is positive.

$x = 4$

Thus, $\cos\alpha = \dfrac{x}{r} = \dfrac{4}{5}$, and

$\tan\alpha = \dfrac{\sin\alpha}{\cos\alpha} = \dfrac{\frac{3}{5}}{\frac{4}{5}} = \dfrac{3}{4}$.

$\sin\beta = \dfrac{5}{13} = \dfrac{y}{r}$

$x^2 + y^2 = r^2$

$x^2 + 5^2 = 13^2$

$x^2 + 25 = 169$

$x^2 = 144$

Because β lies in quadrant II, x is negative.

$x = -12$

Thus, $\cos\beta = \dfrac{x}{r} = \dfrac{-12}{13} = -\dfrac{12}{13}$, and

$\tan\beta = \dfrac{\sin\beta}{\cos\beta} = \dfrac{\frac{5}{13}}{-\frac{12}{13}} = -\dfrac{5}{12}$.

a. $\cos(\alpha + \beta) = \cos\alpha\cos\beta - \sin\alpha\sin\beta$

$= \dfrac{4}{5} \cdot \left(-\dfrac{12}{13}\right) - \dfrac{3}{5} \cdot \dfrac{5}{13} = -\dfrac{63}{65}$

b. $\sin(\alpha + \beta) = \sin\alpha\cos\beta + \cos\alpha\sin\beta$

$= \dfrac{3}{5} \cdot \left(-\dfrac{12}{13}\right) + \dfrac{4}{5} \cdot \dfrac{5}{13} = -\dfrac{16}{65}$

c. $\tan(\alpha + \beta) = \dfrac{\tan\alpha + \tan\beta}{1 - \tan\alpha\tan\beta}$

$= \dfrac{\frac{3}{4} + \left(-\frac{5}{12}\right)}{1 - \frac{3}{4} \cdot \left(-\frac{5}{12}\right)} = \dfrac{\frac{4}{12}}{\frac{63}{48}} = \dfrac{16}{63}$

59. $\tan\alpha = -\dfrac{3}{4} = \dfrac{3}{-4} = \dfrac{y}{x}$

$x^2 + y^2 = r^2$

$(-4)^2 + 3^2 = r^2$

$16 + 9 = r^2$

$25 = r^2$

Because r is a distance, it is positive.

$r = 5$

Thus, $\cos\alpha = \dfrac{x}{r} = \dfrac{-4}{5} = -\dfrac{4}{5}$, and

$\sin\alpha = \dfrac{y}{r} = \dfrac{3}{5}$.

$\cos\beta = \dfrac{1}{3} = \dfrac{x}{r}$

$x^2 + y^2 = r^2$

$1^2 + y^2 = 3^2$

$1 + y^2 = 9$

$y^2 = 8$

Because β lies in quadrant I, y is positive.

$y = \sqrt{8} = 2\sqrt{2}$

Thus, $\sin\beta = \dfrac{y}{r} = \dfrac{2\sqrt{2}}{3}$, and

$\tan\beta = \dfrac{\sin\beta}{\cos\beta} = \dfrac{\frac{2\sqrt{2}}{3}}{\frac{1}{3}} = 2\sqrt{2}$.

a. $\cos(\alpha + \beta) = \cos\alpha\cos\beta - \sin\alpha\sin\beta$

$$= \left(-\frac{4}{5}\right)\cdot\frac{1}{3} - \frac{3}{5}\cdot\frac{2\sqrt{2}}{3}$$

$$= -\frac{4}{15} - \frac{6\sqrt{2}}{15}$$

$$= \frac{-4 - 6\sqrt{2}}{15}$$

$$= -\frac{4 + 6\sqrt{2}}{15}$$

b. $\sin(\alpha + \beta) = \sin\alpha\cos\beta + \cos\alpha\sin\beta$

$$= \frac{3}{5}\cdot\frac{1}{3} + \left(-\frac{4}{5}\right)\cdot\frac{2\sqrt{2}}{3}$$

$$= \frac{3}{15} - \frac{8\sqrt{2}}{15}$$

$$= \frac{3 - 8\sqrt{2}}{15}$$

c. $\tan(\alpha + \beta) = \dfrac{\tan\alpha + \tan\beta}{1 - \tan\alpha\tan\beta}$

$$= \frac{-\frac{3}{4} + 2\sqrt{2}}{1 - \left(-\frac{3}{4}\right)\left(2\sqrt{2}\right)}$$

$$= \frac{\frac{-3 + 8\sqrt{2}}{4}}{\frac{4 + 6\sqrt{2}}{4}}$$

$$= \frac{-3 + 8\sqrt{2}}{4 + 6\sqrt{2}}\cdot\frac{\left(4 - 6\sqrt{2}\right)}{\left(4 - 6\sqrt{2}\right)}$$

$$= \frac{-108 + 50\sqrt{2}}{-56}$$

$$= \frac{54 - 25\sqrt{2}}{28}$$

61. $\cos\alpha = \dfrac{8}{17} = \dfrac{x}{r}$

$$x^2 + y^2 = r^2$$

$$8^2 + y^2 = 17^2$$

$$64 + y^2 = 289$$

$$y^2 = 225$$

Because α lies in quadrant IV, y is negative.

$$y = -15$$

Thus, $\sin\alpha = \dfrac{y}{r} = \dfrac{-15}{17} = -\dfrac{15}{17}$, and

$$\tan\alpha = \frac{\sin\alpha}{\cos\alpha} = \frac{-\frac{15}{17}}{\frac{8}{17}} = -\frac{15}{8}.$$

$$\sin\beta = -\frac{1}{2} = \frac{-1}{2} = \frac{y}{r}$$

$$x^2 + y^2 = r^2$$

$$x^2 + (-1)^2 = 2^2$$

$$x^2 + 1 = 4$$

$$x^2 = 3$$

Because β lies in quadrant III, x is negative.

$$x = -\sqrt{3}$$

Thus, $\cos\beta = \dfrac{x}{r} = \dfrac{-\sqrt{3}}{2} = -\dfrac{\sqrt{3}}{2}$, and

$$\tan\beta = \frac{\sin\beta}{\cos\beta} = \frac{-\frac{1}{2}}{-\frac{\sqrt{3}}{2}} = \frac{1}{\sqrt{3}} = \frac{\sqrt{3}}{3}.$$

a. $\cos(\alpha + \beta) = \cos\alpha\cos\beta - \sin\alpha\sin\beta$

$$= \frac{8}{17}\cdot\left(-\frac{\sqrt{3}}{2}\right) - \left(-\frac{15}{17}\right)\cdot\left(-\frac{1}{2}\right)$$

$$= \frac{-8\sqrt{3} - 15}{34}$$

$$= -\frac{8\sqrt{3} + 15}{34}$$

b. $\sin(\alpha + \beta) = \sin\alpha\cos\beta + \cos\alpha\sin\beta$

$$= \left(-\frac{15}{17}\right)\cdot\left(-\frac{\sqrt{3}}{2}\right) + \frac{8}{17}\cdot\left(-\frac{1}{2}\right)$$

$$= \frac{15\sqrt{3} - 8}{34}$$

63. $\tan\alpha = \dfrac{3}{4} = \dfrac{y}{x}$

Because α lies in quadrant III, x and y are negative.

$r^2 = x^2 + y^2$

$r^2 = (-4)^2 + (-3)^2$

$r^2 = 25$

$r = 5$

$\sin\alpha = \dfrac{y}{r} = \dfrac{-3}{5} = -\dfrac{3}{5}$

$\cos\alpha = \dfrac{x}{r} = \dfrac{-4}{5} = -\dfrac{4}{5}$

$\cos\beta = \dfrac{1}{4} = \dfrac{x}{r}$

Because β lies in quadrant IV, y is negative.

$x^2 + y^2 = r^2$

$1^2 + y^2 = 4^2$

$y^2 = 15$

$y = -\sqrt{15}$

$\sin\beta = \dfrac{y}{r} = \dfrac{-\sqrt{15}}{4} = -\dfrac{\sqrt{15}}{4}$

$\tan\beta = \dfrac{y}{x} = \dfrac{-\sqrt{15}}{1} = -\sqrt{15}$

a. $\cos(\alpha+\beta) = \cos\alpha\cos\beta - \sin\alpha\sin\beta$

$= -\dfrac{4}{5}\cdot\left(\dfrac{1}{4}\right) - \left(-\dfrac{3}{5}\right)\left(-\dfrac{\sqrt{15}}{4}\right)$

$= -\dfrac{4}{20} - \dfrac{3\sqrt{15}}{20}$

$= -\dfrac{4+3\sqrt{15}}{20}$

b. $\sin(\alpha+\beta) = \sin\alpha\cos\beta + \cos\alpha\sin\beta$

$= \left(-\dfrac{3}{5}\right)\left(\dfrac{1}{4}\right) + \left(-\dfrac{4}{5}\right)\left(-\dfrac{\sqrt{15}}{4}\right)$

$= -\dfrac{3}{20} + \dfrac{4\sqrt{15}}{20}$

$= \dfrac{-3+4\sqrt{15}}{20}$

c. $\tan(\alpha+\beta) = \dfrac{\tan\alpha + \tan\beta}{1 - \tan\alpha\tan\beta}$

$= \dfrac{\dfrac{3}{4} + \left(-\sqrt{15}\right)}{1 - \dfrac{3}{4}\left(-\sqrt{15}\right)}$

$= \dfrac{\dfrac{3}{4} - \dfrac{4\sqrt{15}}{4}}{\dfrac{4}{4} + \dfrac{3\sqrt{15}}{4}}$

$= \dfrac{3 - 4\sqrt{15}}{4 + 3\sqrt{15}}$

$= \dfrac{3 - 4\sqrt{15}}{4 + 3\sqrt{15}}\cdot\dfrac{4 - 3\sqrt{15}}{4 - 3\sqrt{15}}$

$= \dfrac{12 - 9\sqrt{15} - 16\sqrt{15} + 180}{16 - 135}$

$= \dfrac{192 - 25\sqrt{15}}{-119}$

$= \dfrac{-192 + 25\sqrt{15}}{119}$

65. a. The graph appears to be the sine curve, $y = \sin x$. It cycles through intercept, maximum, minimum and back to intercept. Thus, $y = \sin x$ also describes the graph.

b. $\sin(\pi - x) = \sin\pi\cos x - \cos\pi\sin x$

$= 0\cdot\cos x - (-1)\cdot\sin x$

$= \sin x$

This verifies our observation that $y = \sin(\pi - x)$ and $y = \sin x$ describe the same graph.

67. a. The graph appears to be 2 times the cosine curve, $y = 2\cos x$. It cycles through maximum, intercept, minimum, intercept and back to maximum. Thus $y = 2\cos x$ also describes the graph.

b. $\sin\left(x + \dfrac{\pi}{2}\right) + \sin\left(\dfrac{\pi}{2} - x\right)$

$= \sin x\cos\dfrac{\pi}{2} + \cos x\sin\dfrac{\pi}{2} + \sin\dfrac{\pi}{2}\cos x$

$\quad - \cos\dfrac{\pi}{2}\sin x$

$= \sin x\cdot 0 + \cos x\cdot 1 + 1\cdot\cos x - 0\cdot\sin x$

$= \cos x + \cos x = 2\cos x$

This verifies our observation that

$y = \sin\left(x + \dfrac{\pi}{2}\right) + \sin\left(\dfrac{\pi}{2} - x\right)$ and $y = 2\cos x$

describe the same graph.

69. $\cos(\alpha+\beta)\cos\beta+\sin(\alpha+\beta)\sin\beta$

$\quad = \cos\left[(\alpha+\beta)-\beta\right]$

$\quad = \cos\alpha$

71. $\dfrac{\sin(\alpha+\beta)-\sin(\alpha-\beta)}{\cos(\alpha+\beta)+\cos(\alpha-\beta)}$

$= \dfrac{\left(\sin\alpha\cos\beta+\cos\alpha\sin\beta\right)-\left(\sin\alpha\cos\beta-\cos\alpha\sin\beta\right)}{\left(\cos\alpha\cos\beta-\sin\alpha\sin\beta\right)+\left(\cos\alpha\cos\beta+\sin\alpha\sin\beta\right)}$

$= \dfrac{\sin\alpha\cos\beta+\cos\alpha\sin\beta-\sin\alpha\cos\beta+\cos\alpha\sin\beta}{\cos\alpha\cos\beta-\sin\alpha\sin\beta+\cos\alpha\cos\beta+\sin\alpha\sin\beta}$

$= \dfrac{\cos\alpha\sin\beta+\cos\alpha\sin\beta}{\cos\alpha\cos\beta+\cos\alpha\cos\beta}$

$= \dfrac{2\cos\alpha\sin\beta}{2\cos\alpha\cos\beta}$

$= \dfrac{\sin\beta}{\cos\beta}$

$= \tan\beta$

73. $\cos\left(\dfrac{\pi}{6}+\alpha\right)\cos\left(\dfrac{\pi}{6}-\alpha\right)-\sin\left(\dfrac{\pi}{6}+\alpha\right)\sin\left(\dfrac{\pi}{6}-\alpha\right)$

$= \cos\left[\left(\dfrac{\pi}{6}+\alpha\right)+\left(\dfrac{\pi}{6}-\alpha\right)\right]$

$= \cos\left[\dfrac{\pi}{6}+\alpha+\dfrac{\pi}{6}-\alpha\right]$

$= \cos\dfrac{\pi}{3}$

$= \dfrac{1}{2}$

75. Conjecture: the left side is equal to $\cos 3x$.

$\cos 2x\cos 5x+\sin 2x\sin 5x$

$= \cos(2x-5x)$

$= \cos(-3x)$

$= \cos 3x$

77. Conjecture: the left side is equal to $\sin\dfrac{x}{2}$.

$\sin\dfrac{5x}{2}\cos 2x-\cos\dfrac{5x}{2}\sin 2x$

$= \sin\left(\dfrac{5x}{2}-2x\right)$

$= \sin\left(\dfrac{5x}{2}-\dfrac{4x}{2}\right)$

$= \sin\dfrac{x}{2}$

79. $\tan\theta=\dfrac{3}{2}=\dfrac{y}{x}$

$\quad x^2+y^2=r^2$

$\quad 2^2+3^3=r^2$

$\quad 4+9=r^2$

$\quad 13=r^2$

Because r is a distance, it is positive.

$\quad r=\sqrt{13}$

Thus, $\sin\theta=\dfrac{y}{r}=\dfrac{3}{\sqrt{13}}$

and $\cos\theta=\dfrac{x}{r}=\dfrac{2}{\sqrt{13}}$

$\sqrt{13}\cos(t-\theta)=\sqrt{13}(\cos t\cos\theta+\sin t\sin\theta)$

$\quad = \sqrt{13}\left(\cos t\cdot\dfrac{2}{\sqrt{13}}+\sin t\cdot\dfrac{3}{\sqrt{13}}\right)$

$\quad = \cos t\cdot 2+\sin t\cdot 3$

$\quad = 2\cos t+3\sin t$

For the equation $y=\sqrt{13}\cos(t-\theta)$, the amplitude is

$\left|\sqrt{13}\right|=\sqrt{13}$, and the period is $\dfrac{2\pi}{1}=2\pi.$

81. – 87. Answers may vary.

89.

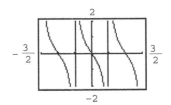

$\tan(\pi-x)=\dfrac{\tan\pi-\tan x}{1+\tan\pi\tan x}$

$\quad\quad = \dfrac{0-\tan x}{1+0\cdot\tan x}$

$\quad\quad = \dfrac{-\tan x}{1}$

$\quad\quad = -\tan x$

91.

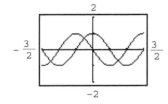

The graphs do not coincide.
Values for *x* may vary.

93.

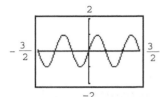

$\sin 1.2x \cos 0.8x + \cos 1.2x \sin 0.8x$

$\sin(1.2x + 0.8x)$

$\sin 2x$

95.   makes sense

97.   makes sense

99. $\cos^{-1}\dfrac{1}{2}$

$x = 1$

$y = \sqrt{3}$

$r = 2$

$\sin^{-1}\dfrac{3}{5}$

$x = 4$

$y = 3$

$r = 5$

$\sin\left(\cos^{-1}\dfrac{1}{2} + \sin^{-1}\dfrac{3}{5}\right) = \sin\cos^{-1}\dfrac{1}{2}\cos\sin^{-1}\dfrac{3}{5} \quad + \cos\cos^{-1}\dfrac{1}{2}\sin\sin^{-1}\dfrac{3}{5}$

$= \dfrac{\sqrt{3}}{2}\cdot\dfrac{4}{5} + \dfrac{1}{2}\cdot\dfrac{3}{5}$

$= \dfrac{4\sqrt{3}+3}{10}$

101. $\tan^{-1}\dfrac{4}{3}$

$x = 3$

$y = 4$

$r = 5$

$\cos^{-1}\dfrac{5}{13}$

$x = 5$

$y = 12$

$r = 13$

$\cos\left(\tan^{-1}\dfrac{4}{3}+\cos^{-1}\dfrac{5}{13}\right)$

$= \cos\tan^{-1}\dfrac{4}{3}\cos\cos^{-1}\dfrac{5}{13}$

$\quad - \sin\tan^{-1}\dfrac{4}{3}\sin\cos^{-1}\dfrac{5}{13}$

$= \dfrac{3}{5}\cdot\dfrac{5}{13}-\dfrac{4}{5}\cdot\dfrac{12}{13}$

$= -\dfrac{33}{65}$

103. Let $\alpha = \sin^{-1}x$, where $-\dfrac{\pi}{2}\le\alpha\le\dfrac{\pi}{2}$. $\sin\alpha = x$

Because x is positive, $\sin\alpha$ is positive. Thus α is in quadrant I. Using a right triangle in quadrant I with $\sin\alpha = x$, the third side can be found using the Pythagorean Theorem.

$a^2 + x^2 = 1^2$

$\quad a^2 = 1 - x^2$

$\quad\quad a = \sqrt{1-x^2}$

Thus $\cos\alpha = \dfrac{\sqrt{1-x^2}}{1}=\sqrt{1-x^2}$

Because y is positive, $\cos\beta$ is positive. Thus β is in quadrant I. Using a right triangle in quadrant I with $\cos\beta = y$, the third side can be found using the Pythagorean Theorem.

$b^2 + y^2 = 1^2$

$\quad b^2 = 1 - x^2$

$\quad\quad a = \sqrt{1-y^2}$

Thus $\cos\alpha = \dfrac{\sqrt{1-y^2}}{1}=\sqrt{1-y^2}$

$\cos(\sin^{-1}x - \cos^{-1}y) = \cos(\alpha - \beta)$

$\quad = \cos\alpha\cos\beta + \sin\alpha\sin\beta$

$\quad = \sqrt{1-x^2}\,y + x\sqrt{1-y^2}$

$\quad = y\sqrt{1-x^2} + x\sqrt{1-y^2}$

105. $\sin^{-1}$

$x = \sqrt{1-x^2}$

$y = x$

$r = 1$

$\cos^{-1}y$

$x = y$

$y = \sqrt{1-y^2}$

$r = 1$

$\tan(\sin^{-1}x + \cos^{-1}y) = \dfrac{\tan\sin^{-1}x + \tan\cos^{-1}y}{1 - \tan\sin^{-1}x\cdot\tan\cos^{-1}y}$

$= \dfrac{\dfrac{x}{\sqrt{1-x^2}}+\dfrac{\sqrt{1-y^2}}{y}}{1 - \dfrac{x}{\sqrt{1-x^2}}\cdot\dfrac{\sqrt{1-y^2}}{y}}$

$= \dfrac{xy+\sqrt{1-y^2}\,\sqrt{1-x^2}}{y\sqrt{1-x^2}-x\sqrt{1-y^2}}$

107. $\sin 30° = \dfrac{1}{2}$

$\cos 30° = \dfrac{\sqrt{3}}{2}$

$\sin 60° = \dfrac{\sqrt{3}}{2}$

$\cos 60° = \dfrac{1}{2}$

108. a. No, they are not equal.
$$\sin(2\cdot30°) \neq 2\sin30°$$
$$\sin60° \neq 2\cdot\frac{1}{2}$$
$$\frac{\sqrt{3}}{2} \neq 1$$

b. Yes, they are equal.
$$\sin(2\cdot30°) = 2\sin30°\cos30°$$
$$\sin60° = 2\cdot\frac{1}{2}\cdot\frac{\sqrt{3}}{2}$$
$$\frac{\sqrt{3}}{2} = \frac{\sqrt{3}}{2}$$

109. a. No, they are not equal.
$$\cos(2\cdot30°) \neq 2\cos30°$$
$$\cos60° \neq 2\cdot\frac{\sqrt{3}}{2}$$
$$\frac{1}{2} \neq \sqrt{3}$$

b. Yes, they are equal.
$$\cos(2\cdot30°) = \cos^2 30° - \sin^2 30°$$
$$\cos60° = \left(\frac{\sqrt{3}}{2}\right)^2 - \left(\frac{1}{2}\right)^2$$
$$\frac{1}{2} = \frac{3}{4} - \frac{1}{4}$$
$$\frac{1}{2} = \frac{1}{2}$$

Section 6.3

Check Point Exercises

1. $\sin\theta = \frac{4}{5} = \frac{y}{r}$

Because θ lies in quadrant II, x is negative.
$$x^2 + y^2 = r^2$$
$$x^2 + 4^2 = 5^2$$
$$x^2 = 5^2 - 4^2 = 9$$
$$x = -\sqrt{9} = -3$$
Now we use values for x, y, and r to find the required values.

a. $\sin2\theta = 2\sin\theta\cos\theta$
$$= 2\left(\frac{4}{5}\right)\left(-\frac{3}{5}\right) = -\frac{24}{25}$$

b. $\cos2\theta = \cos^2\theta - \sin^2\theta$
$$= \left(-\frac{3}{5}\right)^2 - \left(\frac{4}{5}\right)^2 = \frac{9}{25} - \frac{16}{25}$$
$$= -\frac{7}{25}$$

c. $\tan2\theta = \frac{2\tan\theta}{1-\tan^2\theta}$
$$= \frac{2\left(-\frac{4}{3}\right)}{1-\left(-\frac{4}{3}\right)^2} = \frac{-\frac{8}{3}}{1-\frac{16}{9}} = \frac{-\frac{8}{3}}{-\frac{7}{9}}$$
$$= \left(-\frac{8}{3}\right)\left(-\frac{9}{7}\right) = \frac{24}{7}$$

2. The given expression is the right side of the formula for $\cos2\theta$ with $\theta = 15°$.
$$\cos^2 15° - \sin^2 15° = \cos(2\cdot15°) = \cos30° = \frac{\sqrt{3}}{2}$$

3. $\sin3\theta = \sin(2\theta+\theta)$
$$= \sin2\theta\cos\theta + \cos2\theta\sin\theta$$
$$= 2\sin\theta\cos\theta\cos\theta + (2\cos^2\theta-1)\sin\theta$$
$$= 2\sin\theta\cos^2\theta + 2\sin\theta\cos^2\theta - \sin\theta$$
$$= 4\sin\theta\cos^2\theta - \sin\theta$$
$$= 4\sin\theta(1-\sin^2\theta) - \sin\theta$$
$$= 4\sin\theta - 4\sin^3\theta - \sin\theta$$
$$= 3\sin\theta - 4\sin^3\theta$$

By working with the left side and expressing it in a form identical to the right side, we have verified the identity.

4. $\sin^4 x = \left(\sin^2 x\right)^2$

$$= \left(\frac{1-\cos 2x}{2}\right)^2$$

$$= \frac{1 - 2\cos 2x + \cos^2 2x}{4}$$

$$= \frac{1}{4} - \frac{1}{2}\cos 2x + \frac{1}{4}\cos^2 2x$$

$$= \frac{1}{4} - \frac{1}{2}\cos 2x + \frac{1}{4}\left(\frac{1+\cos 2(2x)}{2}\right)$$

$$= \frac{1}{4} - \frac{1}{2}\cos 2x + \frac{1}{8} + \frac{1}{8}\cos 4x$$

$$= \frac{3}{8} - \frac{1}{2}\cos 2x + \frac{1}{8}\cos 4x$$

5.   Because $105°$ lies in quadrant II, $\cos 105° < 0$.

$$\cos 105° = \cos\left(\frac{210°}{2}\right)$$

$$= -\sqrt{\frac{1+\cos 210°}{2}}$$

$$= -\sqrt{\frac{1-\frac{\sqrt{3}}{2}}{2}}$$

$$= -\sqrt{\frac{2-\sqrt{3}}{4}}$$

$$= -\frac{\sqrt{2-\sqrt{3}}}{2}$$

6. $\dfrac{\sin 2\theta}{1+\cos 2\theta} = \dfrac{2\sin\theta\cos\theta}{1+\left(1-2\sin^2\theta\right)}$

$$= \frac{2\sin\theta\cos\theta}{2-2\sin^2\theta}$$

$$= \frac{2\sin\theta\cos\theta}{2\left(1-\sin^2\theta\right)}$$

$$= \frac{2\sin\theta\cos\theta}{2\cos^2\theta}$$

$$= \frac{\sin\theta}{\cos\theta} = \tan\theta$$

The right side simplifies to $\tan\theta$, the expression on the left side. Thus, the identity is verified.

7. $\dfrac{\sec\alpha}{\sec\alpha\csc\alpha+\csc\alpha} = \dfrac{\frac{1}{\cos\alpha}}{\frac{1}{\cos\alpha}\cdot\frac{1}{\sin\alpha}+\frac{1}{\sin\alpha}}$

$$= \frac{\frac{1}{\cos\alpha}}{\frac{1}{\cos\alpha\sin\alpha}+\frac{\cos\alpha}{\cos\alpha\sin\alpha}}$$

$$= \frac{\frac{1}{\cos\alpha}}{\frac{1+\cos\alpha}{\cos\alpha\sin\alpha}}$$

$$= \frac{1}{\cos\alpha}\cdot\frac{\cos\alpha\sin\alpha}{1+\cos\alpha}$$

$$= \frac{\sin\alpha}{1+\cos\alpha}$$

$$= \tan\frac{\alpha}{2}$$

We worked with the right side and arrived at the left side. Thus, the identity is verified.

Concept and Vocabulary Check 6.3

1. $2\sin x\cos x$

2. $\sin^2 A$; $2\cos^2 A$; $2\sin^2 A$

3. $\dfrac{2\tan B}{1-\tan^2 B}$

4. $1-\cos 2\alpha$

5. $1+\cos 2\alpha$

6. $1-\cos 2y$

7. $1-\cos x$

8. $1+\cos y$

9. $1-\cos\alpha$; $1-\cos\alpha$; $1+\cos\alpha$

10. false

11. false

12. false

13. +

14. −

15. +

Exercise Set 6.3

1. $\sin 2\theta = 2\sin\theta\cos\theta = 2\left(\dfrac{3}{5}\right)\left(\dfrac{4}{5}\right) = \dfrac{24}{25}$

3. $\tan 2\theta = \dfrac{2\tan\theta}{1-\tan^2\theta}$

$= \dfrac{2\left(\frac{3}{4}\right)}{1-\left(\frac{3}{4}\right)^2} = \dfrac{\frac{3}{2}}{1-\frac{9}{16}}$

$= \dfrac{\frac{3}{2}}{\frac{7}{16}} = \left(\dfrac{3}{2}\right)\left(\dfrac{16}{7}\right) = \dfrac{24}{7}$

Use this information to solve problems 4, 5, and 6.

$\tan\alpha = \dfrac{7}{24} = \dfrac{y}{x}$

Because r is a distance it is positive.

$x^2 + y^2 = r^2$

$24^2 + 7^2 = r^2$

$576 + 49 = r^2$

$625 = r^2$

$r = 25$

$\sin\alpha = \dfrac{y}{r} = \dfrac{7}{25}$

$\cos\alpha = \dfrac{x}{r} = \dfrac{24}{25}$

5. $\cos 2\alpha = \cos^2\alpha - \sin^2\alpha$

$= \left(\dfrac{24}{25}\right)^2 - \left(\dfrac{7}{25}\right)^2 = \dfrac{576}{625} - \dfrac{49}{625}$

$= \dfrac{527}{625}$

7. $\sin\theta = \dfrac{15}{17} = \dfrac{y}{r}$

Because θ lies in quadrant II, x is negative.

$x^2 + y^2 = r^2$

$x^2 + 15^2 = 17^2$

$x^2 = 17^2 - 15^2 = 64$

$x = -\sqrt{64} = -8$

Now we use values for x, y, and r to find the required values.

a. $\sin 2\theta = 2\sin\theta\cos\theta$

$= 2\left(\dfrac{15}{17}\right)\left(-\dfrac{8}{17}\right) = -\dfrac{240}{289}$

b. $\cos 2\theta = \cos^2\theta - \sin^2\theta$

$= \left(-\dfrac{8}{17}\right)^2 - \left(\dfrac{15}{17}\right)^2 = \dfrac{64}{289} - \dfrac{225}{289}$

$= -\dfrac{161}{289}$

c. $\tan 2\theta = \dfrac{2\tan\theta}{1-\tan^2\theta}$

$= \dfrac{2\left(-\frac{15}{8}\right)}{1-\left(-\frac{15}{8}\right)^2} = \dfrac{-\frac{15}{4}}{1-\frac{225}{64}} = \dfrac{-\frac{15}{4}}{-\frac{161}{64}}$

$= \left(-\dfrac{15}{4}\right)\left(-\dfrac{64}{161}\right) = \dfrac{240}{161}$

9. $\cos\theta = \dfrac{24}{25} = \dfrac{x}{r}$

Because θ lies in quadrant IV, y is negative.

$x^2 + y^2 = r^2$

$24^2 + y^2 = 25^2$

$y^2 = 25^2 - 24^2 = 49$

$y = -\sqrt{49} = -7$

Now we use values for x, y, and r to find the required values.

a. $\sin 2\theta = 2\sin\theta\cos\theta$

$= 2\left(-\dfrac{7}{25}\right)\left(\dfrac{24}{25}\right) = -\dfrac{336}{625}$

b. $\cos 2\theta = \cos^2\theta - \sin^2\theta$

$= \left(\dfrac{24}{25}\right)^2 - \left(-\dfrac{7}{25}\right)^2$

$= \dfrac{576}{625} - \dfrac{49}{625} = \dfrac{527}{625}$

c. $\tan 2\theta = \dfrac{2\tan\theta}{1-\tan^2\theta}$

$= \dfrac{2\left(-\frac{7}{24}\right)}{1-\left(-\frac{7}{24}\right)^2} = \dfrac{-\frac{7}{12}}{1-\frac{49}{576}} = \dfrac{-\frac{7}{12}}{\frac{527}{576}}$

$= \left(-\dfrac{7}{12}\right)\left(\dfrac{576}{527}\right) = -\dfrac{336}{527}$

11. $\cot\theta = 2 = \dfrac{-2}{-1} = \dfrac{x}{y}$

Because r is a distance, it is positive.

$r^2 = x^2 + y^2$

$r^2 = (-2)^2 + (-1)^2$

$r^2 = 5$

$r = \sqrt{5}$

Now we use values for x, y, and r to find the required values.

a. $\sin 2\theta = 2\sin\theta\cos\theta$

$= 2\left(-\dfrac{1}{\sqrt{5}}\right)\left(-\dfrac{2}{\sqrt{5}}\right) = \dfrac{4}{5}$

b. $\cos 2\theta = \cos^2\theta - \sin^2\theta$

$= \left(-\dfrac{2}{\sqrt{5}}\right)^2 - \left(-\dfrac{1}{\sqrt{5}}\right)^2$

$= \dfrac{4}{5} - \dfrac{1}{5} = \dfrac{3}{5}$

c. $\tan 2\theta = \dfrac{2\tan\theta}{1-\tan^2\theta}$

$= \dfrac{2\left(\frac{1}{2}\right)}{1-\left(\frac{1}{2}\right)^2} = \dfrac{1}{1-\frac{1}{4}} = \dfrac{1}{\frac{3}{4}}$

$= (1)\left(\dfrac{4}{3}\right) = \dfrac{4}{3}$

13. $\sin\theta = -\dfrac{9}{41} = \dfrac{-9}{41} = \dfrac{y}{r}$

Because θ lies in quadrant III, x is negative.

$x^2 + y^2 = r^2$

$x^2 + (-9)^2 = 41^2$

$x^2 = 1600$

$x = -\sqrt{1600}$

$x = -40$

Now we use values for x, y, and r to find the required values.

a. $\sin 2\theta = 2\sin\theta\cos\theta$

$= 2\left(-\dfrac{9}{41}\right)\left(-\dfrac{40}{41}\right) = \dfrac{720}{1681}$

b. $\cos 2\theta = \cos^2\theta - \sin^2\theta$

$= \left(-\dfrac{40}{41}\right)^2 - \left(-\dfrac{9}{41}\right)^2$

$= \dfrac{1600}{1681} - \dfrac{81}{1681}$

$= \dfrac{1519}{1681}$

c. $\tan 2\theta = \dfrac{2\tan\theta}{1-\tan^2\theta}$

$= \dfrac{2\left(\frac{9}{40}\right)}{1-\left(\frac{9}{40}\right)^2} = \dfrac{\frac{9}{20}}{1-\frac{81}{1600}} = \dfrac{\frac{9}{20}}{\frac{1519}{1600}}$

$= \left(\dfrac{9}{20}\right)\left(\dfrac{1600}{1519}\right) = \dfrac{720}{1519}$

15. The given expression is the right side of the formula for $\sin 2\theta$ with $\theta = 15°$.

$2\sin 15°\cos 15° = \sin(2\cdot 15°)$

$= \sin 30° = \dfrac{1}{2}$

17. The given expression is the right side of the formula for $\cos 2\theta$ with $\theta = 75°$.

$\cos^2 75° - \sin^2 75° = \cos(2\cdot 75°)$

$= \cos 150° = -\dfrac{\sqrt{3}}{2}$

19. The given expression is the right side of the formula for $\cos 2\theta$ with $\theta = \dfrac{\pi}{8}$.

$2\cos^2\dfrac{\pi}{8} - 1 = \cos\left(2\cdot\dfrac{\pi}{8}\right)$

$= \cos\dfrac{\pi}{4} = \dfrac{\sqrt{2}}{2}$

21. The given expression is the right side of the formula for $\tan 2\theta$ with $\theta = \dfrac{\pi}{12}$.

$\dfrac{2\tan\frac{\pi}{12}}{1-\tan^2\frac{\pi}{12}} = \tan\left(2\cdot\dfrac{\pi}{12}\right) = \tan\dfrac{\pi}{6} = \dfrac{\sqrt{3}}{3}$

23. $\dfrac{2\tan\theta}{1+\tan^2\theta} = \dfrac{2\cdot\frac{\sin\theta}{\cos\theta}}{\frac{\cos^2\theta}{\cos^2\theta}+\frac{\sin^2\theta}{\cos^2\theta}}$

$= \dfrac{\frac{2\sin\theta}{\cos\theta}}{\frac{\cos^2\theta+\sin^2\theta}{\cos^2\theta}}$

$= \dfrac{\frac{2\sin\theta}{\cos\theta}}{\frac{1}{\cos^2\theta}}$

$= \dfrac{2\sin\theta}{\cos\theta}\cdot\dfrac{\cos^2\theta}{1}$

$= 2\sin\theta\cos\theta$

$= \sin 2\theta$

25. $(\sin\theta+\cos\theta)^2 = \sin^2\theta+2\sin\theta\cos\theta+\cos^2\theta$

$= \sin^2\theta+\cos^2\theta+2\sin\theta\cos\theta$

$= 1+2\sin\theta\cos\theta$

$= 1+\sin 2\theta$

27. $\sin^2 x+\cos 2x = \sin^2 x+\cos^2 x-\sin^2 x$

$= \cos^2 x$

29. $\dfrac{\sin 2x}{1-\cos 2x} = \dfrac{2\sin x\cos x}{1-\left(\cos^2 x-\sin^2 x\right)}$

$= \dfrac{2\sin x\cos x}{1-\cos^2 x+\sin^2 x}$

$= \dfrac{2\sin x\cos x}{\sin^2+\sin^2 x}$

$= \dfrac{2\sin x\cos x}{2\sin^2 x}$

$= \dfrac{\cos x}{\sin x}$

$= \cot x$

31. $\tan t\cos 2t = \dfrac{\sin t}{\cos t}\cdot\left(2\cos^2 t-1\right)$

$= \dfrac{2\sin t\cos^2 t}{\cos t}-\dfrac{\sin t}{\cos t}$

$= 2\sin t\cos t-\tan t$

$= \sin 2t-\tan t$

33. $\sin 4t = \sin(2t+2t)$

$= \sin 2t\cos 2t+\cos 2t\sin 2t$

$= \cos 2t(\sin 2t+\sin 2t)$

$= \cos 2t\cdot 2\sin 2t$

$= \left(\cos^2 t-\sin^2 t\right)\cdot 2\cdot 2\sin t\cos t$

$= 4\sin t\cos^3 t-4\sin^3 t\cos t$

35. $6\sin^4 x$

$= 6\left(\dfrac{1-\cos 2x}{2}\right)^2$

$= 6\left(\dfrac{1-2\cos 2x+\cos^2 2x}{4}\right)$

$= \dfrac{6-12\cos 2x+6\cos^2 2x}{4}$

$= \dfrac{3}{4}-3\cos 2x+\dfrac{3}{2}\cos^2 2x$

$= \dfrac{3}{4}-3\cos 2x+\dfrac{3}{2}\left(\dfrac{1+\cos 4x}{2}\right)$

$= \dfrac{3}{4}-3\cos 2x+\dfrac{3}{2}\left(\dfrac{1}{2}+\dfrac{\cos 4x}{2}\right)$

$= \dfrac{3}{4}-3\cos 2x+\dfrac{3}{4}+\dfrac{3}{4}\cos 4x$

$= \dfrac{9}{4}-3\cos 2x+\dfrac{3}{4}\cos 4x$

37. $\sin^2 x\cos^2 x = \left(\dfrac{1-\cos 2x}{2}\right)\left(\dfrac{1+\cos 2x}{2}\right)$

$= \dfrac{1-\cos^2 2x}{4}$

$= \dfrac{1}{4}-\dfrac{1}{4}\cos^2 2x$

$= \dfrac{1}{4}-\dfrac{1}{4}\left(\dfrac{1+\cos(2\cdot 2x)}{2}\right)$

$= \dfrac{1}{4}-\dfrac{1}{8}(1+\cos 4x)$

$= \dfrac{1}{4}-\dfrac{1}{8}-\dfrac{1}{8}\cos 4x$

$= \dfrac{1}{8}-\dfrac{1}{8}\cos 4x$

39. Because $15°$ lies in quadrant I, $\sin 15° > 0$.

$$\sin 15° = \sin \frac{30°}{2}$$

$$= \sqrt{\frac{1 - \cos 30°}{2}} = \sqrt{\frac{1 - \frac{\sqrt{3}}{2}}{2}}$$

$$= \sqrt{\frac{2 - \sqrt{3}}{4}} = \frac{\sqrt{2 - \sqrt{3}}}{2}$$

41. Because $157.5°$ lies in quadrant II, $\cos 157.5° < 0$.

$$\cos 157.5° = \cos \frac{315°}{2} = -\sqrt{\frac{1 + \cos 315°}{2}}$$

$$= -\sqrt{\frac{1 + \frac{\sqrt{2}}{2}}{2}} = -\sqrt{\frac{2 + \sqrt{2}}{4}}$$

$$= -\frac{\sqrt{2 + \sqrt{2}}}{2}$$

43. Because $75°$ lies in quadrant I, $\tan 75° > 0$.

$$\tan 75° = \tan \frac{150°}{2} = \frac{1 - \cos 150°}{\sin 150°}$$

$$= \frac{1 - \left(-\frac{\sqrt{3}}{2}\right)}{\frac{1}{2}} = 2 + \sqrt{3}$$

45. Because $\frac{7\pi}{8}$ lies in quadrant II, $\tan \frac{7\pi}{8} < 0$.

$$\tan \frac{7\pi}{8} = \tan \left(\frac{\frac{7\pi}{4}}{2}\right) = \frac{1 - \cos \frac{7\pi}{4}}{\sin \frac{7\pi}{4}}$$

$$= \frac{1 - \frac{\sqrt{2}}{2}}{-\frac{\sqrt{2}}{2}} = -\frac{2}{\sqrt{2}} + 1$$

$$= -\sqrt{2} + 1$$

47. $\sin \frac{\theta}{2} = \sqrt{\frac{1 - \cos \theta}{2}}$

$$= \sqrt{\frac{1 - \frac{4}{5}}{2}} = \sqrt{\frac{1}{10}}$$

$$= \frac{1}{\sqrt{10}} = \frac{\sqrt{10}}{10}$$

49. $\tan \frac{\theta}{2} = \frac{1 - \cos \theta}{\sin \theta}$

$$= \frac{1 - \frac{4}{5}}{\frac{3}{5}}$$

$$= \frac{1}{3}$$

Use this information to solve problems 50, 51, 52 and 54.

$$\tan \alpha = \frac{7}{24} = \frac{y}{x}$$

Because r is a distance, it is positive.

$$r^2 = x^2 + y^2$$

$$r^2 = 24^2 + 7^2$$

$$r^2 = 625$$

$$r = 25$$

$$\sin \alpha = \frac{y}{r} = \frac{7}{25}$$

$$\cos \alpha = \frac{x}{r} = \frac{24}{25}$$

51. $\cos \frac{\alpha}{2} = \sqrt{\frac{1 + \cos \alpha}{2}} = \sqrt{\frac{1 + \frac{24}{25}}{2}} = \sqrt{\frac{49}{50}}$

$$= \frac{7}{5\sqrt{2}} = \frac{7\sqrt{2}}{10}$$

53. $2 \sin \frac{\theta}{2} \cos \frac{\theta}{2} = 2 \cdot \sqrt{\frac{1 - \cos \theta}{2}} \cdot \sqrt{\frac{1 + \cos \theta}{2}}$

$$= 2 \sqrt{\frac{1 - \frac{4}{5}}{2}} \cdot \sqrt{\frac{1 + \frac{4}{5}}{2}}$$

$$= 2 \cdot \sqrt{\frac{1}{10}} \cdot \sqrt{\frac{9}{10}}$$

$$= 2 \cdot \frac{1}{\sqrt{10}} \cdot \frac{3}{\sqrt{10}}$$

$$= \frac{6}{10} = \frac{3}{5}$$

55. $\tan \alpha = \frac{4}{3} = \frac{-4}{-3} = \frac{y}{x}$

Because r is a distance, it is positive.

$$r^2 = x^2 + y^2$$

$$r^2 = (-4)^2 + (-3)^2$$

$$r^2 = 25$$

$$r = 5$$

Since $180° < \alpha < 270°$, then $90° < \frac{\alpha}{2} < 135°$.

Therefore $\dfrac{\alpha}{2}$ lies in quadrant II.

Thus, $\sin\dfrac{\alpha}{2} > 0$, $\cos\dfrac{\alpha}{2} < 0$, and $\tan\dfrac{\alpha}{2} < 0$.

a. $\sin\dfrac{\alpha}{2} = \sqrt{\dfrac{1-\cos\alpha}{2}} = \sqrt{\dfrac{1-\left(-\dfrac{3}{5}\right)}{2}}$

$= \sqrt{\dfrac{\dfrac{8}{5}}{2}} = \sqrt{\dfrac{4}{5}} = \dfrac{2}{\sqrt{5}} = \dfrac{2\sqrt{5}}{5}$

b. $\cos\dfrac{\alpha}{2} = -\sqrt{\dfrac{1+\cos\alpha}{2}} = -\sqrt{\dfrac{1+\left(-\dfrac{3}{5}\right)}{2}}$

$= -\sqrt{\dfrac{\dfrac{2}{5}}{2}} = -\sqrt{\dfrac{1}{5}} = -\dfrac{1}{\sqrt{5}} = -\dfrac{\sqrt{5}}{5}$

c. $\tan\dfrac{\alpha}{2} = \dfrac{1-\cos\alpha}{\sin\alpha} = \dfrac{1-\left(-\dfrac{3}{5}\right)}{-\dfrac{4}{5}}$

$= \dfrac{\dfrac{8}{5}}{-\dfrac{4}{5}} = \dfrac{8}{-4} = -2$

57. $\sec\alpha = -\dfrac{13}{5} = \dfrac{13}{-5} = \dfrac{r}{x}$

Because α lies in quadrant II, y is positive.

$x^2 + y^2 = r^2$

$(-5)^2 + y^2 = (13)^2$

$y^2 = 144$

$y = 12$

Since $\dfrac{\pi}{2} < \alpha < \pi$, then $\dfrac{\pi}{4} < \dfrac{\alpha}{2} < \dfrac{\pi}{2}$. Therefore $\dfrac{\alpha}{2}$ lies in quadrant I.

Thus, $\sin\dfrac{\alpha}{2} > 0$, $\cos\dfrac{\alpha}{2} > 0$, and $\tan\dfrac{\alpha}{2} > 0$.

a. $\sin\dfrac{\alpha}{2} = \sqrt{\dfrac{1-\cos\alpha}{2}} = \sqrt{\dfrac{1-\left(-\dfrac{5}{13}\right)}{2}}$

$= \sqrt{\dfrac{18}{26}} = \sqrt{\dfrac{9}{13}} = \dfrac{3}{\sqrt{13}}$

$= \dfrac{3\sqrt{13}}{13}$

b. $\cos\dfrac{\alpha}{2} = \sqrt{\dfrac{1+\cos\alpha}{2}} = \sqrt{\dfrac{1+\left(-\dfrac{5}{13}\right)}{2}}$

$= \sqrt{\dfrac{8}{26}} = \sqrt{\dfrac{4}{13}} = \dfrac{2}{\sqrt{13}}$

$= \dfrac{2\sqrt{13}}{13}$

c. $\tan\dfrac{\alpha}{2} = \dfrac{1-\cos\alpha}{\sin\alpha} = \dfrac{1-\left(-\dfrac{5}{13}\right)}{\dfrac{12}{13}}$

$= \dfrac{13+5}{12} = \dfrac{18}{12} = \dfrac{3}{2}$

59. $\sin^2\dfrac{\theta}{2} = \dfrac{1-\cos 2\left(\dfrac{\theta}{2}\right)}{2}$

$= \dfrac{1-\cos\theta}{2} \cdot \dfrac{\dfrac{1}{\cos\theta}}{\dfrac{1}{\cos\theta}}$

$= \dfrac{\dfrac{1-\cos\theta}{\cos\theta}}{\dfrac{2}{\cos\theta}}$

$= \dfrac{\dfrac{1}{\cos\theta}-\dfrac{\cos\theta}{\cos\theta}}{2\cdot\dfrac{1}{\cos\theta}}$

$= \dfrac{\sec\theta-1}{2\sec\theta}$

61. $\cos^2\dfrac{\theta}{2} = \dfrac{1+\cos 2\left(\dfrac{\theta}{2}\right)}{2}$

$= \dfrac{1+\cos\theta}{2}$

$= \dfrac{1+\cos\theta}{2} \cdot \dfrac{\dfrac{\sin\theta}{\cos\theta}}{\dfrac{\sin\theta}{\cos\theta}}$

$= \dfrac{\dfrac{\sin\theta}{\cos\theta}+\sin\theta}{2\cdot\dfrac{\sin\theta}{\cos\theta}}$

$= \dfrac{\tan\theta+\sin\theta}{2\tan\theta}$

$= \dfrac{\sin\theta+\tan\theta}{2\tan\theta}$

63. $\tan\dfrac{\alpha}{2} = \dfrac{\sin\alpha}{1+\cos\alpha}$

$\qquad = \dfrac{\sin\alpha}{1+\cos\alpha} \cdot \dfrac{\frac{1}{\cos\alpha}}{\frac{1}{\cos\alpha}}$

$\qquad = \dfrac{\frac{\sin\alpha}{\cos\alpha}}{\frac{1+\cos\alpha}{\cos\alpha}}$

$\qquad = \dfrac{\tan\alpha}{\frac{1}{\cos\alpha}+\frac{\cos\alpha}{\cos\alpha}}$

$\qquad = \dfrac{\tan\alpha}{\sec\alpha+1}$

65. $\dfrac{\sin x}{1-\cos x} = \dfrac{\sin x}{1-\cos x} \cdot \dfrac{\frac{1}{\sin x}}{\frac{1}{\sin x}}$

$\qquad = \dfrac{\frac{\sin x}{\sin x}}{\frac{1-\cos x}{\sin x}}$

$\qquad = \dfrac{1}{\tan\frac{x}{2}}$

$\qquad = \cot\dfrac{x}{2}$

67. $\tan\dfrac{x}{2}+\cot\dfrac{x}{2} = \dfrac{1-\cos x}{\sin x}+\dfrac{1}{\tan\frac{x}{2}}$

$\qquad = \dfrac{1-\cos x}{\sin x}+\dfrac{1}{\frac{\sin x}{1+\cos x}}$

$\qquad = \dfrac{1-\cos x}{\sin x}+\dfrac{1+\cos x}{\sin x}$

$\qquad = \dfrac{1-\cos x+1+\cos x}{\sin x}$

$\qquad = \dfrac{2}{\sin x} = 2\csc x$

69. Conjecture: The left side is equal to $\cos 2x$.

$\dfrac{\cot x - \tan x}{\cot x + \tan x} = \dfrac{\frac{\cos x}{\sin x}-\frac{\sin x}{\cos x}}{\frac{\cos x}{\sin x}+\frac{\sin x}{\cos x}}$

$\qquad = \dfrac{\frac{\cos x}{\sin x}\cdot\frac{\cos x}{\cos x}-\frac{\sin x}{\cos x}\cdot\frac{\sin x}{\sin x}}{\frac{\cos x}{\sin x}\cdot\frac{\cos x}{\cos x}+\frac{\sin x}{\cos x}\cdot\frac{\sin x}{\sin x}}$

$\qquad = \dfrac{\frac{\cos^2 x}{\sin x\cos x}-\frac{\sin^2 x}{\sin x\cos x}}{\frac{\cos^2 x}{\sin x\cos x}+\frac{\sin^2 x}{\sin x\cos x}}$

$\qquad = \dfrac{\cos^2 x-\sin^2 x}{\cos^2 x+\sin^2 x}$

$\qquad = \dfrac{\cos 2x}{1}$

$\qquad = \cos 2x$

71. Conjecture: The left side is equal to $\sin x + 1$.

$\left(\sin\dfrac{x}{2}+\cos\dfrac{x}{2}\right)^2 = \sin^2\dfrac{x}{2}+2\sin\dfrac{x}{2}\cos\dfrac{x}{2}+\cos^2\dfrac{x}{2}$

$\qquad = 2\sin\dfrac{x}{2}\cos\dfrac{x}{2}+\sin^2\dfrac{x}{2}+\cos^2\dfrac{x}{2}$

$\qquad = \left[2\sin\dfrac{x}{2}\cos\dfrac{x}{2}\right]+\left[\sin^2\dfrac{x}{2}+\cos^2\dfrac{x}{2}\right]$

$\qquad = \sin\left(2\cdot\dfrac{x}{2}\right)+1$

$\qquad = \sin x + 1$

73. Conjecture: The left side is equal to $\sec x$.

$\dfrac{\sin 2x}{\sin x}-\dfrac{\cos 2x}{\cos x} = \dfrac{2\sin x\cos x}{\sin x}-\dfrac{2\cos^2 x-1}{\cos x}$

$\qquad = 2\cos x-\dfrac{2\cos^2 x}{\cos x}+\dfrac{1}{\cos x}$

$\qquad = 2\cos x-2\cos x+\sec x$

$\qquad = \sec x$

75. Conjecture: The left side is equal to $2\csc 2x$.

$$\frac{\csc^2 x}{\cot x} = \frac{\dfrac{1}{\sin^2 x}}{\dfrac{\cos x}{\sin x}}$$

$$= \frac{1}{\sin^2 x} \cdot \frac{\sin x}{\cos x}$$

$$= \frac{1}{\sin x \cos x}$$

$$= \frac{2}{2\sin x \cos x}$$

$$= \frac{2}{\sin 2x}$$

$$= 2\csc 2x$$

77. Conjecture: The left side is equal to $\sin 3x$.

$$\sin x \left(4\cos^2 x - 1\right) = \sin x \left(2\cos^2 x + 2\cos^2 x - 1\right)$$

$$= \sin x \left(2\cos^2 x + \cos 2x\right)$$

$$= 2\sin x \cos^2 x + \sin x \cos 2x$$

$$= 2\sin x \cos x \cos x + \sin x \cos 2x$$

$$= \sin 2x \cos x + \sin x \cos 2x$$

$$= \sin(2x + x)$$

$$= \sin 3x$$

79. **a.** $d = \dfrac{v_o^2}{16}\sin\theta\cos\theta$

$$= \frac{v_o^2}{32} \cdot 2\sin\theta\cos\theta$$

$$= \frac{v_o^2}{32} \cdot \sin 2\theta$$

b. $\sin\alpha$ is at a maximum in the interval $[0, 2\pi]$

when $\alpha = \dfrac{\pi}{2}$, so $\sin 2\theta$ is at a maximum when

$2\theta = \dfrac{\pi}{2}$ or $\theta = \dfrac{\pi}{4}$.

81. $\theta = \dfrac{\pi}{4}$

$$\sin\frac{\theta}{2} = \sqrt{\frac{1 - \cos\theta}{2}}$$

$$= \sqrt{\frac{1 - \cos\frac{\pi}{4}}{2}}$$

$$= \sqrt{\frac{1 - \frac{\sqrt{2}}{2}}{2}}$$

$$= \sqrt{\frac{2 - \sqrt{2}}{4}}$$

$$= \frac{\sqrt{2 - \sqrt{2}}}{2}$$

$$\sin\frac{\theta}{2} = \frac{1}{M}$$

$$\frac{\sqrt{2 - \sqrt{2}}}{2} = \frac{1}{M}$$

$$M = \frac{2}{\sqrt{2 - \sqrt{2}}}$$

$$= \frac{2\sqrt{2 - \sqrt{2}}}{2 - \sqrt{2}}$$

$$= \frac{2\sqrt{2 - \sqrt{2}}}{2 - \sqrt{2}} \cdot \frac{2 + \sqrt{2}}{2 + \sqrt{2}}$$

$$= \frac{4\sqrt{2 - \sqrt{2}} + 2\sqrt{2}\sqrt{2 - \sqrt{2}}}{4 - 2}$$

$$= \frac{2\left(2\sqrt{2 - \sqrt{2}} + \sqrt{2}\sqrt{2 - \sqrt{2}}\right)}{2}$$

$$= 2\sqrt{2 - \sqrt{2}} + \sqrt{2}\sqrt{2 - \sqrt{2}}$$

$$= \sqrt{2 - \sqrt{2}} \cdot \left(2 + \sqrt{2}\right) \approx 2.6$$

83. – 93. Answers may vary.

95.

$$3 - 6\sin^2 x = 3 - 6\left(\frac{1 - \cos 2x}{2}\right)$$

$$= 3 - 3(1 - \cos 2x)$$

$$= 3 - 3 + 3\cos 2x$$

$$= 3\cos 2x$$

97.

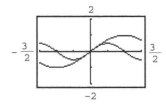

The graphs do not coincide.
Values for x may vary.

99.

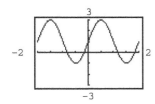

a. The graph appears to be the sum of 1 and 2 times the sine curve, $y = 1 + 2\sin x$. If you subtract 1 from the graph, it cycles through intercept, maximum, intercept, minimum, and back to intercept. Thus, $y = 1 + 2\sin x$ also describes the graph.

b.
$$\frac{1 - 2\cos 2x}{2\sin x - 1}$$
$$= \frac{1 - 2\left(1 - 2\sin^2 x\right)}{2\sin x - 1} = \frac{1 - 2 + 4\sin^2 x}{2\sin x - 1}$$
$$= \frac{4\sin^2 x - 1}{2\sin x - 1} = \frac{(2\sin x - 1)(2\sin x + 1)}{2\sin x - 1}$$
$$= 2\sin x + 1 = 1 + 2\sin x$$

This verifies our observation that
$$y = \frac{1 - 2\cos 2x}{2\sin x - 1} \text{ and } y = 1 + 2\sin x \text{ describe}$$
the same graph.

101.

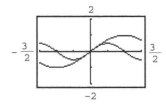

a. The graph appears to be the tangent of half the angle. It cycles from negative infinity through intercept to positive infinity. Thus, $y = \tan\dfrac{x}{2}$ also describes the graph.

b. $\tan\dfrac{x}{2} = \dfrac{1 - \cos x}{\sin x} = \dfrac{1}{\sin x} - \dfrac{\cos x}{\sin x} = \csc x - \cot x$
This verifies our observation that
$$y = \csc x - \cot x \text{ and } y = \tan\frac{x}{2} \text{ describe the}$$
same graph.

103. does not make sense; Explanations will vary. Sample explanation: That procedure is not algebraically sound.

105. does not make sense; Explanations will vary. Sample explanation: That method will not work well because $200°$ is not an angle with known trigonometric values.

107. $\sin\left(2\sin^{-1}\dfrac{\sqrt{3}}{2}\right) = \sin\left(2 \cdot \dfrac{\pi}{3}\right)$
$$= \sin\frac{2\pi}{3}$$
$$= \frac{\sqrt{3}}{2}$$

109. $\cos^2\left[\dfrac{1}{2}\sin^{-1}\dfrac{3}{5}\right]$

Let $\theta = \sin^{-1}\dfrac{3}{5}$, then $\sin\theta = \dfrac{y}{r} = \dfrac{3}{5}$
Since θ is in quadrant I, x is positive.
$$x^2 + y^2 = r^2$$
$$x^2 + 3^2 = 5^2$$
$$x^2 = 16$$
$$x = 4$$
$$\cos^2\left[\frac{1}{2}\sin^{-1}\frac{3}{5}\right] = \cos^2\left[\frac{1}{2}\theta\right]$$
$$= \frac{1 + \cos\left(2 \cdot \frac{1}{2}\theta\right)}{2}$$
$$= \frac{1 + \cos\theta}{2}$$
$$= \frac{1 + \frac{4}{5}}{2}$$
$$= \frac{1 + \frac{x}{r}}{2}$$
$$= \frac{9}{10}$$

111. Let $\alpha = \sin^{-1} x$ where $-\dfrac{\pi}{2} \le \alpha \le \dfrac{\pi}{2}$.

$\sin \alpha = x$

Because x is positive, $\sin \alpha$ is positive. Thus, α is in quadrant I. Using a right triangle in quadrant I with

$\sin \alpha = x = \dfrac{x}{1}$ the third side a can be found using the Pythagorean Theorem.

$a^2 + x^2 = 1^2$

$a^2 = 1 - x^2$

$a = \sqrt{1 - x^2}$

$\cos \alpha = \dfrac{\sqrt{1-x^2}}{1} = \sqrt{1-x^2}$

$\sin(2\sin^{-1} x) = \sin 2\alpha = 2x\sqrt{1-x^2}$

113. $\sin 60° \sin 30° = \dfrac{1}{2}\Big[\cos(60° - 30°) - \cos(60° + 30°) \Big]$

$\quad\quad \dfrac{\sqrt{3}}{2} \cdot \dfrac{1}{2} = \dfrac{1}{2}\Big[\cos 30° - \cos 90° \Big]$

$\quad\quad\quad \dfrac{\sqrt{3}}{4} = \dfrac{1}{2}\Big[\dfrac{\sqrt{3}}{2} - 0 \Big]$

$\quad\quad\quad \dfrac{\sqrt{3}}{4} = \dfrac{\sqrt{3}}{4}$

114. $\cos\dfrac{\pi}{2}\cos\dfrac{\pi}{3} = \dfrac{1}{2}\Big[\cos\Big(\dfrac{\pi}{2} - \dfrac{\pi}{3}\Big) + \cos\Big(\dfrac{\pi}{2} + \dfrac{\pi}{3}\Big) \Big]$

$\quad\quad 0 \cdot \dfrac{1}{2} = \dfrac{1}{2}\Big[\cos\Big(\dfrac{\pi}{6}\Big) + \cos\Big(\dfrac{5\pi}{6}\Big) \Big]$

$\quad\quad\quad 0 = \dfrac{1}{2}\Big[\dfrac{\sqrt{3}}{2} - \dfrac{\sqrt{3}}{2} \Big]$

$\quad\quad\quad 0 = \dfrac{1}{2}[0]$

$\quad\quad\quad 0 = 0$

115. $\sin \pi \cos\dfrac{\pi}{2} = \dfrac{1}{2}\Big[\sin\Big(\pi + \dfrac{\pi}{2}\Big) + \sin\Big(\pi - \dfrac{\pi}{2}\Big) \Big]$

$\quad\quad 0 \cdot 0 = \dfrac{1}{2}\Big[\sin\Big(\dfrac{3\pi}{2}\Big) + \sin\Big(\dfrac{\pi}{2}\Big) \Big]$

$\quad\quad 0 \cdot 0 = \dfrac{1}{2}[-1 + 1]$

$\quad\quad\quad 0 = \dfrac{1}{2}[0]$

$\quad\quad\quad 0 = 0$

Mid-Chapter 6 Check Point

1. $\cos x(\tan x + \cot x)$

$= \cos x \left(\dfrac{\sin x}{\cos x} + \dfrac{\cos x}{\sin x} \right)$

$= \cos x \left(\dfrac{\sin x}{\cos x} \cdot \dfrac{\sin x}{\sin x} + \dfrac{\cos x}{\sin x} \cdot \dfrac{\cos x}{\cos x} \right)$

$= \cos x \left(\dfrac{\sin^2 x}{\sin x \cos x} + \dfrac{\cos^2 x}{\sin x \cos x} \right)$

$= \cos x \left(\dfrac{\sin^2 x + \cos^2 x}{\sin x \cos x} \right)$

$= \cos x \left(\dfrac{1}{\sin x \cos x} \right)$

$= \dfrac{\cos x}{\sin x \cos x}$

$= \dfrac{1}{\sin x}$

$= \csc x$

2. $\dfrac{\sin(x + \pi)}{\cos\left(x + \dfrac{3\pi}{2} \right)} = \dfrac{-\sin x}{\cos\left((x+\pi) + \dfrac{\pi}{2} \right)}$

$= \dfrac{-\sin x}{-\sin(x + \pi)}$

$= \dfrac{-\sin x}{\sin x}$

$= -1$

$= -\left(\sec^2 x - \tan^2 x \right)$

$= \tan^2 x - \sec^2 x$

3. $(\sin\theta + \cos\theta)^2 + (\sin\theta - \cos\theta)^2$

$= \sin^2\theta + 2\sin\theta\cos\theta + \cos^2\theta$

$\quad + \sin^2\theta - 2\sin\theta\cos\theta + \cos^2\theta$

$= \sin^2\theta + \cos^2\theta + \sin^2\theta + \cos^2\theta$

$= 1 + 1$

$= 2$

4. $\dfrac{\sin t - 1}{\cos t} = \dfrac{\sin t - 1}{\cos t} \cdot \dfrac{\cot t}{\cot t}$

$= \dfrac{\sin t \cot t - \cot t}{\cos t \cot t}$

$= \dfrac{\sin t \cdot \dfrac{\cos t}{\sin t} - \cot t}{\cos t \cot t}$

$= \dfrac{\cos t - \cot t}{\cos t \cot t}$

5. $\dfrac{1 - \cos 2x}{\sin 2x} = \dfrac{1 - 2\cos^2 x - 1}{2\sin x \cos x}$

$= \dfrac{2\cos^2 x}{2\sin x \cos x}$

$= \dfrac{\cos x}{\sin x}$

$= \tan x$

6. $\sin\theta\cos\theta + \cos^2\theta$

$= \cos\theta(\sin\theta + \cos\theta)$

$= \cos\theta(\sin\theta + \cos\theta) \cdot \dfrac{\csc\theta}{\csc\theta}$

$= \dfrac{\cos\theta(\sin\theta\csc\theta + \cos\theta\csc\theta)}{\csc\theta}$

$= \dfrac{\cos\theta\left(\sin\theta \cdot \dfrac{1}{\sin\theta} + \cos\theta \cdot \dfrac{1}{\sin\theta} \right)}{\csc\theta}$

$= \dfrac{\cos\theta(1 + \tan\theta)}{\csc\theta}$

7. $\dfrac{\sin x}{\tan x} + \dfrac{\cos x}{\cot x} = \dfrac{\sin x}{\dfrac{\sin x}{\cos x}} + \dfrac{\cos x}{\dfrac{\cos x}{\sin x}}$

$= \sin x \cdot \dfrac{\cos x}{\sin x} + \cos x \cdot \dfrac{\sin x}{\cos x}$

$= \cos x + \sin x$

$= \sin x + \cos x$

8. $\sin^2\dfrac{t}{2} = \left(\sin\dfrac{t}{2}\right)^2$

$\quad = \left(\pm\sqrt{\dfrac{1-\cos t}{2}}\right)^2$

$\quad = \dfrac{1-\cos t}{2}$

$\quad = \dfrac{1-\cos t}{2}\cdot\dfrac{\tan t}{\tan t}$

$\quad = \dfrac{\tan t - \cos t\tan t}{2\tan t}$

$\quad = \dfrac{\tan t - \cos t\cdot\dfrac{\sin t}{\cos t}}{2\tan t}$

$\quad = \dfrac{\tan t - \sin t}{2\tan t}$

9. $\dfrac{1}{2}\big[\sin(\alpha+\beta)+\sin(\alpha-\beta)\big]$

$\quad = \dfrac{1}{2}\big[\sin\alpha\cos\beta+\cos\alpha\sin\beta+$

$\qquad\qquad \sin\alpha\cos\beta-\cos\alpha\sin\beta\big]$

$\quad = \dfrac{1}{2}\big[2\sin\alpha\cos\beta\big]$

$\quad = \sin\alpha\cos\beta$

10. $\dfrac{1+\csc x}{\sec x}-\cot x = \dfrac{1+\dfrac{1}{\sin x}}{\dfrac{1}{\cos x}}-\dfrac{\cos x}{\sin x}$

$\qquad\qquad = \cos x\left(1+\dfrac{1}{\sin x}\right)-\dfrac{\cos x}{\sin x}$

$\qquad\qquad = \cos x+\dfrac{\cos x}{\sin x}-\dfrac{\cos x}{\sin x}$

$\qquad\qquad = \cos x$

11. $\dfrac{\cot x-1}{\cot x+1} = \dfrac{\dfrac{\cot x}{\cot x}-\dfrac{1}{\cot x}}{\dfrac{\cot x}{\cot x}+\dfrac{1}{\cot x}}$

$\qquad\qquad = \dfrac{1-\tan x}{1+\tan x}$

12. $2\sin^3\theta\cos\theta+2\sin\theta\cos^3\theta$

$\quad = 2\sin\theta\cos\theta\left(\sin^2\theta+\cos^2\theta\right)$

$\quad = 2\sin\theta\cos\theta$

$\quad = \sin 2\theta$

13. $\dfrac{\sin t+\cos t}{\sec t+\csc t} = \dfrac{\sin t+\cos t}{\dfrac{1}{\cos t}+\dfrac{1}{\sin t}}$

$\qquad = \dfrac{\sin t+\cos t}{\dfrac{1}{\cos t}\cdot\dfrac{\sin t}{\sin t}+\dfrac{1}{\sin t}\cdot\dfrac{\cos t}{\cos t}}$

$\qquad = \dfrac{\sin t+\cos t}{\dfrac{\sin t+\cos t}{\sin t\cos t}}$

$\qquad = (\sin t+\cos t)\dfrac{\sin t\cos t}{\sin t+\cos t}$

$\qquad = \sin t\cos t$

$\qquad = \sin t\dfrac{1}{\sec t}$

$\qquad = \dfrac{\sin t}{\sec t}$

14. $\dfrac{\sec^2 x}{2-\sec^2 x} = \dfrac{\dfrac{\sec^2 x}{\sec^2 x}}{\dfrac{2}{\sec^2 x}-\dfrac{\sec^2 x}{\sec^2 x}}$

$\qquad = \dfrac{1}{2\cos^2 x-1}$

$\qquad = \dfrac{1}{\cos 2x}$

$\qquad = \sec 2x$

15. $\tan(\alpha+\beta)\tan(\alpha-\beta)$

$\quad = \tan(\alpha+\beta)\tan(\alpha-\beta)$

$\quad = \dfrac{\tan\alpha+\tan\beta}{1-\tan\alpha\tan\beta}\cdot\dfrac{\tan\alpha-\tan\beta}{1+\tan\alpha\tan\beta}$

$\quad = \dfrac{\tan^2\alpha-\tan^2\beta}{1-\tan^2\alpha\tan^2\beta}$

16. $\dfrac{\sin\theta}{1-\cos\theta} = \dfrac{\sin\theta}{1-\cos\theta}\cdot\dfrac{1+\cos\theta}{1+\cos\theta}$

$\qquad = \dfrac{\sin\theta+\sin\theta\cos\theta}{1-\cos^2\theta}$

$\qquad = \dfrac{\sin\theta+\sin\theta\cos\theta}{\sin^2\theta}$

$\qquad = \dfrac{\sin\theta}{\sin^2\theta}+\dfrac{\sin\theta\cos\theta}{\sin^2\theta}$

$\qquad = \dfrac{1}{\sin\theta}+\dfrac{\cos\theta}{\sin\theta}$

$\qquad = \csc\theta+\cot\theta$

17. $\dfrac{1}{\csc 2x} = \sin 2x$

$= 2\sin x \cos x$

$= 2\sin x \cos x \cdot \dfrac{\cos x}{\cos x}$

$= \dfrac{2\sin x \cos^2 x}{\cos x}$

$= \dfrac{2\sin x}{\cos x} \cdot \cos^2 x$

$= 2\tan x \cdot \dfrac{1}{\sec^2 x}$

$= \dfrac{2\tan x}{\sec^2 x}$

$= \dfrac{2\tan x}{1 + \tan^2 x}$

18. $\dfrac{\sec t - 1}{t \sec t} = \dfrac{\sec t - 1}{t \sec t} \cdot \dfrac{\cos t}{\cos t}$

$= \dfrac{\sec t \cos t - \cos t}{t \sec t \cos t}$

$= \dfrac{\dfrac{1}{\cos t} \cos t - \cos t}{t \dfrac{1}{\cos t} \cos t}$

$= \dfrac{1 - \cos t}{t}$

19. Use $\sin \alpha = \dfrac{3}{5} = \dfrac{y}{r}$ to find $\cos \alpha$ and $\tan \alpha$.

Because α is in Quadrant II, x is negative.

$x^2 + y^2 = r^2$

$x^2 + 3^2 = 5^2$

$x^2 = 16$

$x = -\sqrt{16}$

$x = -4$

Thus, $\cos \alpha = \dfrac{-4}{5} = -\dfrac{4}{5}$ and $\tan \alpha = \dfrac{-3}{4} = -\dfrac{3}{4}$.

Use $\cos \beta = \dfrac{-12}{13} = \dfrac{x}{r}$ to find $\sin \beta$ and $\tan \beta$.

Because β is in Quadrant III, x and y are negative.

$x^2 + y^2 = r^2$

$(-12)^2 + y^2 = 13^2$

$y^2 = 25$

$y = -\sqrt{25}$

$y = -5$

Thus, $\sin \beta = \dfrac{-5}{13} = -\dfrac{5}{13}$ and $\tan \beta = \dfrac{-5}{-12} = \dfrac{5}{12}$.

$\cos(\alpha - \beta) = \cos \alpha \cos \beta + \sin \alpha \sin \beta$

$= \left(-\dfrac{4}{5}\right)\left(-\dfrac{12}{13}\right) + \left(\dfrac{3}{5}\right)\left(-\dfrac{5}{13}\right)$

$= \dfrac{33}{65}$

20. In exercise 19 it was shown that

$\tan \alpha = -\dfrac{3}{4}$ and $\tan \beta = \dfrac{5}{12}$. Thus,

$\tan(\alpha + \beta) = \dfrac{\tan \alpha + \tan \beta}{1 - \tan \alpha \tan \beta}$

$= \dfrac{-\dfrac{3}{4} + \dfrac{5}{12}}{1 - \left(-\dfrac{3}{4}\right)\dfrac{5}{12}}$

$= -\dfrac{16}{63}$

21. In exercise 19 it was shown that

$\sin \alpha = \dfrac{3}{5}$ and $\cos \alpha = -\dfrac{4}{5}$.

Thus, $\sin 2\alpha = 2\sin \alpha \cos \alpha$

$= 2 \cdot \dfrac{3}{5}\left(-\dfrac{4}{5}\right)$

$= -\dfrac{24}{25}$

22. $\cos \beta = -\dfrac{12}{13}$.

Since β is in quadrant III, $\dfrac{\beta}{2}$ is in quadrant II.

The cosine is negative in quadrant II.

$\cos \dfrac{\beta}{2} = -\sqrt{\dfrac{1 + \cos \beta}{2}}$

$= -\sqrt{\dfrac{1 + \left(-\dfrac{12}{13}\right)}{2}}$

$= -\sqrt{\dfrac{1}{26}}$

$= -\dfrac{1}{\sqrt{26}}$

$= -\dfrac{\sqrt{26}}{26}$

23. $\sin\left(\dfrac{3\pi}{4} + \dfrac{5\pi}{6}\right)$

$= \sin\dfrac{3\pi}{4}\cos\dfrac{5\pi}{6} + \cos\dfrac{3\pi}{4}\sin\dfrac{5\pi}{6}$

$= \left(\dfrac{\sqrt{2}}{2}\right)\left(-\dfrac{\sqrt{3}}{2}\right) + \left(-\dfrac{\sqrt{2}}{2}\right)\left(\dfrac{1}{2}\right)$

$= -\dfrac{\sqrt{6}}{4} - \dfrac{\sqrt{2}}{4}$

$= -\dfrac{\sqrt{6} + \sqrt{2}}{4}$

24. $\cos^2 15° - \sin^2 15° = \cos(2 \cdot 15°)$

$= \cos 30°$

$= \dfrac{\sqrt{3}}{2}$

25. $\cos\dfrac{5\pi}{12}\cos\dfrac{\pi}{12} + \sin\dfrac{5\pi}{12}\sin\dfrac{\pi}{12} = \cos\left(\dfrac{5\pi}{12} - \dfrac{\pi}{12}\right)$

$= \cos\dfrac{4\pi}{12}$

$= \cos\dfrac{\pi}{3}$

$= \dfrac{1}{2}$

26. $\tan 22.5° = \tan\dfrac{45°}{2}$

$= \dfrac{\sin 45°}{1 + \cos 45°}$

$= \dfrac{\dfrac{\sqrt{2}}{2}}{1 + \dfrac{\sqrt{2}}{2}}$

$= \dfrac{\sqrt{2}}{2 + \sqrt{2}}$

$= \dfrac{\sqrt{2}}{2 + \sqrt{2}} \cdot \dfrac{2 - \sqrt{2}}{2 - \sqrt{2}}$

$= \dfrac{2\sqrt{2} - 2}{4 - 2}$

$= \dfrac{2\sqrt{2} - 2}{2}$

$= \sqrt{2} - 1$

Section 6.4

Check Point Exercises

1. a. $\sin 5x \sin 2x$

$= \dfrac{1}{2}[\cos(5x - 2x) - \cos(5x + 2x)]$

$= \dfrac{1}{2}[\cos 3x - \cos 7x]$

b. $\cos 7x \cos x$

$= \dfrac{1}{2}[\cos(7x - x) + \cos(7x + x)]$

$= \dfrac{1}{2}[\cos 6x + \cos 8x]$

2. a. $\sin 7x + \sin 3x$

$= 2\sin\left(\dfrac{7x + 3x}{2}\right)\cos\left(\dfrac{7x - 3x}{2}\right)$

$= 2\sin\left(\dfrac{10x}{2}\right)\cos\left(\dfrac{4x}{2}\right)$

$= 2\sin 5x \cos 2x$

b. $\cos 3x + \cos 2x$

$= 2\cos\left(\dfrac{3x + 2x}{2}\right)\cos\left(\dfrac{3x - 2x}{2}\right)$

$= 2\cos\left(\dfrac{5x}{2}\right)\cos\left(\dfrac{x}{2}\right)$

3. $\dfrac{\cos 3x - \cos x}{\sin 3x + \sin x} = \dfrac{-2\sin\left(\dfrac{3x + x}{2}\right)\sin\left(\dfrac{3x - 2x}{2}\right)}{2\sin\dfrac{3x + x}{2}\cos\left(\dfrac{3x - x}{2}\right)}$

$= \dfrac{-2\sin 2x \sin x}{2\sin 2x \cos x}$

$= \dfrac{-\sin x}{\cos x}$

$= -\tan x$

We worked with the left side and arrived at the right side. Thus, the identity is verified.

Concept and Vocabulary Check 6.4

1. product; difference

2. product; sum

3. product; sum

4. product; difference

5. sum; product

6. difference; product

7. sum; product

8. difference; product

Exercise Set 6.4

1. $\sin 6x \sin 2x = \dfrac{1}{2}\left[\cos(6x - 2x) - \cos(6x + 2x)\right]$

    $\qquad = \dfrac{1}{2}\left[\cos 4x - \cos 8x\right]$

3. $\cos 7x \cos 3x = \dfrac{1}{2}\left[\cos(7x - 3x) + \cos(7x + 3x)\right]$

    $\qquad = \dfrac{1}{2}\left[\cos 4x + \cos 10x\right]$

5. $\sin x \cos 2x = \dfrac{1}{2}\left[\sin(x + 2x) + \sin(x - 2x)\right]$

    $\qquad = \dfrac{1}{2}\left[\sin 3x + \sin(-x)\right]$

    $\qquad = \dfrac{1}{2}\left[\sin 3x - \sin x\right]$

7. $\cos \dfrac{3x}{2} \sin \dfrac{x}{2} = \dfrac{1}{2}\left[\sin\left(\dfrac{3x}{2} + \dfrac{x}{2}\right) - \sin\left(\dfrac{3x}{2} - \dfrac{x}{2}\right)\right]$

    $\qquad = \dfrac{1}{2}\left[\sin\left(\dfrac{4x}{2}\right) - \sin\left(\dfrac{2x}{2}\right)\right]$

    $\qquad = \dfrac{1}{2}\left[\sin 2x - \sin x\right]$

9. $\sin 6x + \sin 2x = 2\sin\left(\dfrac{6x + 2x}{2}\right)\cos\left(\dfrac{6x - 2x}{2}\right)$

    $\qquad = 2\sin\left(\dfrac{8x}{2}\right)\cos\left(\dfrac{4x}{2}\right)$

    $\qquad = 2\sin 4x \cos 2x$

11. $\sin 7x - \sin 3x = 2\sin\left(\dfrac{7x - 3x}{2}\right)\cos\left(\dfrac{7x + 3x}{2}\right)$

    $\qquad = 2\sin\left(\dfrac{4x}{2}\right)\cos\left(\dfrac{10x}{2}\right)$

    $\qquad = 2\sin 2x \cos 5x$

13. $\cos 4x + \cos 2x = 2\cos\left(\dfrac{4x + 2x}{2}\right)\cos\left(\dfrac{4x - 2x}{2}\right)$

    $\qquad = 2\cos\left(\dfrac{6x}{2}\right)\cos\left(\dfrac{2x}{2}\right)$

    $\qquad = 2\cos 3x \cos x$

15. $\sin x + \sin 2x = 2\sin\left(\dfrac{x + 2x}{2}\right)\cos\left(\dfrac{x - 2x}{2}\right)$

    $\qquad = 2\sin\left(\dfrac{3x}{2}\right)\cos\left(\dfrac{-x}{2}\right)$

    $\qquad = 2\sin\dfrac{3x}{2}\cos\dfrac{x}{2}$

17. $\cos \dfrac{3x}{2} + \cos \dfrac{x}{2} = 2\cos\left(\dfrac{\frac{3x}{2} + \frac{x}{2}}{2}\right)\cos\left(\dfrac{\frac{3x}{2} - \frac{x}{2}}{2}\right)$

    $\qquad = 2\cos\left(\dfrac{4x}{4}\right)\cos\left(\dfrac{2x}{4}\right)$

    $\qquad = 2\cos x \cos\dfrac{x}{2}$

19. $\sin 75° + \sin 15°$

    $\qquad = 2\sin\left(\dfrac{75° + 15°}{2}\right)\cos\left(\dfrac{75° - 15°}{2}\right)$

    $\qquad = 2\sin(45°)\cos(30°)$

    $\qquad = 2\left(\dfrac{\sqrt{2}}{2}\right)\left(\dfrac{\sqrt{3}}{2}\right)$

    $\qquad = \dfrac{\sqrt{6}}{2}$

21. $\sin\dfrac{\pi}{12} - \sin\dfrac{5\pi}{12}$

$= 2\sin\left(\dfrac{\dfrac{\pi}{12}-\dfrac{5\pi}{12}}{2}\right)\cos\left(\dfrac{\dfrac{\pi}{12}+\dfrac{5\pi}{12}}{2}\right)$

$= 2\sin\left(-\dfrac{4\pi}{24}\right)\cos\left(\dfrac{6\pi}{24}\right)$

$= -2\sin\dfrac{\pi}{6}\cos\dfrac{\pi}{4}$

$= -2\left(\dfrac{1}{2}\right)\left(\dfrac{\sqrt{2}}{2}\right)$

$= -\dfrac{\sqrt{2}}{2}$

23. $\dfrac{\sin 3x - \sin x}{\cos 3x - \cos x}$

$= \dfrac{2\sin\left(\dfrac{3x-x}{2}\right)\cos\left(\dfrac{3x+x}{2}\right)}{-2\sin\left(\dfrac{3x+x}{2}\right)\sin\left(\dfrac{3x-x}{2}\right)}$

$= \dfrac{2\sin\left(\dfrac{2x}{2}\right)\cos\left(\dfrac{4x}{2}\right)}{-2\sin\left(\dfrac{4x}{2}\right)\sin\left(\dfrac{2x}{2}\right)}$

$= \dfrac{2\sin x\cos 2x}{-2\sin 2x\sin x}$

$= -\dfrac{\cos 2x}{\sin 2x} = -\cot 2x$

25. $\dfrac{\sin 2x + \sin 4x}{\cos 2x + \cos 4x}$

$= \dfrac{2\sin\left(\dfrac{2x+4x}{2}\right)\cos\left(\dfrac{2x-4x}{2}\right)}{2\cos\left(\dfrac{2x+4x}{2}\right)\cos\left(\dfrac{2x-4x}{2}\right)}$

$= \dfrac{2\sin\left(\dfrac{6x}{2}\right)\cos\left(\dfrac{-2x}{2}\right)}{2\cos\left(\dfrac{6x}{2}\right)\cos\left(\dfrac{-2x}{2}\right)}$

$= \dfrac{2\sin 3x\cos(-x)}{2\cos 3x\cos(-x)}$

$= \dfrac{\sin 3x}{\cos 3x}$

$= \tan 3x$

27. $\dfrac{\sin x - \sin y}{\sin x + \sin y} = \dfrac{2\sin\left(\dfrac{x-y}{2}\right)\cos\left(\dfrac{x+y}{2}\right)}{2\sin\left(\dfrac{x+y}{2}\right)\cos\left(\dfrac{x-y}{2}\right)}$

$= \dfrac{\sin\left(\dfrac{x-y}{2}\right)}{\cos\left(\dfrac{x-y}{2}\right)} \cdot \dfrac{\cos\left(\dfrac{x+y}{2}\right)}{\sin\left(\dfrac{x+y}{2}\right)}$

$= \tan\dfrac{x-y}{2}\cot\dfrac{x+y}{2}$

29. $\dfrac{\sin x + \sin y}{\cos x + \cos y} = \dfrac{2\sin\left(\dfrac{x+y}{2}\right)\cos\left(\dfrac{x-y}{2}\right)}{2\cos\left(\dfrac{x+y}{2}\right)\cos\left(\dfrac{x-y}{2}\right)}$

$= \dfrac{\sin\left(\dfrac{x+y}{2}\right)}{\cos\left(\dfrac{x+y}{2}\right)}$

$= \tan\dfrac{x+y}{2}$

31. a. $y = \cos x$ also describes the graph.

b.
$$\frac{\sin x + \sin 3x}{2\sin 2x} = \frac{2\sin\left(\dfrac{x+3x}{2}\right)\cos\left(\dfrac{x-3x}{2}\right)}{2\sin 2x} = \frac{2\sin\left(\dfrac{4x}{2}\right)\cos\left(\dfrac{-2x}{2}\right)}{2\sin 2x}$$

$$= \frac{2\sin 2x\cos(-x)}{2\sin 2x} = \cos(-x) = \cos x$$

33. a. $y = \tan 2x$ also describes the graph.

b.
$$\frac{\cos x - \cos 5x}{\sin x + \sin 5x} = \frac{-2\sin\left(\dfrac{x+5x}{2}\right)\sin\left(\dfrac{x-5x}{2}\right)}{2\sin\left(\dfrac{x+5x}{2}\right)\cos\left(\dfrac{x-5x}{2}\right)} = \frac{-2\sin\left(\dfrac{6x}{2}\right)\sin\left(\dfrac{-4x}{2}\right)}{2\sin\left(\dfrac{6x}{2}\right)\cos\left(\dfrac{-4x}{2}\right)}$$

$$= \frac{-2\sin 3x\sin(-2x)}{2\sin 3x\cos(-2x)} = \frac{-\sin(-2x)}{\cos 2x} = \frac{\sin 2x}{\cos 2x} = \tan 2x$$

35. a. $y = -\cot 2x$ also describes the graph.

b.
$$\frac{\sin x - \sin 3x}{\cos x - \cos 3x} = \frac{2\sin\left(\dfrac{x-3x}{2}\right)\cos\left(\dfrac{x+3x}{2}\right)}{-2\sin\left(\dfrac{x+3x}{2}\right)\sin\left(\dfrac{x-3x}{2}\right)}$$

$$= \frac{2\sin(-x)\cos 2x}{-2\sin 2x\sin(-x)}$$

$$= \frac{\cos 2x}{-\sin 2x}$$

$$= -\cot 2x$$

37. a.   The low frequency is $l = 852$ cycles per second and the high frequency is $h = 1209$ cycles per second. The sound produced by touching 7 is described by $y = \sin 2\pi(852)t + \sin 2\pi(1209)t$, or $y = \sin 1704\pi t + \sin 2418\pi t$.

b. $y = \sin 1704\pi t + \sin 2418\pi t$

$$= 2\sin\left(\frac{1704\pi t + 2418\pi t}{2}\right) \cdot \cos\left(\frac{1704\pi t - 2418\pi t}{2}\right)$$

$$= 2\sin 2061\pi t \cdot \cos(-357\pi t)$$

$$= 2\sin 2061\pi t \cdot \cos 357\pi t$$

39. – 43.   Answers may vary.

45.
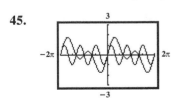

The graphs do not coincide.
Values for x may vary.

47.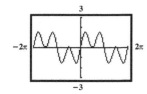

$$\sin x + \sin 3x = 2\sin\left(\frac{x+3x}{2}\right)\cos\left(\frac{x-3x}{2}\right)$$
$$= 2\sin 2x\cos(-x)$$
$$= 2\sin 2x\cos x$$

We worked with the left side and arrived at the right side. Thus, the identity is verified.

49.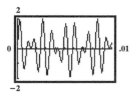

51. a.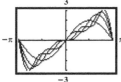

Answers may vary.

b.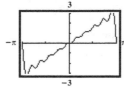

Answers may vary.

c. When $x = \dfrac{\pi}{2}$,

$$\frac{\pi}{2} = 2\left(\frac{\sin\frac{\pi}{2}}{1} - \frac{\sin\left(2\cdot\frac{\pi}{2}\right)}{2} + \frac{\sin\left(3\cdot\frac{\pi}{2}\right)}{3} - \frac{\sin\left(4\cdot\frac{\pi}{2}\right)}{4} + \frac{\sin\left(5\cdot\frac{\pi}{2}\right)}{5} - \frac{\sin\left(6\cdot\frac{\pi}{2}\right)}{6} + \frac{\sin\left(7\cdot\frac{\pi}{2}\right)}{7} - \frac{\sin\left(8\cdot\frac{\pi}{2}\right)}{8} + \cdots\right)$$

$$= 2\left(1 - 0 + \left(-\frac{1}{3}\right) - 0 + \frac{1}{5} - 0 + \left(-\frac{1}{7}\right) + \cdots\right)$$

$$= 2 - \frac{2}{3} + \frac{2}{5} - \frac{2}{7} + \cdots$$

Multiplying both sides by 2 gives: $\pi = 4 - \dfrac{4}{3} + \dfrac{4}{5} - \dfrac{4}{7} + \cdots$

53. makes sense

55. makes sense

57.
$$\sin(\alpha + \beta) = \sin\alpha\cos\beta + \cos\alpha\sin\beta$$
$$\underline{-[\sin(\alpha - \beta) = \sin\alpha\cos\beta - \cos\alpha\sin\beta]}$$
$$\sin(\alpha + \beta) - \sin(\alpha - \beta) = 2\cos\alpha\sin\beta$$

Solve for $\cos\alpha\sin\beta$ by multiplying both sides by $\dfrac{1}{2}$: $\dfrac{1}{2}\big[\sin(\alpha + \beta) - \sin(\alpha - \beta)\big] = \cos\alpha\sin\beta$

59. $2\cos\dfrac{\alpha+\beta}{2}\cos\dfrac{\alpha-\beta}{2} = 2\cdot\dfrac{1}{2}\left[\cos\left(\dfrac{\alpha+\beta}{2}-\dfrac{\alpha-\beta}{2}\right)+\cos\left(\dfrac{\alpha+\beta}{2}+\dfrac{\alpha-\beta}{2}\right)\right]$

$\qquad\qquad\qquad\quad = \cos\left(\dfrac{2\beta}{2}\right)+\cos\left(\dfrac{2\alpha}{2}\right)$

$\qquad\qquad\qquad\quad = \cos\beta+\cos\alpha$

$\qquad\qquad\qquad\quad = \cos\alpha+\cos\beta$

61. $\sin 2x+\sin 4x+\sin 6x = \sin 4x+(\sin 2x+\sin 6x)$

$\qquad\qquad\qquad = \sin 4x+2\sin\left(\dfrac{2x+6x}{2}\right)\cos\left(\dfrac{2x-6x}{2}\right)$

$\qquad\qquad\qquad = \sin 4x+2\sin\left(\dfrac{8x}{2}\right)\cos\left(\dfrac{-4x}{2}\right)$

$\qquad\qquad\qquad = \sin 4x+2\sin 4x\cos(-2x)$

$\qquad\qquad\qquad = \sin 4x+2\sin 4x\cos 2x$

$\qquad\qquad\qquad = \sin(2\cdot 2x)+2\sin 4x\cos 2x$

$\qquad\qquad\qquad = 2\sin 2x\cos 2x+2\sin 4x\cos 2x$

$\qquad\qquad\qquad = 2\cos 2x(\sin 2x+\sin 4x)$

$\qquad\qquad\qquad = 2\cos 2x\left(2\sin\left(\dfrac{2x+4x}{2}\right)\cos\left(\dfrac{2x-4x}{2}\right)\right)$

$\qquad\qquad\qquad = 2\cos 2x\cdot 2\sin\left(\dfrac{6x}{2}\right)\cos\left(\dfrac{-2x}{2}\right)$

$\qquad\qquad\qquad = 2\cos 2x\cdot 2\sin 3x\cos(-x)$

$\qquad\qquad\qquad = 4\cos 2x\sin 3x\cos x$

$\qquad\qquad\qquad = 4\cos x\cos 2x\sin 3x$

63.
$$2(1-u^2)+3u=0$$
$$2-2u^2+3u=0$$
$$2u^2-3u-2=0$$
$$(2u+1)(u-2)=0$$
$$2u+1=0 \quad \text{and} \quad u-2=0$$
$$2u=-1 \qquad\qquad u=2$$
$$u=-\frac{1}{2}$$

The solution set is $\left\{-\dfrac{1}{2},2\right\}$.

64.
$$u^3-3u=0$$
$$u(u^2-3)=0$$
$$u=0 \quad \text{or} \quad u^2-3=0$$
$$u^2=3$$
$$u=\pm\sqrt{3}$$

The solution set is $\left\{-\sqrt{3},0,\sqrt{3}\right\}$.

65.
$$u^2-u-1=0$$
$$a=1, \quad b=-1, \quad c=-1$$
$$x=\frac{-b\pm\sqrt{b^2-4ac}}{2a}$$
$$x=\frac{-(-1)\pm\sqrt{(-1)^2-4(1)(-1)}}{2(1)}$$
$$x=\frac{1\pm\sqrt{5}}{2}$$

The solution set is $\left\{\dfrac{1-\sqrt{5}}{2},\dfrac{1+\sqrt{5}}{2}\right\}$.

Section 6.5

Check Point Exercises

1.
$$5\sin x=3\sin x+\sqrt{3}$$
$$5\sin x-3\sin x=3\sin x-3\sin x+\sqrt{3}$$
$$2\sin x=\sqrt{3}$$
$$\sin x=\frac{\sqrt{3}}{2}$$

Because $\sin\dfrac{\pi}{3}=\dfrac{\sqrt{3}}{2}$, the solutions for $\sin x=\dfrac{\sqrt{3}}{2}$ in $[0,2\pi)$ are

$$x=\frac{\pi}{3}$$

$$x=\pi-\frac{\pi}{3}=\frac{3\pi}{3}-\frac{\pi}{3}=\frac{2\pi}{3}.$$

Because the period of the sine function is 2π, the solutions are given by

$$x=\frac{\pi}{3}+2n\pi \quad \text{or}$$

$$x=\frac{2\pi}{3}+2n\pi$$

where *n* is any integer.

2. The period of the tangent function is π. In the interval $[0,\pi)$, the only value for which the tangent function is $\sqrt{3}$ is $\dfrac{\pi}{3}$. All the solutions to $\tan 2x=\sqrt{3}$ are given by

$$2x=\frac{\pi}{3}+n\pi$$

$$x=\frac{\pi}{6}+\frac{n\pi}{2}$$

where *n* is any integer. In the interval $[0,2\pi)$, we obtain solutions as follows:

Let $n=0.$ $\quad x=\dfrac{\pi}{6}+\dfrac{0\pi}{2}$

$$=\frac{\pi}{6}$$

Let $n=1.$ $\quad x=\dfrac{\pi}{6}+\dfrac{1\pi}{2}$

$$=\frac{\pi}{6}+\frac{3\pi}{6}=\frac{2\pi}{3}$$

Let $n = 2$. $x = \dfrac{\pi}{6} + \dfrac{2\pi}{2}$

$= \dfrac{\pi}{6} + \dfrac{6\pi}{6} = \dfrac{7\pi}{6}$

Let $n = 3$. $x = \dfrac{\pi}{6} + \dfrac{3\pi}{2}$

$= \dfrac{\pi}{6} + \dfrac{9\pi}{6} = \dfrac{5\pi}{3}$

In the interval $[0, 2\pi)$, the solutions are

$\dfrac{\pi}{6}, \dfrac{2\pi}{3}, \dfrac{7\pi}{6},$ and $\dfrac{5\pi}{3}$.

3. The period of the sine function is 2π.

In the interval $[0, 2\pi)$, there are two values at

which the sine function is $\dfrac{1}{2}$. One is $\dfrac{\pi}{6}$. The sine

is positive in quadrant II. Thus, the other value is

$\pi - \dfrac{\pi}{6} = \dfrac{5\pi}{6}$. All the solutions to $\sin \dfrac{x}{3} = \dfrac{1}{2}$ are

given by

$\dfrac{x}{3} = \dfrac{\pi}{6} + 2n\pi$

$x = \dfrac{\pi}{2} + 6n\pi$

or

$\dfrac{x}{3} = \dfrac{5\pi}{6} + 2n\pi$

$x = \dfrac{5\pi}{2} + 6n\pi$

where n is any integer. In the interval $[0, 2\pi)$, we

obtain solutions as follows:

Let $n = 0$. $x = \dfrac{\pi}{2}$ or $x = \dfrac{5\pi}{2}$

The value $\dfrac{5\pi}{2}$ exceeds 2π.

If we let $n = 1$, we are adding 6π to each of these

expressions. These values of x exceed 2π. Thus in

the interval $[0, 2\pi)$, the solution set is $\left\{ \dfrac{\pi}{2} \right\}$

4. The given equation is in quadratic form

$2t^2 - 3t + 1 = 0$ with $t = \sin x$.

$2\sin^2 x - 3\sin x + 1 = 0$

$(2\sin x - 1)(\sin x - 1) = 0$

$2\sin x - 1 = 0$ or $\sin x - 1 = 0$

$2\sin x = 1$ $\sin x = 1$

$\sin x = \dfrac{1}{2}$

$x = \dfrac{\pi}{6}$ $x = \dfrac{\pi}{2}$

$x = \pi - \dfrac{\pi}{6} = \dfrac{5\pi}{6}$

The solutions in the interval $[0, 2\pi)$ are $\dfrac{\pi}{6}, \dfrac{\pi}{2},$

and $\dfrac{5\pi}{6}$.

5. $4\cos^2 x - 3 = 0$

$\cos^2 x = \dfrac{3}{4}$

$\cos x = \pm\sqrt{\dfrac{3}{4}}$

$\cos x = \pm\dfrac{\sqrt{3}}{2}$

$\cos x = \dfrac{\sqrt{3}}{2}$ or $\cos x = -\dfrac{\sqrt{3}}{2}$

$x = \dfrac{\pi}{6}, \dfrac{11\pi}{6}$ $x = \dfrac{5\pi}{6}, \dfrac{7\pi}{6}$

The solutions in the interval $[0, 2\pi)$ are

$\dfrac{\pi}{6}, \dfrac{5\pi}{6}, \dfrac{7\pi}{6},$ and $\dfrac{11\pi}{6}$.

6. $\sin x \tan x = \sin x$

$\sin x \tan x - \sin x = 0$

$\sin x(\tan x - 1) = 0$

$\sin x = 0$ or $\tan x - 1 = 0$

$x = 0$ $x = \pi$ $\tan x = 1$

$x = \dfrac{\pi}{4}$

$x = \pi + \dfrac{\pi}{4} = \dfrac{5\pi}{4}$

The solutions in the interval $[0, 2\pi)$ are

$0, \dfrac{\pi}{4}, \pi,$ and $\dfrac{5\pi}{4}$.

7.
$$2\sin^2 x - 3\cos x = 0$$
$$2(1 - \cos^2 x) - 3\cos x = 0$$
$$2 - 2\cos^2 x - 3\cos x = 0$$
$$-2\cos^2 x - 3\cos x + 2 = 0$$
$$2\cos^2 x + 3\cos x - 2 = 0$$
$$(2\cos x - 1)(\cos x + 2) = 0$$
$$2\cos x - 1 = 0 \quad \text{or} \quad \cos x + 2 = 0$$

$$\cos x = \frac{1}{2} \qquad\qquad \cos x = -2$$

$$\qquad\qquad\qquad\qquad \cancel{\cos x = -2}$$

$$x = \frac{\pi}{3}, \frac{5\pi}{3} \qquad \text{This equation}$$
$$\qquad\qquad\qquad\qquad \text{has no solution.}$$

The solutions in the interval $[0, 2\pi)$ are $\dfrac{\pi}{3}$ and $\dfrac{5\pi}{3}$.

8.
$$\cos 2x + \sin x = 0$$
$$1 - 2\sin^2 x + \sin x = 0$$
$$-2\sin^2 x + \sin x + 1 = 0$$
$$2\sin^2 x - \sin x - 1 = 0$$
$$(2\sin x + 1)(\sin x - 1) = 0$$
$$2\sin x + 1 = 0 \qquad \text{or} \qquad \sin x - 1 = 0$$
$$2\sin x = -1 \qquad\qquad\qquad \sin x = 1$$
$$\sin x = -\frac{1}{2} \qquad\qquad\qquad x = \frac{\pi}{2}$$

$$x = \pi + \frac{\pi}{6} = \frac{7\pi}{6} \text{ or}$$
$$x = 2\pi - \frac{\pi}{6} = \frac{11\pi}{6}$$

The solutions in the interval $[0, 2\pi)$ are
$$\frac{\pi}{2}, \frac{7\pi}{6}, \text{ and } \frac{11\pi}{6}.$$

9. $\sin x \cos x = -\dfrac{1}{2}$

$$2\sin x \cos x = -1$$
$$\sin 2x = -1$$

The period of the sine function is 2π. In the

interval $[0, 2\pi)$, the sine function is -1 at $\dfrac{3\pi}{2}$. All

the solutions to $\sin 2x$ are given by

$$2x = \frac{3\pi}{2} + 2n\pi$$
$$x = \frac{3\pi}{4} + n\pi,$$

where n is any integer. The solutions in the interval $[0, 2\pi)$ are obtained by letting $n = 0$ and $n = 1$. The

solutions are $\dfrac{3\pi}{4}$ and $\dfrac{7\pi}{4}$.

10.
$$\cos x - \sin x = -1$$
$$(\cos x - \sin x)^2 = (-1)^2$$
$$\cos^2 x - 2\cos x \sin x + \sin^2 x = 1$$
$$\cos^2 x + \sin^2 x - 2\cos x \sin x = 1$$
$$1 - 2\cos x \sin x = 1$$
$$-2\cos x \sin x = 0$$
$$\cos x \sin x = 0$$
$$\cos x = 0 \quad \text{or} \quad \sin x = 0$$

$$x = \frac{\pi}{2} \qquad\qquad x = 0$$

$$x = \frac{3\pi}{2} \qquad\qquad x = \pi$$

We check these proposed solutions to see if any are extraneous.

Check 0: $\cos 0 - \sin 0 = -1$
$$1 - 0 = -1$$
$$1 = -1, \text{ false}$$

Check $\dfrac{\pi}{2}$: $\cos\dfrac{\pi}{2} - \sin\dfrac{\pi}{2} = -1$
$$0 - 1 = -1$$
$$-1 = -1, \text{ true}$$

Check π: $\cos\dfrac{\pi}{2} - \sin\dfrac{\pi}{2} = -1$
$$-1 - 0 = -1$$
$$-1 = -1, \text{ true}$$

Check $\dfrac{3\pi}{2}$: $\cos\dfrac{3\pi}{2} - \sin\dfrac{3\pi}{2} = -1$
$$0 - (-1) = -1$$
$$1 = -1, \text{ false}$$

The actual solutions in the interval $[0, 2\pi)$

are $\dfrac{\pi}{2}$ and π.

11. a. $\tan x = 3.1044$

Be sure calculator is in radian mode and find the inverse tangent of 3.1044. This gives the first quadrant reference angle.

$\theta = \tan^{-1} 3.1044 \approx 1.2592$

The tangent is positive in quadrants I and III thus,

$x \approx 1.2592 \quad \text{or} \quad x \approx \pi + 1.2592$

$x \approx 4.4008$

b. $\sin x = -0.2315$

Be sure calculator is in radian mode and find the inverse sine of +0.2315. This gives the first quadrant reference angle.

$\theta = \sin^{-1}(0.2315) \approx 0.2336$

The sine is negative in quadrants III and IV thus,

$x \approx \pi + 0.2336 \quad \text{or} \quad x \approx 2\pi - 1.2592$

$x \approx 3.3752 \qquad\qquad x \approx 6.0496$

12. $\cos^2 x + 5\cos x + 3 = 0$

Use the quadratic formula to solve for $\cos x$.

$\cos x = \dfrac{-b \pm \sqrt{b^2 - 4ac}}{2a}$

$\cos x = \dfrac{-(5) \pm \sqrt{(5)^2 - 4(1)(3)}}{2(1)}$

$\cos x = \dfrac{-5 \pm \sqrt{13}}{2}$

$\cos x \approx -0.6972 \quad \text{or} \quad \cos x \approx -4.3028$

~~$\cos x \approx -4.3028$~~

This equation

has no solution.

Be sure calculator is in radian mode and find the inverse cosine of +0.6972. This gives the first quadrant reference angle.

$\theta = \cos^{-1} 0.6972 \approx 0.7993$

The cosine is negative in quadrants II and III thus,

$x \approx \pi - 0.7993 \quad \text{or} \quad x \approx \pi + 0.7993$

$x \approx 2.3423 \qquad\qquad x \approx 3.9409$

Concept and Vocabulary Check 6.5

1. $\dfrac{3\pi}{4}$; $\dfrac{\pi}{4} + 2n\pi$; $\dfrac{3\pi}{4} + 2n\pi$

2. $\dfrac{2\pi}{3}$; $x = \dfrac{2\pi}{3} + n\pi$

3.   false

4.   true

5.   false

6. $2\cos x + 1$; $\cos x - 5$; $\cos x - 5 = 0$

7. $\cos x$; $2\sin x + \sqrt{2}$

8. $\cos^2 x$; $1 - \sin^2 x$

9.   ; 2π

Exercise Set 6.5

1. $\cos \dfrac{\pi}{4} = \dfrac{\sqrt{2}}{2}$

$\dfrac{\sqrt{2}}{2} = \dfrac{\sqrt{2}}{2}$ is true.

Thus, $\dfrac{\pi}{4}$ is a solution.

3. $\sin \dfrac{\pi}{6} = \dfrac{\sqrt{3}}{2}$

$\dfrac{1}{2} = \dfrac{\sqrt{3}}{2}$ is false.

Thus, $\dfrac{\pi}{6}$ is not a solution.

5. $\cos \dfrac{2\pi}{3} = -\dfrac{1}{2}$

$-\dfrac{1}{2} = -\dfrac{1}{2}$ is true.

Thus, $\dfrac{2\pi}{3}$ is a solution.

7. $\tan\left(2 \cdot \dfrac{5\pi}{12}\right) = -\dfrac{\sqrt{3}}{3}$

$\tan\dfrac{5\pi}{6} = -\dfrac{\sqrt{3}}{3}$

$-\dfrac{\sqrt{3}}{3} = -\dfrac{\sqrt{3}}{3}$ is true.

Thus, $\dfrac{5\pi}{12}$ is a solution.

9. $\cos\dfrac{\pi}{3} = \dfrac{\sqrt{3}}{2}$

$\dfrac{1}{2} = \dfrac{\sqrt{3}}{2}$ is false.

Thus, $\dfrac{\pi}{3}$ is not a solution.

11. $\sin x = \dfrac{\sqrt{3}}{2}$

Because $\sin\dfrac{\pi}{3} = \dfrac{\sqrt{3}}{2}$, the solutions

for $\sin x = \dfrac{\sqrt{3}}{2}$ in $[0, 2\pi)$ are

$x = \dfrac{\pi}{3}$

$x = \pi - \dfrac{\pi}{3} = \dfrac{3\pi}{3} - \dfrac{\pi}{3} = \dfrac{2\pi}{3}$.

Because the period of the sine function is 2π, the solutions are given by

$x = \dfrac{\pi}{3} + 2n\pi$ or $x = \dfrac{2\pi}{3} + 2n\pi$

where n is any integer.

13. $\tan x = 1$

Because $\tan\dfrac{\pi}{4} = 1$, the solution

for $\tan x = 1$ in $[0, \pi)$ is

$x = \dfrac{\pi}{4}$.

Because the period of the tangent function is π, the solutions are given by

$x = \dfrac{\pi}{4} + n\pi$

where n is any integer.

15. $\cos x = -\dfrac{1}{2}$

Because $\cos\dfrac{\pi}{3} = \dfrac{1}{2}$, the solutions

for $\cos x = -\dfrac{1}{2}$ in $[0, 2\pi)$ are

$x = \pi - \dfrac{\pi}{3} = \dfrac{3\pi}{3} - \dfrac{\pi}{3} = \dfrac{2\pi}{3}$

$x = \pi + \dfrac{\pi}{3} = \dfrac{3\pi}{3} + \dfrac{\pi}{3} = \dfrac{4\pi}{3}$.

Because the period of the cosine function is 2π, the solutions are given by

$x = \dfrac{2\pi}{3} + 2n\pi$ or $x = \dfrac{4\pi}{3} + 2n\pi$

where n is any integer.

17. $\tan x = 0$

Because $\tan 0 = 0$, the solution

for $\tan x = 0$ in $[0, \pi)$ is

$x = 0$.

Because the period of the tangent function is π, the solutions are given by

$x = 0 + n\pi = n\pi$

where n is any integer.

19. $2\cos x + \sqrt{3} = 0$

$2\cos x = -\sqrt{3}$

$\cos x = -\dfrac{\sqrt{3}}{2}$

Because $\cos\dfrac{\pi}{6} = \dfrac{\sqrt{3}}{2}$, the solutions

for $\cos x = -\dfrac{\sqrt{3}}{2}$ in $[0, 2\pi)$ are

$x = \pi - \dfrac{\pi}{6} = \dfrac{6\pi}{6} - \dfrac{\pi}{6} = \dfrac{5\pi}{6}$

$x = \pi + \dfrac{\pi}{6} = \dfrac{6\pi}{6} + \dfrac{\pi}{6} = \dfrac{7\pi}{6}$.

Because the period of the cosine function is 2π, the solutions are given by

$x = \dfrac{5\pi}{6} + 2n\pi$ or $x = \dfrac{7\pi}{6} + 2n\pi$

where n is any integer.

21.
$$4\sin\theta - 1 = 2\sin\theta$$
$$4\sin\theta - 2\sin\theta = 1$$
$$2\sin\theta = 1$$
$$\sin\theta = \frac{1}{2}$$

Because $\sin\dfrac{\pi}{6} = \dfrac{1}{2}$, the solutions

for $\sin\theta = \dfrac{1}{2}$ in $[0, 2\pi)$ are

$$\theta = \frac{\pi}{6}$$

$$\theta = \pi - \frac{\pi}{6} = \frac{6\pi}{6} - \frac{\pi}{6} = \frac{5\pi}{6}.$$

Because the period of the sine function is 2π, the solutions are given by

$$\theta = \frac{\pi}{6} + 2n\pi \quad \text{or} \quad \theta = \frac{5\pi}{6} + 2n\pi$$

where n is any integer.

23.
$$3\sin\theta + 5 = -2\sin\theta$$
$$3\sin\theta + 2\sin\theta = -5$$
$$5\sin\theta = -5$$
$$\sin\theta = -1$$

Because $\sin\dfrac{\pi}{2} = 1$, the solutions

for $\sin\theta = -1$ in $[0, 2\pi)$ are

$$\theta = \pi + \frac{\pi}{2} = \frac{2\pi}{2} + \frac{\pi}{2} = \frac{3\pi}{2}$$

$$\theta = 2\pi - \frac{\pi}{2} = \frac{4\pi}{2} - \frac{\pi}{2} = \frac{3\pi}{2}.$$

Because the period of the sine function is 2π, the solutions are given by

$$\theta = \frac{3\pi}{2} + 2n\pi$$

where n is any integer.

25. The period of the sine function is 2π. In the interval $[0, 2\pi)$, there are two values at which the sine

function is $\dfrac{\sqrt{3}}{2}$. One is $\dfrac{\pi}{3}$. The sine is positive in

quadrant II; thus, the other value is $\pi - \dfrac{\pi}{3} = \dfrac{2\pi}{3}$. All

the solutions to $\sin 2x = \dfrac{\sqrt{3}}{2}$ are given by

$$2x = \frac{\pi}{3} + 2n\pi \quad \text{or} \quad 2x = \frac{2\pi}{3} + 2n\pi$$

$$x = \frac{\pi}{6} + n\pi \qquad\qquad x = \frac{\pi}{3} + n\pi$$

Where n is any integer.
The solutions in the interval $[0, 2\pi)$ are obtained by letting $n = 0$ and $n = 1$.

The solutions are $\dfrac{\pi}{6}, \dfrac{\pi}{3}, \dfrac{7\pi}{6}$, and $\dfrac{4\pi}{3}$.

27. The period of the cosine function is 2π. In the interval $[0, 2\pi)$, there are two values at

which the cosine function is $-\dfrac{\sqrt{3}}{2}$. One is $\dfrac{5\pi}{6}$. The

cosine is negative in quadrant III; thus, the other

value is $2\pi - \dfrac{5\pi}{6} = \dfrac{7\pi}{6}$. All the solutions to

$$\cos 4x = -\frac{\sqrt{3}}{2}$$ are given by

$$4x = \frac{5\pi}{6} + 2n\pi \quad \text{or} \quad 4x = \frac{7\pi}{6} + 2n\pi$$

$$x = \frac{5\pi}{24} + \frac{n\pi}{2} \qquad\qquad x = \frac{7\pi}{24} + \frac{n\pi}{2}$$

where n is any integer.
The solutions in the interval $[0, 2\pi)$ are obtained by letting $n = 0$, $n = 1$, $n = 2$, and $n = 3$.

The solutions are $\dfrac{5\pi}{24}, \dfrac{7\pi}{24}, \dfrac{17\pi}{24}, \dfrac{19\pi}{24}$,

$\dfrac{29\pi}{24}, \dfrac{31\pi}{24}, \dfrac{41\pi}{24}$ and $\dfrac{43\pi}{24}$.

29. The period of the tangent function is π. In the interval $[0, \pi)$, the only value for which the tangent function is $\dfrac{\sqrt{3}}{3}$ is $\dfrac{\pi}{6}$.

All the solutions to $\tan 3x = \dfrac{\sqrt{3}}{3}$ are given by

$$3x = \frac{\pi}{6} + n\pi$$

$$x = \frac{\pi}{18} + \frac{n\pi}{3}$$

where n is any integer.
The solutions in the interval $[0, 2\pi)$ are obtained by letting $n = 0, n = 1, n = 2, n = 3, n = 4$, and $n = 5$. The solutions are

$$\frac{\pi}{18}, \frac{7\pi}{18}, \frac{13\pi}{18}, \frac{19\pi}{18}, \frac{25\pi}{18}, \text{ and } \frac{31\pi}{18}.$$

31. The period of the tangent function is π. In the interval $[0, \pi)$, the only value for which the tangent function is $\sqrt{3}$ is $\dfrac{\pi}{3}$.

All the solutions to $\tan \dfrac{x}{2} = \sqrt{3}$ are given by

$$\frac{x}{2} = \frac{\pi}{3} + n\pi$$

$$x = \frac{2\pi}{3} + 2n\pi \text{ where } n \text{ is any integer. The solution}$$

in the interval $[0, 2\pi)$ is obtained by letting $n = 0$.

The only solution is $\dfrac{2\pi}{3}$.

33. The period of the sine function is 2π. In the interval $[0, 2\pi)$, the only value for which the sine function is -1 is $\dfrac{3\pi}{2}$.

All the solutions to $\sin \dfrac{2\theta}{3} = -1$ are given by

$$\frac{2\theta}{3} = \frac{3\pi}{2} + 2n\pi$$

$$\theta = \frac{9\pi}{4} + 3n\pi \text{ where } n \text{ is any integer. All values of}$$

θ exceed 2π or are less than zero.
Thus, in the interval $[0, 2\pi)$ there is no solution.

35. The period of the secant function is 2π. In the interval $[0, 2\pi)$, there are two values at which the secant function is -2. One is $\dfrac{2\pi}{3}$. The secant is negative in quadrant III; thus, the other value is $2\pi - \dfrac{2\pi}{3} = \dfrac{4\pi}{3}$. All the solutions to $\sec \dfrac{3\theta}{2} = -2$ are given by

$$\frac{3\theta}{2} = \frac{2\pi}{3} + 2n\pi \quad \text{or} \quad \frac{3\theta}{2} = \frac{4\pi}{3} + 2n\pi$$

$$\theta = \frac{4\pi}{9} + \frac{4n\pi}{3} \qquad\qquad \theta = \frac{8\pi}{9} + \frac{4n\pi}{3}$$

where n is any integer. The solutions in the interval $[0, 2\pi)$ are obtained by letting $n = 0$ and $n = 1$.

Since $\dfrac{20\pi}{9}$ is not in $[0, 2\pi)$, the solutions are

$$\frac{4\pi}{9}, \frac{8\pi}{9}, \text{ and } \frac{16\pi}{9}.$$

37. The period of the sine function is 2π. In the interval $[0, 2\pi)$, there are two values at which the sine function is $\dfrac{1}{2}$. One is $\dfrac{\pi}{6}$. The sine is positive in quadrant II; Thus, the other value is $\pi - \dfrac{\pi}{6} = \dfrac{5\pi}{6}$.

All the solutions to $\sin\left(2x + \dfrac{\pi}{6}\right) = \dfrac{1}{2}$ are given by

$$2x + \frac{\pi}{6} = \frac{\pi}{6} + 2n\pi$$

$$2x = 2n\pi$$

$$x = n\pi \qquad \text{or}$$

$$2x + \frac{\pi}{6} = \frac{5\pi}{6} + 2n\pi$$

$$2x = \frac{4\pi}{6} + 2n\pi$$

$$x = \frac{2\pi}{6} + n\pi$$

$$x = \frac{\pi}{3} + n\pi$$

where n is any integer. The solutions in the interval $[0, 2\pi)$ are obtained by letting $n = 0$ and $n = 1$.

The solutions are $0, \dfrac{\pi}{3}, \pi,$ and $\dfrac{4\pi}{3}$.

39.
$$2\sin^2 x - \sin x - 1 = 0$$
$$(2\sin x + 1)(\sin x - 1) = 0$$

$2\sin x + 1 = 0$ or $\sin x - 1 = 0$

$2\sin x = -1$ $\sin x = 1$

$$\sin x = -\frac{1}{2}$$

$x = \dfrac{7\pi}{6}$ $x = \dfrac{11\pi}{6}$ $x = \dfrac{\pi}{2}$

The solutions in the interval $[0, 2\pi)$ are $\dfrac{\pi}{2}, \dfrac{7\pi}{6}$, and

$\dfrac{11\pi}{6}$.

41.
$$2\cos^2 x + 3\cos x + 1 = 0$$
$$(2\cos x + 1)(\cos x + 1) = 0$$

$2\cos x + 1 = 0$ or $\cos x + 1 = 0$

$2\cos x = -1$ $\cos x = -1$

$$\cos x = -\frac{1}{2}$$

$x = \dfrac{2\pi}{3}$ $x = \dfrac{4\pi}{3}$ $x = \pi$

The solutions in the interval $[0, 2\pi)$ are $\dfrac{2\pi}{3}, \pi$, and

$\dfrac{4\pi}{3}$.

43.
$$2\sin^2 x = \sin x + 3$$
$$2\sin^2 x - \sin x - 3 = 0$$
$$(2\sin x - 3)(\sin x + 1) = 0$$

$2\sin x - 3 = 0$ or $\sin x + 1 = 0$

$2\sin x = 3$ $\sin x = -1$

$\sin x = \dfrac{3}{2}$ $x = \dfrac{3\pi}{2}$

$\sin x$ cannot be greater than 1.

The solution in the interval $[0, 2\pi)$ is $\dfrac{3\pi}{2}$.

45.
$$\sin^2 \theta - 1 = 0$$
$$(\sin\theta - 1)(\sin\theta + 1) = 0$$

$\sin\theta - 1 = 0$ or $\sin\theta + 1 = 0$

$\sin\theta = 1$ $\sin\theta = -1$

$\theta = \dfrac{\pi}{2}$ $\theta = \dfrac{3\pi}{2}$

The solutions in the interval $[0, 2\pi)$ are $\dfrac{\pi}{2}$ and $\dfrac{3\pi}{2}$.

47.
$$4\cos^2 x - 1 = 0$$
$$(2\cos x + 1)(2\cos x - 1) = 1$$

$2\cos x + 1 = 0$ or $2\cos x - 1 = 0$

$\cos x = -\dfrac{1}{2}$ $\cos x = \dfrac{1}{2}$

$x = \dfrac{2\pi}{3}, \dfrac{4\pi}{3}$ $x = \dfrac{\pi}{3}, \dfrac{5\pi}{3}$

The solutions in the interval $[0, 2\pi)$ are

$\dfrac{\pi}{3}, \dfrac{2\pi}{3}, \dfrac{4\pi}{3}$, and $\dfrac{5\pi}{3}$.

49. $9\tan^2 x - 3 = 0$

$$\tan^2 x = \frac{3}{9}$$

$$\tan x = \pm\sqrt{\frac{3}{9}}$$

$$\tan x = \pm\frac{\sqrt{3}}{3}$$

$\tan x = \dfrac{\sqrt{3}}{3}$ or $\tan x = -\dfrac{\sqrt{3}}{3}$

$x = \dfrac{\pi}{6}, \dfrac{7\pi}{6}$ $x = \dfrac{5\pi}{6}, \dfrac{11\pi}{6}$

The solutions in the interval $[0, 2\pi)$ are

$\dfrac{\pi}{6}, \dfrac{5\pi}{6}, \dfrac{7\pi}{6}$, and $\dfrac{11\pi}{6}$.

51. $\sec^2 x - 2 = 0$

$$\sec^2 x = 2$$

$$\cos^2 x = \frac{1}{2}$$

$$\cos x = \pm\sqrt{\frac{1}{2}}$$

$$\cos x = \pm\frac{\sqrt{2}}{2}$$

$\cos x = \dfrac{\sqrt{2}}{2}$ or $\cos x = -\dfrac{\sqrt{2}}{2}$

$x = \dfrac{\pi}{4}, \dfrac{7\pi}{4}$ $x = \dfrac{3\pi}{4}, \dfrac{5\pi}{4}$

The solutions in the interval $[0, 2\pi)$ are

$\dfrac{\pi}{4}, \dfrac{3\pi}{4}, \dfrac{5\pi}{4}$, and $\dfrac{7\pi}{4}$.

53. $(\tan x - 1)(\cos x + 1) = 0$

$\tan x - 1 = 0$ or $\cos x + 1 = 0$

$\tan x = 1$ $\cos x = -1$

$x = \dfrac{\pi}{4}$ $x = \dfrac{5\pi}{4}$ $x = \pi$

The solutions in the interval $[0, 2\pi)$ are

$\dfrac{\pi}{4}, \pi$, and $\dfrac{5\pi}{4}$

55. $\left(2\cos x + \sqrt{3}\right)(2\sin x + 1) = 0$

$2\cos x + \sqrt{3} = 0$ or $2\sin x + 1 = 0$

$2\cos x = -\sqrt{3}$ $2\sin x = -1$

$\cos x = -\dfrac{\sqrt{3}}{2}$ $\sin x = -\dfrac{1}{2}$

$x = \dfrac{5\pi}{6}$ $x = \dfrac{7\pi}{6}$ $x = \dfrac{7\pi}{6}$ $x = \dfrac{11\pi}{6}$

The solutions in the interval $[0, 2\pi)$ are

$\dfrac{5\pi}{6}, \dfrac{7\pi}{6}$, and $\dfrac{11\pi}{6}$.

57. $\cot x(\tan x - 1) = 0$

$\cot x = 0$ or $\tan x - 1 = 0$

$\tan x = 1$

$x = \dfrac{\pi}{2}$ $x = \dfrac{3\pi}{2}$ $x = \dfrac{\pi}{4}$ $x = \dfrac{5\pi}{4}$

The solutions in the interval $[0, 2\pi)$ are $\dfrac{\pi}{4}$ and $\dfrac{5\pi}{4}$

since tan is undefined for $\dfrac{\pi}{2}$ and $\dfrac{3\pi}{2}$.

59. $\sin x + 2\sin x \cos x = 0$

$\sin x(1 + 2\cos x) = 0$

$\sin x = 0$ or $1 + 2\cos x = 0$

$2\cos x = -1$

$\cos x = -\dfrac{1}{2}$

$x = 0$ $x = \pi$ $x = \dfrac{2\pi}{3}$ $x = \dfrac{4\pi}{3}$

The solutions in the interval $[0, 2\pi)$ are

$0, \dfrac{2\pi}{3}, \pi$, and $\dfrac{4\pi}{3}$.

61. $\tan^2 x \cos x = \tan^2 x$

$\tan^2 x \cos x - \tan^2 x = 0$

$\tan^2 x(\cos x - 1) = 0$

$\tan^2 x = 0$ or $\cos x - 1 = 0$

$\tan x = 0$ $\cos x = 1$

$x = 0$ $x = \pi$ $x = 0$

The solutions in the interval $[0, 2\pi)$ are 0 and π.

63. $2\cos^2 x + \sin x - 1 = 0$

$2\left(1 - \sin^2 x\right) + \sin x - 1 = 0$

$2 - 2\sin^2 x + \sin x - 1 = 0$

$-2\sin^2 x + \sin x + 1 = 0$

$2\sin^2 x - \sin x - 1 = 0$

$(2\sin x + 1)(\sin x - 1) = 0$

$2\sin x + 1 = 0$ or $\sin x - 1 = 0$

$2\sin x = -1$ $\sin x = 1$

$\sin x = -\dfrac{1}{2}$

$x = \dfrac{7\pi}{6}$ $x = \dfrac{11\pi}{6}$ $x = \dfrac{\pi}{2}$

The solutions in the interval $[0, 2\pi)$ are

$\dfrac{\pi}{2}, \dfrac{7\pi}{6}$, and $\dfrac{11\pi}{6}$.

65. $\sin^2 x - 2\cos x - 2 = 0$

$1 - \cos^2 x - 2\cos x - 2 = 0$

$-\cos^2 x - 2\cos x - 1 = 0$

$\cos^2 x + 2\cos x + 1 = 0$

$(\cos x + 1)(\cos x + 1) = 0$

$\cos x + 1 = 0$

$\cos x = -1$

$x = \pi$

The solution in the interval $[0, 2\pi)$ is π.

67.
$$4\cos^2 x = 5 - 4\sin x$$
$$4\cos^2 x + 4\sin x - 5 = 0$$
$$4\left(1 - \sin^2 x\right) + 4\sin x - 5 = 0$$
$$4 - 4\sin^2 x + 4\sin x - 5 = 0$$
$$-4\sin^2 x + 4\sin x - 1 = 0$$
$$4\sin^2 x - 4\sin x + 1 = 0$$
$$(2\sin x - 1)(2\sin x - 1) = 0$$
$$2\sin x - 1 = 0$$
$$2\sin x = 1$$
$$\sin x = \frac{1}{2}$$
$$x = \frac{\pi}{6} \quad x = \frac{5\pi}{6}$$

The solutions in the interval $[0, 2\pi)$ are $\dfrac{\pi}{6}$ and $\dfrac{5\pi}{6}$.

69.
$$\sin 2x = \cos x$$
$$2\sin x \cos x = \cos x$$
$$2\sin x \cos x - \cos x = 0$$
$$\cos x(2\sin x - 1) = 0$$
$$\cos x = 0 \quad \text{or} \quad 2\sin x - 1 = 0$$
$$2\sin x = 1$$
$$\sin x = \frac{1}{2}$$
$$x = \frac{\pi}{2} \quad x = \frac{3\pi}{2} \qquad x = \frac{\pi}{6} \quad x = \frac{5\pi}{6}$$

The solutions in the interval $[0, 2\pi)$ are

$$\frac{\pi}{6}, \frac{\pi}{2}, \frac{5\pi}{6}, \text{and } \frac{3\pi}{2}.$$

71.
$$\cos 2x = \cos x$$
$$2\cos^2 x - 1 = \cos x$$
$$2\cos^2 x - 1 - \cos x = 0$$
$$2\cos^2 x - \cos x - 1 = 0$$
$$(2\cos x + 1)(\cos x - 1) = 0$$
$$2\cos x + 1 = 0 \qquad \text{or} \qquad \cos x - 1 = 0$$
$$2\cos x = -1 \qquad\qquad \cos x = 1$$
$$\cos x = -\frac{1}{2} \qquad\qquad \cos x = 1$$
$$x = \frac{2\pi}{3} \quad x = \frac{4\pi}{3} \qquad\qquad x = 0$$

The solutions in the interval $[0, 2\pi)$ are

$$0, \frac{2\pi}{3}, \text{and } \frac{4\pi}{3}.$$

73.
$$\cos 2x + 5\cos x + 3 = 0$$
$$2\cos^2 x - 1 + 5\cos x + 3 = 0$$
$$2\cos^2 x + 5\cos x + 2 = 0$$
$$(2\cos x + 1)(\cos x + 2) = 0$$
$$2\cos x + 1 = 0 \quad \text{or} \quad \cos x + 2 = 0$$
$$2\cos x = -1 \qquad\qquad \cos x = -2$$
$$\cos x = -\frac{1}{2}$$
$$x = \frac{2\pi}{3} \quad x = \frac{4\pi}{3} \qquad \begin{array}{l} \cos x \text{ cannot} \\ \text{be less than } -1 \end{array}$$

The solutions in the interval $[0, 2\pi)$ are $\dfrac{2\pi}{3}$ and $\dfrac{4\pi}{3}$.

75.
$$\sin x \cos x = \frac{\sqrt{2}}{4}$$
$$2\sin x \cos x = \frac{\sqrt{2}}{2}$$
$$\sin 2x = \frac{\sqrt{2}}{2}$$

The period of the sine function is 2π. In the interval $[0, 2\pi)$, there are two values at which the sine

function is $\dfrac{\sqrt{2}}{2}$. One is $\dfrac{\pi}{4}$. The sine is positive in

quadrant II; thus, the other value is $\pi - \dfrac{\pi}{4} = \dfrac{3\pi}{4}$.

All the solutions to $\sin 2x = \dfrac{\sqrt{2}}{2}$ are given by

$$2x = \frac{\pi}{4} + 2n\pi \quad \text{or} \quad 2x = \frac{3\pi}{4} + 2n\pi$$
$$x = \frac{\pi}{8} + n\pi \qquad\qquad x = \frac{3\pi}{8} + n\pi$$

where n is any integer.
The solutions in the interval $[0, 2\pi)$ are obtained by
letting $n = 0$ and $n = 1$.
The solutions are $\dfrac{\pi}{8}, \dfrac{3\pi}{8}, \dfrac{9\pi}{8}, \text{and } \dfrac{11\pi}{8}$.

77.
$$\sin x + \cos x = 1$$
$$(\sin x + \cos x)^2 = 1^2$$
$$\sin^2 x + 2\sin x \cos x + \cos^2 x = 1$$
$$\sin^2 x + \cos^2 x + 2\sin x \cos x = 1$$
$$1 + 2\sin x \cos x = 1$$
$$2\sin x \cos x = 0$$
$$\sin x \cos x = 0$$
$$\sin x = 0 \quad \text{or} \quad \cos x = 0$$

$$x = 0 \qquad\qquad x = \frac{\pi}{2}$$

$$x = \pi \qquad\qquad x = \frac{3\pi}{2}$$

After checking these proposed solutions, the actual solutions in the interval $[0, 2\pi)$ are 0 and

$$\frac{\pi}{2}.$$

79.
$$\sin\left(x + \frac{\pi}{4}\right) + \sin\left(x - \frac{\pi}{4}\right) = 1$$

$$\frac{1}{2}\left[\sin\left(x + \frac{\pi}{4}\right) + \sin\left(x - \frac{\pi}{4}\right)\right] = 1 \cdot \frac{1}{2}$$

$$\sin x \cos\frac{\pi}{4} = \frac{1}{2}$$

$$\sin x \cdot \frac{\sqrt{2}}{2} = \frac{1}{2}$$

$$\sin x = \frac{1}{\sqrt{2}}$$

$$\sin x = \frac{\sqrt{2}}{2}$$

$$x = \frac{\pi}{4} \quad \text{or} \quad x = \frac{3\pi}{4}$$

The solutions in the interval $[0, 2\pi)$ are $\frac{\pi}{4}$ and $\frac{3\pi}{4}$.

81. $\sin 2x \cos x + \cos 2x \sin x = \dfrac{\sqrt{2}}{2}$

$$\sin(2x + x) = \frac{\sqrt{2}}{2}$$

$$\sin 3x = \frac{\sqrt{2}}{2}$$

The period of the sine function is 2π. In the interval $[0, 2\pi)$, there are two values at which the sine

function is $\dfrac{\sqrt{2}}{2}$. One is $\dfrac{\pi}{4}$. The sine function is

positive in quadrant II; thus, the other value is

$$\pi - \frac{\pi}{4} = \frac{3\pi}{4}.$$

All the solutions to $\sin 3x = \dfrac{\sqrt{2}}{2}$ are given by

$$3x = \frac{\pi}{4} + 2n\pi \quad \text{or} \quad 3x = \frac{3\pi}{4} + 2n\pi$$

$$x = \frac{\pi}{12} + \frac{2n\pi}{3} \qquad\qquad x = \frac{\pi}{4} + \frac{2n\pi}{3}$$

where n is any integer. The solutions in the interval $[0, 2\pi)$ are obtained by letting $n = 0$, $n = 1$, and $n = 2$.

The solutions are $\dfrac{\pi}{12}, \dfrac{\pi}{4}, \dfrac{3\pi}{4}, \dfrac{11\pi}{12}, \dfrac{17\pi}{12}$, and $\dfrac{19\pi}{12}$.

83.
$$\tan x + \sec x = 1$$
$$\tan x - 1 = -\sec x$$
$$(\tan x - 1)^2 = (-\sec x)^2$$
$$\tan^2 x - 2\tan x + 1 = \sec^2 x$$
$$\tan^2 x - 2\tan x + 1 = 1 + \tan^2 x$$
$$-2\tan x = 0$$
$$\tan x = 0$$
$$x = 0 \quad x = \pi$$

We check these proposed solutions to see if any are extraneous.

Check 0: $\tan 0 \ + \ \sec 0 \ \overset{?}{=} \ 1$

$\qquad\qquad 0 \ + \ 1 \ \overset{?}{=} \ 1$ True

Check π: $\tan \pi \ + \ \sec \pi \ \overset{?}{=} \ 1$

$\qquad\qquad 0 \ + \ (-1) \ \overset{?}{=} \ 1$ False

The actual solution in the interval $[0, 2\pi)$ is 0.

85. $\sin x = 0.8246$

Be sure calculator is in radian mode and find the inverse sine of 0.8246. This gives the first quadrant reference angle.

$\theta = \sin^{-1} 0.8246 \approx 0.9695$

The sine is positive in quadrants I and II thus,

$x \approx 0.9695 \quad \text{or} \quad x \approx \pi - 0.9695$

$$x \approx 2.1721$$

87. $\cos x = -\dfrac{2}{5}$

Be sure calculator is in radian mode and find the inverse cosine of $+\dfrac{2}{5}$. This gives the first quadrant reference angle.

$\theta = \cos^{-1}\dfrac{2}{5} \approx 1.1593$

The cosine is negative in quadrants II and III thus,

$x \approx \pi - 1.1593$ or $x \approx \pi + 1.1593$

$x \approx 1.9823$ $x \approx 4.3009$

89. $\tan x = -3$

Be sure calculator is in radian mode and find the inverse tangent of $+3$. This gives the first quadrant reference angle.

$\theta = \tan^{-1}3 \approx 1.2490$

The tangent is negative in quadrants II and IV thus,

$x \approx \pi - 1.2490$ or $x \approx 2\pi - 1.2490$

$x \approx 1.8925$ $x \approx 5.0341$

91. $\cos^2 x - \cos x - 1 = 0$

Use the quadratic formula to solve for $\cos x$.

$\cos x = \dfrac{-b \pm \sqrt{b^2 - 4ac}}{2a}$

$\cos x = \dfrac{-(-1) \pm \sqrt{(-1)^2 - 4(1)(-1)}}{2(1)}$

$\cos x = \dfrac{1 \pm \sqrt{5}}{2}$

$\cos x \approx -0.6180$ or $\cos x \approx 1.6180$

$\phantom{\cos x \approx -0.6180 \text{ or }}$ ~~$\cos x \approx 1.6180$~~

$\phantom{\cos x \approx -0.6180 \text{ or }}$ This equation

$\phantom{\cos x \approx -0.6180 \text{ or }}$ has no solution.

Be sure calculator is in radian mode and find the inverse cosine of $+0.6180$. This gives the first quadrant reference angle.

$\theta = \cos^{-1}0.6180 \approx 0.9046$

The cosine is negative in quadrants II and III thus,

$x \approx \pi - 0.9046$ or $x \approx \pi + 0.9046$

$x \approx 2.2370$ $x \approx 4.0462$

93. $4\tan^2 x - 8\tan x + 3 = 0$

$(2\tan x - 1)(2\tan x - 3) = 0$

$2\tan x - 1 = 0$ or $2\tan x - 3 = 0$

$\tan x = \dfrac{1}{2}$ $\tan x = \dfrac{3}{2}$

$x \approx 0.4636,\ 3.6052$ $x \approx 0.9828,\ 4.1244$

95. $7\sin^2 x - 1 = 0$

$\sin^2 x = \dfrac{1}{7}$

$\sin x = \pm\sqrt{\dfrac{1}{7}}$

$\sin x = \pm\dfrac{\sqrt{7}}{7}$

$\sin x \approx 0.3780$ or $\sin x \approx -0.3780$

$x \approx 0.3876,\ 2.7540$ $x \approx 3.5292,\ 5.8956$

97. $2\cos 2x + 1 = 0$

$\cos 2x = -\dfrac{1}{2}$

The period of the cosine function is 2π. On the interval $[0, 2\pi)$ the cosine function equals $-\dfrac{1}{2}$ at

$\dfrac{2\pi}{3}$ and $\dfrac{4\pi}{3}$. This means that $2x = \dfrac{2\pi}{3}$ or

$2x = \dfrac{4\pi}{3}$. Because the period is 2π, all the solutions of the equation are given by

$2x = \dfrac{2\pi}{3} + 2n\pi$ or $2x = \dfrac{4\pi}{3} + 2n\pi$

$x = \dfrac{\pi}{3} + n\pi$ $x = \dfrac{2\pi}{3} + n\pi$

Use all values of n that result in x values on the interval $[0, 2\pi)$. Thus,

$x = \dfrac{\pi}{3}, \dfrac{5\pi}{3}$ or $x = \dfrac{2\pi}{3}, \dfrac{4\pi}{3}$

99. $\sin 2x + \sin x = 0$

$2\sin x \cos x + \sin x = 0$

$\sin x(2\cos x + 1) = 0$

$\sin x = 0$ or $2\cos x + 1 = 0$

$x = 0, \pi$ $\cos x = -\dfrac{1}{2}$

 $x = \dfrac{2\pi}{3}, \dfrac{4\pi}{3}$

Thus, $x = 0, \dfrac{2\pi}{3}, \pi,$ and $\dfrac{4\pi}{3}$.

101. $3\cos x - 6\sqrt{3} = \cos x - 5\sqrt{3}$

$$2\cos x = \sqrt{3}$$

$$\cos x = \frac{\sqrt{3}}{2}$$

$$x = \frac{\pi}{6}, \frac{11\pi}{6}$$

103. $\tan x = -4.7143$

Be sure calculator is in radian mode and find the inverse tangent of +4.7143. This gives the first quadrant reference angle. $\theta = \tan^{-1} 4.7143 \approx 1.3618$
The tangent is negative in quadrants II and IV thus,

$x \approx \pi - 1.3618$ or $x \approx 2\pi - 1.3618$

$x \approx 1.7798$ $\qquad$ $x \approx 4.9214$

105. $\qquad 2\sin^2 x = 3 - \sin x$

$$2\sin^2 x + \sin x - 3 = 0$$

$$(\sin x - 1)(2\sin x + 3) = 0$$

$\sin x - 1 = 0$ or $2\sin x + 3 = 0$

$\sin x = 1$ $\qquad$ $\sin x = -\dfrac{3}{2}$

$x = \dfrac{\pi}{2}$ $\qquad$ $\cancel{\sin x = -\dfrac{3}{2}}$

107. $\qquad \cos x \csc x = 2\cos x$

$$\cos x \csc x - 2\cos x = 0$$

$$\cos x (\csc x - 2) = 0$$

$\cos x = 0$ $\qquad$ or $\qquad$ $\csc x - 2 = 0$

$\sin x = \dfrac{1}{2}$ $\qquad\qquad\qquad$ $\csc x = 2$

$x = \dfrac{\pi}{2}, \dfrac{3\pi}{2}$ $\qquad\qquad$ $\sin x = \dfrac{1}{2}$

$\qquad\qquad\qquad\qquad\qquad$ $x = \dfrac{\pi}{6}, \dfrac{5\pi}{6}$

Thus, $x = \dfrac{\pi}{6}, \dfrac{\pi}{2}, \dfrac{5\pi}{6},$ and $\dfrac{3\pi}{2}$.

109. $5\cot^2 x - 15 = 0$

$$\cot^2 x = 3$$

$$\tan^2 x = \frac{1}{3}$$

$$\tan x = \pm\sqrt{\frac{1}{3}}$$

$$\tan x = \pm\frac{\sqrt{3}}{3}$$

$\tan x = \dfrac{\sqrt{3}}{3}$ $\qquad$ or $\qquad$ $\tan x = -\dfrac{\sqrt{3}}{3}$

$x = \dfrac{\pi}{6}, \dfrac{7\pi}{6}$ $\qquad\qquad$ $x = \dfrac{5\pi}{6}, \dfrac{11\pi}{6}$

Thus, $x = \dfrac{\pi}{6}, \dfrac{5\pi}{6}, \dfrac{7\pi}{6},$ and $\dfrac{11\pi}{6}$.

111. $\cos^2 x + 2\cos x - 2 = 0$

Use the quadratic formula to solve for $\cos x$.

$$\cos x = \frac{-b \pm \sqrt{b^2 - 4ac}}{2a}$$

$$\cos x = \frac{-(2) \pm \sqrt{(2)^2 - 4(1)(-2)}}{2(1)}$$

$$\cos x = \frac{-2 \pm \sqrt{12}}{2}$$

$$\cos x = \frac{-2 \pm 2\sqrt{3}}{2}$$

$$\cos x = -1 \pm \sqrt{3}$$

$\cos x \approx 0.7321$ $\qquad$ or $\qquad$ $\cos x \approx -2.7321$

$\qquad\qquad\qquad\qquad\qquad$ $\cancel{\cos x \approx -2.7321}$

$\qquad\qquad\qquad\qquad\qquad$ This equation

$\qquad\qquad\qquad\qquad\qquad$ has no solution.

Be sure calculator is in radian mode and find the inverse cosine of 0.7321. This gives the first quadrant reference angle.

$\theta = \cos^{-1} 0.7321 \approx 0.7494$

The cosine is positive in quadrants I and IV thus,

$x \approx 0.7494$ $\qquad$ or $\qquad$ $x \approx 2\pi - 0.7494$

$\qquad\qquad\qquad\qquad\qquad$ $x \approx 5.5338$

113.
$$5\sin x = 2\cos^2 x - 4$$
$$5\sin x = 2\left(1 - \sin^2 x\right) - 4$$
$$5\sin x = 2 - 2\sin^2 x - 4$$
$$2\sin^2 x + 5\sin x + 2 = 0$$
$$\left(2\sin x + 1\right)\left(\sin x + 2\right) = 0$$

$2\sin x + 1 = 0$ or $\sin x + 2 = 0$

$\sin x = -\dfrac{1}{2}$ $\qquad \sin x = -2$

$\qquad\qquad\qquad\qquad$ ~~$\sin x = -2$~~

$x = \dfrac{7\pi}{6},\ \dfrac{11\pi}{6}$

115. $2\tan^2 x + 5\tan x + 3 = 0$

$\left(\tan x + 1\right)\left(2\tan x + 3\right) = 0$

$\tan x + 1 = 0$ or $2\tan x + 3 = 0$

$\tan x = -1$ $\qquad \tan x = -\dfrac{3}{2}$

$x = \dfrac{3\pi}{4},\ \dfrac{7\pi}{4}$ $\qquad x \approx 2.1588,\ 5.3004$

117.

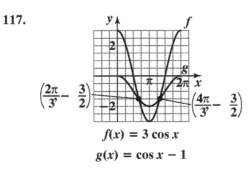

$f(x) = 3\cos x$

$g(x) = \cos x - 1$

$3\cos x = \cos x - 1$

$2\cos x = -1$

$\cos x = -\dfrac{1}{2}$

$x = \dfrac{2\pi}{3},\ \dfrac{4\pi}{3}$

$\left(\dfrac{2\pi}{3}, -\dfrac{3}{2}\right), \left(\dfrac{4\pi}{3}, -\dfrac{3}{2}\right)$

119.

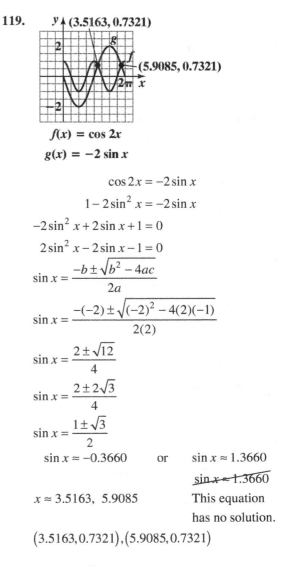

$f(x) = \cos 2x$

$g(x) = -2\sin x$

$$\cos 2x = -2\sin x$$
$$1 - 2\sin^2 x = -2\sin x$$
$$-2\sin^2 x + 2\sin x + 1 = 0$$
$$2\sin^2 x - 2\sin x - 1 = 0$$

$\sin x = \dfrac{-b \pm \sqrt{b^2 - 4ac}}{2a}$

$\sin x = \dfrac{-(-2) \pm \sqrt{(-2)^2 - 4(2)(-1)}}{2(2)}$

$\sin x = \dfrac{2 \pm \sqrt{12}}{4}$

$\sin x = \dfrac{2 \pm 2\sqrt{3}}{4}$

$\sin x = \dfrac{1 \pm \sqrt{3}}{2}$

$\sin x \approx -0.3660$ or $\sin x \approx 1.3660$

$\qquad\qquad\qquad\qquad$ ~~$\sin x \approx 1.3660$~~

$x \approx 3.5163,\ 5.9085$ $\qquad$ This equation has no solution.

$\left(3.5163, 0.7321\right), \left(5.9085, 0.7321\right)$

121. $|\cos x| = \dfrac{\sqrt{3}}{2}$

$\cos x = \dfrac{\sqrt{3}}{2}$ or $\cos x = -\dfrac{\sqrt{3}}{2}$

$x = \dfrac{\pi}{6},\ \dfrac{11\pi}{6}$ $\qquad x = \dfrac{5\pi}{6},\ \dfrac{7\pi}{6}$

123.

$$10\cos^2 x + 3\sin x - 9 = 0$$

$$10\left(1 - \sin^2 x\right) + 3\sin x - 9 = 0$$

$$10 - 10\sin^2 x + 3\sin x - 9 = 0$$

$$-10\sin^2 x + 3\sin x + 1 = 0$$

$$10\sin^2 x - 3\sin x - 1 = 0$$

$$\left(2\sin x - 1\right)\left(5\sin x + 1\right) = 0$$

$$2\sin x - 1 = 0 \qquad \text{or} \quad 5\sin x + 1 = 0$$

$$\sin x = \frac{1}{2} \qquad\qquad \sin x = -\frac{1}{5}$$

$$x = \frac{\pi}{6}, \frac{5\pi}{6} \qquad x = 3.3430, \ 6.0818$$

125.

$$2\cos^3 x + \cos^2 x - 2\cos x - 1 = 0$$

$$\cos^2 x\left(2\cos x + 1\right) - 1\left(2\cos x + 1\right) = 0$$

$$\left(2\cos x + 1\right)\left(\cos^2 x - 1\right) = 0$$

$$\left(2\cos x + 1\right)\left(\cos x + 1\right)\left(\cos x - 1\right) = 0$$

$$\cos x = -\frac{1}{2} \ \text{ or } \ \cos x = -1 \ \text{ or } \ \cos x = 1$$

$$x = 0, \ \frac{2\pi}{3}, \ \pi, \ \frac{4\pi}{3}$$

127. 0.3649, 1.2059, 3.5065, 4.3475

This matches graph a.

129. Substitute $y = 0.3$ into the equation and solve for x:

$$0.3 = 0.6\sin\frac{2\pi}{5}x$$

$$\frac{0.3}{0.6} = \frac{0.6\sin\frac{2\pi}{5}x}{0.6}$$

$$\frac{1}{2} = \sin\frac{2\pi}{5}x$$

$$\sin\frac{2\pi}{5}x = \frac{1}{2}$$

The period of the sine function is 2π. In the interval $[0, 2\pi)$, there are two values at which the sine function is $\frac{1}{2}$. One is $\frac{\pi}{6}$. The sine is positive in quadrant II; thus, the other value is $\pi - \frac{\pi}{6} = \frac{5\pi}{6}$. All of the solutions to $\sin\frac{2\pi}{5}x = \frac{1}{2}$ are given by

$$\frac{2\pi}{5}x = \frac{\pi}{6} + 2n\pi$$

$$x = \frac{5}{12} + 5n$$

or

$$\frac{2\pi}{5}x = \frac{5\pi}{6} + 2n\pi$$

$$x = \frac{25}{12} + 5n$$

where n is any integer. In the interval $[0,5]$ we obtain solutions when $n = 0$. The solutions are $\frac{5}{12}$ and $\frac{25}{12}$.

Therefore, we are inhaling at 0.3 liter per second at $x = \frac{5}{12} \approx 0.4$ second and at $x = \frac{25}{12} \approx 2.1$ seconds.

131. Substitute $y = 10.5$ into the equation and solve for x:

$$10.5 = 3\sin\left[\frac{2\pi}{365}(x - 79)\right] + 12$$

$$-1.5 = 3\sin\left[\frac{2\pi}{365}(x - 79)\right]$$

$$\frac{-1.5}{3} = \frac{3\sin\left[\frac{2\pi}{365}(x - 79)\right]}{3}$$

$$-\frac{1}{2} = \sin\left[\frac{2\pi}{365}(x - 79)\right]$$

$$\sin\left[\frac{2\pi}{365}(x - 79)\right] = -\frac{1}{2}$$

The period of the sine function is 2π. In the interval $[0, 2\pi)$, there are two values at which the sine function is $-\frac{1}{2}$. One is $\pi + \frac{\pi}{6} = \frac{7\pi}{6}$. The other is $2\pi - \frac{\pi}{6} = \frac{11\pi}{6}$. All the solutions to $\sin\left[\frac{2\pi}{365}(x - 79)\right] = -\frac{1}{2}$ are given by

$$\frac{2\pi}{365}(x - 79) = \frac{7\pi}{6} + 2n\pi$$

$$x - 79 = \frac{2555}{12} + 365n$$

$$x = \frac{3503}{12} + 365n$$

or

$$\frac{2\pi}{365}(x-79) = \frac{11\pi}{6} + 2n\pi$$

$$x - 79 = \frac{4015}{12} + 365n$$

$$x = \frac{4963}{12} + 365n$$

where n is any integer.
Substitute various integers for n in the two equations. In the interval $[0,365]$ we obtain values of 49 and 292 days. Thus, Boston has 10.5 hours of daylight 49 and 292 days after January 1.

133. Substitute $d = 2$ into the equation and solve for t:

$$2 = -4\cos\frac{\pi}{3}t$$

$$\frac{2}{-4} = \frac{-4\cos\frac{\pi}{3}t}{-4}$$

$$-\frac{1}{2} = \cos\frac{\pi}{3}t$$

$$\cos\frac{\pi}{3}t = -\frac{1}{2}$$

The period of the cosine function is 2π. In the interval $[0, 2\pi)$, there are two values at which the cosine function is $-\frac{1}{2}$. One is $\frac{2\pi}{3}$. The cosine function is negative in quadrant III; thus, the other value is $2\pi - \frac{2\pi}{3} = \frac{4\pi}{3}$. All solutions to

$$\cos\frac{\pi}{3}t = -\frac{1}{2}$$ are given by

$$\frac{\pi}{3}t = \frac{2\pi}{3} + 2n\pi$$

$$t = 2 + 6n$$

or

$$\frac{\pi}{3}t = \frac{4\pi}{3} + 2n\pi$$

$$t = 4 + 6n$$

where n is any nonnegative integer.

135. Substitute $v_0 = 90$ and $d = 170$, and solve for θ:

$$170 = \frac{90^2}{16}\sin\theta\cos\theta$$

$$\frac{136}{405} = \sin\theta\cos\theta$$

$$2 \cdot \frac{136}{405} = 2\sin\theta\cos\theta$$

$$\frac{272}{405} = \sin 2\theta$$

$$\sin 2\theta = \frac{272}{405}$$

The period of the sine function is $360°$. In the interval $[0, 360°]$, there are two values at which the sine function is $\frac{272}{405}$.

One is $\sin^{-1}\left(\frac{272}{405}\right) \approx 42.19°$. The sine function is positive in quadrant II; Thus, the other value is $180° - 42.19° = 137.81°$. All solutions to

$$\sin 2\theta = \frac{272}{405}$$ are given by

$$2\theta = 42.19° + 360°n$$

$$\theta = 21.095° + 180°n$$

or

$$2\theta = 137.81° + 360°n$$

$$\theta = 68.905° + 180°n$$

where n is any integer.
In the interval $[0, 90°)$ we obtain the solutions by letting $n = 0$. The solutions are approximately $21°$ and $69°$.
Therefore, the angle of elevation should be $21°$ or $69°$.

137. – 145. Answers may vary.

147.

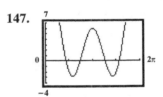

$$x = 1.37, \qquad x = 2.30$$

$$x = 3.98, \quad \text{or} \quad x = 4.91$$

$$x = 0.74$$

149.

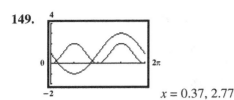

$x = 0.37, 2.77$

151.

$x = 0,$ $x = 1.57,$ $x = 2.09,$

$x = 3.14,$ $x = 4.19,$ or $x = 4.71$

153. makes sense

155. does not make sense; Explanations will vary. Sample explanation: You do not need to solve a trigonometric equation. You nee to find a trigonometric value of an angle ang simplify using arithmetic.

157. false; Changes to make the statement true will vary. A sample change is: The equation has an infinite number of solutions

159. false; Changes to make the statement true will vary. A sample change is: Over this interval, the first equation has two solutions and the second equation has 4 solutions.

161. $\sin 3x + \sin x + \cos x = 0$

$$2\sin\left(\frac{3x+x}{2}\right)\cos\left(\frac{3x-x}{2}\right) + \cos x = 0$$

$$2\sin 2x \cos x + \cos x = 0$$

$$\cos x(2\sin 2x + 1) = 0$$

$\cos x = 0$ or $2\sin 2x + 1 = 0$

$x = \dfrac{\pi}{2}$ $x = \dfrac{3\pi}{2}$ $2\sin 2x = -1$

$$\sin 2x = -\frac{1}{2}$$

The period of the sine function is 2π. In the

interval $[0, 2\pi)$, there are two values at which the sine function is $-\dfrac{1}{2}$. One is $\pi + \dfrac{\pi}{6} = \dfrac{7\pi}{6}$. The other is $2\pi - \dfrac{\pi}{6} = \dfrac{11\pi}{6}$. All the solutions to $\sin 2x = -\dfrac{1}{2}$ are given

$$2x = \frac{7\pi}{6} + 2n\pi \quad \text{or} \quad 2x = \frac{11\pi}{6} + 2n\pi$$

by   where n is

$$x = \frac{7\pi}{12} + n\pi \qquad x = \frac{11\pi}{12} + n\pi$$

any integer. The solutions in the interval $[0, 2\pi)$ are obtained by letting $n = 0$ and $n = 1$. The solutions are

$\dfrac{\pi}{2}, \dfrac{3\pi}{2}, \dfrac{7\pi}{12}, \dfrac{11\pi}{12}, \dfrac{19\pi}{12},$ and $\dfrac{23\pi}{12}$.

163. $\dfrac{a}{\sin 46°} = \dfrac{56}{\sin 63°}$

$a \sin 63° = 56 \sin 46°$

$$a = \frac{56\sin 46°}{\sin 63°}$$

$a \approx 45.2°$

164. $\dfrac{81}{\sin 43°} = \dfrac{62}{\sin B}$

$81 \sin B = 62 \sin 43°$

$$\sin B = \frac{62\sin 43°}{81}$$

$\sin B \approx 0.522023436$

$B \approx \sin^{-1}(0.522023436)$

$B \approx 31.5°$

165. $\dfrac{51}{\sin 75°} = \dfrac{71}{\sin B}$

$51 \sin B = 71 \sin 75°$

$$\sin B = \frac{71\sin 75°}{51}$$

$\sin B \approx 1.344720268$

No solution.

Chapter 6 Review Exercises

1. $\sec x - \cos x = \dfrac{1}{\cos x} - \cos x$

$\qquad\qquad\quad = \dfrac{1}{\cos x} - \dfrac{\cos x}{1} \cdot \dfrac{\cos x}{\cos x}$

$\qquad\qquad\quad = \dfrac{1}{\cos x} - \dfrac{\cos^2 x}{\cos x}$

$\qquad\qquad\quad = \dfrac{1 - \cos^2 x}{\cos x}$

$\qquad\qquad\quad = \dfrac{\sin^2 x}{\cos x}$

$\qquad\qquad\quad = \dfrac{\sin x}{\cos x} \cdot \sin x$

$\qquad\qquad\quad = \tan x \sin x$

2. $\cos x + \sin x \tan x$

$\qquad = \dfrac{\cos x}{\cos x} \cdot \cos x + \sin x \cdot \dfrac{\sin x}{\cos x}$

$\qquad = \dfrac{\cos^2 x}{\cos x} + \dfrac{\sin^2 x}{\cos x}$

$\qquad = \dfrac{\cos^2 x + \sin^2 x}{\cos x}$

$\qquad = \dfrac{1}{\cos x} = \sec x$

3. $\sin^2 \theta (1 + \cot^2 \theta) = \sin^2 \theta + \sin^2 \theta \cot^2 \theta$

$\qquad\qquad\qquad\quad = \sin^2 \theta + \sin^2 \theta \cdot \dfrac{\cos^2 \theta}{\sin^2 \theta}$

$\qquad\qquad\qquad\quad = \sin^2 \theta + \cos^2 \theta$

$\qquad\qquad\qquad\quad = 1$

4. $(\sec \theta - 1)(\sec \theta + 1) = \sec^2 \theta - 1$

$\qquad\qquad\qquad\qquad\quad = 1 + \tan^2 \theta - 1$

$\qquad\qquad\qquad\qquad\quad = \tan^2 \theta$

5. $\dfrac{1 - \tan x}{\sin x} = \dfrac{1}{\sin x} - \dfrac{\tan x}{\sin x}$

$\qquad\qquad\quad = \csc x - \dfrac{\sin x}{\cos x} \cdot \dfrac{1}{\sin x}$

$\qquad\qquad\quad = \csc x - \dfrac{1}{\cos x}$

$\qquad\qquad\quad = \csc x - \sec x$

6. $\dfrac{1}{\sin t - 1} + \dfrac{1}{\sin t + 1}$

$\qquad = \dfrac{1}{\sin t - 1} \cdot \dfrac{\sin t + 1}{\sin t + 1} + \dfrac{1}{\sin t + 1} \cdot \dfrac{\sin t - 1}{\sin t - 1}$

$\qquad = \dfrac{\sin t + 1}{\sin^2 t - 1} + \dfrac{\sin t - 1}{\sin^2 t - 1}$

$\qquad = \dfrac{\sin t + 1 + \sin t - 1}{\sin^2 t - 1}$

$\qquad = \dfrac{2 \sin t}{\sin^2 t - 1}$

$\qquad = \dfrac{2 \sin t}{-\cos^2 t}$

$\qquad = -2 \cdot \dfrac{\sin t}{\cos t} \cdot \dfrac{1}{\cos t}$

$\qquad = -2 \tan t \sec t$

7. $\dfrac{1 + \sin t}{\cos^2 t} = \dfrac{1}{\cos^2 t} + \dfrac{\sin t}{\cos^2 t}$

$\qquad\qquad\quad = \sec^2 t + \dfrac{\sin t}{\cos t} \cdot \dfrac{1}{\cos t}$

$\qquad\qquad\quad = \tan^2 t + 1 + \tan t \sec t$

8. $\dfrac{\cos x}{1 - \sin x} = \dfrac{\cos x}{1 - \sin x} \cdot \dfrac{1 + \sin x}{1 + \sin x}$

$\qquad\qquad\quad = \dfrac{\cos x(1 + \sin x)}{1 - \sin^2 x}$

$\qquad\qquad\quad = \dfrac{\cos x(1 + \sin x)}{\cos^2 x}$

$\qquad\qquad\quad = \dfrac{1 + \sin x}{\cos x}$

9. $1 - \dfrac{\sin^2 x}{1 + \cos x} = 1 - \dfrac{1 - \cos^2 x}{1 + \cos x}$

$\qquad\qquad\qquad\quad = 1 - \dfrac{(1 + \cos x)(1 - \cos x)}{1 + \cos x}$

$\qquad\qquad\qquad\quad = 1 - (1 - \cos x)$

$\qquad\qquad\qquad\quad = 1 - 1 + \cos x$

$\qquad\qquad\qquad\quad = \cos x$

10. $(\tan \theta + \cot \theta)^2$

$\qquad = \tan^2 \theta + 2 \tan \theta \cot \theta + \cot^2 \theta$

$\qquad = \sec^2 \theta - 1 + 2 \dfrac{\sin \theta}{\cos \theta} \cdot \dfrac{\cos \theta}{\sin \theta} + \csc^2 \theta - 1$

$\qquad = \sec^2 \theta - 1 + 2 + \csc^2 \theta - 1$

$\qquad = \sec^2 \theta + \csc^2 \theta$

11. $\dfrac{1}{\sin\theta+\cos\theta}+\dfrac{1}{\sin\theta-\cos\theta}$

$=\dfrac{\sin\theta-\cos\theta}{\sin\theta-\cos\theta}\cdot\dfrac{1}{\sin\theta+\cos\theta}$

$\quad+\dfrac{\sin\theta+\cos\theta}{\sin\theta+\cos\theta}\cdot\dfrac{1}{\sin\theta-\cos\theta}$

$=\dfrac{\sin\theta-\cos\theta}{\sin^2\theta-\cos^2\theta}+\dfrac{\sin\theta+\cos\theta}{\sin^2\theta-\cos^2\theta}$

$=\dfrac{2\sin\theta}{\sin^2\theta-\cos^2\theta}$

$=\dfrac{2\sin\theta}{\sin^2\theta-\cos^2\theta}\cdot\dfrac{\sin^2\theta+\cos^2\theta}{\sin^2\theta+\cos^2\theta}$

$=\dfrac{2\sin\theta\cdot 1}{\sin^4\theta-\cos^4\theta}$

$=\dfrac{2\sin\theta}{\sin^4\theta-\cos^4\theta}$

12. $\dfrac{\cos t}{\cot t-5\cos t}=\dfrac{\cos t}{\cot t-5\cos t}\cdot\dfrac{\frac{1}{\cos t}}{\frac{1}{\cos t}}$

$=\dfrac{\frac{\cos t}{\cos t}}{\frac{\cot t-5\cos t}{\cos t}}$

$=\dfrac{1}{\frac{\cot t}{\cos t}-5}$

$=\dfrac{1}{\frac{\frac{\cos t}{\sin t}}{\cos t}-5}$

$=\dfrac{1}{\frac{\cos t}{\sin t}\cdot\frac{1}{\cos t}-5}$

$=\dfrac{1}{\frac{1}{\sin t}-5}$

$=\dfrac{1}{\csc t-5}$

13. $\dfrac{1-\cos t}{1+\cos t}=\dfrac{1-\cos t}{1+\cos t}\cdot\dfrac{1-\cos t}{1-\cos t}$

$=\dfrac{(1-\cos t)^2}{1-\cos^2 t}$

$=\dfrac{(1-\cos t)^2}{\sin^2 t}$

$=\left(\dfrac{1-\cos t}{\sin t}\right)^2$

$=\left(\dfrac{1}{\sin t}-\dfrac{\cos t}{\sin t}\right)^2$

$=(\csc t-\cot t)^2$

14. $\cos(45°+30°)$

$=\cos 45°\cos 30°-\sin 45°\sin 30°$

$=\dfrac{\sqrt{2}}{2}\cdot\dfrac{\sqrt{3}}{2}-\dfrac{\sqrt{2}}{2}\cdot\dfrac{1}{2}$

$=\dfrac{\sqrt{6}}{4}-\dfrac{\sqrt{2}}{4}$

$=\dfrac{\sqrt{6}-\sqrt{2}}{4}$

15. $\sin 195°=\sin(135°+60°)$

$=\sin 135°\cos 60°+\cos 135°\sin 60°$

$=\dfrac{\sqrt{2}}{2}\cdot\dfrac{1}{2}+\left(-\dfrac{\sqrt{2}}{2}\right)\cdot\dfrac{\sqrt{3}}{2}$

$=\dfrac{\sqrt{2}}{4}-\dfrac{\sqrt{6}}{4}$

$=\dfrac{\sqrt{2}-\sqrt{6}}{4}$

16. $\tan\left(\dfrac{4\pi}{3}-\dfrac{\pi}{4}\right)=\dfrac{\tan\frac{4\pi}{3}-\tan\frac{\pi}{4}}{1+\tan\frac{4\pi}{3}\cdot\tan\frac{\pi}{4}}=\dfrac{\sqrt{3}-1}{1+\sqrt{3}\cdot(1)}$

$=\dfrac{\left(\sqrt{3}-1\right)\left(1-\sqrt{3}\right)}{\left(1+\sqrt{3}\right)\left(1-\sqrt{3}\right)}=\dfrac{-\left(1-\sqrt{3}\right)^2}{1-3}$

$=\dfrac{-\left(1-2\sqrt{3}+3\right)}{-2}=\dfrac{1-2\sqrt{3}+3}{2}$

$=\dfrac{4-2\sqrt{3}}{2}=\dfrac{2\left(2-\sqrt{3}\right)}{2}=2-\sqrt{3}$

17. $\tan\dfrac{5\pi}{12}=\tan\left(\dfrac{2\pi}{12}+\dfrac{3\pi}{12}\right)$

$=\tan\left(\dfrac{\pi}{6}+\dfrac{\pi}{4}\right)$

$=\dfrac{\tan\frac{\pi}{6}+\tan\frac{\pi}{4}}{1-\tan\frac{\pi}{6}\tan\frac{\pi}{4}}$

$=\dfrac{\frac{\sqrt{3}}{3}+1}{1-\frac{\sqrt{3}}{3}\cdot 1}=\dfrac{\frac{\sqrt{3}}{3}+1}{1-\frac{\sqrt{3}}{3}}\cdot\dfrac{\left(1+\frac{\sqrt{3}}{3}\right)}{\left(1+\frac{\sqrt{3}}{3}\right)}$

$=\dfrac{\frac{2\sqrt{3}}{3}+1+\frac{1}{3}}{1-\frac{1}{3}}=\dfrac{\frac{2\sqrt{3}}{3}+\frac{4}{3}}{\frac{2}{3}}$

$=\left(\dfrac{2\sqrt{3}}{3}+\dfrac{4}{3}\right)\cdot\dfrac{3}{2}=\sqrt{3}+2$

18. $\cos 65° \cos 5° + \sin 65° \sin 5°$

$= \cos(65° - 5°)$

$= \cos 60°$

$= \dfrac{1}{2}$

19. $\sin 80° \cos 50° - \cos 80° \sin 50°$

$= \sin(80° - 50°)$

$= \sin 30°$

$= \dfrac{1}{2}$

20. $\sin\left(x + \dfrac{\pi}{6}\right) - \cos\left(x + \dfrac{\pi}{3}\right)$

$= \sin x \cos \dfrac{\pi}{6} + \cos x \sin \dfrac{\pi}{6}$

$\qquad - \left(\cos x \cos \dfrac{\pi}{3} - \sin x \sin \dfrac{\pi}{3}\right)$

$= \sin x \cdot \dfrac{\sqrt{3}}{2} + \cos x \cdot \dfrac{1}{2}$

$\qquad - \left(\cos x \cdot \dfrac{1}{2} - \sin x \cdot \dfrac{\sqrt{3}}{2}\right)$

$= 2 \cdot \dfrac{\sqrt{3}}{2} \cdot \sin x$

$= \sqrt{3} \sin x$

21. $\tan\left(x + \dfrac{3\pi}{4}\right) = \dfrac{\tan x + \tan \frac{3\pi}{4}}{1 - \tan x \tan \frac{3\pi}{4}}$

$= \dfrac{\tan x + (-1)}{1 - \tan x(-1)}$

$= \dfrac{\tan x - 1}{1 + \tan x}$

22. $\sec(\alpha + \beta) = \dfrac{1}{\cos(\alpha + \beta)}$

$= \dfrac{1}{\cos\alpha\cos\beta - \sin\alpha\sin\beta}$

$= \dfrac{1}{\cos\alpha\cos\beta - \sin\alpha\sin\beta} \cdot \dfrac{\frac{1}{\cos\alpha\cos\beta}}{\frac{1}{\cos\alpha\cos\beta}}$

$= \dfrac{\frac{1}{\cos\alpha\cos\beta}}{\frac{\cos\alpha\cos\beta - \sin\alpha\sin\beta}{\cos\alpha\cos\beta}}$

$= \dfrac{\frac{1}{\cos\alpha} \cdot \frac{1}{\cos\beta}}{\frac{\cos\alpha\cos\beta}{\cos\alpha\cos\beta} - \frac{\sin\alpha\sin\beta}{\cos\alpha\cos\beta}}$

$= \dfrac{\sec\alpha\sec\beta}{1 - \frac{\sin\alpha\sin\beta}{\cos\alpha\cos\beta}}$

$= \dfrac{\sec\alpha\sec\beta}{1 - \tan\alpha\tan\beta}$

23. $\dfrac{\cos(\alpha - \beta)}{\cos\alpha\cos\beta} = \dfrac{\cos\alpha\cos\beta + \sin\alpha\sin\beta}{\cos\alpha\cos\beta}$

$= \dfrac{\cos\alpha\cos\beta}{\cos\alpha\cos\beta} + \dfrac{\sin\alpha\sin\beta}{\cos\alpha\cos\beta}$

$= 1 + \tan\alpha\tan\beta$

24. $\cos^4 t - \sin^4 t = \left(\cos^2 t - \sin^2 t\right)\left(\cos^2 t + \sin^2 t\right)$

$= (\cos 2t) \cdot (1)$

$= \cos 2t$

25. $\sin t - \cos 2t = \sin t - \left(1 - 2\sin^2 t\right)$

$= \sin t - 1 + 2\sin^2 t$

$= 2\sin^2 t + \sin t - 1$

$= (2\sin t - 1)(\sin t + 1)$

26. $\dfrac{\sin 2\theta - \sin \theta}{\cos 2\theta + \cos \theta} = \dfrac{2\sin\theta\cos\theta - \sin\theta}{2\cos^2\theta - 1 + \cos\theta}$

$\qquad = \dfrac{\sin\theta(2\cos\theta - 1)}{2\cos^2\theta + \cos\theta - 1}$

$\qquad = \dfrac{\sin\theta(2\cos\theta - 1)}{(2\cos\theta - 1)(\cos\theta + 1)}$

$\qquad = \dfrac{\sin\theta}{\cos\theta + 1}$

$\qquad = \dfrac{\sin\theta}{\cos\theta + 1} \cdot \dfrac{\cos\theta - 1}{\cos\theta - 1}$

$\qquad = \dfrac{\sin\theta(\cos\theta - 1)}{\cos^2\theta - 1}$

$\qquad = \dfrac{\sin\theta(\cos\theta - 1)}{-\sin^2\theta}$

$\qquad = \dfrac{-(\cos\theta - 1)}{\sin\theta}$

$\qquad = \dfrac{1 - \cos\theta}{\sin\theta}$

27. $\dfrac{\sin 2\theta}{1 - \sin^2\theta} = \dfrac{2\sin\theta\cos\theta}{\cos^2\theta}$

$\qquad = \dfrac{2\sin\theta}{\cos\theta} \cdot \dfrac{\cos\theta}{\cos\theta}$

$\qquad = 2\tan\theta$

28. $2\sin t\cos t\sec 2t = \sin 2t \cdot \sec 2t$

$\qquad = \sin 2t \cdot \dfrac{1}{\cos 2t}$

$\qquad = \dfrac{\sin 2t}{\cos 2t}$

$\qquad = \tan 2t$

29. $\cos 4t = \cos(2 \cdot 2t)$

$\qquad = 1 - 2\sin^2 2t$

$\qquad = 1 - 2(\sin 2t)^2$

$\qquad = 1 - 2 \cdot (2\sin t\cos t)^2$

$\qquad = 1 - 2 \cdot 4\sin^2 t\cos^2 t$

$\qquad = 1 - 8\sin^2 t\cos^2 t$

30. $\tan\dfrac{x}{2}(1 + \cos x) = \dfrac{\sin x}{1 + \cos x} \cdot (1 + \cos x) = \sin x$

31. $\tan\dfrac{x}{2} = \dfrac{1 - \cos x}{\sin x}$

$\qquad = \dfrac{1 - \cos x}{\sin x} \cdot \dfrac{\frac{1}{\cos x}}{\frac{1}{\cos x}}$

$\qquad = \dfrac{\frac{1 - \cos x}{\cos x}}{\frac{\sin x}{\cos x}}$

$\qquad = \dfrac{\frac{1}{\cos x} - \frac{\cos x}{\cos x}}{\tan x}$

$\qquad = \dfrac{\sec x - 1}{\tan x}$

32. a. The graph appears to be the cosine curve, $y = \cos x$. It cycles through maximum, intercept, minimum, intercept and back to maximum. Thus, $y = \cos x$ also describes the graph.

b. $\sin\left(x - \dfrac{3\pi}{2}\right) = \sin x\cos\dfrac{3\pi}{2} - \cos x\sin\dfrac{3\pi}{2}$

$\qquad = \sin x \cdot 0 - \cos x \cdot (-1)$

$\qquad = \cos x$

33. a. The graph appears to be the negative of the sine curve, $y = -\sin x$. It cycles through intercept, minimum, intercept, maximum and back to intercept. Thus, $y = -\sin x$ also describes the graph.

b. $\cos\left(x + \dfrac{\pi}{2}\right) = \cos x\cos\dfrac{\pi}{2} - \sin x\sin\dfrac{\pi}{2}$

$\qquad = \cos x \cdot 0 - \sin x \cdot 1$

$\qquad = -\sin x$

34. **a.** The graph appears to be the tangent curve, $y = \tan x$. It cycles through intercept to positive infinity, then from negative infinity through the intercept. Thus, $y = \tan x$ also describes the graph.

b.
$$y = \frac{\tan x - 1}{1 - \cot x}$$
$$= \frac{\frac{\sin x}{\cos x} - 1}{1 - \frac{\cos x}{\sin x}}$$
$$= \frac{\frac{\sin x - \cos x}{\cos x}}{\frac{\sin x - \cos x}{\sin x}}$$
$$= \frac{\sin x - \cos x}{\cos x} \cdot \frac{\sin x}{\sin x - \cos x}$$
$$= \frac{\sin x}{\cos x}$$
$$= \tan x$$

35. $\sin \alpha = \dfrac{3}{5} = \dfrac{y}{r}$

Because α lies in quadrant I, x is positive.
$$x^2 + 3^2 = 5^2$$
$$x^2 = 5^2 - 3^2 = 16$$
$$x = \sqrt{16} = 4$$
Thus, $\cos \alpha = \dfrac{x}{r} = \dfrac{4}{5}$, and $\tan \alpha = \dfrac{y}{x} = \dfrac{3}{4}$.

$\sin \beta = \dfrac{12}{13} = \dfrac{y}{r}$

Because β lies in quadrant II, x is negative.
$$x^2 + 12^2 = 13^2$$
$$x^2 = 13^2 - 12^2 = 25$$
$$x = -\sqrt{25} = -5$$
Thus, $\cos \beta = \dfrac{x}{r} = \dfrac{-5}{13} = -\dfrac{5}{13}$, and

$\tan \beta = \dfrac{y}{x} = \dfrac{12}{-5} = -\dfrac{12}{5}$.

a. $\sin(\alpha + \beta) = \sin \alpha \cos \beta + \cos \alpha \sin \beta$
$$= \frac{3}{5} \cdot \left(-\frac{5}{13}\right) + \frac{4}{5} \cdot \frac{12}{13} = \frac{33}{65}$$

b. $\cos(\alpha - \beta) = \cos \alpha \cos \beta + \sin \alpha \sin \beta$
$$= \frac{4}{5} \cdot \left(-\frac{5}{13}\right) + \frac{3}{5} \cdot \frac{12}{13} = \frac{16}{65}$$

c. $\tan(\alpha + \beta) = \dfrac{\tan \alpha + \tan \beta}{1 - \tan \alpha \tan \beta}$
$$= \frac{\frac{3}{4} + \left(-\frac{12}{5}\right)}{1 - \frac{3}{4}\left(-\frac{12}{5}\right)}$$
$$= \frac{-\frac{33}{20}}{1 + \frac{36}{20}} = \frac{-\frac{33}{20}}{\frac{56}{20}}$$
$$= -\frac{33}{56}$$

d. $\sin 2\alpha = 2 \sin \alpha \cos \alpha = 2 \cdot \dfrac{3}{5} \cdot \dfrac{4}{5} = \dfrac{24}{25}$

e. $\cos \dfrac{\beta}{2} = \sqrt{\dfrac{1 + \cos \beta}{2}} = \sqrt{\dfrac{1 - \frac{5}{13}}{2}}$
$$= \sqrt{\frac{8}{26}} = \sqrt{\frac{4}{13}} = \frac{2}{\sqrt{13}} = \frac{2\sqrt{13}}{13}$$

36. $\tan \alpha = \dfrac{4}{3} = \dfrac{-4}{-3} = \dfrac{y}{x}$

Because r is a distance, it is positive.
$$r^2 = (-4)^2 + (-3)^2 = 25$$
$$r = \sqrt{25} = 5$$
Thus, $\sin \alpha = \dfrac{y}{r} = \dfrac{-4}{5} = -\dfrac{4}{5}$, and

$\cos \alpha = \dfrac{x}{r} = \dfrac{-3}{5} = -\dfrac{3}{5}$.

$\tan \beta = \dfrac{5}{12} = \dfrac{y}{x}$

Because r is a distance, it is positive.
$$r^2 = 5^2 + 12^2 = 169$$
$$r = \sqrt{169} = 13$$
Thus, $\sin \beta = \dfrac{y}{r} = \dfrac{5}{13}$, and $\cos \beta = \dfrac{x}{r} = \dfrac{12}{13}$.

a. $\sin(\alpha + \beta) = \sin \alpha \cos \beta + \cos \alpha \sin \beta$
$$= -\frac{4}{5} \cdot \frac{12}{13} + \left(-\frac{3}{5}\right) \cdot \frac{5}{13}$$
$$= -\frac{63}{65}$$

b. $\cos(\alpha - \beta) = \cos \alpha \cos \beta + \sin \alpha \sin \beta$
$$= -\frac{3}{5} \cdot \frac{12}{13} + \left(-\frac{4}{5}\right) \cdot \frac{5}{13}$$
$$= -\frac{56}{65}$$

c. $\tan(\alpha+\beta) = \dfrac{\tan\alpha+\tan\beta}{1-\tan\alpha\tan\beta}$

$$= \dfrac{\dfrac{4}{3}+\dfrac{5}{12}}{1-\dfrac{4}{3}\cdot\dfrac{5}{12}} = \dfrac{\dfrac{21}{12}}{1-\dfrac{20}{36}} = \dfrac{\dfrac{21}{12}}{\dfrac{16}{36}}$$

$$= \dfrac{21}{12}\cdot\dfrac{36}{16} = \dfrac{63}{16}$$

d. $\sin 2\alpha = 2\sin\alpha\sin\alpha$

$$= 2\left(-\dfrac{4}{5}\right)\left(-\dfrac{4}{5}\right) = \dfrac{24}{25}$$

e. $\cos\dfrac{\beta}{2} = \sqrt{\dfrac{1+\cos\beta}{2}} = \sqrt{\dfrac{1+\dfrac{12}{13}}{2}}$

$$= \sqrt{\dfrac{25}{26}} = \dfrac{5}{\sqrt{26}} = \dfrac{5\sqrt{26}}{26}$$

37. $\tan\alpha = -3 = \dfrac{3}{-1} = \dfrac{y}{x}$

Because r is a distance, it is positive.

$$r^2 = 3^2 + (-1)^2$$
$$r^2 = 10$$
$$r = \sqrt{10}$$

$$\sin\alpha = \dfrac{3}{\sqrt{10}} = \dfrac{3\sqrt{10}}{10}$$

$$\cos\alpha = \dfrac{-1}{\sqrt{10}} = -\dfrac{\sqrt{10}}{10}$$

$$\cot\beta = -3 = \dfrac{3}{-1} = \dfrac{x}{y}$$

Because r is a distance, it is positive.

$$r^2 = 3^2 + (-1)^2$$
$$r^2 = 10$$
$$r = \sqrt{10}$$

$$\sin\beta = \dfrac{-1}{\sqrt{10}} = -\dfrac{\sqrt{10}}{10}$$

$$\cos\beta = \dfrac{3}{\sqrt{10}} = \dfrac{3\sqrt{10}}{10}$$

a. $\sin(\alpha+\beta) = \sin\alpha\cos\beta + \cos\alpha\sin\beta$

$$= \dfrac{3\sqrt{10}}{10}\cdot\dfrac{3\sqrt{10}}{10} + \left(-\dfrac{\sqrt{10}}{10}\right)\left(-\dfrac{\sqrt{10}}{10}\right)$$

$$= \dfrac{90}{100} + \dfrac{10}{100}$$

$$= \dfrac{100}{100}$$

$$= 1$$

b. $\cos(\alpha-\beta)$
$$= \cos\alpha\cos\beta + \sin\alpha\sin\beta$$

$$= \left(-\dfrac{\sqrt{10}}{10}\right)\left(\dfrac{3\sqrt{10}}{10}\right) + \dfrac{3\sqrt{10}}{10}\left(\dfrac{-\sqrt{10}}{10}\right)$$

$$= -\dfrac{60}{100}$$

$$= -\dfrac{3}{5}$$

c. $\tan(\alpha+\beta) = \dfrac{\tan\alpha+\tan\beta}{1-\tan\alpha\tan\beta}$

$$= \dfrac{-3+\left(\dfrac{-1}{3}\right)}{1-(-3)\left(-\dfrac{1}{3}\right)}$$

$$= \dfrac{\dfrac{-10}{3}}{0}$$

Since this value is undefined, the tangent function is undefined at $\alpha+\beta$.

d. $\sin 2\alpha = 2\sin\alpha\cos\alpha$

$$= 2\left(\dfrac{3\sqrt{10}}{10}\right)\left(\dfrac{-\sqrt{10}}{10}\right)$$

$$= -\dfrac{3}{5}$$

e. $\cos\dfrac{\beta}{2} = \sqrt{\dfrac{1+\cos\beta}{2}}$

$$= \sqrt{\dfrac{1+\dfrac{3\sqrt{10}}{10}}{2}}$$

$$= \sqrt{\dfrac{10+3\sqrt{10}}{20}}$$

$$= \dfrac{\sqrt{10+3\sqrt{10}}}{2\sqrt{5}}$$

38. $\sin\alpha = -\dfrac{1}{3} = \dfrac{-1}{3} = \dfrac{y}{r}$

Because α is in quadrant II, x is negative.

$x^2 + (-1)^2 = 3^2$

$\quad x^2 + 1 = 9$

$\quad\quad x^2 = 8$

$\quad\quad\quad x = -\sqrt{8} = -2\sqrt{2}$

$\cos\alpha = \dfrac{-2\sqrt{2}}{3}$

$\tan\alpha = \dfrac{-1}{-2\sqrt{2}} = \dfrac{\sqrt{2}}{4}$

$\cos\beta = -\dfrac{1}{3} = \dfrac{-1}{3} = \dfrac{x}{r}$

Because β is in quadrant III, y is negative.

$(-1)^2 + y^2 = 3^2$

$\quad\quad y^2 = 8$

$\quad\quad\quad y = -\sqrt{8} = -2\sqrt{2}$

$\sin\beta = \dfrac{-2\sqrt{2}}{3}$

$\tan\beta = \dfrac{-2\sqrt{2}}{-1} = 2\sqrt{2}$

a. $\sin(\alpha+\beta)$

$= \sin\alpha\cos\beta + \cos\alpha\sin\beta$

$= -\dfrac{1}{3}\cdot-\dfrac{1}{3} + \left(-\dfrac{2\sqrt{2}}{3}\right)\cdot\left(-\dfrac{2\sqrt{2}}{3}\right)$

$= \dfrac{9}{9} = 1$

b. $\cos(\alpha-\beta)$

$= \cos\alpha\cos\beta + \sin\alpha\sin\beta$

$= -\dfrac{2\sqrt{2}}{3}\cdot\left(-\dfrac{1}{3}\right) + \left(-\dfrac{1}{3}\right)\cdot\left(-\dfrac{2\sqrt{2}}{3}\right)$

$= \dfrac{4\sqrt{2}}{9}$

c. $\tan(\alpha+\beta) = \dfrac{\tan\alpha + \tan\beta}{1 - \tan\alpha\tan B}$

$= \dfrac{\frac{\sqrt{2}}{4} + 2\sqrt{2}}{1 - \left(\frac{\sqrt{2}}{4}\right)\left(2\sqrt{2}\right)}$

$= \dfrac{\frac{9\sqrt{2}}{4}}{0}$

Since this value is undefined, the tangent function is undefined at $\alpha+\beta$.

d. $\sin 2\alpha = 2\sin\alpha\cos\alpha$

$= 2\left(-\dfrac{1}{3}\right)\left(-\dfrac{2\sqrt{2}}{3}\right) = \dfrac{4\sqrt{2}}{9}$

e. $\cos\dfrac{\beta}{2} = -\sqrt{\dfrac{1+\cos\beta}{2}}$

$= -\sqrt{\dfrac{1+\left(-\frac{1}{3}\right)}{2}}$

$= -\sqrt{\dfrac{\frac{2}{3}}{2}} = -\sqrt{\dfrac{1}{3}}$

$= -\dfrac{1}{\sqrt{3}}$

$= -\dfrac{\sqrt{3}}{3}$

39. The given expression is the right side of the formula for $\cos 2\theta$ with $\theta = 15°$.

$\cos^2 15° - \sin^2 15° = \cos(2\cdot 15°)$

$\quad\quad\quad\quad\quad\quad\quad = \cos 30°$

$\quad\quad\quad\quad\quad\quad\quad = \dfrac{\sqrt{3}}{2}$

40. The given expression is the right side of the formula for $\tan 2\theta$ with $\theta = \dfrac{5\pi}{12}$.

$\dfrac{2\tan\frac{5\pi}{12}}{1-\tan^2\frac{5\pi}{12}} = \tan\left(2\cdot\dfrac{5\pi}{12}\right)$

$\quad\quad\quad\quad\quad = \tan\dfrac{5\pi}{6}$

$\quad\quad\quad\quad\quad = -\dfrac{\sqrt{3}}{3}$

41. Because 22.5° lies in quadrant I,
sin 22.5° > 0.

$$\sin 22.5° = \sin \frac{45°}{2}$$

$$= \sqrt{\frac{1 - \cos 45°}{2}} = \sqrt{\frac{1 - \frac{\sqrt{2}}{2}}{2}}$$

$$= \sqrt{\frac{2 - \sqrt{2}}{4}} = \frac{\sqrt{2 - \sqrt{2}}}{2}$$

42. Because $\frac{\pi}{12}$ lies in quadrant I, $\tan \frac{\pi}{12} > 0$.

$$\tan \frac{\pi}{12} = \tan \frac{\frac{\pi}{6}}{2}$$

$$= \frac{1 - \cos \frac{\pi}{6}}{\sin \frac{\pi}{6}} = \frac{1 - \frac{\sqrt{3}}{2}}{\frac{1}{2}}$$

$$= 2 - \sqrt{3}$$

43. $\sin 6x \sin 4x$

$$= \frac{1}{2} \left[\cos(6x - 4x) - \cos(6x + 4x) \right]$$

$$= \frac{1}{2} \left[\cos 2x - \cos 10x \right]$$

44. $\sin 7x \cos 3x$

$$= \frac{1}{2} \left[\sin(7x + 3x) + \sin(7x - 3x) \right]$$

$$= \frac{1}{2} \left[\sin 10x + \sin 4x \right]$$

45. $\sin 2x - \sin 4x$

$$= 2 \sin \left(\frac{2x - 4x}{2} \right) \cos \left(\frac{2x + 4x}{2} \right)$$

$$= 2 \sin(-x) \cos 3x$$

$$= -2 \sin x \cos 3x$$

46. $\cos 75° + \cos 15°$

$$= 2 \cos \left(\frac{75° + 15°}{2} \right) \cos \left(\frac{75° - 15°}{2} \right)$$

$$= 2 \cos 45° \cos 30°$$

$$= 2 \left(\frac{\sqrt{2}}{2} \right) \left(\frac{\sqrt{3}}{2} \right) = \frac{\sqrt{6}}{2}$$

47. $\dfrac{\cos 3x + \cos 5x}{\cos 3x - \cos 5x} = \dfrac{2 \cos \left(\frac{3x + 5x}{2} \right) \cos \left(\frac{3x - 5x}{2} \right)}{-2 \sin \left(\frac{3x + 5x}{2} \right) \sin \left(\frac{3x - 5x}{2} \right)}$

$$= \frac{2 \cos \left(\frac{8x}{2} \right) \cos \left(\frac{-2x}{2} \right)}{-2 \sin \left(\frac{8x}{2} \right) \sin \left(\frac{-2x}{2} \right)}$$

$$= \frac{2 \cos 4x \cos(-x)}{-2 \sin 4x \sin(-x)}$$

$$= \frac{2 \cos 4x \cos x}{2 \sin 4x \sin x}$$

$$= \frac{\cos 4x}{\sin 4x} \cdot \frac{\cos x}{\sin x}$$

$$= \cot 4x \cot x$$

$$= \cot x \cot 4x$$

48. $\dfrac{\sin 2x + \sin 6x}{\sin 2x - \sin 6x} = \dfrac{2 \sin \left(\frac{2x + 6x}{2} \right) \cos \left(\frac{2x - 6x}{2} \right)}{2 \sin \left(\frac{2x - 6x}{2} \right) \cos \left(\frac{2x + 6x}{2} \right)}$

$$= \frac{2 \sin \left(\frac{8x}{2} \right) \cos \left(\frac{-4x}{2} \right)}{2 \sin \left(\frac{-4x}{2} \right) \cos \left(\frac{8x}{2} \right)}$$

$$= \frac{\sin 4x \cos(-2x)}{\sin(-2x) \cos 4x}$$

$$= -\frac{\sin 4x \cos 2x}{\sin 2x \cos 4x}$$

$$= -\frac{\sin 4x}{\cos 4x} \cdot \frac{\cos 2x}{\sin 2x}$$

$$= -\tan 4x \cot 2x$$

49. **a.** The graph appears to be the cotangent curve, $y = \cot x$. It cycles from positive infinity through the intercept to negative infinity. Thus, $y = \cot x$ also describes the graph.

b. $\dfrac{\cos 3x + \cos x}{\sin 3x - \sin x} = \dfrac{2\cos\left(\dfrac{3x+x}{2}\right)\cos\left(\dfrac{3x-x}{2}\right)}{2\sin\left(\dfrac{3x-x}{2}\right)\cos\left(\dfrac{3x+x}{2}\right)}$

$= \dfrac{2\cos\left(\dfrac{4x}{2}\right)\cos\left(\dfrac{2x}{2}\right)}{2\sin\left(\dfrac{2x}{2}\right)\cos\left(\dfrac{4x}{2}\right)}$

$= \dfrac{2\cos 2x\cos x}{2\sin x\cos 2x}$

$= \dfrac{\cos x}{\sin x}$

$= \cot x$

This verifies our observation that

$y = \dfrac{\cos 3x + \cos x}{\sin 3x - \sin x}$ and $y = \cot x$ describe the same graph.

50. $\cos x = -\dfrac{1}{2}$

Because $\cos\dfrac{\pi}{3} = \dfrac{1}{2}$, the solutions for $\cos x = -\dfrac{1}{2}$ in

$[0, 2\pi)$ are $x = \pi - \dfrac{\pi}{3} = \dfrac{2\pi}{3}$ $x = \pi + \dfrac{\pi}{3} = \dfrac{4\pi}{3}$.

Because the period of the cosine function is 2π, the solutions are given by

$x = \dfrac{2\pi}{3} + 2n\pi$ or $x = \dfrac{4\pi}{3} + 2n\pi$ where n is any integer.

51. $\sin x = \dfrac{\sqrt{2}}{2}$

Because $\sin\dfrac{\pi}{4} = \dfrac{\sqrt{2}}{2}$, the solutions for $\sin x = \dfrac{\sqrt{2}}{2}$

in $[0, 2\pi)$ are $x = \dfrac{\pi}{4}$ and $x = \pi - \dfrac{\pi}{4} = \dfrac{3\pi}{4}$. Because

the period of the cosine function is 2π, the solutions

are given by $x = \dfrac{\pi}{4} + 2n\pi$ or $x = \dfrac{3\pi}{4} + 2n\pi$ where n

is any integer.

52. $2\sin x + 1 = 0$

$2\sin x = -1$

$\sin x = -\dfrac{1}{2}$

Because $\sin\dfrac{\pi}{6} = \dfrac{1}{2}$, the solutions for

$\sin x = -\dfrac{1}{2}$ in $[0, 2\pi)$ are

$x = \pi + \dfrac{\pi}{6} = \dfrac{7\pi}{6}$

$x = 2\pi - \dfrac{\pi}{6} = \dfrac{11\pi}{6}$.

Because the period of the sine function is 2π, the solutions are given by

$x = \dfrac{7\pi}{6} + 2n\pi$ or $x = \dfrac{11\pi}{6} + 2n\pi$

where n is any integer

53. $\sqrt{3}\tan x - 1 = 0$

$\sqrt{3}\tan x = 1$

$\tan x = \dfrac{1}{\sqrt{3}}$

Because $\tan\dfrac{\pi}{6} = \dfrac{1}{\sqrt{3}}$, the solution for

$\tan x = \dfrac{1}{\sqrt{3}}$ in $[0, \pi)$ is $x = \dfrac{\pi}{6}$.

Because the period of the tangent function is π, the solutions are given by

$x = \dfrac{\pi}{6} + n\pi$ where n is any integer.

54. The period of the cosine function is 2π. In the interval $[0, 2\pi)$, the only value at which the cosine function is -1 is π. All the solutions to $\cos 2x = -1$ are given by

$2x = \pi + 2n\pi$

$x = \dfrac{\pi}{2} + n\pi$ where n is any integer.

The solutions in the interval $[0, 2\pi)$ are obtained by letting $n = 0$ and $n = 1$.

The solutions are $\dfrac{\pi}{2}$ and $\dfrac{3\pi}{2}$.

55. The period of the sine function is 2π. In the interval $[0, 2\pi)$, the only value at which the sine function is

1 is $\dfrac{\pi}{2}$. All the solutions to $\sin 3x = 1$ are given by

$$3x = \frac{\pi}{2} + 2n\pi$$

$$x = \frac{\pi}{6} + \frac{2n\pi}{3}$$

where n is any integer. The solutions in the interval $[0, 2\pi)$ are obtained by letting $n = 0$, $n = 1$, and $n = 2$.

The solutions are $\dfrac{\pi}{6}, \dfrac{5\pi}{6}$, and $\dfrac{9\pi}{6}$.

56. The period of the tangent function is π. In the interval $[0, \pi)$, the only value for which the tangent

function is -1 is $\dfrac{3\pi}{4}$. All the solutions to

$\tan \dfrac{x}{2} = -1$ are given by

$$\frac{x}{2} = \frac{3\pi}{4} + n\pi$$

$$x = \frac{3\pi}{2} + 2n\pi$$

where n is any integer. The solution in the interval $[0, 2\pi)$ is obtained by letting $n = 0$.

The solution is $\dfrac{3\pi}{2}$.

57.
$$\tan x = 2\cos x \tan x$$
$$\tan x - 2\cos x \tan x = 0$$
$$\tan x (1 - 2\cos x) = 0$$
$$\tan x = 0 \quad \text{or} \quad 1 - 2\cos x = 0$$
$$x = 0 \ \ x = \pi \qquad -2\cos x = -1$$
$$\cos x = \frac{1}{2}$$
$$x = \frac{\pi}{3} \quad x = \frac{5\pi}{3}$$

The solutions in the interval $[0, 2\pi)$ are 0,

$\dfrac{\pi}{3}$, π, and $\dfrac{5\pi}{3}$.

58. The given equation is in quadratic form $t^2 - 2t = 3$ with $t = \cos x$.

$$\cos^2 x - 2\cos x = 3$$
$$\cos^2 x - 2\cos x - 3 = 0$$
$$(\cos x + 1)(\cos x - 3) = 0$$
$$\cos x + 1 = 0 \quad \text{or} \quad \cos x - 3 = 0$$
$$\cos x = -1 \qquad\qquad \cos x = 3$$
$$x = \pi \qquad\qquad \cos x \text{ cannot be}$$
$$\text{greater than 1.}$$

The solution in the interval $[0, 2\pi)$ is π.

59.
$$2\cos^2 x - \sin x = 1$$
$$2\left(1 - \sin^2 x\right) - \sin x = 1$$
$$2 - 2\sin^2 x - \sin x - 1 = 0$$
$$-2\sin^2 x - \sin x + 1 = 0$$
$$2\sin^2 x + \sin x - 1 = 0$$
$$(2\sin x - 1)(\sin x + 1) = 0$$
$$2\sin x - 1 = 0 \quad \text{or} \quad \sin x + 1 = 0$$
$$2\sin x = 1 \qquad\qquad \sin x = -1$$
$$\sin x = \frac{1}{2} \qquad\qquad x = \frac{3\pi}{2}$$
$$x = \frac{\pi}{6}, \frac{5\pi}{6}$$

The solutions in the interval $[0, 2\pi)$ are $\dfrac{\pi}{6}, \dfrac{5\pi}{6}$, and

$\dfrac{3\pi}{2}$.

60. The given equation is in quadratic form $4t^2 = 1$ with $t = \sin x$.

$$4\sin^2 x = 1$$
$$4\sin^2 x - 1 = 0$$
$$(2\sin x - 1)(2\sin x + 1) = 0$$
$$2\sin x - 1 = 0 \quad \text{or} \quad 2\sin x + 1 = 0$$
$$2\sin x = 1 \qquad\qquad 2\sin x = -1$$
$$\sin x = \frac{1}{2} \qquad\qquad \sin x = -\frac{1}{2}$$
$$x = \frac{\pi}{6} \quad x = \frac{5\pi}{6} \qquad x = \frac{7\pi}{6} \quad x = \frac{11\pi}{6}$$

The solutions in the interval $[0, 2\pi)$ are

$\dfrac{\pi}{6}, \dfrac{5\pi}{6}, \dfrac{7\pi}{6}$, and $\dfrac{11\pi}{6}$.

61.
$$\cos 2x - \sin x = 1$$
$$2\cos^2 x - 1 - \sin x = 1$$
$$2(1 - \sin^2 x) - \sin x - 2 = 0$$
$$2 - 2\sin^2 x - \sin x - 2 = 0$$
$$-2\sin^2 x - \sin x = 0$$
$$2\sin^2 x + \sin x = 0$$
$$\sin x (2\sin x + 1) = 0$$

$\sin x = 0 \qquad 2\sin x + 1 = 0$
$x = 0, \pi$
$$\sin x = -\frac{1}{2}$$
$$x = \frac{7\pi}{6}, \frac{11\pi}{6}$$

The solutions in the interval $[0, 2\pi)$ are

$0, \pi, \dfrac{7\pi}{6},$ and $\dfrac{11\pi}{6}$..

62.
$$\sin 2x = \sqrt{3} \sin x$$
$$2\sin x \cos x = \sqrt{3} \sin x$$
$$2\sin x \cos x - \sqrt{3} \sin x = 0$$
$$\sin x (2\cos x - \sqrt{3}) = 0$$

$\sin x = 0 \quad$ or $\quad 2\cos x - \sqrt{3} = 0$
$x = 0 \ \ x = \pi \qquad 2\cos x = \sqrt{3}$
$$\cos x = \frac{\sqrt{3}}{2}$$
$$x = \frac{\pi}{6} \quad x = \frac{11\pi}{6}$$

The solutions in the interval $[0, 2\pi)$ are $0,$

$\dfrac{\pi}{6}, \pi,$ and $\dfrac{11\pi}{6}.$

63.
$$\sin x = \tan x$$
$$\sin x = \frac{\sin x}{\cos x}$$
$$\sin x \bullet \cos x = \sin x$$
$$\sin x \cos x - \sin x = 0$$
$$\sin x (\cos x - 1) = 0$$

$\sin x = 0 \quad$ or $\quad \cos x - 1 = 0$
$x = 0 \ \ x = \pi \qquad \cos x = 1$
$$x = 0$$

The solutions in the interval $[0, 2\pi)$ are 0 and π.

64. $\sin x = -0.6031$
Be sure calculator is in radian mode and find the inverse sine of $+0.6031$. This gives the first quadrant reference angle.
$$\theta = \sin^{-1}(0.6031) \approx 0.6474$$
The sine is negative in quadrants III and IV thus,
$x \approx \pi + 0.6474 \quad$ or $\quad x \approx 2\pi - 0.6474$
$x \approx 3.7890 \qquad\qquad x \approx 5.6358$

65. $5\cos^2 x - 3 = 0$
$$\cos^2 x = \frac{3}{5}$$
$$\cos x = \pm\sqrt{\frac{3}{5}}$$
$$\cos x = \pm\frac{\sqrt{15}}{5}$$

$\cos x \approx 0.7746 \qquad$ or $\quad \cos x \approx -0.7746$
$x \approx 0.6847, \ \ 5.5985 \qquad x \approx 2.4569, \ \ 3.8263$

66.
$$1 + \tan^2 x = 4\tan x - 2$$
$$\tan^2 x - 4\tan x + 3 = 0$$
$$(\tan x - 1)(\tan x - 3) = 0$$
$\tan x = 1 \quad$ or $\quad \tan x = 3$
$x = \dfrac{\pi}{4}, \dfrac{5\pi}{4} \qquad x \approx 1.2490, \ \ 4.3906$

67. $2\sin^2 x + \sin x - 2 = 0$
$$\sin x = \frac{-b \pm \sqrt{b^2 - 4ac}}{2a}$$
$$\sin x = \frac{-(1) \pm \sqrt{(1)^2 - 4(2)(-2)}}{2(2)}$$
$$\sin x = \frac{-1 \pm \sqrt{17}}{4}$$
$\sin x = 0.7808 \qquad$ or $\quad \sin x = -1.2808$
$x = 0.8959, \ \ 2.2457 \qquad \cancel{\sin x = -1.2808}$

68. Substitute $d = -3$ into the equation and solve for t:
$$-3 = -6\cos\frac{\pi}{2}t$$
$$\frac{-3}{-6} = \frac{-6\cos\frac{\pi}{2}t}{-6}$$
$$\frac{1}{2} = \cos\frac{\pi}{2}t$$
$$\cos\frac{\pi}{2}t = \frac{1}{2}$$

The period of the cosine function is 2π. In the interval $[0, 2\pi)$, there are two values at which the cosine function is $\dfrac{1}{2}$. One is $\dfrac{\pi}{3}$. The cosine function is positive in quadrant IV. Thus, the other value is $2\pi - \dfrac{\pi}{3} = \dfrac{5\pi}{3}$.

All solutions to $\cos \dfrac{\pi}{2} t = \dfrac{1}{2}$ are given by

$$\dfrac{\pi}{2} t = \dfrac{\pi}{3} + 2n\pi$$

$$\dfrac{\pi}{2} t = \dfrac{5\pi}{3} + 2n\pi$$

$$t = \dfrac{2}{3} + 4n \text{ or}$$

$$t = \dfrac{10}{3} + 4n$$

where n is any integer.

69. Substitute $v_0 = 90$ and $d = 100$, and solve for θ:

$$100 = \dfrac{90^2}{16} \sin\theta\cos\theta$$

$$\dfrac{16}{81} = \sin\theta\cos\theta$$

$$2 \cdot \dfrac{16}{81} = 2\sin\theta\cos\theta$$

$$\dfrac{32}{81} = \sin 2\theta$$

$$\sin 2\theta = \dfrac{32}{81}$$

The period of the sine function is $360°$. In the interval $[0, 360°)$, there are two values at which the sine function is $\dfrac{32}{81}$. One is $\sin^{-1}\left(\dfrac{32}{81}\right) \approx 23.27°$. The sine function is positive in quadrant II. Thus, the other value is $180° - 23.27° = 156.73°$. All solutions to $\sin 2\theta = \dfrac{32}{81}$ are given by

$$2\theta = 23.27° + 360°n$$

$$\theta = 11.635° + 180°n$$

or

$$2\theta = 156.73° + 360°n$$

$$\theta = 78.365° + 180°n$$

where n is any integer.
In the interval $[0, 90°)$ we obtain the solutions by letting $n = 0$. The solutions are approximately $12°$ and $78°$. Therefore, the angle of elevation should be $12°$ or $78°$.

Chapter 6 Test

For exercises 1–4: $\sin \alpha = \dfrac{4}{5} = \dfrac{y}{r}$

Because α lies in quadrant II, x is negative.
$$x^2 + 4^2 = 5^2$$
$$x^2 = 5^2 - 4^2 = 9$$
$$x = -\sqrt{9} = -3$$

Thus, $\cos \alpha = \dfrac{x}{r} = \dfrac{-3}{5} = -\dfrac{3}{5}$, and

$$\tan \alpha = \dfrac{y}{x} = \dfrac{4}{-3} = -\dfrac{4}{3}.$$

$$\cos \beta = \dfrac{5}{13} = \dfrac{x}{r}$$

Because β lies in quadrant I, y is positive.
$$5^2 + y^2 = 13^2$$
$$y^2 = 13^2 - 5^2 = 144$$
$$y = \sqrt{144} = 12$$

Thus, $\sin \beta = \dfrac{y}{r} = \dfrac{12}{13}$, and $\tan \beta = \dfrac{y}{x} = \dfrac{12}{5}$.

1. $\cos(\alpha + \beta) = \cos\alpha\cos\beta - \sin\alpha\sin\beta$

$$= -\dfrac{3}{5} \cdot \dfrac{5}{13} - \dfrac{4}{5} \cdot \dfrac{12}{13} = -\dfrac{63}{65}$$

2. $\tan(\alpha - \beta) = \dfrac{\tan\alpha - \tan\beta}{1 + \tan\alpha\tan\beta}$

$$= \dfrac{-\dfrac{4}{3} - \dfrac{12}{5}}{1 + \left(-\dfrac{4}{3}\right) \cdot \dfrac{12}{5}} = \dfrac{-\dfrac{56}{15}}{-\dfrac{33}{15}} = \dfrac{56}{33}$$

3. $\sin 2\alpha = 2\sin\alpha\cos\alpha = 2\left(\dfrac{4}{5}\right)\left(-\dfrac{3}{5}\right) = -\dfrac{24}{25}$

4. $\cos\dfrac{\beta}{2} = \sqrt{\dfrac{1 + \cos\beta}{2}} = \sqrt{\dfrac{1 + \frac{5}{13}}{2}} = \sqrt{\dfrac{18}{26}}$

$$= \dfrac{3\sqrt{2}}{\sqrt{26}} = \dfrac{3\sqrt{52}}{26} = \dfrac{3 \cdot 2\sqrt{13}}{26}$$

$$= \dfrac{3\sqrt{13}}{13}$$

5. $\sin 105° = \sin(135° - 30°)$

$\quad\quad = \sin 135° \cos 30° - \cos 135° \sin 30°$

$\quad\quad = \dfrac{\sqrt{2}}{2} \cdot \dfrac{\sqrt{3}}{2} - \left(-\dfrac{\sqrt{2}}{2}\right) \cdot \dfrac{1}{2}$

$\quad\quad = \dfrac{\sqrt{6}}{4} + \dfrac{\sqrt{2}}{4} = \dfrac{\sqrt{6} + \sqrt{2}}{4}$

6. $\cos x \csc x = \cos x \cdot \dfrac{1}{\sin x} = \dfrac{\cos x}{\sin x} = \cot x$

7. $\dfrac{\sec x}{\cot x + \tan x} = \dfrac{\frac{1}{\cos x}}{\frac{\cos x}{\sin x} + \frac{\sin x}{\cos x}}$

$\quad\quad = \dfrac{\frac{1}{\cos x}}{\frac{\cos x}{\sin x} \cdot \frac{\cos x}{\cos x} + \frac{\sin x}{\cos x} \cdot \frac{\sin x}{\sin x}}$

$\quad\quad = \dfrac{\frac{1}{\cos x}}{\frac{\cos^2 x + \sin^2 x}{\sin x \cos x}} = \dfrac{\frac{1}{\cos x}}{\frac{1}{\sin x \cos x}}$

$\quad\quad = \dfrac{1}{\cos x} \cdot \dfrac{\sin x \cos x}{1}$

$\quad\quad = \sin x$

8. $1 - \dfrac{\cos^2 x}{1 + \sin x} = 1 - \dfrac{\left(1 - \sin^2 x\right)}{1 + \sin x}$

$\quad\quad = 1 - \dfrac{(1 + \sin x)(1 - \sin x)}{1 + \sin x}$

$\quad\quad = 1 - (1 - \sin x)$

$\quad\quad = \sin x$

9. $\cos\left(\theta + \dfrac{\pi}{2}\right) = \cos\theta \cos\dfrac{\pi}{2} - \sin\theta \sin\dfrac{\pi}{2}$

$\quad\quad = \cos\theta \cdot 0 - \sin\theta \cdot 1$

$\quad\quad = -\sin\theta$

10. $\dfrac{\sin(\alpha - \beta)}{\sin\alpha \cos\beta} = \dfrac{\sin\alpha \cos\beta - \cos\alpha \sin\beta}{\sin\alpha \cos\beta}$

$\quad\quad = \dfrac{\sin\alpha \cos\beta}{\sin\alpha \cos\beta} - \dfrac{\cos\alpha \sin\beta}{\sin\alpha \cos\beta}$

$\quad\quad = 1 - \cot\alpha \tan\beta$

11. $\sin t \cos t(\tan t + \cot t) = \sin t \cos t\left(\dfrac{\sin t}{\cos t} + \dfrac{\cos t}{\sin t}\right)$

$\quad\quad = \dfrac{\sin^2 t \cos t}{\cos t} + \dfrac{\sin t \cos^2 t}{\sin t}$

$\quad\quad = \sin^2 t + \cos^2 t$

$\quad\quad = 1$

12.   The period of the sine function is 2π. In the interval $[0, 2\pi)$, there are two values at which the sine function is $-\dfrac{1}{2}$.

One is $\pi + \dfrac{\pi}{6} = \dfrac{7\pi}{6}$. The other is $2\pi - \dfrac{\pi}{6} = \dfrac{11\pi}{6}$. All the solutions to $\sin 3x = -\dfrac{1}{2}$ are given by

$3x = \dfrac{7\pi}{6} + 2n\pi$

$x = \dfrac{7\pi}{18} + \dfrac{2n\pi}{3}$

or

$3x = \dfrac{11\pi}{6} + 2n\pi$

$x = \dfrac{11\pi}{18} + \dfrac{2n\pi}{3}$

where n is any integer. The solutions in the interval $[0, 2\pi)$ are obtained by letting $n = 0$, $n = 1$, and $n = 2$.

The solutions are $\dfrac{7\pi}{18}, \dfrac{11\pi}{18}, \dfrac{19\pi}{18}, \dfrac{23\pi}{18},$

$\dfrac{31\pi}{18}$, and $\dfrac{35\pi}{18}$.

13. $\sin 2x + \cos x = 0$

$\quad 2\sin x \cos x + \cos x = 0$

$\quad\quad \cos x(2\sin x + 1) = 0$

$\cos x = 0 \quad$ or $\quad 2\sin x + 1 = 0$

$x = \dfrac{\pi}{2} \; x = \dfrac{3\pi}{2} \quad\quad\quad 2\sin x = -1$

$\quad\quad\quad\quad\quad\quad\quad\quad\quad \sin x = -\dfrac{1}{2}$

$\quad\quad\quad\quad\quad\quad x = \dfrac{7\pi}{6} \; x = \dfrac{11\pi}{6}$

The solutions in the interval $[0, 2\pi)$ are $\dfrac{\pi}{2}$,

$\dfrac{7\pi}{6}, \dfrac{3\pi}{2}$, and $\dfrac{11\pi}{6}$.

14. $2\cos^2 x - 3\cos x + 1 = 0$

$(2\cos x - 1)(\cos x - 1) = 0$

$2\cos x - 1 = 0$ or $\cos x - 1 = 0$

$2\cos x = 1$ $\cos x = 1$

$\cos x = \dfrac{1}{2}$ $x = 0$

$x = \dfrac{\pi}{3}$ $x = \dfrac{5\pi}{3}$

The solutions in the interval $[0, 2\pi)$ are

$0, \dfrac{\pi}{3},$ and $\dfrac{5\pi}{3}$.

15. $2\sin^2 x + \cos x = 1$

$2\left(1 - \cos^2 x\right) + \cos x - 1 = 0$

$2 - 2\cos^2 x + \cos x - 1 = 0$

$-2\cos^2 x + \cos x + 1 = 0$

$2\cos^2 x - \cos x - 1 = 0$

$(2\cos x + 1)(\cos x - 1) = 0$

$2\cos x + 1 = 0$ or $\cos x - 1 = 0$

$2\cos x = -1$ $\cos x = 1$

$\cos x = -\dfrac{1}{2}$ $x = 0$

$x = \dfrac{2\pi}{3}$ $x = \dfrac{4\pi}{3}$

The solutions in the interval

$[0, 2\pi)$ are $0, \dfrac{2\pi}{3},$ and $\dfrac{4\pi}{3}$.

16. $\cos x = -0.8092$

Be sure calculator is in radian mode and find the inverse cosine of $+0.8092$. This gives the first quadrant reference angle.

$\theta = \cos^{-1}(0.8092) \approx 0.6280$

The cosine is negative in quadrants II and III thus,

$x \approx \pi - 0.6280$ or $x \approx \pi + 0.6280$

$x \approx 2.5136$ $x \approx 3.7696$

17. $\tan x \sec x = 3\tan x$

$\tan x \sec x - 3\tan x = 0$

$\tan x \left(\sec x - 3\right) = 0$

$\tan x = 0$ or $\sec x - 3 = 0$

$x = 0, \ \pi$ $\sec x = 3$

$\cos x = \dfrac{1}{3}$

$x \approx 1.2310, \ 5.0522$

18. $\tan^2 x - 3\tan x - 2 = 0$

$\tan x = \dfrac{-b \pm \sqrt{b^2 - 4ac}}{2a}$

$\tan x = \dfrac{-(-3) \pm \sqrt{(-3)^2 - 4(1)(-2)}}{2(1)}$

$\tan x = \dfrac{3 \pm \sqrt{17}}{2}$

$\tan x \approx 3.5616$ or $\tan x \approx -0.5616$

$x \approx 1.2971, \ 4.4387$ $x \approx 2.6299, \ 5.7715$

Cumulative Review Exercises (Chapters 1–6)

1. $x^3 + x^2 - x + 15 = 0$

The possible rational zeros are: $\pm 1, \ \pm 3, \ \pm 5, \ \pm 15$.

Synthetic division shows that -3 is a zero:

$$
\begin{array}{r|rrrr}
-3 & 1 & 1 & -1 & 15 \\
   &   & -3 & 6 & -15 \\
\hline
   & 1 & -2 & 5 & 0
\end{array}
$$

The quotient is $x^2 - 2x + 5$. The remaining zeros are found using the quadratic formula:

$x = \dfrac{-(-2) \pm \sqrt{(-2)^2 - 4(1)(5)}}{2(1)}$

$= \dfrac{2 \pm \sqrt{4 - 20}}{2}$

$= \dfrac{2 \pm \sqrt{-16}}{2}$

$= \dfrac{2 \pm 4i}{2}$

$= 1 \pm 2i$

All solutions are: -3, $1 + 2i$ and $1 - 2i$.

2. $11^{x-1} = 125$

$\log 11^{x-1} = \log 125$

$(x - 1)\log 11 = \log 125$

$x - 1 = \dfrac{\log 125}{\log 11}$

$x = \dfrac{\log 125}{\log 11} + 1$

or $x \approx 3.01$

3. $x^2 + 2x - 8 > 0$

$(x-2)(x+4) > 0$

zero points are $x = 2$ and $x = -4$.

Test Interval	Representative Number	Substitute into $x^2 + 2x - 8 > 0$	Conclusion
$(-\infty, -4)$	-5	$(-5)^2 + 2(-5) - 8 = 25 - 10 - 8 = 7 > 0$	$(-\infty, -4)$ belongs to the solution set.
$(-4, 2)$	0	$0^2 + 2(0) - 8 = -8 > 0$	$(-4, 2)$ does not belong to the solution set.
$(2, \infty)$	3	$3^2 + 2(3) - 8 = 9 + 6 - 8 = 7 > 0$	$(2, \infty)$ belongs to the solution set.

The solution intervals are $(-\infty, -4) \cup (2, \infty)$.

4. $\cos 2x + 3 = 5\cos x$

$2\cos^2 x - 1 + 3 = 5\cos x$

$2\cos^2 x - 5\cos x + 2 = 0$

$(2\cos x - 1)(\cos x - 2) = 0$

$2\cos x - 1 = 0 \quad$ or $\quad \cos x - 2 = 0$

$2\cos x = 1 \qquad\qquad \cos x = 2$

$\cos x = \dfrac{1}{2} \qquad$ $\cos x$ cannot be greater than 1.

$x = \dfrac{\pi}{3} \quad x = \dfrac{5\pi}{3}$

The solutions in the interval $[0, 2\pi)$ are $\dfrac{\pi}{3}$ and $\dfrac{5\pi}{3}$.

5. $\tan x + \sec^2 x = 3$

$\tan x + 1 + \tan^2 x = 3$

$\tan^2 x + \tan x - 2 = 0$

$(\tan x - 1)(\tan x + 2) = 0$

$\tan x - 1 = 0 \quad$ or $\quad \tan x + 2 = 0$

$\tan x = 1 \qquad\qquad \tan x = -2$

$x = \dfrac{\pi}{4}, \dfrac{5\pi}{4} \qquad x \approx 2.0344, \ 5.1761$

6.

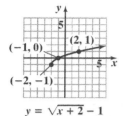

$$y = \sqrt{x+2} - 1$$

Shift the graph of $y = \sqrt{x}$ left 2 units and down 1 unit.

7.

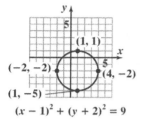

$$(x-1)^2 + (y+2)^2 = 9$$

8.

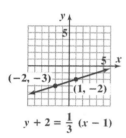

$$y + 2 = \frac{1}{3}(x-1)$$

9.

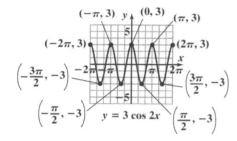

$$y = 3\cos 2x$$

10.

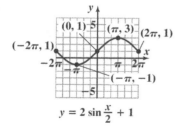

$$y = 2\sin\frac{x}{2} + 1$$

11.

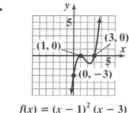

$$f(x) = (x-1)^2(x-3)$$

12. $f(x) = x^2 + 3x - 1$

$$\frac{f(a+h) - f(a)}{h}$$

$$= \frac{(a+h)^2 + 3(a+h) - 1 - \left(a^2 + 3a - 1\right)}{h}$$

$$= \frac{a^2 + 2ah + h^2 + 3a + 3h - 1 - a^2 - 3a + 1}{h}$$

$$= \frac{2ah + h^2 + 3h}{h}$$

$$= 2a + h + 3$$

13. $\sin 225° = \sin(180° + 45°)$

$$= \sin 180° \cos 45° + \cos 180° \sin 45°$$

$$= 0 \cdot \frac{\sqrt{2}}{2} + (-1) \cdot \frac{\sqrt{2}}{2}$$

$$= -\frac{\sqrt{2}}{2}$$

14. $\sec^4 x - \sec^2 x$

$$= \sec^2 x \cdot \sec^2 x - \sec^2 x$$

$$= \left(1 + \tan^2 x\right)\left(1 + \tan^2 x\right) - \left(1 + \tan^2 x\right)$$

$$= 1 + 2\tan^2 x + \tan^4 x - 1 - \tan^2 x$$

$$= \tan^4 x + \tan^2 x$$

We worked with the left side and arrived at the right side. Thus, the identity is verified.

15. $320° \times \dfrac{\pi}{180°} = \dfrac{16}{9}\pi$ or 5.59 radians

16.

$$A = Pe^{rt}$$

$$3P = Pe^{0.0575t}$$

$$3 = e^{0.0575t}$$

$$\ln 3 = \ln e^{0.0575t}$$

$$\ln 3 = 0.0575t$$

$$\frac{\ln 3}{0.0575} = t$$

$$t \approx 19.1 \text{ years}$$

17.
$$f(x) = \frac{2x+1}{x-3}$$
$$y = \frac{2x+1}{x-3}$$
$$x = \frac{2y+1}{y-3}$$
$$x(y-3) = 2y+1$$
$$xy - 3x = 2y+1$$
$$xy - 2y = 3x+1$$
$$y(x-2) = 3x+1$$
$$y = \frac{3x+1}{x-2}$$
$$f^{-1}(x) = \frac{3x+1}{x-2}$$

18. The third angle is:
$B = 180° - 90° - 23° = 67°$.

Since $\sin\theta = \dfrac{\text{opposite}}{\text{hypotenuse}}$,

$$\sin A = \sin 23° = \frac{12}{c}$$

$c = \dfrac{12}{\sin 23°} \approx 30.71$ and

$$\sin B = \sin 67° = \frac{b}{30.71}$$
$b = 30.71 \cdot \sin 67° \approx 28.27$
The angles are 90°, 23°, and 67°.
The sides are 12, 30.71, and 28.27.

19. Solve $8.5 = \dfrac{12}{150} \cdot a$
where a is the adult dose.
$$a = \frac{(8.5) \cdot 150}{12}$$
$$= 106.25 \text{ mg}$$
$$a \approx 106 \text{ mg}$$

20.. Let h be the height of the flagpole.
Then $\tan 53° = \dfrac{h}{12}$
$$h = 12 \cdot \tan 53°$$
$$h \approx 15.9 \text{ feet}$$

Chapter 7
Additional Topics in Trigonometry

Section 7.1

Check Point Exercises

1. Begin by finding B, the third angle of the triangle.
$$A + B + C = 180°$$
$$64° + B + 82° = 180°$$
$$146° + B = 180°$$
$$B = 34°$$
In this problem, we are given c and C:
$c = 14$ and $C = 82°$. Thus, use the ratio
$\dfrac{c}{\sin C}$, or $\dfrac{14}{\sin 82°}$, to find the other two sides. Use
the Law of Sines to find a.
$$\frac{a}{\sin A} = \frac{c}{\sin C}$$
$$\frac{a}{\sin 64°} = \frac{14}{\sin 82°}$$
$$a = \frac{14\sin 64°}{\sin 82°}$$
$$a \approx 12.7 \text{ centimeters}$$
Use the Law of Sines again, this time to find b.
$$\frac{b}{\sin B} = \frac{c}{\sin C}$$
$$\frac{b}{\sin 34°} = \frac{14}{\sin 82°}$$
$$b = \frac{14\sin 34°}{\sin 82°}$$
$$b \approx 7.4 \text{ centimeters}$$
The solution is $B = 34°$, $a \approx 12.7$ centimeters, and $b \approx$ 7.4 centimeters.

2. Begin by finding B.
$$A + B + C = 180°$$
$$40° + B + 22.5° = 180°$$
$$62.5° + B = 180°$$
$$B = 117.5°$$
In this problem, we are given that $b = 12$ and we find that $B = 117.5°$. Thus, use the ratio
$\dfrac{b}{\sin B}$, or $\dfrac{12}{\sin 117.5°}$, to find the other two sides. Use
the Law of Sines to find a.

$$\frac{a}{\sin A} = \frac{b}{\sin B}$$
$$\frac{a}{\sin 40°} = \frac{12}{\sin 117.5°}$$
$$a = \frac{12\sin 40°}{\sin 117.5°} \approx 8.7$$
Use the Law of Sines again, this time to find c.
$$\frac{c}{\sin C} = \frac{b}{\sin B}$$
$$\frac{c}{\sin 22.5°} = \frac{12}{\sin 117.5°}$$
$$c = \frac{12\sin 22.5°}{\sin 117.5°} \approx 5.2$$
The solution is $B = 117.5°$, $a \approx 8.7$, and $c \approx 5.2$.

3. The known ratio is $\dfrac{a}{\sin A}$, or $\dfrac{33}{\sin 57°}$. Because side b
is given, Use the Law of Sines to find angle B.
$$\frac{a}{\sin A} = \frac{b}{\sin B}$$
$$\frac{33}{\sin 57°} = \frac{26}{\sin B}$$
$$33\sin B = 26\sin 57°$$
$$\sin B = \frac{26\sin 57°}{33} \approx 0.6608$$
$$\sin B \approx 0.6608$$
$$B \approx 41°$$
$180° - 41° = 139°$ also has this sine value, but, the
sum of $57°$ and $139°$ exceeds $180°$, so B cannot have
this value.
$C = 180° - B - A = 180° - 41° - 57° = 82°$.
Use the law of sines to find C.
$$\frac{a}{\sin A} = \frac{c}{\sin C}$$
$$\frac{33}{\sin 57°} = \frac{c}{\sin 82°}$$
$$c = \frac{33\sin 82°}{\sin 57°}$$
$$c \approx 39$$
Thus, $B \approx 41°$, $C \approx 82°$, $c \approx 39$.

4. The known ratio is $\dfrac{a}{\sin A}$, or $\dfrac{10}{\sin 50°}$. Because side b is given, Use the Law of Sines to find angle B.

$$\frac{a}{\sin A} = \frac{b}{\sin B}$$

$$\frac{10}{\sin 50°} = \frac{20}{\sin B}$$

$$10\sin B = 20\sin 50°$$

$$\sin B = \frac{20\sin 50°}{10} \approx 1.53$$

Because the sine can never exceed 1, there is no angle B for which $\sin B \approx 1.53$. There is no triangle with the given measurements.

5. The known ratio is $\dfrac{a}{\sin A}$, or $\dfrac{12}{\sin 35°}$. Because side b is given, Use the Law of Sines to find angle B.

$$\frac{a}{\sin A} = \frac{b}{\sin B}$$

$$\frac{12}{\sin 35°} = \frac{16}{\sin B}$$

$$12\sin B = 16\sin 35°$$

$$\sin B = \frac{16\sin 35°}{12} \approx 0.7648$$

There are two angles possible:
$B_1 \approx 50°$, $B_2 \approx 180° - 50° = 130°$

There are two triangles:
$C_1 = 180° - A - B_1 \approx 180° - 35° - 50° = 95°$
$C_2 = 180° - A - B_2 \approx 180° - 35° - 130° = 15°$

Use the Law of Sines to find c_1 and c_2.

$$\frac{c_1}{\sin C_1} = \frac{a}{\sin A}$$

$$\frac{c_1}{\sin 95°} = \frac{12}{\sin 35°}$$

$$c_1 = \frac{12\sin 95°}{\sin 35°} \approx 20.8$$

$$\frac{c_2}{\sin C_2} = \frac{a}{\sin A}$$

$$\frac{c_2}{\sin 15°} = \frac{12}{\sin 35°}$$

$$c_2 = \frac{12\sin 15°}{\sin 35°} \approx 5.4$$

In one triangle, the solution is $B_1 \approx 50°$, $C_1 \approx 95°$, and $c_1 \approx 20.8$. In the other triangle, $B_2 \approx 130°$, $C_2 \approx 15°$, and $c_2 \approx 5.4$.

6. The area of the triangle is half the product of the lengths of the two sides times the sine of the included angle.

$$\text{Area} = \frac{1}{2}(8)(12)(\sin 135°) \approx 34$$

The area of the triangle is approximately 34 square meters.

7.

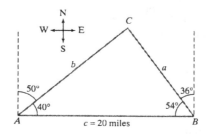

Using a north-south line, the interior angles are found as follows:
$A = 90° - 35° = 55°$

$B = 90° - 49° = 41°$
Find angle C using a 180° angle sum in the triangle.
$C = 180° - A - B = 180° - 55° - 41° = 84°$

The ratio $\dfrac{c}{\sin C}$, or $\dfrac{13}{\sin 84°}$ is now known. Use this ratio and the Law of Sines to find a.

$$\frac{a}{\sin A} = \frac{c}{\sin C}$$

$$\frac{a}{\sin 55°} = \frac{13}{\sin 84°}$$

$$a = \frac{13\sin 55°}{\sin 84°} \approx 11$$

The fire is approximately 11 miles from station B.

Concept and Vocabulary Check 7.1

1. oblique; sides; angles

2. $\dfrac{a}{\sin A} = \dfrac{b}{\sin B} = \dfrac{c}{\sin C}$

3. side; angles

4. false

5. $\dfrac{1}{2}ab\sin C$

Exercise Set 7.1

1. Begin by finding B.
$$A + B + C = 180°$$
$$42° + B + 96° = 180°$$
$$138° + B = 180°$$
$$B = 42°$$

Use the ratio $\dfrac{c}{\sin C}$, or $\dfrac{12}{\sin 96°}$, to find the other two sides. Use the Law of Sines to find a.

$$\frac{a}{\sin A} = \frac{c}{\sin C}$$
$$\frac{a}{\sin 42°} = \frac{12}{\sin 96°}$$
$$a = \frac{12\sin 42°}{\sin 96°}$$
$$a \approx 8.1$$

Use the Law of Sines again, this time to find b.

$$\frac{b}{\sin B} = \frac{c}{\sin C}$$
$$\frac{b}{\sin 42°} = \frac{12}{\sin 96°}$$
$$b = \frac{12\sin 42°}{\sin 96°}$$
$$b \approx 8.1$$

The solution is $B = 42°$, $a \approx 8.1$, and $b \approx 8.1$.

3. Begin by finding A.
$$A + B + C = 180°$$
$$A + 54° + 82° = 180°$$
$$A + 136° = 180°$$
$$A = 44°$$

Use the ratio $\dfrac{a}{\sin A}$, or $\dfrac{16}{\sin 44°}$, to find the other two sides. Use the Law of Sines to find b.

$$\frac{b}{\sin B} = \frac{a}{\sin A}$$
$$\frac{b}{\sin 54°} = \frac{16}{\sin 44°}$$
$$b = \frac{16\sin 54°}{\sin 44°}$$
$$b \approx 18.6$$

Use the Law of Sines again, this time to find c.

$$\frac{c}{\sin C} = \frac{a}{\sin A}$$
$$\frac{c}{\sin 82°} = \frac{16}{\sin 44°}$$
$$c = \frac{16\sin 82°}{\sin 44°}$$
$$c \approx 22.8$$

The solution is $A = 44°$, $b \approx 18.6$, and $c \approx 22.8$.

5. Begin by finding C.
$$A + B + C = 180°$$
$$48° + 37° + C = 180°$$
$$85° + C = 180°$$
$$C = 95°$$

Use the ratio $\dfrac{a}{\sin A}$, or $\dfrac{100}{\sin 48°}$, to find the other two sides. Use the Law of Sines to find b.

$$\frac{b}{\sin B} = \frac{a}{\sin A}$$
$$\frac{b}{\sin 37°} = \frac{100}{\sin 48°}$$
$$b = \frac{100\sin 37°}{\sin 48°}$$
$$b \approx 81.0$$

Use the Law of Sines again, this time to find c.

$$\frac{c}{\sin C} = \frac{a}{\sin A}$$
$$\frac{c}{\sin 95°} = \frac{100}{\sin 48°}$$
$$c = \frac{100\sin 95°}{\sin 48°}$$
$$c \approx 134.1$$

The solution is $C = 95°$, $b \approx 81.0$, and $c \approx 134.1$.

7. Begin by finding B.

$$A + B + C = 180°$$
$$38° + B + 102° = 180°$$
$$B + 140° = 180°$$
$$B = 40°$$

Use the ratio $\dfrac{a}{\sin A}$, or $\dfrac{20}{\sin 38°}$, to find the other two sides. Use the Law of Sines to find b.

$$\frac{b}{\sin B} = \frac{a}{\sin A}$$
$$\frac{b}{\sin 40°} = \frac{20}{\sin 38°}$$
$$b = \frac{20 \sin 40°}{\sin 38°}$$
$$b \approx 20.9$$

Use the Law of Sines again, this time to find c.

$$\frac{c}{\sin C} = \frac{a}{\sin A}$$
$$\frac{c}{\sin 102°} = \frac{20}{\sin 38°}$$
$$c = \frac{20 \sin 102°}{\sin 38°}$$
$$c \approx 31.8$$

The solution is $B = 40°$, $b \approx 20.9$, and $c \approx 31.8$.

9. Begin by finding C.

$$A + B + C = 180°$$
$$44° + 25° + C = 180°$$
$$69° + C = 180°$$
$$C = 111°$$

Use the ratio $\dfrac{a}{\sin A}$, or $\dfrac{12}{\sin 44°}$, to find the other two sides. Use the Law of Sines to find b.

$$\frac{b}{\sin B} = \frac{a}{\sin A}$$
$$\frac{b}{\sin 25°} = \frac{12}{\sin 44°}$$
$$b = \frac{12 \sin 25°}{\sin 44°}$$
$$b \approx 7.3$$

Use the Law of Sines again, this time to find c.

$$\frac{c}{\sin C} = \frac{a}{\sin A}$$
$$\frac{c}{\sin 111°} = \frac{12}{\sin 44°}$$
$$c = \frac{12 \sin 111°}{\sin 44°}$$
$$c \approx 16.1$$

The solution is $C = 111°$, $b \approx 7.3$, and $c \approx 16.1$.

11. Begin by finding A.

$$A + B + C = 180°$$
$$A + 85° + 15° = 180°$$
$$A + 100° = 180°$$
$$A = 80°$$

Use the ratio $\dfrac{b}{\sin B}$, or $\dfrac{40}{\sin 85°}$, to find the other two sides. Use the Law of Sines to find a.

$$\frac{a}{\sin A} = \frac{b}{\sin B}$$
$$\frac{a}{\sin 80°} = \frac{40}{\sin 85°}$$
$$a = \frac{40 \sin 80°}{\sin 85°}$$
$$a \approx 39.5$$

Use the Law of Sines again, this time to find c.

$$\frac{c}{\sin C} = \frac{b}{\sin B}$$
$$\frac{c}{\sin 15°} = \frac{40}{\sin 85°}$$
$$c = \frac{40 \sin 15°}{\sin 85°}$$
$$c \approx 10.4$$

The solution is $A = 80°$, $a \approx 39.5$, and $c \approx 10.4$.

13. Begin by finding B.

$$A + B + C = 180°$$
$$115° + B + 35° = 180°$$
$$B + 150° = 180°$$
$$B = 30°$$

Use the ratio $\dfrac{c}{\sin C}$, or $\dfrac{200}{\sin 35°}$, to find the other two sides. Use the Law of Sines to find a.

$$\frac{a}{\sin A} = \frac{c}{\sin C}$$
$$\frac{a}{\sin 115°} = \frac{200}{\sin 35°}$$
$$a = \frac{200 \sin 115°}{\sin 35°}$$
$$a \approx 316.0$$

Use the Law of Sines again, this time to find b.

$$\frac{b}{\sin B} = \frac{c}{\sin C}$$
$$\frac{b}{\sin 30°} = \frac{200}{\sin 35°}$$
$$b = \frac{200 \sin 30°}{\sin 35°}$$
$$b \approx 174.3$$

The solution is $B = 30°$, $a \approx 316.0$, and $b \approx 174.3$.

15. Begin by finding C.
$$A + B + C = 180°$$
$$65° + 65° + C = 180°$$
$$130° + C = 180°$$
$$C = 50°$$

Use the ratio $\dfrac{c}{\sin C}$, or $\dfrac{6}{\sin 50°}$, to find the other two sides. Use the Law of Sines to find a.
$$\frac{a}{\sin A} = \frac{c}{\sin C}$$
$$\frac{a}{\sin 65°} = \frac{6}{\sin 50°}$$
$$a = \frac{6\sin 65°}{\sin 50°}$$
$$a \approx 7.1$$

Use the Law of Sines to find angle B.
$$\frac{b}{\sin B} = \frac{c}{\sin C}$$
$$\frac{b}{\sin 65°} = \frac{6}{\sin 50°}$$
$$b = \frac{6\sin 65°}{\sin 50°}$$
$$b \approx 7.1$$
The solution is $C = 50°$, $a \approx 7.1$, and $b \approx 7.1$.

17. The known ratio is $\dfrac{a}{\sin A}$, or $\dfrac{20}{\sin 40°}$.
Use the Law of Sines to find angle B.
$$\frac{a}{\sin A} = \frac{b}{\sin B}$$
$$\frac{20}{\sin 40°} = \frac{15}{\sin B}$$
$$20\sin B = 15\sin 40°$$
$$\sin B = \frac{15\sin 40°}{20}$$
$$\sin B \approx 0.4821$$
There are two angles possible:
$B_1 \approx 29°$, $B_2 \approx 180° - 29° = 151°$
B_2 is impossible, since $40° + 151° = 191°$.
We find C using B_1 and the given information $A = 40°$.
$$C = 180° - B_1 - A \approx 180° - 29° - 40° = 111°$$
Use the Law of Sines to find side c.

$$\frac{c}{\sin C} = \frac{a}{\sin A}$$
$$\frac{c}{\sin 111°} = \frac{20}{\sin 40°}$$
$$c = \frac{20\sin 111°}{\sin 40°} \approx 29.0$$
There is one triangle and the solution is
B_1 (or B) $\approx 29°$, $C \approx 111°$, and $c \approx 29.0$.

19. The known ratio is $\dfrac{a}{\sin A}$, or $\dfrac{10}{\sin 63°}$.
Use the Law of Sines to find angle C.
$$\frac{a}{\sin A} = \frac{c}{\sin C}$$
$$\frac{10}{\sin 63°} = \frac{8.9}{\sin C}$$
$$10\sin C = 8.9\sin 63°$$
$$\sin C = \frac{8.9\sin 63°}{10}$$
$$\sin C \approx 0.7930$$
There are two angles possible:
$C_1 \approx 52°$, $C_2 \approx 180° - 52° = 128°$
C_2 is impossible, since $63° + 128° = 191°$.
We find B using C_1 and the given information $A = 63°$.
$$B = 180° - C_1 - A \approx 180° - 52° - 63° = 65°$$
Use the Law of Sines to find side b.
$$\frac{b}{\sin B} = \frac{a}{\sin A}$$
$$\frac{b}{\sin 65°} = \frac{10}{\sin 63°}$$
$$b = \frac{10\sin 65°}{\sin 63°} \approx 10.2$$
There is one triangle and the solution is
C_1 (or C) $\approx 52°$, $B \approx 65°$, and $b \approx 10.2$.

21. The known ratio is $\dfrac{a}{\sin A}$, or $\dfrac{42.1}{\sin 112°}$.

Use the Law of Sines to find angle C.

$$\frac{a}{\sin A} = \frac{c}{\sin C}$$

$$\frac{42.1}{\sin 112°} = \frac{37}{\sin C}$$

$$42.1\sin C = 37\sin 112°$$

$$\sin C = \frac{37\sin 112°}{42.1}$$

$$\sin C \approx 0.8149$$

There are two angles possible:

$C_1 \approx 55°$, $C_2 \approx 180° - 55° = 125°$

C_2 is impossible, since $112° + 125° = 237°$.

We find B using C_1 and the given information $A = 112°$.

$B = 180° - C_1 - A \approx 180° - 55° - 112° = 13°$

Use the Law of Sines to find b.

$$\frac{b}{\sin B} = \frac{a}{\sin A}$$

$$\frac{b}{\sin 13°} = \frac{42.1}{\sin 112°}$$

$$b = \frac{42.1\sin 13°}{\sin 112°} \approx 10.2$$

There is one triangle and the solution is
C_1 (or C) $\approx 55°$, $B \approx 13°$, and $b \approx 10.2$.

23. The known ratio is $\dfrac{a}{\sin A}$, or $\dfrac{10}{\sin 30°}$.

Use the Law of Sines to find angle B.

$$\frac{a}{\sin A} = \frac{b}{\sin B}$$

$$\frac{10}{\sin 30°} = \frac{40}{\sin B}$$

$$10\sin B = 40\sin 30°$$

$$\sin B = \frac{40\sin 30°}{10} = 2$$

Because the sine can never exceed 1, there is no
angle B for which $\sin B = 2$. There is no triangle with
the given measurements.

25. The known ratio is $\dfrac{a}{\sin A}$, or $\dfrac{16}{\sin 60°}$.

Use the Law of Sines to find angle B.

$$\frac{a}{\sin A} = \frac{b}{\sin B}$$

$$\frac{16}{\sin 60°} = \frac{18}{\sin B}$$

$$16\sin B = 18\sin 60°$$

$$\sin B = \frac{18\sin 60°}{16}$$

$$\sin B \approx 0.9743$$

There are two angles possible:

$B_1 \approx 77°$, $B_2 \approx 180° - 77° = 103°$

There are two triangles:

$C_1 = 180° - B_1 - A \approx 180° - 77° - 60° = 43°$

$C_2 = 180° - B_2 - A \approx 180° - 103° - 60° = 17°$ Use the

Law of Sines to find c_1 and c_2.

$$\frac{c_1}{\sin C_1} = \frac{a}{\sin A}$$

$$\frac{c_1}{\sin 43°} = \frac{16}{\sin 60°}$$

$$c_1 = \frac{16\sin 43°}{\sin 60°} \approx 12.6$$

$$\frac{c_2}{\sin C_2} = \frac{a}{\sin A}$$

$$\frac{c_2}{\sin 17°} = \frac{16}{\sin 60°}$$

$$c_2 = \frac{16\sin 17°}{\sin 60°} \approx 5.4$$

In one triangle, the solution is
$B_1 \approx 77°$, $C_1 \approx 43°$, and $c_1 \approx 12.6$.

In the other triangle,
$B_2 \approx 103°$, $C_2 \approx 17°$, and $c_2 \approx 5.4$.

27. The known ratio is $\dfrac{a}{\sin A}$, or $\dfrac{12}{\sin 37°}$.
Use the Law of Sines to find angle B.

$$\frac{a}{\sin A} = \frac{b}{\sin B}$$
$$\frac{12}{\sin 37°} = \frac{16.1}{\sin B}$$
$$12\sin B = 16.1\sin 37°$$
$$\sin B = \frac{16.1\sin 37°}{12}$$
$$\sin B \approx 0.8074$$

There are two angles possible:
$B_1 \approx 54°$, $B_2 \approx 180° - 54° = 126°$
There are two triangles:
$C_1 = 180° - B_1 - A \approx 180° - 54° - 37° = 89°$
$C_2 = 180° - B_2 - A \approx 180° - 126° - 37° = 17°$ Use the
Law of Sines to find c_1 and c_2.

$$\frac{c_1}{\sin C_1} = \frac{a}{\sin A}$$
$$\frac{c_1}{\sin 89°} = \frac{12}{\sin 37°}$$
$$c_1 = \frac{12\sin 89°}{\sin 37°} \approx 19.9$$
$$\frac{c_2}{\sin C_2} = \frac{a}{\sin A}$$
$$\frac{c_2}{\sin 17°} = \frac{12}{\sin 37°}$$
$$c_2 = \frac{12\sin 17°}{\sin 37°} \approx 5.8$$

In one triangle, the solution is
$B_1 \approx 54°$, $C_1 \approx 89°$, and $c_1 \approx 19.9$.
In the other triangle,
$B_2 \approx 126°$, $C_2 \approx 17°$, and $c_2 \approx 5.8$.

29. The known ratio is $\dfrac{a}{\sin A}$, or $\dfrac{22}{\sin 58°}$.
Use the Law of Sines to find angle C.

$$\frac{a}{\sin A} = \frac{c}{\sin C}$$
$$\frac{22}{\sin 58°} = \frac{24.1}{\sin C}$$
$$22\sin C = 24.1\sin 58°$$
$$\sin C = \frac{24.1\sin 58°}{22}$$
$$\sin C \approx 0.9290$$

There are two angles possible:
$C_1 \approx 68°$, $C_2 \approx 180° - 68° = 112°$
There are two triangles:

$B_1 = 180° - C_1 - A \approx 180° - 68° - 58° = 54°$
$B_2 = 180° - C_2 - A \approx 180° - 112° - 58° = 10°$ Use the
Law of Sines to find b_1 and b_2.

$$\frac{b_1}{\sin B_1} = \frac{a}{\sin A}$$
$$\frac{b_1}{\sin 54°} = \frac{22}{\sin 58°}$$
$$b_1 = \frac{22\sin 54°}{\sin 58°} \approx 21.0$$
$$\frac{b_2}{\sin B_2} = \frac{a}{\sin A}$$
$$\frac{b_2}{\sin 10°} = \frac{22}{\sin 58°}$$
$$b_2 = \frac{22\sin 10°}{\sin 58°} \approx 4.5$$

In one triangle, the solution is
$C_1 \approx 68°$, $B_1 \approx 54°$, and $b_1 \approx 21.0$.
In the other triangle,
$C_2 \approx 112°$, $B_2 \approx 10°$, and $b_2 \approx 4.5$.

31. The known ratio is $\dfrac{a}{\sin A}$, or $\dfrac{9.3}{\sin 18°}$.
Use the Law of Sines to find angle B.

$$\frac{a}{\sin A} = \frac{b}{\sin B}$$
$$\frac{9.3}{\sin 18°} = \frac{41}{\sin B}$$
$$9.3\sin B = 41\sin 18°$$
$$\sin B = \frac{41\sin 18°}{9.3} \approx 1.36$$

Because the sine can never exceed 1, there is no
angle B for which $\sin B = 1.36$. There is no triangle
with the given measurements.

33. Area $= \dfrac{1}{2}bc\sin A = \dfrac{1}{2}(20)(40)(\sin 48°) \approx 297$
The area of the triangle is approximately
297 square feet.

35. Area $= \dfrac{1}{2}ac\sin B = \dfrac{1}{2}(3)(6)(\sin 36°) \approx 5$
The area of the triangle is approximately
5 square yards.

37. Area $= \dfrac{1}{2}ab\sin C = \dfrac{1}{2}(4)(6)(\sin 124°) \approx 10$
The area of the triangle is approximately
10 square meters.

39. $\angle ABC = 180° - 67° = 113°$

$\angle ACB = 180° - 43° - 113° = 24°$

Use the law of sines to find $\overline{BC}$.

$$\frac{\overline{BC}}{\sin 43°} = \frac{312}{\sin 24°}$$

$$\overline{BC} = \frac{312 \sin 43°}{\sin 24°}$$

$$\overline{BC} \approx 523.1$$

Use the law of sines to find h.

$$\frac{h}{\sin 67°} = \frac{523.1}{\sin 90°}$$

$$h = \frac{523.1 \sin 67°}{\sin 90°}$$

$$h \approx 481.6$$

41. Begin by finding the six angles inside the two triangles. Then use the law of sines.

$$\frac{a}{\sin 4°} = \frac{\dfrac{450 \sin 145°}{\sin 34°}}{\sin 30°}$$

$$a \approx 64.4$$

43.

$$\frac{a}{\sin A} = \frac{b}{\sin B}$$

$$\frac{300}{\sin 2\theta} = \frac{200}{\sin \theta}$$

$$200 \sin 2\theta = 300 \sin \theta$$

$$400 \sin \theta \cos \theta = 300 \sin \theta$$

$$\cos \theta = \frac{300 \sin \theta}{400 \sin \theta}$$

$$\cos \theta = \frac{3}{4}$$

$$\theta \approx 41°$$

$$2\theta \approx 82°$$

$$A \approx 82°, B \approx 41°, C \approx 57°, c \approx 255.7$$

45.

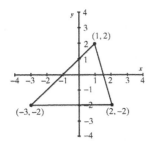

$$A = \frac{1}{2} bh$$

$$= \frac{1}{2}(5)(4)$$

$$= 10$$

47.

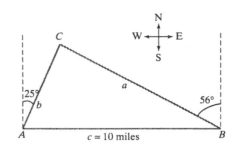

Using a north-south line, the interior angles are found as follows:

$$A = 90° - 25° = 65°$$

$$B = 90° - 56° = 34°$$

Find angle C using a 180° angle sum in the triangle.

$$C = 180° - A - B = 180° - 65° - 34° = 81°$$

The ratio $\dfrac{c}{\sin C}$, or $\dfrac{10}{\sin 81°}$, is now known. Use this ratio and the Law of Sines to find b and a.

$$\frac{b}{\sin B} = \frac{c}{\sin C}$$

$$\frac{b}{\sin 34°} = \frac{10}{\sin 81°}$$

$$b = \frac{10 \sin 34°}{\sin 81°} \approx 5.7$$

Station A is about 5.7 miles from the fire.

$$\frac{a}{\sin A} = \frac{c}{\sin C}$$

$$\frac{a}{\sin 65°} = \frac{10}{\sin 81°}$$

$$a = \frac{10 \sin 65°}{\sin 81°} \approx 9.2$$

Station B is about 9.2 miles from the fire.

49.

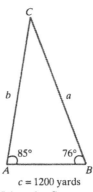

Using the figure,

$$C = 180° - A - B = 180° - 85° - 76° = 19°$$

The ratio $\dfrac{c}{\sin C}$, or $\dfrac{1200}{\sin 19°}$, is now known. Use this ratio and the Law of Sines to find a and b.

$$\frac{a}{\sin A} = \frac{c}{\sin C}$$

$$\frac{a}{\sin 85°} = \frac{1200}{\sin 19°}$$

$$a = \frac{1200 \sin 85°}{\sin 19°} \approx 3671.8$$

$$\frac{b}{\sin B} = \frac{c}{\sin C}$$

$$\frac{b}{\sin 76°} = \frac{1200}{\sin 19°}$$

$$b = \frac{1200 \sin 76°}{\sin 19°} \approx 3576.4$$

The platform is about 3671.8 yards from one end of the beach and 3576.4 yards from the other.

51. According to the figure,
$C = 180° - A - B = 180° - 84.7° - 50° = 45.3°$ The

ratio $\dfrac{c}{\sin C}$, or $\dfrac{171}{\sin 45.3°}$, is now known. Use this

ratio and the Law of Sines to find b.

$$\frac{b}{\sin B} = \frac{c}{\sin C}$$

$$\frac{b}{\sin 50°} = \frac{171}{\sin 45.3°}$$

$$b = \frac{171 \sin 50°}{\sin 45.3°} \approx 184$$

The distance is about 184 feet.

53. The ratio $\dfrac{b}{\sin B}$, or $\dfrac{562}{\sin 85.3°}$, is known.

Use this ratio, the figure, and the Law of Sines to find c.

$$\frac{c}{\sin C} = \frac{b}{\sin B}$$

$$\frac{c}{\sin 5.7°} = \frac{562}{\sin 85.3°}$$

$$c = \frac{562 \sin 5.7°}{\sin 85.3°} \approx 56.0$$

The toss was about 56.0 feet.

55.

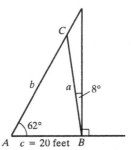

Using the figure,
$B = 90° - 8° = 82°$

$C = 180° - A - B = 180° - 62° - 82° = 36°$

The ratio $\dfrac{c}{\sin C}$, or $\dfrac{20}{\sin 36°}$, is now known. Use this

ratio and the Law of Sines to find a.

$$\frac{a}{\sin A} = \frac{c}{\sin C}$$

$$\frac{a}{\sin 62°} = \frac{20}{\sin 36°}$$

$$a = \frac{20 \sin 62°}{\sin 36°} \approx 30.0$$

The length of the pole is about 30.0 feet.

57. **a.** Using the figure and the measurements shown,
$B = 180° - 44° = 136°$

$C = 180° - B - A = 180° - 136° - 37° = 7°$

The ratio $\dfrac{c}{\sin C}$, or $\dfrac{100}{\sin 7°}$, is now known. Use

this ratio and the Law of Sines to find a.

$$\frac{a}{\sin A} = \frac{c}{\sin C}$$

$$\frac{a}{\sin 37°} = \frac{100}{\sin 7°}$$

$$a = \frac{100 \sin 37°}{\sin 7°} \approx 493.8$$

To the nearest foot, $a = 493.8$ feet.

b. From part a, let $a = 493.8$ be the hypotenuse of the right triangle.
If h represents the height of the tree,

$$\frac{h}{\sin 44°} = \frac{493.8}{\sin 90°}$$

$$h = \frac{493.8 \sin 44°}{\sin 90°} \approx 343.0$$

A typical redwood tree is about 343.0 feet.

59.

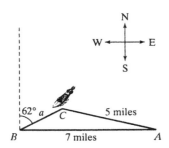

Using the figure,
$B = 90° - 62° = 28°$

The known ratio is $\dfrac{b}{\sin B}$, or $\dfrac{5}{\sin 28°}$.

Use the Law of Sines to find angle C.

$$\frac{b}{\sin B} = \frac{c}{\sin C}$$

$$\frac{5}{\sin 28°} = \frac{7}{\sin C}$$

$$5\sin C = 7\sin 28°$$

$$\sin C = \frac{7\sin 28°}{5} \approx 0.6573$$

There are two angles possible:
$C_1 \approx 41°$, $C_2 \approx 180° - 41° = 139°$

There are two triangles:
$A_1 = 180° - C_1 - B \approx 180° - 41° - 28° = 111°$

$A_2 = 180° - C_2 - B \approx 180° - 139° - 28° = 13°$ Use the

Law of Sines to find a_1 and a_2.

$$\frac{a_1}{\sin A_1} = \frac{b}{\sin B}$$

$$\frac{a_1}{\sin 111°} = \frac{5}{\sin 28°}$$

$$a_1 = \frac{5\sin 111°}{\sin 28°} \approx 9.9$$

$$\frac{a_2}{\sin A_2} = \frac{b}{\sin B}$$

$$\frac{a_2}{\sin 13°} = \frac{5}{\sin 28°}$$

$$a_2 = \frac{5\sin 13°}{\sin 28°} \approx 2.4$$

The boat is either 9.9 miles or 2.4 miles from
lighthouse B, to the nearest tenth of a mile.

61. – 69. Answers may vary.

71. does not make sense; Explanations will vary.
Sample explanation: The law of cosines would be
appropriate for this situation.

73. does not make sense; Explanations will vary.
Sample explanation: The calculator will give you
the acute angle. The obtuse angle is the supplement
of the acute angle.

75. No. Explanations may vary.

77.

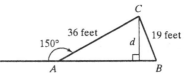

Using the figure, $A = 180° - 150° = 30°$
Using the Law of Sines we have,

$$\frac{d}{\sin A} = \frac{36}{\sin 90°}$$

$$\frac{d}{\sin 30°} = \frac{36}{\sin 90°}$$

$$d = \frac{36\sin 30°}{\sin 90°} = 18$$

$$CC' = 18 + 5 + 18 = 41$$

The wingspan CC' is 41 feet.

78. $\cos B = \dfrac{6^2 + 4^2 - 9^2}{2\cdot 6\cdot 4}$

$\cos B = \dfrac{-29}{48}$

$\cos B = \dfrac{-29}{48}$

$B = \cos^{-1}\left(\dfrac{-29}{48}\right)$

$B \approx 127°$

79. $\sqrt{26(26-12)(26-16)(26-24)}$

$= \sqrt{26(14)(10)(2)}$

$= \sqrt{7280}$

$= 4\sqrt{455}$

≈ 85

80. Diagram:

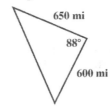

Section 7.2

Check Point Exercises

1. Apply the three-step procedure for solving
    a SAS triangle. Use the Law of Cosines to find the
    side opposite the given angle.
    Thus, we will find *a*.
    $$a^2 = b^2 + c^2 - 2bc\cos A$$
    $$a^2 = 7^2 + 8^2 - 2(7)(8)\cos 120°$$
    $$= 49 + 64 - 112(-0.5)$$
    $$= 169$$
    $$a = \sqrt{169} = 13$$
    Use the Law of Sines to find the angle opposite the
    shorter of the two sides. Thus, we will find acute
    angle *B*.
    $$\frac{b}{\sin B} = \frac{a}{\sin A}$$
    $$\frac{7}{\sin B} = \frac{13}{\sin 120°}$$
    $$13\sin B = 7\sin 120°$$
    $$\sin B = \frac{7\sin 120°}{13} \approx 0.4663$$
    $$B \approx 28°$$
    Find the third angle.
    $$C = 180° - A - B \approx 180° - 120° - 28° = 32°$$
    The solution is $a = 13, B \approx 28°,$ and $C \approx 32°$.

2. Apply the three-step procedure for solving a SSS
    triangle. Use the Law of Cosines to find the angle
    opposite the longest side.
    Thus, we will find angle *B*.
    $$b^2 = a^2 + c^2 - 2ac\cos B$$
    $$2ac\cos B = a^2 + c^2 - b^2$$
    $$\cos B = \frac{a^2 + c^2 - b^2}{2ac}$$
    $$\cos B = \frac{8^2 + 5^2 - 10^2}{2 \cdot 8 \cdot 5} = -\frac{11}{80}$$
    $$\cos^{-1}\left(\frac{11}{80}\right) \approx 82°$$
    B is obtuse, since cos *B* is negative.
    $$B \approx 180° - 82° = 98°$$
    Use the Law of Sines to find either of the two
    remaining acute angles. We will find angle *A*.

$$\frac{a}{\sin A} = \frac{b}{\sin B}$$
$$\frac{8}{\sin A} = \frac{10}{\sin 98°}$$
$$10\sin A = 8\sin 98°$$
$$\sin A = \frac{8\sin 98°}{10} \approx 0.7922$$
$$A \approx 52°$$
Find the third angle.
$$C = 180° - A - B \approx 180° - 52° - 98°$$
$$= 30°$$
The solution is $B \approx 98°, A \approx 52°,$ and $C \approx 30°$

3. The plane flying 400 miles per hour travels
    $400 \cdot 2 = 800$ miles in 2 hours. Similarly, the other
    plane travels 700 miles.

    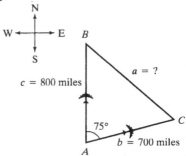

    Use the figure and the Law of Cosines to find *a* in
    this SAS situation.
    $$a^2 = b^2 + c^2 - 2bc\cos A$$
    $$a^2 = 700^2 + 800^2 - 2(700)(800)\cos 75°$$
    $$\approx 840,123$$
    $$a \approx \sqrt{840,123} \approx 917$$
    After 2 hours, the planes are approximately 917 miles
    apart.

4. Begin by calculating one-half the perimeter:
    $$s = \frac{1}{2}(a + b + c) = \frac{1}{2}(6 + 16 + 18) = 20$$
    Use Heron's formula to find the area.
    $$\text{Area} = \sqrt{s(s-a)(s-b)(s-c)}$$
    $$= \sqrt{20(20-6)(20-16)(20-18)}$$
    $$= \sqrt{2240} \approx 47$$
    The area of the triangle is approximately
    47 square meters.

Concept and Vocabulary Check 7.2

1. $b^2 + c^2 - 2bc\cos A$

2. side; Cosines; Sines; acute; 180°

3. Cosines; Sines

4. $\sqrt{s(s-a)(s-b)(s-c)}$; $\frac{1}{2}(a+b+c)$

Exercise Set 7.2

1. Apply the three-step procedure for solving a SAS triangle. Use the Law of Cosines to find the side opposite the given angle.
   Thus, we will find a.
   $a^2 = b^2 + c^2 - 2bc\cos A$
   $a^2 = 4^2 + 8^2 - 2(4)(8)\cos 46°$
   $a^2 = 16 + 64 - 64(\cos 46°)$
   $a^2 \approx 35.54$
   $a \approx \sqrt{35.54} \approx 6.0$
   Use the Law of Sines to find the angle opposite the shorter of the two given sides. Thus, we will find acute angle B.
   $$\frac{b}{\sin B} = \frac{a}{\sin A}$$
   $$\frac{4}{\sin B} = \frac{\sqrt{35.54}}{\sin 46°}$$
   $\sqrt{35.54}\sin B = 4\sin 46°$
   $$\sin B = \frac{4\sin 46°}{\sqrt{35.54}} \approx 0.4827$$
   $B \approx 29°$
   Find the third angle.
   $C = 180° - A - B \approx 180° - 46° - 29° = 105°$
   The solution is $a \approx 6.0, B \approx 29°,$ and $C \approx 105°$.

3. Apply the three-step procedure for solving a SAS triangle. Use the Law of Cosines to find the side opposite the given angle.
   Thus, we will find c.
   $c^2 = a^2 + b^2 - 2ab\cos C$
   $c^2 = 6^2 + 4^2 - 2(6)(4)\cos 96°$
   $c^2 = 36 + 16 - 48(\cos 96°)$
   $c^2 \approx 57.02$
   $c \approx \sqrt{57.02} \approx 7.6$
   Use the Law of Sines to find the angle opposite the shorter of the two given sides. Thus, we will find acute angle B.
   $$\frac{b}{\sin B} = \frac{c}{\sin C}$$
   $$\frac{4}{\sin B} = \frac{\sqrt{57.02}}{\sin 96°}$$
   $\sqrt{57.02}\sin B = 4\sin 96°$
   $$\sin B = \frac{4\sin 96°}{\sqrt{57.02}} \approx 0.5268$$
   $B \approx 32°$
   Find the third angle.
   $A = 180° - B - C \approx 180° - 32° - 96° = 52°$
   The solution is $c \approx 7.6, A \approx 52°,$ and $B \approx 32°$.

5. Apply the three-step procedure for solving a SSS triangle. Use the Law of Cosines to find the angle opposite the longest side. Since two sides have length 8, we can begin by finding angle B or C.
   $b^2 = a^2 + c^2 - 2ac\cos B$
   $\cos B = \dfrac{a^2 + c^2 - b^2}{2ac}$
   $\cos B = \dfrac{6^2 + 8^2 - 8^2}{2 \cdot 6 \cdot 8} = \dfrac{36}{96} = \dfrac{3}{8}$
   $B \approx 68°$
   Use the Law of Sines to find either of the two remaining acute angles. We will find angle A.
   $$\frac{a}{\sin A} = \frac{b}{\sin B}$$
   $$\frac{6}{\sin A} = \frac{8}{\sin 68°}$$
   $8\sin A = 6\sin 68°$
   $$\sin A = \frac{6\sin 68°}{8} \approx 0.6954$$
   $A \approx 44°$
   Find the third angle.
   $C = 180° - B - A \approx 180° - 68° - 44° = 68°$
   The solution is $A \approx 44°, B \approx 68°,$ and $C \approx 68°$.

7. Apply the three-step procedure for solving a SSS triangle. Use the Law of Cosines to find the angle opposite the longest side. Thus, we will find angle A

$$a^2 = b^2 + c^2 - 2bc\cos A$$

$$\cos A = \frac{b^2 + c^2 - a^2}{2bc}$$

$$\cos A = \frac{4^2 + 3^2 - 6^2}{2 \cdot 4 \cdot 3} = -\frac{11}{24}$$

A is obtuse, since $\cos A$ is negative.

$$\cos^{-1}\left(\frac{11}{24}\right) \approx 63°$$

$$A \approx 180° - 63° = 117°$$

Use the Law of Sines to find either of the two remaining acute angles. We will find angle B.

$$\frac{b}{\sin B} = \frac{a}{\sin A}$$

$$\frac{4}{\sin B} = \frac{6}{\sin 117°}$$

$$6\sin B = 4\sin 117°$$

$$\sin B = \frac{4\sin 117°}{6} \approx 0.5940$$

$$B \approx 36°$$

Find the third angle.

$$C = 180° - B - A \approx 180° - 36° - 117° = 27°$$

The solution is $A \approx 117°, B \approx 36°$, and $C \approx 27°$.

9. Apply the three-step procedure for solving a SAS triangle. Use the Law of Cosines to find the side opposite the given angle.

Thus, we will find c.

$$c^2 = a^2 + b^2 - 2ab\cos C$$

$$c^2 = 5^2 + 7^2 - 2(5)(7)\cos 42°$$

$$c^2 = 25 + 49 - 70(\cos 42°)$$

$$c^2 \approx 21.98$$

$$c \approx \sqrt{21.98} \approx 4.7$$

Use the Law of Sines to find the angle opposite the shorter of the two given sides. Thus, we will find acute angle A.

$$\frac{a}{\sin A} = \frac{c}{\sin C}$$

$$\frac{5}{\sin A} = \frac{4.7}{\sin 42°}$$

$$4.7\sin A = 5\sin 42°$$

$$\sin A = \frac{5\sin 42°}{4.7} \approx 0.7118$$

$$A \approx 45°$$

Find the third angle.

$$B = 180° - C - A \approx 180° - 42° - 45° = 93°$$

The solution is $c \approx 4.7, A \approx 45°,$ and $B \approx 93°$.

11. Apply the three-step procedure for solving a SAS triangle. Use the Law of Cosines to find the side opposite the given angle.

Thus, we will find a.

$$a^2 = b^2 + c^2 - 2bc\cos A$$

$$a^2 = 5^2 + 3^2 - 2(5)(3)\cos 102°$$

$$a^2 = 25 + 9 - 30(\cos 102°)$$

$$a^2 \approx 40.24$$

$$a \approx \sqrt{40.24} \approx 6.3$$

Use the Law of Sines to find the angle opposite the shorter of the two given sides. Thus, we will find acute angle C.

$$\frac{c}{\sin C} = \frac{a}{\sin A}$$

$$\frac{3}{\sin C} = \frac{6.3}{\sin 102°}$$

$$6.3\sin C = 3\sin 102°$$

$$\sin C = \frac{3\sin 102°}{6.3} \approx 0.4658$$

$$C \approx 28°$$

Find the third angle.

$$B = 180° - C - A \approx 180° - 28° - 102° = 50°$$

The solution is $a \approx 6.3, C \approx 28°,$ and $B \approx 50°$.

13. Apply the three-step procedure for solving a SAS triangle. Use the Law of Cosines to find the side opposite the given angle.

Thus, we will find b.

$$b^2 = a^2 + c^2 - 2ac\cos B$$

$$b^2 = 6^2 + 5^2 - 2(6)(5)\cos 50°$$

$$b^2 = 36 + 25 - 60(\cos 50°)$$

$$b^2 \approx 22.43$$

$$b \approx \sqrt{22.43} \approx 4.7$$

Use the Law of Sines to find the angle opposite the shorter of the two given sides. Thus, we will find acute angle C.

$$\frac{c}{\sin C} = \frac{b}{\sin B}$$

$$\frac{5}{\sin C} = \frac{4.7}{\sin 50°}$$

$$4.7\sin C = 5\sin 50°$$

$$\sin C = \frac{5\sin 50°}{4.7} \approx 0.8149$$

$$C \approx 55°$$

Find the third angle.

$$A = 180° - C - B \approx 180° - 55° - 50° = 75°$$

The solution is $b \approx 4.7, C \approx 55°,$ and $A \approx 75°$.

15. Apply the three-step procedure for solving a SAS triangle. Use the Law of Cosines to find the side opposite the given angle.

Thus, we will find b.

$$b^2 = a^2 + c^2 - 2ac \cos 90°$$

$$b^2 = 5^2 + 2^2 - 2(5)(2) \cos 90°$$

$$b^2 = 25 + 4 - 20 \cos 90°$$

$$b^2 = 29$$

$$b = \sqrt{29} \approx 5.4$$

(use exact value of b from previous step) Use the Law of Sines to find the angle opposite the shorter of the two given sides. Thus, we will find acute angle C.

$$\frac{c}{\sin C} = \frac{b}{\sin B}$$

$$\frac{2}{\sin C} = \frac{5.4}{\sin 90°}$$

$$5.4 \sin C = 2 \sin 90°$$

$$\sin C = \frac{2 \sin 90°}{5.4} \approx 0.3704$$

$$C \approx 22°$$

Find the third angle.

$$A = 180° - C - B \approx 180° - 22° - 90° = 68°$$

The solution is $b \approx 5.4, C \approx 22°,$ and $A \approx 68°$.

17. Apply the three-step procedure for solving a SSS triangle. Use the Law of Cosines to find the angle opposite the longest side. Thus, we will find C.

$$c^2 = a^2 + b^2 - 2ab \cos C$$

$$\cos C = \frac{a^2 + b^2 - c^2}{2ab}$$

$$\cos C = \frac{5^2 + 7^2 - 10^2}{2 \cdot 5 \cdot 7} = -\frac{13}{35}$$

C is obtuse, since $\cos C$ is negative.

$$\cos^{-1}\left(\frac{13}{35}\right) \approx 68°$$

$$C \approx 180° - 68° = 112°$$

Use the Law of Sines to find either of the two remaining angles. We will find angle A.

$$\frac{a}{\sin A} = \frac{c}{\sin C}$$

$$\frac{5}{\sin A} = \frac{10}{\sin 112°}$$

$$10 \sin A = 5 \sin 112°$$

$$\sin A = \frac{5 \sin 112°}{10} \approx 0.4636$$

$$A \approx 28°$$

Find the third angle.

$$B = 180° - C - A \approx 180° - 112° - 28° = 40°$$

The solution is $C \approx 112°, A \approx 28°,$ and $B \approx 40°$.

19. Apply the three-step procedure for solving a SSS triangle. Use the Law of Cosines to find the angle opposite the longest side. Thus, we will find B.

$$b^2 = a^2 + c^2 - 2ac \cos B$$

$$\cos B = \frac{a^2 + c^2 - b^2}{2ac}$$

$$\cos B = \frac{3^2 + 8^2 - 9^2}{2 \cdot 3 \cdot 8} = -\frac{1}{6}$$

B is obtuse, since $\cos B$ is negative.

$$\cos^{-1}\left(\frac{1}{6}\right) \approx 80°$$

$$B \approx 180° - 80° = 100°$$

Use the Law of Sines to find either of the two remaining angles. We will find angle A.

$$\frac{a}{\sin A} = \frac{b}{\sin B}$$

$$\frac{3}{\sin A} = \frac{9}{\sin 100°}$$

$$9 \sin A = 3 \sin 100°$$

$$\sin A = \frac{3 \sin 100°}{9} \approx 0.3283$$

$$A \approx 19°$$

Find the third angle.

$$C = 180° - B - A \approx 180° - 100° - 19° = 61°$$

The solution is $B \approx 100°, A \approx 19°,$ and $C \approx 61°$.

21. Apply the three-step procedure for solving a SSS triangle. Use the Law of Cosines to find any of the three angles, since each side has the same measure.

$$a^2 = b^2 + c^2 - 2bc \cos A$$

$$\cos A = \frac{b^2 + c^2 - a^2}{2bc}$$

$$\cos A = \frac{3^2 + 3^2 - 3^2}{2 \cdot 3 \cdot 3} = \frac{1}{2}$$

$$A = 60°$$

Use the Law of Sines to find either of the two remaining angles. We will find angle B.

$$\frac{b}{\sin B} = \frac{a}{\sin A}$$

$$\frac{3}{\sin B} = \frac{3}{\sin 60°}$$

$$3 \sin B = 3 \sin 60°$$

$$\sin B = \sin 60°$$

$$B = 60°$$

Find the third angle.

$$C = 180° - A - B = 180° - 60° - 60° = 60°$$

The solution is $A = 60°, B = 60°,$ and $C = 60°$.

23. Apply the three-step procedure for solving a SSS triangle. Use the Law of Cosines to find the angle opposite the longest side. Thus, we will find A.

$$a^2 = b^2 + c^2 - 2bc \cos A$$

$$\cos A = \frac{b^2 + c^2 - a^2}{2bc}$$

$$\cos A = \frac{22^2 + 50^2 - 63^2}{2 \cdot 22 \cdot 50} = -\frac{985}{2200}$$

$$A \approx 117°$$

Use the Law of Sines to find either of the two remaining angles. We will find angle B.

$$\frac{b}{\sin B} = \frac{a}{\sin A}$$

$$\frac{22}{\sin B} = \frac{63}{\sin 117°}$$

$$63 \sin B = 22 \sin 117°$$

$$\sin B = \frac{22 \sin 117°}{63}$$

$$B = 18°$$

Find the third angle.

$$C = 180° - A - B = 180° - 117° - 18° = 45°$$

The solution is $A = 117°, B = 18°$, and $C = 45°$.

25. $s = \dfrac{1}{2}(a + b + c) = \dfrac{1}{2}(4 + 4 + 2) = 5$

$$\text{Area} = \sqrt{s(s-a)(s-b)(s-c)}$$

$$= \sqrt{5(5-4)(5-4)(5-2)}$$

$$= \sqrt{15} \approx 4$$

The area of the triangle is approximately 4 square feet.

27. $s = \dfrac{1}{2}(a + b + c) = \dfrac{1}{2}(14 + 12 + 4) = 15$

$$\text{Area} = \sqrt{s(s-a)(s-b)(s-c)}$$

$$= \sqrt{15(15-14)(15-12)(15-4)}$$

$$= \sqrt{495} \approx 22$$

The area of the triangle is approximately 22 square meters.

29. $s = \dfrac{1}{2}(a + b + c) = \dfrac{1}{2}(11 + 9 + 7) = 13.5$

$$\text{Area} = \sqrt{s(s-a)(s-b)(s-c)}$$

$$= \sqrt{13.5(13.5-11)(13.5-9)(13.5-7)}$$

$$= \sqrt{987.1875} \approx 31$$

The area of the triangle is approximately 31 square yards.

31. $C = 180° - 15° - 35° = 130°$

$$c^2 = b^2 + c^2 - 2bc \cos C$$

$$c^2 = 8^2 + 13^2 - 2(8)(13) \cos 130°$$

$$c^2 \approx 366.6998$$

$$c \approx 19.1$$

Use the law of sines to find the solution is $A \approx 31°, B \approx 19°, C \approx 130°$, and $c \approx 19.1$.

33. Use the given radii to determine that $a = BC = 7.5$, $b = AC = 8.5$, and $c = AB = 9.0$.

$$c^2 = a^2 + b^2 - 2ab \cos C$$

$$9^2 = 7.5^2 + 8.5^2 - 2(7.5)(8.5) \cos C$$

$$\cos C \approx 0.3725$$

$$C \approx 68°$$

Use the law of sines to find the solution is $A \approx 51°, B \approx 61°$, and $C \approx 68°$.

35. Use the distance formula to determine that $a = \sqrt{61} \approx 7.8$, $b = \sqrt{10} \approx 3.2$, and $c = 5$.

$$a^2 = b^2 + c^2 - 2bc \cos A$$

$$\sqrt{61}^2 = \sqrt{10}^2 + 5^2 - 2\left(\sqrt{10}\right)(5) \cos A$$

$$\cos A \approx -0.8222$$

$$A \approx 145°$$

Use the law of sines to find the solution is $A \approx 145°, B \approx 13°$, and $C \approx 22°$.

37. Use the law of cosines.

$$c^2 = a^2 + b^2 - 2ab \cos C$$

$$5.78^2 = 2.9^2 + 3.0^2 - 2(2.9)(3.0) \cos \theta$$

$$\cos \theta \approx -0.9194$$

$$\theta \approx 157°$$

This dinosaur was an efficient walker.

39. Let b = the distance between the ships after three hours. After three hours, the ship traveling 14 miles per hour has gone $3 \cdot 14$ or 42 miles. Similarly, the ship traveling 10 miles per hour has gone 30 miles.

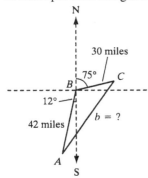

Using the figure,
$$B = 180° - 75° + 12° = 117°$$
$$b^2 = a^2 + c^2 - 2ac \cos B$$
$$b^2 = 30^2 + 42^2 - 2(30)(42) \cos 117° \approx 3808$$
$$b \approx 61.7$$
After three hours, the ships will be about 61.7 miles apart.

41. Let b = the distance across the lake.
$$b^2 = a^2 + c^2 - 2ac \cos B$$
$$b^2 = 160^2 + 140^2 - 2(160)(140) \cos 80°$$
$$\approx 37,421$$
$$b \approx \sqrt{37,421} \approx 193$$
The distance across the lake is about 193 yards.

43. Assume that Island B is due east of Island A. Let A = angle at Island A.
$$a^2 = b^2 + c^2 - 2bc \cos A$$
$$\cos A = \frac{b^2 + c^2 - a^2}{2bc}$$
$$\cos A = \frac{5^2 + 6^2 - 7^2}{2 \cdot 5 \cdot 6} = \frac{1}{5}$$
$$A \approx 78°$$
Since $90° - 78° = 12°$, you should navigate on a bearing of N12°E.

45. a. Using the figure,
$$B = 90° - 40° = 50°$$
$$b^2 = a^2 + c^2 - 2ac \cos B$$
$$b^2 = 13.5^2 + 25^2 - 2(13.5)(25) \cos 50°$$
$$\approx 373$$
$$b \approx \sqrt{373} \approx 19.3$$
You are about 19.3 miles from the pier.

b.
$$\frac{a}{\sin A} = \frac{b}{\sin B}$$
$$\frac{13.5}{\sin A} = \frac{\sqrt{373}}{\sin 50°}$$
$$\sqrt{373} \sin A = 13.5 \sin 50°$$
$$\sin A = \frac{13.5 \sin 50°}{\sqrt{373}} \approx 0.5355$$
$$A \approx 32°$$
Since $90° - 32° = 58°$, the original bearing could have been S58°E.

47.

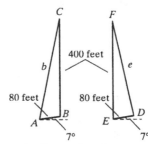

In the figure, b = the guy wire anchored downhill, e = the guy wire anchored uphill.
$$B = 90° + 7° = 97°$$
$$E = 90° - 7° = 83°$$
$$b^2 = a^2 + c^2 - 2ac \cos B$$
$$b^2 = 400^2 + 80^2 - 2(400)(80) \cos 97°$$
$$\approx 174,200$$
$$b \approx \sqrt{174,200} \approx 417.4$$
$$e^2 = d^2 + f^2 - 2df \cos E$$
$$e^2 = 400^2 + 80^2 - 2(400)(80) \cos 83°$$
$$\approx 158,600$$
$$e \approx \sqrt{158.600} \approx 398.2$$
The guy wire anchored downhill is about 417.4 feet long. The one anchored uphill is about 398.2 feet long.

49.

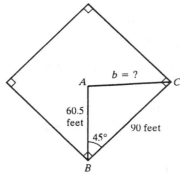

Using the figure,
$B = 90° \div 2 = 45°$ (using symmetry)
$b^2 = a^2 + c^2 - 2ac\cos B$
$b^2 = 90^2 + 60.5^2 - 2(90)(60.5)\cos 45°$
$\quad \approx 4060$
$b \approx \sqrt{4060} \approx 63.7$

It is about 63.7 feet from the pitcher's mound to first base.

51. First, find the area using Heron's formula.
$$s = \frac{1}{2}(a+b+c) = \frac{1}{2}(240+300+420) = 480$$
$$\text{Area} = \sqrt{s(s-a)(s-b)(s-c)}$$
$$= \sqrt{480(480-240)(480-300)(480-420)}$$
$$= \sqrt{1,244,160,000} \approx 35,272.65$$

Now multiply by the price per square foot.
$(35,272.65)(3.50) \approx 123,454$

The cost is $123,454, to the nearest dollar.

53. –59. Answers may vary.

61. does not make sense; Explanations will vary.
Sample explanation: The Law of Cosines is not simply the negative of the Law of Sines.

63. makes sense

65.

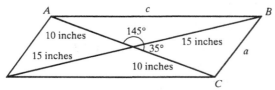

Using the given information and the hint, we arrive at the figure above. Let a = the side opposite the 35° angle, c = the side opposite the 145° angle.

$a^2 = 15^2 + 10^2 - 2(15)(10)\cos 35° \approx 79.3$
$a \approx \sqrt{79.3} \approx 8.9$
$c^2 = 15^2 + 10^2 - 2(15)(10)\cos 145° \approx 570.7$
$b \approx \sqrt{570.7} \approx 23.9$

The lengths of the parallelogram's sides are about 8.9 inches and 23.9 inches.

67.

The angle between the minute and hour hand is $\frac{2}{3}$ of the 90° angle from 9 to 12, or 60°.
Let d = the distance between the tips of the hands.
$d^2 = m^2 + h^2 - 2mh\cos 60°$
$$= m^2 + h^2 - 2mh\left(\frac{1}{2}\right)$$
$$= m^2 + h^2 - mh$$
$$d = \sqrt{m^2 + h^2 - mh}$$

69. $y = 3$ is a horizontal line through (0, 3).

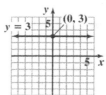

70. $x^2 + (y-1)^2 = 1$ is a circle centered at (0, 1) with a radius of 1.

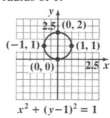

71.

$$x^2 + 6x + y^2 = 0$$

$$x^2 + 6x \qquad + y^2 = 0$$

$$x^2 + 6x + 9 + y^2 = 0 + 9$$

$$(x+3)^2 + y^2 = 9$$

$(x+3)^2 + y^2 = 9$ is a circle centered at $(-3, 0)$ with a radius of 3.

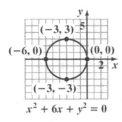

Section 7.3

Check Point Exercises

1. a. $(r, \theta) = (3, 315°)$

Because $315°$ is a positive angle, draw $\theta = 315°$ counterclockwise from the polar axis. Because $r > 0$, plot the point by going out 3 units on the terminal side of θ.

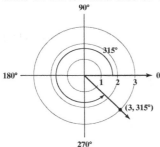

b. $(r, \theta) = (-2, \pi)$

Because π is a positive angle, draw $\theta = \pi$ counterclockwise from the polar axis. Because $r < 0$, plot the point by going out 2 units along the ray opposite the terminal side of θ.

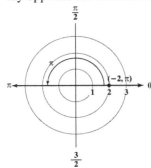

c. $(r, \theta) = \left(-1, -\dfrac{\pi}{2}\right)$

Because $-\dfrac{\pi}{2}$ is a negative angle, draw $\theta = -\dfrac{\pi}{2}$ clockwise from the polar axis. Because $r < 0$, plot the point by going out one unit along the ray opposite the terminal side of θ.

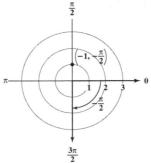

2. a. Add 2π to the angle and do not change r.

$$\left(5, \frac{\pi}{4}\right) = \left(5, \frac{\pi}{4} + 2\pi\right) = \left(5, \frac{\pi}{4} + \frac{8\pi}{4}\right)$$

$$= \left(5, \frac{9\pi}{4}\right)$$

b. Add π to the angle and replace r by $-r$.

$$\left(5, \frac{\pi}{4}\right) = \left(-5, \frac{\pi}{4} + \pi\right) = \left(-5, \frac{\pi}{4} + \frac{4\pi}{4}\right)$$

$$= \left(-5, \frac{5\pi}{4}\right)$$

c. Subtract 2π from the angle and do not change r.

$$\left(5, \frac{\pi}{4}\right) = \left(5, \frac{\pi}{4} - 2\pi\right) = \left(5, \frac{\pi}{4} - \frac{8\pi}{4}\right)$$

$$= \left(5, -\frac{7\pi}{4}\right)$$

3. a. $(r, \theta) = (3, \pi)$

$$x = r\cos\theta = 3\cos\pi = 3(-1) = -3$$

$$y = r\sin\theta = 3\sin\pi = 3(0) = 0$$

The rectangular coordinates of $(3, \pi)$ are $(-3, 0)$.

b. $(r, \theta) = \left(-10, \dfrac{\pi}{6}\right)$

$$x = r\cos\theta = -10\cos\dfrac{\pi}{6} = -10\left(\dfrac{\sqrt{3}}{2}\right)$$

$$= -5\sqrt{3}$$

$$y = r\sin\theta = -10\sin\dfrac{\pi}{6} = -10\left(\dfrac{1}{2}\right) = -5$$

The rectangular coordinates of $\left(-10, \dfrac{\pi}{6}\right)$ are

$\left(-5\sqrt{3}, -5\right)$.

4.

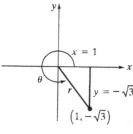

$$r = \sqrt{x^2 + y^2} = \sqrt{1^2 + \left(-\sqrt{3}\right)^2}$$

$$= \sqrt{1 + 3} = \sqrt{4} = 2$$

$$\tan\theta = \dfrac{y}{x} = \dfrac{-\sqrt{3}}{1} = -\sqrt{3}$$

Because $\tan\dfrac{\pi}{3} = \sqrt{3}$ and θ lies in quadrant IV,

$$\theta = 2\pi - \dfrac{\pi}{3} = \dfrac{6\pi}{3} - \dfrac{\pi}{3} = \dfrac{5\pi}{3}$$

The polar coordinates of $\left(1, -\sqrt{3}\right)$ are $(r, \theta) = \left(2, \dfrac{5\pi}{3}\right)$.

5.

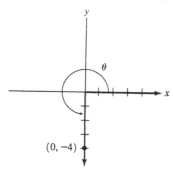

$$r = \sqrt{x^2 + y^2} = \sqrt{(0)^2 + (-4)^2} = \sqrt{16} = 4$$

The point $(0, -4)$ is on the negative y-axis. Thus,

$\theta = \dfrac{3\pi}{2}$. Polar coordinates of $(0, -4)$ are $\left(4, \dfrac{3\pi}{2}\right)$.

6. a.

$$3x - y = 6$$
$$3r\cos\theta - r\sin\theta = 6$$
$$r(3\cos\theta - \sin\theta) = 6$$
$$r = \dfrac{6}{3\cos\theta - \sin\theta}$$

b.

$$x^2 + (y + 1)^2 = 1$$
$$\left(r\cos\theta\right)^2 + \left(r\sin\theta + 1\right)^2 = 1$$
$$r^2\cos^2\theta + r^2\sin^2\theta + 2r\sin\theta + 1 = 1$$
$$r^2 + 2r\sin\theta = 0$$
$$r(r + 2\sin\theta) = 0$$
$$r = 0 \quad \text{or} \quad r + 2\sin\theta = 0$$
$$r = -2\sin\theta$$

7. a. Use $r^2 = x^2 + y^2$ to convert to a rectangular equation.

$$r = 4$$
$$r^2 = 16$$
$$x^2 + y^2 = 16$$

The rectangular equation for $r = 4$ is $x^2 + y^2 = 16$.

b. Use $\tan\theta = \dfrac{y}{x}$ to convert to a rectangular equation in x and y.

$$\theta = \dfrac{3\pi}{4}$$
$$\tan\theta = \tan\dfrac{3\pi}{4}$$
$$\tan\theta = -1$$
$$\dfrac{y}{x} = -1$$
$$y = -x$$

The rectangular equation for $\theta = \dfrac{3\pi}{4}$ is $y = -x$.

c.

$$r = -2\sec\theta$$
$$r = \dfrac{-2}{\cos\theta}$$
$$r\cos\theta = -2$$
$$x = -2$$

d.
$$r = 10\sin\theta$$
$$r^2 = 10r\sin\theta$$
$$x^2 + y^2 = 10y$$
$$x^2 + y^2 - 10y = 0$$
$$x^2 + y^2 - 10y + 25 = 25$$
$$x^2 + (y-5)^2 = 25$$

Concept and Vocabulary Check 7.3

1. pole; polar axis

2. pole; polar axis

3. II

4. IV

5. IV

6. III

7. IV

8. II

9. $-r$

10. $-r$

11. $r\cos\theta$; $r\sin\theta$

12. squaring; $x^2 + y^2$

13. tangent; $\dfrac{y}{x}$

14. multiplying ; r; $x^2 + y^2$; y

Exercise Set 7.3

1. 225° is in the third quadrant.
   C

3. $\dfrac{5\pi}{4} = 225°$ is in the third quadrant. Since r is negative, the point lies along the ray opposite the terminal side of θ, in the first quadrant.
   A

5. $\pi = 180°$ lies on the negative *x*-axis.
   B

7. −135° is measured clockwise 135° from the positive *x*-axis. The point lies in the third quadrant.
   C

9. $-\dfrac{3\pi}{4} = -135°$ is measured clockwise 135° from the positive *x*-axis. Since r is negative, the point lies along the ray opposite the terminal side of θ, in the first quadrant.
   A

11. Draw $\theta = 45°$ counterclockwise, since θ is positive, from the polar axis. Go out 2 units on the terminal side of θ, since $r > 0$.

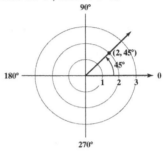

13. Draw $\theta = 90°$ counterclockwise, since θ is positive, from the polar axis. Go out 3 units on the terminal side of θ, since $r > 0$.

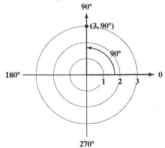

15. Draw $\theta = \dfrac{4\pi}{3} = 240°$ counterclockwise, since θ is positive, from polar axis. Go out 3 units on the terminal side of θ, since $r > 0$.

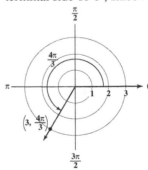

17. Draw $\theta = \pi = 180°$ counterclockwise, since θ is positive, from the polar axis. Go one unit out on the ray opposite the terminal side of θ, since $r < 0$.

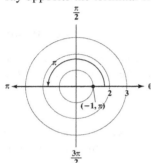

19. Draw $\theta = -\dfrac{\pi}{2} = -90°$ clockwise, since θ is positive,

from the polar axis. Go 2 units out on the ray opposite the terminal side of θ, since $r < 0$.

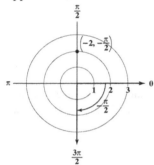

21. Draw $\theta = \dfrac{\pi}{6} = 30°$ counterclockwise, since θ is

positive, from the polar axis. Go 5 units out on the terminal side of θ, since $r > 0$.

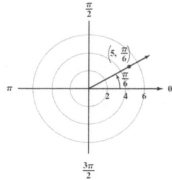

a. Add 2π to the angle and do not change r.

$$\left(5, \frac{\pi}{6}\right) = \left(5, \frac{\pi}{6} + 2\pi\right) = \left(5, \frac{13\pi}{6}\right)$$

b. Add π to the angle and replace r by $-r$.

$$\left(5, \frac{\pi}{6}\right) = \left(-5, \frac{\pi}{6} + \pi\right) = \left(-5, \frac{7\pi}{6}\right)$$

c. Subtract 2π from the angle and do not change r.

$$\left(5, \frac{\pi}{6}\right) = \left(5, \frac{\pi}{6} - 2\pi\right) = \left(5, -\frac{11\pi}{6}\right)$$

23. Draw $\theta = \dfrac{3\pi}{4} = 135°$ counterclockwise, since θ is

positive, from the polar axis. Go out 10 units on the terminal side of θ, since $r > 0$.

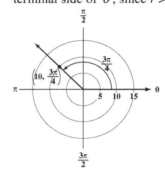

a. Add 2π to the angle and do not change r.

$$\left(10, \frac{3\pi}{4}\right) = \left(10, \frac{3\pi}{4} + 2\pi\right) = \left(10, \frac{11\pi}{4}\right)$$

b. Add π to the angle and replace r by $-r$.

$$\left(10, \frac{3\pi}{4}\right) = \left(-10, \frac{3\pi}{4} + \pi\right) = \left(-10, \frac{7\pi}{4}\right)$$

c. Subtract 2π from the angle and do not change r.

$$\left(10, \frac{3\pi}{4}\right) = \left(10, \frac{3\pi}{4} - 2\pi\right) = \left(10, \frac{-5\pi}{4}\right)$$

25. Draw $\theta = \dfrac{\pi}{2} = 90°$ counterclockwise, since θ is

positive, from the polar axis. Go 4 units out on the terminal side of θ, since $r > 0$.

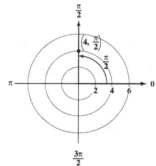

a. Add 2π to the angle and do not change r.

$$\left(4, \frac{\pi}{2}\right) = \left(4, \frac{\pi}{2} + 2\pi\right) = \left(4, \frac{5\pi}{2}\right)$$

b. Add π to the angle and replace r by $-r$.

$$\left(4, \frac{\pi}{2}\right) = \left(-4, \frac{\pi}{2} + \pi\right) = \left(-4, \frac{3\pi}{2}\right)$$

c. Subtract 2π from the angle and do not change r.

$$\left(4, \frac{\pi}{2}\right) = \left(4, \frac{\pi}{2} - 2\pi\right) = \left(4, -\frac{3\pi}{2}\right)$$

27. a, b, d

29. b, d

31. a, b

33. The rectangular coordinates of $(4, 90°)$ are $(0, 4)$.

35. $x = r\cos\theta = 2\cos\dfrac{\pi}{3} = 2\left(\dfrac{1}{2}\right) = 1$

$y = r\sin\theta = 2\sin\dfrac{\pi}{3} = 2\left(\dfrac{\sqrt{3}}{2}\right) = \sqrt{3}$

The rectangular coordinates of $\left(2, \dfrac{\pi}{3}\right)$ are $\left(1, \sqrt{3}\right)$.

37. $x = r\cos\theta = -4\cos\dfrac{\pi}{2} = -4 \cdot 0 = 0$

$y = r\sin\theta = -4\sin\dfrac{\pi}{2} = -4(1) = -4$

The rectangular coordinates of $\left(-4, \dfrac{\pi}{2}\right)$ are $(0, -4)$.

39. $x = r\cos\theta = 7.4\cos 2.5 \approx 7.4(-0.80) \approx -5.9$

$y = r\sin\theta = 7.4\sin 2.5 \approx 7.4(0.60) \approx 4.4$

The rectangular coordinates of $(7.4, 2.5)$ are approximately $(-5.9, 4.4)$.

41. $r = \sqrt{x^2 + y^2} = \sqrt{(-2)^2 + 2^2}$

$ = \sqrt{4 + 4} = \sqrt{8} = 2\sqrt{2}$

$\tan\theta = \dfrac{y}{x} = \dfrac{2}{-2} = -1$

Because $\tan\theta = -1$ and θ lies in quadrant II,

$\theta = \dfrac{3\pi}{4}$.

The polar coordinates of $(-2, 2)$ are

$(r, \theta) = \left(\sqrt{8}, \dfrac{3\pi}{4}\right)$.

43. $r = \sqrt{x^2 + y^2} = \sqrt{(2)^2 + \left(-2\sqrt{3}\right)^2}$

$ = \sqrt{4 + 12} = \sqrt{16} = 4$

$\tan\theta = \dfrac{y}{x} = \dfrac{-2\sqrt{3}}{2} = -\sqrt{3}$

Because $\tan\dfrac{\pi}{3} = \sqrt{3}$ and θ lies in quadrant IV,

$\theta = 2\pi - \dfrac{\pi}{3} = \dfrac{5\pi}{3}$.

The polar coordinates of $\left(2, -2\sqrt{3}\right)$ are

$(r, \theta) = \left(4, \dfrac{5\pi}{3}\right)$.

45. $r = \sqrt{x^2 + y^2} = \sqrt{\left(-\sqrt{3}\right)^2 + (-1)^2}$

$ = \sqrt{3 + 1} = \sqrt{4} = 2$

$\tan\theta = \dfrac{y}{x} = \dfrac{-1}{-\sqrt{3}} = \dfrac{1}{\sqrt{3}}$

Because $\tan\dfrac{\pi}{6} = \dfrac{1}{\sqrt{3}}$ and θ lies in quadrant III,

$\theta = \pi + \dfrac{\pi}{6} = \dfrac{7\pi}{6}$.

The polar coordinates of $\left(-\sqrt{3}, -1\right)$ are

$(r, \theta) = \left(2, \dfrac{7\pi}{6}\right)$.

47. $r = \sqrt{x^2 + y^2} = \sqrt{(5)^2 + (0)^2} = \sqrt{25} = 5$

$\tan\theta = \dfrac{y}{x} = \dfrac{0}{5} = 0$

Because $\tan 0 = 0$ and θ lies on the polar axis, $\theta = 0$.

The polar coordinates of $(5, 0)$ are $(r, \theta) = (5, 0)$.

49.
$$3x + y = 7$$
$$3r\cos\theta + r\sin\theta = 7$$
$$r(3\cos\theta + \sin\theta) = 7$$
$$r = \frac{7}{3\cos\theta + \sin\theta}$$

51.
$$x = 7$$
$$r\cos\theta = 7$$
$$r = \frac{7}{\cos\theta}$$

53.
$$x^2 + y^2 = 9$$
$$r^2 = 9$$
$$r = 3$$

55.
$$(x - 2)^2 + y^2 = 4$$
$$(r\cos\theta - 2)^2 + (r\sin\theta)^2 = 4$$
$$r^2\cos^2\theta - 4r\cos\theta + 4 + r^2\sin^2\theta^2 = 4$$
$$r^2 - 4r\cos\theta = 0$$
$$r^2 = 4r\cos\theta$$
$$r = 4\cos\theta$$

57.
$$y^2 = 6x$$
$$(r\sin\theta)^2 = 6r\cos\theta$$
$$r^2\sin^2\theta = 6r\cos\theta$$
$$r\sin^2\theta = 6\cos\theta$$
$$r = \frac{6\cos\theta}{\sin^2\theta}$$

59.
$$r = 8$$
$$r^2 = 64$$
$$x^2 + y^2 = 64$$

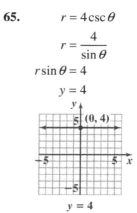

61.
$$\theta = \frac{\pi}{2}$$
$$\tan\theta = \tan\frac{\pi}{2}$$
$$\tan\theta \text{ is undefined}$$
$$\frac{y}{x} \text{ is undefined}$$
$$x = 0$$

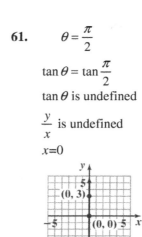

63.
$$r\sin\theta = 3$$
$$y = 3$$

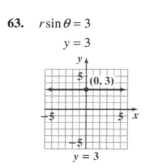

65.
$$r = 4\csc\theta$$
$$r = \frac{4}{\sin\theta}$$
$$r\sin\theta = 4$$
$$y = 4$$

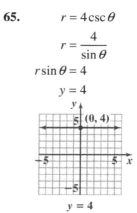

67.
$$r = \sin\theta$$
$$r \cdot r = r \cdot \sin\theta$$
$$r^2 = r\sin\theta$$
$$x^2 + y^2 = y$$

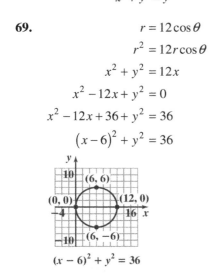

$x^2 + y^2 = y$

69.
$$r = 12\cos\theta$$
$$r^2 = 12r\cos\theta$$
$$x^2 + y^2 = 12x$$
$$x^2 - 12x + y^2 = 0$$
$$x^2 - 12x + 36 + y^2 = 36$$
$$(x-6)^2 + y^2 = 36$$

$(x - 6)^2 + y^2 = 36$

71.
$$r = 6\cos\theta + 4\sin\theta$$
$$r \cdot r = r(6\cos\theta + 4\sin\theta)$$
$$r^2 = 6r\cos\theta + 4r\sin\theta$$
$$x^2 + y^2 = 6x + 4y$$

$x^2 + y^2 = 6x + 4y$

73.
$$r^2 \sin 2\theta = 2$$
$$r^2(2\sin\theta\cos\theta) = 2$$
$$2r\sin\theta\, r\cos\theta = 2$$
$$2yx = 2$$
$$xy = 1$$
$$y = \frac{1}{x}$$

$y = \dfrac{1}{x}$

75.
$$r = a\sec\theta$$
$$r = \frac{a}{\cos\theta}$$
$$r\cos\theta = a$$
$$x = a$$
This is the equation of a vertical line.

77.
$$r = a\sin\theta$$
$$r^2 = ar\sin\theta$$
$$x^2 + y^2 = ay$$
$$x^2 + y^2 - ay = 0$$
$$x^2 + y^2 - ay + \frac{a^2}{4} = \frac{a^2}{4}$$
$$x^2 + \left(y - \frac{a}{2}\right)^2 = \left(\frac{a}{2}\right)^2$$

This is the equation of a circle of radius $\dfrac{a}{2}$ centered at $\left(0, \dfrac{a}{2}\right)$.

79.
$$r\sin\left(\theta - \frac{\pi}{4}\right) = 2$$
$$r\left(\sin\theta\cos\frac{\pi}{4} - \cos\theta\sin\frac{\pi}{4}\right) = 2$$
$$r\sin\theta \cdot \frac{\sqrt{2}}{2} - r\cos\theta \cdot \frac{\sqrt{2}}{2} = 2$$
$$y \cdot \frac{\sqrt{2}}{2} - x \cdot \frac{\sqrt{2}}{2} = 2$$
$$y = x + 2\sqrt{2}$$

$y = x + 2\sqrt{2}$ has slope of 1 and y-intercept of $2\sqrt{2}$.

81. $x_1 = r\cos\theta = 2\cos\dfrac{2\pi}{3} = -1$

$y_1 = r\sin\theta = 2\sin\dfrac{2\pi}{3} = \sqrt{3}$

$\left(-1,\sqrt{3}\right)$

$x_2 = r\cos\theta = 4\cos\dfrac{\pi}{6} = 2\sqrt{3}$

$y_2 = r\sin\theta = 4\sin\dfrac{\pi}{6} = 2$

$\left(2\sqrt{3},2\right)$

$d = \sqrt{\left(x_2 - x_1\right)^2 + \left(y_2 - y_1\right)^2}$

$d = \sqrt{\left(2\sqrt{3}+1\right)^2 + \left(2-\sqrt{3}\right)^2}$

$d = 2\sqrt{5}$

83. The angle is measured counterclockwise from the polar axis.

$\theta = \dfrac{2}{3}(360°) = 240°$ or $\dfrac{4\pi}{3}$.

The distance from the inner circle's center
to the outer circle is
$r = 6 + 3(3) = 6 + 9 = 15$

The polar coordinates are $(r,\theta) = \left(15, \dfrac{4\pi}{3}\right)$.

85. (6.3, 50°) represents a sailing speed of 6.3 knots at an angle of 50° to the wind.

87. Out of the four points in this 10-knot-wind situation, you would recommend a sailing angle of 105°. A sailing speed of 7.5 knots is achieved at this angle.

89. – 95. Answers may vary.

97.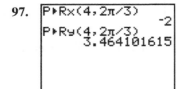

To three decimal places, the rectangular coordinates are (–2, 3.464).

99.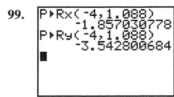

To three decimal places, the rectangular coordinates are (–1.857, –3.543).

101.
```
R▸Pr(√(5),2)
               3
R▸Pθ(√(5),2)
      .7297276562
```

To three decimal places, the polar coordinates are $(r, \theta) = (3, 0.730)$.

103. does not make sense; Explanations will vary. Sample explanation: There are multiple polar representations for a given point.

105. makes sense

107. Use the distance formula for rectangular coordinates, $d = \sqrt{(x_2 - x_1)^2 + (y_2 - y_1)^2}$.

Let $x_1 = r_1 \cos \theta_1$, $y_1 = r_1 \sin \theta_1$,

$x_2 = r_2 \cos \theta_2$, $y_2 = r_2 \sin \theta_2$

$$d = \sqrt{(r_2 \cos \theta_2 - r_1 \cos \theta_1)^2 + (r_2 \sin \theta_2 - r_1 \sin \theta_1)^2}$$

$$= \sqrt{r_2^2 \cos^2 \theta_2 - 2r_1 r_2 \cos \theta_1 \cos \theta_2 + r_1^2 \cos^2 \theta_1 + r_2^2 \sin^2 \theta_2 - 2r_1 r_2 \sin \theta_1 \sin \theta_2 + r_1^2 \sin^2 \theta_1}$$

$$= \sqrt{r_2^2 \left(\cos^2 \theta_2 + \sin^2 \theta_2 \right) + r_1^2 \left(\cos^2 \theta_1 + \sin^2 \theta_1 \right) - 2r_1 r_2 \left(\cos \theta_1 \cos \theta_2 + \sin \theta_1 \sin \theta_2 \right)}$$

$$= \sqrt{r_2^2 (1) + r_1^2 (1) - 2r_1 r_2 \left(\cos(\theta_2 - \theta_1) \right)}$$

$$= \sqrt{r_1^2 + r_2^2 - 2r_1 r_2 \cos(\theta_2 - \theta_1)}$$

109.

θ	0	$\dfrac{\pi}{6}$	$\dfrac{\pi}{3}$	$\dfrac{\pi}{2}$	$\dfrac{2\pi}{3}$	$\dfrac{5\pi}{6}$	π
$r = 1 - \cos\theta$	0	$\dfrac{2-\sqrt{3}}{2}$ ≈ 0.13	$\dfrac{1}{2}$ $= 0.5$	1	$\dfrac{3}{2}$ $= 1.5$	$\dfrac{2+\sqrt{3}}{2}$ ≈ 1.87	0

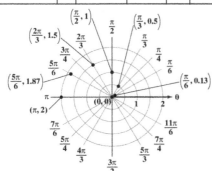

110.

θ	0	$\dfrac{\pi}{6}$	$\dfrac{\pi}{3}$	$\dfrac{\pi}{2}$	$\dfrac{2\pi}{3}$	$\dfrac{5\pi}{6}$	π	$\dfrac{7\pi}{6}$	$\dfrac{4\pi}{3}$	$\dfrac{3\pi}{2}$
$r = 1 + 2\sin\theta$	1	2	$1+\sqrt{3}$ ≈ 2.73	3	$1+\sqrt{3}$ ≈ 2.73	2	1	0	$1-\sqrt{3}$ ≈ -0.73	-1

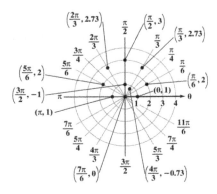

111.

θ	0	$\dfrac{\pi}{6}$	$\dfrac{\pi}{4}$	$\dfrac{\pi}{3}$	$\dfrac{\pi}{2}$	$\dfrac{2\pi}{3}$	$\dfrac{3\pi}{4}$	$\dfrac{5\pi}{6}$	π
$r = 4\sin 2\theta$	0	$2\sqrt{3}$ ≈ 3.46	4	$2\sqrt{3}$ ≈ 3.46	0	$-2\sqrt{3}$ ≈ -3.46	-4	$-2\sqrt{3}$ ≈ -3.46	0

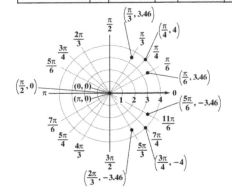

Section 7.4

Check Point Exercises

1. Construct a partial table of coordinates using multiples of $\dfrac{\pi}{6}$. Then plot the points and join them with a smooth curve.

    $r = 4\sin\theta$

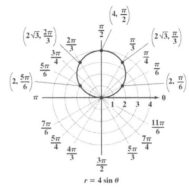

θ	r	(r, θ)
0	$4\sin 0 = 4 \cdot 0 = 0$	$(0, 0)$
$\dfrac{\pi}{6}$	$4\sin\dfrac{\pi}{6} = 4 \cdot \dfrac{1}{2} = 2$	$2, \dfrac{\pi}{6}$
$\dfrac{\pi}{3}$	$4\sin\dfrac{\pi}{3} = 4 \cdot \dfrac{\sqrt{3}}{2} = 2\sqrt{3}$	$2\sqrt{3}, \dfrac{\pi}{3}$
$\dfrac{\pi}{2}$	$4\sin\dfrac{\pi}{2} = 4 \cdot 1 = 4$	$4, \dfrac{\pi}{2}$
$\dfrac{2\pi}{3}$	$4\sin\dfrac{2\pi}{3} = 4 \cdot \dfrac{\sqrt{3}}{2} = 2\sqrt{3}$	$2\sqrt{3}, \dfrac{2\pi}{3}$
$\dfrac{5\pi}{6}$	$4\sin\dfrac{5\pi}{6} = 4 \cdot \dfrac{1}{2} = 2$	$2, \dfrac{5\pi}{6}$
π	$4\sin\pi = 4 \cdot 0 = 0$	$(0, \pi)$

2. **Polar Axis:** Replace θ by $-\theta$ in $r = 1 + \cos\theta$.
    $$r = 1 + \cos(-\theta)$$
    $$r = 1 + \cos\theta$$

    Because the polar equation does not change when θ is replaced by $-\theta$, the graph is symmetric with respect to the polar axis.

    The Line $\theta = \dfrac{\pi}{2}$: Replace (r, θ) by $(-r, -\theta)$ in $r = 1 + \cos\theta$.
    $$-r = 1 + \cos(-\theta)$$
    $$-r = 1 + \cos\theta$$
    $$r = -1 - \cos\theta$$

    Because the polar equation changes when (r, θ) is replaced by $(-r, -\theta)$, the equation fails the symmetry test. The graph

    may or may not be symmetric with respect to the line $\theta = \dfrac{\pi}{2}$.

    The Pole: Replace r by $-r$ in $r = 1 + \cos\theta$.
    $$-r = 1 + \cos\theta$$
    $$r = -1 - \cos\theta$$

    Because the polar equation changes when r is replaced by $-r$, the equation fails the symmetry test. The graph may or may not be symmetric with respect to the pole.

    Because the period of the cosine function is 2π, and the graph is symmetric with respect to the polar axis, begin by finding vales of r for values of θ from 0 to π. then graph $r = 1 + \cos\theta$ for these values and reflect the graph about the polar axis.

θ	0	$\dfrac{\pi}{6}$	$\dfrac{\pi}{3}$	$\dfrac{\pi}{2}$	$\dfrac{2\pi}{3}$	$\dfrac{5\pi}{6}$	π
r	2	1.87	1.5	1	0.5	0.13	0

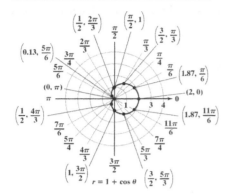

3. $r = 1 - 2\sin\theta$

Check for symmetry:

Polar Axis	The Line $\theta = \dfrac{\pi}{2}$	The Pole
$r = 1 - 2\sin(-\theta)$	$-r = 1 - 2\sin(-\theta)$	$-r = 1 - 2\sin\theta$
$r = 1 - 2(-\sin\theta)$	$-r = 1 + 2\sin\theta$	
$r = 1 + 2\sin\theta$	$r = -1 - 2\sin\theta$	$r = -1 + 2\sin\theta$

There may be no symmetry, since each equation is not equivalent to $r = 1 - 2\sin\theta$. Because the period of the sine function is 2π, we need not consider values of θ beyond 2π.

θ	0	$\dfrac{\pi}{6}$	$\dfrac{\pi}{3}$	$\dfrac{\pi}{2}$	$\dfrac{2\pi}{3}$	$\dfrac{5\pi}{6}$	π	$\dfrac{7\pi}{6}$	$\dfrac{4\pi}{3}$	$\dfrac{3\pi}{2}$	$\dfrac{5\pi}{6}$	$\dfrac{11\pi}{6}$	2π
r	1	0	-0.73	-1	-0.73	0	1	2	2.73	3	2.73	2	1

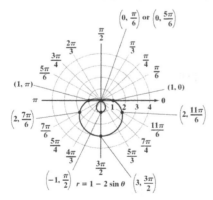

4. $r = 3\cos 2\theta$
Check for symmetry:

Polar Axis	The Line $\theta = \dfrac{\pi}{2}$	The Pole
$r = 3\cos 2(-\theta)$	$-r = 3\cos 2(-\theta)$	$-r = 3\cos 2\theta$
	$-r = 3\cos 2\theta$	
$r = 3\cos 2\theta$	$r = -3\cos 2\theta$	$r = -3\cos 2\theta$

The graph has symmetry with respect to the polar axis. The graph may or may not be symmetric with respect to the line $\theta = \dfrac{\pi}{2}$ or the pole.

Since the graph is symmetric with respect to the polar axis, calculate values of r for θ from 0 to π. Then, graph $r = 3\cos 2\theta$ for these values and reflect the graph about the polar axis.

θ	0	$\dfrac{\pi}{6}$	$\dfrac{\pi}{3}$	$\dfrac{\pi}{2}$	$\dfrac{2\pi}{3}$	$\dfrac{5\pi}{6}$	π
r	3	1.5	-1.5	-3	-1.5	1.5	3

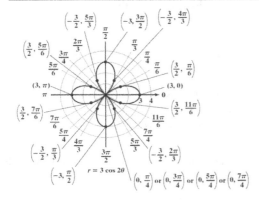

5. $r^2 = 4\cos 2\theta$
Check for symmetry:

Polar Axis	The Line $\theta = \dfrac{\pi}{2}$	The Pole
$r^2 = 4\cos 2(-\theta)$	$(-r)^2 = 4\cos 2(-\theta)$	$(-r)^2 = 4\cos 2\theta$
$r^2 = 4\cos(-2\theta)$	$r^2 = 4\cos(-2\theta)$	
$r^2 = 4\cos 2\theta$	$r^2 = 4\cos 2\theta$	$r^2 = 4\cos 2\theta$

The graph has symmetry with respect to the polar axis, the line $\theta = \dfrac{\pi}{2}$, and the pole.

Calculate values of r for θ from 0 to $\dfrac{\pi}{2}$, and use symmetry to obtain the graph.

θ	0	$\dfrac{\pi}{6}$	$\dfrac{\pi}{4}$	$\dfrac{\pi}{3}$	$\dfrac{\pi}{2}$
r	± 2	± 1.41	0	undef.	undef.

Use symmetry to obtain the graph.

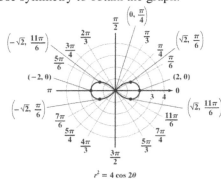

$r^2 = 4\cos 2\theta$

Concept and Vocabulary Check 7.4

1. θ; r

2. circles

3. θ; $-\theta$

4. (r,θ); $(-r,-\theta)$

5. r; $-r$

6. true

7. limaçons; cardioid; loop

8. $2n$; n

9. lemniscates; pole; polar axis; pole; $\theta = \dfrac{\pi}{2}$

Exercise Set 7.4

1. heart-shaped limaçon or cardioid
   $\dfrac{a}{b} = 1$

   $r = 0$ when $\theta = \dfrac{\pi}{2}$
   The polar equation is $r = 1 - \sin\theta$.

3. circle
   $r = 2$ when $\theta = 0$
   The polar equation is $r = 2\cos\theta$.

5. rose curve
   3 petals n 3
   The polar equation is $r = 3\sin 3\theta$.

7. **a.** $r = \sin\theta$
      Replace θ with $-\theta$.
      $r = \sin(-\theta)$

      $r = -\sin\theta$
      The graph may or may not have symmetry with respect to the polar axis.

   b. $r = \sin\theta$
      Replace (r, θ) with $(-r, -\theta)$.
      $-r = \sin(-\theta)$

      $-r = -\sin\theta$

      $r = \sin\theta$
      The graph has symmetry with respect to the line $\theta = \dfrac{\pi}{2}$.

   c. $r = \sin\theta$
      Replace r with $-r$.
      $-r = \sin\theta$

      $r = -\sin\theta$
      The graph may or may not have symmetry about the pole.

9. **a.** $r = 4 + 3\cos\theta$
      Replace θ with $-\theta$.
      $r = 4 + 3\cos(-\theta)$

      $r = 4 + 3\cos\theta$
      The graph has symmetry with respect to the polar axis.

   b. $r = 4 + 3\cos\theta$
      Replace (r, θ) with $(-r, -\theta)$.
      $-r = 4 + 3\cos(-\theta)$

      $-r = 4 + 3\cos\theta$

      $r = -4 - 3\cos\theta$
      The graph may or may not have symmetry with respect to the line $\theta = \dfrac{\pi}{2}$.

   c. $r = 4 + 3\cos\theta$
      Replace r with $-r$.
      $-r = 4 + 3\cos\theta$

      $r = -4 - 3\cos\theta$
      The graph may or may not have symmetry about the pole.

11. a. $r^2 = 16\cos 2\theta$

Replace θ with $-\theta$.

$r^2 = 16\cos 2(-\theta)$

$r^2 = 16\cos(-2\theta)$

$r^2 = 16\cos 2\theta$

The graph has symmetry with respect to the polar axis.

b. $r^2 = 16\cos 2\theta$

Replace (r, θ) with $(-r, -\theta)$.

$(-r)^2 = 16\cos 2(-\theta)$

$r^2 = 16\cos 2\theta$

The graph has symmetry with respect to the line $\theta = \dfrac{\pi}{2}$.

c. $r^2 = 16\cos 2\theta$

Replace r with $-r$.

$(-r)^2 = 16\cos 2\theta$

$r^2 = 16\cos 2\theta$

The graph has symmetry about the pole.

13. $r = 2\cos\theta$

Check for symmetry:

Polar Axis	The Line $\theta = \dfrac{\pi}{2}$	The Pole
$r = 2\cos(-\theta)$	$-r = 2\cos(-\theta)$	$-r = 2\cos\theta$
	$-r = 2\cos\theta$	
$r = 2\cos\theta$	$r = -2\cos\theta$	$r = -2\cos\theta$

The graph is symmetric with respect to the polar axis. The graph may or may not be symmetric with respect to the line

$\theta = \dfrac{\pi}{2}$ or the pole.

Calculate values of r for θ from 0 to π and use symmetry to obtain the graph.

θ	0	$\dfrac{\pi}{6}$	$\dfrac{\pi}{3}$	$\dfrac{\pi}{2}$
r	2	1.73	1	0

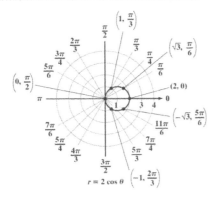

Notice that there are no points in quadrants II or III. Because the cosine is negative in quadrants II and III, r is negative here. This places the points in quadrants IV and I respectively.

15. $r = 1 - \sin\theta$

Check for symmetry:

Polar Axis	The Line $\theta = \dfrac{\pi}{2}$	The Pole
$r = 1 - \sin(-\theta)$	$-r = 1 - \sin(-\theta)$	$-r = 1 - \sin\theta$
	$-r = 1 + \sin\theta$	
$r = 1 + \sin\theta$	$r = -1 - \sin\theta$	$r = -1 + \sin\theta$

There may be no symmetry since each equation is not equivalent to $r = 1 - \sin\theta$. Because the period of the sine function is 2π, we need not consider values of θ beyond 2π.

θ	0	$\dfrac{\pi}{6}$	$\dfrac{\pi}{3}$	$\dfrac{\pi}{2}$	$\dfrac{2\pi}{3}$	$\dfrac{5\pi}{6}$	π	$\dfrac{7\pi}{6}$	$\dfrac{4\pi}{3}$	$\dfrac{3\pi}{2}$	$\dfrac{5\pi}{6}$	$\dfrac{11\pi}{6}$	2π
r	1	0.5	0.13	0	0.13	0.5	1	1.5	1.87	2	1.87	1.5	1

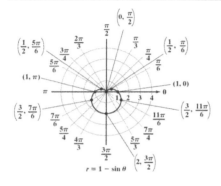

17. $r = 2 + 2\cos\theta$

Check for symmetry:

Polar Axis	The Line $\theta = \dfrac{\pi}{2}$	The Pole
$r = 2 + 2\cos(-\theta)$	$-r = 2 + 2\cos(-\theta)$	$-r = 2 + 2\cos\theta$
	$-r = 2 + 2\cos\theta$	
$r = 2 + 2\cos\theta$	$r = -2 - 2\cos\theta$	$r = -2 - 2\cos\theta$

The graph is symmetric with respect to the polar axis. The graph may or may not be symmetric with respect to the line $\theta = \dfrac{\pi}{2}$ or the pole.

Calculate values of r for θ from 0 to π and use symmetry to obtain the graph.

θ	0	$\dfrac{\pi}{6}$	$\dfrac{\pi}{3}$	$\dfrac{\pi}{2}$	$\dfrac{2\pi}{3}$	$\dfrac{5\pi}{6}$	π
r	4	3.73	3	2	1	0.27	0

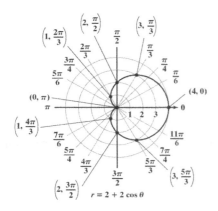

19. $r = 2 + \cos\theta$

Check for symmetry:

Polar Axis	**The Line** $\theta = \dfrac{\pi}{2}$	**The Pole**
$r = 2 + \cos(-\theta)$	$-r = 2 + \cos(-\theta)$	$-r = 2 + \cos\theta$
	$-r = 2 + \cos\theta$	
$r = 2 + \cos\theta$	$r = -2 - \cos\theta$	$r = -2 - \cos\theta$

The graph is symmetric with respect to the polar axis. The graph may or may not be symmetric with respect to the line $\theta = \dfrac{\pi}{2}$ or the pole.

Calculate values of r for θ from 0 to π and use symmetry to obtain the graph.

θ	0	$\dfrac{\pi}{6}$	$\dfrac{\pi}{3}$	$\dfrac{\pi}{2}$	$\dfrac{2\pi}{3}$	$\dfrac{5\pi}{6}$	π
r	3	2.87	2.5	2	1.5	1.13	1

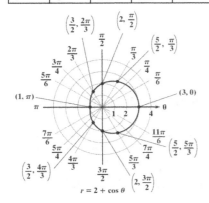

21. $r = 1 + 2\cos\theta$
Check for symmetry:

Polar Axis	The Line $\theta = \dfrac{\pi}{2}$	The Pole
$r = 1 + 2\cos(-\theta)$	$-r = 1 + 2\cos(-\theta)$	$-r = 1 + 2\cos\theta$
	$-r = 1 + 2\cos\theta$	
$r = 1 + 2\cos\theta$	$r = -1 - 2\cos\theta$	$r = -1 - 2\cos\theta$

The graph is symmetric with respect to the polar axis. The graph may or may not be symmetric with respect to the line $\theta = \dfrac{\pi}{2}$ or the pole.

Calculate values of r for θ from 0 to π and use symmetry to obtain the graph.

θ	0	$\dfrac{\pi}{6}$	$\dfrac{\pi}{3}$	$\dfrac{\pi}{2}$	$\dfrac{2\pi}{3}$	$\dfrac{5\pi}{6}$	π
r	3	2.73	2	1	0	-0.73	-1

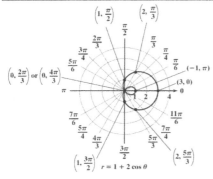

23. $r = 2 - 3\sin\theta$
Check for symmetry:

Polar Axis	The Line $\theta = \dfrac{\pi}{2}$	The Pole
$r = 2 - 3\sin(-\theta)$	$-r = 2 - 3\sin(-\theta)$	$-r = 2 - 3\sin\theta$
	$-r = 2 + 3\sin\theta$	
$r = 2 + 3\sin\theta$	$r = -2 - 3\sin\theta$	$r = -2 + 3\sin\theta$

There may be no symmetry since each equation is not equivalent to $r = 2 - 3\sin\theta$. Because the period of the sine function is 2π, we need not consider values of θ beyond 2π.

θ	0	$\dfrac{\pi}{6}$	$\dfrac{\pi}{3}$	$\dfrac{\pi}{2}$	$\dfrac{2\pi}{3}$	$\dfrac{5\pi}{6}$	π	$\dfrac{7\pi}{6}$	$\dfrac{4\pi}{3}$	$\dfrac{3\pi}{2}$	$\dfrac{5\pi}{6}$	$\dfrac{11\pi}{6}$	2π
r	2	0.5	-0.60	-1	-0.60	0.5	2	3.5	4.6	5	4.6	3.5	2

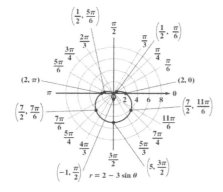

25. $r = 2\cos 2\theta$

Check for symmetry:

Polar Axis	The Line $\theta = \dfrac{\pi}{2}$	The Pole
$r = 2\cos 2(-\theta)$	$-r = 2\cos 2(-\theta)$	$-r = 2\cos 2\theta$
$r = 2\cos(-2\theta)$	$-r = 2\cos(-2\theta)$	
	$-r = 2\cos 2\theta$	
$r = 2\cos 2\theta$	$r = -2\cos 2\theta$	$r = -2\cos 2\theta$

The graph is symmetric with respect to the polar axis. The graph may or may not be symmetric with respect to the line $\theta = \dfrac{\pi}{2}$ or the pole.

Calculate values of r for θ from 0 to π and use symmetry to obtain the graph.

θ	0	$\dfrac{\pi}{6}$	$\dfrac{\pi}{3}$	$\dfrac{\pi}{2}$	$\dfrac{2\pi}{3}$	$\dfrac{5\pi}{6}$	π
r	2	1	-1	-2	-1	1	2

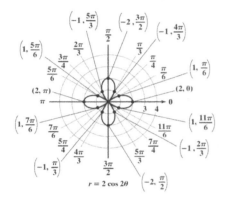

27. $r = 4\sin 3\theta$

Check for symmetry:

Polar Axis	The Line $\theta = \dfrac{\pi}{2}$	The Pole
$r = 4\sin 3(-\theta)$	$-r = 4\sin 3(-\theta)$	$-r = 4\sin 3\theta$
$r = 4\sin(-3\theta)$	$-r = 4\sin(-3\theta)$	
	$-r = -4\sin 3\theta$	
$r = -4\sin 3\theta$	$r = 4\sin 3\theta$	$r = -4\sin 3\theta$

The graph is symmetric with respect to the line $\theta = \dfrac{\pi}{2}$. The graph may or may not be symmetric with respect to the polar axis or the poles.

Calculate values of r for θ from 0 to $\dfrac{\pi}{2}$ and for θ from π to $\dfrac{3\pi}{2}$. Then, use symmetry to obtain the graph.

θ	0	$\dfrac{\pi}{6}$	$\dfrac{\pi}{4}$	$\dfrac{\pi}{3}$	$\dfrac{\pi}{2}$	π	$\dfrac{7\pi}{6}$	$\dfrac{5\pi}{4}$	$\dfrac{4\pi}{3}$	$\dfrac{3\pi}{2}$
r	0	4	2.83	0	-4	0	-4	-2.83	0	4

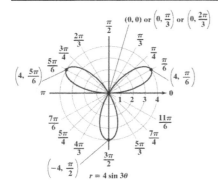

29. $r^2 = 9\cos 2\theta$

Check for symmetry:

Polar Axis	The Line $\theta = \dfrac{\pi}{2}$	The Pole
$r^2 = 9\cos 2(-\theta)$	$(-r)^2 = 9\cos 2(-\theta)$	$(-r)^2 = 9\cos 2\theta$
$r^2 = 9\cos(-2\theta)$	$r^2 = 9\cos(-2\theta)$	
$r^2 = 9\cos 2\theta$	$r^2 = 9\cos 2\theta$	$r^2 = 9\cos 2\theta$

The graph is symmetric with respect to the polar axis, the line $\theta = \dfrac{\pi}{2}$, and the pole.

Note that since $\cos 2\theta$ is negative for $\dfrac{\pi}{4} < \theta < \dfrac{3\pi}{4}$, there is no graph there.

Calculate values of r for θ from 0 to $\dfrac{\pi}{4}$ and use symmetry to obtain the graph.

θ	0	$\dfrac{\pi}{6}$	$\dfrac{\pi}{4}$
r	± 3	± 2.12	0

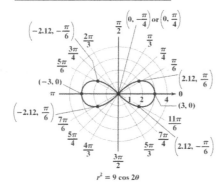

Copyright © 2014 Pearson Education, Inc.

31. $r = 1 - 3\sin\theta$

Check for symmetry:

Polar Axis	The Line $\theta = \dfrac{\pi}{2}$	The Pole
$r = 1 - 3\sin(-\theta)$	$-r = 1 - 3\sin(-\theta)$	$-r = 1 - 3\sin\theta$
	$-r = 1 + 3\sin\theta$	
$r = 1 + 3\sin\theta$	$r = -1 - 3\sin\theta$	$r = -1 + 3\sin\theta$

There may be no symmetry. Since each equation is not equivalent to $r = 1 - 3\sin\theta$. Because the period of the sine function is 2π, we need not consider values of θ beyond 2π.

θ	0	$\dfrac{\pi}{6}$	$\dfrac{\pi}{3}$	$\dfrac{\pi}{2}$	$\dfrac{2\pi}{3}$	$\dfrac{5\pi}{6}$	π	$\dfrac{7\pi}{6}$	$\dfrac{4\pi}{3}$	$\dfrac{3\pi}{2}$	$\dfrac{5\pi}{6}$	$\dfrac{11\pi}{6}$	2π
r	1	-0.5	-1.6	-2	-1.6	-0.5	1	2.5	3.6	4	3.6	2.5	1

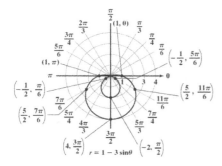

33. $r\cos\theta = -3$

$$r = \frac{-3}{\cos\theta} = -3\sec\theta$$

Check for symmetry:

Polar Axis	The Line $\theta = \dfrac{\pi}{2}$	The Pole
$r = -3\sec(-\theta)$	$-r = -3\sec(-\theta)$	$-r = -3\sec\theta$
	$-r = -3\sec\theta$	
$r = -3\sec\theta$	$r = 3\sec\theta$	$r = 3\sec\theta$

The graph is symmetric with respect to the polar axis. The graph may or may not be symmetric with respect to the line $\theta = \dfrac{\pi}{2}$ or the pole.

Calculate values of r for θ from 0 to π. Then, use symmetry to obtain the graph.

θ	0	$\dfrac{\pi}{6}$	$\dfrac{\pi}{3}$	$\dfrac{\pi}{2}$	$\dfrac{2\pi}{3}$	$\dfrac{5\pi}{6}$	π
r	-3	-3.46	-6	Undef.	6	3.46	3

Note that since $\sec\theta$ is undefined when $\theta=\dfrac{\pi}{2}$ and $\theta=\dfrac{3\pi}{2}$, r increases without bound as θ approaches these angles. $r\cos\theta=-3$ is equivalent to $x=-3$.

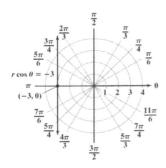

35. $r=\cos\dfrac{\theta}{2}$

Symmetrical about the polar axis: yes

Symmetrical about the line $\theta=\dfrac{\pi}{2}$: yes

Symmetrical about the pole: yes

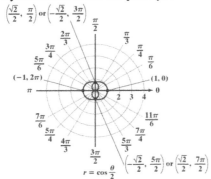

37. $r=\sin\theta+\cos\theta$

Symmetrical about the polar axis: maybe

Symmetrical about the line $\theta=\dfrac{\pi}{2}$: maybe

Symmetrical about the pole: maybe

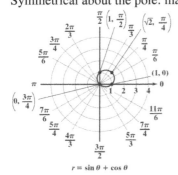

39. $r=\dfrac{1}{1-\cos\theta}$

Symmetrical about the polar axis: yes

Symmetrical about the line $\theta=\dfrac{\pi}{2}$: maybe

Symmetrical about the pole: maybe

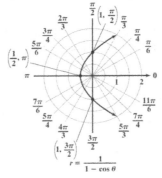

41. $r=\sin\theta\cos^2\theta$

Symmetrical about the polar axis: maybe

Symmetrical about the line $\theta=\dfrac{\pi}{2}$: yes

Symmetrical about the pole: maybe

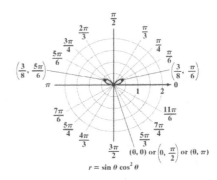

43. $r=2+3\sin 2\theta$

Symmetrical about the polar axis: maybe

Symmetrical about the line $\theta=\dfrac{\pi}{2}$: maybe

Symmetrical about the pole: yes

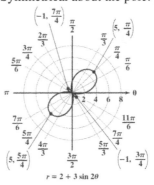

45. Using the graph, sailing at a 60° angle to the wind gives a speed of about 6 knots (to the nearest knot).

47. Using the graph, sailing at a 90° angle to the wind gives a speed of about 8 knots (to the nearest knot).

49. It appears that an angle of 90° gives a maximum speed of about $7\frac{1}{2}$ knots.

51. – 57. Answers may vary.

59.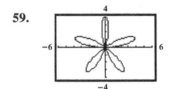

61.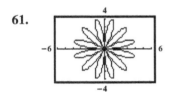

63.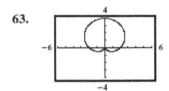

65.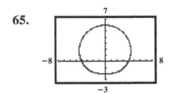

67.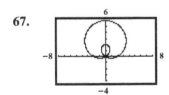

69.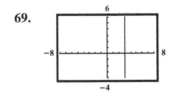

71.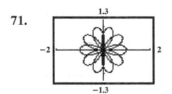

73.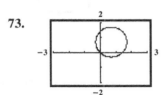

75.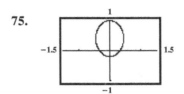

77. If $\theta\max = 2\pi$, the graph is drawn once.

79. θ step $= 0.1$

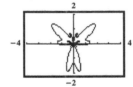

81. θ step $= 0.1$

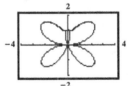

83. As n increases, $\sin n\theta$ increases its number of loops. If n is odd, there are n loops and $\theta\max = \pi$ traces the graph once, while if n is even, there are $2n$ loops and $\theta\max = 2\pi$ traces the graph once.

85. There are n small petals and n large petals for each value of n. For odd values of n, the small petals are inside the large petals. For even n, they are between the large petals.

87. $\theta \min = 0$, $\theta \max = 2\pi$ $\theta \min = 0$, $\theta \max = 4\pi$ $\theta \min = 0$, $\theta \max = 8\pi$

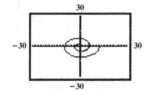 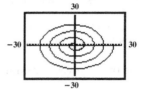

89. does not make sense; Explanations will vary. Sample explanation: If a polar equation fails a symmetry test, its graph may still have that kind of symmetry.

91. makes sense

93.

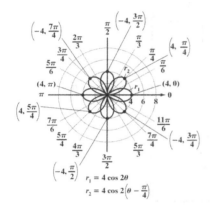

The graph of r_2 is the graph of r_1 rotated $\dfrac{\pi}{4}$ or $45°$.

95. Answers may vary.

96.
$$\begin{aligned}(1+i)(2+2i) &= 2+2i+2i+2i^2 \\ &= 2+4i-2 \\ &= 4i\end{aligned}$$

97.
$$\begin{aligned}(-1+i\sqrt{3})(-1+i\sqrt{3})(-1+i\sqrt{3}) &= (-1+i\sqrt{3})(1-i\sqrt{3}-i\sqrt{3}+3i^2) \\ &= (-1+i\sqrt{3})(1-2i\sqrt{3}-3) \\ &= (-1+i\sqrt{3})(-2-2i\sqrt{3}) \\ &= 2+2i\sqrt{3}-2i\sqrt{3}-6i^2 \\ &= 2+6 \\ &= 8\end{aligned}$$

98.
$$\begin{aligned}\frac{2+2i}{1+i} &= \frac{2+2i}{1+i} \cdot \frac{1-i}{1-i} \\ &= \frac{2-2i+2i-2i^2}{1-i+i-i^2} \\ &= \frac{2+2}{1+1} \\ &= 2\end{aligned}$$

Mid-Chapter 7 Check Point

1. $C = 180° - 32° - 41° = 107°$
Use the Law of Sines to find b.
$$\frac{a}{\sin A} = \frac{b}{\sin B}$$
$$\frac{20}{\sin 32°} = \frac{b}{\sin 41°}$$
$$b = \frac{20\sin 41°}{\sin 32°}$$
$$b \approx 24.8$$
Use the Law of Sines to find c.
$$\frac{a}{\sin A} = \frac{c}{\sin C}$$
$$\frac{20}{\sin 32°} = \frac{c}{\sin 107°}$$
$$c = \frac{20\sin 107°}{\sin 32°}$$
$$c \approx 36.1$$
The solution is $C = 107°$, $b \approx 24.8$, and $c \approx 36.1$.

2. Use the Law of Sines to find B.
$$\frac{a}{\sin A} = \frac{b}{\sin B}$$
$$\frac{63}{\sin 42°} = \frac{57}{\sin B}$$
$$\sin B = \frac{57\sin 42°}{63}$$
$$\sin B \approx 0.6054$$
There are two angles possible:
$B_1 \approx 37°$, $B_2 \approx 180° - 37° = 143°$
B_2 is impossible, since $42° + 143° = 185°$.
$C = 180° - B_1 - A \approx 180° - 37° - 42° = 101°$
Use the Law of Sines to find c.
$$\frac{c}{\sin C} = \frac{a}{\sin A}$$
$$\frac{c}{\sin 101°} = \frac{63}{\sin 42°}$$
$$c = \frac{63\sin 101°}{\sin 42°}$$
$$c \approx 92.4$$
There is one triangle and the solution is
B_1 (or B) $\approx 37°$, $C \approx 101°$, and $c \approx 92.4$.

3. Use the Law of Sines to find angle B.
$$\frac{a}{\sin A} = \frac{b}{\sin B}$$
$$\frac{6}{\sin 65°} = \frac{7}{\sin B}$$
$$\sin B = \frac{7\sin 65°}{6}$$
$$\sin B \approx 1.0574$$
The sine can never exceed 1. There is no triangle with the given measurements.

4. Use the Law of Cosines to find b.
$$b^2 = a^2 + c^2 - 2ac\cos B$$
$$b^2 = 10^2 + 16^2 - 2(10)(16)\cos 110°$$
$$b^2 \approx 465.4464$$
$$b \approx 21.6$$
Use the Law of Sines to find A.
$$\frac{b}{\sin B} = \frac{a}{\sin A}$$
$$\frac{21.6}{\sin 110°} = \frac{10}{\sin A}$$
$$\sin A = \frac{10\sin 110°}{21.6}$$
$$\sin A \approx 0.4350$$
$$A \approx 26°$$
Find the third angle.
$C = 180° - A - B \approx 180° - 26° - 110° = 44°$
The solution is $A \approx 26°$, $C \approx 44°$, and $b \approx 21.6$.

5. Use the Law of Sines to find angle A.
$$\frac{a}{\sin A} = \frac{c}{\sin C}$$
$$\frac{16}{\sin A} = \frac{13}{\sin 42°}$$
$$\sin A = \frac{16\sin 42°}{13}$$
$$\sin A \approx 0.8235$$
There are two angles possible:
$A_1 \approx 55°$, $A_2 \approx 180° - 55° = 125°$
There are two triangles:
$B_1 = 180° - C - A_1 \approx 180° - 42° - 55° = 83°$
$B_2 = 180° - C - A_2 \approx 180° - 42° - 125° = 13°$
Use the Law of Sines to find b_1 and b_2.

$$\frac{b_1}{\sin B_1} = \frac{c}{\sin C}$$

$$\frac{b_1}{\sin 83°} = \frac{13}{\sin 42°}$$

$$b_1 = \frac{13\sin 83°}{\sin 42°} \approx 19.3$$

$$\frac{b_2}{\sin B_2} = \frac{c}{\sin C}$$

$$\frac{b_2}{\sin 13°} = \frac{13}{\sin 42°}$$

$$b_2 = \frac{13\sin 13°}{\sin 42°} \approx 4.4$$

In one triangle, the solution is
$A_1 \approx 55°$, $B_1 \approx 83°$, $b_1 \approx 19.3$.
In the other triangle, $A_2 \approx 125°$, $B_2 \approx 13°$, $b_2 \approx 4.4$.

6. Use the Law of Cosines to find the angle opposite the longest side.
    Thus, find angle C.

    $$c^2 = a^2 + b^2 - 2ab\cos C$$

    $$\cos C = \frac{a^2 + b^2 - c^2}{2ab}$$

    $$\cos C = \frac{5^2 + 7.2^2 - 10.1^2}{2 \cdot 5 \cdot 7.2}$$

    $$\cos C \approx -0.3496$$

    $$C \approx 110°$$

    Use the Law of Sines to find angle A.

    $$\frac{a}{\sin A} = \frac{c}{\sin C}$$

    $$\frac{5}{\sin A} = \frac{10.1}{\sin 110°}$$

    $$\sin A = \frac{5\sin 110°}{10.1}$$

    $$\sin A \approx 0.4652$$

    $$A \approx 28°$$

    Find the third angle.
    $B = 180° - A - C = 180° - 28° - 110° = 42°$
    The solution is $A \approx 28°, B \approx 42°$, and $C \approx 110°$

7. The area of the triangle is half the product of the lengths of the two sides times the sine of the included angle.

    $$\text{Area} = \frac{1}{2}(5)(7)(\sin 36°) \approx 10$$

    The area of the triangle is approximately 10 square feet.

8. Begin by calculating one-half the perimeter:

    $$s = \frac{1}{2}(a + b + c) = \frac{1}{2}(7 + 9 + 12) = 14$$

    Use Heron's formula to find the area.

    $$\begin{aligned} \text{Area} &= \sqrt{s(s-a)(s-b)(s-c)} \\ &= \sqrt{14(14-7)(14-9)(14-12)} \\ &= \sqrt{980} \approx 31 \end{aligned}$$

    The area of the triangle is approximately 31 square meters.

9. The first train traveled 100 miles, the second train traveled 80 miles.
    Use the Law of Cosines to find the distance.

    $$c^2 = a^2 + b^2 - 2ab\cos C$$

    $$c^2 = 100^2 + 80^2 - 2(100)(80)\cos 110°$$

    $$c^2 \approx 21872.32229$$

    $$c \approx 147.9$$

    The two trains are 147.9 miles apart.

10. Let the fire be at point C.
    $A = 90° - 56° = 34°$
    $B = 90° - 23° = 67°$
    $C = 180° - 34° - 67° = 79°$
    Use the law of sines to find b.

    $$\frac{b}{\sin B} = \frac{c}{\sin C}$$

    $$\frac{b}{\sin 67°} = \frac{16}{\sin 79°}$$

    $$b = \frac{16\sin 67°}{\sin 79°}$$

    $$b \approx 15.0$$

    The fire is 15.0 miles from station A

11. Let point A be where the angle of elevation is $66°$.

Let point B be where the angle of elevation is $50°$.

Let point C be at the top of the tree.
$C = 180° - A - B = 180° - 66° - 50° = 64°$
Use the law of sines to find a.

$$\frac{a}{\sin A} = \frac{c}{\sin C}$$

$$\frac{a}{\sin 66°} = \frac{420}{\sin 64°}$$

$$a = \frac{420 \sin 66°}{\sin 64°}$$

$$a \approx 426.9$$

The height of the tree, h, is given by
$h = a \sin B$

$h = 426.9 \sin 50°$

$h \approx 327.0$

The tree is 327.0 feet tall.

12. $x = r \cos\theta = -3\cos\dfrac{5\pi}{4} = \dfrac{3\sqrt{2}}{2}$

$y = r\sin\theta = -3\sin\dfrac{5\pi}{4} = -\dfrac{3\sqrt{2}}{2}$

Ordered pair: $\left(\dfrac{3\sqrt{2}}{2}, -\dfrac{3\sqrt{2}}{2}\right)$

13. $x = r\cos\theta = 6\cos\left(-\dfrac{\pi}{2}\right) = 0$

$y = r\sin\theta = 6\sin\left(-\dfrac{\pi}{2}\right) = -6$

Ordered pair: $(0, -6)$

14. $r = \sqrt{x^2 + y^2} = \sqrt{2^2 + \left(-2\sqrt{3}\right)^2} = 4$

$\tan\theta = \dfrac{y}{x} = \dfrac{-2\sqrt{3}}{2} = -\sqrt{3}$

Because θ lies in quadrant IV, $\theta = 2\pi - \dfrac{\pi}{3} = \dfrac{5\pi}{3}$

Polar coordinates: $\left(4, \dfrac{5\pi}{3}\right)$

15. $r = \sqrt{x^2 + y^2} = \sqrt{(-6)^2 + 0^2} = 6$

$\tan\theta = \dfrac{y}{x} = \dfrac{0}{-6} = 0$

$\theta = \pi$

Polar coordinates: $(6, \pi)$

16.

a. $\left(4, \dfrac{11\pi}{4}\right)$

b. $\left(-4, \dfrac{7\pi}{4}\right)$

c. $\left(4, \dfrac{5\pi}{4}\right)$

17.

a. $\left(\dfrac{5}{2}, \dfrac{5\pi}{2}\right)$

b. $\left(-\dfrac{5}{2}, \dfrac{3\pi}{2}\right)$

c. $\left(\dfrac{5}{2}, -\dfrac{3\pi}{2}\right)$

18. $5x - y = 7$
$5r\cos\theta - r\sin\theta = 7$
$r(5\cos\theta - \sin\theta) = 7$

$$r = \frac{7}{5\cos\theta - \sin\theta}$$

19.
$$y = -7$$
$$r\sin\theta = -7$$
$$r = \frac{-7}{\sin\theta}$$
$$r = -7\csc\theta$$

20.
$$(x+1)^2 + y^2 = 1$$
$$(r\cos\theta + 1)^2 + (r\sin\theta)^2 = 1$$
$$r^2\cos^2\theta + 2r\cos\theta + 1 + r^2\sin^2\theta^2 = 1$$
$$r^2 + 2r\cos\theta = 0$$
$$r^2 = -2r\cos\theta$$
$$r = -2\cos\theta$$

21.
$$r = 6$$
$$r^2 = 36$$
$$x^2 + y^2 = 36$$

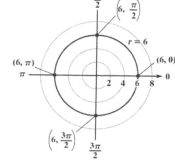

22.
$$\theta = \frac{\pi}{3}$$
$$\tan\theta = \tan\frac{\pi}{3}$$
$$\tan\theta = \sqrt{3}$$
$$\frac{y}{x} = \sqrt{3}$$
$$y = \sqrt{3}x$$

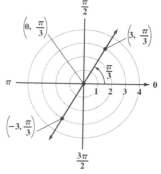

23.
$$r = -3\csc\theta$$
$$r = \frac{-3}{\sin\theta}$$
$$r\sin\theta = -3$$
$$y = -3$$

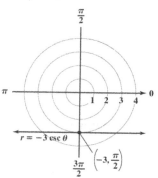

24.
$$r = -10\cos\theta$$
$$r^2 = -10r\cos\theta$$
$$x^2 + y^2 = -10x$$
$$x^2 + 10x + y^2 = 0$$
$$x^2 + 10x + 25 + y^2 = 25$$
$$(x+5)^2 + y^2 = 25$$

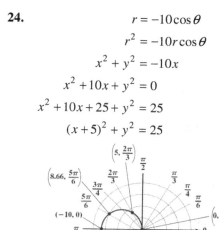
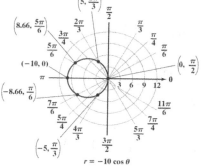

25.
$$r = 4\sin\theta\sec^2\theta$$
$$r = \frac{4\sin\theta}{\cos^2\theta}$$
$$r\cos^2\theta = 4\sin\theta$$
$$r^2\cos^2\theta = 4r\sin\theta$$
$$x^2 = 4y$$
$$y = \frac{1}{4}x^2$$

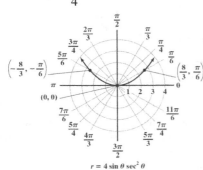

$r = 4\sin\theta\sec^2\theta$

26. $r = 1 - 4\cos\theta$

a. Replace θ with $-\theta$.
$$r = 1 - 4\cos(-\theta)$$
$$r = 1 - 4\cos\theta$$
The graph is symmetric with respect to the polar axis.

b. Replace (r,θ) with $(-r,-\theta)$.
$$-r = 1 - 4\cos(-\theta)$$
$$r = -1 + 4\cos\theta$$
The graph may or may not be symmetric with respect to the line $\theta = \frac{\pi}{2}$.

c. Replace r with $-r$.
$$-r = 1 - 4\cos\theta$$
$$r = -1 + 4\cos\theta$$
The graph may or may not be symmetric with respect to the polar axis.

27. $r^2 = 4\cos 2\theta$

a. Replace θ with $-\theta$.
$$r^2 = 4\cos(-2\theta)$$
$$r^2 = 4\cos 2\theta$$
The graph is symmetric with respect to the polar axis.

b. Replace (r,θ) with $(-r,-\theta)$.
$$(-r)^2 = 4\cos(-2\theta)$$
$$r^2 = 4\cos 2\theta$$
The graph is symmetric with respect to the line $\theta = \frac{\pi}{2}$.

c. Replace r with $-r$.
$$(-r)^2 = 4\cos 2\theta$$
$$r^2 = 4\cos 2\theta$$
The graph is symmetric with respect to the polar axis.

28.

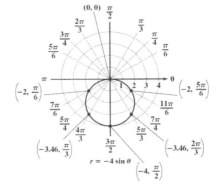

$r = -4\sin\theta$

29.

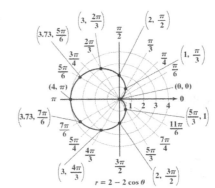

$r = 2 - 2\cos\theta$

30.

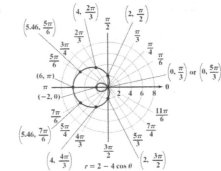

$r = 2 - 4\cos\theta$

31.

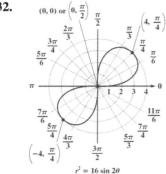

$r = 2 \sin 3\theta$

32.

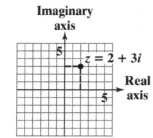

$r^2 = 16 \sin 2\theta$

Section 7.5

Check Point Exercises

1. **a.** $z = 2 + 3i$ corresponds to the point (2, 3). Plot the complex number by moving two units to the right on the real axis and 3 units up parallel to the imaginary axis.

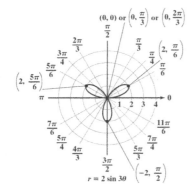

b. $z = -3 - 5i$ corresponds to the point (−3, −5). Plot the complex number by moving three units to the left on the real axis and five units down parallel to the imaginary axis.

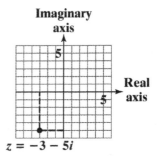

$z = -3 - 5i$

c. Because $z = -4 = -4 + 0i$, this complex number corresponds to the point (−4, 0). Plot the complex number by moving four units to the left on the real axis.

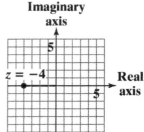

d. Because $z = -i = 0 - i$, this complex number corresponds to the point (0, −1). Plot the complex number by moving one unit down on the imaginary axis.

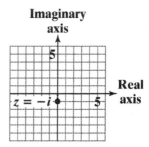

2 **a.** $z = 5 + 12i$

$a = 5, b = 12$

$|z| = \sqrt{5^2 + 12^2} = \sqrt{25 + 144} = \sqrt{169} = 13$

b. $z = 2 - 3i$

$a = 2, b = -3$

$|z| = \sqrt{2^2 + (-3)^2} = \sqrt{4 + 9} = \sqrt{13}$

3. $z = -1 - i\sqrt{3}$ corresponds to the point $\left(-1, -\sqrt{3}\right)$.

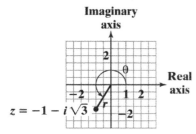

Use $r = \sqrt{a^2 + b^2}$ with $a = -1$ and $b = -\sqrt{3}$ to find r.

$$r = \sqrt{a^2 + b^2} = \sqrt{(-1)^2 + \left(-\sqrt{3}\right)^2}$$
$$= \sqrt{1 + 3} = \sqrt{4} = 2$$

Use $\tan\theta = \dfrac{b}{a}$ with $a = -1$ and $b = -\sqrt{3}$ to find θ.

$$\tan\theta = \frac{b}{a} = \frac{-\sqrt{3}}{-1} = \sqrt{3}$$

Because $\tan\dfrac{\pi}{3} = \sqrt{3}$ and θ lies in

quadrant III, $\theta = \pi + \dfrac{\pi}{3} = \dfrac{3\pi}{3} + \dfrac{\pi}{3} = \dfrac{4\pi}{3}$.

The polar form of $z = -1 - i\sqrt{3}$ is

$$z = r(\cos\theta + i\sin\theta) = 2\left(\cos\frac{4\pi}{3} + i\sin\frac{4\pi}{3}\right).$$

4. The complex number $z = 4(\cos 30° + i\sin 30°)$ is in polar form, with $r = 4$ and $\theta = 30°$. We use exact values for $\cos 30°$ and $\sin 30°$ to write the number in rectangular form.

$$4(\cos 30° + i\sin 30°) = 4\left(\frac{\sqrt{3}}{2} + i\frac{1}{2}\right) = 2\sqrt{3} + 2i$$

The rectangular form of $z = 4(\cos 30° + i\sin 30°)$ is $z = 2\sqrt{3} + 2i$.

5. $z_1 z_2 = [6(\cos 40° + i\sin 40°)][5(\cos 20° + i\sin 20°)] = (6 \cdot 5)[(\cos(40° + 20°) + i\sin(40° + 20°)]$
$ = 30(\cos 60° + i\sin 60°)$

6. $\dfrac{z_1}{z_2} = \dfrac{50\left(\cos\dfrac{4\pi}{3} + i\sin\dfrac{4\pi}{3}\right)}{5\left(\cos\dfrac{\pi}{3} + i\sin\dfrac{\pi}{3}\right)} = \dfrac{50}{5}\left[\cos\left(\dfrac{4\pi}{3} - \dfrac{\pi}{3}\right) + i\sin\left(\dfrac{4\pi}{3} - \dfrac{\pi}{3}\right)\right] = 10(\cos\pi + i\sin\pi)$

7. $[2(\cos 30° + i\sin 30°)]^5 = 2^5\left[\cos(5 \cdot 30°) + i\sin(5 \cdot 30°)\right] = 32(\cos 150° + i\sin 150°) = 32\left(-\dfrac{\sqrt{3}}{2} + i\dfrac{1}{2}\right)$

$$= -16\sqrt{3} + 16i$$

8. Write $1+i$ in $r(\cos\theta + i\sin\theta)$ form.

$$r = \sqrt{a^2 + b^2} = \sqrt{1^2 + 1^2} = \sqrt{2}$$

$$\tan\theta = \frac{b}{a} = \frac{1}{1} = 1 \text{ and } \theta = \frac{\pi}{4}$$

$$1+i = r(\cos\theta + i\sin\theta) = \sqrt{2}\left(\cos\frac{\pi}{4} + i\sin\frac{\pi}{4}\right)$$

Use DeMoivre's Theorem to raise $1 + i$ to the fourth power.

$$(1+i)^4 = \left[\sqrt{2}\left(\cos\frac{\pi}{4} + i\sin\frac{\pi}{4}\right)\right]^4 = \left(\sqrt{2}\right)^4\left[\cos\left(4\cdot\frac{\pi}{4}\right) + i\sin\left(4\cdot\frac{\pi}{4}\right)\right] = 4(\cos\pi + i\sin\pi) = 4(-1+0i) = -4$$

9. From DeMoivre's Theorem for finding complex roots, the fourth roots of $16(\cos 60° + i\sin 60°)$ are

$$z_k = \sqrt[4]{16}\left[\cos\left(\frac{60° + 360°k}{4}\right) + i\sin\left(\frac{60° + 360°k}{4}\right)\right], \; k = 0, 1, 2, 3.$$

Substitute 0, 1, 2, and 3 for k in the above expression for z_k.

$$z_0 = \sqrt[4]{16}\left[\cos\left(\frac{60° + 360°\cdot 0}{4}\right) + i\sin\left(\frac{60° + 360°\cdot 0}{4}\right)\right] = \sqrt[4]{16}\left[\cos\frac{60°}{4} + i\sin\frac{60°}{4}\right] = 2(\cos 15° + i\sin 15°)$$

$$z_1 = \sqrt[4]{16}\left[\cos\left(\frac{60° + 360°\cdot 1}{4}\right) + i\sin\left(\frac{60° + 360°\cdot 1}{4}\right)\right] = \sqrt[4]{16}\left[\cos\frac{420°}{4} + i\sin\frac{420°}{4}\right] = 2(\cos 105° + i\sin 105°)$$

$$z_2 = \sqrt[4]{16}\left[\cos\left(\frac{60° + 360°\cdot 2}{4}\right) + i\sin\left(\frac{60° + 360°\cdot 2}{4}\right)\right] = \sqrt[4]{16}\left[\cos\frac{780°}{4} + i\sin\frac{780°}{4}\right] = 2(\cos 195° + i\sin 195°)$$

$$z_3 = \sqrt[4]{16}\left[\cos\left(\frac{60° + 360°\cdot 3}{4}\right) + i\sin\left(\frac{60° + 360°\cdot 3}{4}\right)\right] = \sqrt[4]{16}\left[\cos\frac{1140°}{4} + i\sin\frac{1140°}{4}\right] = 2(\cos 285° + i\sin 285°)$$

10. First, write 27 in polar form. $27 = r(\cos\theta + i\sin\theta) = 27(\cos 0 + \sin 0)$. From DeMoivre's theorem for finding complex roots, the cube roots of 27 are

$$z_k = \sqrt[3]{27}\left[\cos\left(\frac{0 + 2\pi k}{3}\right) + i\sin\left(\frac{0 + 2\pi k}{3}\right)\right], \; k = 0, 1, 2.$$

$$z_0 = \sqrt[3]{27}\left[\cos\left(\frac{0 + 2\pi\cdot 0}{3}\right) + i\sin\left(\frac{0 + 2\pi\cdot 0}{3}\right)\right] = 3(\cos 0 + i\sin 0) = 3(1 + i\cdot 0) = 3$$

$$z_1 = \sqrt[3]{27}\left[\cos\left(\frac{0 + 2\pi\cdot 1}{3}\right) + i\sin\left(\frac{0 + 2\pi\cdot 1}{3}\right)\right] = 3\left(\cos\frac{2\pi}{3} + i\sin\frac{2\pi}{3}\right) = 3\left(-\frac{1}{2} + i\cdot\frac{\sqrt{3}}{2}\right) = -\frac{3}{2} + i\frac{3\sqrt{3}}{2}$$

$$z_2 = \sqrt[3]{27}\left[\cos\left(\frac{0 + 2\pi\cdot 2}{3}\right) + i\sin\left(\frac{0 + 2\pi\cdot 2}{3}\right)\right] = 3\left(\cos\frac{4\pi}{3} + i\sin\frac{4\pi}{3}\right) = 3\left(-\frac{1}{2} + i\cdot\left(-\frac{\sqrt{3}}{2}\right)\right) = -\frac{3}{2} - i\frac{3\sqrt{3}}{2}$$

Concept and Vocabulary Check 7.5

1. real; imaginary

2. absolute value

3. modulus; argument

4. $\sqrt{a^2 + b^2}$; $\dfrac{b}{a}$

5. $r_1 r_2$; $\theta_1 + \theta_2$; $\theta_1 + \theta_2$; multiplying; adding

6. $\dfrac{r_1}{r_2}$; $\theta_1 - \theta_2$; $\theta_1 - \theta_2$; dividing; subtracting

7. r^n ; $n\theta$; $n\theta$

8. n

Exercise Set 7.5

1. Because $z = 4i = 0 + 4i$, this complex number corresponds to the point (0, 4).

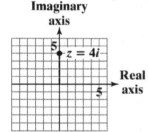

With $a = 0$ and $b = 4$,
$|z| = \sqrt{0^2 + 4^2} = \sqrt{16} = 4.$

3. Because $z = 3 = 3 + 0i$, this complex number corresponds to the point (3, 0).

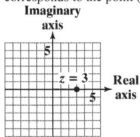

With $a = 3$ and $b = 0$, $|z| = \sqrt{3^2 + 0^2} = \sqrt{9} = 3.$

5. $z = 3 + 2i$ corresponds to the point (3, 2).

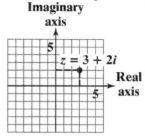

With $a = 3$ and $b = 2$,
$|z| = \sqrt{3^2 + 2^2} = \sqrt{9 + 4} = \sqrt{13}.$

7. $z = 3 - i$ corresponds to the point (3, −1).

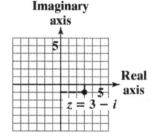

With $a = 3$ and $b = -1$,
$|z| = \sqrt{3^2 + (-1)^2} = \sqrt{9 + 1} = \sqrt{10}.$

9. $z = -3 + 4i$ corresponds to the point (−3, 4).

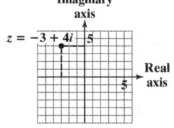

With $a = -3$ and $b = 4$,
$|z| = \sqrt{(-3)^2 + 4^2} = \sqrt{9 + 16} = \sqrt{25} = 5.$

11. $z = 2 + 2i$ corresponds to the point (2, 2).

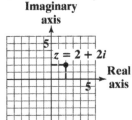

Use $r = \sqrt{a^2 + b^2}$ and $\tan\theta = \dfrac{b}{a}$, with $a = 2$ and

$b = 2$, to find r and θ.

$r = \sqrt{2^2 + 2^2} = \sqrt{4 + 4} = \sqrt{8} = 2\sqrt{2}$

$\tan\theta = \dfrac{2}{2} = 1$

Because $\tan\dfrac{\pi}{4} = 1$ and θ lies in quadrant I, $\theta = \dfrac{\pi}{4}$.

$z = 2 + 2i = r(\cos\theta + i\sin\theta)$

$ = 2\sqrt{2}\left(\cos\dfrac{\pi}{4} + i\sin\dfrac{\pi}{4}\right)$

or $2\sqrt{2}(\cos 45° + i\sin 45°)$

13. $z = -1 - i$ corresponds to the point (–1, –1).

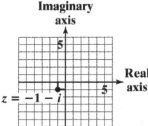

Use $r = \sqrt{a^2 + b^2}$ and $\tan\theta = \dfrac{b}{a}$, with

$a = -1$ and $b = -1$, to find r and θ.

$r = \sqrt{(-1)^2 + (-1)^2} = \sqrt{1 + 1} = \sqrt{2}$

$\tan\theta = \dfrac{-1}{-1} = 1$

Because $\tan\dfrac{\pi}{4} = 1$ and θ lies in

quadrant III, $\theta = \pi + \dfrac{\pi}{4} = \dfrac{5\pi}{4}$.

$z = -1 - i = r(\cos\theta + i\sin\theta)$

$ = \sqrt{2}\left(\cos\dfrac{5\pi}{4} + i\sin\dfrac{5\pi}{4}\right)$

or $\sqrt{2}(\cos 225° + i\sin 225°)$

15. $z = -4i$ corresponds to the point (0, –4).

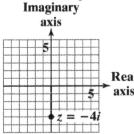

Use $r = \sqrt{a^2 + b^2}$ and $\tan\theta = \dfrac{b}{a}$, with

$a = 0$ and $b = -4$, to find r and θ.

$r = \sqrt{0^2 + (-4)^2} = \sqrt{16} = 4$

$\tan\theta = \dfrac{-4}{0}$ is undefined.

Because $\tan\dfrac{\pi}{2}$ is undefined and θ lies on the

negative *y*-axis, $\theta = \dfrac{\pi}{2} + \pi = \dfrac{3\pi}{2}$.

$z = -4i = r(\cos\theta + i\sin\theta)$

$ = 4\left(\cos\dfrac{3\pi}{2} + i\sin\dfrac{3\pi}{2}\right)$

or $4(\cos 270° + i\sin 270°)$

17. $z = 2\sqrt{3} - 2i$ corresponds to the point $\left(2\sqrt{3}, -2\right)$.

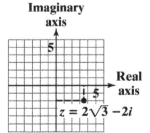

Use $r = \sqrt{a^2 + b^2}$ and $\tan\theta = \dfrac{b}{a}$, with

$a = 2\sqrt{3}$ and $b = -2$, to find r and θ.

$r = \sqrt{\left(2\sqrt{3}\right)^2 + (-2)^2} = \sqrt{12 + 4} = \sqrt{16} = 4$

$\tan\theta = \dfrac{-2}{2\sqrt{3}} = -\dfrac{1}{\sqrt{3}}$

Because $\tan\dfrac{\pi}{6} = \dfrac{1}{\sqrt{3}}$ and θ lies in

quadrant IV, $\theta = 2\pi - \dfrac{\pi}{6} = \dfrac{11\pi}{6}$.

$z = 2\sqrt{3} - 2i = r(\cos\theta + i\sin\theta)$

$\quad = 4\left(\cos\dfrac{11\pi}{6} + i\sin\dfrac{11\pi}{6}\right)$

$\quad$ or $4(\cos 330° + i\sin 330°)$

19. $z = -3$ corresponds to the point $(-3, 0)$.

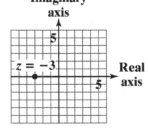

Use $r = \sqrt{a^2 + b^2}$ and $\tan\theta = \dfrac{b}{a}$, with

$a = -3$ and $b = 0$, to find r and θ.

$r = \sqrt{(-3)^2 + 0^2} = \sqrt{9} = 3$

$\tan\theta = \dfrac{0}{-3} = 0$

Because $\tan 0 = 0$ and θ lies on the negative x-axis,

$\theta = 0 + \pi = \pi$.

$z = -3 = r(\cos\theta + i\sin\theta)$

$\quad = 3\left(\cos\pi + i\sin\pi\right)$

$\quad$ or $3(\cos 180° + i\sin 180°)$

21. $z = -3\sqrt{2} - 3i\sqrt{3}$ corresponds to the point $\left(-3\sqrt{2}, -3\sqrt{3}\right)$.

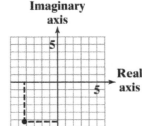

$z = -3\sqrt{2} - 3i\sqrt{3}$

Use $r = \sqrt{a^2 + b^2}$ and $\tan\theta = \dfrac{b}{a}$, with

$a = -3\sqrt{2}$ and $b = -3\sqrt{3}$, to find r and θ.

$r = \sqrt{\left(-3\sqrt{2}\right)^2 + \left(-3\sqrt{3}\right)^2} = \sqrt{18 + 27}$

$\quad = \sqrt{45} = 3\sqrt{5}$

$\tan\theta = \dfrac{-3\sqrt{3}}{-3\sqrt{2}} = \dfrac{\sqrt{3}}{\sqrt{2}} = \dfrac{\sqrt{6}}{2}$

Because θ lies in quadrant III,

$\theta = 180° + \tan^{-1}\left(\dfrac{\sqrt{6}}{2}\right) \approx 180° + 50.8°$

$\quad = 230.8°$

$z = -3\sqrt{2} - 3i\sqrt{3} = r(\cos\theta + i\sin\theta)$

$\quad \approx 3\sqrt{5}(\cos 230.8° + i\sin 230.8°)$

23. $z = -3 + 4i$ corresponds to the point $(-3, 4)$.

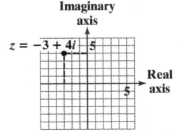

Use $r = \sqrt{a^2 + b^2}$ and $\tan\theta = \dfrac{b}{a}$, with

$a = -3$ and $b = 4$, to find r and θ.

$r = \sqrt{(-3)^2 + (4)^2} = \sqrt{9 + 16} = \sqrt{25} = 5$

$\tan\theta = \dfrac{4}{-3} = -\dfrac{4}{3}$

Because θ lies in quadrant II,

$\theta = 180° - \tan^{-1}\left(\dfrac{4}{3}\right) \approx 180° - 53.1° = 126.9°$.

$z = -3 + 4i = r(\cos\theta + i\sin\theta)$

$\quad \approx 5(\cos 126.9° + i\sin 126.9°)$

25. $z = 2 - i\sqrt{3}$ corresponds to the point $\left(2, -\sqrt{3}\right)$.

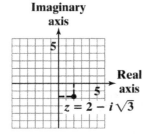

Use $r = \sqrt{a^2 + b^2}$ and $\tan\theta = \dfrac{b}{a}$, with

$a = 2$ and $b = -\sqrt{3}$, to find r and θ.

$r = \sqrt{2^2 + \left(-\sqrt{3}\right)^2} = \sqrt{4+3} = \sqrt{7}$

$\tan\theta = \dfrac{-\sqrt{3}}{2} = -\dfrac{\sqrt{3}}{2}$

Because θ lies in quadrant IV,

$\theta = 360° - \tan^{-1}\left(\dfrac{\sqrt{3}}{2}\right) \approx 360° - 40.9° = 319.1°$

$z = 2 - i\sqrt{3} = r(\cos\theta + i\sin\theta)$

$\approx \sqrt{7}\left(\cos 319.1° + i\sin 319.1°\right)$

27. $6(\cos 30° + i\sin 30°) = 6\left(\dfrac{\sqrt{3}}{2} + i\dfrac{1}{2}\right)$

$= 3\sqrt{3} + 3i$

The rectangular form of

$z = 6(\cos 30° + i\sin 30°)$ is $z = 3\sqrt{3} + 3i$.

29. $4(\cos 240° + i\sin 240°) = 4\left(-\dfrac{1}{2} + i\left(-\dfrac{\sqrt{3}}{2}\right)\right)$

$= -2 - 2i\sqrt{3}$

The rectangular form of

$z = 4(\cos 240° + i\sin 240°)$ is $z = -2 - 2i\sqrt{3}$.

31. $8\left(\cos\dfrac{7\pi}{4} + i\sin\dfrac{7\pi}{4}\right) = 8\left(\dfrac{\sqrt{2}}{2} + i\left(-\dfrac{\sqrt{2}}{2}\right)\right)$

$= 4\sqrt{2} - 4i\sqrt{2}$

The rectangular form of

$8\left(\cos\dfrac{7\pi}{4} + i\sin\dfrac{7\pi}{4}\right)$ is $z = 4\sqrt{2} - 4i\sqrt{2}$.

33. $5\left(\cos\dfrac{\pi}{2} + i\sin\dfrac{\pi}{2}\right) = 5\left(0 + i(1)\right)$

$= 5i$

The rectangular form of

$z = 5\left(\cos\dfrac{\pi}{2} + i\sin\dfrac{\pi}{2}\right)$ is $z = 5i$.

35. $20\left(\cos 205° + i\sin 205°\right)$

$\approx 20\left(-0.91 + i(-0.42)\right) = -18.2 - 8.4i$

The rectangular form of

$z = 20\left(\cos 205° + i\sin 205°\right)$ is $z \approx -18.2 - 8.5i$.

37. $z_1 z_2$

$= \left[6(\cos 20° + i\sin 20°)\right]\left[5(\cos 50° + i\sin 50°)\right]$

$= (6 \cdot 5)\left[\cos(20° + 50°) + i\sin(20° + 50°)\right]$

$= 30(\cos 70° + i\sin 70°)$

39. $z_1 z_2 = \left[3\left(\cos\dfrac{\pi}{5} + i\sin\dfrac{\pi}{5} \right) \right] \left[4\left(\cos\dfrac{\pi}{10} + i\sin\dfrac{\pi}{10} \right) \right] = (3 \cdot 4) \left[\cos\left(\dfrac{\pi}{5} + \dfrac{\pi}{10} \right) + i\sin\left(\dfrac{\pi}{5} + \dfrac{\pi}{10} \right) \right]$

$= 12\left(\cos\dfrac{3\pi}{10} + i\sin\dfrac{3\pi}{10} \right)$

41. $z_1 z_2 = \left[\cos\dfrac{\pi}{4} + i\sin\dfrac{\pi}{4} \right] \left[\cos\dfrac{\pi}{3} + i\sin\dfrac{\pi}{3} \right] = \cos\left(\dfrac{\pi}{4} + \dfrac{\pi}{3} \right) + i\sin\left(\dfrac{\pi}{4} + \dfrac{\pi}{3} \right) = \cos\left(\dfrac{3\pi}{12} + \dfrac{4\pi}{12} \right) + i\sin\left(\dfrac{3\pi}{12} + \dfrac{4\pi}{12} \right)$

$= \cos\dfrac{7\pi}{12} + i\sin\dfrac{7\pi}{12}$

43. Begin by converting $z_1 = 1 + i$ and $z_2 = -1 + i$ to polar form.

For $z_1 : a = 1$ and $b = 1$

$r = \sqrt{a^2 + b^2} = \sqrt{1^2 + 1^2} = \sqrt{2}$

$\tan\theta = \dfrac{b}{a} = \dfrac{1}{1} = 1$ and $\theta = \dfrac{\pi}{4}$.

$z_1 = r(\cos\theta + i\sin\theta) = \sqrt{2}\left(\cos\dfrac{\pi}{4} + i\sin\dfrac{\pi}{4} \right)$

For $z_2 : a = -1$ and $b = 1$

$r = \sqrt{a^2 + b^2} = \sqrt{(-1)^2 + 1^2} = \sqrt{2}$

$\tan\theta = \dfrac{b}{a} = \dfrac{1}{-1} = -1$

Because $\tan\dfrac{\pi}{4} = 1$ and θ lies in quadrant II, $\theta = \pi - \dfrac{\pi}{4} = \dfrac{3\pi}{4}$.

$z_2 = r(\cos\theta + i\sin\theta) = \sqrt{2}\left(\cos\dfrac{3\pi}{4} + i\sin\dfrac{3\pi}{4} \right)$

Now, find the product.

$z_1 z_2 = (1 + i)(-1 + i)$

$= \left[\sqrt{2}\left(\cos\dfrac{\pi}{4} + i\sin\dfrac{\pi}{4} \right) \right] \left[\sqrt{2}\left(\cos\dfrac{3\pi}{4} + i\sin\dfrac{3\pi}{4} \right) \right] = \left(\sqrt{2} \cdot \sqrt{2} \right) \left[\cos\left(\dfrac{\pi}{4} + \dfrac{3\pi}{4} \right) + i\sin\left(\dfrac{\pi}{4} + \dfrac{3\pi}{4} \right) \right]$

$= 2\left(\cos\pi + i\sin\pi \right)$

45. $\dfrac{z_1}{z_2} = \dfrac{20(\cos 75° + i\sin 75°)}{4(\cos 25° + i\sin 25°)} = \dfrac{20}{4}\left[\cos(75° - 25°) + i\sin(75° - 25°) \right]$

$= 5(\cos 50° + i\sin 50°)$

47. $\dfrac{z_1}{z_2} = \dfrac{3\left(\cos\dfrac{\pi}{5} + i\sin\dfrac{\pi}{5} \right)}{4\left(\cos\dfrac{\pi}{10} + i\sin\dfrac{\pi}{10} \right)} = \dfrac{3}{4}\left[\cos\left(\dfrac{\pi}{5} - \dfrac{\pi}{10} \right) + i\sin\left(\dfrac{\pi}{5} - \dfrac{\pi}{10} \right) \right] = \dfrac{3}{4}\left(\cos\dfrac{\pi}{10} + i\sin\dfrac{\pi}{10} \right)$

49. $\dfrac{z_1}{z_2} = \dfrac{\cos 80° + i\sin 80°}{\cos 200° + i\sin 200°} = \cos(80° - 200°) + i\sin(80° - 200°) = \cos(-120°) + i\sin(-120°)$

$= \cos 240° + i\sin 240°$

51. Begin by converting $z_1 = 2 + 2i$ and $z_2 = 1 + i$ to polar form.

For z_1 : $a = 2$ and $b = 2$

$$r = \sqrt{a^2 + b^2} = \sqrt{2^2 + 2^2} = \sqrt{8} = 2\sqrt{2}$$

$$\tan\theta = \frac{b}{a} = \frac{2}{2} = 1 \text{ and } \theta = \frac{\pi}{4}$$

$$z_1 = r(\cos\theta + i\sin\theta) = 2\sqrt{2}\left(\cos\frac{\pi}{4} + i\sin\frac{\pi}{4}\right)$$

For z_2 : $a = 1$ and $b = 1$

$$r = \sqrt{a^2 + b^2} = \sqrt{1^2 + 1^2} = \sqrt{2}$$

$$\tan\theta = \frac{b}{a} = \frac{1}{1} = 1 \text{ and } \theta = \frac{\pi}{4}$$

$$z_2 = r(\cos\theta + i\sin\theta) = \sqrt{2}\left(\cos\frac{\pi}{4} + i\sin\frac{\pi}{4}\right).$$

Now, find the quotient.

$$\frac{z_1}{z_2} = \frac{2+2i}{1+i} = \frac{2\sqrt{2}\left(\cos\dfrac{\pi}{4} + i\sin\dfrac{\pi}{4}\right)}{\sqrt{2}\left(\cos\dfrac{\pi}{4} + i\sin\dfrac{\pi}{4}\right)} = 2\left[\cos\left(\frac{\pi}{4} - \frac{\pi}{4}\right) + i\sin\left(\frac{\pi}{4} - \frac{\pi}{4}\right)\right] = 2(\cos 0 + i\sin 0)$$

53. $\left[4(\cos 15° + i\sin 15°)\right]^3 = (4)^3\left[\cos(3\cdot 15°) + i\sin(3\cdot 15°)\right] = 64(\cos 45° + i\sin 45°) = 64\left(\dfrac{\sqrt{2}}{2} + i\dfrac{\sqrt{2}}{2}\right)$

$$= 32\sqrt{2} + 32i\sqrt{2}$$

55. $\left[2(\cos 80° + i\sin 80°)\right]^3 = (2)^3\left[\cos(3\cdot 80°) + i\sin(3\cdot 80°)\right] = 8(\cos 240° + i\sin 240°)$

$$= 8\left(-\frac{1}{2} + i\left(-\frac{\sqrt{3}}{2}\right)\right) = -4 - 4i\sqrt{3}$$

57. $\left[\dfrac{1}{2}\left(\cos\dfrac{\pi}{12} + i\sin\dfrac{\pi}{12}\right)\right]^6 = \left(\dfrac{1}{2}\right)^6\left[\cos\left(6\cdot\dfrac{\pi}{12}\right) + i\sin\left(6\cdot\dfrac{\pi}{12}\right)\right] = \dfrac{1}{64}\left(\cos\dfrac{\pi}{2} + i\sin\dfrac{\pi}{2}\right) = \dfrac{1}{64}(0 + i) = \dfrac{1}{64}i$

59. $\left[\sqrt{2}\left(\cos\dfrac{5\pi}{6} + i\sin\dfrac{5\pi}{6}\right)\right]^4 = \left(\sqrt{2}\right)^4\left[\cos\left(4\cdot\dfrac{5\pi}{6}\right) + i\sin\left(4\cdot\dfrac{5\pi}{6}\right)\right]$

$$= 4\left(\cos\frac{20\pi}{6} + i\sin\frac{20\pi}{6}\right) = 4\left(\cos\frac{4\pi}{3} + i\sin\frac{4\pi}{3}\right)$$

$$= 4\left(-\frac{1}{2} + i\left(-\frac{\sqrt{3}}{2}\right)\right) = -2 - 2i\sqrt{3}$$

61. Write $1+i$ in $r(\cos\theta + i\sin\theta)$ form.

$$r = \sqrt{a^2 + b^2} = \sqrt{1^2 + 1^2} = \sqrt{2}$$

$$\tan\theta = \frac{b}{a} = \frac{1}{1} = 1 \text{ and } \theta = \frac{\pi}{4}$$

$$1+i = r(\cos\theta + i\sin\theta) = \sqrt{2}\left(\cos\frac{\pi}{4} + i\sin\frac{\pi}{4}\right)$$

Use DeMoivre's Theorem to raise $1+i$ to the fifth power.

$$(1+i)^5 = \left[\sqrt{2}\left(\cos\frac{\pi}{4} + i\sin\frac{\pi}{4}\right)\right]^5 = \left(\sqrt{2}\right)^5\left[\cos\left(5\cdot\frac{\pi}{4}\right) + i\sin\left(5\cdot\frac{\pi}{4}\right)\right]$$

$$= 4\sqrt{2}\left(\cos\frac{5\pi}{4} + i\sin\frac{5\pi}{4}\right) = 4\sqrt{2}\left(-\frac{\sqrt{2}}{2} + i\left(-\frac{\sqrt{2}}{2}\right)\right) = -4 - 4i$$

63. Write $\sqrt{3} - i$ in $r(\cos\theta + i\sin\theta)$ form.

$$r = \sqrt{a^2 + b^2} = \sqrt{\left(\sqrt{3}\right)^2 + (-1)^2} = \sqrt{4} = 2$$

$$\tan\theta = \frac{b}{a} = \frac{-1}{\sqrt{3}} = -\frac{1}{\sqrt{3}}$$

Because $\tan 30° = \frac{1}{\sqrt{3}}$ and θ lies in quadrant IV, $\theta = 360° - 30° = 330°$.

$$\sqrt{3} - i = r(\cos\theta + i\sin\theta) = 2(\cos 330° + i\sin 330°)$$

Use DeMoivre's Theorem to raise $\sqrt{3} - i$ to the sixth power.

$$(\sqrt{3} - i)^6 = \left[2(\cos 330° + i\sin 330°)\right]^6 = (2)^6\left[\cos(6\cdot 330°) + i\sin(6\cdot 330°)\right]$$

$$= 64(\cos 1980° + i\sin 1980°) = 64(\cos 180° + i\sin 180°)$$

$$= 64(-1 + 0i) = -64$$

65. $9(\cos 30° + i\sin 30°)$

$$z_k = \sqrt[2]{9}\left[\cos\left(\frac{30° + 360°k}{2}\right) + i\sin\left(\frac{30° + 360°k}{2}\right)\right], \ k = 0, 1$$

$$z_0 = \sqrt{9}\left[\cos\left(\frac{30° + 360°\cdot 0}{2}\right) + i\sin\left(\frac{30° + 360°\cdot 0}{2}\right)\right] = \sqrt{9}\left[\cos\left(\frac{30°}{2}\right) + i\sin\left(\frac{30°}{2}\right)\right] = 3(\cos 15° + i\sin 15°)$$

$$z_1 = \sqrt{9}\left[\cos\left(\frac{30° + 360°\cdot 1}{2}\right) + i\sin\left(\frac{30° + 360°\cdot 1}{2}\right)\right] = \sqrt{9}\left[\cos\left(\frac{390°}{2}\right) + i\sin\left(\frac{390°}{2}\right)\right] = 3(\cos 195° + i\sin 195°)$$

67. $8(\cos 210° + i \sin 210°)$

$$z_k = \sqrt[3]{8}\left[\cos\left(\frac{210° + 360°k}{3}\right) + i \sin\left(\frac{210° + 360°k}{3}\right)\right], \; k = 0, 1, 2$$

$$z_0 = \sqrt[3]{8}\left[\cos\left(\frac{210° + 360° \cdot 0}{3}\right) + i \sin\left(\frac{210° + 360° \cdot 0}{3}\right)\right] = \sqrt[3]{8}\left[\cos\left(\frac{210°}{3}\right) + i \sin\left(\frac{210°}{3}\right)\right]$$

$$= 2(\cos 70° + i \sin 70°)$$

$$z_1 = \sqrt[3]{8}\left[\cos\left(\frac{210° + 360° \cdot 1}{3}\right) + i \sin\left(\frac{210° + 360° \cdot 1}{3}\right)\right] = \sqrt[3]{8}\left[\cos\left(\frac{570°}{3}\right) + i \sin\left(\frac{570°}{3}\right)\right]$$

$$= 2(\cos 190° + i \sin 190°)$$

$$z_2 = \sqrt[3]{8}\left[\cos\left(\frac{210° + 360° \cdot 2}{3}\right) + i \sin\left(\frac{210° + 360° \cdot 2}{3}\right)\right] = \sqrt[3]{8}\left[\cos\left(\frac{930°}{3}\right) + i \sin\left(\frac{930°}{3}\right)\right]$$

$$= 2(\cos 310° + i \sin 310°)$$

69. $81\left(\cos\frac{4\pi}{3} + i \sin\frac{4\pi}{3}\right)$

$$z_k = \sqrt[4]{81}\left[\cos\left(\frac{\frac{4\pi}{3} + 2\pi k}{4}\right) + i \sin\left(\frac{\frac{4\pi}{3} + 2\pi k}{4}\right)\right], \; k = 0, 1, 2, 3$$

$$z_0 = \sqrt[4]{81}\left[\cos\left(\frac{\frac{4\pi}{3} + 2\pi \cdot 0}{4}\right) + i \sin\left(\frac{\frac{4\pi}{3} + 2\pi \cdot 0}{4}\right)\right] = \sqrt[4]{81}\left(\cos\frac{\pi}{3} + i \sin\frac{\pi}{3}\right) = 3\left(\frac{1}{2} + i\frac{\sqrt{3}}{2}\right) = \frac{3}{2} + \frac{3\sqrt{3}}{2}i$$

$$z_1 = \sqrt[4]{81}\left[\cos\left(\frac{\frac{4\pi}{3} + 2\pi \cdot 1}{4}\right) + i \sin\left(\frac{\frac{4\pi}{3} + 2\pi \cdot 1}{4}\right)\right] = \sqrt[4]{81}\left(\cos\frac{5\pi}{6} + i \sin\frac{5\pi}{6}\right) = 3\left(-\frac{\sqrt{3}}{2} + i\frac{1}{2}\right) = -\frac{3\sqrt{3}}{2} + \frac{3}{2}i$$

$$z_2 = \sqrt[4]{81}\left[\cos\left(\frac{\frac{4\pi}{3} + 2\pi \cdot 2}{4}\right) + i \sin\left(\frac{\frac{4\pi}{3} + 2\pi \cdot 2}{4}\right)\right] = \sqrt[4]{81}\left(\cos\frac{4\pi}{3} + i \sin\frac{4\pi}{3}\right)$$

$$= 3\left(-\frac{1}{2} + i\left(-\frac{\sqrt{3}}{2}\right)\right) = -\frac{3}{2} - \frac{3\sqrt{3}}{2}i$$

$$z_3 = \sqrt[4]{81}\left[\cos\left(\frac{\frac{4\pi}{3} + 2\pi \cdot 3}{4}\right) + i \sin\left(\frac{\frac{4\pi}{3} + 2\pi \cdot 3}{4}\right)\right] = \sqrt[4]{81}\left(\cos\frac{11\pi}{6} + i \sin\frac{11\pi}{6}\right)$$

$$= 3\left(\frac{\sqrt{3}}{2} + i\left(-\frac{1}{2}\right)\right) = \frac{3\sqrt{3}}{2} - \frac{3}{2}i$$

71. $32 = 32(\cos 0° + i \sin 0°)$

$$z_k = \sqrt[5]{32}\left[\cos\left(\frac{0° + 360°k}{5}\right) + i \sin\left(\frac{0° + 360°k}{5}\right)\right], \; k = 0, 1, 2, 3, 4$$

$$z_0 = \sqrt[5]{32}\left[\cos\left(\frac{0° + 360° \cdot 0}{5}\right) + i \sin\left(\frac{0° + 360° \cdot 0}{5}\right)\right] = \sqrt[5]{32}(\cos 0° + i \sin 0°) = 2(1 + 0i) = 2$$

$$z_1 = \sqrt[5]{32}\left[\cos\left(\frac{0° + 360° \cdot 1}{5}\right) + i \sin\left(\frac{0° + 360° \cdot 1}{5}\right)\right] = \sqrt[5]{32}(\cos 72° + i \sin 72°) \approx 2(0.31 + i(0.95))$$

$$\approx 0.6 + 1.9i$$

$$z_2 = \sqrt[5]{32}\left[\cos\left(\frac{0° + 360° \cdot 2}{5}\right) + i\sin\left(\frac{0° + 360° \cdot 2}{5}\right)\right] = \sqrt[5]{32}\left(\cos 144° + i\sin 144°\right) \approx 2(-0.81 + i(0.59))$$

$$\approx -1.6 + 1.2i$$

$$z_3 = \sqrt[5]{32}\left[\cos\left(\frac{0° + 360° \cdot 3}{5}\right) + i\sin\left(\frac{0° + 360° \cdot 3}{5}\right)\right] = \sqrt[5]{32}\left(\cos 216° + i\sin 216°\right) \approx 2(-0.81 + i(-0.59))$$

$$\approx -1.6 - 1.2i$$

$$z_4 = \sqrt[5]{32}\left[\cos\left(\frac{0° + 360° \cdot 4}{5}\right) + i\sin\left(\frac{0° + 360° \cdot 4}{5}\right)\right] = \sqrt[5]{32}\left(\cos 288° + i\sin 288°\right) \approx 2(0.31 + i(-0.95))$$

$$\approx 0.6 - 1.9i$$

73. $1 = 1(\cos 0° + i\sin 0°)$

$$z_k = \sqrt[3]{1}\left[\cos\left(\frac{0° + 360°k}{3}\right) + i\sin\left(\frac{0° + 360°k}{3}\right)\right], \; k = 0, 1, 2$$

$$z_0 = \sqrt[3]{1}\left[\cos\left(\frac{0° + 360° \cdot 0}{3}\right) + i\sin\left(\frac{0° + 360° \cdot 0}{3}\right)\right] = \sqrt[3]{1}\left(\cos 0° + i\sin 0°\right) = 1(1 + 0i) = 1$$

$$z_1 = \sqrt[3]{1}\left[\cos\left(\frac{0° + 360° \cdot 1}{3}\right) + i\sin\left(\frac{0° + 360° \cdot 1}{3}\right)\right] = \sqrt[3]{1}\left(\cos 120° + i\sin 120°\right) = 1\left(-\frac{1}{2} + i\frac{\sqrt{3}}{2}\right) = -\frac{1}{2} + \frac{\sqrt{3}}{2}i$$

$$z_2 = \sqrt[3]{1}\left[\cos\left(\frac{0° + 360° \cdot 2}{3}\right) + i\sin\left(\frac{0° + 360° \cdot 2}{3}\right)\right] = \sqrt[3]{1}\left(\cos 240° + i\sin 240°\right) = 1\left(-\frac{1}{2} + i\left(-\frac{\sqrt{3}}{2}\right)\right)$$

$$= -\frac{1}{2} - \frac{\sqrt{3}}{2}i$$

75. $1 + i = \sqrt{2}\left(\cos 45° + i\sin 45°\right)$

$$z_k = \sqrt[4]{\sqrt{2}}\left[\cos\left(\frac{45° + 360°k}{4}\right) + i\sin\left(\frac{45° + 360°k}{4}\right)\right], \; k = 0, 1, 2, 3$$

$$z_0 = \sqrt[4]{\sqrt{2}}\left[\cos\left(\frac{45° + 360° \cdot 0}{4}\right) + i\sin\left(\frac{45° + 360° \cdot 0}{4}\right)\right] = \sqrt[4]{\sqrt{2}}\left(\cos 11.25° + i\sin 11.25°\right) \approx 1.1 + 0.2i$$

$$z_1 = \sqrt[4]{\sqrt{2}}\left[\cos\left(\frac{45° + 360° \cdot 1}{4}\right) + i\sin\left(\frac{45° + 360° \cdot 1}{4}\right)\right] = \sqrt[4]{\sqrt{2}}\left(\cos 101.25° + i\sin 101.25°\right) \approx -0.2 + 1.1i$$

$$z_2 = \sqrt[4]{\sqrt{2}}\left[\cos\left(\frac{45° + 360° \cdot 2}{4}\right) + i\sin\left(\frac{45° + 360° \cdot 2}{4}\right)\right] = \sqrt[4]{\sqrt{2}}\left(\cos 191.25° + i\sin 191.25°\right) \approx -1.1 - 0.2i$$

$$z_3 = \sqrt[4]{\sqrt{2}}\left[\cos\left(\frac{45° + 360° \cdot 3}{4}\right) + i\sin\left(\frac{45° + 360° \cdot 3}{4}\right)\right] = \sqrt[4]{\sqrt{2}}\left(\cos 281.25° + i\sin 281.25°\right) \approx 0.2 - 1.1i$$

77. $i(2 + 2i)\left(-\sqrt{3} + i\right)$

$$= \left[1(\cos 90° + i\sin 90°)\right]\left[2\sqrt{2}\left(\cos 45° + i\sin 45°\right)\right]\left[2(\cos 150° + i\sin 150°)\right]$$

$$= 4\sqrt{2}\left(\cos 285° + i\sin 285°\right)$$

$$\approx 1.4641 - 5.4641i$$

79. $\dfrac{\left(1+i\sqrt{3}\right)\left(1-i\right)}{2\sqrt{3}-2i}$

$= \dfrac{\left[2\left(\cos 60° + i\sin 60°\right)\right]\left[\sqrt{2}\left(\cos 315° + i\sin 315°\right)\right]}{\left[4\left(\cos 330° + i\sin 330°\right)\right]}$

$= \dfrac{\sqrt{2}}{2}\left(\cos 45° + i\sin 45°\right)$

$= \dfrac{1}{2} + \dfrac{1}{2}i$

81. $x^6 - 1 = 0$

$\qquad x^6 = 1$

$\qquad x = \sqrt[6]{1}$

$\qquad x = \sqrt[6]{1 + 0i}$

$\qquad x = \sqrt[6]{\cos 0° + i\sin 0°}$

$z_k = \sqrt[6]{1}\left[\cos\left(\dfrac{0° + 360° \cdot k}{6}\right) + i\sin\left(\dfrac{0° + 360° \cdot k}{6}\right)\right]$, $k = 0, 1, 2, 3, 4, 5$

$z_0 = \sqrt[6]{1}\left[\cos\left(\dfrac{0° + 360° \cdot 0}{6}\right) + i\sin\left(\dfrac{0° + 360° \cdot 0}{6}\right)\right] = \cos 0° + i\sin 0° = 1 + 0i = 1$

$z_1 = \sqrt[6]{1}\left[\cos\left(\dfrac{0° + 360° \cdot 1}{6}\right) + i\sin\left(\dfrac{0° + 360° \cdot 1}{6}\right)\right] = \cos 60° + i\sin 60° = \dfrac{1}{2} + \dfrac{\sqrt{3}}{2}i$

$z_2 = \sqrt[6]{1}\left[\cos\left(\dfrac{0° + 360° \cdot 2}{6}\right) + i\sin\left(\dfrac{0° + 360° \cdot 2}{6}\right)\right] = \cos 120° + i\sin 120° = -\dfrac{1}{2} + \dfrac{\sqrt{3}}{2}i$

$z_3 = \sqrt[6]{1}\left[\cos\left(\dfrac{0° + 360° \cdot 3}{6}\right) + i\sin\left(\dfrac{0° + 360° \cdot 3}{6}\right)\right] = \cos 180° + i\sin 180° = -1 + 0i = -1$

$z_4 = \sqrt[6]{1}\left[\cos\left(\dfrac{0° + 360° \cdot 4}{6}\right) + i\sin\left(\dfrac{0° + 360° \cdot 4}{6}\right)\right] = \cos 240° + i\sin 240° = -\dfrac{1}{2} + i\left(-\dfrac{\sqrt{3}}{2}\right) = -\dfrac{1}{2} - \dfrac{\sqrt{3}}{2}i$

$z_5 = \sqrt[6]{1}\left[\cos\left(\dfrac{0° + 360° \cdot 5}{6}\right) + i\sin\left(\dfrac{0° + 360° \cdot 5}{6}\right)\right] = \cos 300° + i\sin 300° = \dfrac{1}{2} + i\left(-\dfrac{\sqrt{3}}{2}\right) = \dfrac{1}{2} - \dfrac{\sqrt{3}}{2}i$

83. $x^4 + 16i = 0$

$\qquad x^4 = -16i$

$\qquad x = \sqrt[4]{-16i}$

$\qquad x = \sqrt[4]{0 - 16i}$

$\qquad x = \sqrt[4]{16\left(\cos 270° + i\sin 270°\right)}$

$z_k = \sqrt[4]{16}\left[\cos\left(\dfrac{270° + 360°k}{4}\right) + i\sin\left(\dfrac{270° + 360°k}{4}\right)\right]$, $k = 0, 1, 2, 3$

$z_0 = \sqrt[4]{16}\left[\cos\left(\dfrac{270° + 360° \cdot 0}{4}\right) + i\sin\left(\dfrac{270° + 360° \cdot 0}{4}\right)\right] = 2\left(\cos 67.5° + i\sin 67.5°\right) \approx 0.7654 + 1.8478i$

$$z_1 = \sqrt[4]{16}\left[\cos\left(\frac{270° + 360° \cdot 1}{4}\right) + i\sin\left(\frac{270° + 360° \cdot 1}{4}\right)\right] = 2\left(\cos 157.5° + i\sin 157.5°\right) \approx -1.8478 + 0.7654i$$

$$z_2 = \sqrt[4]{16}\left[\cos\left(\frac{270° + 360° \cdot 2}{4}\right) + i\sin\left(\frac{270° + 360° \cdot 2}{4}\right)\right] = 2\left(\cos 247.5° + i\sin 247.5°\right) \approx -0.7654 - 1.8478i$$

$$z_3 = \sqrt[4]{16}\left[\cos\left(\frac{270° + 360° \cdot 3}{4}\right) + i\sin\left(\frac{270° + 360° \cdot 3}{4}\right)\right] = 2\left(\cos 337.5° + i\sin 337.5°\right) \approx 1.8478 - 0.7654i$$

85. $x^3 - \left(1 + i\sqrt{3}\right) = 0$

$$x^3 = 1 + i\sqrt{3}$$
$$x = \sqrt[3]{1 + i\sqrt{3}}$$
$$x = \sqrt[3]{2(\cos 60° + i\sin 60°)}$$

$$z_k = \sqrt[3]{2}\left[\cos\left(\frac{60° + 360°k}{3}\right) + i\sin\left(\frac{60° + 360°k}{3}\right)\right], \ k = 0, 1, 2$$

$$z_0 = \sqrt[3]{2}\left[\cos\left(\frac{60° + 360° \cdot 0}{3}\right) + i\sin\left(\frac{60° + 360° \cdot 0}{3}\right)\right] = \sqrt[3]{2}\left(\cos 20° + i\sin 20°\right) \approx 1.1839 + 0.4309i$$

$$z_1 = \sqrt[3]{2}\left[\cos\left(\frac{60° + 360° \cdot 1}{3}\right) + i\sin\left(\frac{60° + 360° \cdot 1}{3}\right)\right] = \sqrt[3]{2}\left(\cos 140° + i\sin 140°\right) \approx -0.9652 + 0.8099i$$

$$z_2 = \sqrt[3]{2}\left[\cos\left(\frac{60° + 360° \cdot 2}{3}\right) + i\sin\left(\frac{60° + 360° \cdot 2}{3}\right)\right] = \sqrt[3]{2}\left(\cos 260° + i\sin 260°\right) \approx -0.2188 - 1.2408i$$

87. $e^{\frac{\pi i}{4}} = \cos\frac{\pi}{4} + i\sin\frac{\pi}{4}$

$$= \frac{\sqrt{2}}{2} + \frac{\sqrt{2}}{2}i$$

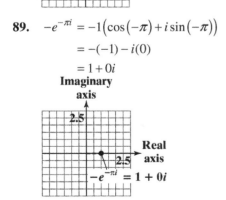

89. $-e^{-\pi i} = -1\left(\cos(-\pi) + i\sin(-\pi)\right)$

$$= -(-1) - i(0)$$
$$= 1 + 0i$$

91. $z = i$

    a. $z_1 = z = i$

$$z_2 = z^2 + z = (i)^2 + i = -1 + i$$

$$z_3 = \left(z^2 + z\right)^2 + z = z_2^2 + z = (-1 + i)^2 + i = -i$$

$$z_4 = \left[\left(z^2 + z\right)^2 + z\right]^2 + z = z_3^2 + z = (-i)^2 + i = -1 + i$$

$$z_5 = z_4^2 + z = (-1 + i)^2 + i = -i$$

$$z_6 = z_5^2 + z = (-i)^2 + i = -1 + i$$

    b. $|-1 + i| = \sqrt{(-1)^2 + 1^2} = \sqrt{2}$

$$|i| = \sqrt{0^2 + 1^2} = 1$$

The absolute values of the terms in the sequence are 1 and $\sqrt{2}$.

Choose a complex number with absolute value less than 1, and another with absolute value greater than $\sqrt{2}$.
Complex numbers may vary.

93. – 105. Answers may vary.

107. does not make sense; Explanations will vary. Sample explanation: This process involves four multiplications.

109. does not make sense; Explanations will vary. Sample explanation: $-1 - i\sqrt{3}$ and $-1 + i\sqrt{3}$ are the other 2 cube roots of 8.

111. $1 = 1\left(\cos 0° + i \sin 0°\right)$

$$z_k = \sqrt[4]{1}\left[\cos\left(\frac{0° + 360°k}{4}\right) + i\sin\left(\frac{0° + 360°k}{4}\right)\right], \; k = 0, 1, 2, 3$$

$$z_0 = \sqrt[4]{1}\left[\cos\left(\frac{0° + 360° \cdot 0}{4}\right) + i\sin\left(\frac{0° + 360° \cdot 0}{4}\right)\right] = \sqrt[4]{1}\left(\cos 0° + i \sin 0\right) = 1(1 + 0i) = 1$$

$$z_1 = \sqrt[4]{1}\left[\cos\left(\frac{0° + 360° \cdot 1}{4}\right) + i\sin\left(\frac{0° + 360° \cdot 1}{4}\right)\right] = \sqrt[4]{1}\left(\cos 90° + i \sin 90°\right) = 1(0 + i(1)) = i$$

$$z_2 = \sqrt[4]{1}\left[\cos\left(\frac{0° + 360° \cdot 2}{4}\right) + i\sin\left(\frac{0° + 360° \cdot 2}{4}\right)\right] = \sqrt[4]{1}\left(\cos 180° + i \sin 180°\right) = 1(-1 + 0i) = -1$$

$$z_3 = \sqrt[4]{1}\left[\cos\left(\frac{0° + 360° \cdot 3}{4}\right) + i\sin\left(\frac{0° + 360° \cdot 3}{4}\right)\right] = \sqrt[4]{1}\left(\cos 270° + i \sin 270°\right) = 1(0 + i(-1)) = -i$$

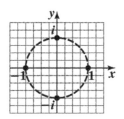

113. Find the distance from $(-3, -3)$ and $(0, 3)$.

$$d = \sqrt{(x_1 - x_2)^2 + (y_1 - y_2)^2}$$
$$= \sqrt{(3+3)^2 + (0+3)^2}$$
$$= \sqrt{45}$$
$$= 3\sqrt{5}$$

Find the distance from $(0, 0)$ and $(3, 6)$.

$$d = \sqrt{(x_1 - x_2)^2 + (y_1 - y_2)^2}$$
$$= \sqrt{(6-0)^2 + (3-0)^2}$$
$$= \sqrt{45}$$
$$= 3\sqrt{5}$$

The line segments have the same length.

114. $(-3, -3)$ and $(0, 3)$

$$m = \frac{y_2 - y_1}{x_2 - x_1}$$
$$= \frac{3 - (-3)}{0 - (-3)}$$
$$= 2$$

$(0, 3)$ and $(3, 6)$

$$m = \frac{y_2 - y_1}{x_2 - x_1}$$
$$= \frac{6 - 0}{3 - 0}$$
$$= 2$$

The lines have the same slope.
Thus, the lines are parallel.

115. $4(5x + 4y) - 2(6x - 9y)$
$$= 20x + 16y - 12x + 18y$$
$$= 8x + 34y$$

Section 7.6

Check Point Exercises

1. First, we show that **u** and **v** have the same magnitude.

$$\| \mathbf{u} \| = \sqrt{(x_2 - x_1)^2 + (y_2 - y_1)^2}$$
$$= \sqrt{(-2 - (-5))^2 + (6 - 2)^2}$$
$$= \sqrt{3^2 + 4^2}$$
$$= \sqrt{9 + 16}$$
$$= \sqrt{25}$$
$$= 5$$

$$\| \mathbf{v} \| = \sqrt{(x_2 - x_1)^2 + (y_2 - y_1)^2}$$
$$= \sqrt{(5 - 2)^2 + (6 - 2)^2}$$
$$= \sqrt{3^2 + 4^2}$$
$$= \sqrt{9 + 16}$$
$$= \sqrt{25}$$
$$= 5$$

Thus, **u** and **v** have the same magnitude: $\| \mathbf{u} \| = \| \mathbf{v} \|$.

Next, we show that **u** and **v** have the same direction. the line on which **u** lies has slope

$$m = \frac{y_2 - y_1}{x_2 - x_1} = \frac{6 - 2}{-2 - (-5)} = \frac{4}{3}.$$

The line on which **v** lies has slope

$$m = \frac{y_2 - y_1}{x_2 - x_1} = \frac{6 - 2}{5 - 2} = \frac{4}{3}.$$

Because **u** and **v** are both directed toward

the upper right on lines having the same slope, $\frac{4}{3}$,

they have the same direction. Thus, **u** and **v** have the same magnitude and direction, and **u** = **v**.

2. For the given vector **v** = 3**i** − 3**j**, $a = 3$ and $b = -3$. The vector's initial point is the origin, (0, 0). The vector's terminal point is $(a, b) = (3, -3)$. We sketch the vector by drawing an arrow from (0, 0) to (3, −3).

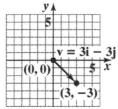

We determine the magnitude of the vector by using the distance formula. Thus, the magnitude is

$$\| \mathbf{v} \| = \sqrt{a^2 + b^2}$$
$$= \sqrt{3^2 + (-3)^2}$$
$$= \sqrt{9 + 9}$$
$$= \sqrt{18}$$
$$= 3\sqrt{2}.$$

3. We identify the values for the variables in the formula.

$$P_1 = (-1, 3) \quad P_2 = (2, 7)$$
$$\uparrow \uparrow \qquad \uparrow \uparrow$$
$$x_1 \; y_1 \qquad x_2 \; y_2$$

Using these values, we write **v** in terms of **i** and **j** as follows:

$$\mathbf{v} = (x_2 - x_1)\mathbf{i} + (y_2 - y_1)\mathbf{j}$$
$$= (2 - (-1))\mathbf{i} + (7 - 3)\mathbf{j}$$
$$= 3\mathbf{i} + 4\mathbf{j}$$

4. a. $\mathbf{v} + \mathbf{w} = (7\mathbf{i} + 3\mathbf{j}) + (4\mathbf{i} - 5\mathbf{j})$
$$= (7 + 4)\mathbf{i} + (3 - 5)\mathbf{j}$$
$$= 11\mathbf{i} - 2\mathbf{j}$$

   b. $\mathbf{v} - \mathbf{w} = (7\mathbf{i} + 3\mathbf{j}) - (4\mathbf{i} - 5\mathbf{j})$
$$= (7 - 4)\mathbf{i} + (3 - (-5))\mathbf{j}$$
$$= 3\mathbf{i} + 8\mathbf{j}$$

5. a. $8\mathbf{v} = 8(7\mathbf{i} + 10\mathbf{j})$
$$= 56\mathbf{i} + 80\mathbf{j}$$

   b. $-5\mathbf{v} = -5(7\mathbf{i} + 10\mathbf{j})$
$$= -35\mathbf{i} - 50\mathbf{j}$$

6. $6\mathbf{v} - 3\mathbf{w} = 6(7\mathbf{i} + 3\mathbf{j}) - 3(4\mathbf{i} - 5\mathbf{j})$
$$= 42\mathbf{i} + 18\mathbf{j} - 12\mathbf{i} + 15\mathbf{j}$$
$$= (42 - 12)\mathbf{i} + (18 + 15)\mathbf{j}$$
$$= 30\mathbf{i} + 33\mathbf{j}$$

7. First, find the magnitude of **v**.

$$\| \mathbf{v} \| = \sqrt{a^2 + b^2}$$
$$= \sqrt{4^2 + (-3)^2}$$
$$= \sqrt{16 + 9}$$
$$= \sqrt{25}$$
$$= 5$$

A unit vector in the same direction as **v** is

$$\frac{\mathbf{v}}{\|\mathbf{v}\|} = \frac{4\mathbf{i} - 3\mathbf{j}}{5} = \frac{4}{5}\mathbf{i} - \frac{3}{5}\mathbf{j}$$

Now, we must verify that the magnitude of the vector

is 1. The magnitude of $\frac{4}{5}\mathbf{i} - \frac{3}{5}\mathbf{j}$ is

$$\sqrt{\left(\frac{4}{5}\right)^2 + \left(-\frac{3}{5}\right)^2} = \sqrt{\frac{16}{25} + \frac{9}{25}} = \sqrt{\frac{25}{25}} = 1.$$

8. $60 \cos 45° \, \mathbf{i} + 60 \sin 45° \, \mathbf{j}$

$$= 60 \cdot \frac{\sqrt{2}}{2}\mathbf{i} + 60 \cdot \frac{\sqrt{2}}{2}\mathbf{j}$$

$$= 30\sqrt{2}\mathbf{i} + 30\sqrt{2}\mathbf{j}$$

9. We need to find $\|\mathbf{F}\|$ and θ.

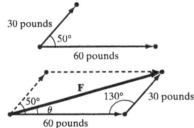

Use the Law of Cosines to find the magnitude of **F**.

$\|\mathbf{F}\|^2 = 60^2 + 30^2 - 2(60)(30)\cos 130° \approx 6814$ The

$\|\mathbf{F}\| \approx \sqrt{6814} \approx 82.5$

magnitude of the resultant force is about 82.5 pounds.

To find θ, the direction of the resultant force, we use the Law of Sines.

$$\frac{82.5}{\sin 130°} = \frac{30}{\sin \theta}$$

$$82.5 \sin \theta = 30 \sin 130°$$

$$\sin \theta = \frac{30 \sin 130°}{82.5}$$

$$\theta = \sin^{-1}\left(\frac{30 \sin 130°}{82.5}\right) \approx 16.2°$$

The direction of the resultant force is approximately 16.2° relative to the 60-pound force, which gives a direction angle of 46.2°.

The two given forces are equivalent to a single force of approximately 82.5 pounds with a direction angle of approximately 46.2°.

Concept and Vocabulary Check 7.6

1. vector

2. scalar

3. **v**; **b**

4. **a**

5. **w**

6. unit; x; y

7. a; b; $\sqrt{a^2 + b^2}$

8. position

9. $x_2 - x_1$; $y_2 - y_1$

10. $a_1 + a_2$; $b_1 + b_2$; $a_1 - a_2$; $b_1 - b_2$; ka_1; kb_1

11. $\frac{\mathbf{v}}{\|\mathbf{v}\|}$; magnitude

12. $\cos \theta$; $\sin \theta$

13. resultant

Exercise Set 7.6

1. **a.** $\|\mathbf{u}\| = \sqrt{(x_2 - x_1)^2 + (y_2 - y_1)^2}$

$$= \sqrt{(4 - (-1))^2 + (6 - 2)^2}$$

$$= \sqrt{5^2 + 4^2}$$

$$= \sqrt{25 + 16}$$

$$= \sqrt{41}$$

   b. $\|\mathbf{v}\| = \sqrt{(x_2 - x_1)^2 + (y_2 - y_1)^2}$

$$= \sqrt{(5 - 0)^2 + (4 - 0)^2}$$

$$= \sqrt{5^2 + 4^2}$$

$$= \sqrt{25 + 16}$$

$$= \sqrt{41}$$

   c. Since $\|\mathbf{u}\| = \|\mathbf{v}\|$, and **u** and **v** have the same direction, we can conclude that **u** = **v**.

3. a. $\|\mathbf{u}\| = \sqrt{(x_2 - x_1)^2 + (y_2 - y_1)^2}$

$= \sqrt{(5 - (-1))^2 + (1 - 1)^2}$

$= \sqrt{6^2 + 0^2}$

$= \sqrt{36 + 0}$

$= \sqrt{36}$

$= 6$

b. $\|\mathbf{v}\| = \sqrt{(x_2 - x_1)^2 + (y_2 - y_1)^2}$

$= \sqrt{(4 - (-2))^2 + (-1 - (-1))^2}$

$= \sqrt{6^2 + 0^2}$

$= \sqrt{36 + 0}$

$= \sqrt{36}$

$= 6$

c. Since $\|\mathbf{u}\| = \|\mathbf{v}\|$, and **u** and **v** have the same direction, we can conclude that
u = v.

5.

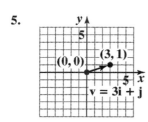

$\|\mathbf{v}\| = \sqrt{3^2 + 1^2} = \sqrt{9 + 1} = \sqrt{10}$

7.

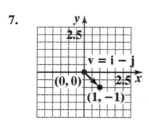

$\|\mathbf{v}\| = \sqrt{1^2 + (-1)^2} = \sqrt{1 + 1} = \sqrt{2}$

9.

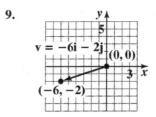

$\|\mathbf{v}\| = \sqrt{(-6)^2 + (-2)^2}$

$= \sqrt{36 + 4}$

$= \sqrt{40}$

$= 2\sqrt{10}$

11.

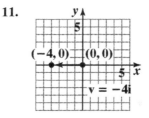

$\|\mathbf{v}\| = \sqrt{(-4)^2 + 0^2} = \sqrt{16 + 0} = \sqrt{16} = 4$

13. $\mathbf{v} = (x_2 - x_1)\mathbf{i} + (y_2 - y_1)\mathbf{j}$
$\mathbf{v} = (6 - (-4))\mathbf{i} + (2 - (-4))\mathbf{j} = 10\mathbf{i} + 6\mathbf{j}$

15. $\mathbf{v} = (x_2 - x_1)\mathbf{i} + (y_2 - y_1)\mathbf{j}$
$\mathbf{v} = (-2 - (-8))\mathbf{i} + (3 - 6)\mathbf{j} = 6\mathbf{i} - 3\mathbf{j}$

17. $\mathbf{v} = (x_2 - x_1)\mathbf{i} + (y_2 - y_1)\mathbf{j}$
$\mathbf{v} = (-7 - (-1))\mathbf{i} + (-7 - 7)\mathbf{j} = -6\mathbf{i} - 14\mathbf{j}$

19. $\mathbf{v} = (x_2 - x_1)\mathbf{i} + (y_2 - y_1)\mathbf{j}$
$\mathbf{v} = (6 - (-3))\mathbf{i} + (4 - 4)\mathbf{j} = 9\mathbf{i} + 0\mathbf{j} = 9\mathbf{i}$

21. $\mathbf{u} + \mathbf{v} = (2\mathbf{i} - 5\mathbf{j}) + (-3\mathbf{i} + 7\mathbf{j})$
$= (2 - 3)\mathbf{i} + (-5 + 7)\mathbf{j}$
$= -\mathbf{i} + 2\mathbf{j}$

23. $\mathbf{u} - \mathbf{v} = (2\mathbf{i} - 5\mathbf{j}) - (-3\mathbf{i} + 7\mathbf{j})$
$= 2\mathbf{i} - 5\mathbf{j} + 3\mathbf{i} - 7\mathbf{j}$
$= (2 + 3)\mathbf{i} + (-5 - 7)\mathbf{j}$
$= 5\mathbf{i} - 12\mathbf{j}$

25. $\mathbf{v} - \mathbf{u} = (-3\mathbf{i} + 7\mathbf{j}) - (2\mathbf{i} - 5\mathbf{j})$
$= -3\mathbf{i} + 7\mathbf{j} - 2\mathbf{i} + 5\mathbf{j}$
$= (-3 - 2)\mathbf{i} + (7 + 5)\mathbf{j}$
$= -5\mathbf{i} + 12\mathbf{j}$

27. $5\mathbf{v} = 5(-3\mathbf{i} + 7\mathbf{j}) = -15\mathbf{i} + 35\mathbf{j}$

29. $-4\mathbf{w} = -4(-\mathbf{i} - 6\mathbf{j}) = 4\mathbf{i} + 24\mathbf{j}$

31. $3\mathbf{w} + 2\mathbf{v} = 3(-\mathbf{i} - 6\mathbf{j}) + 2(-3\mathbf{i} + 7\mathbf{j})$
$= -3\mathbf{i} - 18\mathbf{j} - 6\mathbf{i} + 14\mathbf{j}$
$= (-3 - 6)\mathbf{i} + (-18 + 14)\mathbf{j}$
$= -9\mathbf{i} - 4\mathbf{j}$

33. $3\mathbf{v} - 4\mathbf{w} = 3(-3\mathbf{i} + 7\mathbf{j}) - 4(-\mathbf{i} - 6\mathbf{j})$
$= -9\mathbf{i} + 21\mathbf{j} + 4\mathbf{i} + 24\mathbf{j}$
$= (-9 + 4)\mathbf{i} + (21 + 24)\mathbf{j}$
$= -5\mathbf{i} + 45\mathbf{j}$

35. $\| 2\mathbf{u} \| = \| 2(2\mathbf{i} - 5\mathbf{j}) \|$
$= \| 4\mathbf{i} - 10\mathbf{j} \|$
$= \sqrt{4^2 + (-10)^2}$
$= \sqrt{16 + 100}$
$= \sqrt{116}$
$= 2\sqrt{29}$

37. $\| \mathbf{w} - \mathbf{u} \| = \| (-\mathbf{i} - 6\mathbf{j}) - (2\mathbf{i} - 5\mathbf{j}) \|$
$= \| -\mathbf{i} - 6\mathbf{j} - 2\mathbf{i} + 5\mathbf{j} \|$
$= \| (-1 - 2)\mathbf{i} + (-6 + 5)\mathbf{j} \|$
$= \| -3\mathbf{i} - \mathbf{j} \|$
$= \sqrt{(-3)^2 + (-1)^2}$
$= \sqrt{9 + 1}$
$= \sqrt{10}$

39. $\dfrac{\mathbf{v}}{\| \mathbf{v} \|} = \dfrac{6\mathbf{i}}{\sqrt{6^2 + 0^2}} = \dfrac{6\mathbf{i}}{\sqrt{36}} = \dfrac{6\mathbf{i}}{6} = \mathbf{i}$

41. $\dfrac{\mathbf{v}}{\| \mathbf{v} \|} = \dfrac{3\mathbf{i} - 4\mathbf{j}}{\sqrt{3^2 + (-4)^2}}$
$= \dfrac{3\mathbf{i} - 4\mathbf{j}}{\sqrt{9 + 16}}$
$= \dfrac{3\mathbf{i} - 4\mathbf{j}}{\sqrt{25}}$
$= \dfrac{3\mathbf{i} - 4\mathbf{j}}{5}$
$= \dfrac{3}{5}\mathbf{i} - \dfrac{4}{5}\mathbf{j}$

43. $\dfrac{\mathbf{v}}{\| \mathbf{v} \|} = \dfrac{3\mathbf{i} - 2\mathbf{j}}{\sqrt{3^2 + (-2)^2}}$
$= \dfrac{3\mathbf{i} - 2\mathbf{j}}{\sqrt{9 + 4}}$
$= \dfrac{3\mathbf{i} - 2\mathbf{j}}{\sqrt{13}}$
$= \dfrac{3}{\sqrt{13}}\mathbf{i} - \dfrac{2}{\sqrt{13}}\mathbf{j}$

45. $\dfrac{\mathbf{v}}{\| \mathbf{v} \|} = \dfrac{\mathbf{i} + \mathbf{j}}{\sqrt{1^2 + 1^2}}$
$= \dfrac{\mathbf{i} + \mathbf{j}}{\sqrt{2}}$
$= \dfrac{\mathbf{i}}{\sqrt{2}} + \dfrac{\mathbf{j}}{\sqrt{2}}$
$= \dfrac{\sqrt{2}}{2}\mathbf{i} + \dfrac{\sqrt{2}}{2}\mathbf{j}$

47. $\mathbf{v} = \| \mathbf{v} \| \cos\theta \mathbf{i} + \| \mathbf{v} \| \sin\theta \mathbf{j}$
$= 6\cos 30° \mathbf{i} + 6\sin 30° \mathbf{j}$
$= 6\left(\dfrac{\sqrt{3}}{2} \right)\mathbf{i} + 6\left(\dfrac{1}{2} \right)\mathbf{j}$
$= 3\sqrt{3}\mathbf{i} + 3\mathbf{j}$

49. $\mathbf{v} = \| \mathbf{v} \| \cos\theta \mathbf{i} + \| \mathbf{v} \| \sin\theta \mathbf{j}$
$= 12\cos 225° \mathbf{i} + 12\sin 225° \mathbf{j}$
$= 12\left(-\dfrac{\sqrt{2}}{2} \right)\mathbf{i} + 12\left(-\dfrac{\sqrt{2}}{2} \right)\mathbf{j}$
$= -6\sqrt{2}\mathbf{i} - 6\sqrt{2}\mathbf{j}$

51. $\mathbf{v} = \|\mathbf{v}\|\cos\theta\mathbf{i} + \|\mathbf{v}\|\sin\theta\mathbf{j}$

$\quad = \dfrac{1}{2}\cos 113°\mathbf{i} + \dfrac{1}{2}\sin 113°\mathbf{j}$

$\quad \approx \dfrac{1}{2}(-0.39)\mathbf{i} + \dfrac{1}{2}(0.92)\mathbf{j}$

$\quad \approx -0.20\mathbf{i} + 0.46\mathbf{j}$

53. $4\mathbf{u} - (2\mathbf{v} - \mathbf{w}) = 4(-2\mathbf{i} + 3\mathbf{j}) - \left[2(6\mathbf{i} - \mathbf{j}) - (-3\mathbf{i})\right]$

$\quad = -8\mathbf{i} + 12\mathbf{j} - \left[12\mathbf{i} - 2\mathbf{j} + 3\mathbf{i}\right]$

$\quad = -8\mathbf{i} + 12\mathbf{j} - 12\mathbf{i} + 2\mathbf{j} - 3\mathbf{i})$

$\quad = -23\mathbf{i} + 14\mathbf{j}$

55. $\|\mathbf{u} + \mathbf{v}\|^2 - \|\mathbf{u} - \mathbf{v}\|^2$

$\quad = \|-2\mathbf{i} + 3\mathbf{j} + 6\mathbf{i} - \mathbf{j}\|^2 - \|-2\mathbf{i} + 3\mathbf{j} - (6\mathbf{i} - \mathbf{j})\|^2$

$\quad = \|4\mathbf{i} + 2\mathbf{j}\|^2 - \|-8\mathbf{i} + 4\mathbf{j}\|^2$

$\quad = \left(\sqrt{4^2 + 2^2}\right)^2 - \left(\sqrt{(-8)^2 + 4^2}\right)^2$

$\quad = 16 + 4 - (64 + 16)$

$\quad = 20 - 80$

$\quad = -60$

57. $\mathbf{u} + \mathbf{v} = \mathbf{v} + \mathbf{u}$

$\left(a_1\mathbf{i} + b_1\mathbf{j}\right) + \left(a_2\mathbf{i} + b_2\mathbf{j}\right) = \left(a_2\mathbf{i} + b_2\mathbf{j}\right) + \left(a_1\mathbf{i} + b_1\mathbf{j}\right)$

$\left(a_1 + a_2\right)\mathbf{i} + \left(b_1 + b_2\right)\mathbf{j} = \left(a_1 + a_2\right)\mathbf{i} + \left(b_1 + b_2\right)\mathbf{j}$

This demonstrates the commutative property of vectors.

59. $c(\mathbf{u} + \mathbf{v}) = c\mathbf{u} + c\mathbf{v}$

$c\left(\left(a_1\mathbf{i} + b_1\mathbf{j}\right) + \left(a_2\mathbf{i} + b_2\mathbf{j}\right)\right) = c\left(a_1\mathbf{i} + b_1\mathbf{j}\right) + c\left(a_2\mathbf{i} + b_2\mathbf{j}\right)$

$\left(ca_1 + ca_2\right)\mathbf{i} + \left(cb_1 + cb_2\right)\mathbf{j} = \left(ca_1 + ca_2\right)\mathbf{i} + \left(cb_1 + cb_2\right)\mathbf{j}$

This demonstrates a distributive property of vectors.

61. $\|\mathbf{v}\| = \sqrt{(-10)^2 + 15^2} = \sqrt{325} \approx 18.03$

$\theta = \tan^{-1}\left(\dfrac{15}{-10}\right) \approx 123.7°$

63. $\mathbf{v} = (4\mathbf{i} - 2\mathbf{j}) - (4\mathbf{i} - 8\mathbf{j}) = 6\mathbf{j}$

$\|\mathbf{v}\| = 6$

$\theta = 90°$

65. $\mathbf{v} = \|\mathbf{v}\|\cos\theta\mathbf{i} + \|\mathbf{v}\|\sin\theta\mathbf{j}$

$\quad = 44\cos 30°\mathbf{i} + 44\sin 30°\mathbf{j}$

$\quad = 44\left(\dfrac{\sqrt{3}}{2}\right)\mathbf{i} + 44\left(\dfrac{1}{2}\right)\mathbf{j}$

$\quad = 22\sqrt{3}\mathbf{i} + 22\mathbf{j}$

67. $\mathbf{v} = \|\mathbf{v}\|\cos\theta\mathbf{i} + \|\mathbf{v}\|\sin\theta\mathbf{j}$

$\quad = 150\cos 8°\mathbf{i} + 150\sin 8°\mathbf{j}$

$\quad \approx 148.5\mathbf{i} + 20.9\mathbf{j}$

69. $\mathbf{v} = \|\mathbf{v}\|\cos\theta\mathbf{i} + \|\mathbf{v}\|\sin\theta\mathbf{j}$

$\quad = 1.5\cos 25°\mathbf{i} + 1.5\sin 25°\mathbf{j}$

$\quad \approx 1.4\mathbf{i} + 0.6\mathbf{j}$

The length of the shadow is $|1.4| = 1.4$ inches.

71. $\mathbf{F}_1 = \|\mathbf{F}_1\|\cos\theta\mathbf{i} + \|\mathbf{F}_1\|\sin\theta\,\mathbf{j}$

$\quad = 70\cos 326°\mathbf{i} + 70\sin 326°\,\mathbf{j}$

$\quad = 58\mathbf{i} - 39.1\mathbf{j}$

$\mathbf{F}_2 = \|\mathbf{F}_2\|\cos\theta\mathbf{i} + \|\mathbf{F}_2\|\sin\theta\,\mathbf{j}$

$\quad = 50\cos 18°\mathbf{i} + 50\sin 18°\,\mathbf{j}$

$\quad = 47.6\mathbf{i} + 15.5\mathbf{j}$

$\mathbf{F} = \mathbf{F}_1 + \mathbf{F}_2 = (58\mathbf{i} - 39.1\mathbf{j}) + (47.6\mathbf{i} + 15.5\mathbf{j})$

$\quad\quad = 105.6\mathbf{i} - 23.6\,\mathbf{j}$

$\|F\| = \sqrt{105.6^2 + (-23.6)^2} = 108.2$ pounds

$\cos\theta = \dfrac{a}{\|F\|}$

$\theta = \cos^{-1}\dfrac{105.6}{108.2} = 12.6°$

$360° - 12.6° = 347.4°$

73. $\mathbf{F}_1 = 1610\cos 125°\,\mathbf{i} + 1610\sin 125°\,\mathbf{j}$

$\quad = -923.46\mathbf{i} + 1318.83\mathbf{j}$

$\mathbf{F}_2 = 1250\cos 215°\,\mathbf{i} + 1250\sin 215°\,\mathbf{j}$

$\quad = -1023.94\mathbf{i} - 716.97\mathbf{j}$

$\mathbf{F} = (-923.46 - 1023.94)\,\mathbf{i} + (1318.83 - 716.97)\mathbf{j}$

$\quad = -1947.40\mathbf{i} + 601.86\mathbf{j}$

$\|\mathbf{F}\| = \sqrt{(-1947.40)^2 + 601.86^2} \approx 2038.28$

2038 kilograms

75. $\mathbf{F}_1 = 70\cos 326^\circ \mathbf{i} + 70\sin 326^0 \mathbf{j}$

$\quad = -100\mathbf{j}$

To find the length of the BC: $\quad \cos 18^\circ = \dfrac{a}{100}$

$\qquad\qquad\qquad\qquad\qquad\qquad a \approx 95$

$\mathbf{F}_2 = 95\cos 2888^0 + 95\sin 288^0$

$\quad = 29.4\mathbf{i} - 90.4\mathbf{j}$

$\mathbf{F} = \mathbf{F}_1 - \mathbf{F}_2 = (-100\mathbf{j}) - (29.4\mathbf{i} - 90.4\mathbf{j})$

$\qquad = -29.4\,\mathbf{i} - 9.6\,\mathbf{j}$

$\sqrt{(-29.4)^2 + (-9.6)^2} \approx 30.9$

The force required to pull the weight is 30.9 pounds.

77. a. 335 lb

b. 3484 lb

79. a. $\mathbf{F}_1 + \mathbf{F}_2 = (3 + 6)\mathbf{i} + (-5 + 2)\mathbf{j} = 9\mathbf{i} - 3\mathbf{j}$

b. $-9\mathbf{i} + 3\mathbf{j}$

81. a. $\mathbf{F}_1 = -3\mathbf{i} \quad (-3, 0)$

$\mathbf{F}_2 = -\mathbf{i} + 4\mathbf{j} \quad (-1, 4)$

$\mathbf{F}_3 = 4\mathbf{i} - 2\mathbf{i} \quad (4, -2)$

$\mathbf{F}_4 = -4\mathbf{j} \quad (0, -4)$

$\mathbf{F}_1 + \mathbf{F}_2 + \mathbf{F}_2 + \mathbf{F}_2 = (-3 - 1 + 4)\mathbf{i}$

$+ (4 - 2 - 4)\mathbf{j} = -2\mathbf{j}$

b. $2\mathbf{j}$

83. a. $\mathbf{v} = 180\cos 40^\circ \mathbf{i} + 180\sin 40^\circ \mathbf{j}$

$\quad = 137.88\mathbf{i} + 115.7\mathbf{j}$

$\mathbf{w} = 40\cos 0^\circ \mathbf{i} + 40\sin 0^0 \mathbf{j}$

$\quad = 40\mathbf{i}$

b. $\mathbf{v} + \mathbf{w} = (137.88 + 40)\mathbf{i} + 115.7\mathbf{j}$

$\quad = 177.88\mathbf{i} + 115.7\mathbf{j}$

c. $\sqrt{177.88^2 + 115.7^2} \approx 212$ mph

d. $\cos\theta = \dfrac{177.88}{212}$

$\theta = 33^\circ$

$90^\circ - 33^\circ = \text{N}57^\circ\text{E}$

85. $\mathbf{v} = 320\cos 20^\circ \mathbf{i} + 320\sin 20^\circ \mathbf{j}$

$\quad = 300.7\mathbf{i} + 109.5\mathbf{j}$

$\mathbf{w} = 370\cos 30^\circ \mathbf{i} + 370\sin 30^0 \mathbf{j}$

$\quad = 320.4\mathbf{i} + 185\mathbf{j}$

$\mathbf{w} - \mathbf{v} = (320.4 - 300.7)\mathbf{i}$

$\qquad + (115.7 - 109.5)\mathbf{j}$

$\quad = 19.7\mathbf{i} + 75.6\mathbf{j}$

$\sqrt{19.7^2 + 75.6^2} \approx 78$ mph

$\cos\theta = \dfrac{19.7}{78}$

$\theta = 75.4^\circ$

87. – 103. Answers may vary.

105. does not make sense; Explanations will vary. Sample explanation: A vector represents a distance and a direction. A rate of change does not represent a distance and a direction.

107. does not make sense; Explanations will vary. Sample explanation: The resultant force will have a magnitude less than two pounds unless both forces are in the same direction.

109. true

111. true

113.

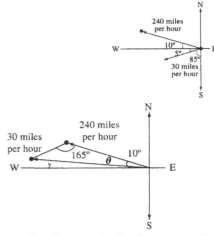

To find the magnitude of $\mathbf{v}$, we use the Law of Cosines.

$\|\mathbf{v}\|^2 = 30^2 + 240^2 - 2(30)(240)\cos 165^\circ$

$\qquad \approx 72,409.3$

$\|\mathbf{v}\| \approx \sqrt{72,409.3} \approx 269$

The plane's true speed relative to the ground is about 269.1 miles per hour. To find the compass heading, relative to the ground, use the Law of Sines.

$$\frac{269}{\sin 165°} = \frac{30}{\sin \theta}$$

$$269 \sin \theta = 30 \sin 165°$$

$$\sin \theta = \frac{30 \sin 165°}{269}$$

$$\theta = \sin^{-1}\left(\frac{30 \sin 165°}{269}\right)$$

$$\theta \approx 1.7$$

The compass heading relative to the ground, is approximately $270° + (10° - 1.7°) = 278.3°$.

115. a. $\mathbf{a} = 310 \cos \theta° \, \mathbf{i} + 310 \sin \theta° \, \mathbf{j}$
$\mathbf{w} = 75 \cos 0° \, \mathbf{i} + 75 \sin 0° \, \mathbf{j}$
$= 75\mathbf{i}$
$310 \cos \theta + 75 = 0$

$$\cos \theta = \frac{-75}{310}$$

$$\theta = 104°$$

b. decrease

116. $\cos \theta = \dfrac{3(-1) + (-2)(4)}{\|\mathbf{v}\|\|\mathbf{w}\|}$

$$\cos \theta = \frac{-3 - 8}{\sqrt{3^2 + (-2)^2}\sqrt{(-1)^2 + 4^2}}$$

$$\cos \theta = \frac{-11}{\sqrt{13}\sqrt{17}}$$

$$\cos \theta = \frac{-11}{\sqrt{221}}$$

$$\theta = \cos^{-1}\left(\frac{-11}{\sqrt{221}}\right)$$

$$\theta \approx 137.7°$$

117. $\dfrac{2(-2) + 4(-6)}{\|\mathbf{w}\|^2}\mathbf{w} = \dfrac{-4 - 24}{\sqrt{(-2)^2 + 6^2}^2}(-2\mathbf{i} + 6\mathbf{j})$

$$= \frac{-28}{40}(-2\mathbf{i} + 6\mathbf{j})$$

$$= \frac{7}{5}\mathbf{i} - \frac{21}{5}\mathbf{j}$$

118. a. $\|\mathbf{u}\|^2 = \|\mathbf{v}\|^2 + \|\mathbf{w}\|^2 - 2\|\mathbf{v}\|\|\mathbf{w}\|\cos \theta$

b. $\|\mathbf{u}\| = \sqrt{(a_1 - a_2)^2 + (b_1 - b_2)^2}$

$\|\mathbf{u}\|^2 = (a_1 - a_2)^2 + (b_1 - b_2)^2$

$\|\mathbf{v}\| = \sqrt{(a_1 - 0)^2 + (b_1 - 0)^2} = \sqrt{a_1^2 + b_1^2}$

$\|\mathbf{v}\|^2 = a_1^2 + b_1^2$

$\|\mathbf{w}\| = \sqrt{(0 - a_2)^2 + (0 - b_2)^2} = \sqrt{a_2^2 + b_2^2}$

$\|\mathbf{w}\|^2 = a_2^2 + b_2^2$

Section 7.7

Check Point Exercises

1. a. $\mathbf{v} \cdot \mathbf{w} = 7(2) + (-4)(-1) = 14 + 4 = 18$

b. $\mathbf{w} \cdot \mathbf{v} = 2(7) + (-1)(-4) = 14 + 4 = 18$

c. $\mathbf{w} \cdot \mathbf{w} = 2(2) + (-1)(-1) = 4 + 1 = 5$

2. $\cos \theta = \dfrac{\mathbf{v} \cdot \mathbf{w}}{\|\mathbf{v}\| \cdot \|\mathbf{w}\|}$

$$= \frac{(4\mathbf{i} - 3\mathbf{j}) \cdot (\mathbf{i} + 2\mathbf{j})}{\sqrt{4^2 + (-3)^2}\sqrt{1^2 + 2^2}}$$

$$= \frac{4(1) + (-3)(2)}{\sqrt{25}\sqrt{5}}$$

$$= \frac{2}{\sqrt{125}}$$

The angle θ between the vectors is

$$\theta = \cos^{-1}\left(-\frac{2}{\sqrt{125}}\right) \approx 100.3°.$$

3. $\mathbf{v} \cdot \mathbf{w} = (6\mathbf{i} - 3\mathbf{j}) \cdot (\mathbf{i} + 2\mathbf{j})$
$= 6(1) + (-3)(2) = 6 - 6 = 0$
The dot product is zero.
Thus, the given vectors are orthogonal.

4. $\text{proj}_\mathbf{w} \mathbf{v} = \dfrac{\mathbf{v} \cdot \mathbf{w}}{\|\mathbf{w}\|^2} \mathbf{w}$

$= \dfrac{(2\mathbf{i} - 5\mathbf{j}) \cdot (\mathbf{i} - \mathbf{j})}{\left(\sqrt{1^2 + (-1)^2}\right)^2} \mathbf{w}$

$= \dfrac{2(1) + (-5)(-1)}{\left(\sqrt{2}\right)^2} \mathbf{w}$

$= \dfrac{7}{2} \mathbf{w}$

$= \dfrac{7}{2} (\mathbf{i} - \mathbf{j})$

$= \dfrac{7}{2}\mathbf{i} - \dfrac{7}{2}\mathbf{j}$

5. $\mathbf{v}_1 = \text{proj}_\mathbf{w} \mathbf{v} = \dfrac{7}{2}\mathbf{i} - \dfrac{7}{2}\mathbf{j}$

$\mathbf{v}_2 = \mathbf{v} - \mathbf{v}_1$

$= (2\mathbf{i} - 5\mathbf{j}) - \left(\dfrac{7}{2}\mathbf{i} - \dfrac{7}{2}\mathbf{j}\right)$

$= -\dfrac{3}{2}\mathbf{i} - \dfrac{3}{2}\mathbf{j}$

6. $W = \|\mathbf{F}\| \left\|\overrightarrow{AB}\right\| \cos\theta = (20)(150)\cos 30°$

≈ 2598

The work done is approximately 2598 foot-pounds.

Concept and Vocabulary Check 7.7

1. dot product; $a_1 a_2 + b_1 b_2$

2. $\|\mathbf{v}\|\|\mathbf{w}\| \cos\theta$

3. orthogonal

4. true

5. false

Exercise Set 7.7

1. $\mathbf{v} \cdot \mathbf{w} = 3(1) + 1(3) = 3 + 3 = 6$
$\mathbf{v} \cdot \mathbf{v} = 3(3) + 1(1) = 9 + 1 = 10$

3. $\mathbf{v} \cdot \mathbf{w} = 5(-2) + (-4)(-1) = -10 + 4 = -6$
$\mathbf{v} \cdot \mathbf{v} = 5(5) + (-4)(-4) = 25 + 16 = 41$

5. $\mathbf{v} \cdot \mathbf{w} = -6(-10) + (-5)(-8) = 60 + 40 = 100$
$\mathbf{v} \cdot \mathbf{v} = -6(-6) + (-5)(-5) = 36 + 25 = 61$

7. $\mathbf{v} \cdot \mathbf{w} = 5(0) + 0(1) = 0 + 0 = 0$
$\mathbf{v} \cdot \mathbf{v} = 5(5) + 0(0) = 25 + 0 = 25$

9. $\mathbf{v} \cdot (\mathbf{v} + \mathbf{w}) = (2\mathbf{i} - \mathbf{j})[(3\mathbf{i} + \mathbf{j}) + (\mathbf{i} + 4\mathbf{j})]$
$= (2\mathbf{i} - \mathbf{j})[(3+1)\mathbf{i} + (1+4)\mathbf{j})]$
$= (2\mathbf{i} - \mathbf{j})(4\mathbf{i} + 5\mathbf{j})$
$= 2(4) + (-1)(5)$
$= 8 - 5$
$= 3$

11. $\mathbf{u} \cdot \mathbf{v} + \mathbf{u} \cdot \mathbf{w}$
$= (2\mathbf{i} - \mathbf{j}) \cdot (3\mathbf{i} + \mathbf{j}) + (2\mathbf{i} - \mathbf{j})(\mathbf{i} + 4\mathbf{j})$
$= (2)(3) + (-1)(1) + 2(1) + (-1)(4)$
$= 6 - 1 + 2 - 4$
$= 3$

13. $(4\mathbf{u}) \cdot \mathbf{v}$
$= [(4(2\mathbf{i} - \mathbf{j})] \cdot (3\mathbf{i} + \mathbf{j})$
$= (8\mathbf{i} - 4\mathbf{j}) \cdot (3\mathbf{i} + \mathbf{j})$
$= (8)(3) + (-4)(1)$
$= 24 - 4$
$= 20$

15. $4(\mathbf{u} \cdot \mathbf{v})$
$= 4[(2\mathbf{i} - \mathbf{j}) \cdot (3\mathbf{i} + \mathbf{j})]$
$= 4[2(3) + (-1)1]$
$= 4[6 - 1]$
$= 4[5]$
$= 20$

17. $\cos\theta \dfrac{\mathbf{v} \cdot \mathbf{w}}{\|\mathbf{v}\| \; \|\mathbf{w}\|}$

$= \dfrac{(2\mathbf{i} - \mathbf{j}) \cdot (3\mathbf{i} + 4\mathbf{j})}{\sqrt{2^2 + (-1)^2}\sqrt{3^2 + 4^2}}$

$= \dfrac{2(3) + (-1)(4)}{\sqrt{5}\sqrt{25}}$

$= \dfrac{6 - 4}{\sqrt{125}}$

$= \dfrac{2}{\sqrt{125}}$

The angle θ between the vectors is

$\theta = \cos^{-1}\left(\dfrac{2}{\sqrt{125}}\right) \approx 79.7°.$

19. $\cos\theta\dfrac{\mathbf{v}\cdot\mathbf{w}}{\|\mathbf{v}\|\ \|\mathbf{w}\|}$

$= \dfrac{(-3\mathbf{i}+2\mathbf{j})\cdot(4\mathbf{i}-\mathbf{j})}{\sqrt{(-3)^2+2^2}\ \sqrt{4^2+(-1)^2}}$

$= \dfrac{-3(4)+2(-1)}{\sqrt{13}\sqrt{17}}$

$= \dfrac{-14}{\sqrt{221}}$

The angle θ between the vectors is

$\theta = \cos^{-1}\left(-\dfrac{14}{\sqrt{221}}\right) \approx 160.3°.$

21. $\cos\theta\dfrac{\mathbf{v}\cdot\mathbf{w}}{\|\mathbf{v}\|\ \|\mathbf{w}\|}$

$= \dfrac{(6\mathbf{i}+0\mathbf{j})\cdot(5\mathbf{i}+4\mathbf{j})}{\sqrt{6^2+0^2}\ \sqrt{5^2+4^2}}$

$= \dfrac{6(5)+0(4)}{\sqrt{36}\sqrt{41}}$

$= \dfrac{30}{\sqrt{1476}}$

The angle θ between the vectors is

$\theta = \cos^{-1}\left(\dfrac{30}{\sqrt{1476}}\right) \approx 38.7°.$

23. $\mathbf{v}\cdot\mathbf{w} = (\mathbf{i}+\mathbf{j})\cdot(\mathbf{i}-\mathbf{j}) = (1)(1)+1(-1) = 1-1 = 0$

The dot product is zero. Thus, the given vectors are orthogonal.

25. $\mathbf{v}\cdot\mathbf{w} = (2\mathbf{i}+8\mathbf{j})\cdot(4\mathbf{i}-\mathbf{j})$

$= 2(4)+(8)(-1)$

$= 8-8$

$= 0$

The dot product is zero. Thus, the given vectors are orthogonal.

27. $\mathbf{v}\cdot\mathbf{w} = (2\mathbf{i}-2\mathbf{j})\cdot(-\mathbf{i}+\mathbf{j})$

$= 2(-1)+(-2)(1)$

$= -2-2$

$= -4$

The dot product is not zero. Thus, the given vectors are not orthogonal.

29. $\mathbf{v}\cdot\mathbf{w} = (3\mathbf{i}+0\mathbf{j})\cdot(-4\mathbf{i}+0\mathbf{j})$

$= 3(-4)+0(0)$

$= -12+0$

$= -12$

The dot product is not zero. Thus, the given vectors are not orthogonal.

31. $\mathbf{v}\cdot\mathbf{w} = (3\mathbf{i}+0\mathbf{j})\cdot(0\mathbf{i}-4\mathbf{j})$

$= 3(0)+(0)(-4)$

$= 0+0$

$= 0$

The dot product is zero. Thus, the given vectors are orthogonal.

33. $\text{proj}_{\mathbf{w}}\mathbf{v} = \dfrac{\mathbf{v}\cdot\mathbf{w}}{\|\mathbf{w}\|^2}\mathbf{w}$

$= \dfrac{(3\mathbf{i}-2\mathbf{j})\cdot(\mathbf{i}-\mathbf{j})}{\left(\sqrt{1^2+(-1)^2}\right)^2}\mathbf{w}$

$= \dfrac{3(1)+(-2)(-1)}{\left(\sqrt{2}\right)^2}$

$= \dfrac{5}{2}\mathbf{w}$

$= \dfrac{5}{2}(\mathbf{i}-\mathbf{j})$

$= \dfrac{5}{2}\mathbf{i}-\dfrac{5}{2}\mathbf{j}$

$\mathbf{v}_1 = \text{proj}_{\mathbf{w}}\mathbf{v} = \dfrac{5}{2}\mathbf{i}-\dfrac{5}{2}\mathbf{j}$

$\mathbf{v}_2 = \mathbf{v}-\mathbf{v}_1 = (3\mathbf{i}-2\mathbf{j})-\left(\dfrac{5}{2}\mathbf{i}-\dfrac{5}{2}\mathbf{j}\right)$

$= \dfrac{1}{2}\mathbf{i}+\dfrac{1}{2}\mathbf{j}$

35. $\text{proj}_{\mathbf{w}} \mathbf{v} = \dfrac{\mathbf{v} \cdot \mathbf{w}}{\| \mathbf{w} \|} \mathbf{w}$

$\qquad = \dfrac{(\mathbf{i} + 3\mathbf{j}) \cdot (-2\mathbf{i} + 5\mathbf{j})}{\sqrt{1^2 + (-1)^2}} \mathbf{w}$

$\qquad = \dfrac{1(-2) + 3(5)}{\left(\sqrt{(-2)^2 + 5^2} \right)^2} \mathbf{w}$

$\qquad = \dfrac{13}{\left(\sqrt{29} \right)^2} \mathbf{w}$

$\qquad = \dfrac{13}{29} \mathbf{w}$

$\qquad = \dfrac{13}{29} (-2\mathbf{i} + 5\mathbf{j})$

$\qquad = \dfrac{-26}{29} \mathbf{i} + \dfrac{65}{29} \mathbf{j}$

$\mathbf{v}_1 = \text{proj}_{\mathbf{w}} \mathbf{v} = -\dfrac{26}{29} \mathbf{i} + \dfrac{65}{29} \mathbf{j}$

$\mathbf{v}_2 = \mathbf{v} - \mathbf{v}_1$

$\qquad = (\mathbf{i} + 3\mathbf{j}) - \left(-\dfrac{26}{29} \mathbf{i} + \dfrac{65}{29} \mathbf{j} \right)$

$\qquad = \dfrac{55}{29} \mathbf{i} + \dfrac{22}{29} \mathbf{j}$

37. $\text{proj}_{\mathbf{w}} \mathbf{v} = \dfrac{\mathbf{v} \cdot \mathbf{w}}{\| \mathbf{w} \|^2} \mathbf{w}$

$\qquad = \dfrac{(\mathbf{i} + 2\mathbf{j}) \cdot (3\mathbf{i} + 6\mathbf{j})}{\left(\sqrt{3^2 + 6^2} \right)^2} \mathbf{w}$

$\qquad = \dfrac{1(3) + 2(6)}{\sqrt{45}}$

$\qquad = \dfrac{15}{45} \mathbf{w}$

$\qquad = \dfrac{1}{3} \mathbf{w}$

$\qquad = \dfrac{1}{3} (3\mathbf{i} + 6\mathbf{j})$

$\qquad = \mathbf{i} + 2\mathbf{j}$

$\mathbf{v}_1 = \text{proj}_{\mathbf{w}} \mathbf{v} = \mathbf{i} + 2\mathbf{j}$

$\mathbf{v}_2 = \mathbf{v} - \mathbf{v}_1$

$\qquad = (\mathbf{i} + 2\mathbf{j}) - (\mathbf{i} + 2\mathbf{j})$

$\qquad = 0\mathbf{i} + 0\mathbf{j}$

$\qquad = \mathbf{0}$

39. $5\mathbf{u} \cdot (3\mathbf{v} - 4\mathbf{w}) = 15\mathbf{u} \cdot \mathbf{v} - 20\mathbf{u} \cdot \mathbf{w}$

$$= 15\big[(-1)(3) + (1)(-2)\big] - 20\big[(-1)(0) + (1)(-5)\big]$$

$$= 15\big[-5\big] - 20\big[-5\big]$$

$$= 25$$

41. $\mathrm{proj}_{\mathbf{u}}(\mathbf{v} + \mathbf{w}) = \dfrac{(\mathbf{v} + \mathbf{w}) \cdot \mathbf{u}}{\|u\|^2}\mathbf{u}$

$$= \frac{(3\mathbf{i} - 2\mathbf{j} - 5\mathbf{j}) \cdot (-\mathbf{i} + \mathbf{j})}{\|-\mathbf{i} + \mathbf{j}\|^2}(-\mathbf{i} + \mathbf{j})$$

$$= \frac{(3\mathbf{i} - 7\mathbf{j}) \cdot (-\mathbf{i} + \mathbf{j})}{\left(\sqrt{(-1)^2 + 1^2}\right)^2}(-\mathbf{i} + \mathbf{j})$$

$$= \frac{-3 - 7}{2}(-\mathbf{i} + \mathbf{j})$$

$$= -5(-\mathbf{i} + \mathbf{j})$$

$$= 5\mathbf{i} - 5\mathbf{j}$$

43. $\cos\theta = \dfrac{\mathbf{v} \cdot \mathbf{w}}{\|\mathbf{v}\|\|\mathbf{w}\|}$

$$\cos\theta = \frac{\left(2\cos\dfrac{4\pi}{3}\right)\left(3\cos\dfrac{3\pi}{2}\right) + \left(2\sin\dfrac{4\pi}{3}\right)\left(3\sin\dfrac{3\pi}{2}\right)}{\sqrt{\left(2\cos\dfrac{4\pi}{3}\right)^2 + \left(2\sin\dfrac{4\pi}{3}\right)^2}\sqrt{\left(3\cos\dfrac{3\pi}{2}\right)^2 + \left(3\sin\dfrac{3\pi}{2}\right)^2}}$$

$$\cos\theta = \frac{3\sqrt{3}}{6}$$

$$\cos\theta = \frac{\sqrt{3}}{2}$$

$$\theta = 30°$$

45. $\cos\theta = \dfrac{\mathbf{v} \cdot \mathbf{w}}{\|\mathbf{v}\|\|\mathbf{w}\|}$

$$\cos\theta = \frac{(3)(6) + (-5)(-10)}{\sqrt{(3)^2 + (-5)^2}\sqrt{(6)^2 + (-10)^2}}$$

$$\cos\theta = \frac{68}{68}$$

$$\cos\theta = 1$$

$$\theta = 0°$$

The vectors are parallel.

$v \cdot w = 68$

$v \cdot w \neq 0$

The vectors are not orthogonal.

47. $\cos\theta = \dfrac{\mathbf{v}\cdot\mathbf{w}}{\|\mathbf{v}\|\|\mathbf{w}\|}$

$\cos\theta = \dfrac{(3)(6)+(-5)(10)}{\sqrt{(3)^2+(-5)^2}\,\sqrt{(6)^2+(10)^2}}$

$\cos\theta = \dfrac{-32}{68}$

$\theta \approx 118°$

The vectors are not parallel.

$\mathbf{v}\cdot\mathbf{w} = -32$

$\mathbf{v}\cdot\mathbf{w} \neq 0$

The vectors are not orthogonal.

49. $\mathbf{v}\cdot\mathbf{w} = (3)(6)+(-5)\left(\dfrac{18}{5}\right) = 0$

The vectors are orthogonal.

51. $\mathbf{v}\cdot\mathbf{w} = (240\mathbf{i}+300\mathbf{j})\cdot(2.90\mathbf{i}+3.07\mathbf{j})$

$\qquad = 240(2.90)+300(3.07)$

$\qquad = 696+921$

$\qquad = 1617$

$\mathbf{v}\cdot\mathbf{w} = 1617$ means \$1617 in revenue
was generated on Monday by the sale of
240 gallons of regular gas at \$2.90 per gallon and 300 gallons of premium gas at \$3.07 per gallon.

53. Since the car is pushed along a level road, the angle between the force and the direction of motion is $\theta = 0$. The work done

$W = \|\mathbf{F}\|\,\|\overrightarrow{AB}\|\cos\theta$

$\quad = (95)(80)\cos 0°$

$\quad = 7600.$

The work done is 7600 foot-pounds.

55. $W = \|\mathbf{F}\|\,\|\overrightarrow{AB}\|\cos\theta$

$\quad = (40)(100)\cos 32°$

$\quad \approx 3392$

The work done is approximately 3392 foot-pounds.

57. $\mathbf{w} = \mathbf{F}\cdot\overrightarrow{AB}$

$\quad = 60(20)\cos(38°-12°)$

$\quad = 1200\cos 26°$

$\quad \approx 1079$ foot-pounds

59. $\mathbf{w} = \mathbf{F}\cdot\overrightarrow{AB}$

$\quad = (3,2)\cdot[(10,20)-(4,9)]$

$\quad = (3,2)\cdot(6,11)$

$\quad = 18+22$

$\quad = 40$ foot-pounds

61. $\mathbf{w} = \mathbf{F} \cdot \overrightarrow{\mathbf{AB}}$

$= (4\cos 50°, 4\sin 50°) \cdot [(8,10) - (3,7)]$

$= (4\cos 50°, 4\sin 50°) \cdot (5,3)$

$= 20\cos 50° + 12\sin 50°$

≈ 22.05 foot-pounds

63. a. $\cos 30° \mathbf{i} + \sin 30° \mathbf{j} = \dfrac{\sqrt{3}}{2}\mathbf{i} + \dfrac{1}{2}\mathbf{j}$

b. $\text{proj}_{\mathbf{u}}\mathbf{F} = \dfrac{(0,-700) \cdot \left(\dfrac{\sqrt{3}}{2}, \dfrac{1}{2}\right)}{\|\mathbf{u}\|^2}\left(\dfrac{\sqrt{3}}{2}, \dfrac{1}{2}\right)$

$= -350\left(\dfrac{\sqrt{3}}{2}, \dfrac{1}{2}\right) = -175\sqrt{3}\mathbf{i} - 175\mathbf{j}$

c. $\sqrt{\left(-175\sqrt{3}\right)^2 + \left(-175\right)^2}$

$= \sqrt{122,500} = 350$

A force of 350 pounds is required to keep the boat from rolling down the ramp.

65. – 73. Answers may vary.

75. makes sense

77. makes sense

79. $\mathbf{u} \cdot \mathbf{v} = (a_1\mathbf{i} + b_1\mathbf{j}) \cdot (a_2\mathbf{i} + b_2\mathbf{j})$

$= a_1 a_2 + b_1 b_2$

$= a_2 a_1 + b_2 b_1$

$= (a_2\mathbf{i} + b_2\mathbf{j}) \cdot (a_1\mathbf{i} + b_1\mathbf{j})$

$= \mathbf{v} \cdot \mathbf{u}$

Thus $\mathbf{u} \cdot \mathbf{v} = \mathbf{v} \cdot \mathbf{u}$.

81. $\mathbf{u} \cdot (\mathbf{v} + \mathbf{w}) = (a_1\mathbf{i} + b_1\mathbf{j}) \cdot [(a_2\mathbf{i} + b_2\mathbf{j}) + (a_3\mathbf{i} + a_3\mathbf{j})]$

$= (a_1\mathbf{i} + b_1\mathbf{j}) \cdot [(a_2 + a_3)\mathbf{i} + (b_2 + b_3)\mathbf{j}]$

$= a_1(a_2 + a_3) + b_1(b_2 + b_3)$

$= a_1 a_2 + a_1 a_3 + b_1 b_2 + b_1 b_3$

$= a_1 a_2 + b_1 b_2 + a_1 a_3 + b_1 b_3$

$= (a_1\mathbf{i} + b_1\mathbf{j}) \cdot (a_2\mathbf{i} + b_2\mathbf{j}) + (a_1\mathbf{i} + b_1\mathbf{j}) \cdot (a_3\mathbf{i} + b_3\mathbf{j})$

$= \mathbf{u} \cdot \mathbf{v} + \mathbf{u} \cdot \mathbf{w}$

83. Let $\mathbf{v} = 15\mathbf{i} - 3\mathbf{j}$ and $\mathbf{w} = -4\mathbf{i} + b\mathbf{j}$. The vectors $\mathbf{v}$ and $\mathbf{w}$ are orthogonal if $\mathbf{u} \cdot \mathbf{w} = 0$.

$\mathbf{v} \cdot \mathbf{w} = (15\mathbf{i} - 3\mathbf{j}) \cdot (-4\mathbf{i} + b\mathbf{j}) = 15(-4) + (-3)b = -60 - 3b$

$\mathbf{v} \cdot \mathbf{w} = 0$ if $-60 - 3b = 0$. Solving the equation for b, we find $b = -20$.

85. We know that $\text{proj}_{\mathbf{w}}\mathbf{v} = \dfrac{\mathbf{v}\cdot\mathbf{w}}{\|\mathbf{w}\|^2}\mathbf{w}$ If the projection of

$\mathbf{v}$ onto $\mathbf{w}$ is $\mathbf{v}$, then $\mathbf{v} = \dfrac{\mathbf{v}\cdot\mathbf{w}}{\|\mathbf{w}\|^2}\mathbf{w}$.

Since $\dfrac{\mathbf{v}\cdot\mathbf{w}}{\|\mathbf{w}\|^2}$ is a scalar for all $\mathbf{v}$ and $\mathbf{w}$, let k

$=\dfrac{\mathbf{v}\cdot\mathbf{w}}{\|\mathbf{w}\|^2}$. Substituting, we have $\mathbf{v} = k\mathbf{w}$.

When one vector can be expressed as a scalar multiple of another, the vectors have the same direction. Thus, the projection of $\mathbf{v}$ onto $\mathbf{w}$ is $\mathbf{v}$ only if $\mathbf{v}$ and $\mathbf{w}$ have the same direction. Thus, any two vectors, $\mathbf{v}$ and $\mathbf{w}$, having the same direction will satisfy the condition that the projection of $\mathbf{v}$ onto $\mathbf{w}$ is $\mathbf{v}$.

87. **a.** $x + 2y = 2$
$4 + 2(-1) = 2$
$2 = 2$, true
Yes, $(4, -1)$ satisfies the equation.

  b. $x - 2y = 6$
$4 - 2(-1) = 6$
$6 = 6$, true
Yes, $(4, -1)$ satisfies the equation.

88. The graphs intersect at $(4, -1)$.

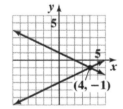

89. $5(2x - 3) - 4x = 9$
$10x - 15 - 4x = 9$
$6x - 15 = 9$
$6x = 24$
$x = 4$
The solution set is $\{4\}$.

Chapter 7 Review Exercises

1. Begin by finding C.
$A + B + C = 180°$
$70° + 55° + C = 180°$
$125° + C = 180°$
$C = 55°$

Use the ratio $\dfrac{a}{\sin A}$, or $\dfrac{12}{\sin 70°}$, to find the other two sides. Use the Law of Sines to find b.
$\dfrac{b}{\sin B} = \dfrac{a}{\sin A}$
$\dfrac{b}{\sin 55°} = \dfrac{12}{\sin 70°}$
$b = \dfrac{12\sin 55°}{\sin 70°} \approx 10.5$

Use the Law of Sines again, this time to find c.
$\dfrac{c}{\sin C} = \dfrac{a}{\sin A}$
$\dfrac{c}{\sin 55°} = \dfrac{12}{\sin 70°}$
$c = \dfrac{12\sin 55°}{\sin 70°} \approx 10.5$
The solution is $C = 55°$, $b \approx 10.5$, $c \approx 10.5$.

2. Begin by finding A.
$A + B + C = 180°$
$A + 107° + 30° = 180°$
$A + 137° = 180°$
$A = 43°$

Use the ratio $\dfrac{c}{\sin C}$, or $\dfrac{126}{\sin 30°}$, to find the other two sides. Use the Law of Sines to find a.
$\dfrac{a}{\sin A} = \dfrac{c}{\sin C}$
$\dfrac{a}{\sin 43°} = \dfrac{126}{\sin 30°}$
$a = \dfrac{126\sin 43°}{\sin 30°} \approx 171.9$

Use the Law of Sines again, this time to find b.

$\dfrac{b}{\sin B} = \dfrac{c}{\sin C}$
$\dfrac{b}{\sin 107°} = \dfrac{126}{\sin 30°}$
$b = \dfrac{126\sin 107°}{\sin 30°} \approx 241.0$
The solution is $A = 43°$, $a \approx 171.9$, and $b \approx 241.0$.

3. Apply the three-step procedure for solving a SAS triangle. Use the Law of Cosines to find the side opposite the given angle. Thus, we will find b.

$$b^2 = a^2 + c^2 - 2ac\cos B$$

$$b^2 = 17^2 + 12^2 - 2(17)(12)\cos 66°$$

$$b^2 = 289 + 144 - 408(\cos 66°)$$

$$b^2 \approx 267.05$$

$$b \approx \sqrt{267.05} \approx 16.3$$

Use the Law of Sines to find the angle opposite the shorter of the two given sides. Thus, we will find acute angle C.

$$\frac{c}{\sin C} = \frac{b}{\sin B}$$

$$\frac{12}{\sin C} = \frac{\sqrt{267.05}}{\sin 66°}$$

$$\sqrt{267.05}\sin C = 12\sin 66°$$

$$\sin C = \frac{12\sin 66°}{\sqrt{267.05}} \approx 0.6708$$

$$C \approx 42°$$

$$A = 180° - B - C = 180° - 66° - 42° = 72°$$

The solution is $b \approx 16.3$, $A \approx 72°$, and $C \approx 42°$.

4. Apply the three-step procedure for solving a SSS triangle. Use the Law of Cosines to find the angle opposite the longest side. Thus, we will find angle C.

$$c^2 = a^2 + b^2 - 2ab\cos C$$

$$\cos C = \frac{a^2 + b^2 - c^2}{2ab}$$

$$= \frac{117^2 + 66^2 - 142^2}{2 \cdot 117 \cdot 66} \approx -0.1372$$

C is obtuse because $\cos C$ is negative.

$$\cos^{-1}(0.1372) \approx 82°$$

$$C \approx 180° - 82° = 98°$$

Use the Law of Sines to find either of the two remaining acute angles. We will find angle A.

$$\frac{a}{\sin A} = \frac{c}{\sin C}$$

$$\frac{117}{\sin A} = \frac{142}{\sin 98°}$$

$$142\sin A = 117\sin 98°$$

$$\sin A = \frac{117\sin 98°}{142} \approx 0.8159$$

$$A \approx 55°$$

$$B = 180° - A - C \approx 180° - 55° - 98° = 27°$$

The solution is $C \approx 98°$, $A \approx 55°$, and $B \approx 27°$.

5. Begin by finding C.

$$A + B + C = 180°$$

$$35° + 25° + C = 180°$$

$$60° + C = 180°$$

$$C = 120°$$

Use the ratio $\dfrac{c}{\sin C}$, or $\dfrac{68}{\sin 120°}$, to find the other two sides. Use the Law of Sines to find a.

$$\frac{a}{\sin A} = \frac{c}{\sin C}$$

$$\frac{a}{\sin 35°} = \frac{68}{\sin 120°}$$

$$a = \frac{68\sin 35°}{\sin 120°} \approx 45.0$$

Use the Law of Sines again, this time to find b.

$$\frac{b}{\sin B} = \frac{c}{\sin C}$$

$$\frac{b}{\sin 25°} = \frac{68}{\sin 120°}$$

$$b = \frac{68\sin 25°}{\sin 120°} \approx 33.2$$

The solution is $C = 120°$, $a \approx 45.0$, and $b \approx 33.2$.

6. The known ratio is $\dfrac{a}{\sin A}$, or $\dfrac{20}{\sin 39°}$. Because side b is given, we used the Law of Sines to find angle B.

$$\frac{b}{\sin B} = \frac{a}{\sin A}$$

$$\frac{26}{\sin B} = \frac{20}{\sin 39°}$$

$$\sin B = \frac{26\sin 39°}{20} \approx 0.8181$$

$$B_1 \approx 55°,\ B_2 \approx 180° - 55° = 125°$$

$$C_1 = 180° - A - B_1 \approx 180° - 39° - 55° = 86°$$

$$C_2 = 180° - A - B_2 \approx 180° - 39° - 125° = 16°$$

Use the Law of Sines to find c_1 and c_2.

$$\frac{c_1}{\sin C_1} = \frac{a}{\sin A}$$

$$\frac{c_1}{\sin 86°} = \frac{20}{\sin 39°}$$

$$c_1 = \frac{20 \sin 86°}{\sin 39°} \approx 31.7$$

$$\frac{c_2}{\sin C_2} = \frac{a}{\sin A}$$

$$\frac{c_2}{\sin 16°} = \frac{20}{\sin 39°}$$

$$c_2 = \frac{20 \sin 16°}{\sin 39°} \approx 8.8$$

There are two triangles. In one triangle, the solution is $B_1 \approx 55°$, $C_1 \approx 86°$, and $c_1 \approx 31.7$. In the other triangle, $B_2 \approx 125°$, $C_2 \approx 16°$, and $c_2 \approx 8.8$.

7. The known ration is $\dfrac{c}{\sin C}$, or $\dfrac{1}{\sin 50°}$. Because side a is given, we used the Law of Sines to find angle A.

$$\frac{a}{\sin A} = \frac{c}{\sin C}$$

$$\frac{3}{\sin A} = \frac{1}{\sin 50°}$$

$$\sin A = \frac{3 \sin 50°}{1} \approx 2.30$$

Because the sine can never exceed 1, there is no triangle with the given measurements.

8. Apply the three-step procedure for solving a SAS triangle. Use the Law of Cosines to find the side opposite the given angle. Thus, we will find a.

$$a^2 = b^2 + c^2 - 2bc \cos A \qquad \text{Use}$$

$$a^2 = (11.2)^2 + (48.2)^2 - 2(11.2)(48.2)\cos 162°$$

$$\approx 3475.5$$

$$a \approx \sqrt{3475.5} \approx 59.0$$

the Law of Sines to find the angle opposite the shorter of the two given sides. Thus, we will find acute angle B.

$$\frac{b}{\sin B} = \frac{a}{\sin A}$$

$$\frac{11.2}{\sin B} = \frac{\sqrt{3475.5}}{\sin 162°}$$

$$\sin B = \frac{11.2 \sin 162°}{\sqrt{3475.5}} \approx 0.0587$$

$$B \approx 3°$$

$$C = 180° - A - B \approx 180° - 162° - 3° = 15°$$

The solution is $a \approx 59.0$, $B \approx 3°$, and $C \approx 15°$.

9. Apply the three-step procedure for solving a SSS triangle. Use the Law of Cosines to find the angle opposite the longest side. Thus, we will find angle B.

$$\cos B = \frac{a^2 + c^2 - b^2}{2ac}$$

$$\cos B = \frac{(26.1)^2 + (36.5)^2 - (40.2)^2}{2 \cdot 26.1 \cdot 36.5}$$

$$\approx 0.2086$$

$$B \approx 78°$$

Use the Law of Sines to find either of the two remaining acute angles. We will find angle A.

$$\frac{a}{\sin A} = \frac{b}{\sin B}$$

$$\frac{26.1}{\sin A} = \frac{40.2}{\sin 78°}$$

$$\sin A = \frac{26.1 \sin 78°}{40.2} \approx 0.6351$$

$$A \approx 39°$$

$$C = 180° - A - B \approx 180° - 39° - 78° = 63°$$

The solution is $B \approx 78°$, $A \approx 39°$, and $C \approx 63°$.

10. The known ratio is $\dfrac{a}{\sin A}$, or $\dfrac{6}{\sin 40°}$. Because side b is given, we used the Law of Sines to find angle B.

$$\frac{b}{\sin B} = \frac{a}{\sin A}$$

$$\frac{4}{\sin B} = \frac{6}{\sin 40°}$$

$$\sin B = \frac{4 \sin 40°}{6} \approx 0.4285$$

$$B_1 \approx 25°, \ B_2 \approx 180° - 25° = 155°$$

B_2 is impossible, since $40° + 155° = 195°$.

$$C = 180° - A - B_1 \approx 180° - 40° - 25° = 115°$$

Use the Law of Sines to find c.

$$\frac{c}{\sin C} = \frac{a}{\sin A}$$

$$\frac{c}{\sin 115°} = \frac{6}{\sin 40°}$$

$$c = \frac{6 \sin 115°}{\sin 40°} \approx 8.5$$

The solution is B_1 (or B) $\approx 25°$, $C \approx 115°$, and $c \approx 8.5$.

11. The known ratio is $\dfrac{b}{\sin B}$, or $\dfrac{8.7}{\sin 37°}$. Because side a is given, we use the Law of Sines to find angle A.

$$\dfrac{a}{\sin A} = \dfrac{b}{\sin B}$$

$$\dfrac{12.4}{\sin A} = \dfrac{8.7}{\sin 37°}$$

$$\sin A = \dfrac{12.4 \sin 37°}{8.7} \approx 0.8578$$

$A_1 \approx 59°$, $A_2 \approx 180° - 59° = 121°$

$C_1 = 180° - A_1 - B$

$\quad \approx 180° - 59° - 37° - 84°$

$C_2 = 180° - A_2 - B$

$\quad \approx 180° - 121° - 37° = 22°$

Use the Law of Sines to find c_1 and c_2.

$$\dfrac{c_1}{\sin C_1} = \dfrac{b}{\sin B}$$

$$\dfrac{c_1}{\sin 84°} = \dfrac{8.7}{\sin 37°}$$

$$c_1 = \dfrac{8.7 \sin 84°}{\sin 37°} \approx 14.4$$

$$\dfrac{c_2}{\sin C_2} = \dfrac{b}{\sin B}$$

$$\dfrac{c_2}{\sin 22°} = \dfrac{8.7}{\sin 37°}$$

$$c_2 = \dfrac{8.7 \sin 22°}{\sin 37°} \approx 5.4$$

There are two triangles. In one triangle, the solution is $A_1 \approx 59°$, $C_1 \approx 84°$, and $c_1 \approx 14.4$. In the other triangle, $A_2 \approx 121°$, $C_2 \approx 22°$, and $c_2 \approx 5.4$.

12. The known ratio is $\dfrac{a}{\sin A}$, or $\dfrac{54.3}{\sin 23°}$. Because side b is given, we used the Law of Sines to find angle B.

$$\dfrac{b}{\sin B} = \dfrac{a}{\sin A}$$

$$\dfrac{22.1}{\sin B} = \dfrac{54.3}{\sin 23°}$$

$$\sin B = \dfrac{22.1 \sin 23°}{54.3} \approx 0.1590$$

$B_1 \approx 9°$, $B_2 \approx 180° - 9° = 171°$

B_2 is impossible, since $23° + 171° = 194°$.

$C = 180° - A - B_1 \approx 180° - 23° - 9° = 148°$

Use the Law of Sines to find c.

$$\dfrac{c}{\sin C} = \dfrac{a}{\sin A}$$

$$\dfrac{c}{\sin 148°} = \dfrac{54.3}{\sin 23°}$$

$$c = \dfrac{54.3 \sin 148°}{\sin 23°} \approx 73.6$$

The solution is B_1 (or B) $\approx 9°$, $C \approx 148°$, and $c \approx 73.6$.

13. Area $= \dfrac{1}{2} ab \sin C$

$\quad = \dfrac{1}{2}(4)(6) \sin 42°$

$\quad \approx 8$

The area of the triangle is approximately 8 square feet.

14. Area $= \dfrac{1}{2} bc \sin A$

$\quad = \dfrac{1}{2}(4)(5) \sin 22°$

$\quad \approx 4$

The area of the triangle is approximately 4 square feet.

15.

$$s = \dfrac{1}{2}(a + b + c) = \dfrac{1}{2}(2 + 4 + 5) = \dfrac{11}{2}$$

$$\text{Area} = \sqrt{s(s-a)(s-b)(s-c)}$$

$$= \sqrt{\dfrac{11}{2}\left(\dfrac{11}{2} - 2\right)\left(\dfrac{11}{2} - 4\right)\left(\dfrac{11}{2} - 5\right)}$$

$$= \sqrt{\dfrac{231}{16}} \approx 4$$

The area of the triangle is approximately 4 square meters.

16.

$$s = \dfrac{1}{2}(a + b + c) = \dfrac{1}{2}(2 + 2 + 2) = 3$$

$$\text{Area} = \sqrt{s(s-a)(s-b)(s-c)}$$

$$= \sqrt{3(3-2)(3-2)(3-2)}$$

$$= \sqrt{3} \approx 2$$

The area of the triangle is approximately 2 square meters.

17.

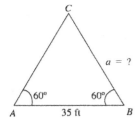

Using the figure, $C = 180° - 60° - 60° = 60°$
Use the Law of Sines to find a.
$$\frac{a}{\sin 60°} = \frac{35}{\sin 60°}$$
$$a = 35$$
The length of the roof is 35 feet.

18. One car travels 60 miles per hour for 30 minutes (half an hour), or $60\left(\frac{1}{2}\right) = 30$ miles. Similarly, the other car travels 25 miles.

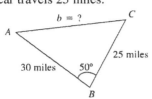

Using the figure,
$$b^2 = a^2 + c^2 - 2ac \cos B$$
$$= 25^2 + 30^2 - 2(25)(30)\cos 80° \approx 1264.53$$
$$b \approx \sqrt{1264.53} \approx 35.6$$
The cars will be about 35.6 miles apart.

19. The first plane travels 325 miles per hour for 2 hours, or $325 \cdot 2 = 650$ miles. Similarly, the other plane travels $300 \cdot 2 = 600$ miles.

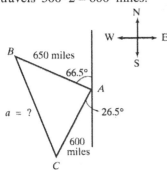

Using the figure,
$A = 180° - 66.5° - 26.5° = 87°$
Use the Law of Cosines to find a.

$$a^2 = b^2 + c^2 - 2bc \cos A$$
$$= 600^2 + 650^2 - 2(600)(650)\cos 87°$$
$$\approx 741,678$$
$$a \approx \sqrt{741,678} \approx 861$$
The planes are about 861 miles apart.

20.

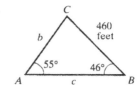

Using the figure,
$C = 180° - A - B = 180° - 55° - 46° = 79°$
Use the Law of Sines to find b.
$$\frac{b}{\sin B} = \frac{a}{\sin A}$$
$$\frac{b}{\sin 46°} = \frac{460}{\sin 55°}$$
$$b = \frac{460 \sin 46°}{\sin 55°} \approx 404$$
Use the Law of Sines again, this time to find c.
$$\frac{c}{\sin C} = \frac{a}{\sin A}$$
$$\frac{c}{\sin 79°} = \frac{460}{\sin 55°}$$
$$c = \frac{460 \sin 79°}{\sin 55°} \approx 551$$
The lengths are about 404 feet and 551 feet.

21. $s = \dfrac{1}{2}(a+b+c) = \dfrac{1}{2}(260+320+450) = 515$
$$\text{Area} = \sqrt{s(s-a)(s-b)(s-c)}$$
$$= \sqrt{515(515-260)(515-320)(515-450)}$$
$$= \sqrt{1,664,544,375} \approx 40,798.83$$
$$\text{cost} \approx (5.25)(40,798.83) \approx 214,194$$
The cost is approximately \$214,194.

22. Draw $\theta = 60°$ counterclockwise, since, θ is positive, from the polar axis. Go 4 units out on the terminal side of θ, since $r > 0$.

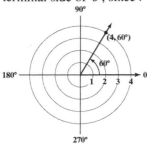

$$x = r\cos\theta = 4\cos 60° = 4\left(\frac{1}{2}\right) = 2$$

$$y = r\sin\theta = 4\sin 60° = 4\left(\frac{\sqrt{3}}{2}\right) = 2\sqrt{3}$$

The rectangular coordinates of $(4, 60°)$ are $\left(2, 2\sqrt{3}\right)$.

23. Draw $\theta = 150°$ counterclockwise, since θ is positive, from the polar axis. Go 3 units out on the terminal side of θ, since $r > 0$.

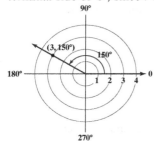

$$x = r\cos\theta = 3\cos 150° = 3\left(-\frac{\sqrt{3}}{2}\right) = -\frac{3\sqrt{3}}{2}$$

$$y = r\sin\theta = 3\sin 150° = 3\left(\frac{1}{2}\right) = \frac{3}{2}$$

The rectangular coordinates of $(3, 150°)$ are $\left(-\frac{3\sqrt{3}}{2}, \frac{3}{2}\right)$.

24. Draw $\theta = \frac{4\pi}{3} = 240°$ counterclockwise, since θ is positive, from the polar axis. Go 4 units out opposite the terminal side of θ, since $r < 0$.

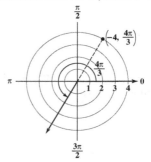

$$x = r\cos\theta = -4\cos\frac{4\pi}{3} = -4\left(-\frac{1}{2}\right) = 2$$

$$y = r\sin\theta = -4\sin\frac{4\pi}{3} = -4\left(-\frac{\sqrt{3}}{2}\right) = 2\sqrt{3}$$

The rectangular coordinates of $\left(-4, \frac{4\pi}{3}\right)$ are $\left(2, 2\sqrt{3}\right)$.

25. Draw $\theta = \frac{5\pi}{4} = 225°$ counterclockwise, since θ is positive from the polar axis. Go 2 units out opposite the terminal side of θ, since $r < 0$.

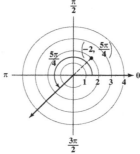

$$x = r\cos\theta = -2\cos\frac{5\pi}{4} = -2\left(-\frac{\sqrt{2}}{2}\right) = \sqrt{2}$$

$$y = r\sin\theta = -2\sin\frac{5\pi}{4} = -2\left(-\frac{\sqrt{2}}{2}\right) = \sqrt{2}$$

The rectangular coordinates of $\left(-2, \frac{5\pi}{4}\right)$ are $\left(\sqrt{2}, \sqrt{2}\right)$.

26. Draw $\theta = -\dfrac{\pi}{2} = -90°$ clockwise, since θ is

negative, from the polar axis. Go 4 units out opposite the terminal side of θ, since $r < 0$.

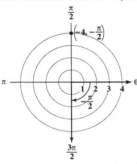

$$x = r\cos\theta = -4\cos\left(-\frac{\pi}{2}\right) = -4(0) = 0$$

$$y = r\sin\theta = -4\sin\left(-\frac{\pi}{2}\right) = -4(-1) = 4$$

The rectangular coordinates of $\left(-4, -\dfrac{\pi}{2}\right)$ are $(0, 4)$.

27. Draw $\theta = -\dfrac{\pi}{4} = -45°$ clockwise, since θ is

negative, from the polar axis. Plot the point out 2 units opposite the terminal side of θ, since $r < 0$.

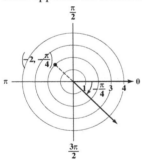

$$x = r\cos\theta = -2\cos\left(-\frac{\pi}{4}\right) = -2\left(\frac{\sqrt{2}}{2}\right) = -\sqrt{2}$$

$$y = r\sin\theta = -2\sin\left(-\frac{\pi}{4}\right) = -2\left(-\frac{\sqrt{2}}{2}\right) = \sqrt{2}$$

The rectangular coordinates of $\left(-2, -\dfrac{\pi}{4}\right)$ are

$\left(-\sqrt{2}, \sqrt{2}\right)$.

28. Draw $\theta = \dfrac{\pi}{6} = 30°$ counterclockwise,

since θ is positive, from the polar axis. Go out 3 units on the terminal side of θ, since $r > 0$.

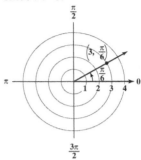

a. $\left(3, \dfrac{\pi}{6}\right) = \left(3, \dfrac{\pi}{6} + 2\pi\right) = \left(3, \dfrac{13\pi}{6}\right)$

b. $\left(3, \dfrac{\pi}{6}\right) = \left(-3, \dfrac{\pi}{6} + \pi\right) = \left(-3, \dfrac{7\pi}{6}\right)$

c. $\left(3, \dfrac{\pi}{6}\right) = \left(3, \dfrac{\pi}{6} - 2\pi\right) = \left(3, -\dfrac{11\pi}{6}\right)$

29. Draw $\theta = \dfrac{2\pi}{3} = 120°$ counterclockwise,

since θ is positive, from the polar axis. Go out 3 units on the terminal side of θ, since $r > 0$.

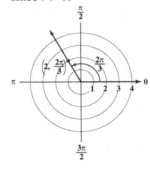

a. $\left(2, \dfrac{2\pi}{3}\right) = \left(2, \dfrac{2\pi}{3} + 2\pi\right) = \left(2, \dfrac{8\pi}{3}\right)$

b. $\left(2, \dfrac{2\pi}{3}\right) = \left(-2, \dfrac{2\pi}{3} + \pi\right) = \left(-2, \dfrac{5\pi}{3}\right)$

c. $\left(2, \dfrac{2\pi}{3}\right) = \left(2, \dfrac{2\pi}{3} - 2\pi\right) = \left(2, -\dfrac{4\pi}{3}\right)$

30.

Draw $\theta = \dfrac{\pi}{2} = 90°$ counterclockwise,

since θ is positive, from the polar axis.
Go out 3.5 units on the terminal side of
θ, since $r > 0$.

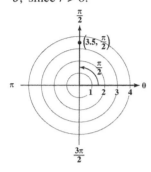

a. $\left(3, \dfrac{\pi}{2} + 2\pi\right) = \left(3, \dfrac{5\pi}{2}\right)$

b. $\left(-3.5, \dfrac{\pi}{2} + \pi\right) = \left(-3.5, \dfrac{3\pi}{2}\right)$

c. $\left(3.5, \dfrac{\pi}{2} - 2\pi\right) = \left(3.5, -\dfrac{3\pi}{2}\right)$

31. $(-4, 4)$

$r = \sqrt{(-4)^2 + (-4)^2} = \sqrt{16 + 16} = \sqrt{32} = 4\sqrt{2}$

$\tan\theta = \dfrac{4}{-4} = -1$

Because $\tan\dfrac{\pi}{4} = 1$ and θ lies in quadrant II,

$\theta = \pi - \dfrac{\pi}{4} = \dfrac{3\pi}{4}$.

The polar coordinates of $(-4, 4)$ are $\left(4\sqrt{2}, \dfrac{3\pi}{4}\right)$.

32. $(3, -3)$

$r = \sqrt{3^2 + (-3)^2} = \sqrt{9 + 9} = \sqrt{18} = 3\sqrt{2}$

$\tan\theta = \dfrac{-3}{3} = -1$

Because $\tan\dfrac{\pi}{4} = 1$, and θ lies in

quadrant IV, $\theta = 2\pi - \dfrac{\pi}{4} = \dfrac{7\pi}{4}$.

The polar coordinates of $(3, -3)$ are $\left(3\sqrt{2}, \dfrac{7\pi}{4}\right)$

33. $(5, 12)$

$r = \sqrt{5^2 + 12^2} = \sqrt{25 + 144} = \sqrt{169} = 13$

$\tan\theta = \dfrac{12}{5}$

Because $\tan^{-1}\left(\dfrac{12}{5}\right) \approx 67°$ and θ lies in quadrant I,

$\theta \approx 67°$.
The polar coordinates of $(5, 12)$ are approximately
$(13, 67°)$.

34. $(-3, 4)$

$r = \sqrt{(-3)^2 + 4^2} = \sqrt{9 + 16} = \sqrt{25} = 5$

$\tan\theta = \dfrac{4}{-3} = -\dfrac{4}{3}$

Because $\tan^{-1}\left(\dfrac{4}{3}\right) \approx 53°$ and θ lies in quadrant II,

$\theta \approx 180° - 53° = 127°$. The polar coordinates of $(-3, 4)$ are $(5, 127°)$.

35. $(0, -5)$

$r = \sqrt{0^2 + (-5)^2} = \sqrt{25} = 5$

$\tan\theta = \dfrac{-5}{0}$ is undefined

Because $\tan\dfrac{\pi}{2}$ is undefined and θ lies on the

negative *y*-axis, $\theta = \dfrac{\pi}{2} + \pi = \dfrac{3\pi}{2}$. The polar

coordinates of $(0, -5)$ are $\left(5, \dfrac{3\pi}{2}\right)$.

36. $(1, 0)$

$r = \sqrt{1^2 + 0^2} = \sqrt{1} = 1$

$\tan\theta = \dfrac{0}{1} = 0$

Because $\tan 0 = 0$ and θ lies on the positive *x*-axis,
$\theta = 0$.
The polar coordinates of $(1, 0)$ are $(1, 0)$.

37.
$$2x + 3y = 8$$
$$2r\cos\theta + 3r\sin\theta = 8$$
$$r(2\cos\theta + 3\sin\theta) = 8$$
$$r = \dfrac{8}{2\cos\theta + 3\sin\theta}$$

38. $x^2 + y^2 = 100$
$$r^2 = 100$$
$$r = 10$$

39.
$$(x-6)^2 + y^2 = 36$$
$$(r\cos\theta - 6)^2 + (r\sin\theta)^2 = 36$$
$$r^2\cos^2\theta - 12r\cos\theta + 36 + r^2\sin^2\theta = 36$$
$$r^2 - 12r\cos\theta = 0$$
$$r^2 = 12r\cos\theta$$
$$r = 12\cos\theta$$

40.
$$r = 3$$
$$r^2 = 3^2$$
$$x^2 + y^2 = 9$$

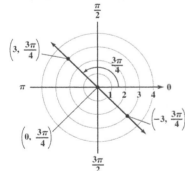

41.
$$\theta = \frac{3\pi}{4}$$
$$\tan\theta = \tan\frac{3\pi}{4}$$
$$\frac{x}{y} = -1$$
$$x = -y$$

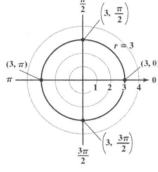

42. $r\cos\theta = -1$
$$x = -1$$

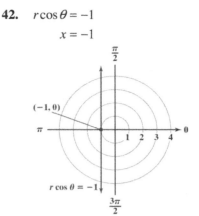

43.
$$r = 5\csc\theta$$
$$r = \frac{5}{\sin\theta}$$
$$r\sin\theta = 5$$
$$y = 5$$

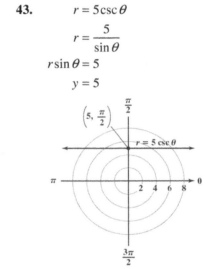

44.
$$r = 3\cos\theta$$
$$r\cdot r = r\cdot 3\cos\theta$$
$$r^2 = 3r\cos\theta$$
$$x^2 + y^2 = 3x$$
$$x^2 - 3x + y^2 = 0$$
$$x^2 - 3x + \frac{9}{4} + y^2 = \frac{9}{4}$$
$$\left(x - \frac{3}{2}\right)^2 + y^2 = \frac{9}{4}$$

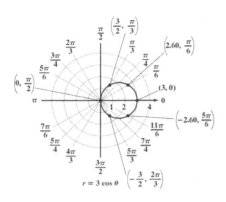

$r = 3\cos\theta$

45. $4r\cos\theta + r\sin\theta = 8$

$$4x + y = 8$$
$$y = -4x + 8$$

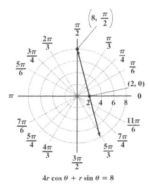

$4r\cos\theta + r\sin\theta = 8$

46. $r^2\sin 2\theta = -2$

$$r^2(2\sin\theta\cos\theta) = -2$$
$$2r\sin\theta r\cos\theta = -2$$
$$2yx = -2$$
$$y = \frac{-2}{2x}$$
$$y = -\frac{1}{x}$$

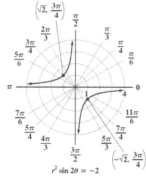

$r^2\sin 2\theta = -2$

47. $r = 5 + 3\cos\theta$

a. $r = 5 + 3\cos(-\theta)$

$r = 5 + 3\cos\theta$

The graph has symmetry about the polar axis.

b. $-r = 5 + 3\cos(-\theta)$

$-r = 5 + 3\cos\theta$

$r = -5 - 3\cos\theta$

The graph may or may not have symmetry with

respect to the line $\theta = \dfrac{\pi}{2}$.

c. $-r = 5 + 3\cos\theta$

$r = -5 - 3\cos\theta$

The graph may or may not have symmetry with
respect to the pole.

48. $r = 3\sin\theta$

a. $r = 3\sin(-\theta)$

$r = -3\sin\theta$

The graph may or may not have symmetry with
respect to the polar axis.

b. $-r = 3\sin(-\theta)$

$-r = -3\sin\theta$

$r = 3\sin\theta$

The graph has symmetry with respect to

line $\theta = \dfrac{\pi}{2}$.

c. $-r = 3\sin\theta$

$r = -3\sin\theta$

The graph may or may not have symmetry with
respect to the pole.

49. $r^2 = 9\cos 2\theta$

a. $r^2 = 9\cos 2(-\theta)$

$r^2 = 9\cos(-2\theta)$

$r^2 = 9\cos 2\theta$

The graph has symmetry with respect to the
polar axis.

b. $(-r)^2 = 9\cos 2(-\theta)$

$r^2 = 9\cos(-2\theta)$

$r^2 = 9\cos 2\theta$

The graph has symmetry with respect to the line

$\theta = \dfrac{\pi}{2}$.

    c. $(-r)^2 = 9\cos 2\theta$

          $r^2 = 9\cos 2\theta$

    The graph has symmetry with respect to
the pole.

50. $r = 3\cos\theta$
Check for symmetry:

Polar Axis	**The Line** $\theta = \dfrac{\pi}{2}$	**The Pole**
$r = 3\cos(-\theta)$	$-r = 3\cos(-\theta)$	
	$-r = 3\cos\theta$	
$r = 3\cos\theta$	$r = -3\cos\theta$	

The graph has symmetry with respect to the polar axis. The graph may or may not be symmetric with respect to the line

$\theta = \dfrac{\pi}{2}$ or the pole. Calculate values of r for θ from 0 to π and use symmetry to obtain the graph.

θ	0	$\dfrac{\pi}{6}$	$\dfrac{\pi}{3}$	$\dfrac{\pi}{2}$	$\dfrac{2\pi}{3}$	$\dfrac{5\pi}{6}$	π
r	3	2.6	1.5	0	-1.5	-2.6	-3

Use symmetry to obtain the graph.

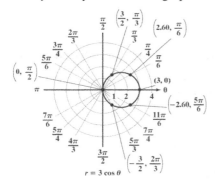

Notice that there are no points in quadrants II or III. Because the cosine is negative in quadrants II and III, r is negative here. This places the points in quadrants IV and I respectively.

51. $r = 2 + 2\sin\theta$
Check for symmetry:

Polar Axis	**The Line** $\theta = \dfrac{\pi}{2}$	**The Pole**
$r = 2 + 2\sin(-\theta)$	$-r = 2 + 2\sin(-\theta)$	$-r = 2 + 2\sin\theta$
	$-r = 2 - 2\sin\theta$	
$r = 2 - 2\sin\theta$	$r = -2 + 2\sin\theta$	$r = -2 - 2\sin\theta$

There may be no symmetry, since each equation is not equivalent to $r = 2 + 2\sin\theta$. Calculate values of r for θ from 0 to 2π.

θ	0	$\dfrac{\pi}{6}$	$\dfrac{\pi}{3}$	$\dfrac{\pi}{2}$	$\dfrac{2\pi}{3}$	$\dfrac{5\pi}{6}$	π	$\dfrac{7\pi}{6}$	$\dfrac{4\pi}{3}$	$\dfrac{3\pi}{2}$	$\dfrac{5\pi}{3}$	$\dfrac{11\pi}{6}$	2π
r	2	3	3.73	4	3.73	3	2	1	0.27	0	0.27	1	2

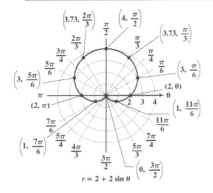

$r = 2 + 2\sin\theta$

52. $r = \sin 2\theta$

Check for symmetry:

Polar Axis	**The Line** $\theta = \dfrac{\pi}{2}$	**The Pole**
$r = \sin 2(-\theta)$	$-r = \sin 2(-\theta)$	$-r = \sin 2\theta$
$r = \sin(-2\theta)$	$-r = \sin(-2\theta)$	
	$-r = -\sin 2\theta$	
$r = -\sin 2\theta$	$r = \sin 2\theta$	$r = -\sin 2\theta$

The graph has symmetry with respect to the line $\theta = \dfrac{\pi}{2}$. The graph may or may not be symmetric with respect to the polar axis or the pole. Calculate values of r for θ from 0 to $\dfrac{\pi}{2}$ and for θ from π to $\dfrac{3\pi}{2}$. Then, use symmetry to obtain the graph.

θ	0	$\dfrac{\pi}{6}$	$\dfrac{\pi}{4}$	$\dfrac{\pi}{3}$	$\dfrac{\pi}{2}$	π	$\dfrac{7\pi}{6}$	$\dfrac{5\pi}{4}$	$\dfrac{4\pi}{3}$	$\dfrac{3\pi}{2}$
r	0	0.87	1	0.87	0	0	0.87	1	0.87	0

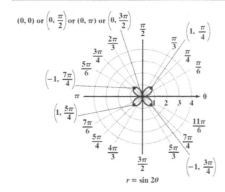

$r = \sin 2\theta$

53. $r = 2 + \cos\theta$

Check for symmetry:

Polar Axis	The Line $\theta = \dfrac{\pi}{2}$	The Pole
$r = 2 + \cos(-\theta)$	$-r = 2 + \cos(-\theta)$	$-r = 2 + \cos\theta$
	$-r = 2 + \cos\theta$	
$r = 2 + \cos\theta$	$r = -2 - \cos\theta$	$r = -2 - \cos\theta$

The graph is symmetric with respect to the polar axis. The graph may or may not be symmetric with respect to the line

$\theta = \dfrac{\pi}{2}$ or the pole. Calculate values of r for θ from 0 to π and use symmetry to obtain the graph.

θ	0	$\dfrac{\pi}{6}$	$\dfrac{\pi}{3}$	$\dfrac{\pi}{2}$	$\dfrac{2\pi}{3}$	$\dfrac{5\pi}{6}$	π
r	3	2.87	2.5	2	1.5	1.13	1

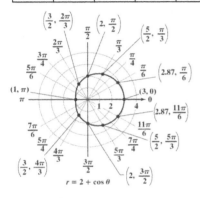

54. $r = 1 + 3\sin\theta$

Check for symmetry:

Polar Axis	The Line $\theta = \dfrac{\pi}{2}$	The Pole
$r = 1 + 3\sin(-\theta)$	$-r = 1 + 3\sin(-\theta)$	$-r = 1 + 3\sin\theta$
	$-r = 1 - 3\sin\theta$	
$r = 1 - 3\sin\theta$	$r = -1 + 3\sin\theta$	$r = -1 - 3\sin\theta$

There may be no symmetry, since each equation is not equivalent to $r = 1 + 3\sin\theta$. Calculate values of r for θ from 0 to 2π.

θ	0	$\dfrac{\pi}{6}$	$\dfrac{\pi}{3}$	$\dfrac{\pi}{2}$	$\dfrac{2\pi}{3}$	$\dfrac{5\pi}{6}$	π	$\dfrac{7\pi}{6}$	$\dfrac{4\pi}{3}$	$\dfrac{3\pi}{2}$	$\dfrac{5\pi}{3}$	$\dfrac{11\pi}{6}$	2π
r	1	2.5	3.6	4	3.6	2.5	1	-0.5	-1.6	-2	-1.6	-0.5	1

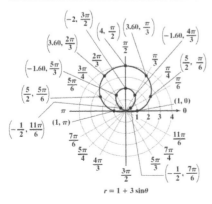

55. $r = 1 - 2\cos\theta$

Check for symmetry:

Polar Axis	The Line $\theta = \dfrac{\pi}{2}$	The Pole
$r = 1 - 2\cos(-\theta)$	$-r = 1 - 2\cos(-\theta)$	$-r = 1 - 2\cos\theta$
	$-r = 1 - 2\cos\theta$	
$r = 1 - 2\cos\theta$	$r = -1 + 2\cos\theta$	$r = -1 + 2\cos\theta$

The graph is symmetric with respect to the polar axis. The graph may or may not be symmetric with respect to the line $\theta = \dfrac{\pi}{2}$ or the pole. Calculate values of r for θ from 0 to π and use symmetry to obtain the graph.

θ	0	$\dfrac{\pi}{6}$	$\dfrac{\pi}{3}$	$\dfrac{\pi}{2}$	$\dfrac{2\pi}{3}$	$\dfrac{5\pi}{6}$	π
r	-1	-0.73	0	1	2	2.73	3

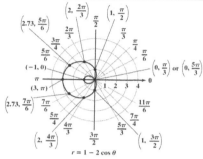

$r = 1 - 2\cos\theta$

56. $r^2 = \cos 2\theta$

Check for symmetry:

Polar Axis	The Line $\theta = \dfrac{\pi}{2}$	The Pole
$r^2 = \cos 2(-\theta)$	$(-r)^2 = \cos 2(-\theta)$	$(-r)^2 = \cos 2\theta$
$r^2 = \cos(-2\theta)$	$r^2 = \cos(-2\theta)$	
$r^2 = \cos 2\theta$	$r^2 = \cos 2\theta$	$r^2 = \cos 2\theta$

The graph has symmetry with respect to the polar axis, the line $\theta = \dfrac{\pi}{2}$, and the pole.

Calculate values of r for θ from 0 to $\dfrac{\pi}{2}$ and use symmetry to obtain the graph.

θ	0	$\dfrac{\pi}{6}$	$\dfrac{\pi}{4}$	$\dfrac{\pi}{3}$	$\dfrac{\pi}{2}$
r	± 1	± 0.71	0	undef	undef

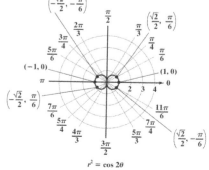

$r^2 = \cos 2\theta$

57. $z = 1 - i$ corresponds to the point $(1, -1)$.

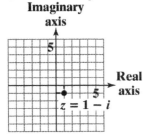

Use $r = \sqrt{a^2 + b^2}$ and $\tan\theta = \dfrac{b}{a}$, with $a = 1$ and $b = -1$, to find r and θ.

$r = \sqrt{1^2 + (-1)^2} = \sqrt{1+1} = \sqrt{2}$

$\tan\theta = \dfrac{-1}{1} = -1$

Because $\tan\dfrac{\pi}{4} = 1$ and θ lies in quadrant IV, $\theta = 2\pi - \dfrac{\pi}{4} = \dfrac{7\pi}{4}$.

$z = 1 - i = r(\cos\theta + i\sin\theta)$

$\quad = \sqrt{2}\left(\cos\dfrac{7\pi}{4} + i\sin\dfrac{7\pi}{4}\right)$

or $\sqrt{2}(\cos 315° + i\sin 315°)$

58. $z = -2\sqrt{3} + 2i$ corresponds to the point $(-2\sqrt{3},\ 2)$.

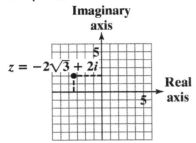

Use $r = \sqrt{a^2 + b^2}$ and $\tan\theta = \dfrac{b}{a}$, with $a = -2\sqrt{3}$ and $b = 2$, to find r and θ.

$r = \sqrt{\left(-2\sqrt{3}\right)^2 + 2^2} = \sqrt{12 + 4} = \sqrt{16} = 4$

$\tan\theta = \dfrac{2}{-2\sqrt{3}} = -\dfrac{1}{\sqrt{3}}$

Because $\tan 30° = \dfrac{1}{\sqrt{3}}$ and θ lies in quadrant II, $\theta = 180° - 30° = 150°$.

$z = -2\sqrt{3} + 2i$

$\quad = r(\cos\theta + i\sin\theta)$

$\quad = 4(\cos 150° + i\sin 150°)$

or $4\left(\cos\dfrac{5\pi}{6} + i\sin\dfrac{5\pi}{6}\right)$

59. $z = -3 - 4i$ corresponds to the point $(-3, -4)$.

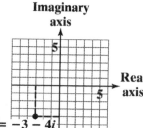

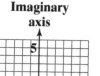

$z = -3 - 4i$

Use $r = \sqrt{a^2 + b^2}$ and $\tan\theta = \dfrac{b}{a}$, with

$a = -3$ and $b = -4$, to find r and θ.

$r = \sqrt{(-3)^2 + (-4)^2} = \sqrt{9 + 16} = \sqrt{25} = 5$

$\tan\theta = \dfrac{-4}{-3} = \dfrac{4}{3}$

Because $\tan^{-1}\left(\dfrac{4}{3}\right) \approx 53°$ and θ lies in quadrant III, $\theta \approx 180° + 53° = 233°$.

$z = -3 - 4i = r(\cos\theta + i\sin\theta)$

$\approx 5(\cos 233° + i\sin 233°)$

60. $z = -5i = 0 - 5i$ corresponds to the point $(0, -5)$.

$z = -5i$

Use $r = \sqrt{a^2 + b^2}$ and $\tan\theta = \dfrac{b}{a}$, with $a = 0$ and $b = -5$, to find r and θ.

$r = \sqrt{0^2 + (-5)^2} = \sqrt{25} = 5$

$\tan\theta = \dfrac{-5}{0}$ is undefined

Because $\tan\dfrac{\pi}{2}$ is undefined and θ lies on the negative y-axis, $\theta = \dfrac{\pi}{2} + \pi = \dfrac{3\pi}{2}$.

$z = -5i = r(\cos\theta + i\sin\theta) = 5\left(\cos\dfrac{3\pi}{2} + i\sin\dfrac{3\pi}{2}\right)$ or $5(\cos 270° + i\sin 270°)$

61. $8(\cos 60° + i\sin 60°) = 8\left(\dfrac{1}{2} + i\dfrac{\sqrt{3}}{2}\right) = 4 + 4i\sqrt{3}$

The rectangular form of $z = 8(\cos 60° + i\sin 60°)$ is $z = 4 + 4i\sqrt{3}$.

62. $4(\cos 210° + i\sin 210°) = 4\left(-\dfrac{\sqrt{3}}{2} + i\left(-\dfrac{1}{2}\right)\right) = -2\sqrt{3} - 2i$

The rectangular form of $z = 4(\cos 210° + i\sin 210°)$ is $z = -2\sqrt{3} - 2i$.

63. $6\left(\cos\dfrac{2\pi}{3}+i\sin\dfrac{2\pi}{3}\right)=6\left(-\dfrac{1}{2}+i\dfrac{\sqrt{3}}{2}\right)=-3+3i\sqrt{3}$

The rectangular form of $z=6\left(\cos\dfrac{2\pi}{3}+i\sin\dfrac{2\pi}{3}\right)$ is $z=-3+3i\sqrt{3}$.

64. $0.6(\cos100°+i\sin100°)\approx0.6\left(-0.17+i(0.98)\right)\approx-0.1+0.6i$

The rectangular form of $z=0.6(\cos100°+i\sin100°)$ is $z\approx-0.1+0.6i$.

65. $z_1z_2=\left[3(\cos40°+i\sin40°)\right]\left[5(\cos70°+i\sin70°)\right]$

$=(3\cdot5)\left[\cos(40°+70°)+i\sin(40°+70°)\right]$

$=15(\cos110°+i\sin110°)$

66. $z_1z_2=\left[\cos210°+i\sin210°\right]\left[\cos55°+i\sin55°\right]$

$=\cos\left(210°+55°\right)+i\sin(210°+55°)$

$=\cos265°+i\sin265°$

67. $z_1z_2=\left[4\left(\cos\dfrac{3\pi}{7}+i\sin\dfrac{3\pi}{7}\right)\right]\left[10\left(\cos\dfrac{4\pi}{7}+i\dfrac{4\pi}{7}\right)\right]$

$=(4\cdot10)\left[\cos\left(\dfrac{3\pi}{7}+\dfrac{4\pi}{7}\right)+i\sin\left(\dfrac{3\pi}{7}+\dfrac{4\pi}{7}\right)\right]$

$=40(\cos\pi+i\sin\pi)$

68. $\dfrac{z_1}{z_2}=\dfrac{10(\cos10°+i\sin10°)}{5(\cos5°+i\sin5°)}=\dfrac{10}{5}\left[\cos(10°-5°)+i\sin(10°-5°)\right]=2(\cos5°+i\sin5°)$

69. $\dfrac{z_1}{z_2}=\dfrac{5\left(\cos\dfrac{4\pi}{3}+i\sin\dfrac{4\pi}{3}\right)}{10\left(\cos\dfrac{\pi}{3}+i\sin\dfrac{\pi}{3}\right)}=\dfrac{5}{10}\left[\cos\left(\dfrac{4\pi}{3}-\dfrac{\pi}{3}\right)+i\sin\left(\dfrac{4\pi}{3}-\dfrac{\pi}{3}\right)\right]=\dfrac{1}{2}(\cos\pi+i\sin\pi)$

70. $\dfrac{z_1}{z_2}=\dfrac{2\left(\cos\dfrac{5\pi}{3}+i\sin\dfrac{5\pi}{3}\right)}{\cos\dfrac{\pi}{2}+i\sin\dfrac{\pi}{2}}=2\left[\cos\left(\dfrac{5\pi}{3}-\dfrac{\pi}{2}\right)+i\sin\left(\dfrac{5\pi}{3}-\dfrac{\pi}{2}\right)\right]$

$=2\left[\cos\left(\dfrac{10\pi}{6}-\dfrac{3\pi}{6}\right)+i\sin\left(\dfrac{10\pi}{6}-\dfrac{3\pi}{6}\right)\right]=2\left(\cos\dfrac{7\pi}{6}+i\sin\dfrac{7\pi}{6}\right)$

71. $\left[2(\cos20°+i\sin20°)\right]^3=(2)^3\left[\cos(3\cdot20°)+i\sin(3\cdot20°)\right]=8(\cos60°+i\sin60°)$

$=8\left(\dfrac{1}{2}+i\dfrac{\sqrt{3}}{2}\right)=4+4\sqrt{3}i$

72. $\left[4(\cos50°+i\sin50°)\right]^3=(4)^3\left[\cos(3\cdot50°)+i\sin(3\cdot50°)\right]=64(\cos150°+i\sin150°)=64\left(-\dfrac{\sqrt{3}}{2}+i\dfrac{1}{2}\right)$

$=-32\sqrt{3}+32i$

73. $\left[\dfrac{1}{2}\left(\cos\dfrac{\pi}{14}+i\sin\dfrac{\pi}{14}\right)\right]^7 = \left(\dfrac{1}{2}\right)^7\left[\cos\left(7\cdot\dfrac{\pi}{14}\right)+i\sin\left(7\cdot\dfrac{\pi}{14}\right)\right] = \dfrac{1}{128}\left(\cos\dfrac{\pi}{2}+i\sin\dfrac{\pi}{2}\right)$

$$= \dfrac{1}{128}(0+i1) = \dfrac{1}{128}i$$

74. Write $1-i\sqrt{3}$ in $r(\cos\theta+i\sin\theta)$ form.

$r = \sqrt{a^2+b^2} = \sqrt{1^2+\left(-\sqrt{3}\right)^2} = \sqrt{1+3} = 2$

$\tan\theta = \dfrac{b}{a} = \dfrac{-\sqrt{3}}{1} = -\sqrt{3}$

Because $\tan 60° = \sqrt{3}$ and θ lies in quadrant IV, $\theta = 360° - 60° = 300°$.

$1-i\sqrt{3} = r(\cos\theta+i\sin\theta) = 2(\cos 300° + i\sin 300°)$

Use DeMoivre's Theorem to raise $1-i\sqrt{3}$ to the seventh power.

$\left(1-i\sqrt{3}\right)^7 = \left[2(\cos 300° + i\sin 300°)\right]^7$

$$= (2)^7\left[\cos(7\cdot 300°)+i\sin(7\cdot 300°)\right]$$

$$= 128(\cos 2100° + i\sin 2100°)$$

$$= 128(\cos 300° + i\sin 300°)$$

$$= 128\left(\dfrac{1}{2}+i\left(-\dfrac{\sqrt{3}}{2}\right)\right)$$

$$= 64 - 64i\sqrt{3}$$

75. Write $-2-2i$ in $r(\cos\theta+i\sin\theta)$ form.

$r = \sqrt{a^2+b^2} = \sqrt{(-2)^2+(-2)^2} = \sqrt{4+4} = 2\sqrt{2}$

$\tan\theta = \dfrac{b}{a} = \dfrac{-2}{-2} = 1$

Because $\tan\dfrac{\pi}{4} = 1$ and θ lies in quadrant III, $\theta = \pi + \dfrac{\pi}{4} = \dfrac{5\pi}{4}$.

$-2-2i = r(\cos\theta+i\sin\theta) = 2\sqrt{2}\left(\cos\dfrac{5\pi}{4}+i\sin\dfrac{5\pi}{4}\right)$

Use DeMoivre's Theorem to raise $-2-2i$ to the fifth power.

$(-2-2i)^5 = \left[2\sqrt{2}\left(\cos\dfrac{5\pi}{4}+i\sin\dfrac{5\pi}{4}\right)\right]^5$

$$= \left(2\sqrt{2}\right)^5\left[\cos\left(5\cdot\dfrac{5\pi}{4}\right)+i\sin\left(5\cdot\dfrac{5\pi}{4}\right)\right]$$

$$= 128\sqrt{2}\left(\cos\dfrac{25\pi}{4}+i\sin\dfrac{25\pi}{4}\right)$$

$$= 128\sqrt{2}\left(\cos\dfrac{\pi}{4}+i\sin\dfrac{\pi}{4}\right)$$

$$= 128\sqrt{2}\left(\dfrac{\sqrt{2}}{2}+i\dfrac{\sqrt{2}}{2}\right)$$

$$= 128 + 128i$$

76. $49(\cos 50° + i \sin 50°)$

$$z_k = \sqrt[2]{49}\left[\cos\left(\frac{50° + 360°k}{2}\right) + i \sin\left(\frac{50° + 360°k}{2}\right)\right],\ k = 0,1$$

$$z_0 = \sqrt{49}\left[\cos\left(\frac{50° + 360° \cdot 0}{2}\right) + i \sin\left(\frac{50° + 360° \cdot 0}{2}\right)\right] = \sqrt{49}\left[\cos\left(\frac{50°}{2}\right) + i \sin\left(\frac{50°}{2}\right)\right]$$

$$= 7(\cos 25° + i \sin 25°)$$

$$z_1 = \sqrt{49}\left[\cos\left(\frac{50° + 360° \cdot 1}{2}\right) + i \sin\left(\frac{50° + 360° \cdot 1}{2}\right)\right] = \sqrt{49}\left[\cos\left(\frac{410°}{2}\right) + i \sin\left(\frac{410°}{2}\right)\right]$$

$$= 7(\cos 205° + i \sin 205°)$$

77. $125(\cos 165° + i \sin 165°)$

$$z_k = \sqrt[3]{125}\left[\cos\left(\frac{165° + 360°k}{3}\right) + i \sin\left(\frac{165° + 360°k}{3}\right)\right],\ k = 0,1,2$$

$$z_0 = \sqrt[3]{125}\left[\cos\left(\frac{165° + 360° \cdot 0}{3}\right) + i \sin\left(\frac{165° + 360° \cdot 0}{3}\right)\right] = \sqrt[3]{125}\left[\cos\left(\frac{165°}{3}\right) + i \sin\left(\frac{165°}{3}\right)\right]$$

$$= 5(\cos 55° + i \sin 55°)$$

$$z_1 = \sqrt[3]{125}\left[\cos\left(\frac{165° + 360° \cdot 1}{3}\right) + i \sin\left(\frac{165° + 360° \cdot 1}{3}\right)\right] = \sqrt[3]{125}\left[\cos\left(\frac{525°}{3}\right) + i \sin\left(\frac{525°}{3}\right)\right]$$

$$= 5(\cos 175° + i \sin 175°)$$

$$z_2 = \sqrt[3]{125}\left[\cos\left(\frac{165° + 360° \cdot 2}{3}\right) + i \sin\left(\frac{165° + 360° \cdot 2}{3}\right)\right] = \sqrt[3]{125}\left[\cos\left(\frac{885°}{3}\right) + i \sin\left(\frac{885°}{3}\right)\right]$$

$$= 5(\cos 295° + i \sin 295°)$$

78. $16\left(\cos\frac{2\pi}{3} + i \sin\frac{2\pi}{3}\right)$

$$z_k = \sqrt[4]{16}\left[\cos\left(\frac{\frac{2\pi}{3} + 2\pi k}{4}\right) + i \sin\left(\frac{\frac{2\pi}{3} + 2\pi k}{4}\right)\right],\ k = 0,1,2,3$$

$$z_0 = \sqrt[4]{16}\left[\cos\left(\frac{\frac{2\pi}{3} + 2\pi \cdot 0}{4}\right) + i \sin\left(\frac{\frac{2\pi}{3} + 2\pi \cdot 0}{4}\right)\right] = \sqrt[4]{16}\left[\cos\left(\frac{\pi}{6}\right) + i \sin\left(\frac{\pi}{6}\right)\right] = 2\left(\frac{\sqrt{3}}{2} + i\frac{1}{2}\right) = \sqrt{3} + i$$

$$z_1 = \sqrt[4]{16}\left[\cos\left(\frac{\frac{2\pi}{3} + 2\pi \cdot 1}{4}\right) + i \sin\left(\frac{\frac{2\pi}{3} + 2\pi \cdot 1}{4}\right)\right] = \sqrt[4]{16}\left(\cos\frac{2\pi}{3} + i \sin\frac{2\pi}{3}\right) = 2\left(-\frac{1}{2} + i\frac{\sqrt{3}}{2}\right) = -1 + i\sqrt{3}$$

$$z_2 = \sqrt[4]{16}\left[\cos\left(\frac{\frac{2\pi}{3} + 2\pi \cdot 2}{4}\right) + i \sin\left(\frac{\frac{2\pi}{3} + 2\pi \cdot 2}{4}\right)\right] = \sqrt[4]{16}\left(\cos\frac{7\pi}{6} + i \sin\frac{7\pi}{6}\right) = 2\left(-\frac{\sqrt{3}}{2} + i\left(-\frac{1}{2}\right)\right) = -\sqrt{3} - i$$

$$z_3 = \sqrt[4]{16}\left[\cos\left(\frac{\frac{2\pi}{3} + 2\pi \cdot 3}{4}\right) + i \sin\left(\frac{\frac{2\pi}{3} + 2\pi \cdot 3}{4}\right)\right] = \sqrt[4]{16}\left(\cos\frac{5\pi}{3} + i \sin\frac{5\pi}{3}\right) = 2\left(\frac{1}{2} + i\left(-\frac{\sqrt{3}}{2}\right)\right) = 1 - i\sqrt{3}$$

79. $8i = 8(\cos 90° + i \sin 90°)$

$$z_k = \sqrt[3]{8}\left[\cos\left(\frac{90° + 360°k}{3}\right) + i\sin\left(\frac{90° + 360°k}{3}\right)\right], \; k = 0,1,2$$

$$z_0 = \sqrt[3]{8}\left[\cos\left(\frac{90° + 360°\cdot 0}{3}\right) + i\sin\left(\frac{90° + 360°\cdot 0}{3}\right)\right] = \sqrt[3]{8}(\cos 30° + i\sin 30°) = 2\left(\frac{\sqrt{3}}{2} + i\frac{1}{2}\right) = \sqrt{3} + i$$

$$z_1 = \sqrt[3]{8}\left[\cos\left(\frac{90° + 360°\cdot 1}{3}\right) + i\sin\left(\frac{90° + 360°\cdot 1}{3}\right)\right] = \sqrt[3]{8}(\cos 150° + i\sin 150°) = 2\left(-\frac{\sqrt{3}}{2} + i\frac{1}{2}\right) = -\sqrt{3} + i$$

$$z_2 = \sqrt[3]{8}\left[\cos\left(\frac{90° + 360°\cdot 2}{3}\right) + i\sin\left(\frac{90° + 360°\cdot 2}{3}\right)\right] = \sqrt[3]{8}(\cos 270° + i\sin 270°) = 2(0 + i(-1)) = -2i$$

80. $-1 = \cos 180° + i \sin 180°$

$$z_k = \sqrt[3]{1}\left[\cos\left(\frac{180° + 360°k}{3}\right) + i\sin\left(\frac{180° + 360°k}{3}\right)\right], \; k = 0,1,2$$

$$z_0 = \sqrt[3]{1}\left[\cos\left(\frac{180° + 360°\cdot 0}{3}\right) + i\sin\left(\frac{180° + 360°\cdot 0}{3}\right)\right] = \sqrt[3]{1}(\cos 60° + i\sin 60°) = 1\left(\frac{1}{2} + i\frac{\sqrt{3}}{2}\right) = \frac{1}{2} + \frac{\sqrt{3}}{2}i$$

$$z_1 = \sqrt[3]{1}\left[\cos\left(\frac{180° + 360°\cdot 1}{3}\right) + i\sin\left(\frac{180° + 360°\cdot 1}{3}\right)\right] = \sqrt[3]{1}(\cos 180° + i\sin 180°) = 1(-1 + i0) = -1$$

$$z_2 = \sqrt[3]{1}\left[\cos\left(\frac{180° + 360°\cdot 2}{3}\right) + i\left(\frac{180° + 360°\cdot 2}{3}\right)\right] = \sqrt[3]{1}(\cos 300° + i\sin 300°) = 1\left(\frac{1}{2} + i\left(-\frac{\sqrt{3}}{2}\right)\right) = \frac{1}{2} - \frac{\sqrt{3}}{2}i$$

81. $-1 - i = \sqrt{2}(\cos 225° + i \sin 225°)$

$$z_k = \sqrt[5]{\sqrt{2}}\left[\cos\left(\frac{225° + 360°k}{5}\right) + i\sin\left(\frac{225° + 360°k}{5}\right)\right], \; k = 0,1,2,3,4$$

$$z_0 = \sqrt[5]{\sqrt{2}}\left[\cos\left(\frac{225° + 360°\cdot 0}{5}\right) + i\sin\left(\frac{225° + 360°\cdot 0}{5}\right)\right] = \sqrt[5]{\sqrt{2}}(\cos 45° + i\sin 45°) = \sqrt[5]{\sqrt{2}}\left(\frac{\sqrt{2}}{2} + i\frac{\sqrt{2}}{2}\right)$$

$$= \frac{\sqrt[5]{8}}{2} + \frac{\sqrt[5]{8}}{2}i$$

$$z_1 = \sqrt[5]{\sqrt{2}}\left[\cos\left(\frac{225° + 360°\cdot 1}{5}\right) + i\sin\left(\frac{225° + 360°\cdot 1}{5}\right)\right] = \sqrt[5]{\sqrt{2}}(\cos 117° + i\sin 117°) \approx -0.49 + 0.95i$$

$$z_2 = \sqrt[5]{\sqrt{2}}\left[\cos\left(\frac{225° + 360°\cdot 2}{5}\right) + i\sin\left(\frac{225° + 360°\cdot 2}{5}\right)\right] = \sqrt[5]{\sqrt{2}}(\cos 189° + i\sin 189°) \approx -1.06 - 0.17i$$

$$z_3 = \sqrt[5]{\sqrt{2}}\left[\cos\left(\frac{225° + 360°\cdot 3}{5}\right) + i\sin\left(\frac{225° + 360°\cdot 3}{5}\right)\right] = \sqrt[5]{\sqrt{2}}(\cos 261° + i\sin 261°) \approx -0.17 - 1.06i$$

$$z_4 = \sqrt[5]{\sqrt{2}}\left[\cos\left(\frac{225° + 360°\cdot 4}{5}\right) + i\sin\left(\frac{225° + 360°\cdot 4}{5}\right)\right] = \sqrt[5]{\sqrt{2}}(\cos 333° + i\sin 333°) \approx 0.95 - 0.49i$$

82.

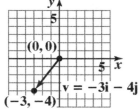

$$\|\mathbf{v}\| = \sqrt{a^2 + b^2}$$
$$= \sqrt{(-3)^2 + (-4)^2}$$
$$= \sqrt{9 + 16}$$
$$= \sqrt{25}$$
$$= 5$$

83.

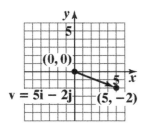

$$\|\mathbf{v}\| = \sqrt{a^2 + b^2}$$
$$= \sqrt{5^2 + (-2)^2}$$
$$= \sqrt{25 + 4}$$
$$= \sqrt{29}$$

84.

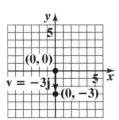

$$\|\mathbf{v}\| = \sqrt{a^2 + b^2}$$
$$= \sqrt{0^2 + (-3)^2}$$
$$= \sqrt{0 + 9}$$
$$= \sqrt{9}$$
$$= 3$$

85. $\mathbf{v} = (x_2 - x_1)\mathbf{i} + (y_2 - y_1)\mathbf{j}$
$$= (5 - 2)\mathbf{i} + [-3 - (-1)]\mathbf{j}$$
$$= 3\mathbf{i} - 2\mathbf{j}$$

86. $\mathbf{v} = (x_2 - x_1)\mathbf{i} + (y_2 - y_1)\mathbf{j}$
$$= [-2 - (-3)\mathbf{i}] + (-2 - 0)\mathbf{j}$$
$$= \mathbf{i} - 2\mathbf{j}$$

87. $\mathbf{v} + \mathbf{w} = (\mathbf{i} - 5\mathbf{j}) + (-2\mathbf{i} + 7\mathbf{j})$
$$= [1 + (-2)]\mathbf{i} + [-5 + 7]\mathbf{j}$$
$$= -\mathbf{i} + 2\mathbf{j}$$

88. $\mathbf{w} - \mathbf{v} = (-2\mathbf{i} + 7\mathbf{j}) - (\mathbf{i} - 5\mathbf{j})$
$$= (-2 - 1)\mathbf{i} + [7 - (-5)]\mathbf{j}$$
$$= -3\mathbf{i} + 12\mathbf{j}$$

89. $6\mathbf{v} - 3\mathbf{w} = 6(\mathbf{i} - 5\mathbf{j}) - 3(-2\mathbf{i} + 7\mathbf{j})$
$$= 6\mathbf{i} - 30\mathbf{j} + 6\mathbf{i} - 21\mathbf{j}$$
$$= 12\mathbf{i} - 51\mathbf{j}$$

90. $\|-2\mathbf{v}\| = |-2| \|\mathbf{v}\|$
$$= 2\|\mathbf{v}\|$$
$$= 2\sqrt{a^2 + b^2}$$
$$= 2\sqrt{1^2 + (-5)^2}$$
$$= 2\sqrt{1 + 25}$$
$$= 2\sqrt{26}$$

91. First, find the magnitude of $\mathbf{v}$.
$$\|\mathbf{v}\| = \sqrt{a^2 + b^2}$$
$$= \sqrt{8^2 + (-6)^2}$$
$$= \sqrt{64 + 36}$$
$$= \sqrt{100}$$
$$= 10$$
A unit vector in the same direction as $\mathbf{v}$ is
$$\frac{\mathbf{v}}{\|\mathbf{v}\|} = \frac{8\mathbf{i} - 6\mathbf{j}}{10} = \frac{4}{5}\mathbf{i} - \frac{3}{5}\mathbf{j}.$$

92. First, find the magnitude of $\mathbf{v}$.
$$\|\mathbf{v}\| = \sqrt{a^2 + b^2} = \sqrt{(-1)^2 + (2)^2} = \sqrt{1 + 4} = \sqrt{5}$$
A unit vector in the same direction as $\mathbf{v}$ is
$$\frac{\mathbf{v}}{\|\mathbf{v}\|} = \frac{-\mathbf{i} + 2\mathbf{j}}{\sqrt{5}} = -\frac{1}{\sqrt{5}}\mathbf{i} + \frac{2}{\sqrt{5}}\mathbf{j}.$$

93. $\mathbf{v} = \|\mathbf{v}\|\cos\theta \, \mathbf{i} + \|\mathbf{v}\|\sin\theta \, \mathbf{j}$

$\qquad = 12\cos 60°\mathbf{i} + 12\sin 60°\mathbf{j}$

$\qquad = 12\left(\dfrac{1}{2}\right)\mathbf{i} + 12\left(\dfrac{\sqrt{3}}{2}\right)\mathbf{j}$

$\qquad = 6\mathbf{i} + 6\sqrt{3}\mathbf{j}$

94. $\quad \mathbf{F}_1 = 100\cos 65°\,\mathbf{i} + 100\sin 65°\,\mathbf{j}$

$\qquad\qquad = 42.3\mathbf{i} + 90.6\mathbf{j}$

$\quad \mathbf{F}_2 = 200\cos 10°\mathbf{i} + 200\sin 10°\mathbf{j}$

$\qquad\quad = 197\mathbf{i} + 34.7\mathbf{j}$

$\mathbf{F}_1 + \mathbf{F}_2 = (42.3 + 197)\mathbf{i} + (90.6 + 34.7)\mathbf{j}$

$\qquad\qquad = 239.3\mathbf{i} + 125.3\mathbf{j}$

$\sqrt{239.3^2 + 125.3^2} \approx 270$ pounds

$\qquad\qquad \cos\theta = \dfrac{239.3}{270}$

$\qquad\qquad\qquad \theta = 27.7°$

95. $\mathbf{v} = 15\cos 25°\,\mathbf{i} + 15\sin 25°\,\mathbf{j} = 13.59\mathbf{i} + 6.34\mathbf{j}$

$\quad \mathbf{w} = 4\cos 270°\,\mathbf{i} + 4\sin 270°\,\mathbf{j} = -4\mathbf{j}$

a. $\quad 13.59\mathbf{i} + (6.34 - 4)\mathbf{j} = 13.59\mathbf{i} + 2.34\mathbf{j}$

b. $\quad \sqrt{13.6^2 + 2.3^2} \approx 14$ mph

c. $\quad \cos\theta = \dfrac{13.59}{14}; \theta = 13.9°$

96. $\mathbf{v}\cdot(\mathbf{v} + \mathbf{w}) = (5\mathbf{i} + 2\mathbf{j})[(\mathbf{i} - \mathbf{j}) + (3\mathbf{i} - 7\mathbf{j})]$

$\qquad\qquad = (5\mathbf{i} + 2\mathbf{j})\cdot[4\mathbf{i} - 8\mathbf{j}]$

$\qquad\qquad = 5(4) + 2(-8)$

$\qquad\qquad = 20 - 16$

$\qquad\qquad = 4$

97. $\quad \mathbf{v}\cdot\mathbf{w} = (2\mathbf{i} + 3\mathbf{j})\cdot(7\mathbf{i} - 4\mathbf{j}) = 2(7) + 3(-4) = 2$

$\cos\theta = \dfrac{2}{\sqrt{2^2 + 3^2}\,\sqrt{7^2 + (-4)^2}}$

$\qquad = \dfrac{2}{\sqrt{13}\sqrt{65}}$

$\qquad = \dfrac{2}{\sqrt{845}}$

The angle θ between the vectors is

$\theta = \cos^{-1}\left(\dfrac{2}{\sqrt{845}}\right) \approx 86.1°$.

98. $\quad \mathbf{v}\cdot\mathbf{w} = (2\mathbf{i} + 4\mathbf{j})\cdot(6\mathbf{i} - 11\mathbf{j}) = 2(6) + 4(-11)$

$\qquad\qquad = 12 - 44$

$\qquad\qquad = -32$

$\cos\theta = \dfrac{-32}{\sqrt{2^2 + 4^2}\,\sqrt{6^2 + (-11)^2}}$

$\qquad = \dfrac{-32}{\sqrt{20}\sqrt{157}}$

$\qquad = \dfrac{-32}{\sqrt{3140}}$

The angle θ between the vectors is

$\theta = \cos^{-1}\left(-\dfrac{32}{\sqrt{3140}}\right) \approx 124.8°$.

99. $\quad \mathbf{v}\cdot\mathbf{w} = (2\mathbf{i} + \mathbf{j})\cdot(\mathbf{i} - \mathbf{j}) = 2(1) + 1(-1)$

$\qquad\qquad = 2 - 1 = 1$

$\cos\theta = \dfrac{1}{\sqrt{2^2 + 1^2}\,\sqrt{1^2 + (-1)^2}}$

$\qquad = \dfrac{1}{\sqrt{5}\sqrt{2}}$

$\qquad = \dfrac{1}{\sqrt{10}}$

The angle θ between the vectors is

$\theta = \cos^{-1}\left(\dfrac{1}{\sqrt{10}}\right) \approx 71.6°$.

100. $\mathbf{v}\cdot\mathbf{w} = (12\mathbf{i} - 8\mathbf{j})\cdot(2\mathbf{i} + 3\mathbf{j})$

$\qquad\qquad = 12(2) + (-8)(3)$

$\qquad\qquad = 24 - 24$

$\qquad\qquad = 0$

The dot product is zero. Thus, the given vectors are orthogonal.

101. $\mathbf{v}\cdot\mathbf{w} = (\mathbf{i} + 3\mathbf{j})\cdot(-3\mathbf{i} - \mathbf{j})$

$\qquad\qquad = 1(-3) + 3(-1)$

$\qquad\qquad = -3 - 3$

$\qquad\qquad = -6$

The dot product is not zero. Thus, the given vectors are not orthogonal.

102. $\text{proj}_{\mathbf{W}}\mathbf{v} = \dfrac{\mathbf{v} \cdot \mathbf{w}}{\|\mathbf{w}\|^2}\mathbf{w}$

$$= \dfrac{(-2\mathbf{i}+5\mathbf{j}) \cdot (5\mathbf{i}+4\mathbf{j})}{\left(\sqrt{5^2+4^2}\right)^2}\mathbf{w}$$

$$= \dfrac{-2(5)+5(4)}{\left(\sqrt{41}\right)^2}\mathbf{w}$$

$$= \dfrac{10}{41}(5\mathbf{i}+4\mathbf{j})$$

$$= \dfrac{50}{41}\mathbf{i}+\dfrac{40}{41}\mathbf{j}$$

$$\mathbf{v}_1 = \text{proj}_{\mathbf{W}}\mathbf{v} = \dfrac{50}{41}\mathbf{i}+\dfrac{40}{41}\mathbf{j}$$

$$\mathbf{v}_2 = \mathbf{v}-\mathbf{v}_1 = (-2\mathbf{i}+5\mathbf{j})-\left(\dfrac{50}{41}\mathbf{i}+\dfrac{40}{41}\mathbf{j}\right)$$

$$= -\dfrac{132}{41}\mathbf{i}+\dfrac{165}{41}\mathbf{j}$$

103. $\text{proj}_{\mathbf{W}}\mathbf{v} = \dfrac{\mathbf{v} \cdot \mathbf{w}}{\|\mathbf{w}\|^2}\mathbf{w}$

$$= \dfrac{(-\mathbf{i}+2\mathbf{j}) \cdot (3\mathbf{i}-\mathbf{j})}{\left(\sqrt{3^2+(-1)^2}\right)^2}\mathbf{w}$$

$$= \dfrac{-1(3)+2(-1)}{\left(\sqrt{10}\right)^2}\mathbf{w}$$

$$= \dfrac{-5}{10}\mathbf{w}$$

$$= -\dfrac{1}{2}(3\mathbf{i}-1\mathbf{j})$$

$$= \dfrac{-3}{2}\mathbf{i}+\dfrac{1}{2}\mathbf{j}$$

$$\mathbf{v}_1 = \text{proj}_{\mathbf{W}}\mathbf{v} = \dfrac{-3}{2}\mathbf{i}+\dfrac{1}{2}\mathbf{j}$$

$$\mathbf{v}_2 = \mathbf{v}-\mathbf{v}_1 = (-\mathbf{i}+2\mathbf{j})-\left(\dfrac{-3}{2}\mathbf{i}+\dfrac{1}{2}\mathbf{j}\right)$$

$$= \dfrac{1}{2}\mathbf{i}+\dfrac{3}{2}\mathbf{j}$$

104. $\mathrm{W} = \|\mathbf{F}\|\left\|\overrightarrow{AB}\right\|\cos\theta$

$$= (30)(50)\cos 42°$$

$$\approx 1115$$

The work done is approximately 1115 foot-pounds.

Chapter 7 Test

1. The known ratio is $\dfrac{a}{\sin A}$, or $\dfrac{4.8}{\sin 34°}$. Because angle B is given, we use the Law of Sines to find side b

$$\dfrac{b}{\sin B} = \dfrac{a}{\sin A}$$

$$\dfrac{b}{\sin 68°} = \dfrac{4.8}{\sin 34°}$$

$$b = \dfrac{4.8\sin 68°}{\sin 34°} \approx 8.0$$

2. Use the Law of Cosines to find c.

$$c^2 = a^2 + b^2 - 2ab\cos C$$

$$c^2 = 5^2 + 6^2 - 2(5)(6)\cos 68°$$

$$= 61 - 60\cos 68°$$

$$\approx 38.52$$

$$c \approx \sqrt{38.52} \approx 6.2$$

3. $s = \dfrac{1}{2}(a+b+c) = \dfrac{1}{2}(17+45+32) = 47$

$$\text{Area} = \sqrt{s(s-a)(s-b)(s-c)}$$

$$= \sqrt{47(47-17)(47-45)(47-32)}$$

$$= \sqrt{42,300} \approx 206$$

The area of the triangle is approximately 206 square inches.

4. Draw $\theta = \dfrac{5\pi}{4} = 225°$ counterclockwise, since θ is positive, from the polar axis. Go 4 units out on the terminal side of θ, since $r > 0$.

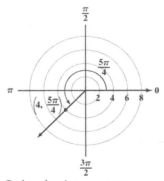

Ordered pairs may vary.

5. $(1, -1)$

$$r = \sqrt{1^2 + (-1)^2} = \sqrt{1+1} = \sqrt{2}$$

$$\tan\theta = \frac{-1}{1} = -1$$

Because $\tan\dfrac{\pi}{4} = 1$ and θ lies in quadrant IV, $\theta = 2\pi - \dfrac{\pi}{4} = \dfrac{7\pi}{4}$.

The polar coordinated of $(1, -1)$ are $(r, \theta) = \left(\sqrt{2}, \dfrac{7\pi}{4}\right)$.

6.
$$x^2 + (y+8)^2 = 64$$
$$(r\cos\theta)^2 + (r\sin\theta + 8)^2 = 64$$
$$r^2\cos^2\theta + r^2\sin^2\theta + 16r\sin\theta + 64 = 64$$
$$r^2 + 16r\sin\theta = 0$$
$$r^2 = -16r\sin\theta$$
$$r = -16\sin\theta$$

7.
$$r = -4\sec\theta$$
$$r = \frac{-4}{\cos\theta}$$
$$r\cos\theta = -4$$
$$x = -4$$

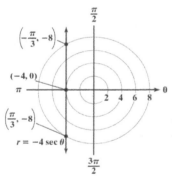

8. $r = 1 + \sin\theta$

Check for symmetry:

Polar Axis	The Line $\theta = \dfrac{\pi}{2}$	The Pole
$r = 1 + \sin(-\theta)$	$-r = 1 + \sin(-\theta)$	$-r = 1 + \sin\theta$
	$-r = 1 - \sin\theta$	
$r = 1 - \sin\theta$	$r = -1 + \sin\theta$	$r = -1 - \sin\theta$

There may be no symmetry, since each equation is not equivalent to $r = 1 + \sin\theta$.
Calculate values of r for θ from 0 to 2π.

θ	0	$\dfrac{\pi}{6}$	$\dfrac{\pi}{3}$	$\dfrac{\pi}{2}$	$\dfrac{2\pi}{3}$	$\dfrac{5\pi}{6}$	π	$\dfrac{7\pi}{6}$	$\dfrac{4\pi}{3}$	$\dfrac{3\pi}{2}$	$\dfrac{5\pi}{3}$	$\dfrac{11\pi}{6}$	2π
r	1	1.5	1.87	2	1.87	1.5	1	0.5	0.13	0	0.13	0.5	1

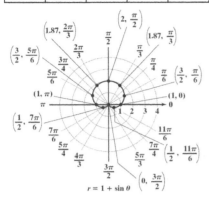

9. $r = 1 + 3\cos\theta$

Check for symmetry:

Polar Axis	**The Line** $\theta = \dfrac{\pi}{2}$	**The Pole**
$r = 1 + 3\cos(-\theta)$	$-r = 1 + 3\cos(-\theta)$	$-r = 1 + 3\cos\theta$
	$-r = 1 + 3\cos\theta$	
$r = 1 + 3\cos\theta$	$r = -1 - 3\cos\theta$	$r = -1 - 3\cos\theta$

The graph has symmetry with respect to the polar axis. The graph may or may not be symmetric with respect to the line

$\theta = \dfrac{\pi}{2}$ or the pole. Calculate values of r for θ from 0 to π and use symmetry to complete the graph.

θ	0	$\dfrac{\pi}{6}$	$\dfrac{\pi}{3}$	$\dfrac{\pi}{2}$	$\dfrac{2\pi}{3}$	$\dfrac{5\pi}{6}$	π
r	4	3.6	2.5	1	–0.5	–1.6	–2

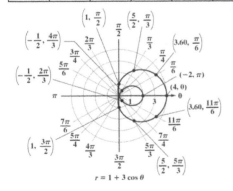

10. Use $r = \sqrt{a^2 + b^2}$ and $\tan \theta = \dfrac{b}{a}$, with $a = -\sqrt{3}$ and $b = 1$, to find r and θ.

$$r = \sqrt{\left(-\sqrt{3}\right)^2 + (1)^2} = \sqrt{3+1} = \sqrt{4} = 2$$

$$\tan \theta = \frac{1}{-\sqrt{3}} = -\frac{1}{\sqrt{3}}$$

Because $\tan 30° = \dfrac{1}{\sqrt{3}}$ and θ lies in quadrant II, $\theta = 180° - 30° = 150°$.

The polar form of $z = -\sqrt{3} + i$ is $z = r(\cos \theta + i \sin \theta) = 2(\cos 150° + i \sin 150°)$ or $2\left(\cos\dfrac{5\pi}{6} + i \sin \dfrac{5\pi}{6}\right)$.

11. $5(\cos 15° + i \sin 15°) \cdot 10(\cos 5° + i \sin 5°) = (5 \cdot 10)\left[\cos(15° + 5°) + i \sin(15° + 5°)\right]$
$$= 50(\cos 20° + i \sin 20°)$$

12. $\dfrac{2\left(\cos\dfrac{\pi}{2} + i \sin \dfrac{\pi}{2}\right)}{4\left(\cos\dfrac{\pi}{3} + i \sin \dfrac{\pi}{3}\right)} = \dfrac{2}{4}\left[\cos\left(\dfrac{\pi}{2} - \dfrac{\pi}{3}\right) + i \sin\left(\dfrac{\pi}{2} - \dfrac{\pi}{3}\right)\right] = \dfrac{2}{4}\left[\cos\left(\dfrac{3\pi}{6} - \dfrac{2\pi}{6}\right) + i \sin\left(\dfrac{3\pi}{6} - \dfrac{2\pi}{6}\right)\right]$

$$= \dfrac{1}{2}\left(\cos\dfrac{\pi}{6} + i \sin \dfrac{\pi}{6}\right)$$

13. $\left[2(\cos 10° + i \sin 10°)\right]^5 = (2)^5\left[\cos(5 \cdot 10°) + i \sin(5 \cdot 10°)\right] = 32(\cos 50° + i \sin 50°)$

14. $27 = 27(\cos 0° + i \sin 0°)$

$$z_k = \sqrt[3]{27}\left[\cos\left(\dfrac{0° + 360°k}{3}\right) + i \sin\left(\dfrac{0° + 360°k}{3}\right)\right], \ k = 0, 1, 2$$

$$z_0 = \sqrt[3]{27}\left[\cos\left(\dfrac{0° + 360° \cdot 0}{3}\right) + i \sin\left(\dfrac{0° + 360° \cdot 0}{3}\right)\right] = \sqrt[3]{27}(\cos 0° + i \sin 0°) = 3(1 + 0i) = 3$$

$$z_1 = \sqrt[3]{27}\left[\cos\left(\dfrac{0° + 360° \cdot 1}{3}\right) + i \sin\left(\dfrac{0° + 360° \cdot 1}{3}\right)\right] = \sqrt[3]{27}(\cos 120° + i \sin 120°) = 3\left(-\dfrac{1}{2} + i\dfrac{\sqrt{3}}{2}\right) = -\dfrac{3}{2} + \dfrac{3\sqrt{3}}{2}i$$

$$z_2 = \sqrt[3]{27}\left[\cos\left(\dfrac{0° + 360° \cdot 2}{3}\right) + i \sin\left(\dfrac{0° + 360° \cdot 2}{3}\right)\right] = \sqrt[3]{27}(\cos 240° + i \sin 240°) = 3\left(-\dfrac{1}{2} - \dfrac{\sqrt{3}}{2}i\right) = -\dfrac{3}{2} - \dfrac{3\sqrt{3}}{2}i$$

15. a. $\mathbf{v} = \left(x_2 - x_1\right)\mathbf{i} + \left(y_2 - y_1\right)\mathbf{j}$
$\mathbf{v} = [-1 - (-2)]\mathbf{i} + (5 - 3)\mathbf{j} = \mathbf{i} + 2\mathbf{j}$

b. $\|\mathbf{v}\| = \sqrt{a^2 + b^2} = \sqrt{1^2 + 2^2} = \sqrt{1 + 4} = \sqrt{5}$

16. $3\mathbf{v} - 4\mathbf{w} = 3(-5\mathbf{i} + 2\mathbf{j}) - 4(2\mathbf{i} - 4\mathbf{j}) = -15\mathbf{i} + 6\mathbf{j} - 8\mathbf{i} + 16\mathbf{j}$
$= (-15 - 8)\mathbf{i} + (6 + 16)\mathbf{j} = -23\mathbf{i} + 22\mathbf{j}$

17. $\mathbf{v} \cdot \mathbf{w} = (-5\mathbf{i} + 2\mathbf{j}) \cdot (2\mathbf{i} - 4\mathbf{j}) = -5(2) + 2(-4) = -10 - 8 = -18$

18. $\cos\theta = \dfrac{\mathbf{v}\cdot\mathbf{w}}{\|\mathbf{v}\|\|\mathbf{w}\|}\mathbf{w}$

$$= \dfrac{(-5\mathbf{i}+2\mathbf{j})\cdot(2\mathbf{i}-4\mathbf{j})}{\sqrt{(-5)^2+2^2}\sqrt{2^2+(-4)^2}}$$

$$= \dfrac{-5(2)+2(-4)}{\sqrt{29}\sqrt{20}}$$

$$= -\dfrac{18}{\sqrt{580}}$$

The angle θ between the vectors is $\theta = \cos^{-1}\left(-\dfrac{18}{\sqrt{580}}\right) \approx 138°$.

19. $\text{proj}_{\mathbf{W}}\,\mathbf{v} = \dfrac{\mathbf{v}\cdot\mathbf{w}}{\|\mathbf{w}\|^2}\mathbf{w}$

$$= \dfrac{(-5\mathbf{i}+2\mathbf{j})\cdot(2\mathbf{i}-4\mathbf{j})}{\left(\sqrt{2^2+(-4)^2}\right)^2}\mathbf{w}$$

$$= \dfrac{-5(2)+2(-4)}{\left(\sqrt{20}\right)^2}\mathbf{w}$$

$$= -\dfrac{18}{20}\mathbf{w}$$

$$= -\dfrac{9}{10}(2\mathbf{i}-4\mathbf{j})$$

$$= -\dfrac{9}{5}\mathbf{i}+\dfrac{18}{5}\mathbf{j}$$

20.

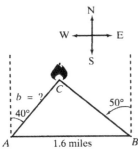

Using the figure,
$B = 90° - 50° = 40°$

$A = 90° - 40° = 50°$

$C = 180° - B - A = 180° - 40° - 50° = 90°$

Use the Law of Sines to find b. $\dfrac{b}{\sin B} = \dfrac{c}{\sin C}$

$$\dfrac{b}{\sin 40°} = \dfrac{1.6}{\sin 90°}$$

$$b = \dfrac{1.6\sin 40°}{\sin 90°} \approx 1.0$$

The fire is about 1.0 mile from the station.

21.
$$\mathbf{F}_1 = 250 \cos 30° \mathbf{i} + 250 \sin 30° \mathbf{j}$$
$$= 216.5\mathbf{i} + 125\mathbf{j}$$

$$\mathbf{F}_2 = 150 \cos 315° \mathbf{i} + 150 \sin 315° \mathbf{j}$$
$$= 106\mathbf{i} - 106\mathbf{j}$$

$$\mathbf{F}_1 + \mathbf{F}_2 = (216.5 + 106)\mathbf{i} + (125 - 106)\mathbf{j}$$
$$= 322.5\mathbf{i} + 19\mathbf{j}$$

$$\| \mathbf{F}_1 + \mathbf{F}_2 \| = \sqrt{322.5^2 + 19^2} \approx 323 \text{ pounds}$$

$$\cos \theta = \frac{322.5}{323} = 3.4°$$

22. $W = \| \mathbf{F} \| \| \overrightarrow{AB} \| \cos \theta$

$$= (40)(60) \cos 35° \approx 1966$$

The work done is approximately 1966 foot-pounds.

Cumulative Review Exercises (Chapters 1–7)

1. $x^4 - x^3 - x^2 - x - 2 = 0$

$$\frac{p}{q} : \pm \frac{2}{1}, \pm \frac{1}{1}$$

$$\begin{array}{r|rrrrr}
-1 & 1 & -1 & -1 & -1 & -2 \\
   &   & -1 & 2 & -1 & 2 \\
\hline
   & 1 & -2 & 1 & -2 & 0
\end{array}$$

$$x^4 - x^3 - x^2 - x - 2 = 0$$
$$(x+1)(x^3 - 2x^2 + x - 2) = 0$$
$$(x+1)[x^2(x-2) + 1(x-2)] = 0$$
$$(x+1)(x-2)(x^2+1) = 0$$

$$x+1 = 0 \quad x-2 = 0 \quad x^2 + 1 = 0$$
$$x = -1 \quad\quad x = 2 \quad\quad x^2 = -1$$
$$x = \pm i$$

The solution set is $\{-1, 2\ i, -i\}$.

2. $2\sin^2 \theta - 3\sin \theta + 1 = 0, \ 0 \le \theta < 2\pi$

$$(2\sin \theta - 1)(\sin \theta - 1) = 0$$

$$2\sin \theta - 1 = 0 \quad \text{or} \quad \sin \theta - 1 = 0$$
$$2\sin \theta = 1 \quad\quad\quad \sin \theta = 1$$
$$\sin \theta = \frac{1}{2}$$

The solutions in the interval $\left[0, 2\pi \right)$ are $\dfrac{\pi}{6}, \dfrac{5\pi}{6}$, and $\dfrac{\pi}{2}$.

3. Begin by solving the related quadratic equation. Thus, we will solve $x^2 + 2x + 3 = 11$.

$$x^2 + 2x + 3 = 11$$
$$x^2 + 2x - 8 = 0$$
$$(x + 4)(x - 2) = 0$$
$$x + 4 = 0 \quad \text{or} \quad x - 2 = 0$$
$$x = -4 \quad \text{or} \quad x = 2$$

The boundary points are –4 and 2. The boundary points divide the number line into three test intervals, namely $(-\infty, -4)$, $(-4, 2)$, and $(2, \infty)$. Take one representative number within each test interval and substitute that number into the original inequality.

Test Interval	Representative Number	Substitute into $x^2 + 2x + 3 > 11$	Conclusion
$(-\infty, -4)$	–5	$(-5)^2 + 2(-5) + 3 > 11$ $18 > 11$ True	$(-\infty, -4)$ belongs to the solution set.
$(-4, -2)$	0	$0^2 + 2(0) + 3 > 11$ $3 > 11$ False	$(-4, -2)$ does not belong to the solution set.
$(2, \infty)$	3	$3^2 + 2(3) + 3 > 11$ $18 > 11$ True	$(2, \infty)$ belongs to the solution set.

The solution is $(-\infty, -4) \cup (2, \infty)$.

4. $\sin\theta\cos\theta = -\dfrac{1}{2}$

$$\frac{\sin 2\theta}{2} = -\frac{1}{2}$$
$$\sin 2\theta = -1$$

The period of the sine function is 2π. In the interval $[0, 2\pi)$, the only value for which the sine function is –1 is $\dfrac{3\pi}{2}$.

This means that $2\theta = \dfrac{3\pi}{2}$. Since the period is 2π, all the solutions to $\sin 2\theta = -1$ are given by $2\theta = \dfrac{3\pi}{2} + 2n\pi$

$$\theta = \frac{3\pi}{4} + n\pi$$

where n is any integer.

The solution in the interval $[0, 2\pi)$ is obtained by letting $n = 0$ and $n = 1$. The solutions are $\dfrac{3\pi}{4}$ and $\dfrac{7\pi}{4}$.

5. The equation $y = 3\sin(2x - \pi)$ is of the form $y = A\sin(Bx - C)$ with $A = 3$, $B = 2$, and $C = \pi$. The amplitude is $|A| = |3| = 3$. The period is $\dfrac{2\pi}{B} = \dfrac{2\pi}{2} = \pi$. The phase shift is $\dfrac{C}{B} = \dfrac{\pi}{2}$. The quarter-period is $\dfrac{\pi}{4}$. The cycle begins at $x = \dfrac{\pi}{2}$. Add quarter-periods to generate x-values for the key points.

$$x = \frac{\pi}{2}$$

$$x = \frac{\pi}{2} + \frac{\pi}{4} = \frac{3\pi}{4}$$

$$x = \frac{3\pi}{4} + \frac{\pi}{4} = \pi$$

$$x = \pi + \frac{\pi}{4} = \frac{5\pi}{4}$$

$$x = \frac{5\pi}{4} + \frac{\pi}{4} = \frac{3\pi}{2}$$

We evaluate the function at each value of x.

x	$y = 3\sin(2x - \pi)$	coordinates
$\dfrac{\pi}{2}$	$y = 3\sin\left(2 \cdot \dfrac{\pi}{2} - \pi\right)$ $= 3\sin(\pi - \pi)$ $= 3\sin 0 = 3 \cdot 0 = 0$	$\left(\dfrac{\pi}{2}, 0\right)$
$\dfrac{3\pi}{4}$	$y = 3\sin\left(2 \cdot \dfrac{3\pi}{4} - \pi\right)$ $= 3\sin\left(\dfrac{3\pi}{2} - \pi\right)$ $= 3\sin\dfrac{\pi}{2} = 3 \cdot 1 = 3$	$\left(\dfrac{3\pi}{4}, 3\right)$
π	$y = 3\sin(2 \cdot \pi - \pi)$ $= 3\sin(2\pi - \pi)$ $= 3\sin \pi = 3 \cdot 0 = 0$	$(\pi, 0)$
$\dfrac{5\pi}{4}$	$y = 3\sin\left(2 \cdot \dfrac{5\pi}{4} - \pi\right)$ $= 3\sin\left(\dfrac{5\pi}{2} - \pi\right)$ $= 3\sin\dfrac{3\pi}{2}$ $= 3(-1) = -3$	$\left(\dfrac{5\pi}{4}, -3\right)$
$\dfrac{3\pi}{2}$	$y = 3\sin\left(2 \cdot \dfrac{3\pi}{2} - \pi\right)$ $= 3\sin(3\pi - \pi)$ $= 3\sin 2\pi = 3 \cdot 0 = 0$	$\left(\dfrac{3\pi}{2}, 0\right)$

Connect the five points with a smooth curve and graph one complete cycle of the given function.

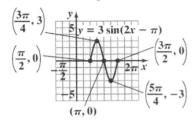

6. The equation $y = -4\cos \pi x$ is of the form $y = A\cos Bx$ with $A = -4$, and $B = \pi$. Thus, the amplitude is
    $|A| = |-4| = 4$.

    The period is $\dfrac{2\pi}{B} = \dfrac{2\pi}{\pi} = 2$.

    Find the x-values for the five key points by dividing the period, 2, by 4, $\dfrac{\text{period}}{4} = \dfrac{2}{4} = \dfrac{1}{2}$, then by adding quarter- periods to the value of x where the cycle begins, $x = 0$. The five x-values are

    $x = 0$

    $x = 0 + \dfrac{1}{2} = \dfrac{1}{2}$

    $x = \dfrac{1}{2} + \dfrac{1}{2} = 1$

    $x = 1 + \dfrac{1}{2} = \dfrac{3}{2}$

    $x = \dfrac{3}{2} + \dfrac{1}{2} = 2$

    We evaluate the function at each value of x.

x	$y = -4\cos \pi x$	coordinates
0	$y = -4\cos(\pi \cdot 0)$ $= -4\cos 0$ $= -4 \cdot 1 = -4$	$(0, -4)$
$\dfrac{1}{2}$	$y = -4\cos\left(\pi \cdot \dfrac{1}{2}\right)$ $= -4\cos\dfrac{\pi}{2}$ $= -4 \cdot 0 = 0$	$\left(\dfrac{1}{2}, 0\right)$
1	$y = -4\cos(\pi \cdot 1)$ $= -4\cos \pi$ $= -4 \cdot (-1) = 4$	$(1, 4)$
$\dfrac{3}{2}$	$y = -4\cos\left(\pi \cdot \dfrac{3}{2}\right)$ $= -4\cos\dfrac{3\pi}{2}$ $= -4 \cdot 0 = 0$	$\left(\dfrac{3}{2}, 0\right)$
2	$y = -4\cos(\pi \cdot 2)$ $= -4\cos 2\pi$ $= -4 \cdot 1 = -4$	$(2, -4)$

Connect the five key points with a smooth curve and graph one complete cycle of the given function.

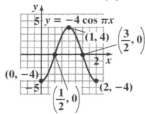

7. $\sin\theta\csc\theta - \cos^2\theta = \sin\theta\left(\dfrac{1}{\sin\theta}\right) - \cos^2\theta$

$$= 1 - \cos^2\theta$$
$$= \sin^2\theta$$

8. $\cos\left(\theta + \dfrac{3\pi}{2}\right) = \cos\theta\cos\dfrac{3\pi}{2} - \sin\theta\sin\dfrac{3\pi}{2}$

$$= \cos\theta(0) - \sin\theta(-1)$$
$$= \sin\theta$$

9. $2x + 4y - 8 = 0$

$$4y = -2x + 8$$
$$\dfrac{4y}{4} = \dfrac{-2x + 8}{4}$$
$$y = -\dfrac{1}{2}x + 2$$

The slope is $-\dfrac{1}{2}$, and the *y*-intercept is 2.

10. $2\sin\dfrac{\pi}{3} - 3\tan\dfrac{\pi}{6} = 2\left(\dfrac{\sqrt{3}}{2}\right) - 3\left(\dfrac{1}{\sqrt{3}}\right)$

$$= \sqrt{3} - \dfrac{3}{\sqrt{3}}$$
$$= \sqrt{3} - \sqrt{3}$$
$$= 0$$

11. Let $\theta = \tan^{-1}\left(\dfrac{1}{2}\right)$, then $\tan\theta = \dfrac{1}{2}$. Because $\tan\theta$ is positive, θ is in the first quadrant.

Use the Pythagorean Theorem to find *r*.
$$r = \sqrt{1^2 + 2^2} = \sqrt{1 + 4} = \sqrt{5}$$

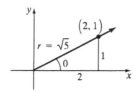

Use the right triangle to find the exact value.
$$\sin\left(\tan^{-1}\dfrac{1}{2}\right) = \sin\theta = \dfrac{1}{\sqrt{5}} = \dfrac{\sqrt{5}}{5}$$

12. $f(x) = \sqrt{5-x}$

$5 - x \geq 0$

$-x \geq -5$

$x \leq 5$

The domain of the function is $(-\infty, 5]$.

13.
$$g(x) = \frac{x-3}{x^2-9}$$

$x^2 - 9 = 0$

$(x-3)(x+3) = 0$

$x - 3 = 0 \quad \text{or} \quad x + 3 = 0$

$x = 3 \qquad\qquad x = -3$

The domain of the function is

$(-\infty, -3) \cup (-3, 3) \cup (3, \infty)$.

14. $s(t) = -16t^2 + 48t + 8$

$= -16\left(t^2 - 3t - \dfrac{1}{2}\right)$

$= -16\left(t^2 - 3t + \dfrac{9}{4} - \dfrac{1}{2} - \dfrac{9}{4}\right)$

$= -16\left[\left(t - \dfrac{3}{2}\right)^2 - \dfrac{1}{2} - \dfrac{9}{4}\right]$

$= -16\left(t - \dfrac{3}{2}\right)^2 + 44$

The ball reaches its maximum height after the first 1.5 seconds. The maximum height is 44 feet.

15. $d = 4\sin 5t$ is of the form $d = a\sin \omega t$ with $a = 4$ and $\omega = 5$.

a. $|a| = |4| = 4$

The maximum displacement is 4 meters.

b. $f = \dfrac{\omega}{2\pi} = \dfrac{5}{2\pi}$

The frequency is $\dfrac{5}{2\pi}$ cycle per second.

c. $\text{period} = \dfrac{2\pi}{\omega} = \dfrac{2\pi}{5}$

$\dfrac{2\pi}{5}$ seconds are required for one cycle.

16. Because $22.5°$ lies in quadrant I, $\cos 22.5° > 0$.

$\cos 22.5° = \cos \dfrac{45°}{2}$

$= \sqrt{\dfrac{1 + \cos 45°}{2}}$

$= \sqrt{\dfrac{1 + \frac{\sqrt{2}}{2}}{2}}$

$= \sqrt{\dfrac{2 + \sqrt{2}}{4}}$

$= \dfrac{\sqrt{\sqrt{2} + 2}}{2}$

17. a. $3\mathbf{v} - \mathbf{w} = 3(2\mathbf{i} + 7\mathbf{j}) - (\mathbf{i} - 2\mathbf{j})$

$= 6\mathbf{i} + 21\mathbf{j} - \mathbf{i} + 2\mathbf{j}$

$= 5\mathbf{i} + 23\mathbf{j}$

b. $\mathbf{v} \cdot \mathbf{w} = (2\mathbf{i} + 7\mathbf{j}) \cdot (\mathbf{i} - 2\mathbf{j})$

$= 2(1) + 7(-2) = 2 - 14$

$= -12$

18. $\dfrac{1}{2}\log_b x - \log_b(x^2 + 1)$

$= \log_b x^{1/2} - \log_b(x^2 + 1)$

$= \log_b \sqrt{x} - \log_b(x^2 + 1)$

$= \log_b \dfrac{\sqrt{x}}{x^2 + 1}$

19. $(4, -1)$ and $(-8, 5)$

$m = \dfrac{5 - (-1)}{-8 - 4} = \dfrac{6}{-12} = -\dfrac{1}{2}$

$y - (-1) = -\dfrac{1}{2}(x - 4)$

$y + 1 = -\dfrac{1}{2}x + 2$

$y = -\dfrac{1}{2}x + 1$

20. $L = A\left(1 - e^{-kt}\right)$

a.
$$20 = 300\left(1 - e^{-k(5)}\right)$$
$$20 = 300 - 300e^{-5k}$$
$$300e^{-5k} = 280$$
$$e^{-5k} = \frac{14}{15}$$
$$\ln\left(e^{-5k}\right) = \ln\left(\frac{14}{15}\right)$$
$$-5k = \ln\left(\frac{14}{15}\right)$$
$$k = -\frac{\ln\left(\frac{14}{15}\right)}{5} \approx 0.014$$

b. $L = 300\left(1 - e^{-0.014(20)}\right) \approx 73$

After 20 minutes, the student will have learned approximately 73 words.

$$260 = 300\left(1 - e^{-0.014t}\right)$$
$$\frac{13}{15} = 1 - e^{-0.014t}$$
$$-\frac{2}{15} = -e^{-0.014t}$$
$$\frac{2}{15} = e^{-0.014t}$$
$$\ln\left(\frac{2}{15}\right) = \ln\left(e^{-0.014t}\right)$$
$$\ln\left(\frac{2}{15}\right) = -0.014t$$
$$t = -\frac{\ln\left(\frac{2}{15}\right)}{0.014} \approx 144$$

It will take about 144 minutes.

Chapter 8
Systems of Equations and Inequalities

Section 8.1

Check Point Exercises

1. a.
$$2x = 3y = -4$$
$$2(1) - 3(2) = -4$$
$$2 - 6 = -4$$
$$-4 = -4 \text{ true}$$
$$2x + y = 4$$
$$2(1) + 2 = 4$$
$$2 + 2 = 4$$
$$4 = 4 \text{ true}$$
$(1, 2)$ is a solution of the system.

b.
$$2x = 3y = -4$$
$$2(7) - 3(6) = -4$$
$$14 - 18 = -4$$
$$-4 = -4 \text{ true}$$
$$2x + y = 4$$
$$2(7) + 6 = 4$$
$$14 + 6 = 4$$
$$20 = 4 \text{ false}$$
$(7, 6)$ is not a solution of the system.

2. $3x + 2y = 4$
$2x + y = 1$
Solve $2x + y = 1$ for y.
$2x + y = 1$
$$y = 1 - 2x$$
Substitute $1 - 2x$ for y in the other equation and solve.
$$3x + 2\overbrace{(1 - 2x)}^{y} = 4$$
$$3x + 2 - 4x = 4$$
$$-x = 2$$
$$x = -2$$
Back-substitute the obtained value:
$$3x + 2y = 4$$
$$3(-2) + 2y = 4$$
$$-6 + 2y = 4$$
$$2y = 10$$
$$y = 5$$
Checking confirms the solution set is $\{(-2, 5)\}$.

3. Rewrite one or both equations:
$$4x + 5y = 3 \xrightarrow{\text{No change}} 4x + 5y = 3$$
$$2x - 3y = 7 \xrightarrow{\text{Mult. by } -2} \underline{-4x + 6y = -14}$$
$$11y = -11$$
$$y = -1$$
Back-substitute into either equation:
$$4x + 5y = 3$$
$$4x + 5(-1) = 3$$
$$4x - 5 = 3$$
$$4x = 8$$
$$x = 2$$
Checking confirms the solution set is $\{(2, -1)\}$.

4. Rewrite both equations in the form $Ax + By = C$:
$$2x = 9 + 3y \quad \rightarrow \quad 2x - 3y = 9$$
$$4y = 8 - 3x \quad \rightarrow \quad 3x + 4y = 8$$
Rewrite with opposite coefficients, then add and solve:
$$2x - 3y = 9 \xrightarrow{\text{Mult. by } 4} 8x - 12y = 36$$
$$3x + 4y = 8 \xrightarrow{\text{Mult. by } 3} \underline{9x + 12y = 24}$$
$$17x = 60$$
$$x = \frac{60}{17}$$
Back-substitute into either equation:
$$4y = 8 - 3x$$
$$4y = 8 - 3\left(\frac{60}{17}\right)$$
$$4y = -\frac{44}{17}$$
$$y = -\frac{11}{17}$$
Checking confirms the solution set is
$$\left\{\left(\frac{60}{17}, -\frac{11}{17}\right)\right\}.$$

5. Rewrite with a pair of opposite coefficients, then add:
$$5x - 2y = 4 \xrightarrow{\text{Mult. by } 2} 10x - 4y = 8$$
$$-10x + 4y = 7 \xrightarrow{\text{No change}} \underline{-10x + 4y = 7}$$
$$0 = 15$$
The statement $0 = 15$ is false which indicates that the system has no solution. The solution set is the empty set, $\varnothing$.

6. Substitute $4y - 8$ for x in the other equation:

$$5(\overbrace{4y-8}^{x}) - 20y = -40$$
$$20y - 40 - 20y = -40$$
$$-40 = -40$$

The statement $-40 = -40$ is true which indicates that the system has infinitely many solutions. The solution set is $\{(x, y) | x = 4y - 8\}$ or $\{(x, y) | 5x - 20y = -40\}$.

7. a. $C(x) = 300,000 + 30x$

b. $R(x) = 80x$

c. $R(x) = C(x)$
$$80x = 300,000 + 30x$$
$$50x = 300,000$$
$$x = 6000$$
$$C(6000) = 300,000 + 30(6000) = 480,000$$

The break-even point is (6000, 480,000). The company will need to make 6000 pairs of shoes and earn \$480,000 to break even.

Concept and Vocabulary Check 8.1

1. satisfies both equations in the system

2. the intersection point

3. $\left\{ \left(\dfrac{1}{3}, -2 \right) \right\}$

4. -2

5. -3

6. $\varnothing$; inconsistent; parallel

7. $\{(x, y) | x = 3y + 2\}$ or $\{(x, y) | 5x - 15y = 10\}$; dependent; are identical or coincide

8. revenue; profit

9. break-even point

Exercise Set 8.1

1. $x + 3y = 11$
$$2 + 3(3) = 11$$
$$2 + 9 = 11$$
$$11 = 11 \text{ true}$$
$$x - 5y = -13$$
$$2 - 5(3) = -13$$
$$2 - 15 = -13$$
$$-13 = -13 \text{ true}$$
(2, 3) is a solution.

3. $2x + 3y = 17$
$$2(2) + 3(5) = 17$$
$$4 + 15 = 17$$
$$19 = 17 \text{ false}$$
(2, 5) is not a solution.

5. $x + y = 4$
$$y = 3x$$
Substitute the expression $3x$ for y in the first equation and solve for x.
$$x + 3x = 4$$
$$4x = 4$$
$$x = 1$$
Substitute 1 for x in the second equation.
$$y = 3(1) = 3$$
The solution set is $\{(1, 3)\}$.

7. $x + 3y = 8$
$$y = 2x - 9$$
Substitute the expression $2x - 9$ for y in the first equation and solve for x.
$$x + 3(2x - 9) = 8$$
$$x + 6x - 27 = 8$$
$$7x = 35$$
$$x = 5$$
Substitute 5 for x in the second equation.
$$y = 2(5) - 9 = 10 - 9 = 1$$
The solution set is $\{(5, 1)\}$.

9. $x = 4y - 2$
$$x = 6y + 8$$
Substitute the expression $4y - 2$ for x in the second equation and solve for y.
$$4y - 2 = 6y + 8$$
$$-10 = 2y$$
$$-5 = y$$
Substitute -5 for y in the equation $x = 4y - 2$.
$$x = 4(-5) - 2 = -22$$
The solution set is $\{(-22, -5)\}$.

11. $5x + 2y = 0$

$x - 3y = 0$

Solve the second equation for x.

$x = 3y$

Substitute the expression $3y$ for x in the first equation and solve for y.

$5(3y) + 2y = 0$

$15y + 2y = 0$

$17y = 0$

$y = 0$

Substitute 0 for y in the equation $x = 3y$

$y = 3(0) = 0$

The solution set is $\{(0, 0)\}$.

13. $2x + 5y = -4$

$3x - y = 11$

Solve the second equation for y.

$-y = -3x + 11$

$y = 3x - 11$

Substitute the expression $3x - 11$ for y in the first equation and solve for x.

$2x + 5(3x - 11) = -4$

$2x + 15x - 55 = -4$

$17x = 51$

$x = 3$

Substitute 3 for x in the equation $y = 3x - 11$.

$y = 3(3) - 11 = 9 - 11 = -2$

The solution set is $\{(3, -2)\}$.

15. $2x - 3y = 8 - 2x$

$2x + 4y = x + 3y + 14$

Solve the second equation for y.

$y = -2x + 14$

Substitute the expression $-2x + 14$ for y in the first equation and solve for x.

$2x - 3(-2x + 14) = 8 - 2x$

$2x + 6x - 42 = 8 - 2x$

$8x - 42 = 8 - 2x$

$10x = 50$

$x = 5$

Substitute 5 for x in the equation $y = -2x + 14$.

$y = -2(5) + 14 = -10 + 14 = 4$

The solution set is $\{(5, 4)\}$.

17. $y = \dfrac{1}{3}x + \dfrac{2}{3}$

$y = \dfrac{5}{7}x - 2$

Substitute the expression $y = \dfrac{1}{3}x + \dfrac{2}{3}$ for y in the second equation and solve for x.

$\dfrac{1}{3}x + \dfrac{2}{3} = \dfrac{5}{7}x - 2$

$7x + 14 = 15x - 42$

$56 = 8x$

$7 = x$

Substitute 7 for x in the equation $y = \dfrac{1}{3}x + \dfrac{2}{3}$ and solve for y.

$y = \dfrac{1}{3}(7) + \dfrac{2}{3} = \dfrac{7}{3} + \dfrac{2}{3} = \dfrac{9}{3} = 3$

The solution set is $\{(7, 3)\}$.

19. Eliminate y by adding the equations.

$x + y = 1$

$\underline{x - y = 3}$

$2x = 4$

$x = 2$

Substitute 2 for x in the first equation.

$2 + y = 1$

$y = -1$

The solution set is $\{(2, -1)\}$.

21. Eliminate y by adding the equations.

$2x + 3y = 6$

$\underline{2x - 3y = 6}$

$4x = 12$

$x = 3$

Substitute 3 for x in the first equation.

$2(3) + 3y = 6$

$6 + 3y = 6$

$3y = 0$

$y = 0$

The solution set is $\{(3, 0)\}$.

23. $x + 2y = 2$

$-4x + 3y = 25$

Eliminate x by multiplying the first equation by 4 and adding the resulting equations.

$4x + 8y = 8$

$\underline{-4x + 3y = 25}$

$11y = 33$

$y = 3$

Substitute 3 for y in the first equation.

$x + 2(3) = 2$

$\quad x + 6 = 2$

$\qquad x = -4$

The solution set is $\{(-4, 3)\}$.

25. $4x + 3y = 15$

$2x - 5y = 1$

Eliminate x by multiplying the second equation by –2 and adding the resulting equations.

$\quad 4x + 3y = 15$

$\underline{-4x + 10y = -2}$

$\qquad 13y = 13$

$\qquad\quad y = 1$

Substitute 1 for y in the second equation.

$2x - 5(1) = 1$

$\quad 2x = 6$

$\qquad x = 3$

The solution set is $\{(3, 1)\}$.

27. $3x - 4y = 11$

$2x + 3y = -4$

Eliminate x by multiplying the first equation by 2 and the second equation by –3. Add the resulting equations.

$\quad 6x - 8y = 22$

$\underline{-6x - 9y = 12}$

$\qquad -17y = 34$

$\qquad\quad y = -2$

Substitute –2 for y in the second equation.

$2x + 3(-2) = -4$

$\quad 2x - 6 = -4$

$\qquad 2x = 2$

$\qquad\quad x = 1$

The solution set is $\{(1, -2)\}$.

29. $3x = 4y + 1$

$3y = 1 - 4x$

Arrange the system so that variable terms appear on the left and constants appear on the right.

$3x - 4y = 1$

$4x + 3y = 1$

Eliminate y by multiplying the first equation by 3 and the second equation by 4. Add the resulting equations.

$\quad 9x - 12y = 3$

$\underline{16x + 12y = 4}$

$\qquad 25x = 7$

$\qquad\quad x = \dfrac{7}{25}$

Substitute $\dfrac{7}{25}$ for x in the second equation.

$3y = 1 - 4\left(\dfrac{7}{25}\right)$

$3y = \dfrac{-3}{25}$

$\ y = \dfrac{-1}{25}$

The solution set is $\left\{\left(\dfrac{7}{25}, -\dfrac{1}{25}\right)\right\}$.

31. The substitution method is used here to solve the system.

$\quad x = 9 - 2y$

$x + 2y = 13$

Substitute the expression $9 - 2y$ for x in the second equation and solve for y.

$9 - 2y + 2y = 13$

$\qquad\quad 9 = 13$

The false statement $9 = 13$ indicates that the system has no solution.

The solution set is the empty set, $\varnothing$.

33. The substitution method is used here to solve the system.

$\quad y = 3x - 5$

$21x - 35 = 7y$

Substitute the expression $3x - 5$ for y in the second equation and solve for x.

$21x - 35 = 7(3x - 5)$

$21x - 35 = 21x - 35$

$\quad -35 = -35$

This true statement indicates that the system has infinitely many solutions.

The solution set is $\left\{(x, y)\,\middle|\, y = 3x - 5\right\}$

35. The elimination method is used here to solve the system.

$3x - 2y = -5$

$4x + y = 8$

Eliminate y by multiplying the second equation by 2 and adding the resulting equations.

$\quad 3x - 2y = -5$

$\underline{8x + 2y = 16}$

$\qquad 11x = 11$

$\qquad\quad x = 1$

Substitute 1 for x in the second equation.

$4(1) + y = 8$

$\qquad\quad y = 4$

The solution set is $\{(1, 4)\}$.

37. The elimination method is used here to solve the system.
$$x + 3y = 2$$
$$3x + 9y = 6$$
Eliminate x by multiplying the first equation by -3 and adding the resulting equations.
$$-3x - 9y = -6$$
$$\underline{3x + 9y = 6}$$
$$0 = 0$$
This true statement indicates that the system has infinitely many solutions.

The solution set is $\left\{ (x, y) \mid x + 3y = 2 \right\}$.

39. First multiply each term in the first equation by 4 to eliminate the fractions.
$$\frac{x}{4} - \frac{y}{4} = -1$$
$$x - y = -4$$
Multiply the first equation by -1 and add to the second equation and solve for y.
$$-x + y = 4$$
$$x + 4y = -9$$
$$5y = -5$$
$$y = -1$$
Substitute -1 for y in the equation $x - y = -4$ and solve for x.
$$x - (-1) = -4$$
$$x + 1 = -4$$
$$x = -5$$
The solution set is $\{(-5, -1)\}$.

41. Rearrange the equations to get in the standard form.
$$2x - 3y = 4$$
$$4x + 5y = 3$$
Multiply the first equation by -2 and add to the second equation. Solve for y.
$$-4x + 6y = -8$$
$$4x + 5y = 3$$
$$11y = -5$$
$$y = -\frac{5}{11}$$

Multiply the first equation by 5 and the second equation by 3 and add the equations. Solve for x.
$$10x - 15y = 20$$
$$12x + 15y = 9$$
$$22x = 29$$
$$x = \frac{29}{22}$$
The solution set is $\left\{ \left(\frac{29}{22}, -\frac{5}{11} \right) \right\}$.

43. Add the equations to eliminate y.
$$x + y = 7$$
$$\underline{x - y = -1}$$
$$2x = 6$$
$$x = 3$$
Substitute 3 for x in the first equation.
$$3 + y = 7$$
$$y = 4$$
The numbers are 3 and 4.

45. $3x - y = 1$
$x + 2y = 12$
Eliminate y by multiplying the first equation by 2 and adding the resulting equations.
$$6x - 2y = 2$$
$$\underline{x + 2y = 12}$$
$$7x = 14$$
$$x = 2$$
Substitute 2 for x in the first equation.
$$3(2) - y = 1$$
$$6 - y = 1$$
$$-y = -5$$
$$y = 5$$
The numbers are 2 and 5.

47. $\dfrac{x+2}{2} - \dfrac{y+4}{3} = 3$

$\dfrac{x+y}{5} = \dfrac{x-y}{2} - \dfrac{5}{2}$

Start by multiplying each equation by its LCD and simplifying to clear the fractions.

$\dfrac{x+2}{2} - \dfrac{y+4}{3} = 3$

$\dfrac{x+y}{5} = \dfrac{x-y}{2} - \dfrac{5}{2}$

Start by multiplying each equation by its LCD and simplifying to clear the fractions.

$6\left(\dfrac{x+2}{2} - \dfrac{y+4}{3}\right) = 6(3)$

$3(x+2) - 2(y+4) = 18$

$3x + 6 - 2y - 8 = 18$

$3x - 2y = 20$

$10\left(\dfrac{x+y}{5}\right) = 10\left(\dfrac{x-y}{2} - \dfrac{5}{2}\right)$

$2(x+y) = 5(x-y) - 5(5)$

$2x + 2y = 5x - 5y - 25$

$3x - 7y = 25$

We now need to solve the equivalent system of equations:

$3x - 2y = 20$

$3x - 7y = 25$

Subtract the two equations:

$3x - 2y = 20$

$\dfrac{-(3x - 7y = 25)}{5y = -5}$

$y = -1$

Back-substitute this value for y and solve for x.

$3x - 2y = 20$

$3x - 2(-1) = 20$

$3x + 2 = 20$

$3x = 18$

$x = 6$

The solution is $(6, -1)$.

49. $5ax + 4y = 17$

$ax + 7y = 22$

Multiply the second equation by -5 and add the equations.

$5ax + 4y = 17$

$\dfrac{-5ax - 35y = -110}{-31y = -93}$

$y = 3$

Back-substitute into one of the original equations to solve for x.

$ax + 7y = 22$

$ax + 7(3) = 22$

$ax + 21 = 22$

$ax = 1$

$x = \dfrac{1}{a}$

The solution is $\left(\dfrac{1}{a}, 3\right)$.

51. $f(-2) = 11 \quad \rightarrow \quad -2m + b = 11$

$f(3) = -9 \quad \rightarrow \quad 3m + b = -9$

We need to solve the resulting system of equations:

$-2m + b = 11$

$3m + b = -9$

Subtract the two equations:

$-2m + b = 11$

$\dfrac{3m + b = -9}{-5m = 20}$

$m = -4$

Back-substitute into one of the original equations to solve for b.

$-2m + b = 11$

$-2(-4) + b = 11$

$8 + b = 11$

$b = 3$

Therefore, $m = -4$ and $b = 3$.

53. The solution to a system of linear equations is the point of intersection of the graphs of the equations in the system. If $(6,2)$ is a solution, then we need to find the lines that intersect at that point.
Looking at the graph, we see that the graphs of $x+3y=12$ and $x-y=4$ intersect at the point $(6,2)$. Therefore, the desired system of equations is

$$x+3y=12 \quad \text{or} \quad y=-\frac{1}{3}x+4$$
$$x-y=4 \qquad\qquad y=x-4$$

55. At the break-even point, $R(x)=C(x)$.
$$10000+30x=50x$$
$$10000=20x$$
$$10000=20x$$
$$500=x$$
Five hundred radios must be produced and sold to break-even.

57. $R(x)=50x$
$$R(200)=50(200)=10000$$
$$C(x)=10000+30x$$
$$C(200)=10000+30(200)$$
$$=10000+6000=16000$$
$$R(200)-C(200)=10000-16000$$
$$=-6000$$
This means that if 200 radios are produced and sold the company will lose \$6,000.

59. a. $P(x)=R(x)-C(x)$
$$=50x-(10,000+30x)$$
$$=50x-10,000-30x$$
$$=20x-10,000$$
$$P(x)=20x-10,000$$

b. $P(10,000)=20(10,000)-10,000$
$$=200,000-10,000=190,000$$
If 10,0000 radios are produced and sold the profit will be \$190,000.

61. a. The cost function is:
$$C(x)=18,000+20x$$

b. The revenue function is:
$$R(x)=80x$$

c. At the break-even point, $R(x)=C(x)$.
$$80x=18,000+20x$$
$$60x=18,000$$
$$x=300$$
$$R(x)=80x$$
$$R(300)=80(300)$$
$$=24,000$$
When approximately 300 canoes are produced the company will break-even with cost and revenue at \$24,000.

63. a. The cost function is:
$$C(x)=30000+2500x$$

b. The revenue function is:
$$R(x)=3125x$$

c. At the break-even point, $R(x)=C(x)$.
$$3125x=30000+2500x$$
$$625x=30000$$
$$x=48$$
After 48 sold out performances, the investor will break-even. (\$150,000)

65. a. Substitute $0.375x+3$ for p in the first equation.
$$p=-0.325x+5.8$$
$$\overbrace{0.375x+3}^{p}=-0.325x+5.8$$
$$0.375x+3=-0.325x+5.8$$
$$0.375x+0.325x+3=-0.325x+0.325x+5.8$$
$$0.7x+3=5.8$$
$$0.7x+3-3=5.8-3$$
$$0.7x=2.8$$
$$\frac{0.7x}{0.7}=\frac{2.8}{0.7}$$
$$x=4$$
Back-substitute to find p.
$$p=-0.325x+5.8$$
$$p=-0.325(4)+5.8=4.5$$

The ordered pair is (4,4.5).
Equilibrium number of workers: 4 million
Equilibrium hourly wage: \$4.50

b. If workers are paid $4.50 per hour, there will be 4 million available workers and 4 million workers will be hired. In this state of market equilibrium, there is no unemployment.

c.
$$p = -0.325x + 5.8$$
$$5.15 = -0.325x + 5.8$$
$$0.65 = -0.325x$$
$$\frac{-0.65}{-0.325} = \frac{-0.325x}{-0.325}$$
$$2 = x$$
At $5.15 per hour, 2 million workers will be hired.

d.
$$p = 0.375x + 3$$
$$5.15 = 0.375x + 3$$
$$2.15 = 0.375x$$
$$\frac{2.15}{0.375} = \frac{0.375x}{0.375}$$
$$x \approx 5.7$$
At $5.15 per hour, there will be about 5.7 million available workers.

e. $5.7 - 2 = 3.7$
At $5.15 per hour, there will be about 3.7 million more people looking for work than employers are willing to hire.

67. a.
$$-3x + 10y = 160$$
$$x + 2y = 142$$
Multiply the second equation by -3 and then add the equations.
$$-3x + 10y = 160$$
$$\underline{3x + 6y = 426}$$
$$16y = 586$$
$$y = \frac{586}{16}$$
$$y = \frac{293}{8}$$
$$y \approx 37$$

Back-substitute $\frac{293}{8}$ for y and find x.
$$x + 2y = 142$$
$$x + 2\left(\frac{293}{8}\right) = 142$$
$$x + \frac{293}{4} = 142$$
$$x = \frac{275}{4}$$
$$x \approx 69$$
About 69 years after 1970, or 2039, the

percentage of never-married adults will be the same as the percentage of married adults, about 37%.

b. The answer from part (a) is represented by the intersection point $(69, 37)$.

69. a. $y = 0.45x + 0.8$

b. $y = 0.15x + 2.6$

c. To find the week in the semester when both groups report the same number of symptoms, we set the two equations equal to each other and solve for x.
$$0.45x + 0.8 = 0.15x + 2.6$$
$$0.3x = 1.8$$
$$x = 6$$
The number of symptoms will be the same in week 6.
$$y = 0.15x + 2.6$$
$$y = 0.15(6) + 2.6$$
$$y = 3.5$$
The number of symptoms in week 6 will be 3.5 for both groups. This is shown in the graph by the intersection point $(6, 3.5)$.

71. a. $y = 0.2x + 5$

b. $y = -0.4x + 47$

c. To project the year in which the percentage of adults living alone will be the same as the percentage of married adults living with kids, we set the two equations equal to each other and solve for x.
$$0.2x + 5 = -0.4x + 47$$
$$0.6x = 42$$
$$x = 70$$
The percentage will be the same 70 years after 1960, or 2030.

$$y = 0.2x + 5$$
$$y = 0.2(70) + 5$$
$$= 19$$
At that time about 19% will belong to each group.

73. Let x = the number of calories in a Mr. Goodbar.
Let y = the number of calories in a Mounds bar.
$$x + 2y = 780$$
$$2x + y = 786$$
Multiply the bottom equation by -2 and then add the equations to eliminate y.
$$x + 2y = 780$$
$$\underline{-4x - 2y = -1572}$$
$$-3x = -792$$
$$x = 264$$
Back-substitute to find y.
$$x + 2y = 780$$
$$264 + 2y = 780$$
$$2y = 516$$
$$y = 258$$
There are 264 calories in a Mr. Goodbar and 258 calories in a Mounds bar.

75. Let x = the number of Mr. Goodbars.
Let y = the number of Mounds bars.
$$x + y = 5$$
$$16.3x + 14.1y - 70 = 7.1$$
Solve the first equation for y in terms of x.
$$x + y = 5$$
$$y = -x + 5$$
Substitute $-x + 5$ for y in the second equation.
$$16.3x + 14.1y - 70 = 7.1$$
$$16.3x + 14.1(\overbrace{-x+5}^{y}) - 70 = 7.1$$
$$16.3x + 14.1(-x+5) - 70 = 7.1$$
$$16.3x - 14.1x + 70.5 - 70 = 7.1$$
$$2.2x + 0.5 = 7.1$$
$$2.2x = 6.6$$
$$x = 3$$
Back-substitute to find y.
$$x + y = 5$$
$$3 + y = 5$$
$$y = 2$$
There are 3 Mr. Goodbars and 2 Mounds bars.

77.
$$x + y = 200$$
$$100x + 80y = 17000$$
Multiply the first equation by -100 and add to the second equation. Solve for y.
$$-100x - 100y = -20000$$
$$100x + 80y = 17000$$
$$-20y = -3000$$
$$y = 150$$
Substitute 150 for y in the first equation and solve for x.
$$x + 150 = 200$$
$$x = 50$$
There are 50 rooms with kitchenettes and 150 rooms without.

79.
$$2x + 2y = 360$$
$$20x + 8(2y) = 3280$$
Multiply the first equation by -10 and add to the second equation. Solve for y.
$$-20x - 20y = -3600$$
$$20x + 16y = 3280$$
$$-4y = -320$$
$$y = 80$$
Substitute 80 for y in the first equation and solve for x.
$$2x + 2(80) = 360$$
$$2x + 160 = 360$$
$$2x = 200$$
$$x = 100$$
The lot is 100 feet long and 80 feet wide.

81.
$$(x + y)2 = 16$$
$$(x - y)2 = 8$$
Multiply to remove the parentheses and then add the two equations together. Solve for x.
$$2x + 2y = 16$$
$$2x - 2y = 8$$
$$4x = 24$$
$$x = 6$$
Substitute 6 for x in the first equation and solve for y.
$$2(6) + 2y = 16$$
$$12 + 2y = 16$$
$$2y = 4$$
$$y = 2$$
The crew rows 6 mph and the current is 2 mph.

83.
$$x + 2y = 180$$
$$(2x - 30) + y = 180$$

Rewrite the second equation in standard form.
$$x + 2y = 180$$
$$2x + y = 210$$

Multiply the first equation by -2 and add the equations.

$$
\begin{array}{r}
-2x - 4y = -360 \\
\underline{2x + y = 210} \\
-3y = -150 \\
y = 50
\end{array}
$$

Back-substitute to solve for x.
$$x + 2y = 180$$
$$x + 2(50) = 180$$
$$x + 100 = 180$$
$$x = 80$$

The three interior angles measure $80°$, $50°$, and $50°$.

85. – 93. Answers will vary.

95. makes sense

97. makes sense

99. $a_1 x + b_1 y = c_1$
$a_2 x + b_2 y = c_2$

Solve the first equation for x.
$$x = \frac{c_1 - b_1 y}{a_1}$$

Substitute the expression $\dfrac{c_1 - b_1 y}{a_1}$ for x in the second

equation and solve for y.

$$a_2 \left(\frac{c_1 - b_1 y}{a_1} \right) + b_2 y = c_2$$

$$a_2 \left(\frac{c_1 - b_1 y}{a_1} \right) + \frac{a_1 b_2 y}{a_1} = c_2$$

$$\frac{a_2 c_1 - a_2 b_1 y + a_1 b_2 y}{a_1} = c_2$$

$$a_2 c_1 - a_2 b_1 y + a_1 b_2 y = a_1 c_2$$

$$y(a_1 b_2 - a_2 b_1) = a_1 c_2 - a_2 c_1$$

$$y = \frac{a_1 c_2 - a_2 c_1}{a_1 b_2 - a_2 b_1}$$

Substitute the expression $\dfrac{a_1 c_2 - a_2 c_1}{a_1 b_2 - a_2 b_1}$ for y in the first

equation and solve for x.

$$a_1 x + b_1 \left(\frac{a_1 c_2 - a_2 c_1}{a_1 b_2 - a_2 b_1} \right) = c_1$$

$$a_1 x + \frac{a_1 b_1 c_2 - a_2 b_1 c_1}{a_1 b_2 - a_2 b_1} = c_1$$

$$a_1 x = c_1 - \frac{a_1 b_1 c_2 - a_2 b_1 c_1}{a_1 b_2 - a_2 b_1}$$

$$= \frac{c_1 (a_1 b_2 - a_2 b_1)}{a_1 b_2 - a_2 b_1} - \frac{a_1 b_1 c_2 - a_2 b_1 c_1}{a_1 b_2 - a_2 b_1}$$

$$= \frac{a_1 b_2 c_1 - a_1 b_1 c_2}{a_1 b_2 - a_2 b_1}$$

$$x = \frac{a_1 b_2 c_1 - a_1 b_1 c_2}{a_1 b_2 - a_2 b_1} \div a_1$$

$$= \frac{a_1 b_2 c_1 - a_1 b_1 c_2}{a_1 (a_1 b_2 - a_2 b_1)} = \frac{a_1 (b_2 c_1 - b_1 c_2)}{a_1 (a_1 b_2 - a_2 b_1)}$$

$$x = \frac{b_2 c_1 - b_1 c_2}{a_1 b_2 - a_2 b_1}$$

101. x = number of hexagons formed
y = number of squares formed
$$6x + y = 52$$
$$x + 4y = 24$$

Eliminate x by multiplying the second equation by -6 and adding the resulting equations.

$$
\begin{array}{r}
6x + y = 52 \\
\underline{-6x - 24y = -144} \\
-23y = -92 \\
y = 4
\end{array}
$$

Substitute 4 for y in the second equation.
$$x + 4(4) = 24$$
$$x + 16 = 24$$
$$x = 8$$

Yes, they should make 8 hexagons and 4 squares.

103.
$$2x - y + 4z = -8$$
$$2(3) - (2) + 4(-3) = -8$$
$$-8 = -8, \text{ true}$$
Yes, the ordered triple satisfies the equation.

104. $5x - 2y - 4z = 3$
$3x + 3y + 2z = -3$
Multiply Equation 2 by 2.
$5x - 2y - 4z = 3$
$6x + 6y + 4z = -6$
Then add to eliminate z.
$$\begin{array}{r} 5x - 2y - 4z = 3 \\ \underline{6x + 6y + 4z = -6} \\ 11x + 4y \quad\quad = -3 \end{array}$$

105. $y = ax^2 + bx + c$
$$1682 = a(4)^2 + b(4) + c$$
$$1682 = 16a + 4b + c$$

Section 8.2

Check Point Exercises

1.
$$x - 2y + 3z = 22$$
$$-1 - 2(-4) + 3(5) = 22$$
$$-1 + 8 + 15 = 22$$
$$22 = 22 \text{ true}$$
$$2x - 3y - z = 5$$
$$2(-1) - 3(-4) - 5 = 5$$
$$-2 + 12 - 5 = 5$$
$$5 = 5 \text{ true}$$
$$3x + y - 5z = -32$$
$$3(-1) - 4 - 5(5) = -32$$
$$-3 - 4 - 25 = -32$$
$$-32 = -32 \text{ true}$$
$(-1, -4, 5)$ is a solution of the system.

2.
$$x + 4y - z = 20$$
$$3x + 2y + z = 8$$
$$2x - 3y + 2z = -16$$
Eliminate z from Equations 1 and 2 by adding Equation 1 and Equation 2.
$$\begin{array}{r} x + 4y - z = 20 \\ \underline{3x + 2y + z = 8} \\ 4x + 6y = 28 \text{ Equation 4} \end{array}$$
Eliminate z from Equations 2 and 3 by multiplying Equation 2 by -2 and adding the resulting equation to Equation 3.
$$\begin{array}{r} -6x - 4y - 2z = -16 \\ \underline{2x - 3y + 2z = -16} \\ -4x - 7y = -32 \text{ Equation 5} \end{array}$$
Solve Equations 4 and 5 for x and y by adding Equation 4 and Equation 5.
$$\begin{array}{r} 4x + 6y = 28 \\ \underline{-4x - 7y = -32} \\ -y = -4 \\ y = 4 \end{array}$$
Substitute 4 for y in Equation 4 and solve for x.
$$4x + 6(4) = 28$$
$$4x + 24 = 28$$
$$4x = 4$$
$$x = 1$$
Substitute 1 for x and 4 for y in Equation 2 and solve for z.
$$3(1) + 2(4) + z = 8$$
$$3 + 8 + z = 8$$
$$11 + z = 8$$
$$z = -3$$
The solution set is $\{(1, 4, -3)\}$.

3.
$$2y - z = 7$$
$$x + 2y + z = 17$$
$$2x - 3y + 2z = -1$$
Eliminate x and z from Equations 2 and 3 by multiplying Equation 2 by -2 and adding the resulting equation to Equation 3.
$$\begin{array}{r} -2x - 4y - 2z = -34 \\ \underline{2x - 3y + 2z = -1} \\ -7y = -35 \\ y = 5 \end{array}$$

Substitute 5 for y in Equation 1 and solve for z.

$2(5) - z = 7$

$10 - z = 7$

$-z = -3$

$z = 3$

Substitute 5 for y and 3 for z in Equation 2 and solve for x.

$x + 2(5) + 3 = 17$

$x + 10 + 3 = 17$

$x + 13 = 17$

$x = 4$

The solution set is $\{(4, 5, 3)\}$.

4. $(1, 4), (2, 1), (3, 4)$

$y = ax^2 + bx + c$

Substitute 1 for x and 4 for y in

$y = ax^2 + bx + c$.

$4 = a(1)^2 + b(1) + c$

$4 = a + b + c \quad \text{Equation 1}$

Substitute 2 for x and 1 for y in

$y = ax^2 + bx + c$.

$1 = a(2)^2 + b(2) + c$

$1 = 4a + 2b + c \quad \text{Equation 2}$

Substitute 3 for x and 4 for y in

$y = ax^2 + bx + c$.

$4 = a(3)^2 + b(3) + c$

$4 = 9a + 3b + c \quad \text{Equation 3}$

Eliminate c from Equations 1 and 2 by multiplying Equation 2 by -1 and adding the resulting equation to Equation 1.

$4 = a + b + c$

$\underline{-1 = -4a - 2b - c}$

$3 = -3a - b \qquad \text{Equation 4}$

Eliminate c from Equation 2 and 3 by multiplying Equation 3 by -1 and adding the resulting equation to Equation 2.

$1 = 4a + 2b + c$

$\underline{-4 = -9a - 3b - c}$

$-3 = -5a - b \qquad \text{Equation 5}$

Solve Equations 4 and 5 for a and b by multiplying Equation 5 by -1 and adding the resulting equation to Equation 4.

$3 = -3a - b$

$\underline{3 = 5a + b}$

$6 = 2a$

$a = 3$

Substitute 3 for a in Equation 4 and solve for b.

$3 = -3(3) - b$

$3 = -9 - b$

$12 = -b$

$b = -12$

Substitute 3 for a and -12 for b in Equation 1 and solve for c.

$4 = 3 - 12 + c$

$4 = -9 + c$

$c = 13$

Substituting 3 for a, -12 for b, and 13 for c in the quadratic equation $y = ax^2 + bx + c$ gives

$y = 3x^2 - 12x + 13$.

Concept and Vocabulary Check 8.2

1. triple; all

2. -2; -4

3. z; add Equations 1 and 3

Exercise Set 8.2

1. $x + y + z = 4$

$2 - 1 + 3 = 4$

$4 = 4 \text{ true}$

$x - 2y - z = 1$

$2(2) - 2(-1) - 3 = 1$

$4 + 2 - 3 = 1$

$1 = 1 \text{ true}$

$2x - y - 2z = -1$

$2(2) - (-1) - 2(3) = -1$

$4 + 1 - 6 = -1$

$-1 = -1 \text{ false}$

$(2, -1, 3)$ is a solution.

3. $x - 2y = 2$

$4 - 2(1) = 2$

$4 - 2 = 2$

$2 = 2 \text{ true}$

$2x + 3y = 11$

$2(4) + 3(1) = 11$

$8 + 3 = 11$

$11 = 11 \text{ true}$

$y - 4z = -7$

$1 - 4(2) = -7$

$1 - 8 = -7$

$-7 = -7 \text{ true}$

$(4, 1, 2)$ is a solution.

5. $x+y+2z=11$

$x+y+3z=14$

$x+2y-z=5$

Eliminate x and y from Equations 1 and 2 by multiplying Equation 2 by -1 and adding the resulting equation to Equation 1.

$-x-y-3z=-14$

$\underline{x+y+2z=11}$

$-z=-3$

$z=3$

Substitute 3 for z in Equations 1 and 3.

$x+y+2(3)=11$

$x+2y-(3)=5$

Simplify:

$x+y=5$　　Equation 4

$x+2y=8$　　Equation 5

Solve Equations 4 and 5 for x and y by multiplying Equation 5 by -1 and adding the resulting equation to Equation 4.

$x+y=5$

$\underline{-x-2y=-8}$

$-y=-3$

$y=3$

Substitute 3 for z and 3 for y in Equation 2 and solve for x.

$x+3+3(3)=14$

$x+12=14$

$x=2$

The solution set is $\{(2, 3, 3)\}$.

7. $4x-y+2z=11$

$x+2y-z=-1$

$2x+2y-3z=-1$

Eliminate y from Equation 1 and 2 by multiplying Equation 1 by 2 and adding the resulting equation to Equation 2 and 3.

$8x-2y+4z=22$

$\underline{x+2y-z=-1}$

$9x+3z=21$　　Equation 4

Eliminate y from Equations 1 and 3 by multiplying Equation 1 by 2 and adding the resulting equation to Equation 3.

$8x-2y+4z=22$

$\underline{2x+2y-3z=-1}$

$10x+z=21$　　Equation 5

Solve Equations 4 and 5 for x and z by multiplying Equation 5 by -3 and adding the resulting equation to Equation 4.

$9x+3z=21$

$\underline{-30x-3z=-63}$

$-21x=-42$

$x=2$

Substitute 2 for x in Equation 5 and solve for z. $10(2)+z=21$

$20+z=21$

$z=1$

Substitute 2 for x and 1 for z in Equation 2 and solve for y.

$2+2y-1=-1$

$2y+1=-1$

$2y=-2$

$y=-1$

The solution set is $\{(2,-1,1)\}$.

9. $3x+2y-3z=-2$

$2x-5y+2z=-2$

$4x-3y+4z=10$

Eliminate z from Equations 1 and 2 by multiplying Equation 1 by 2 and Equation 2 by 3. Add the resulting equations.

$6x+4y-6z=-4$

$\underline{6x-15y+6z=-6}$

$12x-11y=-10$　　Equation 4

Eliminate z from Equations 2 and 3 by multiplying Equation 2 by -2.

$-4x+10y-4z=4$

$\underline{4x-3y+4z=10}$

$7y=14$　　Equation 5

Solve Equation 5 for y

$7y=14$

$y=2$

Solve for x by substituting 7 for y in Equation 4.

$12x-11y=-10$

$12x-11(2)=-10$

$12x-22=-10$

$12x=12$

$x=1$

Substitute 2 for y and 1 for x in Equation 2 and solve for z.

$2x-5y+2z=-2$

$2(1)-5(2)+2z=-2$

$2-10+2z=-2$

$2z=6$

$z=3$

The solution set is $\{(1, 2, 3)\}$.

11. $2x - 4y + 3z = 17$

$x + 2y - z = 0$

$4x - y - z = 6$

Eliminate z from Equations 1 and 2 by multiplying Equation 2 by 3 and adding the resulting equation to Equation 1.

$2x - 4y + 3z = 17$

$\underline{3x + 6y - 3z = 0}$

$5x + 2y = 17$ \quad Equation 4

Eliminate z from Equations 2 and 3 by multiplying Equation 2 by -1 and adding the resulting equation to Equation 3.

$-x - 2y + z = 0$

$\underline{4x - y - z = 6}$

$3x - 3y = 6$ \quad Equation 5

Solve Equations 4 and 5 for x and y by multiplying

Equation 5 by $\dfrac{2}{3}$ and adding the resulting equation to

Equation 4.

$5x + 2y = 17$

$\underline{2x - 2y = 4}$

$7x = 21$

$x = 3$

Substitute 3 for x in Equation 4 and solve for y.

$5(3) + 2y = 17$

$15 + 2y = 17$

$2y = 2$

$y = 1$

Substitute 3 for x and 1 for y in Equation 2 and solve for z.

$3 + 2(1) - z = 0$

$3 + 2 - z = 0$

$5 - z = 0$

$5 = z$

The solution set is $\left\{(3, 1, 5)\right\}$.

13. $2x + y = 2$

$x + y - z = 4$

$3x + 2y + z = 0$

Eliminate z from Equations 2 and 3 by adding Equation 2 and Equation 3.

$x + y - z = 4$

$\underline{3x + 2y + z = 0}$

$4x + 3y = 4$ \quad Equation 4

Solve Equations 1 and 4 for x and y by multiplying Equation 1 by -3 and adding the resulting equation to Equation 4.

$-6x - 3y = -6$

$\underline{4x + 3y = 4}$

$-2x = -2$

$x = 1$

Substitute 1 for x in Equation 1 and solve for y.

$2(1) + y = 2$

$2 + y = 2$

$y = 0$

Substitute 1 for x and 0 for y in Equation 2 and solve for z.

$1 + 0 - z = 4$

$1 - z = 4$

$-z = 3$

$z = -3$

The solution set is $\{(1, 0, -3)\}$.

15. $x + y = -4$

$y - z = 1$

$2x + y + 3z = -21$

Eliminate y from Equations 1 and 2 by multiplying Equation 1 by -1 and adding the resulting equation to Equation 2.

$-x - y = 4$

$\underline{y - z = 1}$

$-x - z = 5$ \quad Equation 4

Eliminate y from Equations 2 and 3 by multiplying Equation 2 by -1 and adding the resulting equation to Equation 3.

$-y + z = -1$

$\underline{2x + y + 3z = -21}$

$2x + 4z = -22$ \quad Equation 5

Solve Equations 4 and 5 for x and z by multiplying Equation 4 by 2 and adding the resulting equation to Equation 5.

$-2x - 2z = 10$

$\underline{2x + 4z = -22}$

$2z = -12$

$z = -6$

Substitute -6 for z in Equation 2 and solve for y.

$y - (-6) = 1$

$y + 6 = 1$

$y = -5$

Substitute -5 for y in Equation 1 and solve for x.

$x + (-5) = -4$

$x = 1$

The solution set is $\{(1, -5, -6)\}$.

17. $3(2x + y) + 5z = -1$

$2(x - 3y + 4z) = -9$

$4(1 + x) = -3(z - 3y)$

Simplify each equation.

$6x + 3y + 5z = -1$ Equation 4

$2x - 6y + 8z = -9$ Equation 5

$4 + 4x = -3z + 9y$

$4x - 9y + 3z = -4$ Equation 6

Eliminate x from Equations 4 and 5 by multiplying Equation 5 by –3 and adding
the resulting equation to Equation 4.

$-6x + 3y + 5z = -1$

$\underline{-6x + 18y - 24z = 27}$

$21y - 19z = 26$ Equation 7

Eliminate x from Equations 5 and 6 by multiplying Equation 5 by –2 and adding
the resulting equation to Equation 6.

$-4x + 12y - 16z = 18$

$\underline{4x - 9y + 3z = -4}$

$3y - 13z = 14$ Equation 8

Solve Equations 7 and 8 for y and z by multiplying Equation 8 by –7 and adding
the resulting equation to Equation 7.

$21y - 19z = 26$

$\underline{-21y + 91z = -98}$

$72z = -72$

$z = -1$

Substitute -1 for z in Equation 8 and solve for y.

$3y - 13(-1) = 14$

$3y + 13 = 14$

$3y = 1$

$y = \dfrac{1}{3}$

Substitute $\dfrac{1}{3}$ for y and –1 for z in Equation 5 and

solve for x.

$2x - 6\left(\dfrac{1}{3}\right) + 8(-1) = -9$

$2x - 2 - 8 = -9$

$2x - 10 = -9$

$2x = 1$

$x = \dfrac{1}{2}$

The solution set is $\left\{\left(\dfrac{1}{2}, \dfrac{1}{3}, -1\right)\right\}$.

19. $(-1, 6), (1, 4), (2, 9)$

$y = ax^2 + bx + c$

Substitute –1 for x and 6 for y in $y = ax^2 + bx + c$.

$6 = a(-1)^2 + b(-1) + c$

$6 = a - b + c$ Equation 1

Substitute 1 for x and 4 for y in $y = ax^2 + bx + c$.

$4 = a(1)^2 + b(1) + c$

$4 = a + b + c$ Equation 2

Substitute 2 for x and 9 for y in
$y = ax^2 + bx + c$.

$9 = a(2)^2 + b(2) + c$

$9 = 4a + 2b + c$ Equation 3

Eliminate b from Equations 1 and 2 by adding Equation 1 and Equation 2.

$6 = a - b + c$

$\underline{4 = a + b + c}$

$10 = 2a + 2c$ Equation 4

Eliminate b from Equations 1 and 3 by multiplying Equation 1 by 2 and adding the resulting equation to Equation 3.

$12 = 2a - 2b + 2c$

$\underline{9 = 4a + 2b + c}$

$21 = 6a + 3c$ Equation 5

Solve Equations 4 and 5 for a and c by multiplying Equation 4 by –3 and adding the resulting equation to Equation 5.

$-30 = -6a - 6c$

$\underline{21 = 6a + 3c}$

$-9 = -3c$

$c = 3$

Substitute 3 for c in Equation 4 and solve for a.

$10 = 2a + 2(3)$

$10 = 2a + 6$

$4 = 2a$

$a = 2$

Substitute 2 for a and 3 for c in Equation 2 and solve
for b.

$4 = 2 + b + 3$

$4 = b + 5$

$b = -1$

Substituting 2 for a, –1 for b, and 3 for c in the
quadratic equation $y = ax^2 + bx + c$ gives

$y = 2x^2 - x + 3$.

21. $(-1, -4), (1, -2), (2, 5)$

Substitute -1 for x and -4 for y in $y = ax^2 + bx + c$.

$-4 = a(-1)^2 + b(-1) + c$

$-4 = a - b + c$ Equation 1

Substitute 1 for x and -2 for y in $y = ax^2 + bx + c$.

$-2 = a(1)^2 + b(1) + c$

$-2 = a + b + c$ Equation 2

Substitute 2 for x and 5 for y in $y = ax^2 + bx + c$.

$5 = a(2)^2 + b(2) + c$

$5 = 4a + 2b + c$ Equation 3

Eliminate a and b from Equations 1 and 2 by multiplying Equation 1 by -1 and adding the resulting equation to Equation 2.

$4 = -a + b - c$

$\underline{-2 = a + b + c}$

$2 = 2b$

$b = 1$

Eliminate c from Equations 1 and 3 by multiplying Equation 1 by -1 and adding the resulting equation to Equation 3.

$4 = -a + b - c$

$\underline{5 = 4a + 2b + c}$

$9 = 3a + 3b$ Equation 4

Substitute 1 for b in Equation 4 and solve for a.

$9 = 3a + 3(1)$

$9 = 3a + 3$

$6 = 3a$

$a = 2$

Substitute 2 for a and 1 for b in Equation 2 and solve for c.

$-2 = 2 + 1 + c$

$-2 = c + 3$

$c = -5$

Substituting 2 for a, 1 for b, and -5 for c in quadratic equation $y = ax^2 + bx + c$ gives $y = 2x^2 + x - 5$.

23. $x + y + z = 16$

$2x + 3y + 4z = 46$

$5x - y = 31$

Eliminate z from Equations 1 and 2 by multiplying Equation 1 by -4 and adding the resulting equation to Equation 2.

$-4x - 4y - 4z = -64$

$\underline{2x + 3y + 4z = 46}$

$-2x - y = -18$ Equation 4

Solve Equations 3 and 4 for x and y by multiplying

Equation 4 by -1 and adding the resulting equation to Equation 3.

$5x - y = 31$

$\underline{2x + y = 18}$

$7x = 49$

$x = 7$

Substitute 7 for x in Equation 3 and solve for y.

$5(7) - y = 31$

$35 - y = 31$

$-y = -4$

$y = 4$

Substitute 7 for x and 4 for y in Equation 1 and solve for z.

$7 + 4 + z = 16$

$z + 11 = 16$

$z = 5$

The numbers are 7, 4 and 5.

25.

$\dfrac{x+2}{6} - \dfrac{y+4}{3} + \dfrac{z}{2} = 0$

$6\left(\dfrac{x+2}{6} - \dfrac{y+4}{3} + \dfrac{z}{2}\right) = 6(0)$

$(x+2) - 2(y+4) + 3z = 0$

$x + 2 - 2y - 8 + 3z = 0$

$x - 2y + 3z = 6$

$\dfrac{x+1}{2} + \dfrac{y-1}{2} - \dfrac{z}{4} = \dfrac{9}{2}$

$4\left(\dfrac{x+1}{2} + \dfrac{y-1}{2} - \dfrac{z}{4}\right) = 4\left(\dfrac{9}{2}\right)$

$2(x+1) + 2(y-1) - z = 18$

$2x + 2 + 2y - 2 - z = 18$

$2x + 2y - z = 18$

$\dfrac{x-5}{4} + \dfrac{y+1}{3} + \dfrac{z-2}{2} = \dfrac{19}{4}$

$12\left(\dfrac{x-5}{4} + \dfrac{y+1}{3} + \dfrac{z-2}{2}\right) = 12\left(\dfrac{19}{4}\right)$

$3(x-5) + 4(y+1) + 6(z-2) = 57$

$3x - 15 + 4y + 4 + 6z - 12 = 57$

$3x + 4y + 6z = 80$

We need to solve the equivalent system:

$$x - 2y + 3z = 6$$
$$2x + 2y - z = 18$$
$$3x + 4y + 6z = 80$$

Add the first two equations together.
$$x - 2y + 3z = 6$$
$$\underline{2x + 2y - z = 18}$$
$$3x + 2z = 24$$

Multiply the second equation by -2 and add it to the third equation.
$$-4x - 4y + 2z = -36$$
$$\underline{3x + 4y + 6z = 80}$$
$$-x + 8z = 44$$

Using the two reduced equations, we solve the system
$$3x + 2z = 24$$
$$-x + 8z = 44$$

Multiply the second equation by 3 and add the equations.

$$3x + 2z = 24$$
$$\underline{-3x + 24z = 132}$$
$$26z = 156$$
$$z = 6$$

Back-substitute to find x.
$$-x + 8(6) = 44$$
$$-x + 48 = 44$$
$$-x = -4$$
$$x = 4$$

Back substitute to find y.
$$x - 2y + 3z = 6$$
$$4 - 2y + 3(6) = 6$$
$$-2y = -16$$
$$y = 8$$

The solution is $(4, 8, 6)$.

27. Selected points may vary, but the equation will be the same.
$$y = ax^2 + bx + c$$
Use the points $(2, -2)$, $(4, 1)$, and $(6, -2)$ to get the system
$$4a + 2b + c = -2$$
$$16a + 4b + c = 1$$
$$36a + 6b + c = -2$$

Multiply the first equation by -1 and add to the second equation.
$$-4a - 2b - c = 2$$
$$\underline{16a + 4b + c = 1}$$
$$12a + 2b = 3$$

Multiply the first equation by -1 and add to the third equation.
$$-4a - 2b - c = 2$$
$$\underline{36a + 6b + c = -2}$$
$$32a + 4b = 0$$

Using the two reduced equations, we get the system
$$12a + 2b = 3$$
$$32a + 4b = 0$$

Multiply the first equation by -2 and add to the second equation.
$$-24a - 4b = -6$$
$$\underline{32a + 4b = 0}$$
$$8a = -6$$
$$a = -\frac{3}{4}$$

Back-substitute to solve for b.
$$12a + 2b = 3$$
$$12\left(-\frac{3}{4}\right) + 2b = 3$$
$$-9 + 2b = 3$$
$$2b = 12$$
$$b = 6$$

Back-substitute to solve for c.
$$4a + 2b + c = -2$$
$$4\left(-\frac{3}{4}\right) + 2(6) + c = -2$$
$$-3 + 12 + c = -2$$
$$c = -11$$

The equation is:
$$y = -\frac{3}{4}x^2 + 6x - 11$$

29. $ax - by - 2cz = 21$
$$ax + by + cz = 0$$
$$2ax - by + cz = 14$$

Add the first two equations.
$$ax - by - 2cz = 21$$
$$\underline{ax + by + cz = 0}$$
$$2ax - cz = 21$$

Multiply the first equation by -1 and add to the

third equation.
$$-ax + by + 2cz = -21$$
$$2ax - by + cz = 14$$
$$\overline{}$$
$$ax + 3cz = -7$$
Use the two reduced equations to get the following system:
$$2ax - cz = 21$$
$$ax + 3cz = -7$$
Multiply the second equation by -2 and add the equations.
$$2ax - cz = 21$$
$$-2ax - 6cz = 14$$
$$\overline{}$$
$$-7cz = 35$$
$$z = -\frac{5}{c}$$
Back-substitute to solve for x.
$$ax + 3cz = -7$$
$$ax + 3c\left(-\frac{5}{c}\right) = -7$$
$$ax - 15 = -7$$
$$ax = 8$$
$$x = \frac{8}{a}$$

Back-substitute to solve for y.
$$ax + by + cz = 0$$
$$a\left(\frac{8}{a}\right) + by + c\left(-\frac{5}{c}\right) = 0$$
$$8 + by - 5 = 0$$
$$by = -3$$
$$y = -\frac{3}{b}$$
The solution is $\left(\dfrac{8}{a}, -\dfrac{3}{b}, -\dfrac{5}{c}\right)$.

31. **a.** Substitute the values for x and y into the quadratic form.
$$224 = a(1)^2 + b(1) + c$$
$$a + b + c = 224$$

$$176 = a(3)^2 + b(3) + c$$
$$9a + 3b + c = 176$$

$$104 = a(4)^2 + b(4) + c$$
$$16a + 4b + c = 104$$

Multiply the first equation by -1 and add to both the second and the third equations to obtain 2 new equations with 2 variables.
$$-a - b - c = -224$$
$$9a + 3b + c = 176$$
$$\overline{}$$
$$8a + 2b = -48$$

$$-a - b - c = -224$$
$$16a + 4b + c = 104$$
$$\overline{}$$
$$15a + 3b = -120$$
Use the two new equations to solve for a and b. Multiply the first equation by -3 and the second equation by 2 and add the results together. Solve for a. Substitute that value in $8a + 2b = -48$ and solve for b.
$$-24a - 6b = 144$$
$$30a + 6b = -240$$
$$\overline{}$$
$$6a = -96$$
$$a = -16$$

$$8(-16) + 2b = -48$$
$$-128 + 2b = -48$$
$$2b = 80$$
$$b = 40$$

Substitute -16 for a and 40 for b into the equation $a + b + c = 224$ and solve for c.
$$-16 + 40 + c = 224$$
$$c = 200$$
The equation is $y = -16x^2 + 40x + 200$.

b. $y = -16(5)^2 + 40(5) + 200 = 0$
The ball hit the ground after 5 seconds.

33. Let x = annual spending in 2010 per person on housing.
Let y = annual spending in 2010 per person on vehicles/gas.
Let z = annual spending in 2010 per person on health care.
$$x + y + z = 13,840$$
$$x - y = 3864$$
$$x - z = 695$$
Solve the second equation for y.
$$x - y = 3864$$
$$-y = -x + 3864$$
$$y = x - 3864$$

Solve the third equation for z.

$$x - z = 695$$
$$-z = -x + 695$$
$$z = x - 695$$

Substitute the expressions for x and z into the first equation and solve for y.

$$x \;+\; y \;+\; z \;= 13{,}840$$

$$x + \overbrace{(x - 3864)}^{y} + \overbrace{(x - 695)}^{z} = 13{,}840$$
$$x + x - 3864 + x - 695 = 13{,}840$$
$$3x - 4559 = 13{,}840$$
$$3x = 18{,}399$$
$$x = 6133$$

Back-substitute to solve for y and z.

$$y = x - 3864$$
$$= 6133 - 3864$$
$$= 2269$$

$$z = x - 695$$
$$= 6133 - 695$$
$$= 5438$$

The annual spending in 2010 per person on housing is \$6133, on vehicles/gas is \$2269, and on health care is \$5438.

35. Let $m = $ the cost per gallon of milk.
Let $w = $ the cost per bottle of water.
Let $c = $ the cost per bag of chips.
The information is represented by the following system of equations.

$$2m + 5w + 6c = 19$$
$$w = 2c$$
$$m = w + 2$$

Use substitution to find m in terms of c:

$$m = \overset{2c}{\overset{\frown}{w}} + 2$$
$$m = 2c + 2$$

Use substitution to find c.

$$2m + 5w + 6c = 19$$
$$2(2c + 2) + 5(2c) + 6c = 19$$
$$4c + 4 + 10c + 6c = 19$$
$$20c + 4 = 19$$
$$20c = 15$$
$$c = 0.75$$

Find w.

$$w = 2c$$
$$w = 2(0.75)$$
$$= 1.50$$

Find m.

$$m = w + 2$$
$$m = 1.50 + 2$$
$$= 3.50$$

The cost per gallon of milk is \$3.50. The cost per bottle of water is \$1.50. The cost per bag of chips is \$0.75.

37. $x = $ number of \$8 tickets sold
$y = $ number of \$10 tickets sold
$z = $ number of \$12 tickets sold
From the given conditions we have the following system of equations.

$$x + y + z = 400$$
$$8x + 10y + 12z = 3700$$
$$x + y = 7z \;\text{ or }\; x + y - 7z = 0$$

Eliminate z from Equations 1 and 2 multiplying Equation 1 by -12 and adding the resulting equation to Equation 2.

$$-12x - 12y - 12z = -4800$$
$$\underline{8x + 10y + 12z = 3700}$$
$$-4x - 2y = -1100 \quad \text{Equation 4}$$

Eliminate z from Equations 1 and 3 by multiplying Equation 1 by 7 and adding the resulting equation to Equation 3.

$$7x + 7y + 7z = 2800$$
$$\underline{x + y - 7z = 0}$$
$$8x + 8y = 2800 \quad \text{Equation 5}$$

Solve Equations 4 and 5 for x and y by multiplying Equation 4 by 2 and adding the resulting equation to Equation 5.

$$-8x - 4y = -2200$$
$$\underline{8x + 8y = 2800}$$
$$4y = 600$$
$$y = 150$$

Substitute 150 for y in Equation 5 and solve for x.

$$8x + 8(150) = 2800$$
$$8x = 2800 - 1200$$
$$8x = 1600$$
$$x = 200$$

Substitute 200 for x and 150 for y in Equation 1 and solve for z.

$$200 + 150 + z = 400$$
$$350 + z = 400$$
$$z = 50$$

The number of \$8 tickets sold was 200.
The number of \$10 tickets sold was 150.
The number of \$12 tickets sold was 50.

39. x = amount of money invested at 10%
y = amount of money invested at 12%
z = amount of money invested at 15%

$$x + y + z = 6700$$
$$0.08x + 0.10y + 0.12z = 716$$
$$z = x + y + 300$$

Arrange Equation 3 so that variable terms appear on the left and constants appear on the right.

$-x - y + z = 300$ Equation 4

Eliminate x and y from Equations 1 and 4 by adding Equations 1 and 4.

$$\begin{aligned} x + y + z &= 6700 \\ -x - y + z &= 300 \\ \hline 2z &= 7000 \\ z &= 3500 \end{aligned}$$

Substitute 3500 for z in Equation 1 and Equation 2 and simplify.

$$x + y + 3500 = 6700$$
$$x + y = 3200 \quad \text{Equation 5}$$
$$0.08x + 0.10y + 0.12(3500) = 716$$
$$0.08x + 0.10y + 420 = 716$$
$$0.08x + 10y = 296 \quad \text{Equation 6}$$

Solve Equations 5 and 6 for x and y by multiplying Equation 5 by –0.10 and adding the resulting equation to Equation 6.

$$\begin{aligned} -0.10x - 0.10y &= -320 \\ 0.08x + 0.10y &= 296 \\ \hline -0.02x &= 24 \\ x &= 1200 \end{aligned}$$

Substitute 1200 for x and 3,500 for z in Equation 1 and solve for y.

$$1200 + y + 3500 = 6700$$
$$y + 4700 = 6700$$
$$y = 2000$$

The person invested $1200 at 8%, $2000 at 10%, and $3500 at 12%.

41. $x + y + z = 180$

$$2x - 5 + z = 180$$
$$2x + z = 185$$

$$2x + 5 + y = 180$$
$$2x + y = 175$$

Multiply the second equation by –1 and add to the first equation. Use the new equation and the third equation to solve for x and z.

$$\begin{aligned} -2x - z &= -185 \\ x + y + z &= 180 \\ \hline -x + y &= -5 \end{aligned}$$

Multiply the new equation by -1.

$$\begin{aligned} x - y &= 5 \\ 2x + y &= 175 \\ \hline 3x &= 180 \\ x &= 60 \end{aligned}$$

$$\begin{aligned} 60 - y &= 5 \\ -y &= -55 \\ y &= 55 \end{aligned}$$

Substitute 60 for x and 55 for y in the first equation and solve for z.

$$60 + 55 + z = 180$$
$$z = 65$$

43. – 47. Answers will vary.

49. does not make sense; Explanations will vary. Sample explanation: A system of linear equations in three variables can contain an equation of the form $y = mx + b$. For this equation, the coefficient of z is 0.

51. makes sense

53. $x =$ number of triangles
$y =$ number of rectangles
$z =$ number of pentagons
$x + y + z = 40$
$3x + 4y + 5z = 153$
$2y + 5z = 72$

Eliminate x from Equations 1 and 2 by multiplying Equation 1 by –3 and adding the resulting equation to Equation 2.
$-3x - 3y - 3z = -120$
$\underline{3x + 4y + 5z = 153}$
$\qquad y + 2z = 33 \qquad$ Equation 4

Solve for z by multiplying Equation 4 by –2 and adding the resulting equation to Equation 3.
$2y + 5z = 72$
$\underline{-2y - 4z = -66}$
$\qquad z = 6$

Substitute 6 for z in Equation 4 and solve for y.
$y + 2(6) = 33$
$\qquad y + 12 = 33$
$\qquad y = 21$

Substitute 21 for y and 6 for z in Equation 1 and solve for x.
$x + 21 + 6 = 40$
$\qquad x + 27 = 40$
$\qquad x = 13$

The painting has 13 triangles, 21 rectangles, and 6 pentagons.

55.
$$\frac{3}{x-4} - \frac{2}{x+2} = \frac{3(x+2)}{(x-4)(x+2)} - \frac{2(x-4)}{(x-4)(x+2)}$$
$$= \frac{3x+6}{(x-4)(x+2)} - \frac{2x-8}{(x-4)(x+2)}$$
$$= \frac{3x+6-2x+8}{(x-4)(x+2)}$$
$$= \frac{x+14}{(x-4)(x+2)}$$

56.
$$\frac{5x-3}{x^2+1} + \frac{2x}{(x^2+1)^2} = \frac{(5x-3)(x^2+1)}{(x^2+1)(x^2+1)} + \frac{2x}{(x^2+1)^2}$$
$$= \frac{5x^3 - 3x^2 + 5x - 3}{(x^2+1)^2} + \frac{2x}{(x^2+1)^2}$$
$$= \frac{5x^3 - 3x^2 + 5x - 3 + 2x}{(x^2+1)^2}$$
$$= \frac{5x^3 - 3x^2 + 7x - 3}{(x^2+1)^2}$$

57.
$$A + B = 3$$
$$2A - 2B + C = 17$$
$$4A - 2C = 14$$
Solving $A + B = 3$ for B gives $B = 3 - A$.
Solving $4A - 2C = 14$ for A gives $C = 2A - 7$.
Use substitution to find f.
$$2A - 2B + C = 17$$
$$2A - 2\overset{B}{\overbrace{(3-A)}} + \overset{C}{\overbrace{2A - 7}} = 17$$
$$2A - 2(3-A) + 2A - 7 = 17$$
$$2A - 6 + 2A + 2A - 7 = 17$$
$$6A - 13 = 17$$
$$6A = 30$$
$$A = 5$$
Find B.
$B = 3 - A$
$B = 3 - 5 = -2$
Find C.
$C = 2A - 7$
$C = 2(5) - 7 = 3$
The solution set is $\{(5, -2, 3)\}$.

Section 8.3

Check Point Exercises

1. $\dfrac{5x-1}{(x-3)(x+4)} = \dfrac{A}{x-3} + \dfrac{B}{x+4}$

Multiply both sides of the equation by the least common denominator $(x - 3)(x + 4)$ and divide out common factors.

$5x-1 = A(x+4) + B(x-3)$

$5x-1 = Ax + 4A + Bx - 3B$

$5x-1 = (A+B)x + 4A - 3B$

Equate coefficients of like powers of x and equate constant terms.

$A + B = 5$

$4A - 3B = -1$

Solving the above system for A and B we find $A = 2$ and $B = 3$.

$\dfrac{5x-1}{(x-3)(x+4)} = \dfrac{2}{x-3} + \dfrac{3}{x+4}$

2. $\dfrac{x+2}{x(x-1)^2} = \dfrac{A}{x} + \dfrac{B}{x-1} + \dfrac{C}{(x-1)^2}$

Multiply both sides of the equation by the least common denominator $x(x-1)^2$ and divide out common factors.

$x+2 = A(x-1)^2 + Bx(x-1) + Cx$

$x+2 = A\left(x^2 - 2x + 1\right) + Bx^2 - Bx + Cx$

$x+2 = Ax^2 - 2Ax + A + Bx^2 - Bx + Cx$

$x+2 = Ax^2 + Bx^2 - 2Ax - Bx + Cx + A$

$x+2 = (A+B)x^2 + (-2A - B + C)x + A$

Equate coefficients of like powers of x and equate constant terms.

$A + B = 0$

$-2A - B + C = 1$

$A = 2$

Since $A = 2$, we find that $B = -2$ and $C = 3$ by substitution.

$\dfrac{x+2}{x(x-1)^2} = \dfrac{2}{x} - \dfrac{2}{x-1} + \dfrac{3}{(x-1)^2}$

3. $\dfrac{8x^2 + 12x - 20}{(x+3)\left(x^2+x+2\right)} = \dfrac{A}{x+3} + \dfrac{Bx+C}{x^2+x+2}$

Multiply both sides of the equation by the least common denominator $(x+3)\left(x^2+x+2\right)$ and divide out common factors.

$8x^2 + 12x - 20 = A\left(x^2+x+2\right) + (Bx+C)(x+3)$

$8x^2 + 12x - 20 = Ax^2 + Ax + 2A + Bx^2 + 3Bx + Cx + 3C$

$8x^2 + 12x - 20 = Ax^2 + Bx^2 + Ax + 3Bx + Cx + 2A + 3C$

$8x^2 + 12x - 20 = (A+B)x^2 + (A+3B+C)x + 2A + 3C$

Equate coefficients of like powers of x and equate constant terms.

$A + B = 8$

$A + 3B + C = 12$

$2A + 3C = -20$

Solving the above system for A, B, and C we find $A = 2$, $B = 6$, and $C = -8$.

$\dfrac{8x^2 + 12x - 20}{(x+3)\left(x^2+x+2\right)} = \dfrac{2}{x+3} + \dfrac{6x-8}{x^2+x+2}$

4. $\dfrac{2x^3 + x + 3}{\left(x^2+1\right)^2} = \dfrac{Ax+B}{x^2+1} + \dfrac{Cx+D}{\left(x^2+1\right)^2}$

Multiply both sides of the equation by the common denominator $\left(x^2+1\right)^2$ and divide out common factors.

$2x^3 + x + 3 = (Ax+B)(x^2+1) + Cx + D$

$2x^3 + x + 3 = Ax^3 + Bx^2 + Ax + B + Cx + D$

$2x^3 + x + 3 = Ax^3 + Bx^2 + Ax + Cx + B + D$

$2x^3 + x + 3 = Ax^3 + Bx^2 + (A+C)x + B + D$

Equate coefficients of like powers of x and equate constant terms.

$A = 2$

$B = 0$

$A + C = 1$

$B + D = 3$

Since $A = 2$ and $B = 0$ we find that $C = -1$ and $D = 3$ by substitution.

$\dfrac{2x^3 + x + 3}{\left(x^2+1\right)^2} = \dfrac{2x}{x^2+1} + \dfrac{-x+3}{\left(x^2+1\right)^2}$ or $\dfrac{2x}{x^2+1} - \dfrac{x-3}{\left(x^2+1\right)^2}$

Concept and Vocabulary Check 8.3

1. correct

2. incorrect

3. incorrect

4. correct

Exercise Set 8.3

1. $\dfrac{11x-10}{(x-2)(x+1)} = \dfrac{A}{x-2} + \dfrac{B}{x+1}$

3. $\dfrac{6x^2-14x-27}{(x+2)(x-3)^2} = \dfrac{A}{x+2} + \dfrac{B}{x-3} + \dfrac{C}{(x-3)^2}$

5. $\dfrac{5x^2-6x+7}{(x-1)(x^2+1)} = \dfrac{A}{x-1} + \dfrac{Bx+C}{x^2+1}$

7. $\dfrac{x^3+x^2}{(x^2+4)^2} = \dfrac{Ax+B}{x^2+4} + \dfrac{Cx+D}{(x^2+4)^2}$

9. $\dfrac{x}{(x-3)(x-2)} = \dfrac{A}{x-3} + \dfrac{B}{x-2}$

Multiply both sides of the equation by the least common denominator $(x-3)(x-2)$ and divide out common factors.

$x = A(x-2) + B(x-3)$

$x = A\,x - 2A + Bx - 3B$

$x = Ax + Bx - 2A - 3B$

$x = (A+B)x - (2A+3B)$

Equate coefficients of like powers of x, and equate constant terms.

$A+B = 1$

$2A+3B = 0$

Solving the above system for A and B, we find $A = 3$ and $B = -2$.

$\dfrac{x}{(x-3)(x-2)} = \dfrac{3}{x-3} - \dfrac{2}{x-2}$

11. $\dfrac{3x+50}{(x-9)(x+2)} = \dfrac{A}{x-9} + \dfrac{B}{x+2}$

Multiply both sides of the equation by the least common denominator $(x-9)(x+2)$ and divide out common factors.

$3x+50 = A(x+2) + B(x-9)$

$3x+50 = Ax + 2A + Bx - 9B$

$3x+50 = Ax + Bx + 2A - 9B$

$3x+50 = (A+B)x + (2A-9B)$

Equate coefficients of like powers of x, and equate constant terms.

$A+B = 3$

$2A-9B = 50$

Solving the above system for A and B, we find $A = 7$ and $B = -4$.

$\dfrac{3x+50}{(x-9)(x+2)} = \dfrac{7}{x-9} - \dfrac{4}{x+2}$

13. $\dfrac{7x-4}{x^2-x-12} = \dfrac{7x-4}{(x-4)(x+3)} = \dfrac{A}{x-4} + \dfrac{B}{x+3}$

Multiply both sides of the last equation by the least common denominator $(x-4)(x-3)$ and divide out common factors.

$7x-4 = A(x+3) + B(x-4)$

$7x-4 = A\,x + 3A + Bx - 4B$

$7x-4 = Ax + Bx + 3A - 4B$

$7x-4 = (A+B)x + (3A-4B)$

Equate coefficients of like powers of x, and equate constant terms.

$A+B = 7$

$3A-4B = -4$

Solving the above system for A and B, we find $A = \dfrac{24}{7}$ and $B = \dfrac{25}{7}$.

$\dfrac{7x-4}{x^2-x-12} = \dfrac{24}{7(x-4)} + \dfrac{25}{7(x+3)}$

15. $\dfrac{4}{(2x+1)(x-3)} = \dfrac{A}{2x+1} + \dfrac{B}{x-3}$

Multiply both sides of the equation by the least common denominator $(2x + 1)(x - 3)$ and divide out common factors.

$4 = A(x-3) + B(2x+1)$

$4 = Ax - A3 + B2x + B$

$4 = (A+2B)x + (-3A+B)$

Equate coefficients of like powers of x and equate the constant terms. Solve for A and B.

$A + 2B = 0$

$-3A + B = 4$

$3A + 6B = 0$

$-3A + B = 4$

$7B = 4$

$B = \dfrac{4}{7}$

$A + 2B = 0$

$6A - 2B = -8$

$7A = -8$

$A = -\dfrac{8}{7}$

$$\dfrac{4}{(2x+1)(x-3)} = \dfrac{-8}{7(2x+1)} + \dfrac{4}{7(x-3)}$$

17. $\dfrac{4x^2 + 13x - 9}{x(x-1)(x+3)} = \dfrac{A}{x} + \dfrac{B}{x-3} + \dfrac{C}{x+3}$

Multiply both sides of the equation by the least common denominator $x(x-1)(x+3)$ and divide out common factors.

$4x^2 + 13x - 9 = A(x\text{-}1)(x+3) + Bx(x+3) + Cx(x-1)$

$4x^2 + 13x - 9 = A(x^2 + 2x - 3) + Bx^2 + 3Bx + Cx^2 - Cx$

$4x^2 + 13x - 9 = Ax^2 + 2Ax - 3A + Bx^2 + 3Bx + Cx^2 - Cx$

$4x^2 + 13x - 9 = Ax^2 + Bx^2 + Cx^2 + 2Ax + 3Bx - Cx - 3A$

$4x^2 + 13x - 9 = (A + B + C)x^2 + (2A + 3B - C)x - 3A$

Equate coefficients of like powers of x, and equate constant terms.

$A + B + C = 4$

$2A + 3B - C = 13$

$-3A = -9$

Solving the above system for A, B, and C, we find $A = 3$ and $B = 2$, and $C = -1$.

$$\dfrac{4x^2 + 13x - 9}{x(x-1)(x+3)} = \dfrac{3}{x} + \dfrac{2}{x-1} - \dfrac{1}{x+3}$$

19. $\dfrac{4x^2 - 7x - 3}{x^3 - x} = \dfrac{4x^2 - 7x - 3}{x(x+1)(x-1)} = \dfrac{A}{x} + \dfrac{B}{x+1} + \dfrac{C}{x-1}$

Multiply both sides of the last equation by the least common denominator $x(x+1)(x-1)$ and divide out common factors.

$4x^2 - 7x - 3 = A(x+1)(x-1) + Bx(x-1) + Cx(x+1)$

$4x^2 - 7x - 3 = A(x^2 - 1) + Bx^2 - Bx + Cx^2 + Cx$

$4x^2 - 7x - 3 = Ax^2 - A + Bx^2 - Bx + Cx^2 + Cx$

$4x^2 - 7x - 3 = Ax^2 + Bx^2 + Cx^2 - Bx + Cx - A$

$4x^2 - 7x - 3 = (A + B + C)x^2 + (-B + C)x - A$

Equate coefficients of like powers of x, and equate constant terms.

$A + B + C = 4$

$-B + C = -7$

$-A = -3$

Solving the above system for A, B, and C, we find $A = 3$ and $B = 4$, and $C = -3$.

$$\dfrac{4x^2 - 7x - 3}{x^3 - x} = \dfrac{3}{x} + \dfrac{4}{x+1} - \dfrac{3}{x-1}$$

21. $\dfrac{6x - 11}{(x-1)^2} = \dfrac{A}{x-1} + \dfrac{B}{(x-1)^2}$

Multiply both sides of the equation by the least common denominator $(x-1)^2$ and divide out common factors.

$6x - 11 = A(x - 1) + B$

$6x - 11 = Ax - A + B$

Equate coefficients of like powers of x, and equate constant terms.

$A = 6$

$-A + B = -11$

Since $A = 6$, we find that $B = -5$ by substitution.

$$\dfrac{6x - 11}{(x-1)^2} = \dfrac{6}{x-1} - \dfrac{5}{(x-1)^2}$$

23. $\dfrac{x^2-6x+3}{(x-2)^3} = \dfrac{A}{x-2} + \dfrac{B}{(x-2)^2} + \dfrac{C}{(x-2)^3}$

Multiply both sides of the equation by the least common denominator $(x-2)^3$ and divide out common factors.

$x^2-6x+3 = A(x-2)^2 + B(x-2) + C$

$x^2-6x+3 = A(x^2-4x+4) + Bx-2B + C$

$x^2-6x+3 = Ax^2-4Ax+4A+Bx-2B+C$

$x^2-6x+3 = Ax^2-4Ax+Bx+4A-2B+C$

$x^2-6x+3 = Ax^2+(-4A+B)x+4A-2B+C$

Equate coefficients of like powers of x, and equate constant terms.

$A = 1$

$-4A + B = -6$

$4A - 2B + C = 3$

Since $A = 1$, we find that $B = -2$ and $C = -5$ by substitution. $\dfrac{x^2-6x+3}{(x-2)^3} = \dfrac{1}{x-2} - \dfrac{2}{(x-2)^2} - \dfrac{5}{(x-2)^3}$

25. $\dfrac{x^2+2x+7}{x(x-1)^2} = \dfrac{A}{x} + \dfrac{B}{x-1} + \dfrac{C}{(x-1)^2}$

Multiply both sides of the equation by the least common denominator $x(x-1)^2$ and divide out common factors.

$x^2+2x+7 = A(x-1)^2 + Bx(x-1) + Cx$

$x^2+2x+7 = A(x^2-2x+1) + Bx^2-Bx + Cx$

$x^2+2x+7 = Ax^2-2Ax+A+Bx^2-Bx+Cx$

$x^2+2x+7 = Ax^2+Bx^2-2Ax-Bx+Cx+A$

$x^2+2x+7 = (A+B)x^2+(-2A-B+C)x+A$

$A + B = 1$

$-2A - B + C = 2$

$A = 7$

Since $A = 7$, we find that $B = -6$ and $C = 10$ by substitution. $\dfrac{x^2+2x+7}{x(x-1)^2} = \dfrac{7}{x} - \dfrac{6}{x-1} + \dfrac{10}{(x-1)^2}$

27. $\dfrac{x^2}{(x+1)(x-1)^2} = \dfrac{A}{x+1} + \dfrac{B}{x-1} + \dfrac{C}{(x-1)^2}$

Multiply both sides of the equation by the least common denominator $(x+1)(x-1)^2$ and divide out common factors.

$x^2 = A(x-1)^2 + B(x+1)(x-1) + C(x+1)$

$x^2 = x^2A-2xA+A+Bx^2-B+Cx+C$

$x^2 = (A+B)x^2+(-2A+C)x+(A-B+C)$

Equate coefficients of like powers of x, and equate constant terms.

$A + B = 1$

$-2A + C = 0$

$A - B + C = 0$

Solving the above system for A, B, and C, we find $A = \dfrac{1}{4}$, $B = \dfrac{3}{4}$, and $C = \dfrac{1}{2}$.

$\dfrac{x^2}{(x+1)(x-1)^2} = \dfrac{1}{4(x+1)} + \dfrac{3}{4(x-1)} + \dfrac{1}{2(x-1)^2}$

29. $\dfrac{5x^2 - 6x + 7}{(x-1)(x^2+1)} = \dfrac{A}{x-1} + \dfrac{Bx+C}{x^2+1}$

Multiply both sides of the equation by the least common denominator $(x-1)(x^2+1)$ and divide out common factors.

$5x^2 - 6x + 7 = A(x^2+1) + (Bx+C)(x-1)$

$5x^2 - 6x + 7 = Ax^2 + A + Bx^2 - Bx + Cx - C$

$5x^2 - 6x + 7 = Ax^2 + Bx^2 - Bx + Cx + A - C$

$5x^2 - 6x + 7 = (A+B)x^2 + (-B+C)x + A - C$

Equate coefficients of like powers of x, and equate constant terms.

$A + B = 5$
$-B + C = -6$
$A - C = 7$

Solving the above system for A, B, and C, we find $A = 3$, $B = 2$, and $C = -4$.

$\dfrac{5x^2 - 6x + 7}{(x-1)(x^2+1)} = \dfrac{3}{x-1} + \dfrac{2x-4}{x^2+1}$

31. $\dfrac{5x^2 + 6x + 3}{(x+1)(x^2+2x+2)} = \dfrac{A}{x+1} + \dfrac{Bx+C}{x^2+2x+2}$

Multiply both sides of the equation by the least common denominator $(x+1)(x^2+2x+2)$ and divide out common factors.

$5x^2 + 6x + 3 = A(x^2+2x+2) + (Bx+C)(x+1)$

$5x^2 + 6x + 3 = Ax^2 + 2Ax + 2A + Bx^2 + Bx + Cx + C$

$5x^2 + 6x + 3 = Ax^2 + Bx^2 + 2Ax + Bx + Cx + 2A + C$

$5x^2 + 6x + 3 = (A+B)x^2 + (2A+B+C)x + 2A + C$

Equate coefficients of like powers of x, and equate constant terms.

$A + B = 5$
$2A + B + C = 6$
$2A + C = 3$

Solving the above system for A, B, and C, we find $A = 2$, $B = 3$, and $C = -1$.

$\dfrac{5x^2 + 6x + 3}{(x+1)(x^2+2x+2)} = \dfrac{2}{x+1} + \dfrac{3x-1}{x^2+2x+2}$

33. $\dfrac{x+4}{x^2(x^2+4)} = \dfrac{A}{x} + \dfrac{B}{x^2} + \dfrac{Cx+D}{x^2+4}$

Multiply both sides of the equation by the least common denominator $x^2(x^2+4)$ and divide out common factors.

$x + 4 = Ax(x^2+4) + B(x^2+4) + (Cx+D)x^2$

$x + 4 = Ax^3 + 4Ax + Bx^2 + 4B + Cx^3 + Dx^2$

$x + 4 = (A+C)x^3 + (B+D)x^2 + 4Ax + 4B$

Equate coefficients of like powers of x, and equate constant terms

$A + C = 0$

$B + D = 0$

$4A = 1$

$4B = 4$

Solving the above system for A, B, and C, we find $A = \dfrac{1}{4}$, $B = 1$, $C = -\dfrac{1}{4}$, and $D = -1$.

$$\frac{x+4}{x^2(x^2+4)} = \frac{1}{4x} + \frac{1}{x^2} + \frac{-1x-4}{4(x^2+4)}$$

35. $\dfrac{6x^2 - x + 1}{x^3 + x^2 + x + 1} = \dfrac{6x^2 - x + 1}{(x+1)(x^2+1)} = \dfrac{A}{x+1} + \dfrac{Bx+C}{x^2+1}$

Multiply both sides of the last equation by the least common denominator $(x+1)(x^2+1)$ and divide out common factors.

$6x^2 - x + 1 = A(x^2+1) + (Bx+C)(x+1)$

$6x^2 - x + 1 = Ax^2 + A + Bx^2 + Bx + Cx + C$

$6x^2 - x + 1 = Ax^2 + Bx^2 + Bx + Cx + A + C$

$6x^2 - x + 1 = (A+B)x^2 + (B+C)x + A + C$

Equate coefficients of like powers of x, and equate constant terms.

$A + B = 6$

$B + C = -1$

$A + C = 1$

Solving the above system for A, B, and C, we find $A = 4$, $B = 2$, and $C = -3$.

$$\frac{6x^2 - x + 1}{x^3 + x^2 + x + 1} = \frac{4}{x+1} + \frac{2x-3}{x^2+1}$$

37. $\dfrac{x^3 + x^2 + 2}{\left(x^2+2\right)^2} = \dfrac{Ax+B}{x^2+2} + \dfrac{Cx+D}{\left(x^2+2\right)^2}$

Multiply both sides of the last equation by the least common denominator $(x^2+2)^2$ and divide out common factors.

$x^3 + x^2 + 2 = \left(Ax+B\right)\left(x^2+2\right) + Cx + D$

$x^3 + x^2 + 2 = Ax^3 + Bx^2 + 2Ax + 2B + Cx + D$

$x^3 + x^2 + 2 = Ax^3 + Bx^2 + 2Ax + Cx + 2B + D$

$x^3 + x^2 + 2 = Ax^3 + Bx^2 + \left(2A+C\right)x + \left(2B+D\right)$

Equate coefficients of like powers of x, and equate constant terms.

$A = 1$

$B = 1$

$2A + C = 0$

$2B + D = 2$

Since $A = 1$ and $B = 1$, we find that $C = -2$ and $D = 0$ by substitution.

$$\frac{x^3 + x^2 + 2}{\left(x^2+2\right)^2} = \frac{x+1}{x^2+2} - \frac{2x}{\left(x^2+2\right)^2}$$

39. $\dfrac{x^3 - 4x^2 + 9x - 5}{(x^2 - 2x + 3)^2} = \dfrac{Ax + B}{x^2 - 2x + 3} + \dfrac{Cx + D}{(x^2 - 2x + 3)^2}$

Multiply both sides of the equation by the least common denominator $(x^2 - 2x + 3)^2$ and divide out common factors.

$x^3 - 4x^2 + 9x - 5 = (Ax + B)(x^2 - 2x + 3) + Cx + D$

$x^3 - 4x^2 + 9x - 5 = Ax^3 - 2Ax^2 + 3Ax + Bx^2 - 2Bx + 3B + Cx + D$

$x^3 - 4x^2 + 9x - 5 = Ax^3 - 2Ax^2 + Bx^2 + 3Ax - 2Bx + Cx + 3B + D$

$x^3 - 4x^2 + 9x - 5 = Ax^3 + (-2A + B)x^2 + (3A - 2B + C)x + 3B + D$

Equate coefficients of like powers of x, and equate constant terms.

$A = 1$
$-2A + B = -4$
$3A - 2B + C = 9$
$3B + D = -5$

Since $A = 1$, we find that $B = -2$, $C = 2$, and $D = 1$ by substitution.

$\dfrac{x^3 - 4x^2 + 9x - 5}{(x^2 - 2x + 3)^2} = \dfrac{x - 2}{x^2 - 2x + 3} + \dfrac{2x + 1}{(x^2 - 2x + 3)^2}$

41. $\dfrac{4x^2 + 3x + 14}{x^3 - 8} = \dfrac{4x^2 + 3x + 14}{(x - 2)(x^2 + 2x + 4)} = \dfrac{A}{x - 2} + \dfrac{Bx + C}{x^2 + 2x + 4}$

Multiply both sides of the last equation by the least common denominator $(x - 2)(x^2 + 2x + 4)$ and divide out common factors.

$4x^2 + 3x + 14 = A(x^2 + 2x + 4) + (Bx + C)(x - 2)$

$4x^2 + 3x + 14 = A^2 + 2Ax + 4A + Bx^2 - 2Bx + Cx - 2C$

$4x^2 + 3x + 14 = Ax^2 + Bx^2 + 2Ax - 2Bx + Cx + 4A - 2C$

$4x^2 + 3x + 14 = (A + B)x^2 + (2A - 2B + C)x + (4A - 2C)$

Equate coefficients of like powers of x, and equate constant terms.

$A + B = 4$

$2A - 2B + C = 3$

$4A - 2C = 14$

Solving the above system for A, B, and C, we find $A = 3$, $B = 1$, and $C = -1$.

$\dfrac{4x^2 + 3x + 4}{x^3 - 8} = \dfrac{3}{x - 2} + \dfrac{x - 1}{x^2 + 2x + 4}$

43. Divide $x^5 + 2$ by $x^2 - 1$.

$$
\begin{array}{r}
x^3 + x \\
x^2 - 1 \overline{) x^5 + 2} \\
\underline{x^5 - x^3} \\
x^3 \\
\underline{x^3 - x} \\
x + 2
\end{array}
$$

$\dfrac{x^5 + 2}{x^2 - 1} = x^3 + x + \dfrac{x + 2}{x^2 - 1}$

Decompose $\dfrac{x+2}{x^2-1}$.

$$\dfrac{x+2}{x^2-1}=\dfrac{x+2}{(x+1)(x-1)}=\dfrac{A}{x+1}+\dfrac{B}{x-1}$$

$$(x+1)(x-1)\dfrac{x+2}{(x+1)(x-1)}=(x+1)(x-1)\left(\dfrac{A}{x+1}+\dfrac{B}{x-1}\right)$$

$$x+2=(x-1)A+(x+1)B$$

$$x+2=Ax-A+Bx+B$$

$$x+2=(A+B)x+(-A+B)$$

Equate coefficients:

$$A+B=1$$
$$-A+B=2$$

Solving this system results in $A=-\dfrac{1}{2}$ and $B=\dfrac{3}{2}$.

$$\dfrac{x^5+2}{x^2-1}=x^3+x+\dfrac{-\frac{1}{2}}{x+1}+\dfrac{\frac{3}{2}}{x-1} \quad\text{or}\quad x^3+x+\dfrac{-1}{2(x+1)}+\dfrac{3}{2(x-1)}$$

45. Divide x^4-x^2+2 by x^3-x^2.

$$
\begin{array}{r}
x+1 \\
x^3-x^2\,\overline{)\,x^4-x^2+2} \\
\underline{x^4-x^3} \\
x^3-x^2 \\
\underline{x^3-x^2} \\
2
\end{array}
$$

$$\dfrac{x^4-x^2+2}{x^3-x^2}=x+1+\dfrac{2}{x^3-x^2}$$

Decompose $\dfrac{2}{x^3-x^2}$.

$$\dfrac{2}{x^3-x^2}=\dfrac{2}{x^2(x-1)}=\dfrac{A}{x}+\dfrac{B}{x^2}+\dfrac{C}{x-1}$$

$$x^2(x-1)\dfrac{2}{x^2(x-1)}=x^2(x-1)\left(\dfrac{A}{x}+\dfrac{B}{x^2}+\dfrac{C}{x-1}\right)$$

$$2=x(x-1)A+(x-1)B+x^2C$$

$$2=Ax^2-Ax+Bx-B+Cx^2$$

$$2=(A+C)x^2+(-A+B)x+(-B)$$

Equate coefficients:

$$A+C=0$$
$$-A+B=0$$
$$-B=2$$

Solving this system results in $A=-2$, $B=-2$, and $C=2$.

$$\dfrac{x^4-x^2+2}{x^3-x^2}=x+1+\dfrac{-2}{x}+\dfrac{-2}{x^2}+\dfrac{2}{x-1}$$

47. $\dfrac{1}{x^2 - c^2} = \dfrac{1}{(x+c)(x-c)} = \dfrac{A}{x+c} + \dfrac{B}{x-c}$

$(x+c)(x-c)\dfrac{1}{(x+c)(x-c)} = (x+c)(x-c)\left(\dfrac{A}{x+c} + \dfrac{B}{x-c}\right)$

$1 = (x-c)A + (x+c)B$

$1 = Ax - Ac + Bx + Bc$

$1 = (A+B)x + (-Ac + Bc)$

Equate coefficients:

$A + B = 0$

$-Ac + Bc = 1$

Solving this system results in $A = \dfrac{-1}{2c}$ and $B = \dfrac{1}{2c}$.

$\dfrac{1}{x^2 - c^2} = \dfrac{\dfrac{-1}{2c}}{x+c} + \dfrac{\dfrac{1}{2c}}{x-c}$

49. $\dfrac{ax+b}{(x-c)^2} = \dfrac{D}{x-c} + \dfrac{E}{(x-c)^2}$

$\dfrac{ax+b}{(x-c)^2} = \dfrac{D}{x-c} + \dfrac{E}{(x-c)^2}$

$(x-c)^2 \dfrac{ax+b}{(x-c)^2} = (x-c)^2\left(\dfrac{D}{x-c} + \dfrac{E}{(x-c)^2}\right)$

$ax + b = (x-c)D + E$

$ax + b = Dx - Dc + E$

$ax + b = (D)x + (-Dc + E)$

Equate coefficients:

$D = a$

$-Dc + E = b$

Solving this system results in $D = a$ and $E = ac + b$.

$\dfrac{ax+b}{(x-c)^2} = \dfrac{a}{x-c} + \dfrac{ac+b}{(x-c)^2}$

51. $\dfrac{1}{x(x+1)} = \dfrac{A}{x} + \dfrac{B}{x+1}$

Multiply both sides of the equation by the least common denominator $x(x+1)$ and divide out common factors.

$1 = A(x+1) + Bx$

$1 = Ax + A + Bx$

$1 = Ax + Bx + A$

$1 = (A+B)x + A$

Equate coefficients of like powers of x, and equate constant terms.

$A + B = 0$

$A = 1$

Since $A = 1$ we find that $B = -1$ by substitution.

$$\frac{1}{x(x+1)} = \frac{1}{x} - \frac{1}{x+1}$$

$$\frac{1}{1 \cdot 2} + \frac{1}{2 \cdot 3} + \frac{1}{3 \cdot 4} + \cdots \frac{1}{99 \cdot 100} = \left(\frac{1}{1} - \frac{1}{2}\right) + \left(\frac{1}{2} - \frac{1}{3}\right) + \left(\frac{1}{3} - \frac{1}{4}\right) + \cdots \left(\frac{1}{99} - \frac{1}{100}\right)$$

$$= \frac{1}{1} - \frac{1}{100}$$

$$= \frac{99}{100}$$

53. – 59. Answers will vary.

61. does not make sense; Explanations will vary. Sample explanation: To perform partial fraction decomposition, the degree in the numerator must be less than the degree in the denominator.

63. does not make sense; Explanations will vary. Sample explanation: The second denominator should be $(x+3)^2$.

65. $\dfrac{4x^2 + 5x - 9}{x^3 - 6x - 9} = \dfrac{4x^2 + 5x - 9}{(x-3)(x^2+3x+3)} = \dfrac{A}{x-3} + \dfrac{Bx+C}{x^2+3x+3}$

Multiply both sides of the last equation by the common denominator $(x-3)(x^2+3x+3)$ and divide out common factors.

$4x^2 + 5x - 9 = A(x^2 + 3x + 3) + (Bx + C)(x-3)$

$4x^2 + 5x - 9 = Ax^2 + 3Ax + 3A + Bx^2 - 3Bx + Cx - 3C$

$4x^2 + 5x - 9 = Ax^2 + Bx^2 + 3Ax - 3Bx + Cx + 3A - 3C$

$4x^2 + 5x - 9 = (A+B)x^2 + (3A - 3B + C)x + 3A - 3C$

Equate coefficients of like powers of x and equate constant terms.

$A + B = 4$

$3A - 3B + C = 5$

$3A - 3C = -9$

Solving the above system for A, B, and C, we find $A = 2$, and $B = 2$, and $C = 5$.

$$\frac{4x^2 + 5x - 9}{x^3 - 6x - 9} = \frac{2}{x-3} + \frac{2x+5}{x^2+3x+3}$$

66. $4x + 3y = 4$

$y = 2x - 7$

Substitute.

$4x + 3y = 4$

$4x + 3\overbrace{(2x-7)}^{y} = 4$

$4x + 3(2x - 7) = 4$

$4x + 6x - 21 = 4$

$10x = 25$

$x = 2.5$

Back-substitute.

$y = 2x - 7$

$y = 2(2.5) - 7 = -2$

The solution set is $\{(2.5, -2)\}$.

67. $2x+4y=-4$

$3x+5y=-3$

Multiply the first equation by –5 and the second equation by 4.

$-10x-20y=20$

$12x+20y=-12$

Add the equations and solve for x.

$2x=8$

$x=4$

Back-substitute to find y.

$2x+4y=-4$

$2(4)+4y=-4$

$8+4y=-4$

$4y=-12$

$y=-3$

The solution set is $\{(4,-3)\}$.

68. The points of intersection are $(0,-3)$ and $(2,-1)$.

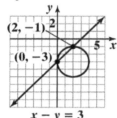

$x-y=3$

$(x-2)^2+(y+3)^2=4$

Point $(0,-3)$:

$x-y=3$

$0-(-3)=3$

$3=3,$ true

$(x-2)^2+(y+3)^2=4$

$(0-2)^2+(-3+3)^2=4$

$(-2)^2+(0)^2=4$

$4=4,$ true

Point $(2,-1)$:

$x-y=3$

$2-(-1)=3$

$3=3,$ true

$(x-2)^2+(y+3)^2=4$

$(2-2)^2+(-1+3)^2=4$

$(0)^2+(2)^2=4$

$4=4,$ true

Section 8.4

Check Point Exercises

1. $x^2=y-1$

$4x-y=-1$

Solve the first equation for y.

$y=x^2+1$

Substitute the expression x^2+1 for y in the second equation and solve for x.

$4x-(x^2+1)=-1$

$4x-x^2-1=-1$

$x^2-4x=0$

$x(x-4)=0$

$x=0$ or $x-4=0$

$x=4$

If $x=0,$ $y=(0)^2+1=1.$

If $x=4,$ $y=(4)^2+1=17.$

The solution set is $\{(0,1),(4,17)\}$.

2. $x+2y=0$

$(x-1)^2+(y-1)^2=5$

Solve the first equation for x.

$x=-2y$

Substitute the expression $-2y$ for x in the second equation and solve for y.

$(-2y-1)^2+(y-1)^2=5$

$4y^2+4y+1+y^2-2y+1=5$

$5y^2+2y-3=0$

$(5y-3)(y+1)=0$

$5y-3=0$ or $y+1=0$

$y=\dfrac{3}{5}$ or $y=-1$

If $y=\dfrac{3}{5},$ $x=-2\left(\dfrac{3}{5}\right)=-\dfrac{6}{5}.$

If $y=-1,$ $x=-2(-1)=2.$

The solution set is $\left\{\left(-\dfrac{6}{5},\dfrac{3}{5}\right),(2,-1)\right\}.$

3. $3x^2 + 2y^2 = 35$

$4x^2 + 3y^2 = 48$

Eliminate the y^2-term by multiplying the first equation by –3 and the second equation by 2. Add the resulting equations.

$-9x^2 - 6y = -105$

$\underline{8x^2 + 6y^2 = 96}$

$-x^2 = -9$

$x^2 = 9$

$x = \pm 3$

If $x = 3$,

$3(3)^2 + 2y^2 = 35$

$y^2 = 4$

$y = \pm 2$

If $x = -3$,

$3(-3)^2 + 2y^2 = 35$

$y^2 = 4$

$y = \pm 2$

The solution set is $\{(3,2),(3,-2),(-3,2),(-3,-2)\}$.

4. $y = x^2 + 5$

$x^2 + y^2 = 25$

Arrange the first equation so that variable terms appear on the left, and constants appear on the right.

Add the resulting equations to eliminate the x^2-terms and solve for y.

$-x^2 + y = 5$

$\underline{x^2 + y^2 = 25}$

$y^2 + y = 30$

$y^2 + y - 30 = 0$

$(y+6)(y-5) = 0$

$y + 6 = 0$ or $y - 5 = 0$

$y = -6$ or $y = 5$

If $y = -6$,

$x^2 + (-6)^2 = 25$

$x^2 = -11$

no real solution

If $y = 5$,

$x^2 + (5)^2 = 25$

$x^2 = 0$

$x = 0$

The solution set is $\{(0, 5)\}$.

5. $2x + 2y = 20$

$xy = 21$

Solve the second equation for x.

$x = \dfrac{21}{7}$

Substitute the expression $\dfrac{21}{y}$ for x in the first equation and solve for y.

$2\left(\dfrac{21}{y}\right) + 2y = 20$

$\dfrac{42}{y} + 2y = 20$

$y^2 - 10y + 21 = 0$

$(y-7)(y-3) = 0$

$y - 7 = 0$ or $y - 3 = 0$

$y = 7$ or $y = 3$

If $y = 7$, $x = \dfrac{21}{7} = 3$.

If $y = 3$, $x = \dfrac{21}{3} = 7$.

The dimensions are 7 feet by 3 feet.

Concept and Vocabulary Check 8.4

1. nonlinear

2. $\{(-4, 3), (0, -1)\}$

3. 3

4. $\{(2,\sqrt{3}), (2,-\sqrt{3}), (-2,\sqrt{3}), (-2,-\sqrt{3})\}$

5. -1; $y^2 + y = 6$

6. $\dfrac{4}{x}$; $\dfrac{4}{x}$; y

Exercise Set 8.4

1. $x + y = 2$

$\quad\quad y = x^2 - 4$

Solve the first equation for y. $y = 2 - x$.

Substitute the expression $2 - x$ for y in the second equation and solve for x.

$\quad\quad 2 - x = x^2 - 4$

$\quad\quad x^2 + x - 6 = 0$

$\quad\quad (x+3)(x-2) = 0$

$\quad\quad x+3 = 0 \quad \text{or} \quad x-2 = 0$

$\quad\quad x = -3 \quad \text{or} \quad x = 2$

If $x = -3$, $y = 2 - (-3) = 5$.

If $x = 2$, $y = 2 - 2 = 0$.

The solution set is $\{(-3, 5), (2, 0)\}$.

3. $x + y = 2$

$\quad\quad y = x^2 - 4x + 4$

Substitute the expression $x^2 - 4x + 4$ for y in the first equation and solve for x.

$\quad x + x^2 - 4x + 4 = 2$

$\quad\quad x^2 - 3x + 2 = 0$

$\quad\quad (x-1)(x-2) = 0$

$\quad\quad\quad x - 1 = 0 \quad\quad x - 2 = 0$

$\quad\quad\quad\quad x = 1 \quad\quad\quad x = 2$

Substitute $x = 1$ and then $x = 2$ into the equation $x + y = 2$ and solve for each value of y.

$\quad 1 + y = 2 \quad 2 + y = 2$

$\quad\quad y = 1 \quad\quad\quad y = 0$

The solution set is $\{(1, 1), (2, 0)\}$.

5. $y = x^2 - 4x - 10$

$\quad\quad y = -x^2 - 2x + 14$

Substitute the expression $x^2 - 4x - 10$ for y in the second equation and solve for x.

$\quad x^2 - 4x - 10 = -x^2 - 2x + 14$

$\quad\quad 2x^2 - 2x - 24 = 0$

$\quad\quad\quad x^2 - x - 12 = 0$

$\quad\quad (x-4)(x+3) = 0$

$\quad x - 4 = 0 \quad \text{or} \quad x + 3 = 0$

$\quad\quad x = 4 \quad \text{or} \quad x = -3$

If $x = 4$, $y = (4)^2 - 4(4) - 10 = -10$.

If $x = -3$, $y = (-3)^2 - 4(-3) - 10 = 11$.

The solution set is $\{(4, -10), (-3, 11)\}$.

7. $x^2 + y^2 = 25$

$\quad\quad x - y = 1$

Solve the second equation for y. $y = x - 1$

Substitute the expression $x - 1$ for y in the first equation and solve for x.

$\quad\quad x^2 + (x-1)^2 = 25$

$\quad x^2 + x^2 - 2x + 1 = 25$

$\quad\quad 2x^2 - 2x - 24 = 0$

$\quad\quad\quad x^2 - x - 12 = 0$

$\quad\quad (x-4)(x+3) = 0$

$\quad x - 4 = 0 \quad \text{or} \quad x + 3 = 0$

$\quad\quad x = 4 \quad \text{or} \quad x = -3$

If $x = 4$, $y = 4 - 1 = 3$.

If $x = -3$, $y = -3 - 1 = -4$.

The solution set is $\{(4, 3), (-3, -4)\}$.

9. $\quad\quad xy = 6$

$\quad 2x - y = 1$

Solve the first equation for y.

$\quad\quad y = \dfrac{6}{x}$

Substitute the expression $\dfrac{6}{x}$ for y in the second equation and solve for x.

$\quad\quad 2x - \dfrac{6}{x} = 1$

$\quad\quad\quad 2x^2 - 6 = x$

$\quad\quad 2x^2 - x - 6 = 0$

$\quad\quad (2x+3)(x-2) = 0$

$\quad 2x + 3 = 0 \quad \text{or} \quad x - 2 = 0$

$\quad\quad x = -\dfrac{3}{2} \quad \text{or} \quad x = 2$

If $x = -\dfrac{3}{2}$, $y = \dfrac{6}{-\frac{3}{2}} = -4$.

If $x = 2$, $y = \dfrac{6}{2} = 3$.

The solution set is $\left\{\left(-\dfrac{3}{2}, -4\right), (2, 3)\right\}$.

11. $y^2 = x^2 - 9$
$2y = x - 3$

Solve the second equation for y.

$y = \dfrac{x-3}{2}$

Substitute the expression $\dfrac{x-3}{2}$ for y in the first

equation and solve for x.

$\left(\dfrac{x-3}{2}\right)^2 = x^2 - 9$

$\dfrac{x^2 - 6x + 9}{4} = x^2 - 9$

$x^2 - 6x + 9 = 4x^2 - 36$

$3x^2 + 6x - 45 = 0$

$x^2 + 2x - 15 = 0$

$(x+5)(x-3) = 0$

$x + 5 = 0$ or $x - 3 = 0$

$x = -5$ or $x = 3$

If $x = -5$, $y = \dfrac{-5-3}{2} = -4$.

If $x = 3$, $y = \dfrac{3-3}{2} = 0$.

The solution set is $\{(-5,-4),(3,0)\}$.

13. $xy = 3$
$x^2 + y^2 = 10$

Solve the second equation for y.

$y = \dfrac{3}{x}$

Substitute the expression $\dfrac{3}{x}$ for y in the second

equation and solve for x.

$x^2 + \left(\dfrac{3}{x}\right)^2 = 10$

$x^2 + \dfrac{9}{x^2} - 10 = 0$

$x^4 - 10x^2 + 9 = 0$

$(x^2 - 9)(x^2 - 1) = 0$

$(x-3)(x+3)(x-1)(x+1) = 0$

$x - 3 = 0$ or $x + 3 = 0$ or $x - 1 = 0$ or $x + 1 = 0$

$x = 3$ or $x = -3$ or $x = 1$ or $x = -1$

If $x = 3$, $y = \dfrac{3}{3} = 1$.

If $x = -3$, $y = \dfrac{3}{-3} = -1$.

If $x = 1$, $y = \dfrac{3}{1} = 3$.

If $x = -1$, $y = \dfrac{3}{-1} = -3$.

The solution set is $\{(3,1),(-3,-1),(1,3),(-1,-3)\}$.

15. $x + y = 1$
$x^2 + xy - y^2 = -5$

Solve the first equation for y. $y = 1 - x$

Substitute the expression $1 - x$ for y in the second
equation and solve for x.

$x^2 + x(1-x) - (1-x)^2 = -5$

$x^2 + x - x^2 - (1 - 2x + x^2) = -5$

$x - 1 + 2x - x^2 = -5$

$x^2 - 3x - 4 = 0$

$(x-4)(x+1) = 0$

$x - 4 = 0$ or $x + 1 = 0$

$x = 4$ or $x = -1$

If $x = 4$, $y = 1 - 4 = -3$.

If $x = -1$, $y = 1 - (-1) = 2$.

The solution set is $\{(4,-3),(-1,2)\}$.

17. $x + y = 1$

$(x-1)^2 + (y+2)^2 = 10$

Solve the first equation for *y*.

$y = 1 - x$

Substitute the expression $1 - x$ for *y* in the second equation and solve for *x*.

$(x-1)^2 + (1-x+2)^2 = 10$

$(x-1)^2 + (3-x)^2 = 10$

$x^2 - 2x + 1 + 9 - 6x + x^2 - 10 = 0$

$2x^2 - 8x = 0$

$x^2 - 4x = 0$

$x(x-4) = 0$

$x = 0$ or $x - 4 = 0$

$x = 4$

If $x = 0, y = 1 - 0 = 1$.

If $x = 4, y = 1 - 4 = -3$.

The solution set is $\{(0,1),(4,-3)\}$.

19. Eliminate the y^2 –terms by adding the equations.

$x^2 + y^2 = 13$

$\underline{x^2 - y^2 = 5}$

$2x^2 = 18$

$x^2 = 9$

$x = \pm 3$

If $x = 3$,

$(3)^2 + y^2 = 13$

$y^2 = 4$

$y = \pm 2$

If $x = -3$,

$(-3)^2 + y^2 = 13$

$y^2 = 4$

$y = \pm 2$

The solution set is

$\{(3, 2), (3, -2), (-3, 2), (-3, -2)\}$.

21. $x^2 - 4y^2 = -7$

$3x^2 + y^2 = 31$

Eliminate the x^2 –terms by multiplying the first equation by –3 and adding the resulting equations.

$-3x^2 + 12y^2 = 21$

$\underline{3x^2 + y^2 = 31}$

$13y^2 = 52$

$y^2 = 4$

$y = \pm 2$

If $y = 2$,

$x^2 - 4(2)^2 = -7$

$x^2 = 9$

$x = \pm 3$

If $y = -2$,

$x^2 - 4(-2)^2 = -7$

$x^2 = 9$

$x = \pm 3$

The solution set is $\{(3,2),(3,-2),(-3,2),(-3,-2)\}$.

23. Arrange the equations so that variable terms appear on the left and constants appear on the right.

$3x^2 + 4y^2 = 16$

$2x^2 - 3y^2 = 5$

Eliminate the y^2 –terms by multiplying the first equation by 3 and the second equation by 4. Add the resulting equations.

$9x^2 + 12y^2 = 48$

$\underline{8x^2 - 12y^2 = 20}$

$17x^2 = 68$

$x^2 = 4$

$x = \pm 2$

If $x = 2$,

$3(2)^2 + 4y^2 = 16$

$y^2 = 1$

$y = \pm 1$

If $x = -2$,

$3(-2)^2 + 4y^2 = 16$

$y = \pm 1$

The solution set is

$\{(2, 1), (2, -1), (-2, 1), (-2, -1)\}$.

25.
$$x^2 + y^2 = 25$$
$$(x-8)^2 + y^2 = 41$$

Expand the second equation and eliminate x^2 and y^2–terms by multiplying the first equation by –1 and adding the resulting equations.

$$x^2 - 16x + 64 + y^2 = 41$$
$$\underline{-x^2 - y^2 = -25}$$
$$-16x + 64 = 16$$
$$-16x = -48$$
$$x = 3$$

If $x = 3$,
$$(3)^2 + y^2 = 25$$
$$y^2 = 16$$
$$y = \pm 4$$

The solution set is $\{(3,4),\ (3,-4)\}$.

27.
$$y^2 - x = 4$$
$$x^2 + y^2 = 4$$

Eliminate the y^2–terms by multiplying the first equation by –1 and adding the resulting equations.

$$x - y^2 = -4$$
$$\underline{x^2 + y^2 = 4}$$
$$x^2 + x = 0$$
$$x(x+1) = 0$$
$$x = 0 \quad \text{or} \quad x + 1 = 0$$
$$x = -1$$

If $x = 0$,
$$y^2 = 4$$
$$y = \pm 2$$

If $x = -1$,
$$y^2 - (-1) = 4$$
$$y^2 = 3$$
$$y = \pm\sqrt{3}$$

The solution set is
$$\left\{(0,2),(0,-2),\left(-1,\sqrt{3}\right),\left(-1,-\sqrt{3}\right)\right\}.$$

$$\left\{(0,-4),\left(2\sqrt{3},2\right),\left(-2\sqrt{3},2\right)\right\}.$$

29. The addition method is used here to solve the system.
$$3x^2 + 4y^2 = 16$$
$$2x^2 - 3y^2 = 5$$

Eliminate the y^2–terms by multiplying the first equation by 3 and the second equation by 4. Add the resulting equations.

$$9x^2 + 12y^2 = 48$$
$$\underline{8x^2 - 12y^2 = 20}$$
$$17x^2 = 68$$
$$x^2 = 4$$
$$x = \pm 2$$

If $x = 2$,
$$3(2)^2 + 4y^2 = 16$$
$$y^2 = 1$$
$$y = \pm 1$$

If $x = -2$,
$$3(-2)^2 + 4y^2 = 16$$
$$y = \pm 1$$

The solution set is
$$\{(2, 1), (2, -1), (-2, 1), (-2, -1)\}.$$

31. The substitution method is used here to solve the system.
$$2x^2 + y^2 = 18$$
$$xy = 4$$

Solve the second equation for y.
$$y = \frac{4}{x}$$

Substitute the expression $\dfrac{4}{x}$ for y in the first equation and solve for x.

$$2x^2 + \left(\frac{4}{x}\right)^2 = 18$$
$$2x^2 + \frac{16}{x^2} = 18$$
$$2x^4 + 16 = 18x^2$$
$$x^4 - 9x^2 + 8 = 0$$
$$\left(x^2 - 8\right)\left(x^2 - 1\right) = 0$$
$$x^2 - 8 = 0 \quad \text{or} \quad x^2 - 1 = 0$$
$$x^2 = 8 \quad \text{or} \quad x^2 = 1$$
$$x = \pm 2\sqrt{2} \quad \text{or} \quad x = \pm 1$$

If $x = 2\sqrt{2}$, $y = \dfrac{4}{2\sqrt{2}} = \sqrt{2}$.

If $x = -2\sqrt{2}$, $y = \dfrac{4}{-2\sqrt{2}} = -\sqrt{2}$.

If $x = 1$, $y = \dfrac{4}{1} = 4$.

If $x = -1$, $y = \dfrac{4}{-1} = -4$.

The solution set is

$$\left\{ \left(2\sqrt{2}, \sqrt{2}\right), \left(-2\sqrt{2}, -\sqrt{2}\right), (1,4), (-1,-4) \right\}.$$

33. The substitution method is used here to solve the system.
$x^2 + 4y^2 = 20$
$x + 2y = 6$
Solve the second equation for x.
$x = 6 - 2y$
Substitute the expression $6 - 2y$ for x in the first equation and solve for y.

$$\left(6 - 2y\right)^2 + 4y^2 = 20$$

$$36 - 24y + 4y^2 + 4y^2 - 20 = 0$$

$$8y^2 - 24y + 16 = 0$$

$$y^2 - 3y + 2 = 0$$

$$\left(y - 2\right)\left(y - 1\right) = 0$$

$y - 2 = 0$ or $y - 1 = 0$
$y = 2$ or $y = 1$
If $y = 2, x = 6 - 2(2) = 2$.
If $y = 1, x = 6 - 2(1) = 4$.

The solution set is $\left\{ (2,2),(4,1) \right\}$.

35. Eliminate y by adding the equations.

$$x^3 + y = 0$$
$$\underline{x^2 - y = 0}$$
$$x^3 + x^2 = 0$$

$$x^2(x+1) = 0$$

$x^2 = 0$ or $x + 1 = 0$
$x = 0$ or $x = -1$
If $x = 0$,
$(0)^3 + y = 0$
$y = 0$
If $x = -1$,
$(-1)^3 + y = 0$
$y = 1$
The solution set is $\left\{ (0,0),(-1,1) \right\}$.

37. The substitution method is used here to solve the system.

$$x^2 + \left(y - 2\right)^2 = 4$$
$$x^2 - 2y = 0$$

Solve the second equation for x^2.
$x^2 = 2y$

Substitute the expression $2y$ for x^2 in the first equation and solve for y.

$$2y + \left(y - 2\right)^2 = 4$$

$$2y + y^2 - 4y + 4 = 4$$

$$y^2 - 2y = 0$$

$$y(y - 2) = 0$$

$y = 0$ or $y - 2 = 0$
$y = 2$
If $y = 0$,
$x^2 = 2(0)$
$x^2 = 0$
$x = 0$
If $y = 2$,
$x^2 = 2(2)$
$x^2 = 4$
$x = \pm 2$
The solution set is $\{(0, 0), (-2, 2), (2, 2)\}$.

39. The substitution method is used here to solve the system.

$$y = \left(x + 3\right)^2$$
$$x + 2y = -2$$

Solve the first equation for x.
$x = -2y - 2$
Substitute the expression $-2y-2$ for x in the first equation and solve for y.

$$y = \left(-2y - 2 + 3\right)^2 = \left(-2y + 1\right)^2$$

$$y = 4y^2 - 4y + 1$$

$$4y^2 - 5y + 1 = 0$$

$$(4y - 1)(y - 1) = 0$$

$4y - 1 = 0$ or $y - 1 = 0$

$$y = \frac{1}{4} \quad \text{or} \quad y = 1$$

If $y = \dfrac{1}{4}$, $x = -2\left(\dfrac{1}{4}\right) - 2 = -\dfrac{5}{2}$.

If $y = 1$, $x = -2(1) - 2 = -4$.

The solution set is $\left\{ (-4, 1), \left(-\dfrac{5}{2}, \dfrac{1}{4}\right) \right\}$.

41. The substitution method is used here to solve the system.

$x^2 + y^2 + 3y = 22$

$2x + y = -1$

Solve the second equation for y.

$y = -2x - 1$

Substitute the expression $-2x-1$ for y in the first equation and solve for x.

$x^2 + (-2x-1)^2 + 3(-2x-1) - 22 = 0$

$x^2 + 4x^2 + 4x + 1 - 6x - 3 - 22 = 0$

$5x^2 - 2x - 24 = 0$

$(5x-12)(x+2) = 0$

$5x - 12 = 0$ or $x + 2 = 0$

$x = \dfrac{12}{5}$ or $x = -2$

If $x = \dfrac{12}{5}$, $y = -2\left(\dfrac{12}{5}\right) - 1 = -\dfrac{29}{5}$.

If $x = -2$, $y = -2(-2) - 1 = 3$.

The solution set is $\left\{\left(\dfrac{12}{5}, -\dfrac{29}{5}\right), (-2, 3)\right\}$.

43. The substitution method is used here to solve the system.

$x + y = 10$

$xy = 24$

Solve the first equation for y.

$y = 10 - x$

Substitute the expression $10 - x$ for y in the second equation and solve for x.

$x(10 - x) = 24$

$10x - x^2 = 24$

$x^2 - 10x + 24 = 0$

$(x-4)(x-6) = 0$

$x - 4 = 0$ or $x - 6 = 0$

$x = 4$ or $x = 6$

If $x = 4$, $y = 10 - 4 = 6$.

If $x = 6$, $y = 10 - 6 = 4$.

The numbers are 4 and 6.

45. Eliminate the y^2–terms by adding the equations.

$x^2 - y^2 = 3$

$\underline{2x^2 + y^2 = 9}$

$3x^2 = 12$

$x^2 = 4$

$x = \pm 2$

If $x = 2$,

$2(2)^2 + y^2 = 9$

$y^2 = 1$

$y = \pm 1$

If $x = -2$,

$2(-2)^2 + y^2 = 9$

$y^2 = 1$

$y = \pm 1$

The numbers are 2 and 1, 2 and –1, –2 and 1, or –2 and –1.

47. $2x^2 + xy = 6$

$x^2 + 2xy = 0$

Multiply the first equation by -2 and add the two equations.

$-4x^2 - 2xy = -12$

$\underline{x^2 + 2xy = 0}$

$-3x^2 = -12$

$x^2 = 4$

$x = \pm 2$

Back-substitute these values for x in the second equation and solve for y.

For $x = -2$: $(-2)^2 + 2(-2)y = 0$

$4 - 4y = 0$

$-4y = -4$

$y = 1$

For $x = 2$: $(2)^2 + 2(2)y = 0$

$4 + 4y = 0$

$4y = -4$

$y = -1$

The solution set is $\{(-2,1),(2,-1)\}$.

49. $-4x + y = 12$

$y = x^3 + 3x^2$

Substitute $x^3 + 3x^2$ for y in the first equation and solve for x.

$-4x + \left(x^3 + 3x^2 \right) = 12$

$x^3 + 3x^2 - 4x - 12 = 0$

$x^2 \left(x + 3 \right) - 4\left(x + 3 \right) = 0$

$\left(x + 3 \right)\left(x^2 - 4 \right) = 0$

$\left(x + 3 \right)\left(x - 2 \right)\left(x + 2 \right) = 0$

$x = -3$, $x = 2$, or $x = -2$

Substitute these values for x in the second equation and solve for y.

For $x = -3$: $y = (-3)^3 + 3(-3)^2 = -27 + 27$

$= 0$

For $x = 2$: $y = (2)^3 + 3(2)^2 = 8 + 12$

$= 20$

For $x = -2$: $y = (-2)^3 + 3(-2)^2 = -8 + 12$

$= 4$

The solution set is $\left\{ (-3,0),(2,20),(-2,4) \right\}$.

51. $\dfrac{3}{x^2} + \dfrac{1}{y^2} = 7$

$\dfrac{5}{x^2} - \dfrac{2}{y^2} = -3$

Multiply the first equation by 2 and add the equations.

$\dfrac{6}{x^2} + \dfrac{2}{y^2} = 14$

$\dfrac{5}{x^2} - \dfrac{2}{y^2} = -3$

────────────

$\dfrac{11}{x^2} = 11$

$x^2 = 1$

$x = \pm 1$

Back-substitute these values for x in the first equation and solve for y.

For $x = -1$:

$\dfrac{3}{(-1)^2} + \dfrac{1}{y^2} = 7$

$3 + \dfrac{1}{y^2} = 7$

$\dfrac{1}{y^2} = 4$

$y^2 = \dfrac{1}{4}$

$y = \pm \dfrac{1}{2}$

For $x = 1$:

$\dfrac{3}{(1)^2} + \dfrac{1}{y^2} = 7$

$3 + \dfrac{1}{y^2} = 7$

$\dfrac{1}{y^2} = 4$

$y^2 = \dfrac{1}{4}$

$y = \pm \dfrac{1}{2}$

The solution set is

$\left\{ \left(-1, -\dfrac{1}{2} \right), \left(-1, \dfrac{1}{2} \right), \left(1, -\dfrac{1}{2} \right), \left(1, \dfrac{1}{2} \right) \right\}$.

53.

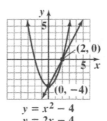

$y = x^2 - 4$
$y = 2x - 4$

55. $16x^2 + 4y^2 = 64$

$y = x^2 - 4$

Substitute the expression $x^2 - 4$ for y in the first equation and solve for x.

$16x^2 + 4\left(x^2 - 4 \right)^2 = 64$

$16x^2 + 4\left(x^4 - 8x^2 + 16 \right) = 64$

$16x^2 + 4x^4 - 32x^2 + 64 = 64$

$4x^4 - 16x^2 = 0$

$x^4 - 4x^2 = 0$

$x^2 \left(x^2 - 4 \right) = 0$

$x^2 = 0$ or $x^2 - 4 = 0$

$x = 0$ or $x^2 = 4$

$x = \pm 2$

If $x = 0$, $y = (0)^2 - 4 = -4$.

If $x = 2$, $y = (2)^2 - 4 = 0$.

If $x = -2$, $y = (-2)^2 - 4 = 0$.

It is possible for the comet to intersect the orbiting body at (0, –4), (–2, 0), (2, 0).

57. $2L + 2W = 36$

$\qquad LW = 77$

Divide each term in the first equation by 2 and solve *L*.

$L + W = 18$

$\qquad L = 18 - W$

Substitute the expression $18 - W$ for *L* in the second equation and solve for *W*.

$$(18 - W)W = 77$$

$$18W - W^2 = 77$$

$$W^2 - 18W + 77 = 0$$

$$(W - 11)(W - 7) = 0$$

$$W - 11 = 0 \quad \text{or} \quad W - 7 = 0$$

$$W = 11 \quad \text{or} \quad W = 7$$

If $W = 11, L = 18 - 11 = 7$.

If $W = 7, L = 18 - 7 = 11$.

The dimensions are 11 feet by 7 feet.

59. $L^2 + W^2 = 10^2 = 100$

$\qquad LW = 48$

Solve the second equation for *L*. $L = \dfrac{48}{W}$

Substitute the expression $\dfrac{48}{W}$ for *L* in the first equation and solve for *W*.

$$\left(\frac{48}{W}\right)^2 + W^2 = 100$$

$$\frac{2304}{W^2} + W^2 - 100 = 0$$

$$2304 + W^4 - 100W^2 = 0$$

$$W^4 - 100W^2 + 2304 = 0$$

$$(W^2 - 36)(W^2 - 64) = 0$$

$$W^2 - 36 = 0 \quad \text{or} \quad W^2 - 64 = 0$$

$$W^2 = 36 \quad \text{or} \quad W^2 = 64$$

$$W = \pm 6 \quad \text{or} \quad W = \pm 8$$

The width cannot be –6 or –8 inches.

If $W = 6$,

$$L = \frac{48}{6} = 8$$

If $W = 8$,

$$L = \frac{48}{8} = 6$$

The dimensions are 8 inches by 6 inches.

61. $x^2 - y^2 = 21$

$4x + 2y = 24$

Divide each term in the second equation by 2 and solve for *y*.

$2x + y = 12$

$y = 12 - 2x$

Substitute the expression $12 - 2x$ for *y* in the first equation and solve for *x*.

$$x^2 - (12 - 2x)^2 = 21$$

$$x^2 - (144 - 48x + 4x^2) = 21$$

$$3x^2 - 48x + 165 = 0$$

$$x^2 - 16x + 55 = 0$$

$$(x - 5)(x - 11) = 0$$

$$x - 5 = 0 \quad \text{or} \quad x - 11 = 0$$

$$x = 5 \quad \text{or} \quad x = 11$$

If $x = 11$, $y = 12 - 2(11) = -10$.

If $x = 5$, $y = 12 - 2(5) = 2$.

The dimensions of the floor are 5 meters by 5 meters and the dimensions of the square that will accommodate the pool are 2 meters by 2 meters.

63. **a.** It appears from the graphs that the percentage of white-collar workers was the same as blue-collar workers between the 1940s and 1960s.

b. $0.5x - y = -18$

$\qquad y = -0.004x^2 + 0.23x + 41$

Substitute $-0.004x^2 + 0.23x + 41$ in the first equation and solve for *x*.

$$0.5x - y = -18$$

$$0.5x - \overbrace{(-0.004x^2 + 0.23x + 41)}^{y} = -18$$

$$0.5x + 0.004x^2 - 0.23x - 41 = -18$$

$$0.004x^2 + 0.27x - 23 = 0$$

Use the quadratic formula.

$$x = \frac{-b \pm \sqrt{b^2 - 4ac}}{2a}$$

$$x = \frac{-(0.27) \pm \sqrt{0.27^2 - 4(0.004)(-23)}}{2(0.004)}$$

$$x \approx 49 \quad \text{or} \quad -117$$

The percentage of white-collar workers was the same as blue-collar workers 49 years after 1900, or 1949.

Let $x = 49$ and solve for y in the white-collar model.

$$0.5x - y = -18$$
$$0.5(49) - y = -18$$
$$24.5 - y = -18$$
$$-y = -42.5$$
$$y \approx 43$$

The percentage of white-collar workers in 1949 was about 43%

Let $x = 49$ and solve for y in the blue-collar model.

$$y = -0.004(49)^2 + 0.23(49) + 41 \approx 43\%$$

The percentage of blue-collar workers in 1949 was about 43%.

c. According to the graph, the percentage of white-collar workers was the same as farmers in 1920. The percentages of white-collar workers and farmers in 1920 were both 28%.

d.
$$0.5x - y = -18$$
$$0.4x + y = 35$$
$$\overline{0.9x = 17}$$
$$x = \frac{17}{0.9}$$
$$x \approx 19$$

According to the models, the percentage of white-collar workers was the same as farmers 19 years after 1900, or 1919.

Let $x = 19$ and solve for y in the white-collar model.

$$0.5x - y = -18$$
$$0.5(19) - y = -18$$
$$9.5 - y = -18$$
$$-y = -27.5$$
$$y = 27.5$$

The percentage of white-collar workers in 1919 was about 27.5%

Let $x = 19$ and solve for y in the farming model.

$$0.4x + y = 35$$
$$0.4(19) + y = 35$$
$$0.4(19) + y = 35$$
$$7.6 + y = 35$$
$$y = 27.4$$

The percentage of farm workers in 1919 was about 27.4%

These answers model the actual data from part c (the graph) fairly well.

65. – 67. Answers will vary.

69. makes sense

71. makes sense

73. false; Changes to make the statement true will vary. A sample change is: A circle and a line can intersect in at most two points, and therefore such a system has at most two real solutions.

75. false; Changes to make the statement true will vary. A sample change is: It is possible that a system of two equations in two variables whose graphs represent circles do not intersect, or intersect in a single point. This means that the system would have no solution, or a single solution, respectively.

77. First determine the solution to the following system of equations.

$$xy = 20$$
$$x^2 + y^2 = 41$$

Solve the first equation for x.

$$x = \frac{20}{y}$$

Substitute the expression $\dfrac{20}{y}$ for x in the second equation and solve for in the second equation and solve for y.

$$\left(\frac{20}{y} \right)^2 + y^2 = 41$$
$$\frac{400}{y^2} + y^2 - 41 = 0$$
$$y^4 - 41y^2 + 400 = 0$$
$$\left(y^2 - 25 \right)\left(y^2 - 16 \right) = 0$$
$$y^2 - 25 = 0 \quad \text{or} \quad y^2 - 16 = 0$$
$$y^2 = 25 \quad \text{or} \quad y^2 = 16$$
$$y = \pm 5 \quad \text{or} \quad y = \pm 4$$

If $y = 5$, $x = \dfrac{20}{5} = 4$.

If $y = -5$, $x = \dfrac{20}{-5} = -4$.

If $y = 4$, $x = \dfrac{20}{4} = 5$.

If $y = -4$, $x = \dfrac{20}{-4} = -5$.

The rectangle formed by joining the points of intersection has sides a and b. The length of a is $\sqrt{(5-4)^2+(4-5)^2}=\sqrt{2}$. The length of b is $\sqrt{(4-(-5))^2+(5-(-4))^2}=9\sqrt{2}$.

The area of the rectangle is $a \cdot b = (\sqrt{2})(9\sqrt{2}) = 18$ square units.

79. $\log_y x = 3$

$\log_y (4x) = 5$

$x = y^3$

$4x = y^5$

Substitute y^3 for x in the equation $4x = y^5$ and solve for y.

$4(y^3) = y^5$

$4 = y^2$

$y = \pm 2$

Keep the positive base.

$x = 2^3 = 8$

The solution set is $\{(8, 2)\}$.

81.

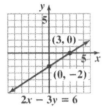

82.

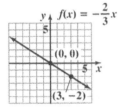

83.

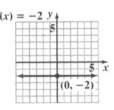

Mid-Chapter 8 Check Point

1. $x = 3y - 7$

$4x + 3y = 2$

Since the first equation is solved for x already, we will use substitution.
Let $x = 3y - 7$ in the second equation and solve for y.

$4(3y - 7) + 3y = 2$

$12y - 28 + 3y = 2$

$15y = 30$

$y = 2$

Substitute this value for y in the first equation.

$x = 3(2) - 7 = 6 - 7 = -1$

The solution is $(-1, 2)$.

2. $3x + 4y = -5$

$2x - 3y = 8$

Multiply the first equation by 3 and the second equation by 4, then add the equations.

$9x + 12y = -15$

$\underline{8x - 12y = 32}$

$17x = 17$

$x = 1$

Back-substitute to solve for y.

$3x + 4y = -5$

$3(1) + 4y = -5$

$3 + 4y = -5$

$4y = -8$

$y = -2$

The solution is $(1, -2)$.

3.
$$\frac{2x}{3}+\frac{y}{5}=6$$
$$\frac{x}{6}-\frac{y}{2}=-4$$

Multiply the first equation by 15 and the second equation by 6 to eliminate the fractions.

$$15\left(\frac{2x}{3}+\frac{y}{5}\right)=15(6)$$
$$10x+3y=90$$
$$6\left(\frac{x}{6}-\frac{y}{2}\right)=6(-4)$$
$$x-3y=-24$$

We now need to solve the equivalent system
$$10x+3y=90$$
$$x-3y=-24$$
Add the two equations to eliminate y.
$$10x+3y=90$$
$$\underline{x-3y=-24}$$
$$11x=66$$
$$x=6$$
Back-substitute to solve for y.
$$x-3y=-24$$
$$6-3y=-24$$
$$-3y=-30$$
$$y=10$$
The solution is $(6,10)$.

4.
$$y=4x-5$$
$$8x-2y=10$$

Since the first equation is already solved for y, we will use substitution.
Let $y=4x-5$ in the second equation and solve for x.
$$8x-2(4x-5)=10$$
$$8x-8x+10=10$$
$$10=10$$
This statement is an identity. The system is dependent so there are an infinite number of solutions. The solution set is $\{(x,y)\,|\,y=4x-5\}$.

5.
$$2x+5y=3$$
$$3x-2y=1$$

Multiply the first equation by 3 and the second equation by -2, then add the equations.
$$6x+15y=9$$
$$\underline{-6x+4y=-2}$$
$$19y=7$$
$$y=\frac{7}{19}$$
Back-substitute to solve for x.
$$2x+5y=3$$
$$2x+5\left(\frac{7}{19}\right)=3$$
$$2x+\frac{35}{19}=3$$
$$2x=\frac{22}{19}$$
$$x=\frac{11}{19}$$
The solution is $\left(\frac{11}{19},\frac{7}{19}\right)$.

6.
$$\frac{x}{12}-y=\frac{1}{4}$$
$$4x-48y=16$$

Solve the first equation for y.
$$\frac{x}{12}-y=\frac{1}{4}$$
$$-y=-\frac{x}{12}+\frac{1}{4}$$
$$y=\frac{x}{12}-\frac{1}{4}$$

Let $y=\frac{x}{12}-\frac{1}{4}$ in the second equation and solve for x.

$$4x-48\left(\frac{x}{12}-\frac{1}{4}\right)=16$$
$$4x-4x+12=16$$
$$12=16$$

This statement is a contradiction. The system is inconsistent so there is no solution. The solution is $\{\ \}$ or $\varnothing$.

7.

$$2x - y + 2z = -8$$
$$x + 2y - 3z = 9$$
$$3x - y - 4z = 3$$

Multiply the first equation by 2 and add to the second equation.

$$4x - 2y + 4z = -16$$
$$\underline{x + 2y - 3z = 9}$$
$$5x + z = -7$$

Multiply the first equation by -1 and add to the third equation.

$$-2x + y - 2z = 8$$
$$\underline{3x - y - 4z = 3}$$
$$x - 6z = 11$$

Use the two reduced equations to get the following system:

$$5x + z = -7$$
$$x - 6z = 11$$

Multiply the first equation by 6 and add to the second equation.

$$30x + 6z = -42$$
$$\underline{x - 6z = 11}$$
$$31x = -31$$
$$x = -1$$

Back-substitute to solve for z.

$$5x + z = -7$$
$$5(-1) + z = -7$$
$$-5 + z = -7$$
$$z = -2$$

Back-substitute to solve for y.

$$2x - y + 2z = -8$$
$$2(-1) - y + 2(-2) = -8$$
$$-2 - y - 4 = -8$$
$$-y = -2$$
$$y = 2$$

The solution is $(-1, 2, -2)$.

8.

$$x \quad\quad - 3z = -5$$
$$2x - y + 2z = 16$$
$$7x - 3y - 5z = 19$$

Multiply the second equation by -3 and add to the third equation.

$$-6x + 3y - 6z = -48$$
$$\underline{7x - 3y - 5z = 19}$$
$$x - 11z = -29$$

Use this reduced equation and the original first equation to obtain the following system:

$$x - 3z = -5$$
$$x - 11z = -29$$

Multiply the second equation by -1 and add to the first equation.

$$x - 3z = -5$$
$$\underline{-x + 11z = 29}$$
$$8z = 24$$
$$z = 3$$

Back-substitute to solve for x.

$$x - 3z = -5$$
$$x - 3(3) = -5$$
$$x - 9 = -5$$
$$x = 4$$

Back-substitute to solve for y.

$$2x - y + 2z = 16$$
$$2(4) - y + 2(3) = 16$$
$$8 - y + 6 = 16$$
$$-y = 2$$
$$y = -2$$

The solution is $(4, -2, 3)$.

9. Solve $x + 2y - 3 = 0$ for x and substitute into the other equation.

$$(\overbrace{-2y + 3}^{x})^2 + y^2 = 9$$
$$4y^2 - 12y + 9 + y^2 = 9$$
$$5y^2 - 12y = 0$$
$$y(5y - 12) = 0$$
$$y = 0 \ \text{ or } \ y = \frac{12}{5}$$

Back-substitute these values to find x.

When $y = 0$, $x = -2(0) + 3 = 3$.

When $y = \frac{12}{5}$, $x = -2\left(\frac{12}{5}\right) + 3 = -\frac{9}{5}$.

The solution set is $\left\{ (3, 0), \left(-\frac{9}{5}, \frac{12}{5} \right) \right\}$.

10. $3x^2 + 2y^2 = 14$

$2x^2 - y^2 = 7$

Multiply the second equation by 2 and add.

$3x^2 + 2y^2 = 14$

$\underline{4x^2 - 2y^2 = 14}$

$\qquad 7x^2 = 28$

$\qquad x^2 = 4$

$\qquad x = \pm 2$

Back-substitute these values to find y.

$3(2)^2 + 2y^2 = 14$

$\quad 12 + 2y^2 = 14$

$\qquad 2y^2 = 2$

$\qquad y^2 = 1$

$\qquad y = \pm 1$

$3(-2)^2 + 2y^2 = 14$

$\quad 12 + 2y^2 = 14$

$\qquad 2y^2 = 2$

$\qquad y^2 = 1$

$\qquad y = \pm 1$

The solution set is $\{(2,1),(2,-1),(-2,1),(-2,-1)\}$.

11. Use the first equation to substitute for y in the second equation.

$$x^2 + \overbrace{(x^2 - 6)^2}^{y} = 8$$

$x^2 + (x^2 - 6)^2 = 8$

$x^2 + x^4 - 12x^2 + 36 = 8$

$x^4 - 11x^2 + 28 = 0$

$(x^2 - 4)(x^2 - 7) = 0$

$x = \pm 2$ or $x = \pm\sqrt{7}$

Back-substitute these values to find x.

$y = (2)^2 - 6 \qquad y = (-2)^2 - 6$

$y = -2 \qquad\qquad y = -2$

$y = (\sqrt{7})^2 - 6 \quad y = (-\sqrt{7})^2 - 6$

$y = 1 \qquad\qquad y = 1$

The solution set is $\left\{(2,-2),(-2,-2),\left(\sqrt{7},1\right),\left(-\sqrt{7},1\right)\right\}$.

12. Use the first equation to substitute for x in the second equation.

$$2y^2 + (\overbrace{2y+4}^{x})y = 8$$

$$2y^2 + (2y+4)y = 8$$

$$2y^2 + 2y^2 + 4y = 8$$

$$4y^2 + 4y - 8 = 0$$

$$y^2 + y - 2 = 0$$

$$(y+2)(y-1) = 0$$

$$y = -2 \text{ or } y = 1$$

Back-substitute these values to find x.

$$x = 2(-2)+4 \quad x = 2(1)+4$$

$$x = 0 \qquad\qquad y = 6$$

The solution set is $\{(0,-2),(6,1)\}$.

13. $\dfrac{x^2-6x+3}{(x-2)^3} = \dfrac{A}{x-2} + \dfrac{B}{(x-2)^2} + \dfrac{C}{(x-2)^3}$

Multiply both sides of the equation by the common denominator $(x-2)^3$.

$$(x-2)^3 \frac{x^2-6x+3}{(x-2)^3} = (x-2)^3 \left(\frac{A}{x-2} + \frac{B}{(x-2)^2} + \frac{C}{(x-2)^3} \right)$$

$$x^2 - 6x + 3 = A(x-2)^2 + B(x-2) + C$$

$$x^2 - 6x + 3 = A(x^2-4x+4) + Bx - 2B + C$$

$$x^2 - 6x + 3 = Ax^2 - 4Ax + 4A + Bx - 2B + C$$

$$x^2 - 6x + 3 = (A)x^2 + (-4A+B)x + (4A-2B+C)$$

Equate coefficients of like powers of x and equate constant terms.

$$A = 1$$

$$-4A + B = -6$$

$$4A - 2B + C = 3$$

Since $A = 1$, we find that $B = -2$ and $C = -5$ by substitution.

$$\frac{2x^2-6x+3}{(x-2)^3} = \frac{1}{x-2} + \frac{-2}{(x-2)^2} + \frac{-5}{(x-2)^3}$$

14. $\dfrac{10x^2+9x-7}{(x+2)(x^2-1)} = \dfrac{10x^2+9x-7}{(x+2)(x+1)(x-1)} = \dfrac{A}{x+2} + \dfrac{B}{x+1} + \dfrac{C}{x-1}$

Multiply both sides of the equation by the least common denominator.

$$(x+2)(x+1)(x-1)\frac{10x^2+9x-7}{(x+2)(x+1)(x-1)} = (x+2)(x+1)(x-1)\left(\frac{A}{x+2} + \frac{B}{x+1} + \frac{C}{x-1} \right)$$

$$10x^2 + 9x - 7 = (x+1)(x-1)A + (x+2)(x-1)B + (x+2)(x+1)C$$

$$10x^2 + 9x - 7 = (x^2-1)A + (x^2+x-2)B + (x^2+3x+2)C$$

$$10x^2 + 9x - 7 = x^2A - A + x^2B + xB - 2B + x^2C + 3xC + 2C$$

$$10x^2 + 9x - 7 = (A+B+C)x^2 + (B+3C)x + (-A-2B+2C)$$

Equate coefficients of like powers of x, and equate constant terms.

$A + B + C = 10$

$B + 3C = 9$

$-A - 2B + 2C = -7$

Solving the above system for A, B, and C, we find $A = 5$, $B = 3$, and $C = 2$.

$$\frac{10x^2 + 9x - 7}{(x+2)(x^2-1)} = \frac{5}{x+2} + \frac{3}{x+1} + \frac{2}{x-1}$$

15. $\dfrac{x^2 + 4x - 23}{(x+3)(x^2+4)} = \dfrac{A}{x+3} + \dfrac{Bx+C}{x^2+4}$

Multiply both sides of the equation by the least common denominator.

$$(x+3)(x^2+4)\frac{x^2+4x-23}{(x+3)(x^2+4)} = (x+3)(x^2+4)\left(\frac{A}{x+3} + \frac{Bx+C}{x^2+4}\right)$$

$$x^2 + 4x - 23 = (x^2+4)A + (x+3)(Bx+C)$$

$$x^2 + 4x - 23 = x^2A + 4A + x^2B + 3xB + xC + 3C$$

$$x^2 + 4x - 23 = (A+B)x^2 + (3B+C)x + (4A+3C)$$

Equate coefficients of like powers of x, and equate constant terms.

$A + B = 1$

$3B + C = 4$

$4A + 3C = -23$

Solving the above system for A, B, and C, we find $A = -2$, $B = 3$, and $C = -5$.

$$\frac{x^2 + 4x - 23}{(x+3)(x^2+4)} = \frac{-2}{x+3} + \frac{3x-5}{x^2+4}$$

16. $\dfrac{x^3}{\left(x^2+4\right)^2} = \dfrac{Ax+B}{x^2+4} + \dfrac{Cx+D}{\left(x^2+4\right)^2}$

Multiply both sides of the equation by the least common denominator.

$$\left(x^2+4\right)^2\frac{x^3}{\left(x^2+4\right)^2} = \left(x^2+4\right)^2\left(\frac{Ax+B}{x^2+4} + \frac{Cx+D}{\left(x^2+4\right)^2}\right)$$

$$x^3 = \left(x^2+4\right)(Ax+B) + (Cx+D)$$

$$x^3 = x^3A + x^2B + 4xA + 4B + Cx + D$$

$$x^3 = (A)x^3 + (B)x^2 + (4A+C)x + (4B+D)$$

Equate coefficients of like powers of x, and equate constant terms.

$A = 1$

$B = 0$

$4A + C = 0$

$4B + D = 0$

Since $A = 1$ and $B = 0$, we find that $C = -4$ and $D = 0$ by substitution.

$$\frac{x^3}{\left(x^2+4\right)^2} = \frac{1x+0}{x^2+4} + \frac{-4x+0}{\left(x^2+4\right)^2} = \frac{x}{x^2+4} + \frac{-4x}{\left(x^2+4\right)^2}$$

17. **a.** $C(x) = 400,000 + 20x$

b. $R(x) = 100x$

c. $P(x) = R(x) - C(x)$

$\qquad = 100x - (400,000 + 20x)$

$\qquad = 80x - 400,000$

d. The break even point is the point where cost and revenue are the same. We need to solve the following system.

$y = 400,000 + 20x$

$y = 100x$

Let $y = 400,000 + 20x$ in the second equation and solve for x.

$400,000 + 20x = 100x$

$\qquad 400,000 = 80x$

$\qquad\quad 5000 = x$

Back-substitute to solve for y.

$y = 100x$

$\quad = 100(5000)$

$\quad = 500,000$

Thus, the break-even point is $(5000, \ 500,000)$. The company will break even when it produces and sells 5000 PDAs. At this level, the cost and revenue will both be $500,000.

18. Let $x =$ the number of roses.

Let $y =$ the number of carnations.

$\quad x + y = 20$

$3x + 1.5y = 39$

Solve the first equation for x.

$x + y = 20$

$\quad x = 20 - y$

Substitute this expression for x in the second equation and solve for y.

$3(20 - y) + 1.5y = 39$

$\quad 60 - 3y + 1.5y = 39$

$\qquad\qquad -1.5y = -21$

$\qquad\qquad\quad y = 14$

Back-substitute to solve for x.

$x = 20 - y = 20 - 14 = 6$

There are 6 roses and 14 carnations in the bouquet.

19. Because the sum of the measures of the angles of any triangle is $180°$, or $x + y + 90 = 180$

$$x + y = 90.$$

Because the angle with measures x and $(3y + 20)$ are supplementary, $x + (3y + 20) = 180$.

Simplify this equation.

$x + 3y + 20 = 180$

$\quad x + 3y = 160$

We now have the system

$x + y = 90$

$x + 3y = 160.$

To solve this system by the addition method, multiply the first equation by -1 and add to the second equation.

$-x - y = -90$

$\underline{x + 3y = 160}$

$\qquad 2y = 70$

$\qquad\; y = 35$

Back-substitute.

$x + y = 90$

$x + 35 = 90$

$\quad x = 55$

Thus, $x = 55°$, $y = 35°$, and $(3y + 20) = 125°$.

20. Using the points $(-1, 0)$, $(1, 4)$, and $(2, 3)$ in the equation $y = ax^2 + bx + c$, we get the following system of equations:

$a - b + c = 0$

$a + b + c = 4$

$4a + 2b + c = 3$

Add the first two equations.

$a - b + c = 0$

$\underline{a + b + c = 4}$

$2a \quad + 2c = 4$

Multiply the first equation by 2 and add to the third equation.

$2a - 2b + 2c = 0$

$\underline{4a + 2b + c = 3}$

$6a \quad\;\; + 3c = 3$

Using the two reduced equations, we get the following system of equations:

$2a + 2c = 4$

$6a + 3c = 3$

Multiply the first equation by -3 and add to the second equation.

$-6a - 6c = -12$

$\underline{6a + 3c = 3}$

$\qquad -3c = -9$

$\qquad\quad c = 3$

Back-substitute to solve for a.

$2a + 2c = 4$

$2a + 2(3) = 4$

$2a + 6 = 4$

$2a = -2$

$a = -1$

Back-substitute to solve for b.

$a + b + c = 4$

$-1 + b + 3 = 4$

$b = 2$

The equation is $y = -x^2 + 2x + 3$.

21. $2l + 2w = 21$

$lw = 20$

Solving $lw = 20$ for l gives $l = \dfrac{20}{w}$.

Substitute into the other equation:

$2\left(\overbrace{\dfrac{20}{w}}^{l}\right) + 2w = 21$

$\dfrac{40}{w} + 2w = 21$

$w\left(\dfrac{40}{w} + 2w\right) = w(21)$

$40 + 2w^2 = 21w$

$2w^2 - 21w + 40 = 0$

$(2w - 5)(w - 8) = 0$

$w = \dfrac{5}{2} = 2.5$ or $w = 8$.

Back-substitute to find l.

When $w = 2.5$, $l = \dfrac{20}{2.5} = 8$.

When $w = 8$, $l = \dfrac{20}{8} = 2.5$.

The dimensions of the rectangle are 2.5 m by 8.

Section 8.5

Check Point Exercises

1. $4x - 2y \geq 8$

    Graph the equation $4x - 2y = 8$ as a solid line.

    Choose a test point that is not on the line.

    Test $(0, 0)$

    $4x - 2y \geq 8$

    $4(0) - 2(0) \geq 8$

    $0 \geq 8$, false

    Since the statement is false, shade the other half-plane.

    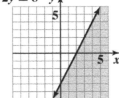

2. $y > -\dfrac{3}{4}x$

    Graph the equation $y = -\dfrac{3}{4}x$ as a dashed line.

    Choose a test point that is not on the line.

    Test $(1, 1)$

    $y > -\dfrac{3}{4}x$

    $1 > -\dfrac{3}{4}(1)$

    $1 > -\dfrac{3}{4}$, true

    Since the statement is true, shade the half-plane containing the point.

    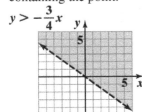

3. **a.** $y > 1$

    Graph the equation $y = 1$ as a dashed line.

    Choose a test point that is not on the line.

    Test $(0, 0)$

    $y > 1$

    $0 > 1$, false

    Since the statement is false, shade the other half-plane.

    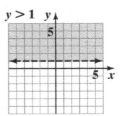

    b. Graph the equation $x = -2$ as a solid line.

    Choose a test point that is not on the line.

    Test $(0, 0)$

    $x \leq -2$

    $0 \leq -2$, false

    Since the statement is false, shade the other half-plane.

    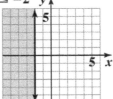

4. Graph $x^2 + y^2 = 16$ as a solid circle with radius 4 and center $(0, 0)$.

    Test $(0, 0)$

    $x^2 + y^2 \geq 16$

    $(0)^2 + (0)^2 \geq 16$

    $0 \geq 16$, false

    Shade the region not containing $(0, 0)$.

    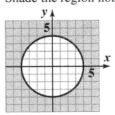

5. Point $B = (66, 130)$

$$4.9x - y \geq 165$$
$$4.9(66) - 130 \geq 165$$
$$193.4 \geq 165, \text{ true}$$
$$3.7x - y \leq 125$$
$$3.7(66) - 130 \leq 125$$
$$114.2 \leq 125, \text{ true}$$

Point B is a solution of the system.

6. Graph the equation $x - 3y = 6$ as a dashed line.

$$\text{Test } (0, 0)$$
$$x - 3y < 6$$
$$(0) - 3(0) < 6$$
$$0 < 6, \text{ true}$$

Since the statement is true, shade the half-plane containing the point.
Graph the equation $2x + 3y = -6$ as a solid line.

$$\text{Test } (0, 0)$$
$$2x + 3y \geq -6$$
$$2(0) + 3(0) \geq -6$$
$$0 \geq -6, \text{ true}$$

Since the statement is true, shade the half-plane containing the point.
For the solution graph, place an open circle at the point of intersection and shade the region that satisfies both inequalities.

$$\begin{cases} x - 3y < 6 \\ 2x + 3y \geq -6 \end{cases}$$

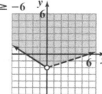

7. Begin by graphing $y = x^2 - 4$ as a solid parabola with vertex $(0, -4)$ and x-intercepts $(-2, 0)$ and $(2, 0)$. Since $(0, 0)$ makes the inequality $y \geq x^2 - 4$ true, shade the region containing $(0, 0)$. Graph $x + y = 2$ as a solid line by using its x-intercept, $(2, 0)$, and its y-intercept $(0, 2)$. Since $(0, 0)$ makes the inequality $x + y \leq 2$ true, shade the region containing $(0, 0)$.

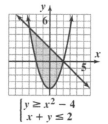

$$\begin{cases} y \geq x^2 - 4 \\ x + y \leq 2 \end{cases}$$

8. Graph the lines $x + y = 2$, $x = 1$, and $y = -3$ with dashed lines.
Graph the line $x = -2$ with a solid line.
Test points indicate that the solution contains the region to the right of -2, to the left of 1, above -3, and below the line $x + y = 2$. The corner points are represented as open circles because none satisfy all three inequalities.

$$\begin{cases} x + y < 2 \\ -2 \leq x < 1 \\ y > -3 \end{cases}$$

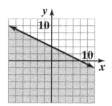

Concept and Vocabulary Check 8.5

1. solution; x; y; $5 > 1$

2. graph

3. half-plane

4. false

5. true

6. false

7. (0, 0) answers will vary

8. $x - y < 1$; $2x + 3y \geq 12$

9. false

Exercise Set 8.5

1. Graph $x + 2y = 8$ as a solid line using its x-intercept, $(8, 0)$, and its y-intercept, $(0, 4)$.
Test $(0, 0)$:
$0 + 2(0) \leq 8$?
$0 \leq 8$ true
Shade the half-plane containing $(0, 0)$.

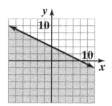

$x + 2y \leq 8$

3. Graph $x - 2y = 10$ as a dashed line using its x-intercept, $(10, 0)$, and its y-intercept, $(0, -5)$.
Test $(0, 0)$:
$0 - 2(0) > 10$?
$\qquad 0 > 10$ false
Shade the half-plane not containing $(0, 0)$.

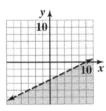

$$x - 2y > 10$$

5. Graph $y = \frac{1}{3}x$ as a solid line using its slope, $\frac{1}{3}$, and its y-intercept $(0, 0)$.
Test $(1, 1)$:
$1 \le \frac{1}{3}(1)$?
$1 \le \frac{1}{3}$ false
Shade the half-plane not containing $(1, 1)$.

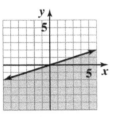

$$y \le \frac{1}{3}x$$

7. Graph $y = 2x - 1$ as a dashed line using its x-intercept, $\left(\frac{1}{2}, 0\right)$ and its y-intercept, $(0, -1)$.
Test $(0,0)$:
$0 > 2(0) - 1$?
$0 > -1$ true
Shade the half-plane containing $(0, 0)$.

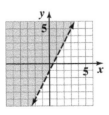

$$y > 2x - 1$$

9. Graph $x = 1$ as a solid vertical line.
Test $(0, 0)$:
$0 \le 1$ true
Shade the half-plane containing $(0, 0)$.

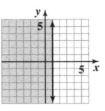

$$x \le 1$$

11. Graph $y = 1$ as a dashed horizontal line.
Test $(0, 0)$:
$0 > 1$ false
Shade the half-plane not containing $(0, 0)$.

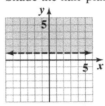

$$y > 1$$

13. Graph $x^2 + y^2 = 1$ as a solid circle with radius 1 and center $(0, 0)$.
Test $(0, 0)$:
$(0)^2 + (0)^2 \le 1$?
$\qquad 0 \le 1$ true
Shade the region containing $(0, 0)$.

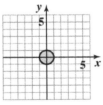

$$x^2 + y^2 \le 1$$

15. Graph $x^2 + y^2 = 25$ as a dashed circle with radius 5 and center (0, 0).
Test (0, 0):
$(0)^2 + (0)^2 > 25?$

$0 > 25$ false
Shade the region not containing (0, 0).

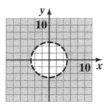

$x^2 + y^2 > 25$

17. Graph $(x-2)^2 + (y+1)^2 = 9$ as a dashed circle.

Test (0,0)

$(x-2)^2 + (y+1)^2 < 9$

$(0-2)^2 + (0+1)^2 < 9$

$5 < 9,$ true
Shade the region containing (0, 0).

$(x - 2)^2 + (y + 1)^2 < 9$

19. Graph $y = x^2 - 1$ as a dashed parabola with vertex (0, −1) and x-intercepts (1, 0) and (−1, 0).
Test (0, 0):
$0 < (0)^2 - 1?$

$0 < -1$ false
Shade the region not containing (0, 0).

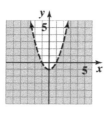

$y < x^2 - 1$

21. Graph $y = x^2 - 9$ as a solid parabola with vertex (0, −9) and x-intercepts (3, 0) and (−3, 0).
Test (0, 0):
$0 \geq (0)^2 - 9?$

$0 \geq -9$ true
Shade the region containing (0, 0).

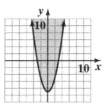

$y \geq x^2 - 9$

23. Graph $y = 2^x$ as a dashed exponential function with base 2 that passes through the point (0, 1).
Test (0, 0):
$0 > 2^0?$

$0 > 1$ false
Shade the region not containing (0, 0).

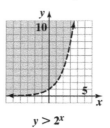

$y > 2^x$

25. Graph $y = \log_2(x+1)$ as a solid logarithmic function.
Test (0, 0):
$y \geq \log_2(x+1)$

$0 \geq \log_2(0+1)$

$0 \geq 0,$ true
Shade the region containing (0, 0).

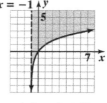

$y \geq \log_2(x + 1)$

27. Begin by graphing $3x + 6y = 6$ as a solid line using its x-intercept, (2, 0), and its y-intercept, (0, 1). Since (0, 0) makes the inequality $3x + 6y \leq 6$ true, shade the half-plane containing (0, 0). Graph $2x + y = 8$ as a solid line using its x-intercept, (4, 0), and its y-intercept, (0, 8). Since (0, 0) makes the inequality $2x + y \leq 8$ true, shade the half-plane containing (0, 0). The solution set of the system is the intersection of the above shaded half-planes, and is shown as the shaded region in the following graph.

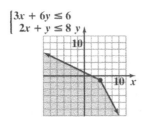

29. Begin by graphing $2x - 5y = 10$ as a solid line using its x-intercept, (5, 0), and its y-intercept, (0, –2). Since (0, 0) makes the inequality $2x - 5y \leq 10$ true, shade the half-plane containing (0, 0). Graph $3x - 2y = 6$ as a dashed line using its x-intercept, (2, 0), and its y-intercept, (0, –3). Since (0, 0) makes the inequality $3x - 2y > 6$ false, shade the half-plane containing (0, 0). The solution set of the system is the intersection of the above shaded half-planes, and is shown as the shaded region in the following graph.

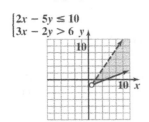

31. Begin by graphing $y = 2x - 3$ as a dashed line using its slope, 2, and its y-intercept, (0, –3). Since (0, 0) makes the inequality $y > 2x - 3$ true, shade the half-plane containing (0, 0). Graph $y = -x + 6$ as a dashed line using its slope, –1, and its y-intercept, (0, 6). Since (0, 0) makes the inequality $y < -x + 6$ true, shade the half-plane containing (0, 0). The solution set of the system is the intersection of the above shaded half-planes, and is shown as the shaded region in the following graph.

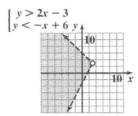

33. Begin by graphing $x + 2y = 4$ as a solid line using its x-intercept, (4, 0), and its y-intercept, (0, 2). Since (0, 0) makes the inequality $x + 2y \leq 4$ true, shade the half-plane containing (0, 0). Graph $y = x - 3$ as a solid line using its slope, 1, and its y-intercept, (0, –3). Since (0, 0) makes the inequality $y \geq x - 3$ true, shade the half-plane containing (0, 0). The solution set of the system is the intersection of the above shaded half-planes, and is shown as the shaded region in the following graph.

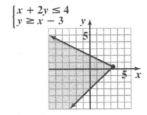

35. Begin by graphing $x = 2$ as a solid vertical line . Since $(0, 0)$ makes the inequality $x \leq 2$ true, shade the half-plane containing $(0, 0)$. Graph $y = 1$ as a solid horizontal line. Since $(0, 0)$ makes the inequality $y \geq -1$ true, shade the half-plane containing $(0, 0)$. The solution set of the system is the intersection of the above shaded half-planes, and is shown as the shaded region in the following graph.

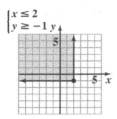

$$\begin{cases} x \leq 2 \\ y \geq -1 \end{cases}$$

37. Graph $x = -2$ as a solid vertical line and $x = 5$ as a dashed vertical line. Since $(0, 0)$ makes the inequality $-2 \leq x < 5$ true, shade the region between the two lines.

$-2 \leq x < 5$

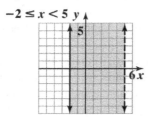

39. Begin by graphing $x - y = 1$ as a solid line using its x-intercept, $(1, 0)$, and its y-intercept $(0, -1)$. Since $(0, 0)$ makes the inequality $x - y \leq 1$ true, shade the half-plane containing $(0, 0)$. Graph $x = 2$ as a solid horizontal line. Since $(0, 0)$ makes the inequality $x \geq 2$ false, shade the half-plane not containing $(0, 0)$. The solution set of the system is the intersection of the above shaded half-planes, and is shown as the shaded region in the following graph.

$$\begin{cases} x - y \leq 1 \\ x \geq 2 \end{cases}$$

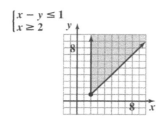

41. $x + y > 4$

$x + y < -1$

Begin by graphing $x + y = 4$ as a dashed line using its x-intercept, $(4, 0)$, and its y-intercept $(0, 4)$. Since $(0, 0)$ makes the inequality $x + y > 4$ false, shade the half-plane not containing $(0, 0)$. Graph $x + y = -1$ as a dashed line using its x-intercept, $(-1, 0)$, and its y-intercept, $(0, -1)$. Since $(0, 0)$ makes the inequality $x + y < -1$ false, shade the half-plane not containing $(0, 0)$. Since these half-planes do not intersect the system has no solution.

43. Begin by graphing $x + y = 4$ as a dashed line using its x-intercept, $(4, 0)$, and its y-intercept, $(0, 4)$. Since $(0, 0)$ makes the inequality $x + y > 4$ false, shade the half-plane not containing $(0, 0)$. Graph $x + y = -1$ as a dashed line using its x-intercept, $(-1, 0)$, and its y-intercept, $(0, -1)$. Since $(0, 0)$ makes the inequality $x + y > -1$ true, shade the half-plane containing $(0, 0)$. The solution set of the system is the intersection of the above half-planes, and is shown as the shaded region in the following graph.

$$\begin{cases} x + y > 4 \\ x + y > -1 \end{cases}$$

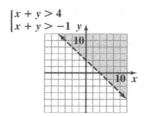

45. Begin by graphing $y = x^2 - 1$ as a solid parabola with vertex $(0, -1)$ and x-intercepts, $(-1, 0)$, and $(1, 0)$. Since $(0, 0)$ makes the inequality $y \geq x^2 - 1$ true, shade the half-plane containing $(0, 0)$. Graph $x - y = -1$ as a solid line using its x-intercept, $(-1, 0)$, and its y-intercept, $(0, 1)$. Since $(0, 0)$ makes the inequality $x - y \geq -1$ true, shade the half-plane containing $(0, 0)$. The solution set of the system is the intersection of the above half-planes, and is shown as the shaded region in the following graph.

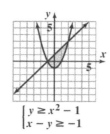

$$\begin{cases} y \geq x^2 - 1 \\ x - y \geq -1 \end{cases}$$

47. Begin by graphing $x^2 + y^2 = 16$ as a solid circle with radius 4 and center, (0, 0). Since (0, 0) makes the inequality $x^2 + y^2 \le 16$ true, shade the half-plane containing (0, 0). Graph $x + y = 2$ as a dashed line using its x-intercept, (2, 0), and its y-intercept, (0, 2). Since (0, 0) makes the inequality $x + y > 2$ false, shade the half-plane not containing (0, 0). The solution set of the system is the intersection of the above half-planes, and is shown as the shaded region in the following graph.

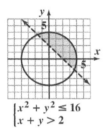

$\begin{cases} x^2 + y^2 \le 16 \\ x + y > 2 \end{cases}$

49. Begin by graphing $x^2 + y^2 = 1$ as a dashed circle with radius 1 and center, (0, 0). Since (0, 0) makes the inequality $x^2 + y^2 > 1$ false, shade the half-plane not containing (0, 0). Graph $x^2 + y^2 = 16$ as a dashed circle with radius 4 and center (0, 0). Since (0, 0) makes the inequality $x^2 + y^2 < 16$ true, shade the half-plane containing (0, 0). The solution set of the system is the intersection of the above half-planes, and is shown as the shaded region in the following graph.

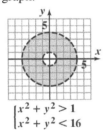

$\begin{cases} x^2 + y^2 > 1 \\ x^2 + y^2 < 16 \end{cases}$

51.

$\begin{cases} (x - 1)^2 + (y + 1)^2 < 25 \\ (x - 1)^2 + (y + 1)^2 \ge 16 \end{cases}$

53.

$\begin{cases} x^2 + y^2 \le 1 \\ y - x^2 > 0 \end{cases}$

55.

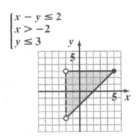

$\begin{cases} x^2 + y^2 < 16 \\ y \ge 2^x \end{cases}$

57. Begin by graphing $x - y = 2$ as a solid line using its x-intercept, (2, 0), and its y-intercept, (0, –2). Since (0, 0) makes the inequality $x - y \le 2$ true, shade the half-plane containing (0, 0). Graph $x = -2$ as a solid vertical line. Since (0, 0) makes the inequality $x \ge -2$ true, shade the half-plane containing (0, 0). Graph $y = 3$ as a solid horizontal line. Since (0, 0) makes the inequality $y \le 3$ true, shade the half-plane containing (0, 0). The solution set of the system is the intersection of the above half-planes, and is shown as the shaded region in the following graph.

$\begin{cases} x - y \le 2 \\ x > -2 \\ y \le 3 \end{cases}$

59. Since $x \geq 0$ and $y \geq 0$ the solution to the system lies in the first quadrant. Graph $2x + 5y = 10$ as a solid line using its x-intercept, (5, 0), and its y-intercept, (0, 2). Since (0, 0) makes the inequality $2x + 5y \leq 10$ true, shade the half-plane containing (0, 0). Graph $3x + 4y = 12$ as a solid line by using its x-intercept, (4, 0), and its y-intercept, (0, 3). Since (0, 0) makes the inequality $3x + 4y \leq 12$ true, shade the half-plane containing (0, 0). The solution set of the system is the intersection of the above half-planes which lies in the first quadrant, and is shown as the shaded region in the following graph.

$$\begin{cases} x \geq 0 \\ y \geq 0 \\ 2x + 5y < 10 \\ 3x + 4y \leq 12 \end{cases}$$

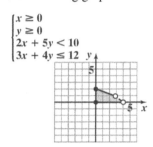

61. Begin by graphing $3x + y = 6$ as a solid line using its x-intercept, (2, 0), and its y-intercept, (0, 6). Since (0, 0) makes the inequality $3x + y \leq 6$ true, shade the half-plane containing (0, 0). Graph $2x - y = -1$ as a solid line using its x-intercept, $\left(-\dfrac{1}{2}, 0\right)$, and its y-intercept, (0, 1). Since (0, 0) makes the inequality $2x - y \leq -1$ false, shade the half-plane not containing (0, 0). Graph $x = -2$ as a solid vertical line. Since (0, 0) makes the inequality $x \geq -2$ true, shade the half-plane containing (0, 0). Graph $y = 4$ as a solid horizontal line. Since (0, 0) makes the inequality $y \leq 4$ true, shade the half-plane containing (0, 0). The solution set of the system is the intersection of the above half-planes, and is shown as the shaded region in the following graph.

$$\begin{cases} 3x + y \leq 6 \\ 2x - y \leq -1 \\ x > -2 \\ y < 4 \end{cases}$$

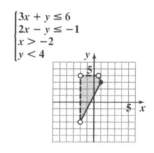

63. $y \geq -2x + 4$
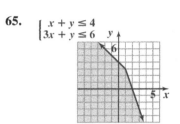

65. $\begin{cases} x + y \leq 4 \\ 3x + y \leq 6 \end{cases}$
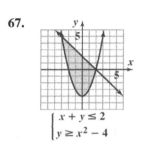

67.
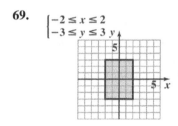
$$\begin{cases} x + y \leq 2 \\ y \geq x^2 - 4 \end{cases}$$

69. $\begin{cases} -2 \leq x \leq 2 \\ -3 \leq y \leq 3 \end{cases}$
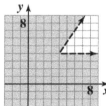

71. Find the union of solutions of
$$y > \frac{3}{2}x - 2 \text{ and } y < 4.$$

73. The system $\begin{aligned} 3x+3y &< 9 \\ 3x+3y &> 9 \end{aligned}$ has no solution.

The number $3x+3y$ cannot both be less than 9 and greater than 9 at the same time.

75. The system has an infinite number of solutions. The solution is the set of points that make up the circle $(x+4)^2+(y-3)^2=9$

77. Point $A = (66,160)$

$$5.3x-y \geq 180$$
$$5.3(66)-160 \geq 180$$
$$189.8 \geq 180, \text{ true}$$

$$4.1x-y \leq 14$$
$$4.1(66)-160 \leq 140$$
$$110.6 \leq 140, \text{ true}$$

Point A is a solution of the system.

79. Point $= (72,205)$

$$5.3x-y \geq 180$$
$$5.3(72)-205 \geq 180$$
$$176.6 \geq 180, \text{ false}$$

$$4.1x-y \leq 14$$
$$4.1(72)-205 \leq 140$$
$$90.2 \leq 140, \text{ true}$$

The data does not satisfy both inequalities. The person is not within the healthy weight region.

81. a. $50x+150y > 2000$

b. Graph $50x+150y$ as a dashed line using its x-intercept, (40, 0), and its y-intercept, $\left(0,\dfrac{40}{3}\right)$.

Test (0, 0):
$50(0)+150(0) > 2000?$
$$0 > 2000 \text{ false}$$
Shade the half-plane not containing (0, 0).

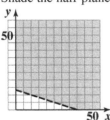

$50x + 150y > 2000$

c. Ordered pairs may vary.

83. a. $\begin{cases} y \geq 0 \\ x+y \geq 5 \\ x \geq 1 \\ 200x+100y \leq 700 \end{cases}$

b. $\begin{cases} y \geq 0 \\ x+y \geq 5 \\ x \geq 1 \\ 200x+100y \leq 700 \end{cases}$

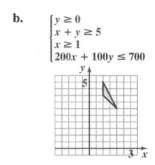

c. 2 nights

85. a. $\text{BMI} = \dfrac{703W}{H^2} = \dfrac{703(200)}{72^2} \approx 27.1$

b. A 20 year old man with a BMI of 27.1 is classified as overweight.

87. – 95. Answers will vary.

97.

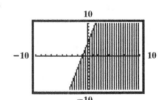

99. $y \geq x^2-4$

101. $2x+y \le 6$

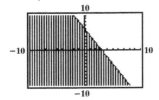

103. – 105. Answers will vary.

107. does not make sense; Explanations will vary. Sample explanation: It is necessary to graph the linear equation with a dashed line to represent its role as a borderline.

109. makes sense

111. $y > x - 3$
   $y \le x$

113. $x + 2y \le 6$ or $2x + y \le 6$

115. $y \ge nx + b$ $(n < 0, b > 0)$
   $y \le mx + b$ $(m > 0, b > 0)$

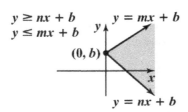

116. a. $x + y \ge 6$
   $x \le 8$
   $y \le 5$

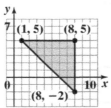

   b. The corner points are $(1,5)$, $(8,5)$, and $(8,-2)$.

   c. At $(1,5)$, $3x + 2y = 3(1) + 2(5) = 13$.
   At $(8,5)$, $3x + 2y = 3(8) + 2(5) = 34$.
   At $(8,-2)$, $3x + 2y = 3(8) + 2(-2) = 20$.

117. a.
$$x \geq 0$$
$$y \geq 0$$
$$3x - 2y \leq 6$$
$$y \leq -x + 7$$

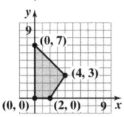

b. The corner points are $(0,0)$, $(2,0)$, $(4,3)$, and $(0,7)$.

c. At $(0,0)$, $2x + 5y = 2(0) + 5(0) = 0$.
At $(2,0)$, $2x + 5y = 2(2) + 5(0) = 4$.
At $(4,3)$, $2x + 5y = 2(4) + 5(3) = 23$.
At $(0,7)$, $2x + 5y = 2(0) + 5(7) = 35$.

118. $20x + 10y \leq 80,000$

Section 8.6

Check Point Exercises

1. The total profit is 25 times the number of bookshelves, x, plus 55 times the number of desks, y. The objective function is $z = 25x + 55y$

2. Not more than a total of 80 bookshelves and desks can be manufactured per day. This is represented by the inequality $x + y \leq 80$.

3. Objective function: $z = 25x + 55y$

Constraints: $\begin{cases} x + y \leq 80 \\ 30 \leq x \leq 80 \\ 10 \leq y \leq 30 \end{cases}$

4. Graph the constraints and find the corners, or vertices, of the region of intersection.

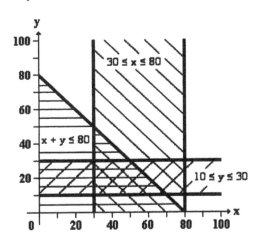 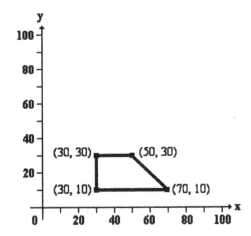

Find the value of the objective function at each corner of the graphed region.

Corner (x, y)	Objective Function $z = 25x + 55y$ z
(30, 10)	$z = 25(30) + 55(10)$ $= 750 + 550 = 1300$
(30, 30)	$z = 25(30) + 55(30)$ $= 750 + 1650 = 2400$
(50, 30)	$z = 25(50) + 55(30)$ $= 1250 + 1650 = 2900 \leftarrow$ Maximum
(70, 10)	$z = 25(70) + 55(10)$ $= 1750 + 550 = 2300$

The maximum value of z is 2900 and it occurs at the point (50, 30).
In order to maximize profit, 50 bookshelves and 30 desks must be produced each day for a profit of $2900.

5. objective function: $z = 3x + 5y$
constraints: $x \geq 0, \ y \geq 0$
$$x + y \geq 1$$
$$x + y \leq 6$$

Evaluate the objective function at the four vertices of the region shown:

$(1, 0) : 3(1) + 5(0) = 3$
$(0, 1) : 3(0) + 5(1) = 5$
$(0, 6) : 3(0) + 5(6) = 30$
$(6, 0) : 3(6) + 5(0) = 18$
The maximum value of z is 30 and this occurs when $x = 0$ and $y = 6$.

Concept and Vocabulary Check 8.6

1. linear programming

2. objective

3. constraints; corner

Exercise Set 8.6

1. $z = 5x + 6y$

(1, 2): 5(1) + 6(2) = 5 + 12 = 17
(2, 10): 5(2) + 6(10) = 10 + 60 = 70
(7, 5): 5(7) + 6(5) = 35 + 30 = 65
(8, 3): 5(8) + 6(3) = 40 + 18 = 58
The maximum value is $z = 70$; the minimum value is $z = 17$.

3. $z = 40x + 50y$

(0, 0): 40(0) + 50(0) = 0 + 0 = 0
(0, 8): 40(0) + 50(8) = 0 + 400 = 400
(4, 9): 40(4) + 50(9) = 160 + 450 = 610
(8, 0): 40(8) + 50(0) = 320 + 0 = 320
The maximum value is $z = 610$; the minimum value is $z = 0$.

5. $z = 3x + 2y$

$x \geq 0, y \geq 0$

$2x + y \leq 8$

$x + y \geq 4$

a.

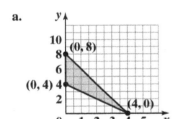

b. $(0, 8): z = 3(0) + 2(8) = 16$

$(0, 4): z = 3(0) + 2(4) = 8$

$(4, 0): z = 3(4) + 2(0) = 12$

c. The maximum value is 16 at $x = 0$ and $y = 8$.

7. $z = 4x + y$

$x \geq 0, y \geq 0$

$2x + 3y \leq 12$

$x + y \geq 3$

a.

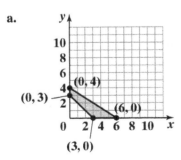

b. (0, 4): $z = 4(0) + 4 = 4$
(0, 3): $z = 4(0) + 3 = 3$
(3, 0): $z = 4(3) + 0 = 12$
(6, 0): $z = 4(6) + 0 = 24$

c. The maximum value is 24 at $x = 6$ and $y = 0$.

9. $z = 3x - 2y$

$1 \leq x \leq 5$

$y \geq 2$

$x - y \geq -3$

a.

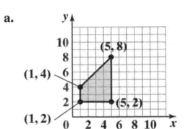

b. (1, 2): $z = 3(1) - 2(2) = -1$
(1, 4): $z = 3(1) - 2(4) = -5$
(5, 8): $z = 3(5) - 2(8) = -1$
(5, 2): $z = 3(5) - 2(2) = 11$

c. Maximum value is 11 at $x = 5$ and $y = 2$.

11. $z = 4x + 2y$

$x \geq 0, y \geq 0$

$2x + 3y \leq 12$

$3x + 2y \leq 12$

$x + y \geq 2$

a.

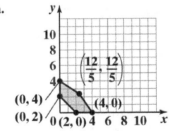

b. (0, 4): $z = 4(0) + 2(4) = 8$
(0, 2): $z = 4(0) + 2(2) = 4$
(2, 0): $z = 4(2) + 2(0) = 8$
(4, 0): $z = 4(4) + 2(0) = 16$

$$\left(\frac{12}{5}, \frac{12}{5} \right): z = 4\left(\frac{12}{5} \right) + 2\left(\frac{12}{5} \right)$$

$$= \frac{48}{5} + \frac{24}{5} = \frac{72}{5}$$

c. The maximum value is 16 at $x = 4$ and $y = 0$.

13. $z = 10x + 12y$

$x \geq 0, y \geq 0$

$x + y \leq 7$

$2x + y \leq 10$

$2x + 3y \leq 18$

a.

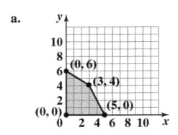

b. $(0, 6): z = 10(0) + 12(6) = 72$

$(0, 0): z = 10(0) + 12(0) = 0$

$(5, 0): z = 10(5) + 12(0) = 50$

$(3, 4): z = 10(3) + 12(4)$

$= 30 + 48 = 78$

c. The maximum value is 78 at $x = 3$ and $y = 4$.

15. a. $z = 125x + 200y$

b. $x \leq 450$

$y \leq 200$

$600x + 900y \leq 360,000$

c. Simplify the third inequality by dividing by 300 to get $2x + 3y \leq 1200$.

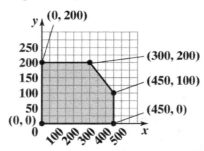

d. $(0, 0): 125(0) + 200(0) = 0 + 0 = 0$

$(0, 200): 125(0) + 200(200)$

$= 0 + 40,000 = 40,000$

$(300, 200): 125(300) + 200(200)$

$= 37,500 + 40,000 = 77,500$

$(450, 100): 125(450) + 200(100)$

$= 56,250 + 20,000 = 76,250$

$(450, 0): 125(450) + 200(0)$

$= 56,250 + 0 = 56,250$

e. The television manufacturer will make the greatest profit by manufacturing <u>300</u> rear-projection televisions each month and <u>200</u> plasma televisions each month. The maximum monthly profit is <u>$77,500</u>.

17. Let x = number of model A bicycles and y = number of model B bicycles.

The constraints are

$5x + 4y \leq 200$

$2x + 3y \leq 108$

Graph these inequalities in the first quadrant, since x and y cannot be negative.

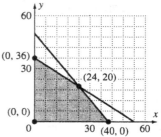

The quantity to be maximized is the profit, which is $25x + 15y$.

$(0, 0): 25(0) + 15(0) = 0 + 0 = 0$

$(0, 36): 25(0) + 15(36) = 0 + 540 = 540$

$(24, 20): 25(24) + 15(20) = 600 + 300 = 900$

$(40, 0): 25(40) + 15(0) = 1000 + 0 = 1000$

40 model A bicycles and 0 model B bicycles should be produced.

19. Let x = the number of cartons of food and y = the number of cartons of clothing.

The constraints are:

$20x + 10y \leq 8,000$ or $2x + y \leq 8000$

$50x + 20y \leq 19,000$ or $5x + 2y \leq 1900$

Graph these inequalities in the first quadrant, since x and y cannot be negative.

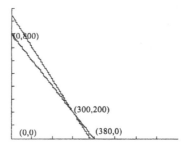

The quantity to be maximized is the number of people helped, which is $12x + 5y$.

$(0, 0): 12(0) + 5(0) = 0 + 0 = 0$

$(0, 800): 12(0) + 5(800) = 0 + 4000 = 4000$

$(300, 200): 12(300) + 5(200) = 4600$

$(380, 0): 12(380) + 5(0) = 4500$

300 cartons of food and 200 cartons of clothing should be shipped. This will help 4600 people.

21. Let x = number of students attending and
y = number of parents attending.
The constraints are
$x + y \leq 150$
$\quad 2x \geq y$
or
$\quad x + y \leq 150$
$2x - y \geq 0$
Graph these inequalities in the first quadrant, since x
and y cannot be negative.

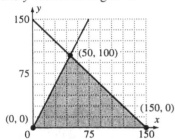

The quantity to be maximized is the amount of
money raised, which is $x + 2y$.
$(0, 0)$: $0 + 2(0) = 0 + 0 = 0$
$(50, 100)$: $50 + 2(100) = 50 + 200 = 250$
$(150, 0)$: $150 + 2(0) = 150 + 0 = 150$
50 students and 100 parents should attend.

23. Let x = number of Boeing 727s, y = number of
Falcon 20s.
Maximize $z = x + y$ with the following constraints:
$1400x + 500y \leq 35,000$ or $14x + 5y \leq 350$
$42,000x + 6000y \geq 672,000$ or $7x + y \geq 112$
$x \leq 20$
$x \geq 0, y \geq 0$

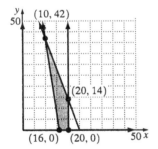

$(16, 0)$: $z = 16$
$(20, 0)$: $z = 20$
$(20, 14)$: $z = 34$
$(10, 42)$: $z = 52$
Federal Express should have purchased 10 Boeing
727s and 42 Falcon 20s.

25. – 27. Answers will vary.

29. does not make sense; Explanations will vary.
Sample explanation: Solving a linear programming
problem does not require graphing the objective
function.

31. makes sense

33. Let x = amount invested in stocks and
y = amount invested in bonds.
The constraints are:
$x + y \leq 10,000$
$\quad y \geq 3000$
$\quad x \geq 2000$
$\quad y \geq x$
Graph these inequalities in the first quadrant, since x
and y cannot be negative.

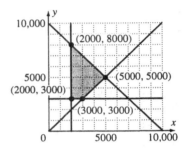

The quantity to be maximized is the return on the
investment, which is $0.12x + 0.08y$.
$(2000, 3000)$:
$0.12(2000) + 0.08(3000) = 240 + 240 = 480$
$(2000, 8000)$:
$0.12(2000) + 0.08(8000) = 240 + 640 = 880$
$(5000, 5000)$:
$0.12(5000) + 0.08(5000) = 600 + 400 = 1000$
$(3000, 3000)$:
$0.12(3000) + 0.08(3000) = 360 + 240 = 600$
The greatest return occurs when \$5000 is invested in
stocks and \$5000 is invested in bonds.

35. Answers will vary.

37. Back-substitute $z = 5$ to find y in the second
equation.
$\quad y + 2z = 13$
$\quad y + 2(5) = 13$
$\quad\quad y + 10 = 13$
$\quad\quad\quad\quad y = 3$
Back-substitute to find x in the first equation.
$\quad x + y + 2z = 19$
$\quad x + 3 + 2(5) = 19$
$\quad\quad\quad x + 13 = 19$
$\quad\quad\quad\quad\quad x = 6$
The solution set is $\{(6, 3, 5)\}$.
Explanations may vary.

38. Back-substitute $z = 3$ to find y in the third equation.

$$y - z = 1$$
$$y - 3 = 1$$
$$y = 4$$

Back-substitute to find x in the second equation.

$$x - \frac{1}{3}y + z = \frac{8}{3}$$
$$x - \frac{1}{3}(4) + 3 = \frac{8}{3}$$
$$x - \frac{4}{3} + \frac{9}{3} = \frac{8}{3}$$
$$x + \frac{5}{3} = \frac{8}{3}$$
$$x = \frac{3}{3}$$
$$x = 1$$

Back-substitute to find w in the first equation.

$$w - x + 2y - 2z = -1$$
$$w - 1 + 2(4) - 2(3) = -1$$
$$w - 1 + 8 - 6 = -1$$
$$w + 1 = -1$$
$$w = -2$$

The solution set is $\{(-2, 1, 4, 3)\}$.
Explanations may vary.

39.
$$\begin{bmatrix} 1 & 2 & -1 \\ 1(-4)+4 & 2(-4)+(-3) & -1(-4)+(-15) \end{bmatrix}$$
$$= \begin{bmatrix} 1 & 2 & -1 \\ 0 & -11 & -11 \end{bmatrix}$$

Chapter 8 Review Exercises

1.
$$y = 4x + 1$$
$$3x + 2y = 13$$

Substitute $4x + 1$ for y in the second equation:

$$3x + 2(4x + 1) = 13$$
$$3x + 8x + 2 = 13$$
$$11x = 11$$
$$x = 1$$
$$y = 4(1) + 1 = 5$$

The solution set is $\{(1, 5)\}$.

2.
$$x + 4y = 14$$
$$2x - y = 1$$

Multiply the second equation by 4 and add to the first equation.

$$x + 4y = 14$$
$$\underline{8x - 4y = 4}$$
$$9x = 18$$
$$x = 2$$
$$2(2) - y = 1$$
$$-y = -3$$
$$y = 3$$

The solution set is $\{(2, 3)\}$.

3.
$$5x + 3y = 1$$
$$3x + 4y = -6$$

Multiply the first equation by 4 and the second equation by -3.
Then add.

$$20x + 12y = 4$$
$$\underline{-9x - 12y = 18}$$
$$11x = 22$$
$$x = 2$$
$$5(2) + 3y = 1$$
$$3y = -9$$
$$y = -3$$

The solution set is $\{(2, -3)\}$.

4.
$$2y - 6x = 7$$
$$3x - y = 9$$

The second equation can be written as $y = 3x - 9$.
Substitute:

$$2(3x - 9) - 6x = 7$$
$$6x - 18 - 6x = 7$$
$$-18 = 7$$

Since this is false, the system has no solution.
The solution set is the empty set, $\varnothing$.

5.
$$4x - 8y = 16$$
$$3x - 6y = 12$$

Divide the first equation by 4 and the second equation by 3.

$$x - 2y = 4$$
$$x - 2y = 4$$

Since these equations are identical, the system has an infinite number of solutions.
The solution set is $\{(x, y) \mid 3x - 6y = 12\}$.

6. a. $C(x) = 60,000 + 200x$

b. $R(x) = 450x$

c. $450x = 60000 + 200x$
$250x = 60000$
$x = 240$
$450(240) = 108,000$
The company must make 240 desks at a cost of $108,000 to break even.

7. Let x = the selling price for Klint's work.
Let y = the selling price for Picasso's work.
$x + y = 239$
$x - y = 31$
Add the equations to eliminate y and solve for x.

$$\begin{aligned} x + y &= 239 \\ x - y &= \ \ 31 \\ \hline 2x \ \ \ \ &= 270 \end{aligned}$$

$x = 135$
Back-substitute to find y.
$x + y = 239$
$135 + y = 239$
$y = 104$
Klint's work sold for $135 million and Picasso's work sold for $104 million.

8. a. $y = 219x + 142$

b. Using substitution,
$95x + y = 770$

$$95x + \overbrace{219x + 142}^{y} = 770$$
$314x + 142 = 770$
$314x = 628$
$x = 2$
Digital track sales caught up with album sales 2 years after 2004, or 2006.
$y = 219x + 142$
$y = 219(2) + 142$
$= 580$
In 2006 there were 580 million digital tracks and 580 million albums sold.

9. Let l = the length of the table.
Let w = the width of the table.
$2l + 2w = 34$
$4l - 3w = 33$
Multiply the first equation by –2 and solve by addition.

$$\begin{aligned} -4l - 4w &= -68 \\ 4l - 3w &= \ \ 33 \\ \hline -7w &= -35 \end{aligned}$$

$w = 5$
Back-substitute 5 for w to find l.
$2l + 2w = 34$
$2l + 2(5) = 34$
$2l + 10 = 34$
$2l = 24$
$l = 12$
The dimensions of the table are 12 feet by 5 feet.

10. Let x = the cost of the hotel
y = the cost of the car
$3x + 2y = 360$
$4x + 3y = 500$
Solve the system.

$$\begin{aligned} 12x + 8y &= 1440 \\ -12x - 9y &= -1500 \\ \hline -y &= -60 \end{aligned}$$

$y = 60$

$3x + 2(60) = 360$
$3x = 240$
$x = 80$
The room costs $80 a day and the car rents for $60 a day.

11. x = number of apples
y = number of avocados
$100x + 350y = 1000$
$24x + 14y = 100$

$$\begin{aligned} 100x + 350y &= 1000 \\ -600x - 350y &= -2500 \\ \hline -500x &= -1500 \end{aligned}$$

$x = 3$
$100(3) + 350y = 1000$
$350y = 700$
$y = 2$
3 apples and 2 avocados supply 1000 calories and 100 grams of carbohydrates.

12.
$$2x - y + z = 1 \,(1)$$
$$3x - 3y + 4z = 5 \,(2)$$
$$4x - 2y + 3z = 4 \,(3)$$
Eliminate y from (1) and (2) by multiplying (1) by -3 and adding the result to (2).
$$-6x + 3y - 3z = -3$$
$$\underline{3x - 3y + 4z = 5}$$
$$-3x + z = 2 \ (4)$$
Eliminate y from (1) and (3) by multiplying (1) by -2 and adding the result to (3).
$$-4x + 2y - 2z = -2$$
$$\underline{4x - 2y + 3z = 4}$$
$$z = 2$$
Substituting $z = 2$ into (4), we get:
$$-3x + 2 = 2$$
$$-3x = 0$$
$$x = 0$$
Substituting $x = 0$ and $z = 2$ into (1), we have:
$$2(0) - y + 2 = 1$$
$$-y = -1$$
$$y = 1$$
The solution set is $\{(0, 1, 2)\}$.

13.
$$x + 2y - z = 5 \ (1)$$
$$2x - y + 3z = 0 \ (2)$$
$$2y + z = 1 \ (3)$$
Eliminate x from (1) and (2) by multiplying (1) by -2 and adding the result to (2).
$$-2x - 4y + 2z = -10$$
$$\underline{2x - y + 3z = 0}$$
$$-5y + 5z = -10$$
$$y - z = 2 \ (4)$$
Adding (3) and (4), we get:
$$2y + z = 1$$
$$\underline{y - z = 2}$$
$$3y = 3$$
$$y = 1$$
Substituting $y = 1$ into (3), we have:
$$2(1) + z = 1$$
$$z = -1$$
Substituting $y = 1$ and $z = -1$ into (1), we obtain:
$$x + 2(1) - (-1) = 5$$
$$x + 3 = 5$$
$$x = 2$$
The solution set is $\{(2, 1, -1)\}$.

14.
$$y = ax^2 + bx + c$$
$$(1, 4): 4 = a + b + c \qquad (1)$$
$$(3, 20): 20 = 9a + 3b + c \qquad (2)$$
$$(-2, 25): 25 = 4a - 2b + c \quad (3)$$
Multiply (1) by -1 and add to (2).
$$20 = 9a + 3b + c$$
$$\underline{-4 = -a - b - c}$$
$$16 = 8a + 2b$$
$$8 = 4a + b$$
$$8 = 4a + b \qquad (4)$$
Multiply (1) by -1 and add to (3).
$$25 = 4a - 2b + c$$
$$\underline{-4 = -a - b - c}$$
$$21 = 3a - 3b$$
$$7 = a - b \qquad (5)$$
Add (4) and (5).
$$8 = 4a + b$$
$$\underline{7 = a - b}$$
$$15 = 5a$$
$$a = 3$$
$$8 = 4(3) + b$$
$$b = -4$$
$$3 - 4 + c = 4$$
$$c = 5$$
Hence, the quadratic function is $y = 3x^2 - 4x + 5$.

15. Let x = number of deaths, in millions, in the 20th century from war.
Let y = number of deaths, in millions, in the 20th century from famine.
Let z = number of deaths, in millions, in the 20th century from tobacco.
$$x + y + z = 306$$
$$x - y = 13$$
$$x - z = 53$$
Solve the second equation for y.
$$x - y = 13$$
$$-y = -x + 13$$
$$y = x - 13$$
Solve the third equation for z.
$$x - z = 53$$
$$-z = -x + 53$$
$$z = x - 53$$
Substitute the expressions for x and z into the first equation and solve for y.

$$x \quad + \quad y \quad + \quad z \quad = 306$$

$$x + \overbrace{(x-13)}^{y} + \overbrace{(x-53)}^{z} = 306$$

$$x + x - 13 + x - 53 = 306$$

$$3x - 66 = 306$$

$$3x = 372$$

$$x = 124$$

Back-substitute to solve for y and z.

$$y = x - 13$$
$$= 124 - 13$$
$$= 111$$

$$z = x - 53$$
$$= 124 - 53$$
$$= 71$$

The number of deaths in the 20th century from war is 124 million, from famine is 111 million, and from tobacco is 71 million.

16. $\dfrac{x}{(x-3)(x+2)} = \dfrac{A}{x-3} + \dfrac{B}{x+2}$

$$x = A(x+2) + B(x-3)$$
$$= (A+B)x + (2A - 3B)$$

$$A + B = 1$$

$$2A - 3B = 0$$

Multiply first equation by 3, then add to second equation.

$$3A + 3B = 3$$
$$\underline{2A - 3B = 0}$$
$$5A = 3$$

$$A = \frac{3}{5}, \ B = \frac{2}{5}$$

$$\frac{x}{(x-3)(x+2)} = \frac{3}{5(x-3)} + \frac{2}{5(x+2)}$$

17. $\dfrac{11x-2}{x^2 - x - 12} = \dfrac{11x-2}{(x-4)(x+3)} = \dfrac{A}{x-4} + \dfrac{B}{x+3}$

$$11x - 2 = A(x+3) + B(x-4)$$
$$= Ax + 3A + Bx - 4B$$
$$= (A+B)x + (3A - 4B)$$

$$A + B = 11$$
$$3A - 4B = -2$$

Multiply first equation by 4, then add to second equation.

$$3A - 4B = -2$$
$$\underline{4A + 4B = 44}$$
$$7A = 42$$

$$A = 6, \ B = 5$$

$$\frac{11x-2}{x^2 - x - 12} = \frac{6}{x-4} + \frac{5}{x+3}$$

18. $\dfrac{4x^2 - 3x - 4}{x^3 + x^2 - 2x} = \dfrac{4x^2 - 3x - 4}{x(x+2)(x-1)}$

$$= \frac{A}{x} + \frac{B}{x+2} + \frac{C}{x-1}$$

$$4x^2 - 3x - 4 = A(x+2)(x-1) + Bx(x-1) + Cx(x+2)$$
$$= A(x^2 + x - 2) + Bx^2 - Bx + Cx^2 + 2Cx$$
$$= Ax^2 + Ax - 2A + Bx^2 - Bx + Cx^2 + 2Cx$$
$$= (A+B+C)x^2 + (A-B+2C)x - 2A$$

$$A + B + C = 4$$
$$A - B + 2C = -3$$
$$-2A = -4$$
$$A = 2$$
$$B + C = 2$$
$$\underline{-B + 2C = -5}$$
$$3C = -3$$
$$C = -1$$
$$B - 1 = 2$$
$$B = 3$$

$$\frac{4x^2 - 3x - 4}{x^3 + x^2 - 2x} = \frac{2}{x} + \frac{3}{x+2} - \frac{1}{x-1}$$

19. $\dfrac{2x+1}{(x-2)^2} = \dfrac{A}{x-2} + \dfrac{B}{(x-2)^2}$

$2x+1 = A(x-2) + B = Ax - 2A + B$

$A = 2$
$-2A + B = 1$
$-2(2) + B = 1$
$B = 5$

$\dfrac{2x+1}{(x-2)^2} = \dfrac{2}{x-2} + \dfrac{5}{(x-2)^2}$

20. $\dfrac{2x-6}{(x-1)(x-2)^2} = \dfrac{A}{x-1} + \dfrac{B}{x-2} + \dfrac{C}{(x-2)^2}$

$2x-6 = A(x-2)^2 + B(x-1)(x-2) + C(x-1)$

$\quad = A(x^2 - 4x + 4) + B(x^2 - 3x + 2) + C(x-1)$

$\quad = Ax^2 - 4Ax + 4A + Bx^2 - 3Bx + 2B + Cx - C$

$\quad = (A+B)x^2 + (-4A - 3B + C)x + (4A + 2B - C)$

$A + B = 0$
$-4A - 3B + C = 2$
$\underline{4A + 2B - C = -6}$
$\quad -B = -4$
$\quad B = 4$
$\quad A = -4$

$4(-4) + 2(4) - C = -6$
$-16 + 8 - C = -6$
$-C - 8 = -6$
$-C = 2$
$C = -2$

$\dfrac{2x-6}{(x-1)(x-2)^2} = -\dfrac{4}{x-1} + \dfrac{4}{x-2} - \dfrac{2}{(x-2)^2}$

21. $\dfrac{3x}{(x-2)(x^2+1)} = \dfrac{A}{x-2} + \dfrac{Bx+C}{x^2+1}$

$3x = A(x^2+1) + (Bx+C)(x-2)$

$\quad = Ax^2 + A + Bx^2 - 2Bx + Cx - 2C$

$\quad = (A+B)x^2 + (-2B+C)x - (2C-A)$

$A + B = 0$
$-2B + C = 3$
$2C - A = 0$
$A = 2C$
$B + 2C = 0$
$\underline{4B - 2C = -6}$
$\quad 5B = -6$

$\quad B = -\dfrac{6}{5}$

$\quad A = \dfrac{6}{5}$

$\quad C = \dfrac{6}{10} = \dfrac{3}{5}$

$\dfrac{3x}{(x-2)(x^2+1)} = \dfrac{6}{5(x-2)} + \dfrac{-6x+3}{5(x^2+1)}$

22. $\dfrac{7x^2 - 7x + 23}{(x-3)(x^2+4)} = \dfrac{A}{x-3} + \dfrac{Bx+C}{x^2+4}$

$7x^2 - 7x + 23 = A(x^2+4) + (Bx+C)(x-3)$

$\quad = Ax^2 + 4A + Bx^2 - 3Bx + Cx - 3C$

$\quad = (A+B)x^2 + (-3B+C)x + (4A - 3C)$

$A + B = 7$
$-3B + C = -7$
$4A - 3C = 23$
$3A + 3B = 21$
$\underline{-3B + C = -7}$
$\quad 3A + C = 14$
$\quad 9A + 3C = 42$
$\underline{4A - 3C = 23}$
$\quad 13A = 65$
$\quad A = 5$
$\quad 5 + B = 7$
$\quad B = 7 - 5 = 2$

$-3(2) + C = -7$
$\quad C = -7 + 6 = -1$

$\dfrac{7x^2 - 7x + 23}{(x-3)(x^2+4)} = \dfrac{5}{x-3} + \dfrac{2x-1}{x^2+4}$

23. $\dfrac{x^3}{(x^2+4)^2} = \dfrac{Ax+B}{x^2+4} + \dfrac{Cx+D}{(x^2+4)^2}$

$x^3 = (Ax+B)(x^2+4)+Cx+D$

$\quad = Ax^3+4Ax+Bx^2+4B+Cx+D$

$\quad = Ax^3+Bx^2+(4A+C)x+(4B+D)$

$\quad A=1$

$\quad B=0$

$4A+C=0$

$4B+D=0$

$\quad C=-4$

$0+D=0, D=0$

$\dfrac{x^2}{(x^2+4)^2} = \dfrac{x}{x^2+4} - \dfrac{4x}{(x^2+4)^2}$

24. $\dfrac{4x^3+5x^2+7x-1}{(x^2+x+1)^2} = \dfrac{Ax+B}{x^2+x+1} + \dfrac{Cx+D}{(x^2+x+1)^2}$

$4x^3+5x^2+7x-1$

$= (Ax+B)(x^2+x+1)+Cx+D$

$= Ax^3+Ax^2+Ax+Bx^2+Bx+B+Cx+D$

$= Ax^3+(A+B)x^2(A+B+C)x+(B+D)$

$\quad A=4$

$\quad A+B=5$

$A+B+C=7$

$\quad B+D=-1$

$4+B=5, B=1$

$4+1+C=7, C=2$

$1+D=-1, D=-2$

$\dfrac{4x^3+5x^2+7x-1}{(x^2+x+1)^2} = \dfrac{4x+1}{x^2+x+1} + \dfrac{2x-2}{(x^2+x+1)^2}$

25. $\qquad 5y = x^2-1$

$\qquad x-y=1$

$\qquad y=x-1$

$\qquad 5(x-1)=x^2-1$

$\qquad 5x-5=x^2-1$

$x^2-5x+4=0$

$(x-4)(x-1)=0$

$\qquad x=4,1$

If $x=4$, $y=4-1=3$.

If $x=1$, $y=1-1=0$.

The solution set is $\{(4,3),(1,0)\}$.

26. $\quad y=x^2+2x+1$

$\quad x+y=1$

$\qquad y=1-x$

$\quad 1-x=x^2+2x+1$

$x^2+3x=0$

$x(x+3)=0$

$\qquad x=0,-3$

If $x=0$, $y=1-0=1$.

If $x=-3$, $y=1-(-3)=4$.

The solution set is $\{(0,1),(-3,4)\}$.

27. $\qquad x^2+y^2=2$

$\qquad x+y=0$

$\qquad\quad x=-y$

$(-y)^2+y^2=2$

$\qquad 2y^2=2$

$\qquad\quad y^2=1$

$\qquad\quad y=1,-1$

If $y=1$, $x=-1$.

If $y=-1$, $x=1$.

The solution set is $\{(1,-1),(-1,1)\}$.

28. $\quad 2x^2+y^2=24$

$\qquad x^2+y^2=15$

$\quad 2x^2+y^2=24$

$\quad \underline{-x^2-y^2=-15}$

$\qquad\quad x^2=9$

$\qquad\quad x=3,-3$

If $x=3$, $3^2+y^2=15$, $y^2=6$ and $y=\pm\sqrt{6}$.

If $x=-3$, $y=\pm\sqrt{6}$.

The solution set is

$\left\{\left(3,\sqrt{6}\right),\left(3,-\sqrt{6}\right),\left(-3,\sqrt{6}\right),\left(-3,-\sqrt{6}\right)\right\}$.

29. $\quad xy-4=0$

$\qquad y-x=0$

$\qquad\quad y=x$

$\qquad\quad xy=4$

$\qquad\quad x^2=4$

$\qquad\quad x=2,-2$

If $x=2$, $y=2$.

If $x=-2$, $y=-2$.

The solution set is $\{(2,2),(-2,-2)\}$.

30.
$$y^2 = 4x$$
$$x - 2y + 3 = 0$$
$$x = \frac{y^2}{4}$$
$$\frac{y^2}{4} - 2y + 3 = 0$$
$$y^2 - 8y + 12 = 0$$
$$(y-6)(y-2) = 0$$
$$y = 6, 2$$

If $y = 6, x = \dfrac{36}{4} = 9.$

If $y = 2, x = \dfrac{4}{4} = 1.$

The solution set is $\{(9,6),(1,2)\}$.

31. $x^2 + y^2 = 10$
$$y = x + 2$$
$$x^2 + (x+2)^2 = 10$$
$$x^2 + x^2 + 4x + 4 - 10 = 0$$
$$2x^2 + 4x - 6 = 0$$
$$x^2 + 2x - 3 = 0$$
$$(x+3)(x-1) = 0$$
$$x = -3, 1$$
If $x = -3, y = -3 + 2 = -1.$
If $x = 1, y = 1 + 2 = 3.$
The solution set is $\{(-3,-1),(1,3)\}$.

32. $xy = 1$
$$y = 2x + 1$$
$$x(2x+1) = 1$$
$$2x^2 + x - 1 = 0$$
$$(2x-1)(x+1) = 0$$
$$x = \frac{1}{2}, -1$$

If $x = \dfrac{1}{2}, y = 2\left(\dfrac{1}{2}\right) + 1 = 2.$

If $x = -1, y = 2(-1) + 1 = -1.$

The solution set is $\left\{\left(\dfrac{1}{2}, 2\right), (-1,-1)\right\}$.

33.
$$x + y + 1 = 0$$
$$x^2 + y^2 + 6y - x = -5$$
$$x = -y - 1$$
$$(-y-1)^2 + y^2 + 6y - (-y-1) + 5 = 0$$
$$y^2 + 2y + 1 + y^2 + 6y + y + 1 + 5 = 0$$
$$2y^2 + 9y + 7 = 0$$
$$(2y+7)(y+1) = 0$$
$$y = -\frac{7}{2}, -1$$

If $y = -\dfrac{7}{2}, x = \dfrac{7}{2} - 1 = \dfrac{5}{2}.$

If $y = -1, x = 1 - 1 = 0.$

The solution set is $\left\{\left(\dfrac{5}{2}, -\dfrac{7}{2}\right), (0,-1)\right\}$.

34. $x^2 + y^2 = 13$
$$x^2 - y = 7$$
$$x^2 + y^2 = 13$$
$$\underline{-x^2 + y = -7}$$
$$y^2 + y = 6$$
$$y^2 + y - 6 = 0$$
$$(y+3)(y-2) = 0$$
$$y = -3, 2$$
If $y = -3, x^2 + 3 = 7$
$$x^2 = 4, x = 2, -2$$
If $y = 2, x^2 - 2 = 7, x^2 = 9, x = 3, -3.$
The solution set is $\{(2,-3),(-2,-3),(3,2),(-3,2)\}$.

35.
$$2x^2 + 3y^2 = 21$$
$$3x^2 - 4y^2 = 23$$
$$8x^2 + 12y^2 = 84$$
$$\underline{9x^2 - 12y^2 = 69}$$
$$17x^2 = 153$$
$$x^2 = \frac{153}{17} = 9$$
$$x = 3, -3$$
If $x = 3, 2(3)^2 + 3y^2 = 21.$
$$3y^2 = 21 - 18 = 3$$
$$y^2 = 1, y = 1, -1$$
If $x = -3, y = 1, -1.$
The solution set is $\{(3,1),(3,-1),(-3,1),(-3,-1)\}$.

36. $2L + 2W = 26$

$LW = 40$

$L = \dfrac{40}{W}$

$2\left(\dfrac{40}{W}\right) + 2W = 26$

$\dfrac{80}{W} + 2W = 26$

$80 + 2W^2 = 26W$

$2W^2 - 26W + 80 = 0$

$W^2 - 13W + 40 = 0$

$(W - 8)(W - 5) = 0$

$W = 8, 5$

If $W = 5, L = \dfrac{40}{5} = 8$

The dimensions are 8 m by 5 m.

37. $xy = 6$

$y = \dfrac{6}{x}$

$2x + y = 8$

$2x + \dfrac{6}{x} = 8$

$2x^2 + 6 = 8x$

$2x^2 - 8x + 6 = 0$

$x^2 - 4x + 3 = 0$

$(x - 1)(x - 3) = 0$

$x = 1, 3$

If $x = 1, y = 6.$
If $x = 3, y = 2.$
The solution set is $\{(1, 6), (3, 2)\}$.

38. $x^2 + y^2 = 2900$

$4x + 2y = 240$

$2x + y = 120$

$y = 120 - 2x$

$x^2 + (120 - 2x)^2 = 2900$

$x^2 + 14{,}400 - 480x + 4x^2 - 2900 = 0$

$5x^2 - 480x + 11{,}500 = 0$

$x^2 - 96x + 2300 = 0$

$(x - 46)(x - 50) = 0$

$x = 46, 50$

If $x = 46, y = 120 - 2(46) = 28.$
If $x = 50, y = 120 - 2(50) = 20.$
$x = 46$ ft and $y = 28$ ft or
$x = 50$ ft and $y = 20$ ft

39.
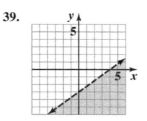

$3x - 4y > 12$

40.
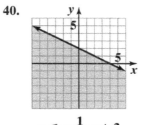

$y \le -\dfrac{1}{2}x + 2$

41.
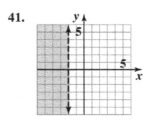

$x < -2$

42.
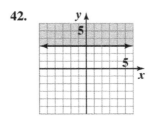

$y \ge 3$

43.
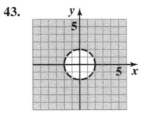

$x^2 + y^2 > 4$

44.

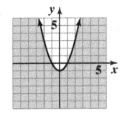

$y \le x^2 - 1$

45.

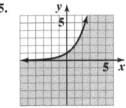

$y \le 2^x$

46.

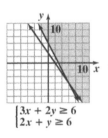

$\begin{cases} 3x + 2y \ge 6 \\ 2x + y \ge 6 \end{cases}$

47.

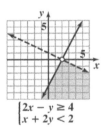

$\begin{cases} 2x - y \ge 4 \\ x + 2y < 2 \end{cases}$

48.

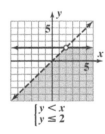

$\begin{cases} y < x \\ y \le 2 \end{cases}$

49.

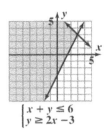

$\begin{cases} x + y \le 6 \\ y \ge 2x - 3 \end{cases}$

50.

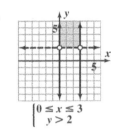

$\begin{cases} 0 \le x \le 3 \\ y > 2 \end{cases}$

51. no solution

52.

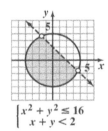

$\begin{cases} x^2 + y^2 \le 16 \\ x + y < 2 \end{cases}$

53.

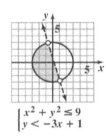

$\begin{cases} x^2 + y^2 \le 9 \\ y < -3x + 1 \end{cases}$

54.

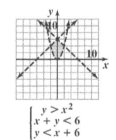

$\begin{cases} y > x^2 \\ x + y < 6 \\ y < x + 6 \end{cases}$

55.

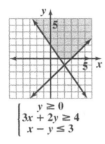

$$\begin{cases} y \geq 0 \\ 3x + 2y \geq 4 \\ x - y \leq 3 \end{cases}$$

56. $z = 2x + 3y$

$(2, 2): z = 2(2) + 3(2) = 10$

$(4, 0): z = 2(4) + 3(0) = 8$

$\left(\dfrac{1}{2}, \dfrac{1}{2}\right): z = 2\left(\dfrac{1}{2}\right) + 3\left(\dfrac{1}{2}\right) = \dfrac{5}{2}$

$(1, 0): z = 2(1) + 3(0) = 2$

The maximum value is 10 and the minimum value is 2.

57. $z = 2x + 3y$

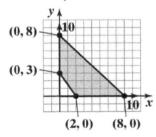

$x \geq 0, \ y \geq 0$

$x + y \leq 8$

$3x + 2y \geq 6$

$(0, 8): z = 2(0) + 3(8) = 24$

$(8, 0): z = 2(8) + 3(0) = 16$

$(0, 3): z = 2(0) + 3(3) = 9$

$(2, 0): z = 2(2) + 3(0) = 6$

Maximum value is 24 at $(0, 8)$.

58. $z = x + 4y$

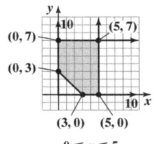

$0 \leq x \leq 5$

$0 \leq y \leq 7$

$x + y \geq 3$

$(0, 3): z = 0 + 4(3) = 12$

$(3, 0): z = 3 + 4(0) = 3$

$(0, 7): z = 0 + 4(7) = 28$

$(5, 0): z = 5 + 4(0) = 5$

$(5, 7): z = 5 + 4(7) = 33$

Maximum value is 33 at $(5,7)$.

59. $z = 5x + 6y$

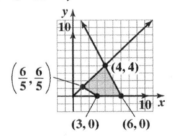

$x \geq 0, y \geq 0$

$y \leq x$

$2x + y \leq 12$

$2x + 3y \geq 6$

$(3, 0): z = 5(3) + 6(0) = 15$

$(6, 0): z = 5(6) + 6(0) = 30$

$\left(\dfrac{6}{5}, \dfrac{6}{5}\right): z = 5\left(\dfrac{6}{5}\right) + 6\left(\dfrac{6}{5}\right) = \dfrac{66}{5} = 13.2$

$(4, 4): 5(4) + 6(4) = 44$

The maximum value is 44.

60. a. $z = 500x + 350y$

b. $x + y \leq 200$
$x \geq 10$
$y \geq 80$

c.

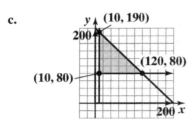

$x + y \leq 200$
$x \geq 10, y \geq 80$

d.

Vertex	Objective Function
	$z = 500x + 350y$
(10, 80)	$z = 500(10) + 350(80)$ $= 33,000$
(10, 190)	$z = 500(10) + 350(190)$ $= 71,500$
(120, 80)	$z = 500(120) + 350(80)$ $= 88,000$

e. The company will make the greatest profit by producing 120 units of writing paper and 80 units of newsprint each day. The maximum daily profit is $88,000.

61. Let x = number of model A tents produced and y = number of model B tents produced.
The constraints are:
$0.9x + 1.8y \leq 864$
$0.8x + 1.2y \leq 672$
$\quad x \geq 0$
$\quad y \geq 0$

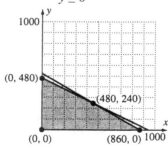

The vertices of the region are (0, 0), (0, 480), (480, 240), and (840, 0).
The objective is to maximize $25x + 40y$.
(0, 0): $25(0) + 40(0) = 0 + 0 = 0$
(0, 480): $25(0) + 40(480) = 0 + 19,200 = 19,200$
(480,240): $25(480) + 40(240) = 12,000 + 9600$
$\qquad\qquad = 21,600$
(840, 0): $25(840) + 40(0) = 21,000 + 0 = 21,000$
The manufacturer should make 480 of model A and 240 of model B.

Chapter 8 Test

1. $x = y + 4$
$3x + 7y = -18$
Substitute $y + 4$ for x into second equation.
$3(y + 4) + 7y = -18$
$3y + 12 + 7y = -18$
$\qquad 10y = -30$
$\qquad\quad y = -3$
$\qquad\quad x = -3 + 4 = 1$
The solution set to the system is $\{(1, -3)\}$.

2. $2x + 5y = -2$
$3x - 4y = 20$
Multiply the first equation by 3 and the second equation by –2 and add the result.
$\quad 6 + 15y = -6$
$\underline{-6x + 8y = -40}$
$\qquad 23y = -46$
$\qquad\quad y = -2$
Substitute $y = -2$ into the first equation:
$2x + 5(-2) = -2$
$\quad 2x - 10 = -2$
$\qquad 2x = 8$
$\qquad\;\; x = 4$
The solution to the system is $\{(4, -2)\}$.

3. $\quad x + y + z = 6\,(1)$
$3x + 4y - 7z = 1\,(2)$
$2x - y + 3z = 5\,(3)$
Eliminate x by multiplying (1) by –3 and adding the result to (2) and by multiplying (1) by –2 and adding the result to (3).
$-3x - 3y - 3z = -18$
$\underline{3x + 4y - 7z = 1}$
$\qquad y - 10z = -17\,(4)$
$-2x - 2y - 2z = -12$
$\underline{2x - y + 3z = 5}$
$\qquad -3y + z = -7\;(5)$

Multiply (4) by 3 and add the result to (5) to eliminate y.
$$3y - 30z = -51$$
$$\underline{-3y + z = -7}$$
$$-29z = -58$$
$$z = 2$$
Substitute $z = 2$ into (5).
$$-3y + 2 = -7$$
$$-3y = -9$$
$$y = 3$$
Substitute $z = 2$ and $y = 3$ into (1).
$$x + 3 + 2 = 6$$
$$x = 1$$
The solution to the system is $\{(1, 3, 2)\}$.

4. $x^2 + y^2 = 25$

   $x + y = 1$

   $y = 1 - x$

   Substitute $1 - x$ for y in the first equation.
   $$x^2 + (1 - x)^2 = 25$$
   $$x^2 + 1 - 2x + x^2 = 25$$
   $$2x^2 - 2x - 24 = 0$$
   $$x^2 - x - 12 = 0$$
   $$(x - 4)(x + 3) = 0$$
   $$x = 4, -3$$
   If $x = 4$, $y = 1 - 4 = -3$.
   If $x = -3$, $y = 1 - (-3) = 4$.
   The solution set is $\{(4, -3), (-3, 4)\}$.

5. $2x^2 - 5y^2 = -2$

   $3x^2 + 2y^2 = 35$

   Multiply first equation by 2 and the second equation by 5. Then add.
   $$4x^2 - 10y^2 = -4$$
   $$\underline{15x^2 + 10y^2 = 175}$$
   $$19x^2 = 171$$
   $$x^2 = 9$$
   $$x = 3, -3$$

If $x = 3, 2(3)^2 - 5y^2 = -2$.
$$18 - 5y^2 = -2$$
$$-5y^2 = -20$$
$$y^2 = 4$$
$$y = 2, -2$$
If $x = -3$, $y = -2$.
The solution to the system is
$\{(3, 2), (3, -2), (-3, 2), (-3, -2)\}$.

6. $$\frac{x}{(x+1)(x^2+9)} = \frac{A}{x+1} + \frac{Bx+C}{x^2+9}$$
   $$x = A(x^2+9) + (Bx+C)(x+1)$$
   $$= Ax^2 + 9A + Bx^2 + Bx + Cx + C$$
   $$= (A+B)x^2 + (B+C)x + (9A+C)$$
   $$A + B = 0 \rightarrow A = -B$$
   $$B + C = 1$$
   $$9A + C = 0$$
   $$-9B + C = 0$$
   $$9B - C = 0$$
   $$\underline{B + C = 1}$$
   $$10B = 1$$
   $$B = \frac{1}{10}$$
   $$A = -\frac{1}{10}$$
   $$\frac{1}{10} + C = 1, \ C = \frac{9}{10}$$
   $$\frac{x}{(x+1)(x^2+9)} = \frac{-1}{10(x+1)} + \frac{x+9}{10(x^2+9)}$$

7.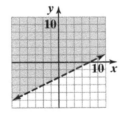

   $x - 2y < 8$

8.

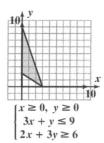

$$\begin{cases} x \geq 0, \ y \geq 0 \\ 3x + y \leq 9 \\ 2x + 3y \geq 6 \end{cases}$$

9.

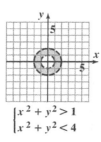

$$\begin{cases} x^2 + y^2 > 1 \\ x^2 + y^2 < 4 \end{cases}$$

10.

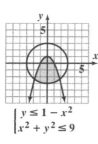

$$\begin{cases} y \leq 1 - x^2 \\ x^2 + y^2 \leq 9 \end{cases}$$

11. $z = 3x + 5y$

$x \geq 0, \ y \geq 0$

$x + y \leq 6$

$x \geq 2$

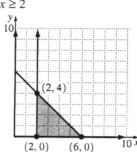

$(2,0): z = 3(2) + 5(0) = 6$

$(6,0): z = 3(6) + 5(0) = 18$

$(2,4): z = 3(2) + 5(4) = 26$

Maximum value is 26.

12. $x =$ mg of cholesterol in one ounce of shrimp

$y =$ mg of cholesterol in one ounce of scallops

$3x + 2y = 156$

$5x + 3y = 255$

Multiply the first equation by -3 and multiply the second equation by 2.

Add the resulting equations together.

$$\begin{array}{r} -9x - 6y = -468 \\ \underline{10x + 6y = 510} \\ x = 42 \end{array}$$

$3(42) + 2y = 156$

$126 + 2y = 156$

$2y = 30$

$y = 15$

$3(42) + 2y = 156$

$126 + 2y = 156$

$2y = 30$

$y = 15$

Shrimp: 42 mg of cholesterol per ounce

Scallops: 15 mg of cholesterol per ounce

13. a. $C(x) = 360,000 + 850x$

b. $R(x) = 1150x$

c. $1150x = 360000 + 850x$

$300x = 360000$

$x = 1200$

$1150(1200) = 1,380,000$

The break-even point is $(1200, 1,380,000)$.

This means that 1200 computers need to be sold to make $1,380,00 and for the company to break even.

14. $y = ax^2 + bx + c$

$(-1, -2): -2 = a - b + c$

$(2, 1): 1 = 4a + 2b + c$

$(-2, 1): 1 = 4a - 2b + c$

$$\begin{array}{r} 4a + 2b + c = 1 \\ \underline{-4a + 2b - c = -1} \\ 4b = 0 \\ b = 0 \end{array}$$

$a + c = -2$

$$\begin{array}{r} 4a + c = 1 \\ \underline{-a - c = 2} \\ 3a = 3 \\ a = 1 \end{array}$$

$$a + c = -2$$

$$c = -3$$

15. $2x + y = 39$

$xy = 180$

$y = 39 - 2x$

$$x(39 - 2x) = 180$$

$$39x - 2x^2 = 180$$

$$2x^2 - 39x + 180 = 0$$

$$(2x - 15)(x - 12) = 0$$

$$x = \frac{15}{2}, 12$$

If $x = \dfrac{15}{2}, \dfrac{15}{2} y = 180$ and $y = 24$.

If $x = 12, 12 y = 180$ and $y = 15$.

The dimensions are 7.5 ft by 24 ft or 12 ft by 15 ft

16. Let x = regular, y = deluxe.

objective function: $z = 200x + 250y$

constraints: $x \geq 50, y \geq 75$

$$x + y \leq 150$$

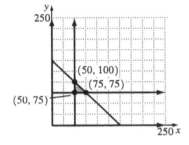

$(50, 75): z = 200(50) + 250(75) = 28,750$

$(50, 100): z = 200(50) + 250(100) = 35,000$

$(75, 75): z = 200(75) + 250(75) = 33,750$

For a maximum profit of $35,000 a week, the company should manufacture 50 regular and 100 deluxe jet skis.

Cumulative Review Exercises (Chapters 1–8)

1. Domain: $(-\infty, \infty)$ Range: $[-\infty, 3)$

2. -1 and 1 are the zeros.

3. The relative maximum is $y = 3$ and it occurs at $x = 0$.

4. $f(x)$ is decreasing on the interval $(0, 2)$.

The quadratic function is $y = x^2 - 3$.

5. At $x = -0.7$, the curve is above the x-axis and thus $f(x)$ is positive.

6. $(f \circ f)(-1) = f(f(-1)) = f(0) = 3$

7. $f(x) \to -\infty$ as $x \to -2^+$ or as $x \to 2^-$.

8. $f(-x) = f(x)$ thus the function is even.

9. The graph of $g(x) = f(x+2) - 1$ can be obtained by shifting $f(x)$ 2 units left and 1 unit down.

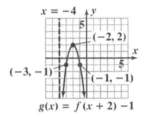

10. The graph of $h(x) = \frac{1}{2} f\left(\frac{1}{2} x\right)$ can be obtained by shrinking the graph of $f(x)$ horizontally by a factor of $\frac{1}{2}$ and vertically by a factor of $\frac{1}{2}$.

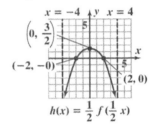

11.

$$\sqrt{x^2 - 3x} = 2x - 6$$

$$x^2 - 3x = 4x^2 - 24x + 36$$

$$3x^2 - 21x + 36 = 0$$

$$x^2 - 7x + 12 = 0$$

$$(x - 3)(x - 4) = 0$$

$$x = 3, 4$$

The solution set is $\{3, 4\}$.

12.
$$4x^2 = 8x - 7$$
$$4x^2 - 8x + 7 = 0$$
$$x = \frac{8 \pm \sqrt{64 - 112}}{8} = \frac{8 \pm \sqrt{-48}}{8}$$
$$= \frac{8 \pm 4\sqrt{3}i}{8} = \frac{2 \pm \sqrt{3}i}{2}$$

The solution set is $\left\{ \frac{2 + i\sqrt{3}}{2}, \frac{2 - i\sqrt{3}}{2} \right\}$.

13. $\left| \frac{x}{3} + 2 \right| < 4$

$$-4 < \frac{x}{3} + 2 < 4$$
$$-6 < \frac{x}{3} < 2$$
$$-18 < x < 6$$

The solution is $\{x \mid -18 < x < 6\}$ or $(-18, 6)$.

14.
$$\frac{x+5}{x-1} > 2$$
$$\frac{x+5}{x-1} - 2 > 0$$
$$\frac{x+5 - 2(x-1)}{x-1} > 0$$
$$\frac{x+5 - 2x + 2}{x-1} > 0$$
$$\frac{-x+7}{x-1} > 0$$

$\frac{-x+7}{x-1} = 0$ when $x = 7$ and is undefined when $x = 1$.

Test $x = 0$:
$$\frac{0+5}{0-1} > 2?$$
$$\frac{5}{-1} > 2?$$
$$-5 \not> 2$$

Test $x = 2$:
$$\frac{2+5}{2-1} > 2?$$
$$\frac{7}{1} > 2?$$
$$7 > 2$$

Test $x = 8$:
$$\frac{8+5}{8-1} > 2?$$
$$\frac{13}{7} > 2?$$
$$\frac{13}{7} \geq \frac{14}{7}$$

The solution is $\{x \mid 1 < x < 7\}$ or $(1, 7)$.

15. $2x^3 + x^2 - 13x + 6 = 0$

$f(x) = 2x^3 + x^2 - 13x + 6$ has 2 sign changes: 2 or 0 positive real roots.

$f(-x) = -2x^3 + x^2 + 13x + 6$ has 1 sign change: 1 negative real root.

p: ±1, ±2, ±3, ±6

q: ±1, ±2

$\frac{p}{q}$: ±1, ±$\frac{1}{2}$, ±2, ±3, ±$\frac{3}{2}$, ±6

$$
\begin{array}{r|rrrr}
-3 & 2 & 1 & -13 & 6 \\
   &   & -6 & 15 & -6 \\
\hline
   & 2 & -5 & 2 & 0
\end{array}
$$

$$2x^3 + x^2 - 13x + 6 = (x+3)(2x^2 - 5x + 2)$$
$$= (x+3)(2x-1)(x-2)$$

$$x = -3, \; x = \frac{1}{2}, \; x = 2$$

The solution set is $\left\{ -3, \frac{1}{2}, 2 \right\}$.

16.
$$6x - 3(5x + 2) = 4(1 - x)$$
$$6x - 15x - 6 = 4 - 4x$$
$$-9x - 6 = 4 - 4x$$
$$-5x = 10$$
$$x = -2$$

The solution set is $\{-2\}$.

17.
$$\log(x+3) + \log x = 1$$
$$\log x(x+3) = 1$$
$$x(x+3) = 10$$
$$x^2 + 3x - 10 = 0$$
$$(x+5)(x-2) = 0$$
$$x = -5 \text{ or } x = 2$$
$$x = -5 \text{ is extraneous.}$$
$$x = 2$$

The solution set is $\{2\}$.

18. $3^{x+2} = 11$

$$\log_3 3^{x+2} = \log_3 11$$

$$x + 2 = \log_3 11$$

$$x = -2 + \log_3 11$$

$$x = -2 + \frac{\log 11}{\log 3} \approx 0.18$$

The solution set is $\{-2 + \log_3 11\}$.

19. $x^{\frac{1}{2}} - 2x^{\frac{1}{4}} - 15 = 0$

$$\sqrt[4]{x}^2 - 2\sqrt[4]{x} - 15 = 0$$

$$\left(\sqrt[4]{x} + 3\right)\left(\sqrt[4]{x} - 5\right) = 0$$

$\sqrt[4]{x} + 3 = 0$ $\qquad$ $\sqrt[4]{x} - 5 = 0$

$\sqrt[4]{x} = -3$ $\qquad$ $\sqrt[4]{x} = 5$

$\qquad$ or

$x = (-3)^4$ $\qquad$ $x = 5^4$

$x = 81$ $\qquad$ $x = 625$

81 does not check. The solution set is $\{625\}$.

20. $3x - y = -2$

$2x^2 - y = 0$

Solve the first equation for y.

$3x - y = -2$

$$y = 3x + 2$$

Use this equation to substitute into the other equation.

$$2x^2 - \overset{y}{\overbrace{(3x+2)}} = 0$$

$$2x^2 - 3x - 2 = 0$$

$$(2x+1)(x-2) = 0$$

$$x = -\frac{1}{2} \quad \text{or} \quad x = 2.$$

Back-substitute to find y.

$y = 3x + 2$ $\qquad$ or $\qquad$ $y = 3x + 2$

$y = 3(2) + 2$

$y = 8$ $\qquad\qquad$ $y = 3\left(-\frac{1}{2}\right) + 2$

$\qquad\qquad\qquad$ $y = \frac{1}{2}$

$y = 3x + 2$

The solution set is $\left\{(2, 8), \left(-\frac{1}{2}, \frac{1}{2}\right)\right\}$.

21. $x + 2y + 3z = -2$

$3x + 3y + 10z = -2$

$2y - 5z = 6$

Multiply equation 1 by -3 and add to equation 2.

$-3x - 6y - 9z = 6$

$\underline{3x + 3y + 10z = -2}$

$-3y + z = 4$ $\quad$ Equation 4

Multiply equation 4 by 5 and add to equation 3 and solve for y.

$-15y + 5z = 20$

$\underline{2y - 5z = 6}$

$-13y = 26$

$y = -2$

Back-substitute to find z.

$-3y + z = 4$

$-3(-2) + z = 4$

$z = -2$

Back-substitute to find x.

$x + 2y + 3z = -2$

$x + 2(-2) + 3(-2) = -2$

$x = 8$

The solution set is $\{8, -2, -2\}$.

22. vertex: $(-2, -4)$

y-intercept:

$$f(0) = (0+2)^2 - 4 = 0$$

x-intercepts:

$$(x+2)^2 - 4 = 0$$

$$x^2 + 4x + 4 - 4 = 0$$

$$x^2 + 4x = 0$$

$$x(x+4) = 0$$

$$x = 0, x - 4$$

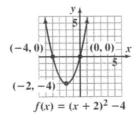

$$f(x) = (x+2)^2 - 4$$

23.

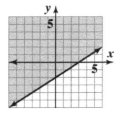

$$2x - 3y \le 6$$

24.

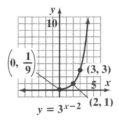

$$y = 3^{x-2}$$

25. vertical asymptote: $x = -1$
horizontal asymptote: $m > n$, none
x-intercepts:
$$x^2 - x - 6 = 0$$
$$(x-3)(x+2) = 0$$
$$x = 3, x = -2$$
y-intercept:
$$f(0) = \frac{0^2 - 0 - 6}{0+1} = -6$$

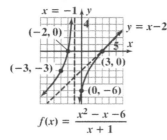

$$f(x) = \frac{x^2 - x - 6}{x + 1}$$

26.

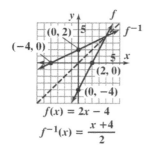

$$f(x) = 2x - 4$$
$$f^{-1}(x) = \frac{x+4}{2}$$

27.

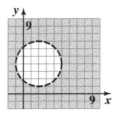

$$(x - 2)^2 + (y - 4)^2 > 9$$

28.

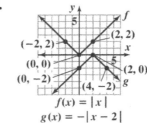

$$f(x) = |x|$$
$$g(x) = -|x - 2|$$

29. $(f \circ g)(x) = f(g(x))$
$$= 2(1-x)^2 - (1-x) - 1$$
$$= 2x^2 - 3x$$

$$(g \circ f) = g(f(x))$$
$$= 1 - (2x^2 - x - 1)$$
$$= 1 - 2x^2 + x + 1$$
$$= -2x^2 + x + 2$$

30. $\dfrac{f(x+h) - f(x)}{h}$

$$= \frac{\left[2(x+h)^2 - (x+h) - 1\right] - \left[2x^2 - x - 1\right]}{h}$$

$$= \frac{2x^2 + 4hx + 2h^2 - x - h - 1 - 2x^2 + x + 1}{h}$$

$$= \frac{4hx + 2h^2 - h}{h}$$

$$= 4x + 2h - 1$$

31. Find slope: $m = \dfrac{4 - (-2)}{2 - 4} = \dfrac{6}{-2} = -3$
Use point slope form to find an equation.
$$y - y_1 = m(x - x_1)$$
$$y - 4 = -3(x - 2)$$
Put in slope-intercept form.
$$y - 4 = -3(x - 2)$$
$$y - 4 = -3x + 6$$
$$y = -3x + 10$$

32. Find the slope of the perpendicular line by putting in slope-intercept form.

$$x + 3y - 6 = 0$$

$$3y = -x + 6$$

$$y = -\frac{1}{3}x + 2$$

The slope of the perpendicular line is $-\frac{1}{3}$ so the slope of the desired line is the negative reciprocal, or 3.

Use point slope form to find an equation.

$$y - y_1 = m(x - x_1)$$

$$y - 0 = 3(x + 1)$$

Put in slope-intercept form.

$$y - 0 = 3(x + 1)$$

$$y = 3x + 3$$

33. Let x = the amount invested at 12%

Let $4000 - x$ = the amount invested at 14%

$$0.12x + 0.14(4000 - x) = 508$$

$$0.12x + 560 - 0.14x = 508$$

$$-0.02x = -52$$

$$x = \frac{-52}{-0.02}$$

$$x = 2600$$

$$4000 - x = 4000 - 2600 = 1400$$

Thus, $2600 was invested at 12% and $1400 was invested at 14%.

34.

$$L = 2W + 1$$

$$LW = 36$$

$$W(2W + 1) = 36$$

$$2W^2 + W - 36 = 0$$

$$(2W + 9)(W - 4) = 0$$

$$W = -\frac{9}{2} \text{ or } 4$$

Length cannot be negative. If $W = 4$, $4L = 36$, $L = 9$. The dimensions are 4 m by 9 m.

35.

$$A = Pe^{rt}$$

$$18{,}000 = 6000e^{10r}$$

$$3 = e^{10r}$$

$$\ln 3 = \ln e^{10r}$$

$$\ln 3 = 10r$$

$$r = \frac{\ln 3}{10} \approx 0.1099$$

$$10.99\%$$

36.

$$\sec\theta - \cos\theta = \frac{1}{\cos\theta} - \cos\theta$$

$$= \frac{1 - \cos^2\theta}{\cos\theta}$$

$$= \frac{\sin^2\theta}{\cos\theta}$$

$$= \frac{\sin\theta}{\cos\theta} \cdot \sin\theta$$

$$= \tan\theta \cdot \sin\theta$$

37.

$$\tan x + \tan y = \frac{\sin x}{\cos x} + \frac{\sin y}{\cos y}$$

$$= \frac{\sin x \cdot \cos y + \sin y \cdot \cos x}{\cos x \cdot \cos y}$$

$$= \frac{\sin(x + y)}{\cos x \cdot \cos y}$$

38.

$$\sin\theta = \tan\theta$$

$$\sin\theta = \frac{\sin\theta}{\cos\theta}$$

$$\sin\theta - \frac{\sin\theta}{\cos\theta} = 0$$

$$\sin\theta\left(1 - \frac{1}{\cos\theta}\right) = 0$$

$$\sin\theta = 0 \quad \text{or} \quad 1 - \frac{1}{\cos\theta} = 0$$

$$\theta = 0, \pi \quad \text{or} \quad 1 = \frac{1}{\cos\theta}$$

$$\cos\theta = 1$$

$$\theta = 0$$

The solutions in the interval $[0, 2\pi)$ are 0 and π.

39.
$$2 + \cos 2\theta = 3\cos\theta$$
$$2 + (2\cos^2\theta - 1) = 3\cos\theta$$
$$2\cos^2\theta + 1 = 3\cos\theta$$
$$2\cos^2\theta - 3\cos\theta + 1 = 0$$
$$(2\cos\theta - 1)(\cos\theta - 1) = 0$$
$$2\cos\theta - 1 = 0 \quad \text{or} \quad \cos\theta - 1 = 0$$
$$2\cos\theta = 1 \qquad\qquad \cos\theta = 1$$
$$\cos\theta = \frac{1}{2}$$
$$\theta = \frac{\pi}{3}, \frac{5\pi}{3} \quad \text{or} \quad \theta = 0$$

The solutions in the interval $[0, 2\pi)$ are

$$0, \frac{\pi}{3}, \text{ and } \frac{5\pi}{3}.$$

40.
$$\frac{b}{\sin B} = \frac{a}{\sin A}$$
$$\frac{b}{\sin 75°} = \frac{20}{\sin 12°}$$
$$b = \frac{20\sin 75°}{\sin 12°}$$
$$b \approx 92.9$$

Chapter 9
Matrices and Determinants

Section 9.1

Check Point Exercises

1. **a.** The notation $R_1 \leftrightarrow R_2$ means to interchange the elements in row 1 and row 2. This results in the row-equivalent matrix

$$\begin{bmatrix} 1 & 6 & -3 & | & 7 \\ 4 & 12 & -20 & | & 8 \\ -3 & -2 & 1 & | & -9 \end{bmatrix}.$$

b. The notation $\frac{1}{4}R_1$ means to multiply each element in row 1 by $\frac{1}{4}$. This results in the row-equivalent matrix

$$\begin{bmatrix} \frac{1}{4}(4) & \frac{1}{4}(12) & \frac{1}{4}(-20) & | & \frac{1}{4}(8) \\ 1 & 6 & -3 & | & 7 \\ -3 & -2 & 1 & | & -9 \end{bmatrix} = \begin{bmatrix} 1 & 3 & -5 & | & 2 \\ 1 & 6 & -3 & | & 7 \\ -3 & -2 & 1 & | & -9 \end{bmatrix}.$$

c. The notation $3R_2 + R_3$ means to add 3 times the elements in row 2 to the corresponding elements in row 3. Replace the elements in row 3 by these sums. First, we find 3 times the elements in row 2:
$3(1) = 3$, $3(6) = 18$, $3(-3) = -9$, $3(7) = 21$. Now we add these products to the corresponding elements in row 3. This results in the row equivalent matrix

$$\begin{bmatrix} 4 & 12 & -20 & | & 8 \\ 1 & 6 & -3 & | & 7 \\ -3+3=0 & -2+18=16 & 1-9=-8 & | & -9+21=12 \end{bmatrix} = \begin{bmatrix} 4 & 12 & -20 & | & 8 \\ 1 & 6 & -3 & | & 7 \\ 0 & 16 & -8 & | & 12 \end{bmatrix}.$$

2.
$$\begin{cases} 2x+y+2z=18 \\ x-y+2z=9 \\ x+2y-z=6 \end{cases} \rightarrow \begin{bmatrix} 2 & 1 & 2 & | & 18 \\ 1 & -1 & 2 & | & 9 \\ 1 & 2 & -1 & | & 6 \end{bmatrix}$$

Interchange row 1 with row 2 to get 1 in the top position of the first column.

$$\begin{bmatrix} 1 & -1 & 2 & | & 9 \\ 2 & 1 & 2 & | & 18 \\ 1 & 2 & -1 & | & 6 \end{bmatrix}$$

Multiply the first row by –2 and add these products to row 2.

$$\begin{bmatrix} 1 & -1 & 2 & | & 9 \\ 2+-2=0 & 1+2=3 & 2+-4=-2 & | & -18+18=0 \\ 1 & 2 & -1 & | & 6 \end{bmatrix} = \begin{bmatrix} 1 & -1 & 2 & | & 9 \\ 0 & 3 & -2 & | & 0 \\ 1 & 2 & -1 & | & 6 \end{bmatrix}$$

Next, multiply the top row by –1 and add these products to row 3.

$$\begin{bmatrix} 1 & -1 & 2 & | & 9 \\ 0 & 3 & -2 & | & 0 \\ 1+-1=0 & 2+1=3 & -1-2=-3 & | & 6-9=-3 \end{bmatrix} = \begin{bmatrix} 1 & -1 & 2 & | & 9 \\ 0 & 3 & -2 & | & 0 \\ 0 & 3 & -3 & | & -3 \end{bmatrix}$$

Next, to obtain a 1 in the second row, second column, multiply 3 by its reciprocal, $\frac{1}{3}$. Therefore, we multiply all the numbers in the second row by $\frac{1}{3}$ to get

$$\begin{bmatrix} 1 & -1 & 2 & | & 9 \\ 0 & 1 & -\frac{2}{3} & | & 0 \\ 0 & 3 & -3 & | & -3 \end{bmatrix}.$$

Next, to obtain a 0 in the third row, second column, multiply the second row by –3 and add the products to row three. The resulting matrix is

$$\begin{bmatrix} 1 & -1 & 2 & | & 9 \\ 0 & 1 & -\frac{2}{3} & | & 0 \\ 0 & 0 & -1 & | & -3 \end{bmatrix}.$$

To get 1 in the third row, third column, multiply –1 by its reciprocal, –1. Multiply all numbers in the third row by –1 to obtain the resulting matrix

$$\begin{bmatrix} 1 & -1 & 2 & | & 9 \\ 0 & 1 & -\frac{2}{3} & | & 0 \\ 0 & 0 & 1 & | & 3 \end{bmatrix}.$$

The system represented by this matrix is:

$x - y + 2z = 9$

$y - \dfrac{2}{3}z = 0$

$z = 3$

Use back substitution to find y and x.

$y - \dfrac{2}{3}(3) = 0 \qquad\qquad x - 2 + 6 = 9$

$\qquad y - 2 = 0 \qquad\qquad\quad x + 4 = 9$

$\qquad\quad y = 2 \qquad\qquad\qquad\quad x = 5$

The solution set for the original system is $\{(5, 2, 3)\}$.

3. $\qquad w - 3x - 2y + z = -3$

$\qquad 2w - 7x - y + 2z = 1$

$\qquad 3w - 7x - 3y + 3z = -5$

$\qquad 5w + x + 4y - 2z = 18$

The augmented matrix is

$$\begin{bmatrix} 1 & -3 & -2 & 1 & | & -3 \\ 2 & -7 & -1 & 2 & | & 1 \\ 3 & -7 & -3 & 3 & | & -5 \\ 5 & 1 & 4 & -2 & | & 18 \end{bmatrix}.$$

Multiply the top row by –2 and add the products to the second row. Multiply the top row by –3 and add the pro-ducts to the third row. Multiply the top row by –5 and add the products to the fourth row. The resulting matrix is

$$\begin{bmatrix} 1 & -3 & -2 & 1 & | & -3 \\ 0 & -1 & 3 & 0 & | & 7 \\ 0 & 2 & 3 & 0 & | & 4 \\ 0 & 16 & 14 & -7 & | & 33 \end{bmatrix}.$$

Next, multiply the second row by –1 to obtain a 1 in the second row, second column.

$$\begin{bmatrix} 1 & -3 & -2 & 1 & | & -3 \\ 0 & 1 & -3 & 0 & | & -7 \\ 0 & 2 & 3 & 0 & | & 4 \\ 0 & 16 & 14 & -7 & | & 33 \end{bmatrix}$$

Next, multiply the second row by –2 and add the products to the third row. Multiply the second row by –16 and add the products to the fourth row. The resulting matrix is

$$\begin{bmatrix} 1 & -3 & -2 & 1 & | & -3 \\ 0 & 1 & -3 & 0 & | & -7 \\ 0 & 0 & 9 & 0 & | & 18 \\ 0 & 0 & 62 & -7 & | & 145 \end{bmatrix}.$$

Next, multiply the third row by $\frac{1}{9}$ to obtain a 1 in the third row, third column. The resulting matrix is

$$\begin{bmatrix} 1 & -3 & -2 & 1 & | & -3 \\ 0 & 1 & -3 & 0 & | & -7 \\ 0 & 0 & 1 & 0 & | & 2 \\ 0 & 0 & 62 & -7 & | & 145 \end{bmatrix}.$$

Multiply the third row by –62 and add the products to the fourth row to obtain the resulting matrix

$$\begin{bmatrix} 1 & -3 & -2 & 1 & | & -3 \\ 0 & 1 & -3 & 0 & | & -7 \\ 0 & 0 & 1 & 0 & | & 2 \\ 0 & 0 & 0 & -7 & | & 21 \end{bmatrix}.$$

Multiply the fourth row by $-\frac{1}{7}$, the reciprocal of –7. The resulting matrix is

$$\begin{bmatrix} 1 & -3 & -2 & 1 & | & -3 \\ 0 & 1 & -3 & 0 & | & -7 \\ 0 & 0 & 1 & 0 & | & 2 \\ 0 & 0 & 0 & 1 & | & -3 \end{bmatrix}.$$

The system of linear equations corresponding to the resulting matrix is
$$w - 3x - 2y + z = -3$$
$$x - 3y = -7$$
$$y = 2$$
$$z = -3$$
Using back-substitution solve for x and w.
$$x - 3(2) = -7$$
$$x = -1$$

$$w - 3(-1) - 2(2) - 3 = -3$$
$$w - 4 = -3$$
$$w = 1$$
The solution set is $\{(1, -1, 2, -3)\}$.

4. The matrix obtained in 3 will be the starting point.

$$\begin{bmatrix} 1 & -1 & 2 & | & 9 \\ 0 & 1 & -\frac{2}{3} & | & 0 \\ 0 & 0 & 1 & | & 3 \end{bmatrix}$$

Next, multiply the third row by $\frac{2}{3}$ and add the products to the second row. Multiply the third row by 2 and add the products to the first row. The resulting matrix is

$$\begin{bmatrix} 1 & -1 & 0 & 3 \\ 0 & 1 & 0 & 2 \\ 0 & 0 & 1 & 3 \end{bmatrix}.$$

Add the second row to the first row and replace the first row.

$$\begin{bmatrix} 1 & 0 & 0 & 5 \\ 0 & 1 & 0 & 2 \\ 0 & 0 & 1 & 3 \end{bmatrix}$$

This matrix corresponds to $x = 5$, $y = 2$ and $z = 3$. The solution set is $\{(5, 2, 3)\}$.

Concept and Vocabulary Check 9.1

1. matrix; elements

2. -3; 2; 3; second; -2; third

3. $\begin{bmatrix} 2 & 1 & 4 & | & -4 \\ 3 & 0 & 1 & | & 1 \\ 4 & 3 & 1 & | & 8 \end{bmatrix}$

4. $\{(1, 3, -2)\}$

5. true

6. false

Exercise Set 9.1

1. $\begin{bmatrix} 2 & 1 & 2 & | & 2 \\ 3 & -5 & -1 & | & 4 \\ 1 & -2 & -3 & | & -6 \end{bmatrix}$

3. $\begin{bmatrix} 1 & -1 & 1 & | & 8 \\ 0 & 1 & -12 & | & -15 \\ 0 & 0 & 1 & | & 1 \end{bmatrix}$

5. $\begin{bmatrix} 5 & -2 & -3 & | & 0 \\ 1 & 1 & 0 & | & 5 \\ 2 & 0 & -3 & | & 4 \end{bmatrix}$

7. $\begin{bmatrix} 2 & 5 & -3 & 1 & | & 2 \\ 0 & 3 & 1 & 0 & | & 4 \\ 1 & -1 & 5 & 0 & | & 9 \\ 5 & -5 & -2 & 0 & | & 1 \end{bmatrix}$

9. $5x + 3z = -11$

$y - 4z = 12$

$7x + 2y = 3$

11. $w + x + 4y + z = 3$

$-w + x - y = 7$

$12w + 5z = 11$

$12y + 4z = 5$

13. $\begin{bmatrix} 2\left(\frac{1}{2}\right) & -6\left(\frac{1}{2}\right) & 4\left(\frac{1}{2}\right) & | & 10\left(\frac{1}{2}\right) \\ 1 & 5 & -5 & | & 0 \\ 3 & 0 & 4 & | & 7 \end{bmatrix} \frac{1}{2}R_1$

$\begin{bmatrix} 1 & -3 & 2 & | & 5 \\ 1 & 5 & -5 & | & 0 \\ 3 & 0 & 4 & | & 7 \end{bmatrix}$

15. $\begin{bmatrix} 1 & -3 & 2 & | & 0 \\ -3(1)+3 & -3(-3)+1 & -3(2)+-1 & | & -3(0)+7 \\ 2 & -2 & 1 & | & 3 \end{bmatrix} -3R_1 + R_2$

$\begin{bmatrix} 1 & -3 & 2 & | & 0 \\ 0 & 10 & -7 & | & 7 \\ 2 & -2 & 1 & | & 3 \end{bmatrix}$

17. $\begin{bmatrix} 1 & -1 & 1 & 1 & | & 3 \\ 0 & 1 & -2 & -1 & | & 0 \\ 2 & 0 & 3 & 4 & | & 11 \\ 5 & 1 & 2 & 4 & | & 6 \end{bmatrix} \begin{matrix} \\ \\ -2R_1 + R_3 \\ -5R_1 + R_4 \end{matrix}$

$\begin{bmatrix} 1 & -1 & 1 & 1 & | & 3 \\ 0 & 1 & -2 & -1 & | & 0 \\ -2(1)+2 & -2(-1)+0 & -2(1)+3 & -2(1)+4 & | & -2(3)+11 \\ -5(1)+5 & -5(-1)+1 & -5(1)+2 & -5(1)+4 & | & -5(3)+6 \end{bmatrix} = \begin{bmatrix} 1 & -1 & 1 & 1 & | & 3 \\ 0 & 1 & -2 & -1 & | & 0 \\ 0 & 2 & 1 & 2 & | & 5 \\ 0 & 6 & -3 & -1 & | & -9 \end{bmatrix}$

19. $\begin{bmatrix} 1 & -1 & 1 & | & 8 \\ 2 & 3 & -1 & | & -2 \\ 3 & -2 & -9 & | & 9 \end{bmatrix}$

$\begin{bmatrix} 1 & -1 & 1 & | & 8 \\ -2(1)+2 & -2(-1)+3 & -2(1)-1 & | & -2(8)-2 \\ -3(1)+3 & -3(-1)-2 & -3(1)-9 & | & -3(8)+9 \end{bmatrix}$

$\begin{bmatrix} 1 & -1 & 1 & | & 8 \\ 0 & 5 & \boxed{-3} & | & \boxed{-18} \\ 0 & 1 & \boxed{-12} & | & \boxed{-15} \end{bmatrix}$

$\begin{bmatrix} 1 & -1 & 1 & | & 8 \\ 0(\frac{1}{5}) & 1(\frac{1}{5}) & -3(\frac{1}{5}) & | & -18(\frac{1}{5}) \\ 0 & 1 & -12 & | & -15 \end{bmatrix}$

$\begin{bmatrix} 1 & -1 & 1 & | & 8 \\ 0 & 1 & \boxed{-\frac{3}{5}} & | & \boxed{-\frac{18}{5}} \\ 0 & 1 & \boxed{-12} & | & \boxed{-15} \end{bmatrix}$

21.
$x + y - z = -2$
$2x - y + z = 5$
$-x + 2y + 2z = 1$

$\begin{bmatrix} 1 & 1 & -1 & | & -2 \\ 2 & -1 & 1 & | & 5 \\ -1 & 2 & 2 & | & 1 \end{bmatrix} -2R_1 + R_2$

$\begin{bmatrix} 1 & 1 & -1 & | & -2 \\ 0 & -3 & 3 & | & 9 \\ -1 & 2 & 2 & | & 1 \end{bmatrix} 1R_1 + R_3$

$\begin{bmatrix} 1 & 1 & -1 & | & -2 \\ 0 & -3 & 3 & | & 9 \\ 0 & 3 & 1 & | & -1 \end{bmatrix} -\frac{1}{3}R_2$

$\begin{bmatrix} 1 & 1 & -1 & | & -2 \\ 0 & 1 & -1 & | & -3 \\ 0 & 3 & 1 & | & -1 \end{bmatrix} -3R_2 + R_3$

$= \begin{bmatrix} 1 & 1 & -1 & | & -2 \\ 0 & 1 & -1 & | & -3 \\ 0 & 0 & 4 & | & 8 \end{bmatrix}$

$4z = 8$
$z = 2$
$y - z = -3$
$y - 2 = -3$
$y = -1$

$x + y - z = -2$
$x - 1 - 2 = -2$
$x - 3 = -2$
$x = 1$
The solution set is $\{(1, -1, 2)\}$.

23.
$x + 3y = 0$
$x + y + z = 1$
$3x - y - z = 11$

$\begin{bmatrix} 1 & 3 & 0 & | & 0 \\ 1 & 1 & 1 & | & 1 \\ 3 & -1 & -1 & | & 11 \end{bmatrix} -1R_1 + R_2$

$\begin{bmatrix} 1 & 3 & 0 & | & 0 \\ 0 & -2 & 1 & | & 1 \\ 3 & -1 & -1 & | & 11 \end{bmatrix} -3R_1 + R_3$

$\begin{bmatrix} 1 & 3 & 0 & | & 0 \\ 0 & -2 & 1 & | & 1 \\ 0 & -10 & -1 & | & 11 \end{bmatrix} -\frac{1}{2}R_2$

$\begin{bmatrix} 1 & 3 & 0 & | & 0 \\ 0 & 1 & -\frac{1}{2} & | & -\frac{1}{2} \\ 0 & -10 & -1 & | & 11 \end{bmatrix} 10R_2 + R_3$

$\begin{bmatrix} 1 & 3 & 0 & | & 0 \\ 0 & 1 & -\frac{1}{2} & | & -\frac{1}{2} \\ 0 & 0 & -6 & | & 6 \end{bmatrix} -\frac{1}{6}R_3$

$\begin{bmatrix} 1 & 3 & 0 & | & 0 \\ 0 & 1 & -\frac{1}{2} & | & -\frac{1}{2} \\ 0 & 0 & 1 & | & -1 \end{bmatrix}$

$z = -1$

$y - \frac{1}{2}z = -\frac{1}{2}$

$y - \frac{1}{2}(-1) = -\frac{1}{2}$

$y + \frac{1}{2} = -\frac{1}{2}$

$y = -1$

Interchange row one and row two.
$x + 3y = 0$
$x + 3(-1) = 0$
$x = 3$
The solution set is $\{(3, -1, -1)\}$.

25. $2x - y - z = 4$
  $x + y - 5z = -4$
  $x - 2y = 4$

$$\begin{bmatrix} 2 & -1 & -1 & 4 \\ 1 & 1 & -5 & -4 \\ 1 & -2 & 0 & 4 \end{bmatrix}$$

Interchange rows one and two.

$$\begin{bmatrix} 1 & 1 & -5 & -4 \\ 2 & -1 & -1 & 4 \\ 1 & -2 & 0 & 4 \end{bmatrix}$$

Replace row two with $-2R_1 + R_2$.
Replace row three with $-R_1 + R_3$.

$$\begin{bmatrix} 1 & 1 & -5 & -4 \\ 0 & -3 & 9 & 12 \\ 0 & -3 & 5 & 8 \end{bmatrix}$$

Replace row two with $-\dfrac{1}{3}R_2$.

$$\begin{bmatrix} 1 & 1 & -5 & -4 \\ 0 & 1 & -3 & -4 \\ 0 & -3 & 5 & 8 \end{bmatrix}$$

Replace row three with $3R_2 + R_3$.

$$\begin{bmatrix} 1 & 1 & -5 & -4 \\ 0 & 1 & -3 & -4 \\ 0 & 0 & -4 & -4 \end{bmatrix}$$

Replace row three with $-\dfrac{1}{4}R_3$.

$$\begin{bmatrix} 1 & 1 & -5 & -4 \\ 0 & 1 & -3 & -4 \\ 0 & 0 & 1 & 1 \end{bmatrix}$$

$z = 1$
  $y - 3z = -4$
  $y - 3(1) = -4$
    $y = -1$
  $x + y - 5z = -4$
  $x - 1 - 5(1) = -4$
    $x - 6 = -4$
      $x = 2$

The solution set is $\{(2, -1, 1)\}$.

27. $x + y + z = 4$
  $x - y - z = 0$
  $x - y + z = 2$

$$\begin{bmatrix} 1 & 1 & 1 & 4 \\ 1 & -1 & -1 & 0 \\ 1 & -1 & 1 & 2 \end{bmatrix}$$

Replace row two with $-R_1 + R_2$.
Replace row three with $-R_1 + R_3$.

$$\begin{bmatrix} 1 & 1 & 1 & 4 \\ 0 & -2 & -2 & -4 \\ 0 & -2 & 0 & -2 \end{bmatrix}$$

Replace row two with $-\dfrac{1}{2}R_2$.

$$\begin{bmatrix} 1 & 1 & 1 & 4 \\ 0 & 1 & 1 & 2 \\ 0 & -2 & 0 & -2 \end{bmatrix}$$

Replace row 3 with $2R_2 + R_3$.

$$\begin{bmatrix} 1 & 1 & 1 & 4 \\ 0 & 1 & 1 & 2 \\ 0 & 0 & 2 & 2 \end{bmatrix}$$

Replace row 3 with $\dfrac{1}{2}R_3$.

$$\begin{bmatrix} 1 & 1 & 1 & 4 \\ 0 & 1 & 1 & 2 \\ 0 & 0 & 1 & 1 \end{bmatrix}$$

$z = 1$
$y + 1 = 2$
  $y = 1$
$x + 1 + 1 = 4$
    $x = 2$
The solution set is $\{(2, 1, 1)\}$.

29. Write the equations in standard form.
  $x + 2y - z = -1$
    $x - y + z = 4$
  $x + y - 3z = -2$

$$\begin{bmatrix} 1 & 2 & -1 & -1 \\ 1 & -1 & 1 & 4 \\ 1 & 1 & -3 & -2 \end{bmatrix}$$

Replace row two with $-R_1 + R_2$.
Replace row three with $-R_1 + R_3$.

$$\begin{bmatrix} 1 & 2 & -1 & -1 \\ 0 & -3 & 2 & 5 \\ 0 & -1 & -2 & -1 \end{bmatrix}$$

Replace row two with $-R_3$.
Replace row three with R_2.

$$\begin{bmatrix} 1 & 2 & -1 & -1 \\ 0 & 1 & 2 & 1 \\ 0 & -3 & 2 & 5 \end{bmatrix}$$

Replace row 3 with $3R_2 + R_3$.

$$\begin{bmatrix} 1 & 2 & -1 & -1 \\ 0 & 1 & 2 & 1 \\ 0 & 0 & 8 & 8 \end{bmatrix}$$

Replace row 3 with $\dfrac{1}{8}R_3$.

$$\begin{bmatrix} 1 & 2 & -1 & -1 \\ 0 & 1 & 2 & 1 \\ 0 & 0 & 1 & 1 \end{bmatrix}$$

$z = 1$
$y + 2(1) = 1$
$\qquad y = -1$
$x + 2(-1) - 1 = -1$
$\qquad x = 2$
The solution set is $\{(2, -1, 1)\}$.

31. $3a - b - 4c = 3$
$\quad 2a - b + 2c = -8$
$\quad a + 2b - 3c = 9$
Interchange equations 1 and 3.

$$\begin{bmatrix} 1 & 2 & -3 & 9 \\ 2 & -1 & 2 & -8 \\ 3 & -1 & -4 & 3 \end{bmatrix}$$

Replace row two with $-2R_1 + R_2$.
Replace row three with $-3R_1 + R_3$.

$$\begin{bmatrix} 1 & 2 & -3 & 9 \\ 0 & -5 & 8 & -26 \\ 0 & -7 & 5 & -24 \end{bmatrix}$$

Replace row two with $-\dfrac{1}{5}R_2$

$$\begin{bmatrix} 1 & 2 & -3 & 9 \\ 0 & 1 & -\dfrac{8}{5} & \dfrac{26}{5} \\ 0 & -7 & 5 & -24 \end{bmatrix}$$

Replace row three with $7R_2 + R_3$.

$$\begin{bmatrix} 1 & 2 & -3 & 9 \\ 0 & 1 & -\dfrac{8}{5} & \dfrac{26}{5} \\ 0 & 0 & -\dfrac{31}{5} & \dfrac{62}{5} \end{bmatrix}$$

Replace row 3 with $-\dfrac{5}{31}R_3$.

$$\begin{bmatrix} 1 & 2 & -3 & 9 \\ 0 & 1 & -\dfrac{8}{5} & \dfrac{26}{5} \\ 0 & 0 & 1 & -2 \end{bmatrix}$$

$z = -2$
$$y - \frac{8}{5}(-2) = \frac{26}{5}$$
$$y + \frac{16}{5} = \frac{26}{5}$$
$$y = 2$$
$x + 2(2) - 3(-2) = 9$
$\quad x + 4 + 6 = 9$
$\qquad x = -1$
The solution set is $\{(-1, 2, -2)\}$.

33. $2x + 2y + 7z = -1$
$\quad 2x + y + 2z = 2$
$\quad 4x + 6y + z = 15$

$$\left[\begin{array}{ccc|c} 2 & 2 & 7 & -1 \\ 2 & 1 & 2 & 2 \\ 4 & 6 & 1 & 15 \end{array}\right]\begin{array}{l} \frac{1}{2}R_1 \\ \\ \\ \end{array}$$

$$\left[\begin{array}{ccc|c} 1 & 1 & \frac{7}{2} & -\frac{1}{2} \\ 2 & 1 & 2 & 2 \\ 4 & 6 & 1 & 15 \end{array}\right]-2R_1 + R_2$$

$$\left[\begin{array}{ccc|c} 1 & 1 & \frac{7}{2} & -\frac{1}{2} \\ 0 & -1 & -5 & 3 \\ 4 & 6 & 1 & 15 \end{array}\right]-4R_1 + R_3$$

$$\left[\begin{array}{ccc|c} 1 & 1 & \frac{7}{2} & -\frac{1}{2} \\ 0 & -1 & -5 & 3 \\ 0 & 2 & -13 & 17 \end{array}\right]-1R_2$$

$$\left[\begin{array}{ccc|c} 1 & 1 & \frac{7}{2} & -\frac{1}{2} \\ 0 & 1 & 5 & -3 \\ 0 & 2 & -13 & 17 \end{array}\right]-2R_2 + R_3$$

$$\left[\begin{array}{ccc|c} 1 & 1 & \frac{7}{2} & -\frac{1}{2} \\ 0 & 1 & 5 & -3 \\ 0 & 0 & -23 & 23 \end{array}\right]-\frac{1}{23}R_3$$

$$\left[\begin{array}{ccc|c} 1 & 1 & \frac{7}{2} & -\frac{1}{2} \\ 0 & 1 & 5 & -3 \\ 0 & 0 & 1 & -1 \end{array}\right]$$

$z = -1$

$$y + 5z = -3$$
$$y + 5(-1) = -3$$
$$y - 5 = -3$$
$$y = 2$$
$$x + y + \frac{7}{2}z = -\frac{1}{2}$$
$$x + 2 + \frac{7}{2}(-1) = -\frac{1}{2}$$
$$x - \frac{3}{2} = -\frac{1}{2}$$
$$x = 1$$

The solution set is $\{(1, 2, -1)\}$.

35.
$$w + x + y + z = 4$$
$$2w + x - 2y - z = 0$$
$$w - 2x - y - 2z = -2$$
$$3w + 2x + y + 3z = 4$$

$$\begin{bmatrix} 1 & 1 & 1 & 1 & | & 4 \\ 2 & 1 & -2 & -1 & | & 0 \\ 1 & -2 & -1 & -2 & | & -2 \\ 3 & 2 & 1 & 3 & | & 4 \end{bmatrix} \begin{matrix} \\ -2R_1 + R_2 \\ \\ \end{matrix}$$

$$\begin{bmatrix} 1 & 1 & 1 & 1 & | & 4 \\ 0 & -1 & -4 & -3 & | & -8 \\ 1 & -2 & -1 & -2 & | & -2 \\ 3 & 2 & 1 & 3 & | & 4 \end{bmatrix} \begin{matrix} \\ \\ -1R_1 + R_3 \\ \end{matrix}$$

$$\begin{bmatrix} 1 & 1 & 1 & 1 & | & 4 \\ 0 & -1 & -4 & -3 & | & -8 \\ 0 & -3 & -2 & -3 & | & -6 \\ 3 & 2 & 1 & 3 & | & 4 \end{bmatrix} \begin{matrix} \\ \\ \\ -3R_1 + R_4 \end{matrix}$$

$$\begin{bmatrix} 1 & 1 & 1 & 1 & | & 4 \\ 0 & -1 & -4 & -3 & | & -8 \\ 0 & -3 & -2 & -3 & | & -6 \\ 0 & -1 & -2 & 0 & | & -8 \end{bmatrix} \begin{matrix} \\ -1R_2 \\ \\ \end{matrix}$$

$$\begin{bmatrix} 1 & 1 & 1 & 1 & | & 4 \\ 0 & 1 & 4 & 3 & | & 8 \\ 0 & -3 & -2 & -3 & | & -6 \\ 0 & -1 & -2 & 0 & | & -8 \end{bmatrix} \begin{matrix} \\ \\ 3R_2 + R_3 \\ \end{matrix}$$

$$\begin{bmatrix} 1 & 1 & 1 & 1 & | & 4 \\ 0 & 1 & 4 & 3 & | & 8 \\ 0 & 0 & 10 & 6 & | & 18 \\ 0 & -1 & -2 & 0 & | & -8 \end{bmatrix} \begin{matrix} \\ \\ \\ 1R_2 + R_4 \end{matrix}$$

$$\begin{bmatrix} 1 & 1 & 1 & 1 & | & 4 \\ 0 & 1 & 4 & 3 & | & 8 \\ 0 & 0 & 10 & 6 & | & 18 \\ 0 & 0 & 2 & 3 & | & 0 \end{bmatrix} \begin{matrix} \\ \\ \frac{1}{10}R_3 \\ \end{matrix}$$

$$\begin{bmatrix} 1 & 1 & 1 & 1 & | & 4 \\ 0 & 1 & 4 & 3 & | & 8 \\ 0 & 0 & 1 & \frac{3}{5} & | & \frac{9}{5} \\ 0 & 0 & 2 & 3 & | & 0 \end{bmatrix} \begin{matrix} \\ \\ \\ -2R_3 + R_4 \end{matrix}$$

$$\begin{bmatrix} 1 & 1 & 1 & 1 & | & 4 \\ 0 & 1 & 4 & 3 & | & 8 \\ 0 & 0 & 1 & \frac{3}{5} & | & \frac{9}{5} \\ 0 & 0 & 0 & \frac{9}{5} & | & -\frac{18}{5} \end{bmatrix} \begin{matrix} \\ \\ \\ \frac{5}{9}R_4 \end{matrix}$$

$$\begin{bmatrix} 1 & 1 & 1 & 1 & | & 4 \\ 0 & 1 & 4 & 3 & | & 8 \\ 0 & 0 & 1 & \frac{3}{5} & | & \frac{9}{5} \\ 0 & 0 & 0 & 1 & | & -2 \end{bmatrix}$$

$$z = -2$$
$$y + \frac{3}{5}z = \frac{9}{5}$$
$$y + \frac{3}{5}(-2) = \frac{9}{5}$$
$$y - \frac{6}{5} = \frac{9}{5}$$
$$y = 3$$
$$x + 4y + 3z = 8$$
$$x + 4(3) + 3(-2) = 8$$
$$x + 6 = 8$$
$$x = 2$$
$$w + x + y + z = 4$$
$$w + 2 + 3 - 2 = 4$$
$$w + 3 = 4$$
$$w = 1$$

The solution set is $\{(1, 2, 3, -2)\}$.

37. $3w - 4x + y + z = 9$

$w + x - y - z = 0$

$2w + x + 4y - 2z = 3$

$-w + 2x + y - 3z = 3$

$$\begin{bmatrix} 3 & -4 & 1 & 1 & | & 9 \\ 1 & 1 & -1 & -1 & | & 0 \\ 2 & 1 & 4 & -2 & | & 3 \\ -1 & 2 & 1 & -3 & | & 3 \end{bmatrix} R_1 \leftrightarrow R_2$$

$$\begin{bmatrix} 1 & 1 & -1 & -1 & | & 0 \\ 3 & -4 & 1 & 1 & | & 9 \\ 2 & 1 & 4 & -2 & | & 3 \\ -1 & 2 & 1 & -3 & | & 3 \end{bmatrix} -3R_1 + R_2$$

$$\begin{bmatrix} 1 & 1 & -1 & -1 & | & 0 \\ 0 & -7 & 4 & 4 & | & 9 \\ 2 & 1 & 4 & -2 & | & 3 \\ -1 & 2 & 1 & -3 & | & 3 \end{bmatrix} -2R_1 + R_3$$

$$\begin{bmatrix} 1 & 1 & -1 & -1 & | & 0 \\ 0 & -7 & 4 & 4 & | & 9 \\ 0 & -1 & 6 & 0 & | & 3 \\ -1 & 2 & 1 & -3 & | & 3 \end{bmatrix} 1R_1 + R_4$$

$$\begin{bmatrix} 1 & 1 & -1 & -1 & | & 0 \\ 0 & -7 & 4 & 4 & | & 9 \\ 0 & -1 & 6 & 0 & | & 3 \\ 0 & 3 & 0 & -4 & | & 3 \end{bmatrix} R_2 \leftrightarrow R_3$$

$$\begin{bmatrix} 1 & 1 & -1 & -1 & | & 0 \\ 0 & -1 & 6 & 0 & | & 3 \\ 0 & -7 & 4 & 4 & | & 9 \\ 0 & 3 & 0 & -4 & | & 3 \end{bmatrix} -R_2$$

$$\begin{bmatrix} 1 & 1 & -1 & -1 & | & 0 \\ 0 & 1 & -6 & 0 & | & -3 \\ 0 & -7 & 4 & 4 & | & 9 \\ 0 & 3 & 0 & -4 & | & 3 \end{bmatrix} 7R_2 + R_3$$

$$\begin{bmatrix} 1 & 1 & -1 & -1 & | & 0 \\ 0 & 1 & -6 & 0 & | & -3 \\ 0 & 0 & -38 & 4 & | & -12 \\ 0 & 3 & 0 & -4 & | & 3 \end{bmatrix} -3R_2 + R_4$$

$$\begin{bmatrix} 1 & 1 & -1 & -1 & | & 0 \\ 0 & 1 & -6 & 0 & | & -3 \\ 0 & 0 & -38 & 4 & | & -12 \\ 0 & 0 & 18 & -4 & | & 12 \end{bmatrix} -\frac{1}{38} R_3$$

$$\begin{bmatrix} 1 & 1 & -1 & -1 & | & 0 \\ 0 & 1 & -6 & 0 & | & -3 \\ 0 & 0 & 1 & -\frac{2}{19} & | & \frac{6}{19} \\ 0 & 0 & 18 & -4 & | & 12 \end{bmatrix} -18R_3 + R_4$$

$$\begin{bmatrix} 1 & 1 & -1 & -1 & | & 0 \\ 0 & 1 & -6 & 0 & | & -3 \\ 0 & 0 & 1 & -\frac{2}{19} & | & \frac{6}{19} \\ 0 & 0 & 0 & -\frac{40}{19} & | & \frac{120}{19} \end{bmatrix} -\frac{19}{40} R_4$$

$$\begin{bmatrix} 1 & 1 & -1 & -1 & | & 0 \\ 0 & 1 & -6 & 0 & | & -3 \\ 0 & 0 & 1 & -\frac{2}{19} & | & \frac{6}{19} \\ 0 & 0 & 0 & 1 & | & -3 \end{bmatrix}$$

$z = -3$

$y - \dfrac{2}{19} z = \dfrac{6}{19}$

$y - \dfrac{2}{19}(-3) = \dfrac{6}{19}$

$y + \dfrac{6}{19} = \dfrac{6}{19}$

$y = 0$

$x - 6y = -3$

$x - 6(0) = -3$

$x = -3$

$w + x - y - z = 0$

$w - 3 + 0 + 3 = 0$

$w = 0$

The solution set is $\{(0, -3, 0, -3)\}$.

39. $f(x) = ax^2 + bx + c$

Use the given function values to find three equations in terms of a, b, and c.

$f(-2) = a(-2)^2 + b(-2) + c = -4$

$4a - 2b + c = -4$

$f(1) = a(1)^2 + b(1) + c = 2$

$a + b + c = 2$

$f(2) = a(2)^2 + b(2) + c = 0$

$4a + 2b + c = 0$

System of equations:

$4a - 2b + c = -4$

$a + b + c = 2$

$4a + 2b + c = 0$

Matrix:

$$\begin{bmatrix} 4 & -2 & 1 & | & -4 \\ 1 & 1 & 1 & | & 2 \\ 4 & 2 & 1 & | & 0 \end{bmatrix}$$

This gives $a = -1$, $b = 1$, and $c = 2$.

Thus, $f(x) = -x^2 + x + 2$.

41. $f(x) = ax^3 + bx^2 + cx + d$

Use the given function values to find four equations in terms of a, b, c, and d.

$f(-1) = a(-1)^3 + b(-1)^2 + c(-1) + d = 0$

$\qquad -a + b - c + d = 0$

$f(1) = a(1)^3 + b(1)^2 + c(1) + d = 2$

$\qquad a + b + c + d = 2$

$f(2) = a(2)^3 + b(2)^2 + c(2) + d = 3$

$\qquad 8a + 4b + 2c + d = 3$

$f(3) = a(3)^3 + b(3)^2 + c(3) + d = 12$

$\qquad 27a + 9b + 3c + d = 12$

System of equations:

$-a + b - c + d = 0$

$a + b + c + d = 2$

$8a + 4b + 2c + d = 3$

$27a + 9b + 3c + d = 12$

Matrix:

$$\begin{bmatrix} -1 & 1 & -1 & 1 & | & 0 \\ 1 & 1 & 1 & 1 & | & 2 \\ 8 & 4 & 2 & 1 & | & 3 \\ 27 & 9 & 3 & 1 & | & 12 \end{bmatrix}$$

This gives $a = 1$, $b = -2$, $c = 0$, and $d = 3$.

Thus, $f(x) = x^3 - 2x^2 + 3$.

43. Let $A = \ln w$, $B = \ln x$, $C = \ln y$, and $D = \ln z$.

System of equations:

$2A + B + 3C - 2D = -6$

$4A + 3B + C - D = -2$

$A + B + C + D = -5$

$A + B - C - D = 5$

Matrix:

$$\begin{bmatrix} 2 & 1 & 3 & -2 & | & -6 \\ 4 & 3 & 1 & -1 & | & -2 \\ 1 & 1 & 1 & 1 & | & -5 \\ 1 & 1 & -1 & -1 & | & 5 \end{bmatrix}$$

This gives $A = -1$, $B = 1$, $C = -3$, and $D = -2$.

Substitute back to find w, x, y, and z.

$A = -1 \qquad\qquad B = 1$

$\ln w = -1 \qquad\quad \ln x = 1$

$w = e^{-1} \qquad\qquad x = e^1$

$w \approx 0.37 \qquad\qquad x \approx 2.72$

$C = -3 \qquad\qquad D = -2$

$\ln y = -3 \qquad\quad \ln z = -2$

$y = e^{-3} \qquad\qquad z = e^{-2}$

$y \approx 0.05 \qquad\qquad z \approx 0.14$

45. a. $s(t) = \dfrac{1}{2}at^2 + v_0 t + s_0$

Use the given function values to find three equations in terms of a, v_0, and s_0.

$s(1) = \dfrac{1}{2}a(1)^2 + v_0(1) + s_0 = 40$

$\qquad \dfrac{1}{2}a + v_0 + s_0 = 40$

$s(2) = \dfrac{1}{2}a(2)^2 + v_0(2) + s_0 = 48$

$\qquad 2a + 2v_0 + s_0 = 48$

$s(3) = \dfrac{1}{2}a(3)^2 + v_0(3) + s_0 = 24$

$\qquad \dfrac{9}{2}a + 3v_0 + s_0 = 24$

System of equations:

$\dfrac{1}{2}a + v_0 + s_0 = 40$

$2a + 2v_0 + s_0 = 48$

$\dfrac{9}{2}a + 3v_0 + s_0 = 24$

Matrix:

$$\begin{bmatrix} \dfrac{1}{2} & 1 & 1 & | & 40 \\ 2 & 2 & 1 & | & 48 \\ \dfrac{9}{2} & 3 & 1 & | & 24 \end{bmatrix}$$

This gives $a = -32$, $v_0 = 56$, and $s_0 = 0$.

Thus, $s(t) = \dfrac{1}{2}(-32)t^2 + (56)t + (0)$

$s(t) = -16t^2 + 56t$

b. $s(t) = -16t^2 + 56t$

$s(3.5) = -16(3.5)^2 + 56(3.5) = 0$

This is the point $(3.5, 0)$.

The ball's height is 0 feet after 3.5 seconds.

This is the point $(3.5, 0)$.

c. The maximum occurs when $x = -\dfrac{b}{2a}$.

$$x = -\frac{b}{2a} = -\frac{v_0}{2a} = -\frac{56}{2(-16)} = 1.75$$

$$s(1.75) = -16(1.75)^2 + 56(1.75) = 49$$

At 1.75 seconds the ball will reach its maximum height of 49 feet.

47. Let x = Food A
Let y = Food B
Let z = Food C
$40x + 200y + 400z = 660$

$\quad 5x + 2y + 4z = 25$
$30x + 10y + 300z = 425$
$\quad 2x + 10y + 20z = 33$
$\quad 5x + 2y + 4z = 25$
$\quad 6x + 2y + 60z = 85$

$\begin{bmatrix} 2 & 10 & 20 & | & 33 \\ 5 & 2 & 4 & | & 25 \\ 6 & 2 & 60 & | & 85 \end{bmatrix} \frac{1}{2}R_1$

$\begin{bmatrix} 1 & 5 & 10 & | & \frac{33}{2} \\ 5 & 2 & 4 & | & 25 \\ 6 & 2 & 60 & | & 85 \end{bmatrix} -5R_1 + R_2$

$\begin{bmatrix} 1 & 5 & 10 & | & \frac{33}{2} \\ 0 & -23 & -46 & | & -\frac{115}{2} \\ 6 & 2 & 60 & | & 85 \end{bmatrix} -6R_1 + R_3$

$\begin{bmatrix} 1 & 5 & 10 & | & \frac{33}{2} \\ 0 & -23 & -46 & | & -\frac{115}{2} \\ 0 & -28 & 0 & | & -14 \end{bmatrix} -\frac{1}{23}R_2$

$\begin{bmatrix} 1 & 5 & 10 & | & \frac{33}{2} \\ 0 & 1 & 2 & | & \frac{5}{2} \\ 0 & -28 & 0 & | & -14 \end{bmatrix} 28R_2 + R_3$

$\begin{bmatrix} 1 & 5 & 10 & | & \frac{33}{2} \\ 0 & 1 & 2 & | & \frac{5}{2} \\ 0 & 0 & 56 & | & 56 \end{bmatrix} \frac{1}{56}R_3$

$\begin{bmatrix} 1 & 5 & 10 & | & \frac{33}{2} \\ 0 & 1 & 2 & | & \frac{5}{2} \\ 0 & 0 & 1 & | & 1 \end{bmatrix}$

$z = 1$

$y + 2z = \dfrac{5}{2}$

$y + 2 = \dfrac{5}{2}$

$2y + 4 = 5$

$\quad 2y = 1$

$\quad\quad y = \dfrac{1}{2}$

$x + 5y + 10z = \dfrac{33}{2}$

$x + \dfrac{5}{2} + 10 = \dfrac{33}{2}$

$2x + 5 + 20 = 33$

$\quad 2x + 25 = 33$

$\quad\quad 2x = 8$

$\quad\quad\quad x = 4$

4 ounces of Food A

$\dfrac{1}{2}$ ounce of Food B

1 ounce of Food C

49. Let w = number of Asians
Let x = number of Africans
Let y = number of Europeans
Let z = number of Americans
Use the variables to model each sentence.
$w + x + y + z = 183$

$\quad w - x - y = 70$

$\quad\quad\quad y - z = 15$

$\quad 2x - y - z = 23$

$\begin{bmatrix} 1 & 1 & 1 & 1 & | & 183 \\ 1 & -1 & -1 & 0 & | & 70 \\ 0 & 0 & 1 & -1 & | & 15 \\ 0 & 2 & -1 & -1 & | & 23 \end{bmatrix} -R_1 + R_2$

$\begin{bmatrix} 1 & 1 & 1 & 1 & | & 183 \\ 0 & -2 & -2 & -1 & | & -113 \\ 0 & 0 & 1 & -1 & | & 15 \\ 0 & 2 & -1 & -1 & | & 23 \end{bmatrix} R_4 + R_2$

$\begin{bmatrix} 1 & 1 & 1 & 1 & | & 183 \\ 0 & -2 & -2 & -1 & | & -113 \\ 0 & 0 & 1 & -1 & | & 15 \\ 0 & 0 & -3 & -2 & | & -90 \end{bmatrix} -\frac{1}{2}R_2$

$\begin{bmatrix} 1 & 1 & 1 & 1 & | & 183 \\ 0 & 1 & 1 & \frac{1}{2} & | & \frac{113}{2} \\ 0 & 0 & 1 & -1 & | & 15 \\ 0 & 0 & -3 & -2 & | & -90 \end{bmatrix} 3R_3 + R_4$

$$\begin{bmatrix} 1 & 1 & 1 & 1 & | & 183 \\ 0 & 1 & 1 & \frac{1}{2} & | & \frac{113}{2} \\ 0 & 0 & 1 & -1 & | & 15 \\ 0 & 0 & 0 & -5 & | & -45 \end{bmatrix} -\frac{1}{5}R_4$$

$$\begin{bmatrix} 1 & 1 & 1 & 1 & | & 183 \\ 0 & 1 & 1 & \frac{1}{2} & | & \frac{113}{2} \\ 0 & 0 & 1 & -1 & | & 15 \\ 0 & 0 & 0 & 1 & | & 9 \end{bmatrix}$$

Back-substitute $z = 9$ to find y.

$y - z = 15$

$y - 9 = 15$

$y = 24$

Back-substitute to find x.

$$x + y + \frac{1}{2}z = \frac{113}{2}$$

$$x + 24 + \frac{1}{2}(9) = \frac{113}{2}$$

$$x + 24 + \frac{1}{2}(9) = \frac{113}{2}$$

$$x + 28\frac{1}{2} = 56\frac{1}{2}$$

$$x = 28$$

Back-substitute to find w.

$w + x + y + z = 183$

$w + 28 + 24 + 9 = 183$

$w + 28 + 24 + 9 = 183$

$w + 61 = 183$

$w = 122$

The number of Asians, Africans, Europeans, and Americans are 122, 28, 24, and 9, respectively.

51. – 57. Answers will vary.

59. makes sense

61. makes sense

63. false; Changes to make the statement true will vary. A sample change is: Multiplying a row by a negative fraction is permitted.

65. false; Changes to make the statement true will vary. A sample change is: When solving a system of three equations in three variables, we use row operations to obtain ones along the diagonal and zeros below the ones.

67.
$$y = ax^2 + bx + c$$
$$5900 = a(30)^2 + b(30) + c$$
$$5900 = 900a + 30b + c$$

$$7500 = a(50)^2 + b(50) + c$$
$$7500 = 2500a + 50b + c$$
$$4500 = a(100)^2 + b(100) + c$$
$$4500 = 10,000a + 100b + c$$
$$900a + 30b + c = 5900$$
$$2500a + 50b + c = 7500$$
$$10,000a + 100b + c = 4500$$

$$\begin{bmatrix} 900 & 30 & 1 & | & 5900 \\ 2500 & 50 & 1 & | & 7500 \\ 10,000 & 100 & 1 & | & 4500 \end{bmatrix} R_1 \leftrightarrow R_2$$

$$\begin{bmatrix} 2500 & 50 & 1 & | & 7500 \\ 900 & 30 & 1 & | & 5900 \\ 10,000 & 100 & 1 & | & 4500 \end{bmatrix} \frac{1}{2500}R_1$$

$$\begin{bmatrix} 1 & \frac{1}{50} & \frac{1}{2500} & | & 3 \\ 900 & 30 & 1 & | & 5900 \\ 10,000 & 100 & 1 & | & 4500 \end{bmatrix} \begin{matrix} -900R_1 + R_2 \\ -10,000R_1 + R_3 \end{matrix}$$

$$\begin{bmatrix} 1 & \frac{1}{50} & \frac{1}{2500} & | & 3 \\ 0 & 12 & \frac{16}{25} & | & 3200 \\ 0 & -100 & -3 & | & -25,500 \end{bmatrix} R_2 \leftrightarrow R_3$$

$$\begin{bmatrix} 1 & \frac{1}{50} & \frac{1}{2500} & | & 3 \\ 0 & -100 & -3 & | & -25,500 \\ 0 & 12 & \frac{16}{25} & | & 3200 \end{bmatrix} \frac{-1}{100}R_2$$

$$\begin{bmatrix} 1 & \frac{1}{50} & \frac{1}{2500} & | & 3 \\ 0 & 1 & \frac{3}{100} & | & 255 \\ 0 & 12 & \frac{16}{25} & | & 3200 \end{bmatrix} \begin{matrix} -12R_2 + R_3 \\ -\frac{1}{50}R_2 + R_1 \end{matrix}$$

$$\begin{bmatrix} 1 & 0 & \frac{-1}{5000} & | & \frac{-21}{10} \\ 0 & 1 & \frac{3}{100} & | & 255 \\ 0 & 0 & \frac{7}{25} & | & 140 \end{bmatrix} \frac{25}{7}R_3$$

$$\begin{bmatrix} 1 & 0 & \frac{-1}{5000} & | & \frac{-21}{10} \\ 0 & 1 & \frac{3}{100} & | & 255 \\ 0 & 0 & 1 & | & 500 \end{bmatrix} \begin{matrix} \frac{-3}{100}R_3 + R_2 \\ \frac{1}{5000}R_3 + R_1 \end{matrix}$$

$$\begin{bmatrix} 1 & 0 & 0 & | & -2 \\ 0 & 1 & 0 & | & 240 \\ 0 & 0 & 1 & | & 500 \end{bmatrix}$$

$y = -2x^2 + 240x + 500$

$y = -2(x^2 - 120x) + 500$

$y = -2(x^2 - 120x + 3600) + 500 + 7200$

$y = -2(x - 60)^2 + 7700$

60 units produce $7700.

68. When $z = 0$, $(12z + 1, 10z - 1, z)$ is equivalent to

$(12(0)+1,\ 10(0)-1,\ 0)$ or $(1,\ -1,\ 0)$.

Check $(1,\ -1,\ 0)$ in each equation.

$$3x-4y+4z=7$$
$$3(1)-4(-1)+4(0)=7$$
$$7=7,\ \text{true}$$

$$x-y-2z=2$$
$$1-(-1)-2(0)=2$$
$$2=2,\ \text{true}$$
$$2x-3y-2z=5$$
$$2(1)-3(-1)+6(0)=5$$
$$5=5,\ \text{true}$$

$(1,\ -1,\ 0)$ satisfies each equation and, therefore, satisfies the system.

69. When $z=1$, $(12z+1,\ 10z-1,\ z)$ is equivalent to $(12(1)+1,\ 10(1)-1,\ 1)$ or $(13,\ 9,\ 1)$.

Check $(13,\ 9,\ 1)$ in each equation.

$$3x-4y+4z=7$$
$$3(13)-4(9)+4(1)=7$$
$$7=7,\ \text{true}$$
$$x-y-2z=2$$
$$13-(9)-2(1)=2$$
$$2=2,\ \text{true}$$
$$2x-3y-2z=5$$
$$2(13)-3(9)+6(1)=5$$
$$5=5,\ \text{true}$$

$(13,\ 9,\ 1)$ satisfies each equation and, therefore, satisfies the system.

70. a. Answers will vary. A sample answer is given selecting $z=10$.

When $z=10$, $(12z+1,\ 10z-1,\ z)$ is equivalent to $(12(10)+1,\ 10(10)-1,\ 10)$ or $(121,\ 99,\ 10)$.

Check $(121,\ 99,\ 10)$ in each equation.

$$3x-4y+4z=7$$
$$3(121)-4(99)+4(10)=7$$
$$7=7,\ \text{true}$$
$$x-y-2z=2$$
$$121-(99)-2(10)=2$$
$$2=2,\ \text{true}$$
$$2x-3y-2z=5$$
$$2(121)-3(99)+6(10)=5$$
$$5=5,\ \text{true}$$

$(121,\ 99,\ 10)$ satisfies each equation and, therefore, satisfies the system.

  b. This system has more than one solution.

Section 9.2

Check Point Exercises

1. $\begin{cases} x-2y-z=5 \\ 2x-3y-z=0 \\ 3x-4y-z=1 \end{cases} \rightarrow \begin{bmatrix} 1 & -2 & -1 & | & -5 \\ 2 & -3 & -1 & | & 0 \\ 3 & -4 & -1 & | & 1 \end{bmatrix}$

$\begin{bmatrix} 1 & -2 & -1 & | & -5 \\ 2 & -3 & -1 & | & 0 \\ 3 & -4 & -1 & | & 1 \end{bmatrix} \begin{matrix} \\ -2R_1+R_2 \\ -3R_1+R_3 \end{matrix}$

$\begin{bmatrix} 1 & -2 & -1 & | & -5 \\ 0 & 1 & 1 & | & 10 \\ 0 & 2 & 2 & | & 16 \end{bmatrix} -2R_2+R_3$

$\begin{bmatrix} 1 & -2 & -1 & | & -5 \\ 0 & 1 & -1 & | & -10 \\ 0 & 0 & 0 & | & -4 \end{bmatrix}$

$0x+0y+0z=-4$ This equation can never be a true statement. Consequently, the system has no solution. The solution set is $\varnothing$, the empty set.

2. $\begin{cases} x-2y-z=5 \\ 2x-5y+3z=16 \\ x-3y+4z=1 \end{cases} \rightarrow \begin{bmatrix} 1 & -2 & -1 & | & 5 \\ 2 & -5 & 3 & | & 6 \\ 1 & -3 & 4 & | & 1 \end{bmatrix}$

$\begin{bmatrix} 1 & -2 & -1 & | & 5 \\ 2 & -5 & 3 & | & 6 \\ 1 & -3 & 4 & | & 1 \end{bmatrix} \begin{matrix} \\ -2R_1+R_2 \\ -1R_1+R_3 \end{matrix}$

$\begin{bmatrix} 1 & -2 & -1 & | & 5 \\ 0 & -1 & 5 & | & -4 \\ 0 & -1 & 5 & | & -4 \end{bmatrix} -1R_2$

$\begin{bmatrix} 1 & -2 & -1 & | & 5 \\ 0 & 1 & -5 & | & 4 \\ 0 & -1 & 5 & | & -4 \end{bmatrix} 1R_2+R_3$

$\begin{bmatrix} 1 & -2 & -1 & | & 5 \\ 0 & 1 & -5 & | & 4 \\ 0 & 0 & 0 & | & 0 \end{bmatrix}$

$0x+0y+0z=0$ or $0=0$

This equation, $0x+0y+0z=0$ is *dependent* on the other two equations. Thus, it can be dropped from the system which can now be expressed in the form

$\begin{bmatrix} 1 & -2 & -1 & | & 5 \\ 0 & 1 & -5 & | & 4 \end{bmatrix}$

The original system is equivalent to the system

$x-2y-z=5$

$y-5z=4$

Solve for x and y in terms of z.

$y=5z+4$

Use back-substitution for y in the previous equation.

$x-2(5z+4)-z=5$

$x-10z-8-z=5$

$x=11z+13$

Finally, letting $z=t$ (or any letter of your choice), the solutions to the system are all of the form $x=11t+13, \quad y=5t+4, \quad z=t,$ where t is a real number. The solution set of the system with dependent equations can be written as $\{(11t+13, 5t+4, t)\}$.

3. $\begin{cases} x+2y+3z=70 \\ x+y+z=60 \end{cases} \rightarrow \begin{bmatrix} 1 & 2 & 3 & | & 70 \\ 1 & 1 & 1 & | & 60 \end{bmatrix}$

$\begin{bmatrix} 1 & 2 & 3 & | & 70 \\ 1 & 1 & 1 & | & 60 \end{bmatrix} -1R_1+R_2$

$\begin{bmatrix} 1 & 2 & 3 & | & 70 \\ 0 & -1 & -2 & | & -10 \end{bmatrix} -1R_2$

$\begin{bmatrix} 1 & 2 & 3 & | & 70 \\ 0 & 1 & 2 & | & 10 \end{bmatrix} \rightarrow \begin{matrix} x+2y+3z=70 \\ y+2z=10 \end{matrix}$

Express x and y in terms of z using back-substitution.

$y=-2z+10$

$x+2(-2z+10)+3z=70$

$x-4z+20+3z=70$

$x=z+50$

With $z=t$, the ordered solution (x, y, z) enables us to express the system's solution set as $\{(t+50, -2t+10, t)\}$.

4. a. I_1: $10+5=15$ cars enter I_1, and $w+z$ cars leave I_1, then $w+z=15$.

   I_2: $20+10=30$ cars enter I_2 and $w+x$ cars leave I_2, then $w+x=30$.

   I_3: $15+30=45$ cars enter I_3 and $x+y$ cars leave I_3, then $x+y=45$.

   I_4: $10+20=30$ cars enter I_4 and $y+z$ cars leave I_4, then $y+z=30$.

   The system of equations that describes this situation is given by

   $w+z=15$

   $w+x=30$

   $x+y=45$

   $y+z=30$

b.

$$\left[\begin{array}{cccc|c} 1 & 0 & 0 & 1 & 15 \\ 1 & 1 & 0 & 0 & 30 \\ 0 & 1 & 1 & 0 & 45 \\ 0 & 0 & 1 & 1 & 30 \end{array}\right] \begin{array}{c} \\ -1R_1 + R_2 \\ \\ \end{array}$$

$$\left[\begin{array}{cccc|c} 1 & 0 & 0 & 1 & 15 \\ 0 & 1 & 0 & -1 & 15 \\ 0 & 1 & 1 & 0 & 45 \\ 0 & 0 & 1 & 1 & 30 \end{array}\right] \begin{array}{c} \\ \\ -1R_2 + R_3 \\ \end{array}$$

$$\left[\begin{array}{cccc|c} 1 & 0 & 0 & 1 & 15 \\ 0 & 1 & 0 & -1 & 15 \\ 0 & 0 & 1 & 1 & 30 \\ 0 & 0 & 1 & 1 & 30 \end{array}\right] \begin{array}{c} \\ \\ -1R_3 + R_4 \\ \end{array}$$

$$\left[\begin{array}{cccc|c} 1 & 0 & 0 & 1 & 15 \\ 0 & 1 & 0 & -1 & 15 \\ 0 & 0 & 1 & 1 & 30 \\ 0 & 0 & 0 & 0 & 0 \end{array}\right]$$

$x + w = 15$

$y - w = 15$

$z + w = 30$

The last row of the matrix shows that the system has dependent equations and infinitely many solutions.

Let z be any real number.

Express w, x and y in terms of z:

$w = 15 - z$

$x = 15 + z$

$y = 30 - z$

With $w = t$, the ordered solution (w, x, y, z) enables us to express the system's solution set as $\{(-t + 15, t + 15, -t + 30, t)\}$

c. If $z = 10$,

$w = 15 - z$

$w = 15 - 10$

$\quad = 5$

$x = 15 + z$

$x = 15 + 10$

$\quad = 25$

$y = 30 - z$

$y = 30 - 10$

$\quad = 20$

If construction limits z to 10 cars per minute, then the number of cars per minute that must pass between the other intersections to keep traffic flowing is $w = 5$, $x = 25$, and $y = 20$.

Concept and Vocabulary Check 9.2

1. no solution

2. one solution

3. infinitely many solutions

4. true

5. $\begin{cases} x - y + z = 5 \\ y - z = -3 \end{cases}$; $z - 3$; 2; $\{(2, z-3, z)\}$

Exercise Set 9.2

1.

$$\left[\begin{array}{ccc|c} 5 & 12 & 1 & 10 \\ 2 & 5 & 2 & -1 \\ 1 & 2 & -3 & 5 \end{array}\right] R_1 \leftrightarrow R_3$$

$$\left[\begin{array}{ccc|c} 1 & 2 & -3 & 5 \\ 2 & 5 & 2 & -1 \\ 5 & 12 & 1 & 10 \end{array}\right] \begin{array}{c} \\ -2R_1 + R_2 \\ -5R_1 + R_3 \end{array}$$

$$\left[\begin{array}{ccc|c} 1 & 2 & -3 & 5 \\ 0 & 1 & 8 & -11 \\ 0 & 2 & 16 & -15 \end{array}\right] \begin{array}{c} \\ \\ -2R_2 + R_3 \end{array}$$

$$\left[\begin{array}{ccc|c} 1 & 2 & 3 & 5 \\ 0 & 1 & 8 & -11 \\ 0 & 0 & 0 & 7 \end{array}\right]$$

From the last row, we see that the system has no solution. The solution set is $\varnothing$, the empty set.

3.

$$\left[\begin{array}{ccc|c} 5 & 8 & -6 & 14 \\ 3 & 4 & -2 & 8 \\ 1 & 2 & -2 & 3 \end{array}\right] R_1 \leftrightarrow R_3$$

$$\left[\begin{array}{ccc|c} 1 & 2 & -2 & 3 \\ 3 & 4 & -2 & 8 \\ 5 & 8 & -6 & 14 \end{array}\right] \begin{array}{c} \\ -3R_1 + R_2 \\ -5R_1 + R_3 \end{array}$$

$$\left[\begin{array}{ccc|c} 1 & 2 & -2 & 3 \\ 0 & -2 & 4 & -1 \\ 0 & -2 & 4 & -1 \end{array}\right] \begin{array}{c} \\ -1R_2 + R_3 \\ \end{array}$$

$$\left[\begin{array}{ccc|c} 1 & 2 & -2 & 3 \\ 0 & -2 & 4 & -1 \\ 0 & 0 & 0 & 0 \end{array}\right] -\frac{1}{2}R_2$$

$$\left[\begin{array}{ccc|c} 1 & 2 & -2 & 3 \\ 0 & 1 & -2 & \frac{1}{2} \\ 0 & 0 & 0 & 0 \end{array}\right]$$

The system $\begin{cases} x+2y-2z=3 \\ \quad\;\; y-2z=\dfrac{1}{2} \end{cases}$ has no unique solution.

Express x and y in terms of z:

$$y = 2z + \frac{1}{2}$$

$$x + 2\left(2z + \frac{1}{2}\right) - 2z = 3$$

$$x + 4z + 1 - 2z = 3$$

$$x + 2z + 1 = 3$$

$$x = -2z + 2$$

With $z = t$, the complete solution to the system is

$$\left\{\left(-2t + 2,\; 2t + \frac{1}{2},\; t\right)\right\}.$$

5. $\begin{bmatrix} 3 & 4 & 2 & | & 3 \\ 4 & -2 & -8 & | & -4 \\ 1 & 1 & -1 & | & 3 \end{bmatrix} R_1 \leftrightarrow R_3$

$\begin{bmatrix} 1 & 1 & -1 & | & 3 \\ 4 & -2 & -8 & | & -4 \\ 3 & 4 & 2 & | & 3 \end{bmatrix} \begin{matrix} \\ -4R_1 + R_2 \\ -3R_1 + R_3 \end{matrix}$

$\begin{bmatrix} 1 & 1 & -1 & | & 3 \\ 0 & -6 & -4 & | & -16 \\ 0 & 1 & 5 & | & -6 \end{bmatrix} R_2 \leftrightarrow R_3$

$\begin{bmatrix} 1 & 1 & -1 & | & 3 \\ 0 & 1 & 5 & | & -6 \\ 0 & -6 & -4 & | & -16 \end{bmatrix} 6R_2 + R_3$

$\begin{bmatrix} 1 & 1 & 1 & | & 3 \\ 0 & 1 & 5 & | & -6 \\ 0 & 0 & 26 & | & -52 \end{bmatrix} \dfrac{1}{26}R_3$

$\begin{bmatrix} 1 & 1 & -1 & | & 3 \\ 0 & 1 & 5 & | & -6 \\ 0 & 0 & 1 & | & -2 \end{bmatrix}$

This corresponds to the system

$$x + y - z = 3$$
$$y + 5z = -6$$
$$z = -2$$

Use back-substitution to find the values of x and y:

$$y + 5(-2) = -6$$
$$y - 10 = -6$$
$$y = 4$$
$$x + 4 + 2 = 3$$
$$x + 6 = 3$$
$$x = -3$$

The solution to the system is $\{(-3, 4, -2)\}$.

7. $\begin{bmatrix} 8 & 5 & 11 & | & 30 \\ -1 & -4 & 2 & | & 3 \\ 2 & -1 & 5 & | & 12 \end{bmatrix} R_1 \leftrightarrow R_2$

$\begin{bmatrix} -1 & -4 & 2 & | & 3 \\ 8 & 5 & 11 & | & 30 \\ 2 & -1 & 5 & | & 12 \end{bmatrix} -1R_1$

$\begin{bmatrix} 1 & 4 & -2 & | & -3 \\ 8 & 5 & 11 & | & 30 \\ 2 & -1 & 5 & | & 12 \end{bmatrix} \begin{matrix} \\ -8R_1 + R_2 \\ -2R_1 + R_3 \end{matrix}$

$\begin{bmatrix} 1 & 4 & -2 & | & -3 \\ 0 & -27 & 27 & | & 54 \\ 0 & -9 & 9 & | & 18 \end{bmatrix} -\dfrac{1}{27}R_2$

$\begin{bmatrix} 1 & 4 & -2 & | & -3 \\ 0 & 1 & -1 & | & -2 \\ 0 & -9 & 9 & | & 18 \end{bmatrix} 9R_2 + R_3$

$\begin{bmatrix} 1 & 4 & -2 & | & -3 \\ 0 & 1 & -1 & | & -2 \\ 0 & 0 & 0 & | & 0 \end{bmatrix}$

The system $\begin{array}{l} x + 4y - 2z = -3 \\ \quad\;\; y - z = -2 \end{array}$ has no unique solution.

Express x and y in terms of z:

$$y = -2 + z$$
$$x + 4(-2 + z) - 2z = -3$$
$$x - 8 + 4z - 2z = -3$$
$$x - 8 + 2z = -3$$
$$x = 5 - 2z$$

With $z = t$, the complete solution to the system is $\{(5 - 2t, -2 + t, t)\}$.

9. $\begin{bmatrix} 1 & -2 & -1 & -3 & | & -9 \\ 1 & 1 & -1 & 0 & | & 0 \\ 3 & 4 & 0 & 1 & | & 6 \\ 0 & 2 & -2 & 1 & | & 3 \end{bmatrix} \begin{matrix} \\ -1R_1 + R_2 \\ -3R_1 + R_3 \\ \\ \end{matrix}$

$\begin{bmatrix} 1 & -2 & -1 & -3 & | & -9 \\ 0 & 3 & 0 & 3 & | & 9 \\ 0 & 10 & 3 & 10 & | & 33 \\ 0 & 2 & -2 & 1 & | & 3 \end{bmatrix} \dfrac{1}{3}R_2$

$\begin{bmatrix} 1 & -2 & -1 & -3 & | & -9 \\ 0 & 1 & 0 & 1 & | & 3 \\ 0 & 10 & 3 & 10 & | & 33 \\ 0 & 2 & -2 & 1 & | & 3 \end{bmatrix} \begin{matrix} \\ \\ -10R_2 + R_3 \\ -2R_2 + R_4 \end{matrix}$

$$\begin{bmatrix} 1 & -2 & -1 & -3 & | & -9 \\ 0 & 1 & 0 & 1 & | & 3 \\ 0 & 0 & 3 & 0 & | & 3 \\ 0 & 0 & -2 & -1 & | & -3 \end{bmatrix} \frac{1}{3}R_3$$

$$\begin{bmatrix} 1 & -2 & -1 & -3 & | & -9 \\ 0 & 1 & 0 & 1 & | & 3 \\ 0 & 0 & 1 & 0 & | & 1 \\ 0 & 0 & -2 & -1 & | & -3 \end{bmatrix} 2R_3 + R_4$$

$$\begin{bmatrix} 1 & -2 & -1 & -3 & | & -9 \\ 0 & 1 & 0 & 1 & | & 3 \\ 0 & 0 & 1 & 0 & | & 1 \\ 0 & 0 & 0 & -1 & | & -1 \end{bmatrix} -1R_4$$

$$\begin{bmatrix} 1 & -2 & -1 & -3 & | & -9 \\ 0 & 1 & 0 & 1 & | & 3 \\ 0 & 0 & 1 & 0 & | & 1 \\ 0 & 0 & 0 & 1 & | & 1 \end{bmatrix}$$

This corresponds to the system
$$w - 2x - y - 3z = -9$$
$$x + z = 3$$
$$y = 1$$
$$z = 1$$

Use back-substitution to find the values of w and x:
$$x + 1 = 3$$
$$x = 2$$
$$w - 2(2) - 1 - 3(1) = -9$$
$$w - 4 - 1 - 3 = -9$$
$$w - 8 = -9$$
$$w = -1$$

The solution to the system is $\{(-1, 2, 1, 1)\}$.

11. $$\begin{bmatrix} 2 & 1 & -1 & 0 & | & 3 \\ 1 & -3 & 2 & 0 & | & -4 \\ 3 & 1 & -3 & 1 & | & 1 \\ 1 & 2 & -4 & -1 & | & -2 \end{bmatrix} R_1 \leftrightarrow R_2$$

$$\begin{bmatrix} 1 & -3 & 2 & 0 & | & -4 \\ 2 & 1 & -1 & 0 & | & 3 \\ 3 & 1 & -3 & 1 & | & 1 \\ 1 & 2 & -4 & -1 & | & -2 \end{bmatrix} \begin{matrix} -2R_1 + R_2 \\ -3R_1 + R_3 \\ -1R_1 + R_4 \end{matrix}$$

$$\begin{bmatrix} 1 & -3 & 2 & 0 & | & -4 \\ 0 & 7 & -5 & 0 & | & 11 \\ 0 & 10 & -9 & 1 & | & 13 \\ 0 & 5 & -6 & -1 & | & 2 \end{bmatrix} \frac{1}{7}R_2$$

$$\begin{bmatrix} 1 & -3 & 2 & 0 & | & -4 \\ 0 & 1 & -\frac{5}{7} & 0 & | & \frac{11}{7} \\ 0 & 10 & -9 & 1 & | & 13 \\ 0 & 5 & -6 & -1 & | & 2 \end{bmatrix} \begin{matrix} -10R_2 + R_3 \\ -5R_2 + R_4 \end{matrix}$$

$$\begin{bmatrix} 1 & -3 & 2 & 0 & | & -4 \\ 0 & 1 & -\frac{5}{7} & 0 & | & \frac{11}{7} \\ 0 & 0 & -\frac{13}{7} & 1 & | & -\frac{19}{7} \\ 0 & 0 & -\frac{17}{7} & -1 & | & -\frac{41}{7} \end{bmatrix} -\frac{7}{13}R_3$$

$$\begin{bmatrix} 1 & -3 & 2 & 0 & | & -4 \\ 0 & 1 & -\frac{5}{7} & 0 & | & \frac{11}{7} \\ 0 & 0 & 1 & -\frac{7}{13} & | & \frac{19}{13} \\ 0 & 0 & -\frac{17}{7} & -1 & | & -\frac{41}{7} \end{bmatrix} \frac{17}{7}R_3 + R_4$$

$$\begin{bmatrix} 1 & -3 & 2 & 0 & | & -4 \\ 0 & 1 & -\frac{5}{7} & 0 & | & \frac{11}{7} \\ 0 & 0 & 1 & -\frac{7}{13} & | & \frac{19}{13} \\ 0 & 0 & 0 & -\frac{30}{13} & | & -\frac{30}{13} \end{bmatrix} -\frac{13}{30}R_4$$

$$\begin{bmatrix} 1 & -3 & 2 & 0 & | & -4 \\ 0 & 1 & -\frac{5}{7} & 0 & | & \frac{11}{7} \\ 0 & 0 & 1 & -\frac{7}{13} & | & \frac{19}{13} \\ 0 & 0 & 0 & 1 & | & 1 \end{bmatrix}$$

This corresponds to the system
$$w - 3x + 2y = -4$$
$$x - \frac{5}{7}y = \frac{11}{7}$$
$$y - \frac{7}{13}z = \frac{19}{13}$$
$$z = 1$$

Use back-substitution to find the values of w, x, and y:
$$y - \frac{7}{13}z = \frac{19}{13}$$
$$y - \frac{7}{13}(1) = \frac{19}{13}$$
$$y = 2$$
$$x - \frac{5}{7}(2) = \frac{11}{7}$$
$$x - \frac{10}{7} = \frac{11}{7}$$
$$x = 3$$
$$w - 3(3) + 2(2) = -4$$
$$w - 9 + 4 = -4$$
$$w - 5 = -4$$
$$w = 1$$

The solution to the system is $\{(1, 3, 2, 1)\}$.

13.
$$\left[\begin{array}{cccc|c} 1 & -3 & 1 & -4 & 4 \\ -2 & 1 & 2 & 0 & -2 \\ 3 & -2 & 1 & -6 & 2 \\ -1 & 3 & 2 & -1 & -6 \end{array}\right] \begin{array}{l} 2R_1 + R_2 \\ -3R_1 + R_3 \\ R_1 + R_4 \end{array}$$

$$\left[\begin{array}{cccc|c} 1 & -3 & 1 & -4 & 4 \\ 0 & -5 & 4 & -8 & 6 \\ 0 & 7 & -2 & 6 & -10 \\ 0 & 0 & 3 & -5 & -2 \end{array}\right] -\tfrac{1}{5}R_2$$

$$\left[\begin{array}{cccc|c} 1 & -3 & 1 & -4 & 4 \\ 0 & 1 & -\tfrac{4}{5} & \tfrac{8}{5} & -\tfrac{6}{5} \\ 0 & 7 & -2 & 6 & -10 \\ 0 & 0 & 3 & -5 & -2 \end{array}\right] -7R_2 + R_3$$

$$\left[\begin{array}{cccc|c} 1 & -3 & 1 & -4 & 4 \\ 0 & 1 & -\tfrac{4}{5} & \tfrac{8}{5} & -\tfrac{6}{5} \\ 0 & 0 & \tfrac{18}{5} & -\tfrac{26}{5} & -\tfrac{8}{5} \\ 0 & 0 & 3 & -5 & -2 \end{array}\right] \tfrac{5}{18}R_3$$

$$\left[\begin{array}{cccc|c} 1 & -3 & 1 & -4 & 4 \\ 0 & 1 & -\tfrac{4}{5} & \tfrac{8}{5} & -\tfrac{6}{5} \\ 0 & 0 & 1 & -\tfrac{13}{9} & -\tfrac{4}{9} \\ 0 & 0 & 3 & -5 & -2 \end{array}\right] -3R_3 + R_4$$

$$\left[\begin{array}{cccc|c} 1 & -3 & 1 & -4 & 4 \\ 0 & 1 & -\tfrac{4}{5} & \tfrac{8}{5} & -\tfrac{6}{5} \\ 0 & 0 & 1 & -\tfrac{13}{9} & -\tfrac{4}{9} \\ 0 & 0 & 0 & -\tfrac{2}{3} & -\tfrac{2}{3} \end{array}\right] -\tfrac{3}{2}R_4$$

$$\left[\begin{array}{cccc|c} 1 & -3 & 1 & -4 & 4 \\ 0 & 1 & -\tfrac{4}{5} & \tfrac{8}{5} & -\tfrac{6}{5} \\ 0 & 0 & 1 & -\tfrac{13}{9} & -\tfrac{4}{9} \\ 0 & 0 & 0 & 1 & 1 \end{array}\right]$$

This corresponds to the system
$$w - 3x + y - 4z = 4$$
$$x - \frac{4}{5}y + \frac{8}{5}z = -\frac{6}{5}$$
$$y - \frac{13}{9}z = -\frac{4}{9}$$
$$z = 1$$

Use back-substitution to find the values of w, z, and y:
$$y - \frac{13}{9}(1) = -\frac{4}{9}$$
$$y = 1$$

$$x - \frac{4}{5}(1) + \frac{8}{5}(1) = -\frac{6}{5}$$
$$x + \frac{4}{5} = -\frac{6}{5}$$
$$x = -2$$
$$w - 3(-2) + 1 - 4 = 4$$
$$w + 6 - 3 = 4$$
$$w = 1$$
The solution to the system is $\{(1, -2, 1, 1)\}$.

15.
$$\left[\begin{array}{ccc|c} 2 & 1 & -1 & 2 \\ 3 & 3 & -2 & 3 \end{array}\right] \tfrac{1}{2}R_1 \left[\begin{array}{ccc|c} 1 & \tfrac{1}{2} & -\tfrac{1}{2} & 1 \\ 3 & 3 & -2 & 3 \end{array}\right] -3R_1 + R_2$$

$$\left[\begin{array}{ccc|c} 1 & \tfrac{1}{2} & -\tfrac{1}{2} & 1 \\ 0 & \tfrac{3}{2} & -\tfrac{1}{2} & 0 \end{array}\right] \tfrac{2}{3}R_2$$

$$\left[\begin{array}{ccc|c} 1 & \tfrac{1}{2} & -\tfrac{1}{2} & 1 \\ 0 & 1 & -\tfrac{1}{3} & 0 \end{array}\right]$$

The system $x + \dfrac{1}{2}y - \dfrac{1}{2}z = 1$ has no unique solution.
$$y - \frac{1}{3}z = 0$$

Express x and y in terms of z:
$$y = \frac{1}{3}z$$
$$x + \frac{1}{2}\left(\frac{1}{3}z\right) - \frac{1}{2}z = 1$$
$$x + \frac{1}{6}z - \frac{1}{2}z = 1$$
$$x - \frac{1}{3}z = 1$$
$$x = 1 + \frac{1}{3}z$$

With $z = t$, the complete solution to the system is
$$\left\{\left(1 + \frac{1}{3}t, \frac{1}{3}t, t\right)\right\}.$$

17. The system $\begin{array}{l} x + 2y + 3z = 5 \\ y - 5z = 0 \end{array}$ has no unique solution.

Express x and y in terms of z:
$$y = 5z$$
$$x + 2(5z) + 3z = 5$$
$$x + 10z + 3z = 5$$
$$x = -13z + 5$$
With $z = t$, the complete solution to the system is
$\{(-13t + 5, 5t, t)\}$.

19.
$$\begin{bmatrix} 1 & 1 & -2 & | & 2 \\ 3 & -1 & -6 & | & -7 \end{bmatrix} -3R_1 + R_2$$

$$\begin{bmatrix} 1 & 1 & -2 & | & 2 \\ 0 & -4 & 0 & | & -13 \end{bmatrix} -\tfrac{1}{4}R_2$$

$$\begin{bmatrix} 1 & 1 & -2 & | & 2 \\ 0 & 1 & 0 & | & \frac{13}{4} \end{bmatrix}$$

The system
$$\begin{aligned} x+y-2z &= 2 \\ y &= \frac{13}{4} \end{aligned}$$
has no unique solution.

Express x in terms of z:

$$x+\frac{13}{4}-2z = 2$$

$$x = 2z - \frac{5}{4}$$

With $z = t$, the complete solution to the system is
$$\left\{\left(2t-\frac{5}{4}, \frac{13}{4}, t\right)\right\}.$$

21.
$$\begin{bmatrix} 1 & 1 & -1 & 1 & | & -2 \\ 2 & -1 & 2 & -1 & | & 7 \\ -1 & 2 & 1 & 2 & | & -1 \end{bmatrix} \begin{matrix} -2R_1 + R_2 \\ 1R_1 + R_3 \end{matrix}$$

$$\begin{bmatrix} 1 & 1 & -1 & 1 & | & -2 \\ 0 & -3 & 4 & -3 & | & 11 \\ 0 & 3 & 0 & 3 & | & -3 \end{bmatrix} R_2 \leftrightarrow R_3$$

$$\begin{bmatrix} 1 & 1 & -1 & 1 & | & -2 \\ 0 & 3 & 0 & 3 & | & -3 \\ 0 & -3 & 4 & -3 & | & 11 \end{bmatrix} \tfrac{1}{3}R_2$$

$$\begin{bmatrix} 1 & 1 & -1 & 1 & | & -2 \\ 0 & 1 & 0 & 1 & | & -1 \\ 0 & -3 & 4 & -3 & | & 11 \end{bmatrix} 3R_2 + R_3$$

$$\begin{bmatrix} 1 & 1 & -1 & 1 & | & -2 \\ 0 & 1 & 0 & 1 & | & -1 \\ 0 & 0 & 4 & 0 & | & 8 \end{bmatrix} \tfrac{1}{4}R_3$$

$$\begin{bmatrix} 1 & 1 & -1 & 1 & | & -2 \\ 0 & 1 & 0 & 1 & | & -1 \\ 0 & 0 & 1 & 0 & | & 2 \end{bmatrix}$$

The system
$$\begin{aligned} x+y-z+w &= -2 \\ x+z &= -1 \\ y &= 2 \end{aligned}$$
has no unique solution. Let $z = t$ and use back substitution to find remaining variables. The complete solution to the system is $\{(1, -t-1, 2, t)\}$.

23.
$$\begin{bmatrix} 1 & 2 & 3 & -1 & | & 7 \\ 0 & 2 & -3 & 1 & | & 4 \\ 1 & -4 & 1 & 0 & | & 3 \end{bmatrix} -1R_1 + R_3$$

$$\begin{bmatrix} 1 & 2 & 3 & -1 & | & 7 \\ 0 & 2 & -3 & 1 & | & 4 \\ 0 & -6 & -2 & 1 & | & -4 \end{bmatrix} \tfrac{1}{2}R_2$$

$$\begin{bmatrix} 1 & 2 & 3 & -1 & | & 7 \\ 0 & 1 & -\frac{3}{2} & \frac{1}{2} & | & 2 \\ 0 & -6 & -2 & 1 & | & -4 \end{bmatrix} 6R_2 + R_3$$

$$\begin{bmatrix} 1 & 2 & 3 & -1 & | & 7 \\ 0 & 1 & -\frac{3}{2} & \frac{1}{2} & | & 2 \\ 0 & 0 & -11 & 4 & | & 8 \end{bmatrix} -\tfrac{1}{11}R_3$$

$$\begin{bmatrix} 1 & 2 & 3 & -1 & | & 7 \\ 0 & 1 & -\frac{3}{2} & \frac{1}{2} & | & 2 \\ 0 & 0 & 1 & -\frac{4}{11} & | & -\frac{8}{11} \end{bmatrix}$$

The system has no unique solution. Let $z = t$ and use back substitution to find remaining variables. The complete solution to the system is
$$\left\{\left(-\frac{2}{11}t+\frac{81}{11}, \frac{1}{22}t+\frac{10}{11}, \frac{4}{11}t-\frac{8}{11}, t\right)\right\}.$$

25. a. System of equations:
$$\begin{aligned} 4w-2x+2y-3z &= 0 \\ 7w-x-y-3z &= 0 \\ w+x-y-z &= 0 \end{aligned}$$

b. Reduced system:
$$\begin{aligned} w-0.5z &= 0 \\ x &= 0 \\ y-0.5z &= 0 \end{aligned}$$
Find w and y in terms of z.
$$\begin{aligned} w-0.5z &= 0 \\ w &= 0.5z \\ y-0.5z &= 0 \\ y &= 0.5z \end{aligned}$$
The complete solution to the system is
$$\{(0.5z, 0, 0.5z, z)\}.$$

27. a. System of equations:
$$\begin{aligned} w+2x+5y+5z &= 3 \\ w+x+3y+4z &= -1 \\ w-x-y+2z &= 3 \end{aligned}$$

b. Reduced system:
$$\begin{aligned} w+y+3z &= 1 \\ x+2y+z &= -2 \end{aligned}$$

Find w and x in terms of y and z.
$$w + y + 3z = 1$$
$$w = 1 - y - 3z$$
$$x + 2y + z = -2$$
$$x = -2 - 2y - z$$
The complete solution to the system is
$$\{(1 - y - 3z, \; -2 - 2y - z, \; y, \; z)\}.$$

29. $z + 12 = x + 6$

31. $x - y = 4$
$x - z = 6$
$y - z = 2$

$$\begin{bmatrix} 1 & -1 & 0 & | & 4 \\ 1 & 0 & -1 & | & 6 \\ 0 & 1 & -1 & | & 2 \end{bmatrix} \begin{matrix} -1R_1 + R_2 \\ \\ \\ \end{matrix}$$

$$\begin{bmatrix} 1 & -1 & 0 & 4 \\ 0 & -1 & 1 & -2 \\ 0 & 1 & -1 & 2 \end{bmatrix} -1R_2$$

$$\begin{bmatrix} 1 & -1 & 0 & 4 \\ 0 & 1 & -1 & 2 \\ 0 & 1 & -1 & 2 \end{bmatrix} \begin{matrix} \\ -1R_2 + R_3 \\ 1R_2 + R_1 \end{matrix}$$

$$\begin{bmatrix} 1 & 0 & -1 & 6 \\ 0 & 1 & -1 & 2 \\ 0 & 0 & 0 & 0 \end{bmatrix}$$

The system has no unique solution. Express x and y in terms of z:
$$x - z = 6$$
$$y - z = 2$$
$$x = z + 6$$
$$y = z + 2$$
With $z = t$, the complete solution to the system is $\{(t + 6, t + 2, t)\}$.

33. a. From left to right along Palm Drive, then along Sunset Drive, we get the equations
$$w + z = 200 + 180 = 380;$$
$$w + x = 400 + 200 = 600;$$
$$z + 70 = y + 20 \text{ or } y - z = 50;$$
$$y + 200 = x + 30 \text{ or } x - y = 170.$$
The system is
$$w + z = 380$$
$$w + x = 600$$
$$y - z = 50$$
$$x - y = 170$$

b.
$$\begin{bmatrix} 1 & 0 & 0 & 1 & | & 380 \\ 0 & 1 & 0 & -1 & | & 220 \\ 0 & 0 & 1 & -1 & | & 50 \\ 0 & 1 & -1 & 0 & | & 170 \end{bmatrix} -1R_2 + R_4$$

$$\begin{bmatrix} 1 & 0 & 0 & 1 & | & 380 \\ 0 & 1 & 0 & -1 & | & 220 \\ 0 & 0 & 1 & -1 & | & 50 \\ 0 & 0 & -1 & 1 & | & -50 \end{bmatrix} 1R_3 + R_4$$

$$\begin{bmatrix} 1 & 0 & 0 & 1 & | & 380 \\ 0 & 1 & 0 & -1 & | & 220 \\ 0 & 0 & 1 & -1 & | & 50 \\ 0 & 0 & 0 & 0 & | & 0 \end{bmatrix}$$

The system has no unique solution. Express x and y and z in terms of z:
$$w = 380 - z$$
$$x = 220 + z$$
$$y = 50 + z$$
With $z = t$, the complete solution to the system is $\{(380 - t, 220 + t, 50 + t, t)\}$.

c. Letting $z = 50$, the solution is
$$w = 380 - 50 = 330$$
$$x = 220 + 50 = 270$$
$$y = 50 + 50 = 100$$

35. Let $x =$ the number of ounces of Food 1,
$y =$ the number of ounces of Food 2, and
$z =$ the number of ounces of Food 3.
The amount of vitamin A is $20x + 30y + 10z$; the amount of iron is $20x + 10y + 10z$; the amount of calcium is $10x + 10y + 30z$.

a. Not having Food 1 means that all x terms are left out. The vitamin A requirement can then be represented by $30y + 10z = 220$; the iron requirement is $10y + 10z = 180$; the calcium requirement is $10y + 30z = 340$.
The corresponding system is
$$30y + 10z = 220$$
$$10y + 10z = 180$$
$$10y + 30z = 340.$$
Dividing all of the numbers by 10, the matrix for this system is

$$\begin{bmatrix} 3 & 1 & | & 22 \\ 1 & 1 & | & 18 \\ 1 & 3 & | & 34 \end{bmatrix} R_1 \leftrightarrow R_2 \begin{bmatrix} 1 & 1 & | & 18 \\ 3 & 1 & | & 22 \\ 1 & 3 & | & 34 \end{bmatrix} \begin{matrix} -3R_1 + R_2 \\ -1R_1 + R_3 \end{matrix}$$

$$\begin{bmatrix} 1 & 1 & | & 18 \\ 0 & -2 & | & -32 \\ 0 & 2 & | & 16 \end{bmatrix} 1R_2 + R_3$$

$$\begin{bmatrix} 1 & 1 & | & 18 \\ 0 & -2 & | & -32 \\ 0 & 0 & | & -16 \end{bmatrix}.$$

From the last row, we see that the system has no solution, so there is no way to satisfy these dietary requirements with no Food 1 available.

b. With Food 1 available, and dropping the vitamin A requirement, the system is
$20x + 10y + 10z = 180$
$10x + 10y + 30z = 340.$
Dividing all of the numbers by 10, the matrix for this system is

$$\begin{bmatrix} 2 & 1 & 1 & | & 18 \\ 1 & 1 & 3 & | & 34 \end{bmatrix} R_1 \leftrightarrow R_2$$

$$\begin{bmatrix} 1 & 1 & 3 & | & 34 \\ 2 & 1 & 1 & | & 18 \end{bmatrix} -2R_1 + R_2$$

$$\begin{bmatrix} 1 & 1 & 3 & | & 34 \\ 0 & -1 & -5 & | & -50 \end{bmatrix} -1R_2$$

$$\begin{bmatrix} 1 & 1 & 3 & | & 34 \\ 0 & 1 & 5 & | & 50 \end{bmatrix}.$$

The system $\begin{matrix} x + y + 3z = 34 \\ y + 5z = 50 \end{matrix}$ has no unique

solution. Express x and y in terms of z:
$y = -5z + 50$
$x + (-5z + 50) + 3z = 34$
$\qquad x - 2z + 50 = 34$
$\qquad\qquad x = 2z - 16$

Now we can choose a value for z, i.e., an amount of Food 3, and find the corresponding values of x and y. Note that negative amounts of food are not realistic, so $z \geq 0$, $y = -5z + 50 \geq 0$, and $x = 2z - 16 \geq 0$. These conditions are equivalent to $8 \leq z \leq 10$.

Using $z = 8$ and $z = 10$, two possibilities are 0 ounces of Food 1, 10 ounces of Food 2, and 8 ounces of Food 3 or 4 ounces of Food 1, 0 ounces of Food 2, and 10 ounces of Food 3. (Other answers are possible.)

37. – 39. Answers will vary.

41. does not make sense; Explanations will vary. Sample explanation: Row 3 indicates that this system has no solution. If eliminated, it would falsely appear that the system has an infinite number of solutions.

43. does not make sense; Explanations will vary. Sample explanation: In a nonsquare system, the number of equations differs from the number of variables.

45.
$$\begin{bmatrix} 1 & 3 & 1 & | & a^2 \\ 2 & 5 & 2a & | & 0 \\ 1 & 1 & a^2 & | & -9 \end{bmatrix} \begin{matrix} \\ -2R_1 + R_2 \\ -1R_1 + R_3 \end{matrix}$$

$$\begin{bmatrix} 1 & 3 & 1 & | & a^2 \\ 0 & -1 & 2a-2 & | & -2a^2 \\ 0 & -2 & a^2-1 & | & -9-a^2 \end{bmatrix} -1R_2$$

$$\begin{bmatrix} 1 & 3 & 1 & | & a^2 \\ 0 & 1 & 2-2a & | & 2a^2 \\ 0 & -2 & a^2-1 & | & -9-a^2 \end{bmatrix} 2R_2 + R_3$$

$$\begin{bmatrix} 1 & 3 & 1 & | & a^2 \\ 0 & 1 & 2-2a & | & 2a^2 \\ 0 & 0 & a^2-4a+3 & | & -9+3a^2 \end{bmatrix}$$

The system will be inconsistent when $a^2 - 4a + 3 = 0$ but $-9 + 3a^2 \neq 0$.
$a^2 - 4a + 3 = (a-1)(a-3) = 0$ when $a = 1$ or $a = 3$. $-9 + 3a^2 = 0$ when $a = \pm\sqrt{3}$.
Thus, the system is inconsistent when $a = 1$ or $a = 3$.

47. $-6 - (-5) = -6 + 5 = -1$

48. $1(-4) + 2(5) + 3(-6) = -4 + 10 - 18 = -12$

49. $\frac{1}{2}[8 - (-8)] = \frac{1}{2}[8+8] = \frac{1}{2}[16] = 8$

Section 9.3

Check Point Exercises

1. **a.** The matrix $A = \begin{bmatrix} 5 & -2 \\ -3 & \pi \\ 1 & 6 \end{bmatrix}$ has 3 rows and 2 columns, so it is of order 3×2.

b. The element a_{12} is in the first row and second column. Thus, $a_{12} = -2$. The element a_{31} is in the third row and first column. Thus, $a_{31} = 1$.

2. **a.** $\begin{bmatrix} -4 & 3 \\ 7 & -6 \end{bmatrix} + \begin{bmatrix} 6 & -3 \\ 2 & -4 \end{bmatrix}$

$= \begin{bmatrix} -4+6 & 3+(-3) \\ 7+2 & -6+(-4) \end{bmatrix} = \begin{bmatrix} 2 & 0 \\ 9 & -10 \end{bmatrix}$

b. $\begin{bmatrix} 5 & 4 \\ -3 & 7 \\ 0 & 1 \end{bmatrix} - \begin{bmatrix} -4 & 8 \\ 6 & 0 \\ -5 & 3 \end{bmatrix}$

$= \begin{bmatrix} 5-(-4) & 4-8 \\ -3-6 & 7-0 \\ 0-(-5) & 1-3 \end{bmatrix}$

$= \begin{bmatrix} 9 & -4 \\ -9 & 7 \\ 5 & -2 \end{bmatrix}$

3. **a.** $-6B = -6 \begin{bmatrix} -1 & -2 \\ 8 & 5 \end{bmatrix}$

$= \begin{bmatrix} -6(-1) & -6(-2) \\ -6(8) & -6(5) \end{bmatrix}$

$= \begin{bmatrix} -6 & 12 \\ -48 & -30 \end{bmatrix}$

b. $3A + 2B = \begin{bmatrix} -4 & 1 \\ 3 & 0 \end{bmatrix} + 2\begin{bmatrix} -1 & -2 \\ 8 & 5 \end{bmatrix}$

$= \begin{bmatrix} 3(-4) & 3(1) \\ 3(3) & 3(0) \end{bmatrix} + \begin{bmatrix} 2(-1) & 2(-2) \\ 2(8) & 2(5) \end{bmatrix}$

$= \begin{bmatrix} -12 & 3 \\ 9 & 0 \end{bmatrix} + \begin{bmatrix} -2 & -4 \\ 16 & 10 \end{bmatrix}$

$= \begin{bmatrix} -12+(-2) & 3+(-4) \\ 9+16 & 0+10 \end{bmatrix}$

$= \begin{bmatrix} -14 & -1 \\ 25 & 10 \end{bmatrix}$

4. $3X + A = B$

$3X = B - A$

$X = \frac{1}{3}(B - A)$

$X = \frac{1}{3}\left(\begin{bmatrix} -10 & 1 \\ -9 & 17 \end{bmatrix} - \begin{bmatrix} 2 & -8 \\ 0 & 4 \end{bmatrix} \right)$

$X = \frac{1}{3}\begin{bmatrix} -12 & 9 \\ -9 & 13 \end{bmatrix}$

$X = \begin{bmatrix} -4 & 3 \\ -3 & \frac{13}{3} \end{bmatrix}$

5. Given $A = \begin{bmatrix} 1 & 3 \\ 2 & 5 \end{bmatrix}$ and $B = \begin{bmatrix} 4 & 6 \\ 1 & 0 \end{bmatrix}$,

$AB = \begin{bmatrix} 1 & 3 \\ 2 & 5 \end{bmatrix} \cdot \begin{bmatrix} 4 & 6 \\ 1 & 0 \end{bmatrix} = \begin{bmatrix} 1(4)+3(1) & 1(6)+3(0) \\ 2(4)+5(1) & 2(6)+5(0) \end{bmatrix}$

$= \begin{bmatrix} 7 & 6 \\ 13 & 12 \end{bmatrix}$

6. If $A = \begin{bmatrix} 2 & 0 & 4 \end{bmatrix}$ and $B = \begin{bmatrix} 1 \\ 3 \\ 7 \end{bmatrix}$, then

$AB = \begin{bmatrix} 2 & 0 & 4 \end{bmatrix}\begin{bmatrix} 1 \\ 3 \\ 7 \end{bmatrix}$

$= \begin{bmatrix} 2(1)+0(3)+4(7) \end{bmatrix}$

$= \begin{bmatrix} 2+0+28 \end{bmatrix}$

$= \begin{bmatrix} 30 \end{bmatrix}$

and

$BA = \begin{bmatrix} 1 \\ 3 \\ 7 \end{bmatrix}\begin{bmatrix} 2 & 0 & 4 \end{bmatrix}$

$= \begin{bmatrix} 1(2) & 1(0) & 1(4) \\ 3(2) & 3(0) & 3(4) \\ 7(2) & 7(0) & 7(4) \end{bmatrix}$

$= \begin{bmatrix} 2 & 0 & 4 \\ 6 & 0 & 12 \\ 14 & 0 & 28 \end{bmatrix}$.

7. a. $\begin{bmatrix} 1 & 3 \\ 0 & 2 \end{bmatrix} \cdot \begin{bmatrix} 2 & 3 & -1 & 6 \\ 0 & 5 & 4 & 1 \end{bmatrix}$

$= \begin{bmatrix} 1(2)+3(0) & 1(3)+3(5) & 1(-1)+3(4) & 1(6)+3(1) \\ 0(2)+2(0) & 0(3)+2(5) & 0(-1)+2(4) & 0(6)+2(1) \end{bmatrix}$

$= \begin{bmatrix} 2 & 18 & 11 & 9 \\ 0 & 10 & 8 & 2 \end{bmatrix}$

b. $\begin{bmatrix} 2 & 3 & -1 & 6 \\ 0 & 5 & 4 & 1 \end{bmatrix}\begin{bmatrix} 1 & 3 \\ 0 & 2 \end{bmatrix}$

The number of columns in the first matrix does not equal the number of rows in the second matrix. Thus, the product of these two matrices is <u>undefined</u>.

8. Because the *L* is dark gray and the background is light gray, the digital photograph can be represented by the matrix

$\begin{bmatrix} 2 & 1 & 1 \\ 2 & 1 & 1 \\ 2 & 2 & 1 \end{bmatrix}$

We can make the *L* light gray by decreasing each 2 in the above matrix to 1. We can make the background black by increasing each 1 in the matrix to 3. This is accomplished using the following matrix addition.

$\begin{bmatrix} 2 & 1 & 1 \\ 2 & 1 & 1 \\ 2 & 2 & 1 \end{bmatrix} + \begin{bmatrix} -1 & 2 & 2 \\ -1 & 2 & 2 \\ -1 & -1 & 2 \end{bmatrix} = \begin{bmatrix} 1 & 3 & 3 \\ 1 & 3 & 3 \\ 1 & 1 & 3 \end{bmatrix}$

9. **a.** $\begin{bmatrix} 0 & 3 & 4 \\ 0 & 5 & 2 \end{bmatrix} + \begin{bmatrix} -3 & -3 & -3 \\ -1 & -1 & -1 \end{bmatrix} = \begin{bmatrix} -3 & 0 & 1 \\ -1 & 4 & 1 \end{bmatrix}$

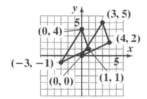

b. $2\begin{bmatrix} 0 & 3 & 4 \\ 0 & 5 & 2 \end{bmatrix} = \begin{bmatrix} 0 & 6 & 8 \\ 0 & 10 & 4 \end{bmatrix}$

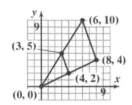

c. $\begin{bmatrix} 1 & 0 \\ 0 & -1 \end{bmatrix}\begin{bmatrix} 0 & 3 & 4 \\ 0 & 5 & 2 \end{bmatrix}$

$= \begin{bmatrix} (1)(0)+0(0) & (1)(3)+0(5) & (1)(4)+0(2) \\ 0(0)-1(0) & 0(3)-1(5) & 0(4)-1(2) \end{bmatrix}$

$= \begin{bmatrix} 0 & 3 & 4 \\ 0 & -5 & -2 \end{bmatrix}$

Multiplication with $\begin{bmatrix} 1 & 0 \\ 0 & -1 \end{bmatrix}$ reflects the

triangle over the *x*-axis.

Concept and Vocabulary Check 9.3

1. third; fourth

2. 1×3

3. square

4. −10; 6

5. true

6. false

7. true

8. *m*; *p*; columns; rows

9. true

10. false

Exercise Set 9.3

1. **a.** 2×3

b. a_{32} does not exist (*A* only has 2 rows).
$a_{23} = -1$

3. **a.** 3×4

b. $a_{32} = \dfrac{1}{2}$; $a_{23} = -6$

5. $\begin{bmatrix} x \\ 4 \end{bmatrix} = \begin{bmatrix} 6 \\ y \end{bmatrix}$
$x = 6$
$y = 4$

7. $\begin{bmatrix} x & 2y \\ z & 9 \end{bmatrix} = \begin{bmatrix} 4 & 12 \\ 3 & 9 \end{bmatrix}$
$x = 4$
$2y = 12$
$y = 6$
$z = 3$

9. **a.** $A + B = \begin{bmatrix} 4+5 & 1+9 \\ 3+0 & 2+7 \end{bmatrix} = \begin{bmatrix} 9 & 10 \\ 3 & 9 \end{bmatrix}$

b. $A - B = \begin{bmatrix} 4-5 & 1-9 \\ 3-0 & 2-7 \end{bmatrix} = \begin{bmatrix} -1 & -8 \\ 3 & -5 \end{bmatrix}$

c. $-4A = \begin{bmatrix} -16 & -4 \\ -12 & -8 \end{bmatrix}$

d. $3A + 2B = \begin{bmatrix} 12+10 & 3+18 \\ 9+0 & 6+14 \end{bmatrix} = \begin{bmatrix} 22 & 21 \\ 9 & 20 \end{bmatrix}$

11. **a.** $A + B = \begin{bmatrix} 1+2 & 3+(-1) \\ 3+3 & 4+(-2) \\ 5+0 & 6+1 \end{bmatrix} = \begin{bmatrix} 3 & 2 \\ 6 & 2 \\ 5 & 7 \end{bmatrix}$

b. $A - B = \begin{bmatrix} 1-2 & 3-(-1) \\ 3-3 & 4-(-2) \\ 5-0 & 6-1 \end{bmatrix} = \begin{bmatrix} -1 & 4 \\ 0 & 6 \\ 5 & 5 \end{bmatrix}$

c. $-4A = \begin{bmatrix} -4 & -12 \\ -12 & -16 \\ -20 & -24 \end{bmatrix}$

d. $3A + 2B = \begin{bmatrix} 3+4 & 9-2 \\ 9+6 & 12-4 \\ 15+0 & 18+2 \end{bmatrix} = \begin{bmatrix} 7 & 7 \\ 15 & 8 \\ 15 & 20 \end{bmatrix}$

13. a. $A + B = \begin{bmatrix} 2+(-5) \\ -4+3 \\ 1+(-1) \end{bmatrix} = \begin{bmatrix} -3 \\ -1 \\ 0 \end{bmatrix}$

b. $A - B = \begin{bmatrix} 2-(-5) \\ -4-3 \\ 1-(-1) \end{bmatrix} = \begin{bmatrix} 7 \\ -7 \\ 2 \end{bmatrix}$

c. $-4A = \begin{bmatrix} -8 \\ 16 \\ -4 \end{bmatrix}$

d. $3A + 2B = \begin{bmatrix} 6-10 \\ -12+6 \\ 3-2 \end{bmatrix} = \begin{bmatrix} -4 \\ -6 \\ 1 \end{bmatrix}$

15. a. $A + B = \begin{bmatrix} 2+6 & -10+10 & -2+(-2) \\ 14+0 & 12+(-12) & 10+(-4) \\ 4+(-5) & -2+2 & 2+(-2) \end{bmatrix}$

$= \begin{bmatrix} 8 & 0 & -4 \\ 14 & 0 & 6 \\ -1 & 0 & 0 \end{bmatrix}$

b. $A - B = \begin{bmatrix} 2-6 & -10-10 & -2-(-2) \\ 14-0 & 12-(-12) & 10-(-4) \\ 4-(-5) & -2-2 & 2-(-2) \end{bmatrix}$

$= \begin{bmatrix} -4 & -20 & 0 \\ 14 & 24 & 14 \\ 9 & -4 & 4 \end{bmatrix}$

c. $-4A = \begin{bmatrix} -8 & 40 & 8 \\ -56 & -48 & -40 \\ -16 & 8 & -8 \end{bmatrix}$

d. $3A + 2B = \begin{bmatrix} 6+12 & -30+20 & -6-4 \\ 42+0 & 36-24 & 30-8 \\ 12-10 & -6+4 & 6-4 \end{bmatrix}$

$= \begin{bmatrix} 18 & -10 & -10 \\ 42 & 12 & 22 \\ 2 & -2 & 2 \end{bmatrix}$

17. $X - A = B$

$\qquad X = A + B$

$$X = \begin{bmatrix} -3 & -7 \\ 2 & -9 \\ 5 & 0 \end{bmatrix} + \begin{bmatrix} -5 & -1 \\ 0 & 0 \\ 3 & -4 \end{bmatrix} = \begin{bmatrix} -8 & -8 \\ 2 & -9 \\ 8 & -4 \end{bmatrix}$$

19. $2X + A = B$

$\qquad 2X = B - A$

$\qquad X = \dfrac{1}{2}(B - A)$

$$X = \frac{1}{2}\left(\begin{bmatrix} -5 & -1 \\ 0 & 0 \\ 3 & -4 \end{bmatrix} - \begin{bmatrix} -3 & -7 \\ 2 & -9 \\ 5 & 0 \end{bmatrix} \right) = \frac{1}{2}\begin{bmatrix} -2 & 6 \\ -2 & 9 \\ -2 & -4 \end{bmatrix} = \begin{bmatrix} -1 & 3 \\ -1 & \frac{9}{2} \\ -1 & -2 \end{bmatrix}$$

21. $3X + 2A = B$

$\qquad 3X = B - 2A$

$\qquad X = \dfrac{1}{3}(B - 2A)$

$$X = \frac{1}{3}\left(\begin{bmatrix} -5 & -1 \\ 0 & 0 \\ 3 & -4 \end{bmatrix} - 2\begin{bmatrix} -3 & -7 \\ 2 & -9 \\ 5 & 0 \end{bmatrix} \right) = \frac{1}{3}\begin{bmatrix} 1 & 13 \\ -4 & 18 \\ -7 & -4 \end{bmatrix} = \begin{bmatrix} \frac{1}{3} & \frac{13}{3} \\ -\frac{4}{3} & 6 \\ -\frac{7}{3} & -\frac{4}{3} \end{bmatrix}$$

23. $\quad B - X = 4A$

$\quad B - 4A = X$

$$X = \begin{bmatrix} -5 & -1 \\ 0 & 0 \\ 3 & -4 \end{bmatrix} - 4\begin{bmatrix} -3 & -7 \\ 2 & -9 \\ 5 & 0 \end{bmatrix} = \begin{bmatrix} -5 & -1 \\ 0 & 0 \\ 3 & -4 \end{bmatrix} + \begin{bmatrix} 12 & 28 \\ -8 & 36 \\ -20 & 0 \end{bmatrix} = \begin{bmatrix} 7 & 27 \\ -8 & 36 \\ -17 & -4 \end{bmatrix}$$

25. $\quad 4A + 3B = -2X$

$\quad -\dfrac{1}{2}(4A + 3B) = X$

$$X = -\frac{1}{2}\left(4\begin{bmatrix} -3 & -7 \\ 2 & -9 \\ 5 & 0 \end{bmatrix} + 3\begin{bmatrix} -5 & -1 \\ 0 & 0 \\ 3 & -4 \end{bmatrix} \right) = -\frac{1}{2}\left(\begin{bmatrix} -12 & -28 \\ 8 & -36 \\ 20 & 0 \end{bmatrix} + \begin{bmatrix} -15 & -3 \\ 0 & 0 \\ 9 & -12 \end{bmatrix} \right) = -\frac{1}{2}\begin{bmatrix} -27 & -31 \\ 8 & -36 \\ 29 & -12 \end{bmatrix} = \begin{bmatrix} \frac{27}{2} & \frac{31}{2} \\ -4 & 18 \\ -\frac{29}{2} & 6 \end{bmatrix}$$

27. a. $\quad AB = \begin{bmatrix} 1 & 3 \\ 5 & 3 \end{bmatrix}\begin{bmatrix} 3 & -2 \\ -1 & 6 \end{bmatrix} = \begin{bmatrix} (1)(3)+(3)(-1) & (1)(-2)+(3)(6) \\ (5)(3)+(3)(-1) & (5)(-2)+(3)(6) \end{bmatrix} = \begin{bmatrix} 3-3 & -2+18 \\ 15-3 & -10+18 \end{bmatrix} = \begin{bmatrix} 0 & 16 \\ 12 & 8 \end{bmatrix}$

b. $\quad BA = \begin{bmatrix} 3 & -2 \\ -1 & 6 \end{bmatrix}\begin{bmatrix} 1 & 3 \\ 5 & 3 \end{bmatrix} = \begin{bmatrix} (3)(1)+(-2)(5) & (3)(3)+(-2)(3) \\ (-1)(1)+(6)(5) & (-1)(3)+(6)(3) \end{bmatrix} = \begin{bmatrix} 3-10 & 9-6 \\ -1+30 & -3+18 \end{bmatrix} = \begin{bmatrix} -7 & 3 \\ 29 & 15 \end{bmatrix}$

29. a. $\quad AB = \begin{bmatrix} 1 & 2 & 3 & 4 \end{bmatrix} \begin{bmatrix} 1 \\ 2 \\ 3 \\ 4 \end{bmatrix} = [(1)(1) + (2)(2) + (3)(3) + (4)(4)] = [1 + 4 + 9 + 16] = [30]$

b. $\quad BA = \begin{bmatrix} 1 \\ 2 \\ 3 \\ 4 \end{bmatrix} \begin{bmatrix} 1 & 2 & 3 & 4 \end{bmatrix} = \begin{bmatrix} (1)(1) & (1)(2) & (1)(3) & (1)(4) \\ (2)(1) & (2)(2) & (2)(3) & (2)(4) \\ (3)(1) & (3)(2) & (3)(3) & (3)(4) \\ (4)(1) & (4)(2) & (4)(3) & (4)(4) \end{bmatrix} = \begin{bmatrix} 1 & 2 & 3 & 4 \\ 2 & 4 & 6 & 8 \\ 3 & 6 & 9 & 12 \\ 4 & 8 & 12 & 16 \end{bmatrix}$

31. a. $\quad AB = \begin{bmatrix} 1 & -1 & 4 \\ 4 & -1 & 3 \\ 2 & 0 & -2 \end{bmatrix} \begin{bmatrix} 1 & 1 & 0 \\ 1 & 2 & 4 \\ 1 & -1 & 3 \end{bmatrix}$

$= \begin{bmatrix} (1)(1) + (-1)(1) + (4)(1) & (1)(1) + (-1)(2) + (4)(-1) & (1)(0) + (-1)(4) + (4)(3) \\ (4)(1) + (-1)(1) + (3)(1) & (4)(1) + (-1)(2) + (3)(-1) & (4)(0) + (-1)(4) + (3)(3) \\ (2)(1) + (0)(1) + (-2)(1) & (2)(1) + (0)(2) + (-2)(-1) & (2)(0) + (0)(4) + (-2)(3) \end{bmatrix}$

$= \begin{bmatrix} 1-1+4 & 1-2-4 & 0-4+12 \\ 4-1+3 & 4-2-3 & 0-4+9 \\ 2+0-2 & 2+0+2 & 0+0-6 \end{bmatrix} = \begin{bmatrix} 4 & -5 & 8 \\ 6 & -1 & 5 \\ 0 & 4 & -6 \end{bmatrix}$

b. $\quad BA = \begin{bmatrix} 1 & 1 & 0 \\ 1 & 2 & 4 \\ 1 & -1 & 3 \end{bmatrix} \begin{bmatrix} 1 & -1 & 4 \\ 4 & -1 & 3 \\ 2 & 0 & -2 \end{bmatrix}$

$= \begin{bmatrix} (1)(1) + (1)(4) + (0)(2) & (1)(-1) + (1)(-1) + (0)(0) & (1)(4) + (1)(3) + (0)(-2) \\ (1)(1) + (2)(4) + (4)(2) & (1)(-1) + (2)(-1) + (4)(0) & (1)(4) + (2)(3) + (4)(-2) \\ (1)(1) + (-1)(4) + (3)(2) & (1)(-1) + (-1)(-1) + (3)(0) & (1)(4) + (-1)(3) + (3)(-2) \end{bmatrix}$

$= \begin{bmatrix} 1+4+0 & -1-1+0 & 4+3+0 \\ 1+8+8 & -1-2+0 & 4+6-8 \\ 1-4+6 & -1+1+0 & 4-3-6 \end{bmatrix} = \begin{bmatrix} 5 & -2 & 7 \\ 17 & -3 & 2 \\ 3 & 0 & -5 \end{bmatrix}$

33. a. $\quad AB = \begin{bmatrix} 4 & 2 \\ 6 & 1 \\ 3 & 5 \end{bmatrix} \begin{bmatrix} 2 & 3 & 4 \\ -1 & -2 & 0 \end{bmatrix} = \begin{bmatrix} (4)(2) + (2)(-1) & (4)(3) + (2)(-2) & (4)(4) + (2)(0) \\ (6)(2) + (1)(-1) & (6)(3) + (1)(-2) & (6)(4) + (1)(0) \\ (3)(2) + (5)(-1) & (3)(3) + (5)(-2) & (3)(4) + (5)(0) \end{bmatrix}$

$= \begin{bmatrix} 8-2 & 12-4 & 16+0 \\ 12-1 & 18-2 & 24+0 \\ 6-5 & 9-10 & 12+0 \end{bmatrix} = \begin{bmatrix} 6 & 8 & 16 \\ 11 & 16 & 24 \\ 1 & -1 & 12 \end{bmatrix}$

b. $\quad BA = \begin{bmatrix} 2 & 3 & 4 \\ -1 & -2 & 0 \end{bmatrix} \begin{bmatrix} 4 & 2 \\ 6 & 1 \\ 3 & 5 \end{bmatrix} = \begin{bmatrix} (2)(4) + (3)(6) + (4)(3) & (2)(2) + (3)(1) + (4)(5) \\ (-1)(4) + (-2)(6) + (0)(3) & (-1)(2) + (-2)(1) + (0)(5) \end{bmatrix}$

$= \begin{bmatrix} 8+18+12 & 4+3+20 \\ -4-12+0 & -2-2+0 \end{bmatrix} = \begin{bmatrix} 38 & 27 \\ -16 & -4 \end{bmatrix}$

35. a. $AB = \begin{bmatrix} 2 & -3 & 1 & -1 \\ 1 & 1 & -2 & 1 \end{bmatrix} \begin{bmatrix} 1 & 2 \\ -1 & 1 \\ 5 & 4 \\ 10 & 5 \end{bmatrix} = \begin{bmatrix} (2)(1)+(-3)(-1)+(1)(5)+(-1)(10) & (2)(2)+(-3)(1)+(1)(4)+(-1)(5) \\ (1)(1)+(1)(-1)+(-2)(5)+(1)(10) & (1)(2)+(1)(1)+(-2)(4)+(1)(5) \end{bmatrix}$

$= \begin{bmatrix} 2+3+5-10 & 4-3+4-5 \\ 1-1-10+10 & 2+1-8+5 \end{bmatrix} = \begin{bmatrix} 0 & 0 \\ 0 & 0 \end{bmatrix}$

37. $4B - 3C = \begin{bmatrix} 20 & 4 \\ -8 & -8 \end{bmatrix} - \begin{bmatrix} 3 & -3 \\ -3 & 3 \end{bmatrix} = \begin{bmatrix} 20-3 & 4-(-3) \\ -8-(-3) & -8-3 \end{bmatrix} = \begin{bmatrix} 17 & 7 \\ -5 & -11 \end{bmatrix}$

39. $BC + CB = \begin{bmatrix} 5-1 & -5+1 \\ -2+2 & 2-2 \end{bmatrix} + \begin{bmatrix} 5+2 & 1+2 \\ -5-2 & -1-2 \end{bmatrix} = \begin{bmatrix} 4 & -4 \\ 0 & 0 \end{bmatrix} + \begin{bmatrix} 7 & 3 \\ -7 & -3 \end{bmatrix} = \begin{bmatrix} 11 & -1 \\ -7 & -3 \end{bmatrix}$

41. $A - C$ is not defined because A is 3 x 2 and C is 2 x 2.

43. $A(BC) = \begin{bmatrix} 4 & 0 \\ -3 & 5 \\ 0 & 1 \end{bmatrix} \begin{bmatrix} 5-1 & -5+1 \\ -2+2 & 2-2 \end{bmatrix} = \begin{bmatrix} 4 & 0 \\ -3 & 5 \\ 0 & 1 \end{bmatrix} \begin{bmatrix} 4 & -4 \\ 0 & 0 \end{bmatrix} = \begin{bmatrix} 16+0 & -16+0 \\ -12+0 & 12+0 \\ 0+0 & 0+0 \end{bmatrix} = \begin{bmatrix} 16 & -16 \\ -12 & 12 \\ 0 & 0 \end{bmatrix}$

45. $(A+B)(C-D) = \left(\begin{bmatrix} 1 & 0 \\ 0 & 1 \end{bmatrix} + \begin{bmatrix} 1 & 0 \\ 0 & -1 \end{bmatrix} \right) \left(\begin{bmatrix} -1 & 0 \\ 0 & 1 \end{bmatrix} - \begin{bmatrix} -1 & 0 \\ 0 & -1 \end{bmatrix} \right) = \left(\begin{bmatrix} 2 & 0 \\ 0 & 0 \end{bmatrix} \right) \left(\begin{bmatrix} 0 & 0 \\ 0 & 2 \end{bmatrix} \right) = \begin{bmatrix} 0 & 0 \\ 0 & 0 \end{bmatrix}$

47. Answers will vary.

49. $BZ = \begin{bmatrix} 1 & 0 \\ 0 & -1 \end{bmatrix} \begin{bmatrix} x \\ y \end{bmatrix} = \begin{bmatrix} x \\ -y \end{bmatrix}$ This reflects the graphic about the *x*-axis because all *y*-coordinates are negated.

51. a. $\begin{bmatrix} 1 & 3 & 1 \\ 3 & 3 & 3 \\ 1 & 3 & 1 \end{bmatrix}$

b. $\begin{bmatrix} 1 & 3 & 1 \\ 3 & 3 & 3 \\ 1 & 3 & 1 \end{bmatrix} + \begin{bmatrix} -1 & -1 & -1 \\ -1 & -1 & -1 \\ -1 & -1 & -1 \end{bmatrix} = \begin{bmatrix} 0 & 2 & 0 \\ 2 & 2 & 2 \\ 0 & 2 & 0 \end{bmatrix}$

c. $\begin{bmatrix} 1 & 3 & 1 \\ 3 & 3 & 3 \\ 1 & 3 & 1 \end{bmatrix} + \begin{bmatrix} 1 & -2 & 1 \\ -2 & -2 & -2 \\ 1 & -2 & 1 \end{bmatrix} = \begin{bmatrix} 2 & 1 & 2 \\ 1 & 1 & 1 \\ 2 & 1 & 2 \end{bmatrix}$

53.
$$\begin{bmatrix} 0 & 3 & 3 & 1 & 1 & 0 \\ 0 & 0 & 1 & 1 & 5 & 5 \end{bmatrix} + \begin{bmatrix} -2 & -2 & -2 & -2 & -2 & -2 \\ -3 & -3 & -3 & -3 & -3 & -3 \end{bmatrix} = \begin{bmatrix} -2 & 1 & 1 & -1 & -1 & -2 \\ -3 & -3 & -2 & -2 & 2 & 2 \end{bmatrix}$$

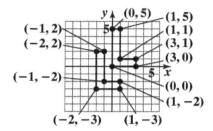

55. $0.5\begin{bmatrix} 0 & 3 & 3 & 1 & 1 & 0 \\ 0 & 0 & 1 & 1 & 5 & 5 \end{bmatrix} + \begin{bmatrix} 0 & 0 & 0 & 0 & 0 & 0 \\ 1 & 1 & 1 & 1 & 1 & 1 \end{bmatrix} = \begin{bmatrix} 0 & 1.5 & 1.5 & 0.5 & 0.5 & 0 \\ 0 & 0 & 0.5 & 0.5 & 2.5 & 2.5 \end{bmatrix} + \begin{bmatrix} 0 & 0 & 0 & 0 & 0 & 0 \\ 1 & 1 & 1 & 1 & 1 & 1 \end{bmatrix}$

$$= \begin{bmatrix} 0 & 1.5 & 1.5 & 0.5 & 0.5 & 0 \\ 1 & 1 & 1.5 & 1.5 & 3.5 & 3.5 \end{bmatrix}$$

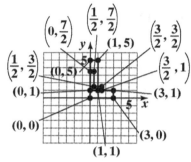

57. a. $AB = \begin{bmatrix} 1 & 0 \\ 0 & -1 \end{bmatrix} \cdot \begin{bmatrix} 0 & 3 & 3 & 1 & 1 & 0 \\ 0 & 0 & 1 & 1 & 5 & 5 \end{bmatrix} = \begin{bmatrix} 0 & 3 & 3 & 1 & 1 & 0 \\ 0 & 0 & -1 & -1 & -5 & -5 \end{bmatrix}$

b.

Rotated L about the *x*-axis.

59. a. $AB = \begin{bmatrix} 0 & -1 \\ 1 & 0 \end{bmatrix} \cdot \begin{bmatrix} 0 & 3 & 3 & 1 & 1 & 0 \\ 0 & 0 & 1 & 1 & 5 & 5 \end{bmatrix} = \begin{bmatrix} 0 & 0 & -1 & -1 & -5 & -5 \\ 0 & 3 & 3 & 1 & 1 & 0 \end{bmatrix}$

b.

Rotated L 90° counterclockwise about the origin.

61. a. $A = \begin{bmatrix} 2 & 6 \\ 31 & 46 \end{bmatrix}$

b. $B = \begin{bmatrix} 9 & 29 \\ 65 & 77 \end{bmatrix}$

c. $B - A = \begin{bmatrix} 9 & 29 \\ 65 & 77 \end{bmatrix} - \begin{bmatrix} 2 & 6 \\ 31 & 46 \end{bmatrix} = \begin{bmatrix} 7 & 23 \\ 34 & 31 \end{bmatrix}$

This matrix represents the percentages of people completing the transition to adulthood in 1960 and 2000 by age and gender.

63. a. System 1: The midterm and final both count for 50% of the course grade.
System 2: The midterm counts for 30% of the course grade and the final counts for 70%.

b. $AB = \begin{bmatrix} 84 & 87.2 \\ 79 & 81 \\ 90 & 88.4 \\ 73 & 68.6 \\ 69 & 73.4 \end{bmatrix}$

System 1 grades are listed first (if different). Student 1: B; Student 2: C or B; Student 3: A or B; Student 4: C or D; Student 5: D or C

65. – 75. Answers will vary.

77. makes sense

79. makes sense

81. Answers will vary.

83. $AB = \begin{bmatrix} 0 & -1 \\ 1 & 0 \end{bmatrix} \begin{bmatrix} 1 & 0 \\ 0 & -1 \end{bmatrix} = \begin{bmatrix} 0 & 1 \\ 1 & 0 \end{bmatrix}$

$-BA = \begin{bmatrix} 1 & 0 \\ 0 & -1 \end{bmatrix} \begin{bmatrix} 0 & -1 \\ 1 & 0 \end{bmatrix} = -\begin{bmatrix} 0 & -1 \\ -1 & 0 \end{bmatrix} = \begin{bmatrix} 0 & 1 \\ 1 & 0 \end{bmatrix}$

$AB = -BA$ so they are anticommutative.

85. $AB = \begin{bmatrix} a_{11} & a_{12} \\ a_{21} & a_{22} \end{bmatrix} \begin{bmatrix} 1 & 0 \\ 0 & 1 \end{bmatrix} = \begin{bmatrix} a_{11} & a_{12} \\ a_{21} & a_{22} \end{bmatrix}$

Nothing happens to the elements in the first matrix.

86. $\left[\begin{array}{ccc|c} -1 & -1 & -1 & 1 \\ 4 & 5 & 0 & 0 \\ 0 & 1 & -3 & 0 \end{array} \right] 4R_1 + R_2$

$\left[\begin{array}{ccc|c} -1 & -1 & -1 & 1 \\ 0 & 1 & -4 & 4 \\ 0 & 1 & -3 & 0 \end{array} \right] \begin{array}{l} R_2 + R_1 \\ \\ -R_2 + R_3 \end{array}$

$\left[\begin{array}{ccc|c} -1 & 0 & -5 & 5 \\ 0 & 1 & -4 & 4 \\ 0 & 0 & 1 & -4 \end{array} \right] -R_1$

$\left[\begin{array}{ccc|c} 1 & 0 & 5 & -5 \\ 0 & 1 & -4 & 4 \\ 0 & 0 & 1 & -4 \end{array} \right] \begin{array}{l} -5R_3 + R_1 \\ 4R_3 + R_2 \end{array}$

$\left[\begin{array}{ccc|c} 1 & 0 & 0 & 15 \\ 0 & 1 & 0 & -12 \\ 0 & 0 & 1 & -4 \end{array} \right]$

The solution set is $\{(15, -12, -4)\}$.

87. $\begin{bmatrix} a_1 & b_1 & c_1 \\ a_2 & b_2 & c_2 \\ a_3 & b_3 & c_3 \end{bmatrix} \begin{bmatrix} x \\ y \\ x \end{bmatrix} = \begin{bmatrix} d_1 \\ d_2 \\ d_3 \end{bmatrix}$

$\begin{bmatrix} a_1x + b_1y + c_1x \\ a_2x + b_2y + c_2z \\ a_3x + b_3y + c_3z \end{bmatrix} = \begin{bmatrix} d_1 \\ d_2 \\ d_3 \end{bmatrix}$

The linear system is written as follows.
$a_1x + b_1y + c_1x = d_1$
$a_2x + b_2y + c_2z = d_2$
$a_3x + b_3y + c_3z = d_3$

Mid-Chapter 9 Check Point

1.
$$\left[\begin{array}{ccc|c} 1 & 2 & -3 & -7 \\ 3 & -1 & 2 & 8 \\ 2 & -1 & 1 & 5 \end{array}\right]\begin{array}{l} \\ -3R_1 + R_2 \\ -2R_1 + R_3 \end{array}$$

$$\left[\begin{array}{ccc|c} 1 & 2 & -3 & -7 \\ 0 & -7 & 11 & 29 \\ 0 & -5 & 7 & 19 \end{array}\right]\begin{array}{l} \\ -\frac{1}{7}R_2 \\ -\frac{1}{5}R_3 \end{array}$$

$$\left[\begin{array}{ccc|c} 1 & 2 & -3 & -7 \\ 0 & 1 & -\frac{11}{7} & -\frac{29}{7} \\ 0 & 1 & -\frac{7}{5} & -\frac{19}{5} \end{array}\right]\begin{array}{l} \\ \\ -R_2 + R_3 \end{array}$$

$$\left[\begin{array}{ccc|c} 1 & 2 & -3 & -7 \\ 0 & 1 & -\frac{11}{7} & -\frac{29}{7} \\ 0 & 0 & \frac{6}{35} & \frac{12}{35} \end{array}\right]\begin{array}{l} \\ \\ \frac{35}{6}R_3 \end{array}$$

$$\left[\begin{array}{ccc|c} 1 & 2 & -3 & -7 \\ 0 & 1 & -\frac{11}{7} & -\frac{29}{7} \\ 0 & 0 & 1 & 2 \end{array}\right]$$

Back-substitute to find y.
$$y - \frac{11}{7}z = -\frac{29}{7}$$
$$y - \frac{11}{7}(2) = -\frac{29}{7}$$
$$y - \frac{22}{7} = -\frac{29}{7}$$
$$y = -\frac{7}{7}$$
$$y = -1$$

Back-substitute to find x.
$$x + 2y - 3z = -7$$
$$x + 2(-1) - 3(2) = -7$$
$$x - 2 - 6 = -7$$
$$x - 8 = -7$$
$$x = 1$$

The solution is $\{(1, -1, 2)\}$.

2.
$$\left[\begin{array}{ccc|c} 2 & 4 & 5 & 2 \\ 1 & 1 & 2 & 1 \\ 3 & 5 & 7 & 4 \end{array}\right]R_1 \leftrightarrow R_3$$

$$\left[\begin{array}{ccc|c} 1 & 1 & 2 & 1 \\ 2 & 4 & 5 & 2 \\ 3 & 5 & 7 & 4 \end{array}\right]\begin{array}{l} \\ -2R_1 + R_2 \\ -3R_1 + R_3 \end{array}$$

$$\left[\begin{array}{ccc|c} 1 & 1 & 2 & 1 \\ 0 & 2 & 1 & 0 \\ 0 & 2 & 1 & 1 \end{array}\right]-R_2 + R_3$$

$$\left[\begin{array}{ccc|c} 1 & 1 & 2 & 1 \\ 0 & 2 & 1 & 0 \\ 0 & 0 & 0 & 1 \end{array}\right]$$

The third row of the matrix is equivalent to
$0x + 0y + 0z = 1$ which is false.

The solution is $\varnothing$.

3.
$$\left[\begin{array}{ccc|c} 1 & -2 & 2 & -2 \\ 2 & 3 & -1 & 1 \end{array}\right]-2R_1 + R_2$$

$$\left[\begin{array}{ccc|c} 1 & -2 & 2 & -2 \\ 0 & 7 & -5 & 5 \end{array}\right]\frac{1}{7}R_2$$

$$\left[\begin{array}{ccc|c} 1 & -2 & 2 & -2 \\ 0 & 1 & -\frac{5}{7} & \frac{5}{7} \end{array}\right]$$

Back-substitute to find y in terms of z.
$$y - \frac{5}{7}z = \frac{5}{7}$$
$$y = \frac{5}{7}z + \frac{5}{7}$$

Back-substitute to find x in terms of z.
$$x - 2y + 2z = -2$$
$$x - 2\left(\frac{5}{7}z + \frac{5}{7}\right) + 2z = -2$$
$$x - \frac{10}{7}z - \frac{10}{7} + 2z = -2$$
$$x + \frac{4}{7}z - \frac{10}{7} = -2$$
$$x = -\frac{4}{7}z - \frac{4}{7}$$

The solution is $\left\{\left(-\frac{4}{7}z - \frac{4}{7}, \frac{5}{7}z + \frac{5}{7}, z\right)\right\}$.

4.
$$\left[\begin{array}{cccc|c} 1 & 1 & 1 & 1 & 6 \\ 1 & -1 & 3 & 1 & -14 \\ 1 & 2 & 0 & -3 & 12 \\ 2 & 3 & 6 & 1 & 1 \end{array}\right]\begin{array}{l} \\ -R_1 + R_2 \\ -R_1 + R_3 \\ -2R_1 + R_4 \end{array}$$

$$\left[\begin{array}{cccc|c} 1 & 1 & 1 & 1 & 6 \\ 0 & -2 & 2 & 0 & -20 \\ 0 & 1 & -1 & -4 & 6 \\ 0 & 1 & 4 & -1 & -11 \end{array}\right]R_2 \leftrightarrow R_3$$

$$\left[\begin{array}{cccc|c} 1 & 1 & 1 & 1 & 6 \\ 0 & 1 & -1 & -4 & 6 \\ 0 & -2 & 2 & 0 & -20 \\ 0 & 1 & 4 & -1 & -11 \end{array}\right]\begin{array}{l} \\ \\ 2R_2 + R_3 \\ -R_2 + R_4 \end{array}$$

$$\begin{bmatrix} 1 & 1 & 1 & 1 & | & 6 \\ 0 & 1 & -1 & -4 & | & 6 \\ 0 & 0 & 0 & -8 & | & -8 \\ 0 & 0 & 5 & 3 & | & -17 \end{bmatrix} R_3 \leftrightarrow R_4$$

$$\begin{bmatrix} 1 & 1 & 1 & 1 & | & 6 \\ 0 & 1 & -1 & -4 & | & 6 \\ 0 & 0 & 5 & 3 & | & -17 \\ 0 & 0 & 0 & -8 & | & -8 \end{bmatrix} \begin{matrix} \\ \\ \frac{1}{5}R_3 \\ -\frac{1}{8}R_4 \end{matrix}$$

$$\begin{bmatrix} 1 & 1 & 1 & 1 & | & 6 \\ 0 & 1 & -1 & -4 & | & 6 \\ 0 & 0 & 1 & \frac{3}{5} & | & -\frac{17}{5} \\ 0 & 0 & 0 & 1 & | & 1 \end{bmatrix}$$

Back-substitute to find y in terms of z.

$$y + \frac{3}{5}z = -\frac{17}{5}$$

$$y + \frac{3}{5}(1) = -\frac{17}{5}$$

$$y + \frac{3}{5} = -\frac{17}{5}$$

$$y = -\frac{20}{5}$$

$$y = -4$$

Back-substitute to find x in terms of z.

$$x - y - 4z = 6$$

$$x - (-4) - 4(1) = 6$$

$$x + 4 - 4 = 6$$

$$x = 6$$

Back-substitute to find w in terms of z.

$$w + x + y + z = 6$$

$$w + (6) + (-4) + (1) = 6$$

$$w + 3 = 6$$

$$w = 3$$

The solution is $\{(3, 6, -4, 1)\}$.

5.
$$\begin{bmatrix} 2 & -2 & 2 & | & 5 \\ 1 & -1 & 1 & | & 2 \\ 2 & 1 & -1 & | & 1 \end{bmatrix} \begin{matrix} -R_2 + R_1 \\ \\ -R_1 + R_3 \end{matrix}$$

$$\begin{bmatrix} 1 & -1 & 1 & | & 3 \\ 1 & -1 & 1 & | & 2 \\ 0 & 3 & -3 & | & -4 \end{bmatrix} -R_1 + R_2$$

$$\begin{bmatrix} 1 & -1 & 1 & | & 3 \\ 0 & 0 & 0 & | & -1 \\ 0 & 3 & -3 & | & -4 \end{bmatrix}$$

The second row of the matrix is equivalent to

$0x + 0y + 0z = -1$ which is false.

The solution is $\varnothing$.

6. $2C - \frac{1}{2}B$

$$= 2\begin{bmatrix} -1 & 0 \\ 0 & 1 \end{bmatrix} - \frac{1}{2}\begin{bmatrix} 4 & 1 \\ -6 & -2 \end{bmatrix}$$

$$= \begin{bmatrix} -2 & 0 \\ 0 & 2 \end{bmatrix} - \begin{bmatrix} 2 & \frac{1}{2} \\ -3 & -1 \end{bmatrix}$$

$$= \begin{bmatrix} -4 & -\frac{1}{2} \\ 3 & 3 \end{bmatrix}$$

7. $A(B + C)$

$$= \begin{bmatrix} 0 & 2 \\ -1 & 3 \\ 1 & 0 \end{bmatrix}\left(\begin{bmatrix} 4 & 1 \\ -6 & -2 \end{bmatrix} + \begin{bmatrix} -1 & 0 \\ 0 & 1 \end{bmatrix}\right)$$

$$= \begin{bmatrix} 0 & 2 \\ -1 & 3 \\ 1 & 0 \end{bmatrix}\left(\begin{bmatrix} 3 & 1 \\ -6 & -1 \end{bmatrix}\right)$$

$$= \begin{bmatrix} -12 & -2 \\ -21 & -4 \\ 3 & 1 \end{bmatrix}$$

8. $A(BC)$

$$= \begin{bmatrix} 0 & 2 \\ -1 & 3 \\ 1 & 0 \end{bmatrix}\left(\begin{bmatrix} 4 & 1 \\ -6 & -2 \end{bmatrix} \cdot \begin{bmatrix} -1 & 0 \\ 0 & 1 \end{bmatrix}\right)$$

$$= \begin{bmatrix} 0 & 2 \\ -1 & 3 \\ 1 & 0 \end{bmatrix}\left(\begin{bmatrix} -4 & 1 \\ 6 & -2 \end{bmatrix}\right)$$

$$= \begin{bmatrix} 12 & -4 \\ 22 & -7 \\ -4 & 1 \end{bmatrix}$$

9. The operation is not defined. Matrices must have the same dimensions in order to be added.

10. $2X - 3C = B$

$\qquad 2X = B + 3C$

$\qquad X = \tfrac{1}{2}(B + 3C)$

$\qquad = \tfrac{1}{2}\left(\begin{bmatrix} 4 & 1 \\ -6 & -2 \end{bmatrix} + 3 \begin{bmatrix} -1 & 0 \\ 0 & 1 \end{bmatrix} \right)$

$\qquad = \tfrac{1}{2}\left(\begin{bmatrix} 4 & 1 \\ -6 & -2 \end{bmatrix} + \begin{bmatrix} -3 & 0 \\ 0 & 3 \end{bmatrix} \right)$

$\qquad = \tfrac{1}{2}\left(\begin{bmatrix} 1 & 1 \\ -6 & 1 \end{bmatrix} \right)$

$\qquad = \begin{bmatrix} \tfrac{1}{2} & \tfrac{1}{2} \\ -3 & \tfrac{1}{2} \end{bmatrix}$

Section 9.4

Check Point Exercises

1. We must show that: $AB = I_2 = \begin{bmatrix} 1 & 0 \\ 0 & 1 \end{bmatrix}$, and

$BA = I_2 = \begin{bmatrix} 1 & 0 \\ 0 & 1 \end{bmatrix}.$

$AB = \begin{bmatrix} 2 & 1 \\ 1 & 1 \end{bmatrix}\begin{bmatrix} 1 & -1 \\ -1 & 2 \end{bmatrix}$

$\quad = \begin{bmatrix} 2(1)+1(-1) & 2(-1)+1(2) \\ 1(1)+1(-1) & 1(-1)+1(2) \end{bmatrix}$

$\quad = \begin{bmatrix} 1 & 0 \\ 0 & 1 \end{bmatrix}$

$BA = \begin{bmatrix} 1 & -1 \\ -1 & 2 \end{bmatrix}\begin{bmatrix} 2 & 1 \\ 1 & 1 \end{bmatrix}$

$\quad = \begin{bmatrix} 1(2)+-1(1) & 1(1)+-1(1) \\ -1(2)+2(1) & -1(1)+2(1) \end{bmatrix}$

$\quad = \begin{bmatrix} 1 & 0 \\ 0 & 1 \end{bmatrix}$

Both products (AB and BA) give the multiplicative identity matrix, I_2. Thus, B is the multiplicative inverse of A.

2. Let us denote the multiplicative inverse of A by $A^{-1} = \begin{bmatrix} w & x \\ y & z \end{bmatrix}$. Because A is a 2×2 matrix, we use the equation $AA^{-1} = I_2$ to find values for w, x, y and z.

$\begin{bmatrix} 5 & 7 \\ 2 & 3 \end{bmatrix}\begin{bmatrix} w & x \\ y & z \end{bmatrix} = \begin{bmatrix} 1 & 0 \\ 0 & 1 \end{bmatrix}$

$\begin{bmatrix} 5w+7y & 5x+7z \\ 2w+3y & 2x+3z \end{bmatrix} = \begin{bmatrix} 1 & 0 \\ 0 & 1 \end{bmatrix}$

$5w + 7y = 1 \qquad 5x + 7z = 0$

$2w + 3y = 0 \qquad 2x + 3z = 1$

Each of these systems can be solved using the addition method.

Multiply by –2: $5w + 7y = 1 \rightarrow -10w - 14y = -2$

Multiply by 5: $2w + 3y = 0 \rightarrow \quad 10w + 15y = 0$

Use back substitution: $w = 3, y = -2$

Multiply by –2: $5x + 7z = 0 \rightarrow -10x - 14z = 0$

Multiply by 5: $2x + 3z = 1 \rightarrow 10x + 15z = 5$

Use back substitution: $x = -7, z = 5$

Using these values, we have

$A^{-1} = \begin{bmatrix} w & x \\ y & z \end{bmatrix} = \begin{bmatrix} 3 & -7 \\ -2 & 5 \end{bmatrix}.$

3. $A^{-1} = \dfrac{1}{ad-bc}\begin{bmatrix} d & -b \\ -c & a \end{bmatrix}$

$\quad = \dfrac{1}{3(1)-(-2)(-1)}\begin{bmatrix} 1 & -(-2) \\ -(-1) & 3 \end{bmatrix}$

$\quad = \dfrac{1}{3-2}\begin{bmatrix} 1 & 2 \\ 1 & 3 \end{bmatrix}$

$\quad = \dfrac{1}{1}\begin{bmatrix} 1 & 2 \\ 1 & 3 \end{bmatrix}$

$\quad = \begin{bmatrix} 1 & 2 \\ 1 & 3 \end{bmatrix}$

4. The augmented matrix $\begin{bmatrix} A & | & I_3 \end{bmatrix}$ is

$\begin{bmatrix} 1 & 0 & 2 & | & 1 & 0 & 0 \\ -1 & 2 & 3 & | & 0 & 1 & 0 \\ 1 & -1 & 0 & | & 0 & 0 & 1 \end{bmatrix}.$

Perform row transformations on $\begin{bmatrix} A & | & I_3 \end{bmatrix}$ to obtain a matrix of the form $\begin{bmatrix} I_3 & | & B \end{bmatrix}$.

$$\begin{bmatrix} 1 & 0 & 2 & | & 1 & 0 & 0 \\ -1 & 2 & 3 & | & 0 & 1 & 0 \\ 1 & -1 & 0 & | & 0 & 0 & 1 \end{bmatrix} 1R_1 / R_2$$

$$= \begin{bmatrix} 1 & 0 & 2 & | & 1 & 0 & 0 \\ 0 & 2 & 5 & | & 1 & 1 & 0 \\ 1 & -1 & 0 & | & 0 & 0 & 1 \end{bmatrix} -1R_3$$

$$= \begin{bmatrix} 1 & 0 & 2 & | & 1 & 0 & 0 \\ 0 & 2 & 5 & | & 1 & 1 & 0 \\ -1 & 1 & 0 & | & 0 & 0 & -1 \end{bmatrix} R_1 + R_3$$

$$= \begin{bmatrix} 1 & 0 & 2 & | & 1 & 0 & 0 \\ 0 & 2 & 5 & | & 1 & 1 & 0 \\ 0 & 1 & 2 & | & 1 & 0 & -1 \end{bmatrix} \tfrac{1}{2}R_2$$

$$= \begin{bmatrix} 1 & 0 & 2 & | & 1 & 0 & 0 \\ 0 & 1 & \tfrac{5}{2} & | & \tfrac{1}{2} & \tfrac{1}{2} & 0 \\ 0 & 1 & 2 & | & 1 & 0 & -1 \end{bmatrix} -1R_2 + R_3$$

$$= \begin{bmatrix} 1 & 0 & 2 & | & 1 & 0 & 0 \\ 0 & 1 & \tfrac{5}{2} & | & \tfrac{1}{2} & \tfrac{1}{2} & 0 \\ 0 & 0 & -\tfrac{1}{2} & | & \tfrac{1}{2} & -\tfrac{1}{2} & -1 \end{bmatrix} -2R_3$$

$$= \begin{bmatrix} 1 & 0 & 2 & | & 1 & 0 & 0 \\ 0 & 1 & \tfrac{5}{2} & | & \tfrac{1}{2} & \tfrac{1}{2} & 0 \\ 0 & 0 & 1 & | & -1 & 1 & 2 \end{bmatrix} \begin{matrix} -2R_3 + R_1 \\ -\tfrac{5}{2}R_3 + R_2 \end{matrix}$$

$$= \begin{bmatrix} 1 & 0 & 0 & | & 3 & -2 & -4 \\ 0 & 1 & 0 & | & 3 & -2 & -5 \\ 0 & 0 & 1 & | & 1 & 1 & 2 \end{bmatrix}$$

Thus, the multiplicative inverse of *A* is

$$A^{-1} = \begin{bmatrix} 3 & -2 & -4 \\ 3 & -2 & -5 \\ -1 & 1 & 2 \end{bmatrix}.$$

5. The linear system can be written as $AX = B$.

$$\begin{bmatrix} 1 & 0 & 2 \\ -1 & 2 & 3 \\ 1 & -1 & 0 \end{bmatrix} \begin{bmatrix} x \\ y \\ z \end{bmatrix} = \begin{bmatrix} 6 \\ -5 \\ 6 \end{bmatrix}.$$

$$X = A^{-1}B = \begin{bmatrix} 3 & -2 & -4 \\ 3 & -2 & -5 \\ -1 & 1 & 2 \end{bmatrix} \begin{bmatrix} 6 \\ -5 \\ 6 \end{bmatrix}$$

$$= \begin{bmatrix} 3(6) + -2(-5) + -4(6) \\ 3(6) + -2(-5) + -5(6) \\ -1(6) + 1(-5) + 2(6) \end{bmatrix}$$

$$= \begin{bmatrix} 18+10-24 \\ 18+10-30 \\ -6-5+12 \end{bmatrix} = \begin{bmatrix} 4 \\ -2 \\ 1 \end{bmatrix}$$

Thus, $x = 4$, $y = -2$, and $z = 1$. The solution set is $\{(4, -2, 1)\}$.

6. The numerical representation of the word BASE is 2, 1, 19, 5. The 2×2 matrix formed is $\begin{bmatrix} 2 & 19 \\ 1 & 5 \end{bmatrix}$.

$$\begin{bmatrix} -2 & -3 \\ 3 & 4 \end{bmatrix} \begin{bmatrix} 2 & 19 \\ 1 & 5 \end{bmatrix}$$

$$= \begin{bmatrix} -2(2) + -3(1) & -2(19) + -3(5) \\ 3(2) + 4(1) & 3(19) + 4(5) \end{bmatrix}$$

$$= \begin{bmatrix} -4-3 & -38-15 \\ 6+4 & 57+20 \end{bmatrix} = \begin{bmatrix} -7 & -53 \\ 10 & 77 \end{bmatrix}$$

The encoded message is –7, 10, –53, 77.

7. Use the multiplicative inverse of the coding matrix. It is $\begin{bmatrix} 4 & 3 \\ -3 & -2 \end{bmatrix}$.

$$\begin{bmatrix} 4 & 3 \\ -3 & -2 \end{bmatrix} \begin{bmatrix} -7 & -53 \\ 10 & 77 \end{bmatrix}$$

$$= \begin{bmatrix} 4(-7) + 3(10) & 4(-53) + 3(77) \\ -3(-7) + -2(10) & -3(-53) + -2(77) \end{bmatrix}$$

$$= \begin{bmatrix} -28+30 & -212+231 \\ 21-20 & 159-154 \end{bmatrix} = \begin{bmatrix} 2 & 19 \\ 1 & 5 \end{bmatrix}$$

The numbers are 2, 1, 19, and 5. Using letters, the decoded message is BASE.

Concept and Vocabulary Check 9.4

1. $\begin{bmatrix} 1 & 0 \\ 0 & 1 \end{bmatrix}$

2. $\begin{bmatrix} 1 & 0 & 0 \\ 0 & 1 & 0 \\ 0 & 0 & 1 \end{bmatrix}$

3. multiplicative inverse

4. true

5. $ad - bc \neq 0$

6. singular

7. false

9. A^{-1}

10. $A^{-1}B$

Exercise Set 9.4

1. $A = \begin{bmatrix} 4 & -3 \\ -5 & 4 \end{bmatrix} \quad B = \begin{bmatrix} 4 & 3 \\ 5 & 4 \end{bmatrix}$

   $AB = \begin{bmatrix} 16-15 & 12-12 \\ -20+20 & -15+16 \end{bmatrix} = \begin{bmatrix} 1 & 0 \\ 0 & 1 \end{bmatrix}$

   $BA = \begin{bmatrix} 16-15 & -12+12 \\ 20-20 & -15+16 \end{bmatrix} = \begin{bmatrix} 1 & 0 \\ 0 & 1 \end{bmatrix}$

   Since $AB = I_2 \; BA = I_2, \; B = A^{-1}$.

3. $AB = \begin{bmatrix} 8+0 & -16+0 \\ -2+0 & 4+3 \end{bmatrix} = \begin{bmatrix} 8 & -16 \\ -2 & 7 \end{bmatrix}$

   $BA = \begin{bmatrix} 8+4 & 0+12 \\ 0+1 & 0+3 \end{bmatrix} = \begin{bmatrix} 12 & 12 \\ 1 & 3 \end{bmatrix}$

   If B is the multiplicative inverse of A, both products (AB and BA) will be the multiplicative identity matrix, I_2. Therefore, B is not the multiplicative inverse of A. That is, $B \neq A^{-1}$.

5. $AB = \begin{bmatrix} -2+3 & -4+4 \\ \frac{3}{2}-\frac{3}{2} & 3-2 \end{bmatrix} = \begin{bmatrix} 1 & 0 \\ 0 & 1 \end{bmatrix}$

   $BA = \begin{bmatrix} -2+3 & 1-1 \\ -6+6 & 3-2 \end{bmatrix} = \begin{bmatrix} 1 & 0 \\ 0 & 1 \end{bmatrix}$

   Since $AB = I_2$ and $BA = I_2, \; B = A^{-1}$.

7. $A = \begin{bmatrix} 0 & 1 & 0 \\ 0 & 0 & 1 \\ 1 & 0 & 0 \end{bmatrix}$ $B = \begin{bmatrix} 0 & 0 & 1 \\ 1 & 0 & 0 \\ 0 & 1 & 0 \end{bmatrix}$

$AB = \begin{bmatrix} 0+1+0 & 0+0+0 & 0+0+0 \\ 0+0+0 & 0+0+1 & 0+0+0 \\ 0+0+0 & 0+0+0 & 1+0+0 \end{bmatrix} = \begin{bmatrix} 1 & 0 & 0 \\ 0 & 1 & 0 \\ 0 & 0 & 1 \end{bmatrix}$ $BA = \begin{bmatrix} 0+0+1 & 0+0+0 & 0+0+0 \\ 0+0+0 & 1+0+0 & 0+0+0 \\ 0+0+0 & 0+0+0 & 0+1+0 \end{bmatrix} = \begin{bmatrix} 1 & 0 & 0 \\ 0 & 1 & 0 \\ 0 & 0 & 1 \end{bmatrix}$

Since $AB = I_3$ and $BA = I_3$, $B = A^{-1}$.

9. $AB = \begin{bmatrix} \frac{7}{2}-1-\frac{3}{2} & -3+0+3 & \frac{1}{2}+1-\frac{3}{2} \\ \frac{7}{2}-\frac{3}{2}-2 & -3+0+4 & \frac{1}{2}+\frac{3}{2}-2 \\ \frac{7}{2}-2-\frac{3}{2} & -3+0+3 & \frac{1}{2}+2-\frac{3}{2} \end{bmatrix} = \begin{bmatrix} 1 & 0 & 0 \\ 0 & 1 & 0 \\ 0 & 0 & 1 \end{bmatrix}$

$BA = \begin{bmatrix} \frac{7}{2}-3+\frac{1}{2} & 7-9+2 & \frac{21}{2}-12+\frac{3}{2} \\ -\frac{1}{2}+0+\frac{1}{2} & -1+0+2 & -\frac{3}{2}+0+\frac{3}{2} \\ -\frac{1}{2}+1-\frac{1}{2} & -1+3-2 & -\frac{3}{2}+4-\frac{3}{2} \end{bmatrix} = \begin{bmatrix} 1 & 0 & 0 \\ 0 & 1 & 0 \\ 0 & 0 & 1 \end{bmatrix}$

Since $AB = I_3$ and $BA = I_3$, $B = A^{-1}$.

11. $AB = \begin{bmatrix} 0+0+0+1 & 0+0-2+2 & 0+0+0+0 & 0+0-2+2 \\ -1+0+0+1 & -2+0+1+2 & 0+0+0+0 & -3+0+1+2 \\ 0+0+0+0 & 0+1-1+0 & 0+1+0+0 & 0+1-1+0 \\ 1+0+0-1 & 2+0+0-2 & 0+0+0+0 & 3+0+0-2 \end{bmatrix} = \begin{bmatrix} 1 & 0 & 0 & 0 \\ 0 & 1 & 0 & 0 \\ 0 & 0 & 1 & 0 \\ 0 & 0 & 0 & 1 \end{bmatrix}$

$BA = \begin{bmatrix} 0-2+0+3 & 0+0+0+0 & -2+2+0+0 & 1+2+0-3 \\ 0-1+0+1 & 0+0+1+0 & 0+1-1+0 & 0+1+0-1 \\ 0-1+0+1 & 0+0+0+0 & 0+1+0+0 & 0+1+0-1 \\ 0-2+0+2 & 0+0+0+0 & -2+2+0+0 & 1+2+0-2 \end{bmatrix} = \begin{bmatrix} 1 & 0 & 0 & 0 \\ 0 & 1 & 0 & 0 \\ 0 & 0 & 1 & 0 \\ 0 & 0 & 0 & 1 \end{bmatrix}$

Since $AB = I_4$ and $BA = I_4$, $B = A^{-1}$.

13. $ad - bc = (2)(2) - (3)(-1) = 4 + 3 = 7$

$A^{-1} = \frac{1}{7}\begin{bmatrix} 2 & -3 \\ 1 & 2 \end{bmatrix} = \begin{bmatrix} \frac{2}{7} & -\frac{3}{7} \\ \frac{1}{7} & \frac{2}{7} \end{bmatrix}$

$AA^{-1} = \begin{bmatrix} \frac{4}{7}+\frac{3}{7} & -\frac{6}{7}+\frac{6}{7} \\ -\frac{2}{7}+\frac{2}{7} & \frac{3}{7}+\frac{4}{7} \end{bmatrix} = \begin{bmatrix} 1 & 0 \\ 0 & 1 \end{bmatrix}$ and $A^{-1}A = \begin{bmatrix} \frac{4}{7}+\frac{3}{7} & \frac{6}{7}-\frac{6}{7} \\ \frac{2}{7}-\frac{2}{7} & \frac{3}{7}+\frac{4}{7} \end{bmatrix} = \begin{bmatrix} 1 & 0 \\ 0 & 1 \end{bmatrix}$

15. $ad - bc = (3)(2) - (-1)(-4) = 6 - 4 = 2$

$A^{-1} = \frac{1}{2}\begin{bmatrix} 2 & 1 \\ 4 & 3 \end{bmatrix} = \begin{bmatrix} 1 & \frac{1}{2} \\ 2 & \frac{3}{2} \end{bmatrix}$

$AA^{-1} = \begin{bmatrix} 3-2 & \frac{3}{2}-\frac{3}{2} \\ -4+4 & -\frac{4}{2}+\frac{6}{2} \end{bmatrix} = \begin{bmatrix} 1 & 0 \\ 0 & 1 \end{bmatrix}$ and $A^{-1}A = \begin{bmatrix} 3-\frac{4}{2} & -1+\frac{2}{2} \\ 6-\frac{12}{2} & -2+\frac{6}{2} \end{bmatrix} = \begin{bmatrix} 1 & 0 \\ 0 & 1 \end{bmatrix}$

17. $ad - bc = (10)(1) - (-2)(-5) = 10 - 10 = 0$

Since division by zero is undefined, A does not have an inverse.

For Problems 19–24, verification that $AA^{-1} = I$ and $A^{-1}A = I$ is left to the student.

For Problems 19–23, verification that $AA^{-1} = I$ and $A^{-1}A = I$ is left to the student.

19. $\begin{bmatrix} 2 & 0 & 0 & 1 & 0 & 0 \\ 0 & 4 & 0 & 0 & 1 & 0 \\ 0 & 0 & 6 & 0 & 0 & 1 \end{bmatrix}$

Divide row 1 by 2, divide row 2 by 4 and divide row 4 by 6.

$\begin{bmatrix} 1 & 0 & 0 & \frac{1}{2} & 0 & 0 \\ 0 & 1 & 0 & 0 & \frac{1}{4} & 0 \\ 0 & 0 & 1 & 0 & 0 & \frac{1}{6} \end{bmatrix}$

$A^{-1} = \begin{bmatrix} \frac{1}{2} & 0 & 0 \\ 0 & \frac{1}{4} & 0 \\ 0 & 0 & \frac{1}{6} \end{bmatrix}$

21. $\begin{bmatrix} 1 & 2 & -1 & 1 & 0 & 0 \\ -2 & 0 & 1 & 0 & 1 & 0 \\ 1 & -1 & 0 & 0 & 0 & 1 \end{bmatrix}$

Replace row 2 with $2R_1 + R_2$.
Replace row 3 with $R_1 - R_3$.

$\begin{bmatrix} 1 & 2 & -1 & 1 & 0 & 0 \\ 0 & 4 & -1 & 2 & 1 & 0 \\ 0 & 3 & -1 & 1 & 0 & -1 \end{bmatrix}$

Replace row 1 with $R_2 - 2R_1$.
Replace row 3 with $-3R_2 + 4R_3$.

$\begin{bmatrix} -2 & 0 & 1 & 0 & 1 & 0 \\ 0 & 4 & -1 & 2 & 1 & 0 \\ 0 & 0 & -1 & -2 & -3 & -4 \end{bmatrix}$

Replace row 1 with $R_3 + R_1$.
Replace row 2 with $R_2 - R_3$.
Replace row 3 with $-R_3$.

$\begin{bmatrix} -2 & 0 & 0 & -2 & -2 & -4 \\ 0 & 4 & 0 & 4 & 4 & 4 \\ 0 & 0 & 1 & 2 & 3 & 4 \end{bmatrix}$

Divide row 1 by –2 and divide row 2 by 4.

$\begin{bmatrix} 1 & 0 & 0 & 1 & 1 & 2 \\ 0 & 1 & 0 & 1 & 1 & 1 \\ 0 & 0 & 1 & 2 & 3 & 4 \end{bmatrix}$

$A^{-1} = \begin{bmatrix} 1 & 1 & 2 \\ 1 & 1 & 1 \\ 2 & 3 & 4 \end{bmatrix}$

23. $\begin{bmatrix} 2 & 2 & -1 & 1 & 0 & 0 \\ 0 & 3 & -1 & 0 & 1 & 0 \\ -1 & -2 & 1 & 0 & 0 & 1 \end{bmatrix} R_1 \leftrightarrow R_3$

$\begin{bmatrix} -1 & -2 & 1 & 0 & 0 & 1 \\ 0 & 3 & -1 & 0 & 1 & 0 \\ 2 & 2 & -1 & 1 & 0 & 0 \end{bmatrix} -1R_1$

$\begin{bmatrix} 1 & 2 & -1 & 0 & 0 & -1 \\ 0 & 3 & -1 & 0 & 1 & 0 \\ 2 & 2 & -1 & 1 & 0 & 0 \end{bmatrix} -2R_1 + R_3$

$\begin{bmatrix} 1 & 2 & -1 & 0 & 0 & -1 \\ 0 & 3 & -1 & 0 & 1 & 0 \\ 0 & -2 & 1 & 1 & 0 & 2 \end{bmatrix} \frac{1}{3}R_2$

$\begin{bmatrix} 1 & 2 & -1 & 0 & 0 & -1 \\ 0 & 1 & -\frac{1}{3} & 0 & \frac{1}{3} & 0 \\ 0 & -2 & 1 & 1 & 0 & 2 \end{bmatrix} \begin{matrix} -2R_2 + R_1 \\ 2R_2 + R_3 \end{matrix}$

$\begin{bmatrix} 1 & 0 & -\frac{1}{3} & 0 & -\frac{2}{3} & -1 \\ 0 & 1 & -\frac{1}{3} & 0 & \frac{1}{3} & 0 \\ 0 & 0 & \frac{1}{3} & 1 & \frac{2}{3} & 2 \end{bmatrix} \begin{matrix} 1R_3 + R_1 \\ 1R_2 + R_1 \end{matrix}$

$\begin{bmatrix} 1 & 0 & 0 & 1 & 0 & 1 \\ 0 & 1 & 0 & 1 & 1 & 2 \\ 0 & 0 & \frac{1}{3} & 1 & \frac{2}{3} & 2 \end{bmatrix} 3R_3$

$\begin{bmatrix} 1 & 0 & 0 & 1 & 0 & 1 \\ 0 & 1 & 0 & 1 & 1 & 2 \\ 0 & 0 & 1 & 3 & 2 & 6 \end{bmatrix}$

$A^{-1} = \begin{bmatrix} 1 & 0 & 1 \\ 1 & 1 & 2 \\ 3 & 2 & 6 \end{bmatrix}$

25. $\begin{bmatrix} 5 & 0 & 2 & 1 & 0 & 0 \\ 2 & 2 & 1 & 0 & 1 & 0 \\ -3 & 1 & -1 & 0 & 0 & 1 \end{bmatrix} \frac{1}{5}R_1$

$\begin{bmatrix} 1 & 0 & \frac{2}{5} & \frac{1}{5} & 0 & 0 \\ 2 & 2 & 1 & 0 & 1 & 0 \\ -3 & 1 & -1 & 0 & 0 & 1 \end{bmatrix} \begin{matrix} -2R_1 + R_2 \\ 3R_1 + R_3 \end{matrix}$

$\begin{bmatrix} 1 & 0 & \frac{2}{5} & \frac{1}{5} & 0 & 0 \\ 0 & 2 & \frac{1}{5} & -\frac{2}{5} & 1 & 0 \\ 0 & 1 & \frac{1}{5} & \frac{3}{5} & 0 & 1 \end{bmatrix} R_2 \leftrightarrow R_3$

$$\begin{bmatrix} 1 & 0 & \frac{2}{5} & \frac{1}{5} & 0 & 0 \\ 0 & 1 & \frac{1}{5} & \frac{3}{5} & 0 & 1 \\ 0 & 2 & \frac{1}{5} & -\frac{2}{5} & 1 & 0 \end{bmatrix} -2R_2 + R_3$$

$$\begin{bmatrix} 1 & 0 & \frac{2}{5} & \frac{1}{5} & 0 & 0 \\ 0 & 1 & \frac{1}{5} & \frac{3}{5} & 0 & 1 \\ 0 & 0 & -\frac{1}{5} & -\frac{8}{5} & 1 & -2 \end{bmatrix} \begin{matrix} 2R_3 + R_1 \\ \\ 1R_3 + R_2 \end{matrix}$$

$$\begin{bmatrix} 1 & 0 & 0 & -3 & 2 & -4 \\ 0 & 1 & 0 & -1 & 1 & -1 \\ 0 & 0 & -\frac{1}{5} & -\frac{8}{5} & 1 & -2 \end{bmatrix} -5R_3$$

$$\begin{bmatrix} 1 & 0 & 0 & -3 & 2 & -4 \\ 0 & 1 & 0 & -1 & 1 & -1 \\ 0 & 0 & 1 & 8 & -5 & 10 \end{bmatrix}$$

$$A^{-1} = \begin{bmatrix} -3 & 2 & -4 \\ -1 & 1 & -1 \\ 8 & -5 & 10 \end{bmatrix}$$

27. $\begin{bmatrix} 1 & 0 & 0 & 0 & 1 & 0 & 0 & 0 \\ 0 & -1 & 0 & 0 & 0 & 1 & 0 & 0 \\ 0 & 0 & 3 & 0 & 0 & 0 & 1 & 0 \\ 1 & 0 & 0 & 1 & 0 & 0 & 0 & 1 \end{bmatrix} -1R_1 + R_4$

$$\begin{bmatrix} 1 & 0 & 0 & 0 & 1 & 0 & 0 & 0 \\ 0 & -1 & 0 & 0 & 0 & 1 & 0 & 0 \\ 0 & 0 & 3 & 0 & 0 & 0 & 1 & 0 \\ 0 & 0 & 0 & 1 & -1 & 0 & 0 & 1 \end{bmatrix} -1R_2$$

$$\begin{bmatrix} 1 & 0 & 0 & 0 & 1 & 0 & 0 & 0 \\ 0 & 1 & 0 & 0 & 0 & -1 & 0 & 0 \\ 0 & 0 & 3 & 0 & 0 & 0 & 1 & 0 \\ 0 & 0 & 0 & 1 & -1 & 0 & 0 & 1 \end{bmatrix} \frac{1}{3}R_3$$

$$\begin{bmatrix} 1 & 0 & 0 & 0 & 1 & 0 & 0 & 0 \\ 0 & 1 & 0 & 0 & 0 & -1 & 0 & 0 \\ 0 & 0 & 1 & 0 & 0 & 0 & \frac{1}{3} & 0 \\ 0 & 0 & 0 & 1 & -1 & 0 & 0 & 1 \end{bmatrix}$$

$$A^{-1} = \begin{bmatrix} 1 & 0 & 0 & 0 \\ 0 & -1 & 0 & 0 \\ 0 & 0 & \frac{1}{3} & 0 \\ -1 & 0 & 0 & 1 \end{bmatrix}$$

29. $\begin{bmatrix} 6 & 5 \\ 5 & 4 \end{bmatrix} \begin{bmatrix} x \\ y \end{bmatrix} = \begin{bmatrix} 13 \\ 10 \end{bmatrix}$

31. $\begin{bmatrix} 1 & 3 & 4 \\ 1 & 2 & 3 \\ 1 & 4 & 3 \end{bmatrix} \begin{bmatrix} x \\ y \\ z \end{bmatrix} = \begin{bmatrix} -3 \\ -2 \\ -6 \end{bmatrix}$

33. $4x - 7y = -3$
$2x - 3y = 1$

35. $2x - z = 6$
$3y = 9$
$x + y = 5$

37. a. $\begin{bmatrix} 2 & 6 & 6 \\ 2 & 7 & 6 \\ 2 & 7 & 7 \end{bmatrix} \begin{bmatrix} x \\ y \\ z \end{bmatrix} = \begin{bmatrix} 8 \\ 10 \\ 9 \end{bmatrix}$

$$\begin{bmatrix} \frac{7}{2} & 0 & -3 \\ -1 & 1 & 0 \\ 0 & -1 & 1 \end{bmatrix} \begin{bmatrix} 8 \\ 10 \\ 9 \end{bmatrix} = \begin{bmatrix} 28+0-27 \\ -8+10+0 \\ 0-10+9 \end{bmatrix} = \begin{bmatrix} 1 \\ 2 \\ -1 \end{bmatrix}$$

The solution to the system is $\{(1, 2, -1)\}$.

39. a. $\begin{bmatrix} 1 & -1 & 1 \\ 0 & 2 & -1 \\ 2 & 3 & 0 \end{bmatrix} \begin{bmatrix} x \\ y \\ z \end{bmatrix} = \begin{bmatrix} 8 \\ -7 \\ 1 \end{bmatrix}$

b. $\begin{bmatrix} 3 & 3 & -1 \\ -2 & -2 & 1 \\ -4 & -5 & 2 \end{bmatrix} \begin{bmatrix} 8 \\ -7 \\ 1 \end{bmatrix}$

$$= \begin{bmatrix} 24-21-1 \\ -16+14+1 \\ -32+35+2 \end{bmatrix} = \begin{bmatrix} 2 \\ -1 \\ 5 \end{bmatrix}$$

The solution to the system is
$\{(2, -1, 5)\}$.

41. a.
$$\begin{bmatrix} 1 & -1 & 2 & 0 \\ 0 & 1 & -1 & 1 \\ -1 & 1 & -1 & 2 \\ 0 & -1 & 1 & -2 \end{bmatrix} \begin{bmatrix} w \\ x \\ y \\ z \end{bmatrix} = \begin{bmatrix} -3 \\ 4 \\ 2 \\ -4 \end{bmatrix}$$

b.
$$\begin{bmatrix} 0 & 0 & -1 & -1 \\ 1 & 4 & 1 & 3 \\ 1 & 2 & 1 & 2 \\ 0 & -1 & 0 & -1 \end{bmatrix} \begin{bmatrix} -3 \\ 4 \\ 2 \\ -4 \end{bmatrix}$$

$$= \begin{bmatrix} 0+0-2+4 \\ -3+16+2-12 \\ -3+8+2-8 \\ 0-4+0+4 \end{bmatrix} = \begin{bmatrix} 2 \\ 3 \\ -1 \\ 0 \end{bmatrix}$$

The solution to the system is
$\{(2, 3, -1, 0)\}$.

43. $A = \begin{bmatrix} e^x & e^{3x} \\ -e^{3x} & e^{5x} \end{bmatrix}$

$$A^{-1} = \frac{1}{ad-bc}\begin{bmatrix} d & -b \\ -c & a \end{bmatrix}$$

$$A^{-1} = \frac{1}{(e^x)(e^{5x})-(e^{3x})(-e^{3x})}\begin{bmatrix} e^{5x} & -e^{3x} \\ -(-e^{3x}) & e^x \end{bmatrix}$$

$$A^{-1} = \frac{1}{e^{6x}+e^{6x}}\begin{bmatrix} e^{5x} & -e^{3x} \\ e^{3x} & e^x \end{bmatrix}$$

$$A^{-1} = \frac{1}{2e^{6x}}\begin{bmatrix} e^{5x} & -e^{3x} \\ e^{3x} & e^x \end{bmatrix}$$

$$A^{-1} = \begin{bmatrix} \frac{e^{5x}}{2e^{6x}} & \frac{-e^{3x}}{2e^{6x}} \\ \frac{e^{3x}}{2e^{6x}} & \frac{e^x}{2e^{6x}} \end{bmatrix}$$

$$A^{-1} = \begin{bmatrix} \frac{1}{2e^x} & -\frac{1}{2e^{3x}} \\ \frac{1}{2e^{3x}} & \frac{1}{2e^{5x}} \end{bmatrix} \text{ or } \begin{bmatrix} \frac{e^{-x}}{2} & -\frac{e^{-3x}}{2} \\ \frac{e^{-3x}}{2} & \frac{e^{-5x}}{2} \end{bmatrix}$$

Check:
$$\begin{bmatrix} e^x & e^{3x} \\ -e^{3x} & e^{5x} \end{bmatrix} \cdot \begin{bmatrix} \frac{1}{2}e^{-x} & -\frac{1}{2}e^{-3x} \\ \frac{1}{2}e^{-3x} & \frac{1}{2}e^{-5x} \end{bmatrix} = \begin{bmatrix} 1 & 0 \\ 0 & 1 \end{bmatrix}$$

45. $A = \begin{bmatrix} 8 & -5 \\ -3 & 2 \end{bmatrix}$

$$I - A = \begin{bmatrix} 1 & 0 \\ 0 & 1 \end{bmatrix} - \begin{bmatrix} 8 & -5 \\ -3 & 2 \end{bmatrix} = \begin{bmatrix} -7 & 5 \\ 3 & -1 \end{bmatrix}$$

$$(I-A)^{-1} = \frac{1}{(-7)(-1)-(5)(3)}\begin{bmatrix} -1 & -(5) \\ -(3) & -7 \end{bmatrix}$$

$$(I-A)^{-1} = \frac{1}{7-15}\begin{bmatrix} -1 & -5 \\ -3 & -7 \end{bmatrix}$$

$$(I-A)^{-1} = \frac{1}{-8}\begin{bmatrix} -1 & -5 \\ -3 & -7 \end{bmatrix}$$

$$(I-A)^{-1} = \begin{bmatrix} \frac{-1}{-8} & \frac{-5}{-8} \\ \frac{-3}{-8} & \frac{-7}{-8} \end{bmatrix}$$

$$(I-A)^{-1} = \begin{bmatrix} \frac{1}{8} & \frac{5}{8} \\ \frac{3}{8} & \frac{7}{8} \end{bmatrix}$$

47. $A = \begin{bmatrix} 2 & 1 \\ 3 & 1 \end{bmatrix}$ $B = \begin{bmatrix} 4 & 7 \\ 1 & 2 \end{bmatrix}$

$$A^{-1} = \begin{bmatrix} -1 & 1 \\ 3 & -2 \end{bmatrix} \quad B^{-1} = \begin{bmatrix} 2 & -7 \\ -1 & 4 \end{bmatrix}$$

$$AB = \begin{bmatrix} 2 & 1 \\ 3 & 1 \end{bmatrix}\begin{bmatrix} 4 & 7 \\ 1 & 2 \end{bmatrix} = \begin{bmatrix} 9 & 16 \\ 13 & 23 \end{bmatrix}$$

$$(AB)^{-1} = \left(\begin{bmatrix} 9 & 16 \\ 13 & 23 \end{bmatrix}\right)^{-1} = \begin{bmatrix} -23 & 16 \\ 13 & -9 \end{bmatrix}$$

$$A^{-1}B^{-1} = \begin{bmatrix} -1 & 1 \\ 3 & -2 \end{bmatrix}\begin{bmatrix} 2 & -7 \\ -1 & 4 \end{bmatrix} = \begin{bmatrix} -3 & 11 \\ 8 & -29 \end{bmatrix}$$

$$B^{-1}A^{-1} = \begin{bmatrix} 2 & -7 \\ -1 & 4 \end{bmatrix}\begin{bmatrix} -1 & 1 \\ 3 & -2 \end{bmatrix} = \begin{bmatrix} -23 & 16 \\ 13 & -9 \end{bmatrix}$$

Observe that $(AB)^{-1} = B^{-1}A^{-1}$.

49. $\begin{bmatrix} a & 0 & 0 \\ 0 & b & 0 \\ 0 & 0 & c \end{bmatrix} \begin{bmatrix} \frac{1}{a} & 0 & 0 \\ 0 & \frac{1}{b} & 0 \\ 0 & 0 & \frac{1}{c} \end{bmatrix}$

$= \begin{bmatrix} (a)(\frac{1}{a})+(0)(0)+(0)(0) & (a)(0)+(0)(\frac{1}{b})+(0)(0) & (a)(0)+(0)(0)+(0)(\frac{1}{c}) \\ (0)(\frac{1}{a})+(b)(0)+(0)(0) & (0)(0)+(b)(\frac{1}{b})+(0)(0) & (0)(0)+(b)(0)+(0)(\frac{1}{c}) \\ (0)(\frac{1}{a})+(0)(0)+(c)(0) & (0)(0)+(0)(\frac{1}{b})+(c)(0) & (0)(0)+(0)(0)+(c)(\frac{1}{c}) \end{bmatrix}$

$= \begin{bmatrix} \frac{a}{a}+0+0 & 0+0+0 & 0+0+0 \\ 0+0+0 & 0+\frac{b}{b}+0 & 0+0+0 \\ 0+0+0 & 0+0+0 & 0+0+\frac{c}{c} \end{bmatrix} = \begin{bmatrix} 1 & 0 & 0 \\ 0 & 1 & 0 \\ 0 & 0 & 1 \end{bmatrix}$

51. The numerical equivalent of HELP is
8, 5, 12, 16.

$\begin{bmatrix} 4 & -1 \\ -3 & 1 \end{bmatrix} \begin{bmatrix} 8 \\ 5 \end{bmatrix} = \begin{bmatrix} 27 \\ -19 \end{bmatrix}$,

$\begin{bmatrix} 4 & -1 \\ -3 & 1 \end{bmatrix} \begin{bmatrix} 12 \\ 16 \end{bmatrix} = \begin{bmatrix} 32 \\ -20 \end{bmatrix}$

The encoded message is 27, –19, 32, –20.

$\begin{bmatrix} 1 & 1 \\ 3 & 4 \end{bmatrix} \begin{bmatrix} 27 \\ -19 \end{bmatrix} = \begin{bmatrix} 8 \\ 5 \end{bmatrix}$, $\begin{bmatrix} 1 & 1 \\ 3 & 4 \end{bmatrix} \begin{bmatrix} 32 \\ -20 \end{bmatrix} = \begin{bmatrix} 12 \\ 16 \end{bmatrix}$

The decoded message is 8, 5, 12, 16 or HELP.

53. $\begin{bmatrix} 1 & -1 & 0 \\ 3 & 0 & 2 \\ -1 & 0 & -1 \end{bmatrix} \begin{bmatrix} 19 & 4 & 1 \\ 5 & 0 & 19 \\ 14 & 3 & 8 \end{bmatrix}$

$= \begin{bmatrix} 19-5+0 & 4+0+0 & 1-19+0 \\ 57+0+28 & 12+0+6 & 3+0+16 \\ -19+0-14 & -4+0-3 & -1+0-8 \end{bmatrix}$

$= \begin{bmatrix} 14 & 4 & -18 \\ 85 & 18 & 19 \\ -33 & -7 & -9 \end{bmatrix}$

The encoded message is 14, 85, –33, 4, 18, –7, –18, 19, –9.

$\begin{bmatrix} 0 & 1 & 2 \\ -1 & 1 & 2 \\ 0 & -1 & -3 \end{bmatrix} \begin{bmatrix} 14 & 4 & -18 \\ 85 & 18 & 19 \\ -33 & -7 & -9 \end{bmatrix}$

$= \begin{bmatrix} 0+85-66 & 0+18-14 & 0+19-18 \\ -14+85-66 & -4+18-14 & 18+19-18 \\ 0-85+99 & 0-18+21 & 0-19+27 \end{bmatrix}$

$= \begin{bmatrix} 19 & 4 & 1 \\ 5 & 0 & 19 \\ 14 & 3 & 8 \end{bmatrix}$

The decoded message is 19, 5, 14, 4, 0, 3, 1, 19, 8 or SEND_CASH

55. – 63. Answers will vary.

65. Enter the matrix $\begin{bmatrix} 3 & -1 \\ -2 & 1 \end{bmatrix}$ as [A], then use [A]$^{-1}$.

Verify this result by showing that $[A][A]^{-1} = I_2$ and $[A]^{-1}[A] = I_2$.

67. Enter the matrix $\begin{bmatrix} -2 & 1 & -1 \\ -5 & 2 & -1 \\ 3 & -1 & 1 \end{bmatrix}$ as [A], then use [A]$^{-1}$.

$$[A]^{-1} = \begin{bmatrix} 1 & 0 & 1 \\ 2 & 1 & 3 \\ -1 & 1 & 1 \end{bmatrix}$$

Verify this result by showing that $[A][A]^{-1} = I_3$ and $[A]^{-1}[A] = I_3$.

69. Enter the matrix $\begin{bmatrix} 7 & -3 & 0 & 2 \\ -2 & 1 & 0 & -1 \\ 4 & 0 & 1 & -2 \\ -1 & 1 & 0 & -1 \end{bmatrix}$ as [A], then use [A]$^{-1}$. $[A]^{-1} = \begin{bmatrix} 0 & -1 & 0 & 1 \\ -1 & -5 & 0 & 3 \\ -2 & -4 & 1 & -2 \\ -1 & -4 & 0 & 1 \end{bmatrix}$

Verify this result by showing that $[A][A]^{-1} = I_4$ and $[A]^{-1}[A] = I_4$.

71. The system is $AX = B$ where

$A = \begin{bmatrix} 1 & -1 & 1 \\ 4 & 2 & 1 \\ 4 & -2 & 1 \end{bmatrix}$, $X = \begin{bmatrix} x \\ y \\ z \end{bmatrix}$, and $B = \begin{bmatrix} -6 \\ 9 \\ -3 \end{bmatrix}$. $X = \begin{bmatrix} 2 \\ 3 \\ -5 \end{bmatrix}$, so the solution to the system is $\{(2, 3, -5)\}$.

73. The system is $AX = B$ where $A = \begin{bmatrix} 3 & -2 & 1 \\ 4 & -5 & 3 \\ 2 & -1 & 5 \end{bmatrix}$, $X = \begin{bmatrix} x \\ y \\ z \end{bmatrix}$, and $B = \begin{bmatrix} -2 \\ -9 \\ -5 \end{bmatrix}$.

$X = \begin{bmatrix} 1 \\ 2 \\ -1 \end{bmatrix}$ so the solution to the system is $\{(1, 2, -1)\}$.

75. The system is $AX = B$ where $A = \begin{bmatrix} 1 & 0 & -3 & 0 & 1 \\ 0 & 1 & 0 & 1 & 0 \\ 0 & 0 & 1 & 0 & 1 \\ 1 & 1 & -1 & 4 & 0 \\ 1 & 1 & 1 & 1 & 1 \end{bmatrix}$, $X = \begin{bmatrix} v \\ w \\ x \\ y \\ z \end{bmatrix}$ and $B = \begin{bmatrix} -3 \\ -1 \\ 7 \\ -8 \\ 8 \end{bmatrix}$. $X = \begin{bmatrix} 2 \\ 1 \\ 3 \\ -2 \\ 4 \end{bmatrix}$, so the solution to the system

is $\{(2, 1, 3, -2, 4)\}$.

77. Answers will vary.

79. does not make sense; Explanations will vary. Sample explanation: Only square matrices have inverses.

81. makes sense

83. false; Changes to make the statement true will vary. A sample change is: Not all square matrices have inverses.

85. false; Changes to make the statement true will vary. A sample change is: You need to multiply the inverse of A and B.

87. false; Changes to make the statement true will vary. A sample change is: $(A+B)^{-1} \neq A^{-1} + B^{-1}$

89. Answers will vary.

91. Using the statement before problems 9–14, we want to find values for a such that
$(1)(4) - (a+1)(a-2) = 0$.
$(1)(4) - (a+1)(a-2) = 4 - (a^2 - a - 2)$

$\qquad = -a^2 + a + 6$
$0 = -a^2 + a + 6$
$0 = a^2 - a - 6$
$0 = (a-3)(a+2)$
$a = 3, -2$

93. $2(-5) - (-3)(4) = -10 + 12 = 2$

94. $\dfrac{2(-5) - 1(-4)}{5(-5) - 6(-4)} = \dfrac{-10 + 4}{-25 + 24} = \dfrac{-6}{-1} = 6$

95. $2(-30 - (-3)) - 3(6 - 9) + (-1)(1 - 15)$
$= 2(-27) - 3(-3) + (-1)(-14)$
$= -54 + 9 + 14$
$= -31$

Section 9.5

Check Point Exercises

1. **a.** $\begin{vmatrix} 10 & 9 \\ 6 & 5 \end{vmatrix} = 10 \cdot 5 - 6 \cdot 9 = 50 - 54 = -4$

b. $\begin{vmatrix} 4 & 3 \\ -5 & -8 \end{vmatrix} = 4 \cdot (-8) - (-5) \cdot (3)$
$\qquad = -32 + 15 = -17$

2. $5x + 4y = 12$
$3x - 6y = 24$
$D = \begin{vmatrix} 5 & 4 \\ 3 & -6 \end{vmatrix} = 5 \cdot (-6) - 3 \cdot 4$
$\qquad = -30 - 12 = -42$
$D_x = \begin{vmatrix} 12 & 4 \\ 24 & -6 \end{vmatrix} = 12(-6) - 24(4)$
$\qquad = -72 - 96 = -168$
$D_y = \begin{vmatrix} 5 & 12 \\ 3 & 24 \end{vmatrix} = 5(24) - 3(12)$
$\qquad = 120 - 36 = 84$
Thus, $x = \dfrac{D_x}{D} = \dfrac{-168}{-42} = 4$
$\qquad y = \dfrac{D_y}{D} = \dfrac{84}{-42} = -2$
The solution set is $\{(4, -2)\}$.

3. $\begin{bmatrix} 2 & 1 & 7 \\ -5 & 6 & 0 \\ -4 & 3 & 1 \end{bmatrix}$

The minor for 2 is $\begin{vmatrix} 6 & 0 \\ 3 & 1 \end{vmatrix}$.

The minor for -5 is $\begin{vmatrix} 1 & 7 \\ 3 & 1 \end{vmatrix}$.

The minor for -4 is $\begin{vmatrix} 1 & 7 \\ 6 & 0 \end{vmatrix}$.

$\begin{bmatrix} 2 & 1 & 7 \\ -5 & 6 & 0 \\ -4 & 3 & 1 \end{bmatrix} = 2 \begin{vmatrix} 6 & 0 \\ 3 & 1 \end{vmatrix} - (-5) \begin{vmatrix} 1 & 7 \\ 3 & 1 \end{vmatrix} - 4 \begin{vmatrix} 1 & 7 \\ 6 & 0 \end{vmatrix}$

$\qquad = 2(6 \cdot 1 - 3 \cdot 0) + 5(1 \cdot 1 - 3 \cdot 7) - 4(1 \cdot 0 - 6 \cdot 7)$
$\qquad = 2(6 - 0) + 5(1 - 21) - 4(0 - 42)$
$\qquad = 12 - 100 + 168$
$\qquad = 80$

4. $\begin{vmatrix} 6 & 4 & 0 \\ -3 & -5 & 3 \\ 1 & 2 & 0 \end{vmatrix} = 0 \begin{vmatrix} -3 & -5 \\ 1 & 2 \end{vmatrix} - 3 \begin{vmatrix} 6 & 4 \\ 1 & 2 \end{vmatrix} + 0 \begin{vmatrix} 6 & 4 \\ -3 & -5 \end{vmatrix}$

$\qquad = 0 - 3(6 \cdot 2 - 1 \cdot 4) + 0$
$\qquad = -3(12 - 4)$
$\qquad = -3(8)$
$\qquad = -24$

5. $3x - 2y + z = 16$

$2x + 3y - z = -9$

$x + 4y + 3z = 2$

$$D = \begin{vmatrix} 3 & -2 & 1 \\ 2 & 3 & -1 \\ 1 & 4 & 3 \end{vmatrix}; \quad D_x = \begin{vmatrix} 16 & -2 & 1 \\ -9 & 3 & -1 \\ 2 & 4 & 3 \end{vmatrix}; \quad D_y = \begin{vmatrix} 3 & 16 & 1 \\ 2 & -9 & -1 \\ 1 & 2 & 3 \end{vmatrix}; \quad D_z = \begin{vmatrix} 3 & -2 & 16 \\ 2 & 3 & -9 \\ 1 & 4 & 2 \end{vmatrix}$$

$$D = \begin{vmatrix} 3 & -2 & 1 \\ 2 & 3 & -1 \\ 1 & 4 & 3 \end{vmatrix} = 3\begin{vmatrix} 3 & -1 \\ 4 & 3 \end{vmatrix} - 2\begin{vmatrix} -2 & 1 \\ 4 & 3 \end{vmatrix} + 1\begin{vmatrix} -2 & 1 \\ 3 & -1 \end{vmatrix}$$

$\quad = 3[(3) \cdot 3 - 4 \cdot (-1)] - 2[(-2) \cdot 3 - 4 \cdot 1] + 1[(-2) \cdot (-1) - (3) \cdot 1]$

$\quad = 3(9+4) - 2(-6-4) + 1(2-3)$

$\quad = 39 + 20 - 1$

$\quad = 58$

$$D_x = \begin{vmatrix} 16 & -2 & 1 \\ -9 & 3 & -1 \\ 2 & 4 & 3 \end{vmatrix} = 1\begin{vmatrix} -9 & 3 \\ 2 & 4 \end{vmatrix} - (-1)\begin{vmatrix} 16 & -2 \\ 2 & 4 \end{vmatrix} + 3\begin{vmatrix} 16 & -2 \\ -9 & 3 \end{vmatrix}$$

$\quad = 1[(-9) \cdot 4 - 2 \cdot (3)] + 1[16 \cdot 4 - 2(-2)] + 3[16 \cdot (3) - (-9) \cdot (-2)]$

$\quad = 1(-36-6) + 1(64+4) + 3(48-18)$

$\quad = -42 + 68 + 90$

$\quad = 116$

$$D_y = \begin{vmatrix} 3 & 16 & 1 \\ 2 & -9 & -1 \\ 1 & 2 & 3 \end{vmatrix} = 3\begin{vmatrix} -9 & -1 \\ 2 & 3 \end{vmatrix} - 2\begin{vmatrix} 16 & 1 \\ 2 & 3 \end{vmatrix} + 1\begin{vmatrix} 16 & 1 \\ -9 & -1 \end{vmatrix}$$

$\quad = 3[(-9) \cdot 3 - 2 \cdot (-1)] - 2[16 \cdot 3 - 2 \cdot 1] + 1[16(-1) - (-9) \cdot 1]$

$\quad = 3(-27+2) - 2(48-2) + 1(-16+9)$

$\quad = -75 - 92 - 7$

$\quad = -174$

$$D_z = \begin{vmatrix} 3 & -2 & 16 \\ 2 & 3 & -9 \\ 1 & 4 & 2 \end{vmatrix} = 3\begin{vmatrix} 3 & -9 \\ 4 & 2 \end{vmatrix} - 2\begin{vmatrix} -2 & 16 \\ 4 & 2 \end{vmatrix} + 1\begin{vmatrix} -2 & 16 \\ 3 & -9 \end{vmatrix}$$

$\quad = 3[(3)2 - 4(-9)] - 2[(-2)2 - 4 \cdot 16] + 1[(-2)(-9) - (3) \cdot 16]$

$\quad = 3(6+36) - 2(-4-64) + 1(18-48)$

$\quad = 126 + 136 - 30$

$\quad = 232$

$x = \dfrac{D_x}{D} = \dfrac{116}{58} = 2$

$y = \dfrac{D_y}{D} = \dfrac{-174}{58} = -3$

$z = \dfrac{D_z}{D} = \dfrac{232}{58} = 4$

The solution to the system is $\{(2, -3, 4)\}$.

6. $|A| = \begin{vmatrix} 0 & 4 & 0 & -3 \\ -1 & 1 & 5 & 2 \\ 1 & -2 & 0 & 6 \\ 3 & 0 & 0 & 1 \end{vmatrix} = (-1)^{2+3} 5 \begin{vmatrix} 0 & 4 & -3 \\ 1 & -2 & 6 \\ 3 & 0 & 1 \end{vmatrix} = -5 \begin{vmatrix} 0 & 4 & -3 \\ 1 & -2 & 6 \\ 3 & 0 & 1 \end{vmatrix}$

Evaluate the third-order determinant to get $|A| = -5(50) = -250$.

Concept and Vocabulary Check 9.5

1. 5; 3; 2; 4; 15; 8; 7; determinant; 7

2. $\dfrac{\begin{vmatrix} 8 & 1 \\ -2 & -1 \end{vmatrix}}{\begin{vmatrix} 1 & 1 \\ 1 & -1 \end{vmatrix}}$; $\dfrac{\begin{vmatrix} 1 & 8 \\ 1 & -2 \end{vmatrix}}{\begin{vmatrix} 1 & 1 \\ 1 & -1 \end{vmatrix}}$

3. $\begin{vmatrix} 3 & 1 \\ 1 & 1 \end{vmatrix}$; $\begin{vmatrix} 2 & 1 \\ 1 & 1 \end{vmatrix}$; $\begin{vmatrix} 2 & 1 \\ 3 & 1 \end{vmatrix}$

4. $\dfrac{\begin{vmatrix} 3 & -8 & 4 \\ 2 & 11 & -2 \\ 1 & 4 & -2 \end{vmatrix}}{\begin{vmatrix} 3 & 1 & 4 \\ 2 & 3 & -2 \\ 1 & -3 & -2 \end{vmatrix}}$

5. column 3/the last column

Exercise Set 9.5

1. $\begin{vmatrix} 5 & 7 \\ 2 & 3 \end{vmatrix} = 5 \cdot 3 - 2 \cdot 7 = 15 - 14 = 1$

3. $\begin{vmatrix} -4 & 1 \\ 5 & 6 \end{vmatrix} = (-4)6 - 5 \cdot 1 = -24 - 5 = -29$

5. $\begin{vmatrix} -7 & 14 \\ 2 & -4 \end{vmatrix} = (-7)(-4) - 2(14) = 28 - 28 = 0$

7. $\begin{vmatrix} -5 & -1 \\ -2 & -7 \end{vmatrix} = (-5)(-7) - (-2)(-1) = 35 - 2 = 33$

9. $\begin{vmatrix} \frac{1}{2} & \frac{1}{2} \\ \frac{1}{8} & -\frac{3}{4} \end{vmatrix} = \frac{1}{2}\left(-\frac{3}{4}\right) - \frac{1}{8} \cdot \frac{1}{2} = -\frac{3}{8} - \frac{1}{16} = -\frac{7}{16}$

11. $D = \begin{vmatrix} 1 & 1 \\ 1 & -1 \end{vmatrix} = -1 - 1 = -2$

$D_x = \begin{vmatrix} 7 & 1 \\ 3 & -1 \end{vmatrix} = -7 - 3 = -10$

$D_y = \begin{vmatrix} 1 & 7 \\ 1 & 3 \end{vmatrix} = 3 - 7 = -4$

$x = \dfrac{D_x}{D} = \dfrac{-10}{-2} = 5$

$y = \dfrac{D_y}{D} = \dfrac{-4}{-2} = 2$

The solution set is $\{(5, 2)\}$.

13. $D = \begin{vmatrix} 12 & 3 \\ 2 & -3 \end{vmatrix} = -36 - 6 = -42$

$D_x = \begin{vmatrix} 15 & 3 \\ 13 & -3 \end{vmatrix} = -45 - 39 = -84$

$D_y = \begin{vmatrix} 12 & 15 \\ 2 & 13 \end{vmatrix} = 156 - 30 = 126$

$x = \dfrac{D_x}{D} = \dfrac{-84}{-42} = 2$

$y = \dfrac{D_y}{D} = \dfrac{126}{-42} = -3$

The solution set is $\{(2, -3)\}$.

15. $D = \begin{vmatrix} 4 & -5 \\ 2 & 3 \end{vmatrix} = 12 - (-10) = 22$

$D_x = \begin{vmatrix} 17 & -5 \\ 3 & 3 \end{vmatrix} = 51 - (-15) = 66$

$D_y = \begin{vmatrix} 4 & 17 \\ 2 & 3 \end{vmatrix} = 12 - 34 = -22$

$x = \dfrac{D_x}{D} = \dfrac{66}{22} = 3$

$y = \dfrac{D_y}{D} = \dfrac{-22}{22} = -1$

The solution set is $\{(3, -1)\}$.

17. $D = \begin{vmatrix} 1 & 2 \\ 5 & 10 \end{vmatrix} = 10 - 10 = 0$

$D_x = \begin{vmatrix} 3 & 2 \\ 15 & 10 \end{vmatrix} = 30 - 30 = 0$

$D_y = \begin{vmatrix} 1 & 3 \\ 5 & 15 \end{vmatrix} = 15 - 15 = 0$

Because all 3 determinants equal zero, the system is dependent.

19. $D = \begin{vmatrix} 3 & -4 \\ 2 & 2 \end{vmatrix} = 6 - (-8) = 14$

$D_x = \begin{vmatrix} 4 & -4 \\ 12 & 2 \end{vmatrix} = 8 - (-48) = 56$

$D_y = \begin{vmatrix} 3 & 4 \\ 2 & 12 \end{vmatrix} = 36 - 8 = 28$

$x = \dfrac{D_x}{D} = \dfrac{56}{14} = 4$

$y = \dfrac{D_y}{D} = \dfrac{28}{14} = 2$

The solution set is $\{(4, 2)\}$.

21. $D = \begin{vmatrix} 2 & -3 \\ 5 & 4 \end{vmatrix} = 8 - (-15) = 23$

$D_x = \begin{vmatrix} 2 & -3 \\ 51 & 4 \end{vmatrix} = 8 - (-153) = 161$

$D_y = \begin{vmatrix} 2 & 2 \\ 5 & 51 \end{vmatrix} = 102 - 10 = 92$

$x = \dfrac{D_x}{D} = \dfrac{161}{23} = 7$

$y = \dfrac{D_y}{D} = \dfrac{92}{23} = 4$

The solution set is $\{(7, 4)\}$.

23. $\begin{vmatrix} 3 & 0 & 0 \\ 2 & 1 & -5 \\ -2 & 5 & -1 \end{vmatrix} = 3\begin{vmatrix} 1 & -5 \\ 5 & -1 \end{vmatrix} - 0\begin{vmatrix} 2 & -5 \\ -2 & -1 \end{vmatrix} + 0\begin{vmatrix} 2 & 1 \\ -2 & 5 \end{vmatrix}$

$= 3[(1)(-1) - (5)(-5)]$
$= 3(-1 + 25) = 3(24)$
$= 72$

25. $\begin{vmatrix} 3 & 1 & 0 \\ -3 & 4 & 0 \\ -1 & 3 & -5 \end{vmatrix} = 0\begin{vmatrix} -3 & 4 \\ -1 & 3 \end{vmatrix} - 0\begin{vmatrix} 3 & 1 \\ -1 & 3 \end{vmatrix} + (-5)\begin{vmatrix} 3 & 1 \\ -3 & 4 \end{vmatrix}$

$= -5[3 \cdot 4 - (-3)(1)]$
$= -5(12 + 3) = -5(15)$
$= -75$

27. $\begin{vmatrix} 1 & 1 & 1 \\ 2 & 2 & 2 \\ -3 & 4 & -5 \end{vmatrix} -2R_1 + R_2$

$\begin{vmatrix} 1 & 1 & 1 \\ 0 & 0 & 0 \\ -3 & 4 & -5 \end{vmatrix} = 0$

29. $D = \begin{vmatrix} 1 & 1 & 1 \\ 2 & -1 & 1 \\ -1 & 3 & -1 \end{vmatrix}$

$= \begin{vmatrix} -1 & 1 \\ 3 & -1 \end{vmatrix} - \begin{vmatrix} 2 & 1 \\ -1 & -1 \end{vmatrix} + \begin{vmatrix} 2 & -1 \\ -1 & 3 \end{vmatrix}$

$= (1 - 3) - [-2 - (-1)] + (6 - 1)$

$= -2 - (-1) + 5 = -2 + 1 + 5 = 4$

$D_x = \begin{vmatrix} 0 & 1 & 1 \\ -1 & -1 & 1 \\ -8 & 3 & -1 \end{vmatrix} = (-1) \begin{vmatrix} -1 & 1 \\ -8 & -1 \end{vmatrix} + \begin{vmatrix} -1 & -1 \\ -8 & 3 \end{vmatrix}$

$= (-1)[1 - (-8)] + (-3 - 8) = (-1)(9) - 11$

$= -20$

$D_y = \begin{vmatrix} 1 & 0 & 1 \\ 2 & -1 & 1 \\ -1 & -8 & -1 \end{vmatrix} = \begin{vmatrix} -1 & 1 \\ -8 & -1 \end{vmatrix} + \begin{vmatrix} 2 & -1 \\ -1 & -8 \end{vmatrix}$

$= 1 - (-8) + (-16 - 1) = 1 + 8 - 17 = -8$

$D_z = \begin{vmatrix} 1 & 1 & 0 \\ 2 & -1 & -1 \\ -1 & 3 & -8 \end{vmatrix} = 1 \begin{vmatrix} -1 & -1 \\ 3 & -8 \end{vmatrix} - 1 \begin{vmatrix} 2 & -1 \\ -1 & -8 \end{vmatrix}$

$= 8 - (-3) - 1(-16 - 1) = 11 + 17 = 28$

$x = \dfrac{D_x}{D} = \dfrac{-20}{4} = -5$

$y = \dfrac{D_y}{D} = \dfrac{-8}{4} = -2$

$z = \dfrac{D_z}{D} = \dfrac{28}{4} = 7$

The solution to the system is $\{(-5, -2, 7)\}$.

31. $D = \begin{vmatrix} 4 & -5 & -6 \\ 1 & -2 & -5 \\ 2 & -1 & 0 \end{vmatrix} = 2 \begin{vmatrix} -5 & -6 \\ -2 & -5 \end{vmatrix} - (-1) \begin{vmatrix} 4 & -6 \\ 1 & -5 \end{vmatrix}$

$= 2(25 - 12) + [-20 - (-6)] = 2(13) + (-14)$

$= 26 - 14 = 12$

$D_x = \begin{vmatrix} -1 & -5 & -6 \\ -12 & -2 & -5 \\ 7 & -1 & 0 \end{vmatrix}$

$= 7 \begin{vmatrix} -5 & -6 \\ -2 & -5 \end{vmatrix} - (-1) \begin{vmatrix} -1 & -6 \\ -12 & -5 \end{vmatrix}$

$= 7(25 - 12) + (5 - 72) = 7(13) - 67$

$= 91 - 67 = 24$

$D_y = \begin{vmatrix} 4 & -1 & -6 \\ 1 & -12 & -5 \\ 2 & 7 & 0 \end{vmatrix} = 2 \begin{vmatrix} -1 & -6 \\ -12 & -5 \end{vmatrix} - 7 \begin{vmatrix} 4 & -6 \\ 1 & -5 \end{vmatrix}$

$= 2(5 - 72) - 7[-20 - (-6)]$

$= 2(-67) - 7(-14) = -134 + 98 = -36$

$D_z = \begin{vmatrix} 4 & -5 & -1 \\ 1 & -2 & -12 \\ 2 & -1 & 7 \end{vmatrix}$

$= 4 \begin{vmatrix} -2 & -12 \\ -1 & 7 \end{vmatrix} - (-5) \begin{vmatrix} 1 & -12 \\ 2 & 7 \end{vmatrix} + (-1) \begin{vmatrix} 1 & -2 \\ 2 & -1 \end{vmatrix}$

$= 4(-14 - 12) + 5[7 - (-24)] - [-1 - (-4)]$

$= 4(-26) + 5(31) - (3) = -104 + 155 - 3 = 48$

$x = \dfrac{D_x}{D} = \dfrac{24}{12} = 2, \ y = \dfrac{D_y}{D} = \dfrac{-36}{12} = -3,$

$z = \dfrac{D_z}{D} = \dfrac{48}{12} = 4$

The solution set is $\{(2, -3, 4)\}$.

33. $D = \begin{vmatrix} 1 & 1 & 1 \\ 1 & -2 & 1 \\ 1 & 3 & 2 \end{vmatrix} = 1 \begin{vmatrix} -2 & 1 \\ 3 & 2 \end{vmatrix} - 1 \begin{vmatrix} 1 & 1 \\ 1 & 2 \end{vmatrix} + 1 \begin{vmatrix} 1 & -2 \\ 1 & 3 \end{vmatrix}$

$= -4 - 3 - (2 - 1) + [3 - (-2)]$

$= -7 - 1 + 5 = -3$

$D_x = \begin{vmatrix} 4 & 1 & 1 \\ 7 & -2 & 1 \\ 4 & 3 & 2 \end{vmatrix} = 4 \begin{vmatrix} -2 & 1 \\ 3 & 2 \end{vmatrix} - 1 \begin{vmatrix} 7 & 1 \\ 4 & 2 \end{vmatrix} + 1 \begin{vmatrix} 7 & -2 \\ 4 & 3 \end{vmatrix}$

$= 4(-4 - 3) - (14 - 4) + [21 - (-8)]$

$= 4(-7) - 10 + 29 = -28 + 19 = -9$

$D_y = \begin{vmatrix} 1 & 4 & 1 \\ 1 & 7 & 1 \\ 1 & 4 & 2 \end{vmatrix} = 1 \begin{vmatrix} 7 & 1 \\ 4 & 2 \end{vmatrix} - 1 \begin{vmatrix} 4 & 1 \\ 4 & 2 \end{vmatrix} + 1 \begin{vmatrix} 4 & 1 \\ 7 & 1 \end{vmatrix}$

$= 14 - 4 - (8 - 4) + (4 - 7) = 10 - 4 - 3 = 3$

$D_z = \begin{vmatrix} 1 & 1 & 4 \\ 1 & -2 & 7 \\ 1 & 3 & 4 \end{vmatrix} = 1 \begin{vmatrix} -2 & 7 \\ 3 & 4 \end{vmatrix} - 1 \begin{vmatrix} 1 & 4 \\ 3 & 4 \end{vmatrix} + 1 \begin{vmatrix} 1 & 4 \\ -2 & 7 \end{vmatrix}$

$= -8 - 21 - (4 - 12) + [7 - (-8)]$

$= -29 + 8 + 15 = -6$

$x = \dfrac{D_x}{D} = \dfrac{-9}{-3} = 3, \ y = \dfrac{D_y}{D} = \dfrac{3}{-3} = -1,$

$z = \dfrac{D_z}{D} = \dfrac{-6}{-3} = 2$

The solution set is $\{3, -1, 2\}$.

35. $D = \begin{vmatrix} 1 & 0 & 2 \\ 0 & 2 & -1 \\ 2 & 3 & 0 \end{vmatrix} = \begin{vmatrix} 2 & -1 \\ 3 & 0 \end{vmatrix} + 2 \begin{vmatrix} 0 & 2 \\ 2 & 3 \end{vmatrix}$

$= 0 - (-3) + 2(0 - 4) = 3 - 8 = -5$

$D_x = \begin{vmatrix} 4 & 0 & 2 \\ 5 & 2 & -1 \\ 13 & 3 & 0 \end{vmatrix} = 4 \begin{vmatrix} 2 & -1 \\ 3 & 0 \end{vmatrix} + 2 \begin{vmatrix} 5 & 2 \\ 13 & 3 \end{vmatrix}$

$= 4[0 - (-3)] + 2(15 - 26)$
$= 4(3) + 2(-11) = 12 - 22 = -10$

$D_y = \begin{vmatrix} 1 & 4 & 2 \\ 0 & 5 & -1 \\ 2 & 13 & 0 \end{vmatrix} = \begin{vmatrix} 5 & -1 \\ 13 & 0 \end{vmatrix} + 2 \begin{vmatrix} 4 & 2 \\ 5 & -1 \end{vmatrix}$

$= 0 - (-13) + 2(-4 - 10)$
$= 13 + 2(-14) = 13 - 28 = -15$

$D_z = \begin{vmatrix} 1 & 0 & 4 \\ 0 & 2 & 5 \\ 2 & 3 & 13 \end{vmatrix} = \begin{vmatrix} 2 & 5 \\ 3 & 13 \end{vmatrix} + 4 \begin{vmatrix} 0 & 2 \\ 2 & 3 \end{vmatrix}$

$= 26 - 15 + 4(0 - 4) = 11 + 4(-4)$
$= 11 - 16 = -5$

$x = \dfrac{D_x}{D} = \dfrac{-10}{-5} = 2, \quad y = \dfrac{D_y}{D} = \dfrac{-15}{-5} = 3,$

$z = \dfrac{D_z}{D} = \dfrac{-5}{-5} = 1$

The solution set is $\{(2, 3, 1)\}$.

37. $\begin{vmatrix} 4 & 2 & 8 & -7 \\ -2 & 0 & 4 & 1 \\ 5 & 0 & 0 & 5 \\ 4 & 0 & 0 & -1 \end{vmatrix} = -2 \begin{vmatrix} -2 & 4 & 1 \\ 5 & 0 & 5 \\ 4 & 0 & -1 \end{vmatrix} + 0 \begin{vmatrix} 4 & 8 & -7 \\ 5 & 0 & 5 \\ 4 & 0 & -1 \end{vmatrix} - 0 \begin{vmatrix} 4 & 8 & -7 \\ -2 & 4 & 1 \\ 4 & 0 & -1 \end{vmatrix} + 0 \begin{vmatrix} 4 & 8 & -7 \\ -2 & 4 & 1 \\ 5 & 0 & 5 \end{vmatrix}$

$= (-2) \left[(-4) \begin{vmatrix} 5 & 5 \\ 4 & -1 \end{vmatrix} + 0 \begin{vmatrix} -2 & 1 \\ 4 & -1 \end{vmatrix} - 0 \begin{vmatrix} -2 & 1 \\ 5 & 5 \end{vmatrix} \right] = (-2)(-4)[5(-1) - 4 \cdot 5] = 8(-5 - 20) = 8(-25) = -200$

39. $\begin{vmatrix} -2 & -3 & 3 & 5 \\ 1 & -4 & 0 & 0 \\ 1 & 2 & 2 & -3 \\ 2 & 0 & 1 & 1 \end{vmatrix} = -1 \begin{vmatrix} -3 & 3 & 5 \\ 2 & 2 & -3 \\ 0 & 1 & 1 \end{vmatrix} + (-4) \begin{vmatrix} -2 & 3 & 5 \\ 1 & 2 & -3 \\ 2 & 1 & 1 \end{vmatrix} - 0 \begin{vmatrix} -2 & -3 & 5 \\ 1 & 2 & -3 \\ 2 & 0 & 1 \end{vmatrix} + 0 \begin{vmatrix} -2 & -3 & 3 \\ 1 & 2 & 2 \\ 2 & 0 & 1 \end{vmatrix}$

$= (-1) \left[0 \begin{vmatrix} 3 & 5 \\ 2 & -3 \end{vmatrix} - 1 \begin{vmatrix} -3 & 5 \\ 2 & -3 \end{vmatrix} + 1 \begin{vmatrix} -3 & 3 \\ 2 & 2 \end{vmatrix} \right] - 4 \left[2 \begin{vmatrix} 3 & 5 \\ 2 & -3 \end{vmatrix} - 1 \begin{vmatrix} -2 & 5 \\ 1 & -3 \end{vmatrix} + 1 \begin{vmatrix} -2 & 3 \\ 1 & 2 \end{vmatrix} \right]$

$= (-1)\{(-1)[(-3)(-3) - 2 \cdot 5] + [(-3)(2) - 2 \cdot 3]\} - 4\{2[3(-3) - 2 \cdot 5] - [(-2)(-3) - 1 \cdot 5] + [(-2)(2) - 1 \cdot 3]\} = 195$

41. $\begin{vmatrix} \begin{vmatrix} 3 & 1 \\ -2 & 3 \end{vmatrix} & \begin{vmatrix} 7 & 0 \\ 1 & 5 \end{vmatrix} \\ \begin{vmatrix} 3 & 0 \\ 0 & 7 \end{vmatrix} & \begin{vmatrix} 9 & -6 \\ 3 & 5 \end{vmatrix} \end{vmatrix} = \begin{vmatrix} 3(3) - (-2)(1) & 7(5) - 1(0) \\ 3(7) - 0(0) & 9(5) - 3(-6) \end{vmatrix} = \begin{vmatrix} 9 + 2 & 35 - 0 \\ 21 - 0 & 45 + 18 \end{vmatrix} = \begin{vmatrix} 11 & 35 \\ 21 & 63 \end{vmatrix}$

$= 11(63) - 21(35) = 693 - 735 = -42$

43. From $D = \begin{vmatrix} 2 & -4 \\ 3 & 5 \end{vmatrix}$ we obtain the coefficients of the variables in our equations:

$2x - 4y = c_1$

$3x + 5y = c_2$

From $D_x = \begin{vmatrix} 8 & -4 \\ -10 & 5 \end{vmatrix}$ we obtain the constant coefficients: 8 and -10

$2x - 4y = 8$

$3x + 5y = -10$

45. $\begin{vmatrix} -2 & x \\ 4 & 6 \end{vmatrix} = 32$

$-2(6) - 4(x) = 32$

$-12 - 4x = 32$

$-4x = 44$

$x = -11$

The solution is -11.

47.
$$\begin{vmatrix} 1 & x & -2 \\ 3 & 1 & 1 \\ 0 & -2 & 2 \end{vmatrix} = -8$$

$$0\begin{vmatrix} x & -2 \\ 1 & 1 \end{vmatrix} - (-2)\begin{vmatrix} 1 & -2 \\ 3 & 1 \end{vmatrix} + 2\begin{vmatrix} 1 & x \\ 3 & 1 \end{vmatrix} = -8$$

$$2[1(1) - 3(-2)] + 2[1(1) - 3(x)] = -8$$

$$2(1 + 6) + 2(1 - 3x) = -8$$

$$2(7) + 2(1 - 3x) = -8$$

$$14 + 2 - 6x = -8$$

$$-6x = -24$$

$$x = 4$$

The solution is 4.

49. Area $= \pm\dfrac{1}{2}\begin{vmatrix} 3 & -5 & 1 \\ 2 & 6 & 1 \\ -3 & 5 & 1 \end{vmatrix} = \pm\dfrac{1}{2}\begin{vmatrix} 3 & -5 & 1 \\ -1 & 11 & 0 \\ -6 & 10 & 0 \end{vmatrix} = \pm\dfrac{1}{2}\begin{vmatrix} -1 & 11 \\ -6 & 10 \end{vmatrix} = \pm\dfrac{1}{2}[-10 - (-66)] = \pm\dfrac{1}{2}(56) = 28$

The area is 28 square units.

The slope of the line through $(3, -5)$ and $(-3, 5)$ is $m = \dfrac{5 - (-5)}{-3 - 3} = \dfrac{10}{-6} = -\dfrac{5}{3}$.

The equation of the line is $y - (-5) = -\dfrac{5}{3}(x - 3)$ or $y = -\dfrac{5}{3}x$.

The line perpendicular to $y = -\dfrac{5}{3}x$ through $(2, 6)$ has equation $y - 6 = \dfrac{3}{5}(x - 2)$ or $y = \dfrac{3}{5}x + \dfrac{24}{5}$.

These lines intersect where $-\dfrac{5}{3}x = \dfrac{3}{5}x + \dfrac{24}{5}$.

$-\dfrac{36}{17} = x$ and $-\dfrac{24}{5} = \dfrac{34}{15}x$ $y = -\dfrac{5}{3}\left(-\dfrac{36}{17}\right) = \dfrac{60}{17}$

Using the side connecting $(3, -5)$ and

$(-3, 5)$ as the base, the height is the distance from $(2, 6)$ to $\left(-\dfrac{36}{17}, \dfrac{60}{17}\right)$.

$$b = \sqrt{[3-(-3)]^2 + (-5-5)^2}$$
$$= \sqrt{36+100} = \sqrt{136} = 2\sqrt{34}$$

$$h = \sqrt{\left[2-\left(-\dfrac{36}{17}\right)\right]^2 + \left(6-\dfrac{60}{17}\right)^2}$$

$$= \sqrt{\dfrac{4900}{289} + \dfrac{1764}{289}} = \dfrac{14\sqrt{34}}{17}$$

$$\dfrac{1}{2}bh = \dfrac{1}{2}\left(2\sqrt{34}\right)\left(\dfrac{14\sqrt{34}}{17}\right) = \dfrac{14(34)}{17}$$

$$= 14(2) = 28 \text{ square units}$$

51. $\begin{vmatrix} 3 & -1 & 1 \\ 0 & -3 & 1 \\ 12 & 5 & 1 \end{vmatrix} = \begin{vmatrix} 3 & -1 & 1 \\ -3 & -2 & 0 \\ 9 & 6 & 0 \end{vmatrix} = \begin{vmatrix} -3 & -2 \\ 9 & 6 \end{vmatrix} =$

$= -18 - (-18) = 0$

Yes, the points are collinear.

53. $\begin{vmatrix} x & y & 1 \\ 3 & -5 & 1 \\ -2 & 6 & 1 \end{vmatrix} = x\begin{vmatrix} -5 & 1 \\ 6 & 1 \end{vmatrix} - y\begin{vmatrix} 3 & 1 \\ -2 & 1 \end{vmatrix} + \begin{vmatrix} 3 & -5 \\ -2 & 6 \end{vmatrix} = x(-5-6) - y[3-(-2)] + (18-10)$

$= -11x - 5y + 8$

The equation of the line is $-11x - 5y + 8 = 0$. The equation of the line in slope-intercept form is $y = -\dfrac{11}{5}x + \dfrac{8}{5}$.

55. – 63. Answers will vary.

65. Input the matrix as $[A]$, then use $\det[A]$ to find the determinant.

$\begin{vmatrix} 8 & 2 & 6 & -1 & 0 \\ 2 & 0 & -3 & 4 & 7 \\ 2 & 1 & -3 & 6 & -5 \\ -1 & 2 & 1 & 5 & -1 \\ 4 & 5 & -2 & 3 & -8 \end{vmatrix} = 13,200$

67. does not make sense; Explanations will vary. Sample explanation: Determinants must be square.

69. does not make sense; Explanations will vary. Sample explanation: The number of determinants needed is one greater than the number of variables.

71. In this exercise, expansions are all done about the first column of the matrix and the resulting products of 0 and a determinant are not shown.

a.
$$\begin{vmatrix} a & a \\ 0 & a \end{vmatrix} = a^2 - 0 = a^2$$

b.
$$\begin{vmatrix} a & a & a \\ 0 & a & a \\ 0 & 0 & a \end{vmatrix} = a\begin{vmatrix} a & a \\ 0 & a \end{vmatrix} - 0 + 0$$
$$= a\left(a^2\right) = a^3$$

c.
$$\begin{vmatrix} a & a & a & a \\ 0 & a & a & a \\ 0 & 0 & a & a \\ 0 & 0 & 0 & a \end{vmatrix} = a\begin{vmatrix} a & a & a \\ 0 & a & a \\ 0 & 0 & a \end{vmatrix} - 0 + 0 - 0$$
$$= a\left(a^3\right) = a^4$$

d. Each determinant has zeros below the main diagonal and a's everywhere else.

e. Each determinant equals a raised to the power equal to the order of the determinant.

73. The sign of the value is changed when 2 columns are interchanged in a 2nd order determinant.

75. Evaluate the determinate and write the equation in slope intercept form.
$$\begin{vmatrix} x & y & 1 \\ x_1 & y_1 & 1 \\ x_2 & y_2 & 1 \end{vmatrix} = 0$$
$$x\begin{vmatrix} y_1 & 1 \\ y_2 & 1 \end{vmatrix} - y\begin{vmatrix} x_1 & 1 \\ x_2 & 1 \end{vmatrix} + 1\begin{vmatrix} x_1 & y_1 \\ x_2 & y_2 \end{vmatrix} = 0$$
$$x\left(y_1 - y_2\right) - y\left(x_1 - x_2\right) + x_1 y_2 - x_2 y_1 = 0$$
$$- y\left(x_1 - x_2\right) = -x\left(y_1 - y_2\right) + x_2 y_1 - x_1 y_2$$
$$y\left(x_2 - x_1\right) = x\left(y_2 - y_1\right) + x_2 y_1 - x_1 y_2$$
$$y = \frac{y_2 - y_1}{x_2 - x_1} x + \frac{x_2 y_1 - x_1 y_2}{x_2 - x_1}$$
$$m = \frac{y_2 - y_1}{x_2 - x_1} \qquad b = \frac{x_2 y_1 - x_1 y_2}{x_2 - x_1}$$

Write the slope-point equation of the line the in point slope form.
$$y - y_1 = \frac{y_2 - y_1}{x_2 - x_1}\left(x - x_1\right)$$
$$y - y_1 = \frac{y_2 - y_1}{x_2 - x_1} x + \frac{-x_1 y_2 + x_1 y_1}{x_2 - x_1}$$
$$y = \frac{y_2 - y_1}{x_2 - x_1} x + \frac{-x_1 y_2 + x_1 y_1}{x_2 - x_1} + y_1$$
$$y = \frac{y_2 - y_1}{x_2 - x_1} x + \frac{-x_1 y_2 + x_1 y_1}{x_2 - x_1} + \frac{x_2 y_1 - x_1 y_1}{x_2 - x_1}$$
$$y = \frac{y_2 - y_1}{x_2 - x_1} x + \frac{x_2 y_1 - x_1 y_2}{x_2 - x_1}$$
$$m = \frac{y_2 - y_1}{x_2 - x_1} \qquad b = \frac{x_2 y_1 - x_1 y_2}{x_2 - x_1}$$

Since both forms give the same slope and y-intercept, the determinant does give the equation of the line.

77. **a.** $\dfrac{x^2}{9} + \dfrac{y^2}{4} = 1$
$$\frac{x^2}{9} + \frac{0^2}{4} = 1$$
$$\frac{x^2}{9} = 1$$
$$x^2 = 9$$
$$x = \pm 3$$

b. $\dfrac{x^2}{9} + \dfrac{y^2}{4} = 1$
$$\frac{0^2}{9} + \frac{y^2}{4} = 1$$
$$\frac{y^2}{4} = 1$$
$$y^2 = 4$$
$$y = \pm 2$$

78. $25x^2 + 16y^2 = 400$
$$\frac{25x^2}{400} + \frac{16y^2}{400} = \frac{400}{400}$$
$$\frac{x^2}{16} + \frac{y^2}{25} = 1$$

79.
$$x^2 + y^2 - 2x + 4y = 4$$
$$x^2 - 2x \quad + y^2 + 4y \quad = 4$$
$$x^2 - 2x + 1 + y^2 + 4y + 4 = 4 + 1 + 4$$
$$(x-1)^2 + (y+2)^2 = 9$$
$$(x-1)^2 + (y+2)^2 = 3^2$$
Center: $(1, -2)$
Radius: 3

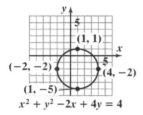

Chapter 9 Review Exercises

1.

2. $\begin{bmatrix} 2 & -2 & 1 & | & -1 \\ 1 & 2 & -1 & | & 2 \\ 6 & 4 & 3 & | & 5 \end{bmatrix} \frac{1}{2}R_1$

$\begin{bmatrix} 1 & -1 & \frac{1}{2} & | & -\frac{1}{2} \\ 1 & 2 & -1 & | & 2 \\ 6 & 4 & 3 & | & 5 \end{bmatrix}$

3. $\begin{bmatrix} 1 & 2 & 3 & | & -5 \\ 2 & 1 & 1 & | & 1 \\ 1 & 1 & -1 & | & 8 \end{bmatrix} \begin{matrix} -2R_1 + R_2 \\ -1R_1 + R_3 \end{matrix}$

$\begin{bmatrix} 1 & 2 & 3 & | & -5 \\ 0 & -3 & -5 & | & 11 \\ 0 & -1 & -4 & | & 13 \end{bmatrix} R_2 \leftrightarrow R_3$

$\begin{bmatrix} 1 & 2 & 3 & | & -5 \\ 0 & -1 & -4 & | & 13 \\ 0 & -3 & -5 & | & 11 \end{bmatrix} -1R_2$

$\begin{bmatrix} 1 & 2 & 3 & | & -5 \\ 0 & 1 & 4 & | & -13 \\ 0 & -3 & -5 & | & 11 \end{bmatrix} 3R_2 + R_3$

$\begin{bmatrix} 1 & 2 & 3 & | & -5 \\ 0 & 1 & 4 & | & -13 \\ 0 & 0 & 7 & | & -28 \end{bmatrix} \frac{1}{7}R_3$

$\begin{bmatrix} 1 & 2 & 3 & | & -5 \\ 0 & 1 & 4 & | & -13 \\ 0 & 0 & 1 & | & -4 \end{bmatrix} -2R_2 + R_1$

$\begin{bmatrix} 1 & 0 & -5 & | & 21 \\ 0 & 1 & 4 & | & -13 \\ 0 & 0 & 1 & | & -4 \end{bmatrix} \begin{matrix} 5R_3 + R_1 \\ -4R_3 + R_2 \end{matrix}$

$\begin{bmatrix} 1 & 0 & 0 & | & 1 \\ 0 & 1 & 0 & | & 3 \\ 0 & 0 & 1 & | & -4 \end{bmatrix}$

The solution set is $\{(1, 3, -4)\}$.

4. $\begin{bmatrix} 1 & -2 & 1 & | & 0 \\ 0 & 1 & -3 & | & -1 \\ 0 & 2 & 5 & | & -2 \end{bmatrix} -2R_2 + R_3$

$\begin{bmatrix} 1 & -2 & 1 & | & 0 \\ 0 & 1 & -3 & | & -1 \\ 0 & 0 & 11 & | & 0 \end{bmatrix} \frac{1}{11}R_3$

$\begin{bmatrix} 1 & -2 & 1 & | & 0 \\ 0 & 1 & -3 & | & -1 \\ 0 & 0 & 1 & | & 0 \end{bmatrix} 2R_2 + R_1$

$\begin{bmatrix} 1 & 0 & -5 & | & -2 \\ 0 & 1 & -3 & | & -1 \\ 0 & 0 & 1 & | & 0 \end{bmatrix} \begin{matrix} 3R_3 + R_2 \\ 5R_3 + R_1 \end{matrix}$

$\begin{bmatrix} 1 & 0 & 0 & -2 \\ 0 & 1 & 0 & -1 \\ 0 & 0 & 1 & 0 \end{bmatrix}$

$x = -2; \ y = -1; \ z = 0$
The solution set is $\{(-2, -1, 0)\}$.

5. $\begin{bmatrix} 3 & 5 & -8 & 5 & | & -8 \\ 1 & 2 & -3 & 1 & | & -7 \\ 2 & 3 & -7 & 3 & | & -11 \\ 4 & 8 & -10 & 7 & | & -10 \end{bmatrix} R_1 \leftrightarrow R_2$

$\begin{bmatrix} 1 & 2 & -3 & 1 & | & -7 \\ 3 & 5 & -8 & 5 & | & -8 \\ 2 & 3 & -7 & 3 & | & -11 \\ 4 & 8 & -10 & 7 & | & -10 \end{bmatrix} \begin{matrix} -3R_1 + R_2 \\ -2R_1 + R_3 \\ -4R_1 + R_4 \end{matrix}$

$\begin{bmatrix} 1 & 2 & -3 & 1 & | & -7 \\ 0 & -1 & 1 & 2 & | & 13 \\ 0 & -1 & -1 & 1 & | & 3 \\ 0 & 0 & 2 & 3 & | & 18 \end{bmatrix} -1R_2$

$$\begin{bmatrix} 1 & 2 & -3 & 1 & | & -7 \\ 0 & 1 & -1 & -2 & | & -13 \\ 0 & -1 & -1 & 1 & | & 3 \\ 0 & 0 & 2 & 3 & | & 18 \end{bmatrix} \begin{matrix} \\ -2R_2 + R_1 \\ 1R_2 + R_3 \\ \\ \end{matrix}$$

$$\begin{bmatrix} 1 & 0 & -1 & 5 & | & 19 \\ 0 & 1 & -1 & -2 & | & -13 \\ 0 & 0 & -2 & -1 & | & -10 \\ 0 & 0 & 2 & 3 & | & 18 \end{bmatrix} \begin{matrix} \\ \\ -\frac{1}{2}R_3 \\ \\ \end{matrix}$$

$$\begin{bmatrix} 1 & 0 & -1 & 5 & | & 19 \\ 0 & 1 & -1 & -2 & | & -13 \\ 0 & 0 & 1 & \frac{1}{2} & | & 5 \\ 0 & 0 & 2 & 3 & | & 18 \end{bmatrix} \begin{matrix} 1R_3 + R_1 \\ 1R_3 + R_2 \\ \\ -2R_3 + R_4 \end{matrix}$$

$$\begin{bmatrix} 1 & 0 & 0 & \frac{11}{2} & | & 24 \\ 0 & 1 & 0 & -\frac{3}{2} & | & -8 \\ 0 & 0 & 1 & \frac{1}{2} & | & 5 \\ 0 & 0 & 0 & 2 & | & 8 \end{bmatrix} \begin{matrix} \\ \\ \frac{1}{2}R_4 \\ \\ \end{matrix}$$

$$\begin{bmatrix} 1 & 0 & 0 & \frac{11}{2} & | & 24 \\ 0 & 1 & 0 & -\frac{3}{2} & | & -8 \\ 0 & 0 & 1 & \frac{1}{2} & | & 5 \\ 0 & 0 & 0 & 1 & | & 4 \end{bmatrix} \begin{matrix} -\frac{11}{2}R_4 + R_1 \\ \frac{3}{2}R_4 + R_2 \\ -\frac{1}{2}R_4 + R_3 \\ \\ \end{matrix}$$

$$\begin{bmatrix} 1 & 0 & 0 & 0 & | & 2 \\ 0 & 1 & 0 & 0 & | & -2 \\ 0 & 0 & 1 & 0 & | & 3 \\ 0 & 0 & 0 & 1 & | & 4 \end{bmatrix}$$

The solution set is $\{(2, -2, 3, 4)\}$.

6. a. The function must satisfy:
$$98 = 4a = 2b + c$$
$$138 = 16a + 4b + c$$
$$162 = 100a + 10b + c.$$

$$\begin{bmatrix} 4 & 2 & 1 & | & 98 \\ 16 & 4 & 1 & | & 138 \\ 100 & 10 & 1 & | & 162 \end{bmatrix} \begin{matrix} \frac{1}{4}R_1 \\ \\ \end{matrix}$$

$$\begin{bmatrix} 1 & \frac{1}{2} & \frac{1}{4} & | & \frac{49}{2} \\ 16 & 4 & 1 & | & 138 \\ 100 & 10 & 1 & | & 162 \end{bmatrix} \begin{matrix} \\ -16R_1 + R_2 \\ -100R_1 + R_3 \end{matrix}$$

$$\begin{bmatrix} 1 & \frac{1}{2} & \frac{1}{4} & | & \frac{49}{2} \\ 0 & -4 & -3 & | & -254 \\ 0 & -40 & -24 & | & -2288 \end{bmatrix} \begin{matrix} \\ -\frac{1}{4}R_2 \\ \\ \end{matrix}$$

$$\begin{bmatrix} 1 & \frac{1}{2} & \frac{1}{4} & | & \frac{49}{2} \\ 0 & 1 & \frac{3}{4} & | & \frac{127}{2} \\ 0 & -40 & -24 & | & -2288 \end{bmatrix} 40R_2 + R_3$$

$$\begin{bmatrix} 1 & \frac{1}{2} & \frac{1}{4} & | & \frac{49}{2} \\ 0 & 1 & \frac{3}{4} & | & \frac{127}{2} \\ 0 & 0 & 6 & | & 252 \end{bmatrix} \frac{1}{6}R_3$$

$$\begin{bmatrix} 1 & \frac{1}{2} & \frac{1}{4} & | & \frac{49}{2} \\ 0 & 1 & \frac{3}{4} & | & \frac{127}{2} \\ 0 & 0 & 1 & | & 42 \end{bmatrix} \begin{matrix} -\frac{1}{4}R_3 + R_1 \\ -\frac{3}{4}R_3 + R_2 \\ \\ \end{matrix}$$

$$\begin{bmatrix} 1 & \frac{1}{2} & 0 & | & 14 \\ 0 & 1 & 0 & | & 32 \\ 0 & 0 & 1 & | & 42 \end{bmatrix} -\frac{1}{2}R_3 + R_1$$

$$\begin{bmatrix} 1 & 0 & 0 & | & -2 \\ 0 & 1 & 0 & | & 32 \\ 0 & 0 & 1 & | & 42 \end{bmatrix}$$

The function is $y = -2x^2 + 32x + 42$ and $a = -2$, $b = 32$ and $c = 42$.

b. $y = -2x^2 + 32x + 42$ is a parabola.

The maximum occurs when
$$x = \frac{-32}{2(-2)} = \frac{-32}{-4} = 8.$$

The air pollution level is a maximum 8 hours after 6 A.M., which is 2 P.M.

When $x = 8$, $y = -2(64) + 32(8) + 42$
$$= -128 + 256 + 42.$$
$$= 170.$$

The maximum level is 170 parts per million at 2 P.M.

7. Write the equations.

$$w + x + y + z = 80$$
$$y - w - x = 18$$
$$y - z = 4$$
$$3x - w - y = 10$$

Rewrite the equations with terms in consistent order.

$$w + x + y + z = 80$$
$$-w - x + y = 18$$
$$y - z = 4$$
$$-w + 3x - y = 10$$

Write as a matrix and solve.

$$\begin{bmatrix} 1 & 1 & 1 & 1 & | & 80 \\ -1 & -1 & 1 & 0 & | & 18 \\ 0 & 0 & 1 & -1 & | & 4 \\ -1 & 3 & -1 & 0 & | & 10 \end{bmatrix} \begin{matrix} \\ R_1 + R_2 \\ \\ R_1 + R_4 \end{matrix}$$

$$\begin{bmatrix} 1 & 1 & 1 & 1 & | & 80 \\ 0 & 0 & 2 & 1 & | & 98 \\ 0 & 0 & 1 & -1 & | & 4 \\ 0 & 4 & 0 & 1 & | & 90 \end{bmatrix} R_2 \leftrightarrow R_4$$

$$\begin{bmatrix} 1 & 1 & 1 & 1 & | & 80 \\ 0 & 4 & 0 & 1 & | & 90 \\ 0 & 0 & 1 & -1 & | & 4 \\ 0 & 0 & 2 & 1 & | & 98 \end{bmatrix} \tfrac{1}{4} R_2$$

$$\begin{bmatrix} 1 & 1 & 1 & 1 & | & 80 \\ 0 & 1 & 0 & \tfrac{1}{4} & | & \tfrac{45}{2} \\ 0 & 0 & 1 & -1 & | & 4 \\ 0 & 0 & 2 & 1 & | & 98 \end{bmatrix} -2R_3 + R_4$$

$$\begin{bmatrix} 1 & 1 & 1 & 1 & | & 80 \\ 0 & 1 & 0 & \tfrac{1}{4} & | & \tfrac{45}{2} \\ 0 & 0 & 1 & -1 & | & 4 \\ 0 & 0 & 0 & 3 & | & 90 \end{bmatrix} \tfrac{1}{3} R_4$$

$$\begin{bmatrix} 1 & 1 & 1 & 1 & | & 80 \\ 0 & 1 & 0 & \tfrac{1}{4} & | & \tfrac{45}{2} \\ 0 & 0 & 1 & -1 & | & 4 \\ 0 & 0 & 0 & 1 & | & 30 \end{bmatrix}$$

Back-substitute $z = 30$ to find y.

$$y - z = 4$$
$$y - 30 = 4$$
$$y = 34$$

Back-substitute to find x.

$$x + \frac{1}{4}z = \frac{45}{2}$$
$$x + \frac{1}{4}(30) = \frac{45}{2}$$
$$x + \frac{15}{2} = \frac{45}{2}$$
$$x = 15$$

Back-substitute to find x.

$$w + x + y + z = 80$$
$$w + 15 + 34 + 30 = 80$$
$$w + 79 = 80$$
$$w = 1$$

Capitalist: 1%
Upper Middle: 15%
Lower Middle: 34%
Working: 30%

8. $$\begin{bmatrix} 2 & -3 & 1 & | & 1 \\ 1 & -2 & 3 & | & 2 \\ 3 & -4 & -1 & | & 1 \end{bmatrix} R_1 \leftrightarrow R_2$$

$$\begin{bmatrix} 1 & -2 & 3 & | & 2 \\ 2 & -3 & 1 & | & 1 \\ 3 & -4 & -1 & | & 1 \end{bmatrix} \begin{matrix} \\ -2R_1 + R_2 \\ -3R_1 + R_3 \end{matrix}$$

$$\begin{bmatrix} 1 & -2 & 3 & | & 2 \\ 0 & 1 & -5 & | & -3 \\ 0 & 2 & -10 & | & -5 \end{bmatrix} -2R_2 + R_3$$

$$\begin{bmatrix} 1 & -2 & 3 & | & 2 \\ 0 & 1 & -5 & | & -3 \\ 0 & 0 & 0 & | & 1 \end{bmatrix}$$

From the last line, we see that the system has no solution. The solution set is $\varnothing$.

9. $$\begin{bmatrix} 1 & -3 & 1 & | & 1 \\ -2 & 1 & 3 & | & -7 \\ 1 & -4 & 2 & | & 0 \end{bmatrix} \begin{matrix} 2R_1 + R_2 \\ -1R_1 + R_3 \end{matrix}$$

$$\begin{bmatrix} 1 & -3 & 1 & | & 1 \\ 0 & -5 & 5 & | & -5 \\ 0 & -1 & 1 & | & -1 \end{bmatrix} -\tfrac{1}{5}R_2$$

$$\begin{bmatrix} 1 & -3 & 1 & | & 1 \\ 0 & 1 & -1 & | & 1 \\ 0 & -1 & 1 & | & -1 \end{bmatrix} 1R_2 + R_3$$

Copyright © 2014 Pearson Education, Inc.

$$\begin{bmatrix} 1 & -3 & 1 & | & 1 \\ 0 & 1 & -1 & | & 1 \\ 0 & 0 & 0 & | & 0 \end{bmatrix}$$

The system $\begin{aligned} x - 3y + z &= 1 \\ y - z &= 1 \end{aligned}$ has no unique solution.

Express x and y in terms of z:

$y = z + 1$

$x - 3(z + 1) + z = 1$

$x - 3z - 3 + z = 1$

$x = 2z + 4$

With $z = t$, the complete solution to the system is $\{(2t + 4, t + 1, t)\}$.

10. $\begin{bmatrix} 1 & 4 & 3 & -6 & | & 5 \\ 1 & 3 & 1 & -4 & | & 3 \\ 2 & 8 & 7 & -5 & | & 11 \\ 2 & 5 & 0 & -6 & | & 4 \end{bmatrix} \begin{aligned} &-1R_1 + R_2 \\ &-2R_1 + R_3 \\ &-2R_1 + R_4 \end{aligned}$

$\begin{bmatrix} 1 & 4 & 3 & -6 & | & 5 \\ 0 & -1 & -2 & 2 & | & -2 \\ 0 & 0 & 1 & 7 & | & 1 \\ 0 & -3 & -6 & 6 & | & -6 \end{bmatrix} -1R_2$

$\begin{bmatrix} 1 & 4 & 3 & -6 & | & 5 \\ 0 & 1 & 2 & -2 & | & 2 \\ 0 & 0 & 1 & 7 & | & 1 \\ 0 & -3 & -6 & 6 & | & -6 \end{bmatrix} 3R_2 + R_4$

$\begin{bmatrix} 1 & 4 & 3 & -6 & | & 5 \\ 0 & 1 & 2 & -2 & | & 2 \\ 0 & 0 & 1 & 7 & | & 1 \\ 0 & 0 & 0 & 0 & | & 0 \end{bmatrix}$

The system $\begin{aligned} x_1 + 4x_2 + 3x_3 - 6x_4 &= 5 \\ x_2 + 2x_3 - 2x_4 &= 2 \\ x_3 + 7x_4 &= 1 \end{aligned}$

does not have a unique solution.

Express x_1, x_2, and x_3 in terms of x_4:

$x_3 = -7x_4 + 1$

$x_2 + 2(-7x_4 + 1) - 2x_4 = 2$

$x_2 - 14x_4 + 2 - 2x_4 = 2$

$x_2 = 16x_4$

$x_1 + 4(16x_4) + 3(-7x_4 + 1) - 6x_4 = 5$

$x_1 + 64x_4 - 21x_4 + 3 - 6x_4 = 5$

$x_1 = -37x_4 + 2$

With $x_4 = t$, the complete solution to the system is $\{(-37t + 2, 16t, -7t + 1, t)\}$.

11. $\begin{bmatrix} 2 & 3 & -5 & | & 15 \\ 1 & 2 & -1 & | & 4 \end{bmatrix} R_1 \leftrightarrow R_2$

$\begin{bmatrix} 1 & 2 & -1 & | & 4 \\ 2 & 3 & -5 & | & 15 \end{bmatrix} -2R_1 + R_2$

$\begin{bmatrix} 1 & 2 & -1 & | & 4 \\ 0 & -1 & -3 & | & 7 \end{bmatrix} -1R_2$

$\begin{bmatrix} 1 & 2 & -1 & | & 4 \\ 0 & 1 & 3 & | & -7 \end{bmatrix}$

The system $\begin{aligned} x + 2y - z &= 4 \\ y + 3z &= -7 \end{aligned}$ has no unique solution.

Express x and y in terms of z:

$y = -3z - 7$

$x + 2(-3z - 7) - z = 4$

$x - 6z - 14 - z = 4$

$x = 7z + 18$

With $z = t$, the complete solution to the system is $\{(7t + 18, -3t - 7, t)\}$.

12. a. $350 + 400 = x + z$

$450 + z = y + 700$

$x + y = 300 + 200$

or

$x + z = 750$

$y - z = -250$

$x + y = 500$

b. $\begin{bmatrix} 1 & 0 & 1 & | & 750 \\ 0 & 1 & -1 & | & -250 \\ 1 & 1 & 0 & | & 500 \end{bmatrix} -1R_1 + R_3$

$\begin{bmatrix} 1 & 0 & 1 & | & 750 \\ 0 & 1 & -1 & | & -250 \\ 0 & 1 & -1 & | & -250 \end{bmatrix} -1R_2 + R_3$

$\begin{bmatrix} 1 & 0 & 1 & | & 750 \\ 0 & 1 & -1 & | & -250 \\ 0 & 0 & 0 & | & 0 \end{bmatrix}$

The system $\begin{aligned} x + z &= 750 \\ y - z &= -250 \end{aligned}$ has no unique solution.

Express x and y in terms of z:

$y = z - 250$

$x = -z + 750$

With $z = t$, the complete solution to the system is $\{(-t + 750, t - 250, t)\}$.

c. $x = -400 + 750 = 350$

$y = 400 - 250 = 150$

13.
$$2x = -10$$
$$x = -5$$
$$y + 7 = 13$$
$$y = 6$$
$$z = 6$$
$$x = -5; \ y = 6; \ z = 6$$

14. $A + D = \begin{bmatrix} 2-2 & -1+3 & 2+1 \\ 5+3 & 3-2 & -1+4 \end{bmatrix} = \begin{bmatrix} 0 & 2 & 3 \\ 8 & 1 & 3 \end{bmatrix}$

15. $2B = \begin{bmatrix} 2(0) & 2(-2) \\ 2(3) & 2(2) \\ 2(1) & 2(-5) \end{bmatrix} = \begin{bmatrix} 0 & -4 \\ 6 & 4 \\ 2 & -10 \end{bmatrix}$

16. $D - A = \begin{bmatrix} -2-2 & 3+1 & 1-2 \\ 3-5 & -2-3 & 4+1 \end{bmatrix}$

$= \begin{bmatrix} -4 & 4 & -1 \\ -2 & -5 & 5 \end{bmatrix}$

17. Not possible since B is 3×2 and C is 3×3.

18. $3A + 2D = \begin{bmatrix} 6 & -3 & 6 \\ 15 & 9 & -3 \end{bmatrix} + \begin{bmatrix} -4 & 6 & 2 \\ 6 & -4 & 8 \end{bmatrix}$

$= \begin{bmatrix} 2 & 3 & 8 \\ 21 & 5 & 5 \end{bmatrix}$

19.
$$-2A + 4D = \begin{bmatrix} -4 & 2 & -4 \\ -10 & -6 & 2 \end{bmatrix} + \begin{bmatrix} -8 & 12 & 4 \\ 12 & -8 & 16 \end{bmatrix}$$

$= \begin{bmatrix} -12 & 14 & 0 \\ 2 & -14 & 18 \end{bmatrix}$

20. $-5(A+D) = -5\left(\begin{bmatrix} 0 & 2 & 3 \\ 8 & 1 & 3 \end{bmatrix} \right) = \begin{bmatrix} 0 & -10 & -15 \\ -40 & -5 & -15 \end{bmatrix}$

21. $AB = \begin{bmatrix} 0-3+2 & -4-2-10 \\ 0+9-1 & -10+6+5 \end{bmatrix} = \begin{bmatrix} -1 & -16 \\ 8 & 1 \end{bmatrix}$

22. $BA = \begin{bmatrix} 0-10 & 0-6 & 0+2 \\ 6+10 & -3+6 & 6-2 \\ 2-25 & -1-15 & 2+5 \end{bmatrix} = \begin{bmatrix} -10 & -6 & 2 \\ 16 & 3 & 4 \\ -23 & -16 & 7 \end{bmatrix}$

23. $BD = \begin{bmatrix} 0-6 & 0+4 & 0-8 \\ -6+6 & 9-4 & 3+8 \\ -2-15 & 3+10 & 1-20 \end{bmatrix} = \begin{bmatrix} -6 & 4 & -8 \\ 0 & 5 & 11 \\ -17 & 13 & -19 \end{bmatrix}$

24. $DB = \begin{bmatrix} 0+9+1 & 4+6-5 \\ 0-6+4 & -6-4-20 \end{bmatrix} = \begin{bmatrix} 10 & 5 \\ -2 & -30 \end{bmatrix}$

25. Not possible since AB is 2 x 2 and BA is 3 x 3.

26. $(A-D)C = \begin{bmatrix} 4 & -4 & 1 \\ 2 & 5 & -5 \end{bmatrix} \begin{bmatrix} 1 & 2 & 3 \\ -1 & 1 & 2 \\ -1 & 2 & 1 \end{bmatrix}$

$= \begin{bmatrix} 4+4-1 & 8-4+2 & 12-8+1 \\ 2-5+5 & 4+5-10 & 6+10-5 \end{bmatrix} = \begin{bmatrix} 7 & 6 & 5 \\ 2 & -1 & 11 \end{bmatrix}$

27. $B(AC) = \begin{bmatrix} 0 & -2 \\ 3 & 2 \\ 1 & -5 \end{bmatrix} \begin{bmatrix} 2+1-2 & 4-1+4 & 6-2+2 \\ 5-3+1 & 10+3-2 & 15+6-1 \end{bmatrix}$

$= \begin{bmatrix} 0 & -2 \\ 3 & 2 \\ 1 & -5 \end{bmatrix} \begin{bmatrix} 1 & 7 & 6 \\ 3 & 11 & 20 \end{bmatrix}$

$= \begin{bmatrix} 0-6 & 0-22 & 0-40 \\ 3+6 & 21+22 & 18+40 \\ 1-15 & 7-55 & 6-100 \end{bmatrix}$

$= \begin{bmatrix} -6 & -22 & -40 \\ 9 & 43 & 58 \\ -14 & -48 & -94 \end{bmatrix}$

28. $3X + A = B$
$$3X = B - A$$
$$X = \frac{1}{3}(B - A)$$
$$X = \frac{1}{3}\left(\begin{bmatrix} -2 & -12 \\ 4 & 1 \end{bmatrix} - \begin{bmatrix} 4 & 6 \\ -5 & 0 \end{bmatrix} \right)$$
$$X = \frac{1}{3} \begin{bmatrix} -6 & -18 \\ 9 & 1 \end{bmatrix}$$
$$X = \begin{bmatrix} -2 & -6 \\ 3 & \frac{1}{3} \end{bmatrix}$$

29. $\begin{bmatrix} 2 & 2 & 2 \\ 1 & 2 & 1 \\ 1 & 2 & 1 \end{bmatrix}$

30. $\begin{bmatrix} 2 & 2 & 2 \\ 1 & 2 & 1 \\ 1 & 2 & 1 \end{bmatrix} + \begin{bmatrix} 1 & 1 & 1 \\ -1 & 1 & -1 \\ -1 & 1 & -1 \end{bmatrix} = \begin{bmatrix} 3 & 3 & 3 \\ 0 & 3 & 0 \\ 0 & 3 & 0 \end{bmatrix}$

$B = \begin{bmatrix} 3 & 3 & 3 \\ 0 & 3 & 0 \\ 0 & 3 & 0 \end{bmatrix}$

31. $\begin{bmatrix} 0 & 2 & 2 \\ 0 & 0 & -4 \end{bmatrix} + \begin{bmatrix} -2 & -2 & -2 \\ 1 & 1 & 1 \end{bmatrix} = \begin{bmatrix} -2 & 0 & 0 \\ 1 & 1 & -3 \end{bmatrix}$

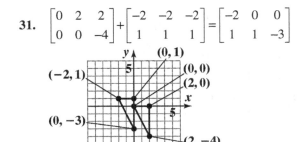

32. $\dfrac{1}{2}\begin{bmatrix} 0 & 2 & 2 \\ 0 & 0 & -4 \end{bmatrix} + \begin{bmatrix} 0 & 0 & 0 \\ -2 & -2 & -2 \end{bmatrix} = \begin{bmatrix} -2 & -1 & -1 \\ -2 & -2 & -4 \end{bmatrix}$

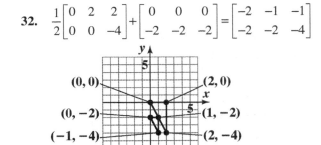

33. $\begin{bmatrix} 1 & 0 \\ 0 & -1 \end{bmatrix}\begin{bmatrix} 0 & 2 & 2 \\ 0 & 0 & -4 \end{bmatrix} = \begin{bmatrix} 0 & 2 & 2 \\ 0 & 0 & 4 \end{bmatrix}$

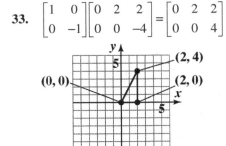

The triangle is reflected about the *x*-axis.

34. $\begin{bmatrix} -1 & 0 \\ 0 & 1 \end{bmatrix}\begin{bmatrix} 0 & 2 & 2 \\ 0 & 0 & -4 \end{bmatrix} = \begin{bmatrix} 0 & -2 & -2 \\ 0 & 0 & -4 \end{bmatrix}$

The triangle is reflected about the *y*-axis.

35. $\begin{bmatrix} 0 & -1 \\ 1 & 0 \end{bmatrix}\begin{bmatrix} 0 & 2 & 2 \\ 0 & 0 & -4 \end{bmatrix} = \begin{bmatrix} 0 & 0 & 4 \\ 0 & 2 & 2 \end{bmatrix}$

The triangle is rotated 90° counterclockwise about the origin.

36. $\begin{bmatrix} 2 & 0 \\ 0 & 1 \end{bmatrix}\begin{bmatrix} 0 & 2 & 2 \\ 0 & 0 & -4 \end{bmatrix} = \begin{bmatrix} 0 & 4 & 4 \\ 0 & 0 & -4 \end{bmatrix}$

The triangle is stretched by a factor of 2 horizontally.

37. $AB = \begin{bmatrix} 8-7 & -14+21 \\ 4-4 & -7+12 \end{bmatrix} = \begin{bmatrix} 1 & 7 \\ 0 & 5 \end{bmatrix}$

$BA = \begin{bmatrix} 8-7 & 28-28 \\ -2+3 & -7+12 \end{bmatrix} = \begin{bmatrix} 1 & 0 \\ 1 & 5 \end{bmatrix}$

If *B* is the multiplicative inverse of *A*, both products (*AB* and *BA*) will be the multiplicative identity matrix, I_2. Therefore, *B* is not the multiplicative inverse of *A*.

38. $AB = \begin{bmatrix} 1 & 0 & 0 \\ 0 & 2 & -7 \\ 0 & -1 & 4 \end{bmatrix}\begin{bmatrix} 1 & 0 & 0 \\ 0 & 4 & 7 \\ 0 & 1 & 2 \end{bmatrix} = \begin{bmatrix} 1 & 0 & 0 \\ 0 & 1 & 0 \\ 0 & 0 & 1 \end{bmatrix}$

$BA = \begin{bmatrix} 1 & 0 & 0 \\ 0 & 4 & 7 \\ 0 & 1 & 2 \end{bmatrix}\begin{bmatrix} 1 & 0 & 0 \\ 0 & 2 & -7 \\ 0 & -1 & 4 \end{bmatrix} = \begin{bmatrix} 1 & 0 & 0 \\ 0 & 1 & 0 \\ 0 & 0 & 1 \end{bmatrix}$

If *B* is the multiplicative inverse of *A*, both products (*AB* and *BA*) will be the multiplicative identity matrix, I_3. Therefore, *B* is the multiplicative inverse of *A*.

39. $A^{-1} = \dfrac{1}{3-2}\begin{bmatrix} 3 & 1 \\ 2 & 1 \end{bmatrix} = \begin{bmatrix} 3 & 1 \\ 2 & 1 \end{bmatrix}$

$AA^{-1} = \begin{bmatrix} 1 & -1 \\ -2 & 3 \end{bmatrix}\begin{bmatrix} 3 & 1 \\ 2 & 1 \end{bmatrix}$

$= \begin{bmatrix} 3-2 & 1-1 \\ -6+6 & -2+3 \end{bmatrix}$

$= \begin{bmatrix} 1 & 0 \\ 0 & 1 \end{bmatrix}$

$A^{-1}A = \begin{bmatrix} 3 & 1 \\ 2 & 1 \end{bmatrix}\begin{bmatrix} 1 & -1 \\ -2 & 3 \end{bmatrix}$

$= \begin{bmatrix} 3-2 & 3+3 \\ 2-2 & -2+3 \end{bmatrix}$

$= \begin{bmatrix} 1 & 0 \\ 0 & 1 \end{bmatrix}$

40. $A^{-1} = \dfrac{1}{0-5}\begin{bmatrix} 3 & -1 \\ -5 & 0 \end{bmatrix}$

$= \dfrac{-1}{5}\begin{bmatrix} 3 & -1 \\ -5 & 0 \end{bmatrix}$

$= \begin{bmatrix} -\frac{3}{5} & \frac{1}{5} \\ 1 & 0 \end{bmatrix}$

$AA^{-1} = \begin{bmatrix} 0 & 1 \\ 5 & 3 \end{bmatrix}\begin{bmatrix} -\frac{3}{5} & \frac{1}{5} \\ 1 & 0 \end{bmatrix}$

$= \begin{bmatrix} 0+1 & 0+0 \\ -3+3 & 1+0 \end{bmatrix}$

$= \begin{bmatrix} 1 & 0 \\ 0 & 1 \end{bmatrix}$

$A^{-1}A = \begin{bmatrix} -\frac{3}{5} & \frac{1}{5} \\ 1 & 0 \end{bmatrix}\begin{bmatrix} 0 & 1 \\ 5 & 3 \end{bmatrix}$

$= \begin{bmatrix} 0+1 & -\frac{3}{5}+\frac{3}{5} \\ 0+0 & 1+0 \end{bmatrix}$

$= \begin{bmatrix} 1 & 0 \\ 0 & 1 \end{bmatrix}$

41. $\left[\begin{array}{ccc|ccc} 1 & 0 & -2 & 1 & 0 & 0 \\ 2 & 1 & 0 & 0 & 1 & 0 \\ 1 & 0 & -3 & 0 & 0 & 1 \end{array}\right]\begin{array}{l} \\ -2R_1+R_2 \\ -1R_1+R_3 \end{array}$

$\left[\begin{array}{ccc|ccc} 1 & 0 & -2 & 1 & 0 & 0 \\ 0 & 1 & 4 & -2 & 1 & 0 \\ 0 & 0 & -1 & -1 & 0 & 1 \end{array}\right]-1R_3$

$\left[\begin{array}{ccc|ccc} 1 & 0 & -2 & 1 & 0 & 0 \\ 0 & 1 & 4 & -2 & 1 & 0 \\ 0 & 0 & 1 & 1 & 0 & -1 \end{array}\right]\begin{array}{l} 2R_3+R_1 \\ \\ -4R_3+R_2 \end{array}$

$\left[\begin{array}{ccc|ccc} 1 & 0 & 0 & 3 & 0 & -2 \\ 0 & 1 & 0 & -6 & 1 & 4 \\ 0 & 0 & 1 & 1 & 0 & -1 \end{array}\right]$

$A^{-1} = \begin{bmatrix} 3 & 0 & -2 \\ -6 & 1 & 4 \\ 1 & 0 & -1 \end{bmatrix}$

$AA^{-1} = \begin{bmatrix} 1 & 0 & -2 \\ 2 & 1 & 0 \\ 1 & 0 & -3 \end{bmatrix}\begin{bmatrix} 3 & 0 & -2 \\ -6 & 1 & 4 \\ 1 & 0 & -1 \end{bmatrix}$

$= \begin{bmatrix} 3+0-2 & 0+0+0 & -2+0+2 \\ 6-6+0 & 0+1+0 & -4+4+0 \\ 3+0-3 & 0+0+0 & -2+0+3 \end{bmatrix}$

$= \begin{bmatrix} 1 & 0 & 0 \\ 0 & 1 & 0 \\ 0 & 0 & 1 \end{bmatrix}$

$A^{-1}A = \begin{bmatrix} 3 & 0 & -2 \\ -6 & 1 & 4 \\ 1 & 0 & -1 \end{bmatrix}\begin{bmatrix} 1 & 0 & -2 \\ 2 & 1 & 0 \\ 1 & 0 & -3 \end{bmatrix}$

$= \begin{bmatrix} 3+0-2 & 0+0+0 & -6+0+6 \\ -6+2+4 & 0+1+0 & 12+0-12 \\ 1+0-1 & 0+0+0 & -2+0+3 \end{bmatrix}$

$= \begin{bmatrix} 1 & 0 & 0 \\ 0 & 1 & 0 \\ 0 & 0 & 1 \end{bmatrix}$

42. $\begin{bmatrix} 1 & 3 & -2 & | & 1 & 0 & 0 \\ 4 & 13 & -7 & | & 0 & 1 & 0 \\ 5 & 16 & -8 & | & 0 & 0 & 1 \end{bmatrix} \begin{matrix} -4R_1 + R_2 \\ -5R_1 + R_3 \end{matrix}$

$\begin{bmatrix} 1 & 3 & -2 & | & 1 & 0 & 0 \\ 0 & 1 & 1 & | & -4 & 1 & 0 \\ 0 & 1 & 2 & | & -5 & 0 & 1 \end{bmatrix} \begin{matrix} -1R_2 + R_3 \\ -3R_2 + R_1 \end{matrix}$

$\begin{bmatrix} 1 & 0 & -5 & | & 13 & -3 & 0 \\ 0 & 1 & 1 & | & -4 & 1 & 0 \\ 0 & 0 & 1 & | & -1 & -1 & 1 \end{bmatrix} \begin{matrix} -1R_3 + R_2 \\ 5R_3 + R_1 \end{matrix}$

$\begin{bmatrix} 1 & 0 & 0 & | & 8 & -8 & 5 \\ 0 & 1 & 0 & | & -3 & 2 & -1 \\ 0 & 0 & 1 & | & -1 & -1 & 1 \end{bmatrix}$

$A^{-1} = \begin{bmatrix} 8 & -8 & 5 \\ -3 & 2 & -1 \\ -1 & -1 & 1 \end{bmatrix}$

$AA^{-1} = \begin{bmatrix} 1 & 3 & -2 \\ 4 & 13 & -7 \\ 5 & 16 & -8 \end{bmatrix} \begin{bmatrix} 8 & -8 & 5 \\ -3 & 2 & -1 \\ -1 & -1 & 1 \end{bmatrix} = \begin{bmatrix} 8-9+2 & -8+6+2 & 5-3-2 \\ 32-39+7 & -32+26+7 & 20-13-7 \\ 40-48+8 & -40+32+8 & 25-16-8 \end{bmatrix} = \begin{bmatrix} 1 & 0 & 0 \\ 0 & 1 & 0 \\ 0 & 0 & 1 \end{bmatrix}$

$A^{-1}A = \begin{bmatrix} 8 & -8 & 5 \\ -3 & 2 & -1 \\ -1 & -1 & -1 \end{bmatrix} \begin{bmatrix} 1 & 3 & -2 \\ 4 & 13 & -7 \\ 5 & 16 & -8 \end{bmatrix} = \begin{bmatrix} 8-32+25 & 24-104+80 & -16+56-40 \\ -3+8-5 & -9+26-16 & 6-14+8 \\ -1-4+5 & -3-13+16 & 2+7-8 \end{bmatrix} = \begin{bmatrix} 1 & 0 & 0 \\ 0 & 1 & 0 \\ 0 & 0 & 1 \end{bmatrix}$

43. a. $\begin{bmatrix} 1 & 1 & 2 \\ 0 & 1 & 3 \\ 3 & 0 & -2 \end{bmatrix} \begin{bmatrix} x \\ y \\ z \end{bmatrix} = \begin{bmatrix} 7 \\ -2 \\ 0 \end{bmatrix}.$

b. $A^{-1}B = \begin{bmatrix} -2 & 2 & 1 \\ 9 & -8 & -3 \\ -3 & 3 & 1 \end{bmatrix} \begin{bmatrix} 7 \\ -2 \\ 0 \end{bmatrix} = \begin{bmatrix} -14-4+0 \\ 63+16+0 \\ -21-6+0 \end{bmatrix} = \begin{bmatrix} -18 \\ 79 \\ -27 \end{bmatrix}$

The solution to the system is $\{(-18, 79, -27)\}$.

44. a. $\begin{bmatrix} 1 & -1 & 2 \\ 0 & 1 & -1 \\ 1 & 0 & 2 \end{bmatrix} \begin{bmatrix} x \\ y \\ z \end{bmatrix} = \begin{bmatrix} 12 \\ -5 \\ 10 \end{bmatrix}$

b. $A^{-1}B = \begin{bmatrix} 2 & 2 & -1 \\ -1 & 0 & 1 \\ -1 & -1 & 1 \end{bmatrix} \begin{bmatrix} 12 \\ -5 \\ 10 \end{bmatrix} = \begin{bmatrix} 24-10-10 \\ -12+10 \\ -12+5+10 \end{bmatrix} = \begin{bmatrix} 4 \\ -2 \\ 3 \end{bmatrix}$

The solution to the system is $\{(4, -2, 3)\}$.

45. R U L E has a numerical equivalent of 18, 21, 12, 5.

$$\begin{bmatrix} 3 & 2 \\ 4 & 3 \end{bmatrix}\begin{bmatrix} 18 & 12 \\ 21 & 5 \end{bmatrix} = \begin{bmatrix} 54+42 & 36+10 \\ 72+63 & 48+15 \end{bmatrix} = \begin{bmatrix} 96 & 46 \\ 135 & 63 \end{bmatrix}$$

The encoded message is 96, 135, 46, 63.

$$\begin{bmatrix} 3 & -2 \\ -4 & 3 \end{bmatrix}\begin{bmatrix} 96 & 46 \\ 135 & 63 \end{bmatrix} = \begin{bmatrix} 288-270 & 138-126 \\ -384+405 & -184+189 \end{bmatrix} = \begin{bmatrix} 18 & 12 \\ 21 & 5 \end{bmatrix}$$

The decoded message is 18, 21, 12, 5 or RULE.

46. $\begin{vmatrix} 3 & 2 \\ -1 & 5 \end{vmatrix} = 15 - (-2) = 17$

47. $\begin{vmatrix} -2 & -3 \\ -4 & -8 \end{vmatrix} = 16 - 12 = 4$

48. $\begin{vmatrix} 2 & 4 & -3 \\ 1 & -1 & 5 \\ -2 & 4 & 0 \end{vmatrix} = -2\begin{vmatrix} 4 & -3 \\ -1 & 5 \end{vmatrix} - 4\begin{vmatrix} 2 & -3 \\ 1 & 5 \end{vmatrix} + 0\begin{vmatrix} 2 & 4 \\ 1 & -1 \end{vmatrix}$

$\qquad\qquad = -2(20 - 3) - 4[10 - (-3)] + 0$
$\qquad\qquad = -2(17) - 4(13)$
$\qquad\qquad = -34 - 52$
$\qquad\qquad = -86$

49. $\begin{vmatrix} 4 & 7 & 0 \\ -5 & 6 & 0 \\ 3 & 2 & -4 \end{vmatrix} = 4\begin{vmatrix} 6 & 0 \\ 2 & -4 \end{vmatrix} + 5\begin{vmatrix} 7 & 0 \\ 2 & -4 \end{vmatrix} + 3\begin{vmatrix} 7 & 0 \\ 6 & 0 \end{vmatrix}$

$\qquad\qquad = 4(-24 - 0) + 5(-28 - 0) + 3(0 - 0)$
$\qquad\qquad = 4(-24) + 5(-28) + 0$
$\qquad\qquad = -236$

50. $\begin{vmatrix} 1 & 1 & 0 & 2 \\ 0 & 3 & 2 & 1 \\ 0 & -2 & 4 & 0 \\ 0 & 3 & 0 & 1 \end{vmatrix} = \begin{vmatrix} 3 & 2 & 1 \\ -2 & 4 & 0 \\ 3 & 0 & 1 \end{vmatrix}$

$\qquad\qquad = 3\begin{vmatrix} 2 & 1 \\ 4 & 0 \end{vmatrix} + \begin{vmatrix} 3 & 2 \\ -2 & 4 \end{vmatrix}$
$\qquad\qquad = 3(0 - 4) + [12 - (-4)]$
$\qquad\qquad = 3(-4) + 16$
$\qquad\qquad = -12 + 16$
$\qquad\qquad = 4$

51. $\begin{vmatrix} 2 & 2 & 2 & 2 \\ 0 & 2 & 2 & 2 \\ 0 & 0 & 2 & 2 \\ 0 & 0 & 0 & 2 \end{vmatrix} = 2\begin{vmatrix} 2 & 2 & 2 \\ 0 & 2 & 2 \\ 0 & 0 & 2 \end{vmatrix}$

$$= 2(2)\begin{vmatrix} 2 & 2 \\ 0 & 2 \end{vmatrix}$$

$$= 2(2)(4)$$

$$= 16$$

52. $D = \begin{vmatrix} 1 & -2 \\ 3 & 2 \end{vmatrix} = 2 - (-6) = 2 + 6 = 8$

$D_x = \begin{vmatrix} 8 & -2 \\ -1 & 2 \end{vmatrix} = 16 - 2 = 14$

$D_y = \begin{vmatrix} 1 & 8 \\ 3 & -1 \end{vmatrix} = -1 - 24 = -25$

$x = \dfrac{D_x}{D} = \dfrac{14}{8} = \dfrac{7}{4}, \ y = \dfrac{D_y}{D} = \dfrac{-25}{8} = -\dfrac{25}{8}$

The solution to the system is $\left\{ \left(\dfrac{7}{4}, -\dfrac{25}{8} \right) \right\}$.

53. $D = \begin{vmatrix} 7 & 2 \\ 2 & 1 \end{vmatrix} = 7 - 4 = 3$

$D = \begin{vmatrix} 7 & 2 \\ 2 & 1 \end{vmatrix} = 7 - 4 = 3$

$D_x = \begin{vmatrix} 0 & 2 \\ -3 & 1 \end{vmatrix} = 0 - (-6) = 6$

$D_y = \begin{vmatrix} 7 & 0 \\ 2 & -3 \end{vmatrix} = -21 - 0 = -21$

$x = \dfrac{D_x}{D} = \dfrac{6}{3} = 2$

$y = \dfrac{D_y}{D} = \dfrac{-21}{3} = -7$

The solution to the system is $\{(2, -7)\}$.

54. $D = \begin{vmatrix} 1 & 2 & 2 \\ 2 & 4 & 7 \\ -2 & -5 & -2 \end{vmatrix} = \begin{vmatrix} 1 & 2 & 2 \\ 0 & 0 & 3 \\ 0 & -1 & 2 \end{vmatrix}$

$= \begin{vmatrix} 0 & 3 \\ -1 & 2 \end{vmatrix}$

$= 0 - (-3)$

$= 3$

$D_x = \begin{vmatrix} 5 & 2 & 2 \\ 19 & 4 & 7 \\ 8 & -5 & -2 \end{vmatrix}$

$= 5\begin{vmatrix} 4 & 7 \\ -5 & -2 \end{vmatrix} - 2\begin{vmatrix} 19 & 7 \\ 8 & -2 \end{vmatrix} + 2\begin{vmatrix} 19 & 4 \\ 8 & -5 \end{vmatrix}$

$= 5[-8 - (-35)] - 2(-38 - 56) + 2(-95 - 32)$

$= 5(27) - 2(-94) - 2(127)$

$= 135 + 188 - 254$

$= 69$

$D_y = \begin{vmatrix} 1 & 5 & 2 \\ 2 & 19 & 7 \\ -2 & 8 & -2 \end{vmatrix}$

$= \begin{vmatrix} 1 & 5 & 2 \\ 0 & 9 & 3 \\ 0 & 18 & 2 \end{vmatrix}$

$= \begin{vmatrix} 9 & 3 \\ 18 & 2 \end{vmatrix}$

$= 18 - 54$

$= -36$

$D_z = \begin{vmatrix} 1 & 2 & 5 \\ 2 & 4 & 19 \\ -2 & -5 & 8 \end{vmatrix}$

$= \begin{vmatrix} 1 & 2 & 5 \\ 0 & 0 & 9 \\ 0 & -1 & 18 \end{vmatrix}$

$= \begin{vmatrix} 0 & 9 \\ -1 & 18 \end{vmatrix}$

$= 0 - (-9)$

$= 9$

$x = \dfrac{D_x}{D} = \dfrac{69}{3} = 23, \ y = \dfrac{D_y}{D} = \dfrac{-36}{3} = -12,$

$z = \dfrac{D_z}{D} = \dfrac{9}{3} = 3$

The solution to the system is $\{(23, -12, 3)\}$.

55.

$$D = \begin{vmatrix} 2 & 1 & 0 \\ 0 & 1 & -2 \\ 3 & 0 & -2 \end{vmatrix}$$

$$= 2\begin{vmatrix} 1 & -2 \\ 0 & -2 \end{vmatrix} + 3\begin{vmatrix} 1 & 0 \\ 1 & -2 \end{vmatrix}$$

$$= 2(-2-0) + 3(-2-0)$$

$$= 2(-2) + 3(-2)$$

$$= -4 - 6$$

$$= -10$$

$$D_x = \begin{vmatrix} -4 & 1 & 0 \\ 0 & 1 & -2 \\ -11 & 0 & -2 \end{vmatrix}$$

$$= -1\begin{vmatrix} 0 & -2 \\ -11 & -2 \end{vmatrix} + 1\begin{vmatrix} -4 & 0 \\ -11 & -2 \end{vmatrix}$$

$$= -1(0-22) + 1(8-0)$$

$$= 22 + 8$$

$$= 30$$

$$D_y = \begin{vmatrix} 2 & -4 & 0 \\ 0 & 0 & -2 \\ 3 & -11 & -2 \end{vmatrix}$$

$$= 2\begin{vmatrix} 0 & -2 \\ -11 & -2 \end{vmatrix} + 3\begin{vmatrix} -4 & 0 \\ 0 & -2 \end{vmatrix}$$

$$= 2(0-22) + 3(8-0)$$

$$= 2(-22) + 3(8)$$

$$= -44 + 24$$

$$= -20$$

$$D_z = \begin{vmatrix} 2 & 1 & -4 \\ 0 & 1 & 0 \\ 3 & 0 & -11 \end{vmatrix}$$

$$= 2\begin{vmatrix} 1 & 0 \\ 0 & -11 \end{vmatrix} + 3\begin{vmatrix} 1 & -4 \\ 1 & 0 \end{vmatrix}$$

$$= 2(-11-0) + 3(0+4)$$

$$= 2(-11) + 3(+4) = -22 + 12$$

$$= -10$$

$$x = \frac{D_x}{D} = \frac{30}{-10} = -3$$

$$y = \frac{D_y}{D} = \frac{-20}{-10} = 2$$

$$z = \frac{D_z}{D} = \frac{-10}{-10} = 1$$

The solution to the system is $\{(-3, 2, 1)\}$.

56. The quadratic function must satisfy
$$f(20) = 400 = 400a + 20b + c$$
$$f(40) = 150 = 1600a + 40b + c$$
$$f(60) = 400 = 3600a + 60b + c$$

$$D = \begin{vmatrix} 400 & 20 & 1 \\ 1600 & 40 & 1 \\ 3600 & 60 & 1 \end{vmatrix}$$

$$= (400)(20)\begin{vmatrix} 1 & 1 & 1 \\ 4 & 2 & 1 \\ 9 & 3 & 1 \end{vmatrix}$$

$$= 8000\begin{vmatrix} 1 & 1 & 1 \\ 3 & 1 & 0 \\ 8 & 2 & 0 \end{vmatrix} = 8000\begin{vmatrix} 3 & 1 \\ 8 & 2 \end{vmatrix}$$

$$= 8000(6-8)$$

$$= 8000(-2)$$

$$= -16,000$$

$$D_a = \begin{vmatrix} 400 & 20 & 1 \\ 150 & 40 & 1 \\ 400 & 60 & 1 \end{vmatrix}$$

$$= (50)(20)\begin{vmatrix} 8 & 1 & 1 \\ 3 & 2 & 1 \\ 8 & 3 & 1 \end{vmatrix}$$

$$= 1000\begin{vmatrix} 8 & 1 & 1 \\ -5 & 1 & 0 \\ 0 & 2 & 0 \end{vmatrix}$$

$$= 1000\begin{vmatrix} -5 & 1 \\ 0 & 2 \end{vmatrix}$$

$$= 1000(-10-0)$$

$$= -10,000$$

$$D_b = \begin{vmatrix} 400 & 400 & 1 \\ 1600 & 150 & 1 \\ 3600 & 400 & 1 \end{vmatrix}$$

$$= (400)(50)\begin{vmatrix} 1 & 8 & 1 \\ 4 & 3 & 1 \\ 9 & 8 & 1 \end{vmatrix}$$

$$= 20,000\begin{vmatrix} 1 & 8 & 1 \\ 3 & -5 & 0 \\ 8 & 0 & 0 \end{vmatrix}$$

$$= 20,000\begin{vmatrix} 3 & -5 \\ 8 & 0 \end{vmatrix}$$

$$= 20,000[0-(-40)]$$

$$= 20,000(40)$$

$$= 800,000$$

$$D_c = \begin{vmatrix} 400 & 20 & 400 \\ 1600 & 40 & 150 \\ 3600 & 60 & 400 \end{vmatrix}$$

$$= (400)(20)(50)\begin{vmatrix} 1 & 1 & 8 \\ 4 & 2 & 3 \\ 9 & 3 & 8 \end{vmatrix}$$

$$= 400,000 \begin{vmatrix} 1 & 0 & 0 \\ 4 & -2 & -29 \\ 2 & -6 & -64 \end{vmatrix}$$

$$= 400,000 \begin{vmatrix} -2 & -29 \\ -6 & -64 \end{vmatrix}$$

$$= 400,000(128 - 174)$$
$$= 400,000(-46)$$
$$= -18,400,000$$

$$a = \frac{D_a}{D} = \frac{-10,000}{-16,000} = \frac{5}{8},$$

$$b = \frac{D_b}{D} = \frac{800,000}{-16,000} = -50,$$

$$c = \frac{D_c}{D} = \frac{-18,400,000}{-16,000} = 1150$$

The model is $f(x) = \frac{5}{8}x^2 - 50x + 1150$.

$$f(30) = \frac{5}{8}(900) - 50(30) + 1150$$
$$= 562.5 - 1500 + 1150$$
$$= 212.5$$

$$f(50) = \frac{5}{8}(2500) - 50(50) + 1150$$
$$= 1562.8 - 2500 + 1150$$
$$= 212.5$$

30- and 50-year-olds are involved in an average of 212.5 automobile accidents per day.

Chapter 9 Test

1. $\begin{bmatrix} 1 & 2 & -1 & | & -3 \\ 2 & -4 & 1 & | & -7 \\ -2 & 2 & -3 & | & 4 \end{bmatrix} \begin{matrix} \\ -2R_1 + R_2 \\ 2R_1 + R_3 \end{matrix}$

$\begin{bmatrix} 1 & 2 & -1 & | & -3 \\ 0 & -8 & 3 & | & -1 \\ 0 & 6 & -5 & | & -2 \end{bmatrix} -\frac{1}{8}R_2$

$\begin{bmatrix} 1 & 2 & -1 & | & -3 \\ 0 & 1 & -\frac{3}{8} & | & \frac{1}{8} \\ 0 & 6 & -5 & | & -2 \end{bmatrix} -6R_2 + R_3$

$\begin{bmatrix} 1 & 2 & -1 & | & -3 \\ 0 & 1 & -\frac{3}{8} & | & \frac{1}{8} \\ 0 & 0 & -\frac{11}{4} & | & -\frac{11}{4} \end{bmatrix} -\frac{4}{11}R_3$

$\begin{bmatrix} 1 & 2 & -1 & | & -3 \\ 0 & 1 & -\frac{3}{8} & | & \frac{1}{8} \\ 0 & 0 & 1 & | & 1 \end{bmatrix}$

$$x + 2y - z = -3$$
$$y - \frac{3}{8}z = \frac{1}{8}$$
$$z = 1$$

Using back substitution,

$$y - \frac{3}{8}(1) = \frac{1}{8} \text{ and } x + 2\left(\frac{1}{2}\right) - 1 = -3.$$

$$y = \frac{1}{2} \qquad x + 1 - 1 = -3$$
$$x = -3$$

The solution to the system is $\left\{ \left(-3, \frac{1}{2}, 1\right) \right\}$.

2. $\begin{bmatrix} 1 & -2 & 1 & | & 2 \\ 2 & -1 & -1 & | & 1 \end{bmatrix} -2R_1 + R_2$

$\begin{bmatrix} 1 & -2 & 1 & | & 2 \\ 0 & 3 & -3 & | & -3 \end{bmatrix} \frac{1}{3}R_2$

$\begin{bmatrix} 1 & -2 & 1 & | & 2 \\ 0 & 1 & -1 & | & -1 \end{bmatrix}$

The system $\begin{matrix} x - 2y + z = 2 \\ y - z = -1 \end{matrix}$ has no unique solution.

Express x and y in terms of z:
$$y = z - 1$$
$$x - 2(z - 1) + z = 2$$
$$x - 2z + 2 + z = 2$$
$$x = z$$

With $z = t$, the complete solution to the system is $\{(t, t - 1, t)\}$.

3. $2B + 3C = \begin{bmatrix} 2 & -2 \\ 4 & 2 \end{bmatrix} + \begin{bmatrix} 3 & 6 \\ -3 & 9 \end{bmatrix} = \begin{bmatrix} 5 & 4 \\ 1 & 11 \end{bmatrix}$

4. $AB = \begin{bmatrix} 3+2 & -3+1 \\ 1+0 & -1+0 \\ 2+2 & -2+1 \end{bmatrix} = \begin{bmatrix} 5 & -2 \\ 1 & -1 \\ 4 & -1 \end{bmatrix}$

5. $C^{-1} = \frac{1}{(1)(3) - (2)(-1)} \begin{bmatrix} 3 & -2 \\ 1 & 1 \end{bmatrix}$

$= \frac{1}{3+2} \begin{bmatrix} 3 & -2 \\ 1 & 1 \end{bmatrix} = \begin{bmatrix} \frac{3}{5} & -\frac{2}{5} \\ \frac{1}{5} & \frac{1}{5} \end{bmatrix}$

6. $BC = \begin{bmatrix} 1+1 & 2-3 \\ 2-1 & 4+3 \end{bmatrix} = \begin{bmatrix} 2 & -1 \\ 1 & 7 \end{bmatrix}$

$BC - 3B = \begin{bmatrix} 2 & -1 \\ 1 & 7 \end{bmatrix} - \begin{bmatrix} 3 & -3 \\ 6 & 3 \end{bmatrix} = \begin{bmatrix} -1 & 2 \\ -5 & 4 \end{bmatrix}$

7. $AB = \begin{bmatrix} -3+14-10 & 2-8+6 & 0+2-2 \\ -6+21-15 & 4-12+9 & 0+3-3 \\ -3-7+10 & 2+4-6 & 0-1+2 \end{bmatrix}$

$= \begin{bmatrix} 1 & 0 & 0 \\ 0 & 1 & 0 \\ 0 & 0 & 1 \end{bmatrix} = I_3$

$BA = \begin{bmatrix} -3+4+0 & -6+6+0 & -6+6+0 \\ 7-8+1 & 14-12-1 & 14-12-2 \\ -5+6-1 & -10+9+1 & -10+9+2 \end{bmatrix}$

$= \begin{bmatrix} 1 & 0 & 0 \\ 0 & 1 & 0 \\ 0 & 0 & 1 \end{bmatrix} = I_3$

8. a. $\begin{bmatrix} 3 & 5 \\ 2 & -3 \end{bmatrix} \begin{bmatrix} x \\ y \end{bmatrix} = \begin{bmatrix} 9 \\ -13 \end{bmatrix}$

b. $A^{-1} = \dfrac{1}{(3)(-3)-(5)(2)} \begin{bmatrix} -3 & -5 \\ -2 & 3 \end{bmatrix}$

$= \dfrac{1}{-19} \begin{bmatrix} -3 & -5 \\ -2 & 3 \end{bmatrix} = \begin{bmatrix} \frac{3}{19} & \frac{5}{19} \\ \frac{2}{19} & -\frac{3}{19} \end{bmatrix}$

c. $A^{-1}B = \begin{bmatrix} \frac{3}{19} & \frac{5}{19} \\ \frac{2}{19} & -\frac{3}{19} \end{bmatrix} \begin{bmatrix} 9 \\ -13 \end{bmatrix} = \begin{bmatrix} \frac{27}{19} - \frac{65}{19} \\ \frac{18}{19} + \frac{39}{19} \end{bmatrix} = \begin{bmatrix} -2 \\ 3 \end{bmatrix}$

The solution to the system is $\{(-2, 3)\}$.

9. $\begin{vmatrix} 4 & -1 & 3 \\ 0 & 5 & -1 \\ 5 & 2 & 4 \end{vmatrix} = 4 \begin{vmatrix} 5 & -1 \\ 2 & 4 \end{vmatrix} + 5 \begin{vmatrix} -1 & 3 \\ 5 & -1 \end{vmatrix}$

$= 4[20-(-2)]+5(1-15)$

$= 4(22)+5(-14)$

$= 88-70$

$= 18$

10. $D = \begin{vmatrix} 3 & 1 & -2 \\ 2 & 7 & 3 \\ 4 & -3 & -1 \end{vmatrix} = 3 \begin{vmatrix} 7 & 3 \\ -3 & -1 \end{vmatrix} - 1 \begin{vmatrix} 2 & 3 \\ 4 & -1 \end{vmatrix} - 2 \begin{vmatrix} 2 & 7 \\ 4 & -3 \end{vmatrix}$

$= 3[-7-(-9)]-1(-2-12)-2(-6-28)$

$= 3(2)-1(-14)-2(-34)$

$= 6+14+68$

$= 88$

$D_x = \begin{vmatrix} -3 & 1 & -2 \\ 9 & 7 & 3 \\ 7 & -3 & -1 \end{vmatrix} = -3 \begin{vmatrix} 7 & 3 \\ -3 & -1 \end{vmatrix} - 1 \begin{vmatrix} 9 & 3 \\ 7 & -1 \end{vmatrix} - 2 \begin{vmatrix} 9 & 7 \\ 7 & -3 \end{vmatrix}$

$= -3[-7-(-9)]-1(-9-21)-2(-27-49)$

$= -3(2)-1(-30)-2(-76)$

$= -6+30+152$

$= 176$

$x = \dfrac{D_x}{D} = \dfrac{176}{88} = 2$

Cumulative Review Exercises (Chapters 1–9)

1. $2x^2 = 4 - x$

$2x^2 + x - 4 = 0$

$x = \dfrac{-1 \pm \sqrt{1^2 - 4(2)(-4)}}{2(2)}$

$x = \dfrac{-1 \pm \sqrt{1 - 32}}{4}$

$x = \dfrac{-1 \pm \sqrt{33}}{4}$

The solution set is $\left\{ \dfrac{-1+\sqrt{33}}{4}, \dfrac{-1-\sqrt{33}}{4} \right\}$.

2. $5x + 8 \le 7(1+x)$

$5x + 8 \le 7 + 7x$

$-2x \le -1$

$x \ge \dfrac{1}{2}$

The solution set is $\left\{ x \mid x \ge \dfrac{1}{2} \right\}$ or $\left[\dfrac{1}{2}, \infty \right)$.

3. $\sqrt{2x+4} - \sqrt{x+3} - 1 = 0$

$\sqrt{2x+4} = \sqrt{x+3} + 1$

$2x + 4 = x + 3 + 2\sqrt{x+3} + 1$

$x = 2\sqrt{x+3}$

$x^2 = 4(x+3)$

$x^2 = 4x + 12$

$x^2 - 4x - 12 = 0$

$(x-6)(x+2) = 0$

$x = 6$ or $x = -2$

$x = -2$ does not check. The solution set is $\{6\}$.

4. $3x^3 + 8x^2 - 15x + 4 = 0$

$p = \pm 1, \pm 2, \pm 4$

$q = \pm 1, \pm 3$

$\dfrac{p}{q} = \pm 1, \pm \dfrac{1}{3}, \pm 2, \pm \dfrac{2}{3}, \ \pm 4, \pm \dfrac{4}{3}$

$$\begin{array}{r|rrrr}
-4 & 3 & 8 & -15 & 4 \\
 & & -12 & 16 & -4 \\
\hline
 & 3 & -4 & 1 & 0
\end{array}$$

$(x+4)(3x^2 - 4x + 1) = 0$

$(x+4)(3x-1)(x-1) = 0$

$x = -4, \ x = \dfrac{1}{3}, \ x = 1$

The solution set is $\left\{ -4, \dfrac{1}{3}, 1 \right\}$.

5. $e^{2x} - 14e^x + 45 = 0 \quad let \ t = e^x$

$t^2 - 14t + 45 = 0$

$(t-5)(t-9) = 0$

$t = 5 \quad t = 9$

$e^x = 5 \quad e^x = 9$

$\ln e^x = \ln 5 \quad \ln e^x = \ln 9$

$x = e^x = \ln 5 \quad x = \ln 9$

The solution set is $\{\ln 5, \ln 9\}$.

6. $\log_3 x + \log_3(x+2) = 1$

$\log_3 x^2 + 2x = 1$

$3^1 = x^2 + 2x$

$x^2 + 2x - 3 = 0$

$(x-1)(x+3) = 0$

$x = 1, \ x = -3$

$x = -3$ does not check. The solution set is $\{1\}$.

7. $\left[\begin{array}{rrr|r} 1 & -1 & 1 & 17 \\ 2 & 3 & 1 & 8 \\ -4 & 1 & 5 & -2 \end{array}\right] \begin{array}{l} \\ -2R_1 + R_2 \\ 4R_1 + R_3 \end{array}$

$\left[\begin{array}{rrr|r} 1 & -1 & 1 & 17 \\ 0 & 5 & -1 & -26 \\ 0 & -3 & 9 & 66 \end{array}\right] \begin{array}{l} \\ \\ -\frac{1}{3}R_3 \end{array}$

$\left[\begin{array}{rrr|r} 1 & -1 & 1 & 17 \\ 0 & 1 & -3 & -22 \\ 0 & 5 & -1 & -26 \end{array}\right] \begin{array}{l} \\ -5R_2 + R_3 \\ 1R_2 + R_1 \end{array}$

$\left[\begin{array}{rrr|r} 1 & 0 & -2 & -5 \\ 0 & 1 & -3 & -22 \\ 0 & 0 & 14 & 84 \end{array}\right] \frac{1}{14}R_3$

$\left[\begin{array}{rrr|r} 1 & 0 & -2 & -5 \\ 0 & 1 & -3 & -22 \\ 0 & 0 & 1 & 6 \end{array}\right] \begin{array}{l} \\ 3R_3 + R_2 \\ 2R_3 + R_1 \end{array}$

$\left[\begin{array}{rrr|r} 1 & 0 & 0 & 7 \\ 0 & 1 & 0 & -4 \\ 0 & 0 & 1 & 6 \end{array}\right]$

$x = 7 \ y = -4 \ z = 6$

The solution set is $\{(7, -4, 6)\}$.

8. $D = \begin{vmatrix} 1 & -2 & 1 \\ 2 & 1 & -1 \\ 3 & 2 & -2 \end{vmatrix}$

$= 1\begin{vmatrix} 1 & -1 \\ 2 & -2 \end{vmatrix} - 2\begin{vmatrix} -2 & 1 \\ 2 & -2 \end{vmatrix} + 3\begin{vmatrix} -2 & 1 \\ 1 & 1 \end{vmatrix}$

$= 1(-2 + 2) - 2(4 - 2) + 3(2 - 1)$

$= 0 - 4 + 3$

$= -1$

$D_y = \begin{vmatrix} 1 & 7 & 1 \\ 2 & 0 & -1 \\ 3 & -2 & -2 \end{vmatrix} = 7\begin{vmatrix} 2 & -1 \\ 3 & -2 \end{vmatrix} - 2\begin{vmatrix} 1 & 1 \\ 2 & -1 \end{vmatrix}$

$= 7(-4 + 3) - 2(-1 - 2)$

$= -7 + 6 = 1$

$y = \dfrac{D_y}{D} = \dfrac{1}{-1} = -1$

$y = -1$

9. $y = \sqrt{4x - 7}$

$x = \sqrt{4y - 7}$

$x^2 = 4y - 7$

$x^2 + 7 - 4y$

$\dfrac{x^2 + 7}{4} = y$

$f^{-1}(x) = \dfrac{x^2 + 7}{4} \ (x \geq 0)$

10. $f(x) = \dfrac{x}{x^2 - 16}$

$f(0) = \dfrac{0}{-16} = 0$

y-intercept at 0

$0 = \dfrac{x}{x^2 - 16}$

$0 = x$

x-intercept at 0

$f(x) = \dfrac{x}{(x+4)(x-4)}$

vertical asymptotes at 4, –4
horizontal asymptote at 0

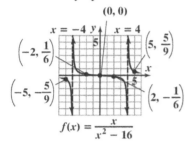

$f(x) = \dfrac{x}{x^2 - 16}$

11. $f(x) = 4x^4 - 4x^3 - 25x^2 + x + 6$

$$\begin{array}{r|rrrrr}
-2 & 4 & -4 & -25 & 1 & 6 \\
 & & -8 & 24 & 2 & -6 \\
\hline
3 & 4 & -12 & -1 & 3 & 0 \\
 & & 12 & 0 & -3 & \\
\hline
 & 4 & 0 & -1 & 0 &
\end{array}$$

$f(x) = (x+2)(x-3)(4x^2 - 1)$

$f(x) = (x+2)(x-3)(2x+1)(2x-1)$

12. $y = \log_2 x$

$2^y = x$

x	y
1	0
2	1
$\frac{1}{2}$	–1

$y = \log_2(x+1)$

Shift the graph of $y = \log_2{}^x$ left one unit.

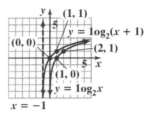

13. a. $A = A_0 e^{kt}$

$450 = 900 e^{k(40)}$

$\dfrac{1}{2} = e^{40k}$

$\ln\left(\dfrac{1}{2}\right) = \ln\left(e^{40k}\right)$

$\ln\left(\dfrac{1}{2}\right) = 40k$

$k = \dfrac{\ln\left(\dfrac{1}{2}\right)}{40} \approx -0.017$

$A = 900 e^{-0.017t}$

b. $A = 900 e^{-0.017(10)}$

$A = 900 e^{-0.17}$

$A \approx 759.30$ grams

14. $\begin{bmatrix} 1 & -1 & 0 \\ 2 & 1 & 3 \end{bmatrix} \begin{bmatrix} 4 & -1 \\ 2 & 0 \\ 1 & 1 \end{bmatrix} = \begin{bmatrix} 4-2+0 & -1+0+0 \\ 8+2+3 & -2+0+3 \end{bmatrix}$

$= \begin{bmatrix} 2 & -1 \\ 13 & 1 \end{bmatrix}$

15. $\dfrac{3x^2 + 17x - 38}{(x-3)(x-2)(x+2)} = \dfrac{A}{x-3} + \dfrac{B}{x-2} + \dfrac{C}{x+2}$

$3x^2 + 17x - 38$

$= A(x^2 - 4) + B(x^2 - x - 6) + C(x^2 - 5x + 6)$

$3x^2 + 17x - 38$

$= Ax^2 - 4A + Bx^2 - Bx - 6B + Cx^2 - 5Cx + 6c$

$3x^2 + 17x - 38$

$= (A+B+C)x^2 + (-B-5C)x - (4A+6B-6C)$

$A + B + C = 3$

$-B - 5C = 17$

$4A + 6B - 6C = 38$

$$\begin{bmatrix} 1 & 1 & 1 & | & 3 \\ 0 & -1 & -5 & | & 17 \\ 4 & 6 & -6 & | & 38 \end{bmatrix} \begin{matrix} \\ \\ -4R_1 + R_3 \end{matrix}$$

$$\begin{bmatrix} 1 & 1 & 1 & | & 3 \\ 0 & -1 & -5 & | & 17 \\ 0 & 2 & -10 & | & 26 \end{bmatrix} \begin{matrix} \\ -1R_2 \\ \end{matrix}$$

$$\begin{bmatrix} 1 & 1 & 1 & | & 3 \\ 0 & 1 & 5 & | & -17 \\ 0 & 2 & -10 & | & 26 \end{bmatrix} \begin{matrix} \\ -2R_2 + R_3 \\ -1R_2 + R_1 \end{matrix}$$

$$\begin{bmatrix} 1 & 0 & -4 & | & 20 \\ 0 & 1 & 5 & | & -17 \\ 0 & 0 & -20 & | & 60 \end{bmatrix} \begin{matrix} \\ \\ -\frac{1}{20}R_3 \end{matrix}$$

$$\begin{bmatrix} 1 & 0 & -4 & | & 20 \\ 0 & 1 & 5 & | & -17 \\ 0 & 0 & 1 & | & -3 \end{bmatrix} \begin{matrix} \\ -5R_3 + R_2 \\ 4R_3 + R_1 \end{matrix}$$

$$\begin{bmatrix} 1 & 0 & 0 & | & 8 \\ 0 & 1 & 0 & | & -2 \\ 0 & 0 & 1 & | & -3 \end{bmatrix}$$

$A = 8, \quad B = -2, \quad C = -3$

$$\frac{3x^2 + 17x - 38}{(x-3)(x-2)(x+2)} = \frac{8}{x-3} + \frac{-2}{x-2} + \frac{-3}{x+2}$$

16. $y = -\dfrac{2}{3}x - 1$

x	y
0	−1
3	−3
−3	1

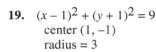

17. $3x - 5y < 15$
$-5y < -3x + 15$
$y > \dfrac{3}{5}x - 3$

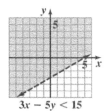

$3x - 5y < 15$

18. $f(x) = x^2 - 2x - 3$
$f(x) = (x^2 - 2x + 1) - 3 - 1$
$f(x) = (x-1)^2 - 4$

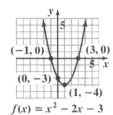

$f(x) = x^2 - 2x - 3$

19. $(x-1)^2 + (y+1)^2 = 9$
center $(1, -1)$
radius $= 3$

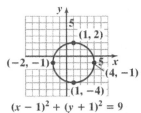

$(x-1)^2 + (y+1)^2 = 9$

20.

$$\begin{array}{r|rrrr} 2 & 1 & 0 & -6 & 4 \\ & & 2 & 4 & -4 \\ \hline & 1 & 2 & -2 & 0 \end{array}$$

$$\frac{x^3 - 6x + 4}{x - 2} = x^2 + 2x - 2$$

21. $y = 2\sin 2\pi x, \ 0 \le x \le 2$

Amplitude: $|A| = |2| = 2$

Period: $\dfrac{2\pi}{B} = \dfrac{2\pi}{2\pi} = 1$

x-intercepts:

$(0, 0), \left(\dfrac{1}{2}, 0\right), (1, 0), \left(\dfrac{3}{2}, 0\right), (2, 0)$

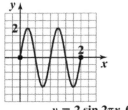

$y = 2 \sin 2\pi x, 0 \le x \le 2$

22. $\cos\left[\tan^{-1}\left(-\dfrac{4}{3} \right) \right]$

If $\tan\theta = -\dfrac{4}{3}$, θ lies in quadrant IV.

$\tan\theta = -\dfrac{4}{3} = \dfrac{y}{x} = \dfrac{-4}{3}$

$r = \sqrt{(3)^2 + (-4)^2} = \sqrt{9 + 16} = \sqrt{25} = 5$

$\cos\left[\tan^{-1}\left(-\dfrac{4}{3} \right) \right] = \dfrac{x}{r} = \dfrac{3}{5}$

23. $\dfrac{\cos 2x}{\cos x - \sin x} = \dfrac{\cos^2 x - \sin^2 x}{\cos x - \sin x}$

$\qquad\qquad = \dfrac{(\cos x - \sin x)(\cos x + \sin x)}{\cos x - \sin x}$

$\qquad\qquad = \cos x + \sin x$

24. $\qquad \cos^2 x + \sin x + 1 = 0$

$\qquad (1 - \sin^2 x) + \sin x + 1 = 0$

$\qquad -\sin^2 x + \sin x + 2 = 0$

$\qquad \sin^2 x - \sin x - 2 = 0$

$\qquad (\sin x - 2)(\sin x + 1) = 0$

$\sin x - 2 = 0 \quad \text{or} \quad \sin x + 1 = 0$

$\qquad \sin x = 2 \qquad\qquad \sin x = -1$

no solution $\quad$ or $\qquad x = \dfrac{3\pi}{2}$

The solution in the interval $[0, 2\pi)$ is $\dfrac{3\pi}{2}$.

25. $4\mathbf{w} - 5\mathbf{v} = 4(-7\mathbf{i} + 3\mathbf{j}) - 5(-6\mathbf{i} + 5\mathbf{j})$

$\qquad\qquad = -28\mathbf{i} + 12\mathbf{j} + 30\mathbf{i} - 25\mathbf{j}$

$\qquad\qquad = 2\mathbf{i} - 13\mathbf{j}$

Chapter 10
Conic Sections and Analytic Geometry

Check Point Exercises

1. $\dfrac{x^2}{36} + \dfrac{y^2}{9} = 1$

   The center is $(0, 0)$.

   $a^2 = 36,\ a = 6$

   $b^2 = 9,\ b = 3$

   $c^2 = a^2 - b^2$

   $\quad = 36 - 9$

   $\quad = 27$

   $c = \sqrt{27}$

   $\quad = 3\sqrt{3}$

   The foci are located at $(-3\sqrt{3},\ 0)$ and $(3\sqrt{3},\ 0)$.

   The vertices are $(0 \pm 6,\ 0)$ which gives $(-6,\ 0)$ and $(6,\ 0)$.

   The endpoints of the minor axis are $(0,\ 0 \pm 3)$ which gives $(0, -3)$ and $(0, 3)$.

   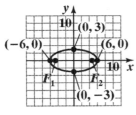

   $$\frac{x^2}{36} + \frac{y^2}{9} = 1$$

2. $16x^2 + 9y^2 = 144$

   $\dfrac{16x^2}{144} + \dfrac{9y^2}{144} = \dfrac{144}{144}$

   $\dfrac{x^2}{9} + \dfrac{y^2}{16} = 1$

   The center is $(0, 0)$.

   $a^2 = 16,\ a = 4$

   $b^2 = 9,\ b = 3$

   $c^2 = a^2 - b^2$

   $\quad = 16 - 9$

   $\quad = 7$

   $c = \sqrt{7}$

   The foci are located at $(0, -\sqrt{7})$ and $(0, \sqrt{7})$.

   The vertices are $(0,\ 0 \pm 4)$ which gives $(0, -4)$ and $(0, 4)$.

   The endpoints of the minor axis are $(0 \pm 3,\ 0)$ which gives $(-3, 0)$ and $(3, 0)$.

   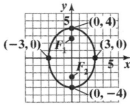

   $$16x^2 + 9y^2 = 144$$

3. Because the foci are located at $(-2, 0)$ and $(2, 0)$, on the x-axis, the major axis is horizontal.
   The center of the ellipse is midway between the foci, located at $(0, 0)$.

   Thus, the form of the equation is $\dfrac{x^2}{a^2} + \dfrac{y^2}{b^2} = 1$.

   We need to determine the values for a^2 and b^2.
   The distance from the center, $(0, 0)$, to either vertex is 3. Thus, $a = 3$ and $a^2 = 9$.
   The distance from the center, $(0, 0)$, to either focus is 2. Thus, $c = 2$ and $c^2 = 4$.

   $b^2 = a^2 - c^2$

   $\quad = 9 - 4$

   $\quad = 5$

   The equation is $\dfrac{x^2}{9} + \dfrac{y^2}{5} = 1$.

4. $\dfrac{(x+1)^2}{9} + \dfrac{(y-2)^2}{4} = 1$

   The center is $(-1, 2)$.

   $a^2 = 9,\ a = 3$

   $b^2 = 4,\ b = 2$

   $c^2 = a^2 - b^2$

   $\quad = 9 - 4$

   $\quad = 5$

   $c = \sqrt{5}$

   The foci are located at $(-1 - \sqrt{5},\ 2)$ and $(-1 + \sqrt{5},\ 2)$.

   The vertices are $(-1 \pm 3,\ 2)$ which gives $(-4, 2)$ and $(2, 2)$.

The endpoints of the minor axis are $(-1, 2\pm 2)$ which gives $(-1, 0)$ and $(-1, 4)$.

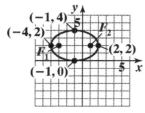

$$\frac{(x+1)^2}{9} + \frac{(y-2)^2}{4} = 1$$

5.
$$\frac{x^2}{20^2} + \frac{y^2}{10^2} = 1$$
$$\frac{x^2}{400} + \frac{y^2}{100} = 1$$
Since the truck is 12 feet wide, substitute $x = 6$ into the equation to find y.
$$\frac{6^2}{400} + \frac{y^2}{100} = 1$$
$$400\left(\frac{36}{400} + \frac{y^2}{100}\right) = 400(1)$$
$$36 + 4y^2 = 400$$
$$4y^2 = 364$$
$$y^2 = 91$$
$$y = \sqrt{91}$$
$$y \approx 9.54$$

6 feet from the center, the height of the archway is 9.54 feet. Since the truck's height is 9 feet, it will fit under the archway.

Concept and Vocabulary Check 10.1

1. ellipse; foci; center

2. 25; −5; 5; (−5, 0); (5, 0); 9; −3; 3; (0,−3); (0, 3)

3. 25; −5; 5; (0, −5); (0, 5); 9; −3; 3; (−3, 0); (3, 0)

4. 5; $(0, -\sqrt{5})$; $(0, \sqrt{5})$

5. (−1, 4)

6. (−2,−2); (8,−2)

7. (1, 9)

9. 4; 1; 16

Exercise Set 10.1

1.
$$\frac{x^2}{16} + \frac{y^2}{4} = 1$$
$$a^2 = 16, a = 4$$
$$b^2 = 4, b = 2$$
$$c^2 = a^2 - b^2 = 16 - 4 = 12$$
$$c = \sqrt{12} = 2\sqrt{3}$$
The foci are located at $(-2\sqrt{3}, 0)$ and $(2\sqrt{3}, 0)$.

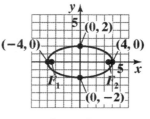

$$\frac{x^2}{16} + \frac{y^2}{4} = 1$$

3.
$$a^2 = 36, a = 6$$
$$b^2 = 9, b = 3$$
$$c^2 = a^2 - b^2 = 36 - 9 = 27$$
$$c = \sqrt{27} = 3\sqrt{3}$$
The foci are located at $(0, -3\sqrt{3})$ and $(0, 3\sqrt{3})$.

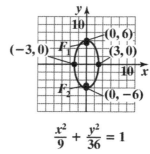

$$\frac{x^2}{9} + \frac{y^2}{36} = 1$$

5. $a^2 = 64,\ a = 8$

$b^2 = 25,\ b = 5$

$c^2 = a^2 - b^2 = 64 - 25 = 39$

$c = \sqrt{39}$

The foci are located at $(0, -\sqrt{39})$ and $(0, \sqrt{39})$.

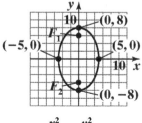

$$\frac{x^2}{25} + \frac{y^2}{64} = 1$$

7. $a^2 = 81,\ a = 9$

$b^2 = 49,\ b = 7$

$c^2 = a^2 - b^2 = 81 - 49 = 32$

$c = \sqrt{32} = 4\sqrt{2}$

The foci are located at $(0, -4\sqrt{2})$ and $(0, 4\sqrt{2})$.

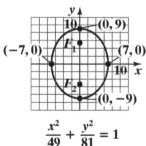

$$\frac{x^2}{49} + \frac{y^2}{81} = 1$$

9. $\dfrac{x^2}{\frac{9}{4}} + \dfrac{y^2}{\frac{25}{4}} = 1$

$c^2 = \dfrac{25}{4} - \dfrac{9}{4}$

$c^2 = \dfrac{16}{4}$

$c^2 = 4$

$c = 2$

The foci are located at $(0, 2)$ and $(0, -2)$.

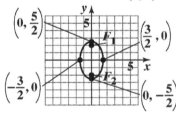

11. $x^2 = 1 - 4y^2$

$x^2 + 4y^2 = 1$

$x^2 + \dfrac{y^2}{\frac{1}{4}} = 1$

$c^2 = 1 - \dfrac{1}{4}$

$c^2 = \dfrac{3}{4}$

$c = \pm\dfrac{\sqrt{3}}{2}$

$c \approx \pm 0.9$

The foci are located at $\left(\dfrac{\sqrt{3}}{2}, 0\right)$ and $\left(-\dfrac{\sqrt{3}}{2}, 0\right)$.

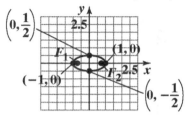

13. $25x^2 + 4y^2 = 100$

$\dfrac{25x^2}{100} + \dfrac{4y^2}{100} = \dfrac{100}{100}$

$\dfrac{x^2}{4} + \dfrac{y^2}{25} = 1$

$a^2 = 25,\ a = 5$

$b^2 = 4,\ b = 2$

$c^2 = a^2 = b^2 = 25 - 4 = 21$

The foci are located at $(0, -\sqrt{21})$ and $(0, \sqrt{21})$.

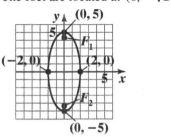

15. $4x^2 + 16y^2 = 64$

$$\frac{x^2}{16} + \frac{y^2}{4} = 1$$

$a^2 = 16, a = 4$

$b^2 = 4, b = 2$

$c^2 = 16 - 4$

$c^2 = 12$

$c = \pm\sqrt{12}$

$c = \pm 2\sqrt{3}$

$c \approx \pm 3.5$

The foci are located at $(2\sqrt{3}, 0)$ and $(-2\sqrt{3}, 0)$.

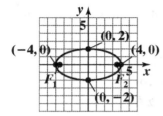

17. $7x^2 = 35 - 5y^2$

$7x^2 + 5y^2 = 35$

$$\frac{x^2}{5} + \frac{y^2}{7} = 1$$

$a^2 = 7, a = \sqrt{7}$

$b^2 = 5, b = \sqrt{5}$

$c^2 = 7 - 5$

$c^2 = 2$

$c = \pm\sqrt{2}$

$c \approx \pm 1.4$

The foci are located at $(0, \sqrt{2})$ and $(0, -\sqrt{2})$.

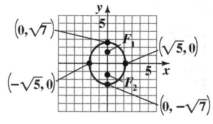

19. $a^2 = 4, b^2 = 1$, center at $(0, 0)$

$$\frac{x^2}{4} + \frac{y^2}{1} = 1$$

$c^2 = a^2 - b^2 = 4 - 1 = 3$

$c = \sqrt{3}$

The foci are at $(-\sqrt{3}, 0)$ and $(\sqrt{3}, 0)$.

21. $a^2 = 4, b^2 = 1$,

center: $(0, 0)$

$$\frac{x^2}{1} + \frac{y^2}{4} = 1$$

$c^2 = a^2 - b^2 = 4 - 1 = 3$

$c = \sqrt{3}$

The foci are at $(0, \sqrt{3})$ and $(0, -\sqrt{3})$.

23. $\dfrac{(x+1)^2}{4} + \dfrac{(y-1)^2}{1} = 1$

$a^2 = 4, \quad b^2 = 1$

$c^2 = 4 - 1$

$c^2 = 3$

$c = \pm\sqrt{3}$

The foci are located at $(-1+\sqrt{3}, 1)$ and $(-1-\sqrt{3}, 1)$.

25. $c^2 = 25, a^2 = 64$

$b^2 = a^2 - c^2 = 64 - 25 = 39$

$$\frac{x^2}{64} + \frac{y^2}{39} = 1$$

27. $c^2 = 16, a^2 = 49$

$b^2 = a^2 - c^2 = 49 - 16 = 33$

$$\frac{x^2}{33} + \frac{y^2}{49} = 1$$

29. $c^2 = 4, b^2 = 9$

$a^2 = b^2 + c^2 = 9 + 4 = 13$

$$\frac{x^2}{13} + \frac{y^2}{9} = 1$$

31. $2a = 8, a = 4, a^2 = 16$

$2b = 4, b = 2, b^2 = 4$

$$\frac{x^2}{16} + \frac{y^2}{4} = 1$$

33. $2a = 10,\ a = 5,\ a^2 = 25$

$2b = 4,\ b = 2,\ b^2 = 4$

$$\frac{(x+2)^2}{4} + \frac{(y-3)^2}{25} = 1$$

35. length of the major axis = 9 − 3 = 6
$2a = 6,\ a = 3$ major axis is vertical
length of the minor axis = 9 − 5 = 4
$2b = 4,\ b = 2$
Center is at (7, 6).

$$\frac{(x-7)^2}{4} + \frac{(y-6)^2}{9} = 1$$

37. $a^2 = 9,\ a = 3$

$b^2 = 4,\ b = 2$

center: (2, 1)

$c^2 = a^2 - b^2 = 9 - 4 = 5$

$c = \sqrt{5}$

The foci are at $(2 - \sqrt{5},\ 1)$ and $(2 + \sqrt{5},\ 1)$.

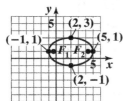

$$\frac{(x-2)^2}{9} + \frac{(y-1)^2}{4} = 1$$

39. $\dfrac{(x+3)^2}{16} + \dfrac{4(y-2)^2}{16} = \dfrac{16}{16}$

$$\frac{(x+3)^2}{16} + \frac{(y-2)^2}{4} = 1$$

$a^2 = 16,\ a = 4$

$b^2 = 4,\ b = 2$

center: (−3, 2)

$c^2 = a^2 - b^2 = 16 - 4 = 12$

$c = \sqrt{12} = 2\sqrt{3}$

The foci are at $(-3 - 2\sqrt{3},\ 2)$ and $(-3 + 2\sqrt{3},\ 2)$.

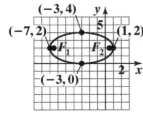

$$(x + 3)^2 + 4(y - 2)^2 = 16$$

41. $a^2 = 25,\ a = 5$

$b^2 = 9,\ b = 3$

center: (4, −2)

$c^2 = a^2 - b^2 = 25 - 9 = 16$

$c = 4$

The foci are at (4, 2) and (4, −6).

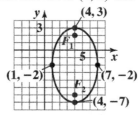

$$\frac{(x-4)^2}{9} + \frac{(y+2)^2}{25} = 1$$

43. $a^2 = 36,\ a = 6$

$b^2 = 25,\ b = 5$

center: (0, 2)

$c^2 = a^2 - b^2 = 36 - 25 = 11$

$c = \sqrt{11}$

The foci are at $(0,\ 2 + \sqrt{11})$ and $(0,\ 2 - \sqrt{11})$.

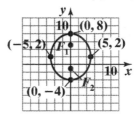

$$\frac{x^2}{25} + \frac{(y-2)^2}{36} = 1$$

45. $a^2 = 9,\ a = 3$
$b^2 = 1,\ b = 1$
center: (−3, 2)
$c^2 = a^2 - b^2 = 9 - 1 = 8$
$c = \sqrt{8} = 2\sqrt{2}$

The foci are at $(-3 - 2\sqrt{2},\ 2)$ and
$(-3 + 2\sqrt{2},\ 2)$.

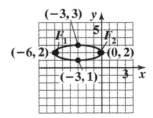

$$\frac{(x+3)^2}{9} + (y-2)^2 = 1$$

47. $c^2 = 5 - 2$

$c^2 = 3$

$c = \pm\sqrt{3}$

$c \approx \pm 1.7$

The foci are located at $(1, -3 + \sqrt{3})$ and $(1, -3 - \sqrt{3})$.

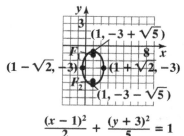

$$\frac{(x-1)^2}{2} + \frac{(y+3)^2}{5} = 1$$

49. $\dfrac{9(x-1)^2}{36} + \dfrac{4(y+3)^2}{36} = \dfrac{36}{36}$

$$\frac{(x-1)^2}{4} + \frac{(y+3)^2}{9} = 1$$

$a^2 = 9, a = 3$

$b^2 = 4, b = 2$

center: $(1, -3)$

$c^2 = a^2 - b^2 = 9 - 4 = 5$

$c = \sqrt{5}$

The foci are at $(1, -3 + \sqrt{5})$ and $(1, -3 - \sqrt{5})$.

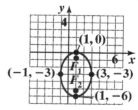

$9(x - 1)^2 + 4(y + 3)^2 = 36$

51. $(9x^2 - 36x) + (25y^2 + 50y) = 164$

$9(x^2 - 4x) + 25(y^2 + 2y) = 164$

$9(x^2 - 4x + 4) + 25(y^2 + 2y + 1)$

$= 164 + 36 + 25$

$9(x - 2)^2 + 25(y + 1)^2 = 225$

$\dfrac{9(x-2)^2}{225} + \dfrac{25(y+1)^2}{225} = \dfrac{225}{225}$

$$\frac{(x-2)^2}{25} + \frac{(y+1)^2}{9} = 1$$

center: $(2, -1)$

$a^2 = 25, a = 5$

$b^2 = 9, b = 3$

$c^2 = a^2 - b^2 = 25 - 9 = 16$

$c = 4$

The foci are at $(-2, -1)$ and $(6, -1)$.

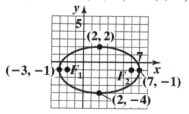

$9x^2 + 25y^2 - 36x + 50y - 164 = 0$

53. $(9x^2 - 18x) + (16y^2 + 64y) = 71$

$9(x^2 - 2x) + 16(y^2 + 4y) = 71$

$9(x^2 - 2x + 1) + 16(y^2 + 4y + 4)$

$= 71 + 9 + 64$

$9(x - 1)^2 + 16(y + 2)^2 = 144$

$\dfrac{9(x-1)^2}{144} + \dfrac{16(y+2)^2}{144} = \dfrac{144}{144}$

$$\frac{(x-1)^2}{16} + \frac{(y+2)^2}{9} = 1$$

center: $(1, -2)$

$a^2 = 16, a = 4$

$b^2 = 9, b = 3$

$c^2 = a^2 - b^2 = 16 - 9 = 7$

$c = \sqrt{7}$

The foci are at

$(1 - \sqrt{7}, -2)$ and $(1 + \sqrt{7}, -2)$.

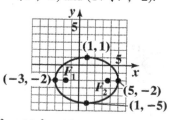

$9x^2 + 16y^2 - 18x + 64y - 71 = 0$

55.
$$(4x^2 + 16x) + (y^2 - 6y) = 39$$
$$4(x^2 + 4x) + (y^2 - 6y) = 39$$
$$4(x^2 + 4x + 4) + (y^2 - 6y + 9) = 39 + 16 + 9$$
$$4(x + 2)^2 + (y - 3)^2 = 64$$
$$\frac{4(x + 2)^2}{64} + \frac{(y - 3)^2}{64} = \frac{64}{64}$$
$$\frac{(x + 2)^2}{16} + \frac{(y - 3)^2}{64} = 1$$

center: (–2, 3)

$a^2 = 64$, $a = 8$

$b^2 = 16$, $b = 4$

$c^2 = a^2 - b^2 = 64 - 16 = 48$

$c = \sqrt{48} = 4\sqrt{3}$

The foci are at $(-2, 3 + 4\sqrt{3})$ and $(-2, 3 - 4\sqrt{3})$.

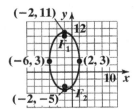

$$4x^2 + y^2 + 16x - 6y - 39 = 0$$

57. $x^2 + y^2 = 1$ $\qquad$ $x^2 + 9y^2 = 9$
$$\frac{x^2}{9} + \frac{9y^2}{9} = \frac{9}{9}$$
$$\frac{x^2}{9} + \frac{y^2}{1} = 1$$

The first equation is that of a circle with center at the origin and $r = 1$. The second equation is that of an ellipse with center at the origin, horizontal major axis of length 6 units $(a = 3)$, and vertical minor axis of length 2 units $(b = 1)$.

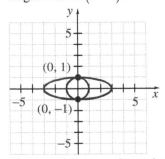

Check each intersection point.

The solution set is $\{(0, -1), (0, 1)\}$.

59. $\dfrac{x^2}{25} + \dfrac{y^2}{9} = 1$ $\qquad$ $y = 3$

The first equation is for an ellipse centered at the origin with horizontal major axis of length 10 units and vertical minor axis of length 6 units. The second equation is for a horizontal line with a y-intercept of 3.

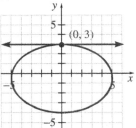

Check the intersection point.

The solution set is $\{(0,3)\}$.

61. $4x^2 + y^2 = 4$ $\qquad$ $2x - y = 2$
$$\frac{4x^2}{4} + \frac{y^2}{4} = \frac{4}{4} \qquad -y = -2x + 2$$
$$\frac{x^2}{1} + \frac{y^2}{4} = 1 \qquad y = 2x - 2$$

The first equation is for an ellipse centered at the origin with vertical major axis of length 4 units $(b = 2)$ and horizontal minor axis of length 2 units $(a = 1)$. The second equation is for a line with slope 2 and y-intercept -2.

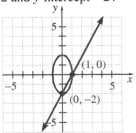

Check the intersection points.

The solution set is $\{(0, -2), (1, 0)\}$.

63. $$y^2 = \left(-\sqrt{16-4x^2}\right)^2$$
$$y^2 = 16 - 4x^2$$
$$4x^2 + y^2 = 16$$
$$\frac{x^2}{4} + \frac{y^2}{16} = 1$$

We want to graph the bottom half of an ellipse centered at the origin with a vertical major axis of length 8 units ($b = 4$) and horizontal minor axis of length 4 units $(a = 2)$.

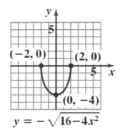

$$y = -\sqrt{16-4x^2}$$

65. $a = 15$, $b = 10$
$$\frac{x^2}{225} + \frac{y^2}{100} = 1$$
Let $x = 4$
$$\frac{4^2}{225} + \frac{y^2}{100} = 1$$
$$900\left(\frac{16}{225} + \frac{y^2}{100}\right) = 900(1)$$
$$64 + 9y^2 = 900$$
$$9y^2 = 836$$
$$y = \sqrt{\frac{836}{9}} \approx 9.64$$

Yes, the truck only needs 7 feet so it will clear.

67. a. $a = 48$, $a^2 = 2304$
$b = 23$, $b^2 = 529$
$$\frac{x^2}{2304} + \frac{y^2}{529} = 1$$

b. $c^2 = a^2 - b^2 = 2304 - 529 = 1775$
$c = \sqrt{1775} \approx 42.13$
He situated his desk about 42 feet from the center of the ellipse, along the major axis.

69. – 77. Answers will vary.

79. does not make sense; Explanations will vary.
Sample explanation: The foci are on the major axis.

81. does not make sense; Explanations will vary. Sample explanation: We must also know the other vertices.

83. $a = 6$, $a^2 = 36$
$$\frac{x^2}{b^2} + \frac{y^2}{36} = 1$$
When $x = 2$ and $y = -4$,
$$\frac{2^2}{b^2} + \frac{(-4)^2}{36} = 1$$
$$\frac{4}{b^2} + \frac{16}{36} = 1$$
$$\frac{4}{b^2} = \frac{5}{9}$$
$$36 = 5b^2$$
$$b^2 = \frac{36}{5}$$
$$\frac{x^2}{\frac{36}{5}} + \frac{y^2}{36} = 1$$

85. The large circle has radius 5 with center $(0, 0)$. Its equation is $x^2 + y^2 = 25$. The small circle has radius 3 with center $(0, 0)$. Its equation is $x^2 + y^2 = 9$

87. $4x^2 - 9y^2 = 36$
$$\frac{4x^2}{36} - \frac{9y^2}{36} = \frac{36}{36}$$
$$\frac{x^2}{9} - \frac{y^2}{4} = 1$$
The terms are separated by subtraction rather than by addition.

88. $\dfrac{x^2}{16} - \dfrac{y^2}{9} = 1$

a. Substitute 0 for y.
$$\frac{x^2}{16} - \frac{0^2}{9} = 1$$
$$\frac{x^2}{16} = 1$$
$$x^2 = 16$$
$$x = \pm 4$$
The x-intercepts are -4 and 4.

b. $\dfrac{0^2}{16} - \dfrac{y^2}{9} = 1$
$$-\frac{y^2}{9} = 1$$
$$y^2 = -9$$
The equation $y^2 = -9$ has no real solutions.

89. $\dfrac{y^2}{9} - \dfrac{x^2}{16} = 1$

a. Substitute 0 for x.

$$\dfrac{y^2}{9} - \dfrac{0^2}{16} = 1$$

$$\dfrac{y^2}{9} = 1$$

$$y^2 = 9$$

$$y = \pm 3$$

The y-intercepts are -3 and 3.

b. $\dfrac{0^2}{9} - \dfrac{x^2}{16} = 1$

$$-\dfrac{x^2}{16} = 1$$

$$x^2 = -16$$

The equation $x^2 = -16$ has no real solutions.

Section 10.2

Check Point Exercises

1. **a.** $\dfrac{x^2}{25} - \dfrac{y^2}{16} = 1$

$a^2 = 25,\ a = 5$

vertices: $(5, 0)$ and $(-5, 0)$

$b^2 = 16$

$c^2 = a^2 + b^2$

$\quad = 25 + 16$

$\quad = 41$

$c = \sqrt{41}$

The foci are at $(\sqrt{41},\ 0)$ and $(-\sqrt{41},\ 0)$.

b. $\dfrac{y^2}{25} - \dfrac{x^2}{16} = 1$

$a^2 = 25,\ a = 5$

vertices: $(0, 5)$ and $(0, -5)$

$b^2 = 16$

$c^2 = a^2 + b^2$

$\quad = 25 + 16$

$\quad = 41$

$c = \sqrt{41}$

The foci are at $(0,\ \sqrt{41})$ and $(0,\ -\sqrt{41})$.

2. Because the foci are located at $(0, -5)$ and $(0, 5)$, on the y-axis, the transverse axis lies on the y-axis. The center of the hyperbola is midway between the foci, located at $(0, 0)$.

Thus, the form of the equation is $\dfrac{y^2}{a^2} - \dfrac{x^2}{b^2} = 1$.

We need to determine the values for a^2 and b^2. The distance from the center, $(0, 0)$, to either vertex is 3. Thus, $a = 3$ and $a^2 = 9$. The distance from the center, $(0, 0)$, to either focus is 5. Thus, $c = 5$ and $c^2 = 25$.

$b^2 = c^2 - a^2$

$\quad = 25 - 9$

$\quad = 16$

The equation is $\dfrac{y^2}{9} - \dfrac{x^2}{16} = 1$.

3. $\dfrac{x^2}{36} - \dfrac{y^2}{9} = 1$

$a^2 = 36,\ a = 6$

The vertices are $(6, 0)$ and $(-6, 0)$.

$b^2 = 9,\ b = 3$

asymptotes: $y = \pm\dfrac{b}{a}x = \pm\dfrac{3}{6}x = \pm\dfrac{1}{2}x$

$c^2 = a^2 + b^2$

$\quad = 36 + 9$

$\quad = 45$

$c = \sqrt{45}$

$\quad = 3\sqrt{5}$

The foci are at $(-3\sqrt{5},\ 0)$ and $(3\sqrt{5},\ 0)$.

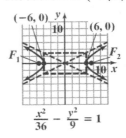

$$\dfrac{x^2}{36} - \dfrac{y^2}{9} = 1$$

4. $y^2 - 4x^2 = 4$

$$\frac{y^2}{4} - \frac{4x^2}{4} = \frac{4}{4}$$

$$\frac{y^2}{4} - x^2 = 1$$

$a^2 = 4, \ a = 2$

The vertices are (0, 2) and (0, –2).

$b^2 = 1, \ b = 1$

asymptotes: $y = \pm \dfrac{a}{b} x = \pm 2x$

$c^2 = a^2 + b^2$

$\quad = 4 + 1$

$\quad = 5$

$c = \sqrt{5}$

The foci are at $(0, \sqrt{5})$ and $(0, -\sqrt{5})$.

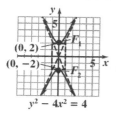

5. $\dfrac{(x-3)^2}{4} - \dfrac{(y-1)^2}{4} = 1$

center at (3, 1)

$a^2 = 4, \ a = 2$

$b^2 = 1, \ b = 1$

The vertices are (1, 1) and (5, 1).

asymptotes: $y - 1 = \pm \dfrac{1}{2}(x - 3)$

$c^2 = a^2 + b^2$

$\quad = 4 + 1$

$\quad = 5$

$c = \sqrt{5}$

The foci are at $(3 - \sqrt{5}, 1)$ and $(3 + \sqrt{5}, 1)$.

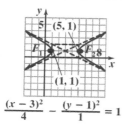

6. $4x^2 - 24x - 9y^2 - 90y - 153 = 0$

$4\left(x^2 - 6x \quad\right) - 9\left(y^2 + 10y \quad\right) = 153$

$4\left(x^2 - 6x + 9\right) - 9\left(y^2 + 10y + 25\right) = 153 + 36 + (-225)$

$4(x - 3)^2 - 9(y + 5)^2 = -36$

$\dfrac{4(x-3)^2}{-36} - \dfrac{9(y+5)^2}{-36} = \dfrac{-36}{-36}$

$-\dfrac{(x-3)^2}{9} + \dfrac{(y+5)^2}{4} = 1$

$\dfrac{(y+5)^2}{4} - \dfrac{(x-3)^2}{9} = 1$

center at (3, –5)

$a^2 = 4, \ a = 2$

$b^2 = 9, \ b = 3$

The vertices are (3, –3) and (3, –7).

asymptotes: $y + 5 = \pm \dfrac{2}{3}(x - 3)$

$c^2 = a^2 + b^2$

$\quad = 4 + 9$

$\quad = 13$

$c = \sqrt{13}$

The foci are at $(3, -5 - \sqrt{13})$ and $(3, -5 + \sqrt{13})$.

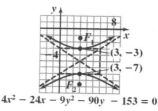

7. $c = 5280$

$2a = 3300, a = 1650$

$b^2 = c^2 - a^2 = 5280^2 - 1650^2 = 25{,}155{,}900$

The explosion occurred somewhere at the right branch of the hyperbola given by

$$\frac{x^2}{2{,}722{,}500} - \frac{y^2}{25{,}155{,}900} = 1.$$

Concept and Vocabulary Check 10.2

1. hyperbola; foci; vertices; transverse

2. $(-5, 0)$; $(5, 0)$; $(-\sqrt{34}, 0)$; $(\sqrt{34}, 0)$

3. $(0, -5)$; $(0, 5)$; $(0, -\sqrt{34})$; $(0, \sqrt{34})$

4. asymptotes; center

5. dividing; 36

6. $y = -\dfrac{3}{2}x$; $y = \dfrac{3}{2}x$

7. $y = -2x$; $y = 2x$

8. $(-3, 3)$; $(7, 3)$

9. $(7, -2)$

10. 16; 1; 128

Exercise Set 10.2

1. $a^2 = 4$, $a = 2$
   The vertices are $(2, 0)$ and $(-2, 0)$.
   $b^2 = 1$
   $c^2 = a^2 + b^2 = 4 + 1 = 5$
   $c = \sqrt{5}$
   The foci are located at $(\sqrt{5}, 0)$ and $(-\sqrt{5}, 0)$.
   graph (b)

3. $a^2 = 4$, $a = 2$
   The vertices are $(0, 2)$ and $(0, -2)$.
   $b^2 = 1$
   $c^2 = a^2 + b^2 = 4 + 1 = 5$
   $c = \sqrt{5}$
   The foci are located at
   $(0, \sqrt{5})$ and $(0, -\sqrt{5})$.
   graph (a)

5. $a = 1$, $c = 3$
   $b^2 = c^2 - a^2 = 9 - 1 = 8$
   $y^2 - \dfrac{x^2}{8} = 1$

7. $a = 3$, $c = 4$
   $b^2 = c^2 - a^2 = 16 - 9 = 7$
   $\dfrac{x^2}{9} - \dfrac{y^2}{7} = 1$

9. $2a = 6 - (-6)$
   $2a = 12$
   $a = 6$
   $\dfrac{a}{b} = 2$
   $\dfrac{6}{b} = 2$
   $6 = 2b$
   $3 = b$
   Transverse axis is vertical.
   $\dfrac{y^2}{36} - \dfrac{x^2}{9} = 1$

11. $a = 2$, $c = 7 - 4 = 3$
    $2^2 + b^2 = 3^2$
    $4 + b^2 = 9$
    $b^2 = 5$
    Transverse axis is horizontal.
    $\dfrac{(x-4)^2}{4} - \dfrac{(y+2)^2}{5} = 1$

13. $a^2 = 9$, $a = 3$
    $b^2 = 25$, $b = 5$
    vertices: $(3, 0)$ and $(-3, 0)$
    asymptotes: $y = \pm\dfrac{b}{a}x = \pm\dfrac{5}{3}x$
    $c^2 = a^2 + b^2 = 9 + 25 = 34$
    $c = \sqrt{34}$ on x-axis
    The foci are at $(\sqrt{34}, 0)$ and $(-\sqrt{34}, 0)$.

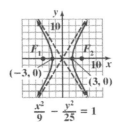

$\dfrac{x^2}{9} - \dfrac{y^2}{25} = 1$

15. $a^2 = 100, a = 10$
$b^2 = 64, b = 8$
vertices: $(10, 0)$ and $(-10, 0)$

asymptotes: $y = \pm\dfrac{b}{a}x = \pm\dfrac{8}{10}x$

or $y = \pm\dfrac{4}{5}x$
$c^2 = a^2 + b^2 = 100 + 64 = 164$
$c = \sqrt{164} = 2\sqrt{41}$ on x-axis
The foci are at $(2\sqrt{41},\ 0)$ and $(-2\sqrt{41},\ 0)$.

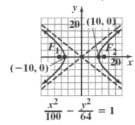

$$\frac{x^2}{100} - \frac{y^2}{64} = 1$$

17. $a^2 = 16, a = 4$
$b^2 = 36, b = 6$
vertices: $(0, 4)$ and $(0, -4)$

asymptotes: $y = \pm\dfrac{a}{b}x = \pm\dfrac{4}{6}x = \pm\dfrac{2}{3}x$

or $y = \pm\dfrac{2}{3}x$
$c^2 = a^2 + b^2 = 16 + 36 = 52$
$c = \sqrt{52} = 2\sqrt{13}$ on y-axis
The foci are at $(0,\ 2\sqrt{13})$ and $(0,\ -2\sqrt{13})$.

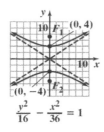

$$\frac{y^2}{16} - \frac{x^2}{36} = 1$$

19. $\dfrac{y^2}{\frac{1}{4}} - x^2 = 1$

$a^2 = \dfrac{1}{4}, a = \dfrac{1}{2}$

$b^2 = 1, b = 1$

$c^2 = a^2 + b^2$

$c^2 = \dfrac{1}{4} + 1$

$c^2 = \dfrac{5}{4}$

$c = \pm\dfrac{\sqrt{5}}{2}$

$c \approx \pm 1.1$

The foci are located at $\left(0, \dfrac{\sqrt{5}}{2}\right)$ and $\left(0, -\dfrac{\sqrt{5}}{2}\right)$.

asymptotes: $y = \pm\dfrac{\frac{1}{2}}{1}x$

$y = \pm\dfrac{1}{2}x$

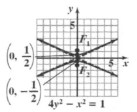

$$4y^2 - x^2 = 1$$

21. $\dfrac{9x^2}{36} - \dfrac{4y^2}{36} = \dfrac{36}{36}$

$\dfrac{x^2}{4} - \dfrac{y^2}{9} = 1$
$a^2 = 4, a = 2$
$b^2 = 9, b = 3$
vertices: $(2, 0)$ and $(-2, 0)$

asymptotes: $y = \pm\dfrac{b}{a}x = \pm\dfrac{3}{2}x$
$c^2 = a^2 + b^2 = 4 + 9 = 13$
$c = \sqrt{13}$ on x-axis
The foci are at $(\sqrt{13},\ 0)$ and $(-\sqrt{13},\ 0)$.

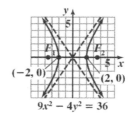

$$9x^2 - 4y^2 = 36$$

23. $\dfrac{9y^2}{225} - \dfrac{25x^2}{225} = \dfrac{225}{225}$

$\dfrac{y^2}{25} - \dfrac{x^2}{9} = 1$

$a^2 = 25, a = 5$

$b^2 = 9, b = 3$

vertices: $(0, 5)$ and $(0, -5)$

asymptotes: $y = \pm\dfrac{a}{b}x = \pm\dfrac{5}{3}x$

$c^2 = a^2 + b^2 = 25 + 9 = 34$

$c = \sqrt{34}$ on y-axis

The foci are at $(0, \sqrt{34})$ and $(0, -\sqrt{34})$.

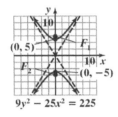

25. $y^2 = x^2 - 2$

$2 = x^2 - y^2$

$1 = \dfrac{x^2}{2} - \dfrac{y^2}{2}$

$a^2 = 2, a = \sqrt{2}$

$b^2 = 2, b = \sqrt{2}$

$c^2 = 2 + 2$

$c^2 = 4$

$c = 2$

The foci are located at $(2, 0)$ and $(-2, 0)$.

asymptotes: $y = \pm\dfrac{\sqrt{2}}{\sqrt{2}}x$

$y = \pm x$

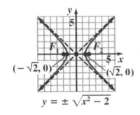

27. $a = 3, b = 5$

$\dfrac{x^2}{9} - \dfrac{y^2}{25} = 1$

29. $a = 2, b = 3$

$\dfrac{y^2}{4} - \dfrac{x^2}{9} = 1$

31. Center $(2, -3), a = 2, b = 3$

$\dfrac{(x-2)^2}{4} - \dfrac{(y+3)^2}{9} = 1$

33. center: $(-4, -3)$

$a^2 = 9, a = 3$

$b^2 = 16, b = 4$

vertices: $(-7, -3)$ and $(-1, -3)$

asymptotes: $y + 3 = \pm\dfrac{4}{3}(x+4)$

$c^2 = a^2 + b^2 = 9 + 16 = 25$

$c = \pm 5$ parallel to x-axis

The foci are at $(-9, -3)$ and $(1, -3)$.

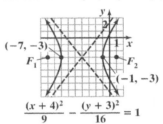

35. center: $(-3, 0)$

$a^2 = 25, a = 5$

$b^2 = 16, b = 4$

vertices: $(2, 0)$ and $(-8, 0)$

asymptotes: $y = \pm\dfrac{4}{5}(x+3)$

$c^2 = a^2 + b^2 = 25 + 16 = 41$

$c = \sqrt{41}$

The foci are at $(-3+\sqrt{41}, 0)$ and $(-3-\sqrt{41}, 0)$.

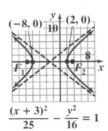

37. center: $(1, -2)$
$a^2 = 4, a = 2$
$b^2 = 16, b = 4$
vertices: $(1, 0)$ and $(1, -4)$

asymptotes: $y + 2 = \pm\dfrac{1}{2}(x - 1)$

$c^2 = a^2 + b^2 = 4 + 16 = 20$

$c = \sqrt{20} = 2\sqrt{5}$ parallel to y-axis

The foci are at $(1, -2 + 2\sqrt{5})$ and $(1, -2 - 2\sqrt{5})$.

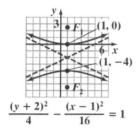

$$\dfrac{(y+2)^2}{4} - \dfrac{(x-1)^2}{16} = 1$$

39. $\dfrac{(x-3)^2}{4} - \dfrac{4(y+3)^2}{4} = \dfrac{4}{4}$

$\dfrac{(x-3)^2}{4} - (y+3)^2 = 1$

center: $(3, -3)$

$a^2 = 4, a = 2$

$b^2 = 1, b = 1$

vertices: $(1, -3)$ and $(5, -3)$

asymptotes: $y + 3 = \pm\dfrac{1}{2}(x - 3)$

$c^2 = a^2 + b^2 = 4 + 1 = 5$

$c = \sqrt{5}$

The foci are at $(3 + \sqrt{5}, -3)$ and $(3 - \sqrt{5}, -3)$.

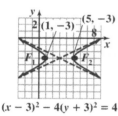

$(x - 3)^2 - 4(y + 3)^2 = 4$

41. $\dfrac{(x-1)^2}{3} - \dfrac{(y-2)^2}{3} = 1$

center: $(1, 2)$

$a^2 = 3, a = \sqrt{3}$

$b^2 = , b = \sqrt{3}$

vertices: $(-1, 2)$ and $(3, 2)$

asymptotes: $y - 2 = \pm(x - 1)$

$c^2 = a^2 + b^2 = 3 + 3 = 6$

$c = \sqrt{6}$ parallel to y-axis

The foci are at $(1 + \sqrt{6}, 2)$ and $(1 - \sqrt{6}, 2)$.

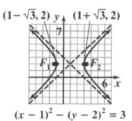

$(x - 1)^2 - (y - 2)^2 = 3$

43. $(x^2 - 2x) - (y^2 + 4y) = 4$
$(x^2 - 2x + 1) - (y^2 + 4y + 4) = 4 + 1 - 4$
$(x - 1)^2 - (y + 2)^2 = 1$

center: $(1, -2)$

$a^2 = 1, a = 1$

$b^2 = 1, b = 1$

$c^2 = a^2 + b^2 = 1 + 1 = 2$

$c = \sqrt{2}$

asymptotes: $y + 2 = \pm(x - 1)$

The foci are at $(1 + \sqrt{2}, -2)$ and $(1 - \sqrt{2}, -2)$.

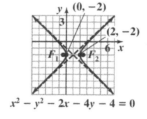

$x^2 - y^2 - 2x - 4y - 4 = 0$

45.
$$(16x^2 + 64x) - (y^2 + 2y) = -67$$
$$16(x^2 + 4x + 4) - (y^2 + 2y + 1) = -67 + 64 - 1$$
$$16(x + 2)^2 - (y + 1)^2 = -4$$
$$\frac{16(x+2)^2}{-4} - \frac{(y+1)^2}{-4} = \frac{-4}{-4}$$
$$\frac{(y+1)^2}{4} - \frac{(x+2)^2}{\frac{1}{4}} = 1$$

center: $(-2, -1)$
$a^2 = 4, a = 2$
$b^2 = \dfrac{1}{4}, b = \dfrac{1}{2}$
$c^2 = a^2 + b^2 = 4 + \dfrac{1}{4} = \dfrac{17}{4}$
$c = \sqrt{\dfrac{17}{4}} = \sqrt{4.25}$

asymptotes:
$$(y + 1) = \pm\frac{2}{\frac{1}{2}}(x + 2)$$
$$y + 1 = \pm 4(x + 2)$$

The foci are at $\left(-2, -1 + \sqrt{4.25}\right)$ and $\left(-2, -1 - \sqrt{4.25}\right)$.

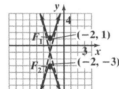

$$16x^2 - y^2 + 64x - 2y + 67 = 0$$

47.
$$(4x^2 - 16x) - (9y^2 - 54y) = 101$$
$$4(x^2 - 4x + 4) - 9(y^2 - 6y + 9) = 101 + 16 - 81$$
$$4(x - 2)^2 - 9(y - 3)^2 = 36$$
$$\frac{(x-2)^2}{9} - \frac{(y-3)^2}{4} = 1$$

center: $(2, 3)$
$a^2 = 9, a = 3$
$b^2 = 4, b = 2$
$c^2 = a^2 + b^2 = 9 + 4 = 13$
$c = \sqrt{13}$

asymptotes: $y - 3 = \pm\dfrac{2}{3}(x - 2)$

The foci are at $(2 + \sqrt{13}, 3)$ and $(2 - \sqrt{13}, 3)$.

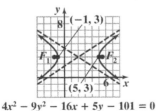

$$4x^2 - 9y^2 - 16x + 5y - 101 = 0$$

49.
$$(4x^2 - 32x) - 25y^2 = -164$$
$$4(x^2 - 8x + 16) - 25y^2 = -164 + 64$$
$$4(x - 4)^2 - 25y^2 = -100$$
$$\frac{4(x-4)^2}{-100} - \frac{25y^2}{-100} = \frac{-100}{-100}$$
$$\frac{y^2}{4} - \frac{(x-4)^2}{25} = 1$$

center: $(4, 0)$
$a^2 = 4, a = 2$
$b^2 = 25, b = 5$
$c^2 = a^2 + b^2 = 4 + 25 = 29$
$c = \sqrt{29}$

asymptotes: $y = \pm\dfrac{2}{5}(x - 4)$

The foci are at $(4, \sqrt{29})$ and $(4, -\sqrt{29})$.

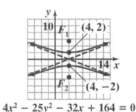

$$4x^2 - 25y^2 - 32x + 164 = 0$$

51. $\dfrac{x^2}{9} - \dfrac{y^2}{16} = 1$

The equation is for a hyperbola in standard form with the transverse axis on the x-axis. We have $a^2 = 9$ and $b^2 = 16$, so $a = 3$ and $b = 4$.

Therefore, the vertices are at $(\pm a, 0)$ or $(\pm 3, 0)$.

Using a dashed line, we construct a rectangle using the ± 3 on the x-axis and ± 4 on the y-axis. Then use dashed lines to draw extended diagonals for the rectangle. These represent the asymptotes of the graph.

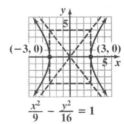

$$\frac{x^2}{9} - \frac{y^2}{16} = 1$$

From the graph we determine the following:
Domain: $\{x \mid x \le -3 \text{ or } x \ge 3\}$ or
$(-\infty, -3] \cup [3, \infty)$

Range: $\{y \mid y \text{ is a real number}\}$ or $(-\infty, \infty)$

53.

$$\frac{x^2}{9} + \frac{y^2}{16} = 1$$

The equation is for an ellipse in standard form with major axis along the y-axis. We have $a^2 = 16$ and $b^2 = 9$, so $a = 4$ and $b = 3$. Therefore, the vertices are $(0, \pm a)$ or $(0, \pm 4)$. The endpoints of the minor axis are $(\pm b, 0)$ or $(\pm 3, 0)$.

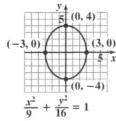

From the graph we determine the following:

Domain: $\{x \mid -3 \le x \le 3\}$ or $[-3, 3]$

Range: $\{y \mid -4 \le y \le 4\}$ or $[-4, 4]$.

55.

$$\frac{y^2}{16} - \frac{x^2}{9} = 1$$

The equation is in standard form with the transverse axis on the y-axis. We have $a^2 = 16$ and $b^2 = 9$, so $a = 4$ and $b = 3$. Therefore, the vertices are at $(0, \pm a)$ or $(0, \pm 4)$. Using a dashed line, we construct a rectangle using the ± 4 on the y-axis and ± 3 on the x-axis. Then use dashed lines to draw extended diagonals for the rectangle. These represent the asymptotes of the graph.

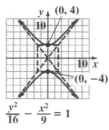

From the graph we determine the following:

Domain: $\{x \mid x \text{ is a real number}\}$ or $(-\infty, \infty)$

Range: $\{y \mid y \le -4 \text{ or } y \ge 4\}$ or $(-\infty, -4] \cup [4, \infty)$

57.

$$x^2 - y^2 = 4$$
$$x^2 + y^2 = 4$$

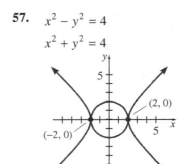

Check $(-2, 0)$:

$$(-2)^2 - 0^2 = 4 \qquad (-2)^2 + 0^2 = 4$$
$$4 - 0 = 4 \qquad\qquad 4 + 0 = 4$$
$$4 = 4 \text{ true} \qquad\qquad 4 = 4 \text{ true}$$

Check $(2, 0)$:

$$(2)^2 - 0^2 = 4 \qquad (2)^2 + 0^2 = 4$$
$$4 - 0 = 4 \qquad\qquad 4 + 0 = 4$$
$$4 = 4 \text{ true} \qquad\qquad 4 = 4 \text{ true}$$

The solution set is $\{(-2, 0), (2, 0)\}$.

59.

$$9x^2 + y^2 = 9 \quad \text{or} \quad \frac{x^2}{1} + \frac{y^2}{9} = 1$$
$$y^2 - 9x^2 = 9 \qquad\qquad \frac{y^2}{9} - \frac{x^2}{1} = 1$$

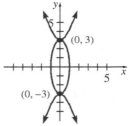

Check $(0, -3)$:

$$9(0)^2 + (-3)^2 = 9 \qquad (-3)^2 - 9(0)^2 = 9$$
$$0 + 9 = 9 \qquad\qquad\qquad 9 - 0 = 9$$
$$9 = 9 \qquad\qquad\qquad\quad 9 = 9$$
$$\text{true} \qquad\qquad\qquad\quad \text{true}$$

Check $(0, 3)$:

$$9(0)^2 + (3)^2 = 9 \qquad (3)^2 - 9(0)^2 = 9$$
$$0 + 9 = 9 \qquad\qquad\qquad 9 - 0 = 9$$
$$9 = 9 \qquad\qquad\qquad\quad 9 = 9$$
$$\text{true} \qquad\qquad\qquad\quad \text{true}$$

The solution set is $\{(0, -3), (0, 3)\}$.

61. $|d_2 - d_1| = 2a = (2 \text{ s})(1100 \text{ ft / s}) = 2200 \text{ ft}$

$a = 1100 \text{ ft}$

$2c = 5280 \text{ ft}, \ c = 2640 \text{ ft}$

$b^2 = c^2 - a^2 = (2640)^2 - (1100)^2$

$= 5,759,600$

$\dfrac{x^2}{(1100)^2} - \dfrac{y^2}{5,759,600} = 1$

$\dfrac{x^2}{1,210,000} - \dfrac{y^2}{5,759,600} = 1$

If M_1 is located 2640 feet to the right of the origin on the *x*-axis, the explosion is located on the right branch of the hyperbola given by the equation above.

63. $625y^2 - 400x^2 = 250,000$

$\dfrac{625y^2}{250,000} - \dfrac{400x^2}{250,000} = \dfrac{250,000}{250,000}$

$\dfrac{y^2}{400} - \dfrac{x^2}{625} = 1$

$a^2 = 400, \ a = \sqrt{400} = 20$

$2a = 40$

The houses are 40 yards apart at their closest point.

65. **a.** ellipse

b. $x^2 + 4y^2 = 4$

67. – 75. Answers will vary.

77. $\dfrac{x^2}{4} - \dfrac{y^2}{9} = 0$

$y^2 = \dfrac{9}{4}x^2$

$y = \pm\dfrac{3}{2}x$

No; in general, the graph is two intersecting lines.

79. $4x^2 - 6xy + 2y^2 - 3x + 10y - 6 = 0$

$2y^2 + (10 - 6x)y + (4x^2 - 3x - 6) = 0$

$y = \dfrac{6x - 10 \pm \sqrt{(10 - 6x)^2 - 8(4x^2 - 3x - 6)}}{4}$

$y = \dfrac{6x - 10 \pm \sqrt{4(x^2 - 24x + 37)}}{4}$

$y = \dfrac{3x - 5 \pm \sqrt{x^2 - 24x + 37}}{2}$

The *xy*-term rotates the hyperbola. Separation of terms into ones containing only *x* or only *y* would not be possible.

81. does not make sense; Explanations will vary. Sample explanation: This would change the ellipse to a hyperbola.

83. makes sense

85. false; Changes to make the statement true will vary. A sample change is: If a hyperbola has a transverse axis along the *x*–axis and one of the branches is removed, the remaining branch does not define a function of *x*.

87. true

89. $\dfrac{c}{a}$ will be large when *a* is small. When this happens, the asymptotes will be nearly vertical.

91. If the asymptotes are perpendicular, then their slopes are negative reciprocals. For the hyperbola $\dfrac{x^2}{a^2} - \dfrac{y^2}{b^2} = 1$, the asymptotes are $y = \pm\dfrac{b}{a}x$. The slopes are negative reciprocals when $\dfrac{b}{a} = \dfrac{a}{b}$ (since one is already the negative of the other). This happens when $b^2 = a^2$, so $a = b$. Any hyperbola where $a = b$, such as $\dfrac{x^2}{4} - \dfrac{y^2}{4} = 1$, has perpendicular asymptotes.

92. $y = x^2 + 4x - 5$

Since $a = 1$ is positive, the parabola opens upward. The x-coordinate of the vertex is

$x = -\dfrac{b}{2a} = -\dfrac{4}{2(1)} = -2$. The y-coordinate of the

vertex is $y = (-2)^2 + 4(-2) - 5 = -9$.

Vertex: $(-2, -9)$.

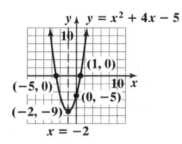

93. $y = -3(x - 1)^2 + 2$

Since $a = -3$ is negative, the parabola opens downward. The vertex of the parabola is $(h, k) = (1, 2)$.

The y–intercept is -1.

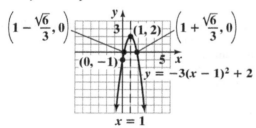

94. $y^2 + 2y + 12x - 23 = 0$

$$y^2 + 2y \qquad\quad = -12x + 23$$
$$y^2 + 2y + 1 = -12x + 23 + 1$$
$$(y + 1)^2 = -12x + 24$$

Section 10.3

Check Point Exercises

1. $y^2 = 8x$

$4p = 8$

$p = 2$

focus: $(p,\ 0) = (2,\ 0)$

directrix: $x = -p;\ \ x = -2$

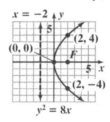

2. $x^2 = -12y$

$4p = -12$

$p = -3$

focus: $(0, p) = (0, -3)$

directrix: $y = -p;\ \ y = 3$

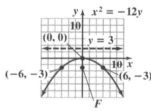

3. The focus is $(8, 0)$. Thus, the focus is on the x-axis. We use the standard form of the equation in which there is x-axis symmetry, namely, $y^2 = 4px$.

The focus is 8 units to the right of the vertex, $(0, 0)$. Thus, p is positive and $p = 8$.

$$y^2 = 4px$$
$$y^2 = 4 \cdot 8x$$
$$y^2 = 32x$$

4. $(x-2)^2 = 4(y+1)$

From the equation we have $h = 2$ and $k = -1$.

vertex: $(2, -1)$

Find p: $4p = 4$

$p = 1$

focus: $(h, k+p) = (2, -1+1)$

$= (2, 0)$

directrix: $y = k - p$

$y = -1 - 1$

$y = -2$

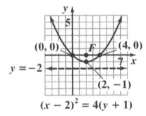

5. $y^2 + 2y = -4x + 7$

$y^2 + 2y + 1 = -4x + 7 + 1$

$(y+1)^2 = -4(x-2)$

From the equation we have $h = 2$ and $k = -1$.

vertex: $(2, -1)$

Find p: $4p = -4$

$p = -1$

focus: $(h+p, k) = (2-1, -1)$

$= (1, -1)$

directrix: $x = h - p$

$x = 2 - (-1)$

$x = 3$

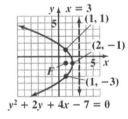

6. $x^2 = 4py$

Let $x = 3$ and $y = 4$.

$3^2 = 4p \cdot 4$

$9 = 16p$

$p = \dfrac{9}{16}$

$x^2 = \dfrac{9}{4}y$

The light should be placed at $\left(0, \dfrac{9}{16}\right)$ or $\dfrac{9}{16}$ inch above the vertex.

Concept and Vocabulary Check 10.3

1. parabola; directrix; focus

2. a

3. $(-7, 0)$

4. $x = 7$

5. 28; $(-7, -14)$; $(-7, 14)$

6. d

7. $(-2, 0)$

8. $y = -2$

9. 4; $(-4, 0)$; $(0, 0)$

Exercise Set 10.3

1. $y^2 = 4x$
$4p = 4, p = 1$
vertex: $(0, 0)$
focus: $(1, 0)$
directrix: $x = -1$
graph (c)

3. $x^2 = -4y$
$4p = -4, p = -1$
vertex: $(0, 0)$
focus: $(0, -1)$
directrix: $y = 1$
graph (b)

5. $4p = 16$, $p = 4$
vertex: $(0, 0)$
focus: $(4, 0)$
directrix: $x = -4$

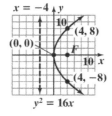

7. $4p = -8$, $p = -2$
vertex: $(0, 0)$
focus: $(-2, 0)$
directrix: $x = 2$

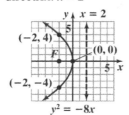

9. $4p = 12$, $p = 3$
vertex: $(0, 0)$
focus: $(0, 3)$
directrix: $y = -3$

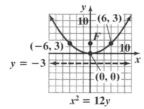

11. $4p = -16$, $p = -4$
vertex: $(0, 0)$
focus: $(0, -4)$
directrix: $y = 4$

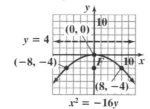

13. $\qquad y^2 = 6x$

$4p = 6$, $p = \dfrac{6}{4} = \dfrac{3}{2}$

vertex: $(0, 0)$

focus: $\left(\dfrac{3}{2}, 0\right)$

directrix: $x = -\dfrac{3}{2}$

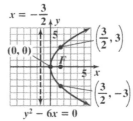

15. $8x^2 = -4y$

$x^2 = -\dfrac{1}{2}y$

$4p = -\dfrac{1}{2}$

$p = -\dfrac{1}{8}$

focus: $\left(0, -\dfrac{1}{8}\right)$

directrix: $y = \dfrac{1}{8}$

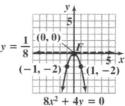

17. $p = 7$, $4p = 28$
$y^2 = 28x$

19. $p = -5$, $4p = -20$
$y^2 = -20x$

21. $p = 15$, $4p = 60$
$x^2 = 60y$

23. $p = -25$, $4p = -100$
$x^2 = -100y$

25. $p = -5 - (-3) = -2$ Vertex, $(2, -3)$
$(x - 2)^2 = -8(y + 3)$

27. vertex: $(1, 2)$ $p = 2$
$(y - 2)^2 = 8(x - 1)$

29. vertex: $(-3, 3)$, $p = 1$
$(x + 3)^2 = 4(y - 3)$

31. $(y - 1)^2 = 4(x - 1)$
$4p = 4, p = 1$
vertex: $(1, 1)$
focus: $(2, 1)$
directrix: $x = 0$
graph (c)

33. $(x + 1)^2 = -4(y + 1)$
$4p = -4, p = -1$
vertex: $(-1, -1)$
focus: $(-1, -2)$
directrix: $y = 0$
graph (d)

35. $4p = 8, p = 2$
vertex: $(2, 1)$
focus: $(2, 3)$
directrix: $y = -1$

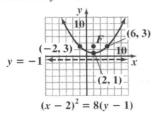

$(x - 2)^2 = 8(y - 1)$

37. $4p = -8, p = -2$
vertex: $(-1, -1)$
focus: $(-1, -3)$
directrix: $y = 1$

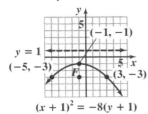

$(x + 1)^2 = -8(y + 1)$

39. $4p = 12, p = 3$
vertex: $(-1, -3)$
focus: $(2, -3)$
directrix: $x = -4$

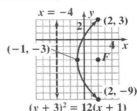

$(y + 3)^2 = 12(x + 1)$

41. $(y + 1)^2 = -8(x - 0)$
$4p = -8, p = -2$
vertex: $(0, -1)$
focus: $(-2, -1)$
directrix: $x = 2$

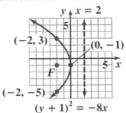

$(y + 1)^2 = -8x$

43. $x^2 - 2x + 1 = 4y - 9 + 1$
$(x - 1)^2 = 4y - 8$
$(x - 1)^2 = 4(y - 2)$
$4p = 4, p = 1$
vertex: $(1, 2)$
focus: $(1, 3)$
directrix: $y = 1$

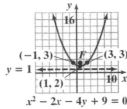

$x^2 - 2x - 4y + 9 = 0$

45.
$$y^2 - 2y + 1 = -12x + 35 + 1$$
$$(y - 1)^2 = -12x + 36$$
$$(y - 1)^2 = -12(x - 3)$$
$$4p = -12, \ p = -3$$
vertex: (3, 1)
focus: (0, 1)
directrix: $x = 6$

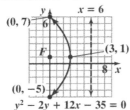

47.
$$x^2 + 6x = 4y - 1$$
$$x^2 + 6x + 9 = 4y - 1 + 9$$
$$(x + 3)^2 = 4(y + 2)$$
$$4p = 4, \ p = 1$$
vertex: (-3, -2)
focus: (-3, -1)
directrix: $y = -3$

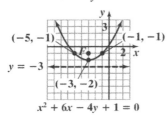

49. The y-coordinate of the vertex is
$$y = -\frac{b}{2a} = -\frac{6}{2(1)} = -3$$
The x-coordinate of the vertex is
$$x = (-3)^2 + 6(-3) + 5$$
$$= 9 - 18 + 5$$
$$= -4$$
The vertex is $(-4, -3)$.
Since the squared term is y and $a > 0$, the graph opens to the right.
Domain: $\{x \mid x \geq -4\}$ or $[-4, \infty)$
Range: $\{y \mid y \text{ is a real number}\}$ or $(-\infty, \infty)$
The relation is not a function.

51. The x-coordinate of the vertex is
$$x = -\frac{b}{2a} = -\frac{(4)}{2(-1)} = 2$$
The y-coordinate of the vertex is
$$y = -(2)^2 + 4(2) - 3$$
$$= -4 + 8 - 3$$
$$= 1$$
The vertex is $(2, 1)$.
Since the squared term is x and $a < 0$, the graph opens down.
Domain: $\{x \mid x \text{ is a real number}\}$ or $(-\infty, \infty)$
Range: $\{y \mid y \leq 1\}$ or $(-\infty, 1]$
The relation is a function.

53. The equation is in the form $x = a(y - k)^2 + h$
From the equation, we can see that the vertex is $(3, 1)$.
Since the squared term is y and $a < 0$, the graph opens to the left.
Domain: $\{x \mid x \leq 3\}$ or $(-\infty, 3]$
Range: $\{y \mid y \text{ is a real number}\}$ or $(-\infty, \infty)$
The relation is not a function.

55.

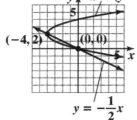

Check $(-4, 2)$:
$$-4 = (2 - 2)^2 - 4 \qquad 2 = -\frac{1}{2}(-4)$$
$$-4 = 0 - 4 \qquad\qquad 2 = 2$$
$$-4 = -4 \qquad\qquad\qquad \text{true}$$
$$\text{true}$$
Check $(0, 0)$:
$$0 = (0 - 2)^2 - 4 \qquad 0 = -\frac{1}{2}(0)$$
$$0 = 4 - 4 \qquad\qquad 0 = 0$$
$$0 = 0 \qquad\qquad\qquad \text{true}$$
$$\text{true}$$
The solution set is $\{(-4, 2), (0, 0)\}$.

57.

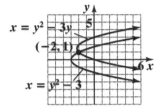

Check $(-2,1)$:

$-2 = (1)^2 - 3$ $-2 = (1)^2 - 3(1)$

$-2 = 1 - 3$ $-2 = 1 - 3$

$-2 = -2$ true $-2 = -2$ true

The solution set is $\{(-2,1)\}$.

59.

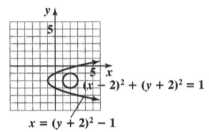

The two graphs do not cross. Therefore, the solution set is the empty set, $\{\ \ \}$ or $\varnothing$.

61. $x^2 = 4py$

$2^2 = 4p(1)$

$4 = 4$

$p = 1$

The light bulb should be placed 1 inch above the vertex.

63. $x^2 = 4py$

$6^2 = 4p(2)$

$36 = 8p$

$p = \dfrac{36}{8} = \dfrac{9}{2} = 4.5$

The receiver should be located 4.5 feet from the base of the dish.

65. $x^2 = 4py$

$(640)^2 = 4p(160)$

$p = \dfrac{(640)^2}{640} = 640$

$x = 640 - 200 = 440$

$(440)^2 = 4(640)y$

$y = \dfrac{(440)^2}{4(640)} = 75.625$

The height is 76 meters.

67. $x^2 = 4py$

$\left(\dfrac{200}{2}\right)^2 = 4p(-50)$

$\dfrac{10,000}{-50} = 4p$

$4p = -200$

$x^2 = -200y$

$(30)^2 = -200y$

$y = \dfrac{900}{-200} = -4.5$

(height of bridge) $= 50 - 4.5 = 45.5$ feet.

Yes, the boat will clear the arch.

69. – 75. Answers will vary.

77. $y^2 + 2y - 6x + 13 = 0$

$y^2 + 2y + (-6x + 13) = 0$

$y = \dfrac{-2 \pm \sqrt{2^2 - 4(-6x+13)}}{2}$

$y = \dfrac{-2 \pm \sqrt{24x - 48}}{2}$

$y = -1 \pm \sqrt{6x - 12}$

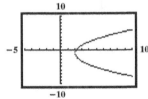

79. $16x^2 - 24xy + 9y^2 - 60x - 80y + 100 = 0$

$9y^2 - (24x + 80)y + (16x^2 - 60x + 100) = 0$

$y = \dfrac{24x + 80 \pm \sqrt{(24x+80)^2 - 36(16x^2 - 60x + 100)}}{18}$

$y = \dfrac{24x + 80 \pm \sqrt{6000x + 2800}}{18}$

$y = \dfrac{24x + 80 \pm 20\sqrt{15x + 7}}{18}$

$y = \dfrac{12x + 40 \pm 10\sqrt{15x + 7}}{9}$

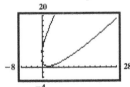

81. does not make sense; Explanations will vary. Sample explanation: Horizontal parabolas will rise without limit.

83. makes sense

85. false; Changes to make the statement true will vary. A sample change is: Because $a = -1$, the parabola will open to the left.

87. false; Changes to make the statement true will vary. A sample change is: If a parabola defines y as a function of x, it will open up or down.

89. $Ax^2 + Ey = 0$

$Ax^2 = -Ey \qquad 4p = -\dfrac{E}{A}y$

$x^2 = -\dfrac{E}{A}y \qquad p = -\dfrac{E}{4A}y$

focus: $\left(0, -\dfrac{E}{4A}\right)$,

directrix: $y = \dfrac{E}{4A}$

91. Answers will vary.

92. $\left[\dfrac{\sqrt{2}}{2}(x' - y')\right]\left[\dfrac{\sqrt{2}}{2}(x' + y')\right] = 1$

$\dfrac{\sqrt{2}}{2}\dfrac{\sqrt{2}}{2}(x' - y')(x' + y') = 1$

$\dfrac{2}{4}\left((x')^2 - (y')^2\right) = 1$

$\dfrac{x'^2}{2} - \dfrac{y'^2}{2} = 1$

$x'^2 - y'^2 = 2$

93. **a.**

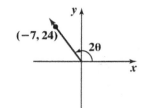

b. $\cos 2\theta = -\dfrac{7}{25}$

c. $\sin \theta = \sqrt{\dfrac{1 - \cos 2\theta}{2}}$

$\sin \theta = \sqrt{\dfrac{1 - \left(-\dfrac{7}{25}\right)}{2}}$

$\sin \theta = \sqrt{\dfrac{16}{25}}$

$\sin \theta = \dfrac{4}{5}$

$\cos \theta = \sqrt{\dfrac{1 + \cos 2\theta}{2}}$

$\cos \theta = \sqrt{\dfrac{1 + \left(-\dfrac{7}{25}\right)}{2}}$

$\cos \theta = \sqrt{\dfrac{9}{25}}$

$\cos \theta = \dfrac{3}{5}$

d. Since $90° < 2\theta < 180°$, we have $45° < \theta < 90°$. Both $\sin \theta$ and $\cos \theta$ are positive when $45° < \theta < 90°$.

94. $B^2 - 4AC = (-2\sqrt{3})^2 - 4(3)(1)$

$\qquad\qquad\quad = 12 - 12$

$\qquad\qquad\quad = 0$

Mid-Chapter 10 Check Point

1. Center: $(0,0)$

Because the denominator of the x^2 – term is greater than the denominator of the y^2 – term, the major axis is horizontal. Since $a^2 = 25$, $a = 5$ and the vertices are $(-5,0)$ and $(5,0)$. Since $b^2 = 4$, $b = 2$ and endpoints of the minor axis are $(0,-2)$ and $(0,2)$.

Foci: $\left(\pm\sqrt{21},0\right)$

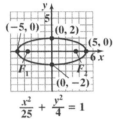

$$\frac{x^2}{25} + \frac{y^2}{4} = 1$$

2. Divide both sides by 36 to get the standard form:

$$\frac{x^2}{4} + \frac{y^2}{9} = 1$$

Center: $(0,0)$

Because the denominator of the y^2 – term is greater than the denominator of the x^2 – term, the major axis is vertical. Since $a^2 = 9$, $a = 3$ and the vertices are $(0,-3)$ and $(0,3)$. Since $b^2 = 4$, $b = 2$ and endpoints of the minor axis are $(-2,0)$ and $(2,0)$.

Foci: $\left(0,\pm\sqrt{5}\right)$

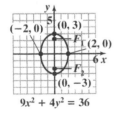

$$9x^2 + 4y^2 = 36$$

3. Center: $(2,-1)$

Because the denominator of the y^2 – term is greater than the denominator of the x^2 – term, the major axis is vertical. We have $a^2 = 25$ and $b^2 = 16$, so $a = 5$ and $b = 4$. The vertices lie 5 units above and below the center. The endpoints of the minor axis lie 4 units to the left and right of the center.

Vertices: $(2,4)$ and $(2,-6)$

Minor endpoints: $(-2,-1)$ and $(6,-1)$

Foci: $(2,2)$, $(2,-4)$

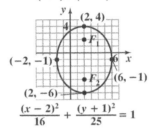

$$\frac{(x-2)^2}{16} + \frac{(y+1)^2}{25} = 1$$

4. Center: $(-2,1)$

Because the denominator of the x^2 – term is greater than the denominator of the y^2 – term, the major axis is horizontal. We have $a^2 = 25$ and $b^2 = 16$, so $a = 5$ and $b = 4$. The vertices lie 5 units to the left and right of the center. The endpoints of the minor axis lie 4 units above and below the center.

Vertices: $(-7,1)$ and $(3,1)$

Minor endpoints: $(-2,5)$ and $(-2,-3)$

Foci: $(-5,1)$, $(1,1)$

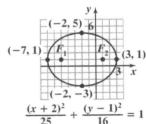

$$\frac{(x+2)^2}{25} + \frac{(y-1)^2}{16} = 1$$

5.
$$x^2 - 4x\ \ + 9y^2 + 54y\ \ = -49$$
$$\left(x^2 - 4x\ \ \right) + 9\left(y^2 + 6y\ \ \right) = -49$$
$$\left(x^2 - 4x + 4\right) + 9\left(y^2 + 6y + 9\right) = -49 + 4 + 81$$
$$\left(x - 2\right)^2 + 9\left(y + 3\right)^2 = 36$$
$$\frac{\left(x-2\right)^2}{36} + \frac{9\left(y+3\right)^2}{36} = \frac{36}{36}$$
$$\frac{\left(x-2\right)^2}{36} + \frac{\left(y+3\right)^2}{4} = 1$$

Center: $\left(2, -3\right)$

Foci: $\left(2 \pm 4\sqrt{2}, -3\right)$

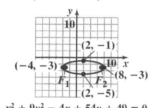

6. The equation is for a hyperbola in standard form with the transverse axis on the *x*-axis. We have $a^2 = 9$ and $b^2 = 1$, so $a = 3$ and $b = 1$. Therefore, the vertices are at $\left(\pm a, 0\right)$ or $\left(\pm 3, 0\right)$.

Using a dashed line, we construct a rectangle using the ± 3 on the *x*-axis and ± 1 on the *y*-axis. Then use dashed lines to draw extended diagonals for the rectangle. These represent the asymptotes of the graph.

Foci: $\left(\pm\sqrt{10}, 0\right)$

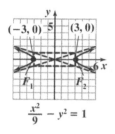

7. The equation is in the form $\dfrac{y^2}{a^2} - \dfrac{x^2}{b^2} = 1$ with $a^2 = 9$, and $b^2 = 1$. We know the transverse axis lies on the *y*-axis and the vertices are $\left(0, -3\right)$ and $\left(0, 3\right)$. Because $a^2 = 9$ and $b^2 = 1$, $a = 3$ and $b = 1$. Construct a rectangle using -1 and 1 on the *x*-axis, and -3 and 3 on the *y*-axis. Draw extended diagonals to obtain the asymptotes.

Foci: $\left(0, \pm\sqrt{10}\right)$

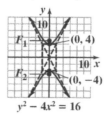

8. $$\frac{y^2}{16} - \frac{x^2}{4} = 1$$

The equation is in the form $\dfrac{y^2}{a^2} - \dfrac{x^2}{b^2} = 1$ with $a^2 = 16$, and $b^2 = 4$. We know the transverse axis lies on the *y*-axis and the vertices are $\left(0, -4\right)$ and $\left(0, 4\right)$. Because $a^2 = 16$ and $b^2 = 4$, $a = 4$ and $b = 2$. Construct a rectangle using -2 and 2 on the *x*-axis, and -4 and 4 on the *y*-axis. Draw extended diagonals to obtain the asymptotes.

Foci: $\left(0, \pm 2\sqrt{5}\right)$

9.

$$\frac{x^2}{49} - \frac{y^2}{4} = 1$$

The equation is for a hyperbola in standard form with the transverse axis on the *x*-axis. We have $a^2 = 49$ and $b^2 = 4$, so $a = 7$ and $b = 2$.

Therefore, the vertices are at $(\pm a, 0)$ or $(\pm 7, 0)$.

Using a dashed line, we construct a rectangle using the ± 7 on the *x*-axis and ± 2 on the *y*-axis. Then use dashed lines to draw extended diagonals for the rectangle. These represent the asymptotes of the graph.

Foci: $\left(\pm\sqrt{53}, 0\right)$

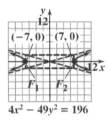

$4x^2 - 49y^2 = 196$

10. The equation is for a hyperbola in standard form with center $(2, -2)$. We have $a^2 = 9$ and $b^2 = 16$, so $a = 3$ and $b = 4$.

Asymptotes: $y + 2 = \pm\frac{4}{3}(x - 2)$

Foci: $(-3, -2)$, $(7, -2)$

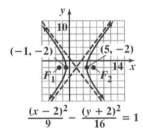

$\dfrac{(x-2)^2}{9} - \dfrac{(y+2)^2}{16} = 1$

11. Write the equation for the hyperbola in standard form:

$$4x^2 - y^2 + 8x + 6y + 11 = 0$$

$$4x^2 + 8x \quad - y^2 + 6y \quad = -11$$

$$4\left(x^2 + 2x \quad\right) - \left(y^2 - 6y \quad\right) = -11$$

$$4\left(x^2 + 2x + 1\right) - \left(y^2 - 6y + 9\right) = -11 + 4 - 9$$

$$4(x+1)^2 - (y-3)^2 = -16$$

$$\frac{4(x+1)^2}{-16} - \frac{(y-3)^2}{-16} = \frac{-16}{-16}$$

$$\frac{(y-3)^2}{16} - \frac{(x+1)^2}{4} = 1$$

Center $(-1, 3)$.

Asymptotes: $y - 3 = \pm 2(x + 1)$

Foci: $\left(-1, 3 \pm 2\sqrt{5}\right)$, $(7, -2)$

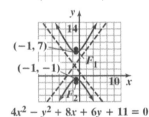

$4x^2 - y^2 + 8x + 6y + 11 = 0$

12. $(x - 2)^2 = -12(y + 1)$

$h = 2$

$k = -1$

$4p = -12$

$p = -3$

Vertex: (h, k)

$\qquad (2, -1)$

Focus: $(h, k + p)$

$\qquad (2, -1 - 3)$

$\qquad (2, -4)$

Directrix: $y = k - p$

$\qquad y = -1 - (-3)$

$\qquad y = 2$

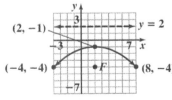

$(x - 2)^2 = -12(y + 1)$

13. $y^2 - 2x - 2y - 5 = 0$

$$y^2 - 2y \quad = 2x + 5$$
$$y^2 - 2y + 1 = 2x + 5 + 1$$
$$(y - 1)^2 = 2x + 6$$
$$(y - 1)^2 = 2(x + 3)$$

$h = -3$

$k = 1$

$4p = 2$

$p = \dfrac{1}{2}$

Vertex: (h, k)

$(-3, 1)$

Focus: $(h + p, k)$

$$\left(-3 + \frac{1}{2}, 1\right)$$

$$\left(-\frac{5}{2}, 1\right)$$

Directrix: $y = h - p$

$$y = -3 - \left(\frac{1}{2}\right)$$

$$y = -\frac{7}{2}$$

$$x = -\frac{7}{2}$$

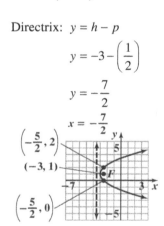

14. This is the equation of a circle centered at the origin with radius $r = \sqrt{4} = 2$.
We can plot points that are 2 units to the left, right, above, and below the origin and then graph the circle. The points are $(-2, 0)$, $(2, 0)$, $(0, 2)$, and $(0, -2)$.

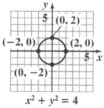

15. $x + y = 4$

$y = -x + 4$

This is the equation of a line with slope $m = -1$ and a y-intercept of 4. We can plot the point $(0, 4)$, use the slope to get an additional point, connect the points with a straight line and then extend the line to represent the graph of the equation.

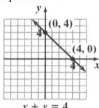

16. $x^2 - y^2 = 4$

$$\frac{x^2}{4} - \frac{y^2}{4} = 1$$

The equation is for a hyperbola in standard form with the transverse axis on the x-axis. We have $a^2 = 4$ and $b^2 = 4$, so $a = 2$ and $b = 2$.
Therefore, the vertices are at $(\pm a, 0)$ or $(\pm 2, 0)$.

Using a dashed line, we construct a rectangle using the ± 2 on the x-axis and ± 2 on the y-axis. Then use dashed lines to draw extended diagonals for the rectangle. These represent the asymptotes of the graph.

Graph the hyperbola.

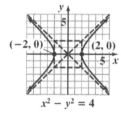

17. $x^2 + 4y^2 = 4$

$\dfrac{x^2}{4} + \dfrac{y^2}{1} = 1$

Center: $(0,0)$

Because the denominator of the x^2 – term is greater than the denominator of the y^2 – term, the major axis is horizontal. We have $a^2 = 4$ and $b^2 = 1$, so $a = 2$ and $b = 1$. The vertices lie 2 units to the left and right of the center. The endpoints of the minor axis lie 1 unit above and below the center.

Vertices: $(-2,0)$ and $(2,0)$

Minor endpoints: $(0,-1)$ and $(0,1)$

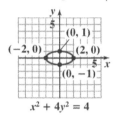

18. Center: $(-1,1)$

Radius: $r = \sqrt{4} = 2$

We plot the points that are 2 units to the left, right, above and below the center.

These points are $(-3,1)$, $(1,1)$, $(-1,3)$, and $(-1.-1)$.

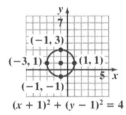

19. $x^2 + 4(y-1)^2 = 4$

$\dfrac{x^2}{4} + \dfrac{(y-1)^2}{1} = 1$

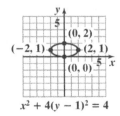

20. $(x-1)^2 - (y-1)^2 = 4$

$\dfrac{(x-1)^2}{4} - \dfrac{(y-1)^2}{4} = 1$

The equation is for a hyperbola in standard form centered at (1, 1). We have $a^2 = 4$ and $b^2 = 4$, so $a = 2$ and $b = 2$.

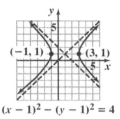

21. $(y+1)^2 = 4(x-1)$

$h = 1$

$k = -1$

$4p = 4$

$p = 1$

Vertex: (h,k)

$\qquad (1,-1)$

Focus: $(h+p,k)$

$\qquad (1+1,-1)$

$\qquad (2,-1)$

Directrix: $y = h - p$

$\qquad y = 1 - 1$

$\qquad y = 0$

$\qquad x = 0$

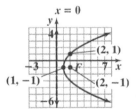

22. The foci and vertices show that c is 4 and a is 5.

$c^2 = a^2 - b^2$

$4^2 = 5^2 - b^2$

$b^2 = 25 - 16$

$b^2 = 9$

$\dfrac{x^2}{25} + \dfrac{y^2}{9} = 1$

23. The endpoints show that the center is (1, 2).

Since $2a = 18$, $a = 9$ and $a^2 = 81$.

Since $2c = 10$, $c = 5$ and $c^2 = 25$.

$$c^2 = a^2 - b^2$$

$$25 = 81 - b^2$$

$$b^2 = 81 - 25$$

$$b^2 = 56$$

$$\frac{(x-1)^2}{81} + \frac{(y-2)^2}{56} = 1$$

24. The foci and vertices show that c is 3 and a is 2.

$$b^2 = c^2 - a^2$$

$$b^2 = 3^2 - 2^2$$

$$b^2 = 9 - 4$$

$$b^2 = 5$$

$$\frac{x^2}{4} - \frac{y^2}{5} = 1$$

25. The endpoints show that the center is (–1, 5).

Since $2a = 4$, $a = 2$ and $a^2 = 4$.

Since $2c = 6$, $c = 3$ and $c^2 = 9$.

$$b^2 = c^2 - a^2$$

$$b^2 = 9 - 4$$

$$b^2 = 5$$

$$\frac{(x+1)^2}{4} - \frac{(y-5)^2}{5} = 1$$

26. Focus: $(\overset{4}{h}, \overset{5}{k+p})$

Directrix: $y = \overset{-1}{k-p}$

$$(k+p) + (k-p) = (5) + (-1)$$

$$2k = 4$$

$$k = 2$$

$$k + p = 5$$

$$p = 5 - k$$

$$p = 5 - 2$$

$$p = 3$$

$$(x-h)^2 = 4p(y-k)$$

$$(x-4)^2 = 4(3)(y-2)$$

$$(x-4)^2 = 12(y-2)$$

27. Focus: $(\overset{-2}{h+p}, \overset{6}{k})$

Directrix: $x = \overset{8}{h-p}$

$$(h+p) + (h-p) = (-2) + (8)$$

$$2h = 6$$

$$h = 3$$

$$h + p = -2$$

$$p = -2 - h$$

$$p = -2 - 3$$

$$p = -5$$

$$(y-k)^2 = 4p(x-h)$$

$$(y-6)^2 = 4(-5)(x-3)$$

$$(y-6)^2 = -20(x-3)$$

28. $a = 15, b = 10$

$$\frac{x^2}{15^2} + \frac{y^2}{10^2} = 1$$

$$\frac{x^2}{225} + \frac{y^2}{100} = 1$$

Since the truck is 10 feet wide, substitute $x = 5$ into the equation to find y.

$$\frac{5^2}{225} + \frac{y^2}{100} = 1$$

$$\frac{25}{225} + \frac{y^2}{100} = 1$$

$$\frac{1}{9} + \frac{y^2}{100} = 1$$

$$900\left(\frac{1}{9} + \frac{y^2}{100}\right) = 900(1)$$

$$100 + 9y^2 = 900$$

$$9y^2 = 800$$

$$y^2 = 88.8889$$

$$y = \sqrt{88.8889}$$

$$y \approx 9.43$$

5 feet from the center, the height of the archway is 9.43 feet. Since the truck's height is 9.5 feet, it will not fit under the archway.

29. Find the distance between the foci.

Since $2a = 40$, $a = 20$ and $a^2 = 400$.

Since $2b = 20$, $b = 10$ and $b^2 = 100$.

$$c^2 = a^2 - b^2$$
$$c^2 = 400 - 100$$
$$c^2 = 300$$
$$c = \sqrt{300}$$
$$= 10\sqrt{3}$$
$$2c = 20\sqrt{3}$$
$$2c \approx 34.64$$

The kidney stone should be 34.64 cm from the electrode that sends the ultrasound waves.

30. a. Since $2c = 6$, $c = 3$ and $c^2 = 9$.

The ranger at the primary station heard the explosion 6 seconds before the other ranger. This means that the explosion occurred $6 \times 0.35 = 2.1$ miles closer to the primary station.

Since $2a = 2.1$, $a = 1.05$ and $a^2 = 1.1025$.

$$b^2 = c^2 - a^2$$
$$b^2 = 9 - 1.1025$$
$$b^2 = 7.8975$$
$$\frac{x^2}{1.1025} - \frac{y^2}{7.8975} = 1$$

b.

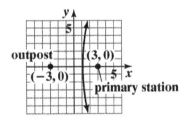

31. Consider the peak of the dome as the point $(0, 10)$. Since the width is 15 meters, the points $(\pm 7.5, 0)$ are on the parabola. Use standard form to find p.

$$(x - h)^2 = 4p(y - k)$$
$$(7.5 - 0)^2 = 4p(0 - 10)$$
$$7.5^2 = -40p$$
$$\frac{56.25}{-40} = p$$
$$p \approx -1.4$$

The light should be about 1.4 meters below the peak of the ceiling.

Section 10.4

Check Point Exercises

1. **a.** $A = 3$ and $C = 2$.
   $AC = 3(2) = 6$. Since $A \neq C$ and $AC > 0$, the graph is an ellipse.

   b. $A = 1$ and $C = 1$. Since $A = C$, the graph is a circle.

   c. $A = 0$ and $C = 1$. Since $AC = 0$, the graph is a parabola.

   d. $A = 9$ and $C = -16$. Since $AC < 0$, the graph is a hyperbola.

2.
$$x = x'\cos 45° - y'\sin 45° = x'\left(\frac{\sqrt{2}}{2}\right) - y'\left(\frac{\sqrt{2}}{2}\right) = \frac{\sqrt{2}}{2}(x' - y')$$

$$y = x'\sin 45° + y'\cos 45° = x'\left(\frac{\sqrt{2}}{2}\right) + y'\left(\frac{\sqrt{2}}{2}\right) = \frac{\sqrt{2}}{2}(x' + y')$$

Substitute into the equation: $xy = 2$

$$\left[\frac{\sqrt{2}}{2}(x' - y')\right]\left[\frac{\sqrt{2}}{2}(x' + y')\right] = 2$$

$$\frac{1}{2}\left(x'^2 - y'^2\right) = 2$$

$$\frac{x'^2}{4} - \frac{y'^2}{4} = 1$$

$(x', y') = (2, 0)$

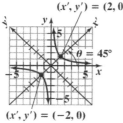

$\theta = 45°$

$(x', y') = (-2, 0)$

3. $2x^2 + \sqrt{3}xy + y^2 - 2 = 0$
   Step 1
   $A = 2, B = \sqrt{3},$ and $C = 1$.

   $$\cot 2\theta = \frac{A - C}{B} = \frac{2 - 1}{\sqrt{3}} = \frac{1}{\sqrt{3}} = \frac{\sqrt{3}}{3}$$

   Step 2
   Since $\cot 2\theta = \frac{\sqrt{3}}{3}, 2\theta = 60°$. Thus, $\theta = 30°$.

   Step 3
   $$x = x'\cos 30° - y'\sin 30° = x'\left(\frac{\sqrt{3}}{2}\right) - y'\left(\frac{1}{2}\right) = \frac{\sqrt{3}x' - y'}{2}$$

   $$y = x'\sin 30° + y'\cos 30° = x'\left(\frac{1}{2}\right) + y'\left(\frac{\sqrt{3}}{2}\right) = \frac{x' + \sqrt{3}y'}{2}$$

Step 4

Substitute into the equation:

$2x^2 + \sqrt{3}xy + y^2 - 2 = 0$

$2\left(\dfrac{\sqrt{3}x' - y'}{2}\right)^2 + \sqrt{3}\left(\dfrac{\sqrt{3}x' - y'}{2}\right)\left(\dfrac{x' + \sqrt{3}y'}{2}\right) + \left(\dfrac{x' + \sqrt{3}y'}{2}\right)^2 - 2 = 0$

$2\left(\dfrac{3x'^2 - 2\sqrt{3}x'y' + y'^2}{4}\right) + \sqrt{3}\left(\dfrac{\sqrt{3}x'^2 + 2x'y' - \sqrt{3}y'^2}{4}\right) + \dfrac{x'^2 + 2\sqrt{3}x'y' + 3y'^2}{4} = 2$

$6x'^2 - 4\sqrt{3}x'y' + 2y'^2 + 3x'^2 + 2\sqrt{3}x'y' - 3y'^2 + x'^2 + 2\sqrt{3}x'y' + 3y'^2 = 8$

$10x'^2 + 2y'^2 = 8$

$\dfrac{10x'^2}{8} + \dfrac{2y'^2}{8} = \dfrac{8}{8}$

$\dfrac{x'^2}{\frac{4}{5}} + \dfrac{y^2}{4} = 1$

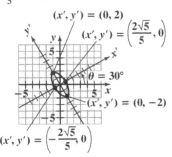

4. $4x^2 - 4xy + y^2 - 8\sqrt{5}x - 16\sqrt{5}y = 0$

Step 1:

$A = 4, B = -4,$ and $C = 1.$

$\cot 2\theta = \dfrac{A - C}{B} = \dfrac{4 - 1}{-4} = \dfrac{-3}{4}$

Step 2: Since θ is always acute, and $\cot 2\theta$ is negative, 2θ is in quadrant II.

    The third side of the right triangle is found using the Pythagorean theorem:

$(-3)^2 + 4^2 = r^2$

$\qquad 25 = r^2$

$\qquad r = 5$

So, $\cos 2\theta = \dfrac{\text{adjacent}}{\text{hypotenuse}} = \dfrac{-3}{5}.$

$\sin\theta = \sqrt{\dfrac{1 - \cos 2\theta}{2}} = \sqrt{\dfrac{1 - \left(-\frac{3}{5}\right)}{2}} = \sqrt{\dfrac{8}{10}} = \sqrt{\dfrac{4}{5}} = \dfrac{2\sqrt{5}}{5}$ and

$\cos\theta = \sqrt{\dfrac{1 + \cos 2\theta}{2}} = \sqrt{\dfrac{1 + \left(-\frac{3}{5}\right)}{2}} = \sqrt{\dfrac{1}{5}} = \dfrac{\sqrt{5}}{5}$

Step 3:

$$x = x'\cos\theta - y'\sin\theta = x'\left(\frac{\sqrt{5}}{5}\right) - y'\left(\frac{2\sqrt{5}}{5}\right) = \sqrt{5}\left(\frac{x' - 2y'}{5}\right)$$

$$y = x'\sin\theta + y'\cos\theta = x'\left(\frac{2\sqrt{5}}{5}\right) + y'\left(\frac{\sqrt{5}}{5}\right) = \sqrt{5}\left(\frac{2x' + y'}{5}\right)$$

Step 4: Substitute into the equation: $4x^2 - 4xy + y^2 - 8\sqrt{5}x - 16\sqrt{5}y = 0$

$$4\left[\sqrt{5}\left(\frac{x' - 2y'}{5}\right)\right]^2 - 4\left[\sqrt{5}\left(\frac{x' - 2y'}{5}\right)\right]\left[\sqrt{5}\left(\frac{2x' + y'}{5}\right)\right] + \left[\sqrt{5}\left(\frac{2x' + y'}{5}\right)\right]^2 - 8\sqrt{5}\left[\sqrt{5}\left(\frac{x' - 2y'}{5}\right)\right]$$

$$-16\sqrt{5}\left[\sqrt{5}\left(\frac{2x' + y'}{5}\right)\right] = 0$$

$$20\left(\frac{x'^2 - 4x'y' + 4y'^2}{25}\right) - 20\left(\frac{2x'^2 - 3x'y' - 2y'^2}{25}\right) + 5\left(\frac{4x'^2 + 4x'y' + y'^2}{25}\right) - 40\left(\frac{x' - 2y'}{5}\right)$$

$$-80\left(\frac{2x' + y'}{5}\right) = 0$$

Multiply both sides by 25:

$$20x'^2 - 80x'y' + 80y'^2 - 40x'^2 + 60x'y' + 40y'^2 + 20x'^2 + 20x'y' + 5y'^2 - 200x' + 400y'$$

$$-800x' - 400y' = 0$$

$$125y'^2 - 1000x' = 0$$

$$y'^2 - 8x' = 0$$

Step 5: This is a parabola, since it has only the y' squared.

$$y'^2 - 8x' = 0$$

$$y'^2 = 8x'$$

The vertex of the parabola, relative to the $x'y'$ system, is (0, 0). Using a calculator to solve $\sin\theta = \frac{2\sqrt{5}}{5}$, we find $\theta = \sin^{-1}\left(\frac{2\sqrt{5}}{5}\right) \approx 63°$. Rotate the axes through approximately 63°.

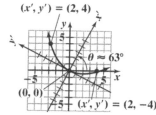

5. $3x^2 - 2\sqrt{3}xy + y^2 + 2x + 2\sqrt{3}y = 0$

$A = 3, B = -2\sqrt{3},$ and $C = 1$.

$B^2 - 4AC = \left(-2\sqrt{3}\right)^2 - 4(3)(1) = 12 - 12 = 0$

Because $B^2 - 4AC = 0$, the graph of the equation is a parabola.

Concept and Vocabulary Check 10.4

1. circle; parabola; ellipse; hyperbola

2. $\cot 2\theta = \dfrac{A - C}{B}$

3. parabola; ellipse; circle; hyperbola

Exercise Set 10.4

1. $A = 0$ and $C = 1$. Since $AC = 0$, the graph is a parabola.

3. $A = 4$ and $C = -9$. Since $AC = -36 < 0$, the graph is a hyperbola.

5. $A = 4$ and $C = 4$. Since $A = C$, the graph is a circle.

7. $A = 100$ and $C = -7$. Since $AC = -700 < 0$, the graph is a hyperbola.

9. $x = x' \cos 45° - y' \sin 45° = x'\left(\dfrac{\sqrt{2}}{2}\right) - y'\left(\dfrac{\sqrt{2}}{2}\right) = \dfrac{\sqrt{2}}{2}(x' - y')$

    $y = x' \sin 45° + y' \cos 45° = x'\left(\dfrac{\sqrt{2}}{2}\right) + y'\left(\dfrac{\sqrt{2}}{2}\right) = \dfrac{\sqrt{2}}{2}(x' + y')$

    Substitute into the equation: $xy = -1$

    $\left[\dfrac{\sqrt{2}}{2}(x' - y')\right]\left[\dfrac{\sqrt{2}}{2}(x' + y')\right] = -1$

    $\dfrac{1}{2}\left(x'^2 - y'^2\right) = -1$

    $\dfrac{y'^2}{2} - \dfrac{x'^2}{2} = 1$

11. $x = x' \cos 45° - y' \sin 45° = x'\left(\dfrac{\sqrt{2}}{2}\right) - y'\left(\dfrac{\sqrt{2}}{2}\right) = \dfrac{\sqrt{2}}{2}(x' - y')$

    $y = x' \sin 45° + y' \cos 45° = x'\left(\dfrac{\sqrt{2}}{2}\right) + y'\left(\dfrac{\sqrt{2}}{2}\right) = \dfrac{\sqrt{2}}{2}(x' + y')$

    Substitute into the equation: $x^2 - 4xy + y^2 - 3 = 0$

    $\left[\dfrac{\sqrt{2}}{2}(x' - y')\right]^2 - 4\left[\dfrac{\sqrt{2}}{2}(x' - y')\right]\left[\dfrac{\sqrt{2}}{2}(x' + y')\right] + \left[\dfrac{\sqrt{2}}{2}(x' + y')\right]^2 - 3 = 0$

    $\dfrac{1}{2}(x' - y')^2 - 4\left[\dfrac{1}{2}\left(x'^2 - y'^2\right)\right] + \dfrac{1}{2}(x' + y')^2 = 3$

    $\dfrac{1}{2}\left(x'^2 - 2x'y' + y'^2\right) - 2x'^2 + 2y'^2 + \dfrac{1}{2}\left(x'^2 + 2x'y' + y'^2\right) = 3$

    $\dfrac{1}{2}x'^2 - x'y' + \dfrac{1}{2}y'^2 - 2x'^2 + 2y'^2 + \dfrac{1}{2}x'^2 + x'y' + \dfrac{1}{2}y'^2 = 3$

    $-x'^2 + 3y'^2 = 3$

    $\dfrac{-x'^2}{3} + \dfrac{3y'^2}{3} = \dfrac{3}{3}$

    $\dfrac{y'^2}{1} - \dfrac{x'^2}{3} = 1$

13. $x = x'\cos 30° - y'\sin 30° = x'\left(\dfrac{\sqrt{3}}{2}\right) - y'\left(\dfrac{1}{2}\right) = \dfrac{\sqrt{3}x' - y'}{2}$

$y = x'\sin 30° + y'\cos 30° = x'\left(\dfrac{1}{2}\right) + y'\left(\dfrac{\sqrt{3}}{2}\right) = \dfrac{x' + \sqrt{3}y'}{2}$

Substitute into the equation: $23x^2 + 26\sqrt{3}xy - 3y^2 - 144 = 0$

$$23\left(\dfrac{\sqrt{3}x' - y'}{2}\right)^2 + 26\sqrt{3}\left(\dfrac{\sqrt{3}x' - y'}{2}\right)\left(\dfrac{x' + \sqrt{3}y'}{2}\right) - 3\left(\dfrac{x' + \sqrt{3}y'}{2}\right)^2 = 144$$

$$23\left(\dfrac{3x'^2 - 2\sqrt{3}x'y' + y'^2}{4}\right) + 26\sqrt{3}\left(\dfrac{\sqrt{3}x'^2 + 2x'y' - \sqrt{3}y'^2}{4}\right) - 3\left(\dfrac{x'^2 + 2\sqrt{3}x'y' + 3y'^2}{4}\right) = 144$$

$$69x'^2 - 46\sqrt{3}x'y' + 23y'^2 + 78x'^2 + 52\sqrt{3}x'y' - 78y'^2 - 3x'^2 - 6\sqrt{3}x'y' - 9y'^2 = 576$$

$$144x'^2 - 64y'^2 = 576$$

$$\dfrac{144x'^2}{576} - \dfrac{64y'^2}{576} = \dfrac{576}{576}$$

$$\dfrac{x'^2}{4} - \dfrac{y'^2}{9} = 1$$

15. $x^2 + xy + y^2 - 10 = 0$

$A = 1, B = 1,$ and $C = 1.$

$\cot 2\theta = \dfrac{A - C}{B} = \dfrac{1 - 1}{1} = 0$

$\quad 2\theta = 90°$

$\quad\ \theta = 45°$

$x = x'\cos 45° - y'\sin 45° = x'\left(\dfrac{\sqrt{2}}{2}\right) - y'\left(\dfrac{\sqrt{2}}{2}\right) = \dfrac{\sqrt{2}}{2}(x' - y')$

$y = x'\sin 45° + y'\cos 45° = x'\left(\dfrac{\sqrt{2}}{2}\right) + y'\left(\dfrac{\sqrt{2}}{2}\right) = \dfrac{\sqrt{2}}{2}(x' + y')$

17. $3x^2 - 10xy + 3y^2 - 32 = 0$

$A = 3, B = -10,$ and $C = 3.$

$\cot 2\theta = \dfrac{A - C}{B} = \dfrac{3 - 3}{-10} = 0$

$\quad 2\theta = 90°$

$\quad\ \theta = 45°$

$x = x'\cos 45° - y'\sin 45° = x'\left(\dfrac{\sqrt{2}}{2}\right) - y'\left(\dfrac{\sqrt{2}}{2}\right) = \dfrac{\sqrt{2}}{2}(x' - y')$

$y = x'\sin 45° + y'\cos 45° = x'\left(\dfrac{\sqrt{2}}{2}\right) + y'\left(\dfrac{\sqrt{2}}{2}\right) = \dfrac{\sqrt{2}}{2}(x' + y')$

19. $11x^2 + 10\sqrt{3}xy + y^2 - 4 = 0$

$A = 11, B = 10\sqrt{3},$ and $C = 1.$

$$x = x'\cos 30° - y'\sin 30°$$

$$x = x'\left(\frac{\sqrt{3}}{2}\right) - y'\left(\frac{1}{2}\right)$$

$$\cot 2\theta = \frac{A-C}{B} = \frac{11-1}{10\sqrt{3}} = \frac{10}{10\sqrt{3}} = \frac{1}{\sqrt{3}} = \frac{\sqrt{3}}{3} \qquad x = \frac{\sqrt{3}x' - y'}{2}$$

$$2\theta = 60° \qquad\qquad\qquad y = x'\sin 30° + y'\cos 30°$$

$$\theta = 30° \qquad\qquad\qquad y = x'\left(\frac{1}{2}\right) + y'\left(\frac{\sqrt{3}}{2}\right)$$

$$y = \frac{x' + \sqrt{3}y'}{2}$$

21. $10x^2 + 24xy + 17y^2 - 9 = 0$

$A = 10, B = 24,$ and $C = 17.$

$$\cot 2\theta = \frac{A-C}{B} = \frac{10-17}{24} = \frac{-7}{24}$$

Since θ is always acute, and $\cot 2\theta$ is negative, 2θ is in quadrant II.

The third side of the right triangle is found by using the Pythagorean theorem:

$$(-7)^2 + 24^2 = r^2$$

$$625 = r^2$$

$$r = 25$$

So, $\cos 2\theta = \frac{-7}{25}.$

$$\sin\theta = \sqrt{\frac{1-\cos 2\theta}{2}} = \sqrt{\frac{1-\left(\frac{-7}{25}\right)}{2}} = \frac{4}{5} \text{ and}$$

$$\cos\theta = \sqrt{\frac{1+\cos 2\theta}{2}} = \sqrt{\frac{1+\left(\frac{-7}{25}\right)}{2}} = \frac{3}{5}$$

$$x = x'\cos\theta - y'\sin\theta$$

$$x = x'\left(\frac{3}{5}\right) - y'\left(\frac{4}{5}\right)$$

$$x = \frac{3x' - 4y'}{5}$$

$$y = x'\sin\theta + y'\cos\theta$$

$$y = x'\left(\frac{4}{5}\right) + y'\left(\frac{3}{5}\right)$$

$$y = \frac{4x' + 3y'}{5}$$

23. $x^2 + 4xy - 2y^2 - 1 = 0$

$A = 1, B = 4,$ and $C = -2.$

$$\cot 2\theta = \frac{A - C}{B} = \frac{1 - (-2)}{4} = \frac{3}{4}$$

Since θ is always acute, and $\cot 2\theta$ is positive, 2θ is in quadrant I.
The third side of the right triangle is found using the Pythagorean theorem:

$$3^2 + 4^2 = r^2$$

$$25 = r^2$$

$$r = 5$$

So, $\cos 2\theta = \dfrac{3}{5}.$

$$\sin\theta = \sqrt{\frac{1 - \cos 2\theta}{2}} = \sqrt{\frac{1 - \frac{3}{5}}{2}} = \frac{\sqrt{5}}{5} \text{ and}$$

$$\cos\theta = \sqrt{\frac{1 + \cos 2\theta}{2}} = \sqrt{\frac{1 + \frac{3}{5}}{2}} = \frac{2\sqrt{5}}{5}$$

$$x = x'\cos\theta - y'\sin\theta$$

$$x = x'\left(\frac{2\sqrt{5}}{5}\right) - y'\left(\frac{\sqrt{5}}{5}\right)$$

$$x = \sqrt{5}\left(\frac{2x' - y'}{5}\right)$$

$$y = x'\sin\theta + y'\cos\theta$$

$$y = x'\left(\frac{\sqrt{5}}{5}\right) + y'\left(\frac{2\sqrt{5}}{5}\right)$$

$$y = \sqrt{5}\left(\frac{x' + 2y'}{5}\right)$$

25. $34x^2 - 24xy + 41y^2 - 25 = 0$

$A = 34, B = -24,$ and $C = 41.$

$$\cot 2\theta = \frac{A - C}{B} = \frac{34 - 41}{-24} = \frac{-7}{-24} = \frac{7}{24}$$

Since θ is always acute, and $\cot 2\theta$ is positive, 2θ is in quadrant I.
The third side of the right triangle is found using the Pythagorean theorem:

$$7^2 + 24^2 = r^2$$

$$625 = r^2$$

$$r = 25$$

So, $\cos 2\theta = \dfrac{7}{25}.$

$$\sin\theta = \sqrt{\frac{1 - \cos 2\theta}{2}} = \sqrt{\frac{1 - \frac{7}{25}}{2}} = \frac{3}{5} \text{ and } \cos\theta = \sqrt{\frac{1 + \cos 2\theta}{2}} = \sqrt{\frac{1 + \frac{7}{25}}{2}} = \frac{4}{5}$$

$$x = x'\cos\theta - y'\sin\theta = x'\left(\frac{4}{5}\right) - y'\left(\frac{3}{5}\right) = \frac{4x' - 3y'}{5}$$

$$y = x'\sin\theta + y'\cos\theta = x'\left(\frac{3}{5}\right) + y'\left(\frac{4}{5}\right) = \frac{3x' + 4y'}{5}$$

27. a. From Exercise 15,

$$x = \frac{\sqrt{2}}{2}(x' - y') \text{ and } y = \frac{\sqrt{2}}{2}(x' + y').$$

Substitute into the equation: $x^2 + xy + y^2 - 10 = 0$

$$\left[\frac{\sqrt{2}}{2}(x' - y')\right]^2 + \left[\frac{\sqrt{2}}{2}(x' - y')\right]\left[\frac{\sqrt{2}}{2}(x' + y')\right] + \left[\frac{\sqrt{2}}{2}(x' + y')\right]^2 - 10 = 0$$

$$\frac{1}{2}\left(x'^2 - 2x'y' + y'^2\right) + \frac{1}{2}\left(x'^2 - y'^2\right) + \frac{1}{2}\left(x'^2 + 2x'y' + y'^2\right) = 10$$

Multiply both sides by 2.

$$x'^2 - 2x'y' + y'^2 + x'^2 - y'^2 + x'^2 + 2x'y' + y'^2 = 20$$

$$3x'^2 + y'^2 = 20$$

b. $\dfrac{x'^2}{\frac{20}{3}} + \dfrac{y'^2}{20} = 1$

c.

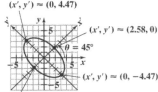

$(x', y') \approx (0, 4.47)$

$(x', y') \approx (2.58, 0)$

$\theta = 45°$

$(x', y') \approx (0, -4.47)$

$(x', y') \approx (-2.58, 0)$

29. a. From Exercise 17, $x = \dfrac{\sqrt{2}}{2}(x' - y')$ and $y = \dfrac{\sqrt{2}}{2}(x' + y')$.

Substitute into the equation: $3x^2 - 10xy + 3y^2 - 32 = 0$

$$3\left[\frac{\sqrt{2}}{2}(x' - y')\right]^2 - 10\left[\frac{\sqrt{2}}{2}(x' - y')\right]\left[\frac{\sqrt{2}}{2}(x' + y')\right] + 3\left[\frac{\sqrt{2}}{2}(x' + y')\right]^2 - 32 = 0$$

$$3\left[\frac{1}{2}\left(x'^2 - 2x'y' + y'^2\right)\right] - 10\left[\frac{1}{2}\left(x'^2 - y'^2\right)\right] + 3\left[\frac{1}{2}\left(x'^2 + 2x'y' + y'^2\right)\right] = 32$$

Multiply both sides by 2.

$$3x'^2 - 6x'y' + 3y'^2 - 10x'^2 + 10y'^2 + 3x'^2 + 6x'y' + 3y'^2 = 64$$

$$-4x'^2 + 16y'^2 = 64$$

b. $\dfrac{y'^2}{4} - \dfrac{x'^2}{16} = 1$

c.

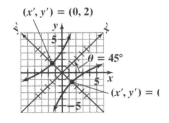

$(x', y') = (0, 2)$

$\theta = 45°$

$(x', y') = ($

31. a. From Exercise 19,

$$x = \frac{\sqrt{3}x' - y'}{2} \text{ and } y = \frac{x' + \sqrt{3}y'}{2}.$$

Substitute into the equation: $11x^2 + 10\sqrt{3}xy + y^2 - 4 = 0$

$$11\left(\frac{\sqrt{3}x' - y'}{2}\right)^2 + 10\sqrt{3}\left(\frac{\sqrt{3}x' - y'}{2}\right)\left(\frac{x' + \sqrt{3}y'}{2}\right) + \left(\frac{x' + \sqrt{3}y'}{2}\right)^2 - 4 = 0$$

$$11\left(\frac{3x'^2 - 2\sqrt{3}x'y' + y'^2}{4}\right) + 10\sqrt{3}\left(\frac{\sqrt{3}x'^2 + 2x'y' - \sqrt{3}y'^2}{4}\right) + \frac{x'^2 + 2\sqrt{3}x'y' + 3y'^2}{4} = 4$$

Multiply both sides by 4:

$$33x'^2 - 22\sqrt{3}x'y' + 11y'^2 + 30x'^2 + 20\sqrt{3}x'y' - 30y'^2 + x'^2 + 2\sqrt{3}x'y' + 3y'^2 = 16$$

$$64x'^2 - 16y'^2 = 16$$

b. $\dfrac{x'^2}{\frac{1}{4}} - \dfrac{y'^2}{1} = 1$

c.

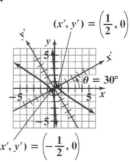

33. a. From Exercise 21, $x = \dfrac{3x' - 4y'}{5}$ and $y = \dfrac{4x' + 3y'}{5}$.

Substitute into the equation: $10x^2 + 24xy + 17y^2 - 9 = 0$

$$10\left(\frac{3x' - 4y'}{5}\right)^2 + 24\left(\frac{3x' - 4y'}{5}\right)\left(\frac{4x' + 3y'}{5}\right) + 17\left(\frac{4x' + 3y'}{5}\right)^2 - 9 = 0$$

$$10\left(\frac{9x'^2 - 24x'y' + 16y'^2}{25}\right) + 24\left(\frac{12x'^2 - 7x'y' - 12y'^2}{25}\right) + 17\left(\frac{16x'^2 + 24x'y' + 9y'^2}{25}\right) = 9$$

Multiply both sides by 25:

$$90x'^2 - 240x'y' + 160y'^2 + 288x'^2 - 168x'y' - 288y'^2 + 272x'^2 + 408x'y' + 153y'^2 = 225$$

$$650x'^2 + 25y'^2 = 225$$

b. $\dfrac{x'^2}{\frac{9}{26}} + \dfrac{y'^2}{9} = 1$

c.

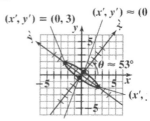

The axes are rotated by $\theta = \sin^{-1}\left(\dfrac{4}{5}\right) \approx 53°$.

35. a. From Exercise 23,

$$x = \sqrt{5}\left(\frac{2x' - y'}{5}\right) \text{ and } y = \sqrt{5}\left(\frac{x' + 2y'}{5}\right).$$

Substitute into the equation: $x^2 + 4xy - 2y^2 - 1 = 0$

$$\left[\sqrt{5}\left(\frac{2x' - y'}{5}\right)\right]^2 + 4\left[\sqrt{5}\left(\frac{2x' - y'}{5}\right)\right]\left[\sqrt{5}\left(\frac{x' + 2y'}{5}\right)\right] - 2\left[\sqrt{5}\left(\frac{x' + 2y'}{5}\right)\right]^2 = 1$$

$$5\left(\frac{4x'^2 - 4x'y' + y'^2}{25}\right) + 20\left(\frac{2x'^2 + 3x'y' - 2y'^2}{25}\right) - 10\left(\frac{x'^2 + 4x'y' + 4y'^2}{25}\right) = 1$$

Multiply both sides by 25:

$$20x'^2 - 20x'y' + 5y'^2 + 40x'^2 + 60x'y' - 40y'^2 - 10x'^2 - 40x'y' - 40y'^2 = 25$$

$$50x'^2 - 75y'^2 = 25$$

b. $\dfrac{x'^2}{\frac{1}{2}} - \dfrac{y'^2}{\frac{1}{3}} = 1$

c.

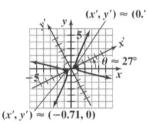

The axes are rotated by $\theta = \sin^{-1}\left(\dfrac{\sqrt{5}}{5}\right) \approx 27°$.

37. a. From Exercise 25,

$$x = \frac{4x' - 3y'}{5} \text{ and } y = \frac{3x' + 4y'}{5}.$$

Substitute into the equation: $34x^2 - 24xy + 41y^2 - 25 = 0$

$$34\left(\frac{4x' - 3y'}{5}\right)^2 - 24\left(\frac{4x' - 3y'}{5}\right)\left(\frac{3x' + 4y'}{5}\right) + 41\left(\frac{3x' + 4y'}{5}\right)^2 = 25$$

$$34\left(\frac{16x'^2 - 24x'y' + 9y'^2}{25}\right) - 24\left(\frac{12x'^2 + 7x'y' - 12y'^2}{25}\right) + 41\left(\frac{9x'^2 + 24x'y' + 16y'^2}{25}\right) = 25$$

Multiply both sides by 25:

$$544x'^2 - 816x'y' + 306y'^2 - 288x'^2 - 168x'y' + 288y'^2 + 369x'^2 + 984x'y' + 656y'^2 = 625$$

$$625x'^2 + 1250y'^2 = 625$$

b. $\dfrac{625x'^2 + 1250y'^2}{625} = \dfrac{625}{625}$

$$\dfrac{x'^2}{1} + \dfrac{y'^2}{\frac{1}{2}} = 1$$

c.

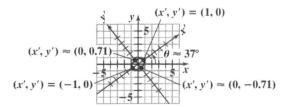

The axes are rotated by $\theta = \sin^{-1}\left(\dfrac{3}{5}\right) \approx 37°$.

39. $5x^2 - 2xy + 5y^2 - 12 = 0$

$A = 5, B = -2,$ and $C = 5$.

$B^2 - 4AC = (-2)^2 - 4(5)(5) = -96$.

Since $B^2 - 4AC < 0$, the graph is an ellipse or a circle.

41. $24x^2 + 16\sqrt{3}xy + 8y^2 - x + \sqrt{3}y - 8 = 0$

$A = 24, B = 16\sqrt{3},$ and $C = 8$.

$B^2 - 4AC = \left(16\sqrt{3}\right)^2 - 4(24)(8) = 768 - 768 = 0$

Since $B^2 - 4AC = 0$, the graph is a parabola.

43. $23x^2 + 26\sqrt{3}xy - 3y^2 - 144 = 0$

$A = 23, B = 26\sqrt{3},$ and $C = -3$.

$B^2 - 4AC = \left(26\sqrt{3}\right)^2 - 4(23)(-3) = 2028 + 276 = 2304$

Since $B^2 - 4AC > 0$, the graph is a hyperbola.

45. Find θ:

$\cot 2\theta = \dfrac{A - C}{B} = \dfrac{5 - 5}{-6} = 0$

$2\theta = 90°$

$\theta = 45°$

Substitute θ into rotation formulas:

$x = x'\cos\theta - y'\sin\theta = x'\cos 45° - y'\sin 45° = x'\dfrac{\sqrt{2}}{2} - y'\dfrac{\sqrt{2}}{2} = \dfrac{x'\sqrt{2} - y'\sqrt{2}}{2}$

$y = x'\sin\theta + y'\cos\theta = x'\sin 45° + y'\cos 45° = x'\dfrac{\sqrt{2}}{2} + y'\dfrac{\sqrt{2}}{2} = \dfrac{x'\sqrt{2} + y'\sqrt{2}}{2}$

Substitute into equation:

$$5x^2 - 6xy + 5y^2 - 8 = 0$$

$$5\left(\dfrac{x'\sqrt{2} - y'\sqrt{2}}{2}\right)^2 - 6\left(\dfrac{x'\sqrt{2} - y'\sqrt{2}}{2}\right)\left(\dfrac{x'\sqrt{2} + y'\sqrt{2}}{2}\right) + 5\left(\dfrac{x'\sqrt{2} + y'\sqrt{2}}{2}\right)^2 - 8 = 0$$

$$\dfrac{x'^2}{4} + \dfrac{y'^2}{1} = 1$$

$\dfrac{x'^2}{4} + \dfrac{y'^2}{1} = 1$ is an ellipse with minor axis vertices of $(0, -1)$ and $(0, 1)$.

47. Find $\sin\theta$ and $\cos\theta$:

$$\cot 2\theta = \frac{A-C}{B} = \frac{1-4}{-4} = \frac{3}{4}$$

$$\cot 2\theta = \frac{x}{y} = \frac{3}{4}$$

$$r = \sqrt{x^2+y^2} = \sqrt{3^2+4^2} = 5$$

$$\sin\theta = \sqrt{\frac{1-\cos 2\theta}{2}} = \sqrt{\frac{1-\left(\frac{3}{5}\right)}{2}} = \frac{\sqrt{5}}{5} \quad\text{and}\quad \cos\theta = \sqrt{\frac{1+\cos 2\theta}{2}} = \sqrt{\frac{1+\left(\frac{3}{5}\right)}{2}} = \frac{2\sqrt{5}}{5}$$

Substitute into rotation formulas:

$$x = x'\cos\theta - y'\sin\theta = x'\frac{2\sqrt{5}}{5} - y'\frac{\sqrt{5}}{5} = \frac{x'2\sqrt{5} - y'\sqrt{5}}{5}$$

$$y = x'\sin\theta + y'\cos\theta = x'\frac{\sqrt{5}}{5} + y'\frac{2\sqrt{5}}{5} = \frac{x'\sqrt{5} - y'2\sqrt{5}}{5}$$

Substitute into equation:

$$x^2 - 4xy + 4y^2 + 5\sqrt{5}y - 10 = 0$$

$$\left(\frac{x'2\sqrt{5}-y'\sqrt{5}}{5}\right)^2 - 4\left(\frac{x'2\sqrt{5}-y'\sqrt{5}}{5}\right)\left(\frac{x'\sqrt{5}+y'2\sqrt{5}}{5}\right) + 4\left(\frac{x'\sqrt{5}+y'2\sqrt{5}}{5}\right)^2 + 5\sqrt{5}\left(\frac{x'\sqrt{5}+y'2\sqrt{5}}{5}\right) - 10 = 0$$

$$x' + y'^2 + 2y' - 2 = 0$$

$$x' = -y'^2 - 2y' + 2$$

$$x' = -\left(y'^2 + 2y'\right) + 2$$

$$x' = -\left(y'^2 + 2y' + 1\right) + 2 + 1$$

$$x' = -\left(y'+1\right)^2 + 3$$

$x' = -\left(y'+1\right)^2 + 3$ is a parabola with vertex $(3,-1.)$

49. – 53. Answers may vary.

55. $y_1 = \dfrac{-(Bx+E)+\sqrt{(Bx+E)^2 - 4C\left(Ax^2+Dx+F\right)}}{2C}$ and $y_2 = \dfrac{-(Bx+E)-\sqrt{(Bx+E)^2 - 4C\left(Ax^2+Dx+F\right)}}{2C}$

for equations of the form $Ax^2 + Bxy + Cy^2 + Dx + Ey + F = 0$.

$A = 7$, $B = 8$, $C = 1$, $D = 0$, $E = 0$, and $F = -1$.

Graph $y_1 = \dfrac{-8x+\sqrt{64x^2 - 4\left(7x^2-1\right)}}{2}$ and $y_2 = \dfrac{-8x-\sqrt{64x^2 - 4\left(7x^2-1\right)}}{2}$.

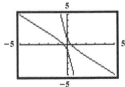

57. $y_1 = \dfrac{-(Bx+E)+\sqrt{(Bx+E)^2 - 4C\left(Ax^2 + Dx + F\right)}}{2C}$ and $y_2 = \dfrac{-(Bx+E)-\sqrt{(Bx+E)^2 - 4C\left(Ax^2 + Dx + F\right)}}{2C}$

for equations of the form $Ax^2 + Bxy + Cy^2 + Dx + Ey + F = 0$.

$A = 3$, $B = -6$, $C = 3$, $D = 10$, $E = -8$, and $F = -2$.

Graph $y_1 = \dfrac{-(-6x-8)+\sqrt{(-6x-8)^2 - 12\left(3x^2 + 10x - 2\right)}}{6}$ and $y_2 = \dfrac{-(-6x-8)-\sqrt{(-6x-8)^2 - 12\left(3x^2 + 10x - 2\right)}}{6}$.

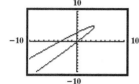

59. $y_1 = \dfrac{-(Bx+E)+\sqrt{(Bx+E)^2 - 4C\left(Ax^2 + Dx + F\right)}}{2C}$ and $y_2 = \dfrac{-(Bx+E)-\sqrt{(Bx+E)^2 - 4C\left(Ax^2 + Dx + F\right)}}{2C}$

for equations of the form $Ax^2 + Bxy + Cy^2 + Dx + Ey + F = 0$.

$A = 1$, $B = 4$, $C = 4$, $D = 10\sqrt{5}$, $E = 0$, and $F = -9$.

Graph $y_1 = \dfrac{-4x+\sqrt{16x^2 - 16\left(x^2 + 10\sqrt{5}x - 9\right)}}{8}$ and $y_2 = \dfrac{-4x-\sqrt{16x^2 - 16\left(x^2 + 10\sqrt{5}x - 9\right)}}{8}$.

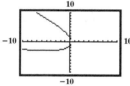

$x = 3\cos t$ and $y = 5\sin t$

$\dfrac{x}{3} = \cos t$ $\dfrac{y}{5} = \sin t$

61. does not make sense; Explanations will vary. Sample explanation: This is not necessary because there is no *xy* term.

63. makes sense

65. $A = 3, B = 2,$ and $C = 3.$

$$\cot 2\theta = \frac{A - C}{B} = \frac{3 - 3}{2} = 0$$

$$2\theta = 90°$$

$$\theta = 45°$$

$$x = x' \cos 45° - y' \sin 45° = x' \left(\frac{\sqrt{2}}{2} \right) - y' \left(\frac{\sqrt{2}}{2} \right) = \frac{\sqrt{2}}{2}(x' - y')$$

$$y = x' \sin 45° + y' \cos 45° = x' \left(\frac{\sqrt{2}}{2} \right) + y' \left(\frac{\sqrt{2}}{2} \right) = \frac{\sqrt{2}}{2}(x' + y')$$

Substitute into the equation: $3x^2 - 2xy + 3y^2 + 2 = 0$

$$3 \left[\frac{\sqrt{2}}{2}(x' - y') \right]^2 - 2 \left[\frac{\sqrt{2}}{2}(x' - y') \right] \left[\frac{\sqrt{2}}{2}(x' + y') \right] + 3 \left[\frac{\sqrt{2}}{2}(x' + y') \right]^2 + 2 = 0$$

$$\frac{3}{2}\left(x'^2 - 2x'y' + y'^2 \right) - \left(x'^2 - y'^2 \right) + \frac{3}{2}\left(x'^2 + 2x'y' + y'^2 \right) = -2$$

Multiply both sides by 2:

$$3x'^2 - 6x'y' + 3y'^2 - 2x'^2 + 2y'^2 + 3x'^2 + 6x'y' + 3y'^2 = -4$$

$$4x'^2 + 8y'^2 = -4$$

$$\frac{4x'^2 + 8y'^2}{-4} = \frac{-4}{-4}$$

$$-x'^2 - 2y'^2 = 1$$

There are no solutions to this equation since the left side of the equation is negative or 0 for all values of x' and y'. Thus, there are no points on the graph of this equation, just as one hand clapping makes no sound.

67.

$$A' = A\cos^2 \theta + B \sin \theta \cos \theta + C \sin^2 \theta$$

$$C' = A\sin^2 \theta - B \sin \theta \cos \theta + C \cos^2 \theta$$

$$A' + C' = A\cos^2 \theta + B \sin \theta \cos \theta + C \sin^2 \theta + A \sin^2 \theta - B \sin \theta \cos \theta + C \cos^2 \theta$$

$$= A \left(\cos^2 \theta + \sin^2 \theta \right) + B \left(\sin \theta \cos \theta - \sin \theta \cos \theta \right) + C \left(\sin^2 \theta + \cos^2 \theta \right)$$

$$= A(1) + B(0) + C(1)$$

$$= A + C$$

68.
$$A' = A\cos^2\theta + B\sin\theta\cos\theta + C\sin^2\theta$$

$$B' = B\left(\cos^2\theta - \sin^2\theta\right) + 2(C - A)(\sin\theta\cos\theta)$$

$$C' = A\sin^2\theta - B\sin\theta\cos\theta + C\cos^2\theta$$

$$B'^2 = \left[B\left(\cos^2\theta - \sin^2\theta\right) + 2(C - A)(\sin\theta\cos\theta)\right]^2$$

$$= B^2\left(\cos^2\theta - \sin^2\theta\right)^2 + 4B(C - A)(\sin\theta\cos\theta)\left(\cos^2\theta - \sin^2\theta\right) + 4(C - A)^2(\sin\theta\cos\theta)^2$$

$$= B^2\left(\cos^4\theta - 2\cos^2\theta\sin^2\theta + \sin^4\theta\right) + 4BC(\sin\theta\cos\theta)\left(\cos^2\theta - \sin^2\theta\right) - 4AB(\sin\theta\cos\theta)\left(\cos^2\theta - \sin^2\theta\right)$$

$$+ 4C^2(\sin\theta\cos\theta)^2 - 8AC(\sin\theta\cos\theta)^2 + 4A^2(\sin\theta\cos\theta)^2$$

$$4A'C' = 4\left[\left(A\cos^2\theta + B\sin\theta\cos\theta + C\sin^2\theta\right)\left(A\sin^2\theta - B\sin\theta\cos\theta + C\cos^2\theta\right)\right]$$

$$= 4\left[A^2(\sin\theta\cos\theta)^2 - AB\cos^2\theta(\sin\theta\cos\theta) + AC\cos^4\theta\right]$$

$$+ 4\left[AB\sin^2\theta(\sin\theta\cos\theta) - B^2(\sin\theta\cos\theta)^2 + BC\cos^2\theta(\sin\theta\cos\theta)\right]$$

$$+ 4\left[AC\sin^4\theta - BC\sin^2\theta(\sin\theta\cos\theta) + C^2(\sin\theta\cos\theta)^2\right]$$

$$= 4A^2(\sin\theta\cos\theta)^2 - 4B^2(\sin\theta\cos\theta)^2 + 4C^2(\sin\theta\cos\theta)^2 + 4AB\left(\sin^2\theta(\sin\theta\cos\theta) - \cos^2\theta(\sin\theta\cos\theta)\right)$$

$$+ 4AC\left(\cos^4\theta + \sin^4\theta\right) + 4BC\left(\cos^2\theta(\sin\theta\cos\theta) - \sin^2\theta(\sin\theta\cos\theta)\right)$$

Subtracting these results,

$$B'^2 - 4A'C'$$

$$= 4A^2\left[(\sin\theta\cos\theta)^2 - (\sin\theta\cos\theta)^2\right] + B^2\left[\cos^4\theta + 2\cos^2\theta\sin^2\theta + \sin^4\theta\right] + 4C^2\left[(\sin\theta\cos\theta)^2 - (\sin\theta\cos\theta)^2\right]$$

$$+ 4AB\left[\cos^2\theta(\sin\theta\cos\theta) - \sin^2\theta(\sin\theta\cos\theta) - (\sin\theta\cos\theta)\left(\cos^2\theta - \sin^2\theta\right)\right]$$

$$- 4AC\left[\cos^4\theta + 2(\sin\theta\cos\theta)^2 + \sin^4\theta\right] + 4BC\left[(\sin\theta\cos\theta)\left(\cos^2\theta - \sin^2\theta\right) - (\sin\theta\cos\theta)\left(\cos^2\theta - \sin^2\theta\right)\right]$$

$$= 4A^2[0] + B^2\left[\left(\cos^2\theta + \sin^2\theta\right)^2\right] + 4C^2[0] + 4AB\left[\sin\theta\cos\theta\left(\cos^2\theta - \sin^2\theta - \cos^2\theta + \sin^2\theta\right)\right]$$

$$- 4AC\left[\left(\cos^2\theta + \sin^2\theta\right)^2\right] + 4BC\left[\sin\theta\cos\theta\left(\cos^2\theta - \sin^2\theta - \cos^2\theta + \sin^2\theta\right)\right]$$

$$= B^2\left[(1)^2\right] + 4AB[0] - 4AC\left[(1)^2\right] + 4BC[0] = B^2 - 4AC$$

69. Answers may vary.

70. parabola

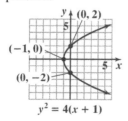

$y^2 = 4(x + 1)$

Section 10.5

Check Point Exercises

1.

t	$x = t^2 + 1$	$y = 3t$	(x, y)
-2	$(-2)^2 + 1 = 5$	$3(-2) = -6$	$(5, -6)$
-1	$(-1)^2 + 1 = 2$	$3(-1) = -3$	$(2, -3)$
0	$0^2 + 1 = 1$	$3(0) = 0$	$(1, 0)$
1	$1^2 + 1 = 2$	$3(1) = 3$	$(2, 3)$
2	$2^2 + 1 = 5$	$3(2) = 6$	$(5, 6)$

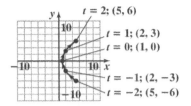

2. $x = \sqrt{t} \implies t = x^2$

Using $t = x^2$ and $y = 2t - 1$, $y = 2x^2 - 1$.

Since $t \geq 0$, x is nonnegative.

3. $x = 6\cos t$ and $y = 4\sin t$, $\pi \leq t \leq 2\pi$

$$\frac{x}{6} = \cos t \qquad \frac{y}{4} = \sin t$$

Square and add the equations:

$$\frac{x^2}{36} = \cos^2 t$$

$$+ \quad \frac{y^2}{16} = \sin^2 t$$

$$\overline{\frac{x^2}{36} + \frac{y^2}{16} = \cos^2 t + \sin^2 t}$$

$$\frac{x^2}{36} + \frac{y^2}{16} = 1$$

Since t is in the interval $[\pi, 2\pi]$, we use $t = \pi$, $t = \dfrac{3\pi}{2}$, and $t = 2\pi$:

$t = \pi$: $\quad x = 6\cos \pi = -6$

$\qquad\qquad y = 4\sin \pi = 0$

$t = \dfrac{3\pi}{2}$: $\quad x = 6\cos\dfrac{3\pi}{2} = 0$

$\qquad\qquad y = 4\sin\dfrac{3\pi}{2} = -4$

$t = 2\pi$: $\quad x = 6\cos 2\pi = 6$

$\qquad\qquad y = 4\cos 2\pi = 0$

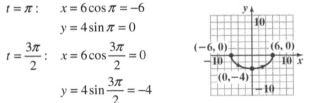

4. $y = x^2 - 25$

Let $x = t$. Then $y = t^2 - 25$.

The parametric equations are $x = t$ and $y = t^2 - 25$.

Concept and Vocabulary Check 10.5

1. parametric; parameter; plane curve

2. t; x; y

3. $\sin t$; $\cos t$; $\sin^2 t + \cos^2 t = 1$

4. true

Exercise Set 10.5

1. $x = 3 - 5(1) = -2$
$y = 4 + 2(1) = 6$;
   $(-2, 6)$

3. $x = 2^2 + 1 = 5$
$y = 5 - 2^3 = 5 - 8 = -3$;
   $(5, -3)$

5. $x = 4 + 2\cos\dfrac{\pi}{2} = 4 + 2(0) = 4$

$y = 3 + 5\sin\dfrac{\pi}{2} = 3 + 5(1) = 8$;

   $(4, 8)$

7. $x = (60\cos 30°)(2) = \left(60 \cdot \dfrac{\sqrt{3}}{2}\right)(2) = 60\sqrt{3}$

$y = 5 + (60\sin 30°)(2) - 16(2)^2$

   $= 5 + \left(60 \cdot \dfrac{1}{2}\right)(2) - 16 \cdot 4$

   $= 5 + 60 - 64 = 1$;

   $\left(60\sqrt{3},\, 1\right)$

9.

t	$x = t + 2$	$y = t^2$	(x, y)
-2	$-2 + 2 = 0$	$(-2)^2 = 4$	$(0, 4)$
-1	$-1 + 2 = 1$	$(-1)^2 = 1$	$(1, 1)$
0	$0 + 2 = 2$	$0^2 = 0$	$(2, 0)$
1	$1 + 2 = 3$	$1^2 = 1$	$(3, 1)$
2	$2 + 2 = 4$	$2^2 = 4$	$(4, 4)$

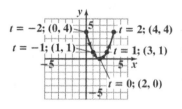

11.

t	$x = t - 2$	$y = 2t + 1$	(x, y)
−2	$-2 - 2 = -4$	$2(-2) + 1 = -3$	$(-4, -3)$
−1	$-1 - 2 = -3$	$2(-1) + 1 = -1$	$(-3, -1)$
0	$0 - 2 = -2$	$2(0) + 1 = 1$	$(-2, 1)$
1	$1 - 2 = -1$	$2(1) + 1 = 3$	$(-1, 3)$
2	$2 - 2 = 0$	$2(2) + 1 = 5$	$(0, 5)$
3	$3 - 2 = 1$	$2(3) + 1 = 7$	$(1, 7)$

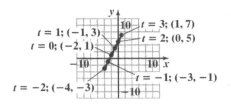

13.

t	$x = t + 1$	$y = \sqrt{t}$	(x, y)
0	$0 + 1 = 1$	$\sqrt{0} = 0$	$(1, 0)$
1	$1 + 1 = 2$	$\sqrt{1} = 1$	$(2, 1)$
4	$4 + 1 = 5$	$\sqrt{4} = 2$	$(5, 2)$
9	$9 + 1 = 10$	$\sqrt{9} = 3$	$(10, 3)$

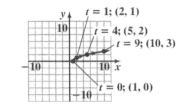

15.

t	$x = \cos t$	$y = \sin t$	(x, y)
0	$\cos 0 = 1$	$\sin 0 = 0$	$(1, 0)$
$\dfrac{\pi}{2}$	$\cos \dfrac{\pi}{2} = 0$	$\sin \dfrac{\pi}{2} = 1$	$(0, 1)$
π	$\cos \pi = -1$	$\sin \pi = 0$	$(-1, 0)$
$\dfrac{3\pi}{2}$	$\cos \dfrac{3\pi}{2} = 0$	$\sin \dfrac{3\pi}{2} = -1$	$(0, -1)$
2π	$\cos 2\pi = 1$	$\sin 2\pi = 0$	$(1, 0)$

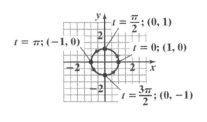

17.

t	$x = t^2$	$y = t^3$	(x, y)
−2	$(-2)^2 = 4$	$(-2)^3 = -8$	$(4, -8)$
−1	$(-1)^2 = 1$	$(-1)^3 = -1$	$(1, -1)$
0	$0^2 = 0$	$0^3 = 0$	$(0, 0)$
1	$1^2 = 1$	$1^3 = 1$	$(1, 1)$
2	$2^2 = 4$	$2^3 = 8$	$(4, 8)$

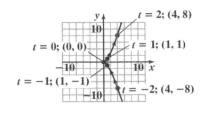

19.

t	$x = 2t$	$y = |t - 1|$	(x, y)
−2	$2(-2) = -4$	$|-2 - 1| = 3$	$(-4, 3)$
−1	$2(-1) = -2$	$|-1 - 1| = 2$	$(-2, 2)$
0	$2(0) = 0$	$|0 - 1| = 1$	$(0, 1)$
1	$2(1) = 2$	$|1 - 1| = 0$	$(2, 0)$
2	$2(2) = 4$	$|2 - 1| = 1$	$(4, 1)$

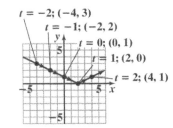

21. $x = t \implies y = 2x$

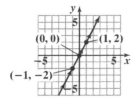

23. $x = 2t - 4$

$$\frac{x+4}{2} = t$$

Substitute into y:

$$y = 4\left(\frac{x+4}{2}\right)^2 = (x+4)^2$$

$$y = (x+4)^2$$

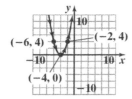

25. $x = \sqrt{t}$

$$x^2 = t$$

Substitute into y:

$$y = x^2 - 1$$

Since $t \geq 0$ in $x = \sqrt{t}$, $x \geq 0$.

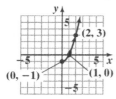

27. $x = 2\sin t$ and $y = 2\cos t$

$$\frac{x}{2} = \sin t \qquad \frac{y}{2} = \cos t$$

Square and add the equations:

$$\frac{x^2}{4} = \sin^2 t$$

$$+ \quad \frac{y^2}{4} = \cos^2 t$$

$$\overline{\frac{x^2}{4} + \frac{y^2}{4} = \sin^2 t + \cos^2 t}$$

$$\frac{x^2}{4} + \frac{y^2}{4} = 1$$

$$x^2 + y^2 = 4$$

The circle centered at $(0, 0)$ with radius 2.

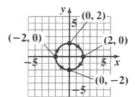

29. $x = 1 + 3\cos t$ and $y = 2 + 3\sin t$

$$\frac{x-1}{3} = \cos t \qquad \frac{y-2}{3} = \sin t$$

Square and add the equations:

$$\frac{(x-1)^2}{9} = \cos^2 t$$

$$+ \quad \frac{(y-2)^2}{9} = \sin^2 t$$

$$\overline{\frac{(x-1)^2}{9} + \frac{(y-2)^2}{9} = \cos^2 t + \sin^2 t}$$

$$\frac{(x-1)^2}{9} + \frac{(y-2)^2}{9} = 1$$

$$(x-1)^2 + (y-2)^2 = 9$$

This is a circle centered at $(1, 2)$ with radius 3.

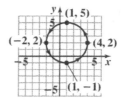

31. $x = 2\cos t$ and $y = 3\sin t$

$\dfrac{x}{2} = \cos t \qquad\qquad \dfrac{y}{3} = \sin t$

Square and add the equations:

$\dfrac{x^2}{4} = \cos^2 t$

$+ \quad \dfrac{y^2}{9} = \sin^2 t$

$\overline{\dfrac{x^2}{4} + \dfrac{y^2}{9} = \cos^2 t + \sin^2 t}$

$\dfrac{x^2}{4} + \dfrac{y^2}{9} = 1$

This is an ellipse centered at (0, 0).

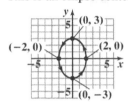

33. $x = 1 + 3\cos t$ and $y = -1 + 2\sin t$

$\dfrac{x-1}{3} = \cos t \qquad\qquad \dfrac{y+1}{2} = \sin t$

Square and add the equations:

$\dfrac{(x-1)^2}{9} = \cos^2 t$

$+ \quad \dfrac{(y+1)^2}{4} = \sin^2 t$

$\overline{\dfrac{(x-1)^2}{9} + \dfrac{(y+1)^2}{4} = \cos^2 t + \sin^2 t}$

$\dfrac{(x-1)^2}{9} + \dfrac{(y+1)^2}{4} = 1$

Since $0 \le t \le \pi$, $-1 \le \cos t \le 1$ and $0 \le \sin t \le 1$.

Thus, $-1 \le \dfrac{x-1}{3} \le 1$ and $0 \le \dfrac{y+1}{2} \le 1$. Hence,

$-2 \le x \le 4$ and $-1 \le y \le 1$. This is the upper half of an ellipse centered at (−1, 1).

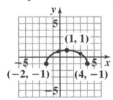

35. $x = \sec t$ and $y = \tan t$

Square and subtract the equations:

$x^2 = \sec^2 t$

$- \quad (y^2 = \tan^2 t)$

$\overline{x^2 - y^2 = \sec^2 t - \tan^2 t}$

$x^2 - y^2 = 1$

This is a hyperbola centered at (0, 0).

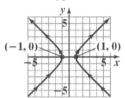

37. $x = t^2 + 2$ and $y = t^2 - 2$

$x - 2 = t^2$

Substitute $x - 2$ into $y = t^2 - 2$ for t^2:

$y = (x - 2) - 2$

$y = x - 4$

Since $t^2 \ge 0$ for all t, $x \ge 2$ and $y \ge -2$.

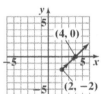

39. $x = 2^t$ and $y = 2^{-t}$

$y = \left(2^t\right)^{-1}$

Substitute x in $y = \left(2^t\right)^{-1}$ for 2^t:

$y = (x)^{-1}$

$y = \dfrac{1}{x}$

Since $t \ge 0$, $x \ge 1$ and $y \ge 0$.

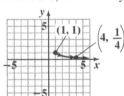

41. $x = h + r\cos t$ and $y = k + r\sin t$

$\dfrac{x-h}{r} = \cos t$ $\dfrac{y-k}{r} = \sin t$

Square and add the equations:

$$\dfrac{(x-h)^2}{r^2} = \cos^2 t$$

$$+ \quad \dfrac{(y-k)^2}{r^2} = \sin^2 t$$

$$\dfrac{(x-h)^2}{r^2} + \dfrac{(y-k)^2}{r^2} = \cos^2 t + \sin^2 t$$

$$\dfrac{(x-h)^2}{r^2} + \dfrac{(y-k)^2}{r^2} = 1$$

$$(x-h)^2 + (y-k)^2 = r^2$$

43. $x = h + a\sec t$ and $y = k + b\tan t$

$\dfrac{x-h}{a} = \sec t$ $\dfrac{y-k}{b} = \tan t$

Square and subtract the equations:

$$\dfrac{(x-h)^2}{a^2} = \sec^2 t$$

$$- \quad \left(\dfrac{(y-k)^2}{b^2} = \tan^2 t \right)$$

$$\dfrac{(x-h)^2}{a^2} - \dfrac{(y-k)^2}{b^2} = \sec^2 t - \tan^2 t$$

$$\dfrac{(x-h)^2}{a^2} - \dfrac{(y-k)^2}{b^2} = 1$$

45. $h = 3, k = 5,$ and $r = 6$

$x = h + r\cos t$ and $y = k + r\sin t$

$x = 3 + 6\cos t$ $y = 5 + 6\sin t$

47. $h = -2, k = 3, a = 5, b = 2$

$x = h + a\cos t$ and $y = k + b\sin t$

$x = -2 + 5\cos t$ $y = 3 + 2\sin t$

49. $h = 0, k = 0, a = 4, c = 6$

$$c^2 = a^2 + b^2$$

$$6^2 = 4^2 + b^2$$

$$36 - 16 = b^2$$

$$b = \sqrt{20} = 2\sqrt{5}$$

$x = h + a\sec t$ and $y = k + b\tan t$

$x = 0 + 4\sec t$ $y = 0 + 2\sqrt{5}\tan t$

$x = 4\sec t$ $y = 2\sqrt{5}\tan t$

51. $x_1 = -2, \ y_1 = 4, \ x_2 = 1, \ y_2 = 7$

53. Answers may vary.
Sample answer:
$x = t$ and $y = 4t - 3$; $x = t + 1$ and $y = 4t + 1$

55. Answers may vary.
Sample answer:
$x = t$ and $y = t^2 + 4$; $x = t + 1$ and $y = t^2 + 2t + 5$

57. a.

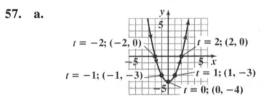

b.

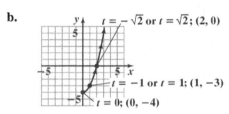

c.

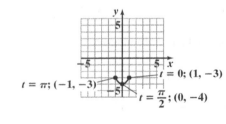

d.

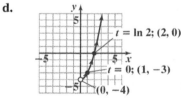

Explanations of how the curves differ from each other may vary.

59.

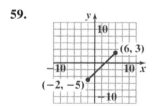

domain: $[-2, 6]$;
range: $[-5, 3]$

61.

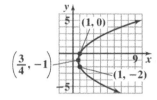

domain: $\left[\dfrac{3}{4}, \infty\right)$;

range: $[-\infty, \infty]$

63.

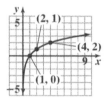

a. increasing: $[-\infty, \infty]$
b. no maximum or minimum

65.

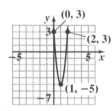

a. decreasing: $[-\infty, 1]$; increasing: $(1, \infty)$
b. minimum of -5 at $x = 1$

67.

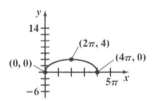

a. increasing: $(1, 2\pi)$; decreasing: $(2\pi, 4\pi)$
b. maximum of 4 at $x = 2\pi$,
minimum of 0 at $x = 0$ and $x = 4\pi$

69. a. $x = (180\cos 40°)t$

$y = 3 + (180\sin 40°)t - 16t^2$

b. After 1 second:
$x = (180\cos 40°) \cdot 1$

≈ 137.9 feet in distance

$y = 3 + (180\sin 40°)1 - 16 \cdot 1^2$

≈ 102.7 feet in height

After 2 seconds:
$x = (180\cos 40°) \cdot 2$

≈ 275.8 feet in distance

$y = 3 + (180\sin 40°) \cdot 2 - 16 \cdot 2^2$

≈ 170.4 feet in height

After 3 seconds:
$x = (180\cos 40°) \cdot 3$

≈ 413.7 feet in distance

$y = 3 + (180\sin 40°) \cdot 3 - 16 \cdot 3^2$

≈ 206.1 feet in height

The points on the curve are (137.9, 102.7), (275.8, 170.4), (413.7, 206.1).

c. The ball is no longer in flight when its height above ground is zero:

$0 = 3 + (180\sin 40°)t - 16t^2$

$0 = -16t^2 + (180\sin 40°)t + 3$

$t = \dfrac{-(180\sin 40°) \pm \sqrt{(180\sin 40°)^2 - 4 \cdot (-16)(3)}}{2(-16)}$

$t \approx -.03$ or $t \approx 7.3$

Since we cannot use the negative time, the ball hits the ground at $t \approx 7.3$ seconds.
The total horizontal distance is:
$x = (180\cos 40°) \cdot (7.3) \approx 1006.6$ feet

d. Answers may vary.

71. – 77. Answers may vary.

79.

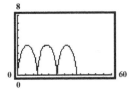

81.

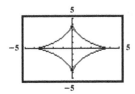

83. $x = (v_0 \cos\theta)t$ and $y = h + (v_0 \sin\theta)t - 16t^2$ where v_0 is the initial velocity, θ is the angle from horizontal, h is the height above the ground, and t is the time, in seconds.

$x = (200\cos 55°)t$ and $y = (200\sin 55°)t - 16t^2$

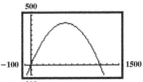

Window: $[-100, 1500] \times [-100, 500]$

The maximum height is 419.4 feet at a time of 5.1 seconds. The range of the projective is 1174.6 feet horizontally. It hits the ground at 10.2 seconds.

85. a. $x = (v_0 \cos\theta)t$ and $y = h + (v_0 \sin\theta)t - 16t^2$ where v_0 is the initial velocity, θ is the angle from horizontal, h is the height above the ground, and t is the time, in seconds.

$x = (140\cos 22°)t$ and $y = 5 + (140\sin 22°)t - 16t^2$

b.

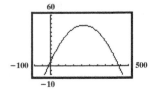

Window: $[-100, 500] \times [-10, 60]$

c. The maximum height is 48.0 feet. It occurs at 1.6 seconds.

d. The ball is in the air for 3.4 seconds.

e. The ball travels 437.5 feet.

87. makes sense

89. makes sense

91. $x = 3\sin t$ and $y = 3\cos t$

93. $r = \dfrac{2}{1 + \frac{1}{2}\cos\theta}$

94.

θ	0	$\dfrac{\pi}{2}$	$\dfrac{2\pi}{3}$	$\dfrac{3\pi}{4}$	$\dfrac{5\pi}{6}$	π
$r = \dfrac{4}{2 + \cos\theta}$	$\dfrac{4}{3}$ ≈ 1.33	2	$\dfrac{8}{3}$ ≈ 2.66	$\dfrac{16 + 4\sqrt{2}}{7}$ ≈ 3.09	$\dfrac{32 + 8\sqrt{3}}{13}$ ≈ 3.53	4

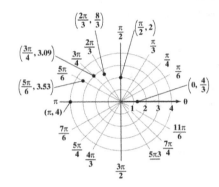

95. **a.**
$$r = \frac{1}{3 - 3\cos\theta}$$
$$r(3 - 3\cos\theta) = 1$$
$$3r - 3r\cos\theta = 1$$
$$3r = 1 + 3r\cos\theta$$
$$(3r)^2 = (1 + 3r\cos\theta)^2$$
$$9r^2 = (1 + 3r\cos\theta)^2$$

b.
$$9r^2 = (1 + 3r\cos\theta)^2$$
$$9(x^2 + y^2) = (1 + 3x)^2$$
$$9x^2 + 9y^2 = 1 + 6x + 9x^2$$
$$9y^2 = 1 + 6x$$

This is the equation of a parabola.

Section 10.6

Check Point Exercises

1. Graph $r = \dfrac{4}{2 - \cos \theta}$.

   Step 1: Divide numerator and denominator by 2 to write the equation in standard form:

   $$r = \dfrac{2}{1 - \frac{1}{2}\cos \theta}$$

   Step 2: $e = \dfrac{1}{2}$ and $ep = \dfrac{1}{2}p = 2$, so $p = 4$. Since $e < 1$, the graph is an ellipse.

   Step 3: The graph has symmetry with respect to the polar axis. One focus is at the pole and the directrix is $x = -4$.
   Find the vertices by selecting $\theta = 0$ and $\theta = \pi$:

   $(4, 0)$ and $\left(\dfrac{4}{3}, \pi \right)$.

   Sketch the upper half by plotting some points, then use the symmetry of the graph to sketch the lower half.

   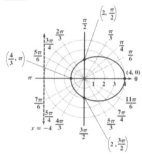

2. Graph $r = \dfrac{8}{4 + 4\sin \theta}$.

   Step 1: Divide numerator and denominator by 4 to write the equation in standard form:

   $$r = \dfrac{2}{1 + \sin \theta}$$

   Step 2: $e = 1$ and $ep = 1$ $p = 2$, so $p = 2$. Since $e = 1$, the graph is parabola.

Step 3: The graph has symmetry with respect to $\theta = \dfrac{\pi}{2}$. The focus is at the pole and the directrix is $y = 2$. Since the vertex is on the line $\theta = \dfrac{\pi}{2}$ (y-axis) the vertex is at $\left(1, \dfrac{\pi}{2} \right)$. To find the intercepts on the polar axis, select $\theta = 0$ and $\theta = \pi$: $(2, 0)$ and $(2, \pi)$.

Sketch the right half by plotting some points, then use symmetry to sketch the left half.

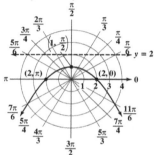

3. Graph $r = \dfrac{9}{3 - 9\cos \theta}$.

   Step 1: Divide numerator and denominator by 3 to write the equation in standard form:

   $$r = \dfrac{3}{1 - 3\cos \theta}$$

   Step 2: $e = 3$ and $ep = 3p = 3$, so $p = 1$. Since $e > 1$, the graph is a hyperbola.

   Step 3: The graph is symmetric with respect to the polar axis. One focus is at the pole and the directrix is at $x = -1$. The transverse axis is horizontal and the vertices lie on the polar axis. Find them by selecting $\theta = 0$ and $0 = \pi$:

   $\left(-\dfrac{3}{2}, 0 \right)$ and $\left(\dfrac{3}{4}, \pi \right)$.

   Sketch the upper half of the hyperbola by plotting some points, then use symmetry to sketch the lower half.

   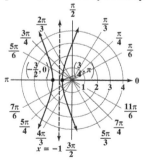

Concept and Vocabulary Check 10.6

1. focus; directrix; eccentricity; parabola; ellipse; hyperbola

2. pole; eccentricity

3. 3; hyperbola; 1; perpendicular; 1; right

4. 1; parabola; 2; parallel; 2; below

5. 1; dividing the numerator and denominator by 4

Exercise Set 10.6

1. $r = \dfrac{3}{1 + \sin \theta}$
   $e = 1$ and $ep = 3$, so $p = 3$.

   a. The graph is a parabola.

   b. The directrix is 3 units above the pole, at $y = 3$.

3. $r = \dfrac{6}{3 - 2 \cos \theta}$
   Divide numerator and denominator

   by 3: $r = \dfrac{2}{1 - \frac{2}{3} \cos \theta}$.

   $e = \dfrac{2}{3}$ and $ep = 2$, so $p = 3$.

   a. The graph is an ellipse.

   b. The directrix is 3 units to the left of the pole, at $x = -3$.

5. $r = \dfrac{8}{2 + 2 \sin \theta}$
   Divide numerator and denominator

   by 2: $r = \dfrac{4}{1 + \sin \theta}$.

   $e = 1$ and $ep = 4$, so $p = 4$.

   a. The graph is a parabola.

   b. The directrix is 4 units above the pole, at $y = 4$.

7. $r = \dfrac{12}{2 - 4 \cos \theta}$
   Divide numerator and denominator

   by 2: $r = \dfrac{6}{1 - 2 \cos \theta}$.

   $e = 2$ and $ep = 6$, so $p = 3$.

   a. The graph is a hyperbola.

   b. The directrix is 3 units to the left of the pole, at $x = -3$.

9. $r = \dfrac{1}{1 + \sin \theta}$
   $e = 1$ and $ep = 1$, so $p = 1$.
   Since $e = 1$, the graph is a parabola. It is symmetric with respect to the y-axis and has a directrix at $y = 1$.

   The vertex is at $\left(\dfrac{1}{2}, \dfrac{\pi}{2} \right)$.

   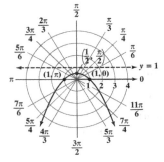

11. $r = \dfrac{2}{1 - \cos \theta}$
    $e = 1$ and $ep = 2$, so $p = 2$.
    Since $e = 1$, the graph is a parabola. It is symmetric with respect to the polar axis and has a directrix at $x = -2$. The vertex is at $(1, \pi)$.

    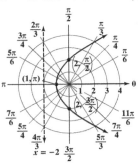

13. $r = \dfrac{12}{5 + 3\cos\theta}$

Write in standard form:

$$r = \dfrac{\frac{12}{5}}{1 + \frac{3}{5}\cos\theta}$$

$e = \dfrac{3}{5}$ and $ep = \dfrac{12}{5}$, so $p = 4$. Since $e < 1$, the graph is an ellipse. It is symmetric with respect to the polar axis and has a directrix at $x = 4$.

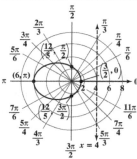

15. $r = \dfrac{6}{2 - 2\sin\theta}$

Write in standard form: $r = \dfrac{3}{1 - \sin\theta}$

$e = 1$ and $ep = 3$, so $p = 3$. Since $e = 1$, the graph is a parabola. It is symmetric with respect to the y-axis and has a directrix at $y = -3$. The vertex is at $\left(\dfrac{3}{2}, \dfrac{3\pi}{2}\right)$.

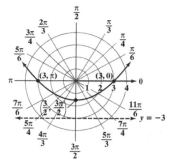

17. $r = \dfrac{8}{2 - 4\cos\theta}$

Write in standard form:

$$r = \dfrac{4}{1 - 2\cos\theta}$$

$e = 2$ and $ep = 4$, so $p = 2$. Since $e > 1$, the graph is a hyperbola. It is symmetric with respect to the polar axis and it has a directrix at $x = -2$. The transverse axis is horizontal and the vertices lie on the polar axis.

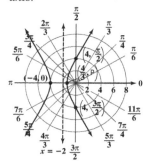

19. $r = \dfrac{12}{3 - 6\cos\theta}$

Write in standard form:

$$r = \dfrac{4}{1 - 2\cos\theta}$$

$e = 2$ and $ep = 4$, so $p = 2$. Since $e > 2$, the graph is a hyperbola. It is symmetric with respect to the polar axis and has a directrix at $x = -2$. The transverse axis is horizontal and the vertices lie on the polar axis.

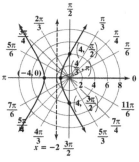

21. $[-3, 15, 1]$ by $[-7, 7, 1]$

23. $[-4, 2, 1]$ by $[-10, 10, 1]$

25. $[-2, 5, 1]$ by $[-10, 10, 1]$

27. $[-4, 4, 1]$ by $[-10, 0.4, 1]$

29. The shortest distance from the sun occurs on the positive *y*-axis, at $\theta = \dfrac{\pi}{2}$.

When $\theta = \dfrac{\pi}{2}$, $r = \dfrac{1.069}{1 + 0.967 \sin \frac{\pi}{2}} = \dfrac{1.069}{1.967}$

≈ 0.54 astronomical units or about 51 million miles.

31. His greatest distance from Earth's center occurred when $\theta = 0$:

$$r = \frac{4090.76}{1 - 0.0076 \cos 0} = \frac{4090.76}{0.9924}$$

≈ 4122 miles from the center of the earth.
Assuming the earth to be perfectly spherical, he was
$4122 - 3960 = 162$ miles from the surface of the earth.

33. – 39. Answers may vary.

41. Write the equation in standard form: $r = \dfrac{3}{1 + \frac{5}{4} \sin \theta}$

Since $e = \dfrac{5}{4} > 1$, the graph is a hyperbola.

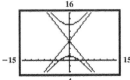

43.

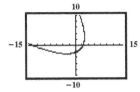

The graph appears to be rotated counter-clockwise through an angle of $\dfrac{\pi}{4}$ radians.

45. Mercury: $r = \dfrac{\left(1 - 0.2056^2\right)\left(36.0 \times 10^6\right)}{1 - 0.2056 \cos \theta}$

Earth: $r = \dfrac{\left(1 - 0.0167^2\right)\left(92.96 \times 10^6\right)}{1 - 0.0167 \cos \theta}$

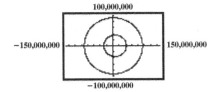

$r : -150,000,000 - 150,000,000$

$y : -100,000,000, -100,000,000$

Observations may vary.

47. does not make sense; Explanations will vary. Sample explanation: This form is not symmetrical with respect to the *y*-axis.

49. does not make sense; Explanations will vary. Sample explanation: A knowledge of conic sections is necessary to graph such equations.

51. Since the equation is an ellipse with a vertex at $(4, 0)$, the polar axis is the major axis. Since $\dfrac{PF}{PD} = e$, then at the point $(4, 0)$,

$\dfrac{4}{x} = \dfrac{1}{2}$, where *x* is the distance between the point $(4, 0)$ and the directrix. Thus, $x = 8$ and the distance between the point $(4, 0)$ and the directrix is 8. The directrix is either $x = -4$ or $x = 12$. There are two polar equations that meet the given conditions. If the directrix is $x = -4$,

$$r = \frac{\frac{1}{2}(4)}{1 - \frac{1}{2}\cos\theta} = \frac{2}{1 - \frac{1}{2}\cos\theta}.$$

If the directrix is $x = 12$,

$$r = \frac{\frac{1}{2}(12)}{1 + \frac{1}{2}\cos\theta} = \frac{6}{1 + \frac{1}{2}\cos\theta}.$$

53. $r = \dfrac{1}{2 - 2\cos\theta}$

Write the equation in standard form: $r = \dfrac{\frac{1}{2}}{1 - \cos\theta}$

Since $e = 1$, the graph is a parabola.
Write in rectangular coordinates:

$$r = \frac{\frac{1}{2}}{1 - \cos\theta}$$

$$r(1 - \cos\theta) = \frac{1}{2}$$

$$r - r\cos\theta = \frac{1}{2}$$

$$r = r\cos\theta + \frac{1}{2}$$

$$r = x + \frac{1}{2}$$

Substitution: $x = r\cos\theta$

$$r^2 = \left(x + \frac{1}{2}\right)^2$$

Square both sides

$$x^2 + y^2 = \left(x + \frac{1}{2}\right)^2$$

Substitution: $r^2 = x^2 + y^2$

$$x^2 + y^2 = x^2 + x + \frac{1}{4}$$

$$y^2 = x + \frac{1}{4}$$

55. For $n = 1$; $\dfrac{(-1)^n}{3^n - 1} = \dfrac{(-1)^1}{3^1 - 1} = \dfrac{-1}{3 - 1} = -\dfrac{1}{2}$

For $n = 2$; $\dfrac{(-1)^n}{3^n - 1} = \dfrac{(-1)^2}{3^2 - 1} = \dfrac{1}{9 - 1} = \dfrac{1}{8}$

For $n = 3$; $\dfrac{(-1)^n}{3^n - 1} = \dfrac{(-1)^3}{3^3 - 1} = \dfrac{-1}{27 - 1} = -\dfrac{1}{26}$

For $n = 4$; $\dfrac{(-1)^n}{3^n - 1} = \dfrac{(-1)^4}{3^4 - 1} = \dfrac{1}{81 - 1} = \dfrac{1}{80}$

56. $5 \cdot 4 \cdot 3 \cdot 2 \cdot 1 = 120$

57. For $i = 1$; $i^2 + 1 = 1^2 + 1 = 1 + 1 = 2$

For $i = 2$; $i^2 + 1 = 2^2 + 1 = 4 + 1 = 5$

For $i = 3$; $i^2 + 1 = 3^2 + 1 = 9 + 1 = 10$

For $i = 4$; $i^2 + 1 = 4^2 + 1 = 16 + 1 = 17$

For $i = 5$; $i^2 + 1 = 5^2 + 1 = 25 + 1 = 26$

For $i = 6$; $i^2 + 1 = 6^2 + 1 = 36 + 1 = 37$

$2 + 5 + 10 + 17 + 26 + 37 = 97$

Chapter 10 Review Exercises

1. $a^2 = 36, a = 6$
$b^2 = 25, b = 5$
$c^2 = a^2 - b^2 = 36 - 25 = 11$
$c = \sqrt{11}$
The foci are at $(\sqrt{11}, 0)$ and $(-\sqrt{11}, 0)$

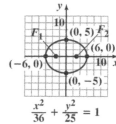

$\dfrac{x^2}{36} + \dfrac{y^2}{25} = 1$

2. $a^2 = 25, a = 5$
$b^2 = 16, b = 4$
$c^2 = a^2 - b^2$
$c^2 = 25 - 16$
$c^2 = 9$
$c = 3$
The foci are $(0, 3)$ and $(0, -3)$.

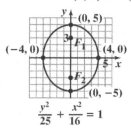

$\dfrac{y^2}{25} + \dfrac{x^2}{16} = 1$

3. $\dfrac{4x^2}{16} + \dfrac{y^2}{16} = \dfrac{16}{16}$

$\dfrac{x^2}{4} + \dfrac{y^2}{16} = 1$

$b^2 = 4, b = 2$
$a^2 = 16, a = 4$
$c^2 = a^2 - b^2 = 16 - 4 = 12$
$c = \sqrt{12} = 2\sqrt{3}$
The foci are at $(0, 2\sqrt{3})$ and $(0, -2\sqrt{3})$.

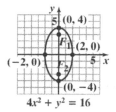

$4x^2 + y^2 = 16$

4. $\dfrac{4x^2}{36} + \dfrac{9y^2}{36} = \dfrac{36}{36}$

$\dfrac{x^2}{9} + \dfrac{y^2}{4} = 1$

$a^2 = 9, a = 3$
$b^2 = 4, b = 2$
$c^2 = a^2 - b^2 = 9 - 4 = 5, c = \sqrt{5}$
The foci are at $(\sqrt{5}, 0)$ and $(-\sqrt{5}, 0)$.

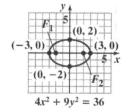

$4x^2 + 9y^2 = 36$

5. $a^2 = 16 \; a = 4$
$b^2 = 9 \; b = 3$
$c^2 = 16 - 9 = 7, \; c = \sqrt{7}$
center: $(1, -2)$
The foci are at $(1+\sqrt{7}, -2)$ and $(1-\sqrt{7}, -2)$.

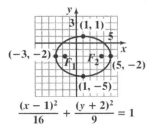

$$\frac{(x-1)^2}{16} + \frac{(y+2)^2}{9} = 1$$

6. $a^2 = 16, \; a = 4$
$b^2 = 9, \; b = 3$
$c^2 = a^2 - b^2 = 16 - 9 = 7, \; c = \sqrt{7}$
center: $(-1, 2)$
The foci are at $(-1, 2+\sqrt{7})$ and $(-1, 2-\sqrt{7})$.

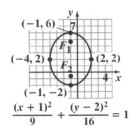

$$\frac{(x+1)^2}{9} + \frac{(y-2)^2}{16} = 1$$

7. $4x^2 + 24x + 9y^2 - 36y = -36$
$4(x^2 + 6x + 9) + 9(y^2 - 4y + 4)$
$= -36 + 36 + 36$
$= 4(x+3)^2 + 9(y-2)^2 = 36$
$$\frac{(x+3)^2}{9} + \frac{(y-2)^2}{4} = 1$$
$c^2 = a^2 - b^2 = 5, \; c = \sqrt{5}$
center: $(-3, 2)$
The foci are at $(-3+\sqrt{5}, 2)$ and $(-3-\sqrt{5}, 2)$.

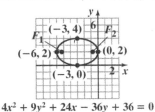

$4x^2 + 9y^2 + 24x - 36y + 36 = 0$

8. $9x^2 - 18x + 4y^2 + 8y = 23$
$9(x^2 - 2x + 1) + 4(y^2 + 2y + 1) = 23 + 9 + 4$
$9(x-1)^2 + 4(y+1)^2 = 36$
$$\frac{(x-1)^2}{4} + \frac{(y+1)^2}{9} = 1$$
$c^2 = a^2 - b^2 = 9 - 4 = 5, \; c = \sqrt{5}$
center: $(1, -1)$
The foci are at $(1, -1+\sqrt{5})$ and $(1, -1-\sqrt{5})$.

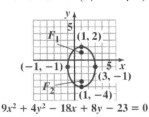

$9x^2 + 4y^2 - 18x + 8y - 23 = 0$

9. $c = 4, \; c^2 = 16$
$a = 5, \; a^2 = 25$
$b^2 = a^2 - c^2 = 25 - 16 = 9$
$$\frac{x^2}{25} + \frac{y^2}{9} = 1$$

10. $c = 3, \; c^2 = 9$
$a = 6, \; a^2 = 36$
$b^2 = a^2 - c^2 = 36 - 9 = 27$
$$\frac{x^2}{27} + \frac{y^2}{36} = 1$$

11. $2a = 12, \; a = 6, \; a^2 = 36$
$2b = 4, \; b = 2, \; b^2 = 4$
$$\frac{(x+3)^2}{36} + \frac{(y-5)^2}{4} = 1$$

12. $2a = 20, \; a = 10, \; a^2 = 100$
$b = 6, \; b^2 = 36$
$$\frac{x^2}{100} + \frac{y^2}{36} = 1$$

13. $2a = 50, \; a = 25$
$b = 15$
$$\frac{x^2}{625} + \frac{y^2}{225} = 1$$
Let $x = 14$.
$$\frac{(14)^2}{625} + \frac{y^2}{225} = 1$$
$$y^2 = 225\left(1 - \frac{196}{625}\right)$$
$y \approx 15(0.8285) \approx 12.4 > 12$
Yes, the truck can drive under the archway.

14. The hit ball will collide with the other ball.

15. $c^2 = a^2 + b^2 = 16 + 1 = 17, c = \sqrt{17}$

The foci are at $(\sqrt{17}, 0)$ and $(-\sqrt{17}, 0)$.

Asymptotes: $y = \pm \dfrac{1}{4} x$

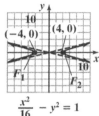

$\dfrac{x^2}{16} - y^2 = 1$

16. $c^2 = a^2 + b^2 = 16 + 1 = 17$

$c = \sqrt{17}$

The foci are at $(0, \sqrt{17})$ and $(0, -\sqrt{17})$.

Asymptotes: $y = \pm 4x$

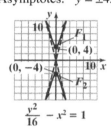

$\dfrac{y^2}{16} - x^2 = 1$

17. $\dfrac{x^2}{16} - \dfrac{y^2}{9} = 1$

$c^2 = a^2 + b^2 = 16 + 9 = 25, c = 5$

The foci are at $(5, 0)$ and $(-5, 0)$.

Asymptotes: $y = \pm \dfrac{3}{4} x$

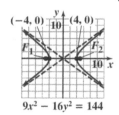

$9x^2 - 16y^2 = 144$

18. $\dfrac{y^2}{4} - \dfrac{x^2}{16} = 1$

$c^2 = a^2 + b^2 = 4 + 16 = 20$

$c = \sqrt{20} = 2\sqrt{5}$

The foci are at $(0, 2\sqrt{5})$ and $(0, -2\sqrt{5})$.

Asymptotes: $y = \pm \dfrac{1}{2} x$

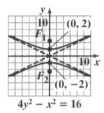

$4y^2 - x^2 = 16$

19. $c^2 = a^2 + b^2 = 25 + 16 = 41, c = \sqrt{41}$

center: $(2, -3)$

The foci are at $(2 + \sqrt{41}, -3)$ and $(2 - \sqrt{41}, -3)$.

Asymptotes: $y + 3 = \pm \dfrac{4}{5} (x - 2)$

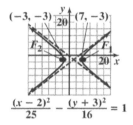

$\dfrac{(x-2)^2}{25} - \dfrac{(y+3)^2}{16} = 1$

20. $c^2 = a^2 + b^2 = 25 + 16 = 41, c = \sqrt{41}$

center: $(3, -2)$

The foci are at $(3, -2 + \sqrt{41})$ and $(3, -2 - \sqrt{41})$.

Asymptotes: $y + 2 = \pm \dfrac{5}{4} (x - 3)$

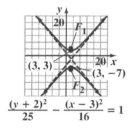

$\dfrac{(y+2)^2}{25} - \dfrac{(x-3)^2}{16} = 1$

21.
$$y^2 - 4y - 4x^2 + 8x - 4 = 0$$
$$(y^2 - 4y + 4) - 4(x^2 - 2x + 1) = 4 + 4 - 4$$
$$(y - 2)^2 - 4(x - 1)^2 = 4$$
$$\frac{(y-2)^2}{4} - (x-1)^2 = 1$$
$$c^2 = a^2 + b^2 = 4 + 1 = 5, \ c = \sqrt{5}$$
center: (1, 2)
The foci are at $(1, 2+\sqrt{5})$ and $(1, 2-\sqrt{5})$.
Asymptotes: $y - 2 = \pm 2(x - 1)$

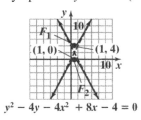

$y^2 - 4y - 4x^2 + 8x - 4 = 0$

22.
$$x^2 - 2x - y^2 - 2y = 1$$
$$(x^2 - 2x + 1) - (y^2 + 2y + 1) = 1 + 1 - 1$$
$$(x - 1)^2 - (y + 1)^2 = 1$$
$$c^2 = a^2 + b^2 = 1 + 1 = 2, \ c = \sqrt{2}$$
center: (1, −1)
The foci are at $(1+\sqrt{2}, -1)$ and $(1-\sqrt{2}, -1)$.
asymptotes: $y + 1 = \pm(x - 1)$

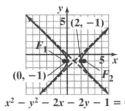

$x^2 - y^2 - 2x - 2y - 1 = 0$

23. $c = 4, \ c^2 = 16$
$$a = 2, \ a^2 = 4$$
$$b^2 = c^2 - a^2 = 16 - 4 = 12$$
$$\frac{y^2}{4} - \frac{x^2}{12} = 1$$

24. $c = 8, \ c^2 = 64$
$$a = 3, \ a^2 = 9$$
$$b^2 = c^2 - a^2 = 64 - 9 = 55$$
$$\frac{x^2}{9} - \frac{y^2}{55} = 1$$

25. If the foci are at (0, −2) and (0, 2), then $c = 2$. If the vertices are at (0, −3) and (0, 3) then $a = 3$. This is not possible since c must be greater than a.

26. foci: (±100, 0), $c = 100$
$$\left|d_1 - d_2\right| = \left(0.186\frac{\text{mi}}{\mu s}\right)(500\mu s) = 93 \text{ mi} = 2a$$
$$a = \frac{93}{2}$$
$$b^2 = c^2 - a^2 = (100)^2 - \left(\frac{93}{2}\right)^2 = 7837.75$$
$$\frac{x^2}{\left(\frac{93}{2}\right)^2} - \frac{y^2}{7837.75} = 1$$
$$\frac{x^2}{2162.25} - \frac{y^2}{7837.75} = 1$$

27. $4p = 8, \ p = 2$
vertex: (0, 0)
focus: (2, 0)
directrix: $x = -2$

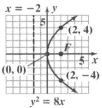

$y^2 = 8x$

28.
$$x^2 + 16y = 0$$
$$x^2 = -16y$$
$$4p = -16$$
$$p = -4$$
vertex: (0, 0)
focus: (0, −4)
directrix: $y = 4$

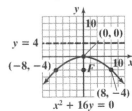

$x^2 + 16y = 0$

29. $4p = -16$
$p = -4$
vertex: $(0, 2)$
focus: $(-4, 2)$
directrix: $x = 4$

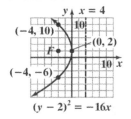

$(y - 2)^2 = -16x$

30. $4p = 4, p = 1$
vertex: $(4, -1)$
focus: $(4, 0)$
directrix: $y = -2$

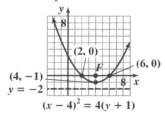

$(x - 4)^2 = 4(y + 1)$

31. $x^2 = -4y + 4$
$x^2 = -4(y - 1)$
$4p = -4, p = -1$
vertex: $(0, 1)$
focus: $(0, 0)$
directrix: $y = 2$

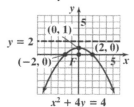

$x^2 + 4y = 4$

32.
$$y^2 - 10y = 4x - 21$$
$$y^2 - 10y + 25 = 4x - 21 + 25$$
$$(y - 5)^2 = 4(x + 1)$$
$4p = 4, p = 1$
vertex: $(-1, 5)$
focus: $(0, 5)$
directrix: $x = -2$

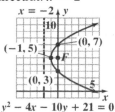

$y^2 - 4x - 10y + 21 = 0$

33. $x^2 - 4x - 2y = 0$
$$x^2 - 4x = 2y$$
$$(x^2 - 4x + 4) = 2y + 4$$
$$(x - 2)^2 = 2(y + 2)$$
$4p = 2, p = \dfrac{1}{2}$
vertex: $(2, -2)$
focus: $\left(2, -\dfrac{3}{2}\right)$
directrix: $y = -\dfrac{5}{2}$

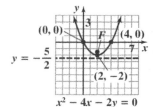

$x^2 - 4x - 2y = 0$

34. $p = 12$
$y^2 = 48x$

35. $p = -11$
$x^2 = -44y$

36. $x^2 = 4py$
$$(6)^2 = 4p(3)$$
$$p = 3$$
$$x^2 = 12y$$
Place the light 3 inches from the vertex at $(0, 3)$.

37. $x^2 = 4py$

$(1750)^2 = 4p(316)$

$4p \approx 9691$

$x^2 = 9691y$

Let $x = 1750 - 1000 = 750$.

$y = \dfrac{x^2}{9691} = \dfrac{(750)^2}{9691} \approx 58$

The height is approximately 58 feet.

38. $x^2 = 4py$

$(150)^2 = 4p(44)$

$22,500 = 176p$

$p \approx 128$

The receiver should be placed approximately 128 feet from the base of the dish.

39. $A = 0, C = 1$.

$AC = 0$, so the graph is a parabola.

40. $A = 1, C = 16$.

$AC = 16 > 0$ and $A \ne C$, so the graph is an ellipse.

41. $A = 16, C = 9$.

$AC = 16 \cdot 9 = 144 > 0$ and $A \ne C$, so the graph is an ellipse.

42. $A = 4, C = -9$.

$AC = 4(-9) = 36 < 0$, so the graph is a hyperbola.

43. $A = 5, B = 2\sqrt{3}, C = 3$.

$B^2 - 4AC = \left(2\sqrt{3}\right)^2 - 4(5)(3) = 12 - 60 = -48$ Since $B^2 - 4AC < 0$, the graph is an ellipse or a circle.

44. $A = 5, B = -8, C = 7$.

$B^2 - 4AC = (-8)^2 - 4(5)(7) = 64 - 140 = -76$. Since $B^2 - 4AC < 0$, the graph is an ellipse or a circle.

45. $A = 1, B = 6, C = 9$.

$B^2 - 4AC = 6^2 - 4(1)(9) = 36 - 36 = 0$. Since $B^2 - 4AC = 0$, the graph is a parabola.

46. $A = 1, B = -2, C = 3$.

$B^2 - 4AC = (-2)^2 - 4(1)(3) = 4 - 12 = -8$ Since $B^2 - 4AC < 0$, the graph is an ellipse or a circle.

47. $xy - 4 = 0$

a. $A = 0, B = 1, C = 0$.

$\cot 2\theta = \dfrac{A - C}{B} = \dfrac{0 - 0}{1} = 0$

$2\theta = 90°$

$\theta = 45°$

$x = x'\cos 45° - y'\sin 45° = x'\left(\dfrac{\sqrt{2}}{2}\right) - y'\left(\dfrac{\sqrt{2}}{2}\right) = \dfrac{\sqrt{2}}{2}(x' - y')$

$y = x'\sin 45° + y'\cos 45° = x'\left(\dfrac{\sqrt{2}}{2}\right) + y'\left(\dfrac{\sqrt{2}}{2}\right) = \dfrac{\sqrt{2}}{2}(x' + y')$

Substitute into the equation: $xy - 4 = 0$

$$\left[\frac{\sqrt{2}}{2}(x' - y')\right]\left[\frac{\sqrt{2}}{2}(x' - y')\right] - 4 = 0$$

$$\frac{1}{2}\left(x'^2 - y'^2\right) - 4 = 0$$

$$x'^2 - y'^2 = 8$$

b. $\dfrac{x'^2}{8} - \dfrac{y'^2}{8} = 1$

c.

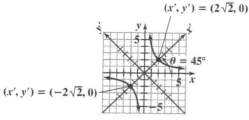

48. $x^2 + xy + y^2 - 1 = 0$

a. $A = 1, B = 1, C = 1.$

$$\cot 2\theta = \frac{A - C}{B} = \frac{1 - 1}{1} = \frac{0}{1} = 0$$

$$2\theta = 90°$$

$$\theta = 45°$$

$$x = x'\cos 45° - y'\sin 45° = \frac{\sqrt{2}}{2}(x' - y')$$

$$y = x'\sin 45° + y'\cos 45° = \frac{\sqrt{2}}{2}(x' + y')$$

Substitute into the equation: $x^2 + xy + y^2 - 1 = 0$

$$\left[\frac{\sqrt{2}}{2}(x' - y')\right]^2 + \left[\frac{\sqrt{2}}{2}(x' - y')\right]\left[\frac{\sqrt{2}}{2}(x' + y')\right] + \left[\frac{\sqrt{2}}{2}(x' + y')\right]^2 - 1 = 0$$

$$\frac{1}{2}\left(x'^2 - 2x'y' + y'^2\right) + \frac{1}{2}\left(x'^2 - y'^2\right) + \frac{1}{2}\left(x'^2 + 2x'y' + y'^2\right) = 1$$

Multiply both sides by 2 and simplify: $3x'^2 + y'^2 = 2$

b. $\dfrac{x'^2}{\dfrac{2}{3}} + \dfrac{y'^2}{2} = 1$

c.

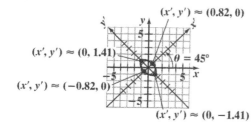

49. $4x^2 + 10xy + 4y^2 - 9 = 0$

a. $A = 4$, $B = 10$, $C = 4$.

$$\cot 2\theta = \frac{A - C}{B} = \frac{4 - 4}{10} = \frac{0}{10} = 0$$

$$2\theta = 90°$$

$$\theta = 45°$$

$$x = x'\cos 45° - y'\sin 45° = \frac{\sqrt{2}}{2}(x' - y')$$

$$y = x'\sin 45° + y'\cos 45° = \frac{\sqrt{2}}{2}(x' + y')$$

Substitute into the equation: $4x^2 + 10xy + 4y^2 - 9 = 0$

$$4\left[\frac{\sqrt{2}}{2}(x' - y')\right]^2 + 10\left[\frac{\sqrt{2}}{2}(x' - y')\right]\left[\frac{\sqrt{2}}{2}(x' + y')\right] + 4\left[\frac{\sqrt{2}}{2}(x' + y')\right]^2 - 9 = 0$$

$$4 \cdot \frac{1}{2}\left(x'^2 - 2x'y' + y'^2\right) + 10 \cdot \frac{1}{2}\left(x'^2 - y'^2\right) + 4 \cdot \frac{1}{2}\left(x'^2 + 2x'y' + y'^2\right) = 9$$

Multiply both sides by 2 and simplify: $18x'^2 - 2y'^2 = 18$

b. $\dfrac{x'^2}{1} - \dfrac{y'^2}{9} = 1$

c.

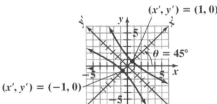

50. $6x^2 - 6xy + 14y^2 - 45 = 0$

a. $A = 6$, $B = -6$ $C = 14$

$$\cot 2\theta = \frac{A - C}{B} = \frac{6 - 14}{-6} = \frac{-8}{-6} = \frac{4}{3}$$

Since θ is always acute, and $\cot 2\theta$ is positive, 2θ lies in quadrant I. The third side of the right triangle is found using the Pythagorean Theorem.

$$4^2 + 3^2 = r^2$$

$$r = 5$$

So, $\cos 2\theta = \dfrac{4}{5}$.

$$\sin\theta = \sqrt{\frac{1 - \cos 2\theta}{2}} = \sqrt{\frac{1 - \frac{4}{5}}{2}} = \frac{\sqrt{10}}{10} \quad \text{and} \quad \cos\theta = \sqrt{\frac{1 + \cos 2\theta}{2}} = \sqrt{\frac{1 + \frac{4}{5}}{2}} = \frac{3\sqrt{10}}{10}.$$

So, $x = x'\cos\theta - y'\sin\theta = x'\left(\dfrac{3\sqrt{10}}{10}\right) - y'\left(\dfrac{\sqrt{10}}{10}\right) = \dfrac{\sqrt{10}}{10}\left(3x' - y'\right)$

and $y = x'\sin\theta - y'\cos\theta = x'\left(\dfrac{\sqrt{10}}{10}\right) + y'\left(\dfrac{3\sqrt{10}}{10}\right) = \dfrac{\sqrt{10}}{10}\left(x' + 3y'\right)$.

Substitute into the equation: $6x^2 - 6xy + 14y^2 - 45 = 0$

$$6\left[\frac{\sqrt{10}}{10}(3x' - y')\right]^2 - 6\left[\frac{\sqrt{10}}{10}(3x' - y')\right]\left[\frac{\sqrt{10}}{10}(x' + 3y')\right] + 14\left[\frac{\sqrt{10}}{10}(x' + 3y')\right]^2 - 45 = 0$$

$$6\left[\frac{1}{10}(9x'^2 - 6x'y' + y'^2)\right] - 6\left[\frac{1}{10}(3x'^2 + 8x'y' - 3y'^2)\right] + 14\left[\frac{1}{10}(x'^2 + 6x'y' + 9y'^2)\right] = 45$$

Multiply both sides by 10 and simplify: $50x'^2 + 150y'^2 = 450$

b. $\dfrac{x'^2}{9} + \dfrac{y'^2}{3} = 1$

c.

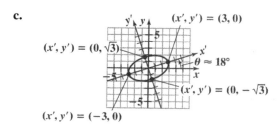

$(x', y') = (0, \sqrt{3})$ $(x', y') = (3, 0)$

$(x', y') = (0, -\sqrt{3})$

$(x', y') = (-3, 0)$

$\theta \approx 18°$

The axes are rotated by $\theta = \sin^{-1}\left(\dfrac{\sqrt{10}}{10}\right) \approx 18°$.

51. $x^2 + 2\sqrt{3}xy + 3y^2 - 12\sqrt{3}x + 12y = 0$

a. $A = 1, B = 2\sqrt{3}, C = 3$

$$\cot 2\theta = \frac{A - C}{B} = \frac{1 - 3}{2\sqrt{3}} = \frac{-2}{2\sqrt{3}} = \frac{-\sqrt{3}}{3}$$

$$2\theta = 120°$$
$$\theta = 60°$$

$$x = x'\cos 60° - y'\sin 60° = x'\left(\frac{1}{2}\right) - y'\left(\frac{\sqrt{3}}{2}\right) = \frac{1}{2}\left(x' - \sqrt{3}y'\right)$$

and $y = x'\sin 60° + y'\cos 60° = x'\left(\dfrac{\sqrt{3}}{2}\right) + y'\left(\dfrac{1}{2}\right) = \dfrac{1}{2}\left(\sqrt{3}x' + y'\right)$

Substitute into the equation: $x^2 + 2\sqrt{3}xy + 3y^2 - 12\sqrt{3}x + 12y = 0$

$$\left[\frac{1}{2}(x' - \sqrt{3}y')\right]^2 + 2\sqrt{3}\left[\frac{1}{2}(x' - \sqrt{3}y')\right]\left[\frac{1}{2}(\sqrt{3}x' + y')\right] + 3\left[\frac{1}{2}(\sqrt{3}x' + y')\right]^2 - 12\sqrt{3}\left[\frac{1}{2}(x' - \sqrt{3}y')\right]$$

$$+ 12\left[\frac{1}{2}(\sqrt{3}x' + y')\right] = 0$$

$$\frac{1}{4}(x'^2 - 2\sqrt{3}x'y' + 3y'^2) + 2\sqrt{3} \cdot \frac{1}{4}(\sqrt{3}x'^2 - 2x'y' - \sqrt{3}y'^2) + 3 \cdot \frac{1}{4}(3x'^2 + 2\sqrt{3}x'y' + y'^2) - 6\sqrt{3}x' + 18y'$$

$$+ 6\sqrt{3}x' + 6y' = 0$$

Multiply both sides by 4 and simplify: $16x'^2 + 96y' = 0$

b. $16x'^2 = -96y'$

$x'^2 = -6y'$

c.

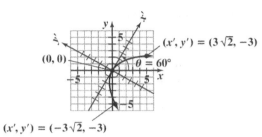

52. $x = 2t - 1$ and $y = 1 - t$; $-\infty < t < \infty$

$\dfrac{x+1}{2} = t$

Substitute into y: $y = 1 - \left(\dfrac{x+1}{2}\right)$

$y = -\dfrac{1}{2}x + \dfrac{1}{2}$

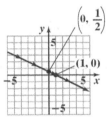

53. $x = t^2$ and $y = t - 1$; $-1 \le t \le 3$

$y + 1 = t$

Substitute into x:

$x = (y+1)^2$

$(y+1)^2 = x$

$0 \le x \le 9, \ -2 \le y \le 2$

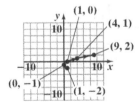

54. $x = 4t^2$ and $y = t + 1$; $-\infty < t < \infty$

$y - 1 = t$

Substitute into x:

$x = 4(y-1)^2$

$\dfrac{1}{4}x = (y-1)^2$

$(y-1)^2 = \dfrac{1}{4}x$

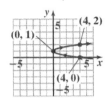

55. $x = 4\sin t$, $y = 3\cos t$; $0 \le t < \pi$

$\dfrac{x}{4} = \sin t$ $\dfrac{y}{3} = \cos t$

Square and add the equations:

$\dfrac{x^2}{16} = \sin^2 t$

$+ \quad \dfrac{y^2}{9} = \cos^2 t$

$\rule{4cm}{0.4pt}$

$\dfrac{x^2}{16} + \dfrac{y^2}{9} = \sin^2 t + \cos^2 t$

$\dfrac{x^2}{16} + \dfrac{y^2}{9} = \sin^2 t + \cos^2 t$

$\dfrac{x^2}{16} + \dfrac{y^2}{9} = 1$

$0 \le x \le 4, \ -3 \le y \le 3$

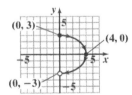

56. $x = 3 + 2\cos t, \ y = 1 + 2\sin t; \ 0 \le t < 2\pi$

$\dfrac{x-3}{2} = \cos t \quad \dfrac{y-1}{2} = \sin t$

Square and add the equations:

$$\dfrac{(x-3)^2}{4} = \cos^2 t$$

$$+ \ \dfrac{(y-1)^2}{4} = \sin^2 t$$

$$\dfrac{(x-3)^2}{4} + \dfrac{(y-1)^2}{4} = \cos^2 t + \sin^2 t$$

$$\dfrac{(x-3)^2}{4} + \dfrac{(y-1)^2}{4} = 1$$

or $(x-3)^2 + (y-1)^2 = 4$

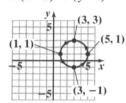

57. $x = 3\sec t, \ y = 3\tan t; \ 0 \le t \le \dfrac{\pi}{4}$

$\dfrac{x}{3} = \sec t \quad \dfrac{y}{3} = \tan t$

Square and subtract the equations:

$$\dfrac{x^2}{9} = \sec^2 t$$

$$- \left(\dfrac{y^2}{9} = \tan^2 t \right)$$

$$\dfrac{x^2}{9} - \dfrac{y^2}{9} = \sec^2 t - \tan^2 t$$

$$\dfrac{x^2}{9} - \dfrac{y^2}{9} = 1$$

$3 \le x \le 3\sqrt{2}, \ 0 \le y \le 3$

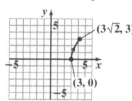

58. Answers may vary. Sample answer:

$x = t$ and $y = t^2 + 6; \ x = t+1$ and $y = t^2 + 2t + 7$

59. a. $x = (100\cos 40°)t$

$y = 6 + (100\sin 40°)t - 16t^2$

b. After 1 second:

$x = (100\cos 40°) \cdot 1$

≈ 76.6 feet in distance

$y = 6 + (100\sin 40°) \cdot 1 - 16(1)^2$

≈ 54.3 feet in height

After 2 seconds:

$x = (100\cos 40°) \cdot 2$

≈ 153.2 feet in distance

$y = 6 + (100\sin 40°) \cdot 2 - 16(2)^2$

≈ 70.6 feet in height

After 3 seconds:

$x = (100\cos 40°) \cdot 3$

≈ 229.8 feet in distance

$y = 6 + (100\sin 40°) \cdot 3 - 16(3)^2$

≈ 54.8 feet in height

c. $0 = 6t(100\sin 40°)t - 16t^2$

Using the quadratic formula with

$a = -16, \ b = 100\sin 40°, \text{ and } c = 6,$

$$t = \dfrac{-100\sin 40° \pm \sqrt{(100\sin 40°)^2 - 4(-16)(6)}}{2(-16)}$$

$t \approx -0.1$ or $t \approx 4.1$

Since t cannot be negative, we discard $t \approx -0.1$.

At $t \approx 4.1, \ x = (100\cos 40°)(4.1) \approx 314.1$

The ball is in flight for 4.1 seconds. It travels a total horizontal distance of 314.1 feet.

d.

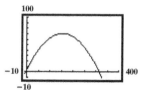

The ball is at its maximum height at 2.0 seconds. The maximum height is 70.6 feet.

60. a. $r = \dfrac{4}{1-\sin\theta}$

b. $e = 1$ and $ep = 4$, so $p = 4$. Since $e = 1$, the graph is a parabola.

c.

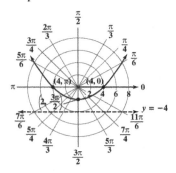

61. a. $r = \dfrac{6}{1+\cos\theta}$

b. $e = 1$ and $ep = 6$, so $p = 6$. Since $e = 1$, the graph is a parabola.

c.

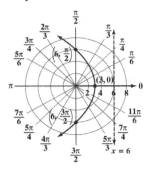

62. a. Divide numerator and denominator by 2:

$$r = \dfrac{3}{1+\frac{1}{2}\sin\theta}$$

b. $e = \dfrac{1}{2}$ and $ep = 3$, so $p = 6$. Since $e < 1$, the graph is an ellipse.

c.

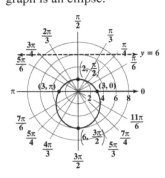

63. a. Divide the numerator and denominator by 3:

$$r = \dfrac{\frac{2}{3}}{1-\frac{2}{3}\cos\theta}$$

b. $e = \dfrac{2}{3}$ and $ep = \dfrac{2}{3}$, so $p = 1$. Since $e < 1$, the graph is an ellipse.

c.

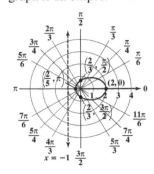

64. a. Divide the numerator and denominator by 3:

$$r = \dfrac{2}{1+2\sin\theta}$$

b. $e = 2$ and $ep = 2$, so $p = 1$. Since $e > 1$, the graph is a hyperbola.

c.

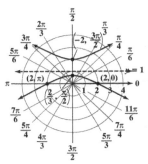

65. a. Divide the numerator and denominator by 4:

$$r = \frac{2}{1 + 4\cos\theta}$$

b. $e = 4$ and $ep = 2$, so $p = \frac{1}{2}$. Since $e > 1$, the graph is a hyperbola.

c.

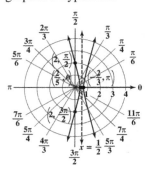

Chapter 10 Test

1. $\frac{x^2}{4} - \frac{y^2}{9} = 1$

$c^2 = a^2 + b^2 = 4 + 9 = 13$, $c = \sqrt{13}$

hyperbola

The foci are at $\left(\sqrt{13}, 0\right)$ and $\left(-\sqrt{13}, 0\right)$.

Asymptotes: $y = \pm\frac{3}{2}x$

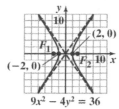

2. $4p = -8$, $p = -2$
parabola
vertex: $(0, 0)$
focus: $(0, -2)$
directrix: $y = 2$

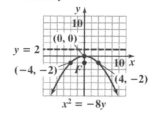

3. The center is at $(-2, 5)$.
$c^2 = a^2 - b^2 = 25 - 9 = 16$, $c = 4$
ellipse
The foci are at $(-6, 5)$ and $(2, 5)$.

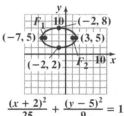

$$\frac{(x+2)^2}{25} + \frac{(y-5)^2}{9} = 1$$

4.
$$4x^2 - y^2 + 8x + 2y + 7 = 0$$
$$\left(4x^2 + 8x\right) - \left(y^2 - 2y\right) = -7$$
$$4\left(x^2 + 2x + 1\right) - \left(y^2 - 2y + 1\right) = -7 + 4 - 1$$
$$4(x+1)^2 - (y-1)^2 = -4$$
$$(y-1)^2 - 4(x+1)^2 = 4$$
$$\frac{(y-1)^2}{4} - (x+1)^2 = 1$$
$c^2 = a^2 + b^2 = 4 + 1 = 5$, $c = \sqrt{5}$
The center is at $(-1, 1)$.
Asymptotes: $y - 1 = \pm 2(x+1)$
hyperbola
The foci are at $\left(-1, 1+\sqrt{5}\right)$ and $\left(-1, 1-\sqrt{5}\right)$.

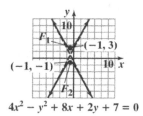

5. $4p = 8$, $p = 2$
parabola
vertex: $(-5, 1)$
focus: $(-5, 3)$
directrix: $y = -1$

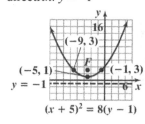

6. $c = 7, c^2 = 49$

$a = 10, a^2 = 100$

$b^2 = a^2 - c^2 = 100 - 49 = 51$

$\dfrac{x^2}{100} + \dfrac{y^2}{51} = 1$

7. $c = 10, c^2 = 100$

$a = 7, a^2 = 49$

$b^2 = c^2 - a^2 = 100 - 49 = 51$

$\dfrac{y^2}{49} - \dfrac{x^2}{51} = 1$

8. $p = 50$

$y^2 = 4px$

$y^2 = 200x$

9. $b = 24, b^2 = 576$

$2a = 80, a = 40, a^2 = 1600$

$c^2 = a^2 - b^2 = 1600 - 576 = 1024$

$c = \sqrt{1024} = 32$

The two people should each stand 32 feet from the center of the room, along the major axis.

10. a. $x^2 = 4py$

when $x = \pm 3, y = 3$

$9 = 4p(3)$

$3 = 4p$

$\dfrac{3}{4} = p$

$x^2 = 3y$

b. focus: $\left(0, \dfrac{3}{4}\right)$

The light is placed $\dfrac{3}{4}$ inch above the vertex.

11. $A = 1, C = 9$

$AC = 1 \cdot 9 = 9 > 0$, so the graph is an ellipse.

12. $A = 1, B = 1, C = 1$

$B^2 - 4AC = 1^2 - 4(1)(1) = -3$.

Since $B^2 - 4AC < 0$, the graph is an ellipse or circle.

13. $7x^2 - 6\sqrt{3}xy + 13y^2 - 16 = 0$

$A = 7, B = -6\sqrt{3}, C = 13$

$\cot 2\theta = \dfrac{A - C}{B} = \dfrac{7 - 13}{-6\sqrt{3}}$

$= \dfrac{-6}{-6\sqrt{3}} = \dfrac{1}{\sqrt{3}} = \dfrac{\sqrt{3}}{3}$

$2\theta = 60°$

$\theta = 30°$

14. $x = t^2, \quad y = t - 1; -\infty < t < \infty$

$y + 1 = t$

Substitute into x:

$x = (y + 1)^2$

$(y + 1)^2 = x$

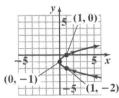

15. $x = 1 + 3\sin t, \quad y = 2\cos t; 0 \le t < 2\pi$

$\dfrac{x - 1}{3} = \sin t \qquad \dfrac{y}{2} = \cos t$

Square and add the equations:

$\dfrac{(x-1)^2}{9} = \sin^2 t$

$+ \quad \dfrac{y^2}{4} = \cos^2 t$

$\rule{3cm}{0.4pt}$

$\dfrac{(x-1)^2}{9} + \dfrac{y^2}{4} = \sin^2 t + \cos^2 t$

$\dfrac{(x-1)^2}{9} + \dfrac{y^2}{4} = 1$

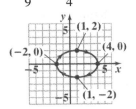

16. $r = \dfrac{2}{1 - \cos\theta}$

$e = 1$ and $ep = 2$, so $p = 2$.

Since $e = 1$, the graph is a parabola.

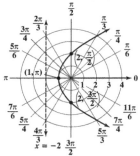

17. $r = \dfrac{4}{2 + \sin\theta}$

Divide the numerator and denominator by 2:

$r = \dfrac{2}{1 + \dfrac{1}{2}\sin\theta}$

$e = \dfrac{1}{2}$ and $ep = 2$, so $p = 4$. Since $e < 1$, the graph is an ellipse.

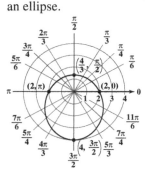

Cumulative Review Exercises (Chapters 1–10)

1. $2(x - 3) + 5x = 8(x - 1)$

$2x - 6 + 5x = 8x - 8$

$7x - 6 = 8x - 8$

$-x = -2$

$x = 2$

The solution set is $\{2\}$.

2. $-3(2x - 4) > 2(6x - 12)$

$-6x + 12 > 12x - 24$

$-18x > -36$

$x < 2$

The solution set is $\{x \mid x < 2\}$.

3. $x - 5 = \sqrt{x + 7}$

$(x - 5)^2 = x + 7$

$x^2 - 10x + 25 = x + 7$

$x^2 - 11x + 18 = 0$

$(x - 2)(x - 9) = 0$

$x = 2$ or $x = 9$

The solution $x = 2$ is extraneous, so the only solution is $x = 9$.

The solution set is $\{9\}$.

4. $(x - 2)^2 = 20$

$x - 2 = \pm\sqrt{20}$

$x - 2 = \pm 2\sqrt{5}$

$x = 2 \pm 2\sqrt{5}$

The solution set is $\left\{2 + 2\sqrt{5},\, 2 - 2\sqrt{5}\right\}$.

5. $|2x - 1| \geq 7$

$2x - 1 \geq 7$ or $2x - 1 \leq -7$

$2x \geq 8$ $\qquad 2x \leq -6$

$x \geq 4$ or $\qquad x \leq -3$

The solution set is $\left\{x \mid x \leq -3 \text{ or } x \geq 4\right\}$

6. $3x^3 + 4x^2 - 7x + 2 = 0$

$p : \pm 1, \pm 2$

$q : \pm 1, \pm 3$

$\dfrac{p}{q} : \pm 1, \pm 2, \pm\dfrac{1}{3}, \pm\dfrac{2}{3}$

Let $f(x) = 3x^3 + 4x^2 - 7x + 2$.

Evaluate f at the possible rational zeros to find $f\left(\dfrac{2}{3}\right) = 0$.

$$
\begin{array}{r|rrrr}
\frac{2}{3} & 3 & 4 & -7 & 2 \\
  &   & 2 & 4 & -2 \\
\hline
  & 3 & 6 & -3 & 0
\end{array}
$$

$\left(x - \dfrac{2}{3}\right)(3x^2 + 6x - 3) = 0$

$(3x - 2)(x^2 + 2x - 1) = 0$

$x = \dfrac{2}{3}$ or $x = \dfrac{-2 \pm \sqrt{(2)^2 - 4(1)(-1)}}{2}$

$x = \dfrac{-2 \pm \sqrt{8}}{2}$

$x = -1 \pm \sqrt{2}$

The solution set is $\left\{\dfrac{2}{3},\, -1 + \sqrt{2},\, -1 - \sqrt{2}\right\}$.

7. $\log_2(x+1) + \log_2(x-1) = 3$

$$\log_2(x^2 - 1) = 3$$
$$x^2 - 1 = 2^3$$
$$x^2 = 9$$
$$x = \pm 3$$

$x = -3$ is not a solution of the original equation. The solution set is $\{3\}$.

8. $3x + 4y = 2$

$2x + 5y = -1$

$6x + 8y = 4$
$\underline{-6x - 15y = 3}$
$-7y = 7$
$y = -1$

$$3x + 4(-1) = 2$$
$$3x = 6$$
$$x = 2$$

The solution set is $\{(2, -1)\}$.

9. $2x^2 - y^2 = -8$

$x - y = 6$

$x - y = 6$
$x = y + 6$

$$x^2 = (y+6)^2 = y^2 + 12y + 36$$

Substitute into first equation.

$$2(y^2 + 12y + 36) - y^2 = -8$$
$$2y^2 + 24y + 72 - y^2 = -8$$
$$y^2 + 24y + 80 = 0$$
$$(y+4)(y+20) = 0$$
$$y = -4 \text{ or } y = -20$$
$$x = 2 \qquad x = -14$$

The solution set is $\{(2, -4), (-14, -20)\}$.

10. Set up the augmented matrix and use Gauss-Jordan reduction.

$$\begin{bmatrix} 1 & -1 & 1 & | & 17 \\ -4 & 1 & 5 & | & -2 \\ 2 & 3 & 1 & | & 8 \end{bmatrix}$$

$$\begin{bmatrix} 1 & -1 & 1 & | & 17 \\ 0 & -3 & 9 & | & 66 \\ 0 & 5 & -1 & | & -26 \end{bmatrix} \begin{matrix} \\ 4R_1 + R_2 \\ -2R_1 + R_3 \end{matrix}$$

$$\begin{bmatrix} 1 & -1 & 1 & | & 17 \\ 0 & 1 & -3 & | & -22 \\ 0 & 5 & -1 & | & -26 \end{bmatrix} -\frac{1}{3}R_2$$

$$\begin{bmatrix} 1 & 0 & -2 & | & -5 \\ 0 & 1 & -3 & | & -22 \\ 0 & 0 & 14 & | & 84 \end{bmatrix} \begin{matrix} R_2 + R_1 \\ \\ -5R_2 + R_3 \end{matrix}$$

$$\begin{bmatrix} 1 & 0 & -2 & | & -5 \\ 0 & 1 & -3 & | & -22 \\ 0 & 0 & 1 & | & 6 \end{bmatrix} \frac{1}{14}R_3$$

$$\begin{bmatrix} 1 & 0 & 0 & | & 7 \\ 0 & 1 & 0 & | & -4 \\ 0 & 0 & 1 & | & 6 \end{bmatrix} \begin{matrix} 2R_3 + R_1 \\ 3R_3 + R_2 \\ \end{matrix}$$

$x = 7, y = -4, z = 6$
The solution set is $\{(7, -4, 6)\}$.

11. Parabola with vertex at $(1, -4)$.

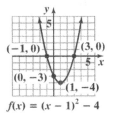

$f(x) = (x - 1)^2 - 4$

12. Ellipse with center at $(0, 0)$ and vertices at $(3, 0)$ and $(-3, 0)$.

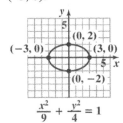

$$\frac{x^2}{9} + \frac{y^2}{4} = 1$$

13. $5x + y \le 10 \qquad\qquad y \ge \frac{1}{4}x + 2$

$y \le -5x + 10$

Graph with solid line $y = -5x + 10$ and $y = \frac{1}{4}x + 2$. Shade the region that is below the line $y = -5x + 10$ and above the line $y = \frac{1}{4}x + 2$. Then dash the solid lines that do not contain the solution set.

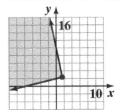

14. a. $p : \pm 1, \pm 3$

$q : \pm 1, \pm 2, \pm 4, \pm 8, \pm 16, \pm 32$

$\dfrac{p}{q} : \pm 1, \ \pm 3, \ \pm \dfrac{1}{2}, \ \pm \dfrac{3}{2}, \ \pm \dfrac{1}{4}, \ \pm \dfrac{3}{4}, \ \pm \dfrac{1}{8},$

$\pm \dfrac{3}{8}, \pm \dfrac{1}{16}, \pm \dfrac{3}{16}, \pm \dfrac{1}{32}, \pm \dfrac{3}{32}$

b. $x = 1$ appears to be a root.

$$
\begin{array}{r|rrrr}
1 & 32 & -52 & 17 & 3 \\
  &   & 32 & -20 & -3 \\
\hline
  & 32 & -20 & -3 & 0
\end{array}
$$

$32x^3 - 52x^2 + 17x + 3 = 0$

$(x - 1)(32x^2 - 20x - 3) = 0$

$(x - 1)(4x - 3)(8x + 1) = 0$

$x = 1$ or $x = \dfrac{3}{4}$ or $x = -\dfrac{1}{8}$

The solution set is $\left\{ -\dfrac{1}{8}, \dfrac{3}{4}, 1 \right\}$.

15. a. domain: $(-2, 2)$

range: $[-3, \infty)$

b. the relative minimum of -3 occurs at $x = 0$.

c. increasing: $(0, 2)$

d. $f(-1) - f(0) = 0 - (-3) = 3$

e. $(f \circ f)(1) = f\big(f(1)\big) = f(0) = -3$

f. $f(x) \to \infty$ as $x \to -2^+$ or as $x \to 2^-$

g.

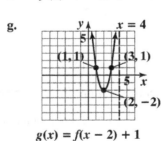

$g(x) = f(x - 2) + 1$

h.

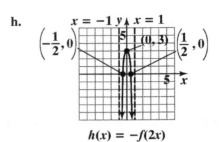

$h(x) = -f(2x)$

16. $f(x) = x^2 - 4, \ g(x) = x + 2$

$(g \circ f)(x) = g\big(x^2 - 4\big) = \big(x^2 - 4\big) + 2 = x^2 - 2$

17. $\log_5 \dfrac{x^3 \sqrt{y}}{125} = \log_5 x^3 \sqrt{y} - \log_5 125$

$= \log_5 x^3 + \log_5 \sqrt{y} - 3$

$= 3 \log_5 x + \dfrac{1}{2} \log_5 y - 3$

18. $m = \dfrac{y_2 - y_1}{x_2 - x_1} = \dfrac{8 - (-4)}{-5 - 1} = \dfrac{12}{-6} = -2$

$y - y_1 = m(x - x_1)$

$y + 4 = -2(x - 1)$

$y = -2x - 2$

19. Let R = the cost of a rental at Rent-a-Truck and let A = the cost of a rental at Ace Truck Rentals.

$R = 39 + 0.16m$

$A = 25 + 0.24m$

where m is the number of miles.

$39 + 0.16m = 25 + 0.24m$

$14 = 0.08m$

$m = 175$

$R = 39 + 0.16(175) = 67$

The cost will be the same when the number of miles driven is 175 miles. The cost will be $67.

20. Let x = cost of basic cable,

Let y = cost of movie channel.

$x + y = 35$

$x + 2y = 45$

Multiply the first equation by -1 and then add the two equations.

$-x - y = -35$

$\underline{x + 2y = 45}$

$y = 10$

Use back-substitution to find x.

$x + 10 = 35$

$x = 25$

Basic cable costs $25 and one movie channel costs $10.

21. $\dfrac{\csc\theta-\sin\theta}{\sin\theta}=\dfrac{\dfrac{1}{\sin\theta}-\sin\theta}{\sin\theta}\cdot\dfrac{\sin\theta}{\sin\theta}$

$\qquad=\dfrac{1-\sin^2\theta}{\sin^2\theta}$

$\qquad=\dfrac{\cos^2\theta}{\sin^2\theta}$

$\qquad=\left(\dfrac{\cos\theta}{\sin\theta}\right)^2$

$\qquad=\cot^2\theta$

22. $y=2\cos(2x+\pi)$

$A=2,\ B=2,\ C=-\pi$

Amplitude: $|A|=|2|=2$

Period: $\dfrac{2\pi}{B}=\dfrac{2\pi}{2}=\pi$

Phase Shift: $\dfrac{C}{B}=\dfrac{-\pi}{2}=-\dfrac{\pi}{2}$

$(0,-2),\left(\dfrac{\pi}{4},0\right),\left(\dfrac{\pi}{2},2\right),\left(\dfrac{3\pi}{4},0\right),(\pi,-2)$

$y=2\cos(2x+\pi)$

23. $(\mathbf{v}\cdot\mathbf{w})\mathbf{w}=\left[(3\mathbf{i}-6\mathbf{j})\cdot(\mathbf{i}+\mathbf{j})\right](\mathbf{i}+\mathbf{j})$

$\qquad=\left[3(1)-6(1)\right](\mathbf{i}+\mathbf{j})$

$\qquad=(3-6)(\mathbf{i}+\mathbf{j})$

$\qquad=-3(\mathbf{i}+\mathbf{j})$

$\qquad=-3\mathbf{i}-3\mathbf{j}$

24. $\qquad\sin2\theta=\sin\theta,\ 0\le\theta<2\pi$

$\qquad 2\sin\theta\cos\theta=\sin\theta$

$\qquad 2\sin\theta\cos\theta-\sin\theta=0$

$\qquad\sin\theta(2\cos\theta-1)=0$

$\sin\theta=0\qquad$ or $\quad 2\cos\theta-1=0$

$\quad\theta=0,\ \pi\qquad\qquad 2\cos\theta=1$

$\qquad\qquad\qquad\qquad\quad\cos\theta=\dfrac{1}{2}$

$\qquad\qquad\qquad\qquad\quad\theta=\dfrac{\pi}{3},\dfrac{5\pi}{3}$

The solutions in the interval $[0,2\pi)$ are $0,\ \pi,\dfrac{\pi}{3}$,

and $\dfrac{5\pi}{3}$.

25. $\quad A+B+C=180°$

$\quad 64°+72°+C=180°$

$\qquad\quad 136°+C=180°$

$\qquad\qquad\qquad C=44°$

$\dfrac{b}{\sin B}=\dfrac{a}{\sin A}$

$\dfrac{b}{\sin72°}=\dfrac{13.6}{\sin64°}$

$\qquad b=\dfrac{13.6\sin72°}{\sin64°}\approx14.4$

$\dfrac{c}{\sin C}=\dfrac{a}{\sin A}$

$\dfrac{c}{\sin44°}=\dfrac{13.6}{\sin64°}$

$\qquad c=\dfrac{13.6\sin44°}{\sin64°}\approx10.5$

The solution is $C=44°,\ b\approx14.4$, and $c\approx10.5$.

Chapter 11
Sequences, Induction, and Probability

Section 11.1

Check Point Exercises

1. a. $a_n = 2n + 5$

$a_1 = 2(1) + 5 = 7$

$a_2 = 2(2) + 5 = 9$

$a_3 = 2(3) + 5 = 11$

$a_4 = 2(4) + 5 = 13$

The first four terms are 7, 9, 11, and 13.

b.
$$a_n = \frac{(-1)^n}{2^n + 1}$$

$$a_1 = \frac{(-1)^1}{2^1 + 1} = \frac{-1}{3} = -\frac{1}{3}$$

$$a_2 = \frac{(-1)^2}{2^2 + 1} = \frac{1}{5}$$

$$a_3 = \frac{(-1)^3}{2^3 + 1} = \frac{-1}{9} = -\frac{1}{9}$$

$$a_4 = \frac{(-1)^4}{2^4 + 1} = \frac{1}{17}$$

The first four terms are $-\frac{1}{3}, \frac{1}{5}, -\frac{1}{9},$ and $\frac{1}{17}$.

2. $a_1 = 3$ and $a_n = 2a_{n-1} + 5$ for $n \geq 2$

$a_2 = 2a_1 + 5$

$\quad = 2(3) + 5 = 11$

$a_3 = 2a_2 + 5$

$\quad = 2(11) + 5 = 27$

$a_4 = 2a_3 + 5$

$\quad = 2(27) + 5 = 59$

The first four terms are 3, 11, 27, and 59.

3. $a_n = \dfrac{20}{(n+1)!}$

$$a_1 = \frac{20}{(1+1)!} = \frac{20}{2!} = 10$$

$$a_2 = \frac{20}{(2+1)!} = \frac{20}{3!} = \frac{20}{6} = \frac{10}{3}$$

$$a_3 = \frac{20}{(3+1)!} = \frac{20}{4!} = \frac{20}{24} = \frac{5}{6}$$

$$a_4 = \frac{20}{(4+1)!} = \frac{20}{5!} = \frac{20}{120} = \frac{1}{6}$$

The first four terms are $10, \frac{10}{3}, \frac{5}{6},$ and $\frac{1}{6}$.

4. a. $\dfrac{14!}{2! \, 12!} = \dfrac{14 \cdot 13 \cdot 12!}{2! \, 12!} = \dfrac{14 \cdot 13}{2 \cdot 1} = 91$

b. $\dfrac{n!}{(n-1)!} = \dfrac{n \cdot (n-1)!}{(n-1)!} = n$

5. a. $\displaystyle\sum_{i=1}^{6} 2i^2$

$= 2(1)^2 + 2(2)^2 + 2(3)^2$

$\quad + 2(4)^2 + 2(5)^2 + 2(6)^2$

$= 2 + 8 + 18 + 32 + 50 + 72$

$= 182$

b. $\displaystyle\sum_{k=3}^{5} \left(2^k - 3\right)$

$= \left(2^3 - 3\right) + \left(2^4 - 3\right) + \left(2^5 - 3\right)$

$= (8 - 3) + (16 - 3) + (32 - 3)$

$= 5 + 13 + 29$

$= 47$

c. $\displaystyle\sum_{i=1}^{5} 4$

$= 4 + 4 + 4 + 4 + 4$

$= 20$

6. a. The sum has nine terms, each of the form i^2, starting at $i = 1$ and ending at $i = 9$.

$$1^2 + 2^2 + 3^2 + \cdots + 9^2 = \sum_{i=1}^{9} i^2$$

b. The sum has n terms, each of the form $\frac{1}{2^{i-1}}$, starting at $i = 1$ and ending at $i = n$.

$$1 + \frac{1}{2} + \frac{1}{4} + \frac{1}{8} + \cdots + \frac{1}{2^{n-1}} = \sum_{i=1}^{n} \frac{1}{2^{i-1}}$$

Concept and Vocabulary Check 11.1

1. sequence; integers; terms

2. general

3. 4

4. $\dfrac{1}{15}$

5. 2

6. factorial; 5; 1; 1

7. $n+3$

8. a_1; a_2; a_3; a_n; index; upper limit; lower limit

Exercise Set 11.1

1. $a_n = 3n+2$

   $a_1 = 3(1)+2 = 5$

   $a_2 = 3(2)+2 = 8$

   $a_3 = 3(3)+2 = 11$

   $a_4 = 3(4)+2 = 14$

   The first four terms are 5, 8, 11, and 14.

3. $a_n = 3^n$

   $a_1 = 3^1 = 3$

   $a_2 = 3^2 = 9$

   $a_3 = 3^3 = 27$

   $a_4 = 3^4 = 81$

   The first four terms are 3, 9, 27, and 81.

5. $a_n = (-3)^n$

   $a_1 = (-3)^1 = -3$

   $a_2 = (-3)^2 = 9$

   $a_3 = (-3)^3 = -27$

   $a_4 = (-3)^4 = 81$

   The first four terms are −3, 9, −27, and 81.

7. $a_n = (-1)^n (n+3)$

   $a_1 = (-1)^1 (1+3) = -4$

   $a_2 = (-1)^2 (2+3) = 5$

   $a_3 = (-1)^3 (3+3) = -6$

   $a_4 = (-1)^4 (4+3) = 7$

   The first four terms are −4, 5, −6, and 7.

9. $a_n = \dfrac{2n}{n+4}$

   $a_1 = \dfrac{2(1)}{1+4} = \dfrac{2}{5}$

   $a_2 = \dfrac{2(2)}{2+4} = \dfrac{4}{6} = \dfrac{2}{3}$

   $a_3 = \dfrac{2(3)}{3+4} = \dfrac{6}{7}$

   $a_4 = \dfrac{2(4)}{4+4} = \dfrac{8}{8} = 1$

   The first four terms are $\frac{2}{5}, \frac{2}{3}, \frac{6}{7},$ and 1.

11. $a_n = \dfrac{(-1)^{n+1}}{2^n - 1}$

    $a_1 = \dfrac{(-1)^{1+1}}{2^1 - 1} = \dfrac{1}{1} \quad n = 1$

    $a_2 = \dfrac{(-1)^{2+1}}{2^2 - 1} = -\dfrac{1}{3}$

    $a_3 = \dfrac{(-1)^{3+1}}{2^3 - 1} = \dfrac{1}{7}$

    $a_4 = \dfrac{(-1)^{4+1}}{2^4 - 1} = -\dfrac{1}{15}$

    The first four terms are $1, -\frac{1}{3}, \frac{1}{7},$ and $-\frac{1}{15}$.

13. $a_1 = 7$ and $a_n = a_{n-1} + 5$ for $n \geq 2$

    $a_2 = a_1 + 5 = 7 + 5 = 12$

    $a_3 = a_2 + 5 = 12 + 5 = 17$

    $a_4 = a_3 + 5 = 17 + 5 = 22$

    The first four terms are 7, 12, 17, and 22.

15. $a_1 = 3$ and $a_n = 4a_{n-1}$ for $n \geq 2$

    $a_2 = 4a_1 = 4(3) = 12$

    $a_3 = 4a_2 = 4(12) = 48$

    $a_4 = 4a_3 = 4(48) = 192$

    The first four terms are 3, 12, 48, and 192.

17. $a_1 = 4$ and $a_n = 2a_{n-1} + 3$

$a_2 = 2(4) + 3 = 11$

$a_3 = 2(11) + 3 = 25$

$a_4 = 2(25) + 3 = 53$

The first four terms are 4, 11, 25, and 53.

19. $a_n = \dfrac{n^2}{n!}$

$a_1 = \dfrac{1^2}{1!} = 1$

$a_2 = \dfrac{2^2}{2!} = 2$

$a_3 = \dfrac{3^2}{3!} = \dfrac{9}{6} = \dfrac{3}{2}$

$a_4 = \dfrac{4^2}{4!} = \dfrac{16}{24} = \dfrac{2}{3}$

The first four terms are $1,\ 2,\ \frac{3}{2},$ and $\frac{2}{3}$.

21. $a_n = 2(n+1)!$

$a_1 = 2(1+1)! = 2(2) = 4$

$a_2 = 2(2+1)! = 2(6) = 12$

$a_3 = 2(3+1)! = 2(24) = 48$

$a_4 = 2(4+1)! = 2(120) = 240$

The first four terms are 4, 12, 48, and 240.

23. $\dfrac{17!}{15!} = \dfrac{17 \cdot 16 \cdot 15!}{15!} = 17 \cdot 16 = 272$

25. $\dfrac{16!}{2! \, 14!} = \dfrac{16 \cdot 15 \cdot 14!}{2! \, 14!} = \dfrac{16 \cdot 15}{2 \cdot 1} = \dfrac{8 \cdot 15}{1} = 120$

27. $\dfrac{(n+2)!}{n!} = \dfrac{(n+2)(n+1)n!}{n!} = (n+2)(n+1)$

29. $\displaystyle\sum_{i=1}^{6} 5i = 5 \cdot 1 + 5 \cdot 2 + 5 \cdot 3 + 5 \cdot 4 + 5 \cdot 5 + 5 \cdot 6$

$\qquad = 5 + 10 + 15 + 20 + 25 + 30$

$\qquad = 105$

31. $\displaystyle\sum_{i=1}^{4} 2i^2 = 2 \cdot 1^2 + 2 \cdot 2^2 + 2 \cdot 3^2 + 2 \cdot 4^2$

$\qquad = 2 + 8 + 18 + 32$

$\qquad = 60$

33. $\displaystyle\sum_{k=1}^{5} k(k+4) = 1(5) + 2(6) + 3(7) + 4(8) + 5(9)$

$\qquad = 5 + 12 + 21 + 32 + 45$

$\qquad = 115$

35. $\displaystyle\sum_{i=1}^{4} \left(\dfrac{-1}{2}\right)^i = \left(-\dfrac{1}{2}\right)^1 + \left(-\dfrac{1}{2}\right)^2 + \left(-\dfrac{1}{2}\right)^3 + \left(-\dfrac{1}{2}\right)^4$

$\qquad = -\dfrac{1}{2} + \dfrac{1}{4} + -\dfrac{1}{8} + \dfrac{1}{16}$

$\qquad = -\dfrac{5}{16}$

37. $\displaystyle\sum_{i=5}^{9} 11 = 11 + 11 + 11 + 11 + 11 = 55$

39. $\displaystyle\sum_{i=0}^{4} \dfrac{(-1)^i}{i!}$

$\qquad = \dfrac{(-1)^0}{0!} + \dfrac{(-1)^1}{1!} + \dfrac{(-1)^2}{2!} + \dfrac{(-1)^3}{3!} + \dfrac{(-1)^4}{4!}$

$\qquad = 1 - 1 + \dfrac{1}{2} - \dfrac{1}{6} + \dfrac{1}{24}$

$\qquad = \dfrac{9}{24} = \dfrac{3}{8}$

41. $\displaystyle\sum_{i=1}^{5} \dfrac{i!}{(i-1)!} = \dfrac{1!}{0!} + \dfrac{2!}{1!} + \dfrac{3!}{2!} + \dfrac{4!}{3!} + \dfrac{5!}{4!}$

$\qquad = 1 + 2 + 3 + 4 + 5 = 15$

43. $1^2 + 2^2 + 3^2 + \cdots + 15^2 = \displaystyle\sum_{i=1}^{15} i^2$

45. $2 + 2^2 + 2^3 + 2^4 + \cdots + 2^{11} = \displaystyle\sum_{i=1}^{11} 2^i$

47. $1 + 2 + 3 + \cdots + 30 = \displaystyle\sum_{i=1}^{30} i$

49. $\dfrac{1}{2} + \dfrac{2}{3} + \dfrac{3}{4} + \cdots + \dfrac{14}{14+1} = \displaystyle\sum_{i=1}^{14} \dfrac{i}{i+1}$

51. $4 + \dfrac{4^2}{2} + \dfrac{4^3}{3} + \cdots + \dfrac{4^n}{n} = \displaystyle\sum_{i=1}^{n} \dfrac{4^i}{i}$

53. $1 + 3 + 5 + \cdots + (2n-1) = \displaystyle\sum_{i=1}^{n} (2i-1)$

55. $5 + 7 + 9 + \cdots + 31$

Possible answer: $\displaystyle\sum_{k=1}^{14} (2k+3)$

57. $a + ar + ar^2 + \cdots + ar^{12}$

Possible answer: $\displaystyle\sum_{k=0}^{12} ar^k$

59. $a + (a+d) + (a+2d) + \cdots + (a+nd)$

Possible answer: $\displaystyle\sum_{k=0}^{n} (a+kd)$

61. $\displaystyle\sum_{i=1}^{5}(a_i^2 + 1) = \left((-4)^2 + 1\right) + \left((-2)^2 + 1\right) + \left((0)^2 + 1\right) + \left((2)^2 + 1\right) + \left((4)^2 + 1\right)$

$= 17 + 5 + 1 + 5 + 17$

$= 45$

63. $\displaystyle\sum_{i=1}^{5}(2a_i + b_i) = \left(2(-4) + 4\right) + \left(2(-2) + 2\right) + \left(2(0) + 0\right) + \left(2(2) + (-2)\right) + \left(2(4) + (-4)\right)$

$= -4 + (-2) + 0 + 2 + 4 = 0$

65. $\displaystyle\sum_{i=4}^{5}\left(\frac{a_i}{b_i}\right)^2 = \left(\frac{2}{-2}\right)^2 + \left(\frac{4}{-4}\right)^2 = (-1)^2 + (-1)^2 = 1 + 1 = 2$

67. $\displaystyle\sum_{i=1}^{5}a_i^2 + \sum_{i=1}^{5}b_i^2 = \left((-4)^2 + (-2)^2 + 0^2 + 2^2 + 4^2\right) + \left(4^2 + 2^2 + 0^2 + (-2)^2 + (-4)^2\right)$

$= (16 + 4 + 0 + 4 + 16) + (16 + 4 + 0 + 4 + 16) = 80$

69.   **a.** $\displaystyle\sum_{i=1}^{8}a_i = 100 + 120 + 145 + 170 + 200 + 220 + 260 + 300 = 1515$

A total of 1515 thousand, or 1,515,000, autism cases were diagnosed in the United States from 2001 through 2008.

b. $\displaystyle\sum_{i=1}^{8}a_i = (28 \cdot 1 + 63) + (28 \cdot 2 + 63) + (28 \cdot 3 + 63) + (28 \cdot 4 + 63) + (28 \cdot 5 + 63) + (28 \cdot 6 + 63) + (28 \cdot 7 + 63) + (28 \cdot 8 + 63)$

$= 91 + 119 + 147 + 175 + 203 + 231 + 259 + 287 = 1512$

The model underestimates the actual sum by 3 thousand.

71. $a_n = 6000\left(1 + \dfrac{0.06}{4}\right)^n, n = 1, 2, 3, \cdots$

$a_{20} = 6000\left(1 + \dfrac{0.06}{4}\right)^{20} \approx 8081.13$

After five years, the balance is $8081.13.

73. – 79.   Answers will vary.

81.   Most calculators give error message if the expression is entered directly.

However, $\dfrac{200!}{198!} = \dfrac{200 \cdot 199 \cdot 198!}{198!} = 200 \cdot 199 = 39,800$

83. $\dfrac{20!}{300} = 8,109,673,360,588,800$

However, most calculators give a rounded answer in scientific notation.

85. $\dfrac{54!}{(54-3)!3!} = 24,804$

87. Answers will vary.

89. $a_n = \dfrac{n}{n+1}$

As *n* gets larger, a_n approaches 1.

91. $a_n = \dfrac{2n^2 + 5n - 7}{n^3}$

As *n* gets larger, a_n approaches 0.

93. does not make sense; Explanations will vary. Sample explanation: There is nothing that implies that there is a negative number of sheep.

95. makes sense

97. false; Changes to make the statement true will vary. A sample change is: $\dfrac{n!}{(n-1)!} = \dfrac{n \cdot (n-1)!}{(n-1)!} = n$

99. false; Changes to make the statement true will vary. A sample change is:

$$\sum_{i=1}^{2} (-1)^i 2^i = (-1)^1 2^1 + (-1)^2 2^2 = -1(2) + 1(4) = -2 + 4 = 2$$

101. $a_n = \begin{cases} \dfrac{a_{n-1}}{2} & \text{if } a_{n-1} \text{ is even.} \\ 3a_n + 5 & \text{if } a_{n-1} \text{ is odd} \end{cases}$

for $n \geq 2$.

$a_1 = 9$

Since 9 is odd, $a_2 = 3(9) + 5 = 32$.

Since 32 is even, $a_3 = \dfrac{32}{2} = 16$.

Similarly, $a_4 = \dfrac{16}{2} = 8$, $a_5 = \dfrac{8}{2} = 4$.

The first five terms of the sequence are 9, 32, 16, 8, and 4.

103. $a_2 - a_1 = 3 - 8 = -5$

$a_3 - a_2 = -2 - 3 = -5$

$a_4 - a_3 = -7 - (-2) = -5$

$a_5 - a_4 = -12 - (-7) = -5$

The difference between consecutive terms is always -5.

104. $a_2 - a_1 = (4(2) - 3) - (4(1) - 3) = 4$

$a_3 - a_2 = (4(3) - 3) - (4(2) - 3) = 4$

$a_4 - a_3 = (4(4) - 3) - (4(3) - 3) = 4$

$a_5 - a_4 = (4(5) - 3) - (4(4) - 3) = 4$

The difference between consecutive terms is always 4.

105. $a_n = 4 + (n-1)(-7)$

$a_8 = 4 + (8-1)(-7) = 4 + (7)(-7) = 4 - 49 = -45$

Section 11.2

Check Point Exercises

1. $a_1 = 100$

$a_2 = a_1 - 30 = 100 - 30 = 70$

$a_3 = a_2 - 30 = 70 - 30 = 40$

$a_4 = a_3 - 30 = 40 - 30 = 10$

$a_5 = a_4 - 30 = 10 - 30 = -20$

$a_6 = a_5 - 30 = -20 - 30 = -50$

The first five terms are 100, 70, 40, 10, -20, -50.

2. $a_1 = 6$, $d = -5$

To find the ninth term, a_9, replace n in the formula with 9, a_1 with 6, and d with -5.

$a_n = a_1 + (n-1)d$

$a_9 = 6 + (9-1)(-5)$

$\quad = 6 + 8(-5)$

$\quad = 6 + (-40)$

$\quad = -34$

3. a. $a_n = a_1 + (n-1)d$

$\quad = 16 + (n-1)0.35$

$\quad = 0.35n + 15.65$

b. $a_n = 0.35n + 15.65$

$a_{21} = 0.35(21) + 15.65$

$\quad = 23$

In 2030, it is projected that 23% of the U.S. population will be Latino.

4. 3, 6, 9, 12, ...

To find the sum of the first 15 terms, S_{15}, replace n in the formula with 15.

$$S_n = \frac{n}{2}(a_1 + a_n)$$

$$S_{15} = \frac{15}{2}(a_1 + a_{15})$$

Use the formula for the general term of a sequence to find a_{15}. The common difference, d, is 3, and the first term, a_1, is 3.

$a_n = a_1 + (n-1)d$

$a_{15} = 3 + (15-1)(3)$

$\quad = 3 + 14(3)$

$\quad = 3 + 42$

$\quad = 45$

Thus, $S_{15} = \frac{15}{2}(3 + 45) = \frac{15}{2}(48) = 360$.

5. $\sum\limits_{i=1}^{30}(6i - 11) = (6 \cdot 1 - 11) + (6 \cdot 2 - 11)$

$\qquad\qquad\qquad + (6 \cdot 3 - 11) + \ldots + (6 \cdot 30 - 11)$

$\qquad\qquad = -5 + 1 + 7 + \ldots + 169$

So the first term, a_1, is -5; the common difference, d, is $1 - (-5) = 6$; the last term, a_{30}, is 169.

Substitute $n = 30$, $a_1 = -5$, and $a_{30} = 169$ in the formula $S_n = \frac{n}{2}(a_1 + a_n)$.

$S_{30} = \frac{30}{2}(-5 + 169) = 15(164) = 2460$

Thus, $\sum\limits_{i=1}^{30}(6i - 11) = 2460$

6. $a_n = 1800n + 64,130$

$a_1 = 1800(1) + 64,130 = 65,930$

$a_{10} = 1800(10) + 64,130 = 82,130$

$$S_n = \frac{n}{2}(a_1 + a_n)$$

$$S_{10} = \frac{10}{2}(a_1 + a_{10})$$

$\quad = 5(65,930 + 82,130)$

$\quad = 5(148,060)$

$\quad = \$740,300$

It would cost \$740,300 for the ten-year period beginning in 2014.

Concept and Vocabulary Check 11.2

1. arithmetic; common difference

2. $a_1 + (n-1)d$; first term; common difference

3. $\dfrac{n}{2}(a_1 + a_2)$; first term; nth term

4. 2; 116

5. 8; 13; 18; 5

Exercise Set 11.2

1. $a_1 = 200,\ d = 20$
   The first six terms are 200, 220, 240, 260, 280, and 300.

3. $a_1 = -7,\ d = 4$
   The first six terms are –7, –3, 1, 5, 9, and 13.

5. $a_1 = 300,\ d = -90$
   The first six terms are 300, 210, 120, 30, –60, and –150.

7. $a_1 = \dfrac{5}{2},\ d = -\dfrac{1}{2}$

   The first six terms are $\dfrac{5}{2}, 2, \dfrac{3}{2}, 1, \dfrac{1}{2},$ and 0.

9. $a_n = a_{n-1} + 6,\ a_1 = -9$
   The first six terms are –9, –3, 3, 9, 15, and 21.

11. $a_n = a_{n-1} - 10,\ a_1 = 30$
    The first six terms are 30, 20, 10, 0, –10, and –20.

13. $a_n = a_{n-1} - 0.4,\ a_1 = 1.6$
    The first six terms are 1.6, 1.2, 0.8, 0.4, 0, and –0.4.

15. $a_1 = 13,\ d = -4$
    $a_n = 13 + (n-1)4$
    $a_6 = 13 + 5(4) = 13 + 20 = 33$

17. $a_1 = 7,\ d = 5$
    $a_n = 7 + (n-1)2$
    $a_{50} = 7 + 49(5) = 252$

19. $a_1 = -40,\ d = 5$
    $a_n = -40 + (n-1)5$
    $a_{200} = -40 + (199)5 = 955$

21. $a_1 = 35,\ d = -3$
    $a_n = 35 - 3(n-1)$
    $a_{60} = 35 - 3(59) = -142$

23. 1, 5, 9, 13, ...
    $d = 5 - 1, = 4$
    $a_n = 1 + (n-1)4 = 1 + 4n - 4$
    $a_n = 4n - 3$
    $a_{20} = 4(20) - 3 = 77$

25. 7, 3, –1, –5, ...
    $d = 3 - 7 = -4$
    $a_n = 7 + (n-1)(-4) = 7 - 4n + 4$
    $a_n = 11 - 4n$
    $a_{20} = 11 - 4(20) = -69$

27. $a_1 = 9,\ d = 2$
    $a_n = 9 + (n-1)(2)$
    $a_n = 7 + 2n$
    $a_{20} = 7 + 2(20) = 47$

29. $a_1 = -20,\ d = -4$
    $a_n = -20 + (n-1)(-4)$
    $a_n = -20 - 4n + 4$
    $a_n = -16 - 4n$
    $a_{20} = -16 - 4(20) = -96$

31. $a_n = a_{n-1} + 3,\ a_1 = 4$
    $d = 3$
    $a_n = 4 + (n-1)(3)$
    $a_n = 1 + 3n$
    $a_{20} = 1 + 3(20) = 61$

33. $a_n = a_{n-1} - 10,\ a_1 = 30,\ d = -10$
    $a_n = 30 - 10(n-1) = 30 - 10n + 10$
    $a_n = 40 - 10n$
    $a_{20} = 40 - 10(20) = -160$

35. 4, 10, 16, 22, . . .
    $d = 10 - 4 = 6$
    $a_n = 4 + (n-1)(6)$
    $a_{20} = 4 + (19)(6) = 118$
    $S_{20} = \dfrac{20}{2}(4 + 118) = 1220$

37. $-10, -6, -2, 2, \ldots$

$$d = -6 - (-10) = -6 + 10 = 4$$

$$a_n = -10 + (n-1)4$$

$$a_{50} = -10 + (49)4 = 186$$

$$S_{50} = \frac{50}{2}(-10 + 186) = 4400$$

39. $1 + 2 + 3 + 4 + \cdots + 100$

$$S_{100} = \frac{100}{2}(1 + 100) = 5050$$

41. $2 + 4 + 6 + \cdots + 120$

$$S_{60} = \frac{60}{2}(2 + 120) = 3660$$

43. even integers between 21 and 45;
$22 + 24 + 26 + \cdots + 44$

$$S_{12} = \frac{12}{2}(22 + 44) = 396$$

45. $\displaystyle\sum_{i=1}^{17}(5i + 3) = (5 + 3) + (10 + 3) + (15 + 3) + \cdots + (85 + 3) = 8 + 13 + 18 + \cdots + 88$

$$S_{17} = \frac{17}{2}(8 + 88) = 816$$

47. $\displaystyle\sum_{i=1}^{30}(-3i + 5) = (-3 + 5) + (-6 + 5) + (-9 + 5) + \cdots + (-90 + 5) = 2 - 1 - 4 - \cdots - 85$

$$S_{30} = \frac{30}{2}(2 - 85) = -1245$$

49. $\displaystyle\sum_{i=1}^{100} 4i = 4 + 8 + 12 + \cdots + 400$

$$S_{100} = \frac{100}{2}(4 + 400) = 20,200$$

51. First find a_{14} and b_{12}:

$$a_{14} = a_1 + (n-1)d$$
$$= 1 + (14 - 1)(-3 - 1) = -51$$
$$b_{12} = b_1 + (n-1)d$$
$$= 3 + (12 - 1)(8 - 3) = 58$$

So, $a_{14} + b_{12} = -51 + 58 = 7$.

53. $a_n = a_1 + (n-1)d$

$-83 = 1 + (n-1)(-3-1)$

$-83 = 1 + -4(n-1)$

$-84 = -4n + 4$

$-88 = -4n$

$n = 22$

There are 22 terms.

55. $S_n = \dfrac{n}{2}(a_1 + a_n)$

For $\{a_n\}$: $S_{14} = \dfrac{14}{2}(a_1 + a_{14}) = 7(1 + (-51)) = -350$

For $\{b_n\}$: $S_{14} = \dfrac{14}{2}(b_1 + b_{14}) = 7(3 + 68) = 497$

So $\displaystyle\sum_{n=1}^{14} b_n - \sum_{n=1}^{14} a_n = 497 - (-350) = 847$

57. Two points on the graph are $(1, 1)$ and $(2, -3)$.

Finding the slope of the line;

$m = \dfrac{y_2 - y_1}{x_2 - x_2} = \dfrac{-3-1}{2-1} = \dfrac{-4}{1} = -4$

Using the point-slope form of an equation of a line;

$y - y_2 = m(x - x_2)$

$y - 1 = -4(x-1)$

$y - 1 = -4x + 4$

$y = -4x + 5$

Thus, $f(x) = -4x + 5$.

59. Using $a_n = a_1 + (n-1)d$ and $a_2 = 4$:

$a_2 = a_1 + (2-1)d$

$4 = a_1 + d$

And since $a_6 = 16$:

$a_6 = a_1 + (6-1)d$

$16 = a_1 + 5d$

The system of equations is

$4 = a_1 + d$

$16 = a_1 + 5d$

Solving the first equation for a_1:

$a_1 = 4 - d$

Substituting the value into the second equation and solving for d:

$16 = (4-d) + 5d$

$16 = 4 + 4d$

$12 = 4d$

$3 = d$

Back-substitute:

$a_1 = 4 - d$

$a_1 = 4 - 3)$

$a_1 = 1$

Then $a_n = a_1 + (n-1)d$

$a_n = 1 + (n-1)3$

$a_n = 1 + 3n - 3$

$a_n = 3n - 2$

61. a. $a_n = a_1 + (n-1)d$

$a_n = 11.0 + (n-1)0.5$

$= 11.0 + 0.5n - 0.5$

$= 0.5n + 10.5$

b. $a_n = 0.5n + 10.5$

$= 0.5(50) + 10.5$

$= 35.5$

The percentage is projected to be 35.5% in 2019.

63. Company A

$a_n = 24000 + (n-1)1600$

$= 24000 + 1600n - 1600$

$= 1600n + 22400$

$a_{10} = 1600(10) + 22400$

$= 16000 + 22400 = 38400$

Company B

$a_n = 28000 + (n-1)1000$

$= 28000 + 1000n - 1000$

$= 1000n + 27000$

$a_{10} = 1000(10) + 27000$

$= 10000 + 27000 = 37000$

Company A will pay $1400 more in year 10.

65. a. Total cost:

$\$5836 + \$6185 + \$6585 + \$7020 = \$25,626$

b. $a_1 = 395(1) + 5419 = 5814$

$a_4 = 395(4) + 5419 = 6999$

$S_n = \dfrac{n}{2}(a_1 + a_n)$

$S_4 = \dfrac{4}{2}(5814 + 6999) = 2(12,813) = \$25,626$

The model gives actual sum of $25,626 obtained in part (a).

67. Answers will vary.

69. Company A:
$$a_n = 19,000 + (n-1)2600$$
$$a_{10} = 19,000 + (9)2600 = \$42,200$$
$$S_{10} = \frac{10}{2}(19000 + 42400) = \$307,000$$
Company B:
$$a_n = 27,000 + (n-1)1200$$
$$a_{10} = 27,000 + (9)1200 = \$37,800$$
$$S_{10} = \frac{10}{2}(27,000 + 37,800) = \$324,000$$
Company B pays the greater total amount.

71. $a_n = 20 + (n+1)3$
$$a_{38} = 20 + (37)3 = 131$$
$$S_{38} = \frac{38}{2}(20 + 131) = 2869$$
The theater has 2869 seats.

73. – 77. Answers will vary.

79. makes sense

81. makes sense

83. Degree days: 23, 25, 27, …
$$a_1 = 23, d = 2$$
$$a_{10} = 23 + 9(2) = 41$$
$$S_{10} = \frac{10}{2}(a_1 + a_{10})$$
$$S_{10} = \frac{10}{2}(23 + 41) = 320$$
There are 320 degree-days.

85. $\dfrac{a_2}{a_1} = \dfrac{-2}{1} = -2$

$\dfrac{a_3}{a_2} = \dfrac{4}{-2} = -2$

$\dfrac{a_4}{a_3} = \dfrac{-8}{4} = -2$

$\dfrac{a_5}{a_4} = \dfrac{16}{-8} = -2$

The ratio of a term to the term that directly precedes it is always −2.

86. $\dfrac{a_2}{a_1} = \dfrac{3 \cdot 5^2}{3 \cdot 5^1} = 5$

$\dfrac{a_3}{a_2} = \dfrac{3 \cdot 5^3}{3 \cdot 5^2} = 5$

$\dfrac{a_4}{a_3} = \dfrac{3 \cdot 5^4}{3 \cdot 5^3} = 5$

$\dfrac{a_5}{a_4} = \dfrac{3 \cdot 5^5}{3 \cdot 5^4} = 5$

The ratio of a term to the term that directly precedes it is always 5.

87. $a_n = a_1 3^{n-1}$
$$a_7 = 11 \cdot 3^{7-1} = 11 \cdot 3^6 = 11 \cdot 729 = 8019$$

Section 11.3

Check Point Exercises

1. $a_1 = 12, r = \dfrac{1}{2}$

$a_2 = 12\left(\dfrac{1}{2}\right)^1 = 6$

$a_3 = 12\left(\dfrac{1}{2}\right)^2 = \dfrac{12}{4} = 3$

$a_4 = 12\left(\dfrac{1}{2}\right)^3 = \dfrac{12}{8} = \dfrac{3}{2}$

$a_5 = 12\left(\dfrac{1}{2}\right)^4 = \dfrac{12}{16} = \dfrac{3}{4}$

$a_6 = 12\left(\dfrac{1}{2}\right)^5 = \dfrac{12}{32} = \dfrac{3}{8}$

The first six terms are 12, 6, 3, $\dfrac{3}{2}$, $\dfrac{3}{4}$, and $\dfrac{3}{8}$.

2. $a_1 = 5, r = -3$
$$a_n = 5r^{n-1}$$
$$a_7 = 5(-3)^{7-1} = 5(-3)^6 = 5(729) = 3645$$
The seventh term is 3645.

3. 3, 6, 12, 24, 48, ...

$$r = \frac{6}{3} = 2, \; a_1 = 3$$

$$a_n = 3(2)^{n-1}$$

$$a_8 = 3(2)^{8-1} = 3(2)^7 = 3(128) = 384$$

The eighth term is 384.

4. $a_1 = 2, \; r = \frac{-6}{2} = -3$

$$S_n = \frac{a_1(1 - r^r)}{1 - r}$$

$$S_9 = \frac{2\left(1 - (-3)^9\right)}{1 - (-3)} = \frac{2(19,684)}{4} = 9842$$

The sum of the first nine terms is 9842.

5. $\displaystyle\sum_{i=1}^{8} 2 \cdot 3^i$

$$a_1 = 2 \cdot (3)^1 = 6, \; r = 3$$

$$S_n = \frac{a_1(1 - r^n)}{1 - r}$$

$$S_8 = \frac{6\left(1 - 3^8\right)}{1 - 3} = \frac{6(-6560)}{-2} = 19,680$$

Thus, $\displaystyle\sum_{i=1}^{8} 2 \cdot 3^i = 19,680.$

6. $a_1 = 30,000, \; r = 1.06$

$$S_n = \frac{a_1(1 - r^n)}{1 - r}$$

$$S_{30} = \frac{30,000\left(1 - (1.06)^{30}\right)}{1 - 1.06} \approx 2,371,746$$

The total lifetime salary is $2,371,746.

7. **a.** $A = \dfrac{P\left[\left(1 + \frac{r}{n}\right)^{nt} - 1\right]}{\frac{r}{n}}$

$$P = 100, \; r = 0.095, \; n = 12, \; t = 35$$

$$A = \frac{100\left[\left(1 + \dfrac{0.095}{12}\right)^{12 \cdot 35} - 1\right]}{\dfrac{0.095}{12}} \approx 333,946$$

The value of the IRA will be $333,946.

b. Interest = Value of IRA − Total deposits

$$\approx \$333,946 - \$100 \cdot 12 \cdot 35$$

$$\approx \$333,946 - \$42,000$$

$$\approx \$291,946$$

8. $3 + 2 + \dfrac{4}{3} + \dfrac{8}{9} + \cdots$

$$a_1 = 3, \; r = \frac{2}{3}$$

$$S = \frac{a_1}{1 - r}$$

$$S = \frac{3}{1 - \frac{2}{3}} = \frac{3}{\frac{1}{3}} = 9$$

The sum of this infinite geometric series is 9.

9. $0.\overline{9} = 0.9999\cdots = \dfrac{9}{10} + \dfrac{9}{100} + \dfrac{9}{1000} + \cdots$

$$a_1 = \frac{9}{10}, r = \frac{1}{10}$$

$$S = \frac{\frac{9}{10}}{1 - \frac{1}{10}} = \frac{\frac{9}{10}}{\frac{9}{10}} = 1$$

An equivalent fraction for $0.\overline{9}$ is 1.

10. $a_1 = 1000(0.8) = 800, \; r = 0.8$

$$S = \frac{800}{1 - 0.8} = 4000$$

The total amount spent is $4000.

Concept and Vocabulary Check 11.3

1. geometric; common ratio

2. $a_1 r^{n-1}$; first term; common ratio

3. $\dfrac{a_1(1 - r^n)}{1 - r}$; first term; common ratio

4. annuity; P; r; n

5. infinite geometric series; 1; $\dfrac{a_1}{1 - r}$; $|r| \geq 1$

6. 2; 4; 8; 16; 2

7. arithmetic

8. geometric

9. geometric

10. arithmetic

Exercise Set 11.3

1. $a_1 = 5,\ r = 3$
First five terms: 5, 15, 45, 135, 405.

3. $a_1 = 20,\ r = \dfrac{1}{2}$
First five terms: 20, 10, 5, $\frac{5}{2}$, $\frac{5}{4}$ $\frac{5}{4}$.

5. $a_n = -4a_{n-1},\ a_1 = 10$
First five terms: 10, –40, 160, –640, 2560.

7. $a_n = -5a_{n-1},\ a_1 = -6$
First five terms: –6, 30, –150, 750, –3750.

9. $a_1 = 6,\ r = 2$
$a_n = 6 \cdot 2^{n-1}$
$a_8 = 6 \cdot 2^7 = 768$

11. $a_1 = 5,\ r = -2$
$a_n = 5 \cdot (-2)^{n-1}$
$a_{12} = 5 \cdot (-2)^{11} = -10,240$

13. $a_1 = 1000,\ r = -\dfrac{1}{2}$
$a_n = 1000\left(-\dfrac{1}{2}\right)^{n-1}$
$a_{40} = 1000\left(-\dfrac{1}{2}\right)^{39}$
≈ 0.000000002

15. $a_1 = 1,000,000,\ r = 0.1$
$a_n = 1,000,000(0.1)^{n-1}$
$a_8 = 1,000,000(0.1)^7 = 0.1$

17. 3, 12, 48, 192, . . .
$r = \dfrac{12}{3} = 4$
$a_n = 3(4)^{n-1}$
$a_7 = 3(4)^6 = 12,288$

19. $19, 6, 2, \dfrac{2}{3}, \cdots$ $r = \dfrac{6}{18} = \dfrac{1}{3}$
$a_n = 18\left(\dfrac{1}{3}\right)^{n-1}$
$a_7 = 18\left(\dfrac{1}{3}\right)^6 = \dfrac{2}{81}$

21. 1.5, –3, 6, –12, . . .
$r = \dfrac{6}{-3} = -2$
$a_n = 1.5(-2)^{n-1}$
$a_7 = 1.5(-2)^6 = 96$

23. 0.0004, –0.004, 0.04, –0.4, . . .
$r = \dfrac{-0.004}{0.0004} = -10$
$a_n = 0.0004(-10)^{n-1}$
$a_7 = 0.0004(-10)^6 = 400$

25. 2, 6, 18, 54, . . .
$r = \dfrac{6}{2} = 3$
$S_{12} = \dfrac{2\left(1-3^{12}\right)}{1-3} = \dfrac{2(-531,440)}{-2} = 531,440$

27. 3, –6, 12, –24, . . .
$r = \dfrac{-6}{3} = -2$
$S_{11} = \dfrac{3\left[1-(-2)^{11}\right]}{1-(-2)} = \dfrac{3(2049)}{3} = 2049$

29. $-\dfrac{3}{2}, 3, -6, 12, \cdots$
$r = \dfrac{3}{-\dfrac{3}{2}} = -2$
$S_{14} = \dfrac{-\dfrac{3}{2}\left[1-(-2)^{14}\right]}{1-(-2)} = \dfrac{-\dfrac{3}{2}(-16,383)}{3} = \dfrac{16,383}{2}$

31. $\displaystyle\sum_{i=1}^{8} 3^i$
$r = 3,\ a_1 = 3$
$S_8 = \dfrac{3\left(1-3^8\right)}{1-3} = \dfrac{3(-6560)}{-2} = 9840$

33. $\displaystyle\sum_{i=1}^{10} 5 \cdot 2^i$

$r = 2, \quad a_1 = 10$

$S_{10} = \dfrac{10\left(1 - 2^{10}\right)}{1 - 2} = \dfrac{10(-1023)}{-1} = 10,230$

35. $\displaystyle\sum_{i=1}^{6} \left(\dfrac{1}{2}\right)^{i+1}$

$r = \dfrac{1}{2}, \quad a_1 = \dfrac{1}{4}$

$S_6 = \dfrac{\dfrac{1}{4}\left(1 - \left(\dfrac{1}{2}\right)^6\right)}{1 - \dfrac{1}{2}} = \dfrac{\dfrac{1}{4}\left(\dfrac{63}{64}\right)}{\dfrac{1}{2}} = \dfrac{63}{128}$

37. $r = \dfrac{1}{3}$

$S_\infty = \dfrac{1}{1 - \dfrac{1}{3}} = \dfrac{1}{\dfrac{2}{3}} = \dfrac{3}{2}$

39. $r = \dfrac{1}{4}$

$S_\infty = \dfrac{3}{1 - \dfrac{1}{4}} = \dfrac{3}{\dfrac{3}{4}} = 4$

41. $r = -\dfrac{1}{2}$

$S_\infty = \dfrac{1}{1 - \left(-\dfrac{1}{2}\right)} = \dfrac{1}{\dfrac{3}{2}} = \dfrac{2}{3}$

43. $r = -0.3$

$S_\infty = \dfrac{8}{1 - (-0.3)} = \dfrac{8}{1.3} \approx 6.15385$

45. $r = \dfrac{1}{10}$

$S_\infty = \dfrac{\dfrac{5}{10}}{1 - \dfrac{1}{10}} = \dfrac{\dfrac{5}{10}}{\dfrac{9}{10}} = \dfrac{5}{9}$

47. $r = \dfrac{1}{100}$

$S_\infty = \dfrac{\dfrac{47}{100}}{1 - \dfrac{1}{100}} = \dfrac{\dfrac{47}{100}}{\dfrac{99}{100}} = \dfrac{47}{99}$

49. $0.\overline{257} = \dfrac{257}{1000} + \dfrac{257}{10^6} + \dfrac{257}{10^9} + \cdots$

$r = \dfrac{1}{1000}$

$S_\infty = \dfrac{\dfrac{257}{1000}}{1 - \dfrac{1}{1000}} = \dfrac{\dfrac{257}{1000}}{\dfrac{999}{1000}} = \dfrac{257}{999}$

51. $a_n = n + 5$

arithmetic, $d = 1$

53. $a_n = 2^n$

geometric, $r = 2$

55. $a_n = n^2 + 5$

neither

57. First find a_{10} and b_{10}:

$a_{10} = a_1 r^{n-1}$

$\qquad = (-5)\left(\dfrac{10}{-5}\right)^{10-1} = (-5)(-2)^9$

$\qquad = 2560$

$b_{10} = b_1 + (n-1)d$

$\qquad = 10 + (10 - 1)(-5 - 10)$

$\qquad = 10 + (9)(-15) = -125$

So, $a_{10} + b_{10} = 2560 + (-125) = 2435$.

59. For $\{a_n\}$, $r = \dfrac{10}{-5} = -2$ and

$S_{10} = \dfrac{a_1(1 - r^n)}{1 - r} = \dfrac{(-5)\left(1 - (-2)^{10}\right)}{1 - (-2)}$

$\qquad = \dfrac{(-5)(-1023)}{3} = 1705$

For $\{b_n\}$,

$$b_{10} = b_1 + (n-1)d$$
$$= 10 + (10-1)(-5-10)$$
$$= 10 + (9)(-15) = -125$$

$$S_n = \frac{n}{2}(b_1 + b_n)$$

$$S_{10} = \frac{n}{2}(b_1 + b_{10})$$

$$= \frac{10}{2}(10 + (-125))$$

$$= 5(-115) = -575$$

61. For $\{a_n\}$,

$$S_6 = \frac{a_1(1-r^n)}{1-r} = \frac{(-5)\left(1-(-2)^6\right)}{1-(-2)}$$

$$= \frac{(-5)(-63)}{3} = 105$$

For $\{c_n\}$,

$$S = \frac{a_1}{1-r} = \frac{-2}{1-\frac{1}{-2}} = \frac{-2}{\frac{3}{2}} = -\frac{4}{3}$$

So, $S_6 \cdot S = 105\left(-\frac{4}{3}\right) = -140$

63. It is given that $a_4 = 27$. Using the formula

$a_n = a_1 r^{n-1}$ when $n = 4$ we have

$$27 = 8r^{4-1}$$

$$\frac{27}{8} = r^3$$

$$r = \sqrt[3]{\frac{27}{8}} = \frac{3}{2}$$

Thus,

$$a_n = a_1 r^{n-1}$$

$$a_2 = 8\left(\frac{3}{2}\right)^{2-1} = 8\left(\frac{3}{2}\right) = 12$$

$$a_3 = 8\left(\frac{3}{2}\right)^{3-1} = 8\left(\frac{3}{2}\right)^2 = 8\left(\frac{9}{4}\right) = 18$$

65. $1, 2, 4, 8, \ldots$

$r = 2$

$$a_n = 2^{n-1}$$

$$a_{15} = 2^{14} = \$16,384$$

67. $a_1 = 3,000,000$

$r = 1.04$

$$a_n = 3,000,000(1.04)^{n-1}$$

$$a_7 = 3,000,000(1.04)^6 = \$3,795,957$$

69. **a.** $r_{2000 \text{ to } 2001} = \frac{34.21}{33.87} \approx 1.01$

$r_{2001 \text{ to } 2002} = \frac{34.55}{34.21} \approx 1.01$

$r_{2002 \text{ to } 2003} = \frac{34.90}{34.55} \approx 1.01$

$r_{2003 \text{ to } 2004} = \frac{35.25}{34.90} \approx 1.01$

$r_{2004 \text{ to } 2005} = \frac{35.60}{35.25} \approx 1.01$

$r_{2005 \text{ to } 2006} = \frac{36.00}{35.60} \approx 1.01$

$r_{2006 \text{ to } 2007} = \frac{36.36}{36.00} \approx 1.01$

$r_{2007 \text{ to } 2008} = \frac{36.72}{36.36} \approx 1.01$

$r_{2008 \text{ to } 2009} = \frac{37.09}{36.72} \approx 1.01$

$r_{2009 \text{ to } 2010} = \frac{37.25}{37.09} \approx 1.01$

r is approximately 1.01 for all but one division.

b. $a_n = a_1 r^{n-1}$

$$a_n = 33.87(1.01)^{n-1}$$

c. Since year 2020 is the 21th term, find a_{21}.

$$a_n = 33.87(1.01)^{n-1}$$

$$a_{21} = 33.87(1.01)^{21-1} \approx 41.33$$

The population of California will be approximately 41.33 million in 2020.

71. $1, 2, 4, 8, \ldots$

$r = 2$

$$S_{15} = \frac{1(1-2^{15})}{1-2} = 32,767$$

The total savings is \$32,767.

73. $a_1 = 24,000, \ r = 1.05$

$$S_{20} = \frac{24,000\left[1-(1.05)^{20}\right]}{1-1.05} = 793,582.90$$

The total salary is \$793,583.

75. $r = 0.9$

$$S_{10} = \frac{20(1 - 0.9^{10})}{1 - 0.9} \approx 130.26$$

The total length is 130.26 inches.

77. **a.** $A = \dfrac{P\left[\left(1 + \dfrac{r}{n}\right)^{nt} - 1\right]}{\dfrac{r}{n}} = \dfrac{2000\left[\left(1 + \dfrac{0.075}{1}\right)^{5} - 1\right]}{\dfrac{0.075}{1}} \approx \$11,617$

 b. $\$11,617 - 5 \times \$2000 = \$1617$

79. **a.** $A = \dfrac{P\left[\left(1 + \dfrac{r}{n}\right)^{nt} - 1\right]}{\dfrac{r}{n}} = \dfrac{50\left[\left(1 + \frac{0.055}{12}\right)^{12 \times 40} - 1\right]}{\frac{0.055}{12}} \approx \$87,052$

 b. $\$87,052 - \$50 \cdot 12 \cdot 40 = \$63,052$

81. **a.** $A = \dfrac{P\left[\left(1 + \frac{r}{n}\right)^{nt} - 1\right]}{\frac{r}{n}} = \dfrac{10,000\left[\left(1 + \frac{0.105}{4}\right)^{4 \times 10} - 1\right]}{\frac{0.105}{4}} \approx \$693,031$

 b. $\$693,031 - \$10,000 \cdot 4 \cdot 10 = \$293,031$

83. Find the total value of the lump-sum investment.

$$A = P(1 + r)^t = 30,000(1 + 0.05)^{20} \approx 79,599$$

Find the total value of the annuity.

$$A = \dfrac{P\left[\left(1 + \dfrac{r}{n}\right)^{nt} - 1\right]}{\dfrac{r}{n}} = \dfrac{1500\left[\left(1 + \dfrac{0.05}{1}\right)^{20} - 1\right]}{\dfrac{0.05}{1}} \approx 49,599$$

$\$79,599 - \$49,599 = \$30,000$

You will have $30,000 more from the lump-sum investment.

85. $r = 0.6$

$$S_\infty = \frac{6(0.6)}{1 - 0.6} = 9$$

The total economic impact is $9 million.

87. $r = \dfrac{1}{4}$

$$S_\infty = \frac{\frac{1}{4}}{1 - \frac{1}{4}} = \frac{1}{4} \cdot \frac{4}{3} = \frac{1}{3}$$

89. – 97. Answers will vary.

99. $f(x) = \dfrac{2\left[1-\left(\frac{1}{3}\right)^x\right]}{1-\frac{1}{3}}$

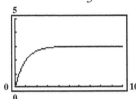

The sum of the series is 3 and the asymptote of the function is $y = 3$.

101. makes sense

103. makes sense

105. false; Changes to make the statement true will vary. A sample change is: The sequence is not geometric. There is not a common ratio.

107. false; Changes to make the statement true will vary. A sample change is: The sum of the sequence is $\dfrac{10}{1-\left(-\frac{1}{2}\right)}$.

109. Let a_1 equal the number of flies released each day.

On any day, the total number of flies is the number released that day, plus 90% of those released the day before, plus 90% of 90% of those released two days before, etc.:

$$S = \frac{a_1}{1-r}$$

$$20{,}000 = \frac{a_1}{1-.9}$$

$$20{,}000 = \frac{a_1}{.1}$$

$$2000 = a_1$$

2000 flies to be released each day.

111. Answers will vary.

112. $1+2+3 = \dfrac{3(3+1)}{2}$

$\quad\quad 6 = \dfrac{3(4)}{2}$

$\quad\quad 6 = 6$

113. $1+2+3+4+5 = \dfrac{5(5+1)}{2}$

$\quad\quad 15 = \dfrac{5(6)}{2}$

$\quad\quad 15 = 15$

114. $\dfrac{k(k+1)(2k+1)}{6} + (k+1)^2$

$= \dfrac{k(k+1)(2k+1)}{6} + \dfrac{6(k+1)^2}{6}$

$= \dfrac{k(k+1)(2k+1) + 6(k+1)^2}{6}$

$= \dfrac{(k+1)[k(2k+1) + 6(k+1)]}{6}$

$= \dfrac{(k+1)[2k^2 + k + 6k + 6]}{6}$

$= \dfrac{(k+1)[2k^2 + 7k + 6]}{6}$

$= \dfrac{(k+1)(k+2)(2k+3)}{6}$

Mid-Chapter 11 Check Point

1. $a_n = (-1)^{n+1}\dfrac{n}{(n-1)!}$

$a_1 = (-1)^{1+1}\dfrac{1}{(1-1)!} = (-1)^2\dfrac{1}{0!} = 1\cdot 1 = 1$

$a_2 = (-1)^{2+1}\dfrac{2}{(2-1)!} = (-1)^3\dfrac{2}{1!} = (-1)(2) = -2$

$a_3 = (-1)^{3+1}\dfrac{3}{(3-1)!} = (-1)^4\dfrac{3}{2!} = 1\cdot\dfrac{3}{2} = \dfrac{3}{2}$

$a_4 = (-1)^{4+1}\dfrac{4}{(4-1)!} = (-1)^5\dfrac{4}{3!} = (-1)\dfrac{4}{6} = -\dfrac{2}{3}$

$a_5 = (-1)^{5+1}\dfrac{5}{(5-1)!} = (-1)^6\dfrac{5}{4!} = 1\cdot\dfrac{5}{24} = \dfrac{5}{24}$

2. Using $a_n = a_1 + (n-1)d$;

$a_1 = 5$

$a_2 = 5 + (2-1)(-3) = 5 + 1(-3) = 5 - 3 = 2$

$a_3 = 5 + (3-1)(-3) = 5 + 2(-3) = 5 - 6 = -1$

$a_4 = 5 + (4-1)(-3) = 5 + 3(-3) = 5 - 9 = -4$

$a_5 = 5 + (5-1)(-3) = 5 + 4(-3) = 5 - 12 = -7$

3. Using $a_n = a_1 r^{n-1}$;

$a_1 = 5$

$a_2 = 5(-3)^{2-1} = 5(-3)^1 = 5(-3) = -15$

$a_3 = 5(-3)^{3-1} = 5(-3)^2 = 5(9) = 45$

$a_4 = 5(-3)^{4-1} = 5(-3)^3 = 5(-27) = -135$

$a_5 = 5(-3)^{5-1} = 5(-3)^4 = 5(81) = 405$

4. $a_n = -a_{n-1} + 4$

$a_1 = 3$

$a_2 = -a_1 + 4 = -3 + 4 = 1$

$a_3 = -a_2 + 4 = -1 + 4 = 3$

$a_4 = -a_3 + 4 = -3 + 4 = 1$

$a_5 = -a_4 + 4 = -1 + 4 = 3$

5. $d = a_2 - a_1 = 6 - 2 = 4$

$\begin{aligned}
a_n &= a_1 + (n-1)d \\
&= 2 + (n-1)4 \\
&= 2 + 4n - 4 \\
&= 4n - 2
\end{aligned}$

$a_{20} = 4(20) - 2 = 78$

6. $r = \dfrac{a_2}{a_1} = \dfrac{6}{3} = 2$

$\begin{aligned}
a_n &= a_1 r^{n-1} \\
&= 3(2)^{n-1}
\end{aligned}$

$\begin{aligned}
a_{10} &= 3(2)^{10-1} \\
&= 3(2)^9 \\
&= 1536
\end{aligned}$

7. $d = a_2 - a_1 = 1 - \dfrac{3}{2} = -\dfrac{1}{2}$

$a_n = a_1 + (n-1)d$

$\quad = \dfrac{3}{2} + (n-1)\left(-\dfrac{1}{2}\right)$

$\quad = \dfrac{3}{2} - \dfrac{1}{2}n + \dfrac{1}{2}$

$\quad = -\dfrac{1}{2}n + 2$

$a_{30} = -\dfrac{1}{2}(30) + 2$

$\quad = -15 + 2$

$\quad = -13$

8. $S_n = \dfrac{a_1(1-r^n)}{1-r}; \; r = \dfrac{a_2}{a_1} = \dfrac{10}{5} = 2$

$S_{10} = \dfrac{5(1-2^{10})}{1-2} = \dfrac{5(-1023)}{-1} = 5115$

9. First find a_{10};

$d = a_2 - a_1 = 0 - (-2) = 2$

$a_{50} = a_1 + (n-1)d = -2 + (50-1)(2) = -2 + 49(2) = 96$

$S_{50} = \dfrac{n}{2}(a_1 + a_n) = \dfrac{50}{2}(-2+96) = 25(94) = 2350$

10. $r = \dfrac{a_2}{a_1} = \dfrac{40}{-20} = -2$

$S_{10} = \dfrac{a_1(1-r^n)}{1-r} = \dfrac{-20(-1-(-2)^{10})}{1-(-2)} = \dfrac{-20(-1023)}{3} = \dfrac{20460}{3} = 6820$

11. First find a_{100};

$d = a_2 - a_1 = -2 - 4 = -6$

$a_{100} = a_1 + (n-1)d = 4 + (100-1)(-6) = 4 + 99(-6) = -590$

$S_{100} = \dfrac{n}{2}(a_1 + a_n) = \dfrac{100}{2}(4-590) = 50(-586) = -29,300$

12. $\displaystyle\sum_{i=1}^{4}(i+4)(i-1) = (1+4)(1-1) + (2+4)(2-1) + (3+4)(3-1) + (4+4)(4-1)$

$\quad = 5(0) + 6(1) + 7(2) + 8(3) = 0 + 6 + 14 + 24 = 44$

13. $\sum_{i=1}^{50}(3i-2) = (3\cdot1-2)+(3\cdot2-2)+(3\cdot3-3)+...+(3\cdot50-2)$

$$= (3-2)+(6-2)+(9-3)+...+(150-2)$$
$$= 1+4+6+...+148$$

The sum of this arithmetic sequence is given by $S_n = \dfrac{n}{2}(a_1+a_n)$;

$$S_{50} = \frac{50}{2}(1+148) = 25(149) = 3725$$

14. $\sum_{i=1}^{6}\left(\dfrac{3}{2}\right)^i = \left(\dfrac{3}{2}\right)^1 + \left(\dfrac{3}{2}\right)^2 + \left(\dfrac{3}{2}\right)^3 + \left(\dfrac{3}{2}\right)^4 + \left(\dfrac{3}{2}\right)^5 + \left(\dfrac{3}{2}\right)^6 = \dfrac{3}{2} + \dfrac{9}{4} + \dfrac{27}{8} + \dfrac{81}{16} + \dfrac{243}{32} + \dfrac{729}{64} = \dfrac{1995}{64}$

15. $\sum_{i=1}^{\infty}\left(-\dfrac{2}{5}\right)^{i-1} = \left(-\dfrac{2}{5}\right)^{1-1} + \left(-\dfrac{2}{5}\right)^{2-1} + \left(-\dfrac{2}{5}\right)^{3-1} + ... = \left(-\dfrac{2}{5}\right)^0 + \left(-\dfrac{2}{5}\right)^1 + \left(-\dfrac{2}{5}\right)^2 + ... = 1 + \left(-\dfrac{2}{5}\right) + \dfrac{4}{25} + ...$

This is an infinite geometric sequence with $r = \dfrac{a_2}{a_1} = \dfrac{-\frac{2}{5}}{1} = -\dfrac{2}{5}$.

$$S = \frac{a_1}{1-r} = \frac{1}{1-\left(-\frac{2}{5}\right)} = \frac{1}{\frac{7}{5}} = \frac{5}{7}$$

16. $0.\overline{45} = \dfrac{a_1}{1-r} = \dfrac{\frac{45}{100}}{1-\frac{1}{100}} = \dfrac{\frac{45}{100}}{\frac{99}{100}}$

$$= \frac{45}{100} \div \frac{99}{100} = \frac{45}{100} \cdot \frac{100}{99} = \frac{45}{99} = \frac{5}{11}$$

17. Answers will vary. An example is $\sum_{i=1}^{18}\dfrac{i}{i+2}$.

18. The arithmetic sequence is 16, 48, 80, 112, ...
First find a_{15} where $d = a_2 - a_1 = 48-16 = 32$.
$$a_{15} = a_1 + (n-1)d = 16 + (15-1)(32) = 16 + 14(32) = 16 + 448 = 464$$
The distance the skydiver falls during the 15th second is 464 feet.
$$S_{15} = \frac{n}{2}(a_1+a_n) = \frac{15}{2}(16+464) = 7.5(480) = 3600$$
The total distance the skydiver falls in 15 seconds is 3600 feet.

19. $r = 0.10$
$$A = P(1+r)^t$$
$$= 120000(1+0.10)^{10}$$
$$\approx 311249$$
The value of the house after 10 years is $311,249.

Section 11.4

Check Point Exercises

1. **a.** $S_1 : 2 = 1(1+1)$

$S_k : 2+4+6+\cdots 2k = k(k+1)$

$S_{k+1} : 2+4+6+\cdots+2(k+1) = (k+1)(k+2)$

b. $S_1 : 1^3 = \dfrac{1^2(1+1)^2}{4}$

$S_k = 1^3 + 2^3 + 3^3 + \cdots + k^3 = \dfrac{k^2(k+1)^2}{4}$

$S_{k+1} = 1^3 + 2^3 + 3^3 + \cdots + (k+1)^3 = \dfrac{(k+1)^2(k+2)^2}{4}$

2. $S_1 : 2 = 1(1+1)$

$2 = 2$ is true.

$S_k : 2+4+6+\cdots+2k = k(k+1)$

$S_{k+1} : 2+4+6+\cdots+2k+2(k+1) = (k+1)(k+2)$

Add $2(k+1)$ to both sides of S_k :

$2+4+6+\cdots+2k+2(k+1) = k(k+1)+2(k+1)$

Simplify the right-hand side:

$k(k+1)+2(k+1) = (k+1)(k+2)$

If S_k is true, then S_{k+1} is true. The statement is true for all *n*.

3. $S_1 : 1^3 = \dfrac{1^2(1+1)^2}{4}$

$1 = \dfrac{4}{4}$

$1 = 1$ is true.

$S_k : 1^3 + 2^3 + 3^3 + \cdots + k^3 = \dfrac{k^2(k+1)^2}{4}$

$S_{k+1} : 1^3 + 2^3 + 3^3 + \cdots + k^3 + (k+1)^3 = \dfrac{(k+1)^2(k+2)^2}{4}$

Add $(k + 1)^3$ to both sides of S_k :

$1^3 + 2^3 + 3^3 + \cdots + k^3 + (k+1)^3 = \dfrac{k^2(k+1)^2}{4} + (k+1)^3$

Simplify the right hand side:

$\dfrac{k^2(k+1)^2}{4} + (k+1)^3 = \dfrac{k^2(k+1)^2 + 4(k+1)^3}{4} = \dfrac{(k+1)^2\left[k^2 + 4(k+1)\right]}{4} = \dfrac{(k+1)^2(k^2+4k+4)}{4}$

$= \dfrac{(k+1)^2(k+2)^2}{4}$

If S_k is true, then S_{k+1} is true. The statement is true for all *n*.

4. S_1: 2 is a factor of $1^2 + 1 = 2$, since $2 = 2 \cdot 1$.

S_k: 2 is a factor of $k^2 + k$

S_{k+1}: 2 is a factor of $(k+1)^2 + (k+1)$
Simplify:
$$(k+1)^2 + (k+1) = k^2 + 2k + 1 + k + 1$$
$$= k^2 + 3k + 2$$
$$= k^2 + k + 2k + 2$$
$$= (k^2 + k) + 2(k+1)$$

Because we assume S_k is true, we know 2 is a factor of $k^2 + k$. Since 2 is a factor of $2(k+1)$, we conclude 2 is a factor of the sum $(k^2 + k) + 2(k+1)$. If S_k is true, then S_{k+1} is true. The statement is true for all n.

Concept and Vocabulary Check 11.4

1. induction; 1; $k+1$

2. $2 + 4 + 6 = 3(3+1)$;
$2 + 4 + 6 + \cdots + 2(k+1) = (k+1)(k+2)$

3. $3 + 7 + 11 = 3(6+1)$;
$3 + 7 + 11 + \cdots + [4(k+1) - 1] = (k+1)[2(k+1) + 1]$

4. 4; 10; 18; $(k+1)^2 + 3(k+1)$; $k^2 + 5k + 4$

5. $k+1$

Exercise Set 11.4

1. $S_n = 1 + 3 + 5 + \cdots + (n-1) = n^2$
$S_1 : 1 = 1^2$
$1 = 1$ true
$S_2 : 1 + 3 = 2^2$
$4 = 4$ true
$S_3 : 1 + 3 + 5 = 3^2$
$9 = 9$ true

3. S_n : 2 is a factor of $n^2 - n$

     S_1 : 2 is a factor of $1^2 - 1 = 0$

       $0 = 0 \cdot 2$ so 2 is a factor of 0 is true.

     S_2 : 2 is a factor of $2^2 - 2 = 2$

       $2 = 1 \cdot 2$ so 2 is a factor of 2 is true.

     S_3 : 2 is a factor of $3^2 - 3 = 6$

       $6 = 3 \cdot 2$ so 2 is a factor of 6 is true.

5. $S_n : 4 + 8 + 12 + \cdots + 4n = 2n(n+1)$

        $S_k : 4 + 8 + 12 + \cdots + 4k = 2k(k+1)$

  $S_{k+1} : 4 + 8 + 12 + \cdots + 4(k+1) = 2(k+1)(k+1+1)$

         $4 + 8 + 12 + \cdots + 4(k+1) = 2(k+1)(k+2)$

7. $S_n : 3 + 7 + 11 + \cdots + (4n - 1) = n(2n+1)$

         $S_k : 3 + 7 + 11 + \cdots + (4k - 1) = k(2k+1)$

  $S_{k+1} : 3 + 7 + 11 + \cdots + [4(k+1) - 1] = (k+1)[2(k+1) + 1]$

         $3 + 7 + 11 + \cdots + (4k + 3) = (k+1)(2k+3)$

9. S_n : 2 is a factor of $n^2 - n + 2$

      S_k : 2 is a factor of $k^2 - k + 2$

  S_{k+1} : 2 is a factor of $(k+1)^2 - (k+1) + 2$

       $k^2 + 2k + 1 - k - 1 + 2 = k^2 + k + 2$

  S_{k+1} : 2 is a factor of $k^2 + k + 2$.

11. $S_1 : 4 = 2(1)(1+1)$

     $4 = 2(2)$

     $4 = 4$ is true.

       $S_k : 4 + 8 + 12 + \cdots + 4k = 2k(k+1)$

  $S_{k+1} : 4 + 8 + 12 + \cdots 4(k+1) = 2(k+1)(k+1+1)$

     Add $4(k+1)$ to both sides of S_k:

     $4 + 8 + 12 + \cdots + 4(k+1) = 2k(k+1) + 4(k+1)$

     Simplify the right-hand side:

     $= 2k(k+1) + 4(k+1) = (2k+4)(k+1)$

     $= 2(k+2)(k+1)$

     $= 2(k+1)(k+1+1)$

     If S_k is true, then S_{k+1} is true. The statement is true for all n.

13. $S_1 : 1 = 1^2$

     $1 = 1$ is true.

     $S_k : 1 + 3 + 5 + \cdots + (2k - 1) = k^2$

  $S_{k+1} : 1 + 3 + 5 + \cdots + (2k - 1) + [2(k+1) - 1] = (k+1)^2$

       $1 + 3 + 5 + \cdots + (2k - 1) + (2k + 1) = (k+1)^2$

Add $(2k + 1)$ to both sides of S_k:

$1+3+5+\cdots+(2k-1)+(2k+1) = k^2 + (2k+1)$

Simplify the right-hand side:

$= k^2 + (2k+1)$

$= (k+1)^2$

If S_k is true, then S_{k+1} is true. The statement is true for all n.

15. $S_1 : 3 = 1[2(1)+1)]$

$3 = 3$ is true.

$S_k : 3+7+11+\cdots+(4k-1) = k(2k+1)$

$S_{k+1} : 3+7+11+\cdots+(4k-1)+[4(k+1)-1] = (k+1)[2(k+1)+1]$

$3+7+11+\cdots+(4k-1)+(4k+3) = (k+1)(2k+3)$

Add $(4k + 3)$ to both sides of S_k:

$3 + 7 + 11 + \ldots + (4k-1) + (4k+3) = k(2k+1) + 4(k+3)$

Simplify the right-hand side:

$= k(2k+1)+(4k+3) = 2k^2 + k + 4k + 3$

$= 2k^2 + 5k + 3$

$= (k+1)(2k+3)$

If S_k is true, then S_{k+1} is true. The statement is true for all n.

17. $S_1 : 1 = 2^1 - 1$

$1 = 1$ is true.

$S_k : 1+2+2^2 +\cdots+2^{k-1} = 2^k - 1$

$S_{k+1} : 1+2+2^2 +\cdots+2^{k-1}+2^{k+1-1} = 2^{k+1} - 1$

$1+2+2^2 +\cdots+2^{k-1}+2^k = 2^{k+1} - 1$

Add 2^k to both sides of S_k:

$1+2+2^2 +\cdots+2^{k-1}+2^k = 2^k + 2^k - 1$

Simplify the right-hand side:

$= 2^k + 2^k - 1 = 2(2^k) - 1$

$= 2^{k+1} - 1$

If S_k is true, then S_{k+1} is true. The statement is true for all n.

19. $S_1 : 2 = 2^{1+1} - 2$

$2 = 4 - 2$

$2 = 2$ is true.

$S_k : 2+4+8+\cdots+2^k = 2^{k+1} - 2$

$S_{k+1} : 2+4+8+\cdots+2^k+2^{k+1} = 2^{k+2} - 2$

Add 2^{k+1} to both sides of S_k:

$2+4+8+\cdots+2^k+2^{k+1} = 2^{k+1}+2^{k+1} - 2$

Simplify the right-hand side:

$= 2^{k+1}+2^{k+1} - 1 = 2(2^{k+1}) - 2$

$= 2^{k+2} - 2$

If S_k is true, then S_{k+1} is true. The statement is true for all n.

21. $S_1 : 1 \cdot 2 = \dfrac{1(1+1)(1+2)}{3}$

$2 = \dfrac{6}{3}$

$2 = 2$ is true.

$S_k : 1 \cdot 2 + 2 \cdot 3 + 3 \cdot 4 + \cdots + k(k+1) = \dfrac{k(k+1)(k+2)}{3}$

$S_{k+1} : 1 \cdot 2 + 2 \cdot 3 + 3 \cdot 4 + \cdots + k(k+1) + (k+1)(k+2) = \dfrac{(k+1)(k+2)(k+3)}{3}$

Add $(k+1)(k+2)$ to both sided of S_k:

$1 \cdot 2 + 2 \cdot 3 + 3 \cdot 4 + \cdots + k(k+1) + (k+1)(k+2) = \dfrac{k(k+1)(k+2)}{3} + (k+1)(k+2)$

Simplify the right-hand side:

$= \dfrac{k(k+1)(k+2)}{3} + (k+1)(k+2) = \dfrac{k(k+1)(k+2) + 3(k+1)(k+2)}{3}$

$= \dfrac{(k+1)(k+2)(k+3)}{3}$

If S_k is true, then S_{k+1} is true. The statement is true for all n.

23. $S_1 : \dfrac{1}{1 \cdot 2} = \dfrac{1}{1+1}$

$\dfrac{1}{2} = \dfrac{1}{2}$ is true.

$S_k : \dfrac{1}{1 \cdot 2} + \dfrac{1}{2 \cdot 3} + \dfrac{1}{3 \cdot 4} + \cdots + \dfrac{1}{k(k+1)} = \dfrac{k}{k+1}$

$S_{k+1} : \dfrac{1}{1 \cdot 2} + \dfrac{1}{2 \cdot 3} + \dfrac{1}{3 \cdot 4} + \cdots + \dfrac{1}{k(k+1)} + \dfrac{1}{(k+1)(k+2)} = \dfrac{k+1}{k+2}$

Add $\dfrac{1}{(k+1)(k+2)}$ to both sides of S_k:

$\dfrac{1}{1 \cdot 2} + \dfrac{1}{2 \cdot 3} + \dfrac{1}{3 \cdot 4} + \cdots + \dfrac{1}{k(k+1)} + \dfrac{1}{(k+1)(k+2)} = \dfrac{k}{k+1} + \dfrac{1}{(k+1)(k+2)}$

Simplify the right-hand side:

$\dfrac{k}{(k+1)} + \dfrac{1}{(k+1)(k+2)} = \dfrac{k(k+2)+1}{(k+1)(k+2)}$

$= \dfrac{k^2 + 2k + 1}{(k+1)(k+2)}$

$= \dfrac{(k+1)(k+1)}{(k+1)(k+2)}$

$= \dfrac{k+1}{k+2}$

If S_k is true, then S_{k+1} is true. The statement is true for all n.

25. S_1 : 2 is a factor of $1^2 - 1 = 0$, since $0 = 2 \cdot 0$.

   S_k : 2 is a factor of $k^2 - k$

S_{k+1} : 2 is a factor of $(k+1)^2 - (k+1)$

$(k+1)^2 - (k-1) = k^2 + 2k + 1 - k - 1$

$$= k^2 + k$$

$$= k^2 - k + 2k$$

$$= (k^2 - k) + 2k$$

Because we assume S_k is true, we know 2 as a factor of $k^2 - k$. Since 2 is a factor of 2*k*, we conclude 2 is factor of the sum $(k^2 + k) + 2k$. If S_k is true, then S_{k+1} is true. The statement is true for all *n*.

27. S_1: 6 is a factor of $1(1+1)(1+2) = 6$, since $6 = 6 \cdot 1$.

 S_k: 6 is a factor of $k(k+1)(k+2)$

 S_{k+1}: 6 is a factor of $(k+1)(k+2)(k+3)$

$(k+1)(k+2)(k+3) = k(k+1)(k+2) + 3(k+1)(k+2)$

Because we assume S_k is true, we know 6 as a factor of $k(k+1)(k+2)$. Since either *k* + 1 or *k* + 2 must be even, the product $(k+1)(k+2)$ is even. Thus 2 is a factor of $(k+1)(k+2)$, and we can conclude that 6 is factor of $3(k+1)(k+2)$ If S_k is true, then S_{k+1} is true.

The statement is true for all *n*.

29. $\displaystyle\sum_{i=1}^{n} 5 \cdot 6^i = 6\left(6^n - 1\right)$

Show that S_1 is true: $\displaystyle\sum_{i=1}^{1} 5 \cdot 6^i = 6\left(6^1 - 1\right)$

$$5 \cdot 6^1 = 6(6-1)$$

$$5 \cdot 6 = 6 \cdot 5, \text{ True}$$

Show that if S_k is true, then S_{k+1} is true:

Assume $S_k : \displaystyle\sum_{i=1}^{k} 5 \cdot 6^i = 6\left(6^k - 1\right)$ is true. Then,

$$\sum_{i=1}^{k} 5 \cdot 6^i + 5 \cdot 6^{k+1} = 6\left(6^k - 1\right) + 5 \cdot 6^{k+1}$$

$$\sum_{i=1}^{k+1} 5 \cdot 6^i = 6^{k+1} - 6 + 5 \cdot 6^{k+1}$$

$$\sum_{i=1}^{k+1} 5 \cdot 6^i = 6 \cdot 6^{k+1} - 6$$

$$\sum_{i=1}^{k+1} 5 \cdot 6^i = 6\left(6^{k+1} - 1\right)$$

The final statement is S_{k+1} . Thus, by mathematical induction, we have proven that $\displaystyle\sum_{i=1}^{n} 5 \cdot 6^i = 6\left(6^n - 1\right)$.

31. $n + 2 > n$

Show that S_1 *is true:* $1 + 2 > 1$

$$3 > 1, \ \text{True}$$

Show that if S_k *is true, then* S_{k+1} *is true:*

Assume $S_k : k + 2 > k$ is true. Then,

$$k + 2 + 1 > k + 1$$
$$(k + 1) + 2 > k + 1$$

The final statement is S_{k+1}. Thus, by mathematical induction, we have proven that $n + 2 > n$.

33. $S_1 : (ab)^1 = a^1 b^1$

$ab = ab$ is true.

$S_k : (ab)^k = a^k b^k$

$S_{k+1} : (ab)^{k+1} = a^{k+1} b^{k+1}$

Multiply both sides of S_k by ab:

$$(ab)^k (ab) = a^k b^k (ab)$$
$$(ab)^{k+1} = a^{k+1} b^{k+1}$$

If S_k is true, then S_{k+1} is true.
The statement is true for all n.

35. Answers will vary.

37. does not make sense; Explanations will vary. Sample explanation: We use mathematical induction to prove statements involving positive integers.

39. does not make sense; Explanations will vary. Sample explanation: It is necessary for all the dominoes to topple.

41. $n^2 > 2n + 1$ for $n \geq 3$

$S_3 : 3^2 > 2 \cdot 3 + 1$

$$9 > 7$$

$S_k : k^2 > 2k + 1$ for $k \geq 3$

$S_{k+1} : (k + 1)^2 > 2k + 3$.

Add $2k + 1$ to both sides of S_k.

$$k^2 + (2k + 1) > 2k + 1 + (2k + 1)$$

Write the left side of the inequalities as the square of a binomial and simplify the right side. $(k + 1)^2 > 4k + 2$

Since $4k + 2 > 2k + 3$ for $k \geq 3$, we can conclude that $(k + 1)^2 > 4k + 2 > 2k + 3$.

By the transitive property,

$$(k + 1)^2 > 2k + 3$$
$$(k + 1)^2 > 2(k + 1) + 1$$

If S_k is true, then S_{k+1} is true.
The statement is true for all n.

43. $S_1 = \dfrac{1}{4} = \dfrac{1}{4}$

$S_2 = \dfrac{1}{4} + \dfrac{1}{12} = \dfrac{1}{3}$

$S_3 = \dfrac{1}{4} + \dfrac{1}{12} + \dfrac{1}{24} = \dfrac{3}{8}$

$S_4 = \dfrac{1}{4} + \dfrac{1}{12} + \dfrac{1}{24} + \dfrac{1}{40} = \dfrac{2}{5}$

$S_5 = \dfrac{1}{4} + \dfrac{1}{12} + \dfrac{1}{24} + \dfrac{1}{40} + \dfrac{1}{60} = \dfrac{5}{12}$

$S_n = \dfrac{1}{4} + \dfrac{1}{12} + \dfrac{1}{24} + \cdots + \dfrac{1}{2n(n+1)} = \dfrac{n}{2n+2}$

$S_k = \dfrac{1}{4} + \dfrac{1}{12} + \dfrac{1}{24} + \cdots + \dfrac{1}{2k(k+1)} = \dfrac{k}{2k+2}$

$S_{k+1} = \dfrac{1}{4} + \dfrac{1}{12} + \dfrac{1}{24} + \cdots + \dfrac{1}{2k(k+1)} + \dfrac{1}{2(k+1)(k+2)} = \dfrac{k+1}{2k+4}$

Add $\dfrac{1}{2(k+1)(k+2)}$ to both sides of S_k:

$\dfrac{1}{4} + \dfrac{1}{12} + \dfrac{1}{24} + \cdots + \dfrac{1}{2k(k+1)} + \dfrac{1}{2(k+1)(k+2)} = \dfrac{k}{2k+2} + \dfrac{1}{2(k+1)(k+2)}$

Simplify the right-hand side:

$\dfrac{k}{2k+2} + \dfrac{1}{2(k+1)(k+2)} = \dfrac{k(k+2)+1}{2(k+1)(k+2)} = \dfrac{k^2+2k+1}{2(k+1)(k+2)} = \dfrac{(k+1)^2}{2(k+1)(k+2)} = \dfrac{k+1}{2k+4}$

If S_k is true, then S_{k+1} is true. The conjecture is proven.

45. Answers will vary.

46. The exponents begin with the exponent on $a+b$ and decrease by 1 in each successive term.

47. The exponents begin with 0, increase by 1 in each successive term, and end with the exponent on $a+b$.

48. The sum of the exponents is the exponent on $a+b$.

Section 11.5

Check Point Exercises

1. **a.** $\dbinom{6}{3} = \dfrac{6!}{3!(6-3)!}$

$= \dfrac{6!}{3!3!}$

$= \dfrac{5 \cdot 4}{1}$

$= 20$

b. $\dbinom{6}{0} = \dfrac{6!}{0!(6-0)!}$

$\qquad = \dfrac{6!}{6!}$

$\qquad = 1$

c. $\dbinom{8}{2} = \dfrac{8!}{2!(8-2)!}$

$\qquad = \dfrac{8!}{2!6!}$

$\qquad = \dfrac{8 \cdot 7}{2}$

$\qquad = 28$

d. $\dbinom{3}{3} = \dfrac{3!}{3!(3-3)!}$

$\qquad = \dfrac{3!}{3!0!}$

$\qquad = \dfrac{3!}{3!}$

$\qquad = 1$

2. $(x+1)^4 = \dbinom{4}{0}x^4 + \dbinom{4}{1}x^3 + \dbinom{4}{2}x^2 + \dbinom{4}{1}x + \dbinom{4}{0} = x^4 + 4x^3 + 6x^2 + 4x + 1$

3. $(x-2y)^5 = \dbinom{5}{0}x^5(-2y)^0 + \dbinom{5}{1}x^4(-2y)^1 + \dbinom{5}{2}x^3(-2y)^2 + \dbinom{5}{3}x^2(-2y)^3 + \dbinom{5}{4}x(-2y)^4 + \dbinom{5}{5}x^0(-2y)^5$

$\qquad = x^5 - 5x^4(2y) + 10x^3(4y^2) - 10x^2(8y^3) + 5x(16y^4) - 32y^5$

$\qquad = x^5 - 10x^4 y + 40x^3 y^2 - 80x^2 y^3 + 80xy^4 - 32y^5$

4. $(2x+y)^9$

$\qquad$ fifth term $= \dbinom{9}{4}(2x)^5 y^4$

$\qquad\qquad\quad = \dfrac{9!}{4!5!}(32x^5)y^4$

$\qquad\qquad\quad = 4032x^5 y^4$

Concept and Vocabulary Check 11.5

1. binomial

2. $\dfrac{8!}{2!6!}$

3. $\dfrac{n!}{r!(n-r)!}$

4. $\dbinom{5}{1}; \dbinom{5}{2}; \dbinom{5}{3}; \dbinom{5}{4}; \dbinom{5}{5}$

5. $\dbinom{n}{1}; \dbinom{n}{2}; \dbinom{n}{3}; \dbinom{n}{n}; n$

6. Binomial

7. $a^{n-r}b^r$

Exercise Set 11.5

1. $\dbinom{8}{3} = \dfrac{8!}{3!(8-3)!} = \dfrac{8\cdot7\cdot6}{3\cdot2\cdot1} = 56$

3. $\dbinom{12}{1} = \dfrac{12!}{1!11!} = 12$

5. $\dbinom{6}{6} = \dfrac{6!}{0!6!} = 1$

7. $\dbinom{100}{2} = \dfrac{100!}{2!98!} = \dfrac{100\cdot99}{2} = 4950$

9. $(x+2)^3 = \dbinom{3}{0}x^3 + \dbinom{3}{1}2x^2 + \dbinom{3}{2}4x + \dbinom{3}{3}8 = x^3 + 3x^2\cdot2 + 3x\cdot4 + 8 = x^3 + 6x^2 + 12x + 8$

11. $(3x+y)^3 = \dbinom{3}{0}27x^3 + \dbinom{3}{1}9x^2y + \dbinom{3}{2}3xy^2 + \dbinom{3}{3}y^3 = 27x^3 + 27x^2y + 9xy^2 + y^3$

13. $(5x-1)^3 = \dbinom{3}{0}125x^3 - \dbinom{3}{1}25x^2 + \dbinom{3}{2}5x - \dbinom{3}{3} = 125x^3 - 75x^2 + 15x - 1$

15. $(2x+1)^4 = \dbinom{4}{0}16x^4 - \dbinom{4}{1}8x^3 + \dbinom{4}{2}4x^2 + \dbinom{4}{3}2x + \dbinom{4}{4} = 16x^4 + 32x^3 + 24x^2 + 8x + 1$

17. $(x^2+2y)^4 = \dbinom{4}{0}(x^2)^4 + \dbinom{4}{1}(x^2)^3(2y) + \dbinom{4}{2}(x^2)^2(2y)^2 + \dbinom{4}{3}(x^2)^1(2y)^3 + \dbinom{4}{4}(2y)^4$

$\qquad = 1(x^8) + 4(x^6)(2y) + 6(x^4)(4y^2) + 4x^2(8y^3) + 1(16y^4)$

$\qquad = x^8 + 8x^6y + 24x^4y^2 + 32x^2y^3 + 16y^4$

19. $(y-3)^4 = \dbinom{4}{0}y^4 + \dbinom{4}{1}y^3(-3) + \dbinom{4}{2}y^2(-3)^2 + \dbinom{4}{3}y(-3)^3 + \dbinom{4}{4}(-3)^4$

$\qquad = y^4 + 4(y^3)(-3) + 6(y^2)(9) + 4(y)(-27) + 81$

$\qquad = y^4 - 12y^3 + 54y^2 - 108y + 81$

21. $(2x^3-1)^4 = \dbinom{4}{0}(2x^3)^4 + \dbinom{4}{1}(2x^3)^3(-1) + \dbinom{4}{2}(2x^3)^2(-1)^2 + \dbinom{4}{3}(2x^3)(-1)^3 + \dbinom{4}{4}(-1)^4$

$\qquad = 16x^{12} - 4(8x^9) + 6(4x^6) - 4(2x^3) + 1$

$\qquad = 16x^{12} - 32x^9 + 24x^6 - 8x^3 + 1$

23. $(c+2)^5 = \binom{5}{0}c^5 + \binom{5}{1}c^4(2) + \binom{5}{2}c^3(2^2) + \binom{5}{3}c^2(2^3) + \binom{5}{4}c(2^4) + \binom{5}{5}(2^5)$

$\qquad = c^5 + 5c^4(2) + 10c^3(4) + 10c^2(8) + 5c(16) + 32$

$\qquad = c^5 + 10c^4 + 40c^3 + 80c^2 + 80c + 32$

25. $(x-1)^5 = \binom{5}{0}x^5 - \binom{5}{1}x^4 + \binom{5}{2}x^3 - \binom{5}{3}x^2 + \binom{5}{4}x - \binom{5}{5} = x^5 - 5x^4 + 10x^3 - 10x^2 + 5x - 1$

27. $(3x-y)^5 = \binom{5}{0}(3x)^5 - \binom{5}{1}(3x)^4 y + \binom{5}{2}(3x)^3 y^2 - \binom{5}{3}(3x)^2 y^3 + \binom{5}{4}3xy^4 - \binom{5}{5}y^5$

$\qquad = (1)243x^5 - 5(81x^4)y + 10(27x^3)y^2 - 10(9x^2)y^3 + 5(3x)y^4 - (1)y^5$

$\qquad = 243x^5 - 405x^4 y + 270x^3 y^2 - 90x^2 y^3 + 15xy^4 - y^5$

29. $(2a+b)^6 = \binom{6}{0}(2a)^6 + \binom{6}{1}(2a)^5 b + \binom{6}{2}(2a)^4 b^2 + \binom{6}{3}(2a)^3 b^3 + \binom{6}{4}(2a)^2 b^4 + \binom{6}{5}(2a)b^5 + \binom{6}{6}b^6$

$\qquad = 64a^6 + 6(32a^5)b + 15(16a^4)b^2 + 20(8a^3)b^3 + 15(4a^2)b^4 + 6(2a)b^5 + b^6$

$\qquad = 64a^6 + 192a^5 b + 240a^4 b^2 + 160a^3 b^3 + 60a^2 b^4 + 12ab^5 + b^6$

31. $(x+2)^8 = \binom{8}{0}x^8 + \binom{8}{1}x^7 2 + \binom{8}{3}x^6(2)^2 + \cdots$

$\qquad = x^8 + 16x^7 + 112x^6 + \cdots$

33. $(x-2y)^{10} = \binom{10}{0}x^{10} - \binom{10}{1}x^9(2y) + \binom{10}{2}x^8(2y)^2 - \cdots$

$\qquad = x^{10} - 20x^9 y + 180x^8 y^2 - \cdots$

35. $(x^2+1)^{16} = \binom{16}{0}(x^2)^{16} + \binom{16}{1}(x^2)^{15} + \binom{16}{2}(x^2)^{14} + \cdots$

$\qquad = x^{32} + 16x^{30} + 120x^{28} + \cdots$

37. $(y^3-1)^{20} = \binom{20}{0}(y^3)^{20} - \binom{20}{1}(y^3)^{19} + \binom{20}{2}(y^3)^{18} - \cdots$

$\qquad = y^{60} - 20y^{57} + 190y^{54} - \cdots$

39. $(2x+y)^6$;   third term $= \binom{6}{2}(2x)^4(y)^2 = 15(16x^4 y^2) = 240x^4 y^2$

41. $(x-1)^9$;   fifth term $= \binom{9}{4}x^5(-1)^4 = 126x^5$

43. $(x^2+y^3)^8$;   sixth term $= \binom{8}{5}(x^2)^3(y^3)^5 = 56x^6 y^{15}$

45. $(x-\frac{1}{2})^9$;   fourth term $= \binom{9}{3}x^6\left(-\frac{1}{2}\right)^3 = 84x^6\left(-\frac{1}{8}\right) = -\frac{21}{2}x^6$

47. $\dbinom{22}{14}(x^2)^8 y^{14} = 319,770x^{16}y^{14}$

49.

$$\left(x^3 + x^{-2}\right)^4 = \dbinom{4}{0}\left(x^3\right)^4 + \dbinom{4}{1}\left(x^3\right)^3\left(x^{-2}\right) + \dbinom{4}{2}\left(x^3\right)^2\left(x^{-2}\right)^2 + \dbinom{4}{3}\left(x^3\right)^1\left(x^{-2}\right)^3 + \dbinom{4}{4}\left(x^{-2}\right)^4$$

$$= \frac{4!}{0!(4-0)!}x^{12} + \frac{4!}{1!(4-1)!}x^9 x^{-2} + \frac{4!}{2!(4-2)!}x^6 x^{-4} + \frac{4!}{3!(4-3)!}x^3 x^{-6} + \frac{4!}{4!(4-4)!}x^{-8}$$

$$= \frac{\cancel{4!}}{0!\,\cancel{4!}}x^{12} + \frac{4\cdot\cancel{3!}}{1!\,\cancel{3!}}x^7 + \frac{4\cdot 3\cdot\cancel{2!}}{2\cdot 1\cdot\cancel{2!}}x^2 + \frac{4\cdot\cancel{3!}}{\cancel{3!}\cdot 1!}x^{-3} + \frac{\cancel{4!}}{\cancel{4!}\cdot 0!}x^{-8}$$

$$= x^{12} + 4x^7 + 6x^2 + \frac{4}{x^3} + \frac{1}{x^8}$$

51.

$$\left(x^{\frac{1}{3}} - x^{-\frac{1}{3}}\right)^3 = \left(x^{\frac{1}{3}} + \left(-x^{-\frac{1}{3}}\right)\right)^3 = \dbinom{3}{0}\left(x^{\frac{1}{3}}\right)^3 + \dbinom{3}{1}\left(x^{\frac{1}{3}}\right)^2\left(-x^{-\frac{1}{3}}\right) +$$

$$+ \dbinom{3}{2}\left(x^{\frac{1}{3}}\right)^1\left(-x^{-\frac{1}{3}}\right)^2 + \dbinom{3}{3}\left(-x^{-\frac{1}{3}}\right)^3$$

$$= \frac{3!}{0!(3-0)!}x^1 + \frac{3!}{1!(3-1)!}x^{\frac{2}{3}}\cdot -x^{-\frac{1}{3}} + \frac{3!}{2!(3-2)!}x^{\frac{1}{3}}x^{-\frac{2}{3}} + \frac{3!}{3!(3-3)!}\cdot -x^{-1}$$

$$= \frac{\cancel{3!}}{0!\,\cancel{3!}}x + \frac{3\cdot\cancel{2!}}{1!\,\cancel{2!}}\cdot -x^{\frac{1}{3}} + \frac{3\cdot\cancel{2!}}{\cancel{2!}\cdot 1!}x^{-\frac{1}{3}} + \frac{\cancel{3!}}{\cancel{3!}\cdot 0!}\cdot -x^{-1}$$

$$= x - 3x^{\frac{1}{3}} + \frac{3}{x^{\frac{1}{3}}} - \frac{1}{x}$$

53. $f(x) = x^4 + 7$;

$$\frac{f(x+h)-f(x)}{h}$$

$$= \frac{(x+h)^4 + 7 - \left(x^4 + 7\right)}{h}$$

$$= \frac{\dbinom{4}{0}x^4 + \dbinom{4}{1}x^3 h + \dbinom{4}{2}x^2 h^2 + \dbinom{4}{3}xh^3 + \dbinom{4}{4}h^4 + 7 - x^4 - 7}{h}$$

$$= \frac{\dfrac{4!}{0!(4-0)!}x^4 + \dfrac{4!}{1!(4-1)!}x^3 h + \dfrac{4!}{2!(4-2)!}x^2 h^2 + \dfrac{4!}{3!(4-3)!}xh^3 + \dfrac{4!}{4!(4-4)!}h^4 - x^4}{h}$$

$$= \frac{\dfrac{\cancel{4!}}{0!\,\cancel{4!}}x^4 + \dfrac{4\cdot\cancel{3!}}{1!\,\cancel{3!}}x^3 h + \dfrac{4\cdot 3\cdot\cancel{2!}}{\cancel{2!}\cdot 2\cdot 1}x^2 h^2 + \dfrac{4\cdot\cancel{3!}}{\cancel{3!}1!}xh^3 + \dfrac{\cancel{4!}}{\cancel{4!}0!}h^4 - x^4}{h}$$

$$= \frac{\cancel{x^4} + 4x^3 h + 6x^2 h^2 + 4xh^3 + h^4 - \cancel{x^4}}{h}$$

$$= \frac{h(4x^3 + 6x^2 h + 4xh^2 + h^3)}{h}$$

$$= 4x^3 + 6x^2 h + 4xh^2 + h^3$$

55. Find the $(5+1) = 6^{th}$ term.

$$\binom{n}{r}a^{n-r}b^r = \binom{10}{5}\left(\frac{3}{x}\right)^{10-5}\left(\frac{x}{3}\right)^5 = \frac{10!}{5!(10-5)!}\left(\frac{3}{x}\right)^5\left(\frac{x}{3}\right)^5$$

$$= \frac{10 \cdot 9 \cdot 8 \cdot 7 \cdot 6 \cdot \cancel{5!}}{\cancel{5!} \cdot \cancel{5} \cdot 4 \cdot 3 \cdot \cancel{2} \cdot 1}\left(\frac{3}{x}\right)^5\left(\frac{x}{3}\right)^5 = 252 \cdot \frac{3^5}{x^5} \cdot \frac{x^5}{3^5} = 252$$

57. $(0.28 + 0.72)^5$

Third Term $(r = 2)$: $\binom{n}{r}a^{n-r}b^r = \binom{5}{2}0.28^{5-2}0.72^2 = \frac{5!}{2!(5-2)!}0.28^{5-2}0.72^2 = \frac{5!}{2!3!}0.28^3 0.72^2 \approx 0.1138$

59. – 67. Answers will vary.

69. $f_1(x) = (x+1)^4$

$f_2(x) = x^4$

$f_3(x) = x^4 + 4x^3$

$f_4(x) = x^4 + 4x^3 + 6x^2$

$f_5(x) = x^4 + 4x^3 + 6x^2 + 4x$

$f_6(x) = x^4 + 4x^3 + 6x^2 + 4x + 1$

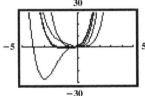

f_2, f_3, f_4, and f_5 are approaching $f_1 = f_6$.

71. $f_1(x) = (x-2)^4$

$$= \binom{4}{0}x^4 + \binom{4}{1}x^3(-2) + \binom{4}{2}x^2(-2)^2 + \binom{4}{3}x(-2)^3 + \binom{4}{4}(-2)^4$$

$$= x^4 + 4x^3(-2) + 6x^2(4) + 4x(-8) + 16$$

$$= x^4 - 8x^3 + 24x^2 - 32x + 16$$

73. makes sense

75. does not make sense; Explanations will vary. Sample explanation: $\binom{n}{0}$ and $\binom{n}{1}$ are the coefficients of the first and second term.

77. false; Changes to make the statement true will vary. A sample change is: The binomial expansion for $(a+b)^n$ contains $n+1$ terms.

79. false; Changes to make the statement true will vary. A sample change is: The sum of the binomial coefficients in $(a+b)^n$ is 2^n.

81. $\left(x^2+x+1\right)^3 = \left[x^2+(x+1)\right]^3$

$$= \binom{3}{0}\left(x^2\right)^3 + \binom{3}{1}\left(x^2\right)^2(x+1) + \binom{3}{2}x^2(x+1)^2 + \binom{3}{3}(x+1)^3$$

$$= x^6 + 3x^4(x+1) + 3x^2\left(x^2+2x+1\right) + x^3 + 3x^2 + 3x + 1$$

$$= x^6 + 3x^5 + 3x^4 + 3x^4 + 6x^3 + 3x^2 + x^3 + 3x^2 + 3x + 1$$

$$= x^6 + 3x^5 + 6x^4 + 7x^3 + 6x^2 + 3x + 1$$

83. $\dbinom{n}{r} = \dfrac{n!}{r!(n-r)!} = \dfrac{n!}{(n-r)!r!} = \dfrac{n!}{(n-r)!\left[n-(n-r)\right]!} = \dbinom{n}{n-r}$

85. **a.** $S_1 : (a+b)^1 = \dbinom{1}{0}a^1 + \dbinom{1}{1}a^{1-1}b = a+b$

b. $S_k : (a+b)^k = \dbinom{k}{0}a^k + \dbinom{k}{1}a^{k-1}b + \dbinom{k}{2}a^{k-2}b^2 + \cdots + \dbinom{k}{k-1}ab^{k-1} + \dbinom{k}{k}b^k$

$S_{k+1} : (a+b)^{k+1} = \dbinom{k+1}{0}a^{k+1} + \dbinom{k+1}{1}a^k b + \dbinom{k+1}{2}a^{k-1}b^2 + \cdots + \dbinom{k+1}{k}ab^k + \dbinom{k+1}{k+1}b^{k+1}$

c. $(a+b)(a+b)^k$

$(a+b)^{k+1} = \dbinom{k}{0}a^{k+1} + \dbinom{k}{0}a^k b + \dbinom{k}{1}a^k b + \dbinom{k}{1}a^{k-1}b^2 + \dbinom{k}{2}a^{k-1}b^2 + \dbinom{k}{2}a^{k-2}b^3 + \cdots$

$= \dbinom{k}{k-1}a^2 b^{k-1} + \dbinom{k}{k-1}ab^k + \dbinom{k}{k}ab^k + \dbinom{k}{k}b^{k+1}$

d. $(a+b)^{k+1} = \dbinom{k}{0}a^{k+1} + \left[\dbinom{k}{0}+\dbinom{k}{1}\right]a^k b + \left[\dbinom{k}{1}+\dbinom{k}{2}\right]a^{k-1}b^2 + \left[\dbinom{k}{2}+\dbinom{k}{3}\right]a^{k-2}b^3 + \cdots$

$+ \left[\dbinom{k}{k-1}+\dbinom{k}{k}\right]ab^k + \dbinom{k}{k}b^{k+1}$

e. $(a+b)^{k+1} = \dbinom{k}{0}a^{k+1} + \dbinom{k+1}{1}a^k b + \dbinom{k+1}{2}a^{k-1}b^2 + \dbinom{k+1}{3}a^{k-2}b^3 + \cdots + \dbinom{k+1}{k}ab^k + \dbinom{k}{k}b^{k+1}$

f. $\dbinom{k}{0} = \dbinom{k+1}{0}$ because both equal 1. $\dbinom{k}{k} = \dbinom{k+1}{k+1}$ also because both equal 1.

$S_{k+1} : (a+b)^{k+1} = \dbinom{k+1}{0}a^{k+1} + \dbinom{k+1}{1}a^k b + \dbinom{k+1}{2}a^{k-1}b^2 + \cdots + \dbinom{k+1}{k}ab^k + \dbinom{k+1}{k+1}b^{k+1}$

86. $\dfrac{n!}{(n-r)!} = \dfrac{20!}{(20-3)!} = \dfrac{20!}{17!} = \dfrac{20 \cdot 19 \cdot 18 \cdot 17!}{17!} = 20 \cdot 19 \cdot 18 = 6840$

87. $\dfrac{n!}{(n-r)!r!} = \dfrac{8!}{(8-3)!3!} = \dfrac{8!}{5!3!} = \dfrac{8 \cdot 7 \cdot 6 \cdot 5!}{3 \cdot 2 \cdot 1 \cdot 5!} = \dfrac{8 \cdot 7 \cdot 6}{3 \cdot 2 \cdot 1} = 56$

88. true

Section 11.6

Check Point Exercises

1. We use the Fundamental Counting Principal to find the number of ways a one-topping pizza can be ordered.

$\underline{\text{Size}} : \underline{\text{Crust}} : \underline{\text{Topping}}:$

$\quad 3 \;\times\; 4 \;\times\; 6 \;\;= 72$

There are 72 different ways of ordering a one-topping pizza.

2. We use the Fundamental Counting Principal to find the number of ways we can answer the questions.

$\underline{\text{Question \#1:}}\; \underline{\text{Question \#2:}}\; \underline{\text{Question \#3:}}\; \underline{\text{Question \#4:}}\; \underline{\text{Question \#5:}}\; \underline{\text{Question \#6:}}$

$\quad 3 \;\;\times\;\; 3 \;\;\times\;\; 3 \;\;\times\;\; 3 \;\;\times\;\; 3 \;\;\times\;\; 3 \;\;= 3^6 = 729$

There are 729 ways of answering the questions.

3. We use the Fundamental Counting Principal to find the number of different license plates that can be manufactured. Multiply the number of different letters, 26, for the first two places and the number of different digits, 10, for the next three places.

$26 \cdot 26 \cdot 10 \cdot 10 \cdot 10 = 26^2 \cdot 1000 = 676,000 \text{ plates}$

There are 676,000 different license plates possible.

4. Your group is choosing $r = 4$ officers from a group of $n = 7$ people.

The order in which the officers are chosen matters because the four officers to be chosen have different responsibilities. Thus, we are looking for the number of permutations of 7 things taken 4 at a time.

We use the formula $_nP_r = \dfrac{n!}{(n-r)!}$ with $n = 7$ and $r = 4$. $\;_7P_4 = \dfrac{7!}{(7-4)!} = \dfrac{7!}{3!} = 840.$

Thus, there are 840 different ways of filling the four offices.

5. Because you are using all six of your books in every possible arrangement, you are arranging $r = 6$ books from a group of $n = 6$ books. Thus, we are looking for the number of permutations of 6 things taken 6 at a time. We use the formula

$_nP_r = \dfrac{n!}{(n-r)!}$ with $n = 6$ and $r = 6$.

$_6P_6 = \dfrac{6!}{(6-6)!} = \dfrac{6!}{0!} = 6! = 720.$

There are 720 different possible permutations.

Thus, you can arrange the books in 720 ways.

6. **a.** The order does not matter; this problem involves combinations.

    b. Since what place each runner finishes matters, this problem involves permutations.

7. The order in which the four people are selected does not matter. This is a problem of selecting $r = 4$ people from a group of $n = 10$ people. We are looking for the number of combinations of 10 things taken 4 at a time. We use the formula

$_nC_r = \dfrac{n!}{(n-r)!\,r!}$ with $n = 10$ and $r = 4$.

$_{10}C_4 = \dfrac{10!}{(10-4)!4!} = \dfrac{10!}{6!4!} = \dfrac{10 \cdot 9 \cdot 8 \cdot 7 \cdot 6!}{6! \cdot 4 \cdot 3 \cdot 2 \cdot 1} = \dfrac{10 \cdot 9 \cdot 8 \cdot 7}{4 \cdot 3 \cdot 2 \cdot 1} = 210$

Thus, 210 committees of 4 people each can be found from 10 people at the conference on acupuncture.

8. Because the order in which the 4 cards are dealt does not matter, this is a problem involving combinations. We are looking for the number of combinations of $n = 16$ cards drawn $r = 4$ at a time. We use the formula $_nC_r = \dfrac{n!}{(n-r)!\ r!}$ with $n = 16$ and $r = 4$.

$$_{16}C_4 = \frac{16!}{(16-4)!4!} = \frac{16!}{12!4!} = \frac{16 \cdot 15 \cdot 14 \cdot 13 \cdot 12!}{12! \cdot 4 \cdot 3 \cdot 2 \cdot 1} = 1820$$

Thus, there are 1820 different 4-card hands possible.

Concept and Vocabulary Check 11.6

1. $M \cdot N$

2. multiplying; Fundamental Counting

3. $\dfrac{n!}{(n-r)!}$

4. $\dfrac{n!}{(n-r)!\ r!}$

5. $r!$

Exercise Set 11.6

1. $_9P_4 = \dfrac{9!}{5!} = 3024$

3. $_8P_5 = \dfrac{8!}{3!} = 8 \cdot 7 \cdot 6 \cdot 5 \cdot 4 = 6720$

5. $_6P_6 = \dfrac{6!}{0!} = 720$

7. $_8P_0 = \dfrac{8!}{8!} = 1$

9. $_9C_5 = \dfrac{9!}{4!5!} = \dfrac{9 \cdot 8 \cdot 7 \cdot 6}{4 \cdot 3 \cdot 2 \cdot 1} = \dfrac{3 \cdot 7 \cdot 6}{1} = 126$

11. $_{11}C_4 = \dfrac{11!}{7!4!} = \dfrac{11 \cdot 10 \cdot 9 \cdot 8}{4 \cdot 3 \cdot 2 \cdot 1} = \dfrac{11 \cdot 10 \cdot 3}{1} = 330$

13. $_7C_7 = \dfrac{7!}{0!7!} = 1$

15. $_5C_0 = \dfrac{5!}{5!0!} = 1$

17. combinations; The order in which the volunteers are chosen does not matter.

19. permutations; The order of the letters matters because ABCD is not the same as BADC.

21. $\dfrac{{}_7P_3}{3!} - {}_7C_3 = \dfrac{\frac{7!}{(7-3)!}}{3!} - \dfrac{7!}{(7-3)!3!} = \dfrac{\frac{7!}{4!}}{3!} - \dfrac{7!}{4!3!} = \dfrac{7!}{4!3!} - \dfrac{7!}{4!3!} = 0$

23. $1 - \dfrac{{}_3P_2}{{}_4P_3} = 1 - \dfrac{\frac{3!}{(3-2)!}}{\frac{4!}{(4-3)!}} = 1 - \dfrac{\frac{3!}{1!}}{\frac{4!}{1!}} = 1 - \dfrac{3!}{4!} = 1 - \dfrac{3!}{4 \cdot 3!} = 1 - \dfrac{1}{4} = \dfrac{3}{4}$

25. $\dfrac{{}_7C_3}{{}_5C_4} - \dfrac{98!}{96!} = \dfrac{\frac{7!}{(7-3)!3!}}{\frac{5!}{(5-4)!4!}} - \dfrac{98 \cdot 97 \cdot 96!}{96!} = \dfrac{\frac{7!}{4!3!}}{\frac{5!}{1!4!}} - 95067 = \dfrac{\frac{7 \cdot 6 \cdot 5 \cdot 4!}{4!3 \cdot 2 \cdot 1}}{\frac{5 \cdot 4!}{1!4!}} - 9506 = \dfrac{35}{5} - 9506 = 7 - 9506 = -9499$

27. $\dfrac{{}_4C_2 \cdot {}_6C_1}{{}_{18}C_3} = \dfrac{\frac{4!}{(4-2)!2!} \cdot \frac{6!}{(6-1)!1!}}{\frac{18!}{(18-3)!3!}} = \dfrac{\frac{4!}{2!2!} \cdot \frac{6!}{5!1!}}{\frac{18!}{15!3!}} = \dfrac{\frac{4 \cdot 3 \cdot 2!}{2!2 \cdot 1} \cdot \frac{6 \cdot 5!}{5!1!}}{\frac{18 \cdot 17 \cdot 16 \cdot 15!}{15!3 \cdot 2 \cdot 1}} = \dfrac{36}{816} = \dfrac{3}{68}$

29. $9 \cdot 3 = 27$ ways

31. $2 \cdot 4 \cdot 5 = 40$ ways

33. $3^5 = 243$ ways

35. $8 \cdot 2 \cdot 9 = 144$ area codes

37. $5 \cdot 4 \cdot 3 \cdot 2 \cdot 1 \cdot 1 = 120$ ways

39. $1 \cdot 3 \cdot 2 \cdot 1 \cdot 1 = 6$ paragraphs

41. ${}_{10}P_3 = \dfrac{10!}{7!3!} = 10 \cdot 9 \cdot 8 = 720$ ways

43. ${}_{13}P_7 = \dfrac{13!}{6!} = 13 \cdot 12 \cdot 11 \cdot 10 \cdot 9 \cdot 8 \cdot 7$
 $= 8,648,640$ ways

45. ${}_6P_3 = \dfrac{6!}{3!} = 6 \cdot 5 \cdot 4 = 120$ ways

47. ${}_9P_5 = \dfrac{9!}{4!} = 9 \cdot 8 \cdot 7 \cdot 6 \cdot 5 = 15,120$ lineups

49. ${}_6C_3 = \dfrac{6!}{3!3!} = \dfrac{6 \cdot 5 \cdot 4}{3 \cdot 2 \cdot 1} = 20$ ways

51. $_{12}C_4 = \dfrac{12!}{8!4!} = \dfrac{12\cdot11\cdot10\cdot9}{4\cdot3\cdot2\cdot1}$
$=495$ collections

53. $_{17}C_8 = \dfrac{17!}{9!8!} = \dfrac{17\cdot16\cdot15\cdot14\cdot13\cdot12\cdot11\cdot10}{8\cdot7\cdot6\cdot5\cdot4\cdot3\cdot2\cdot1}$
$=24,310$ groups

55. $_{53}C_6 = \dfrac{53!}{47!6!} = 22,957,480$ selections

57. $_6P_4 = \dfrac{6!}{2!} = 6\cdot5\cdot4\cdot3 = 360$ ways

59. $_{13}C_6 = \dfrac{13!}{7!6!} = \dfrac{13\cdot12\cdot11\cdot10\cdot9\cdot8}{6\cdot5\cdot4\cdot3\cdot2\cdot1}$
$=1716$ ways

61. $_{20}C_3 = \dfrac{20!}{17!3!} = \dfrac{20\cdot19\cdot18}{3\cdot2\cdot1} = 1140$ ways

63. $_7P_4 = \dfrac{7!}{3!} = 840$ passwords

65. $_{15}P_3 = \dfrac{15!}{12!} = 15\cdot14\cdot13 = 2730$ cones

67. $_6P_6 = \dfrac{6!}{0!} = 6! = 720$ rankings

69. $_6C_3 = \dfrac{6!}{3!3!} = \dfrac{6\cdot5\cdot4}{3\cdot2\cdot1} = 20$ ways

71. $_4P_4 = \dfrac{4!}{0!} = 4! = 24$ ways

73. – 81. Answers will vary.

83. makes sense

85. does not make sense; Explanations will vary. Sample explanation: Since order matters use permutations.

87. false; Changes to make the statement true will vary. A sample change is: The number of ways to choose four questions out of ten questions is $_{10}C_4$.

89. true

91. $5\cdot5\cdot4\cdot4\cdot3\cdot3\cdot2\cdot2\cdot1\cdot1 = 14,400$ ways

93. $_{10}C_8 \cdot_5 C_3 = \dfrac{10!}{(10-8)!8!}\cdot\dfrac{5!}{(5-3)!3!}$
$= 45\cdot10 = 450$
They can be chosen in 450 ways.

95. $\dfrac{4}{6}$ or $\dfrac{2}{3}$ are less than 5.

96. $\dfrac{2}{6}$ or $\dfrac{1}{3}$ are not less than 5.

97. 2, 4, 5, and 6 are even or greater than three. This is a fraction of $\dfrac{4}{6}$ or $\dfrac{2}{3}$.

Section 11.7

Check Point Exercises

1. **a.** $P(\text{positive test}) = \dfrac{\text{\# women w/positive test}}{\text{total number of women}}$

    $= \dfrac{7664}{100,000}$

    $= \dfrac{479}{6250}$

    ≈ 0.077

    b. $P(\text{positive test}) = \dfrac{\text{\# women w/breast cancer and positive test}}{\text{total number of women w/breast cancer}}$

    $= \dfrac{720}{800}$

    $= \dfrac{9}{10}$

    $= 0.9$

    c. $P(\text{breast cancer}) = \dfrac{\text{\# women w/breast cancer and positive test}}{\text{total number of women w/positive test}}$

    $= \dfrac{720}{7664}$

    $= \dfrac{45}{479}$

    $= 0.094$

2. The sample space of equally likely outcomes is $S = \{1, 2, 3, 4, 5, 6\}$. There are six outcomes in the sample space, so $n(S) = 6$. The event of getting a number greater than 4 can be represented by $E = \{5, 6\}$. There are two outcomes in this event, so $n(E) = 2$.

    The probability of rolling a number greater than 4 is $P(E) = \dfrac{n(E)}{n(S)} = \dfrac{2}{6} = \dfrac{1}{3}$.

3. We have $n(S) = 36$. The phrase "getting a sum of 5" describes the event $E = \{(1,4),(2,3),(3,2),(4,1)\}$. This event has 4 outcomes, so $n(E) = 4$. Thus, the probability of getting a sum of 5 is $P(E) = \dfrac{n(E)}{n(S)} = \dfrac{4}{36} = \dfrac{1}{9}$.

4. Let E be the event of being dealt a king. Because there are 4 kings in the deck, the event of being dealt a king can occur in 4 ways, i.e., $n(E) = 4$. With 52 cards in the deck, $n(S) = 52$.

    The probability of being dealt a king is $P(E) = \dfrac{n(E)}{n(S)} = \dfrac{4}{52} = \dfrac{1}{13}$.

5. Because the order of the six numbers does not matter, this is a situation involving combinations. With one lottery ticket, there is only one way of winning so $n(E) = 1$. Using the combinations formula $_nC_r = \dfrac{n!}{(n-r)!\,r!}$ to find the number of outcomes in the sample space, we are selecting $r = 6$ numbers from a collection of $n = 49$ numbers.

$_{49}C_6 = \dfrac{49!}{43!\,6!} = 13{,}983{,}816$ So $n(S) = 13{,}983{,}816$. If a person bought one lottery ticket, the probability of winning was

$$P(E) = \frac{n(E)}{n(S)} = \frac{1}{13{,}983{,}816}$$

The probability of winning the state lottery was 0.0000000715.

6. $P(\text{does not live in North America}) = 1 - P(\text{lives in North America})$

$$= 1 - \frac{550}{7000}$$
$$= \frac{7000}{7000} - \frac{550}{7000}$$
$$= \frac{6450}{7000}$$
$$= \frac{129}{140} \approx 0.921$$

7. We find the probability that either of these mutually exclusive events will occur by adding their individual probabilities.

$$P(4 \text{ or } 5) = P(4) + P(5) = \frac{1}{6} + \frac{1}{6} = \frac{2}{6} = \frac{1}{3}$$

The probability of selecting a 4 or a 5 is $\dfrac{1}{3}$.

8. It is possible for the pointer to land on a number that is odd and less than 5. Two of the numbers, 1 and 3, are odd and less than 5. These events are not mutually exclusive. The probability of landing on a number that is odd and less than 5 is
$P(\text{odd or less than } 5) = P(\text{odd}) + P(\text{less than } 5) - P(\text{odd and less than } 5)$

$$= \frac{4}{8} + \frac{4}{8} - \frac{2}{8}$$
$$= \frac{6}{8}$$
$$= \frac{3}{4}$$

The probability that the pointer will stop on an odd number or a number less than 5 is $\dfrac{3}{4}$.

9.   **a.** These events are not mutually exclusive.
$P(\text{married or female}) = P(\text{married}) + P(\text{female}) - P(\text{married and female})$

$$= \frac{130}{242} + \frac{124}{242} - \frac{65}{242}$$
$$= \frac{189}{242} \approx 0.78$$

   b. These events are mutually exclusive.
$P(\text{divorced or widowed}) = P(\text{divorced}) + P(\text{widowed})$

$$= \frac{24}{242} + \frac{14}{242}$$
$$= \frac{38}{242} \approx 0.16$$

10. The wheel has 38 equally likely outcomes and 2 are green.

Thus, the probability of a green occurring on a play is $\dfrac{2}{38}$, or $\dfrac{1}{19}$.

The result that occurs on each play is independent of all previous results.

Thus, P (green and green) $= P$ (green) $\cdot P$ (green) $= \dfrac{1}{19} \cdot \dfrac{1}{19} = \dfrac{1}{361} \approx 0.00277$.

The probability of green occurring on two consecutive plays is $\dfrac{1}{361}$ or 0.00277..

11. If two or more events are independent, we can find the probability of them all occurring by multiplying the probabilities.

The probability of a baby boy is $\dfrac{1}{2}$.

Thus, the probability of having four boys in a row is $P\left(4 \text{ boys in a row}\right) = \dfrac{1}{2} \cdot \dfrac{1}{2} \cdot \dfrac{1}{2} \cdot \dfrac{1}{2}$

$$= \dfrac{1}{16}.$$

Concept and Vocabulary Check 11.7

1. empirical

2. sample space

3. $P(E)$; number of outcomes in E; total number of possible outcomes

4. 52; Hearts; diamonds; clubs; spades

5. 1; combinations

6. $1 - P(E)$; $1 - P(\text{not } E)$

7. mutually exclusive; $P(A) + P(B)$

8. $P(A) + P(B) - P(A \text{ and } B)$

9. independent; $P(A) \cdot P(B)$

Exercise Set 11.7

1. $P(\text{divorced}) = \dfrac{\text{number of persons divorced}}{\text{total number of U.S. adults}} = \dfrac{24}{242} = \dfrac{12}{121} \approx 0.10$

3. $P(\text{female}) = \dfrac{\text{number of females}}{\text{total number of U.S. adults}} = \dfrac{124}{242} = \dfrac{62}{141} \approx 0.51$

5. $P(\text{widowed male}) = \dfrac{\text{number of widowed males}}{\text{total number of U.S. adults}} = \dfrac{3}{242} \approx 0.01$

7. $P(\text{selecting a woman from the divorced population}) = \dfrac{\text{number of divorced women}}{\text{number of persons divorced}} = \dfrac{14}{24} = \dfrac{7}{14} \approx 0.58$

9. $P(\text{selecting a married man from the adult male population}) = \dfrac{\text{number of married men}}{\text{number of males}} = \dfrac{65}{118} \approx 0.55$

11. $P(R) = \dfrac{n(E)}{n(S)} = \dfrac{1}{6}$

13. $P(E) = \dfrac{n(E)}{n(S)} = \dfrac{3}{6} = \dfrac{1}{2}$

15. $P(E) = \dfrac{n(E)}{n(S)} = \dfrac{2}{6} = \dfrac{1}{3}$

17. $P(E) = \dfrac{n(E)}{n(S)} = \dfrac{4}{52} = \dfrac{1}{13}$

19. $P(E) = \dfrac{n(E)}{n(S)} = \dfrac{12}{52} = \dfrac{3}{13}$

21. $P(E) = \dfrac{n(E)}{n(S)} = \dfrac{1}{4}$

23. $P(E) = \dfrac{n(E)}{n(S)} = \dfrac{7}{8}$

25. $P(E) = \dfrac{n(E)}{n(S)} = \dfrac{3}{36} = \dfrac{1}{12}$

27. Buying 1 ticket:

    $P(E) = \dfrac{n(E)}{n(S)} = \dfrac{1}{{}_{51}C_6} = \dfrac{1}{18,009,460}$

    Buying 100 tickets:

    $P(E) = \dfrac{100}{18,009,460} = \dfrac{5}{900,473}$

29. a. $\begin{aligned}{}_{52}C_5 &= \dfrac{52!}{47!5!}\\ &= \dfrac{52 \cdot 51 \cdot 50 \cdot 49 \cdot 48}{5 \cdot 4 \cdot 3 \cdot 2 \cdot 1} = 2,598,960\end{aligned}$

    b. ${}_{13}C_5 = \dfrac{13!}{8!5!} = \dfrac{13 \cdot 12 \cdot 11 \cdot 10 \cdot 9}{5 \cdot 4 \cdot 3 \cdot 2 \cdot 1} = 1287$

    c. $P(E) = \dfrac{n(E)}{n(S)} = \dfrac{1287}{2,598,960} \approx 0.0005$

31. $P(\text{not completed 4 years of college}) = 1 - P(\text{completed 4 years of college}) = 1 - \dfrac{45}{174} = \dfrac{43}{58}$

33. $P(\text{completed H.S. or less than 4 yrs college}) = P(\text{completed H.S}) + P(\text{less than 4 yrs college})$

$$= \frac{56}{174} + \frac{44}{174} = \frac{100}{174} = \frac{50}{87}$$

35. $P(\text{completed 4 yrs H.S. or man}) = P(\text{completed 4 yrs H.S.}) + P(\text{man}) - P(\text{man who completed 4 yrs H.S})$

$$= \frac{56}{174} + \frac{82}{174} - \frac{25}{174} = \frac{113}{174}$$

37. $P(\text{not king}) = 1 - P(\text{king}) = 1 - \frac{4}{52} = 1 - \frac{1}{13} = \frac{12}{13}$

39. $P(2 \text{ or } 3) = P(2) + P(3) = \frac{4}{52} + \frac{4}{52} = \frac{8}{52} = \frac{2}{13}$

41. $P(7 \text{ or red card}) = P(7) + P(\text{red card}) - P(7 \text{ and red}) = \frac{4}{52} + \frac{26}{52} - \frac{2}{52} = \frac{28}{52} = \frac{7}{13}$

43. $P(\text{odd or less than 6}) = P(\text{odd}) + P(\text{less than 6}) - P(\text{odd \# less than 6}) = \frac{4}{8} + \frac{5}{8} - \frac{3}{8} = \frac{6}{8} = \frac{3}{4}$

45. $P(\text{professor or male}) = P(\text{professor}) + P(\text{male}) - P(\text{male professor}) = \frac{19}{40} + \frac{22}{40} - \frac{8}{40} = \frac{33}{40}$

47. $P(2 \text{ and } 3) = P(2) \cdot P(3) = \frac{1}{6} \cdot \frac{1}{6} = \frac{1}{36}$

49. $P(\text{even and greater than 2}) = P(\text{even}) \cdot P(\text{greater than 2}) = \frac{3}{6} \cdot \frac{4}{6} = \frac{1}{3}$

51. $P(\text{all heads}) = \frac{1}{2} \cdot \frac{1}{2} \cdot \frac{1}{2} \cdot \frac{1}{2} \cdot \frac{1}{2} \cdot \frac{1}{2} = \left(\frac{1}{2}\right)^6 = \frac{1}{64}$

53. **a.** $P(\text{hit 2 yrs in a row}) = P(\text{hit in 1}^{st} \text{ year and 2}^{nd} \text{ year}) = P(\text{hit 1}^{st} \text{ year}) \cdot P(\text{hit 2}^{nd} \text{ year}) = \frac{1}{16} \cdot \frac{1}{16} = \frac{1}{256}$

    b. $P(\text{hit 3 yrs in a row}) = \frac{1}{16} \cdot \frac{1}{16} \cdot \frac{1}{16} = \frac{1}{4096}$

    c. First find the probability that South Florida will not be hit by a major hurricane in a single year:

$$P(\text{not hit}) = 1 - \frac{1}{16} = \frac{15}{16}$$

$$P(\text{not hit in next 10 years}) = \left(\frac{15}{16}\right)^{10} \approx 0.524$$

    d. $P(\text{hit at least once}) = 1 - P(\text{hit none}) = 1 - \left(\frac{15}{16}\right)^{10} \approx 0.476$

55. – 63. Answers will vary.

65. does not make sense; Explanations will vary. Sample explanation: The probability of a Democrat winning and the probability of a Republican winning are not necessarily equal. For instance, it is possible that the probability of a Democrat winning is 0.6 and the probability of a Republican winning is 0.4.

67. makes sense

69. Answers will vary.

71. **a.** P(Democrat who is not a business major)

$$= \frac{\text{\# of students who are Democrats but not business majors}}{\text{\# of students}}$$

$$= \frac{29-5}{50} = \frac{24}{50} = \frac{12}{25}$$

b. P(neither Democrat nor business major)

$= 1 - P$(Democrat or business major)

$= 1 - \left(P(\text{Democrat}) + P(\text{business major}) - P(\text{Democrat and business major}) \right)$

$= 1 - \left(\frac{29}{50} + \frac{11}{50} - \frac{5}{50} \right) = 1 - \frac{35}{50} = \frac{15}{50} = \frac{3}{10}$

73. **a.** The first person can have any birthday in the year. The second person can have all but one birthday.

b. $\dfrac{365}{365} \cdot \dfrac{364}{365} \cdot \dfrac{363}{365} \approx 0.99$

c. $100\% - 99\% = 0.01$

d. $\dfrac{365}{365} \cdot \dfrac{364}{365} \cdot \dfrac{363}{365} \cdots \dfrac{346}{365} \approx 0.59$

$1 - 0.59 = 0.41$

e. With 23 people, the probability that at least two people have the same birthday is

$$P(E) = 1 - \frac{365}{365} \cdot \frac{364}{365} \cdot \frac{363}{365} \cdots \frac{342}{365}$$
$$\approx 1 - 0.4927 \approx 0.5073$$

Chapter 11 Review Exercises

1. $a_n = 7n - 4$

   $a_1 = 7 - 4 = 3$

   $a_2 = 14 - 4 = 10$

   $a_3 = 21 - 4 = 17$

   $a_4 = 28 - 4 = 24$

   First four terms: 3, 10, 17, 24.

2. $a_n = (-1)^n \dfrac{n+2}{n+1}$

   $a_1 = (-1)^1 \dfrac{1+2}{1+1} = -\dfrac{3}{2}$

   $a_2 = (-1)^2 \dfrac{2+2}{2+1} = \dfrac{4}{3}$

   $a_3 = (-1)^3 \dfrac{3+2}{3+1} = -\dfrac{5}{4}$

   $a_4 = (-1)^4 \dfrac{4+2}{4+1} = \dfrac{6}{5}$

   First four terms: $-\dfrac{3}{2}, \dfrac{4}{3}, -\dfrac{5}{4}, \dfrac{6}{5}$.

3. $a_n = \dfrac{1}{(n-1)!}$

   $a_1 = \dfrac{1}{0!} = 1$

   $a_2 = \dfrac{1}{1!} = 1$

   $a_3 = \dfrac{1}{2!} = \dfrac{1}{2}$

   $a_4 = \dfrac{1}{3!} = \dfrac{1}{6}$

   First four terms: $1, 1, \dfrac{1}{2}, \dfrac{1}{6}$.

4. $a_n = \dfrac{(-1)^{n+1}}{2^n}$

   $a_1 = \dfrac{(-1)^2}{2^1} = \dfrac{1}{2}$

   $a_2 = \dfrac{(-1)^3}{2^2} = -\dfrac{1}{4}$

   $a_3 = \dfrac{(-1)^4}{2^3} = \dfrac{1}{8}$

   $a_4 = \dfrac{(-1)^5}{2^4} = -\dfrac{1}{16}$

   First four terms: $\dfrac{1}{2}, -\dfrac{1}{4}, \dfrac{1}{8}, -\dfrac{1}{16}$.

5. $a_1 = 9$ and $a_n = \dfrac{2}{3a_{n-1}}$

   $a_1 = 9$

   $a_2 = \dfrac{2}{3 \cdot 9} = \dfrac{2}{27}$

   $a_3 = \dfrac{2}{3} \cdot \dfrac{27}{2} = \dfrac{54}{6} = 9$

   $a_4 = \dfrac{2}{3 \cdot 9} = \dfrac{2}{27}$

   First four terms: $9, \dfrac{2}{27}, 9, \dfrac{2}{27}$.

6. $a_1 = 4$ and $a_n = 2a_{n-1} + 3$

   $a_1 = 4$

   $a_2 = 2 \cdot 4 + 3 = 8 + 3 = 11$

   $a_3 = 2 \cdot 11 + 3 = 22 + 3 = 25$

   $a_4 = 2 \cdot 25 + 3 = 50 + 3 = 53$

   First four terms: 4, 11, 25, and 53.

7. $\dfrac{40!}{4!\,38!} = \dfrac{40 \cdot 39 \cdot 38!}{4 \cdot 3 \cdot 2 \cdot 1 \cdot 38!} = 65$

8. $\displaystyle\sum_{i=1}^{5}\left(2i^2-3\right)=(2-3)+\left(2\cdot2^2-3\right)+\left(2\cdot3^2-3\right)+\left(2\cdot4^2-3\right)+\left(2\cdot5^2-3\right)$

$$=-1+5+15+29+47$$
$$=95$$

9. $\displaystyle\sum_{i=0}^{4}(-1)^{i+1}i!=(-1)^1 0!+(-1)^2 1!+(-1)^3 3!+(-1)^4 4!$

$$=-1+1-2+6-24$$
$$=-20$$

10. $\dfrac{1}{3}+\dfrac{2}{4}+\dfrac{3}{5}+\cdots+\dfrac{15}{17}=\displaystyle\sum_{i=1}^{15}\dfrac{i}{i+2}$

11. $4^3+5^3+6^3+\cdots+13^3=\displaystyle\sum_{i=1}^{10}(i+3)^3$

12. $a_1=7,\ d=4$
First six terms: 7, 11, 15, 19, 23, 27.

13. $a_1=-4,\ d=-5$
First six terms: $-4,-9,-14,-19,-24,-29$.

14. $a_1=\dfrac{3}{2},\ d=-\dfrac{1}{2}$

First six terms: $\dfrac{3}{2},\ 1,\ \dfrac{1}{2},\ 0,\ -\dfrac{1}{2},\ -1$.

15. $a_{n+1}=a_n+5,\ a_1=-2$
First six terms: $-2,\ 3,\ 8,\ 13,\ 18,\ 23$.

16. $a_1=5,\ d=3$
$a_n=5+(n-1)3$
$a_6=5+(5)3=20$

17. $a_1=-8,\ d=-2$
$a_n=-8+(n-1)(-2)$
$a_{12}=-8+11(-2)=-30$

18. $a_1=14,\ d=-4$
$a_n=14+(n-1)(-4)$
$a_{14}=14+(13)(-4)=-38$

19. $-7,-3,1,5,\ldots$
$d=-3-(-7)=4$
$a_n=-7+(n-1)(4)$
$a_n=4n-11$
$a_{20}=4(20)-11$
$a_{20}=69$

20. $a_1 = 200, d = -20$

$a_n = 200 + (n-1)(-20)$

$a_n = 220 - 20n$

$a_{20} = 220 - 20(20)$

$a_{20} = -180$

21. $a_n = a_{n-1} - 5, a_1 = 3$

$d = -5$

$a_n = 3 + (n-1)(-5) = 3 - 5n + 5$

$a_n = 8 - 5n$

$a_{20} = 8 - 5(20) = -92$

22. 5, 12, 19, 26, ...

$d = 7$

$a_n = 5 + (n-1)(7)$

$a_{22} = 5 + 21(7) = 152$

$S_{22} = \dfrac{22}{2}(5 + 152) = 1727$

23. $-6, -3, 0, 3, ...$

$d = 3$

$a_n = -6 + (n-1)3$

$a_{15} = -6 + (14)3 = 36$

$S_{15} = \dfrac{15}{2}(-6 + 36) = 225$

24. $3 + 6 + 9 + \ldots + 300$

$S_{100} = \dfrac{100}{2}(3 + 300) = 15{,}150$

25. $\displaystyle\sum_{i=1}^{16}(3i + 2)$

$a_1 = 3 + 2 = 5$

$a_{16} = 3(16) + 2 = 50$

$S_{16} = \dfrac{16}{2}(5 + 50) = 440$

26. $\displaystyle\sum_{i=1}^{25}(-2i + 6)$

$a_1 = -2 + 6 = 4$

$a_{25} = -2(25) + 6 = -44$

$S_{25} = \dfrac{25}{2}(4 - 44) = -500$

27. $\displaystyle\sum_{i=1}^{30} -5i$

$a_1 = -5$

$a_{30} = -5(30) = -150$

$S_{30} = \dfrac{30}{2}(-5 - 150) = -2325$

28. **a.** $a_n = 34.5 + (n-1)(-0.3)$

$= 34.5 - 0.3n + 0.3$

$= 34.8 - 0.3n$

b. $a_n = 34.8 - 0.3n$

$a_{56} = 34.8 - 0.3(56)$

$= 18$

The model projects wives will devote 18 hours per week in 2020.

29. $a_n = 31{,}500 + (n-1)2300$

$a_{10} = 31{,}500 + (9)2300 = 52{,}200$

$S_{10} = \dfrac{10}{2}(31{,}500 + 52{,}200) = 418{,}500$

The total salary is $418, 500.

30. $a_n = 25 + (n-1)$

$a_{35} = 25 + 34 = 59$

$S_{35} = \dfrac{35}{2}(25 + 59) = 1470$

There are 1470 seats.

31. $a_1 = 3, r = 2$

First five terms: 3, 6, 12, 24, 48.

32. $a_1 = \dfrac{1}{2}, r = \dfrac{1}{2}$

First five terms: $\dfrac{1}{2}, \dfrac{1}{4}, \dfrac{1}{8}, \dfrac{1}{16}, \dfrac{1}{32}$.

33. $a_1 = 16, r = -\dfrac{1}{2}$

First five terms: 16, -8, 4, -2, 1.

34. $a_n = -5a_{n-1}, a_1 = -1$

First five terms: -1, 5, -25, 125, -625.

35. $a_1 = 2, r = 3$

$a_n = 2 \cdot 3^{n-1}$

$a_7 = 2 \cdot 3^6 = 1458$

36. $a_1 = 16,\ r = \dfrac{1}{2}$

$a_n = 16\left(\dfrac{1}{2}\right)^{n-1}$

$a_6 = 16\left(\dfrac{1}{2}\right)^5 = \dfrac{16}{32} = \dfrac{1}{2}$

37. $a_1 = -3,\ r = 2$

$a_n = -3 \cdot 2^{n-1}$

$a_5 = -3 \cdot 2^4 = -48$

38. $1, 2, 4, 8, \dots$

$a_1 = 1,\ r = \dfrac{2}{1} = 2$

$a_n = 2^{n-1}$

$a_8 = 2^7 = 128$

39. $100, 10, 1, \dfrac{1}{10}, \dots$

$a_1 = 100,\ r = \dfrac{10}{100} = \dfrac{1}{10}$

$a_n = 100\left(\dfrac{1}{10}\right)^{n-1}$

$a_8 = 100\left(\dfrac{1}{10}\right)^7 = \dfrac{1}{100,000}$

40. $12, -4, \dfrac{4}{3}, -\dfrac{4}{9} \dots$

$a_1 = 12,\ r = -\dfrac{4}{12} = -\dfrac{1}{3}$

$a_n = 12\left(-\dfrac{1}{3}\right)^{n-1}$

$a_8 = 12\left(-\dfrac{1}{3}\right)^7 = -\dfrac{4}{729}$

41. $5, -15, 45, -135, \dots$

$r = \dfrac{-15}{5} = -3$

$S_{15} = \dfrac{5\left[1 - (-3)^{15}\right]}{1 - (-3)} = 17,936,135$

42. $r = \dfrac{1}{2},\ a_1 = 8$

$S_{78} = \dfrac{8\left[1 - \left(\dfrac{1}{2}\right)^7\right]}{1 - \dfrac{1}{2}} = -16\left(1 - \dfrac{1}{128}\right)$

$\phantom{S_{78}} = -16\left(-\dfrac{127}{128}\right) = \dfrac{127}{8}$

43. $S_6 = \dfrac{5\left(1 - 5^6\right)}{1 - 5} = \dfrac{5(-15624)}{-4} = 19,530$

44. $\displaystyle\sum_{i=1}^{7} 3(-2)^i$

$a_1 = -6,\ r = -2$

$S_7 = \dfrac{-6\left[1 - (-2)^7\right]}{1 - (-2)} = \dfrac{-6(129)}{3} = -258$

45. $\displaystyle\sum_{i=1}^{5} 2\left(\tfrac{1}{4}\right)^{i-1}$

$a_1 = 2,\ r = \dfrac{1}{4}$

$S_5 = \dfrac{2\left[1 - \left(\dfrac{1}{4}\right)^5\right]}{1 - \dfrac{1}{4}} = \dfrac{2\left(\dfrac{1023}{1024}\right)}{\dfrac{3}{4}} = \dfrac{341}{128}$

46. $a_1 = 9,\ r = \dfrac{1}{3}$

$S_\infty = \dfrac{9}{1 - \dfrac{1}{3}} = \dfrac{9}{\dfrac{2}{3}} = 9 \cdot \dfrac{3}{2} = \dfrac{27}{2}$

47. $a_1 = 2,\ r = -\dfrac{1}{2}$

$S_\infty = \dfrac{2}{1 - \left(-\dfrac{1}{2}\right)} = \dfrac{2}{\dfrac{3}{2}} = \dfrac{4}{3}$

48. $a_1 = -6,\ r = -\dfrac{2}{3}$

$S_\infty = \dfrac{-6}{1 - \left(-\dfrac{2}{3}\right)} = \dfrac{-6}{\dfrac{5}{3}} = -\dfrac{18}{5}$

49. $r = 0.8$

$S_\infty = \dfrac{4}{1 - 0.8} = 20$

50. $0.\overline{6} = 0.6 + 0.06 + 0.006 + \cdots$

$a_1 = \dfrac{6}{10}, \ r = \dfrac{1}{10}$

$S_\infty = \dfrac{\dfrac{6}{10}}{1 - \dfrac{1}{10}} = \dfrac{\dfrac{6}{10}}{\dfrac{9}{10}} = \dfrac{6}{9} = \dfrac{2}{3}$

51. $0.\overline{47} = 0.47 + 0.0047 + 0.000047 + \cdots$

$a_1 = \dfrac{47}{100}, \ r = \dfrac{1}{100}$

$S_\infty = \dfrac{\dfrac{47}{100}}{1 - \dfrac{1}{100}} = \dfrac{\dfrac{47}{100}}{\dfrac{99}{100}} = \dfrac{47}{99}$

52.   **a.**   Divide each value by the previous value:

$\dfrac{18.80}{18.44} \approx 1.02$

$\dfrac{18.44}{18.15} \approx 1.02$

$\dfrac{18.15}{17.86} \approx 1.02$

$\dfrac{17.86}{17.58} \approx 1.02$

$\dfrac{17.58}{17.30} \approx 1.02$

$\dfrac{17.30}{17.03} \approx 1.02$

$\dfrac{17.03}{16.76} \approx 1.02$

$\dfrac{16.76}{16.50} \approx 1.02$

$\dfrac{16.50}{16.24} \approx 1.02$

$\dfrac{16.24}{15.98} \approx 1.02$

The population is increasing geometrically with $r = 1.02$.

b. $a_n = 15.98(1.02)^n$

c.   2080 is 8 decades after 2000 so $n = 8$.

$a_n = 15.98(1.02)^{n-1}$

$a_{31} = 15.98(1.02)^{31-1}$

$\quad \approx 28.95$

In 2030, the model predicts Florida's population will be 28.95 million.

53. $a_1 = 32,000, \ r = 1.06$

$a_6 = 32,000(1.06)^5 \approx \$42,823$

The sixth year salary is \$42, 823.

$S_6 = \dfrac{32,000\left(1 - 1.06^6\right)}{1 - 1.06}$

$= \dfrac{32,000\left(1 - 1.06^6\right)}{-0.06}$

$\approx 223,210$

The total salary paid is \$223, 210.

54. **a.** $A = \dfrac{P\left[\left(1 + \frac{r}{n}\right)^{nt} - 1\right]}{\frac{r}{n}}$

$P = \$520, \ r = 0.06, \ n = 1, \ t = 20$

$A = \dfrac{\$520\left[\left(1 + \frac{0.06}{1}\right)^{1 \cdot 20} - 1\right]}{\frac{0.06}{1}} = \dfrac{\$520\left[(1.06)^{20} - 1\right]}{0.06} \approx \$19,129$

The value of the annuity will be \$19,129.

b. Interest = Value of annuity − Total deposits

$\approx \ \$19,129 \ - \ \$520 \cdot 20$

$\approx \ \$8729$

55. **a.** $A = \dfrac{P\left[\left(1 + \frac{r}{n}\right)^{nt} - 1\right]}{\frac{r}{n}}$

$P = 100, \ r = 0.055, \ n = 12, \ t = 30$

$A = \dfrac{\$100\left[\left(1 + \frac{0.055}{12}\right)^{12 \cdot 30} - 1\right]}{\frac{0.055}{12}} \approx \$91,361$

The value of the IRA will be \$91,361.

b. Interest = Value of IRA − Total deposits

$\approx \ \$91,361 \ - \ \$100 \cdot 12 \cdot 30$

$\approx \ \$55,361$

56. $4(0.7) + 4(0.7)^2 + \cdots; \ r = 0.7$

$S_\infty = \dfrac{4(0.7)}{1 - 0.7} = 9.\overline{3}$

The total spending is $\$9\frac{1}{3}$ million.

57. $S_1 : 5 = \dfrac{5(1)(1+1)}{2}$

$5 = \dfrac{5(2)}{2}$

$5 = 5$ is true.

$S_k : 5 + 10 + 15 + \cdots + 5k = \dfrac{5k(k+1)}{2}$

$S_{k+1} : 5 + 10 + 15 + \cdots + 5k + 5(k+1)$

$\qquad = \dfrac{5(k+1)(k+2)}{2}$

Add $5(k+1)$ to both sides of S_k :

$5 + 10 + 15 + \cdots + 5k + 5(k+1)$

$\qquad = \dfrac{5k(k+1)}{2} + 5k(k+1)$

Simplify the right-hand side:

$\dfrac{5k(k+1)}{2} + 5(k+1) = \dfrac{5k(k+1) + 10(k+1)}{2}$

$\qquad\qquad = \dfrac{(5k+10)(k+1)}{2}$

$\qquad\qquad = \dfrac{5(k+1)(k+2)}{2}$

If S_k is true, then S_{k+1} is true.

The statement is true for all n.

58. $S_1 : 1 = \dfrac{4^1 - 1}{3}$

$1 = \dfrac{3}{3}$

$1 = 1$ is true.

$S_k : 1 + 4 + 4^2 + \cdots + 4^{k-1} = \dfrac{4^k - 1}{3}$

$S_{k+1} : 1 + 4 + 4^2 + \cdots + 4^{k-1} + 4^k = \dfrac{4^{k+1} - 1}{3}$ Add 4^k to both sides of S_k :

$S_k : 1 + 4 + 4^2 + \cdots + 4^{k-1} = \dfrac{4^k - 1}{3}$

$1 + 4 + 4^2 + \cdots + 4^{k-1} + 4^k = \dfrac{4^k - 1}{3} + 4^k$

Simplify the right-hand side:

$\dfrac{4^k - 1}{3} + 4^k = \dfrac{4^k - 1 + 3 \cdot 4^k}{3}$

$\qquad\qquad = \dfrac{4 \cdot 4^k - 1}{3}$

$\qquad\qquad = \dfrac{4^{k+1} - 1}{3}$

If S_k is true, then S_{k+1} is true.

The statement is true for all n.

59. $S_1 : 2 = 2(1)^2$

$2 = 2$ is true.

$S_k : 2 + 6 + 10 + \cdots + (4k - 2) = 2k^2$

$S_{k+1} : 2 + 6 + 10 + \cdots + (4k - 2) + (4k + 2) = 2(k + 1)^2$

Add $(4k + 2)$ to both sides of S_k:

$2 + 6 + 10 + \cdots + (4k - 2) + (4k + 2) = 2k^2 + (4k + 2)$

Simplify the right-hand side:

$2k^2 + 4k + 2 = 2(k^2 + 2k + 1)$

$\qquad\qquad\quad = 2(k + 1)^2$

If S_k is true, then S_{k+1} is true. The statement is true for all n.

60. $S_1 : 1 \cdot 3 = \dfrac{1(1 + 1)[2(1) + 7]}{6}$

$3 = \dfrac{2 \cdot 9}{6}$

$3 = \dfrac{18}{6}$

$3 = 3$ is true.

$S_k : 1 \cdot 3 + 2 \cdot 4 + 3 \cdot 5 + \cdots + k(k + 2) = \dfrac{k(k + 1)(2k + 7)}{6}$

$S_{k+1} : 1 \cdot 3 + 2 \cdot 4 + 3 \cdot 5 + \cdots + k(k + 2) + (k + 1)(k + 3) = \dfrac{(k + 1)(k + 2)(2k + 9)}{6}$

Add $(k + 1)(k + 3)$ to both sides of S_k:

$1 \cdot 3 + 2 \cdot 4 + 3 \cdot 5 + \cdots + k(k + 2) + (k + 1)(k + 3) = \dfrac{k(k + 1)(2k + 7)}{6} + (k + 1)(k + 3)$

Simplify the right-hand side:

$= \dfrac{k(k + 1)(2k + 7)}{6} + (k + 1)(k + 3)$

$= \dfrac{k(k + 1)(2k + 7) + 6(k + 1)(k + 3)}{6}$

$= \dfrac{(k + 1)[k(2k + 7) + 6(k + 3)]}{6}$

$= \dfrac{(k + 1)(2k^2 + 13k + 18)}{6}$

$= \dfrac{(k + 1)(k + 2)(2k + 9)}{6}$

If S_k is true, then S_{k+1} is true. The statement is true for all n.

61. S_1: 2 is a factor of $1^2 + 5(1) = 6$ since $6 = 2 \cdot 3$.

S_k : 2 is a factor of $k^2 + 5k$.

S_{k+1} : 2 is a factor of $(k+1)^2 + 5(k+1)$.

$$
\begin{aligned}
(k+1)^2 + 5(k+1) &= k^2 + 2k + 1 + 5k + 5 \\
&= k^2 + 7k + 6 \\
&= k^2 + 5k + 2(k+3) \\
&= (k^2 + 5k) + 2(k+3)
\end{aligned}
$$

Because we assume S_k is true, we know 2 is a factor of $k^2 + 5k$. Since 2 is a factor of $2(k+3)$, we conclude 2 is a factor of the sum $(k^2 + 5k) + 2(k+3)$. If S_k is true, then S_{k+1} is true. The statement is true for all n.

62. $\dbinom{11}{8} = \dfrac{11!}{3!8!} = \dfrac{11 \cdot 10 \cdot 9}{3 \cdot 2 \cdot 1} = 165$

63. $\dbinom{90}{2} = \dfrac{90!}{88!2!} = \dfrac{90 \cdot 89}{2 \cdot 1} = 4005$

64. $(2x+1)^3 = \dbinom{3}{0}(2x)^3 + \dbinom{3}{1}(2x)^2 \cdot 1 + \dbinom{3}{2}(2x)1^2 + \dbinom{3}{3}1^3$

$\qquad = 8x^3 + 3(4x^2) + 3(2x) + 1$

$\qquad = 8x^3 + 12x^2 + 6x + 1$

65. $(x^2 - 1)^4 = \dbinom{4}{0}(x^2)^4 + \dbinom{4}{1}(x^2)^3(-1) + \dbinom{4}{2}(x^2)^2(-1)^2 + \dbinom{4}{3}x^2(-1)^3 + \dbinom{4}{4}(-1)^4$

$\qquad = x^8 - 4x^6 + 6x^4 - 4x^2 + 1$

66. $(x+2y)^5 = \dbinom{5}{0}x^5 + \dbinom{5}{1}x^4(2y) + \dbinom{5}{2}x^3(2y)^2 + \dbinom{5}{3}x^2(2y)^3 + \dbinom{5}{4}x(2y)^4 + \dbinom{5}{5}(2y)^5$

$\qquad = x^5 + 5(2)x^4y + 10(4)x^3y^2 + 10(8)x^2y^3 + 5(16)xy^4 + 32y^5$

$\qquad = x^5 + 10x^4y + 40x^3y^2 + 80x^2y^3 + 80xy^4 + 32y^5$

67. $(x-2)^6 = \dbinom{6}{0}x^6 + \dbinom{6}{1}x^5(-2) + \dbinom{6}{2}x^4(-2)^2 + \dbinom{6}{3}x^3(-2)^3 + \dbinom{6}{4}x^2(-2)^4 + \dbinom{6}{5}x(-2)^5 \dbinom{6}{6}(-2)^6$

$\qquad = x^6 + 6x^5(-2) + 15x^4(4) + 20x^3(-8) + 15x^2(16) + 6x(-32) + 64$

$\qquad = x^6 - 12x^5 + 60x^4 - 160x^3 + 240x^2 - 192x + 64$

68. $(x^2 + 3)^8 = \dbinom{8}{0}(x^2)^8 + \dbinom{8}{1}(x^2)^7 3 + \dbinom{8}{2}(x^2)^6 3^2 + \cdots$

$\qquad = x^{16} + 8x^{14} 3 + 28x^{12} 9 + \cdots$

$\qquad = x^{16} + 24x^{14} + 252x^{12} + \cdots$

69. $(x-3)^9 = \binom{9}{0}x^9 + \binom{9}{1}x^8(-3) + \binom{9}{2}x^7(-3)^2 - \cdots$

$\qquad = x^9 + 9(-3)x^8 + 36(9)x^7 - \cdots$

$\qquad = x^9 - 27x^8 + 324x^7 - \cdots$

70. $(x+2)^5$

fourth term $= \binom{5}{3}x^2(2)^3$

$\qquad = 10(8)x^2 = 80x^2$

71. $(2x-3)^6$

fifth term $= \binom{6}{4}(2x)^2(-3)^4$

$\qquad = 15(4x^2)(81) = 4860x^2$

72. $_8P_3 = \dfrac{8!}{5!} = 8 \cdot 7 \cdot 6 = 336$

73. $_9P_5 = \dfrac{9!}{4!} = 9 \cdot 8 \cdot 7 \cdot 6 \cdot 5 = 15,120$

74. $_8C_3 = \dfrac{8!}{5!3!} = \dfrac{8 \cdot 7 \cdot 6}{3 \cdot 2 \cdot 1} = 56$

75. $_{13}C_{11} = \dfrac{13!}{2!11!} = \dfrac{13 \cdot 12}{2 \cdot 1} = 78$

76. $4 \cdot 5 = 20$ choices

77. $3^5 = 243$ possibilities

78. $_{15}P_4 = \dfrac{15!}{11!} = 15 \cdot 14 \cdot 13 \cdot 12 = 32,760$ ways

79. $_{20}C_4 = \dfrac{20!}{16!4!} = \dfrac{20 \cdot 19 \cdot 18 \cdot 17}{4 \cdot 3 \cdot 2 \cdot 1} = 4845$ ways

80. $_{20}C_3 = \dfrac{20!}{17!3!} = \dfrac{20 \cdot 19 \cdot 18}{3 \cdot 2 \cdot 1} = 1140$ sets

81. $_{20}P_4 = \dfrac{20!}{16!}$

$\qquad = 20 \cdot 19 \cdot 18 \cdot 17$

$\qquad = 116,280$ ways

82. $5! = 120$ ways

83. $P(\text{public college}) = \dfrac{252}{350} = \dfrac{18}{25}$

84. $P(\text{not from high-income family}) = 1 - P(\text{from high-income family}) = 1 - \dfrac{50}{350} = \dfrac{350}{350} - \dfrac{50}{350} = \dfrac{300}{350} = \dfrac{6}{7}$

85. $P(\text{from middle-income family or high-income family}) = \dfrac{160 + 50}{350} = \dfrac{210}{350} = \dfrac{3}{5}$

86. $P(\text{attended private college or is from a high income family})$

$= P(\text{private college}) + P(\text{high income family}) - P(\text{attended private college and is from a high income family})$

$= \dfrac{98}{350} + \dfrac{50}{350} - \dfrac{28}{350} = \dfrac{120}{350} = \dfrac{12}{35}$

87. $P(\text{selecting a student from a low-income family from among those attending public college}) = \dfrac{120}{252} = \dfrac{10}{21}$

88. $P(\text{selecting a student that attends a private college from among those in middle-income families}) = \dfrac{50}{160} = \dfrac{5}{16}$

89. $P(\text{less than 5}) = P(\text{rolling a 1, 2, 3, 4})$

$= \dfrac{4}{6} = \dfrac{2}{3}$

90. $P(\text{less than 3 or greater than 4})$

$= P(\text{less than 3}) + P(\text{greater than 4})$

$= P(\text{rolling a 1, 2}) + P(\text{rolling a 5, 6})$

$= \dfrac{2}{6} + \dfrac{2}{6} = \dfrac{4}{6} = \dfrac{2}{3}$

91. $P(E) = \dfrac{4}{52} + \dfrac{4}{52} = \dfrac{8}{52} = \dfrac{2}{13}$

92. $P(E) = \dfrac{4}{52} + \dfrac{26}{52} - \dfrac{2}{52} = \dfrac{28}{52} = \dfrac{7}{13}$

93. $P(\text{not yellow}) = 1 - P(\text{yellow}) = 1 - \dfrac{1}{6} = \dfrac{5}{6}$

94. $P(\text{red or greater than 3}) = \dfrac{3}{6} + \dfrac{3}{6} - \dfrac{1}{6} = \dfrac{5}{6}$

95. $P(\text{green, then less than 4}) = \dfrac{2}{6} \cdot \dfrac{3}{6} = \dfrac{6}{36} = \dfrac{1}{6}$

96. a. $P(E) = \dfrac{n(E)}{n(S)} = \dfrac{1}{{}_{20}C_5} = \dfrac{1}{15,504}$

b. $P(E) = \dfrac{100}{15,504} = \dfrac{25}{3876}$

97. $P(E) = \left(\dfrac{1}{2}\right)^5 = \dfrac{1}{32}$

98. a. $(0.2)^2 = 0.04$

b. $(0.2)^3 = 0.008$

c. $(1 - 0.2)^4 = (0.8)^4 = 0.4096$

Chapter 11 Test

1. $a_n = \dfrac{(-1)^{n+1}}{n^2}$

$a_1 = \dfrac{(-1)^2}{1^2} = 1$

$a_2 = \dfrac{(-1)^3}{2^2} = -\dfrac{1}{4}$

$a_3 = \dfrac{(-1)^4}{3^2} = \dfrac{1}{9}$

$a_4 = \dfrac{(-1)^5}{4^2} = -\dfrac{1}{16}$

$a_5 = \dfrac{(-1)^6}{5^2} = \dfrac{1}{25}$

First five terms: $1, -\dfrac{1}{4}, \dfrac{1}{9}, -\dfrac{1}{16}, \dfrac{1}{25}$.

2. $\displaystyle\sum_{i=1}^{5} \left(i^2 + 10\right) = 11 + 14 + 19 + 26 + 35 = 105$

3. $\displaystyle\sum_{i=1}^{20} (3i - 4)$

$a_1 = 3 - 4 = -1$

$d = 3$

$a_n = -1 + (n-1)3$

$a_{20} = -1 + (19)3 = 56$

$S_{20} = \dfrac{20}{2}(-1 + 56) = 550$

4. $\sum\limits_{i=1}^{15}(-2)^i$

$a_1 = -2, \; r = -2$

$S_{15} = \dfrac{-2\left[1-(-2)^{15}\right]}{1-(-2)} = -21{,}846$

5. $\dbinom{9}{2} = \dfrac{9!}{7!2!} = \dfrac{9 \cdot 8}{2 \cdot 1} = 36$

6. $_{10}P_3 = \dfrac{10!}{7!} = 10 \cdot 9 \cdot 8 = 720$

7. $_{10}C_3 = \dfrac{10!}{7!3!} = \dfrac{10 \cdot 9 \cdot 8}{3 \cdot 2 \cdot 1} = 120$

8. $\dfrac{2}{3}+\dfrac{3}{4}+\dfrac{4}{5}+\cdots+\dfrac{21}{22} = \sum\limits_{i=1}^{20}\dfrac{i+1}{i+2}$

9. $4, 9, 14, 19, \ldots$

$a_1 = 4, \; d = 5$

$a_n = 4+(n-1)\cdot 5 = 4+5n-1$

$a_n = 5n-1$

$a_{12} = 5(12)-1 = 59$

10. $16, 4, 1, \dfrac{1}{4}, \cdots$

$a_1 = 16, \; r = \dfrac{1}{4}$

$a_n = 16\left(\dfrac{1}{4}\right)^{n-1}$

$a_{12} = 16\left(\dfrac{1}{4}\right)^{11} = \dfrac{1}{262{,}144}$

11. $7, -14, 28, -56, \ldots$

$a_1 = 7, \; r = -2$

$S_{10} = \dfrac{7\left[1-(-2)^{10}\right]}{1-(-2)} = \dfrac{7(-1023)}{3} = -2387$

12. $-7, -14, -21, -28, \ldots$

$a_1 = -7, \; d = -7$

$a_n = -7+(n-1)(-7)$

$a_{10} = -7+9(-7) = -70$

$S_{10} = \dfrac{10}{2}(-7-70) = -385$

13. $4 + \dfrac{4}{2} + \dfrac{4}{2^2} + \dfrac{4}{2^3} + \cdots$

$r = \dfrac{1}{2}$

$S_\infty = \dfrac{4}{1 - \dfrac{1}{2}} = 8$

14. $0.\overline{73} = 0.73 + 0.0073 + 0.000073 + \cdots$

$a_1 = \dfrac{73}{100},\ r = \dfrac{1}{100}$

$S_\infty = \dfrac{\dfrac{73}{100}}{1 - \dfrac{1}{100}} = \dfrac{\dfrac{73}{100}}{\dfrac{99}{100}} = \dfrac{73}{99}$

15. $a_1 = 30{,}000,\ r = 1.04$

$S_8 = \dfrac{30{,}000\left[1 - (1.04)^8\right]}{1 - 1.04} \approx 276{,}426.79$

The total salary is \$276,427.

16. $S_1 : 1 = \dfrac{1[3(1) - 1]}{2}$

$1 = \dfrac{2}{2}$

$1 = 1$ is true.

$S_k : 1 + 4 + 7 + \cdots + (3k - 2) = \dfrac{k(3k - 1)}{2}$

$S_{k+1} : 1 + 4 + 7 + \cdots + (3k - 2) + (3k + 1) = \dfrac{(k + 1)(3k + 2)}{2}$

Add $(3k + 1)$ to both sides of S_k:

$1 + 4 + 7 + \cdots + (3k - 2) + (3k + 1) = \dfrac{k(3k - 1)}{2} + (3k + 1)$

Simplify the right-hand side:

$\dfrac{k(3k - 1)}{2} + (3k + 1) = \dfrac{k(3k - 1) + 2(3k + 1)}{2}$

$= \dfrac{3k^2 + 5k + 2}{2}$

$= \dfrac{(k + 1)(3k + 2)}{2}$

If S_k is true, then S_{k+1} is true. The statement is true for all n.

17. $\left(x^2 - 1\right)^5 = \binom{5}{0}\left(x^2\right)^5 + \binom{5}{1}\left(x^2\right)^4(-1) + \binom{5}{2}\left(x^2\right)^3(-1)^2 + \binom{5}{3}\left(x^2\right)^2(-1)^3 + \binom{5}{4}x^2(-1)^4 + \binom{5}{5}(-1)^5$

$= x^{10} - 5x^8 + 10x^6 - 10x^4 + 5x^2 - 1$

18. $\left(x+y^2\right)^8$

First Term $\quad \dbinom{n}{r-1}a^{n-r+1}b^{r-1} = \dbinom{8}{1-1}x^{8-1+1}\left(y^2\right)^{1-1} = \dbinom{8}{0}x^8\left(y^2\right)^0 = \dfrac{8!}{0!(8-0)!}x^8 \cdot 1 = \dfrac{\cancel{8!}}{0!\cancel{8!}}x^8$

$\qquad\qquad\qquad = x^8$

Second Term $\quad \dbinom{n}{r-1}a^{n-r+1}b^{r-1} = \dbinom{8}{2-1}x^{8-2+1}\left(y^2\right)^{2-1} = \dbinom{8}{1}x^7\left(y^2\right)^1 = \dfrac{8!}{1!(8-1)!}x^7y^2 = \dfrac{8\cdot\cancel{7!}}{1\cdot\cancel{7!}}x^7y^2$

$\qquad\qquad\qquad = 8x^7y^2$

Third Term $\quad \dbinom{n}{r-1}a^{n-r+1}b^{r-1} = \dbinom{8}{3-1}x^{8-3+1}\left(y^2\right)^{3-1} = \dbinom{8}{2}x^6\left(y^2\right)^2 = \dfrac{8!}{2!(8-2)!}x^6y^4 = \dfrac{8\cdot7\cdot\cancel{6!}}{2\cdot1\cdot\cancel{6!}}x^6y^4$

$\qquad\qquad\qquad = 28x^6y^4$

$x^8 + 8x^7y^2 + 28x^6y^4 + \cdots$

19. $\quad _{11}P_3 = \dfrac{11!}{8!} = 11\cdot10\cdot9 = 990$ ways

20. $\quad _{10}C_4 = \dfrac{10!}{6!4!} = \dfrac{10\cdot9\cdot8\cdot7}{4\cdot3\cdot2\cdot1} = 210$ sets

21. Four digits are open: $10^4 = 10,000$

22. $P(\text{not brown eyes}) = 1 - P(\text{brown eyes})$

$\qquad\qquad\qquad\quad = 1 - \dfrac{40}{50}$

$\qquad\qquad\qquad\quad = \dfrac{60}{100} = \dfrac{3}{5}$

23. $P(\text{brown eyes or blue eyes})$

$\quad = P(\text{brown eyes}) + P(\text{blue eyes})$

$= \dfrac{40}{100} + \dfrac{38}{100} = \dfrac{78}{100} = \dfrac{39}{50}$

24. $P(\text{female or green eyes})$

$\quad = P(\text{female}) + P(\text{green eyes})$

$\qquad\quad - P(\text{female with green eyes})$

$= \dfrac{50}{100} + \dfrac{22}{100} - \dfrac{12}{100} = \dfrac{60}{100} = \dfrac{3}{5}$

25. $P(\text{male, given blue eyes}) = \dfrac{18}{38} = \dfrac{9}{19}$

26. $\quad _{15}C_6 = \dfrac{15!}{9!6!} = \dfrac{15\cdot14\cdot13\cdot12\cdot11\cdot10}{6\cdot5\cdot4\cdot3\cdot2} = 5005$

$P(E) = \dfrac{50}{5005} = \dfrac{10}{1001}$

27. $P(E) = \dfrac{26}{52} + \dfrac{12}{52} - \dfrac{6}{52} = \dfrac{32}{52} = \dfrac{8}{13}$

28. $P(E) = \dfrac{25}{50} + \dfrac{20}{50} - \dfrac{15}{50} = \dfrac{30}{50} = \dfrac{3}{5}$

29. $P(E) = \left(\dfrac{1}{4}\right)^4 = \dfrac{1}{256}$

30. $P(E) = \dfrac{2}{8} \cdot \dfrac{2}{8} = \dfrac{1}{16}$

Cumulative Review Exercises (Chapters 1–11)

1. domain: $[-4,1)$; range: $(-\infty, 2]$

2. maximum of 2 at $x = -2$

3. decreasing interval: $(-2,1)$

4. neither

5. $f(-3) = 1$ and $f(-1) = 1$

6. $(f \circ f)(-4) = f(f(-4)) = f(0) = 0$

7. $f(x) \to -\infty$ as $x \to 1^-$

8.

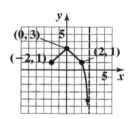

$g(x) = f(x - 2) + 1$

9.

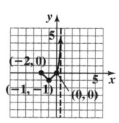

$h(x) = -f(2x)$

10. $-2(x - 5) + 10 = 3(x + 2)$
$-2x + 10 + 10 = 3x + 6$
$14 = 5x$
$x = \dfrac{14}{5}$

The solution set is $\left\{\dfrac{14}{5}\right\}$.

11. $3x^2 - 6x + 2 = 0$

$x = \dfrac{6 \pm \sqrt{36 - 24}}{6}$

$= \dfrac{6 \pm \sqrt{12}}{6}$

$= \dfrac{6 \pm 2\sqrt{3}}{6}$

$= \dfrac{3 \pm \sqrt{3}}{3}$

The solution set is $\left\{\dfrac{3 + \sqrt{3}}{3}, \dfrac{3 - \sqrt{3}}{3}\right\}$.

12. $\log_2 x + \log_2 (2x - 3) = 1$
$\log_2 x(2x - 3) = 1$
$x(2x - 3) = 2$
$2x^2 - 3x - 2 = 0$
$(2x + 1)(x - 2) = 0$
$2x + 1 = 0 \qquad$ or $\qquad x - 2 = 0$
$x = -\dfrac{1}{2} \qquad\qquad\qquad x = 2$

$x = -\dfrac{1}{2}$ does not check since $\log_2\left(-\dfrac{1}{2}\right)$ does not exist.
The solution set is $\{2\}$.

13. $x^{1/2} - 6x^{1/4} + 8 = 0$
Let $t = x^{1/4}$.
$t^2 - 6t + 8 = 0$
$(t - 2)(t - 4) = 0$
$t - 2 = 0 \quad$ or $\quad t - 4 = 0$
$t = 2 \qquad\qquad t = 4$
$x^{1/4} = 2 \qquad\quad x^{1/4} = 4$
$x = 16 \qquad\qquad x = 256$
The solution set is $\{16, 256\}$.

14. $\sqrt{2x+4} - \sqrt{x+3} - 1 = 0$

$$\left(\sqrt{2x+4}\right)^2 = \left(\sqrt{x+3}+1\right)^2$$
$$2x+4 = (x+3)+2\sqrt{x+3}+1$$
$$x = 2\sqrt{x+3}$$
$$x^2 = 4(x+3)$$
$$x^2 - 4x - 12 = 0$$
$$(x-6)(x+2) = 0$$
$$x - 6 = 0 \quad \text{or} \quad x+2 = 0$$
$$x = 6 \qquad\qquad x = -2$$

$x = -2$ does not check.
The solution set is $\{6\}$.

15. $|2x+1| \le 1$
$$-1 \le 2x+1 \le 1$$
$$-2 \le 2x \le 0$$
$$-1 \le x \le 0 \text{ or } [-1, 0]$$
The solution set is $\{x \mid -1 \le x \le 0\}$ or $[-1, 0]$.

16.
$$6x^2 - 6 < 5x$$
$$6x^2 - 5x - 6 < 0$$
$$6x^2 - 5x - 6 = 0$$
$$(3x+2)(2x-3) = 0$$
$$3x+2 = 0 \text{ or } 2x-3 = 0$$
$$x = -\frac{2}{3} \qquad x = \frac{3}{2}$$

The test intervals are $\left(-\infty, -\frac{2}{3}\right), \left(-\frac{2}{3}, \frac{3}{2}\right)$, and $\left(\frac{3}{2}, \infty\right)$. Testing a point in each interval shows that the solution is $\left(-\frac{2}{3}, \frac{3}{2}\right)$.

17. $\dfrac{x-1}{x+3} \le 0$

The test intervals are $(-\infty, -3), (-3, 1)$ and $(1, \infty)$.
Testing a point in each interval shows that the solution is $(-3, 1]$.

18. $30e^{0.7x} = 240$
$$e^{0.7x} = 8$$
$$\ln e^{0.7x} = \ln 8$$
$$0.7x = \ln 8$$
$$x = \frac{\ln 8}{0.7} \approx 2.9706$$

The solution set is $\left\{\dfrac{\ln 8}{0.7}\right\}$ or $\{2.9706\}$.

19. $2x^3 + 3x^2 - 8x + 3 = 0$
$p: \pm 1, \pm 3$
$q: \pm 1, \pm 2$
$\dfrac{p}{q}: \pm 1, \pm 3, \pm\dfrac{1}{2}, \pm\dfrac{3}{2}$

$$\begin{array}{r|rrrr} 1 & 2 & 3 & -8 & 3 \\ & & 2 & 5 & -3 \\ \hline & 2 & 5 & -3 & 0 \end{array}$$

$$(x-1)(2x^2 + 5x - 3) = 0$$
$$(x-1)(2x-1)(x+3) = 0$$
$$x = 1 \text{ or } x = \frac{1}{2} \text{ or } x = -3$$

The solution set is $\left\{-3, \dfrac{1}{2}, 1\right\}$.

20. $4x^2 + 3y^2 = 48$
$3x^2 + 2y^2 = 35$
Multiply equation 1 by –2.
Multiply equation 2 by 3.
$$-8x^2 - 6y^2 = -96$$
$$\underline{9x^2 + 6y^2 = 105}$$
Add: $\qquad x^2 = 9$
$$x = \pm 3$$
Let $x = -3$:
$$4(-3)^2 + 3y^2 = 48$$
$$36 + 3y^2 = 48$$
$$3y^2 = 12$$
$$y^2 = 4$$
$$y = \pm 2$$
Let $x = 3$:
$$4(3)^2 + 3y^2 = 48$$
$$36 + 3y^2 = 48$$
$$3y^2 = 12$$
$$y^2 = 4$$
$$y = \pm 2$$
The solution set is
$\{(3, 2), (3, -2), (-3, 2), (-3, -2)\}$.

21.
$$x - 2y + z = 16$$
$$2x - y - z = 14$$
$$3x + 5y - 4z = -10$$

$$\begin{bmatrix} 1 & -2 & 1 & | & 16 \\ 0 & -1 & -1 & | & 14 \\ 3 & 5 & -4 & | & -10 \end{bmatrix} \begin{matrix} \\ -2R_1 + R_2 \\ -3R_1 + R_3 \end{matrix}$$

$$\begin{bmatrix} 1 & -2 & 1 & | & 16 \\ 0 & 3 & -3 & | & -18 \\ 0 & 11 & -7 & | & -58 \end{bmatrix} \begin{matrix} \\ \frac{1}{3}R_2 \\ \\ \end{matrix}$$

$$\begin{bmatrix} 1 & -2 & 1 & | & 16 \\ 0 & 1 & -1 & | & -6 \\ 0 & 11 & -7 & | & -58 \end{bmatrix} \begin{matrix} \\ 2R_2 + R_1 \\ -11R_2 + R_3 \end{matrix}$$

$$\begin{bmatrix} 1 & 0 & -1 & | & 4 \\ 0 & 1 & -1 & | & -6 \\ 0 & 0 & 4 & | & 8 \end{bmatrix} \begin{matrix} \\ \frac{1}{4}R_3 \\ \\ \end{matrix}$$

$$\begin{bmatrix} 1 & 0 & -1 & | & 4 \\ 0 & 1 & -1 & | & -6 \\ 0 & 0 & 1 & | & 2 \end{bmatrix} \begin{matrix} R_3 + R_1 \\ R_2 + R_1 \\ \\ \end{matrix}$$

$$\begin{bmatrix} 1 & 0 & 0 & | & 6 \\ 0 & 1 & 0 & | & -4 \\ 0 & 0 & 1 & | & 2 \end{bmatrix} \begin{matrix} R_3 + R_1 \\ R_2 + R_1 \\ \\ \end{matrix}$$

The solution set is $\{(6, -4, 2)\}$.

22.
$$\begin{cases} x - y = 1 \\ x^2 - x - y = 1 \end{cases}$$

Solving $x - y = 1$ for y gives $y = x - 1$.
Substitute:
$$x^2 - x - y = 1$$
$$x^2 - x - \overbrace{(x-1)}^{y} = 1$$
$$x^2 - x - x + 1 = 1$$
$$x^2 - 2x + 1 = 1$$
$$x^2 - 2x = 0$$
$$x(x - 2) = 0$$
$x = 0$ or $x = 2$
If $x = 0$, $0 - y = 1$ so $y = -1$.
If $x = 2$, $2 - y = 1$ so $y = 1$.
The solution set is $\{(0, -1), (2, 1)\}$.

23.
$$100x^2 + y^2 = 25$$
$$4x^2 + \frac{y^2}{25} = 1$$
$$\frac{x^2}{\left(\frac{1}{4}\right)} + \frac{y^2}{25} = 1$$
Ellipse
Foci on the *y*-axis

$a^2 = 25$ and $b^2 = \dfrac{1}{4}$, so $\dfrac{1}{4} = 25 - c^2$.

$$c^2 = \frac{99}{4}$$
$$c = \frac{3\sqrt{11}}{2}$$

Foci: $\left(0, -\dfrac{3\sqrt{11}}{2}\right), \left(0, \dfrac{3\sqrt{11}}{2}\right)$

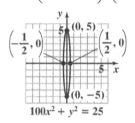

$100x^2 + y^2 = 25$

24.
$$4x^2 - 9y^2 - 16x + 54y - 29 = 0$$
$$4(x^2 - 4x) - 9(y^2 - 6y) = 29$$
$$4(x^2 - 4x + 4) - 9(y^2 - 6y + 9) = 16 - 81 + 29$$
$$4(x - 2)^2 - 9(y - 3)^2 = -36$$
$$\frac{(y - 3)^2}{4} - \frac{(x - 2)^2}{9} = 1$$
Hyperbola with center at (2, 3)
Transverse axis vertical
$a^2 = 4$ and $b^2 = 9$, so $9 = c^2 - 4$.
$$c^2 = 13$$
$$c = \sqrt{13}$$
Foci: $\left(2, 3 - \sqrt{13}\right), \left(2, 3 + \sqrt{13}\right)$

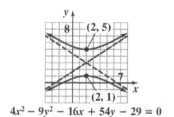

$4x^2 - 9y^2 - 16x + 54y - 29 = 0$

Subject Index

A page number in **boldface** indicates mention of the subject in a figure or figure legend.

in proteinoids, 128
racemic mixture, **117**
relationship between observed and
    expected frequencies, **151**
sequence homologies. *See also* Molecular
    phylogenies
stereoisomers, **117**
Strecker synthesis, 123
substitutions, 257–66. *See also* Mutation,
    missense
    rates, 272, 273, 284, 559
       dependence on protein function, 273
       proposed constancy, 560. *See also*
          Evolutionary clocks
synthesis from glycolytic substrates, **161**
Amino group
    in amino acids, 109
    definition, 632
Aminoacyl adenylates, 130
    structural formula, 130
Aminoamide, synthesis, 123
γ-Aminobutyric acid, in Murchison
    meteorite, 122
Aminoimidazole-5-carboxamidine, abiotic
    synthesis, 124
α-Aminoisobutyric acid
    in Murchison meteorite, 122
    synthesis in Miller electric spark discharge
       experiment, 122
β-Aminoisobutyric acid, in Murchison
    meteorite, 122
Aminomalonodiamidine, abiotic synthesis,
    124
Aminomalononitrile, abiotic synthesis, 124
α-Amino-n-butyric acid
    in Murchison meteorite, 122
    synthesis in Miller electric spark discharge
       experiment, 122
β-Amino-n-butyric acid, in Murchison
    meteorite, 122
Aminonitrile, synthesis, 123
Ammocoete, lamprey larva, 341, 396
Ammonia
    in amphibian excretion, 417
    from hydrothermal deep-sea vents, 120
    interstellar density, **120**
    in Miller electric spark discharge
       experiment, 121, 122
    from nitrogenase activity, 167
    outgassing
       from Earth, 88
       from meteorites, 120
    reducing agent, 161
    synthesis, 123
Ammonites (fossil cephalopod mollusks),
    94, 377
    Late Cretaceous extinctions, 430
    in molluskan phylogeny, **378**
Ammonium cyanide, in abiotic purine
    synthesis, 124
Amniotes (Amniota)
    definition, 419
    as proposed monophyletic taxon, 416
    urogenital duct systems, **418**
Amniotic egg, 416, 417, 419
    description, 632
    membranes, 417, **418**
    preceding evolutionary stages, 419
Amoeba
    aggregating, 329
    protistan phylogeny, **330**

Amphibians, 412
    anthracosaurs, 412, 413, **414, 415**
    anurans, 286, 414, **415**. *See also* Frogs and
       toads
       adaptations, 416
       DNA nucleotide numbers, **268**
       early fossils, **416**
    apodans (caecilians), 414, **415**
       DNA nucleotide numbers, **268**
       early fossils, **416**
    aquatic dependence, 417
    cell-type numbers, **544**
    distinctions from reptiles, 417
    early evolution, 410–14
       number of digits, 414
       proposed osteolepiform ancestry, **406**
          limb bone comparisons, **411**
       reconstructions, **410**
       skull bone comparisons with
          rhipidistians, **411**
       vertebrae, **412**
    ecogeographical rules, 585
    eggs, 417, **418**
    excretory system, 417
    heart, **417**
    ichthyostegids. *See* Ichthyostegids
    labyrinthodonts, **411,** 412, **414,** 419. *See*
       also Ichthyostegids
       limb girdles, **413**
    Late Cretaceous extinctions, 430
    lepospondyls, 413, **415**
    lissamphibians (modern forms), 414–16,
       414, 415
       in amphibian phylogeny, **415**
       characteristics, 414
       DNA nucleotide numbers, **268**
       fossil appearance, **416**
       proposed polyphyletic origin, 416
       reproductive patterns, 414
       urogenital duct systems, **418**
    low energy requirements, 421
    number of species, 416
    Permian extinctions, 420
    phylogeny, **415**
       proposed polyphyletic origin, 414
    protein polymorphism, 231
    reptilian transitional changes, 419
    stereospondyls, 412, **415,** 420
    temnospondyls, 412, **415**
    urodeles (newts and salamanders). *See*
       Salamanders; Urodeles
    vertebral evolution, **414**
Amphidiploids, 206, **209**
Amphilestids (Jurassic-Cretaceous
    mammals), in phylogeny of
    mammalian molar teeth, **449**
Amphioxus (cephalochordate), 395
    development, zootype relationship, **355**
    homeoboxes, **354, 396**
    morphology and characteristics, **396**
Amphipathic molecules, **131,** 132
Amylase, **134**
Anabaena (cyanobacterium), **172,** 283
Anacystis (cyanobacterium), 5S ribosomal
    RNA phylogeny, **281**
Anaerobic
    definition, 632
    glycolysis, 158–62
       Entner-Doudoroff pathway, 161
       enzymatic sequence, **160**
       proposed endosymbiotic transfer, 283

universality, 158, 159
heterotrophs, 281
metabolism, 158–62
    in ectotherms, 421
    in phylogeny of prokaryotes, **283**
photosynthesizers, 281
respiration, 632
Anagenesis
    definition, 632
    horses, **48**
    horseshoe crabs, **242**
Analogous organs, 37
    definition, 632
Anapsids (unfenestrated reptiles), 419
    classification, 421
    in phylogeny of reptiles, **422**
    skull structure, **420**
Anaspids (fossil agnathan fish), 399
    as proposed ancestral cyclostome, 400
    structure, **400**
Anatosaurus (duck-billed hadrosaur), **427**
Ancient DNA, 276–79
    contamination problems, 279
    deterioration and fragmentation, 277
    Neanderthal, 279, 481
    organisms, 277
Aneuploidy, 206
    definition, 632
    hexasomics, 210
    nondisjunctional origin, 208
    trisomics, 210, 616
    types, 210
Angiosperms, 312, 452
    characteristics, 308–10, 633
    competitive advantages, 313
    dicotyledons, **318**
    double fertilization, 310
    endosperm, 310, 311, 313
    evolutionary relationships among major
       groups, **318**
    flowers, **315**
       animal pollination, 308, **316**
       incomplete color dominance, 197
       proposed early forms, 312
    fossil pollen and leaves, **317**
    homeodomain proteins, 355
    monocotyledons, **318**
    number of species, 307
    origins, 310–14
       Cretaceous period, 307
       polyphyletic proposals, 248, 313
    seeds, selective factors, 310
Angstrom, definition, 633
Anhidrotic ectodermal dysplasia, 203
Animals. *See also* Metazoa
    acquisition of cellular organelles,
       181, 279, 281, **283.** *See also*
       Mitochondria
    ancient DNA from fossils, 277
    aneuploidy, 208
    classification problems, 9
    conserved histone proteins, 565
    cytochrome *c* phylogeny, **266**
    evolutionary clock comparisons, 560
    germ line protection, 302
    neuronal variations, 356
    polyploidy, 206
    speciation by hybridization, 596
Animism, 55
Aniridia, 224
Anisomycin, 173

Bacteria—*Cont.*
  symbiosis with early eukaryotes, **181,** 279,
    281, 283
    in origin of eukaryotic organelles, 179,
      180
  synonymous nucleotide substitution
    rates, **286**
  test for random mutation, 227, **228**
  universal genetic code, 147
Bacteriophages. *See also* Viruses
  definition, 633
  DNA melting temperature, **271**
  mutation rates, 224
  in test for random mutation, **228**
Bactrian (Asian) camel, 210
Balanced genetic load, 558, 559. *See also*
    Polymorphism; Genetic loads
  definition, 633
  proposed evolutionary cost, 559
Balanced lethals, segregational load, 559
Balanced polymorphism, **538, 539.** *See also*
    Polymorphism
  definition, 539, 633
Balanoglossus (acorn worm), 397. *See also*
    Hemichordates
  chordate similarities, 395
  coelomic organization, 340
  larva, **394,** 397
  morphology, **394**
  in proposed phylogeny of vertebrates, **399**
Banded iron formations
  description, 633
  proposed biological origin, 172
Barnacles (crustacean arthropods), 383
  Late Cretaceous extinctions, 430
  preadaptations, 33
Barnard's star, 80
Basalts, 90. *See also* Rocks, igneous
  definition, 633
  oceanic, 98
Bases (nucleic acid), 111, **112, 149, 216**
  abiotic synthesis, 121
  changes, 215, **216.** *See also* Mutation, base
      substitutions
    back mutations, 264
    tautomeric, 215, **216**
    transitions and transversions, 215, **216**
  complementary pairing, 113, **115**
    definition, 636
    stereochemical fit, **114,** 144
  description, 633
  hypoxanthine, 149
  proposed reason for limited kinds, 146
  purines, 112, 120
    abiotic synthesis, 112, 121, 124
    adenine, **112**
    guanine, **112**
    structural formulas, **112**
  pyrimidines, 112, 120
    abiotic synthesis, 121, 124, 125
    cytosine, **112**
    orotic acid, 124
    structural formulas, **112**
    thymine, **112**
    uracil, **112,** 124
  unusual tRNA forms, **149**
Basic (alkaline), definition, 633
  pH scale, 651
Basidiomycetes (fungi), 315
  in fungal phylogeny, **319**

Batesian mimicry, 540
  butterflies, **540**
  definition, 633
Batrachosauria (fossil amphibian order), 413
Bats, 358, 431, 468
  absence in Australia, 102
  in Cenozoic mammalian radiation, **456**
  forelimb vertebrate homologies, **41**
  as pollination vectors, **316**
  wing analogies, 37
Bauplan (baupläne). *See also* Archetypes
  definition, 633
Bdellostoma (hagfish), **400**
Beagle HMS. *See* Darwin: the *Beagle* voyage
Bears, 249
Bees, 383
  allodapine, 248
  instinctive dancing patterns, 585
  as pollination vectors, **316,** 516, 595
  social organization, 384
    and haplodiploidy, **385,** 570
Behavior, 10. *See also* Social behavior;
    Sociobiology
  adaptations, 585–87
  altruism, 386, 498, 499, 566, 570, 587,
      610, 611
    definition, 632
  in animal-plant pollination, **316**
  circadian rhythm, 385
  consciousness, 500, 504
  convergences, 243
  density dependent, 586
  disorders, 586
  emotions, 505–6
    advantages, 506
  evolutionarily stable strategies, 587
  genes, 585, **587**
  haplodiploid fratricide, **385**
  hormonal influence, 385
  instincts, 585, 645
  intelligence. *See* Intelligence
  as isolating mechanism, 590
  learned, 585, 586
    definition, 646
    Lamarckian pattern, 606–7, 608
  as level of complexity, 612
  mystical explanations, 14
  parental investment, 475
  phylogenetic limitations, 256
  self-awareness, 497, 498
  sexual bonding, 474, 475, 490
  territoriality, 572
  tropisms, 585
Belding's ground squirrels, 570
Belemnites (fossil cephalopod mollusks), 94,
    378
  Late Cretaceous extinctions, 430
  in molluskan phylogeny, **378**
Bennettitales (fossil plant), 312
Benthic
  algae, 299
  definition, 634
  early metazoans, 336
Benthosuchus (amphibian labyrinthodont), **411**
Benzaldehyde, 125
Bergmann's rule, 585
Beta rays in abiotic organic synthesis, 121
Bible. *See also* Judeo-Christian religion
  age of the Earth, 13, 14, 36
  Book of Daniel, 56

Ecclesiastes, 75
  as edited documents, 62
  Exodus, 29
  Genesis, 13, 29, 36
    creation myths, 63, 64
    as literal truth, 62
  incompatibility with Darwinism, 54
  Noachian flood, 13, 14, 94
  reinterpretations, 61
Big Bang theory of the universe, 76, 77, **78**
  definition, 634
Bilateral body organization (Bilateria), 338,
    373
  acoelan turbellarians, **335**
  in classification of phyla, 366
  definition, 634
  in metazoan phylogeny, **368**
  proposed origin, 336
Binary fission (cell division), 173
  definition, 634
Binomial expansion, 522
  definition, 634
Binomial nomenclature, 10, 634. *See also*
    Systematics
Biogenetic Law. *See* Haeckel's biogenetic law
Biogeography, 634
Biometricians, 515, 519
Biosphere, definition, 634
Biota, definition, 634
Biotic, definition, 634
Bipedalism, 424
  archosaurian reptiles, 424
  definition, 634
  dinosaurs, 425
    pelvic changes, **425**
  hominids, 466, 474–77
    anatomy and evolution, 475–77, **476,**
      477, 478, 479
    australopithecines, 472, **473**
    early forms, 473
Birds, 357
  adaptations
    bills and feet, **438**
    for flight, **435**
    Lamarckian concepts, 24
  ancient DNA from fossils, 277
    moa-kiwi relationship, 279
  arboreal-cursorial origin controversy, 436
  Archaeopteryx. See Archaeopteryx
  brain, **40**
    lateralization, 503
  cell-type numbers, **544**
  in cladistic classification, 248
  convergent molecular evolution
    (lysozyme), 265
  Cretaceous, 436
  "Darwin's finches", 22, 23, 37. *See also*
    "Darwin's finches"
  Dendroica, resource partitioning, **572**
  ecogeographical rules, 585
  egg, **418**
  evolution, 436–37
  flightless, 437
    giant forms, 436, **437**
  forelimb vertebrate homologies, **41**
  Hawaiian honeycreepers, 251
  helpers, 202
  imaginary hybrids, 10
  Late Cretaceous survival, 429
  lens crystallins, 262

male ornamentation, 588, **589**
nucleotide substitution rates, 285
number of species, 437
parental investment, 588
phylogenies from DNA-DNA
    hybridization, 271
in phylogeny of reptiles, **422**
as pollination vectors, 308, **316**
polymorphism
    beaks, 516
      proteins, 231
race formation, 589
recognition of species distinctions, 590
sex chromosomes, 200
sexual isolation, 590
sexual selection, 588
skewed sex ratios, 202
skull structures, **434**
theropod ancestry, 436
1, 3 Bisphosphate glycerate, in progenote
    metabolic pathways, **178**
Biston betularia (peppered moth), **541**, 565,
    580, 597
darwin rate measurement, 598
industrial melanism, 541
Bithorax gene complex, 352. *See also*
    Homeoboxes; Homeotic mutations
Bivalves, 377
fossil borings, 16
Late Cretaceous extinctions, 430
in molluskan phylogeny, **378**
Blastocoele, **332**, 336
definition, 634
Blastoids (echinoderms), 387
phylogeny, **387**
Blastopore, **334**, 336, 365, 376
definition, 634
deuterostomes, 393
protostomes, 393, 394
Blastula, **334**, 335, 376
definition, 634
Blending inheritance, 28, 32, 194, 521, 634
Blood group genes (humans), 583
absence of environmental effects, 357
MN system, 527
polymorphisms, 582
Rh, 546
Bone, 400
advantages, 400
    for preserving ancient DNA, 277
    skeleton, 401
dermal, 401
endotherm-ectotherm differences, 428
haversian canals, 428
Bonellia (marine echiuroid)
male progenesis, 398
non-chromosomal sex determination, 198
Bootstrapping, 264
Borhyaenids (South American fossil
    marsupial carnivores), 455, **458**
Bos taurus. See Cattle
Bottleneck effects, 549, 593. *See also* Founder
    Principle; Random genetic drift
definition, 634
proposed cause for macroevolution, 599
Bowfin (neopterygian fish), 404, **406**
Brachiation, description, 466, 634
Brachiopods
classification among metazoan phyla, 366
coelomic organization, 340

Late Cretaceous extinctions, 430
Lingula ("living fossil"), 49
in metazoan phylogenies, **332, 368**
    18*S* ribosomal RNA sequences, **333**
proposed chordate relationship, 398
stratigraphic (fossil) appearance, 91, **324**
Brachiosaurus (saurischian dinosaur), in
    dinosaur phylogeny, **426**
Brackish (water), 400
definition, 634
Brains, 56, 500. *See also* Intelligence
allometry in hominids, 500
cephalopod, 377
cerebral cortex
    function, 505
    human-primate prefrontal
      comparisons, **505**
    language capacity, 503
cranial nerves, 42, **45**, 402, **403**
cross-modal associations, 497
emotional centers, 506
encephalization quotients, **455**
    hominoids, 500
endocranial casts, 471, **501, 502**
hominids, 467
human
    evolution, 499–505
    obstetric effects, **476**, 502
    speech centers, **498**. *See also* Broca's
      area; Wernicke's area
insect (*Drosophila*), **587**
lateralization, 503
mammals
    Cenozoic increase, 453, **455**
    neocortex, 442
    therapsid comparison, 447
vertebrate comparisons, **40**
volumes (hominoids), 479
    australopithecines, 471, 472, 500
    chimpanzees, 500
    gorillas, 500
    Homo erectus, 479, 500
    Homo habilis, 500
    Homo sapiens, 500
    Neanderthals, 480
Branchiostoma (amphioxus), 395, **396**. *See
    also* Amphioxus; Cephalochordates
Broca's area (speech center), 497, **498,**
    502, 503
Brontosaurus (saurischian herbivore
    dinosaur), 426
Brownian movement, 338
Bryophytes, 300
alternation of generations, 300, **301**
Anthoceros, 300, 305
characteristics, 300
description, 634
DNA nucleotide numbers, **268**
fossil appearance, 300
life cycle, **301**
number of species, 300
proposed origins, 300, **314**
    polyphyletic proposals, 248, 300
Bryozoans
classification among metazoan phyla, 366
Late Cretaceous extinctions, 430
in metazoan phylogeny, **368**
time of first fossil appearance, **324**
Buccal, definition, 634
Burgess Shale fossils, **328**

Burrowing, 336, 339, 340, 374
definition, 634
earthworms, 379
peristaltic, **339, 341**
Butterflies and moths, 597
Biston betularia, **541**, 565, 580, 597
Colias, 588
Danais plexippus, **540**
Euphydras editha, 547
Heliconius eucrate, 595, **540**
industrial melanism, **541**, 565, 580
isolating mechanisms, 593
Limenitis archippus, **540**
Lycorea halia, **540**
mimicry, 597
    as convergent evolution, **540**
Papilio, 597
as pollinators, 308, **316**
sex chromosomes, 200

C

Cacops (amphibian temnospondyl), 412
Cacti, convergent evolution, 313, **317**
Caecilians (apodan amphibians), 414, **415**
DNA nucleotide numbers, **268**
early fossils, **416**
Caenorhabditis elegans (nematode), 200, 374
developmental homologies, **351**, 354
genome sequencing, 276
homeoboxes, **354**
number of genes, 269
social gene, 585
Calamites (Carboniferous tree), **306**
Calcarea (sponges), **370**
Callitrichidae (family of New World
    monkeys), 466
in primate classification system, 464
Calorie, definition, 634
Calvin cycle, **164**, 165, 283, 634
as the dark reaction in photosynthesis, 165
self-catalytic feature, 167
Camarhynchus (a genus of "Darwin's
    finches"), 23, **38**
    crassirostris, 23, **38**
    heliobates, 23, **38**
    pallidus, 23, **38**
    parvulus, 23, **38**
    pauper, 23, **38**
    psittacula, 23, **38**
Cambrian period, 94, 182
"arms race", 382
Burgess Shale, **328**
definition, 634
increased atmospheric oxygen
    concentration, 165
metazoan radiation, 323–26, **324**
naming of, 91
Camels, 210
chromosomal conservatism, 210
cytochrome *c* relationships, 264, **266**
lens crystallins, 262
Pleistocene North American extinction, 454
Canalization (developmental), 357, 358,
    359, 542, 644
selection for, 359, 598
Candida (yeast species), genetic code
    exceptions, **154**
Canis lupus (North American wolf), **458**. *See
    also* Wolves

Cannibalism, 58
Capitalism, 55, 611
  Industrial Revolution, 8, 26
  revolutionary aspects, 54
Captorhinomorphs (stem reptiles), 419, 424
  in amphibian phylogeny, **415**
  classification, 421
  fossil appearance, 420
  in phylogeny of reptiles, **422**
  skull structure, **420**
Capuchins (New World monkeys), in
  primate classification system, 464
Carbamoyl phosphate, in progenote
  metabolic pathways, **178**
  synthase, **178**
Carbohydrates
  definition, 634
  in meteorites, 117
  from photosynthesis, 165
  plant and algal storage, 298
Carbon, 75, 122
  chemical attributes, 119
  fusion temperature, 82
  interstellar compounds, 119, 122
  isotopes, 172
  primitive sources for photosynthesizers, 164
  in radioactive dating, 94, 95
  reduction, early forms, 161, 162
  synthesis in stellar fusion reactions, 80
Carbon dioxide
  "greenhouse" effect, 130
  carbon source for photosynthesis, 165
  fixation in Calvin cycle, **164,** 165
  in Krebs cycle, 167
  outgassing
    from Earth, 88
    from meteorites, 120
  present atmospheric composition, 89
  primitive reduction of, 161
Carbon monosulfide, interstellar density,
  120
Carbon monoxide
  in early Earth atmosphere, 88
  interstellar density, **120**
  meteorite outgassing, 120
  present atmospheric composition, 89
Carbonaceous
  chondrites (meteorites), 117, 122, **123,**
    125, 634
  definition, 634
Carbonic anhydrase, chimpanzee-human
  relationship, 467
Carboniferous period, **308**
  amphibian increase, 412
  coal forests (seams), 304, 306
  Gondwana system, 96
  gymnosperm appearance, 307
  insect expansion, 386
  lycopod abundance, 304
  reptilian emergence, 416
    captorhinomorphs, 419
    pelycosaurs (synapsids), 443
  tree forms, **306**
Carbonyl sulfide, interstellar density, **120**
Carboxyl group, in amino acids, 109
Carboxylic acids
  definition, 634
  nomenclature, **161**
Carnivora (mammals)
  Cenozoic radiation, **456**
  generic evolutionary rates, **598**

Carnivores, 336
  "arms race". *See* "Arms race", predator-
    prey
  brain size comparisons between fossil and
    modern mammalian forms, **455**
  carnosaurs (theropod dinosaurs), 426,
    584
    in dinosaur phylogeny, **426**
  definition, 634
  plants, 329
  Precambrian-Cambrian diversification, 325
  predator-to-prey ratios, 428
  saber-toothed, 455, **458**
Carotenoids (plant pigments), **163,** 165,
  167
  similarity in green algae and higher
    plants, 298
Carp, hemoglobin chain substitutions, 284,
  560
Carpoids (echinoderms), 395
  phylogeny, **387**
  as proposed vertebrate ancestor, 394
Carrying capacity (populations), 554, 556
  definition, 635
  predation effects, 573
Cartilage, 400, 403
Catalase
  antioxidant, 166
  hydrogen peroxide degradation, **140**
Catalpa (woody plant), geographical species
  distinctions, 240
Catalysts, **140**
  advantages, 139
  auto-, 140, 142, 145
  definition, 635
  enzyme, 111. *See also* Enzymes
  laboratory evolution, 142
  organic. *See* Enzymes
  RNA, 140, 142
  "ribozymes", 140, 142
Catarrhines (Old World anthropoids). *See*
  also Hominoids; Old World monkeys
  description, 466
  in primate classification system, 464
  questions of origin, 468
Catastrophism concept, 14, 15, 16
  definition, 635
Catostomus clarkii (freshwater fish), enzyme
  polymorphism, 561
Cats, 204, **610**
  artificial selection, 50
  brain, **40**
  classification problems, 248
  forelimb vertebrate homologies, **41**
  marsupial, **39**
  phylogeny, **243**
  sex-linked traits, 203
Cattle, 204
  in agricultural expansion, 608
  artificial selection, 50
  cloning, 624
  convergent molecular evolution
    (lysozyme), 265
  cytochrome *c* relationships, 264, **266**
  embryology, **44**
  linkage relationships, **206**
  "mad cow" disease, 144
  nucleotide substitution rate comparisons,
    273
  number of repetitive DNA copies, 270
  sex-linked traits, 203

Caulerpa (algae), 298
  structure, **299**
Cave animals, 33, **34,** 40
  common attributes, 15
Cavies, lens crystallins, 262
Cebidae (family, New World monkeys), 466.
  See also New World monkeys
  in primate classification system, 464
Ceboidea (superfamily, New World
  monkeys). *See also* New World
  monkeys
  in primate classification system, 464
  South American radiation, 459
Cell, 565
  carbon isotope ratios, 172
  cytoplasm, 109, **110**
  cytoskeletal protein, 349
  deacidification by proton gradient, 164
  division. *See* Cell division
  eukaryotic, **110.** *See also* Eukaryotes
    cytoskeleton, 173, 179, 182
    early fossils, **182**
    hydrogenosome, 181
    microsphere similarities, 133, **135**
    microtubules, 173, 179, 182, 329
      proposed endosymbiotic origin, 180,
      181
    substructure organization, 329
  necessity for cell division, 189
  nucleus, **110**
    proposed endosymbiotic origin, 181
    RNA processing, **176,** 177
  organelles, **110,** 175, 202. *See also*
    Chloroplasts; Mitochondria
    codon distinctions, 148
    prokaryotic origin, 177, 180, 279, 281,
      329
  plasma membrane, **131, 350**
  prokaryotic, **110.** *See also* Prokaryotes
    early fossils, 171, **172**
    isotope ratios, 172
  proposed minimum number of genes,
    276
  protocells, 134, 135
  signaling systems, 180, 336, 349
    signal transduction, **350**
  specialization ("division of labor"), 296,
    336, 342, 345
  structure, 109, **110**
  types (morphological complexity), **544**
  walls, 635
    superkingdom distinctions, 173
Cell division
  binary fission, 173
  cell plate (plants), 296
  cellular apparatus, 175, 182
  chromosome disjunction, 191
  in early eukaryotic fossil cells, **182**
  evolution, 182
  meiosis, 191, 647. *See also* Meiosis
    stages, **192**
  mitosis, 191, 648. *See also* Mitosis
    stages, **190**
  prokaryotes, 182, 191
  protistan origin and diversification, **330**
  selection for, 189
  spindle fibers, 182, 191, 329
Cenozoic era
  bird expansion, 436
  climatological changes, 429
    vegetative landscapes, **453**

definition, 635
Indian-Asian juncture, 99
mammals
brain size changes, **455**
dispersion, **106**
radiation, 431, 452–59, **456**
Centigrade scale (ºC), definition, 635
Centrifugal selection. *See* Selection, disruptive
Centripetal selection. *See* Selection, stabilizing
Centromeres, 182, **190**, 191
acrocentric, **190**
definition, 635
heterochromatin, 271
metacentric, **190**
pericentric inversion effects, 209
telocentric, **190**
Cephalochordates, 354
Branchiostoma (amphioxus), 395, **396**
DNA nucleotide numbers, **268**
in metazoan phylogenies, **368**
as vertebrate ancestor, 395
common traits, 395
Cephalopods, 377
adaptations for hunting, 378
cell-type numbers, **544**
eyes, 34
analogies, 37
lens crystallins, 262
in molluskan phylogeny, **378**
Ceratopsians (ornithischian dinosaurs), 427
in dinosaur phylogeny, **426**
Cercopithecidae (family, Old World monkeys), in primate classification system, 464. *See also* Old World monkeys
Cercopithecoidea (superfamily, Old World monkeys), 466. *See also* Old World monkeys
in primate classification system, 464
Certhidea olivacea ("Darwin's finches"), **23, 38**
Cestodes (platyhelminth tapeworms), reproductive patterns, 374
Cetaceans
as artiodactyl subtaxon, 252
in Cenozoic mammalian radiation, **456**
convergent evolution in marine predators, **42**
Chaetognathans
classification among metazoan phyla, 366
in metazoan phylogeny, **368**
time of first fossil appearance, **324**
Chagas disease, 20
Chambers (pre-Darwinian) evolutionary concepts, 26
"Chaos", 68, 118, 119, 122, 135
Character
character state, **244, 245**
definition, 635
displacement, 572, 575
in "Darwin's finches", 572, **573**
definition, 635
phenetic, 651
quantitative, 637, 652
Charophyceae (stoneworts), 296
Cheek teeth, 443, 444. *See also* Teeth, molars
definition, 635
Chelicerates (arthropods), characteristics, 383

Chemical elements
abundance in universe, **82**
early terrestrial distribution, 89
solar distribution, 88
synthesis in stellar fusion reactions, 80, 82, 83
Chemiosmosis, **166**
definition, 635
Chemoautotroph (chemolithotroph), definition, 635
Cherts (microcrystalline quartz), 171, **172**
definition, 635
Chiasma (chromosomal), 191, **192, 193**
Chickens
conserved amino acid sequences
aspartate transaminase, 256, **257**
triosephosphate isomerase, 159
cytochrome *c* mutations, 273
embryology, 44
hemoglobin chain substitutions, 284
number of repetitive DNA copies, 270
optimum phenotypes, 542
5*S* ribosomal RNA phylogeny, **281**
silent/replacement mutation rate comparisons, 273
skeletal structure, **435**
Chicxulub crater (Cretaceous-Tertiary boundary), 433
Chimaera (cartilaginous fish), **406**
Chimpanzees
active teaching, 497
animal hunting, 466, 489, 490
body contours, **467**
bonobo (pigmy)
in hominoid phylogeny, **470**
language aptitude ("Kanzi"), 496
brain
encephalization quotient, 500
endocranial cast, **501**
volume, 500
cannibalism, 466, 490
chromosome evolution, 211
banding comparisons, **214**
comparisons with humans
chromosomal, **214**
gene linkages, 203
molecular, 211, 257, 264, 271, **272, 275**
morphological, 287
skull development, **502**
"culture", 607
deceptive strategy, 496
description, 466
illustration, **465**
facial expressions, **492**
foot, **479**
karyotype, **214**
knuckle-walking, 466
language abilities, 495–96
mitochondrial DNA restriction enzyme map, **275**
molecular relationship with humans, 467
pelvic structure, **476**
phylogeny
DNA-DNA hybridization, **272**
immunological methods, 257
mitochondrial DNA hominoid relationships, **470**
in primate classification system, 464
proposed thought processes, 506
protein polymorphism, 231

seated pose, **468**
self-awareness, 498
sex-linked traits, 203
skull, **471**
tactile communication, 492
terrestrial stance, **477**
toolmaking abilities
nut cracking, 497
termite fishing, 496
vocalization
anatomical structures, **494**
attempts, 495
limitations, 493, 495
Chiroptera. *See* Bats
Chlamydomonas (algae), 296, 302
dominance in diploids, 546
mating type evolution, 595
number of repetitive DNA copies, 270
possible phylogeny, **297**
Chloramphenicol, 173
Chlorella (green alga), 5*S* ribosomal RNA phylogeny, **281**
Chlorobium limicola (green photosynthetic bacteria), ferredoxin amino acid sequence, **162**
Chlorophyll, 125, 163
chemical structure, **163**
in coacervate systems, 133
in cyclic photosynthesis, **163**
membrane association, 164
in non-cyclic photosynthesis (photosystems I and II), 165, **166**
phytol groups, 165
resonance forms, **163**
similarity in green algae and higher plants, 298
Chlorophyta (algae), 296, **331**
in plant phylogeny, **314**
Chloroplasts
definition, 635
DNA, 180
maternal inheritance, 202
nuclear genes for, 180, 283
numbers of genomes per cell, 180
prokaryotic origin, 177, 180, **181**, 279, 281, **283**, 296, 329
from secondary invasions, 180, 281
thylakoid membranes, 165, **166**
Choanoflagellates (protistans)
metazoan relationship, 329
as planula ancestor, **335**
in proposed phylogeny, **332**
in protistan phylogeny, **331**
resemblance to sponge choanocytes, 367
Chondrichthyes (cartilaginous fish), 402–3
DNA nucleotide numbers, **268**
in phylogeny of fishes, **406**
Chondrosteans (primitive ray-finned fish), 404
DNA nucleotide numbers, **268**
in phylogeny of fishes, **406**
Chordates, **340**. *See also* Vertebrates
characteristics, 391
classification among metazoan phyla, 366
eucoelomate body plan, **337**
filter feeding in early forms, 396, **399**
hemichordates, **340,** 366, 394, 397
metamerism, 337
in metazoan phylogenies, **332**
18*S* ribosomal RNA sequences, **333**
mutational distance (histones) from echinoderms, 273

Clover, self-sterility alleles, 541
Cnidarians, 365, 369. *See also* Coelenterates
    cell-type numbers, **544**
    classes, 369
        phylogeny, **372**
    classification among metazoan phyla, 366
    in metazoan phylogeny, **368**
    polyp and medusa forms, 369, **371**
    possible Ediacaran fossils, **327**
    possible origin, **331**
Coacervates, 133, 134, 136, 139
    description, 636
    formation, **133**
    glucose-starch reaction system, **134**
    properties, 133
    self-assembly, 134
Coadaptation
    definition, 636
    gene complexes, 525, 593
    mutations, 182, 451
Coal seams (Carboniferous origins),
    304, 306
Coalescence, 636
Coarse-grained environment, 541
    definition, 636
Coccoliths (calcified planktonic algae), Late
    Cretaceous extinctions, 430
Codominance (alleles), 197
    definition, 197, 636
    Hardy–Weinberg equilibrium values, 527
Codons. *See also* Genetic code
    anti-, **116,** 147, 148, **149**. *See also*
        Anticodons
        description, 633
    chain termination (stop, nonsense), **116,**
        147, 148, 216, 649, 655
        codon designations, 148
    definitions, 147, 636
    dictionary, 148
    duet, 148
    evolution, **153**
    for hydrophobic and hydrophilic amino
        acids, 152
    initiation, **116**
    mutations, **217**
        substitutions necessary for amino acid
            conversions, 259, 263
    proposed affinity to amino acids, 150
    quartet families, 148, 149, 150, 151, 152
    synonymous, 147, 217, 562
        substitution rates among taxonomic
            groups, 285, **286**
    in translation process, **115**
    triplet nucleotide sequence, 148
        proposed earlier stages, 150
    universality
        definition, 657
        exceptions, 148, **154**
        frozen accident hypothesis, 150
    wobble pairing, **149**
        extreme forms, 152
Coelacanths, 49
    crossopterygian relationship, 405
    Mesozoic fossils, **49**
    in phylogeny of fishes, **406**
Coelenterates. *See also* Cnidarians;
    Ctenophorans
    characteristics, 369–71
    diploblastic body plan, **337**
    DNA nucleotide numbers, **268**
    endoderm formation, 334

in Hadzi hypothesis of metazoan origin,
    333
    Haeckel "gastraea" hypothesis, 334
    locomotion, 369
    metamerism, 337
    in metazoan phylogenies, **332, 368**
        18*S* ribosomal RNA sequences, **333**
    nematocysts, 369
    number of species, 369
    planula-like larvae, 336, 369
    proposed ancestries
        planula, **335**
        protistan, **331**
    time of first fossil appearance, **324**
Coelolepids (agnathan fish), 399
    structure, **400**
Coelom, 336–37, 378, 381
    acoelomates, 335, **337,** 366
    advantages
        hydrostatic organ, 336, 374, 379
        metabolic transport, 374
    coelomates. *See* Coelomates
    definition, 636
    embryological development
        deuterostomes, 393, 394
        protostomes, 393
    eucoelomates, 336, **337**
        hemichordate organization, **340**
    hypotheses for origin
        gonocoel, 337
        Haeckel "gastraea" hypothesis, 334
        nephrocoel, 337
        polyphyletic, 337
        schizocoel, 337
    locomotory functions, 339–40
    pseudocoelomates, 336, **337,** 366. *See also*
        Aschelminthes; Pseudocoelomates
        convergent evolution, 374
        segmentation patterns, 340
Coelomates, 336, 375–88
    advantages of peristalsis, 339
    Haeckel "gastraea" hypothesis, 334
    phyla, 366
    Precambrian (Ediacaran strata), 376
Coelophysis (early dinosaur), in dinosaur
    phylogeny, **426**
Coelurosaurs (theropod dinosaurs), 426, 428
    in dinosaur phylogeny, **426**
Coenzymes, 159, **164,** 167, **169**
    acetyl-coenzyme A, 142, 167, **168,**
        169, **178**
    definition, 125, 636
    early selection for, 145
    flavine adenine dinucleotide (FAD), 142,
        145, 162, **166, 168**
    nicotinamide adenine dinucleotide
        (NAD), 125, 142, 145, 159, **160,** 161,
        162, 167, **168, 169**
    nicotinamide adenine dinucleotide
        phosphate (NADP), 133, **164,** 165,
        166, 167, **168**
    Q, 165, **169**
    succinyl-coenzyme A, 167, **168,** 169
Coevolution, 349, 574–75
    "arms race" (predator-prey), 324, 336,
        426, 453, 574. *See also* "Arms race"
    brain size, **455**
    constraints, 451
    definition, 636
    flower pollination, 308
    sexual antagonistic genes, 199, 589

Cofactor (enzymatic), definition, 636
Cohort, 252, 555, 655
    definition, 636
Colchicine, effect on chromosome
    disjunction, 206
Coleochaete (algae), 296, **297,** 298
Colias (butterfly), sexual selection, 588
Collagen, 324
Collision theory of Earth's origin, 85
Colobidae (Old World monkeys), in
    primate classification system, 464
Colonization, 325
    founder effects, 548, **549,** 598
    Hawaiian Drosophilidae, 251, 548, **549,**
        593
    Hawaiian honeycreepers, 251
    as source of evolutionary radiation, 36
Columba (pigeon), skull structure, **434**
Comets, **86**
    early terrestrial impacts, 432
    as possible cause for Late Cretaceous
        extinctions, 429
Commensalism, 571
    definition, 636
Communication, 491–99. *See also* Language;
    Speech; Vocalization
    impact of cooperative hunting, 490
    sensory channels, 491
    social behavior and interaction, 498, 503
    symbolic, 492
    warning signals, 492, 499
Comparative anatomy, 37. *See also*
    Vertebrates, comparative anatomy
    vertebrates, **37**
        brains, **40**
        embryos, **44**
        forelimbs, **37, 41**
        gill arches, **44**
Competition, 566, 571. *See also* Population,
    interactions
    advantages of sex, 564
    "arms race". *See* Coevolution, "arms race"
    cropping principle of diversification,
        325–26, 573
    definitions, 571, 636
    density dependent, 556
    evolutionary consequences, 571–73
    interactions, 574
        coevolution, 574–75
    principle of competitive exclusion, 572, **573**
    resource partitioning, 572
    sexual interactions, 587–89
        intersexual, 588, 589
        intrasexual, 588, 589
    "species selection", 571
Complexity, 378
    adaptive, 451
    cell types, **544**
    cellular, 109
    chemical, 119
    cosmological, 78
    C-value paradox, 269
    definition, 636
    developmental, 356, 361
    differences between prokaryotes and
        eukaryotes, 173–77
    and entropy, 118
    hierarchies
        mental abilities, 505
        selection, 571
    levels, 612

Complexity—Cont.
  metazoan innovations, **332**
  morphological, 451
    cell types, 544
  nucleic acids, 112
  parasite life cycle, 373, 374
  and "progress", 544
  reductionist explanations, 612
  regulatory systems, 219, 361
  triploblastic evolution, 371
Compsognathus (theropod dinosaur)
  possible feathers, 436
  skeleton, **435**
  skull structure, **434**
Concerted evolution, 271
  definition, 636
Condensation (gravitational) theory of
  Earth's origin, 85, **86**. *See also* Solar
  system, gravitational condensation
Condensation reactions (chemical
  dehydration), 124, 126, **127**
  in abiotic purine synthesis, 124
  condensing agents, 126
  definition, 636
  in montmorillonite clays, 130
Condylarths (extinct mammalian
  herbivores), **456**, 458
  brain size, **455**
  in Cenozoic mammalian radiation, **456**
  description, 636
Conodonts, 398
Consciousness, 500, 504
Constraints
  adaptive, 358, 598
  cultural, 14
  definition, 636
  developmental, 61, 355, 356, 565
  environmental, 358
  evolutionary effects, 598
  genetic, cause for extinctions, 452
  in "streetcar theory of evolution", 600
Continental drift, 102, 103. *See also* Continents;
  Tectonic movements and plates
  biological effects, 452, 454
    distribution of lungfish, 405
    mammalian distribution, 102, 104, **106**
  convergent events, **104**
  definition, 637
  in dispersal of New World monkeys, 468
  evidence for, 96
  geographical changes, **101**
  Gondwana supercontinent. *See also*
    Gondwana (ancient southern
    continent)
  magnetic pole wanderings, 97, 98, **99, 102**
  major plates, **103**
  paleomagnetism. *See* Paleomagnetism
  as possible cause for Cambrian radiation,
    325
  Precambrian, 100, **102**
  sea floor spreading, 99, **100**, 654
  Wegener's concept, 96
Continents, 97
  "Atlantis", 96
  crustal structure, **90**
  Gondwana system, 96, 97, **98**, 99, 100,
    101, 102, 107
  Laurasia system, 99, 103
  matched fit, 96, **97**
  Pangaea supercontinent, 96, 99, 102, 104,
    650

Convergent evolution, 249, 313, **317,** 374.
  See also Parallel evolution
  among marine predators, **42,** 245
  analogy, 37
  as evolutionary "repetitiveness", 243
  aschelminthes, 280
  australopithecines, 472
  cephalopod and vertebrate vision, **34,** 378
  cichlid fishes, 280, **282**
  definition, 242, 637
  distinctions from homology, **34,** 37, 243,
    244, 245, 280, 360. *See also*
    Homology
  fiddler crab behavior, 243
  fungi and algae, 315
  genotypic basis, **244,** 452
  homoplasy, 243, 249
  mammalian anteaters, **244**
  mammalian triconodonts, **449**
  marsupials and placentals, **39**
    carnivores, 455, **458**
    "horses", 458, **459**
  mimicry, **540**
  molecular, 265, 280, 290, 561
  in phylogenetic trees, **245**
  plants
    plant-algal conducting cells, 300
  segmentation, 381
  visual pigments, 561
Convoluta (platyhelminth), **373**
Cooksonia (fossil plant), 304
  fossil reconstruction, **304**
Cooperation, 575. *See also* Altruism;
  Mutualism; Social behavior
  advantages, 587
  gene networks, 146
  homonid hunting, 490
  selection for, 612
  social insects, 384–86
  through kin selection, 499, 570
Cope's rule, 637
Corn (*Zea mays*)
  agricultural expansion, 608
  conserved triosephosphate isomerase
    sequence, 159
  directional selection, **544**
  evolution, 516, **517**
  hybrid vigor, 538
  inbreeding depression, 530
  mutation rates, 224
  phenotype, **517**
  polyploidy, 206
  selection for sexual isolation, 593
  teosinte relationship, 69, 516, **517,** 595
Correlation
  definition, 637
  in evolutionary clock determination, 285
  in numerical taxonomy, 237
Corythosaurus (duck-billed hadrosaur), **426,**
  427
Cosmic rays
  as energy source, 119
  interplanetary hazards, 118
Cosmology. *See also* Universe (cosmology)
  definition, 637
Cotylosaurs (stem reptiles), 421
  in phylogeny of reptiles, **422**
Covalent bonds, 119
  definition, 637
Cranial nerves (vertebrates), 42, **45,**
  402, **403**

Creation, 6, 10, 13, 19
  definition, 637
  in idealism, 5
  multicultural myths, 63, 113
  "science", 63, 64
    anti-evolution arguments and
      evolutionists' responses, 65–70
    religious prerequisites, 64, 66
  as source of fossils, 13
  U. S. creationist movement, 14, 54, 62–63
Creodonts (fossil mammalian carnivores),
  456
  brain size, **455**
  in Cenozoic mammalian radiation, **456**
  description, 637
Crepuscular, description, 464, 637
Crinoids (echinoderms), phylogeny, **387**
Crocodiles, 428
  in cladistic taxonomy, 248
  lens crystallins, 262
    lactate dehydrogenase, 260
  in phylogeny of reptiles, **422**
  in reptilian classification system, 421
  spontaneous generation myth, 113
  thecodont ancestry, 424
Cro-Magnon, 481. *See also* Humans
  proposed climatological adaptations,
    585
  reconstruction, **481**
Cropping principle of diversification,
  325–26, 573
Cross fertilization. *See also* Sex
  as a source of genetic variability, 199
Crossing over, 191. *See also* Genetic
  recombination; Linkage
  absence in male *Drosophila*, 203
  definition, 637, 653
  four-strand stage, **204**
  inversion effects, 209
  selection against, 199
  sex-linked genes, **204**
  unequal, 209, 260
Crossopterygians (lobe-finned fish), 409
  coelacanths, **49,** 405, **406**
  in phylogeny of fishes, **406**
  rhipidistians, **406**
Crustaceans (arthropods), 248. *See also*
  Barnacles
  characteristics, 383
  convergent behavior (fiddler crabs), 243
  developmental genetic homologies
    between phyla, 353
  gill homologies to insect wings, 360, 382
  homeoboxes, **360**
  satellite DNA, 271
  vestigial organs, 40
Cryptic, definition, 637
Ctenophorans, 369. *See also* Coelenterates
  characteristics, 371
  classification among metazoan phyla, 366
  in coelenterate phylogeny, **372**
  in metazoan phylogeny, **368**
  planula, 369
  time of first fossil appearance, **324**
Cubozoans (cnidarians), 371
  in coelenterate phylogeny, **372**
Culture, 14, 62
  agriculture, 607
    historical expansion, 608
    populational effects, 607
  attitudes toward hunting, 490

and biology
 biological limitations, 613–14
 connections, 611
 distinctions, 607, 609, 610
concepts of death, 14
creation myths, 63
definitions, 606, 607, 637
evolution of religion, 55
rate of change, 607–8
and science, 14
and social behavior, 503
sociobiology, 610–13
transmission patterns, 606–7, 608
universal behaviors, 610
Cursorial, 436
 definition, 637
Cusps (teeth), description, 443, 637. *See also*
 Teeth
Cyanamide
 hydrolysis, 126
 structural formula, 126
Cyanic acid, 126
 structural formula, 126
Cyanides
 hydrolysis, 123
 synthesis, 123
Cyanoacetylene
 in abiotic pyrimidine synthesis, 124, 125
 interstellar density, **120**
 structural formula, 125, 126
Cyanobacteria, 163, 165
 Anabaena, **172,** 283
 Anacystis, **281**
 description, 173, 637
 Krebs cycle precursors, 167
 molecular phylogeny, **283**
 in nitrogen fixation, 167
 proto-, 172
 stromatolite organisms, 171
 as symbionts in evolution of eukaryotic
  organelles, 179, **181,** 279, 281, **283,**
  329
 thylakoid membranes, 165
Cyanogen
 interstellar density, **120**
 structural formula, 126
Cycads
 fossil form, **311**
 in plant phylogeny, **314**
Cyclic adenosine monophosphate (cAMP),
 351
Cyclic guanosine monophosphate (cGMP),
 351
Cyclorana (desert toads), 414
Cyclostomes (jawless fish), **400.** *See also*
 Fish, agnathans (jawless); Hagfish;
 Lampreys
 in phylogeny of fishes, **406**
 proposed anaspid (agnathan) origin, 400
 urogenital duct system, **418**
Cynognathus (therapsid reptile), 97, **98,** 443
Cysteine
 codon designations, 148
 in ferredoxin, **162**
 structural formula, **111**
 synthesis from glycolytic substrates, **161**
Cystic fibrosis, 528, 618
 incidence, 616
Cystinuria (excessive cystine excretion), 528
Cystoids (echinoderms), phylogeny, **387**
Cytochromes, 272, 487

c, 286
 chimpanzee-human relationship, 467
 eukaryote phylogeny, 264
 in evolutionary tree, 280, 281, **283**
 nucleotide substitution rates in
  artiodactyls, **273**
 phylogenies
  eukaryotes, **266**
  vertebrates, **267**
 rate of amino acid replacement, 273
 sequence identities in artiodactyls, 264
cyclic photosynthesis, **163**
description, 637
electron transfer, 169
non-cyclic photosynthesis, **166**
oxidation-reduction pathways, 162
plant-bacterial homologies, 281
respiratory pathway, **169**
Cytology, definition, 637
Cytoplasm, 202, 298
 composition, 109
 definition, 637
Cytoplasmic inheritance, 202. *See*
 Extranuclear inheritance
Cytosine
 abiotic synthesis, 124, 125
 methylation, 28, 270
 in nucleic acids, 112
 structural formula, **112**
 tautomeric changes, **216**
Cytoskeleton, protein, 349
Cytula, in Haeckel's "gastraea" hypothesis,
 334

D

Damselflies, isolating barriers, 590
Danais plexippus (monarch butterfly), **540**
Darwin. *See also* C. Darwin (Author's Index)
 biographical information, 19–20
 the "darwin" (evolutionary rate
  measurement), 598
 the *Beagle* voyage, 20, **21, 22**–24, 23
"Darwin's finches", 37
 adaptive peaks, 567
 character displacement, 572, **573**
 Galapagos species, 22, **23, 38**
 hybridization effects, 596
Darwinism, 3, 53. *See also* Neo-Darwinian
 (modern) synthesis
 basic hypotheses, 27, 36
 concept of variation, 515
 conflict with religion, 60
 definition, 637
 historical background, 3–17
 impact on biology, 36
 impact on nonbiological fields, 53
 in the nineteenth century
  scientific objections, 32–36
  scientific support, 36–50
 social. *See* Social Darwinism
Dasycercus (marsupial mouse), **39**
Dasyurus (marsupial cat), **39**
DDT (dichloro-diphenyl-trichloroethane),
 229, **231, 565**
Death, 56, 57
 cultural concepts, 14
 definitions, 132
 genetic, 557
 religious concepts, 57, 58
 as a selected trait, 556, 614

Deccan Traps (volcanic lava), 433
Deciduous (teeth), definition, 637
Deficiency. *See* Deletions
Degenerate code. *See also* Genetic code
 definition, 637
Dehydrogenases, **160,** 167, **168, 178,** 203,
 260, 262, **539,** 561, 562
 definition, 637
Deinonychus (theropod dinosaur), 426
 in dinosaur phylogeny, **426**
Deleterious alleles. *See also* Gene, deleterious
 definition, 637
 in haploidy and diploidy, 302
 replacement cost, 558
Deletions
 as cause for regulatory changes, 286, **287**
 chromosomal, 208
  pairing in heterozygotes, **211**
  "buckles", 209
 definition, 637
 deleterious efffects, 209
 nucleotides, 147, 215, 216
 sex chromosome incidence (humans),
  616
 viral origin, 225
Demes, 519
 definition, 637
 in Wright's shifting balance hypothesis,
  566, 567
Demospongiae (sponges), **370**
Dendroica (warblers), resource partitioning,
 572
Density dependent. *See also* Ecological
  factors; Population growth
 behavioral gene response, 586
 competition, 556
 definition, 638
 growth, 554
Density independent. *See also* Ecological
  factors; Population growth
 definition, 638
 growth, 554
Dentin, 401
 definition, 638
Deoxyribose sugar, **112**
 in nucleic acids, 111
 structural formula, **112**
Derived characters, 249, **250**
 definition, 638
Dermopterans, in Cenozoic mammalian
  radiation, **456**
Desulfovibrio (sulfate-reducing bacteria)
 in evolutionary tree, **283**
 ferredoxin amino acid sequence, **162**
Deuterium, description, 638
Deuteromycetes (fungi), 315, **319**
Deuterostomes, 394
 cleavage patterns, 376, 393
 definition, 638
 distinguishing characteristics, 376, 393
 embryological development of coelom, 393
 fate of blastopore, 365, 376, 393
 in metazoan phylogeny, **368**
 phyla, 366
 proposed monophyletic origin, 395
 segmentation, 376
Development, 336. *See also* Embryology
 aging, 556. *See also* Aging
 allometry, 360
  description, 632
  hominid brain size, 500

Development—*Cont.*
  amniotic egg membranes, **418**
  bacterial sporulation, **348**
  "biogenetic law" (Haeckel), 40, 41, 634
  canalization, 357, 358, 359, 542
  cleavage patterns, **376**
    radial, 393
    spiral, 393, 394
  complexity, 361
  constraints, 61, 355, 356, 565
    adaptive, 358
    from "design", 584
    evolutionary effects, 598
    mutational, 357
  description, 638
  determinate, 376, 393, 394
  by duplicated genes, 353, **354**, 355
  early concepts, 7
    emboitement theory, 12
  epigenesis, 13, 356
    definition, 639
  eutely, 374
  evolutionary innovations, 360–61
  genetic assimilation, 358
  gradients, 351, **352**
  growth coordinates, 219, **222**
  heterochrony, 360, 397
  homeoboxes, 353
    metazoan homologies, 338, 353, **354**
      the "zootype", 354, 355
  homeostasis, 358
  homeotic mutations, **352**, 353, **355**, 382
  "hopeful monsters", 599
  immune system antibodies, 356
  indeterminate, 376, 393
  Lambda (λ) virus, 347
  metamorphosis, 383, 397
    definition, 648
    tunicates, **398**
  modularity, 361, 382
  morphogen, 351
  "mosaic", 376
  neoteny, 41, 397, 649
  "norm of reaction", 357
  notochord genes, 398
  ontogeny, 40, 649
  orthogenetic concepts, 516
  paedomorphosis, 397, **399**
    definition, 650
  as phenotypic interaction, 356
  phylogenetic recapitulation. *See* Haeckel's
    biogenetic law
  phylotypic stage, 41, **355**
    definition, 651
    pharyngula, 40, **44**, **355**, 356, 391, 412
  positional information, 351, **352**
  preformationism, 12, 652
  "regulative", 376
  regulatory effects, 351
    gill to wing transition, 360, 382
  segmentation, 337–38, 340, **353**, 654
    in proposed metazoan phylogenies, **332**
  selective effects, 349
  sex determination, 200
  shared vertebrate patterns, 42, **44**
    as community of descent, 40
  signal transduction, 349, **350**, 357
  stability, 588
  T4 virus, 345, **346**
  totipotency, 624
  transcriptional innovations, 360

  von Baer's law, 40
  zootype, **355**, 657
Diadectes (amphibian anthracosaur), 413,
  414, 415
Diapsids (fenestrated reptiles), 419
  archosaurs, 423
  cell-type numbers, **544**
  classification, 421
  fossil appearance, 424
  in phylogeny of reptiles, **422**
  pterosaur succession, 431
  skull temporal openings, **420**
Diarthrognatus (Triassic therapsid), 444
Diatryma (flightless early Cenozoic bird),
  436, **437**
Dicotyledons. *See also* Angiosperms
  description, 638
  proposed phylogenies, 318
Dicyanamide, structural formula, 126
Dicyandiamide, structural formula, 126
Didelphis. See Opossum
Didelphodus (early Cenozoic mammal)
  molar teeth, **448**
Differentiation, **352**. *See also* Development;
  Embryology
  description, 345, 638
Digenea (flukes), parasitic platyhelminths,
  374
Dihydroxyacetone, abiotic synthesis, 124
Dihydroxyacetone phosphate
  in anaerobic glycolysis, **160**
  in Calvin cycle, **164**
Dimetrodon (Permian pelycosaur), 421. *See
  also* Pelycosaurs
  reconstruction, **424**
  skull structure, **420**
Dimorphism
  birds, 588, **589**
  Bonellia, 198
  definition, 638
  Dinophilus, 198
  elephant seals, 588
  fish, 198
  hominids, 472
  Irish "elk", **589**
  "runaway" selection, 588
Dinoflagellates (planktonic protistans)
  Late Cretaceous extinctions, 430
  in protistan phylogeny, **330**
  single division meiosis, 191
Dinophilus (annelid sea worm), non-
  chromosomal sex determination, 198
Dinornis maximus (giant moa), **437**
Dinosaurs, 429. *See also* individual taxa
  "arms race", 426, 574
  bird similarities, 431, 436
  Cretaceous extinctions, 427, 428–29
  endothermy-ectothermy controversy,
    427–29
  hadrosaurs, **426, 427**
  ornithischians, 425, **426, 427**
    classification, 421
  pelvic differences, **425**
  phylogeny, **426**
    reptilian groups, **422**
  predator-to-prey ratios, 428
  prevailing plant flora, **311**
  proposed as new vertebrate class, 428
  in reptilian taxonomy, 421
  saurischians, 425, **426**, 427, 428
    classification, 421

  thecodont ancestry, 425
  theropods, 426, 584
Diodon (puffer fish), changed growth
  coordinates, **222**
Dioecious, **199**
  definition, 198, 638
Dipicolinic acid synthetase, **348**
Diploblastic, 333, **337**
  definition, 638
Diplocaulus (amphibian lepospondyl), **415**
Diploids
  advantages, 302
    in multicellular organisms, 302
  amphi-, 206, **209**
  definition, 191, 208, 638
  effects of selection, 536–50
  genotypic equilibrium. *See* Hardy-
    Weinberg principle
  genotypic variability, 197
  meiosis, **193**
  plant sporophytes, 297
Dipnoans (lungfish), 405
  DNA nucleotide content, **268**, 269
  "living fossil", 49
  in phylogeny of fishes, **406**
  possible ancestor to tetrapods, 405
  preadaptations for terrestrial existence,
    409
Directional selection, 290, 359, 542. *See also*
  Selection
  definition, 638
Disassortative mating, 529
Disequilibrium (linkage), 524–25, **526**, 561,
  647
Disruptive selection, 542, 596, 597
  definition, 638
Distorter genes (meiotic drive), 202, 566
Distribution, normal, 219, **222, 223**
Diurnal, definition, 638
Divergence
  in adaptive radiation, **37, 39**. *See also*
    Adaptive radiation
  definition, 638
  in establishing phylogeny, 263
  in gene duplications, 260
  "isolation by distance", 488
  in speciation, **592**
DNA, **216**, 275, 349, 353, **622**
  aberrations. *See also* Mutation
    deletions, 215, 216
    duplications, 215
    insertions, 215, 216
    inversions, 215
    transpositions, 215
  absence of catalytic activity, 146
  advantages
    as genetic material, 146
    in information storage, 146
  ancient, 276–79
    contamination problems, 279
    deterioration and fragmentation, 277
    mummies, 277
    Neanderthal, 279, 481
  biased optical rotation, 150
  "cDNA", 269
  chloroplast, 180, 202
  and chromosome numbers, 269
  complementary organization
    base pairing, 113, **114**
    strands, **114, 115**
  "CpG islands", 269

Electron transport systems
    in cyclic photosynthesis, **163**
    in non-cyclic photosynthesis, **166**
    oxidative (respiratory) phosphorylation,
        169
    primitive forms, 162
Electrophoresis
    description, 639
    detection of allozyme variations, **230**
Elements, early concepts, 4. *See also*
        Chemical elements
Elephants, 7
    Cenozoic proboscidean radiation, **456**
    evolutionary rate, 599
    Pleistocene mammoth extinctions, 454
Ellis-van Creveld syndrome, 528
Embden-Meyerhof glycolytic pathway, 159,
    160. *See also* Anaerobic, glycolysis
Embelomeres, in amphibian phylogeny, **415**
Emboitement theory, 12
Embryology, 13, **354.** *See also* Development
    archenteron, **332, 334,** 336
    blastopore, **334,** 336, 365, 376, 393, 394
    blastula, **334,** 335, 376
    cleavage patterns, 376, 394
        radial, **376,** 393
        spiral, **376,** 393, 394
    gastrula, **334**
    Haeckel's biogenetic law, 40, 41, 634
    phylotypic stage, 355
        definition, 651
        pharyngula, 40, **44, 355,** 356, 391, 412
    teleological concept, 5
    vertebrate gills, 391, 412
        homologies, 40, **44**
Emotions, 505–6
    advantages, 506
Enation theory of leaf origin, 305, **309**
Encephalization quotients, **455**
    hominoids, 500
Endemic, definition, 639
Endocranial casts, 471, **501,** 502
Endocytosis
    definition, 639
    in eukaryotic organelle evolution, 179,
        181. *See also* Endosymbiosis
Endoderm, **332, 334**
    definition, 639
    diploblastic, **334**
Endonucleases, definition, 639
Endoplasmic reticulum, **110,** 175
Endosperm, 310, 311, 313, **315**
Endosymbiosis. *See also* Symbiosis
    cryptic, **178**
    definition, 639
    in *Drosophila,* "sex ratio" condition, 201
    for origin of eukaryotic organelles, 180,
        181, 279, 281, 329
        Gram-negative bacterial proposal, 182
        possible glycolytic enzyme, 283
    proposed origin of eukaryotic nucleus, 181
    secondary invasions, 180, 281, 329
Endothermy
    advantages, 421
    definition, 639
    dinosaur controversy, 427–29
    energy requirements, 421
    mammals, 442, 447, 450
        factor in adaptive radiation, 452
    as possible cause for Late Cretaceous
        survival, 429

predator-to-prey ratios, 428
pterosaurs, 434
therapsids, 421, 443, 451
turbinal respiratory membranes, **444**
    absence in dinosaurs, 428
Energy, 119, 126
    activation, **140**
    chemical, 158, 159. *See also* Metabolism
        from monosaccharide breakdown, 158
        from oxidative phosphorylation, 170
        phosphate bonds, **116,** 126, 128
        photosynthetic, 163
        in proton gradients, 162
    cosmic rays, 119
    electrical discharges, 119
    gamma rays, 125
    photons, quantum organic yield, 125
    in proton gradients, **166**
    radioactivity, 119
    solar, 119, 163
    sources for abiotic synthesis, 119, 121
    thermal, 139
        shift to chemical, 139
    thermodynamic, 118
    ultraviolet, 119
    volcanoes, 119
Enolase
    in anaerobic glycolysis, **160**
    in lens crystallins, 260, 262
Entamoeba histolytica (mitochondrionless
        protistan), **178**
Entoprocts (invertebrate phylum)
    classification among metazoan phyla, 366
    time of first fossil appearance, **324**
Entropy
    definition, 639
    in mass action equilibrium, 132
    origin of life problem, 118, 119
        decrease in metabolic systems, 132
Environment, 554, 571, **581.** *See also*
        Ecological factors; Ecological niche;
        Habitats
    and adaptation, 27
    carrying capacity, 554
    coarse-grained, 541, 636
    constraints, 358
    definition, 639
    desiccation resistance
        plants, 298, 299
        reptiles, 417–19
    deterioration, human responsibility, 431
    effect on intelligence, 618
    effect on sex determination, 198, 202
        sex ratio, 385
    eutrophication, 404
    fine-grained, 541, 640
    genetic assimilation, 358
    homeostatic responses to, 358, **359**
    hypoxia, 404
    Lamarckian concept, 25
    "patchy", 474, 489
    resource partitioning, 45, **572**
    variable features, 553
Enzymes
    acetoacetyl-Co A reductase, **348**
    acid phosphatase, 528, 562
    aconitase, **168, 348**
    activation energy, **140**
    active site, 159
    adenosine deaminase, **348**
    adenylate cyclase, 582

advantages, 139
alcohol dehydrogenases, **160,** 260, 262,
    561, 562
aldehyde dehydrogenase, 262
aldolase, 561
allozymes, 228, **230,** 275, 561. *See also*
    Allozymes
    definition, 632
amino acid-activating, 153
amylase, **134**
in antibiotic synthesis in absence of
    mRNA, 144
argininosuccinase, 178
aspartate transaminase, **178,** 256, **257**
association between function and
    polymorphism, 561
ATPase, **166**
autocatalysts, 140, 142
in bacterial sporulation, **348**
in biasing optical rotation, 150
carbamoyl phosphate synthase, **178**
carbonic anhydrase, 467
catalase, **140,** 166
citrate synthase, **168, 178**
in coacervates, 133, **134**
coenzymes, 125. *See also* Coenzymes
in DDT insecticide breakdown, 230
definition, 639
dehydrogenases, 262, 561, 562
dipicolinic acid synthetase, **348**
DNA polymerase, 142, 215, 291, 570
in DNA repair, 225
in dominance-recessive relationships, 197
electrophoretic detection, **230**
endonucleases
    definition, 639
enolase, **160,** 260, 262
esterase, 561, 562, 582
evolution in laboratory experiments, 290
evolutionary lengthening, 144, 146
fructose bisphosphate aldolase, **160, 164**
fumarase, **168, 348,** 562
α galactosidase deficiency, 203
β galactosidase, **220,** 290, 347
glucose dehydrogenase, **348**
glucose-6-phosphate dehydrogenase, 562
glutamate dehydrogenase, 262
glutathione transferase, 260, 262
glyceraldehyde-3-phosphate
    dehydrogenase, **160, 164, 178,** 283
glycerol-3- phosphate dehydrogenase, 560
glyoxylase, 582
hexokinase, **160**
hexose diphosphatase, **164**
human polymorphisms, 582
hydrogenase, 161
in information transfer, **141**
    protein translation, 113, 140
isocitrate dehydrogenase, **168**
isomerase, definition, 646
isozyme, **230**
kinases, **350**
lactate dehydrogenase, **160,** 260, 262
ligase, **621**
luciferase, 166
lysozyme, 257, 260
    convergent evolution, 265
lysyloxidase, 325
malate dehydrogenase, **168, 178,** 262, 562
mitogen-activated protein kinases
    (MAPKs), 349, **350**

proposed phylogenies, **266**
   metazoans, **332, 333**
   protistans, 330
proto-, **181**
regulatory nucleotide sequence, **221**
some 16*S* rRNA sequences, 177
sexual reproduction, 182. *See also*
   Reproduction
split genes, 175, **176**
   RNA processing, 177
sporulation, 347
superkingdom relationships, 177, **178,**
   247, 276
distinctions, 173
Eupantotheres (Jurassic mammals)
   in phylogeny of mammalian teeth, **449**
   tribosphenic molar, **448**
Euparkeria (Triassic thecodont), 424
   reconstruction, **425**
   skull structure, **420, 434**
Euphorbia (desert plants), convergent
   evolution, 313, **317**
Euphydras editha (butterfly), 547
Euploidy, 204. *See also* Polyploidy
   categories (types), 208
   definition, 640
   possible causes, 206
   triploidy, 208
      angiosperm endosperm, 310, 313
Euplotes (ciliate protozoan), genetic code
   exceptions, **154**
Euryapsids (fenestrated reptiles), skull
   temporal openings, **420**
Eurypterids (fossil arthropods), 382, 383
   predators on early vertebrates, 401
Eusthenopteron (Devonian osteolepiform fish)
   ichthyostegid vertebral comparisons, **412**
   limb bone comparisons with fossil
      amphibians, **411**
   preadapted limb girdles, **413**
   reconstruction, **405**
Eutelegenesis, 618
   definition, 640
Eutely, 374
   definition, 640
Eutherians, 102, 103, 104, **105, 106.** *See also*
   Mammals; Placentals
   definition, 651
Eutrophication, 404
   definition, 640
Evolution. *See also* Phylogeny; Natural
   selection; Speciation
   adaptive constraint, 358
   adaptive radiation, 36, 431. *See* Adaptive
      radiation
   amphibians, 410–16
   anagenesis, 241, **242**
      description, 632
   anti-evolution arguments and
      evolutionists' responses, 65–70
   anti-evolution laws, 63, 69–70
   "arms race". *See* "Arms race"
   basic concepts, 3, 29
   behavior. *See* Behavior
   birds, 436–37
   Buffon's proposals, 10
   caused by changes in gene regulation,
      286–87
   cell division, 182
   Chambers' proposals, 26
   chemical versus biological, 136

chromosomal, 211, 212, **213**
   change in number, 210, **212,** 228
   change in structure, 209, **214, 215**
cladogenesis, 241, **243,** 599
co-, 574–75
complexity. *See* Complexity
concerted, 271, 636
contrast between populations and
   individuals, 517, 563, 566, 570
convergent. *See* Convergent evolution
cost of selection, 558–62. *See also*
   Genetic load
cultural, 606–13. *See also* Culture
definitions, 3, **27,** 518, 534, 600, 640
divergent. *See* Divergence
emotions, 505–6
enzymes, 144, 146, 262, 290
etymology, 3
eukaryotic organelles, 177, 180, **181**
as "fact" and "theory", 65–66
fish, 398–405, **406**
fungi, 315–16
genetic code, 146–54
grade, 245, 247, 382, 452
   coordinated changes, 451
Haeckel's biogenetic law, 40, 41, 634
hemoglobin polypeptide chains,
   258–60
herbicide resistance, 565
hominids. *See* Hominids
homology, 241, **244**
horses, **48**
human. *See* Humans
impact on religion, 53, 54, 58, 60. *See also*
   Religion
impact on social sciences and the
   humanities, 53
innovations, 39, 306, **332,** 336, 356, 360,
   365, 412, 416, 424, 451, 452. *See also*
   Adaptation
insect segmentation, **353**
insecticide resistance, 229, 565
invertebrates, 365–88
Krebs cycle, 167
laboratory-controlled, 290–91. *See also*
   Genetic engineering
   enzymes, 290
   use of genetic recombination, 291
Lamarckian. *See* Lamarckianism
locomotion. *See* Locomotion
macro-. *See* Macroevolution
mammals, 442–59
   jaw articulation, 444, **445**
   middle ear ossicles, **445**
man-made polyploidy species, 206
Mesozoic lessons, 451–52
metabolism, 159, 170
metazoan locomotion, 338–40
micro-. *See* Microevolution
modular, 382
molecular clocks. *See* Molecular
   phylogenies, evolutionary clocks
molecular techniques for determining
   phylogenies, 256–86. *See also*
   Molecular phylogenies
monophyletic, 245, **246.** *See also*
   Monophyletic evolution
   arguments for, 451
mosaic, 382
"mutationists", 516
Neo-Darwinian synthesis, 516–18

neutral mutation theory, 559–60, 569. *See*
   also Neutral mutations
   definition, 649
in the nineteenth century
   scientific objections, 32–36
   scientific support, 36–50
non-Darwinian, 559
objections against, 11, 24
   philosophical criticisms, 14
   religious arguments, 54. *See also* Judeo-
      Christian religion
orthogenetic concepts, 45, 49, 429, 518
   definition, 649
   in development, 516
parallel. *See* Parallel evolution
phyletic, 35, 241, **242,** 250, 599, 651
plants. *See* Plants
polyphyletic, 245, **246,** 247, 248, 308, 313,
   329, 334, 337, 338, 374, 382, 398,
   414, 416, 447. *See also* Polyphyletic
   evolution
   definition, 652
   genotypic basis, 451
   reasons for rarity, 451
polyploidy, 206, **208**
   man-made species, 206, **209**
postadaptive concepts, 227
preadaptive concepts, 227, 229. *See also*
   Preadaptation
Precambrian events, **174**
predictions, 15, 308
primates, 468–88
progenote, 177, 329, 652
   metabolic pathways, **178**
"progress", 451. *See also* Progress
protein, 257–66
   synthesis, 144–46
   translation, 145
radiation. *See also* Adaptive radiation
rates. *See* Evolutionary rates
Red Queen hypothesis, 543, 544, 553, 653
repetitiveness, 243
reptiles, 416–36
reticulate, **246,** 249
retrograde (backward), 159
ribosomes, 145
rice, 349
from RNA to DNA, 146
saltation theory, 516, 518, 599, 654
sex chromosomes, dosage compensation,
   200
sex determination, 199
sex ratio, 202
signaling pathways, 349, **351**
social behavior. *See* Social behavior
split genes, 175
   "introns early" hypothesis, 175
   "introns late" hypothesis, 175, 177
"streetcar theory", 600
"struggle for existence", 25, 26, 36
"survival of the fittest" concept, 27, 60,
   324, 534, 609
"survival of the luckiest" concept, **324**
theistic, 61
"tinkering", 382
transcription, 146
transfer RNA, **149**
transitional forms, 44. *See also* Phylogeny;
   and individual taxa
   ape-hominid (*Ardipithecus ramidus*),
      506

Evolution—*Cont.*
  fish-amphibian, **410**
  horses, 45, **48**
  reasons for rarity, 44, 378, 379
  reptile-bird (*Archaeopteryx*), 44, 47, 50,
    431, **435, 436**
  reptile-mammal. *See* Therapsids
  response to creationist claims, 66–67
  uniformitarian concepts, 15, 16, 25, 657
  unpredictability, 451–52
  vertebrates. *See* Vertebrates
  vestigial organs, 39, 40, **43**
    definition, 657
  vitalistic interpretations. *See* Vitalism
  Wright's shifting balance hypothesis,
    566–69
Evolutionarily stable strategies, 587
Evolutionary classification, 251–52
Evolutionary clocks, 281–86. *See also*
  Molecular phylogenies, evolutionary
  clocks
  definition, 640
  differences among taxonomic groups, 286
  hemoglobin changes, 284, 560
Evolutionary rates, 597–99. *See also*
  Molecular phylogenies
  categories, 598
    bradytelic, 598, 634
    horotelic, 598, 644
    quantum, 598, 652
    tachytelic, 598, 656
  the "darwin" measurement, 598
    in guppy populations, 600
  mammalian genera, **598**
  nucleotide substitutions
    artiodactyls, **273**
    synonymous codons, **286**
  possible causes, 598–99
  problems of determination, 597
"Evolutionary psychology", 506
Exons, 142, 175
  definition, 640
  hemoglobin, **176,** 260
  polymorphism, 562
  and polypeptide domains, 175
Extension (appendage), definition, 640
Extinctions
  Cambrian period, **324**
  caused by adaptive and genetic
    constraints, 358, 452
    high selection coefficients, 559
  caused by meiotic drive, 566
  concept of "racial senescence", 429
  from continental drift, 104
    South American mammals, 103, 104
  definition, 640
  dinosaurs, 428–29
  early amphibians, 412
  as gaps in the "Ladder of Nature", 7
  human responsibility, 431
  from inbreeding, 530
  large scale events, 432
    Late Cretaceous
      Chicxulub crater, 433
      extraterrestrial impact hypothesis, 433
      numbers of genera, 430
      possible causes, 429
    Late Pleistocene, 454
    modern period, 431
  organismic replacement opportunities,
    452

periodicity, **433**
possible extraterrestrial causes, 432–33
sexual/asexual comparisons, 301
therapsids, 424
Extranuclear inheritance, 202. *See also*
  Chloroplasts; Mitochondria
Eyes
  analogies among taxa, 37, 378
  convergences, **34**
    visual pigments, 561
  evolutionary sequence, 35
    mollusks, **34**
  homologous developmental factors, 251,
    354
  lens crystallins, 251, 260
  pineal, 399, 405
  preadaptations, 33
  as vestigial organs, 15, 40

evolution, 398–405, **406**
    rapid rates, **282,** 597, 600
fins. *See* Fins (fish)
gill arches, 42, **45**
    proposal for jaw ancestry, **402**
heart, 42
heterocercal tails, **400, 403**
hypocercal tails, **400**
imaginary hybrids, 10
jawed forms, 400–405
    advantages, 401
    fossils, **401,** 403
        early appearance, 401
    proposed origins, **402**
    urogenital duct systems, **418**
labyrinthodont rhipidistian dentition, **411**
Late Cretaceous extinctions, 430
lateral line system, 401, 412
lobe-finned (sarcopterygians), **49,** 404,
        405, 409, 410
    characteristics, 404
    in phylogeny of fishes, **406**
lungs, 404
    preadaptations, 33
        laryngeal valves, 493
non-chromosomal sex determination, 198
osteichthyans, 403–5, **406**
osteolepiforms, **405,** 409
    limb bone comparisons with fossil
        amphibians, **411**
    preadapted limb girdles, **413**
    vertebrae, **412**
pax-6 gene homologies, 354
phylogeny, 405, **406**
placoderms, **401,** 402
protein polymorphism, 231
ray-finned (actinopterygians), 404
    chondrosteans, 404
    neopterygians, 404
    teleosts, **222,** 280, **282, 403,** 404, **406**
        enzyme polymorphism, 561
        hemoglobin chain substitutions, 284,
            560
    urogenital duct system, **418**
school defense, 570
selection for terrestrial incursions, 409
    modern teleosts, 410
silent/replacement mutation rate
    comparisons, 273
stickleback (premating isolation), 593
stratigraphic (fossil) appearance, 91
swim bladder, 33, 403, 405
vagus nerve, 42, **45**
Fisher's fundamental theorem of natural
    selection, 542, 569
Fission-track dating, 94
Fitness. *See also* Adaptive value; Selection
    adaptive peaks, 565–69, **567,** 569
    altruism effect, 570
    definitions, 534, 640, 653
    evolutionary value, 361
    Haldane's rule, 200, 591
    in hypothesis testing, 15
    inclusive, 610, 645. *See also* Kin selection
    optimum genotypes, 565
        limitations, 558
        linkage effects, 565
    optimum phenotypes, 542, 565
    overdominance, 537, 538, 539, 558, 561
        proposed evolutionary cost, 559
        sickle cell gene, 539

and segregation distortion, 566
and sexual interaction, 587–89
thresholds, 560
variable features, 553
Fixation (gene, gene frequencies),
    definition, 641
Fixity of species concept, 12, 13, 53, 62
    definition, 641
Flavine adenine dinucleotide (FAD), 142, 145
    electron acceptor, 162
    in Krebs cycle, **168**
    in non-cyclic photosynthesis, **166**
    in respiratory pathway, **169**
Flavines (plant pigments), **163**
Flavoprotein, in respiratory pathway, **169**
Flexion (appendage), definition, 641
Flight
    advantages, 431
    insects, proposed origin, 382
    vertebrates, 433–37. *See also* Birds,
        Pterosaurs
Flora, definition, 641
Food chain, proposed increase during
    Cambrian radiation, 325
Foraminiferans (marine protistans)
    Late Cretaceous extinctions, 430
    in protistan phylogeny, **330**
Forces (physical)
    electromagnetic, 79
    gravity, 15, 16, 57, 76, 79
    nuclear, 79
Formaldehyde
    interstellar density, **120**
    structural formula, 123, 124
    synthesis, 123
Formamidine, abiotic synthesis, 124
Formic acid, synthesis in Miller electric
    spark discharge experiment, 122
Formylmethionine, 173
Fossils, 13, 15, 91, 240, **304, 319.** *See also*
        individual taxa
    advantages, 237
    algae, 296, **297**
    angiosperm leaves, 311
    Aysheaia (Cambrian "arthropod"), **382**
    Cambrian metazoa, **328**
    Carboniferous
        insect, **382**
        in tree stumps, 419
    challenges to the Great Chain of Being, 13
    definition, 641
    DNA. *See* Ancient DNA
    early cells, 171, 172
        eukaryotes, **182,** 323
    early plant seed, 308
    early reptilian egg, 419
    estimated rates of fossilization, 237
    evidence for continental drift, **98**
    evolutionary rates, 598
    fish, **400**
    fossilization process, **46**
    fungi, **319**
    in geological dating, 91, **92**
    Gondwana system associations, 96, **98**
    gymnosperms, 307
    historical concepts, 13
    hominids, 470–74
    Homo, 477–82. *See also Homo* species
        numbers, 68
        skulls, **480**
    horses, 44, **48,** 49. *See also* Horses

Lamarckian concepts, 24
"living". *See* Living fossils
"Lusi naturae", 13
mass extinction events, 432
Megalonix jeffersoni (North American
    ground sloth), 13, **16**
Megatherium (giant ground sloth), **16,** 22
    metabolic, 139
"missing links". *See also* Evolution,
    transitional forms
    Archaeopteryx (reptile-bird), 44, 47, 50
    Ardipithecus ramidus (ape-hominid),
        506
mollusks
    ammonites and belemnites, 377
    Paludina species, **37**
    obstacles to durability, 42
Pikaia, possible vertebrate ancestor, **399**
Piltdown forgery, **471,** 472
plants, 303–8
Precambrian metazoa (Ediacaran strata),
    326, **327,** 376
primates, 468–74, **469**
protistan acritarchs, 182
punctuated equilibrium. *See* Punctuated
    Equilibrium
rarity, 91, 237
    because of disarticulation, 445, 468
    because of fragility and erosion, 42, 237
    because of limited accessability, 237
    molecular compounds, 118
    Precambrian metazoans, 323
    transitional forms, 67, 379
as real species, 19, 49
saltation concept, 516, 599
South American mammals, 103
stratigraphic appearance, 13, 91
stromatolites, **170, 171, 172,** 325, 655
systematic sequence, **37**
transitional. *See* Evolution, transitional
    forms
Founder effects, 548, **549,** 593. *See also*
    Bottleneck effect; Random genetic drift
    bottlenecks, 549, 634
        arguments against, 594–95
    definition, 641
    Hawaiian Drosophilidae, 548, **549**
    proposed cause for macroevolution, 599
    proposed rapid evolutionary rate, 598
Fox's proteinoids. *See* Proteinoids
Frameshifts (codon reading frame), 147, 216
    mutations, **217**
Fredericella (ectoproct), **340**
Frequency-dependent selection, 540, 560
    definition, 641
Fritschiella (algae), 297, 298
    structure, **297**
Frogs and toads, **288,** 414, 415
    adaptations, 416
    African clawed toad. *See* Xenopus
    characteristics, 415
    desert toads, 414
    DNA nucleotide numbers, **268**
    evolutionary stability, 286, **288**
    forelimb vertebrate homologies, **41**
    fossil forms, **416**
    hybrid inviability, 591
    nuclear transplantation, 624
    number of repetitive DNA copies, 270
    number of species, 286
        developmental divergence, 599

sex determination. *See* Sex determination
sex-linked. *See* Sex linkage
sexual antagonistic, 199, 589
split, 175, **176,** 655
structural, 219, **220, 287,** 655
"supergene", 209
Gene conversion, 271
Gene duplications, 209. *See also*
    Duplications
in development, 352, 353
in evolution of transfer RNA, 152, **153**
evolutionary advantages, 146, 208, 242
ferredoxin, 262
frequency, 276
glutamate dehydrogenase, 262
haptoglobin, 262
hemoglobin chains, 260
histones, 271
homeobox sequences, **354,** 355
immunoglobulin G, 262
insulin, 260
α lactalbumin, 260
lactate dehydrogenase, 260
lysozyme, 260
in metazoan segmentation, 352
myoglobin, 260
in polyploids, 208
ribosomal RNA, 209, 271
selection for divergence, 146, 562, 565
in serial homology, 242
serine proteases, 260
transfer RNA, 271
from unequal crossing over, 260
vertebrate nerve growth factor, 260
Gene flow. *See also* Migration
definition, 546, 641
impediment to race formation, 589
in proposed human multiple origins, 487
Gene frequencies, 519–32, 519, 520, 527,
    538
binomial expansion, 522
changes, 533–50
    bottleneck effects, 549, 593
        proposed cause for macroevolution,
          599
    by migration, 546–47. *See also*
        Migration
    by mutation, 533–34. *See also* Mutation
    nondirectional, 547
    by random genetic drift, 547–48. *See*
        also Random genetic drift
        definition, 653
        sampling error, 547, 654
    by selection, 534. *See also* Selection
        in diploids, 536–50
        dominance effects, 536–37
        in haploids, 536
conservation, 520–28. *See also* Hardy-
    Weinberg principle
contrast between populations and
    individuals, 517
definition, 641
disequilibrium, 524–25, **526**
elimination, 537, 538, **547**
equilibrium. *See* Equilibrium, genotypic
fixation, 537, 538, **547, 548, 549**
    definition, 641
    neutral mutations, 559
multinomial expansion, 523, 524, 528
in Neo-Darwinian synthesis, 517
as a racial distinction, 581, 583

sampling error. *See* Random genetic drift
sex linkage, 525–27
trinomial expansion, 523
unpredictable events, 549
various human traits, 528
Gene manipulation, 619, 620–23. *See also*
    Genetic engineering
Gene pair, definition, 641
Gene pool
definition, 519, 641
in Neo-Darwinian synthesis, 517
Gene therapy, 620. *See also* Genetic
    engineering
definition, 641
transforming vectors, 620
Genealogy, definition, 641
Genetic assimilation, 358
Genetic background, 358
Genetic code. *See also* Codons
absence of ambiguity, 147, 148
basic features, 148–49
commaless, 147, 148
definition, 637, 642
dictionary, 147
disproportionate codon numbers, 151
doublet, 150
evolution, 146–54
    hypothesis of anticodon duplication,
        153
    hypothesis of complementary amino
        acid-codon pairing, 150
expansion of, 152
reading frame, 147, 148, 216
    frameshifts, 147, 216
redundancy ("degeneracy"), 147, 148,
    149, 151
degeneracy, definition 637
    for minimizing translational errors, 151
    in wobble pairing, 152
restricted alphabet, 146, 147
sense word, 147
singlet, 150
terms used for, 147
triplet codon selection, 150
universality, 147, 149, 154, 177
    definition, 657
    exceptions, 148, **154**
    frozen accident hypothesis, 150
wobble pairing, **149,** 152
Genetic constancy, 189. *See also* Gene
Genetic crossing over. *See* Crossing over;
    Genetic recombination; Linkage
Genetic death, 557. *See also* Genetic loads
definition, 642
evolutionary cost, 558–62
    estimated number, 558
Genetic distance, 487. *See also* Phylogeny,
    distance measurements
definition, 642
human races, 582, **584**
"isolation by distance", 488
Nei's index, 583, **595**
    calculation, 583
Genetic drift. *See* Random genetic drift
Genetic engineering, 619, 620–23
chronology of scientific advances, 623
cloning (gene), 277, **622**
cloning (organisms). *See* Clones
    (organisms)
definition, 642
ethical problems, 623

vectors, 620, **622,** 623
    definition, 657
Genetic loads
advantages, 558
in asexual populations, 564
balanced, 558, 559, 633. *See also*
    Polymorphism
definitions, 557, 642
for deleterious recessives, 557
human, 615
mutational, 336, 558, 648
    relationship to mutation rate, 558
proposals for minimizing, 560
recombinational, 558
segregational, 558
    balanced lethals, 559
    calculation, 558
    proposed evolutionary cost, 559
threshold effects, 560
in Wright's shifting balance theory,
    567
Genetic material
advantages of DNA, 146
alphabet, 146, 147
organismic quantities, **268**
possible early forms, 141
properties, 113
in the "RNA world", 143
selection for increase in, 267
source of homology, 243
Genetic polymorphism. *See also*
    Polymorphism
definition, 642, 652
as source of preadaptation, 229
Genetic recombination, 191. *See also*
    Crossing over; Linkage
definition, 653
in DNA repair systems, 225
in *Drosophila,* **204**
effect on epistasis, 567
effect on linkage disequilibrium, 525,
    526, 564
and evolution of sex, 564
evolutionary value, 565
    advantages, 191, 204, 301, **563,** 564
    disadvantages, 301, 563
inversion effects, 209, **229,** 564
in laboratory controlled evolution, 291
in measuring linkage, 203
mitochondrial proposal, 483
recombinational load, 558
reduction in Y chromosome, **199**
selection against, 199
selection for change in rate, 564
source of variability, 204, 563
unequal crossing over, 209, 260, 657
via transposons, 225
X-Y sex chromosomes, 199
Genetic variability, 33, 189, 229. *See also*
    Gene; Mutation; Polymorphism;
    Variation
absence in pure lines, 542
for aging, 556
behavior, 586
chromosomal, 228
    number-dependence hypothesis, 210
contrast between populations and
    individuals, 517
effect of random genetic drift, 548
estimates from protein polymorphism,
    231, 582

Genetic variability—*Cont.*
  evolutionary effects, 230
  Fisher's fundamental theorem of natural
    selection, 542
  hybridization effect, 596
  from multicellular meiosis, 302
  numerical possibilities, 197, 566
  polymorphic. *See* Polymorphism
  quantitative, 219
  by recessive gene accumulation, 228
  regulatory genes, 361
  sources
    allelic differences, 197
    cross fertilization, 199
    disruptive selection, 596
    meiotic chromosome distribution, 194,
      300
    recombination, 191, 204, 563, 653
Genome
  adaptive constraints, 452
  advantages of increased size, 267
  definition, 642
  sequence analysis, 276
  superkingdom organization, 173
Genotype
  definition, 113, 189, 194, 642
  frequencies. *See* Equilibrium; Gene
    frequencies; Hardy-Weinberg
    principle
  haplotype, 275
  hemizygote, 203, 643
  heterozygotes, 195, 643. *See also*
    Heterozygotes
  homozygotes, 195, 644. *See also*
    Homozygotes
  hypotheses on precedence of function or
    information, 140–44
  as object of selection, 534
  "optimum", 557, 565, 576
    in adaptive peaks, **567**
    limitations, 558
    linkage effects, 565
  as "prime mover", 356
Genus
  definitions, 10, 247, 642
  problems of classification ("lumpers" and
    "splitters"), 248
Geographic isolation, 589, 590. *See also*
  Isolating mechanisms (barriers)
  definition, 642
Geographic speciation, **592, 593,** 596. *See
  also* Speciation
  definition, 642
Geology. *See also* Earth; Rocks
  banded iron formations, 172
  catastrophism concept, 14, 15, 16
  continental drift, 96, **101,** 102, 103, **106,**
    325. *See also* Continental drift
    biological effects, 102, 104
  crosscutting relationships, **96**
  dating methods
    for sedimentary rocks, **96**
    use of fossils, 91, **92**
    use of radioactivity, 94, **95,** 98, 653
    validity, 68
  early classification system, 94
  earthquakes, 22
    plate tectonics, 100, 102
    seismic waves, 89
  Ediacaran strata, **324,** 326, **327,** 376
  intrusions, **96,** 645

mountain formation, 102, **104**
"overthrusting", 69, **104**
paleomagnetism, 96, 97. *See also*
  Paleomagnetism
Phanerozoic time scales and divisions, 94
philosophical criticisms, 14
Precambrian events, **174**
rock cycle, **91**
stratification, 15, **92**
  definition, 642
stromatolite formations, **170, 171, 172,**
  325
superposition, law of, 91, **96**
tectonic plates, 102, **103,** 656. *See also*
  Tectonic movements and plates
  boundary events, 100, 101, **104**
  Hawaiian Islands formation, **549**
time scales, definition, 642
uniformitarian concept, 15, 16, 657
uplifting events, 22
Geospiza (a genus of "Darwin's finches"),
  23, 38
  conirostris, **23, 38**
  difficilis, **23, 38**
  fortis, **23, 38,** 572, **573**
  fuliginosa, **23, 38,** 572, **573**
  magnirostris, **23, 38,** **573**
  scandens, **23, 38**
Germ plasm
  advantages, 302
  animal–plant differences, 302
  definition, 642
  theory, **28,** 515
Germinal choice, 618. *See also* Eugenics
  definition, 642
Gerrothorax (amphibian stereospondyl),
  412
Geum (perennial herb), cross fertility
  differences among species, **240**
Giardia (diplomonad protistan), **178,** 181
Gibbons (*Hylobates*)
  body contours, **467**
  description, 466
  illustration, **465**
  mitochondrial DNA restriction enzyme
    map, **275**
  phylogeny
    DNA-DNA hybridization, **272**
    immunological methods, **258**
    mitochondrial DNA hominoid
      relationships, **470**
  in primate classification system, 464
  sex-linked traits, 203
Gilia (plant genus, Phlox family)
  mode of speciation, 593
  sibling species, 239
Gills (vertebrates), 391, 402
  fish skeletal structures, **403**
  gill arches, 42, **44, 45**
    innervation, **403**
    proposal for jaw ancestry, **402**
  retention in early amphibians, **410**
Ginkgo (gymnosperm)
  "living fossil", 49, **311**
  in plant phylogeny, **314**
  proposed phylogeny, cytochrome *c*
    sequences, **266**
Giraffes
  cervical vertebrae, 356
  laryngeal nerve, 42, 356
Glaciation, Gondwana events, 97

Gloger's rule, 585
  definition, 642
Glossopteris (Permian fossil tree), 97, **98**
Glucose, 161
  abiotic synthesis, 124
  aerobic metabolism, 167
    ATP production, 169
    anaerobic glycolysis, 158, **160.** *See also*
      Anaerobic, glycolysis
  availability for glycolysis, 159
  biosynthesis, 159
  in Calvin cycle, **164,** 165
  environmental stability, 124
  in progenote metabolic pathways, **178**
  starch polymerization in coacervates, **134**
  structural formula, 124
  in synthesis of maltose, **127**
Glucose dehydrogenase, **348**
Glucose-1-phosphate, in coacervate reaction
  system, **134**
Glucose-6-phosphate
  in anaerobic glycolysis, **160**
  in Entner-Doudoroff pathway, 161
Glucose-6-phosphate dehydrogenase, 562
  deficiency (anti-malarial effect), **539**
  polymorphism, 562
Glutamate dehydrogenase, gene
  duplication, 262
Glutamic acid
  codon designations, 148
  in Murchison meteorite, 122
  structural formula, **111**
  synthesis in Miller electric spark discharge
    experiment, 122
Glutamine
  codon designations, 148
  structural formula, **111**
Glutathione transferase, in lens crystallins,
  260, 262
Glyceraldehyde
  abiotic synthesis, 124
  in primitive anaerobic pathway, 159
Glyceraldehyde-3-phosphate
  in anaerobic glycolysis, **160**
  in Calvin cycle, **164**
  in progenote metabolic pathways, **178**
Glyceraldehyde-3-phosphate dehydrogenase
  in anaerobic glycolysis, **160**
  in Calvin cycle, **164**
  in progenote metabolic pathways, **178**
  proposed endosymbiotic origin, 283
Glycerate-1,3-bisphosphate, in Calvin
  cycle, **164**
Glycerate-3-phosphate, in Calvin cycle, **164**
Glycerol-3-phosphate dehydrogenase,
  evolutionary rate, 560
Glycine
  abiotic synthesis, 123
  codon designations, 148
  in Murchison meteorite, 122
  polymerization in flow reactor, 131
  structural formula, **111**
  synthesis from glycolytic substrates, **161**
  synthesis in Miller electric spark discharge
    experiment, 122
Glycoaldehyde, abiotic synthesis, 124
Glycolic acid, synthesis in Miller electric
  spark discharge experiment, 122
Glycolysis, 561. *See also* Embden-Meyerhof
  glycolytic pathway
  aerobic, 170

anaerobic, 158–62
  enzymatic sequence, **161**
conserved enzyme sequence, 159
definition, 642
as level of complexity, 612
universality, 158, 159
Glyoxylase, 582
Glyptodon (South American fossil armadillo), 22, **459**
Gnathostomata (jawed fish), 401. *See also* Fish, jawed forms
Gnathostomulids (invertebrate phylum), classification among metazoan phyla, 366
Gnetales, in plant phylogeny, **314**
God concepts, **4**, 5, 6, 14
Golgi apparatus, 175
Gombe (Tanzanian national park), 490
Gondwana (ancient southern continent), 96, **98**, 99, 100, 102
  breakup, 99
    separation from Laurasia, **101**
  description, 642
  distribution of fossil plants and reptiles, **98**
  distribution of lungfish, 405
  glaciation, 96, 97
  Precambrian wanderings, **102**
Gonium (algae), possible phylogeny, **297**
Gorillas
  body contours, **467**
  brain volume, 500
  chromosome evolution, 211
    banding comparisons, **214**
  description, 466
    illustration, **465**
  encephalization quotient, 500
  β globin gene family relationships, **260**
  hind limb anatomy, **478**
  karyotype, **214**
  knuckle-walking, 466
  language abilities, 495
  mitochondrial DNA restriction enzyme map, **275**
  phylogeny
    DNA-DNA hybridization, **272**
    immunological methods, 257
    mitochondrial DNA hominoid relationships, **470**
  in primate classification system, 464
  sex-linked traits, 203
  terrestrial stance, **477**
Gossypium (cotton), 546
Grade, as a level of organization
  arthropods, 382
  definition, 642
  mammals, 245, 451
Gradient, definition, 642
Gramicidin (antibiotic), 144
Granite, 91. *See also* Rocks, igneous
  description, 642
Gravity, 15, 16, 57, 76, 79
  in planetary condensation, 86
  in sedimentary rock formation, 90
  in stellar contractions, 79, 80, 82
Great Chain of Being, 6, 8, 11
  definition, 642
  fossil challenges, 13
Green algae, 296. *See also* Algae
Grooming

definition, 642
  primates, 491
Group selection, 569–71
  for cooperative behavior, 386
  definition, 642
  hypercycles, 143
  for language, 504
  for senescence and death, 556
  "species selection", 571
Guanine
  abiotic synthesis, 124
  in nucleic acids, 112
  structural formula, **112**
  tautomeric changes, **216**
Guanosine diphosphate (GDP)
  energy transfer, 139
  phosphorylation, **350**
  in signal transduction, **350**
Guanosine triphosphate (GTP)
  energy transfer, 139
  phosphorylation of GDP, **350**
  in protein synthesis, **116**
  in signal transduction, **350**
Guppies (*Poecilia reticulata*), rapid evolution, 600
Gymnomycota (slime molds), **319**
Gymnosperms
  characteristics, 642
  DNA nucleotide numbers, **268**
  fossils
    first appearance, 307
    Glossopteris (Permian), 97, **98**
    reconstruction, **311**
  in plant phylogeny, **314**
  polyphyletic proposals, 248
  progymnosperms, 307, 308
    origin, 308
    in plant phylogeny, **314**
    reconstruction, **311**

H

Habitats, **39**, 375, 383, 447. *See also* Ecological niche; Environment
  arboreal, 431, 463, 633
    early primates, 468
  benthic, 299, 336
    definition, 634
  brackish water, 400
  burrowing, 336, **339**, 340, **341**, 374, 379
  definition, 642
  as isolating mechanisms, 590
  nocturnal, 464
    early mammals, 444, 447
      retinal accommodations, 447
  pelagic, 336
    definition, 650
  planktonic, 299
Hadean (earliest Precambrian era), 93, 102
  early extraterrestrial impacts, 432
Hadrosaurs (duck-billed ornithischians), **427**
Haeckel's "gastraea" hypothesis, 334–35
  definition, 643
Haeckel's biogenetic law, 40, 41
  criticisms, 41
  description, 634
Hagfish
  in phylogeny of fishes, **406**
  proposed ancestry, 400
  structure, **400**
Haldane's rule, 200, 591

Half life (radioactivity), 94, **95**
  definition, 643
Hamamelidae (angiosperm dicotyledon), **318**
Hamsters, sex-linked traits, 203
Haplodiploidy, 570, 643
  fratricide, **385**
  genetic basis, 385
  and social organization, 385
    kin selection, 385. *See also* Kin selection
Haploids
  advantages, 302
  definition, 191, 208, 643
  effects of selection, 535
  monoploid synonym, 206
  plant gametophytes, 297
  spores. *See* Spores
Haplorhini (proposed primate suborder), 464
Haplotype, 275
  definition, 643
Haptoglobins
  gene duplication, 262
  human polymorphism, 582
Haramyids (Triassic-Jurassic mammals), in phylogeny of mammalian molar teeth, **449**
Hardy–Weinberg principle (equilibrium), 517
  algebraic derivation, 521–523
  assumptions and conditions, **522**
  in calculating gene frequency changes, 537
  codominant alleles, 527
  definition, 643
  disequilibrium (linkage), 524–25, **526**, 561, 647
  effect of inbreeding, 529–30
  multiple alleles, **524**, 528
    equilibrium calculations, 523
  multiple loci, 523–25
  in natural populations, 527–28
  sex-linked genes, **526**, 525–27
  two alleles, 522, **523**
    varied frequencies, **523**
Hare (family Leporidae)
  effects on population size, 574
  lens crystallins, 262
  sex-linked traits, 203
Hawaii
  Drosophilidae, 251
  honeycreepers, 251
Heart (vertebrate)
  amphibian, **417**
  ductus arteriosis, 42
  mammalian, 442
  ontological stages, 42
  reptilian, **417**
Hedgehogs, convergent molecular evolution (apolipoprotein), 265
Helianthus (sunflower), hybrid speciation, 596
Heliconius (butterfly), **540**, 595
Heliozoans, protistan phylogeny, **330**
Helium
  distribution in the universe, 76
  fusion
    reaction product, 75, 76, 79
    temperature, 80
  helium flash, 80
  initial formation, 78
  isotopes, 79
  present atmospheric composition, 89
Helobdella (hirudinean annelid), **381**

Heme, 125
    catalytic activity, **140**
    in cytochrome *c*, 264
    in electron transfer, 169
    in hemoglobin, 257
Hemichordates, 397
    Balanoglossus. See Balanoglossus
    classification among metazoan phyla, 366
    coelom partitioning, 340
    proposed as vertebrate ancestors, 394
    pterobranchs, 334, **340**, 398
    Rhabdopleura, **340**, 398
    time of first fossil appearance, **324**
Hemicyclaspis (osteostracan), **400**
Hemizygote, definition, 203, 643
Hemoglobins, 15, 257
    ape-human relationships, 467
    evolutionary clock proposals, 284
    exons and introns, **176**, 260
    gene duplications, 209, 260
    β globin gene
        primate phylogeny, **261**
        structure, **176**
    heme component, 257
    human-ape similarities, 264, 286
    nucleotide substitution rates in
        artiodactyls, **273**
    polypeptide chains, 217
        α chain, 217, 257–60
            amino acid substitution rates, 284,
                560
            human-chimpanzee identity, 264
            phylogenies, **265, 267**
        β chain, **176**, 217, **218**, 257–60, **260**,
            273, 560
            phylogenies, **267**
        δ chain, 257, 259–260, **260**
        ε chain, 258, **260**
        γ chain, 257–60, **260**
        ζ chain, **260**
        evolutionary relationships, 258–60
        organizational advantages, 260
        serial homologies, 242
    in proposed Early Cambrian animals, 325
    pseudogenes, **261**
    rate of amino acid replacement, 273, 284,
        559
    sickle-cell mutation, 217, **218**. *See also*
        Sickle-cell disease
    thalassemia regulatory mutations, 217
Hemophilia, 202, 203
    incidence, 616
    transgenic proteins, 625
Hepatophytes (liverworts and hornworts),
    300
Herbicides, 565
Herbivores
    definition, 643
    Precambrian-Cambrian diversification, 325
Heritability, 223
    definition, 643
Hermaphrodites, 198, 529
    definition, 643, 648
    plants, **303**
    sex determining genes, **199**
Herring (teleost), **406**
Hertzsprung-Russell diagram (stellar
    evolution), **80**
Hesperornis (Cretaceous bird), 436
Heterocercal tails, **400**, 403
    description, 643

Heterochromatin, 200
    definition, 640, 643
    satellite DNA localization, 271
Heterochrony, 360, 397, **502**
    definition, 643
Heterodont dentition, 443, 450
    definition, 643
Heterogametic sex, 201
    definition, 643
    evolutionary stages, **199**
    Haldane's rule, 591
Heterosis, 538. *See also* Heterozygotes,
    superiority
    definition, 643
Heterostracans (jawless fossil fish), 399
    structure, **400**
Heterotrophs, 164, 172
    description, 643
    metabolism, 158
Heterozygotes
    decreased frequency on inbreeding, 529,
        530
    definition, 195, 643
    frequency estimates
        for carriers of rare alleles, 527–28, 535
        from protein polymorphism, 229, 231,
            582
    lethality, 538
    pairing in chromosomal aberrations, **211**
    selection against, 538
    superiority, 538, 616
        cause for diploidy, 302
        definition, 643, 644
        genetic load, 558, 559
            proposed evolutionary cost, 559
        in hybrid vigor ("overdominance"),
            538
        selection coefficients, 538, 558, 559
        sickle cell gene, 217, 561
Hexactinellid ("glass" sponge), **370**
Hexaploids, definition, 208
Hexasomics, 210
Hexokinase, in anaerobic glycolysis, **160**
Hexose diphosphatase, in Calvin cycle, **164**
Hierarchy, 600, 612
    definition, 644
Hirudineans (annelids), 379. *See also*
    Leeches
    external form and cross sections, **380**
Histidine
    codon designations, 148
    structural formula, **111**
Histones
    in coacervates, **133**
    description, 644
    effect on transcription, 270
    in eukaryotes, 175
    gene duplications, 271
    rate of amino acid replacement, 273, 565
Hitchhiking, 525, 561. *See also* Linkage,
    disequilibrium
    definition, 644
    proposed effect on reproductive
        isolation, 595
Hoatzin (bird), lysozyme evolution, 265
Holophyletic, **246**
Holosteans, DNA nucleotide numbers, **268**
Holothurians (echinoderms)
    auricularia larva, **394**, 397
    phylogeny, **387**
Home range, of hunting hominids, 490

Homeoboxes, 353
    Antennapedia complex, 352, 353, **355**
    Bithorax complex, 352
    cephalochordates, 395, **396**
    change in regulatory function, 360
    definition, 644
    diploblastic animals, 372
    echinoderms, 387
    gene duplications, **354**, 355
    hemichordates, 395
    homeodomains. *See* Homeodomains
    homologies between organisms, 353, **354**,
        360, **396**
    proposed origins, 355
        Early Cambrian animals, 325
    vertebrates, **396**
    zootype, 354, **355**
Homeodomains (DNA binding factors),
    353, **644**
    duplications, 355
    in plants and fungi, 355
Homeostasis (developmental), 358
    definition, 644
Homeotic mutations, 352, 382
    Antennapedia, **352**, 353, **355**
    bithorax, 219, **222**, 352, 353
    definition, 644
    regulatory molecules, 351
    Ultrabithorax, 352, **355**
Hominids, 15. *See also* Australopithecines;
    Homo species; Humans; Neanderthals
    age of origin, 470
    animal hunting
        advantages, 489
        evolutionary effects, 489–91
    ape-hominid "missing link", 473
    bipedalism
        advantages, 474–77
        anatomy and evolution, 475–77
        skeletal structures, **473**
    brains
        encephalization quotients, 500
        endocranial casts, **501**
    cell-type numbers, **544**
    characteristics, 466–68, 644
    fossils, 470–74
        early appearance, **469**
        skulls, **471**
    "home base", 474, 490
    pelvic modifications, 475
    population size, 484
    in primate classification system, 464
    proposed phylogeny, 481, **483**
    selection for toolmaking skills, 490
    selection in "patchy" environment, 489
    sexual bonding, 474, 475, 490
        proposed evolution of female breasts
            and buttocks, 475
Homininae (proposed primate group), 464
    age and monophyletic origin, 470
Hominoids. *See also* Apes; Hominids
    ape-human comparisons, 203, 211, 212, **214**,
        257, 264, **272**, 275, **467**, 467–68, **476**
    description, 466, 644
    early fossil appearance, 468, **469**
    intellectual preadaptations, 497
    molecular phylogeny, 470
    pelves, **473**, 476
    phylogenies
        DNA-DNA hybridization, **272**
        immunological methods, 258

Piltdown fossil forgery, **471,** 472
in primate classification system, 464
self-awareness, 498
time of ape-human divergence, 470
Homo
disputed relationships, 481
fossils, 477–82
skulls, **480**
Homo antecessor, **483**
Homo erectus
brain volume, 500
Broca's area, 502
characteristics, 479, 482
encephalization quotient, 500
in proposed hominid phylogeny, **483**
replacement hypothesis, **486,** 504
skull structure, **480**
stone tools, **480**
vocal limitations, 494
Homo ergaster, **483**
Homo habilis
brain volume, 500
Broca's area, 502
characteristics, 482
encephalization quotient, 500
in proposed hominid phylogeny, **483**
relative size and cranial capacity, 477
skull structure, **480**
stone tools, 477, **480,** 490
tree-climbing abilities, 474
Homo heidelbergensis, **483**
Homo rhodesiensis, **483**
Homo rudolfensis, **483**
Homo sapiens neanderthalensis. See
Neanderthals
Homo sapiens sapiens, 479. *See also*
Humans
"archaic", 482
brain volume, 500
characteristics, 482
classification, 247
encephalization quotient, 500
geographic migrations, **488,** 504
proposed origin
African locality, 483, 485, **486–87**
multiple localities, **486,** 487, 488
stone tools, **480**
Homogametic sex
definition, 644
evolutionary stages, **199**
Homologous chromosomes, 191
crossing over, 203
definition, 644
nondisjunction, 208
pairing, 191, **192**
for aberrations, **211**
Homologous organs. *See* Homology
Homology, 37
conceptual difficulties, 242, 360
definitions, 241, 644
in determining phylogeny, 241–45
developmental, 353–55, 360, 382
fish-amphibian
limb bone comparisons, **411**
skull bone comparisons, **411**
genetic (genotypic) basis, 242, **244,** 360
mammalian chromosomes
hominoid banding patterns, **214**
linkage relationships (synteny), **204, 206**
sex-linked genes, **203**
metazoan appendages, 354

molecular sequences. *See* Molecular
phylogenies
partial, 245
serial, 242, 654
in signal pathways, **351**
vertebrates
brains, **40**
embryos, 40, **44**
forelimbs, 37, **41**
Homoplasy, 243, 249, 251, 264. *See also*
Convergent evolution
definition, 644
Homozygotes
definition, 195, 644
deleterious recessives, 528
reasons for persistence, 528
increased frequency on inbreeding,
529–30
"pure lines", 516, 542
selection effects, 536–50
"Hopeful monsters", 356
macromutational concept, 599
rarity, 451
Horizontal transmission, **181,** 225, 226, 240,
242, 249, 252, 264, 276, 575, 607
Hormones, 589
Horses, 15, 357
agricultural expansion, 608
ancient DNA, 277
artificial selection, 50
Cenozoic perissodactyl radiation, **456**
conserved aspartate transaminase
sequence, 256, **257**
Equus, 45, **48**
evolution, 44, 45, **48**
anagenesis and phylogenesis, **48,** 241
habitats, 45
hemoglobin chain substitutions, 560
hind limb structure, **454**
hybrid (mule) sterility, 24, 591
Hyracotherium (*Eohippus*), 44, 45, **48**
parallel-convergent evolution with South
American litopterns, 458
Pleistocene North American extinction,
454
quagga, 277
Horseshoe crabs. *See Limulus*
Hot spots (mutation), 215
Hottentots, presumed "missing links", 8
Houseflies (*Musca domestica*)
DDT resistance, **565**
sexual selection, 588
Howler monkeys (New World monkeys), in
primate classification system, 464
Hox genes. *See* Homeoboxes;
Homeodomains (DNA binding factors)
Hubble constant, **77**
definition, 644
Humans, 6. *See also* Hominids; Hominoids;
Homo sapiens
achondroplasia, 224, 616
achromatopsia, 528, 549
acid phosphatase, 528
adrenogenital syndrome, 616
agammaglobulinemia, 203
agriculture, 608. *See also* Agriculture
Malthusian limitations, 26
populational impacts, 607
albinism, 527, 528, 616
alkaptonuria, 528
allometry, **501**

Alu repeated nucleotide sequences, 226,
272, 277
aminoacidurias, 616
anencephaly, 616
anhidrotic ectodermal dysplasia, 203
aniridia, 224
behavior
disorders, 586, 611
hunting, 489, 490
social. *See* Culture; Social behavior;
Sociobiology
Bergmann's rule, **585**
biological limitations, 613–14
bipedalism. *See also* Hominids,
bipedalism
foot adaptations, **479**
functional utility, 41
stance, **477**
birth defects, 615–17, 616
birth weight, optima, **542**
blindness, 616
blood group genes, 583
ABO, 582, 583
absence of environmental effects, 357
Duffy, 582
MN, 527, 582
polymorphisms, 582
Rh, 546, 582
body contour comparisons with apes, **467**
brachydactyly, 519
brain, **40**
anatomic and metabolic costs, 502
obstetric effects, **476**
Broca's area, 497, **498,** 502, 503
cerebral dominance, 503
development, 500
emotional centers, 506
encephalization quotient, 500
endocranial cast, **501**
evolution, 499–505
neotenous development, **502**
volume, 500
stability, 608
Wernicke's area, 497, **498,** 502
cell-type numbers (complexity), **544**
characteristics, 466–68
illustrations, **465, 481**
chromosomal disorders, 616
chromosome evolution, 211
banding comparisons, **214**
early stages, **215**
cleft lip and palate, 616
clubfoot, 616
congenital heart defects, 616
conserved aspartate transaminase amino
acid sequence, 256, **257**
convergent evolution
apolipoprotein, 265
visual pigments, 561
copper transport deficiency, 203
Creutzfeldt-Jakob disease, 144
Cro-Magnon type
proposed climatological adaptations,
585
reconstruction, **481**
culture. *See* Culture
cystic fibrosis, 528, 616, 618
cystinuria, 528
deafness, 616
defects in DNA repair mechanisms, 225
development, **105**

Humans—*Cont.*
 embryonic, **44**
 self-awareness, 497
DNA
 "ancient", 277, 279
 length of, **268**
 noncoding sequences, 226, 270
 nucleotide numbers, **268**
 protein-coding sequences, 226
Down syndrome, 506
Duchenne/Becker muscular dystrophy, 177, 203, 616
Ellis-van Creveld syndrome, 528, 549
emotions, 505–6
epiloia, 224
estimated number of genes, 224
eugenic proposals, 617–27. *See also* Eugenics
 eutelegenesis, 618
 genetic engineering, 623
facial expressions, **492**
fecundity, 614
 birth control, 615
forelimb vertebrate homologies, **41**
fossils, earliest appearance, 481. *See also Homo* species
frequencies of genetic disorders, 528
galactosemia, 528, 616
α-galactosidase deficiency, 203
gene numbers mapped, **276,** 623
genetic disorders (incidence), 616
genetic load, 615
genome project, 276
gestation and extrauterine development period, 500
gill arches, 42
β globin gene
 exon-intron structure, **176**
 family relationships, **260**
glucose-6-phosphate dehydrogenase deficiency, 203
 geographical distribution, **539**
haptoglobins, 582
heart, ontological stages, 42
height distribution, **222**
hemoglobin chains, 257. *See also* Hemoglobins
 amino acid substitutions, 284, 560
 evolutionary relationships, **260**
hemophilia, 202, 616
 A, 203
 B, 203
hind limb anatomy, **478**
homeoboxes, **354**
hunter-gatherer societies, 489
 replacement by agriculture, 607
 survivorship curve, **615**
hunting, adaptational effects, 490–91
Huntington chorea, 224, 620
intelligence, 606
 preadaptations, 497
karyotype, **214**
language, 493–98
Lesch-Nyhan syndrome, 203
linkage
 homologies with other mammals (synteny), **204, 206**
 using somatic cell hybridization, 204
longevity, 614, **615**
malaria, 539, 565

geographical distribution and resistant genes, **539.** *See also* Sickle-cell disease
selective advantage of sickle cell heterozygotes, 217, 561
Marfan syndrome, 616
mating preferences, 611
menopause, 614
microsatellite loci, 280
migrations
 geographic, 279, **488,** 607
 rates, 546
mitochondria
 DNA restriction enzyme map, **275**
 genome, **484**
  numbers, 180
 "Mitochondrial Eve", 484, 487
mucopolysaccharidoses, 616
multifactorial disorders, 616
mummies, 277
mutation rates, 224, 225
myoglobin chain, 258, **260**
myotonic dystrophy, 224, 616
nearsightedness, 617
neoteny, 41, **502**
neurofibromatosis, 224, 616
"norms of reaction", 357
number of genes, 269
number of nucleotides, 229
number of repetitive DNA copies, 270
olfactory signals, 491
optimum phenotypes, **542**
origin controversy
 African locality, 483, 485, **486–87**
 multiple localities, **486,** 487, 488
ornithine transcarbamylase deficiency, 203
pax-6 gene homologies, 354
pelvic structure, **476**
 vestigial vertebrae, **43**
peramorphosis, **502**
personality-soul concepts, 56
phenylketonuria, 528, 616
phenylthiocarbamide tasting, 519–21
phosphoglycerate kinase, 203
phylogenies
 fossil hominids, **483**
 molecular
  DNA-DNA hybridization, **272**
  immunological methods, **257**
  mitochondrial DNA hominoid relationships, **470**
  5*S* rRNA sequences, **281**
polymorphism, 582
 blood groups genes, 582
 nucleotides, 229
 proteins, 231, 582
 quantitative estimate, 229
Polynesian migrations, 279
population explosion, 614
presumed "missing links", 8, **9**
 in Ladder of Nature, 11
in primate classification system, 464
in proposed hominid phylogeny, **483**
provincialisms, 619
pyloric stenosis, 616
races
 adaptational responses, **586**
 ecogeographical rules, 585
  Bergmann's, **585**
 gene frequencies

comparisons, 581–83
phylogeny, **584**
and intelligence, 617
"isolation by distance", 488
mitochondrial DNA phylogeny, 482–87, **485**
polymorphism, 582
proposed major divisions, 582
 estimated times of divergence, 582
ranked categories in classification, **247**
replacement of Neanderthals, 481, 504
retention of deleterious recessive genes, 528, 618
retinoblastoma, 224
5*S* rRNA molecular structure, **280**
sex-linked traits, 203, 616
 incidence of abnormalities, 616
sickle-cell disease. *See* Sickle-cell disease
similarities to apes, 467–68
 anatomical, **467**
 molecular, 211, 257, 264, **272,** 275, 286, 467
 postural, **468**
skeleton, **473**
skull structure, **480**
 vestigial muscles, **43**
speech, 493–97. *See also* Language
 cerebral centers, **498**
spina bifida, 616
steroid sulfatase deficiency, 203
stone ages and stone tools. *See* Stone ages
Tay-Sachs disease, 528
teeth, vestigial molars, **43**
testicular feminization syndrome, 198, 203
thalassemia, 217, **539,** 623
trisomies, 616
tuberous sclerosis, 616
ulcer belt syndrome, 614, 615
vestigial structures, 40, **43**
vitamin D-resistant rickets, 203
vocalization
 anatomical structures, **494**
 vowel sounds, **493**
Williams syndrome, 490, 506
xeroderma pigmentosum, 225
Xg blood cell antigen, 203
"Y-chromosome Adam", 487
Hummingbirds
 flower pollination, 516, 595
 sexual dimorphism, **589**
Hunting
 adaptational effects, 490–91
 advantages, 489
 baboons, 489
 cephalopod adaptations, 378
 chimpanzees, 466, 489, 490
 cooperative, 490
 Homo erectus, 479
 Homo habilis, 477
 hunter-gatherer societies, 489
 "hunting hypothesis" (hominids), 489–91
Huntington chorea, 224, 620
Hybrid
 breakdown, 591, 645
 definition, 645
 DNA-DNA hybridization, 271–73
  phylogenetic trees
   anthropoids, **272**
   artiodactyls, **273**

dysgenesis, 203
early concepts, 10
Haldane's rule, 591
introgressive, 596, 645
Mendelian, 195
somatic cell, 204
sterility and/or inviability, 11, 24, 591
in allopatric laboratory populations,
593
allopolyploidy override, **209**
as an isolating mechanism, 591
male sensitivity, 201
in speciation, 35
numbers of genes, 287
test for species distinctions, 10, 239, **240**
zones, 596
Hybrid vigor, 538, 616. *See also*
Heterozygotes, superiority
definition, 643
Hydration, 126
Hydrochloric acid, Earth outgassing, 88
Hydrogen, 123
distribution in the universe, 76
early terrestrial abundance, 120
as energy source, 162
fusion temperature, 79
initial formation, 78
interstellar density, **120**
membrane transfer. *See* Proton gradient
in Miller electric spark discharge
experiment, 121, 122
outgassing
from Earth, 88, 120
from meteorites, 85, 120
present atmospheric composition, 89
reducing agent, 88, 120, 136, 161
thermonuclear fuel, 75, 76
Hydrogen bond, **114**
definition, 645
Hydrogen cyanide
in abiotic purine synthesis, 124
interstellar density, **120**
synthesis, 123
Hydrogen ion, 161
definition, 645
Krebs cycle product, **168**
in non-cyclic photosynthesis, **166**
in oxidative phosphorylation, **169**
proton pump, 132, 164, **166**
Hydrogen peroxide
catalysis, **140**
cellular detoxification, 165
Hydrogen sulfide
early source of hydrogen for
photosynthesis, 165
electron donor, 161, 165
formation of pyrite, 161
interstellar density, **120**
outgassing
from Earth, 88
from meteorites, 120
Hydrogenase, 161
in breakdown of hydrogen sulfide, 161
Hydrogenosome, 181
Hydrolimax (platyhelminth), **373**
Hydrolysis, 123, 126
in coacervates, 133
definition, 645
Hydrophilic, **131, 132**
definition, 131, 645
Hydrophobic, **131, 132**

definition, 131, 645
Hydrostatic pressure, 339, 340
definition, 645
Hydrothermal plumes, possible role in
origin of life, 120, 125, 129, 130
Hydroxyapatite, 400
bone constituent, 277
as condensing agent, 126, 130
for extraction of DNA, 271
α-Hydroxybutyric acid, synthesis in Miller
electric spark discharge experiment, 122
Hydroxyl groups, in nucleic acid sugars,
111, 146
Hydroxylamine, mutagen, 215
Hydroxyproline, structural formula, **111**
Hydrozoans (cnidarians), 333, 371
in coelenterate phylogeny, **372**
Hylonomus (early fossil reptile), **413**
Hymenoptera (insects)
pollination, **316,** 516, 595
social organization, 384
and haplodiploidy, 385, 570
genetic basis, **385**
Hypercycles, 142, 291, 613
Hypermutation, 227
Hypocercal tails, **400**
Hypoxanthine, 149
Hypoxia, 404, 409
definition, 645
Hypsilophodon (ornithischian dinosaur), in
dinosaur phylogeny, **426**
Hyracoids, in Cenozoic mammalian
radiation, **456**
Hyracotherium (early fossil horse), 44,
45, **48**

I

Ichthyosaurs (marine fossil reptilian group,
order Ichthyopterygia), 358, 421
classification, 421
convergent evolution in marine
predators, **42,** 245
in phylogeny of reptiles, **422**
skull temporal opening pattern, **420**
Ichthyostegids (early fossil amphibians),
410, 414
in amphibian phylogeny, **415**
fossil reconstruction, **410**
relationship to rhipidistians, 410
skull bone comparisons, **411**
vertebral comparisons, **412**
Idealism, 3, 4, 5, 6, 10
archetypes, 6, **8,** 19, 61
definition, 645
the philosopher-state, 617
Igneous rocks, 89
crustal composition, 91
in dating sedimentary strata, **96**
definition, 645
earliest age, 96
lava, 90, 100
paleomagnetism, 97
magnetite deposits, 97
in radioactive dating, 94
in rock cycle, **91**
Iguanodon (ornithischian dinosaur), in
dinosaur phylogeny, **426**
Iminoaceticpropionic acid, synthesis in
Miller electric spark discharge
experiment, 122

Iminoacetonitrile, abiotic synthesis, 124
Iminodiacetic acid, synthesis in Miller
electric spark discharge experiment,
122
Immunoglobulin G, gene duplication, 262
Immunology
antibodies, 257
description, 633
genetic potentialities, 356
for parasitic protection, 373
antigens, 257
antigenic distance, 257
description, 633
microcomplement fixation, 257
primate phylogenetic tree, 257, **258**
Implantation (mammals), definition, 645
Inbreeding, 528–30, 529
coefficient, **529,** 645
definition, 528, 645
depression, 529–30, 645
effect on genotypic frequencies, 529–30
limited genetic variability, 199
mating systems, **529**
Inclusive fitness, 610. *See also* Kin selection
definition, 645, 646
Independent assortment, **195, 196**
definition, 645
Mendelian ratios, 196
translocation effect, 210
India
Gondwana breakup, 99
Himalayan Mountains formation, 99
Inducers (regulation), 219
Industrial melanism, 541, 565
definition, 645
Industrial Revolution, 8, 26
Inheritance
of acquired characters, 24, 25, 28, 29, 357,
645. *See* also Lamarckianism
cultural transmission, 608
blending, 28, 32, 194, 521, 634
extranuclear (cytoplasmic), 202
germ plasm theory, **28,** 515
horizontal transmission, **181,** 225, 226,
240, 242, 249, 252, 264, 276, 575, 607
maternal, 180, 202
Mendelian. *See* Gene; Mendelian
principles
pangenesis theory, **28,** 33
refutation, 515
primitive forms, 135
quantitative, 219, **223,** 516
spontaneous generation concept, 11, **12,**
25, 26, 113, 655
vertical transmission, 225, 575, 607
Inosine (I), 149
Inosinic acid, 149
Insecticides
DDT (dichloro-diphenyl-
trichloroethane), 229, 230, **565**
dieldrin, 565
resistance, 69, 230, **231,** 516
Insectivores
in Cenozoic mammalian radiation, **456**
Cretaceous fossil, **469**
definition, 645
polyphyletic proposals, 248, 454
stem reptiles, 420
Insects (Hexapoda), 231, 248. *See also*
Drosophila species; and other insect
taxa

social manipulation of altruism, 611
in social primates, 499
Kinetochore, 182
Kingdom (classification), 329
definition, 646
difficulties, **247**
Kinorhynchs (invertebrate phylum)
characteristics, 375
classification among metazoan phyla, 366
time of first fossil appearance, **324**
Knuckle-walking, 466
definition, 646
Krebs cycle
amino acid synthesis, 167
description, 646
evolution, 167
metabolic pathway, 167, **168**
proposed selection for oxidative steps,
169
self-catalytic feature, 167
Kritosaurus (duck-billed hadrosaur), **427**
Krypton, atmospheric composition, 89
Kuehneotheriids (Triassic mammals), 447
in phylogeny of mammalian teeth, **449**
tribosphenic molar, **448**

L

Labyrinthodonts, 412, 419. *See also*
Ichthyostegids
in amphibian phylogeny, **415**
limb girdles, **413**
teeth, **411**
vertebrae, **414**
α Lactalbumin, gene duplication, 260
Lactate dehydrogenase, 262
in anaerobic glycolysis, **160**
gene duplication, 260
and lens crystallins, 260, 262
Lactation (mammals), 450
definition, 646
Lactic acid
in anaerobic glycolysis, 159, **160**
fermentation reaction, **160**
synthesis in Miller electric spark discharge
experiment, 122
Lactose
allolactose, 219
enzyme regulation, 219, **220,** 347
evolution of lactose-hydrolyzing enzyme,
290
Ladder of Nature, 6, 7, 8, **9.** *See also* Great
Chain of Being
definition, 646
Laetoli (African hominid fossil site), 472
bipedal footprints, 473
Lagomorphs. *See also* Rabbits
in Cenozoic mammalian radiation, **456**
Lamarckianism, 24, 25
"arms race" explanation, 356
classification system, **11**
in cultural transmission, 28, 606–7, 608
Herbert Spencer, 609
impact of Neo-Darwinian synthesis, 518
inheritance concepts, 646
acquired characters, 24, 25, 28, 29, 608,
645
use and disuse, 24, 25, 28, 29, 33, 657
modern proponents, 356, 357
postadaptive mutation concepts, 227
in the Soviet Union, 24

species concepts, 11, 24
summary views on variation and heredity,
29
Lambda (λ), development, 347
Lambeosaurus (duck-billed hadrosaur), **427**
Lampreys (*Petromyzon*)
ammocoete larva
segmentation pattern, 341
similarity to cephalochordates, 396
in phylogeny of fishes, **406**
proposed ancestry, 400
structure, **400**
Language, 471, 493–98. *See also*
Communication; Speech
chimpanzee abilities, 495–96
cross-modal associations, 497
definition, 646
evolution, 504
human speech centers, **498**
sequencing rules, 497
social influence, 497, 498, 612
as symbolic communication, 4
syntax, 497, 504
Langur monkeys
convergent molecular evolution
(lysozyme), 265
in primate classification system, 464
Lariophagus distinguendus (wasp), 385
Larvacea (planktonic tunicates), 397
Larvae
auricularia (echinoderm holothurians),
394, 397
definition, 646
life style advantages, 383
paedomorphosis, 397, **399**
tornaria (*Balanoglossus*), **394,** 397
trochophore (annelids and mollusks),
378, **379, 394**
tunicates, **398, 399**
Larynx, laryngeal nerves, 42, 356, 357
Latimeria chalumnae (coelacanth), **49,** 405,
598. *See also* Coelacanths
Laurasia (ancient northern continent), 99, 103
description, 646
separation from Gondwana, **101**
Law of superposition (geology), 91, **96**
Lead, 89
isotopes, 94
in radioactive dating, 94, 95
relative universal abundance, **82**
Lecithin, **131**
Leech (hirudinean annelid)
development, zootype relationship, **355**
external form and cross sections, **380**
locomotion, **381**
parasitism, 381
Lemming, sex-linked traits, 203
Lemurs
description, 464
illustration, **465**
in primate classification system, 464
proposed adapid origin, 468
Lens crystallins, 251, 260, **267**
Lepidodendron (lycopod), 304
fossil reconstruction, **305**
Lepidoptera. *See* Butterflies and moths
Leptospondyls, 413
in amphibian phylogeny, **415**
Lethals
balanced, segregational load, 559
definition, 646

developmental interactions, 356, 357
dominant, 646
selection against, 535
equilibrium values under heterozygote
superiority, 539
"equivalents" (humans), 615
haploid, 535
recessive
definition, 653
selection against, 537
sickle-cell homozygotes, 219
thalassemia homozygotes, 219
Sexlethal gene, 200
Leucine
codon designations, 148
structural formula, **111**
Lichens, algal-fungal symbiosis, **541**
Life. *See also* Origin of life
definitions, 114, 133, 356, 646
as an autocatalytic tautology, 145
as a selection game, 136
as a system of decreasing positive
entropy, 132
in extrasolar possibilities, 86–88
primitive properties, 119
spontaneous generation concept, 11, 12,
13, 655
the "universal" phylogenetic tree, **178**
Life cycle (stages), 556
action of selection, **535**
animal-plant distinctions, 302
definition, 646
hypothetical age structure, 555
parasitism, 373
plants, 297, **303**
alga, **298**
bryophyte, **301**
fern, **307**
Life history, definition, 646
Ligase, **621**
Light year, definition, 647
Limenitis archippus (viceroy butterfly),
540
Limestone (calcium carbonate), formation,
90, 91
Limulus (horseshoe crabs)
characteristics, 383
evolutionary rate, 599
"living fossil", 49
phyletic evolution, **242**
Lineage, definition, 647
Lingual, definition, 647
Lingula (brachiopod), "living fossil", 49, 250
Linkage, 203
calculation of distance by recombination
frequency, 203
conserved relationships in mammals, **206**
definition, 647
disequilibrium, 524–25, **526,** 561, 564,
647
definition, 644
effect on selection, 560, 565
equilibrium (definition), 647
hitchhiking, 561, 644
homeobox conservation in metazoans,
354
maps, 204
definition, 647
Drosophila melanogaster, **205**
Escherichia coli, **207**
restriction enzyme, **275**

Muntiacus species—*Cont.*
   DNA distribution, 269
   evolutionary changes, 210, **212**
Muraenosaurus (reptilian plesiosaur), skull
   structure, **420**
Muramic acid, eubacterial cell walls, 173
Murchison meteorite, 122, **123**
   amino acids, 122
      racemic mixture, 122
   fatty acids, 125
Mus musculus. See Mice
Muscles
   ape-hominid hind limb transitions,
      475–77
   body wall interaction, 338
   circular, 336, 338, 339, **341**, 374, **380**, 381
      effects of parapodia, 381
   hominoid hind limbs, **478**
   longitudinal, 336, 338, 339, **341**, 374, **380**
      effects of parapodia, 381
   mammalian jaw, 444
   parapodial, 381
Mutation, 27, **59**, 189, 224
   "adaptive", 227, 357
   antimutators, 225
   base substitutions, 215, **216**. *See also* Bases
      (nucleic acid), changes
      back mutations, 264, **266**
      number necessary for amino acid
         conversions, 259, 263
      parallel mutations, 264, **266**
      rates, 272, **273**, 284, **285**
      transitions, 215, **216**, **217**
      transversions, 215, **217**
   behavioral, **587**
   beneficial, **227**, 539
   calculation of distance between proteins,
      263–65
   as cause for aging, 556
   as cause for gene frequency changes,
      533–34
   as cause for genetic load, 558
   coadaptive, 182, 451
   definition, 648
   deleterious, 227. *See also* Gene,
      deleterious
      evolutionary cost of replacement, 558
      in gene duplications, **260**
      in haploidy and diploidy, 302
      Y chromosome accumulation, 199
   developmental limitations, 358, 565
   elimination probability, 533
   in evolution of genetic code, 153
   evolutionary value, 564–65
   fixation, 283
   frameshift, 147, 216, **217**
   frequencies. *See* Mutation rates
   gene, 212–25
   heat shock protein effect, 225
   hitchhiking, 561
   homeotic, **352**, 382
      definition, 644
   "hopeful monsters", 356, 599
   "hopeless monsters", 565
   "hot spots", 215, 227
   hyper-, 227
   in laboratory-controlled evolution, 290–91
   lethals. *See* Lethals
   macro-, 516, 599
   missense, 147, 216, **217**, **218**
      error reduction, 151

molecular clock, 281–86. *See also* Molecular
   phylogenies, evolutionary clocks
   from monoecy to dioecy, 199
   "Muller's ratchet", 199, 564
   mutagens, 215
   mutator genes, 225, 227
   neutral, 227, 284, 285, 559, 649
      estimated frequency, 559
   nonrandom, 227
   nonsense, **217**, 649
   post- versus pre-adaptive concepts, 227
   rates. *See* Mutation rates
      reading frame, **217**
   regulatory, 217–19, **220**, 287, 361. *See also*
      Regulation
   repair mechanisms, 225, 227, 270, 285,
      484
   -selection equilibrium, 544, 545
      hydraulic model, **545**
   sex determination, **199**
   site-directed mutagenesis, 262
   spontaneous, 224
   "sports", 32, 515
   "stop" (chain termination, nonsense),
      147, 649, 655
   synonymous (silent), 147, **217**, 285, **286**,
      562
      detection, 275
      relation to evolutionary time, 273
   test for randomness, 227
   transposon effects, 225
   in vestigialization, 437
   viral causes, 225
Mutation rates, 223–25, 534
   deleterious genes, 225
   effect of mutator genes, 225, 227
   effect of nucleotide sequence
      composition, 558
   equilibrium, 534
   frequencies, 224
   and gene frequencies, 533–34
   in limiting the number of nucleic acid
      bases, 146
   "meltdown", 291
   mitochondrial DNA, 485
   "optimal", 217, 291, 570
   relationship to genetic load, 558
   RNA/DNA comparisons, 142, 146, 177, 224
Mutationists, 516
Mutualism, 575
   in animal-plant pollination, **316**
   definition, 648
   population interaction, 571
Mycoplasmas
   codon distinctions, 148
   genetic code exceptions, **154**
   Mycoplasma genitalium, 276
   number of genes, 276
   translation and replication requirements,
      143
Myoglobin, 15, 258
   chimpanzee-human relationship, 467
   evolutionary globin chain relationships, **260**
   folded chain, **115**
   gene duplication, 260
   human-ape similarities, 286
   in vertebrate phylogeny, **267**
Myriapods (arthropods)
   in arthropod lineage, **382**
      segmentation, **353**
   characteristics, 383

Myrmecobius (marsupial anteater), **39**
   parallel evolution, **244**
Myrmecophaga (placental anteater), parallel
   evolution, **244**
Mysticism, 4, 14, 36, 612. *See also* Religion
   in developmental concepts, 13, 357
   in the evolution of religion, 55
   in explaining behavior, 14
   Lamarck's "feu éthéré", 24, 516
   in origin of life concept, 118. *See also*
      Creation, multicultural myths
   in orthogenesis, 227
   in postadaptive mutation concepts, 227
   Spencerian concept of evolution, 609
   teleology, 5, 656
   vitalism, 13, 14, 24, 613
      definition, 657
Myxiniformes (hagfishes)
   proposed ancestry, 400
Myxoma virus, 575
Myxomycetes (acellular slime molds), 315
Myxotricha paradoxa, protistan symbiosis
   with spirochetes, 180
Myxozoa, metazoans, 366

N

Natural selection, 27, 32, 33, 36, 135, 227.
   See also Selection
   versus chance events, 68
   in choosing among hypotheses, 15
   compared to artificial, 50
   on continuous variation, 515
   contrast with cultural evolution, 608
   creative ability, 58, 59
   in Darwin's evolutionary sequence, 27
   definitions, 25, **27**, **648**
   early concepts, 25
   effect of modern medicine, 616
   evolutionary cost, 558–62
   Fisher's fundamental theorem, 542, 569
   illustration for a sequence of alphabetical
      letters, **59**
   versus Lamarckianism, 25
      in cultural evolution, 608
   Malthusian sociological basis, 25, 26, 554
   number of documented cases, 543
   objections against, 32
      tautology argument, 534
   rates. *See* Evolutionary rates
   Red Queen hypothesis, 543, 544, 553, 653
   replacement by human-directed
      evolution, 620
   Social Darwinism concept, 609
   as "struggle for existence", 25, 26
   ubiquity, 543
   Wallace proposal, 26
Naturphilosophie school, 6
Nautiloids (mollusks), 378
   extinctions, 430
   phylogeny, **378**
Neanderthals
   "ancient DNA", 279, 481
   brain volumes, 480
   characteristics, 480, 482
      proposed climatological adaptations,
         481, **585**
   disputed species status, 481
   fossil appearance, 480
   Mousterian stone tool industry, 481
   proposed vocal limitations, 495

O

Placoderms (Devonian jawed fish), **401**, 402
  in phylogeny of fishes, **406**
Placodontia (fossil reptiles), 421
Placozoans, **371**
  characteristics, 367
  classification among metazoan phyla, 366
  in metazoan phylogeny, **368**
Planaria (platyhelminths), locomotion,
    339
Planets, 86
  Earthlike, 88
  formation, **86**
  galactic life-bearing, 118
  origins
    collision theory, 85
    condensation theory, 85
  in other solar systems, 86, 87
  protoplanets, 85
Plankton, algae, 299
Plants, 6, 12. *See also* Vascular plants;
    individual taxa
  acquisition of cellular organelles, **181,**
      279, 281, **283**
  algal ancestors, 296–300
  alternation of generations. *See* Alternation
      of generations
  ancient (fossil) DNA, 277
  aneuploidy, 208
  angiosperms. *See* Angiosperms
  apogamy, 303
  apospory, 303
  bryophytes. *See* Bryophytes
  carnivorous, 313, 329
  Cenozoic distribution of landscapes, **453**
  chloroplasts. *See* Chloroplasts
  classification problems, 9
  conserved amino acid sequences
    aspartate transaminase, 256, **257**
      histone proteins, 565
      triosephosphate isomerase, 159
  convergent evolution, 313, **317**
  cytochrome *c* phylogeny, **266**
  desiccation-resistant mechanisms, 303
  Devonian radiation, 409
  ecological adaptations, 580, **581**
  embryophytes, 300
  endosymbiotic cyanobacterial origin, 180
  estimated time of divergence from
      animals, 276, **281**
  ethylene, 349
  evolution, 296–320
  ferns. *See* Ferns
  gametes, dependence on aqueous transfer,
      303, 306
  gametophytes. *See* Gametophytes
  genome sequencing, 276
  germ line absence, 302
  gymnosperms. *See* Gymnosperms
  hemoglobin-like molecules, 257
  herbicide resistance, 565
  Late Cretaceous extinctions, 430
  leaf origins
    enation theory, 305, **309**
    telome theory, 304, **309**
  life cycle, **303**
  lycopods. *See* Lycopsida
  man-made species, **209**
  mechanisms for resistance to desiccation,
      298, 307
  meristematic cell division, 299
  pigments

carotenoids, **163**, 165, **167**, 298
    chlorophyll. *See* Chlorophyll
    flavines, **163**
    phycobilins, **163**
  polyphyletic proposals, 248
  polyploidy, 206, **209**, 310, 596
    advantages, 206
    man-made species, 206
  progymnosperms. *See* Progymnosperms
  proposed phylogeny, **314**
    5*S* ribosomal RNA phylogeny, **281**
  protein polymorphism, 231
  pteridosperms (seed ferns), **98**, 308,
      312, 313
  18*S* ribosomal RNA phylogeny, **333**
  root variations, 356
  seeds. *See also* Seeds
    ancient DNA, 277
    evolution, **313**
    selective factors, 310, 312
  self-sterility alleles, 310, 529, 541
  sex determination, 198
    non-chromosomal, 198
  signal transduction, 349
  speciation by hybridization, 596
  spores. *See* Spores
  sporophytes. *See* Sporophytes
  thylakoid membranes, 165
  tropical diversity, 312
  urpflanze concept, 6, **8**
  vascular. *See* Vascular plants
Planula, 335, 336
  locomotion, 338
  as metazoan ancestor, **335,** 338, 342, 651
  in sexual reproduction, 369
Plasma membrane
  definition, 651
  eukaryotic cell, **110**
Plasmids
  definition, 651
  in horizontal transmission, 225, 240
  in origin of sex, 564
  source of antibiotic resistance factors, 225
  vectors in genetic engineering, 620, **622,** 657
Plastocyanin, in non-cyclic photosynthesis, **166**
Plastoquinones, in non-cyclic
    photosynthesis, **166**
Plate tectonics, 102. *See also* Tectonic
    movements and plates
  boundary events, 100, 101, **104**
  definition, 651
  Earthquake belts, 100, 102, **103,** 107
  major plates, **103**
  theories of origin, 102
Platyhelminths, 335, 336
  acoelan turbellarians. *See* Acoelan
      turbellarians
  Cambrian absence, **324**
  characteristics, 373–74
  classification among metazoan phyla, 366
  metamerism, 337
  in metazoan phylogenies, **332, 368**
    18*S* ribosomal RNA sequences, **333**
  morphology, **373**
  number of species, 373
  parasitism, 373, 374
    reproductive patterns, 374
  pedal locomotion, 339
  Planaria, **339**
  time of first fossil appearance, **324**
  triploblastic body plan, **337**

Platynereis (polychaete annelid), **341**
Platypus (*Ornithorhynchus*), 102, 455. *See
    also* Prototherians (monotremes)
  characteristics, 446
  development, **105**
Platyrrhines. *See* New World monkeys
Pleiotropy
  definition, 651
  enzymes and lens crystallins, 260
  proposed effect on reproductive
      isolation, 595
  relation to aging, 556, 614
  sickle-cell gene, 217, **218,** 358
  in speciation model, 595
Plesiadapiformes (early primates), 468
Plesiomorphic characters, 249, **250,** 651
Plesiosaurs (fossil marine reptiles), 421
  classification, 421
  skull temporal opening pattern, **420**
Pogonophorans, 336
  classification among metazoan phyla,
      366
  in metazoan phylogenies, **332**
  time of first fossil appearance, **324**
Polemoniaceae. *See* Phlox
Pollen, 307
  angiosperm, **315**
    fossils, 311, **317**
  animal pollination, 308, **316,** 516, 595
  double fertilization, 310, **315**
  as isolating mechanism, 591
  in seed evolution, **313**
  self-sterility alleles, 310
Polychaetes (annelids), 340, **341,** 379, 381
  external form and cross sections, **380**
  metamerism, 337
  parapodia, 381
  trochophore larva, **379**
Polygenes, 33, 219, 542
  definition, 651
  in insecticide resistance, 230
Polymerase, 219, **220, 278**
  definition, 651
  early selection for, 145
  mutational errors, 215
Polymerase chain reaction (PCR) technique,
    277, 291, 623, 651
  illustration, **278**
Polymerization
  by condensation reactions, 126, **127**
  early enzymes, 144
  head-growth process, 144
  proposed coupling with organic
      degradation, 145
Polymers, 123
  definition, 651
  proteinoids, 128, 652
  self-assembly, 134
Polymorphism, 539, 562. *See also* Genetic
    variability; Variation
  adaptive advantage, 228
  balanced, **538,** 539
    definition, 539, 633
    proposed evolutionary cost, 559
  because of frequency dependent
      selection, 540
  chromosomal, 228, **229,** 580, 583
    balanced, adaptive values, 539, **540**
  in coding and noncoding DNA
      sequences, 562
  definition, 652

peristaltic locomotion, **339**
time of first fossil appearance, **324**
Primates, 15. *See also* Anthropoids;
    Humans; Monkeys; Prosimians; and
    individual taxa
adapids, 468
Alu repeated nucleotide sequences, 226
animal hunting, 489
behavior
    altruism, 498
    emotions, 499
    self-awareness, 498
    social interaction, 498
brain
    cerebral dominance, 497
    prefrontal effects, **505**
in Cenozoic mammalian radiation, **456**
characteristics, 463–66
    illustrations, **465**
chromosome evolution, 211
    banding comparisons, **214**
    early stages, **215**
classification, 463–66
    cladistic, 251, 252
    controversies, 464
communication, 491–99
    olfactory signals, 491
    tactile signals, 491
    visual signals, 492
    vocalization, 492
        symbolic, 492
        warning signals, 492
cytochrome *c* phylogeny, **266**
evolution, 468–88
fossils
    difficulties in fossilization, 468
    Fayum deposits, 468
    Piltdown forgery, 472
    sequence, **469**
hind limb structure, **454**
nucleotide substitution rates, **285**
number of species, 463
pelvic modifications, 475
phylogenetic tree
    DNA-DNA hybridization, 271, **272**
    β globin gene relationships, **260**
    immunological methods, 257, **258**
Piltdown fossil forgery, **471,** 472
protein polymorphism, 231
restriction enzyme map comparisons of
    mitochondrial DNA, **275**
social behavior and interaction, 498–99
South American radiation, 458
time of ape-human divergence, 470
Principle of use and disuse (Lamarck). *See*
    Lamarckianism
Proboscideans, in Cenozoic mammalian
    radiation, **456**. *See also* Elephants
Proconsul (early hominoid), 468
    reconstruction, **469**
Progenesis (heterochrony), 398, 643
Progenote, 177, 329
    definition, 652
    presumed metabolic pathways, **178**
Progress, 544
    evolutionary, 451
    Lamarckian concept, 24
    Leibnizian philosophy, 7
Progymnosperms, 307, 308
    fossil reconstruction, **311**
    in plant phylogeny, **314**

Prokaryotes
    cell division, 182, 191
        binary fission, 173
    cell structure, **110**
    descriptions, 173–77, 652
    DNA repair systems, 225
    estimated time of divergence from
        eukaryotes, 276, **281**
    feature comparisons with eukaryotes,
        173–77
    gene regulation, 177
        lactose metabolism, 219, **220,** 347
        operators, 219
        promoters, 219
    genetic code exceptions, **154**
    phylogenetic trees, **178, 281, 283**
    proposed eocyta kingdom, 173
    stromatolite organisms, 171, **172**
    as symbionts in origin of eukaryotic
        organelles, 179, 279, 281, **283,** 296
    thylakoid membranes, 165
Proline
    codon designations, 148
    in Murchison meteorite, 122
    structural formula, **111**
Promoters (gene regulation), 219, **220,**
    270, 347
    definition, 652
    eukaryotic DNA sequence, **221**
    in new functions, 360
    in signal transduction, 349
Propionic acid, synthesis in Miller electric
    spark discharge experiment, 122
Prosauropods (early saurischian dinosaurs),
    in dinosaur phylogeny, **426**
Prosimians. *See also* Lemurs; Lorises
    characteristics, 463
        illustrations, **465**
    early fossil appearance, **469**
    β globin gene family relationships, **260**
    in primate classification system, 464
Protein synthesis. *See also* Translation
    chain termination codons, 147, 148, 216
    complexity, **141,** 143
    early adaptor molecules, 145
    evolution, 144–46
        conserved components, 276
        early stages, 150
        hypotheses, 146
        precedence of function or
            information?, 140–44
    initiation codon, 148
    on messenger RNA template, **115, 116, 141**
    by proteins, 144
    reading frame, **217**
    system universality, 177
    termination codons, **217**
Proteinoids, 128
    amino acid compositions, 128
    catalyzed reactions, 129
    description, 652
    microsphere formation, 133, **135,** 648
    mode of synthesis, 128
    properties, 128
        shared with proteins, 130
    sequential reactions catalyzed by, **130**
Proteins, 219, 263, 561. *See also* Enzymes;
    Polypeptides
    accumulation in membranous droplets,
        132
    actin, 349

allosteric, **220,** 347
amino acids
    composition, 109
        structural varieties, **111**
        substitution rates, 284, 559
            proposed constancy, 560. *See also*
                Evolutionary clocks
catalytic activity, **140**
cellular membranes, 109
"chaperones", 181
coenzymes, 125. *See also* Coenzymes
convergent molecular evolution, 265. *See*
    also Convergent evolution,
    molecular
crystallins, 251, 260
definition, 652
domains, 175, 638
early primitive forms, 119
electrophoresis, 228, **230,** 275
enzymes. *See* Enzymes
evolution (amino acid sequences), 257–66
    aspartate transaminase, 256, **257**
    triosephosphate isomerase, 159
Fox's "proteinoids", 128. *See also*
    Proteinoids
functional variety, 141, 144
    relationship to structure, 113
heat shock (hsp), 181, 225
histones, 175, 270, 271, 273, 565
human-ape similarities, 286
nucleotide substitution rates, 284
    in artiodactyls, **273**
polymorphism, 229, 231
    human, 582
post-translational modification, 349
"prions", 144
repressor, 219, **220**
    definition, 653
in the "RNA world", 143
self-replicating proposals, 141, 144
silent mutations, 273
site-directed mutagenesis, 262
structure, 111
    numbers known, **276**
    relationship to function, 113
substitution rates, 273
synthesis. *See* Protein synthesis
translation. *See* Translation
Proterozoic era, 93
    continental drift, 100
    fossil cells, **182,** 279
Prothylacynus patagonicus (South American
    fossil marsupial wolf), **458**
Protistans, 173, 182. *See also* individual taxa
    acritarch fossils, 182
        Late Cretaceous extinctions, 430
    codon distinctions, 148
    definition, 652
    first fossil appearance, 329
    general features, 328–29
    proposed phylogeny, **330**
        18*S* ribosomal RNA sequences, **333**
    protoflagellate ancestor, 329, **330**
    radiation, 329
    symbiotic acquisition of cellular
        organelles, **180, 181**. *See also*
        Symbiosis
    time of first fossil appearance, 329
    transition to multicellular forms, 182,
        329–36
    widely separate origins, **178**

Protocells, 134, 135
Protoceratops (ceratopsian dinosaur), in
    dinosaur phylogeny, **426**
Protoctista. *See* Protistans
Protoflagellates, 329
Protogalaxies, 78
Proton gradient
    as an early cellular deacidifier, 164
    in membranes, 132, 162, **163**, 164, **166**
    in oxidative phosphorylation, **169**
Proton pump, 132, 164, **166, 169**
Protoplanets, 85
Protoplasm, definition, 652
Protororsauria (Permian-Triassic archosaur
    reptiles), 421
Protostars, 79
Protostomes
    cleavage patterns, 376, 393, 394
    definition, 652
    distinguishing characteristics, 376, 393
    embryological development of coelom,
        393
    fate of blastopore, 365, 376, 393, 394
    in metazoan phylogeny, **368**
    phyla, 366
Prototherians (monotremes), 102, 104, **106,**
    450. *See also* Echidna; Platypus
    Australian presence, 455
    in Cenozoic mammalian radiation, **456**
    characteristics, 446–47, 648
    DNA nucleotide numbers, **268**
    early tooth patterns, 446
    effects of continental drift, 102
    fossil rarity, 455
    mammary glands, **105**
    molar teeth, in mammalian phylogeny, **449**
    parallel evolution, **244**
    proposed therian origin, 447
Protozoans, 142, 329. *See also* Protistans
    DNA nucleotide numbers, **268**
    in Hadzi metazoan origin hypothesis,
        332, **333**
    as metazoan ancestor, 329, **331, 368**
        Haeckel "gastraea" hypothesis, 334
    number of genes, 269
    protistan phylogeny, **178, 330**
    symbiosis with algae, **180**
Psaronius (tree fern), fossil reconstruction, **308**
Pseudocoelomates, 336
    body plan, **337**
    characteristics, 374, 652
    convergent evolution, 374
    phyla, 366
        proposed relationships, 374
Pseudogenes, 260
Psilophyton (Devonian plant), 304
    fossil reconstruction, 304
    in plant phylogeny, **314**
Psilopsida. *See* Psilotum
Psilotum (fern), 303, 304
    DNA nucleotide numbers, **268**
    resemblance to early fossil plants, **304**
Psychology. *See also* Behavior; Brain;
    Consciousness; Intelligence
    emotions, 505–06
    "evolutionary", 506
Pteranodon (large flying pterodactyloid), 434
Pteraspis (heterostracan), **400**
Ptericthyodes (fossil jawed fish), **401**
Pteridophyta. *See* Ferns

Pteridosperms. *See* Seed ferns
Pterobranchs (hemichordates), 334, **340,** 398
    in proposed phylogeny of vertebrates, **399**
Pterodactyloids (pterosaur suborder)
    characteristics, 431
    reconstruction, **435**
Pterophyta, Pteropsida. *See* Ferns
Pterosaurs, 421
    characteristics, 433–36
    fossil appearance, 431
    Late Cretaceous extinction, 436
    in phylogeny of reptiles, **422**
    proposed endothermy, 434
    reconstructions, **435**
    in reptilian classification system, 421
    skull structure, **434**
    thecodont ancestry, 431
Pulsars, 83
Punctuated equilibria, 287, 599–600
    criticisms, 599–600
    definition, 652
    DeVriesian version, 571
Pure lines, 516, 542
Purines, 120
    abiotic synthesis, 121, 124
    adenine, **112, 114**, 124
    description, 652
    guanine, **112, 114**, 124
    hypoxanthine, 149
    in nucleic acids, 112
    structural formulas, **112, 114**
    synthesis from glycolytic substrates, **161**
    transitions and transversions, 215, **216**
Pygmies (Mbuti), 489
Pyrimidines, 112, 120
    abiotic synthesis, 121, 124, 125
    cytosine, **112, 114**
        abiotic synthesis, 124, 125
    description, 652
    in nucleic acids, 112
    orotic acid, 124
    structural formulas, **112, 114**
    thymine, **112, 114**
        abiotic synthesis, 124
    transitions and transversions, 215, **216**
    uracil, **112,** 124
        abiotic synthesis, 124
Pyrite (FeS$_2$), as proposed primitive energy
    source, 161
Pyrophosphate, 126, 139
Pyrroles, 125. *See also* Porphyrins
    structural formula, 125
Pyruvate dehydrogenase, 167, 169
    in Krebs cycle, **168**
Pyruvate kinase, in anaerobic glycolysis, **160**
Pyruvic acid, **161**
    in aerobic metabolism, 167
    from anaerobic glycolysis, 159, **160**
    in Entner-Doudoroff pathway, 161
    fermentation reaction, **160**
    in Krebs cycle, 167, **168**
    in proteinoid-catalyzed reactions, 129,
        130
Python snake, vestigial organs, **43**

Q (coenzyme)
    antioxidant, 165
    in respiratory pathway, **169**

Qβ virus
    evolved secondary structure, **291**
    laboratory evolution, 290–91
    replicase enzyme, 142, 290
Quadrupedal, definition, 652
Quantitative inheritance, 219, **223**, 516
    character, 652
    effect of different modes of selection, **543**
    effect of inbreeding depression, 530
    heritability, 223
    normal distribution, **223**
Quantitative trait loci (QTLs), 219
Quantum evolution, 598
    definition, 652
Quetzalcoatlus (pterydactyloid flying
    reptile), 434
Quinones, in cyclic photosynthesis, **163**

r selection, 556, 575
    characteristics, 557
    definition, 653
Rabbits (family Leporidae)
    coevolution with viral pathogen, 575
    conserved triosephosphate isomerase
        sequence, 159
    embryology, **44**
    β globin gene family relationships, **260**
    lagomorph Cenozoic radiation, **456**
    lens crystallins, 262
    quantitative white spotting, 219, **223**
    sex-linked traits, 203
    silent/replacement mutation rate
        comparisons, 273
Racemic mixture, 117, 122
    definition, 652
Races, 280, 610
    adaptational patterns, 583–87
    in anti-evolution argument, 35
    behavioral genes, 586
    in cladistic species concept, 249
    Darwin's concept, 26
    definitions, 518, 581, 583, 652, 655
    ecogeographical rules, 585
    formation, 589
    human. *See* Humans, races
    "isolation by distance", 488
    Lamarckian concept, 24
    migration rate estimates, 546
    molecular distinctions, 280
    as a natural classification unit, 5
    Nei's index of genetic distance, 582, 583
    in polytypic species, 239
    "purity", 582
    "senescence" concept, 429
Radial cleavage, **376**, 393
Radial symmetry
    in classification of phyla, 366
    coelenterates, 369
    echinoderms, 387, 395
Radiata. *See* Coelenterates
Radiation (energy). *See* Cosmic rays;
    Radioactivity; Sun, solar radiation;
    Ultraviolet; X-rays
Radiation (organisms). *See* Adaptive
    radiation
Radioactivity
    definition, 653
    as early energy source, 119

South America
immigration by "rafting", 103
radiation, 458
Rotiferans, 564
characteristics, 375
classification among metazoan phyla, 366
in metazoan phylogenies, **368**
time of first fossil appearance, **324**
Rubidium, in radioactive dating, 94, 95
Rubus (woodland fleshy-fruited plant genus), different modes of classification, 237

S

Saccharomyces. See Yeast
Salamanders, 414
embryology, **44**
hemoglobin chain substitutions, 284
neoteny, 397
Salmonella enterica (Gram-negative enterobacteria), DNA repair systems, 225
Salmonella typhimurium (Gram-negative enterobacteria)
mutation rates, 224
5*S* ribosomal RNA phylogeny, **281**
similarities to other bacteria, 240
Saltation concept, 516, 518, 599
definition, 654
Samarium, in radioactive dating, 95
Sampling error. *See also* Random genetic drift
as cause for Muller's ratchet, 199
as cause for random genetic drift, 547
definition, 654
San Blas Indians (albinism), 528
Saprophytes
definition, 654
fungi, 315
Sarcopterygians (lobe-finned fish), 404, 409, 410. *See also* Fish
characteristics, 404
crossopterygians, 405, **406**. *See also* Rhipidistians
dipnoans, 405
lungs, 405
in phylogeny of fishes, **406**
preadaptations for terrestrial existence, 409
Sarcosine
in Murchison meteorite, 122
synthesis in Miller spark electric discharge experiment, 122
Saurischians (lizard-hipped dinosaurs), 421, 425, 426, 427, 428. *See also* Dinosaurs
in dinosaur phylogeny, **426**
pelvis, **425**
Saurolophus (duck-billed dinosaur), **427**
Sauropods (saurischian dinosaurs), 426, 427, 428
in dinosaur phylogeny, **426**
Sauropterygians (diapsid reptilian superorder), 421
in phylogeny of reptiles, **422**
Scale of Nature, 6, **7**, 11, 25. *See also* Great Chain of Being
Scandentia (tree shrew order), 468
Scaphiopus (desert toads), 414
Scaphopods, in molluskan phylogeny, **378**
Scavenger, definition, 654
Schistosoma (platyhelminth trematodes), 374

Science
creationist claims and evolutionist responses, 65–70
differences from religion, 65
Sclerospongiae (sponges), **370**
Scyphozoans (cnidarians), 369
in coelenterate phylogeny, **372**
Sea floor spreading, 96, 99, **100**. *See also* Tectonic movements and plates
definition, 654
Second Law of Thermodynamics, 66, 118, 132
definition, 639
Sedimentary rocks, 90, 98. *See also* Rocks
crustal composition, 91
dating, 94, **96**
stratigraphic, 91
description, 654
first appearance, 96, 120
fossil fragility and erosion, 42, 237
limestone, 90, 91
in rock cycle, **91**
tertiary origin, 94
Seed ferns (pteridosperms), 98, 308, 312
fossil reconstruction, **312**
Glossopteris (Permian fossil), **98**
in plant phylogeny, **314**
as possible angiosperm ancestor, 312
seed evolution, **313**
Seeds, 307
ancient DNA, 277
angiosperm, 310
animal dispersal, 308
in Carboniferous gymnosperms, 307
Devonian fossil, 308
evolution, **313**
megaspore origin, 305
selective factors, 310, 312
seed dispersal and protection, 310
Segmentation, 337–38, 340, 360. *See also* Metamerism
arthropod evolution, **353**
in burrowing locomotion, **341**
convergences, 381
definition, 654
developmental genes, 353, 376
metazoan
homologies, 353, 376
phylogenies, **332**
in "short-term" and "long-term" insects, 360
vertebrates, 340
Segregation (meiotic), **195, 196**
abnormal ("distortion"), 201. *See also* Meiotic drive
ratios, 202, 566, 654
alternate, 210
description, 654
effect of colchicine, 206
Mendel's principle, **194**
Segregational load, 558. *See also* Genetic load
balanced lethals, 559
proposed evolutionary cost, 559
Seismic waves, 89, **90**
Moho discontinuity, **90**
Selection, 33, 227, 557. *See also* Sexual selection
as adaptive constraint, 357, 358
in adaptive radiation, 452
for aerial flight in insects, 382

against heterozygotes, 538
for aging, 556
for allozymes, 561
for amino acid replacements, 273
DNA coding regions, 562
interspecific comparisons, 562
for animal flower-pollination, 308
in "arms race". *See* "Arms race"
artificial. *See* Artificial selection
balancing. *See* Balance polymorphism
for basic chromosome structure, 273
for biochemical pathways, 158, 159
for bipedalism in archosaurian reptiles, 424
for bipedalism in hominids, 474–77
for body size in dinosaur "arms race", 426
for canalization, 358, 359, 598
for cell division, 189
centrifugal, 542
centripetal, 542
coefficient. *See* Selection coefficient
for coenzymes, 145
for coevolution, 308, 349
for coiling in mollusks, 377
for conserved amino acid sequences
aspartate transaminase, **257**
triosephosphate isomerase, 159
constraints, 452
on continuous variation, 33, 516
for convergence, 37, **39**, 242, 245, 313, 455
for cooperation, 490, 499, 587, 612
as creative force, 59, 68, 136
dampened response, 359
definitions, 534, 654
for density dependent behavior, 586
in development, 349
constraints, 357, 598
in diploids, 536–50
directional, 290, 359, 542, **543, 544**
definition, 638
on discontinuous variation, 516
disruptive, 542, **543**, 595, 596, 597
definition, 638
Drosophila bristle number, 597
for divergence
in duplicated genes, 146, 562, 565
in sex-related genes, 562
dominance modifiers and alleles, 197, 545
for dominance expression, 197
for dosage compensation, 200
in early land tetrapods, 409
laryngeal valves, 493
for ecogeographical patterns, 585
effect on genetic drift, 548
for enzymatic pleiotropy, 260
evolution of, 135
origin, 134–36
effect on evolutionary clocks, 285
and evolutionary rates, 599
for extended secondary mammalian palate, 443
for fertility, 614
for flightlessness in birds, 437
for fetal development (mammals), 450
frequency-dependent, 540, 560
definition, 641
for gametic size differences, 198
for genetic assimilation, 358
in genetic codon evolution, 151
germ plasm advantages, 302
group. *See* Group selection

Stromatopteris (fern), 303
Strongylocentrotus purpuratus (echinoderm, sea urchin)
    actin protein regulation, 349
    modified larval development, 360
    nucleotide substitution rates, 285
        silent/replacement comparisons, 273
    number of repetitive DNA copies, 270
Strontium, in radioactive dating, 94, 95
"Struggle for existence", 25, 26, 36, 554. *See also* Natural selection
    in Darwin's evolutionary sequence, **27**
    in Social Darwinism proposals, 609
Sturgeon (chondrostean fish), 404, **406**
    "living fossil", 49
    urogenital duct system, **418**
Subspecies, 239. *See also* Races
    definitions, 583, 655
Succinate dehydrogenase, in Krebs cycle, **168**
Succinic acid
    electron donor, 165
    in Krebs cycle, 167, **168**
    synthesis in Miller electric spark discharge experiment, 122
Succinyl thiokinase, in Krebs cycle, **168**
Succinyl-coenzyme A, 167, 169
    in Krebs cycle, 167, **168**
Sugars, 120
    abiotic synthesis, 124
    aldose, 124
    biased optical rotation, 150
    deoxyribose. *See* Deoxyribose
    glucose. *See* Glucose
    ketose, 124
    lactose. *See* Lactose
        enzyme regulation, 219
    maltose, **127, 134**
    monosaccharides, 158, 159
    in nucleic acids, 111, **112**
    polysaccharides, 159
        synthesis by condensation reactions, 126, **127**
    ribose. *See* Ribose
Sulfur
    bacteria, 163, 165, **172**
    in Earth's geological structure, 89
    as energy source, 315
    in ferredoxin, **162**
    sulfur environments (hot sulfur springs), 173
    synthesis in stellar fusion reactions, 82
    thiol compounds, 144
Sun. *See also* Solar system
    angular momentum, 69, 86
    condensation and thermonuclear reactions, 85
    diameter, 82
    estimated age, 80
    geocentric concept, **4**
    location in Milky Way, 78, **79**
    planet formation, **86**
    position on main sequence, **80,** 119
    possible companion star, 432
    as second- or third-generation star, 83
    solar radiation, 36, 119
        effect on interplanetary organismic spores, 118
Sunflowers, hybrid speciation, 596
"Supergene", 209
Superkingdom (classification), 173, **247,** 638

Supernovae, 83
    possible cause for Late Cretaceous extinctions, 429
Superoxide dismutase
    antioxidant, 166
    evolutionary rate, 560
Superoxide radicals, 165, 277
Superphylum, 366
Superposition (geology), 91, **96**
"Survival of the fittest" concept, 27, 60, **324**
    in Social Darwinism proposals, 609
    tautology argument, 534
"Survival of the luckiest" concept, **324**
Survivorship, 555
    ancient and modern human populations, **615**
    definition, 655
    measurements, 555
Sylvilagus brasilensis (rabbit), 575
Symbiont, definition, 655
Symbiosis, 575, 655. *See also* Endocytosis; Endosymbiosis
    between algae and fungi (lichens), **319, 541**
    cnidarian-algal (zooxanthellae), 369
    in the hypercycle, 142, 291
    hypothesis for origin of eukaryotic organelles, 179, **181, 281, 283**
        multiple events, 279
        proposed microtubular origin, 180
    hypothesis for origin of RNA-protein interdependence, 145
    protistan-algal, **180**
Symmetrodonts (Jurassic-Cretaceous mammals), in phylogeny of mammalian molar teeth, **449**
Sympatric, 655
    isolating mechanisms, 590–91
    population, 239, 590
    speciation, **592,** 596–97, 655
        experiments, 597
Symplesiomorphic characters, 249, **250**
Synapomorphic characters, 249, **250**
    definition, 655
Synapsids (fenestrated reptiles), 419
    classification, 421
        problems, 443
    as mammalian ancestors, 443
        proposed mammalian taxon, 248
    in phylogeny of reptiles, **422**
    skull temporal openings, **420**
    sound transmission, 444
Synteny, 204, **206**
    definition, 656
Systematics, 9, 10, 249. *See also* Classification; Taxonomy
    binomial nomenclature, 10, 634
    definition, 236, 656
    holophyletic taxa, **246**
    paraphyletic taxa, 249, 251, 650
        reptiles, 419, 443
    polyphyletic taxa, 245, **246,** 247, 248, 249, 251, 252, 308, 313, 414, 416, 447, 652. *See also* Polyphyletic evolution
    support for Darwin, 36, **37**
    typology, 10, 657

T4 virus
    development, 345, **346**
    DNA melting temperature, **271**

self-assembly, 134
Tachyglossus. See Echidna
Tamarins (New World monkeys), 466
    in primate classification system, 464
Tapeworms (cestodes), parasitic platyhelminths, 373, 374
Tardigrades
    classification among metazoan phyla, 366
    in metazoan phylogeny, **368**
    time of first fossil appearance, **324**
Tarsiers
    description, 465
    illustration, **465**
    in primate classification system, 464
    proposed omomyid origin, 468
Tautology
    in description of life, 145
    "survival of the fittest" argument, 534
Tautomers, 215, **216**. *See also* Mutation, base substitutions
    definition, 656
Taxes (behavior), 585
Taxon, definition, 247, 656
Taxonomy. *See also* Classification; Systematics
    classical, 238
    definition, 236, 656
    family (definition), 640
    genus (definition), 642
    "lumpers" and "splitters", 237, 248, 597
    numerical (phenetic), 237, 238, 248
        difficulties, 248
    order (definition), 649
    paraphyletic taxa, 249, 251, 419
        definition, 650
        reptiles, 419, 443
    phylum (definition), 651
    polyphyletic taxa. *See* Polyphyletic evolution
    species (definition), 655. *See also* Species
Tay-Sachs disease, 528
    incidence, 616
Tectonic movements and plates, 102. *See also* Continental drift; Plate tectonics
    boundary events, 100, 101
    description, 637, 656
    Gondwana (ancient southern continent), 96, **98,** 99, 100, **102**
    Hawaiian Islands formation, **549**
    Laurasia (ancient northern continent), 99, 103
    paleomagnetism, 96, 97. *See also* Paleomagnetism
    Pangaea (ancient supercontinent), 96, 99, 102, 104, 452
    theories of origin, 102
Teeth, 277
    closure, 446
        morganucodontids, **446**
    cusps, description, 443, 637
    deciduous, 444
        definition, 637
        marsupials, 450
    dental comb, 464
    dentin, 401
    heritable patterns, 445
    heterodont, 443, 450
    horses, 45
    human vestigial, **43**
    labyrinthodont, 410, **411,** 419

Tree shrews, 468
pelvis, 475
Trematodes (flukes), parasitic
plathyhelminths, 373, 374
reproductive patterns, 374
Trentepohlia (algae), 298
Triadobatrachus (Triassic frog), **416**
Tribolium castaneum (flour beetle),
homeoboxes, **354**
Tribolium confusum (flour beetle), 570
Tricarboxylic acid cycle. *See* Krebs cycle
Triceratops (ceratopsian dinosaur), in
dinosaur phylogeny, **426**
Trichoplax (placozoan), 367, **371**
Triconodonts (Jurassic-Cretaceous
mammals), 446
in phylogeny of mammalian molar
teeth, **449**
Trilobites (fossil arthropods), 248, 382
characteristics, 383
stratigraphic appearance, 91
Trilophosauria (Triassic archosaur
reptiles), 421
Trinomial expansion, 523, **524**
Triosephosphate isomerase
in anaerobic glycolysis, **160**
in Calvin cycle, **164**
conserved amino acid sequence, 159
Triploblastic, 333, 334, **337**
definition, 656
Triploids
definition, 208
sexual abnormalities, 198
Trisomics
definition, 210
human, 616
Tritium, definition, 657
Trochophore larva (annelids and mollusks),
378, **379, 394**
Tropisms, 585
Trout, silent/replacement mutation rate
comparisons, 273
Tryptophan
codon designation, 148
structural formula, **111**
Tryptophan synthetase, fusion event, 260
Tuatara (New Zealand sphenodont
reptile), 420
Tuberous sclerosis (neuroectodermal
lesions), 616
Tubulidentates (aardvarks), in Cenozoic
mammalian radiation, **456**
Tudinatus (amphibian lepospondyl), **415**
Tunicates (sea squirts), 397
adult morphological characteristics, **397**
DNA nucleotide numbers, **268**
larval features, 397, **398**
paedomorphosis, 397, **399**
pax-6 gene homologies, 354
in proposed phylogeny of vertebrates, **399**
Turtles, 421, 423, 428
non-chromosomal sex determination, 198
in phylogeny of reptiles, **422**
in reptilian classification system, 421
Tympanic membrane, 444, **445**
Typhlodromus occidentalis (mite), **574**
Typhlonectes (amphibian caecilian), 415
Typology, 10
definition, 657
Tyrannosaurus (theropod dinosaur),
426, 584

in dinosaur phylogeny, **426**
endotherm metabolism proposal, 428
Tyrocidin (antibiotic), 144
Tyrosine
codon designations, 148
structural formula, **111**
Tyrosine kinase, **350**

U

Ultraviolet
in abiotic organic synthesis, 121, 124
action on water, 89
decarboxylating agent, 124
degradative effects, 126, 170
description, 657
in detecting DNA melting temperature,
271
early energy source, 119
effect on ammonia, 165, 167
interplanetary organismic hazard, 118
in Lambda (λ) virus regulation, 347
mutagenic activity, 225
ozone screening, 170, 299
as possible cause for Late Cretaceous
extinctions, 429
quantum organic yield, 125
Ulva (algae), 298
life cycle, **298**
Unequal crossing over, 209, 260, 657
Ungulates, 500, 657
brain size comparisons of fossil and
modern forms, **455**
condylarth ancestors, **456**, 458
South American
extinctions, 459
radiation, 458, **459**
Toxodon fossil, 22
Uniformitarianism, 15, 16, 25
definition, 657
Uniramians (arthropods), 383. *See also*
Insects
"Universal" phylogenetic tree, **178**, 279
Universe (cosmology), 78
Big Bang hypothesis, 76, 77, **78**, 634
energy/matter origin, 78
Black Body radiation, 76
change in concept from static to
dynamic, 17
creation myths, 63, 113
"dark matter", 76
Doppler shift, 76, **77**
elements
chemical. *See* Chemical elements
early concepts, 4
entropy, increase and decrease, 132. *See
also* Entropy
estimated number of atoms, 264
estimated number of protons and
neutrons, 246, 566
expansion, 76
Hubble constant, **77**
geocentric concept, **4**
idealistic concepts, 5
"inflationary", 76
as "Ladder of Nature", 7
Oscillating Big Bang hypothesis, 76, **78**
possibilities for life, 86–88
"singularity" origin, 78
Steady State hypothesis, 76, **78**
Upper Paleolithic stone age, **480**, 481

Uracil, 124
abiotic synthesis, 124
in nucleic acids, 112
structural formula, **112**
Uranium
in radioactive dating, 94, 95
relative universal abundance, **82**
synthesis in supernovae, 83
Urea, 13
in abiotic pyrimidine synthesis, 124,
125
cycle for nitrogen excretion, 151
in reptilian excretion, 417
structural formula, 125, 126
synthesis in Miller electric spark discharge
experiment, 122
Ureidoacrylonitrile, abiotic synthesis, 125
Urochordates. *See also* Tunicates
in metazoan phylogenies, **368**
paedomorphosis, 397
time of first fossil appearance, **324**
in vertebrate ancestry, 396
Urodeles (newts and salamanders), 414
characteristics, 415
DNA nucleotide numbers, **268**
fossil appearance, **416**
proposed origin, 416
Urogenital duct systems, vertebrates, **418**
Urpflanze, 6, **8**. *See also* Archetypes
Urskeleton, 7, **8**. *See also* Archetypes
Use and disuse concept (Lamarck), 25, 28,
29, 33
definition, 657

V

Vagus nerve (fish, mammals), 42, **45**
Valine
codon designations, 148
in Murchison meteorite, 122
structural formula, **111**
Variation, 25. *See also* Genetic variability;
Polymorphism
continuous, 33, 219, 515, 516. *See also*
Quantitative inheritance
definition, 637
proponents, 515
description, 657
discontinuous, 515
definition, 638
proponents, 515–16
intraspecies, 10
Lamarckian concept, 25
polygenic, 33, 219, 230. *See also* Polygenes
presumed limits, 33
Variety, 10. *See also* Race
in artificial selection, 50
Vascular plants. *See also* Angiosperms;
Ferns; Gymnosperms; Lycopsida;
Sphenopsida
definition, 657
distinctions from bryophytes, 300
evolution of conductive tissues, 305, **310**
fossil appearance, 300, 303–8
Carboniferous, **306, 308, 312**
Devonian, **311**
Jurassic, **311**
Silurian, **304**
hybridization estimates, 596
polyphyletic proposals, 248

Vascular plants—*Cont.*
proposed phylogeny, **314**
resemblances to algae, 298
seeds. *See* Seeds
Vectors (genetic engineering), 620,
622, 623
definition, 657
Vendian (Precambrian period of metazoan
fossils), 326
Vendozoa (Precambrian metazoan taxon), **327**
Ventral, definition, 657
Vertebrae
cervical numbers, 356, 358
distinctive Vertebrata subphylum
feature, 391
evolution in early tetrapods, 412, **414**
osteolepiform-amphibian comparisons,
412
Vertebrata. *See* Vertebrates
Vertebrates, 44. *See also* Chordates
ancient DNA from fossils, 277
axial segmentation pattern, 340
characteristics, 391, **392,** 393–94, 393
comparative anatomy, 37
brains, **40**
cranial nerves, 402, **403**
forelimbs, 37, **41**
gill arches, 42, **44,** 403
ductus arteriosis, 42
proposal for jaw ancestry, **402**
heart, 42, **417**
urogenital duct systems, **418**
convergent evolution. *See* Convergent
evolution
cranial nerves, 42, **45**
cytochrome *c* phylogeny, **266**
development
constraints, 355
gene duplications, **354**
genetic homologies between phyla, 353
shared patterns, 42
ectotherms, 421
endotherms, 421
gills, 42, 391, **410**
hemoglobin-like molecules, 257
amino acid substitution rates, 284
α chain phylogeny, **265**
homologies, 37
brains, **40**
embryos, 40, **44**
forelimbs, **41,** 241, 245
homeoboxes, 360, **396**
pax-6 gene sequences, **34,** 354
imaginary hybrids, 10
initial gene sequence ("CpG island"), 269
jaws
advantages, 401
proposed origins, **402**
lens crystallins, 260
lung preadaptations, 33, 409
in metazoan phylogenies, **368**
nerve growth factor, gene duplication, 260
nucleotide substitution rates, 284
differences among taxonomic groups,
273, 285
optical preadaptations, 33
parallel evolution. *See* Parallel evolution
phylogeny, **265, 267**
phylotypic stage, **44,** 356
pharyngula, 40, **44, 355,** 356, 391, 412

proposals for evolutionary origin,
392–98
annelids/arthropods, dorso-ventral
inversion, 360, 361, 392–93, **393**
cephalochordates, 395
echinoderms, 393–95, 397
expansion in gene numbers, 270
gene cluster duplication, **354**
hemichordates, 394
lophophorate ancestor, 398
phylogeny, **399**
marine or estuarian localities, 400
Pikaia fossil (Cambrian), **399**
proposed rate of amino acid
substitutions, 559, 560
protein polymorphism, 231
serial homology, 242
skeletal advantages, 401
urskeleton concept, 7, **8**
vestigial structures, **43**
Vertical transmission, 225, 575, 607
Vervet monkeys
in primate classification system, 464
vocalizations, 492
kin selection, 570
Vestigial organs, 39, 357
birds, 437
cave animals, 15, 33, 40
definition, 657
humans, 40, **43**
snakes, 40, **43**
whales, 40, **43**
Vicariance, 590
Viruses, 575
bacteriophages, 224, 240, **271,** 345, **346,** 347
coevolution with hosts, 575
description, 657
human immunodeficiency virus (HIV),
575
laboratory evolution, 290–91
Lambda (λ), 347
linkage maps, 204
modes of transmission, 240, 575
mutagenic activity in host, 225
mutation rates, 224
myxoma, 575
nucleotide numbers, **268**
Qβ. *See* Qβ virus
retro-, 619, 620, 623
RNA. *See* RNA viruses
selection for virulence, 575
self-assembly, 134
T4, 134, 345, **346**
temperate, 575
tobacco mosaic, 134
vectors in genetic engineering, 619, 620,
623, 657
Vitalism, 13, 14, 24, 613. *See also*
Lamarckianism; Mysticism;
Orthogenesis
definition, 657
Vitamins (A, E, K), antioxidants, 165, **167**
Viviparous reproduction
amphibians, 414
definition, 657
mammals, 442
proposed early forms, 450
Vocalization
advantages, 492
anatomical structures, **494**

basic mode, 493
chimpanzee attempts, 495
larynx, 42, 493, **494**
vowel sounds, **493**
nasalized, 495
Volcanoes, 90
amino acid condensation, 129
as early energy source, 119
Late Cretaceous lava (Deccan Traps), 433
outgassing, 88
Volvox (algae), 296, 334
possible phylogeny, **297**

W

Wasps, 371
chalcid species, 308
fig pollination, 308
sex ratio determination in *Lariophagus*, 385
social organization, 384
and haplodiploidy, 385, 570
proposed evolutionary sequence, 386
Water
hydrolysis, 123
hydrophilic interactions, **131, 132**
hydrophobic interactions, **131, 132,** 145
from meteorites, 85, 120, **123**
in Miller electric spark discharge
experiment, 122
outgassing from Earth, 88, 120
in photosynthesis
electron source, 165, **166**
hydrogen source, 165
present atmospheric composition, 89
production in oxidative respiratory
chain, **169**
properties, 119
removal in condensation reactions, 126,
127
ultraviolet dissociation, 89
Wernicke's area (speech center), 497, **498,** 502
Whales
in artiodactyl classification, 252, 454
cervical vertebrae, 356
convergent evolution, 245
cytochrome *c* relationships, 264, **266**
developmental constraints, 357
forelimb vertebrate homologies, **41**
vestigial hind limbs, 40, **43**
Wheat
agricultural expansion, 608
allopolyploidy, 206
as material for spontaneous generation,
11
quantitative inheritance, **223**
White Dwarf star, 82
Wild type, definition, 657
Williamsonia (Jurassic cycad), fossil
reconstruction, **311**
Wings, analogies among taxa, 37
Wobble pairing, **149**
extreme forms, 152
Wolves
marsupial, **39,** 458
placental (*Canis lupus*), **458**
relationship to dogs, **50**
Wombat, in marsupial radiation, **39**
Woodrats, Bergmann's rule, 585
Worm
in annelid-arthropod evolution, **382**

body organization, 338
peristaltic locomotion, **339**
Wright's shifting balance process of
evolution, 566–69
group selection, 569

Part and Chapter Opener Photo Credits:

25. Symmetry:

$$f(-x) = \frac{x^2 - 1}{-x - 2}$$

No symmetry since $f(-x) \neq f(x)$ and $f(-x) \neq -f(-x)$.

x-intercepts:

$$x^2 - 1 = 0$$
$$x = \pm 1$$

y-intercept:

$$f(0) = \frac{1}{2}$$
$$y = \frac{1}{2}$$

Vertical asymptote:

$$x - 2 = 0$$
$$x = 2$$

Horizontal asymptote:

$n > m$, so no horizontal asymptote.

Slant asymptote: $n = m + 1$

$$f(x) = x + 2 + \frac{3}{x - 2}$$
$$y = x + 2$$

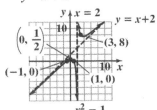

26.

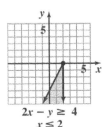

$$2x - y \geq 4$$
$$x \leq 2$$

27. $f(x) = x^2 - 4x - 5$

$$x = \frac{-b}{2a} = \frac{4}{2} = 2$$

$$f(2) = 2^2 - 8 - 5 = -9$$

vertex: $(2, -9)$

x-intercepts:

$$x^2 - 4x - 5 = 0$$
$$(x - 5)(x + 1) = 0$$
$$x = 5, -1$$

y-intercept: $f(0) = -5$

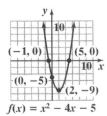

28.

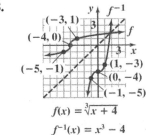

$$f(x) = \sqrt[3]{x + 4}$$
$$f^{-1}(x) = x^3 - 4$$

29.

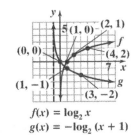

$$f(x) = \log_2 x$$
$$g(x) = -\log_2(x + 1)$$

30. $f(x) = -x^2 - 2x + 1$, $\quad g(x) = x - 1$

$$(f \circ g)(x) = f(g(x))$$
$$= -(x - 1)^2 - 2(x - 1) + 1$$
$$= -x^2 + 2x - 1 - 2x + 2 + 1$$
$$= -x^2 + 2$$

$$(g \circ f)(x) = g(f(x))$$
$$= (-x^2 - 2x + 1) - 1$$
$$= -x^2 - 2x$$

31. $f(x) = -x^2 - 2x + 1$

$$\frac{f(x+h) - f(x)}{h} = \frac{\left(-(x+h)^2 - 2(x+h) + 1\right) - \left(-x^2 - 2x + 1\right)}{h}$$

$$= \frac{-x^2 - 2xh - h^2 - 2x - 2h + 1 + x^2 + 2x - 1}{h}$$

$$= \frac{-2xh - h^2 - 2h}{h}$$

$$= \frac{h(-2x - h - 2)}{h}$$

$$= -2x - h - 2$$

32. $AB - 4A = \begin{bmatrix} 4 & 2 \\ 1 & -1 \\ 0 & 5 \end{bmatrix} \begin{bmatrix} 2 & 4 \\ 3 & 1 \end{bmatrix} - 4 \begin{bmatrix} 4 & 2 \\ 1 & 1 \\ 0 & 5 \end{bmatrix}$

$$= \begin{bmatrix} 14 & 18 \\ -1 & 3 \\ 15 & 5 \end{bmatrix} - \begin{bmatrix} 16 & 8 \\ 4 & -4 \\ 0 & 20 \end{bmatrix} = \begin{bmatrix} -2 & 10 \\ -5 & 7 \\ 15 & -15 \end{bmatrix}$$

33. $\dfrac{2x^2 - 10x + 2}{(x-2)(x^2 + 2x + 2)} = \dfrac{A}{x-2} + \dfrac{Bx + C}{x^2 + 2x + 2}$

$2x^2 - 10x + 2 = A(x^2 + 2x + 2) + (Bx + C)(x - 2)$

$$= Ax^2 + 2Ax + 2A + Bx^2 - 2Bx + Cx - 2C$$

$$= (A + B)x^2 + (2A - 2B + C)x + 2A - 2C$$

Thus we have the following system of equations.

$A + B = 2$
$2A - 2B + C = -10$
$2A - 2C = 2$

Add twice the first equation to the second equation.

$2A + 2B = 4$

$\underline{2A - 2B + C = -10}$

$4A + C = -6$

Add twice the resulting equation to the third equation.

$8A + 2C = -12$

$\underline{2A - 2C = 2}$

$10A = -10$

$A = -1$

Back-substitute to find B and C.

$2(-1) - 2C = 2$
$-2 - 2C = 2$
$-2C = 4$
$C = -2$
$-1 + B = 2$
$B = 3$

$\dfrac{-1}{x - 2} + \dfrac{3x - 2}{x^2 + 2x + 2}$

34. $(x^3 + 2y)^5$ $= \binom{5}{0}(x^3)^5 + \binom{5}{1}(x^3)^4(2y) + \binom{5}{2}(x^3)^3(2y)^2 + \binom{5}{3}(x^3)^2(2y)^3 + \binom{5}{4}(x^3)(2y)^4 + \binom{5}{5}(2y)^5$

$= x^{15} + 5x^{12}(2y) + 10x^9(4y^2) + 10x^6(8y^3) + 5x^3(16y^4) + 32y^5$

$= x^{15} + 10x^{12}y + 40x^9y^2 + 80x^6y^3 + 80x^3y^4 + 32y^5$

35. $\displaystyle\sum_{i=1}^{50}(4i - 25)$

$a_1 = 4(1) - 25 = -21$

$a_{50} = 4(50) - 25 = 175$

$S_{50} = \dfrac{50}{2}(-21 + 175) = 3850$

36. Find slope:

$m = \dfrac{y_2 - y_1}{x_2 - x_1} = \dfrac{1 - 3}{-2 - 6} = \dfrac{-2}{-8} = \dfrac{1}{4}$

Find equation:

$y - y_1 = m(x - x_1)$

$y - 3 = \dfrac{1}{4}(x - 6)$

$y - 3 = \dfrac{1}{4}x - \dfrac{3}{2}$

$y = \dfrac{1}{4}x + \dfrac{3}{2}$

37. Find the slope:

$x - 5y - 20 = 0$

$-5y = -x + 20$

$\dfrac{-5y}{-5} = \dfrac{-x}{-5} + \dfrac{20}{-5}$

$y = \dfrac{1}{5}x - 4$

$m = -\dfrac{1}{\frac{1}{5}} = -5$

Find the equation:

$y - y_1 = m(x - x_1)$

$y - (-2) = -5(x - 0)$

$y + 2 = -5x$

$y = -5x - 2$

38. $200 + 0.05x = 0.15x$

$200 = 0.1x$

$\dfrac{200}{0.1} = \dfrac{0.1x}{0.1}$

$2000 = x$

At \$2000 in sales, the two earnings will be the same.

39. $2L + 2W = 300$
$\qquad\quad L = W + 50$
Rearrange the equations and add:
$L + W = 150$
$\underline{L - W\ = 50}$
$\qquad 2L = 200$
$\qquad\quad L = 100$
$\qquad\quad W = 50$
length: 100 yards, width 50 yards

40. $10x + 12y = 42$
$\quad\ \ 5x + 10y = 29$
Multiply second equation by –2 and add:
$10x + 12y = 42$
$\underline{-10x - 20y = -58}$
$\qquad\quad -8y = -16$
$\qquad\qquad\ y = 2$
Back substitute:
$5x + 10(2) = 29$
$\qquad\ \ 5x = 9$
$\qquad\qquad x = 1.8$
pen: \$1.80, pad: \$2

41. $s(t) = -16t^2 + 80t + 96$

    a. $-16t^2 + 80t + 96 = 0$
$\qquad\quad t^2 - 5t - 6 = 0$
$\qquad (t + 1)(t - 6) = 0$
$\qquad\qquad\qquad t = -1 \text{ or } t = 6$
    The ball will strike the ground after 6 seconds.

    b. $t = \dfrac{-b}{2a} = \dfrac{-80}{-32} = \dfrac{5}{2} \text{ or } 2.5$
$\qquad S(2.5) = -16(2.5)^2 + 80(2.5) + 96 = 196$
    The ball reaches a maximum height of 196 feet, 2.5 seconds after it is thrown.

42. $I = \dfrac{k}{R}$

$5 = \dfrac{k}{22}$

$k = 110$

$I = \dfrac{110}{10} = 11$

11 amperes

43. Let x represent the number of years after 2010.
The data from 2010 and 2011 are represented as $(0, 68.4)$ and $(1, 80.2)$

Find slope: $m = \dfrac{80.2 - 68.4}{1 - 0} \approx 11.8$

Thus, $f(x) = 11.8x + 68.4$
Projections will vary.

44. $d = 10\sin\dfrac{3\pi}{4}t$

    a. $|a| = |10| = 10 \; 2a = 20$

         The maximum displacement is 20 inches.

    b. $f = \dfrac{\omega}{2\pi} = \dfrac{\frac{3\pi}{4}}{2\pi} = \dfrac{3}{8}$

         The frequency is $\dfrac{3}{8}$ cycle per second.

    c.   period $= \dfrac{2\pi}{\omega} = \dfrac{2\pi}{\frac{3\pi}{4}} = \dfrac{8}{3}$

         The time required for one oscillation is $\dfrac{8}{3}$ seconds.

45. $\tan x + \dfrac{1}{\tan x} = \dfrac{\sin x}{\cos x} + \dfrac{1}{\frac{\sin x}{\cos x}} = \dfrac{\sin x}{\cos x} + \dfrac{\cos x}{\sin x}$

$$= \dfrac{\sin^2 x + \cos^2 x}{\cos x \cdot \sin x}$$

$$= \dfrac{1}{\cos x \cdot \sin x}$$

46. $\dfrac{1 - \tan^2 x}{1 + \tan^2 x} = \dfrac{1 - \frac{\sin^2 x}{\cos^2 x}}{1 + \frac{\sin^2 x}{\cos^2 x}} \cdot \dfrac{\cos^2 x}{\cos^2 x}$

$$= \dfrac{\cos^2 x - \sin^2 x}{\cos^2 x + \sin^2 x}$$

$$= \dfrac{\cos 2x}{1} = \cos 2x$$

47. $y = -2\cos(3x - \pi)$

Amplitude: $|A| = |-2| = 2$

Period: $\dfrac{2\pi}{B} = \dfrac{2\pi}{3}$

Phase shift: $\dfrac{C}{B} = \dfrac{\pi}{3}$

$\dfrac{\pi}{3}, -2$, $\dfrac{\pi}{2}, 0$, $\dfrac{2\pi}{3}, 2$, $\dfrac{5\pi}{6}, 0$,

$(\pi, -2)$

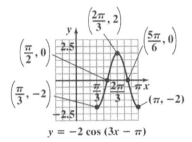

$y = -2\cos(3x - \pi)$

48. $4\cos^2 x = 3$

$$\cos^2 x = \frac{3}{4}$$

$$\cos x = \pm\sqrt{\frac{3}{4}} = \pm\frac{\sqrt{3}}{2}$$

$$x = \frac{\pi}{6}, \frac{5\pi}{6}, \frac{7\pi}{6}, \frac{11\pi}{6}$$

The solutions in the interval $[0, 2\pi)$ are $\frac{\pi}{6}, \frac{5\pi}{6}, \frac{7\pi}{6}$, and $\frac{11\pi}{6}$.

49. $2\sin^2 x + 3\cos x - 3 = 0$

$$2(1 - \cos^2 x) + 3\cos x - 3 = 0$$

$$2 - 2\cos^2 x + 3\cos x - 3 = 0$$

$$2\cos^2 x - 3\cos x + 1 = 0$$

$$(2\cos x - 1)(\cos x - 1) = 0$$

$$2\cos x - 1 = 0 \quad \text{or} \quad \cos x - 1 = 0$$

$$2\cos x = 1 \qquad\qquad \cos x = 1$$

$$\cos x = \frac{1}{2}$$

$$x = \frac{\pi}{3}, \frac{5\pi}{3} \quad \text{or} \qquad x = 0$$

The solutions in the interval $[0, 2\pi)$ are $0, \frac{\pi}{3}$, and $\frac{5\pi}{3}$.

50. $\cot\left(\cos^{-1}\left(-\frac{5}{6}\right)\right)$

If $\cos\theta = -\frac{5}{6}$, θ lies in quadrant II.

$$\cos\theta = -\frac{5}{6} = \frac{x}{r} = \frac{-5}{6}$$

$$x^2 + y^2 = r^2$$

$$(-5)^2 + y^2 = 6^2$$

$$25 + y^2 = 36$$

$$y^2 = 11$$

$$\cot\left(\cos^{-1}\left(-\frac{5}{6}\right)\right) = \frac{x}{y} = \frac{-5}{\sqrt{11}} = -\frac{5\sqrt{11}}{11}$$

51. $r = 1 + 2\cos\theta$
Check for symmetry:

Polar Axis	**The Line** $\theta = \dfrac{\pi}{2}$	**The Pole**
$r = 1 + 2\cos(-\theta)$	$-r = 1 + 2\cos(-\theta)$	$-r = 1 + 2\cos\theta$
$r = 1 + 2\cos\theta$	$r = -1 - 2\cos\theta$	$r = -1 - 2\cos\theta$

Graph is symmetric with respect to the polar axis.

θ	0	$\dfrac{\pi}{6}$	$\dfrac{\pi}{3}$	$\dfrac{\pi}{2}$	$\dfrac{2\pi}{3}$	$\dfrac{5\pi}{6}$	π
r	3	2.73	2	1	0	−0.73	−1

Use symmetry to obtain the graph.

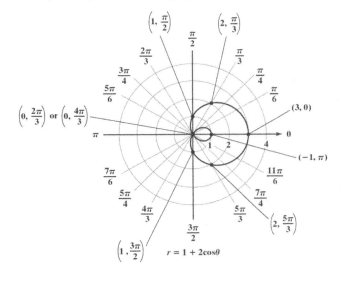

52. The known ratio is $\dfrac{a}{\sin A}$, or $\dfrac{22}{\sin 34°}$. Because side b is given, Use the Law of Sines to find angle B.

$$\frac{a}{\sin A} = \frac{b}{\sin B}$$
$$\frac{22}{\sin 34°} = \frac{32}{\sin B}$$
$$22\sin B = 32\sin 34°$$
$$\sin B = \frac{32\sin 34°}{22}$$
$$\sin B \approx 0.8134$$
$$B \approx \sin^{-1} 0.8134$$
$$B \approx 54°$$

There are two angles possible:
$B_1 \approx 54°$, $B_2 \approx 180° - 54° = 126°$

There are two triangles:

$C_1 = 180° - A - B_1 \approx 180° - 34° - 54° = 92°$

$C_2 = 180° - A - B_2 \approx 180° - 34° - 126° = 20°$

Use the Law of Sines to find c_1 and c_2.

$$\frac{c_1}{\sin C_1} = \frac{a}{\sin A}$$

$$\frac{c_1}{\sin 92°} = \frac{22}{\sin 34°}$$

$$c_1 = \frac{22\sin 92°}{\sin 34°} \approx 39.5$$

$$\frac{c_2}{\sin C_2} = \frac{a}{\sin A}$$

$$\frac{c_2}{\sin 20°} = \frac{22}{\sin 34°}$$

$$c_2 = \frac{22\sin 20°}{\sin 34°} \approx 13.7$$

In one triangle, the solution is $B_1 \approx 54°$, $C_1 \approx 92°$, and $c_1 \approx 39.5$.

In the other triangle, $B_2 \approx 126°$, $C_2 \approx 20°$, and $c_2 \approx 13.7$.

53. Use $\cos^2 t + \sin^2 t = 1$.

Since $x = \sin t$, then $\sin^2 t = x^2$.

Since $y = 1 - \cos^2 t$, then $\cos^2 t = y - 1$.

Substitute these values into $\cos^2 \overbrace{t}^{y-1} + \sin^2 \overbrace{t}^{x^2} = 1$.

$\cos^2 t + \sin^2 t = 1$

$y - 1 + x^2 = 1$

$y = 2 - x^2$

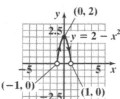